Construction Technology

Trainee Guide
Third Edition

Upper Saddle River, New Jersey
Columbus, OH

Contren® Learning Series

National Center for Construction Education and Research
President: Don Whyte
Director of Product Development: Daniele Stacey
Construction Technology Project Manager: Angela Jonas
Production Manager: Tim Davis
Quality Assurance Coordinator: Debie Ness
Desktop Publishing Coordinator: James McKay
Editors: Rob Richardson and Matt Tischler
Production Asssistant: Laura Wright

Pearson Education, Inc.
Product Manager: Lori Cowen
Text Designer: Kristina D. Holmes
Cover Designer: Kristina D. Holmes

This book was set in Palatino and Helvetica by NCCER. It was printed and bound by Courier Kendallville, Inc. The cover was printed by Phoenix Color Corp.

This information is general in nature and intended for training purposes only. Actual performance of activities described in this manual requires compliance with all applicable operating, service, maintenance, and safety procedures under the direction of qualified personnel. References in this manual to patented or proprietary devices do not constitute a recommendation of their use.

Pearson Prentice Hall™ is a trademark of Pearson Education, Inc.
Pearson® is a registered trademark of Pearson plc
Prentice Hall® is a registered trademark of Pearson Education, Inc.

Pearson Education Ltd.
Pearson Education Singapore Pte. Ltd.
Pearson Education Canada, Ltd.
Pearson Education—Japan

Pearson Education Australia Pty. Limited
Pearson Education North Asia Ltd.
Pearson Educación de Mexico, S.A. de C.V.
Pearson Education Malaysia Pte. Ltd.

10 9 8 7 6 5 4 3 2
ISBN 0-13-609951-3
ISBN 13 978-0-13-609951-2

Contren® Learning Series

PREFACE

TO THE TRAINEE

This curriculum will ground you in the basic knowledge and principles of carpentry, masonry, concrete finishing, electrical work, HVAC, and plumbing. You will become skilled in different phases of a project from start to finish. When you complete this course, you will be able to interpret construction drawings, perform quality concrete and brickwork, frame walls, ceilings, and floors of a structure, and install the proper wiring and piping for electrical and plumbing systems.

If this is your first step towards a future in the construction industry, this training is the ideal starting point for a wide variety of possible career paths. Postsecondary and apprenticeship training provide excellent opportunities to develop your skills in a specific craft and establish a secure and prosperous career. Hourly earnings for craft workers continue to grow, with opportunities for continuing education and advancement. Many in the construction industry work for larger companies or contractors, but a good number are in business for themselves. According to the U.S. Bureau of Labor Statistics, about 10% of electricians, 25% of masons, and 30% of carpenters are self-employed.

Whether or not you become an independent craft worker, your skills will be needed in most regions of the U.S. over the coming decades as the country embarks on a construction building boom. A large portion of the existing workforce is also reaching retirement age, so there will be increased opportunity for new craft workers entering the field. Electricians in particular are expected to enjoy the most job growth over the next decade as more workers retire and newer technologies are required for more projects.

Other possible career development directions open to people trained in the construction field include college (in civil engineering or design, for example), the military, or positions of leadership in an organization, such as project manager and senior manager. If you progress on a professional path in construction, be sure to take advantage of NCCER's national training and certification programs for over 60 different crafts to ensure that your skills are documented and portable for anywhere you want to work. NCCER wishes you success in your skills and career development.

We invite you to visit the NCCER website at www.nccer.org for the latest releases, training information, newsletter, and much more. You can also reference the Contren® product catalog online at www.crafttraining.com. Your feedback is welcome. You may email your comments to curriculum@nccer.org or send general comments and inquiries to info@nccer.org.

CONTREN® LEARNING SERIES

The National Center for Construction Education and Research (NCCER) is a not-for-profit 501(c)(3) education foundation established in 1995 by the world's largest and most progressive construction companies and national construction associations. It was founded to address the severe workforce shortage facing the industry and to develop a standardized training process and curricula. Today, NCCER is supported by hundreds of leading construction and maintenance companies, manufacturers, and national associations. The Contren® Learning Series was developed by NCCER in partnership with Pearson Education, Inc., the world's largest educational publisher.

Some features of NCCER's Contren® Learning Series are as follows:

- An industry-proven record of success
- Curricula developed by the industry for the industry
- National standardization providing portability of learned job skills and educational credits
- Compliance with Office of Apprenticeship requirements for related classroom training (CFR 29:29)
- Well-illustrated, up-to-date, and practical information

NCCER also maintains a National Registry that provides transcripts, certificates, and wallet cards to individuals who have successfully completed modules of NCCER's Contren® Learning Series. *Training programs must be delivered by an NCCER Accredited Training Sponsor in order to receive these credentials.*

Contents

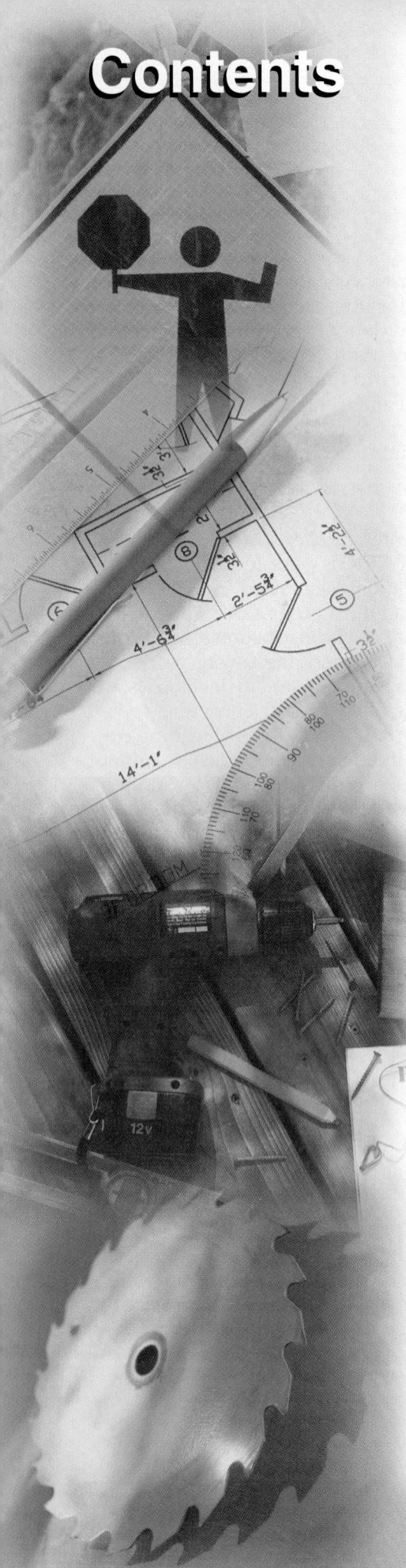

68101-09 Site Layout One – Distance Measurement and Leveling

Covers the equipment, principles, and methods used to perform distance measurement and leveling. Also covers the layout responsibilities of surveyors, field engineers, and carpenters; interpretation and use of site/plot plan drawings; and methods used for on-site communication. **(22.5 Hours)**

68102-09 Introduction to Concrete, Reinforcing Materials, and Forms

Describes the ingredients of concrete, discusses the various types of concrete, and explains how to mix concrete. The module also covers basic job-built footing, edge, and wall forms and form ties and describes the types and uses of concrete reinforcing materials. **(10 Hours)**

68103-09 Handling and Placing Concrete

Covers tools, equipment, and procedures for handling, placing, and finishing concrete. Also covers joints made in concrete structures, the use of joint sealants, and form removal procedures. Emphasizes safety procedures for handling, placing, and finishing concrete. **(22.5 Hours)**

68104-09 Introduction to Masonry

Introduces the trainee to the historic and current materials and processes used in the masonry trade and covers safety concerns specific to the trade. Explains the uses of brick and concrete block, along with basic techniques for mixing mortar and laying masonry units. Covers opportunities in the trade. Allows the trainee to mix mortar and perform basic bricklaying. **(20 Hours)**

68105-09 Masonry Units and Installation Techniques

Covers all types of concrete and clay masonry units and their applications. Explains the use of ties and reinforcing materials. Covers the processes used in placing masonry units, including layout and setup, spreading mortar, cutting brick and block, laying to the line, making corners, tooling joints, patching, and cleanup. **(60 Hours)**

68106-09 Floor Systems

Covers framing basics as well as the procedures for laying out and constructing a wood floor using common lumber as well as engineered building materials. **(25 Hours)**

68107-09 Wall and Ceiling Framing

Describes the procedures for laying out and framing walls and ceilings, including roughing-in door and window openings, constructing corners and partition Ts, bracing walls and ceilings, and applying sheathing. **(20 Hours)**

68108-09 Roof Framing

Describes the various kinds of roofs and contains instructions for laying out rafters for gable roofs, hip roofs, and valley intersections. Covers both stick-built and truss-built roofs. **(37.5 Hours)**

68109-09 Roofing Applications

Covers the common materials used in residential and light commercial roofing, along with the safety practices and application methods for these materials. Includes shingles, roll roofing, shakes, tiles, and metal and membrane roofs, as well as the selection and installation of roof vents. **(25 Hours)**

68110-09 Exterior Finishing

Covers the types of exterior siding used in residential construction and their installation procedures. Includes wood, metal, vinyl, and cement board siding. **(35 Hours)**

68111-09 Basic Stair Layout

Introduces types of stairs and common building code requirements related to stairs. Focuses on the techniques for measuring and calculating rise, run, and stairwell openings, laying out stringers, and fabricating basic stairways. **(12.5 Hours)**

68112-09 Electrical Safety

Covers safety rules and regulations for electricians. Teaches the necessary precautions for electrical hazards on the job. Also covers the OSHA-mandated lockout/tagout procedure. **(10 Hours)**

68113-09 Residential Electrical Services

Covers the electrical devices and wiring techniques common to residential construction and maintenance. Allows trainees to practice making service calculations. Stresses the appropriate NEC® requirements. **(15 Hours)**

68114-09 Introduction to HVAC

Covers the basic principles of heating, ventilating, and air conditioning, career opportunities in HVAC, and apprenticeship programs. **(7.5 Hours)**

68115-09 Introduction to Drain, Waste, and Vent (DWV) Systems

Explains how DWV systems remove waste. Discusses how system components such as pipe, drains, traps, and vents work. Reviews drain and vent sizing, grade, and waste treatment. Also discusses how building sewers and sewer drains connect the DWV system to the public sewer system. **(10 Hours)**

68116-09 Plastic Pipe and Fittings

Introduces the types of plastic pipe and fittings used in plumbing applications, including ABS, PVC, CPVC, PE, PEX, and PB. Describes how to measure, cut, join, and support plastic pipe according to manufacturer's instructions and applicable codes. Also discusses pressure testing of plastic pipe once installed. **(10 Hours)**

68117-09 Copper Pipe and Fittings

Discusses sizing, labeling, and applications of copper pipe and fittings and reviews the types of valves that can be used on copper pipe systems. Explains proper methods for cutting, joining, and installing copper pipe. Also addresses insulation, pressure testing, seismic codes, and handling and storage requirements. **(10 Hours)**

CONTREN® CURRICULA

NCCER's training programs comprise over 60 construction, maintenance, and pipeline areas and include skills assessments, safety training, and management education.

Boilermaking
Cabinetmaking
Carpentry
Concrete Finishing
Construction Craft Laborer
Construction Technology
Core Curriculum:
   Introductory Craft Skills
Drywall
Electrical
Electronic Systems Technician
Heating, Ventilating, and
   Air Conditioning
Heavy Equipment Operations
Highway/Heavy Construction
Hydroblasting
Industrial Maintenance
   Electrical and Instrumentation
   Technician
Industrial Maintenance Mechanic
Instrumentation
Insulating
Ironworking
Masonry
Millwright
Mobile Crane Operations
Painting
Painting, Industrial
Pipefitting
Pipelayer
Plumbing
Reinforcing Ironwork
Rigging
Scaffolding
Sheet Metal
Site Layout
Sprinkler Fitting
Welding

Pipeline
Control Center Operations,
   Liquid
Corrosion Control
Electrical and Instrumentation
Field Operations, Liquid
Field Operations, Gas
Maintenance
Mechanical

Safety
Field Safety
Safety Orientation
Safety Technology

Management
Introductory Skills for the
   Crew Leader
Project Management
Project Supervision

Spanish Translations
Albañilería
Andamios
Currículo Básico
   Habilidades Introductorias
   del Oficio
Instalación de Rociadores
   Nivel Uno
Orientación de Seguridad
Principios Básicos de Maniobras
Seguridad de Campo

Supplemental Titles
Applied Construction Math
Careers in Construction
Your Role in the Green
   Environment

ACKNOWLEDGMENTS

This curriculum was revised as a result of the farsightedness and leadership of the following sponsors:

Frenship I.S.D.
Miami Valley CTC
Ohio University, Facilities Management
Santa Cruz Valley Union High School
Windham School District

This curriculum would not exist were it not for the dedication and unselfish energy of those volunteers who served on the Authoring Team. A sincere thanks is extended to the following:

Gregory Bauer
Larry Beets
Charles Buscher
Mick Harris
John Hoyle

NCCER PARTNERING ASSOCIATIONS

American Fire Sprinkler Association
Associated Builders and Contractors, Inc.
Associated General Contractors of America
Association for Career and Technical Education
Association for Skilled and Technical Sciences
Carolinas AGC, Inc.
Carolinas Electrical Contractors Association
Center for the Improvement of Construction
    Management and Processes
Construction Industry Institute
Construction Users Roundtable
Design Build Institute of America
Green Advantage
Merit Contractors Association of Canada
Metal Building Manufacturers Association
NACE International
National Association of Minority Contractors
National Association of Women in Construction

National Insulation Association
National Ready Mixed Concrete Association
National Systems Contractors Association
National Technical Honor Society
National Utility Contractors Association
NAWIC Education Foundation
North American Crane Bureau
North American Technician Excellence
Painting & Decorating Contractors of America
Portland Cement Association
SkillsUSA
Steel Erectors Association of America
Texas Gulf Coast Chapter, ABC
U.S. Army Corps of Engineers
University of Florida
Women Construction Owners & Executives,
    USA

Site Layout One – Distance Measurement and Leveling
68101-09

68101-09

Site Layout One: Distance Measurement and Leveling

Topics to be presented in this module include:

1.0.0	Introduction	1.2
2.0.0	Building Plan Drawings	1.2
3.0.0	Characteristics of Contour Lines	1.5
4.0.0	Site Layout Control Points	1.7
5.0.0	Communicating with Hand Signals	1.9
6.0.0	Distance Measurement Tools and Equipment	1.11
7.0.0	Measuring Distances by Taping	1.13
8.0.0	Estimating Distances by Pacing	1.19
9.0.0	Electronic Distance Measurements	1.19
10.0.0	Differential Leveling Tools and Equipment	1.20
11.0.0	Basics of Differential Leveling	1.30
12.0.0	Field Notes	1.36
13.0.0	Leveling Applications	1.37
14.0.0	Batter Boards	1.38
15.0.0	3-4-5 Rule	1.40

Site layout involves the accurate placement of structures on a work site. On large jobs, a senior carpenter may work with field engineers or site layout technicians to lay out building foundations, retaining systems, and other types of structures. On small projects, a senior carpenter may be required to do many of the tasks of a field engineer. This module is the first of two that focus on site layout. In this module, you will learn about surveyors' benchmarks and electronic and manual measuring equipment, as well as how to make accurate linear measurements and determine site elevations. This information, coupled with hands-on experience you will gain in laboratory training sessions, will help you to perform common site layout tasks.

Objectives

When you have completed this module, you will be able to do the following:

1. Describe the major responsibilities of the carpenter relative to site layout.
2. Convert measurements stated in feet and inches to equivalent measurements stated in decimal feet, and vice versa.
3. Use and properly maintain tools and equipment associated with taping.
4. Use manual or electronic equipment and procedures to make distance measurements and perform site layout tasks.
5. Determine approximate distances by pacing.
6. Recognize, use, and properly care for tools and equipment associated with differential leveling.
7. Use a builder's level and differential leveling procedures to determine site and building elevations.
8. Record site layout data and information in field notes using accepted practices.
9. Check and/or establish 90-degree angles using the 3-4-5 rule.

Trade Terms

Backsight (BS)	Fill
Breaking the tape	Foresight (FS)
Control points	Height of instrument (HI)
crosshairs	Parallax
Cut	Peg test
Differential leveling	Station
Earthwork	Temporary benchmark
Field notes	Turning point (TP)

Required Trainee Materials

1. Pencil and paper
2. Appropriate personal protective equipment

Prerequisites

Before you begin this module, it is recommended that you successfully complete the *Core Curriculum*.

This course map shows all of the modules in the *Construction Technology* curriculum. The suggested training order begins at the bottom and proceeds up. Skill levels increase as you advance on the course map. The local Training Program Sponsor may adjust the training order.

68117-09
Copper Pipe and Fittings

68116-09
Plastic Pipe and Fittings

68115-09
Introduction to Drain, Waste, and Vent (DWV) Systems

68114-09
Introduction to HVAC

68113-09
Residential Electrical Services

68112-09
Electrical Safety

68111-09
Basic Stair Layout

68110-09
Exterior Finishing

68109-09
Roofing Applications

68108-09
Roof Framing

68107-09
Wall and Ceiling Framing

68106-09
Floor Systems

68105-09
Masonry Units and Installation Techniques

68104-09
Introduction to Masonry

68103-09
Handling and Placing Concrete

68102-09
Introduction to Concrete, Reinforcing Materials, and Forms

68101-09
Site Layout One: Distance Measurement and Leveling

CONSTRUCTION TECHNOLOGY

CORE CURRICULUM:
Introductory Craft Skills

101CMAP.EPS

1.0.0 ◆ INTRODUCTION

Depending on the size of the project, the lead carpenter may do many of the same layout tasks at the job site as a field engineer. On very large construction jobs, the carpenter may work with the field engineers. On smaller projects, the carpenter may be responsible for the layout of the entire project.

Some of the tasks that can be performed by a carpenter include:

- Interpretation of site/contour plans
- Layout of secondary and working **control points** through leveling
- Site work
- Layout of retaining systems
- Layout of footings and foundations
- Column layout
- Layout of embedded items, sleeves, and block-outs

This is the first of two modules concerned with site layout. The focus of this module is on the principles, equipment, and basic methods used to perform the site layout tasks of distance measurement and **differential leveling**. The information presented in this module, along with the hands-on experience gained in the laboratory training sessions related to this module, will enable you to perform these layout tasks.

2.0.0 ◆ BUILDING PLAN DRAWINGS

The content and layout of plot/site drawings were covered in detail in the *Reading Plans and Elevations* module in Carpentry Level One. You may want to review that module before you continue with this module to get the most from this new information.

When you are working with the drawing set, keep in mind that the drawing set is a legal document and must be followed exactly. When it is necessary to deviate from the plans, the change must be recorded on the drawings and initialed by the responsible party. On some projects, construction changes are not permitted until a change order is approved, so be sure to know the policy on your projects.

Because all crafts on the project use the project drawing set, many diagrams will contain information that you will not use. When you are using any plans, be careful not to damage or deface them. Never doodle or scribble on the plans. Never make personal notes on plans that are used by other crafts. When you need to make notes, use a separate piece of paper. When using a drawing that has been assigned to your crew,

make relevant notes legible and initial the notes in case there is a question later.

The drawings that you will be using will most likely include civil plans, architectural drawings, and structural drawings. You have already worked with architectural drawings and structural drawings in previous levels. This module and the following one will discuss civil plans.

2.1.0 Civil Plans

Civil plans are those involving the earth. They are also called site plans, survey plans, or plot plans. *Figures 1*, *2*, and *3* are basic examples. Civil plans show man-made and topographic (natural) features, along with other relevant project information, including the information needed to correctly locate the structure on the site. Man-made features include roads, sidewalks, utilities, and buildings. Topographical features include trees, streams, springs, and existing contours. Sometimes, civil plans include a large-scale map of the overall area that indicates where the project is located on the site. Civil plans, just as any other plan, are referred to in the contract, and therefore are legal documents.

Civil plans, like other drawings, are usually drawn to scale, which is displayed on the drawing. The scale used depends on the size of the project. A project covering a large area typically will have a small scale, such as 1" = 100', while a project on a small site might have a large scale, such as 1" = 10'.

A prominently displayed north direction arrow appears on civil plans for orientation purposes. Project information includes the building outline, general utility information, proposed sidewalks, parking areas, roads, landscape information, proposed contours, and any other information that will convey what is to be constructed or changed on the site.

It is important to note that civil drawings show elevations, including the elevation of the finished first floor, in relation to sea level. Other drawings often use the elevation of the finished first floor as a reference point. Typically, site plans show the following types of detailed information:

- Coordinates of control points or property corners
- Direction and length of property lines or control lines
- Description, or reference to a description, for all control and property monuments
- Location, dimensions, and elevation (relative to sea level) of the structure on site
- Finish and existing grade contour lines

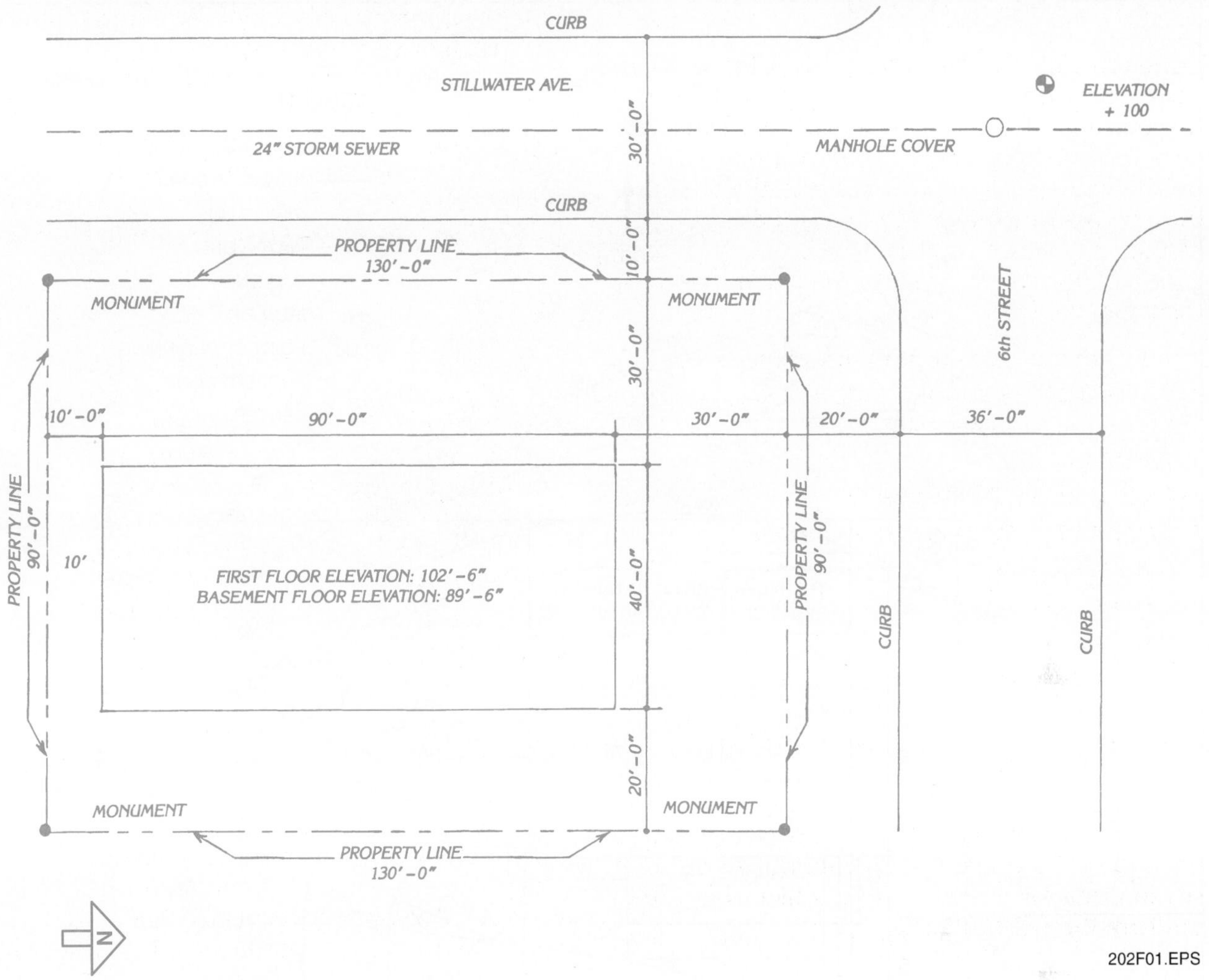

CURB

STILLWATER AVE.

30'-0"

⊕ ELEVATION
+ 100

24" STORM SEWER

○ MANHOLE COVER

CURB

10'-0"

PROPERTY LINE
130'-0"

MONUMENT

MONUMENT

6th STREET

30'-0"

10'-0"

90'-0"

30'-0"

20'-0"

36'-0"

PROPERTY LINE
90'-0"

10'

FIRST FLOOR ELEVATION: 102'-6"
BASEMENT FLOOR ELEVATION: 89'-6"

40'-0"

PROPERTY LINE
90'-0"

CURB

CURB

20'-0"

MONUMENT

MONUMENT

PROPERTY LINE
130'-0"

N

202F01.EPS

Figure 1 ◆ Typical site/plot plan.

- Finished elevations (relative to sea level) of building floors
- Location of utilities
- Location of existing elements such as trees and other structures
- Location and dimensions of roadways, driveways, and sidewalks
- Names of all roads shown on the plan
- Locations and dimensions of any easements

While working with civil plans, keep the following in mind:

- Civil plans are legal documents that must be followed exactly.
- Should changes be needed after the start of construction, they must be officially approved by the owner or owner's representative before the changes are executed.
- Any deviations from the plans must be carefully documented.

In addition to using the civil plan to get information about the building project, you will also be using the plan to get information about what not to do. Follow these rules while you are working:

- Do not cross property boundaries unless you know that the owner has given the managers of the project easement. When easement has been granted, drive on the agreed path to avoid damaging terrain and underground structures such as drainage pipes and culverts.
- Do not operate a vehicle in the location of a benchmark, monument, or control point until you are sure of its location. Damaging these references can cause costly delays.
- Do not work near any utilities unless you are certain of their locations. Hitting an underground gas line or power cable can be fatal.

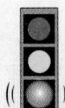 NOTE

References to utilities on drawings are general locations. The utilities must be located in the field.

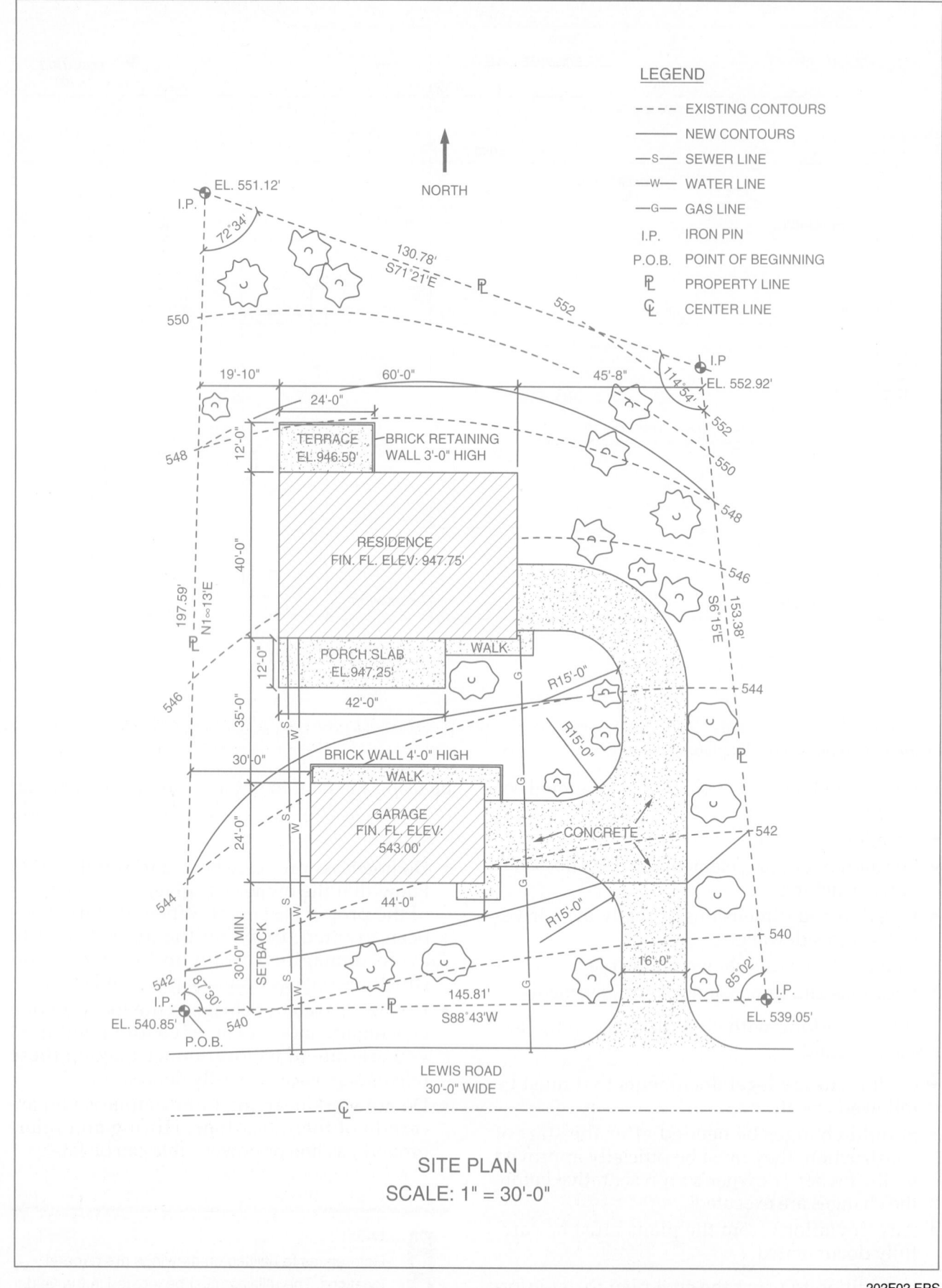

Figure 2 ◆ Typical site/plot plan showing topographical features.

202F02.EPS

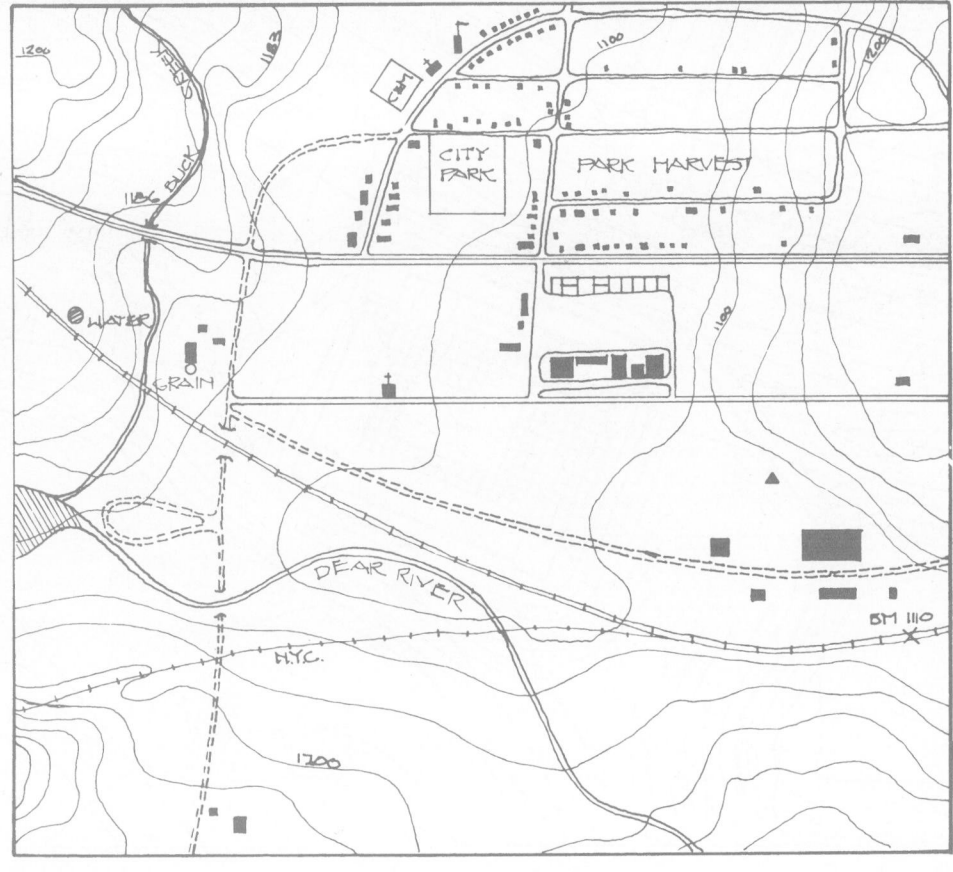

202F03.EPS

Figure 3 ◆ Example of a topographical map.

3.0.0 ◆ CHARACTERISTICS OF CONTOUR LINES

The topography of a job site or other area can be represented by contour lines (*Figures 3* and *4*). Contour lines show changes in the elevation and contour of the land. The lines may be dashed or solid. Generally, dashed lines are used to show the natural or existing grade, and solid lines show the finished grade to be achieved during construction.

Each contour line across the plot of land represents a line of constant elevation relative to some point such as sea level or a local feature. *Figure 5* shows an example of a contour map for a hill. As shown, contour lines are drawn in uniform elevation intervals called contour intervals. Commonly used intervals are 1', 2', 5', and 10'.

On some plans and surveys, every fifth contour line is drawn using a heavier-weight line and is labeled with its elevation to help the user more easily determine the contour. This method of drawing contour lines is called indexing contours. The elevation is marked above the contour line, or the line is interrupted for it.

As shown in *Figure 5*, contour lines form a closed loop within the map. If you start at any

point on the contour and follow its path, you will eventually return to the starting point. A contour may close on a site plan or map or it may be discontinued at any two points at the borders of the plan or map. Examples of this are shown on the topographical survey map in *Figure 3*. Such points mark the ends of the contour on the map, but the contour does not end at these points. The contour is continued on a plan or map of the adjacent land. Some rules for interpreting contours include the following:

- Contour lines do not cross.
- Contour lines crossing a stream point upstream.
- The horizontal distance between contour lines represents the degree of slope. Closely spaced contour lines represent steep ground and widely spaced contour lines represent nearly level ground with a gradual slope. Uniform spacing indicates a uniform slope.
- Contour lines are at right angles to the slope. Therefore, water flow is perpendicular to contour lines.
- Straight contour lines parallel to each other represent man-made features such as terracing.

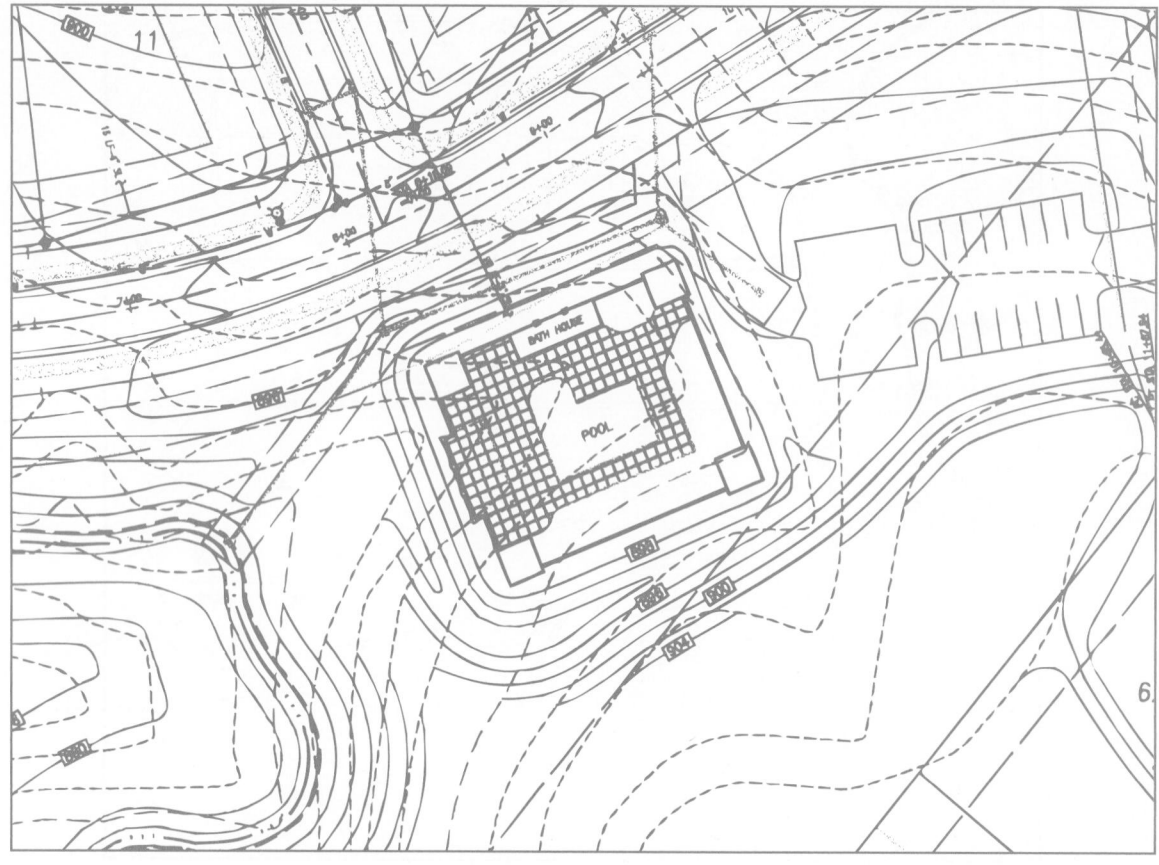

Figure 4 ◆ Example of a topographical survey drawing.

202F04.EPS

CONTOUR MAP OF HILL WITH 5' CONTOUR INTERVAL

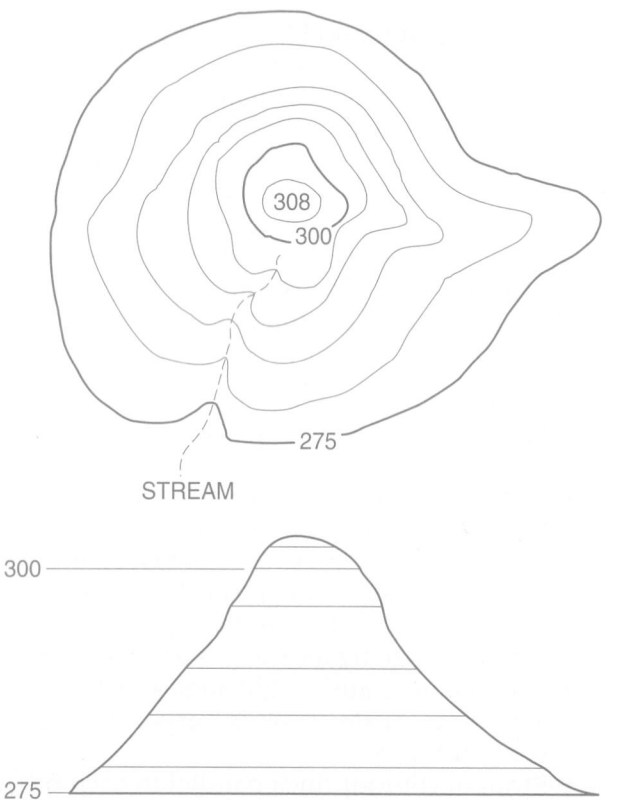

Figure 5 ◆ Contour map of a hill.

202F05.EPS

4.0.0 ◆ SITE LAYOUT CONTROL POINTS

Site plans show the locations of property corners and the direction and length of property lines or control lines. In most states, registered surveyors are required to perform any layout work that establishes legal property lines or boundaries. This is because the surveyor assumes the liability for any mistake in the surveying work. The surveyor is legally responsible if the building ends up on the wrong property or at some location that violates setback requirements or other regulations. Because of the tremendous liability involved, the carpenter should never make any layout measurements that relate to property lines or boundaries. This is a task for the professional land surveyor.

4.1.0 Types of Control Points

Site layout involves establishing a network of control points on a site that serve as a common reference for all construction. The exact locations of these control points are marked at the site and recorded in the **field notes** as they are made. Annotating control point location reference data in the field notes is important for two reasons. First, it makes it possible to locate a point should it become covered up or otherwise hidden. Second, it makes it possible to reestablish a point accurately if the marker is damaged or removed. There are three basic categories of control points:

- *Primary control points* – These points are used as the basis for locating secondary control points and other points on the site. Primary control points are located where they are accessible and protected from damage for the duration of the job. Primary vertical control points can be located and marked on many kinds of permanent and immovable objects such as fire hydrants and power poles. When no suitable permanent objects are available for use as a primary control point marker, iron stakes driven into the ground, a concrete monument dug and poured into the ground (*Figure 6*), etc., can be used. If a poured concrete monument is used, it must be dug a foot deeper than the frost line to prevent freezing and thawing from moving it. It must also have a distinct high point. This can be a rounded brass cap, rebar, etc., that sticks up out of the top of the concrete. Primary control point markers are typically established by a registered surveyor. They are commonly referred to as monuments or benchmarks.

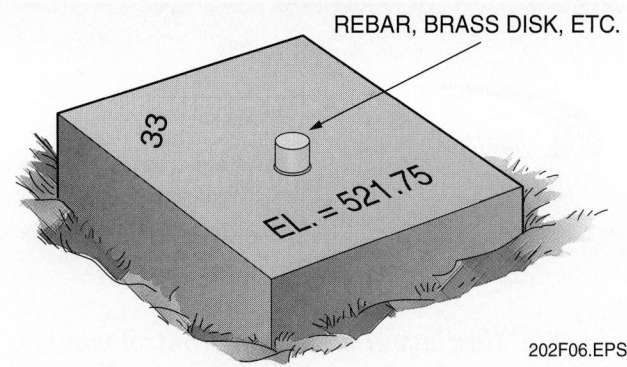

REBAR, BRASS DISK, ETC.

33

EL. = 521.75

202F06.EPS

Figure 6 ◆ Typical control point concrete monument.

- *Secondary control points* – These are additional control points located within the job site to aid in the construction of the individual structures on the site. Secondary control points typically are marked by a hub stake surrounded by protective laths (*Figure 7*), posts, or fencing. The hub stake is typically a 1½" square piece of wood pointed on one end. Its length is normally determined by the hardness of the ground it must be driven into, with lengths between 8" and 12" being typical. The hub stake is driven into the ground until flush or nearly so. A surveyor's tack, with a depression in the center of the head, is driven into the top of the hub stake to locate the exact point.

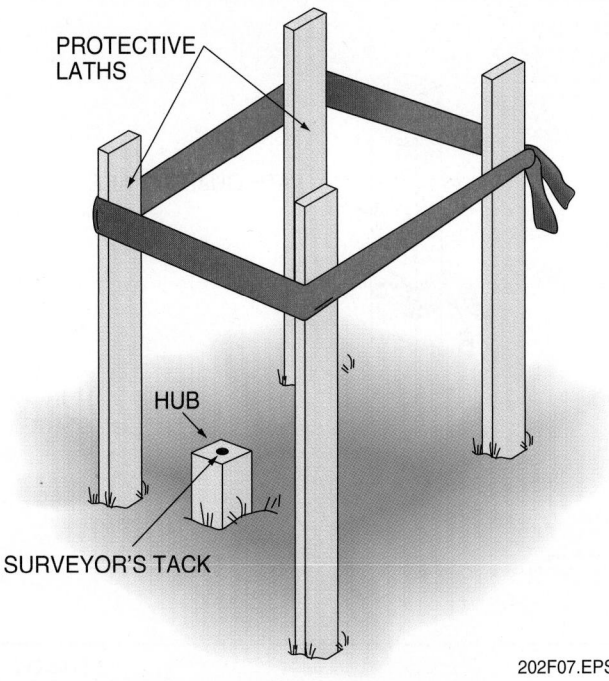

PROTECTIVE LATHS

HUB

SURVEYOR'S TACK

202F07.EPS

Figure 7 ◆ Typical secondary control point marker.

Hubs

Hubs are commonly made by sawing 2 × 4s in half. When necessary to drive wooden hubs into hard-packed ground, the job can be made easier by driving a tempered steel pin called a gad into the ground first to start a pilot hole.

• *Building layout or working control points –* These points are usually located with reference to the secondary control points. These are the points from which actual measurements for construction are taken. Building layout points are used to locate the corners of buildings and building lines. They usually are marked with a hub and a related marker stake (*Figure 8*). The marker stake is typically a ¾" × 1½" piece of wood that varies in length, with 24" to 36" being typical. In addition to serving as hub markers, these stakes are also used to mark line or grade and other information for center lines, offset lines, slope stakes, etc.

4.2.0 Placement of Control Points and Other Markers

The placement of the numerous on-site control point markers depends on the nature of the job, the terrain, other work in progress, the sequence

of work, and many other factors. Accuracy in site layout work requires that benchmarks, control points, and other important markers be referenced in a way that ensures they can be easily located at a later date. These points should be referenced to permanent objects in the surrounding area. Guidelines for establishing such references are summarized here.

• Establish several (three or more) definable permanent or semi-permanent references for each point. Some widely used reference objects include:

  – A bonnet bolt located on a nearby fire hydrant
  – Wooden power poles
  – Sidewalks
  – Trees
  – Fences
  – Building corners
  – Signposts

• To locate a point more accurately, reference it in all directions (north, south, east, west) from the point rather than in the same direction. This is because when arcs are swung from the reference points, there is a very small, distinct area where they intercept.

• If possible, stay within one tape length of the point you are using as a reference.

• Draw a clear, complete sketch in the field notes.

The placement and positioning of stakes at the construction site must be done properly. Some guidelines for performing this task include:

• Face the stakes so that they can be read from the direction of use.

• Offset the stakes as required for their protection.

• Set the stakes within tolerances.

• Place the stakes solidly in the ground.

• Place the stakes and laths so that they are plumb.

• Center the hubs and stakes.

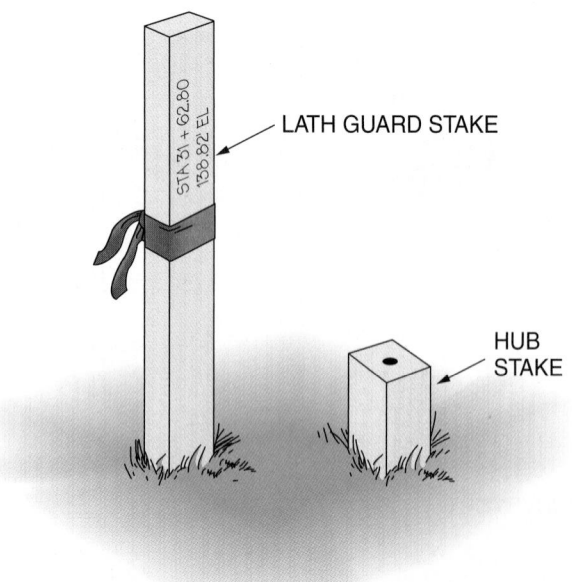

LATH GUARD STAKE

STA 31 + 62.80
136.82 EL

HUB STAKE

202F08.EPS

Figure 8 ◆ Typical working control point marker.

Placement of Control Points

Good practice dictates that a control point should always be placed so that three other control points are visible from it at all times. This is done in case the control point should become covered up or destroyed for some reason. If this were to happen, two tapes can be stretched from the other points in order to relocate the control point.

4.3.0 Communicating Information on Control Markers and Other Markers

Hubs, stakes, and laths must be legibly and accurately marked to correctly communicate location, elevation, and other pertinent construction information. Some guidelines for marking stakes are:

- Use a permanent marker.
- Print neatly. Start from the top of the stake and work toward the bottom. Use all capital letters, being careful not to crowd your words or numbers.
- Avoid the use of abbreviations or use standard abbreviations. The *Appendix* lists some of the common abbreviations used for layout tasks.
- All sides of a stake can be marked with information; however, the main information should always be marked on the direction of use.
- Mark only pertinent information. Too much or too little information can be confusing.
- Mark the following types of information on a stake, as applicable:

  - Alignment information
  - Center line
  - **Cut** or **fill** data
  - Elevation data
  - General information description
  - Grade information
  - Offset
  - Reference information
  - Slope
  - Specific information

In addition to marking stakes, it is often necessary to mark reference lines and points on walls, curbs, foundations, etc. When doing so, the same general guidelines should be followed that were given for marking information on stakes. In addition, any such reference lines must be marked straight, level, and/or plumb. *Figure 9* shows an example of a reference line marked on a wall 4' above a finished floor.

202F09.EPS

Figure 9 ◆ Example of a point marked 4' above the finished floor.

4.4.0 Color-Coding Control Markers

Control points and other markers are identified by color-coding them both to identify their purpose and so that they can be easily seen and recognized. Color-coding can be done by applying paint to monuments, hubs, or other field markers, applying ribbons on stakes, attaching flags on wire markers, etc. Note that color-coding of field markers is not standardized; however, many construction or field engineering organizations have established their own color-coding systems. When performing layout work at a site, you should ask your supervisor what color-coding scheme to use and follow it.

5.0.0 ◆ COMMUNICATING WITH HAND SIGNALS

Two-way radios are commonly used to maintain voice communication between members of a site layout crew. However, because of equipment noises, blasting, non-availability of radios, or other reasons, the use of hand signals may be required. *Figure 10* shows some common hand signals for use when performing distance measurement and differential leveling tasks.

MOVE IN THIS DIRECTION

MOVE UP

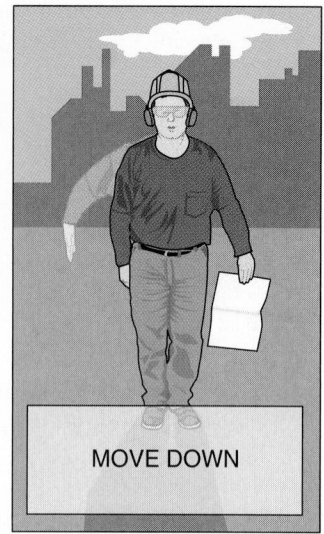

MOVE DOWN

TURNING POINT

WAVE ROD FROM SIDE TO SIDE;
ROCK BACK AND FORTH

OBSERVATION COMPLETED;
MOVE ON; UNDERSTAND

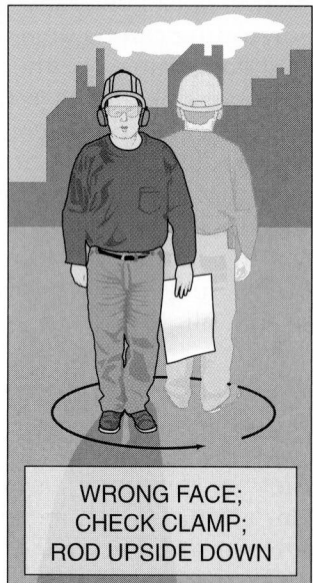

WRONG FACE;
CHECK CLAMP;
ROD UPSIDE DOWN

USE LONG ROD

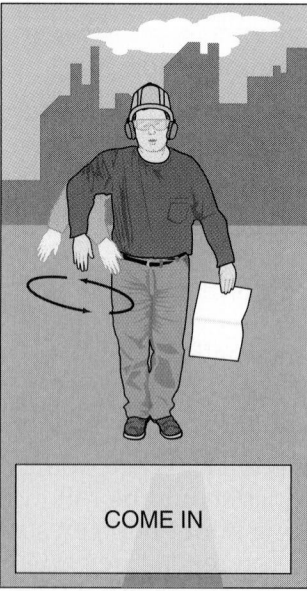

COME IN

202F10.EPS

Figure 10 ◆ Common hand signals.

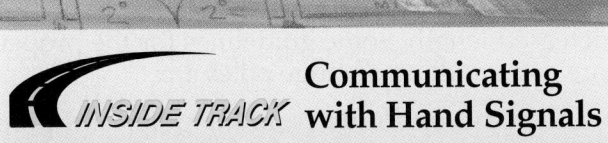
6.0.0 ◆ DISTANCE MEASUREMENT TOOLS AND EQUIPMENT

Site layout involves making horizontal and vertical distance measurements by a process commonly called taping or chaining. The term chaining is derived from past surveying practices when a metal chain was used to measure distances. Today, the terms taping and chaining are used interchangeably in the trade when referring to the measurement of distances using a steel tape or other type of tape. In this module, we will use the term taping when referring to the distance measurement process and the term tape (instead of chain) when referring to the measuring tape. The task of taping involves two people working together and communicating with each other. The major items of equipment (_Figure 11_) needed to perform taping typically include:

- 100' or longer steel tape
- Range poles
- Plumb bobs and gammon reels
- Hand sight levels
- Chaining (taping) pins
- Accessories and stakes

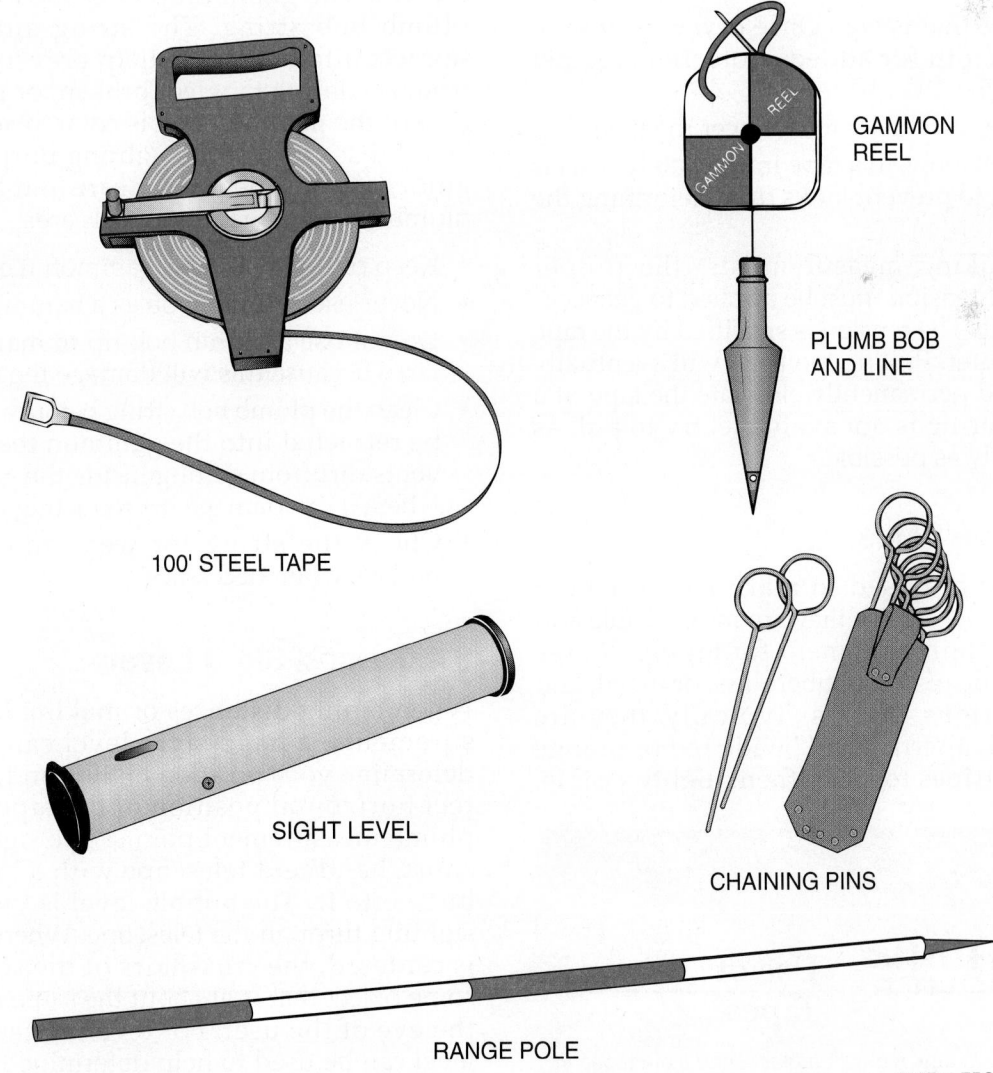

GAMMON REEL

PLUMB BOB AND LINE

100' STEEL TAPE

SIGHT LEVEL

CHAINING PINS

RANGE POLE

202F11.EPS

Figure 11 ◆ Common taping equipment.

6.1.0 Tapes

Tapes can be made of cloth, fiberglass, or steel. However, steel tapes are the most widely used for precision measuring tasks because they stretch less than non-metallic tapes. Steel tapes are made in a variety of graduated lengths, with 100' being common. Typically, the tape is mounted on a reel for ease of handling and storage. Tapes are made with graduations in feet and inches or feet and tenths of a foot. Metric versions are also available. Some steel tapes are nylon-coated to increase their durability. The ends of steel tapes are equipped with heavy loops that provide a place to attach leather thongs or tension handles. These allow the user to tighten or tension the tape firmly. The physical construction of a tape requires that it be handled and used properly.

Some guidelines for the proper care and handling of tapes are:

- Keep the tape on the reel and rolled up when not in use.
- Dry a tape that is wet. Once dry, wipe it with an oiled cloth for added protection against rust.
- Do not allow vehicles to run over the tape.
- Remove all loops in a tape immediately. This is important to prevent kinks from deforming the tape.
- When making measurements, the proper amount of tension must be applied to get accurate results. The tension is specified by the tape manufacturer. Pulling too hard will eventually stretch and permanently elongate the tape. If a tension spring is not available, try to pull as consistently as possible.

6.2.0 Range Poles

Range poles are used to mark measurement points at the site so that they are more visible and to help maintain alignment for taping. Range poles are made of wood, fiberglass, or metal, and come in various lengths. Typically, they are painted with alternating 1" wide red or orange and white stripes to make them highly visible.

Range poles that come in sections are also available, which can be connected together to obtain increased length. Some guidelines for the proper care and handling of range poles are:

- Clean dirt and mud from the poles and pole tips after each use.
- Store poles in a protective case when not in use.
- Repaint poles when necessary.
- Maintain a sharp point. Replace or sharpen the point when necessary.

6.3.0 Plumb Bobs and Gammon Reels

When suspended vertically from a string, a plumb bob is used to create a plumb vertical line that can be used to position yourself or an instrument directly over a reference point. Plumb bobs are usually made of brass and have a replaceable tip. Depending on the model, they can weigh between 8 and 24 ounces, with 16 ounces being typical. The gammon reel is used to store the plumb bob string. The string automatically retracts into the reel to help prevent the string from becoming tangled, broken, or muddy. The case of the gammon reel is colored so that it can be used as a target for sighting purposes. Some guidelines for the proper care and handling of plumb bobs and gammon reels are:

- Keep plumb bobs and gammon reels clean.
- Never use a plumb bob as a hammer.
- Do not use a plumb bob tip to mark hard surfaces because this will damage the tip.
- Clean the plumb bob string before allowing it to be retracted into the gammon reel. This prevents dirt from getting inside the gammon reel where it can damage the retracting mechanism.
- Check the string for wear and knots and replace when necessary.

6.4.0 Hand Sight Levels

When taping distances or making layout measurements, a hand sight level can be used to determine your position on line and/or the correct horizontal position of the tape needed to plumb measurement points. The sight level is a short, handheld telescope with a bubble level built into it. The bubble level is visible when sighting through the telescope. When the bubble is centered, the crosshairs of the scope fall on some object that is at about the same elevation as the eye of the user. For leveling tasks, a hand level can be used to help determine where to set up the leveling instrument so that its line of sight will intercept the leveling rod. This is done by

THINK ABOUT IT Use of Tapes

How would you rank steel, fiberglass, and cloth tapes in terms of their use to make precise measurements?

using a hand level to sight on a benchmark or other object to see if the height is above or below eye level. Some guidelines for the proper care and handling of a sight level are:

- Do not drop the level.
- Wipe the lens with a clean cloth as needed.
- Keep the level in its protective case when it is not being used.
- Check its calibration frequently.

6.5.0 Chaining Pins

Chaining pins are used to mark temporary positions during the taping process. A set consists of 11 metal pins. The tip of each pin is hardened and pointed so that it can easily be implanted into the ground. Chaining pins are usually painted in alternating red and white stripes to make viewing them easier. Chaining pins must be cleaned regularly during use and should be repainted when necessary. When marking a point with a chaining pin, it should be angled into the ground to allow more accurate measurements to the exact point in the ground when using a plumb bob.

6.6.0 Other Equipment

Other types of equipment that may be used in conjunction with the taping process include a tension spring, a thermometer, and clamps. A tension spring is a spring scale device that can be attached to the end of the tape for the purpose of measuring the amount of pull or tension that is applied to the tape. Its use takes the guesswork out of how much tension is being applied to the tape during a measurement. Its repeated use over successive measurements helps yield more accurate measurements by allowing the user to apply the same amount of tension to the tape each time it is used. Clamps are used to grip a tape anywhere along its length to protect both the user's hands from cuts and the tape from bending or breaking. Thermometers are used to take temperature readings of the tape when it is necessary to correct the values measured with a tape to compensate for temperature variations. Carpenters are not responsible for determining tape correction values due to temperature variations.

7.0.0 ◆ MEASURING DISTANCES BY TAPING

To achieve both accuracy and precision when taping, you must have a good understanding of the measurement process and its principles.

7.1.0 Accuracy and Tolerances

Accuracy in measurement is determined by how close the measured distance is to the actual distance. For example, when a crew measures a property boundary line that is known to be 212' as 202', that crew has a poor accuracy. When another crew measures the same boundary line as 212.01', that crew has a high degree of accuracy. In construction, it is important to maintain a high degree of accuracy to avoid costly errors. Errors exist in every measurement that is made because of both human and instrumentation limits, and while errors can never be eliminated, they can be significantly decreased.

Human errors typically result from differences in eyesight, sense of touch and feel, physical strength, and other factors. Human errors can be decreased by ensuring that all members of the carpentry crew approach the task in an organized manner and practice proper measurement procedures. Mistakes can be detected by checking and rechecking your work.

One easy way to decrease errors is to perform the measurements twice, comparing the second set of measurements to the first. A rule of thumb is that for every 100' measured, the difference should be no more than $\frac{1}{100}$ of a foot. When the entire distance has been measured twice, the two sets of measurements are averaged to decrease any errors. This procedure is discussed in detail in the next section.

Instrumentation errors are usually a result of instruments or equipment being damaged, out of calibration, or incapable of measuring within tolerances. These types of errors are compensated for by making sure that you always use well-maintained and calibrated measuring instruments. Other examples of instrumentation errors are a tape's length being different from its marked length because of stretching or temperature variations, or the angle of the tape not being horizontal during the measurement. These types of errors are compensated for by making corrections to the measured values via a mathematical formula.

Tolerances are used in construction layout to define how far off the exact design location something can be and still be acceptable. You must always meet or exceed the specified tolerances. Several factors determine the degree of accuracy required for a specific job or different control points within a job site. There is no one specific standard that defines required tolerances, so the specifications for the specific job must always be consulted. Once the specific tolerances are known, you must select measurement equipment for use that is capable of measuring to these tolerances.

7.2.0 Taping Guidelines

Guidelines that must be followed in order to achieve accuracy when taping are summarized in *Figure 12* and briefly explained below.

- *Use a calibrated tape* – Manufacturers furnish a data sheet with each precision tape they make. This sheet will tell you exactly how accurate the tape is and to which standard it is compared.
- *Know where zero (0) is on the tape* – Determine where the exact 0 point of the tape is and use it. It may be at the end of a loop or other fitting at the end of the tape, or it can be offset from the end of the tape.
- *Maintain good alignment* – Distances must be measured as a straight line. When measuring distances that are longer than the tape, intermediate measurement points must be used. Any such intermediate points should be directly in line between the beginning and end points being measured.
- *Apply correct tape tension* – Apply the correct tension to the tape during measurements. This is necessary to overcome any sag in the tape. For the correct amount of tension to use with a specific tape, use the tension value specified by the manufacturer in the product literature supplied with the tape. As mentioned earlier, a tension spring can be attached to the end of the tape to aid you in applying the right tension.
- *Measure horizontally* – The tape must be read when it is in a horizontal position. On flat ground, this is not much of a problem. On a slope or incline, tape readings can be taken in smaller increments. This allows the downhill person to comfortably hold the tape in a horizontal position while still applying the right tension to the tape. Typically, this puts the end of the tape about chest high.
- *Repeat measurements* – To avoid mistakes that can easily occur when taping, make all measurements at least twice. Reversing the direction of the measurements greatly reduces the chance of repeating a mistake. Greater accuracy can be achieved by averaging the two sets of readings to reduce errors.
- *Make mathematical corrections as necessary* – Mathematical corrections must be made to measurement data to compensate for conditions such as tape length differences resulting from calibration, the expansion or contraction of the tape's length due to temperature variations, the tape positioned on a slope rather than horizontal during a measurement, etc. All of these conditions can affect the accuracy of your measurements. These conditions are described in more detail later in this section.
- *Record readings immediately* – The reading for each measurement should be recorded in the field notes immediately after it is taken in order to avoid omissions that can contribute to errors. Also, make sure to check that you have recorded each entry correctly. It is easy to transpose numbers or misplace a decimal point.
- *Use well-maintained equipment* – Accurate measurements can only be made with well-maintained taping equipment.

7.3.0 Taping a Distance

The task of taping involves two people working together and communicating with each other. The following procedure outlines one method for measuring a distance between two existing points, such as two control monuments. The procedure for measuring a distance between known and unknown points, such as when laying out a

- USE A CALIBRATED TAPE
- KNOW WHERE ZERO IS ON THE TAPE
- MAINTAIN GOOD ALIGNMENT
- APPLY CORRECT TAPE TENSION
- MEASURE HORIZONTALLY
- REPEAT MEASUREMENTS
- MAKE MATHEMATICAL CORRECTIONS AS NECESSARY
- RECORD READINGS IMMEDIATELY
- USE WELL-MAINTAINED EQUIPMENT

202F12.EPS

Figure 12 ◆ Guidelines for distance measurements by taping.

building, would be performed in basically the same manner. In this procedure, the two people involved are designated as the rear tape person and the head tape person (*Figure 13*). For the purpose of explanation, it is assumed that a 100' tape is being used, the overall distance to be measured is greater than 100', and the terrain is relatively flat, allowing for horizontal measurements to be made with the tape on the ground.

Step 1   Determine the straight path for the overall measurement between the start and end points of the line to be measured. This path should be cleared of any brush, rocks, or other obstacles that will hamper making the measurements.

Step 2   Once the measurement path has been determined, the location of the start and end points of the overall measurement path are marked with range poles or laths with flagging.

Step 3   While the rear tape person holds onto the tape reel at the starting point, the head tape person takes the 0' end of the tape and advances along the measurement line to the location of the first intermediate measuring point, presumably at 100' from the beginning point. Some companies prefer that the head tape person advance with the reel while the rear tape person holds onto the 0' end of the tape. Either way is acceptable.

Step 4   The head tape person applies the proper tension to the tape as the rear tape person aligns the 100' mark on the tape exactly on the starting point. Once positioned over the starting point, the rear tape person signals the head tape person, who then marks the position of the 0' point of the tape with a chaining pin or other marker. If a chaining pin is used, the pin should be slanted slightly in a direction away from the tape.

Step 5   Following this, both workers advance along the measurement line in preparation for the next measurement. The starting point for this measurement is at the location of the chaining pin established in Step 4.

Step 6   Steps 3 and 4 are repeated as required until the line has been measured from the original starting point to the end.

Step 7   The overall distance measured is equal to the total number of full tape lengths measured plus the reading of the last measurement. For example, if three 100' tape measurements were made and the last measurement recorded by the rear tape person is 30.25', then the total distance measured is 330.25'.

Step 8   Repeat Steps 1 through 7 and record the second set of measurements.

Step 9   Add the total distances obtained in Steps 7 and 8, then divide the sum by two, using the result as the measurement. For example, in Step 7 the total measured distance was 330.25'. Assume the distance measured in Step 8 was 330.27'. The sum of the two measurements is 660.52' (330.25' + 330.27' = 660.52'), so the average is 330.26' (660.52' ÷ 2 = 330.26').

The general procedure for measuring a distance over terrain with an excessive slope or incline is basically the same as described above.

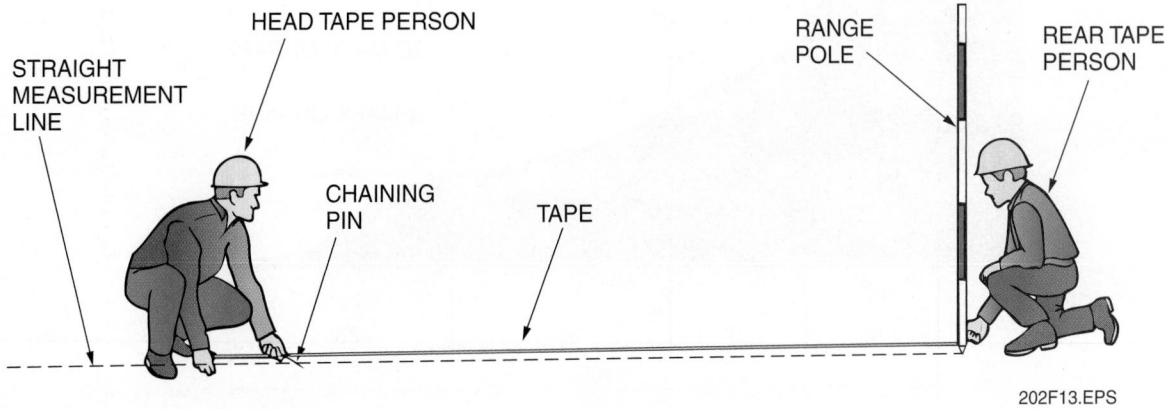

STRAIGHT MEASUREMENT LINE

HEAD TAPE PERSON

CHAINING PIN

TAPE

RANGE POLE

REAR TAPE PERSON

202F13.EPS

Figure 13 ◆ Taping procedure.

However, to aid in maintaining the tape in the horizontal (level) position when making measurements, a method called **breaking the tape** is used. This means making measurements using a portion of the full tape's length in a series of steps until the full tape length has been traversed.

For example, if moving down the slope of the hill, the head tape person advances with the end of the tape along the line for a distance equal to the tape length. Then, leaving the tape on the ground, the head tape person returns as far along the tape as necessary to reach a point that will allow him or her to comfortably hold the tape in the horizontal position and still be able to apply the proper tension during the measurement (*Figure 14*). Typically, this is about chest or waist high.

The head tape person holds a plumb bob string over a convenient whole foot mark, and when the tape is properly tensioned and determined to be horizontal, and both the head tape person and rear tape person agree that a measurement should be taken, the head tape person then marks the location of the whole foot mark onto the ground below.

Because the tape is elevated above ground level during the measurement, this is accomplished by lowering a plumb bob suspended from the whole foot mark on the tape to a point just above the ground. When the plumb bob is stationary, the pressure on the plumb bob string is released, allowing the tip of the plumb bob to come into contact with the ground. The point where the plumb bob tip strikes the ground is then marked with a chaining pin.

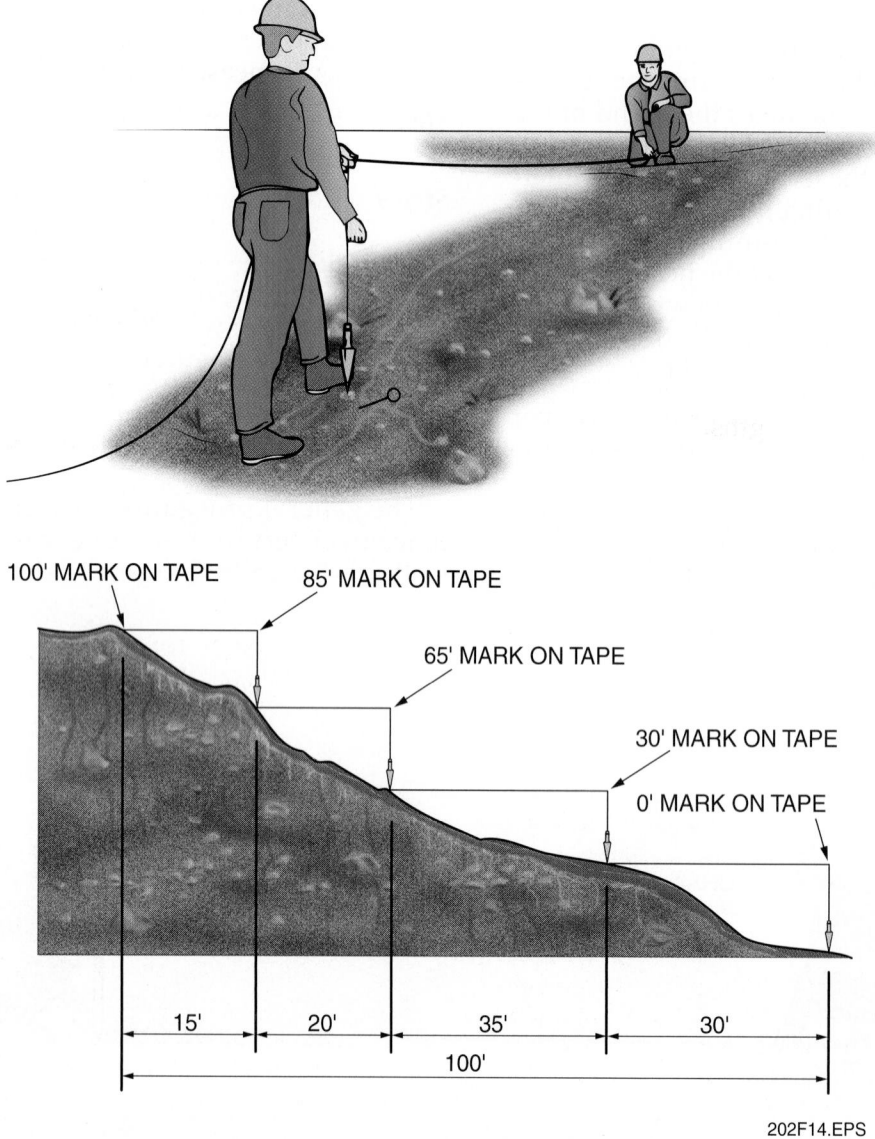

202F14.EPS

Figure 14 ◆ Breaking the tape to measure a distance on a steep slope.

The head tape person continues to hold the intermediate foot mark on the tape until the rear tape person arrives, at which time the head tape person hands the tape to the rear tape person while still maintaining the foot mark position on the tape that he or she has been holding. This procedure is repeated as often as necessary while moving down the hill until the end of the tape is reached and the required total distance is measured or laid out.

When taping, it is easy to make mistakes. However, mistakes can be eliminated or minimized by constantly checking and rechecking your work. Some common mistakes to avoid when taping are:

- Using incorrect measuring tools for the job being performed
- Having too much slack (sag) in the tape during measurement
- Tape not being held horizontal
- Tape twisted or kinked
- Tape not aligned correctly on measurement reference points
- Making measurements using a wrong reference point or points
- Reading tape graduations incorrectly
- Recording incorrect numbers
- Miscounting the number of full tape lengths measured
- Making errors in mathematical computations

It is sometimes desirable to express measured distances in terms of tape lengths. This is done by dividing the distance measured by the length of the tape used to make the measurement. For example, when using a 100' tape, a distance that measured 825.45' would be expressed as 8.2545 tape lengths. If using a 50' tape, the same distance would be expressed as 16.5090 tape lengths.

7.4.0 Converting Between Distance Measurement Systems

Construction project drawings can show dimensions both in feet and inches, in feet and fractions of a foot expressed as a decimal (decimal feet), or both. For example, the dimensions of structures are usually shown in feet and inches. Land measurements, ground elevations, etc., are typically shown in decimal feet. For this reason, it is often desirable (or necessary) to convert between the two measurement systems. Conversion tables are available in many trade-related reference books

that can be used for this purpose. However, you should be familiar with the methods used to make the conversions mathematically in case conversion tables are not readily available.

7.4.1 Converting Feet and Inches to Decimal Feet

To convert values given in feet and inches (and inch-fractions) into equivalent decimal feet values, use the following procedure. For our example, we will convert 45'-4⅜" to decimal feet.

Step 1 Convert the inch-fraction ⅜" to a decimal. This is done by dividing the numerator of the fraction (top number) by the denominator of the fraction (bottom number). For our example, ⅜" = 0.375.

Step 2 Add the 0.375 to 4" to obtain 4.375".

Step 3 Divide 4.375" by 12 to obtain 0.3646' = 0.36' (rounded off).

Step 4 Add 0.36' to 45' to obtain 45.36'.

7.4.2 Converting Decimal Feet to Feet and Inches

To convert values given in decimal feet into equivalent decimal feet and inches, use the following procedure. For our example, we will convert 45.3646' to feet and inches (convert to the nearest ⅛").

Step 1 Subtract 45' from 45.3646' = 0.3646'.

Step 2 Convert 0.3646' to inches by multiplying 0.3646' by 12 = 4.3752".

Step 3 Subtract 4" from 4.3752" = 0.3752".

Step 4 Convert 0.3752" into eighths of an inch by multiplying 0.3752" by 8 = 3.0016 eighths or, when rounded off, ⅜". Therefore, 45.3646' = 45'-4⅜".

NOTE

When making measurement conversions, it is a good practice to use at least four or five decimal places during your calculations, then round off at the end to avoid compounding rounding errors. Also, using an appropriate calculator makes the conversion process easier with less chance of making mistakes.

7.5.0 Making Corrections to Measured Readings

In addition to using a calibrated tape and applying the proper tape tension when taping, it is necessary to make corrections to your readings and to measurements made with a tape in order to compensate for errors resulting from variations in tape length, tape temperature, or both. Failure to make these corrections can greatly affect the accuracy of your work. This is because these errors are repetitive and accumulate in proportion to the number of times the tape is used to measure a distance.

7.5.1 Corrections for Tape Length

The lengths of tapes in tape manufacturers' catalogs and other product literature are listed as being 100', 200', 300', and so on. These lengths are nominal lengths, meaning in name only, not in fact. Tape manufacturers calibrate each precision tape they make against a standard to determine the exact amount of deviation that exists between the actual length of the tape and the standard. The result is that most tapes are actually shorter or longer than their stated nominal length. Manufacturers supply a data sheet with each new tape that records the accuracy of the tape.

In field use, tapes can become stretched, kinked, or broken. Carpenters should replace these tapes rather than repair them.

7.5.2 Other Corrections

Other corrections to measurements that may be necessary to make when taping to increase accuracy can include corrections for tape sag, for tape alignment, and for making measurements when on a slope. Detailed information on the methods for calculating correction factors to compensate for these errors are beyond the scope of this module, but can be found in any of the readily available texts or reference books on surveying or field engineering. This section provides a brief description of measurement errors caused by sag, misalignment, and slope.

Manufacturers calibrate their tapes under a certain amount of tension with the tape supported along its entire length. When the tape is used in the field and is supported at both ends only, the natural sag that results in the middle of the tape has the same effect as if the tape is too long. One way to correct for sag is to increase the tension on the tape. This stretches the tape and reduces the sag to a point where it is inconsequential. This method is usually sufficient for construction layout work. The amount of tension applied to the tape should be within the limits recommended by the tape manufacturer (normal tension). A spring-balanced tension spring can be used for this purpose, but many experienced carpenters rely on feel. When distances shorter than one full tape length are being measured, the tension applied to the tape should be proportionally less than the normal tension. For example, if a distance of 50' is being measured with a 100' tape that has a normal tension of 20 pounds, then the applied tension should be 10 pounds.

Generally, tape alignment is not a problem on a construction site because either all obstacles on a line have been removed or the line has been moved to avoid any obstacles. If it is necessary to tape long distances where trees or other obstacles are directly in the line of the measurements, the tape must be moved offline as needed to avoid the obstacle causing the resulting measured distance to be too long. This requires that a correction factor be calculated and applied to the measured distance.

When making tape measurements on terrain that slopes with the tape resting on the ground rather than being horizontal, the distance measured along the slope will be too long. A slope correction must be made to the measured distance to determine the true horizontal distance.

Errors

Errors resulting from tape length, tape temperature, or both have a cumulative effect. The more measurements you make, the larger the error can be at the completion of your measuring. Good practice is to take your time and repeat a measurement one or more times to make sure you are correct before moving on to the next measurement.

By using trigonometric relationships, the slope correction can be calculated when either the angle of the slope or the difference in elevations between the bottom and top of the slope are known.

8.0.0 ◆ ESTIMATING DISTANCES BY PACING

The ability to pace a distance with reasonable accuracy can be very helpful. Pacing can be used to check measurements that have been made by others or to estimate an unknown distance. You can determine your average pace length by walking a known distance that has been previously measured accurately with a steel tape and dividing that length by the number of paces taken. When pacing the distance, you should walk naturally with a consistent pace length. Some people count each step as a pace. Others only count full strides (two paces) instead of paces when stepping with their right or left foot. It does not matter how you count—just be consistent. Also, for the last pace in the measurement, which is normally less than a full pace, record to the nearest ½ or even ¼ pace, if possible. The procedures for finding your average pace length and determining an unknown distance by pacing are briefly outlined here.

Step 1 Use a tape to lay out a level distance of 100'.

Step 2 Starting at the beginning point, walk naturally to pace the 100' distance. Record the number of paces required to travel the distance.

Step 3 Repeat Step 2 a minimum of four more times and record the number of paces required for each time.

Step 4 Calculate your average number of paces per 100'. For example, assume the total number of paces for the five trips equals 199 paces (40.5 + 39 + 40 + 39.5 + 40 = 199). In this case, the average number of paces per 100' equals 39.8 paces (199 total paces ÷ 5 trips).

Step 5 Calculate your length of pace in feet by dividing the average number of paces into the distance traveled. For our example, the average pace length is 2.51' (100' ÷ 39.8 paces).

Once the average length of your pace is known, other distances can be determined by pacing the distance and calculating its length by multiplying your pace length by the number of paces needed to travel the distance. For example, if it takes 60 paces to travel a distance and your average pace length is 2.51', then the distance is approximately 151' (60 paces × 2.51' = 150.6' = 151' rounded off).

9.0.0 ◆ ELECTRONIC DISTANCE MEASUREMENTS

Electronic distance measurement (EDM) is a widely used technology that provides a fast and extremely accurate method for making long distance measurements, including measurements over obstacles such as lakes, ravines, and roadways. EDM involves the use of an electronic distance measurement instrument (EDMI). There are two classes of EDMIs: electro-optical instruments (*Figure 15*) and microwave instruments. The difference is in the wavelength of the distance measurement signal transmitted by the device. Most site layout work performed by carpenters is done using the electro-optical type; therefore, the remainder of this discussion will focus on this type.

The basic electro-optical measurement system consists of an EDMI and a reflector. The EDMI is set up at one end of the line to be measured and the reflector at the other end (*Figure 16*). The reflector consists of one or more prisms mounted on a tripod or range pole. The number of prisms used is determined by the length of the distance to be measured. The longer the distance, the more prisms used. Depending on the instrument's design, the EDMI transmits either a modulated, visible, low-power laser light signal or an invisible infrared light signal.

Pacing

Can you think of a way to determine the height of a tree by pacing?

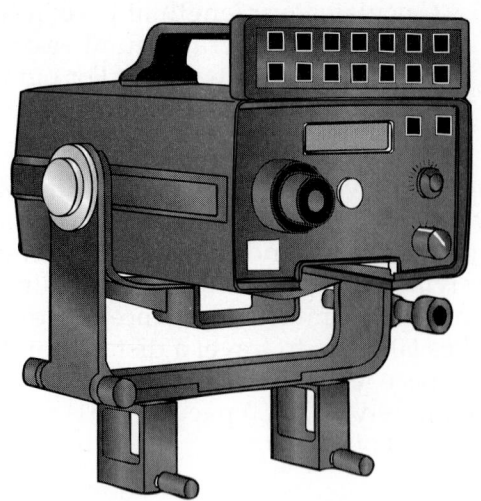

SINGLE TILTING PRISM
ASSEMBLY WITH
ATTACHED TARGET

TRIPLE NON-TILTING
PRISM ASSEMBLY

202F18.EPS

Figure 15 ◆ Typical electronic distance measuring instrument and prisms.

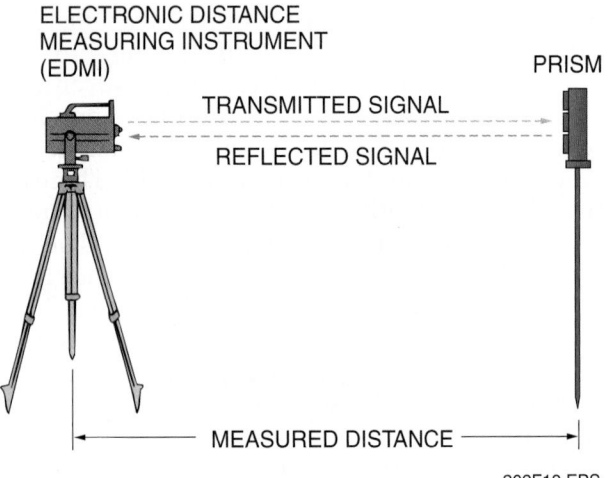

ELECTRONIC DISTANCE
MEASURING INSTRUMENT
(EDMI)

PRISM

TRANSMITTED SIGNAL

REFLECTED SIGNAL

MEASURED DISTANCE

202F19.EPS

Figure 16 ◆ Simplified electronic distance measuring
(EDM) system.

During a distance measurement, this directional signal is aimed at the reflector. When the signal strikes the reflector's prism(s), it is reflected back to the EDMI in the same direction from which it came. However, a phase shift is imparted to this reflected signal relative to the phase of the transmitted signal. Within the EDMI, the time and phase relationships between the transmitted and received reflected signals are compared and the differences are processed electronically to produce the resultant distance measurement value. Many EDMIs can also provide an electronic record of the work done.

Because of the wide variety of EDMIs and prisms that are available, they should be set up, aligned, and operated in accordance with the manufacturer's operating instructions. This procedure may involve entering several pieces of data into the instrument before or during the

measurement. It should be emphasized that the common use of EDMIs to make distance measurements does not eliminate the need to make distance measurements by taping. Because of the time required to set up the EDMI and prisms for a measurement, it is common practice to make shorter distance measurements by taping because it is often quicker and more convenient.

10.0.0 ◆ DIFFERENTIAL LEVELING TOOLS AND EQUIPMENT

Differential leveling is the process used to determine or establish elevations such as those needed for setting slope stakes, grade stakes, footings, anchor bolts, slabs, decks, and sidewalks. This section describes the equipment used to perform differential leveling tasks.

10.1.0 Leveling Instruments

A wide variety of leveling instruments can be used to perform leveling and other on-site layout tasks. The procedural data relating to differential leveling given later in this module emphasizes the use of conventional leveling instruments, such as the builder's level.

10.1.1 Builder's Level

The builder's level (*Figure 17*) is an instrument that can be used to check and establish grades and elevations, and to set up level points over long distances. It consists of a telescope, a bubble spirit level (leveling vial) mounted parallel with the telescope, and a leveling head mounted on a circular base with a horizontal circle scale graduated in degrees. The telescope can be rotated 360 degrees for measuring horizontal angles. A builder's level is mounted on a tripod when used.

Depending on the model, builder's levels are made with telescope powers ranging from 12 power (12×) to 32 power (32×), with 20 power (20×) being the most common. The power of a telescope determines how much closer an object will appear when viewed through the telescope.

There are two types of leveling head systems used in builder's levels: a four-screw system, commonly called a dumpy level, and a three-screw system. The advantage of the three-screw system is that it allows the instrument to be leveled more quickly. Four-screw systems are more common in older models. Note that coverage of the builder's level in this module focuses on its use in determining elevations. Its use for making rough horizontal angular measurements is covered in detail later in your training.

A worker needs a great deal of skill to accurately use a builder's level. Like any other instrument that does not have automatic leveling capabilities, it must be manually leveled and then its level must be checked by measuring the elevation of two benchmarks before it can be used to measure an unknown elevation. This ensures that the builder's level is set up correctly.

10.1.2 Automatic Leveling Instruments

Automatic levels (*Figure 18*) are the most commonly used levels today because they are accurate and easy to use. Automatic levels are used to perform the same measurements and operations as described for the builder's level. These instruments have a built-in compensator mechanism that works to automatically maintain a true level line of sight. Compensator instruments still have to be leveled within the range of the compensator by three screws located on the base. Automatic leveling instruments must be kept upright and should never be carried over your shoulder. They contain a prism that can be damaged if the level is carelessly handled.

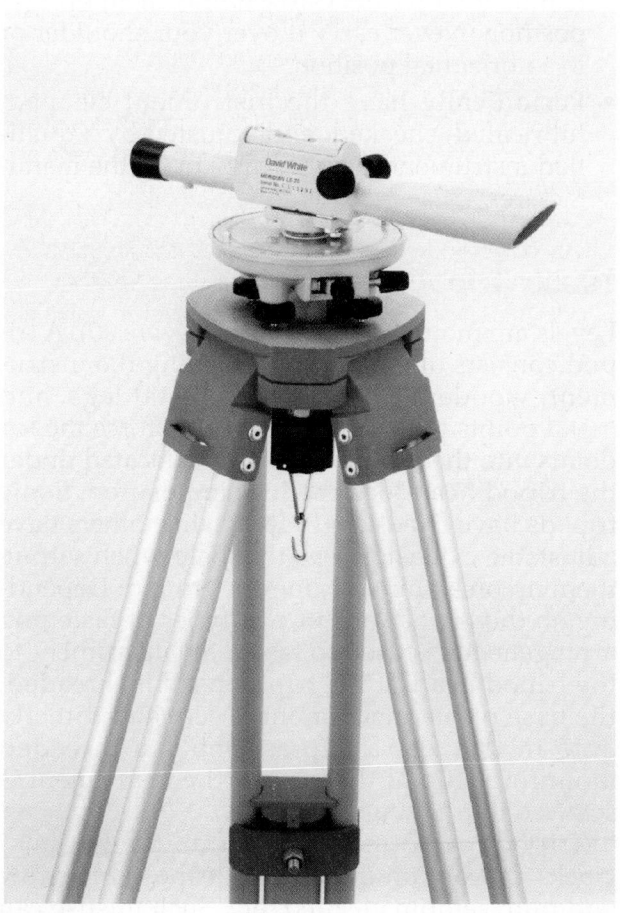

202F20.EPS

Figure 17 ◆ Typical builder's level.

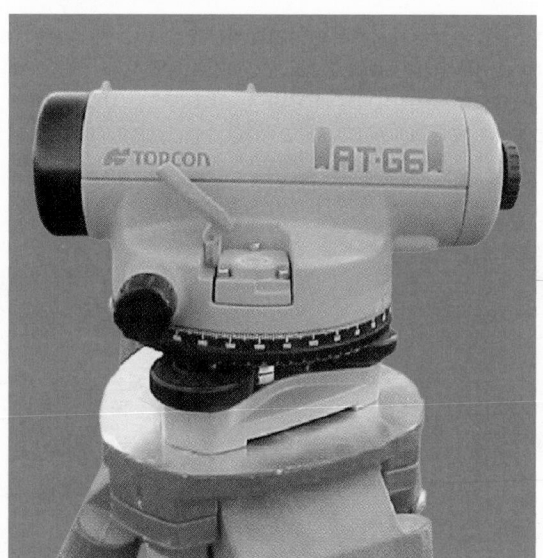

202F22.EPS

Figure 18 ◆ Typical automatic level.

10.1.3 Laser Leveling Instruments

Laser beam levels (*Figure 19*) can be used to perform all of the tasks that can be performed with a conventional leveling instrument. The laser level does not depend on the human eye. Instead, it emits a high-intensity light beam, which is detected by an electronic sensor (target) at distances up to 1,000'. Both fixed and rotating laser models are available. Rotating models enable one person, instead of two, to perform any layout operation. When a rotating laser is operated in the sweep mode, the head rotates through 360 degrees, allowing the laser beam to sweep multiple sensors placed at different locations. The use of laser instruments is covered in detail later in your training.

CAUTION

You must be trained before you can operate a laser beam instrument. Government regulations also require that manufacturers provide a warning in their literature regarding the hazards associated with the use of laser instruments.

10.1.4 Care and Handling of Leveling Instruments

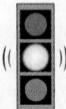

CAUTION

Levels, transits, and other optical instruments are sensitive devices. They provide accurate readings only if in good condition, properly calibrated, and used according to the procedures recommended by the manufacturer.

Leveling instruments should always be maintained and handled in accordance with the manufacturer's instructions. General guidelines for the proper use, care, and handling of leveling instruments are as follows:

202F23.EPS

Figure 19 ◆ Typical laser beam level and detector.

- Only use an instrument if you know how to operate it.
- Keep the instrument in its closed carrying case when not in use.
- Handle the instrument by its base when removing it from the case or attaching it to the tripod.
- Never force any parts of the instrument. All moving parts should turn freely and easily by hand.
- Keep the instrument clean and free of dust and dirt. Clean the objective and eyepiece lenses using a soft brush or lens tissue. Rubbing with a cloth may scratch the lens coating and impair the view. Clean the instrument with a soft, non-abrasive cloth and mild detergent.
- Do not disassemble the instrument.
- Keep the equipment as dry as possible. If it gets wet, dry it before you return it to its case. It may be necessary to leave it out of its case overnight to dry.
- When moving the instrument over a long distance, by foot or by vehicle, remove it from the tripod and place it in its protective case.
- When moving a tripod-mounted instrument, handle it with care. Carry it only in an upright position. Never carry it over your shoulder or in a horizontal position.
- Periodically have the instrument cleaned, lubricated, checked, and adjusted by a qualified instrument repair facility or by the manufacturer.

10.2.0 Tripods

Levels are mounted on a tripod (*Figure 20*). A tripod consists of a head for attaching the instrument, wooden, fiberglass, or metal legs, and metal points with foot pads to help force the leg points into the ground. Wing nuts located under the tripod head lock the legs in position. Some tripods have fixed-length legs, while others have adjustable extension legs that help when setting them up on sloping or uneven ground. Depending on the tripod, one of two types of fastening arrangements is used to fasten the instrument to the tripod head. If the tripod head is threaded, the base of the instrument is screwed directly onto it. If it has a cup assembly, a threaded mounting stud at the base of the instrument is screwed into the cup assembly.

Tripods are often thrown into the backs of trucks, left out on the ground, exposed to snow and rain, seldom cleaned, etc. Such misuse can result in damage to or instability of the tripod that can contribute to measurement errors.

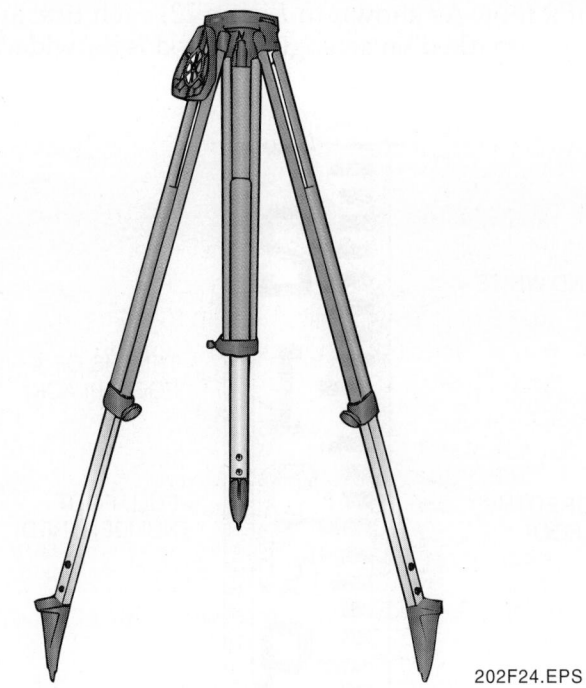

Figure 20 ◆ Tripod.

202F24.EPS

Guidelines for the use and proper care and handling of tripods are:

- When setting up the tripod, position the tripod legs properly. The legs should have about a 3' spread positioned so that the top of the tripod head is horizontal.
- If the tripod's legs are adjustable, make sure that the leg levers are securely tightened.
- If setting up on dirt, make sure that the tripod points are well into the ground. Apply your full weight to each leg to prevent settlement.
- When setting up on a smooth floor or paved surface, secure the points of the legs by attaching chains between the legs or putting a brick or similar object in front of each leg.
- Attach the instrument to the tripod securely. Do not overtighten the attaching hardware.
- Frequently lubricate the joints and adjustable legs of the tripod using an appropriate lubricant.
- When not in use, protect the head of the tripod from damage.
- When transporting a tripod in a vehicle, never pile other materials on top of the tripod. Make sure to protect it from damage that can be caused by shifting equipment or materials.
- Keep the tripod clean and dry.
- When not in use, store the tripod in its protective case.

10.3.0 Leveling Rods

Two people are required when a conventional leveling instrument is used; the first operates the instrument and the second holds a vertical measuring device, called a leveling rod (*Figure 21*), in the area where the grade or elevation is being checked. Leveling rods are made in many sizes, shapes, and colors. They can be made of wood, fiberglass, metal, or a combination of these materials. Leveling rods consist of two or more movable sections, allowing the rod to be adjusted to different lengths. Telescoping rods are also available.

Many styles of leveling rods are given geographic names, such as Philadelphia rods, Chicago rods, San Francisco rods, and Florida rods. The Philadelphia rod is a two-section rod with scales on both the front and back, which can be extended to about 13'. The Chicago and San Francisco rods consist of three sliding sections with the Chicago rod being 12' long and the San Francisco rod available in several lengths. The Florida rod is a 10' long rod graduated with alternating 0.10' wide red and white stripes.

In the United States, the two most commonly used rods are architect's rods and engineer's rods. Metric rods are also available. An architect's

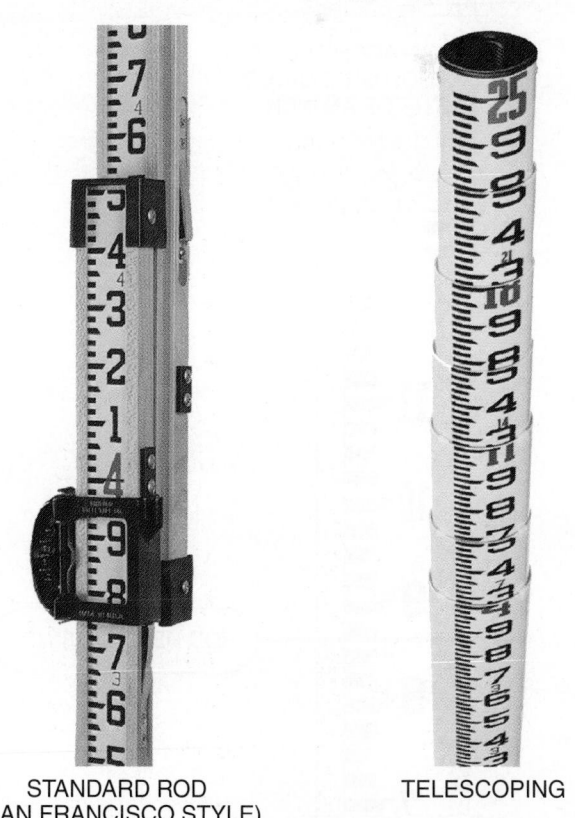

STANDARD ROD
(SAN FRANCISCO STYLE)   TELESCOPING

202F25.EPS

Figure 21 ◆ Leveling rods.

rod is graduated in feet, inches, and eighths of an inch (*Figure 22*). As shown, each line and space on an architect's rod is ⅛" wide. An engineer's rod is marked in feet, tenths of a foot, and hundredths of a foot. As shown in *Figure 22*, each line and space marked on an engineer's rod is ¹⁄₁₀₀' wide.

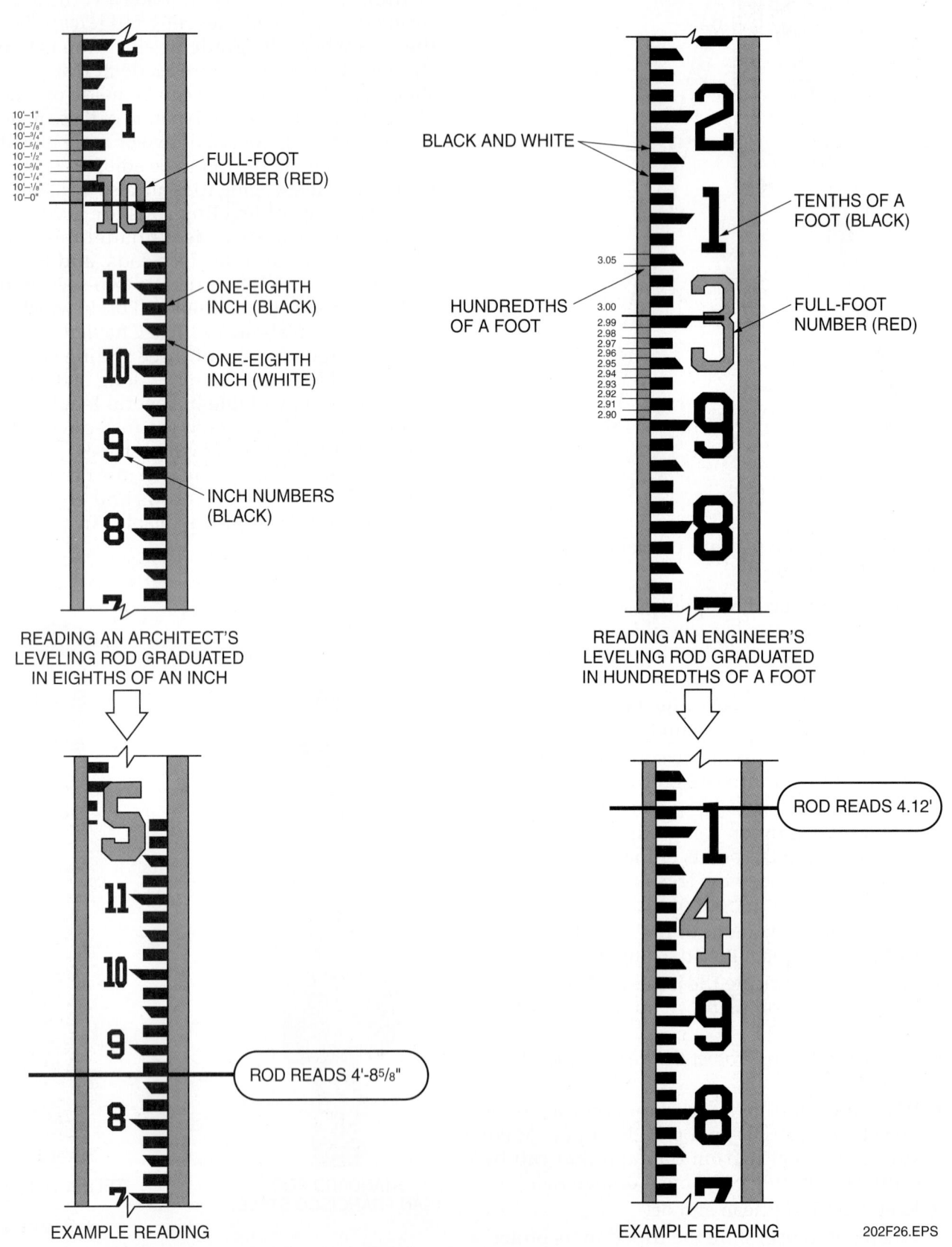

Figure 22 ◆ Reading a leveling rod.

THINK ABOUT IT — Leveling Rods

What is a quick way to tell if a leveling rod is a feet and inches rod (architect's rod) or decimal-foot rod (engineer's rod)?

There are several accessories used with leveling rods. A movable red and white metal disk called a target (*Figure 23*) is used to help make more precise rod readings. The target's vernier scale is set parallel to and beside the primary scale of the leveling rod. Its use enables readings to the nearest sixty-fourth of a foot (architect's rod) or nearest thousandth of a foot (engineer's rod). The target is moved up or down on the rod until the 0 on the vernier scale is lined up with the crosshairs of the leveling instrument. The target is then clamped in place. To read the vernier scale, count the number of vernier divisions up from the 0 (index mark) until one of the vernier divisions lines up exactly with a division on the rod scale itself. This number is added to the last division on the rod, just below the vernier's index mark.

INSIDE TRACK — Engineer's Leveling Rod

Some people who have difficulty reading an engineer's leveling rod find that thinking in terms of money helps them. For example, the red foot numbers can be thought of as dollars; the black tenths numbers as 10 cents, 20 cents, 30 cents, and so on; and each space or black line width as one cent. The point on the longer black line midway between the black numbers is 5 cents.

A bull's-eye rod level is normally attached to a leveling rod for use in keeping the rod plumb for sighting and while the reading is being taken.

NOTE

When sighting short distances, carpenters often use an ordinary measuring tape or wood rule held against a board instead of a leveling rod. Some also use an unmarked wood rod to establish grades. A line of sight is marked on the wood rod as it is held over an established point, then the rod is moved to other locations to establish the grade.

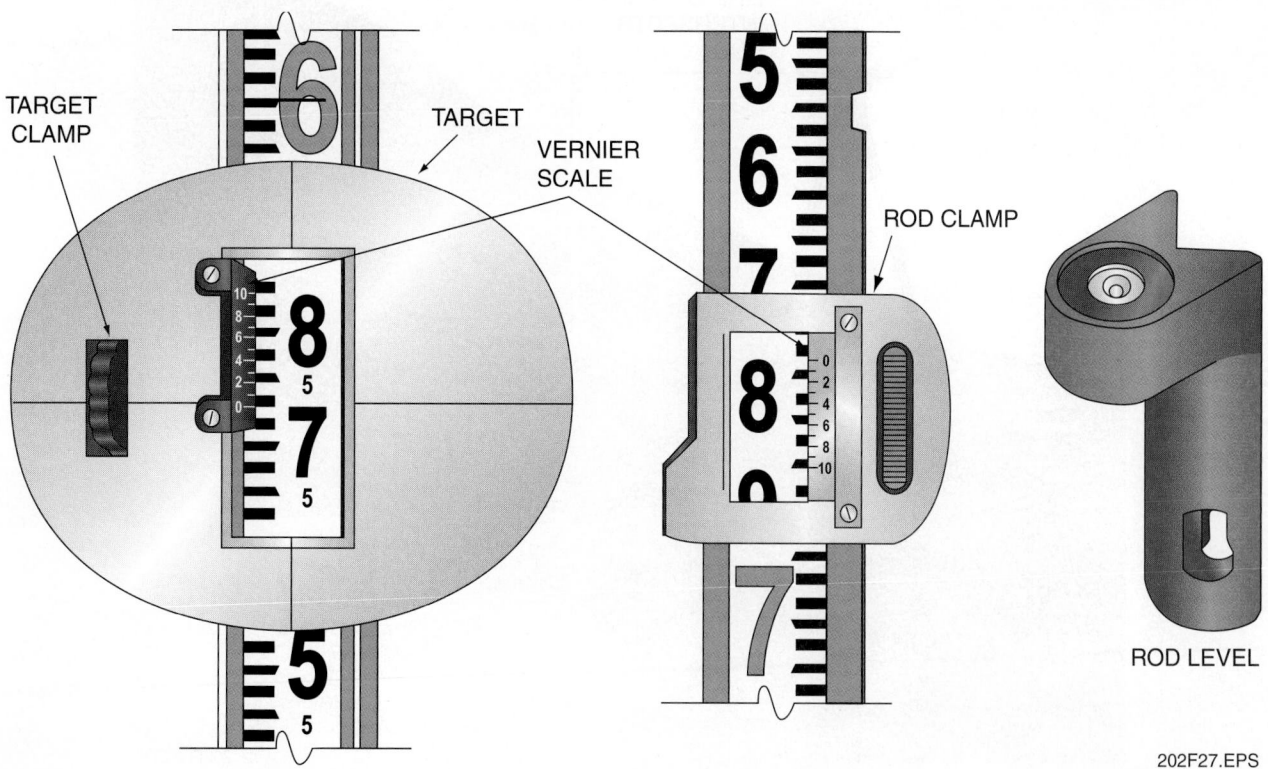

Figure 23 ◆ Rod accessories.

202F27.EPS

Leveling rods are made to withstand the severity of everyday use, but they must be handled, stored, and used properly to avoid unnecessary damage. A damaged rod can contribute to errors.

Guidelines for the proper care and handling of leveling rods are as follows:

- Clean the face, joints, and bottom of the rod frequently during use.
- Avoid touching the face of the rod. Over time, this can cause the numbers and markings to be worn off.
- Make sure all the rod hardware is securely fastened.
- When using a telescoping-type leveling rod, make sure that it is fully extended. Failure to extend a rod fully will result in major errors.
- Never throw a leveling rod into the back of a truck or leave it sticking out from a truck. When not in use, store it in its protective case.

10.4.0 Initial Setup and Adjustment of a Leveling Instrument

The initial setup of a leveling instrument such as a builder's level or automatic level is completed as follows:

Step 1 Select a location to set up the instrument so that its horizontal line of sight will be at a correct height to intercept the level rod, as shown in *Figure 24*.

Step 2 Set up a tripod, making sure to spread its legs wide enough (at least 3' between the legs) to provide a firm foundation for the instrument. Push the legs firmly into the ground and fasten them securely. If setting up on sloping ground, make sure to place one leg of the tripod into the slope. Also, make sure the head of the tripod is horizontal. If the tripod head is too far out of level, there is little chance of correctly leveling the instrument on top of it.

Step 3 Carefully remove the leveling instrument from its case and loosen its horizontal clamp screw (*Figure 25*).

NOTE
When using an automatic level, skip Steps 4 and 5.

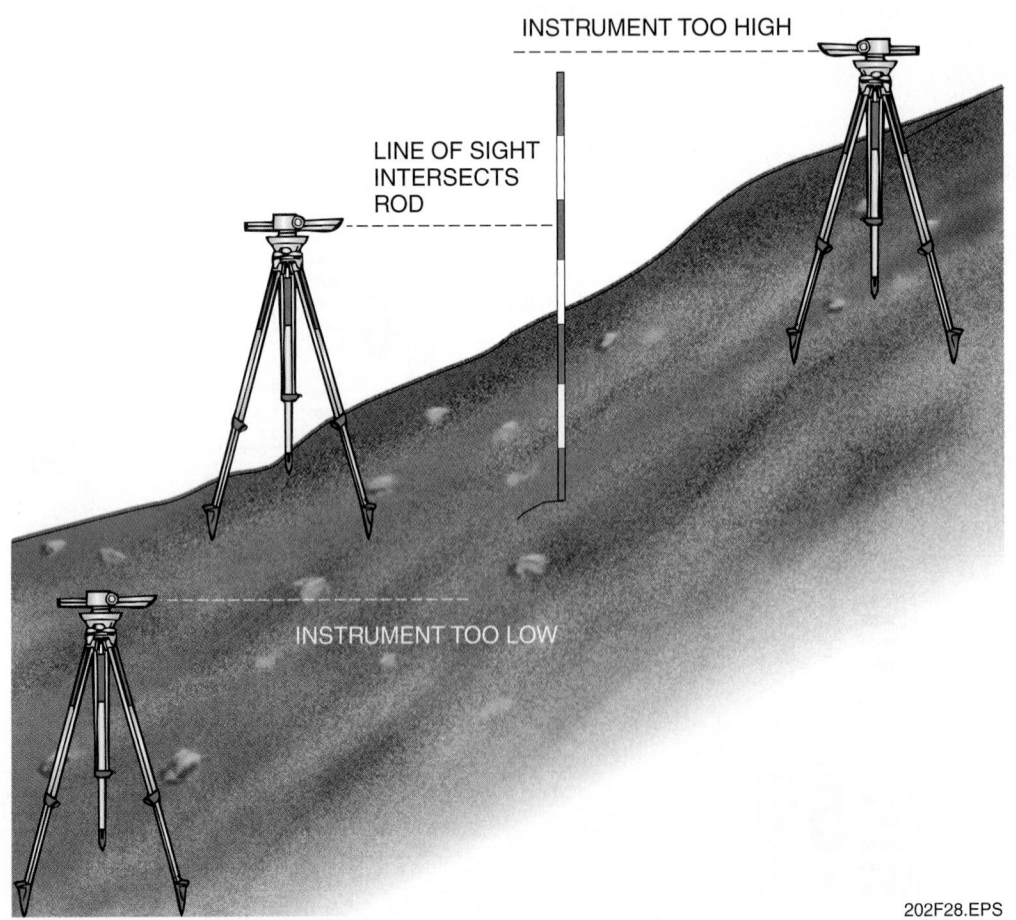

INSTRUMENT TOO HIGH

LINE OF SIGHT INTERSECTS ROD

INSTRUMENT TOO LOW

202F28.EPS

Figure 24 ◆ Set up the instrument so that the line of sight is at the correct level.

Removing a Leveling Instrument from Its Case

When removing a leveling instrument from its case, pay close attention to how it sits in the case. The more moving parts an instrument has, the harder it will be to fit it back into the case if all the parts are not aligned properly.

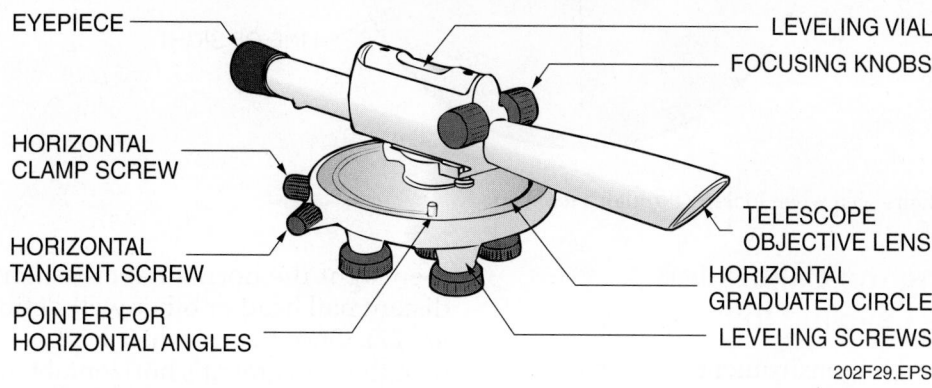

EYEPIECE

LEVELING VIAL
FOCUSING KNOBS

HORIZONTAL CLAMP SCREW

HORIZONTAL TANGENT SCREW

POINTER FOR HORIZONTAL ANGLES

TELESCOPE OBJECTIVE LENS

HORIZONTAL GRADUATED CIRCLE

LEVELING SCREWS

202F29.EPS

Figure 25 ◆ Typical builder's level operator controls.

Step 4   Level the instrument according to the manufacturer's instructions. Avoid over-tightening the leveling screws.

Step 5   Sight along the top of the telescope tube to aim the telescope in the direction of a distant leveling rod or other target, then look through the telescope and adjust the focus. When the crosshairs (*Figure 26*) are positioned on or near the target, tighten the horizontal clamp screw and make the final settings with the horizontal tangent screw to bring the crosshairs exactly on point. Focus the crosshairs by turning the eyepiece one way or another until the crosshairs are as dark and crisp as they can possibly be. Then, adjust the telescope's focusing knob until the graduations on the rod are legible, sharp, and crisp. Keep both eyes open. This eliminates squinting, does not tire the eyes, and gives the best view through the telescope. Failure to focus the telescope crosshairs properly will cause **parallax**.

NOTE

When turning two leveling screws simultaneously (as is required when leveling four-screw and three-screw instruments), always rotate them in opposite directions (turn one in a counterclockwise direction and the other in a clockwise direction when viewed from above) and turn them at the same rate. When rotating the two leveling screws, the spirit level bubble will always follow the direction of the left-hand thumb. That is, if the left thumb is turning the leveling screw in a counterclockwise direction, the bubble will move towards the right; if turning it in a clockwise direction, the bubble will move towards the left. Note that the left-thumb rule also applies if the left hand is used to adjust a single leveling screw, such as is necessary when leveling a three-screw system.

NOTE

Parallax occurs when there is an apparent movement of the crosshairs on the rod or object being viewed as the eye moves. If this occurs when reading a level rod, major errors can occur. You can easily check for parallax by looking at the rod or object being viewed and moving your head slightly while looking at the crosshairs. If they stay on the same spot, no parallax exists.

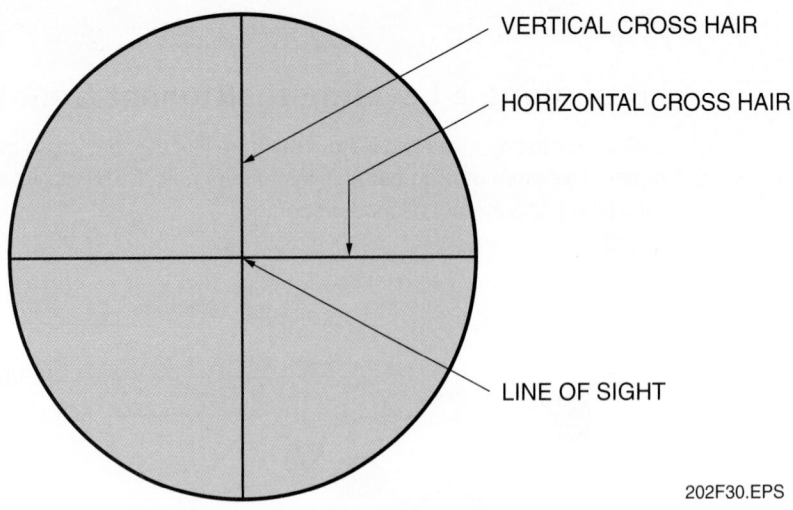

VERTICAL CROSS HAIR

HORIZONTAL CROSS HAIR

LINE OF SIGHT

202F30.EPS

Figure 26 ◆ Crosshairs seen when looking through a telescope.

10.5.0 Testing the Calibration of the Leveling Instrument

Field testing a leveling instrument for correct calibration and adjustment should be done when using an instrument for the first time or if the instrument is suspected of being out of adjustment. Two tests are recommended: a horizontal crosshair test and a line-of-sight test. Each test should be repeated several times to make sure of your results.

10.5.1 Horizontal Crosshair Test

The object of the horizontal crosshair test is to ensure that the instrument's horizontal crosshair is in a plane that is perpendicular to the vertical axis of the instrument. With a properly adjusted instrument, you should be able to place any part of the horizontal crosshair on the object or point being viewed with the telescope and still get an accurate reading. The horizontal crosshair test is simple to perform. First, level the instrument,

then sight the horizontal crosshair reticule on a distant nail head or other well-defined point (*Figure 27*). Once the crosshair is placed on the point, turn the instrument's horizontal tangent screw so that the instrument slowly rotates about its vertical axis. The crosshair should stay fixed on the point as the instrument is rotated. If any part of the crosshair moves above or below the reference point, the instrument needs adjustment and should be returned to a repair facility.

10.5.2 Line-of-Sight Check (Levels and Transits)

The line-of-sight check, commonly called a **peg test**, determines if the instrument's telescope line of sight is horizontal. This means that the line of sight is parallel to the barrel of the telescope and the axis of the telescope bubble tube. The line of sight is checked as follows:

Step 1 In a fairly level and clear area, place two stakes at a distance of about 200' apart.

REFERENCE POINT

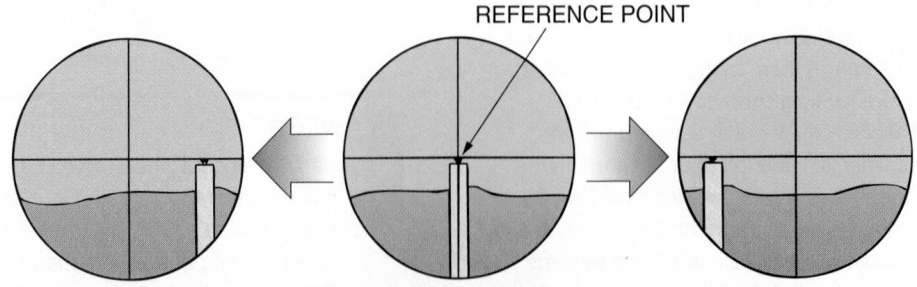

IF THE HORIZONTAL CROSS HAIR MOVES OFF THE
REFERENCE POINT, THE RETICULE NEEDS ADJUSTMENT

202F31.EPS

Figure 27 ◆ Horizontal crosshair test.

Step 2 Set up and level the instrument at a point exactly midway between the two stakes (*Figure 28A*). While working with a rod person, take several elevation rod readings at both stake locations (points A and B). Average the set of **backsight (BS)** readings and set of **foresight (FS)** readings, then subtract the averages to determine the actual difference in elevation between the two stake points. Record this difference.

> **NOTE**
>
> With the instrument placed midway between the two points, any error that may be caused by a line-of-sight problem will result in an identical amount of error in both the rod readings. Because the errors are identical, the calculated difference in elevation between the two points is the true difference in elevation.

Step 3 Move and set up the instrument as close as possible (within a foot) of the stake at point A (*Figure 28B*).

Step 4 While the rod person holds a rod plumb on stake A, sight backwards through the objective lens of the telescope at a pencil point that is being held and moved slowly up and down the rod by the rod person. When the pencil point is exactly centered in your view, read the rod and record the backsight elevation value.

Step 5 Rotate the telescope and take a rod reading on stake B 200 feet away in a normal manner. Record the elevation as a foresight and subtract from the elevation obtained in Step 4. Do this several times and record the average as the difference in elevation readings.

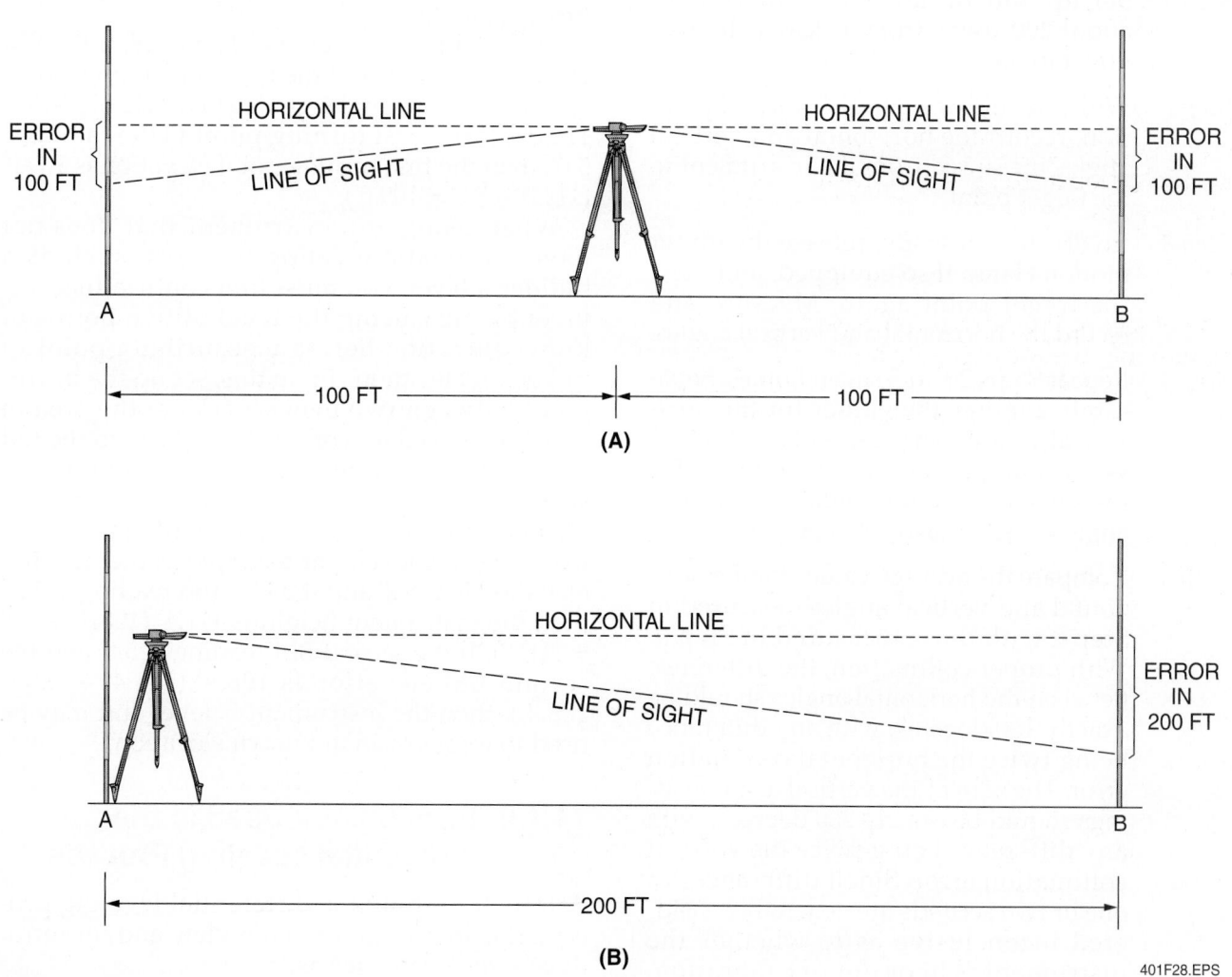

Figure 28 ◆ Peg test.

401F28.EPS

Step 6 Compare the difference in the elevation readings obtained in Steps 1 and 5. Ideally, the difference in elevations should be the same. This indicates no adjustment is required. Any difference greater than two or three thousandths of a foot should be considered significant enough to require adjustment of the instrument. This is done by adjusting the instrument telescope level vial per the manufacturer's instructions.

10.5.3 Line-of-Sight Check (Theodolites and Total Stations)

Horizontal and vertical collimation checks make sure the line of sight of the theodolite or total **station** telescope, as represented by the vertical and horizontal crosshairs and the barrel of the telescope, are aligned. These checks can be done in several different ways. One simple way is described here:

Step 1 Set up and initialize the instrument about 200' away from a clearly defined target point.

Step 2 With the telescope in the normal position, record the horizontal and vertical angles measured from the instrument to the target point.

Step 3 Invert the telescope, release the upper motion clamp if so equipped, and view the target point again. Measure and record the horizontal and vertical angles.

Step 4 Repeat Steps 2 and 3 several times. Separately average the values for the horizontal angles and the values for the vertical angles measured in Step 2. Do the same for the horizontal and vertical angle values measured in Step 3.

Step 5 Compare the average values for the horizontal and vertical angles measured in Step 2 with those measured in Step 3. With proper collimation, the difference between the horizontal angles should be exactly 180 degrees, with any difference being twice the horizontal collimation error. The sum of the vertical angle readings should be exactly 360 degrees, with any difference being twice the vertical collimation error. Small differences of one or two seconds are generally considered inconclusive as to whether the instrument is in or out of calibration.

Larger differences indicate that the instrument should be calibrated before using it. The definition or magnitude of larger differences depends on the instrument and the type of measurements for which it will be used.

11.0.0 ◆ BASICS OF DIFFERENTIAL LEVELING

The process of differential leveling is based on the measurement of vertical distances from a level line. Elevations are transferred from one point to another by using a leveling instrument with auto-leveling features to first read a rod held vertically on a point of known elevation, then to read a rod held on a point of unknown elevation (*Figure 29*). Following this, the unknown elevation is calculated by adding or subtracting the readings. To determine elevations between two or more widely separated points or points on a sloping terrain, several repetitions of the same basic differential leveling process are performed.

For example, as shown in *Figure 29*, if the BM elevation is 112.8' and the first rod reading is 1.2', then the instrument height is 114.0' (112.8' + 1.2' = 114.0'). If the first **turning point (TP)** reading is 5.0', then the first turning point elevation is 109.0' (114.0' − 5.0' = 109.0').

When using any instrument that does not have automatic leveling features, such as a builder's level, you must first confirm that it is level by measuring the level of two points of know elevation before measuring a point of unknown elevation. To do this, set up the instrument between two benchmarks in the area in which you need to work (*Figure 30*). Read the rod set over the first benchmark and then do the same over the second benchmark. Perform the same calculation as above to confirm that the instrument is level. For example, if the first BM elevation is 108.2' and the first rod reading is 3.2', then the instrument height is 111.4' (108.2' + 3.2' = 111.4'). If the second BM reading is 5.0' and the second BM elevation is 106.4' (111.4' − 5.0' = 106.4'), then the instrument is level and may be used to measure an unknown elevation.

11.1.0 Terminology Used in the Differential Leveling Process

Before describing the differential leveling process, it is important to first review and/or introduce some related terms.

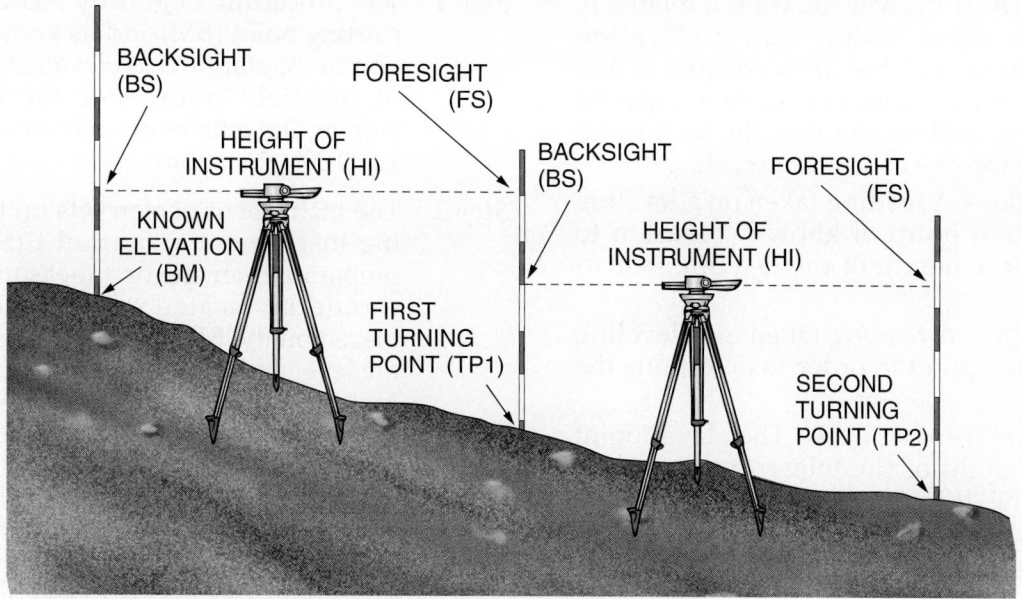

KNOWN ELEVATION (BM) + BACKSIGHT (BS) = HEIGHT OF INSTRUMENT (HI)
HEIGHT OF INSTRUMENT (HI) – FORESIGHT (FS) = TURNING POINT (TP) ELEVATION

202F32.EPS

Figure 29 ◆ Differential leveling relationships.

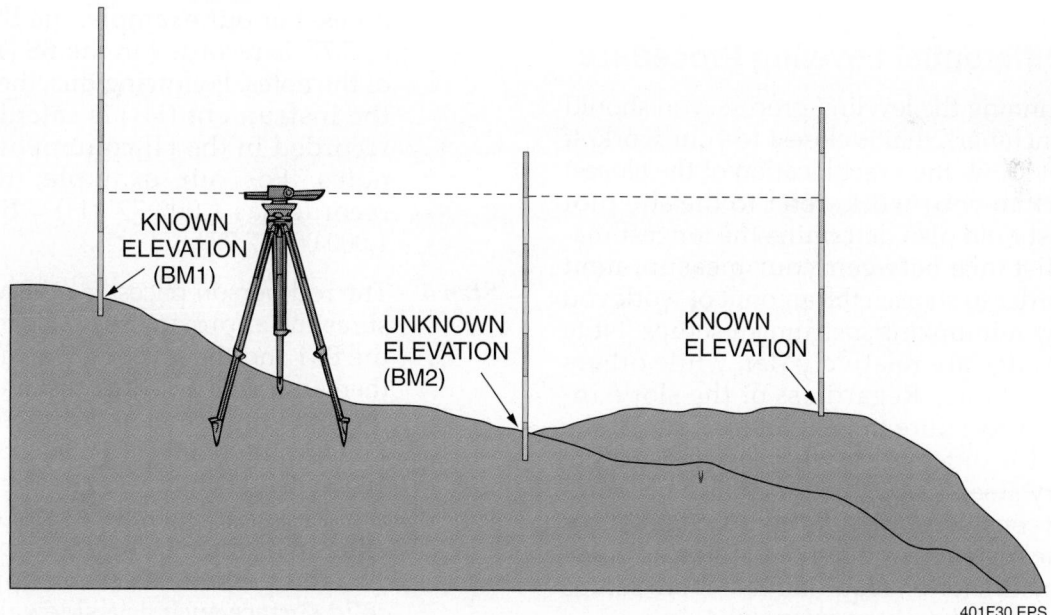

401F30.EPS

Figure 30 ◆ Measuring an unknown elevation.

- *Elevation* – Elevation is the vertical distance above a datum point. For leveling purposes, a datum is normally based on the ocean's mean sea level (MSL). At numerous locations throughout the United States, the government has installed monuments marked with known elevations referenced to MSL. When readily available, such a monument would be used as the elevation reference. When no monument is readily available, such as at most construction sites, a benchmark must be established by a surveyor.

- *Benchmark (BM)* – A benchmark is a relatively permanent object with a known elevation located near or on a site. It can be iron stakes driven into the ground, a concrete monument with a brass disk in the middle, a chiseled mark at the top of a concrete curb, etc.
- *Backsight (BS)* – A reading taken on a leveling rod held on a point of known elevation to determine the height of the leveling instrument.
- *Foresight (FS)* – A reading taken on a leveling rod held on a point in order to determine the elevation.
- *Height of instrument (HI)* – The elevation of the line of sight of the telescope above the datum plane. It is determined by adding the backsight elevation to the known elevation.
- *Turning point (TP)* – A temporary point whose elevation is determined by differential leveling. The turning point elevation is determined by subtracting the foresight reading from the height of the instrument elevation.
- *Closed loop* – Making a traverse consisting of a series of differential measurements that return to the point from which they began.

11.2.0 Differential Leveling Procedure

Before beginning the leveling process, you should select a benchmark that is closest to your work. If you do not know the exact location of the closest benchmark to your work, refer to the site plot plan. You should also determine the longest reasonable distance between your measurement points in order to shorten the amount of work you must do by minimizing instrument setups. Note that some sites are relatively flat, while others have steep slopes. Regardless of the slope involved, the procedure for leveling is done in the same way. The difference is when leveling at a site with a very steep slope, the procedure becomes more time consuming. This is because the line of sight of the instrument relative to intercepting a leveling rod is shorter, requiring that more setups be used to cover the distance involved.

The differential leveling procedure generally involves two people working together and communicating with each other. One person is designated as the rod person and the other the instrument person. Depending on the complexity of the task, recording of the collected measurement data in the field notes may be done by either person, both, or sometimes by a third person. An example of a typical differential elevation procedure is described next, and its path (traverse) is shown in *Figure 31*.

Step 1   The procedure begins by recording the starting point (BM) and its known elevation in the station and elevation columns of the field notes. For the example shown, the entries are BM (station) and 1,000.00' (elevation).

Step 2   The instrument person sets up the leveling instrument at Station 1 (STA 1) in preparation for the first measurement. It should be located so that a level rod placed on the BM is in the line of sight of the level and the rod can be clearly read. Note that this same point should also allow the line of sight of the level to intercept a level rod held on the proposed location of the first turning point (TP1). Set this point equally distant between the two points and no farther away than 150' to 200' from either point of measurement. This reduces the possibility of error if the instrument is out of calibration.

Step 3   While the rod person holds the level rod plumb on the BM, a backsight rod reading is taken, then recorded in the field notes. For our example, the BS reading of 7.77' is recorded in the BS (1) column of the notes. Following this, the height of the instrument (HI) is calculated and recorded in the HI column of the field notes. For our example, the HI is recorded as 1,007.77' (HI = BM + BS = 1,000.00 + 7.77').

Step 4   The rod person paces or otherwise measures the approximate distance between the BM and the leveling instrument and then advances an equal distance beyond the level in the desired direction of the first turning point (TP1). This point must be located such that when the level rod is placed on it, the line of sight of the leveling instrument will intercept the rod. The rod person selects an appropriate solid surface such as a sidewalk or large rock for the turning point. Note that an unmarked point on grass or soil should never be used as a turning point. If no natural solid object is available, a metal turning pin, railroad spike, or wooden stake driven in the ground can serve as a turning point. When a turning point on a solid surface such as a sidewalk or pavement is used, the point should be marked and identified by the turning point number.

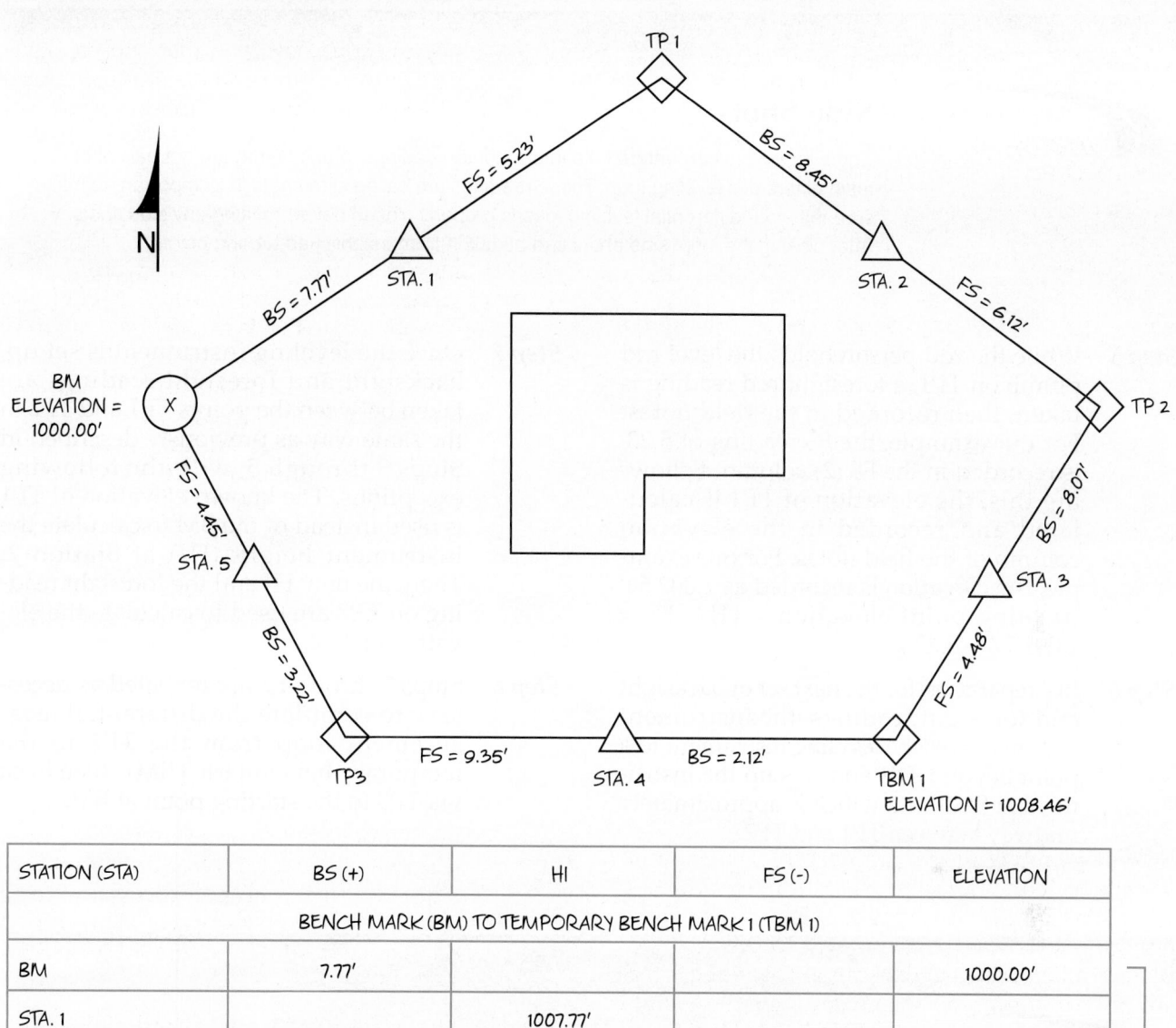

STATION (STA)	BS (+)	HI	FS (–)	ELEVATION
BENCH MARK (BM) TO TEMPORARY BENCH MARK 1 (TBM 1)				
BM	7.77'			1000.00'
STA. 1		1007.77'		
TP 1	8.45'		5.23'	1002.54'
STA. 2		1010.99'		
TP 2	8.07'		6.12'	1004.87'
STA. 3		1012.94'		
TBM 1	2.12'		4.48'	1008.46'
STA. 4		1010.58'		
TP 3	3.22'		9.35'	1001.23'
STA. 5		1004.45'		
BM			4.45'	1000.00'
Σ CHECK	29.63'		29.63'	

DIFFERENCE = 0.00'

DIFFERENCE = 0.00'

202F33.EPS

Figure 31 ◆ Differential leveling traverse and related field notes data.

Side Shots

Some surveyors/carpenters take intermediate readings to points that are not part of the main differential leveling loop. These readings are called side shots. It is important to make sure that your differential leveling loop is properly closed before making any side shots. After closing the loop, side shots can be taken from established turning points.

Step 5 While the rod person holds the level rod plumb on TP1, a foresight rod reading is taken, then recorded in the field notes. For our example, the FS reading of 5.23' is recorded in the FS (2) column. Following this, the elevation of TP1 is calculated and recorded in the elevation column of the field notes. For our example, the elevation is recorded as 1,002.54' (turning point elevation = HI − FS = 1,007.77' − 5.23').

Step 6 In preparation for the next set of backsight and foresight readings, the instrument person moves the leveling instrument to a point beyond TP1 and sets up the instrument at Station 2, which is approximately midway between TP1 and TP2.

Step 7 Once the leveling instrument is set up, backsight and foresight readings are taken between the points TP1 and TP2 in the same way as previously described in Steps 3 through 5, with the following exceptions. The known elevation of TP1 is used instead of the BM to calculate the instrument height (HI) at Station 2. Then, the new HI and the foresight reading on TP2 are used to calculate the elevation of TP2.

Step 8 Steps 3 through 7 are repeated as necessary to complete the differential measurement loop from the TP2 to the temporary benchmark TBM1, then back via TP3 to the starting point at BM.

Differential Leveling

These photos show site layout trainees practicing differential leveling. The left photo shows the instrument person, and the right photo shows the rod person.

202P0201.EPS

202P0202.EPS

In the example shown, the leveling traverse is run back to the starting point at BM. This is called closing the loop or a closed level loop. Any leveling survey should close back either on the starting benchmark or on some other point of known elevation in order to provide a check of the measurements taken.

Leveling notes should always be checked for arithmetic or calculator input errors. This is done by simply summing the backsight (BS) and foresight (FS) columns and comparing the difference between them with the starting and ending elevations. As shown in *Figure 31*, the difference between the BS sum and the FS sum is 0.00'. Also, the difference between the starting elevation of 1,000.00' and the ending elevation of 1,000.00' is 0.00'. Since the differences are equal, the arithmetic checks and the loop is properly closed. An error would exist if the differences were not equal or were not within the established accuracy standard or tolerances specified for the project. Using the same example, the calculations for the traverse

between the BM and TBM1 can be checked in the same manner. This is shown in *Figure 32*.

When performing differential leveling, it is easy to make mistakes. However, mistakes can be eliminated by constantly checking and rechecking your work. Some common mistakes to avoid when performing differential leveling are:

- Backsight and foresight distances not equal
- Instrument not leveled
- Rod not plumb (if not using a level, the rod should be rocked forward and backward, then the smallest reading recorded)
- Sections of an extended leveling rod not adjusted properly
- Dirt, ice, etc., accumulated on the base of the rod
- Misreading the rod
- Recording incorrect values in the field notes
- Moving the position of a turning point between backsight and foresight readings

STATION (STA)	BS (+)	HI	FS (-)	ELEVATION
BENCH MARK (BM) TO TEMPORARY BENCH MARK 1 (TBM 1)				
BM	7.77'			1000.00'
STA. 1		1007.77'		
TP 1	8.45'		5.23'	1002.54'
STA. 2		1010.99'		
TP 2	8.07'		6.12'	1004.87'
STA. 3		1012.94'		
TBM 1			4.48'	1008.46'
	24.29'		15.83'	

MATH CHECK: 1000.00
            +   24.29
            ―――――――――
              1024.29
            ‑   15.83
            ―――――――――
              1008.46

202F34.EPS

Figure 32 ◆ Example of a math check.

12.0.0 ◆ FIELD NOTES

In the taping and differential leveling procedures described earlier, constant reference has been made to recording measurement information in field notes. Writing a legible and accurate set of notes in a field book (*Figure 33*) is just as important as doing the leveling or layout work itself. This is because field notes provide a historical record of the work performed. They serve as a reference should there ever be a question about the correctness or integrity of your work, especially in a court of law. Field notes should leave no room for misinterpretation. Your notes should be written so that others can understand your work. General guidelines for writing and keeping field notes are as follows:

- All field notebooks should contain the name, address, and phone number of the owner.
- All pages should be numbered, and there should be a table of contents page.
- Make neatly printed entries in the book using a suitable sharp pencil with hard lead (3H or 4H). Never use cursive script in a field book.
- Begin each new task on a new page. The left-hand pages are generally used for entering numerical data and the right-hand pages are for making sketches and notes.
- Always record the date, time, weather conditions, names of crew members and their assignments, and a list of the equipment used.
- Record each measurement in the field book immediately after it is taken. Do not trust it to memory.

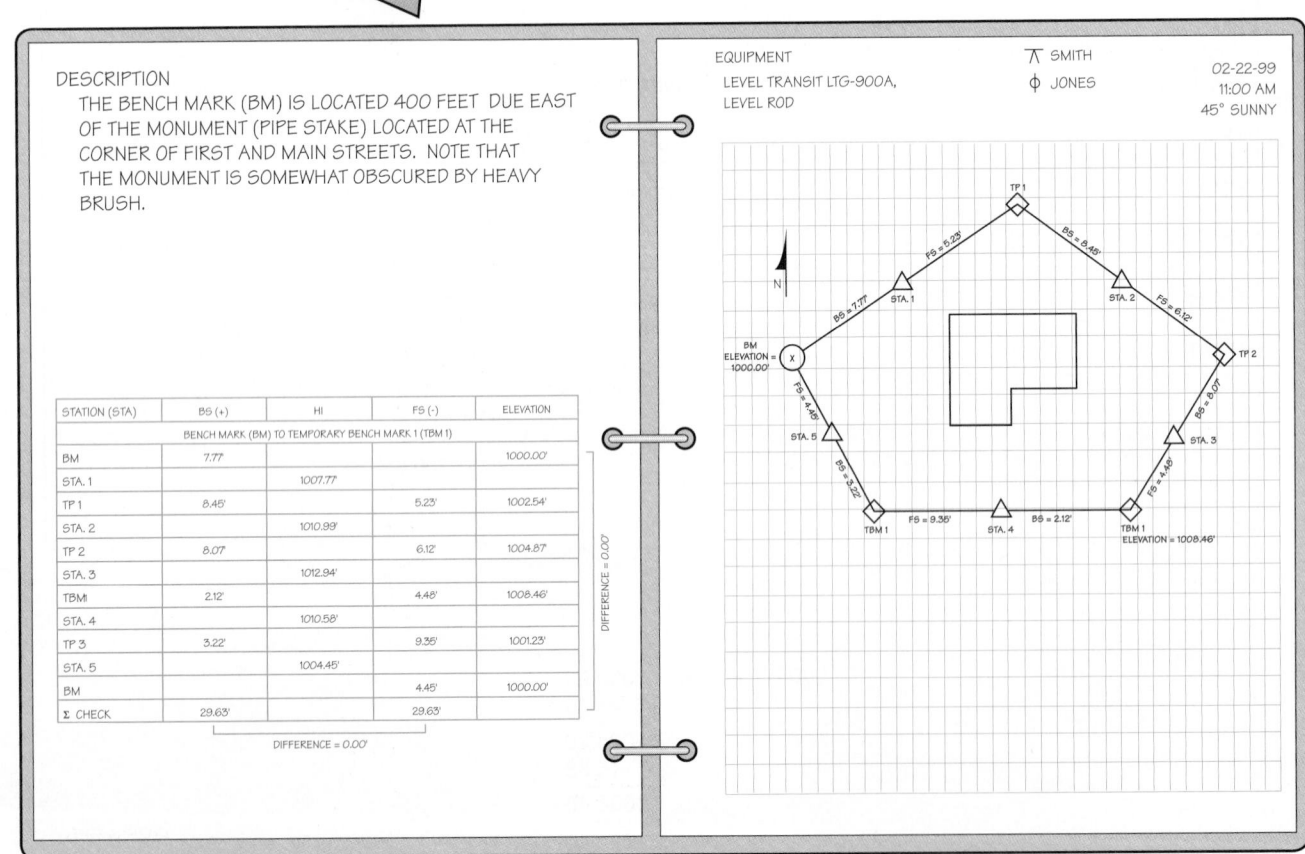

Figure 33 ◆ Example of field notes.

- Record data exactly. Ideally, the data should be checked by two crew members at the time it is recorded.
- Make liberal use of sketches if needed for clarity. They should be neat and clearly labeled, including the approximate north direction. Do not crowd the sketches.
- Never erase. If a mistake is made, draw a single line through the incorrect entry and write the correct data above it.
- Draw a diagonal across the page and mark the word VOID on the tops of pages that, for one reason or another, are invalid. When marking the page, be careful not to make the voided information unreadable. The date and name of the person voiding the page should also be recorded.
- Mark the word COPY on the top of copied pages. Refer to the name and page number of the original document.
- Always keep the field book in a safe place on the job site. At night, lock it up in a fireproof safe. Original field books should never be destroyed, even if copied for one reason or another.

13.0.0 ◆ LEVELING APPLICATIONS

In addition to setting benchmarks, grade stakes, etc., there are many applications involving leveling, including:

- Transferring elevations up a structure
- Profile leveling
- Cross-section leveling
- Grid leveling

13.1.0 Transferring Elevations Up a Structure

When constructing multistory buildings and other tall structures, ground elevations frequently need to be transferred vertically up the structure as it is being built to maintain the design grades. One method for accomplishing this involves the use of both differential leveling and taping skills. The process begins by first establishing a temporary benchmark (see TBM1 in *Figure 34*) with a known elevation at the base of the structure by using differential leveling methods. Following this, a tape is used to measure up from TBM1 the vertical distance needed to establish a second temporary benchmark (TBM2) on the floor or level of the structure on which the elevation(s) are needed. Once TBM2 has been established on the upper level, a leveling instrument can be set up and a height of instrument (HI) calculated in the normal way by backsighting on and reading a rod held on TBM2 (HI = BS + TBM2 elevation). Note that the elevation of TBM2 is equal to the elevation of TBM1 plus the tape distance. Following this, any subsequent leveling tasks are performed on the upper floor or level just as if the instrument were placed on the ground.

ELEVATION OF TBM 2 = ELEVATION TBM 1 + THE TAPE DISTANCE

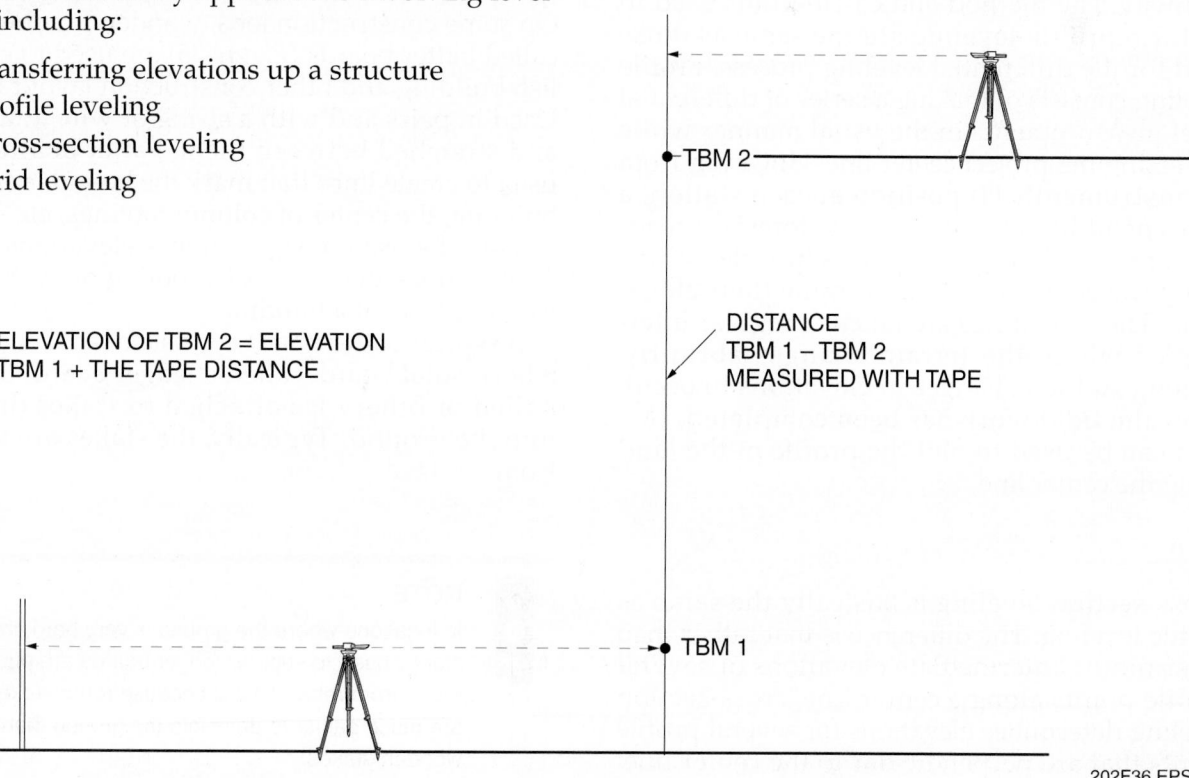

TBM 2

DISTANCE TBM 1 – TBM 2 MEASURED WITH TAPE

TBM 1

202F36.EPS

Figure 34 ◆ Transferring elevations up a structure.

Sometimes, points that need elevations may be above the line of sight of the instrument, such as with elevations for the bottom of a beam and ceiling levels. Taking the elevation in these instances requires that the level rod be held upside down and placed against the beam, ceiling, etc. This results in a positive foresight reading that must be added to the HI rather than subtracted as is normally done.

13.2.0 Profile, Cross-Section, and Grid Leveling

Profile, cross-section, and grid leveling are all methods used to determine the profile of a terrain or surface. These methods are briefly described here. However, procedures for performing these leveling methods are beyond the scope of this module. Such procedures can be found in most surveying or field engineering texts or reference books, some of which are referenced in the back of this module.

13.2.1 Profile Leveling

Profile leveling is the process of determining the elevation of a series of points along the ground at approximately uniform intervals along a continuous center line, such as when determining the profile of the ground along the center line of a highway. The method and calculations used to perform profile leveling are the same as those used for the differential leveling process. Profile leveling consists of making a series of differential level measurements in the usual manner while traversing the project center line. However, from the instrument's HI position at each station, a series of additional intermediate foresight readings are taken on several points (profile points) along the center line to determine their elevations. These readings are taken at regular intervals or where the terrain changes abruptly, causing sudden changes in elevations to occur. After the field work has been completed, this data can be used to plot the profile of the land along the center line.

13.2.2 Cross-Section Leveling

Cross-section leveling is basically the same as profile leveling. The difference is that rather than determining intermediate elevations of several profile points along a center line, cross-section leveling determines elevations for several profile points that are perpendicular to the center line. Note that for a specific project, there is only one center line profile but there can be numerous cross sections. Cross-section profile plots derived from cross-section leveling data are used for estimating quantities of **earthwork** to be performed.

13.2.3 Grid Leveling

Grid leveling is one process that can be used to determine the existing topography of a building lot or other land area. It is also used when necessary to determine earthwork quantities related to an excavation (pit) or a mound. This is normally done when it is necessary to calculate the volume of material that has been excavated or placed. Basically, this method requires that a rectangular profile grid be laid out on the building lot with grid intersections occurring at regular intervals spaced about 50' or 100' apart. Following this, differential leveling is done in a similar manner as for profile leveling, except that more intermediate elevation readings can be taken from one instrument position.

When performed in conjunction with earthwork, grid leveling is normally done both before and after the earthwork is accomplished. The difference between the original and final elevations is then used in a volume formula to calculate the volume of material excavated or filled.

14.0.0 ◆ BATTER BOARDS

On some construction jobs, wooden frameworks called batter boards (*Figure 35*) are used to establish building and other construction layout lines. Used in pairs and with a string or wire attached and stretched between them, batter boards are used to create lines that mark the boundaries of a building, the center of column footings, etc. They can also be used to set reference elevations such as elevations to the top of a footing or to the finish floor level of a building.

A typical batter board consists of a 2 × 4 or 2 × 6 horizontal board, called a ledger board, that is nailed or otherwise attached to stakes driven into the ground. Typically, the stakes are made from 2 × 4s.

 **NOTE**

In locations where the ground is very hard, the stakes used to support ledger boards are often made from rebar. This is because rebar stakes are much easier to drive into the ground than wooden stakes.

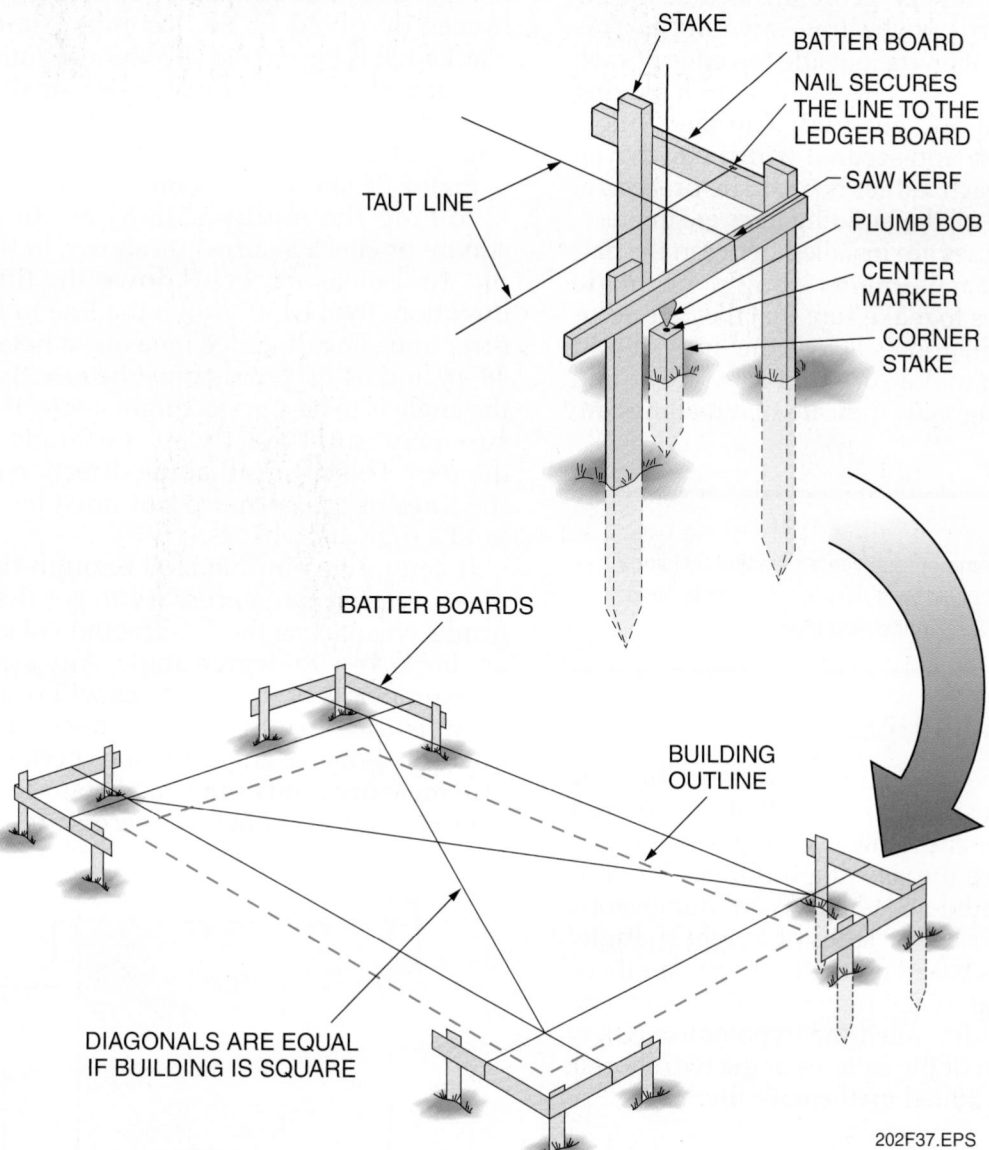

STAKE

BATTER BOARD

NAIL SECURES THE LINE TO THE LEDGER BOARD

SAW KERF

PLUMB BOB

CENTER MARKER

CORNER STAKE

TAUT LINE

BATTER BOARDS

BUILDING OUTLINE

DIAGONALS ARE EQUAL IF BUILDING IS SQUARE

202F37.EPS

Figure 35 ◆ Typical use of batter boards.

The placement of batter boards is normally done after the exact locations of the building corners have been established. Placement involves driving the ledger board support stakes firmly into the ground behind each building corner at a distance that allows enough working room between the batter boards and the immediate construction area. Depending on the type of job and the excavation equipment used, this distance could be anywhere between 4' and 20'. If it is necessary to drive the stakes in soft soil, or if the stakes extend 3' or more out of the ground, they should be braced to prevent any movement. Following this, a leveling instrument is used to sight and mark the stakes at the required elevation.

Then, each ledger board is fastened to the outside of its support stakes so that its top edge is on the elevation mark. It is important to make sure that when all the related batter boards have been installed, the tops of the ledger boards are level with one another.

Once the batter boards are installed, the building corners can be transferred to the batter boards. One method for doing this involves the use of a plumb bob and nylon string. This is done by stretching a nylon string (line) between two opposite batter boards and directly over the building corner stakes. The plumb bob is used to locate the exact position of the line by suspending it directly over the center marker on each corner stake. When the line is accurately located on the two batter boards, a shallow saw cut (kerf) is made at this point on the outside top edge of each ledger board. This prevents the line from moving when stretched and secured. The taut line is placed in the kerf and secured with a nail driven on the back of each ledger board. The procedure is repeated until all the building lines are in place.

After all the lines are installed between the batter boards, measurements should be made between the lines to make sure that they are accurate. Also, the diagonals across the lines should be measured to make sure that they are equal. Equal length diagonals indicate that the lines are square.

NOTE

This is a common practice for smaller buildings. On larger buildings, a different method is used. This is covered in the next module.

15.0.0 ◆ 3-4-5 RULE

The 3-4-5 rule has been used in construction for centuries. It is a simple method that can be used for laying out or checking a 90-degree angle that does not require the use of a builder's level or transit. The numbers 3-4-5 represent dimensions in feet that describe the sides of a right triangle. The 3-4-5 rule is based on the Pythagorean theorem. It states that in any right triangle, the square of the longest side, called the hypotenuse (C), is equal to the sum of the squares of the two shorter sides (A and B). Stated mathematically:

$$C^2 = A^2 + B^2$$

Accordingly, for the 3-4-5 right triangle:

$$5^2 = 3^2 + 4^2$$
$$25 = 9 + 16$$
$$25 = 25$$

This theorem also applies if you multiply each number (3, 4, and 5) by the same number. For example, if multiplied by the constant 3, it becomes a 9-12-15 triangle. The Pythagorean theorem was discussed in more detail in an earlier module.

For most construction layout and checking, right triangles that are multiples of the 3-4-5 triangle are used (such as 9-12-15, 12-16-20, 15-20-25, and 30-40-50). The specific multiple used is determined mainly by the relative distances involved in the job being laid out or checked. It is best to use the highest multiple that is practical. This is because when smaller multiples are used, any error made in measurement will result in a much greater angular error.

Figure 36 shows an example of the 3-4-5 rule involving the multiple 48-64-80. In order to square or check a corner as shown in the example, first measure 48'-0" down the line in one direction, then 64'-0" down the line in the other direction. The distance measured between the 48'-0" and 64'-0" points must be exactly 80'-0" if the angle is to be a perfect right angle. If the measurement is not exactly 80', the angle is not 90 degrees. This means that the direction of one of the lines or the corner point must be adjusted until a right angle exists.

It cannot be emphasized enough that exact measurements are necessary to get the desired results when using the 3-4-5 method of laying out or checking a 90-degree angle. Any error in the measurements of the distances will result in not establishing a right angle as desired, or if an existing 90-degree angle is being checked, inaccurate measurements may cause you to make an adjustment that might be unnecessary.

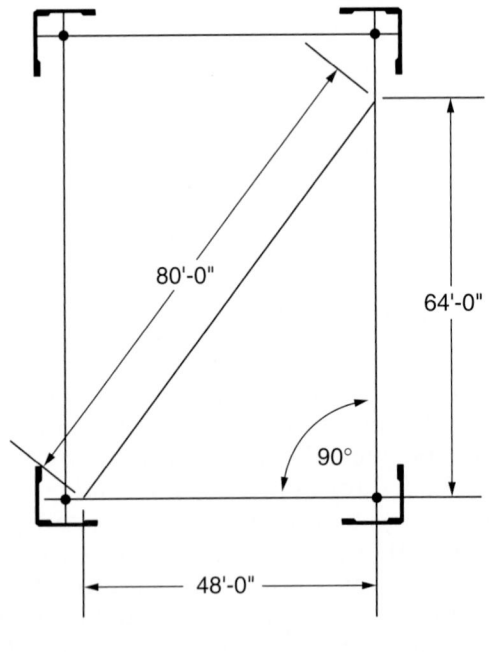

202F38.EPS

Figure 36 ◆ Example of checking lines for square using the 48-64-80 multiple of a 3-4-5 triangle.

1. Contour lines are used on site/plot plans to show _____ elevations.
   a. natural grade
   b. finish grade
   c. both natural and finish grade
   d. finish floor

2. Carpenters should *never* make any measurements that relate to _____.
   a. footings and foundations
   b. working control points
   c. layout of embedded items
   d. property lines

3. When referencing a point such as a benchmark or control point, objects located in the same direction from the point should be referenced from that point rather than from different directions.
   a. True
   b. False

4. All of the following are recommended practices for marking information on stakes *except* _____.
   a. use all capital letters
   b. do not crowd words and numbers
   c. use abbreviations whenever possible
   d. mark the main information on the direction of use

5. While working with a tape to lay out the foundation of a building, you are caught in heavy rain that wets the tape. You know the proper procedure is to _____.
   a. roll up the tape loosely to allow it air circulation to dry
   b. carefully dry the tape as you are rolling it onto the reel
   c. send the tape in for precision equipment maintenance
   d. dispose of the tape. A wet tape stretches so it is inaccurate

6. Pulling too hard on a steel tape can stretch it.
   a. True
   b. False

7. You want to be sure that your instrument is directly over a reference point, so you use a _____.
   a. plumb bob
   b. range pole
   c. hand level
   d. chaining pin

8. When comparing two measurements, the rule of thumb is that for every 100' measured, the difference between the two readings should be no more than _____.
   a. $\frac{1}{100}$ of a foot
   b. $\frac{1}{10}$ of a foot
   c. 1 inch
   d. 1 foot

9. Taping requires _____.
   a. three people; two to measure and one to record the measurements
   b. two people if possible; otherwise, one person can accurately do it
   c. three people; two to measure and one to level the measuring tape
   d. two people working together and communicating with each other

10. When measuring a distance over an incline, you will mark the ground where a measurement was taken with a _____.
    a. plumb bob and chaining pin
    b. grade stake and chaining pin
    c. chaining pin and tensioning spring
    d. tensioning spring and plumb bob

11. Land measurements on a drawing are usually shown in feet and inches.
    a. True
    b. False

12. Convert 3'-4¼" to decimal feet.
    a. 3.066'
    b. 3.333'
    c. 3.354'
    d. 3.510'

13. Convert 6.875' to feet and inches.
    a. 6'-2¼"
    b. 6'-8¼"
    c. 6'-10½"
    d. 6'-12½"

14. To calculate the length of your pace, you need to record the number of paces you take to travel a measured 100' distance at least _____ time(s).
    a. one
    b. five
    c. six
    d. ten

15. If your average pace length is 2.51' and it takes 90 paces to travel a distance, then its approximate length is _____.
    a. 150.6'
    b. 190.9'
    c. 225.9'
    d. 251.0'

16. It takes you a total number of 200 paces to cover a known distance of 100' five times. After first determining your average pace length, solve the following problem. If it takes you 60 paces to cover an unknown distance between two points, the distance is approximately _____.
    a. 120'
    b. 130'
    c. 140'
    d. 150'

17. The best instrument to use for checking elevations is _____.
    a. an electronic distance measuring instrument
    b. an automatic leveling instrument
    c. theodolite
    d. a tripod and range pole

18. As indicated at the pencil point in *Figure 1*, the architect's leveling rod shown below reads _____.
    a. 3.14'
    b. 3'-1⅛"
    c. 3'-1⅜"
    d. 3.05'

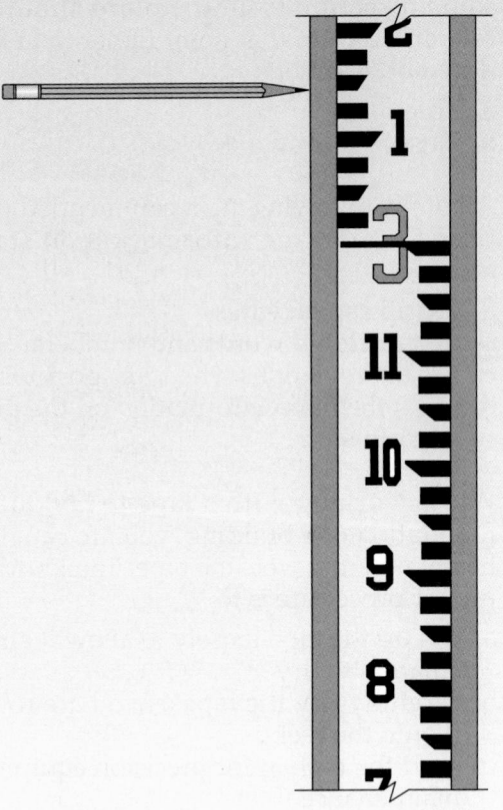

202E01.EPS

Figure 1

19. As indicated at the pencil point in *Figure 2*, the engineer's leveling rod shown below reads _____.
    a. 8'-9⅞"
    b. 9'-⅞"
    c. 8.98'
    d. 8.99'

20. In the differential leveling procedure, the _____ is determined by _____ the backsight elevation _____ the known elevation.
    a. height of instrument (HI); adding; to
    b. turning point (TP) elevation; subtracting; from
    c. height of instrument (HI); subtracting; from
    d. turning point (TP) elevation; adding; to

21. What is the elevation of the anchor bolt (A-bolt) shown in *Figure 3*?
    a. 781.22'
    b. 784.99'
    c. 787.04'
    d. 792.86'

22. You are part way through laying out a building foundation and realize that you started from the wrong benchmark. To correct your field notes, you _____.
    a. erase the entries. This is why field notes are written in pencil
    b. tear out any pages with incorrect information and discard them
    c. draw a diagonal line across the entry and write void on the page
    d. cross out the entries so it can't be read and write void on the page

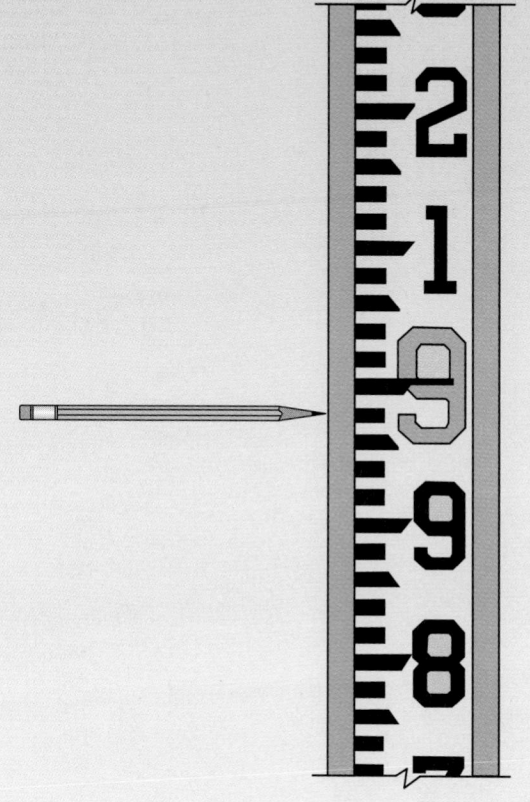

202E02.EPS

Figure 2

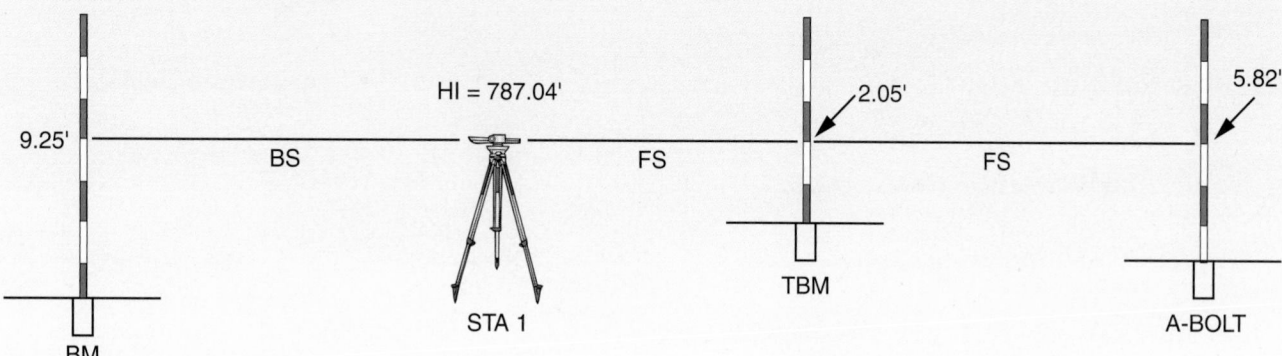

202E03.EPS

Figure 3

---

23. Original field notes can be destroyed if you have made a copy.

    a. True
    b. False

24. If it is necessary to correct any measurement data, neatly erase it and write in the correct data.

    a. True
    b. False

25. When checking the angle formed by the corner of a 15' wall and a 20' wall for square using the 3-4-5 rule, the best angular accuracy is obtained if the multiple of _____ is used.

    a. 9-12-15
    b. 12-16-20
    c. 15-20-25
    d. 30-40-50

Summary

Site preparation and layout tasks can be performed by different individuals, including surveyors, field engineers, and carpenters. However, because of the tremendous liability involved, carpenters should never make any measurements that relate to property lines. This is a task for the professional land surveyor.

The carpenter must have the knowledge to perform standard surveying measurements on the job site such as measuring distances, angles, and elevations. It is important to eliminate mistakes and reduce the size of errors in measurement. This is achieved by having a good understanding of measurement principles and by rechecking your work several times. It is also achieved by making sure to always use calibrated measuring instruments. Do not use instruments that are damaged or otherwise incapable of being able to measure within tolerances.

The task of taping involves two people working together and communicating with each other. To achieve accurate distance measurement by taping, the tape must be held straight between points, held horizontal between points, properly tensioned, held exactly on zero, and read correctly. To prevent errors, it is important that you be familiar with the tape being used and the markings on the tape.

In differential leveling, elevations are transferred from one point to another by using a leveling instrument to first read a rod held vertically on a point of known elevation (backsight) in order to determine the height of the instrument (HI), then a rod held on a point of unknown elevation is read (foresight). Following this, the unknown elevation is calculated by adding or subtracting the foresight reading from the HI.

Writing a legible and accurate set of notes in a field book is just as important as doing the leveling or layout work itself. This is because field notes provide a historical record of the work performed. They serve as a reference should there ever be a question about the correctness or integrity of your work, especially in a court of law.

Notes

Trade Terms
Introduced in This Module

Backsight (BS): A reading taken on a leveling rod held on a point of known elevation to determine the height of the leveling instrument.

Breaking the tape: Making measurements using a portion of a full tape's length in a series of steps.

Control points: A series of horizontal and/or vertical points established in the field to serve as a known framework for all points on the site.

Crosshairs: A set of lines, typically horizontal and vertical, placed in a telescope used for sighting purposes.

Cut: Removing soil or rock on site to achieve a required elevation.

Differential leveling: A method of leveling used to determine the difference in elevation between two points.

Earthwork: All construction operations connected with excavating (cutting) or filling earth.

Field notes: A permanent record of field measurement data and related information.

Fill: Adding soil or rock on site to achieve a required elevation.

Foresight (FS): A reading taken on a leveling rod held on a point in order to determine a new elevation.

Height of instrument (HI): The elevation of the line of sight of the telescope relative to a known elevation. It is determined by adding the backsight elevation to the known elevation.

Parallax: The apparent movement of the crosshairs in a surveying instrument caused by movement of the eyes.

Peg test: A procedure used to check for an out-of-adjustment bubble vial on levels and other instruments.

Station: Instrument setting locations in differential leveling.

Temporary benchmark: A point of known (reference) elevation determined from benchmarks through leveling, and permanent enough to last for the duration of a project.

Turning point (TP): A temporary point within an open or closed differential leveling circuit whose elevation is determined by differential leveling. It is normally the leveling rod location. Its elevation is determined by subtracting the foresight elevation from the height of the instrument elevation.

Common Construction Abbreviations

Above mean sea level	ABMSL
Abutment	abt.
Approximate	approx.
At	@
Avenue	Ave.
Average	avg.
Back of sidewalk	BSW
Back of walk	BW
Backsight	BS
Begin curb return	BCR
Benchmark	BM
Between	betw.
Bottom	bot.
Boulevard	Blvd.
Boundary	bndry.
Bridge	br.
Calculated	calc.
Cast-iron pipe	CIP
Catch basin	CB
Catch point	CP
Cement-treated base	CTB
Concrete block wall	CBW
Construction	const.
Control point	CP
County	Co.
Court	Ct.
Creek	cr.
Curb	cb.
Curb and gutter	C&G
Cut	C
Description	desc.
Destroyed	dest.
Detour	det.
Direct	D
Distance	dist.
Distance	D
Distance, horizontal	Dh
District	Dist.
Ditch	dit.
Drive	Dr.
Driveway	drwy.
Drop inlet	DI

Edge of gutter	EG
Edge of pavement	EP
Edge of shoulder	ES
Elevation	el.
End wall	EW
Equation	eqn.
Existing	exist.
Expressway	Exwy.
Fahrenheit	F
Fence	fe.
Fence post	FP
Feet	ft.
Field book	FB
Fill	f
Finish grade	FG
Fire hydrant	FH
Flow line	FL
Foot	ft.
Footing	ftg.
Foresight	FS
Found	fd.
Foundation	fdn.
Freeway	Fwy.
Galvanized	galv.
Galvanized steel pipe	GSP
Gas line	GL
Gas valve	GV
Geodetic	geod.
Grid	grd.
Ground	grnd.
Gutter	gtr.
Head wall	hdwl.
Height	ht.
Height of instrument	HI
Highway	Hwy.
Hub & tack	H&T
Inch	in.
Inside diameter	ID
Instrument	inst.
Intersection	int
Iron pipe	IP
Irrigation pipe	irr.P

Junction	jct.	Reference monument	RM
Kilometer	km	Reference point	RP
Lane	ln.	Reinforced concrete pipe	RCP
Left	lt.	Retaining wall	ret.W
Manhole	MH	Right	rt.
Marker	mkr.	Right of way	R/W
Maximum	max.	River	Riv.
Measured	meas.	Road	rd.
Median	med.	Roadway	rdwy.
Mile	mi.	Rock	rk.
Millimeter	mm	Route	Rte.
Minimum	min.	Section	S
Minute	min.	Sewer line (sanitary)	SS
Monument	mon.	Shoulder	shldr.
Nail	N	Sidewalk	SW
North	N	Slope stake	SS
Number	# or no.	South	S
Offset	O/S	Spike	spk.
Original ground	OG	Stake	stk.
Outside diameter	OD	Stand pipe	SP
Overhead	OH	Station	sta.
Page	p.	Steel	stl.
Pages	pp.	Storm drain	SDr.
Party chief	PC	Street	St.
Pavement	pvmt.	Structure	str.
Perforated metal pipe	PMP	Subdivision	subd.
Pipe	P	Subgrade	SG
Place	pl.	Tack	tk.
Plastic	plas.	Telephone cable	tel.C.
Point	pt.	Telephone pole	tel.P.
Point of intersection	PI	Temperature	temp.
Portland cement concrete	PCC	Temporary benchmark	TBM
Power pole	PP	Top back of curb	TBC
Pressure	press.	Top of bank	TB
Private	pvt.	Top of curb	TC
Project control survey	PCS	Township	T
Property line	PL	Tract	tr.
Punch mark	PM	Transmission tower	TT
Railroad	RR	Turning point	TP
Railroad spike	RRspk.	Water line	WL
Read head nail	RH	Water valve	WV
Record	rec.	Wing wall	WW
Reference	ref.		

Additional Resources and References

Additional Resources

This module is intended to be a thorough resource for task training. The following reference works are suggested for further study. These are optional materials for continued education rather than for task training.

Construction Surveying and Layout: A Step-by-Step Engineering Methods Manual, Wesley G. Crawford. West Lafayette, IN: Creative Construction Publishing, 1995.

Surveying, Jack McCormac. New York, NY: John Wiley & Sons, 1999.

NCCER makes every effort to keep these textbooks up-to-date and free of technical errors. We appreciate your help in this process. If you have an idea for improving this textbook, or if you find an error, a typographical mistake, or an inaccuracy in NCCER's Contren® textbooks, please write us, using this form or a photocopy. Be sure to include the exact module number, page number, a detailed description, and the correction, if applicable. Your input will be brought to the attention of the Technical Review Committee. Thank you for your assistance.

Instructors – If you found that additional materials were necessary in order to teach this module effectively, please let us know so that we may include them in the Equipment/Materials list in the Annotated Instructor's Guide.

Write: Product Development and Revision
National Center for Construction Education and Research
3600 NW 43rd St, Bldg G, Gainesville, FL 32606

Fax: 352-334-0932

E-mail: curriculum@nccer.org

Craft _____ Module Name _____

Copyright Date _____ Module Number _____ Page Number(s) _____

Description _____
_____
_____
_____
_____

(Optional) Correction _____
_____
_____
_____

(Optional) Your Name and Address _____
_____
_____

Introduction to Concrete, Reinforcing Materials, and Forms
68102-09

In the Associated Builders and Contractors 2005 National Craft Championships, Carpentry contestants are expected to be able to perform several masonry and concrete finishing tasks, such as mixing concrete; building footings, edges, and wall forms; and using concrete reinforcing materials.

68102-09
Introduction to Concrete, Reinforcing Materials, and Forms

Topics to be presented in this module include:

1.0.0   Introduction 2.2
2.0.0   Concrete and Concrete Materials 2.2
3.0.0   Normal Concrete Mix Proportions and Measurements . . .2.5
4.0.0   Special Types of Concrete 2.6
5.0.0   Curing Methods and Materials 2.8
6.0.0   Concrete Slump Testing 2.8
7.0.0   Estimating Concrete Volume 2.9
8.0.0   Concrete Reinforcement Materials 2.12
9.0.0   Concrete Forms 2.17

Overview

Though the finishing of footings, foundations, and floor slabs is typically done by subcontractors, there will be occasions when the residential carpenter is called on to build basic forms and place reinforcing materials. A residential carpenter might also have to mix a batch of concrete simply to pour into a hole to support deck pillars and other vertical supports.

Carpenters doing commercial and industrial construction can expect to spend a lot of time building, bracing, and stripping concrete forms for walls, columns, slabs, beams, and other structures. Such work requires knowledge of form ties and form support structures. In this module, you will learn about concrete, reinforcing materials, and basic formwork.

Objectives

When you have completed this module, you will be able to do the following:

1. Identify the properties of cement.
2. Describe the composition of concrete.
3. Perform volume estimates for concrete quantity requirements.
4. Identify types of concrete reinforcement materials and describe their uses.
5. Identify various types of footings and explain their uses.
6. Identify the parts of various types of forms.
7. Explain the safety procedures associated with the construction and use of concrete forms.
8. Erect, plumb, and brace a simple concrete form with reinforcement.

Trade Terms

ACI	Plastic concrete
Admixtures	Plyform
Aggregates	Pozzolan
Axle-steel	Rail-steel
Brace	Rebars
Cured concrete	Screeding
Flatwork	Shoring
Footing	Slab-on-grade (slab-at-
Forms	grade)
Green concrete	Slag
Hydration	Slump
Kip	Studs
Monolithic slab	Subgrade
(monolithic pour)	Walers
Pascal	Water-cement ratio
Piles	

Required Trainee Materials

1. Pencil and paper
2. Appropriate personal protective equipment

Prerequisites

Before you begin this module, it is recommended that you successfully complete *Core Curriculum*, and *Construction Technology*, Module 68101-09.

This course map shows all of the modules in *Construction Technology*. The suggested training order begins at the bottom and proceeds up. Skill levels increase as you advance on the course map. The local Training Program Sponsor may adjust the training order.

CONSTRUCTION TECHNOLOGY

68117-09
Copper Pipe and Fittings

68116-09
Plastic Pipe and Fittings

68115-09
Introduction to Drain, Waste, and Vent (DWV) Systems

68114-09
Introduction to HVAC

68113-09
Residential Electrical Services

68112-09
Electrical Safety

68111-09
Basic Stair Layout

68110-09
Exterior Finishing

68109-09
Roofing Applications

68108-09
Roof Framing

68107-09
Wall and Ceiling Framing

68106-09
Floor Systems

68105-09
Masonry Units and Installation Techniques

68104-09
Introduction to Masonry

68103-09
Handling and Placing Concrete

68102-09
Introduction to Concrete, Reinforcing Materials, and Forms

68101-09
Site Layout One: Distance Measurement and Leveling

CORE CURRICULUM: Introductory Craft Skills

102CMAP.EPS

1.0.0 ◆ INTRODUCTION

This module introduces the trainee to the types of cement, aggregates, and additives used in concrete. It describes general types of concrete, concrete mixing information, and various concrete tests. It also covers concrete quantity estimating procedures for various job applications as well as compression specimen casting and slump (consistency) testing of freshly mixed concrete.

This module also introduces the trainee to some types of concrete reinforcement material such as steel reinforcement bars, welded-wire mesh, and various reinforcement bar supports.

Because it is in a semiliquid state, concrete must be contained in forms until it hardens. Forms can be built on the job using lumber and plywood, or they can be prefabricated form systems that the contractor buys or rents. In this module, we will introduce the basic job-built forms for concrete walls, footings, and floor slabs.

2.0.0 ◆ CONCRETE AND CONCRETE MATERIALS

Concrete is a mixture of four basic materials: portland cement, fine aggregates, coarse aggregates, and water. When first mixed, concrete is in a semiliquid state and is referred to as plastic concrete. When the concrete hardens, but has not yet gained structural strength, it is called green concrete. After concrete has hardened and gained its structural strength, it is called cured concrete. Various types of concrete can be obtained by varying the basic materials and/or by adding other materials to the mix. These added materials are called admixtures.

2.1.0 Portland Cement

Portland cement is a finely ground powder consisting of varying amounts of lime, silica, alumina, iron, and other trace components. While dry, it may be moved in bulk or can be bagged in moisture-resistant sacks and stored for relatively long periods of time. Portland cement is a hydraulic cement because it will set and harden by reacting with water with or without the presence of air. This chemical reaction is called hydration, and it can occur even when the concrete is submerged in water. The reaction creates a calcium silicate hydrate gel and releases heat. Hydration

begins when water is mixed with the cement and continues as the mixture hardens and cures. The reaction occurs rapidly at first, depending on how finely the cement is ground and what admixtures are present. Then, after its initial cure and strength are achieved, the cement mixture continues to slowly cure over a longer period of time until its ultimate strength is attained.

Today, portland cement is manufactured by heating lime mixed with clay, shale, or slag to about 3000°F. The material that is produced is pulverized, and gypsum is added to regulate the hydration process. Manufacturers use their own brand names, but nearly all portland cements are manufactured to meet American Society for Testing and Materials (ASTM) or American National Standards Institute (ANSI) specifications.

 **WARNING!**

Those working with dry cement or wet concrete should be aware that it is harmful. Dry cement dust can enter open wounds and cause blood poisoning. The cement dust, when it comes in contact with body fluids, can cause chemical burns to the membranes of the eyes, nose, mouth, throat, or lungs. It can also cause a fatal lung disease known as silicosis. Wet cement or concrete can also cause chemical burns to the eyes and skin. Make sure that appropriate personal protective equipment is worn when working with dry cement or wet concrete. If wet concrete enters waterproof boots from the top, remove the boots and rinse your legs, feet, boots, and clothing with clear water as soon as possible. Repeated contact with cement or wet concrete can also cause an allergic skin reaction known as cement dermatitis.

2.2.0 Aggregates for Concrete

Natural sand, gravel, crushed stone, blast furnace slag, and manufactured sand (from crushed stone, gravel, or slag) are the most commonly used aggregates in concrete. In some cases, recycled crushed concrete is used for low-strength concrete applications. There are other types of special aggregates for concrete.

Aggregates make up 60 percent to 80 percent of the volume of concrete and function not only as filler material, but also provide rigidity to greatly restrain volume shrinkage as the concrete mass

The History of Portland Cement

Portland cement is a successor to a hydraulic lime that was first developed by John Smeaton in 1756 when he built a lighthouse in the English Channel. He used a burned mixture of limestone and clay capable of setting and hardening under water as well as in air. The next developments took place around 1800 in England and France where a cement material was made by burning nodules of clayey limestone. About the same time in the United States, a similar material was obtained by burning a naturally-occurring substance called cement rock. These materials belong to a class known as natural cement, similar to portland cement but more lightly burned and not of a controlled composition.

In 1824, Joseph Aspdin, another Englishman, patented a manufacturing process for making a hydraulic cement out of limestone and clay that he called portland cement because when set it resembled portland stone, a type of limestone from the Isle of Portland, England. In 1850, this was followed by a more heavily burned portland cement product developed by Isaac Charles Johnson in southeastern England. Since then, the manufacture of portland cement has spread rapidly throughout the world accompanied by numerous new developments and improvements in its uses and manufacture.

Care of Portland Cement

In the United States, portland cement is normally sold in paper bags that hold one cubic foot of cement by volume and weigh 94 pounds. As shown here, these bags of cement should always be stored off the ground and in a dry place to prevent the cement from absorbing moisture that causes lumps to form in the cement powder. Cement powder should be free flowing. Any cement with lumps that cannot be broken up easily into powder by squeezing them in your hand should not be used.

108SA01.EPS

Concrete, Mortar, and Grout

What is the difference between concrete, mortar, and grout?

cures. At the usual aggregate content of about 75 percent by volume, shrinkage of concrete is only one-tenth that of pure cement paste. The presence of aggregates in concrete provides an enormous contact area for an intimate bond between the paste and aggregate surfaces.

Since aggregates are such a large portion of the concrete volume, they have a substantial effect on the quality of the finished product. In addition to concrete volume stability, they influence concrete weight, strength, and resistance to environmental destruction. Therefore, aggregates must be strong, clean, free of chemicals and coatings, and of the proper size and weight.

2.3.0 Water for Concrete

Generally, any drinkable water (unless it is extremely hard or contains too many sulfates) may be used to make concrete. If the quality of the water is unknown or questionable, the water should be analyzed, or mortar cubes should be made and tested against control cubes made with drinkable water. If the water is satisfactory, the test cubes should show the same compressive strength as the control cubes after a 28-day cure time.

High concentrations of chlorides, sulfates, alkalis, salts, and other contaminants have corrosive effects on concrete and/or metal reinforcing rods, mesh, and cables. Sulfates can cause disintegration of concrete, while alkalis, sodium carbonate, and bicarbonates may affect the hardening times as well as the strength of concrete.

2.4.0 Admixtures for Concrete

Admixtures are materials added to a concrete mix before or during mixing to modify the characteristics of the final concrete. They may be used to improve workability during placement, increase strength, retard or accelerate strength development,

and increase frost resistance. Usually, an admixture will affect more than one characteristic of concrete. Therefore, its effect on all the properties of the concrete must be considered. Admixtures may increase or decrease the cost of concrete work by lowering cement requirements, changing the volume of the mixture, or lowering the cost of handling and placement.

Control of concrete setting time may reduce costs by decreasing waiting time for floor finishing and form removal. Extending the setting time may reduce costs by keeping the concrete plastic,

108SA02.EPS

thereby eliminating bulkheads and construction joints. In practice, it is desirable to fully pretest all admixtures with the specific concrete mix to be used since the mix may affect the admixture efficiency and, ultimately, the final concrete.

There are a number of other admixtures available that are not in general use, such as gas-forming, air-entraining, grouting, expansion-producing or expansion-reducing, bonding, corrosion-inhibiting, fungicidal, germicidal, and insecticidal. These admixtures are for very specialized work, generally under the supervision of the manufacturer or a specialist.

3.0.0 ◆ NORMAL CONCRETE MIX PROPORTIONS AND MEASUREMENTS

Project specifications for concrete can generally be divided into two classes: prescription specifications and performance specifications, with some including features of both. Typical of the prescription type is the outmoded volume-proportion

requirement such as 1:2:3, meaning one part cement, two parts sand, and three parts aggregate by volume with water only being limited by the maximum amount of slump allowed. Most prescription requirements are now stated by weight, such as kilograms per cubic meter or pounds per cubic yard, and often cover only minimum cement content along with a water-cement ratio, usually by weight. The water-cement ratio includes all cementitious components of the concrete, including fly ash and pozzolan, as well as portland cement. *Figure 1* is an example of the typical strength range of a Type I portland cement at various water-cement ratios. In these types of specifications, the strength of the concrete will vary over a certain range due to variations in the cementitious materials, along with additional variations that can be caused by the quality of the aggregates. Numerous prescription-type tables have been prepared and presented in various publications by the ACI (American Concrete Institute). The recommendations of these tables can be combined to produce a calculated concrete mix

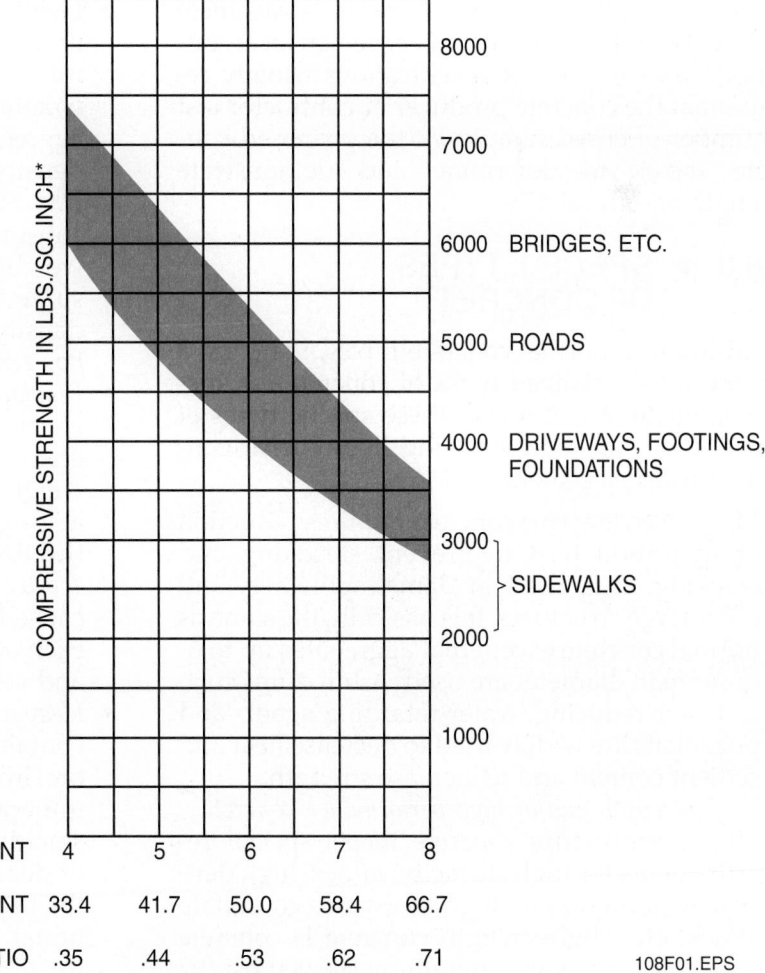

* COMPRESSION STRENGTH AFTER 28 DAYS OF MOIST CURING AT 70°F (21°C)

U.S. GAL. WATER/BAG (94 LB.) CEMENT	4	5	6	7	8
EQUIVALENT LBS. WATER/BAG (94 LB.) CEMENT	33.4	41.7	50.0	58.4	66.7
EQUIVALENT WATER-CEMENT RATIO	.35	.44	.53	.62	.71

108F01.EPS

Figure 1 ◆ Typical Type I portland cement strength for various water-cement ratios.

American Concrete Institute

The American Concrete Institute (ACI) is an international organization established in 1905 whose purpose is to further engineering and technical education, scientific investigation and research, and the development of standards for design and construction incorporating concrete and related materials. Its mission is to be the world's premier developer and disseminator of information on the use of concrete structures and facilities. It does this through the work of thousands of professional volunteers at local, national, and international levels, assisted and guided by a professional staff located at the ACI headquarters in Farmington Hills, Michigan.

design that will be on the side of safety in both strength and durability.

Concrete performance specifications usually state the strength required, the minimum cement content or maximum water-cement ratio, consistency as measured by slump, air-entrainment (if required), and usually some limitation on aggregate sizes and properties. By keeping the limitations to a minimum, concrete producers or contractors can use prescription tables as a starting point to design (with laboratory assistance) the most effective concrete mix based on local conditions. These types of specifications usually require that the concrete producer or contractor test a number of cured samples of the proposed concrete mixes to determine and demonstrate strength and durability.

4.0.0 ◆ SPECIAL TYPES OF CONCRETE

In addition to normal concrete mixes, there are a number of specialized types of concrete for specific applications. A few of these special types of concrete do not look anything like normal concrete during placement.

- *Mass concrete* – This concrete requires reduction of hydration heat to prevent shrinking and cracking. It is used in dams, spillways, and other large structures. It is basically the same as normal concrete except that aggregates up to 6" or more in diameter are used. Admixtures such as water-reducing/water-retarding agents and pozzolans are widely used to decrease heat and cement content and to increase strength.
- *High-strength and/or high-performance concrete* – High-performance concrete meets special requirements for high elasticity values, high density, watertightness, high resistance to sulfate attack, etc. High-strength concrete is concrete with a compressive strength over 6,000 psi (92 megapascals or MPa).

- *Roller-compacted concrete (RCC)* – This is used for concreting large areas that contain little or no reinforcement. One type is a low-cement content type used for dams or other very large structures. The other type is a high-cement content type used for pavements. Both are very dry, stiff mixes that are placed without forms and spread with bulldozers or pavers. After spreading, the concrete is immediately compacted by 10-ton vibratory rollers.
- *Lightweight concrete* – As shown in *Figure 2*, this concrete is classified as either low-density, structural lightweight, or moderate-strength lightweight concrete and uses various density aggregates depending on its classification. Low-density concrete is used primarily for insulation. Structural lightweight concrete is used as a substitute for normal concrete and has reduced weight as well as reduced strength. Moderate-strength lightweight concrete falls between low-density and structural lightweight concrete and has enough strength for use in roofs.
- *Preplaced aggregate (prepacked) concrete* – This concrete is widely used for repair of large concrete structures or in the placement of heavy-weight concrete. It involves placing dry, coarse aggregate in forms and then pumping a highly fluidized cement mixture from the bottom to the top of the forms to make the concrete. Because the coarse aggregates are in contact with each other, the concrete has very low shrinkage and very good strength.
- *Heavyweight (high-density) concrete* – A concrete containing very heavy aggregates such as iron ore, iron shot or punchings, iron slag, and other minerals. It is used primarily for radiation shielding or in construction of counterweights or deadweights. This concrete can be placed by the preplaced-aggregate method or by conventional methods if heavy-duty equipment is used. Forms for this concrete require increased strength.

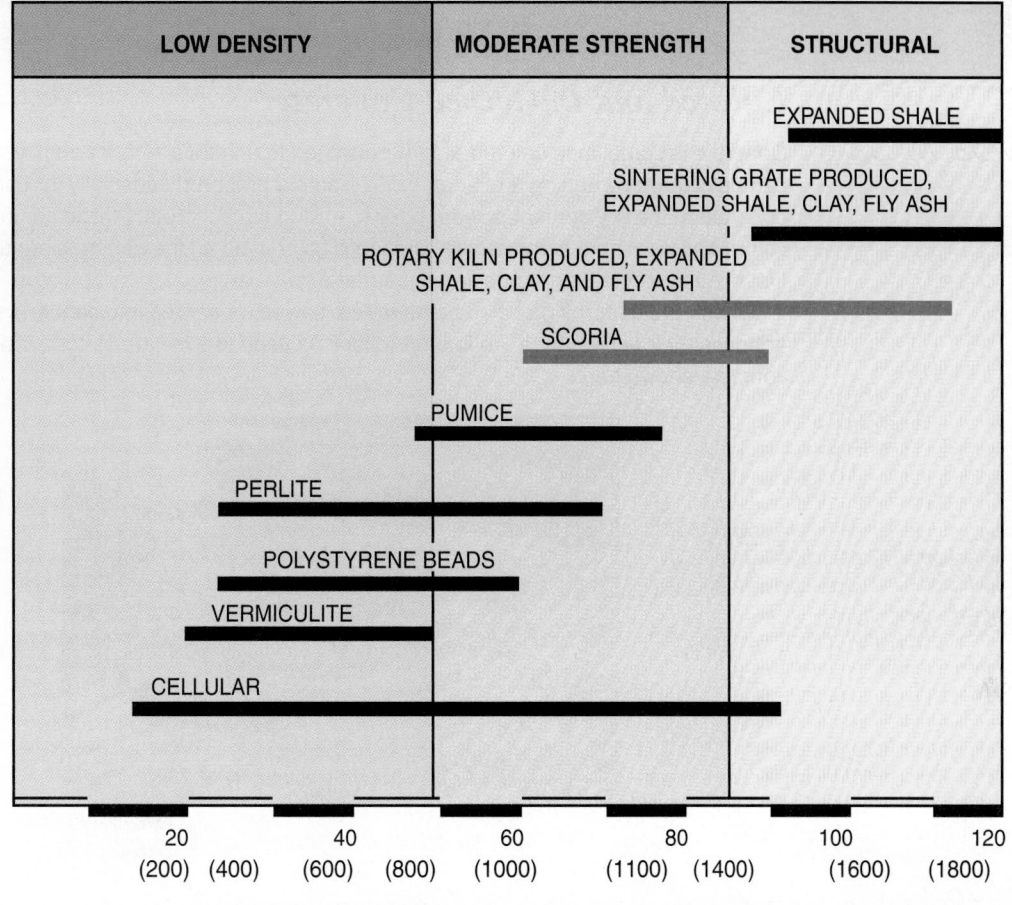

LOW DENSITY	MODERATE STRENGTH	STRUCTURAL

EXPANDED SHALE

SINTERING GRATE PRODUCED,
EXPANDED SHALE, CLAY, FLY ASH

ROTARY KILN PRODUCED, EXPANDED
SHALE, CLAY, AND FLY ASH

SCORIA

PUMICE

PERLITE

POLYSTYRENE BEADS

VERMICULITE

CELLULAR

20 40 60 80 100 120
(200) (400) (600) (800) (1000) (1100) (1400) (1600) (1800)

28-DAY AIR DRY UNIT WEIGHT IN LBS./CU. FT. (KG/CU. METER)

108F02.EPS

Figure 2 ◆ Lightweight concrete aggregates and unit weights.

INSIDE TRACK

RCC

High-strength roller-compacted concrete (RCC) is commonly used for roads where log-handling vehicles can have axle loads exceeding 100,000 pounds.

108SA03.EPS

Colored and Patterned Concrete

Concrete is sometimes colored and/or patterned to enhance its appearance. Concrete can be colored by adding a mineral-oxide pigment prepared especially for use with concrete to the dry cement powder prior to mixing the concrete. Plastic cement can be colored by working a commercially prepared color dye into the concrete surface after first floating, edging, and grooving the surface.

Geometric patterns can also be stamped, sawed, or scored into concrete surfaces to enhance their appearance. An imprint roller was used to simulate a stone pattern in the concrete shown here.

108SA04.EPS

5.0.0 ◆ CURING METHODS AND MATERIALS

Concrete gains strength by the process of hydration, which is a cement-water reaction. If concrete is allowed to dry prematurely, the reaction ceases, and the strength is severely reduced and cannot be recovered even if the concrete is re-wetted. This is especially true on the surface of concrete. Re-wetting dried concrete also causes cracking, especially on the surface.

Since any loss of moisture from the concrete will result in a loss of strength or surface cracking, proper concrete curing, regardless of the method used, should commence as soon as forms are removed or as soon as the surface will not be damaged by the curing method.

• *Water* – Moist curing with water is ideal and is accomplished by keeping the concrete surface covered in water by the use of sprinklers, fog sprayers, or dikes. Water used for making concrete is suitable for curing use.
• *Curing compounds* – Liquid membrane-forming compounds of various types can be applied to fresh concrete to help ensure that the concrete will not dry out due to neglect. Some of these compounds will discolor the concrete, and

others will seal the surface for subsequent application of paint or floor coverings.
• *Curing paper or plastic sheeting* – These products perform the same function as curing compounds and do not discolor the concrete. However, they must be weighted down and sealed at the edges to prevent water evaporation and wind damage.
• *Mats and blankets* – Wet curing with cotton curing mats or moving blankets eliminates the need for continuous spraying. For winter concreting, commercial insulating blankets, fiberglass batts, or expanded plastic cell slabs can be used to protect the concrete from freezing; however, they generally must be protected from snow and water penetration.

6.0.0 ◆ CONCRETE SLUMP TESTING

A slump test is used to determine the consistency of concrete, referred to as its flowability or workability. The test is the distance, in inches, that the top of a conical pile of fresh concrete will sag after a standardized cone mold is removed (*Figure 3*). Concrete slump must conform to the structure specifications.

Figure 3 ◆ Measurement of concrete slump.

7.0.0 ◆ ESTIMATING CONCRETE VOLUME

Accurate measuring and estimating of concrete quantities is required for concrete work. Fortunately, most concrete structures can be divided into rectangular or circular shapes, individually estimated, and the results added together to obtain the required volume of concrete. For instance, a floor is a rectangular horizontal slab. A footing and a wall can be divided into a long rectangular (nearly square) shape representing the footing and a vertical slab representing the wall. The volume of each can be calculated and the results added together for the total volume. Blueprints for a project will provide the dimensions for the various portions of a concrete structure, and these dimensions are used to calculate the volume of

concrete required for the structure. For reference purposes, the *Appendix* provides area or volume formulas for various geometric shapes and a table for conversion of inches to fractions of a foot or a decimal equivalent.

7.1.0 Rectangular Volume Calculations

A number of methods can be used to determine the volume of concrete required for a rectangular object. One method uses the following formula:

Cubic yards of concrete
(rounded up to next $\frac{1}{4}$ yard) =

$$\frac{\text{width or heigh (ft)} \times \text{length (ft)} \times \text{thickness (ft)}}{27(\text{cubic ft/yard})}$$

To use the formula, all dimensions in inches must be converted to feet and/or fractions of a foot and then into a decimal equivalent. For example:

$$7" = \frac{7}{12}' \text{ or } 7 \div 12 = 0.58'$$
$$8" = \frac{8}{12}' \text{ or } \frac{2}{3}' \text{ or } 2 \div 3 = 0.66'$$
$$23" = \frac{23}{12}' \text{ or } 23 \div 12 = 1.92'$$

The other two methods involve knowing the width or height and length of an area in feet, along with the thickness in inches, and then using a concrete calculator (*Figure 4*) or a concrete table (*Figure 5*) and a simple formula to determine the volume. The concrete calculator cannot be used when the volume of concrete required is less than $\frac{1}{2}$ yard. Also, the results of the calculator must be estimated by sight between rule marks, which makes the results subject to some variation.

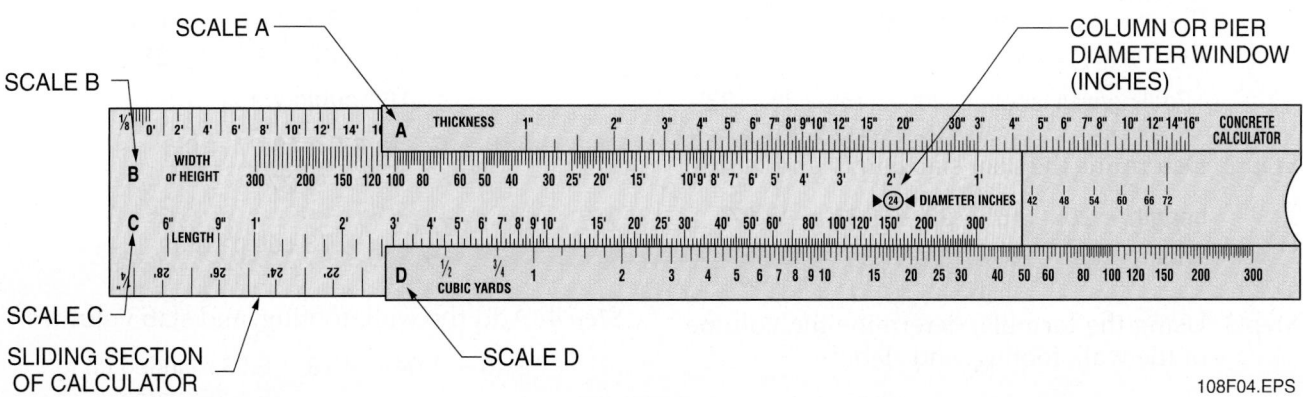

Figure 4 ◆ Typical concrete calculator.

ONE CUBIC YARD OF CONCRETE WILL PLACE:					
THICKNESS	SQ FT	THICKNESS	SQ FT	THICKNESS	SQ FT
1"	324	5"	65	9"	36
1¼"	259	5¼"	62	9¼"	35
1½"	216	5½"	59	9½"	34
1¾"	185	5¾"	56	9¾"	33
2"	162	6"	54	10"	32.5
2¼"	144	6¼"	52	10¼"	31.5
2½"	130	6½"	50	10½"	31
2¾"	118	6¾"	48	10¾"	30
3"	108	7"	46	11"	29.5
3¼"	100	7¼"	45	11¼"	29
3½"	93	7½"	43	11½"	28
3¾"	86	7¾"	42	11¾"	27.5
4"	81	8"	40	12"	27
4¼"	76	8¼"	39	15"	21.5
4½"	72	8½"	38	18"	18
4¾"	68	8¾"	37	24"	13.5

108F05.EPS

Figure 5 ◆ Portion of typical concrete table.

7.1.1 Example Calculation Using the Formula

Using the calculation formula, determine the amount of concrete required for the partial wall, footing, and floor slab plan shown in *Figure 6*.

Step 1 The entire footing and wall length must be determined. Since the wall is centered on the footing, the wall length is the same as the footing length:

Footing/wall length = 20' + (15' − 2') = 33'

Step 2 Determine the floor slab length and width:

Length = 20' − 16" = 20' − 1.33' = 18.67'
Width = 15' − 16" = 15' − 1.33' = 13.67'

Step 3 Using the formula, determine the volume of the wall, footing, and slab:

$$\text{Volume} = \frac{\text{width or height (ft)} \times \text{length (ft)} \times \text{thickness (ft)}}{27(\text{cubic ft} / \text{yard})}$$

$$\text{Wall} = \frac{3 \times 33 \times 0.67}{27} = \frac{66.33}{27} = $$

2.46 cubic yards

$$\text{Footing} = \frac{2 \times 33 \times 0.67}{27} = \frac{44.22}{27} = $$

1.64 cubic yards

$$\text{Slab} = \frac{13.67 \times 18.67 \times 0.5}{27} = $$

$$\frac{127.6}{27} = 4.73 \text{ cubic yards}$$

Step 4 Add the wall, footing, and slab volumes:

2.46 + 1.64 + 4.73 = 8.83 rounded up to 9 cubic yards

For the plan shown, 9 cubic yards of concrete will be required.

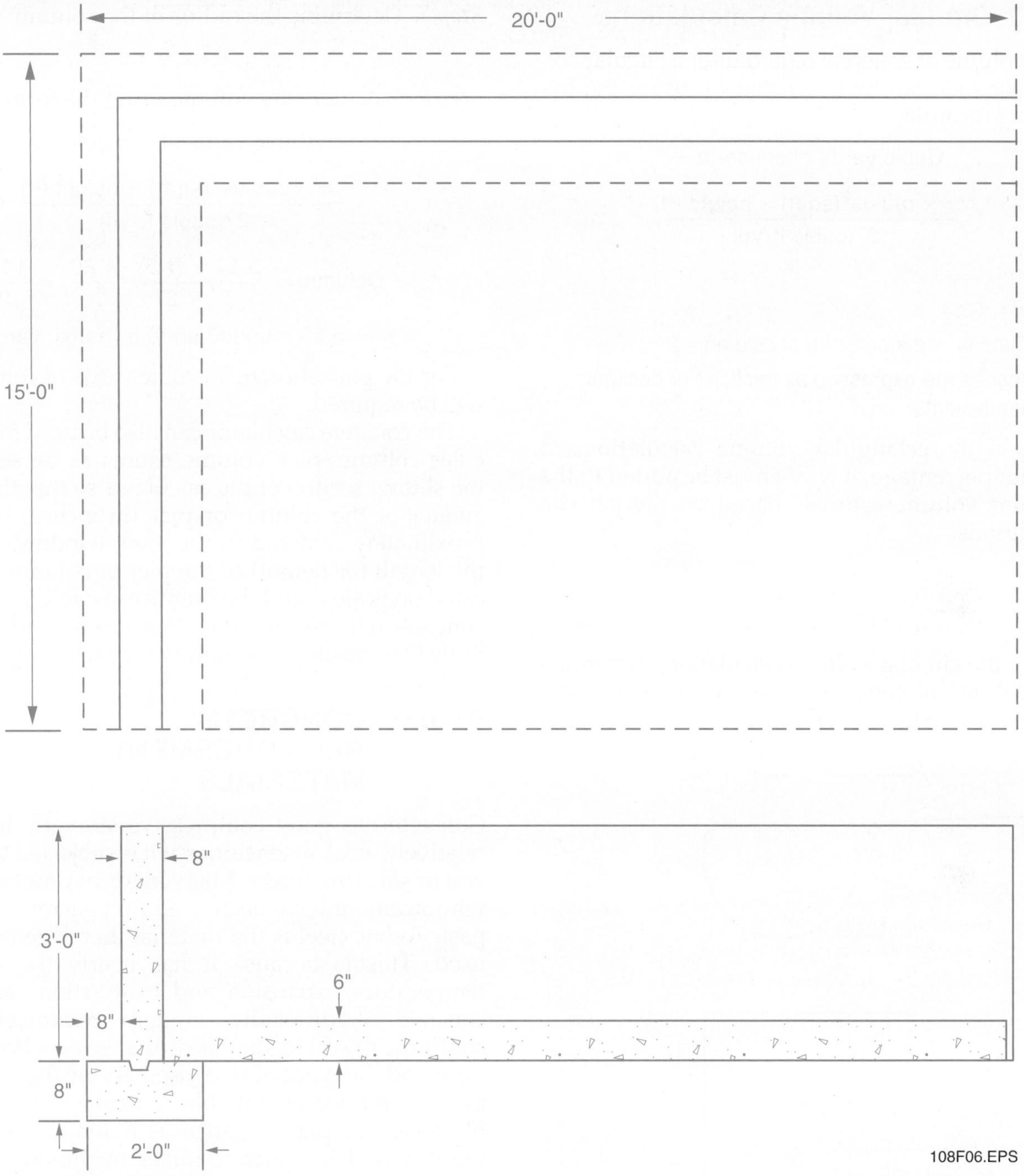

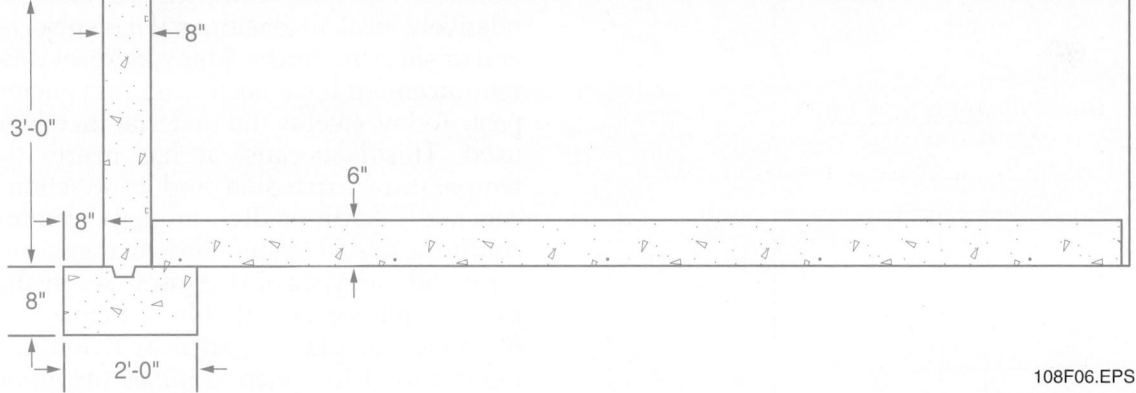

Figure 6 ◆ Partial wall, footing, and floor slab plan.

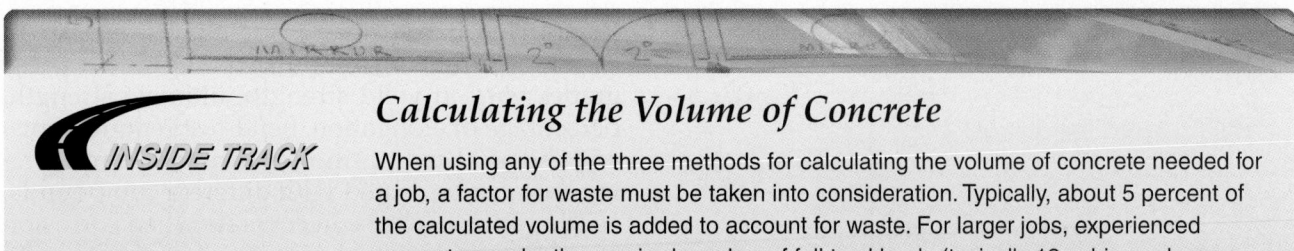

Calculating the Volume of Concrete

When using any of the three methods for calculating the volume of concrete needed for a job, a factor for waste must be taken into consideration. Typically, about 5 percent of the calculated volume is added to account for waste. For larger jobs, experienced carpenters order the required number of full truckloads (typically 12 cubic yards per load), then specify the required amount needed for the last truckload once the total volume of concrete needed is known.

7.2.0 Circular Volume Calculations

The volume of concrete required for a circular column or pier can be calculated by using the following formula:

$$\text{Cubic yards of concrete} =$$
$$\frac{\pi \times \text{radius}^2(\text{sq ft}) \times \text{height (ft)}}{27(\text{cubic ft/yd})}$$

Where:

- $\pi = 3.14$
- Radius = diameter (ft) of column ÷ 2
- Inches are expressed as fractions or decimal equivalents

As with rectangular volume calculations, a certain percentage of waste must be added to the circular volume estimate based on the job site conditions.

7.2.1 Example Calculation of Circular Volume

Using the circular volume calculation, determine the volume of concrete required for the circular column plan shown in *Figure 7*.

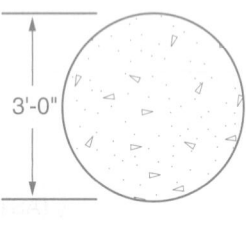

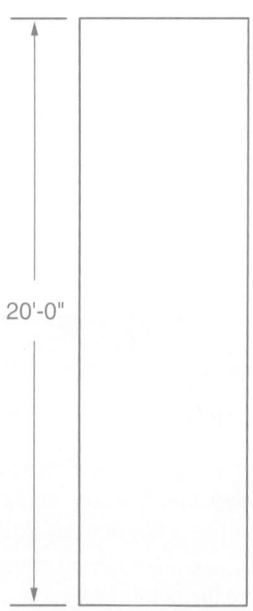

3'-0"

20'-0"

108F07.EPS

Figure 7 ◆ Typical circular column plan.

Step 1 Determine the radius of the column:

$$3' \div 2 = 1.5'$$

Step 2 Calculate the volume using the formula:

$$\text{Cubic yards of concrete} =$$
$$\frac{\pi \times \text{radius}^2(\text{sq ft}) \times \text{height (ft)}}{27(\text{cubic ft/yd})}$$

$$\text{Column} = \frac{3.14 \times (1.5)^2 \times 20}{27} = \frac{141.3}{27}$$

$$= 5.23 \text{ rounded up to } 5\frac{1}{4} \text{ cubic yards}$$

For the plan shown, 5¼ cubic yards of concrete will be required.

The concrete calculator can also be used for circular column/pier volume estimates by setting the sliding section of the calculator so that the diameter of the column or pier (in inches) is approximately centered in the view window. Then the length (or height) of the pier or column is located on Scale C and the number of cubic yards of concrete required for the structure is read from Scale D opposite the length value on Scale C.

8.0.0 ◆ CONCRETE REINFORCEMENT MATERIALS

Concrete has good compressive strength, but is relatively weak in tension or if it is subjected to lateral or shearing forces. Many kinds of proprietary reinforcement have been used for concrete in the past. Today, steel is the material that is generally used. This is because it has nearly the same temperature expansion and contraction rate as concrete. Additionally, modern reinforcement conforms to ASTM standards that govern both its form and the types of steel used. As an alternative to steel reinforcement, fibers made from steel, fiberglass, or plastics, such as nylon, are sometimes added to concrete mixes to provide reinforcement.

8.1.0 Reinforcing Bars

Reinforcing bars (**rebars**), sometimes referred to as rerods, are available in several grades. These grades vary in yield strength, ultimate strength, percentage of elongation, bend-test requirements, and chemical composition. Furthermore, reinforcing bars can be coated with different compounds, such as epoxy, for use in concrete when corrosion could be a problem. In order to obtain uniformity throughout the United States, the ASTM has established standard specifications for these bars. These grades will appear on bar-bundle tags, in color coding, in rolled-on markings on the bars, and/or on bills of materials.

The specifications are:

- *A615–Standard Specification for Deformed and Plain Carbon-Steel Bars for Concrete Reinforcement*
- *A996–Standard Specification for Rail-Steel and Axle-Steel Deformed Bars for Concrete Reinforcement* (This standard replaces *A616* and *A617*.)
- *A706–Standard Specification for Low-Alloy Steel Deformed Bars and Plain Bars for Concrete Reinforcement*

The standard configuration for reinforcing bars is the deformed bar. Different patterns may be impressed upon the bars depending on which mill manufactured them, but all are rolled to conform to ASTM specifications. The deformation improves the bond between the concrete and the bar and prevents the bar from moving in the concrete.

Plain bars are smooth and round without deformations on them and are used for special purposes, such as for dowels at expansion joints where the bars must slide in a sleeve, for expansion and contraction joints in highway pavement, and for column spirals.

Deformed bars are designated by a number in eleven standard sizes (metric or inch-pound), as shown in *Table 1*. The number denotes the approximate diameter of the bar in eighths of an inch or metric (mm). For example, a #5 bar has an approximate diameter of ⅝". The nominal dimension of a deformed bar (nominal does not include the deformation) is equivalent to that of a plain bar having the same weight per foot.

As shown in *Figure 8*, bar identification is accomplished by ASTM specifications, which require that each bar manufacturer roll the following information onto the bar:

- A letter or symbol to indicate the manufacturer's mill
- A number corresponding to the size number of the bar (*Table 1*)
- A symbol or marking to indicate the type of steel (*Table 2*)
- A marking to designate the grade (*Table 3*)

The grade represents the minimum yield (tension strength) measured in **kips** per square inch (ksi) or megapascals (MPa) that the type of steel used will withstand before it permanently stretches (elongates) and will not return to its original length. Today, Grade 420 is the most commonly used rebar. Bars are normally supplied from the mill bundled in 60' lengths.

Bar fabrication is accomplished for straight bars by cutting them to specified lengths from the 60' stock. Bent bars are cut to length the same as straight bars and then they are assigned to a bending machine that is best suited for the type of bend and size of the bar. *Table 1* provides size and weight information for various bars so that

Table 2 Reinforcement Bar Steel Types

Symbol/Marking	Type of Steel
A	Axle (ASTM A996)
S or N	Billet (ASTM A615)
I or IR	Rail (ASTM A996)
W	Low-alloy (ASTM A706) (for welded lap, butt joints, etc.)

Table 1 ASTM Standard Metric and Inch-Pound Reinforcing Bars

		Nominal Characteristics*					
Bar Size		Diameter		Cross-Sectional Area		Weight	
Metric	[in-lb]	mm	[in]	mm	[in]	kg/m	[lbs/ft]
#10	[#3]	9.5	[0.375]	71	[0.11]	0.560	[0.376]
#13	[#4]	12.7	[0.500]	129	[0.20]	0.944	[0.668]
#16	[#5]	15.9	[0.625]	199	[0.31]	1.552	[1.043]
#19	[#6]	19.1	[0.750]	284	[0.44]	2.235	[1.502]
#22	[#7]	22.2	[0.875]	387	[0.60]	3.042	[2.044]
#25	[#8]	25.4	[1.000]	510	[0.79]	3.973	[2.670]
#29	[#9]	28.7	[1.128]	645	[1.00]	5.060	[3.400]
#32	[#10]	32.3	[1.270]	819	[1.27]	6.404	[4.303]
#36	[#11]	35.8	[1.410]	1006	[1.56]	7.907	[5.313]
#43	[#14]	43.0	[1.693]	1452	[2.25]	11.380	[7.650]
#57	[#18]	57.3	[2.257]	2581	[4.00]	20.240	[13.600]

*The equivalent nominal characteristics of inch-pound bars are the values enclosed within the brackets.

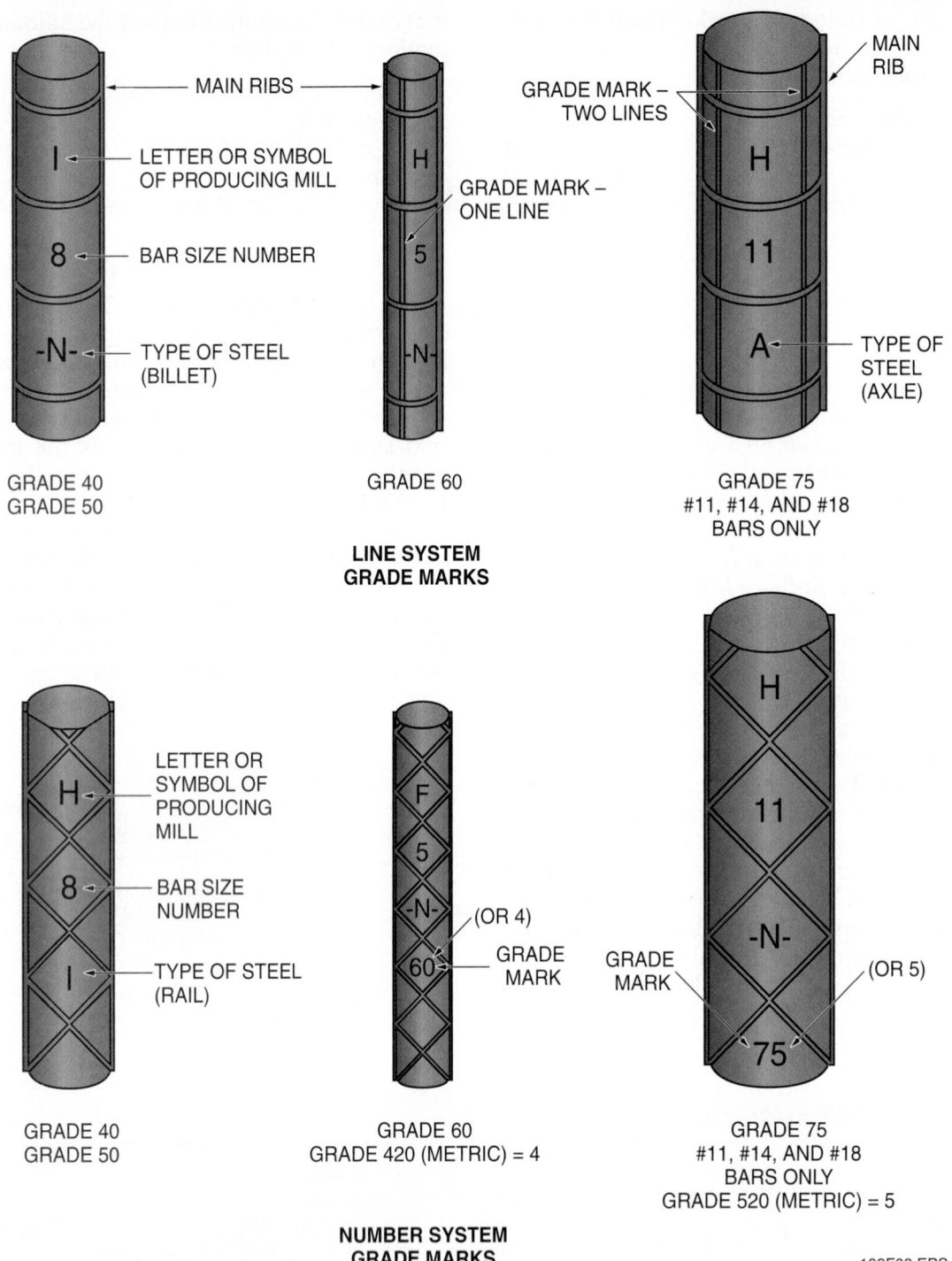

Figure 8 ◆ Reinforcement bar identification.

Table 3	Reinforcement Bar Grades	
Grade	**Identification**	**Minimum Yield Strength**
40 and 50	None	40,000 to 50,000 psi (40 to 50 ksi)
60	One line or the number 60	60,000 psi (60 ksi)
70	Two lines or the number 70	70,000 psi (70 ksi)
420	The number 4	420 MPa (60,000 psi or 60 ksi)
520	The number 5	520 MPa (75,000 psi or 70 ksi)

proper handling and bending equipment can be selected.

Uncoated reinforcement steel that has not been contaminated by oil, grease, or preservatives will normally rust when stored—even for short lengths of time under cover. A number of studies, some conducted over 70 years ago, have shown that rust and tight mill scale actually improve the bond between the steel and the concrete. Other studies have shown that normal handling (moving, bending, etc.) of extremely rusted reinforcement steel prepares it sufficiently for proper bonding with concrete without additional effort to remove the rust.

Bender and Cutter

The portable rebar cutter and bender shown here are typical of those used in the field for cutting and bending rebar in sizes up to #7 (⅞").

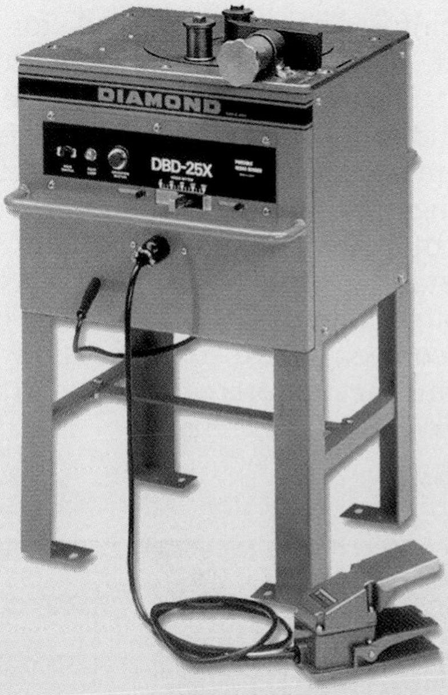

BENDER

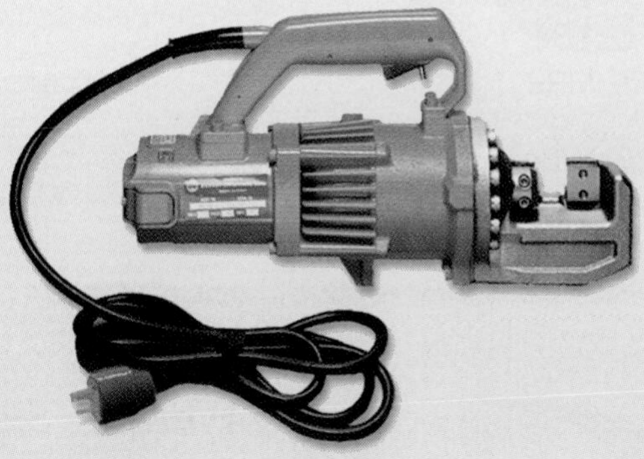

CUTTER

108SA05.EPS

Steel Reinforcement

Steel reinforcement of any kind must be covered by enough concrete to be adequately protected; otherwise, the steel will rust, causing damage to the concrete. Some examples of minimum concrete coverage are:

- Footings – 3"
- Concrete surface exposed to weather –2" for bars larger than #5, 1½" for bars #5 and smaller
- Slabs, walls, joists – ¾"
- Beams and girders – 1½"

8.2.0 Bar Supports

Bar supports are used to support, hold, and space reinforcing bars and mats or wire fabric before and during concrete placement. Bar supports are made from steel, concrete, or plastic. When used with coated reinforcement steel, the supports should also be coated with the same material or made of concrete or plastic to prevent corrosion. It is also important that a sufficient number of supports be used to prevent the reinforcement from shifting out of position or deforming when the concrete is placed.

Figure 9 shows a standard roll of tie wire and a wire tie (pigtail) used to secure lengths of reinforcing steel to each other or to various supports.

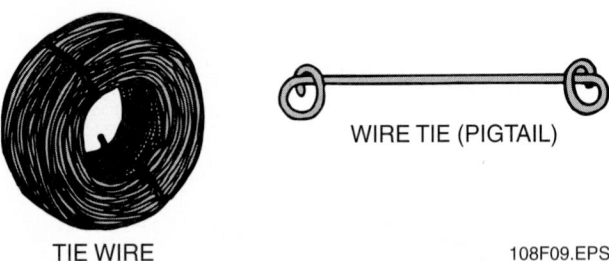

WIRE TIE (PIGTAIL)

TIE WIRE

108F09.EPS

Figure 9 ◆ Tie wire and pigtail.

8.3.0 Splicing Reinforcing Bars

Because in most situations it is impossible to provide full-length bars that run continuously throughout a structure, making splices in reinforcing bars is a common occurrence. The placing drawings will show the location and type of splice to use. Sometimes several methods of splicing will be listed by the placing drawings, and the best method may be chosen from among them. No other type of splice should be used without consulting the proper authority.

There are three basic types of splices used in reinforcing steel work. These are the lap splice, the welded splice, and the mechanical-coupling splice.

8.4.0 Welded-Wire Fabric

When reinforcement is required for concrete pavement, parking lots, driveways, floor slabs, etc., welded-wire fabric (WWF) can be used instead of individual rebars. WWF consists of longitudinal and transverse steel wires electrically welded together to form a square or rectangular mesh or mat. Depending on the wire diameter, which can range up to ¾" or more, WWF is available in roll form or in 4' × 8' flat mats. *Figure 10* illustrates some standard sizes of plain wire WWF in roll form. Bar supports can be used to space and secure WWF as well as rebar.

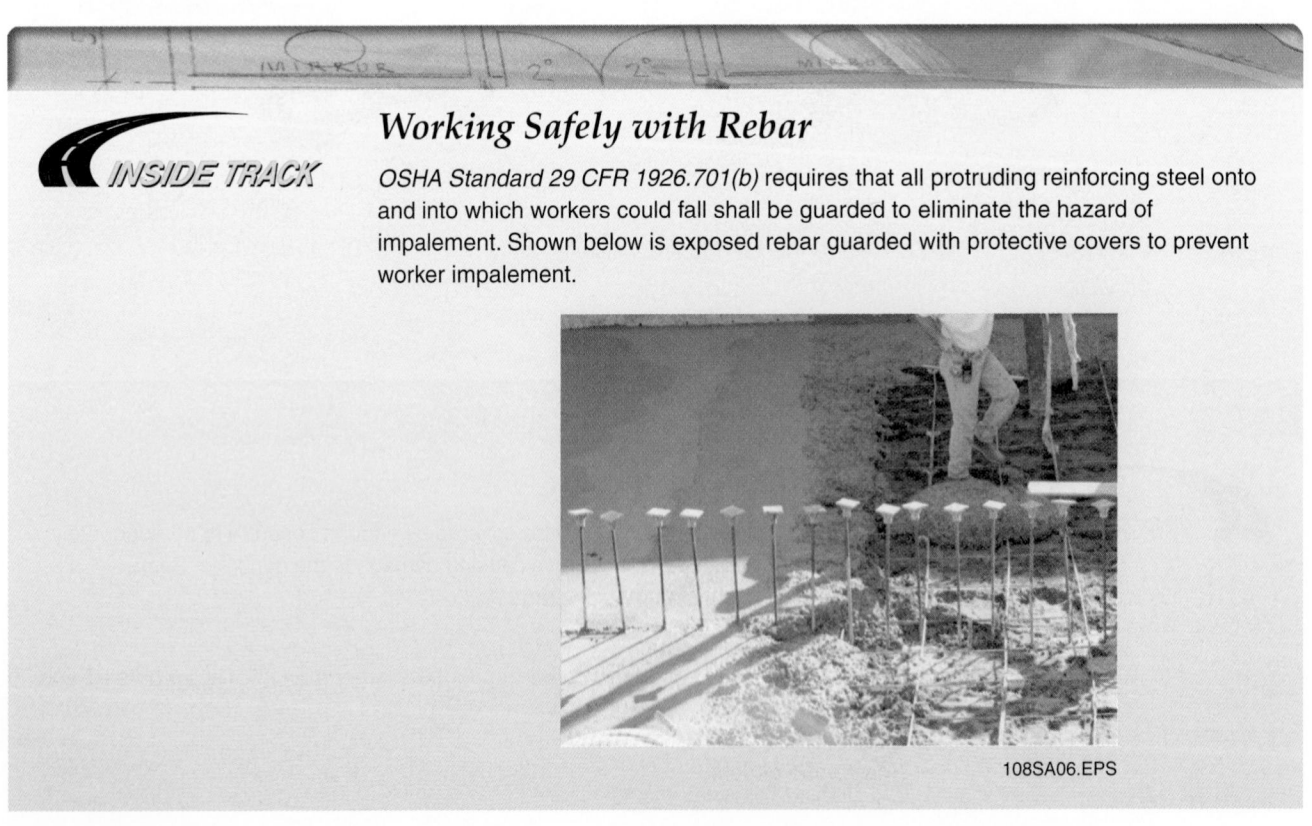

Working Safely with Rebar

INSIDE TRACK

OSHA Standard 29 CFR 1926.701(b) requires that all protruding reinforcing steel onto and into which workers could fall shall be guarded to eliminate the hazard of impalement. Shown below is exposed rebar guarded with protective covers to prevent worker impalement.

108SA06.EPS

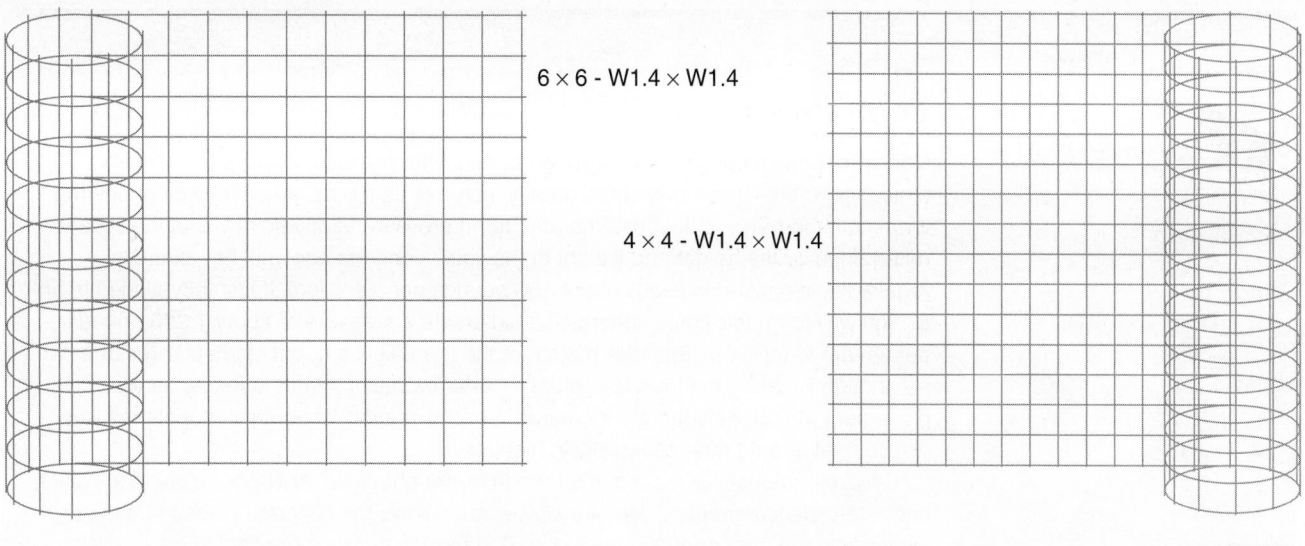

6 × 6 - W1.4 × W1.4

4 × 4 - W1.4 × W1.4

108F10.EPS

Figure 10 ◆ Typical rolled, plain wire welded-wire fabric.

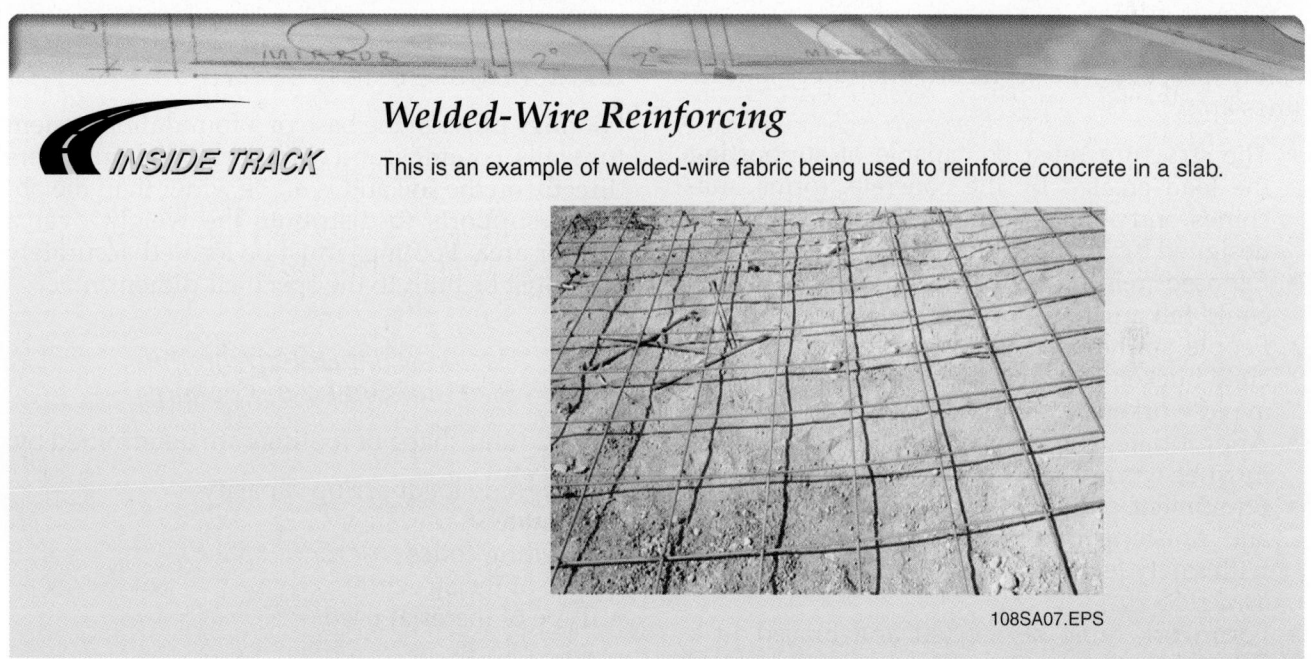

INSIDE TRACK

Welded-Wire Reinforcing

This is an example of welded-wire fabric being used to reinforce concrete in a slab.

108SA07.EPS

9.0.0 ◆ CONCRETE FORMS

A concrete form is a temporary structure or mold used to support concrete until the hardening process gives it sufficient strength to be self-supporting. Carpenters generally assemble the job-built concrete forms that are used to construct foundations, walls, columns, and other concrete structures.

9.1.0 Form Safety

A concrete form is subjected to tremendous pressure from wet concrete. The pressure increases as the height of the wall increases. The rate at which the concrete is poured also affects the pressure on the form. The faster the rate of pour, the greater the pressure.

If the forms are not properly constructed and supported, or if the concrete is poured too quickly, the forms will not be able to withstand the pressure of the wet concrete. They will fracture, releasing the concrete. This can be very dangerous to workers, notwithstanding the mess it will make at the job site. Therefore, it is very important that the form builder construct the form in accordance with established standards and

Form Pressures

Concrete forms must be strong enough to resist the pressures created by plastic (liquid) concrete. These pressures greatly increase as the height of the wall or column form increases. This is because the amount of pressure at any point within the form is determined by the height and weight of the liquid concrete above it. For example, regular liquid concrete weighs about 150 pounds per cubic foot. If instantly placed in an 8' high wall form, this liquid concrete would create a pressure of about 1,200 pounds per square foot (150 × 8) at the bottom of the form. This amount of lateral pressure can be enough to cause the form to rupture. Pressures within a form increase as the placement rate of the concrete increases. For this reason, concrete is not placed in a wall form at a rapid rate, especially in high forms.

Concrete normally is placed in a form in layers about 12" to 18" deep using a slower, consistent placement rate. Use of a slower rate allows the concrete placed in a lower layer within the form enough time to begin to harden before more liquid concrete is placed on top of it. As the lower layer sets up and starts to harden, it becomes stable and ceases to exert pressure on the form even though liquid concrete continues to be placed above it.

safety practices. Basic formwork safety guidelines are:

- The structure must be capable of supporting the load created by the concrete, forms, machines, and people. Forms and shoring must be designed by qualified designers.
- Exposed reinforcing steel on which people could fall must be guarded.
- People are not permitted to work under concrete buckets. Buckets must be routed so as to avoid workers.
- Appropriate personal protective equipment must be worn.
- Equipment and tools used for the placement and finishing of concrete must be equipped with safety features as specified by *OSHA Standard 1926.702*.
- Formwork must be erected and braced in a manner that ensures it will support all vertical and lateral loads.
- Shoring equipment must be inspected and properly maintained. Damaged or defective shoring equipment must be repaired or replaced.
- Forms and shoring equipment may not be removed until the concrete has been verified as strong enough to support its weight and that of its loads.
- Reinforcing steel must be adequately supported to prevent it from falling or collapsing.
- Measures must be taken to prevent uncoiled wire mesh from recoiling.
- Measures must be taken to avoid skin contact with concrete, which is a hazardous material.

9.2.0 Footings

Footings provide the base of a foundation system for walls, columns, and chimneys. A footing bears directly on the soil and is made wider than the object it supports to distribute the weight over a greater area. Footings must be located accurately and must be built to the specified dimensions.

9.2.1 Factors That Determine the Size and Shape of Footings

The size and shape of footings are determined by:

- Subgrade loadbearing capacity
- Climate
- Building codes
- Size of the structure
- Type of material used

When designing the foundation for a structure, an engineer must consider the soil conditions of the building site. Weak or unstable soils may require larger footings or engineered foundations. Some soils are so weak that a slab may be required to spread the weight over the greatest possible area. Building a structure on soil where the soil conditions can cause a large amount of uneven settlement to occur can result in cracks in the foundation and structural damage to the rest of the building. Typically, the architect consults a soil engineer who makes test bores of the soil on the building site and analyzes these samples. Based on the results, the architect designs suitable footings and a foundation for the structure.

INSIDE TRACK

Footings

This is an example of concrete placed in a continuous footing form.

108SA08.EPS

INSIDE TRACK

Reinforced Footings

Steel reinforcing rods, commonly called rebar, are embedded in footings to make them stronger. Additional reinforcement of footings is typically done over weak spots in the soil, such as where there have been excavations for sewer, gas, or other connections.

On sites where the soil is firm and not too porous, the construction of a footing form is often unnecessary. A trench is dug into the soil to the width and depth of the footing and the concrete is placed into the trench. Note that some problems can occur when using this method, such as the soil being porous enough to absorb too much water from the concrete, or the soil from the trench walls falling into the concrete.

WARNING

Any time you are working in or near a trench, you must follow applicable safety precautions.

Frost in the ground also affects foundation design. During the winter months in colder regions, moisture from the rain and snow penetrates the ground and freezes. The depth to which the soil freezes in a region is called the frost line. Freezing and thawing of the soil above the frost line causes the soil to expand and contract. This would cause foundations with footings placed above the frost line to heave and buckle. For this reason, foundation footings must always be installed below the frost line. Local building codes normally specify how deep the foundation footings should be in the area. In extreme northern climates, footings may be as much as 6' to 8' below the surface. In tropical climates, footings usually only need to reach solid soil, with no regard for a frost line.

Building codes can be restrictive when it comes to foundations. Codes regulate the size and thickness of footings and foundation walls, the depth of the footing, and other construction factors. In areas subject to earthquakes (seismic risk zones), the foundation will receive the forces of the quake and therefore needs to be built much stronger than foundations in non-seismic risk zones. In seismic risk zones, local building codes normally require that a foundation be designed to withstand greater stress. This is accomplished by embedding reinforcing steel rods (rebar) in all concrete or masonry foundation walls constructed in these zones.

9.2.2 Types of Footings

There are four basic types of footings:

• Continuous or spread
• Stepped continuous
• Pier
• Grade beam

Continuous footings – Continuous footings (*Figure 11*), also called spread footings, are commonly used to support poured concrete or concrete block foundation walls. They can be formed on the ground or partially in or out of the ground. The method used depends on the soil conditions and specifications defined in the working drawings. For a residential building with normal soil conditions, the width of a continuous footing is typically equal to twice the thickness of the foundation wall, and its height or depth is equal to the wall thickness. Both sides of the footing project out from the wall by one-half of the wall thickness. For commercial and industrial buildings, the dimensions of the footings vary depending on the type of structure and the building loads. Also, the foundation wall is not always centered on the footing.

When a foundation is to be built on hilly or steeply sloped land, it is necessary to step the footing at intervals (*Figure 12*) to achieve grade levels. When building stepped footing forms, the thickness of the footing must be maintained. Building codes normally govern the construction of stepped footings. Many building codes specify that for residential buildings the distance between one horizontal step and another must be no less than 2'. The vertical footing must be no higher than ¾ of the distance between one horizontal step and another and it must be at least 6" thick. Vertical boards are placed between the forms to retain the concrete at each step.

Piers – Piers are isolated footings set in soil to directly support posts, columns, or grade beams. The pier size, which is normally specified in the working drawings, is determined by soil conditions, the weight of the structure, and the loads supported by the structure. Piers are made in different shapes (*Figure 13*) including square, rectangular, tapered (battered), and round. Regardless of pier type, the ground must be dug so that the base of the pier rests on firm soil and is below the frost line, or is supported by a proper footing.

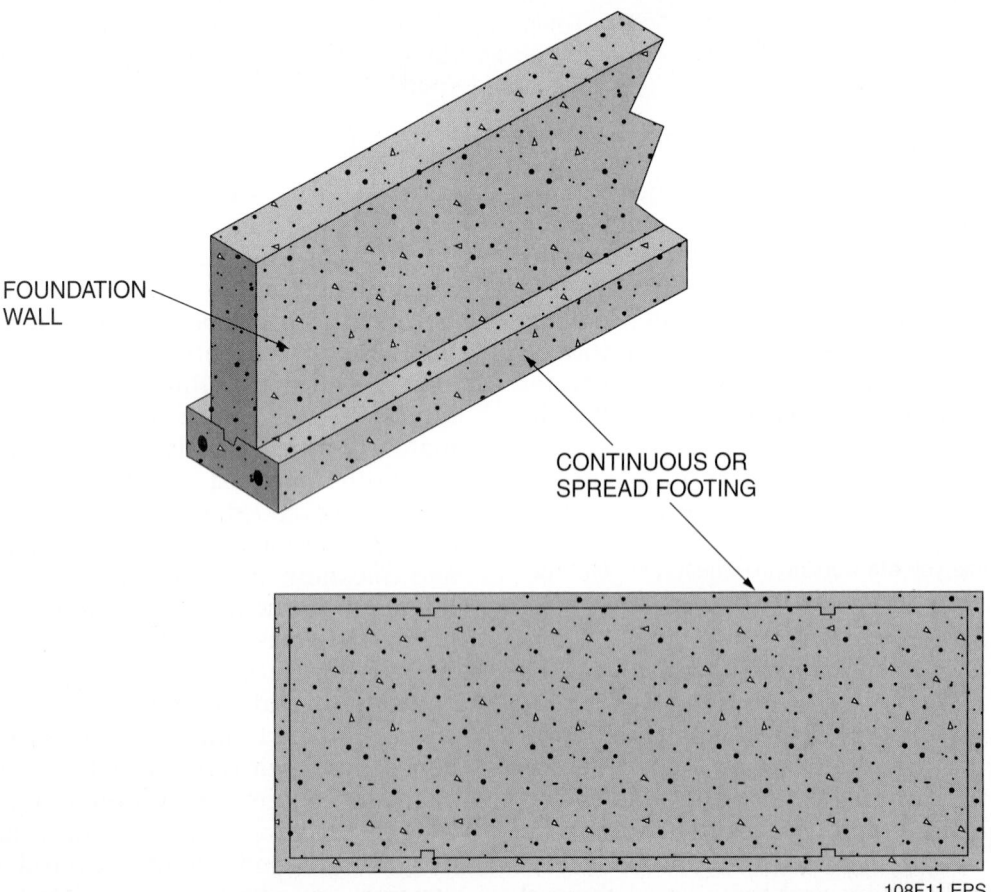

FOUNDATION WALL

CONTINUOUS OR SPREAD FOOTING

108F11.EPS

Figure 11 ◆ Continuous or spread footing.

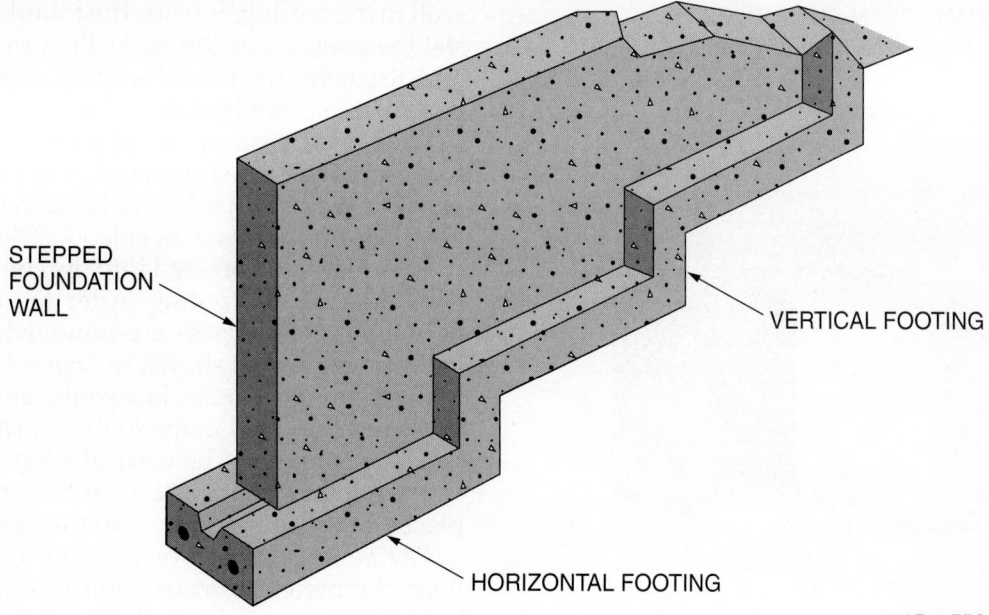

STEPPED
FOUNDATION
WALL

VERTICAL FOOTING

HORIZONTAL FOOTING

108F12.EPS

Figure 12 ◆ Continuous stepped footing.

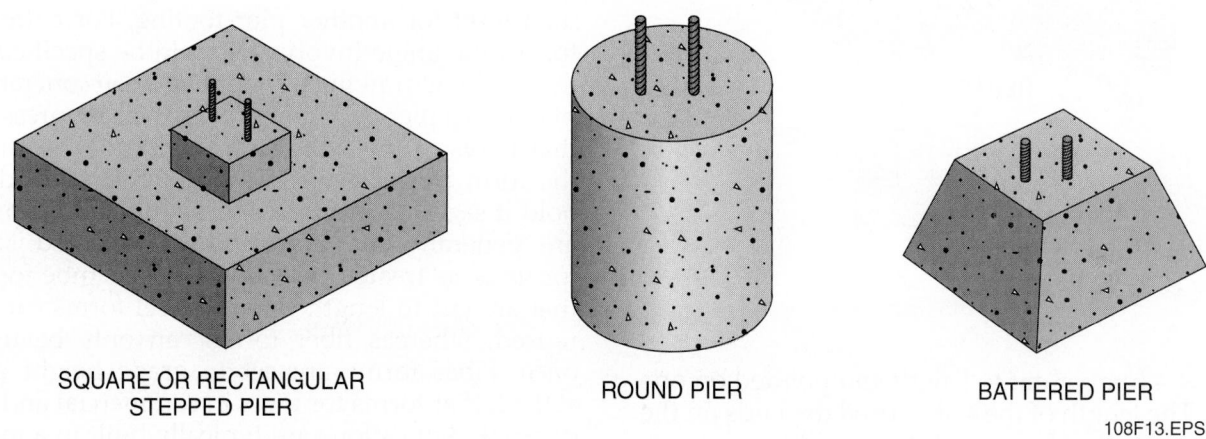

SQUARE OR RECTANGULAR
STEPPED PIER

ROUND PIER

BATTERED PIER

108F13.EPS

Figure 13 ◆ Typical piers.

Grade beams – Grade beams are foundation walls that receive their main support from concrete piers that extend deep into the ground (*Figure 14*). They are often used with stepped or ramped foundations on steep hillsides. Grade beams are also used on level sites where unstable soil conditions exist. In heavy construction, it may not be practical to excavate deep enough to reach loadbearing soil. Instead, **piles** are driven and capped with a grade beam.

9.2.3 Components of Footing and Pier Forms

The components of footing and pier forms are basically the same with some minor exceptions. Both types are described here.

Footing forms – Continuous or spread footing forms (*Figure 15*) are typically made from construction lumber (1×4s, 2×4s, 2×6s, etc.) selected to match the approximate width of the footing thickness. The use of thicker boards allows the form stakes to be placed farther apart and requires less bracing. The side (edge form) boards are placed on edge and held in place by the form stakes, braces, and cross spreaders. The form stakes are driven into the soil to provide primary support for the form.

The form members are fastened together with double-headed (duplex) nails or screws. The use of duplex nails allows for easy nail removal when the form is taken apart after use. The stakes, spreaders, and braces may be made of wood or metal. Wood stakes are typically made from 1×4

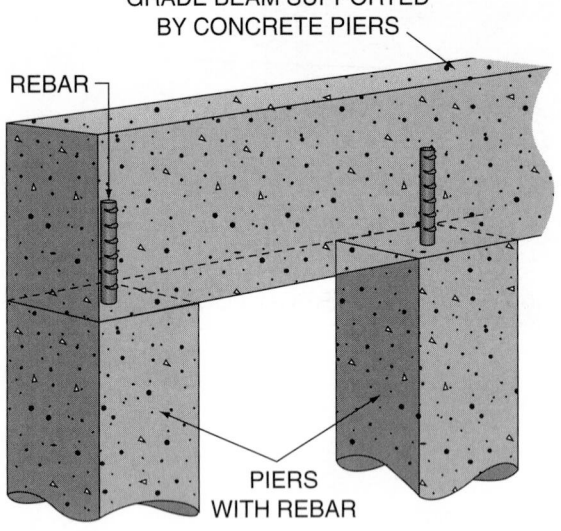

GRADE BEAM SUPPORTED
BY CONCRETE PIERS

REBAR

PIERS
WITH REBAR

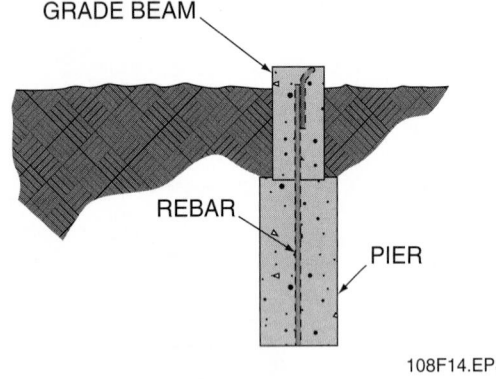

GRADE BEAM

REBAR

PIER

108F14.EPS

Figure 14 ◆ Grade beams.

or 2 × 4 lumber cut to length and pointed on one end. The length of the stakes used depends on the height of the form and the condition of the ground. Soft ground requires the use of longer stakes. Normally, a stake should be driven into the ground to a depth of 8" to 10". Metal stakes are commonly used when the ground is especially hard. These are made with a series of predrilled holes used for fastening them to the edge form boards with nails or screws.

A groove in the footing, called a keyway (*Figure 15*), shapes the footing concrete for interlocking with the future foundation wall pour. This joint acts to lock or secure the bottom of the foundation wall to the footing. It helps the foundation wall resist the pressure of the back-filled earth against it and also helps to prevent seepage of water into the basement or crawlspace.

A keyway strip is not always hung from the form spreaders, as shown in *Figure 15(A)*. Instead, after the concrete has been placed in the form and before it sets, pieces of beveled 2 × 4s (or metal or plastic strips) are pressed into the concrete toward the center of the footing. After the concrete has hardened, these pieces are removed, leaving the keyway groove, as shown in *Figure 15(B)*.

In seismic risk areas, horizontal and vertical rebars are positioned in the footing forms before the concrete is placed. The vertical rebars sticking out of the footing forms are later tied to the rebars placed in the foundation wall formwork.

Pier forms – Forms for piers in residential and light commercial construction typically are built by nailing plywood together in square, rectangular, or tapered (battered) shapes to the specified size. The form is built in such a way that it can be easily removed after the concrete is cured, and then reset for another pier footing. For battered forms, the angle involved should be specified in the working drawings. Note that when concrete is placed in a pier or column form, it exerts pressure that tends to push the form upward. Therefore, the form must be weighted down or staked to hold it securely in place. Forms for circular piers are generally manufactured as patented steel forms or as treated, waterproof, fiber tube forms that are cut to length on site. Steel forms can be reused, whereas fiber forms can only be used once. Fiber forms are set to grade height and staked. Pier forms for piers in commercial and industrial applications are typically built in a manner similar to the one shown in *Figure 16*. The sides of the form can be made from almost any exterior-type plywood sheathing.

The **walers** shown are horizontal members placed on the outside of the form to brace or stiffen the form and to which the form ties are fastened. Form ties are wire or metal cross ties used to hold the pressure of wet concrete and hold the form together to maintain its thickness. There are many types of form ties available, including washer-type, strap-type, cone-type, and she-bolt.

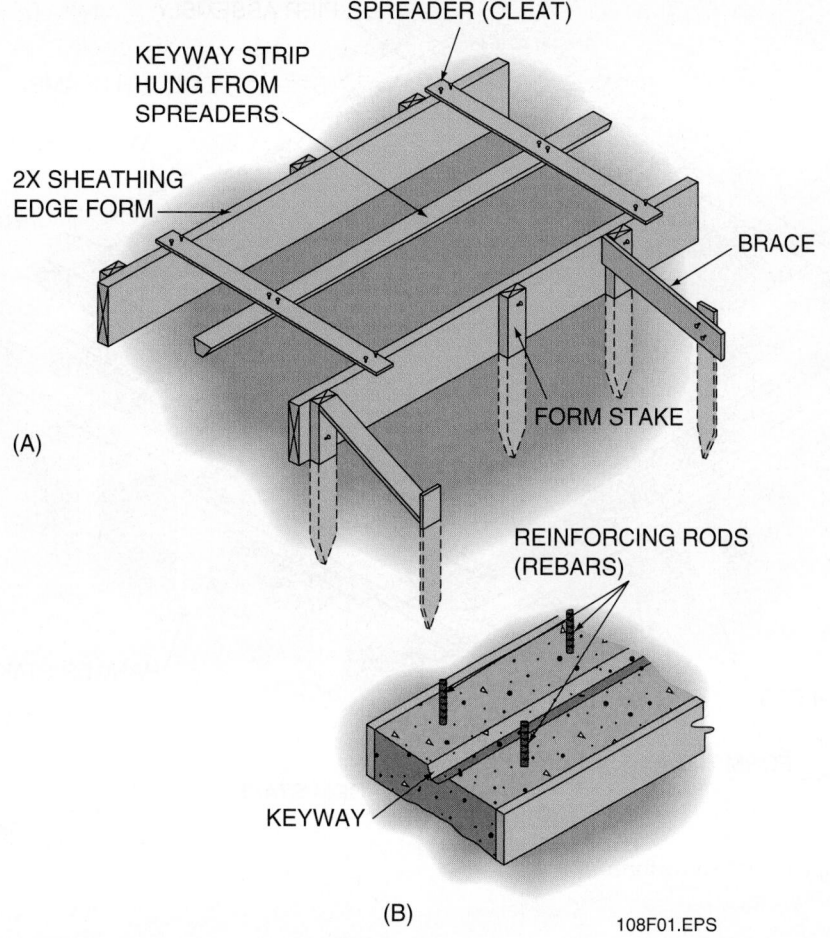

(A)

SPREADER (CLEAT)

KEYWAY STRIP
HUNG FROM
SPREADERS

2X SHEATHING
EDGE FORM

BRACE

FORM STAKE

REINFORCING RODS
(REBARS)

KEYWAY

(B)

108F01.EPS

Figure 15 ◆ Parts of a continuous or spread footing form.

Plywood Forms

The plywood industry makes a reusable plywood, called plyform, which is specifically designed for use in building concrete forms. It is an exterior-type plywood made from special wood and veneer grades to ensure high performance. Two classes of plyform are available. Plyform Class I is stronger and stiffer than Class II. Both can be purchased with a high-density surface overlaid on each side. This provides a very smooth, grainless surface that resists abrasion and moisture penetration.

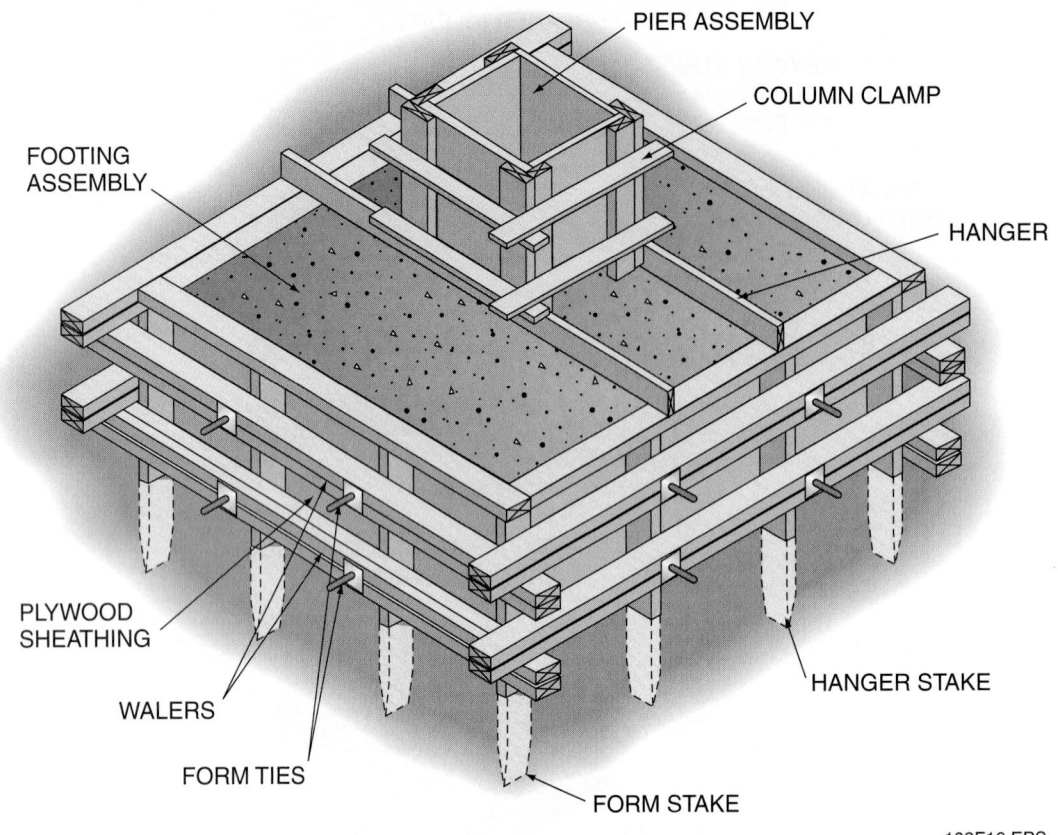

PIER ASSEMBLY

COLUMN CLAMP

FOOTING ASSEMBLY

HANGER

PLYWOOD SHEATHING

WALERS

FORM TIES

HANGER STAKE

FORM STAKE

108F16.EPS

Figure 16 ◆ Parts of a stepped pier footing form.

9.2.4 Laying Out and Constructing Forms for a Continuous Footing

Laying out and constructing forms for a continuous footing involves the following tasks:

- Establishing the exact location of the building on site
- Establishing building lines
- Performing required excavation and/or trenching, observing applicable safety precautions
- Laying out and constructing footing forms

Before any footing or foundation layout can be done, the dimensions of the building and its location on the site must be determined from the site (plot) plan. Following this, the exact location of the four corners of the building must be determined and staked out. This task can be accomplished in different ways. One method involves taping and the use of a builder's level or transit level. You will learn this method later in your training.

9.3.0 Wall Forms

A wall form consists of two sets of panels separated by the thickness of the concrete wall. The following are different types of wall forms:

- Forms that are constructed primarily from wood panels, lumber, and nails, with some specialized hardware
- Forms that are constructed using wood panels and lumber support members secured with patented attaching hardware
- Patented form systems that use metal panels and factory-built attaching hardware

While there may be variations of the above, these represent the basic types. In this module, we will cover the forms that are constructed primarily from wood panels, lumber, and nails, with some specialized hardware.

9.3.1 Components of a Wall Form

See *Figure 17*. The main components of a wall form panel include the following:

- *Top and bottom plates* – The horizontal members of a wall frame panel.
- *Studs* – The upright members that support the sheathing. They are nailed to the top and bottom plates. Studs are usually 2 × 4 or 2 × 6 lumber. Heavier studs may be needed, depending on the pressure that will be applied to the forms by the concrete. In job-built wooden forms, studs are required. Many patented forms have the upright supports built into the frame and therefore do not require separate studs.
- *Sheathing* – Gives the wall surface its shape and texture. It keeps the concrete in place.
- *Walers (wales)* – Horizontal members used to support the form and the studs. Like the studs, walers usually are made of 2 × 4 or 2 × 6 lumber, but heavier lumber may be required by the particular job.
- *Strongbacks* – Upright members used to stiffen or reinforce the form.
- *Form ties or spreaders* – Used to keep the form from being spread apart by the weight of the wet concrete and to keep the walls from shifting while the concrete is being poured. Spreaders must be installed above the level of the concrete.

- *Braces* – Fastened to one side of the form and nailed to stakes driven into the ground. Braces are usually placed 8' to 10' apart. They support the form and are also used to plumb the form. Adjustable metal braces, which make it easier to plumb the form, are available (*Figure 18*). Forms must be properly braced to ensure that they do not collapse under the weight of the plastic concrete.

Sheathing can be made of exterior plywood sheets. Just about any type of exterior plywood may be used. However, a special product called **plyform** is popular because it is specifically made for use in concrete forms. Plyform is an exterior plywood that is made only from certain grades and types of wood. Plyform panels are sanded on both sides and are usually oiled at the mill. The oil reduces moisture penetration and aids in releasing the form from the concrete. Plyform edges exposed during construction must be treated with oil or another release agent before use. There are two basic grades of plyform (Class I and Class II), plus a structural grade that is designed to withstand greater pressures.

Planks may also be used to form walls, in some cases. The planks are nailed to studs.

Quarter-inch tempered hardboard that is specially treated and coated may also be used as

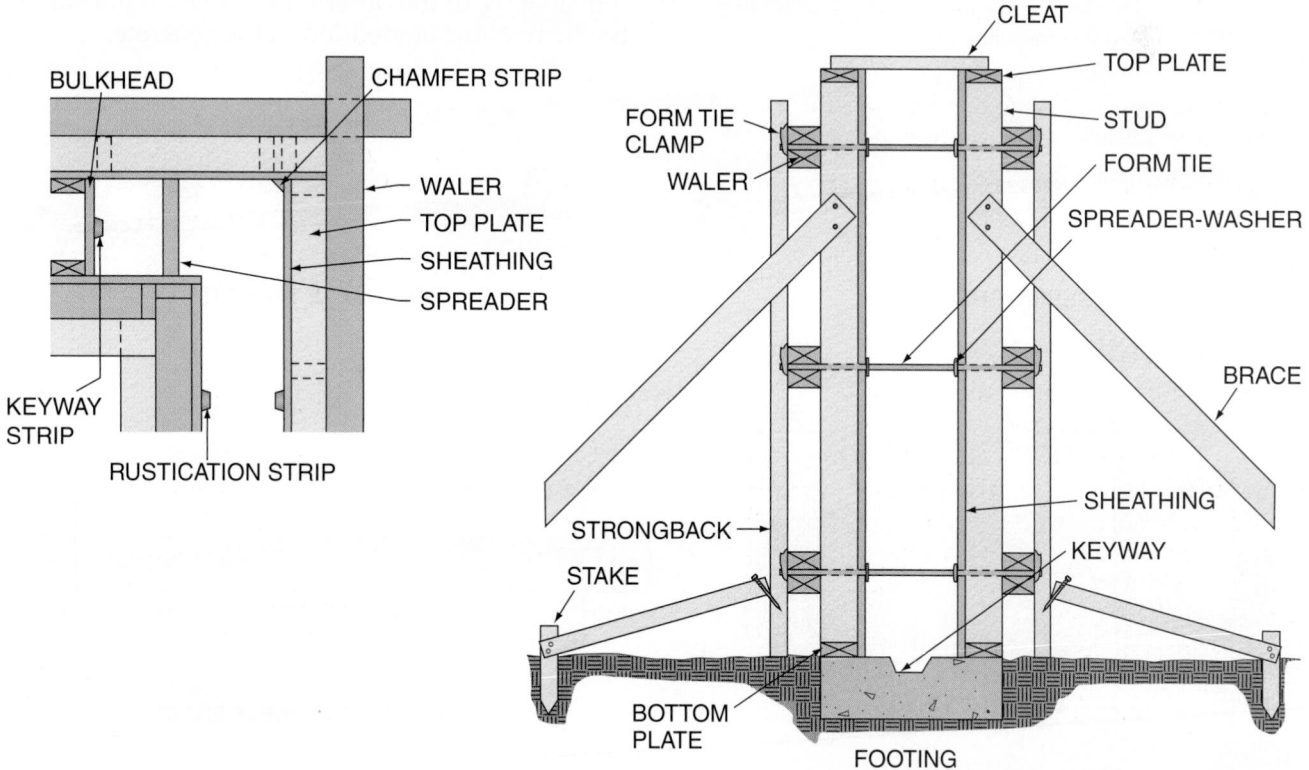

Figure 17 ◆ Parts of a wall form panel.

108F17.EPS

INSIDE TRACK

Integrity of Wall Forms

Always make sure to use lumber that is free of defects when building wall forms, especially lumber used for braces and supports. Before placing concrete in a wall form or any other type of form, the form must be inspected for structural adequacy and compliance with the job specifications by a competent, authorized person.

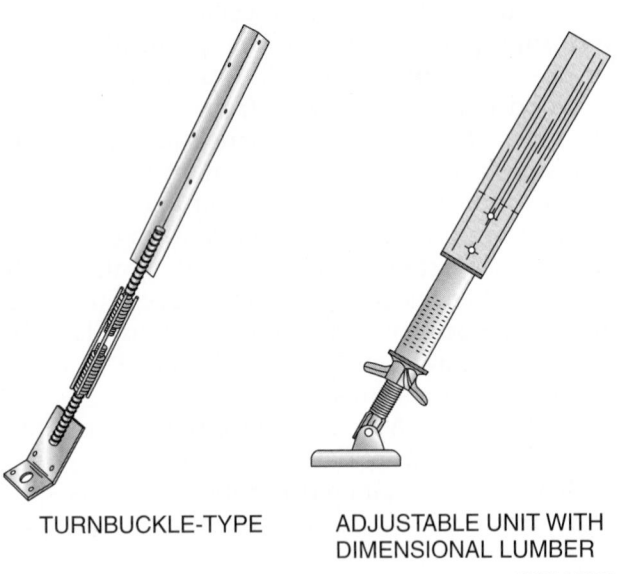

TURNBUCKLE-TYPE　　ADJUSTABLE UNIT WITH DIMENSIONAL LUMBER

108F18.EPS

Figure 18 ◆ Adjustable form braces.

sheathing. If hardboard is used, however, it must be backed with lumber. Hardboard used in this manner acts as a liner to obtain a smooth concrete finish.

Form ties are metal rods or metal straps that hold the two sides of a wall form in position (*Figure 19*). Spreaders, which serve the same purpose, are often made from pieces of wood cut to the thickness of the wall and nailed in place between the two wall form sections above the concrete. Ties generally remain in the concrete; wood spreaders are removed as the concrete is poured to their level. Some ties are equipped with cones or other devices that allow them to act as spreaders.

When the forms are stripped, ties will project out from the hardened concrete. They are broken off, twisted off, disconnected, or removed, depending on the type of tie. The tapered tie is removable. With the other types of ties, a portion of the tie remains embedded in the concrete.

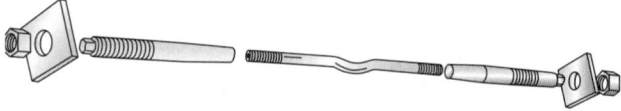

LEAVE-BEHIND SHE-BOLT

REMOVABLE TAPER TIE

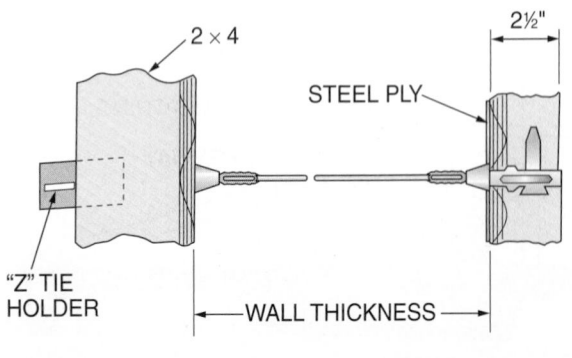

BREAK-OFF LINE TIE

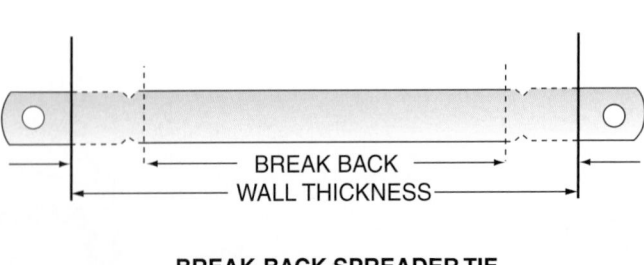

BREAK-BACK SPREADER TIE

108F19.EPS

Figure 19 ◆ Ties and spreaders.

Ties go through both panels and either through a single waler or between double walers. There are several different systems for securing the ties to the form (*Figure 20*). In some systems, slotted metal wedges are driven behind the buttons at the ends of the ties. In other cases, bolts and washers are used.

Additional support is provided by horizontal members known as walers and vertical members known as strongbacks. Diagonal braces are also used to support the forms.

9.4.0 Edge Forms

Edge forms are low-height perimeter forms constructed to contain concrete poured for flat surfaces (flatwork) such as on-grade building slabs with or without a foundation, or for outdoor slabs such as parking lots, driveways, streets, sidewalks, and approaches. A slab is defined as a section of concrete that is larger in its horizontal dimensions than in its thickness. A slab-on-grade (slab-at-grade) is a concrete slab supported by the ground.

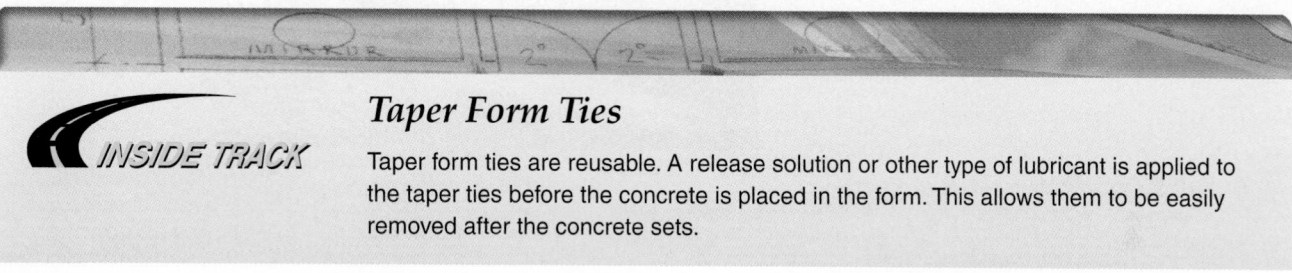

Taper Form Ties

INSIDE TRACK

Taper form ties are reusable. A release solution or other type of lubricant is applied to the taper ties before the concrete is placed in the form. This allows them to be easily removed after the concrete sets.

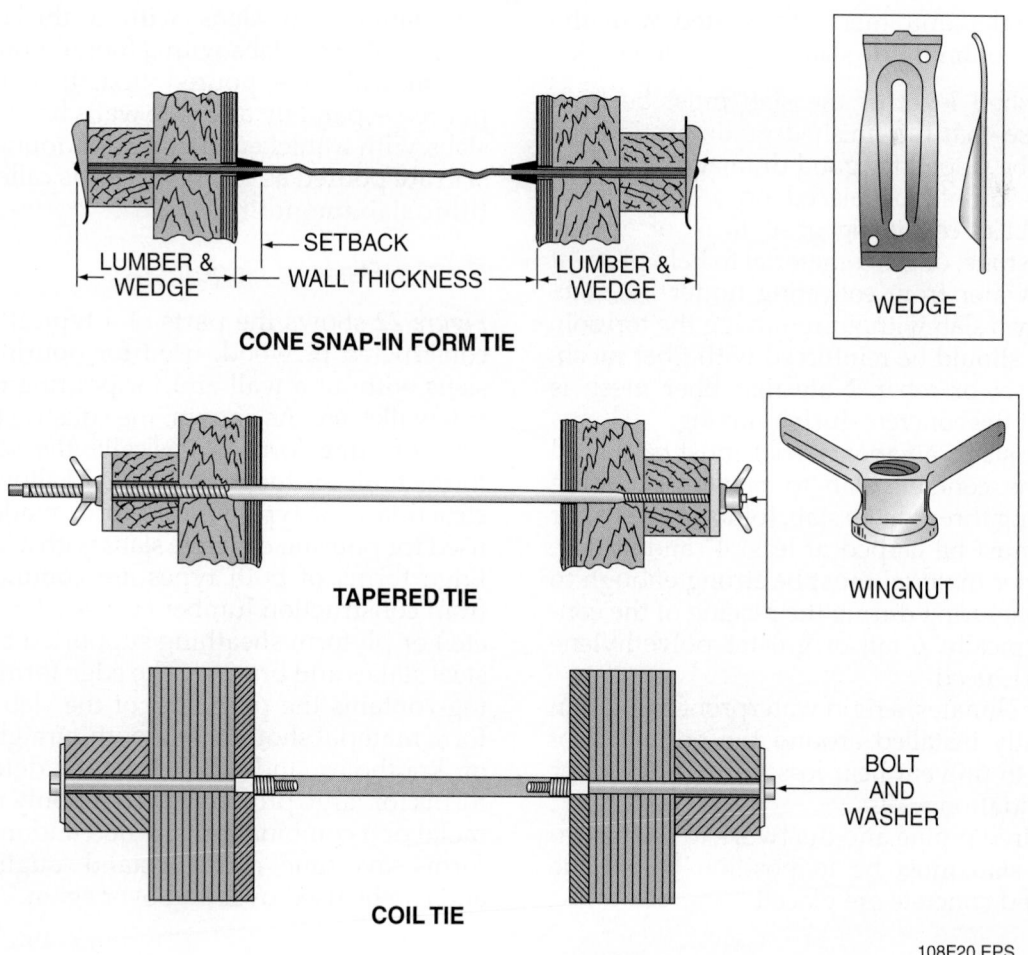

CONE SNAP-IN FORM TIE

WEDGE

TAPERED TIE

WINGNUT

COIL TIE

BOLT AND WASHER

108F20.EPS

Figure 20 ◆ Methods of securing wall ties.

Built-in-Place Form

This is an example of a partially constructed built-in-place form for a foundation wall.

108SA09.EPS

9.4.1 General Requirements for On-Grade Slabs

Some general requirements associated with the construction of on-grade slabs are as follows:

- The finished level of the slab must be high enough so that the finished grade around the slab can be sloped for good drainage.
- The slab should be placed on a solid (compacted) base consisting of 4" to 6" of gravel, crushed stone, or other material to help prevent ground water from collecting under the slab. Never lay a slab without removing the top soil.
- The slab should be reinforced with fiber mesh, wire mesh, or rebar. Note that fiber mesh is placed in the concrete during mixing.
- For floor slabs, a vapor barrier must be placed under the concrete slab to prevent moisture from rising through the slab. Joints in the vapor barrier must be lapped at least 4" and sealed. The barrier material must be strong enough to resist puncturing during the placing of the concrete. Typically, 6 mil or greater polyethylene sheeting is used.
- In colder climates, a rigid waterproof insulation is typically installed around the perimeter of the slab to prevent heat loss through the floor and foundation walls.
- All required piping and ductwork to be run under the slab must be in position before the gravel and concrete are placed.

9.4.2 Types of Slabs

There are two basic kinds of slabs: slabs with a foundation and slabs with a thickened edge (*Figure 21*). For slabs with a foundation, the foundation walls are poured first, then the slab is poured separately after the walls have set up. For slabs with a thickened edge, the foundation and slab are poured as one unit. This is called a **monolithic slab (monolithic pour)**.

9.4.3 Parts of Edge Forms

Figure 22 shows the parts of a typical edge form constructed of wood, used for pouring concrete slabs without a wall and for pouring driveways, sidewalks, etc. As shown, the construction of this type of edge form is basically the same as for forms built for footings. *Figure 23* shows the construction of a typical edge form made of wood used for pouring concrete slabs with a foundation. Edge forms of both types are commonly made from construction lumber (1 × 4s, 2 × 4s, 2 × 6s, etc.) or plyform sheathing supported by wood or steel stakes and bracing. The edge form or sheathing contains the perimeter of the slab. The edge form material should be smooth, straight, and free of knotholes and other surface defects. Edge forms for large projects are commonly made from metal or a combination of wood and metal. Metal forms save time, can withstand rough handling, and can be used over and over again.

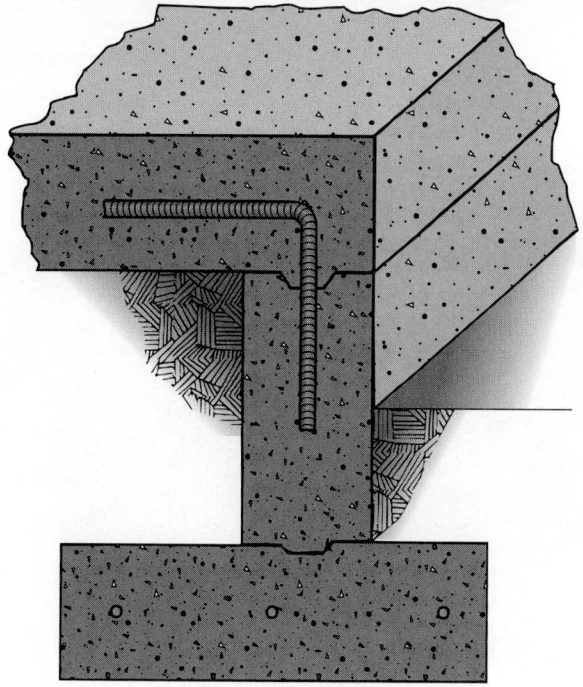

SLAB WITH FOUNDATION

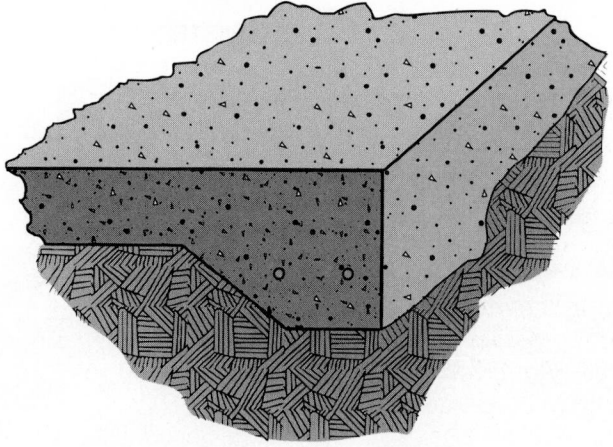

SLAB WITH THICKENED EDGE

108F21.EPS

Figure 21 ◆ Types of slabs.

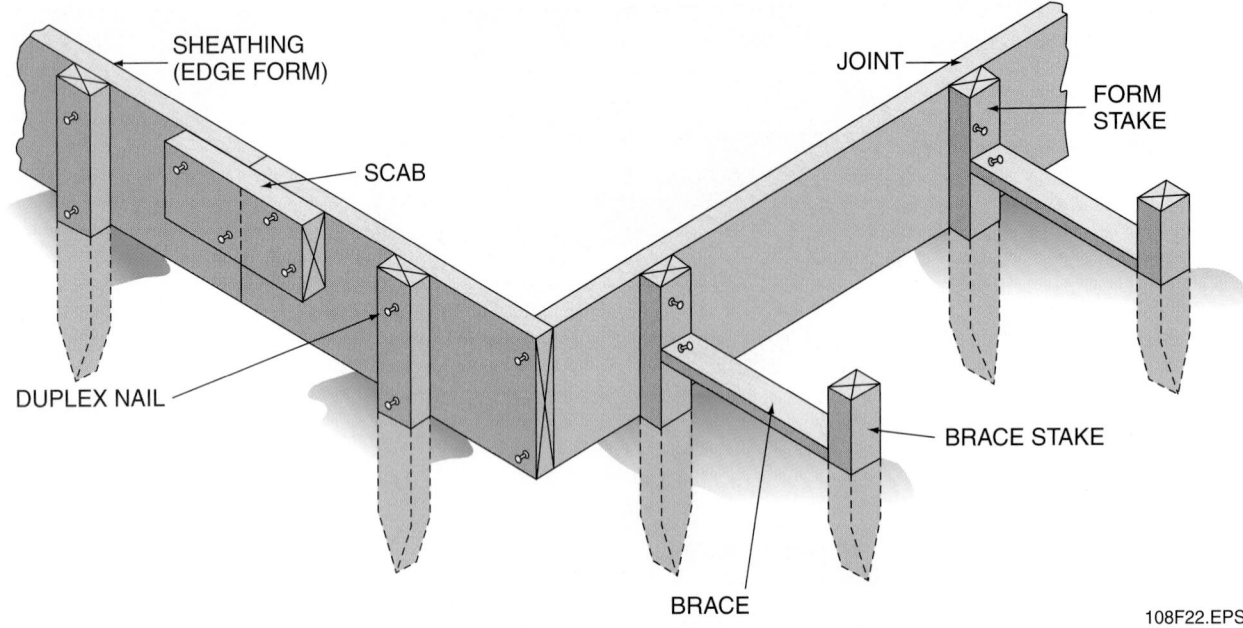

Figure 22 ◆ Parts of an edge form for a slab without a wall.

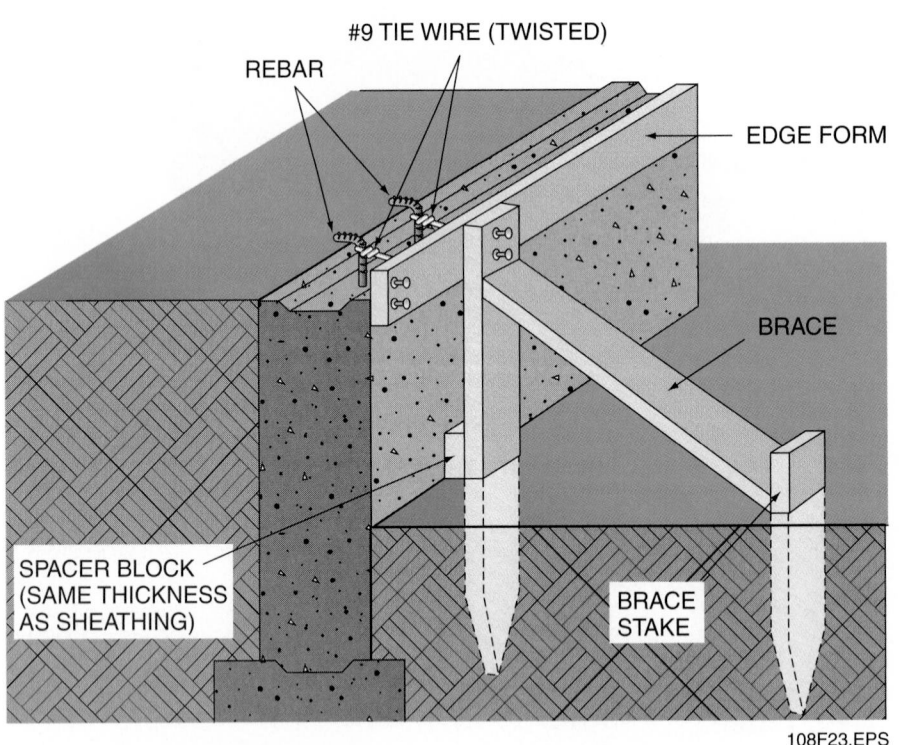

Figure 23 ◆ Parts of an edge form for a slab with a wall.

The form stakes driven into the soil provide primary support for the edge form. When used, brace stakes and the related braces provide additional form support. Braces and brace stakes should be used when the subgrade is too loose or when the form stakes cannot be driven deep enough. Wooden stakes can be made from 1×2, 1×4, 2×2, or 2×4 material. Stakes are normally spaced about 4' apart when using 2" edge form lumber. When using 1" edge form lumber, they are placed about 2' apart. Steel stakes are also made for use with wood or metal forms. They are easier to drive and last longer. Stakes should be driven straight and plumb so that the form is true. They should be driven slightly below the top of the forms to help in **screeding** and finishing the concrete slab. The edge form is attached to the stakes using duplex nails driven through the stakes, then into the edge form. Scabs are pieces of wood or other appropriate material fastened across a splice or joint in the form edge to strengthen and hold the pieces together. Another method of reinforcing a joint in the form edge is to make the joint at a form stake.

9.5.0 Removing Forms

Depending on strength and deflection considerations, forms for floor and similar kinds of slabs may have to be left in place for several days before removal. The time of removal should be determined by a structural engineer. On large construction jobs, the recommended times for removing various types of forms are often spelled out in the specifications.

Forms should be constructed in a manner that allows them to be easily taken apart and removed. Form surfaces that come in contact with the concrete should be coated with form oil before the concrete is placed to prevent the concrete from adhering to the forms. Generally, forms are taken apart in reverse order of the way in which they were built. Removal of forms must be done carefully to avoid damaging the concrete. If the forms are to be used again, they should be carefully removed using tools and equipment that will not damage either the concrete or the forms. Wooden wedges, not steel tools, should be used to separate the forms from the concrete.

Once removed, reusable forms should be cleaned, repaired, handled, and stored in such a way that they are not damaged. Wooden forms can be cleaned using a wood scraper and stiff fiber brush. Holes in forms can be patched with metal plates, corks, or plastic inserts. When the forms are coated with form oil, they should be allowed to dry before stacking or storing them. Reusable form hardware (such as form ties, nut washers, and wedges) should be sorted, cleaned, and stored for reuse. All damaged hardware should be thrown away.

Summary

In some instances, a carpenter will be called upon to mix concrete and build concrete forms. Concrete is a mixture of portland cement, sand, and water in specified proportions. When mixed, concrete is in a semiliquid (plastic) state and must be contained in forms until it hardens. Concrete will support a great deal of weight on its own, but must usually be reinforced with steel bars or steel mesh so it can resist lateral or shearing forces. Job-built concrete forms can be made entirely out of lumber and wood sheathing. In practice, however, many of the attaching components, such as form ties, are made of metal. Plyform is the most common sheathing used in forms because it is specially made for that purpose. Plyform is sanded on both sides and treated with oil, which makes it easier to strip. Only certain kinds of wood are used to make plyform.

The most important thing to remember about building or assembling forms is safety. If forms are improperly built or are not adequate for the concrete pressure they must handle, people can be killed or injured as a result of structural failure. Therefore it is very important to follow the plans and specifications established by the building designers, especially when it comes to supporting and bracing the forms.

Notes

1. The basic ingredients of concrete are _____.
   a. sand, water, and aggregates
   b. aggregates, water, and portland cement
   c. admixtures, aggregates, and water
   d. portland cement, water, and admixtures

2. Aggregates make up about _____ percent of the volume of concrete.
   a. 15
   b. 30
   c. 50
   d. 75

3. High-strength concrete is defined as concrete with a compressive strength over _____ psi.
   a. 4,000
   b. 6,000
   c. 8,000
   d. 10,000

4. A concrete calculator *cannot* be used when the volume required is under _____ yard.
   a. ¼
   b. ½
   c. ¾
   d. ⅞

5. Steel is used as concrete reinforcement because of its _____.
   a. weight
   b. strength
   c. expansion/contraction rate
   d. resistance to sulfate attack

6. Reinforcement bars are identified by a line or number system that designates their _____.
   a. weight
   b. grade
   c. manufacturer
   d. type of steel

7. Footings provide the base of a foundation system for _____.
   a. patios
   b. walkways
   c. driveways
   d. chimneys

8. Construction of forms for a footing is sometimes *not* needed at sites _____
   a. in tropical climates with firm surface topsoil conditions
   b. where the soil is firm and not too porous
   c. where the frost line is less than 2" to 3" below the surface
   d. where the frost line is more than 6" to 8" below the surface

9. A _____ is a type of footing that is commonly used at level sites where unstable soil conditions exist.
   a. continuous footing
   b. stepped continuous footing
   c. pier
   d. grade beam

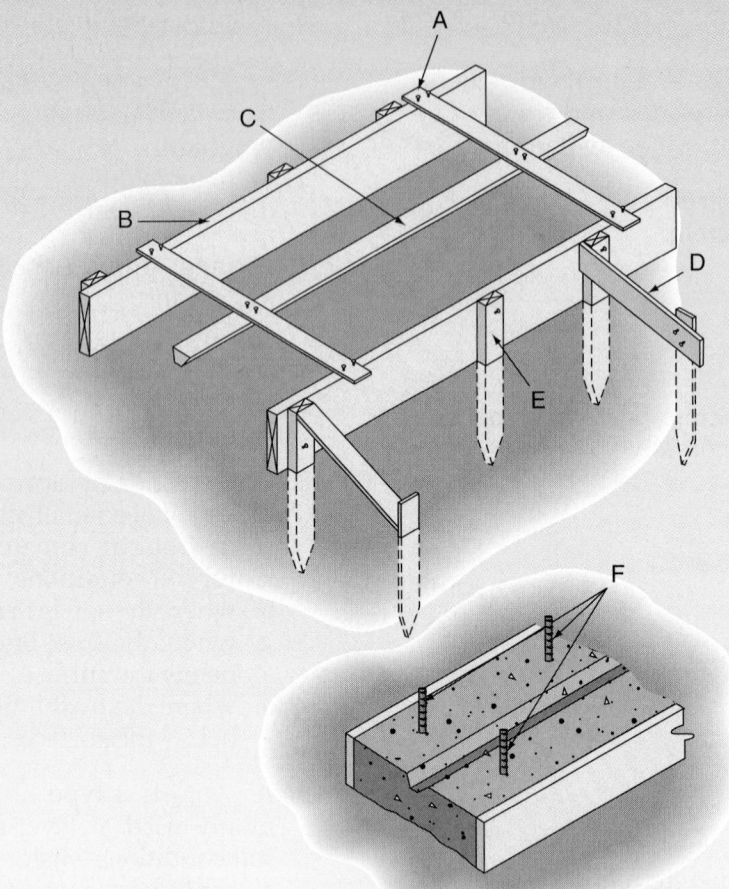

Figure 1

108RQ01.EPS

10. The letter D in *Figure 1* is pointing to _____.
  a. a stake
  b. a spreader
  c. a brace
  d. sheathing

11. The letter C in *Figure 1* is pointing to _____.
  a. a keyway strip
  b. a spreader
  c. a brace
  d. rebar

12. Which of the following requirements applies when using tempered hardboard as form sheathing?
  a. It must be sanded on both sides.
  b. It must be backed with lumber.
  c. It may not have any special treatment or coating.
  d. It may only be used as a form liner in combination with plywood or plyform sheets.

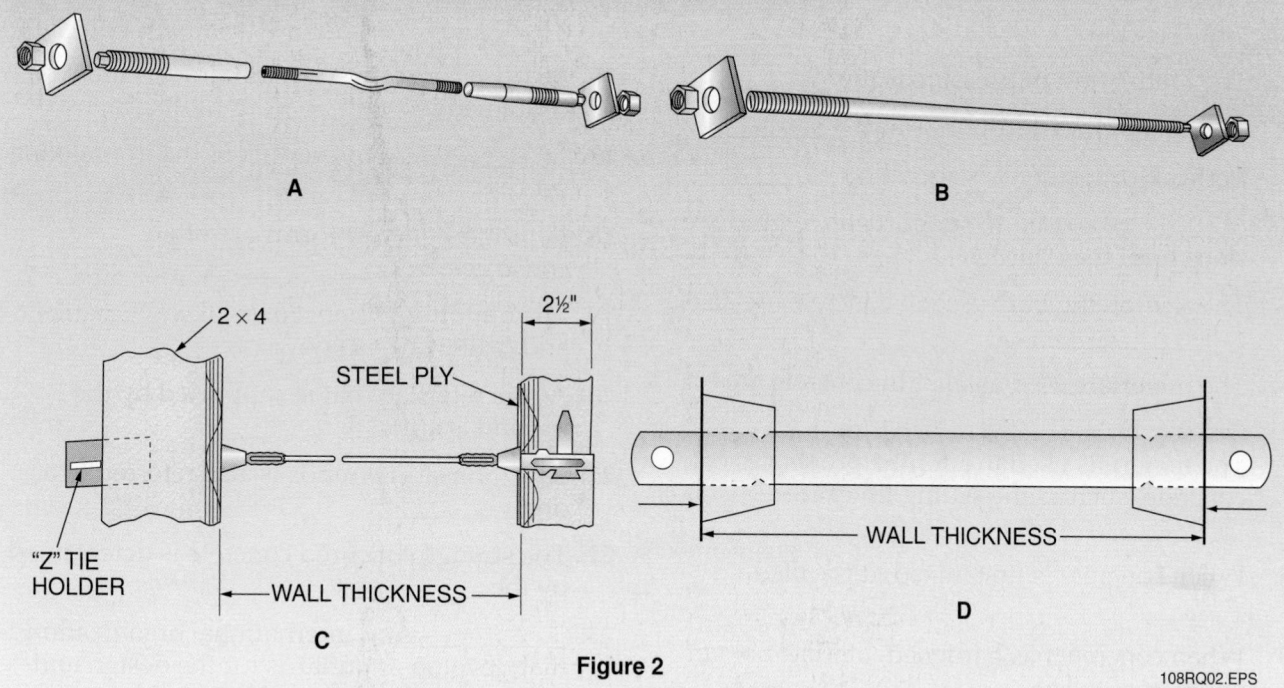

Figure 2

108RQ02.EPS

13. Which item in *Figure 2* shows a form tie that can be completely removed after the concrete has hardened?
a. A
b. B
c. C
d. D

14. A piece of wood placed inside a form to hold the sides of the form the proper distance apart is known as a _____.
a. spandrel
b. cleat
c. ledger
d. spreader

15. A slab poured as a single unit of concrete is called a(n) _____.
a. above-grade slab
b. monolithic slab
c. slab with a foundation
d. slab without a foundation

Trade Terms Quiz

1. The metric unit of pressure is the _____.

2. A diagonal form support is a(n) _____.

3. Vertical form supports are called _____.

4. _____ is used to support above-grade slabs until they harden.

5. The reinforcing bars used in concrete are also called _____.

6. The materials used as filler in concrete are _____.

7. The materials used to alter the properties of concrete, such as the setting time, are _____.

8. When concrete is first mixed, it is called _____.

9. When concrete has hardened, but has not yet cured, it is called _____.

10. When concrete has gained its full structural strength, it is called _____.

11. _____ provide support for grade beams or columns that carry the structural load of a building.

12. Freshly mixed concrete should be tested for the proper _____ before it is placed.

13. _____ is the chemical reaction that causes cement to harden.

14. Reinforcing bars that are rolled from carbon-steel axles used on railroad cars are known as _____.

15. Reinforcing bars that are rolled from used railroad rails are known as _____.

16. Because concrete is a semiliquid, it must be contained in _____ until it hardens.

17. _____ is a byproduct of the ironmaking process.

18. Reusable concrete forms are often constructed of _____.

19. The building of slabs for walkways, patios, and driveways is known as _____.

20. A concrete slab that is supported by the ground is a(n) _____.

21. One thousand pounds is also referred to as one _____.

22. The strength of cured concrete is determined by the _____.

23. _____ is an international organization that develops standards for the design and construction of concrete structures.

24. A wall rests on a foundation known as a(n) _____.

25. The ancient Romans used a type of volcanic ash cement known as _____.

26. Form ties are attached to horizontal form supports known as _____.

27. After the concrete is placed, it is leveled to an established grade using a process known as _____.

28. Building a foundation directly on an unstable _____ can result in cracks and structural damage.

29. A(n) _____ is a continuous pour without construction joints.

Trade Terms

ACI	Forms	Plastic concrete	Slab-on-grade (slab-at-grade)
Admixtures	Green concrete	Plyform	Slag
Aggregates	Hydration	Pozzolan	Slump
Axle-steel	Kip	Rail-steel	Studs
Brace	Monolithic slab (monolithic pour)	Rebars	Subgrade
Cured concrete		Screeding	Walers
Flatwork	Pascal	Shoring	Water-cement ratio
Footing	Piles		

John Payne

Craft Training Adminstrator
Brasfield and Gorrie

When he was a young man, John worked with his dad on odd jobs around the house. In high school, he had a summer job as a builder's assistant. His main occupation is as a field foreman or field engineer. Now he works for Brasfield and Gorrie, which was ranked by *Engineering News-Record* among the top 400 contractors in the United States.

What kinds of work have you done in your career?
I went to college for industrial education, but that only lasted one semester. I preferred working with my hands. I got a job as a surveyor's rodman. Through the years, I always worked on carpentry crews. For eight years, I was a superintendent doing mid-rise commercial buildings. Currently, I am in craft training administration and will be moving up to become a general superintendent soon.

My most mentally challenging project was a nine-story domed structure. I spent many late nights figuring out how to lay it out. It was all radial math. The most physically challenging project I did was a seven-story office building. We poured 248,000 square feet of concrete in four months.

What are some of the things you do in your job?
I run our in-house training program for carpenters, surveyors, and foremen. Our 19 operating divisions maintain a combined staff of more than 2,600 skilled professionals. In my company, our greatest strength is our people and their performance. This job gave me the opportunity to be in one place and not travel as much as I did doing field work.

What do you like about the work you do?
The construction business is a tangible occupation. At the end of the day, you can see and touch what you have built. In my present job, I like to talk about what

I do. I support our organization through longevity, training the new people, building the future.

What do you think it takes to be a success in your trade?
It's the old adage, show up on time and be enthusiastic. This industry does not lend itself to the wishy-washy. You have to stay after it, keep your eyes wide open, and learn every day. There are many opportunities to learn new things. They are always developing new products, new materials, and new methods of doing things.

What advice would you give someone just starting out?
Explore all areas of the trade and the construction industry. What you enjoy doing will become evident. It may be carpentry or it could be another trade. You only know if you try them out. If you enjoy your work, you will put in the time to learn your trade.

Study your math skills. There is a direct correlation between your ability to think rationally and your ability to do simple math. It is impossible to advance without a good math background and reasoning skills. I see my students in their late teens and early twenties who realize that they should have paid more attention in math class. Anyone can cut to the mark, but the one who can tell where to make the mark gets paid more.

Trade Terms
Introduced in This Module

ACI: American Concrete Institute.

Admixtures: Materials that are added to a concrete mix to change certain properties of the concrete such as retarding setting time, reducing water requirements, or making the concrete easier to work with.

Aggregates: Materials used as filler in concrete; may include mixtures of sand, gravel, crushed stone, crushed gravel, or blast-furnace slag.

Axle-steel: Deformed reinforcing bars that are rolled from carbon-steel axles used on railroad cars.

Brace: A diagonal supporting member used to reinforce a form against the weight of the concrete.

Cured concrete: Concrete that has hardened and gained its structural strength.

Flatwork: Work connected with concrete slabs used for walks, driveways, patios, and floors.

Footing: The base of a foundation system for a wall, column, and chimney. It bears directly on the undisturbed soil and is made wider than the object it supports to distribute the weight over a greater area.

Forms: Wood or metal structures built to contain plastic concrete until it hardens.

Green concrete: Concrete that has hardened but has not yet gained its structural strength.

Hydration: The catalytic action water has in transforming the chemicals in portland cement into a hard solid. The water interacts with the chemicals to form calcium silicate hydrate gel.

Kip: An informal unit of force that equals one thousand (kilo) pounds.

Monolithic slab (monolithic pour): Concrete placed in forms in a continuous pour without construction joints.

Pascal: A metric measurement of pressure.

Piles: Column-like structural members that penetrate through unstable, nonbearing soil to lower levels of loadbearing soil. They provide support for grade beams or columns that carry the structural load of a building.

Plastic concrete: Concrete when it is first mixed and is in a semiliquid and moldable state.

Plyform: American Plywood Association's tradename for a reusable material for constructing concrete forms.

Pozzolan: The name given by the ancient Romans to describe the volcanic ash they used as a type of cement. Today, the term is used for natural or calcined materials (including fly ash and silica fume) or air-cooled blast furnace slag.

Rail-steel: Deformed reinforcing bars that are rolled from selected used railroad rails.

Rebars: Abbreviation for reinforcing bars. Also called rerod.

Screeding: Leveling newly placed concrete to an established grade. Also called striking off.

Shoring: Temporary bracing used to support above-grade concrete slabs while they harden.

Slab-on-grade (slab-at-grade): A ground-supported concrete slab 3½" or thicker that is used as a foundation system. It combines concrete foundation walls with a concrete floor slab that rests directly on an approved base that has been placed over the ground.

Slag: The ash produced during the reduction of iron ore to iron in a blast furnace.

Slump: The distance a standard-sized cone made of freshly mixed concrete will sag. This is known as a slump test.

Studs: Vertical members of a form panel used to support sheathing.

Subgrade: Soil prepared and compacted to support a structure or pavement system.

Walers: Horizontal pieces placed on the outsides of the form walls to strengthen and stiffen the walls. The form ties are also fastened to the walers.

Water-cement ratio: The ratio of water to cement, usually by weight (water weight divided by cement weight), in a concrete mix. The water-cement ratio includes all cementitious components of the concrete, including fly ash and pozzolans, as well as portland cement.

Formulas for Geometric Shapes and Conversion Tables

Conversion Table for Changing Measurements (in Inches) to Fractions and Decimal Parts of a Foot		
Inches	**Fractional Part of Foot**	**Decimal Part of Foot**
1	$\frac{1}{12}$	0.08
2	$\frac{1}{6}$	0.17
3	$\frac{1}{4}$	0.25
4	$\frac{1}{3}$	0.33
5	$\frac{5}{12}$	0.42
6	$\frac{1}{2}$	0.50
7	$\frac{7}{12}$	0.58
8	$\frac{2}{3}$	0.67
9	$\frac{3}{4}$	0.75
10	$\frac{5}{6}$	0.83
11	$\frac{11}{12}$	0.92
12	1	1.00

Metric Conversion of Pounds to Grams or Kilograms
1 pound (16 ounces) = 453.6 grams or 0.4536 kilogram (kg)

AREAS OF PLANE FIGURES

NAME FORMULA	SHAPE
(A = Area) Parallelogram $A = B \times h$	
Trapezoid $A = \dfrac{B + C}{2} \times h$	
Triangle $A = \dfrac{B \times h}{2}$	
Trapezium (Divide into 2 triangles) A = Sum of the 2 triangles (See above)	
Regular Polygon $A = \dfrac{\text{Sum of sides (s)}}{2}$ x inside Radius (R)	
Circle $\pi = 3.14$ $A = \begin{cases}(1)\ \pi R^2 \\ (2)\ .784 \times D^2\end{cases}$	
Sector $A = \begin{cases}(1)\ \dfrac{a^2}{360°} \times \pi R^2 \\ (2)\ \text{Length of arc} \times \dfrac{R}{2}\end{cases}$ $(\pi = 3.14)$	
Segment A = Area of sector minus triangle (see above)	
Ellipse $A = M \times m \times .7854$	
Parabola $A = B \times \dfrac{2h}{3}$	

VOLUMES OF SOLID FIGURES

NAME FORMULA	SHAPE
(V - volume) Cube $V = a^3$ (in cubic units)	
Rectangular Solids $V = L \times W \times h$	
Prisms $V(1) = \dfrac{B \times A}{2} \times h$ $V(2) = \dfrac{s \times R}{2} \times 6 \times h$ V = Area of end x h	
Cylinder $V = \pi R^2 \times h$ $(\pi = 3.14)$	
Cone $V = \dfrac{\pi R^2 \times h}{3}$ $(\pi = 3.14)$	
Pyramids $V(1) = L \times W \times \dfrac{h}{3}$ $V(2) = \dfrac{B \times A}{2} \times \dfrac{h}{3}$ V = Area of Base x $\dfrac{h}{3}$	
Sphere $V = \dfrac{1}{6} \pi D^3$	
Circular Ring (Torus) $V = 2\pi^2 \times Rr^2$ V = Area of section x $2\pi R$	

108A01.EPS

METRIC CONVERSION CHART

inches (fractions)	inches (decimals)	mm	inches (fractions)	inches (decimals)	mm	inches (fractions)	inches (decimals)	mm	inches (fractions)	inches (decimals)	mm
–	.0004	.01	25/32	.781	19.844	–	2.165	55.	3-11/16	3.6875	93.663
–	.004	.10	–	.7874	20.	2-3/16	2.1875	55.563	–	3.7008	94.
–	.01	.25	51/64	.797	20.241	–	2.2047	56.	3-23/32	3.719	94.456
1/64	.0156	.397	13/16	.8125	20.638	2-7/32	2.219	56.356	–	3.7401	95.
–	.0197	.50	–	.8268	21.	–	2.244	57.	3-3/4	3.750	95.250
–	.0295	.75	53/64	.828	21.034	2-1/4	2.250	57.150	–	3.7795	96.
1/32	.03125	.794	27/32	.844	21.431	2-9/32	2.281	57.944	3-25/32	3.781	96.044
–	.0394	1.	55/64	.859	21.828	–	2.2835	58.	3-13/16	3.8125	96.838
3/64	.0469	1.191	–	.8661	22.	2-5/16	2.312	58.738	–	3.8189	97.
–	.059	1.5	7/8	.875	22.225	–	2.3228	59.	3-27/32	3.844	97.631
1/16	.062	1.588	57/64	.8906	22.622	2-11/32	2.344	59.531	–	3.8583	98.
5/64	.0781	1.984	–	.9055	23.	–	2.3622	60.	3-7/8	3.875	98.425
–	.0787	2.	29/32	.9062	23.019	2-3/8	2.375	60.325	–	3.8976	99.
3/32	.094	2.381	59/64	.922	23.416	–	2.4016	61.	3-29/32	3.9062	99.219
–	.0984	2.5	15/16	.9375	23.813	2-13/32	2.406	61.119	–	3.9370	100.
7/64	.109	2.778	–	.9449	24.	2-7/16	2.438	61.913	3-15/16	3.9375	100.013
–	.1181	3.	61/64	.953	24.209	–	2.4409	62.	3-31/32	3.969	100.806
1/8	.125	3.175	31/32	.969	24.606	2-15/32	2.469	62.706	–	3.9764	101.
–	.1378	3.5	–	.9843	25.	–	2.4803	63.	4	4.000	101.600
9/64	.141	3.572	63/64	.9844	25.003	2-1/2	2.500	63.500	4-1/16	4.062	103.188
5/32	.156	3.969	1	1.000	25.400	–	2.5197	64.	4-1/8	4.125	104.775
–	.1575	4.	–	1.0236	26.	2-17/32	2.531	64.294	–	4.1338	105.
11/64	.172	4.366	1-1/32	1.0312	26.194	–	2.559	65.	4-3/16	4.1875	106.363
–	.177	4.5	1-1/16	1.062	26.988	2-9/16	2.562	65.088	4-1/4	4.250	107.950
3/16	.1875	4.763	–	1.063	27.	2-19/32	2.594	65.881	4-5/16	4.312	109.538
–	.1969	5.	1-3/32	1.094	27.781	–	2.5984	66.	–	4.3307	110.
13/64	.203	5.159	–	1.1024	28.	2-5/8	2.625	66.675	4-3/8	4.375	111.125
–	.2165	5.5	1-1/8	1.125	28.575	–	2.638	67.	4-7/16	4.438	112.713
7/32	.219	5.556	–	1.1417	29.	2-21/32	2.656	67.469	4-1/2	4.500	114.300
15/64	.234	5.953	1-5/32	1.156	29.369	–	2.6772	68.	–	4.5275	115.
–	.2362	6.	–	1.1811	30.	2-11/16	2.6875	68.263	4-9/16	4.562	115.888
1/4	.250	6.350	1-3/16	1.1875	30.163	–	2.7165	69.	4-5/8	4.625	117.475
–	.2559	6.5	1-7/32	1.219	30.956	2-23/32	2.719	69.056	4-11/16	4.6875	119.063
17/64	.2656	6.747	–	1.2205	31.	2-3/4	2.750	69.850	–	4.7244	120.
–	.2756	7.	1-1/4	1.250	31.750	–	2.7559	70.	4-3/4	4.750	120.650
9/32	.281	7.144	–	1.2598	32.	2-25/32	2.781	70.6439	4-13/16	4.8125	122.238
–	.2953	7.5	1-9/32	1.281	32.544	–	2.7953	71.	4-7/8	4.875	123.825
19/64	.297	7.541	–	1.2992	33.	2-13/16	2.8125	71.4376	–	4.9212	125.
5/16	.312	7.938	1-5/16	1.312	33.338	–	2.8346	72.	4-15/16	4.9375	125.413
–	.315	8.	–	1.3386	34.	2-27/32	2.844	72.2314	5	5.000	127.000
21/64	.328	8.334	1-11/32	1.344	34.131	–	2.8740	73.	–	5.1181	130.
–	.335	8.5	1-3/8	1.375	34.925	2-7/8	2.875	73.025	5-1/4	5.250	133.350
11/32	.344	8.731	–	1.3779	35.	2-29/32	2.9062	73.819	5-1/2	5.500	139.700
–	.3543	9.	1-13/32	1.406	35.719	–	2.9134	74.	–	5.5118	140.
23/64	.359	9.128	–	1.4173	36.	2-15/16	2.9375	74.613	5-3/4	5.750	146.050
–	.374	9.5	1-7/16	1.438	36.513	–	2.9527	75.	–	5.9055	150.
3/8	.375	9.525	–	1.4567	37.	2-31/32	2.969	75.406	6	6.000	152.400
25/64	.391	9.922	1-15/32	1.469	37.306	–	2.9921	76.	6-1/4	6.250	158.750
–	.3937	10.	–	1.4961	38.	3	3.000	76.200	–	6.2992	160.
13/32	.406	10.319	1-1/2	1.500	38.100	3-1/32	3.0312	76.994	6-1/2	6.500	165.100
–	.413	10.5	1-17/32	1.531	38.894	–	3.0315	77.	–	6.6929	170.
27/64	.422	10.716	–	1.5354	39.	3-1/16	3.062	77.788	6-3/4	6.750	171.450
–	.4331	11.	1-9/16	1.562	39.688	–	3.0709	78.	7	7.000	177.800
7/16	.438	11.113	–	1.5748	40.	3-3/32	3.094	78.581	–	7.0866	180.
29/64	.453	11.509	1-19/32	1.594	40.481	–	3.1102	79.	–	7.4803	190.
15/32	.469	11.906	–	1.6142	41.	3-1/8	3.125	79.375	7-1/2	7.500	190.500
–	.4724	12.	1-5/8	1.625	41.275	–	3.1496	80.	–	7.8740	200.
31/64	.484	12.303	–	1.6535	42.	3-5/32	3.156	80.169	8	8.000	203.200
–	.492	12.5	1-21/32	1.6562	42.069	3-3/16	3.1875	80.963	–	8.2677	210.
1/2	.500	12.700	1-11/16	1.6875	42.863	–	3.1890	81.	8-1/2	8.500	215.900
–	.5118	13.	–	1.6929	43.	3-7/32	3.219	81.756	–	8.6614	220.
33/64	.5156	13.097	1-23/32	1.719	43.656	–	3.2283	82.	9	9.000	228.600
17/32	.531	13.494	–	1.7323	44.	3-1/4	3.250	82.550	–	9.0551	230.
35/64	.547	13.891	1-3/4	1.750	44.450	–	3.2677	83.	–	9.4488	240.
–	.5512	14.	–	1.7717	45.	3-9/32	3.281	83.344	9-1/2	9.500	241.300
9/16	.563	14.288	1-25/32	1.781	45.244	–	3.3071	84.	–	9.8425	250.
–	.571	14.5	–	1.8110	46.	3-5/16	3.312	84.1377	10	10.000	254.001
37/64	.578	14.684	1-13/16	1.8125	46.038	3-11/32	3.344	84.9314	–	10.2362	260.
–	.5906	15.	1-27/32	1.844	46.831	–	3.3464	85.	–	10.6299	270.
19/32	.594	15.081	–	1.8504	47.	3-3/8	3.375	85.725	11	11.000	279.401
39/64	.609	15.478	1-7/8	1.875	47.625	–	3.3858	86.	–	11.0236	280.
5/8	.625	15.875	–	1.8898	48.	3-13/32	3.406	86.519	–	11.4173	290.
–	.6299	16.	1-29/32	1.9062	48.419	–	3.4252	87.	–	11.8110	300.
41/64	.6406	16.272	1-15/16	1.9291	49.	3-7/16	3.438	87.313	12	12.000	304.801
–	.6496	16.5	1-15/16	1.9375	49.213	–	3.4646	88.	13	13.000	330.201
21/32	.656	16.669	–	1.9685	50.	3-15/32	3.469	88.106	–	13.7795	350.
–	.6693	17.	1-31/32	1.969	50.006	3-1/2	3.500	88.900	14	14.000	355.601
43/64	.672	17.066	2	2.000	50.800	–	3.5039	89.	15	15.000	381.001
11/16	.6875	17.463	–	2.0079	51.	3-17/32	3.531	89.694	–	15.7480	400.
45/64	.703	17.859	2-1/32	2.03125	51.594	–	3.5433	90.	16	16.000	406.401
–	.7087	18.	–	2.0472	52.	3-9/16	3.562	90.4877	17	17.000	431.801
23/32	.719	18.256	2-1/16	2.062	52.388	–	3.5827	91.	–	17.7165	450.
–	.7283	18.5	–	2.0866	53.	3-19/32	3.594	91.281	18	18.000	457.201
47/64	.734	18.653	2-3/32	2.094	53.181	–	3.622	92.	19	19.000	482.601
–	.7480	19.	2-1/8	2.125	53.975	3-5/8	3.625	92.075	–	19.6850	500.
3/4	.750	19.050	–	2.126	54.	3-21/32	3.656	92.869	20	20.000	508.001
49/64	.7656	19.447	2-5/32	2.156	54.769	–	3.6614	93.			

108A02.EPS

This module is intended to present thorough resources for task training. The following reference works are suggested for further study. These are optional materials for continuing education rather than for task training.

Concrete Masonry Handbook for Architects, Engineers, and Builders, Fifth Edition. W.C. Panarese, S.H. Kosmatka, and F.A. Randall, Jr. Portland Cement Association.

The Homeowner's Guide to Building with Concrete, Brick, and Stone. The Portland Cement Association.

NCCER makes every effort to keep these textbooks up-to-date and free of technical errors. We appreciate your help in this process. If you have an idea for improving this textbook, or if you find an error, a typographical mistake, or an inaccuracy in NCCER's Contren® textbooks, please write us, using this form or a photocopy. Be sure to include the exact module number, page number, a detailed description, and the correction, if applicable. Your input will be brought to the attention of the Technical Review Committee. Thank you for your assistance.

Instructors – If you found that additional materials were necessary in order to teach this module effectively, please let us know so that we may include them in the Equipment/Materials list in the Annotated Instructor's Guide.

Write: Product Development and Revision
National Center for Construction Education and Research
3600 NW 43rd St, Bldg G, Gainesville, FL 32606

Fax: 352-334-0932

E-mail: curriculum@nccer.org

Craft

Module Name

Copyright Date

Module Number

Page Number(s)

Description

(Optional) Correction

(Optional) Your Name and Address

Handling and Placing
Concrete
68103-09

68103-09
Handling and Placing Concrete

Topics to be presented in this module include:

1.0.0   Introduction 3.2
2.0.0   Joints in Concrete Structures 3.2
3.0.0   Moving and Handling Concrete 3.5
4.0.0   Placing Concrete in Forms 3.14
5.0.0   Consolidating Concrete 3.16
6.0.0   Finishing Concrete 3.18
7.0.0   Curing Concrete 3.27
8.0.0   Joint Sealants 3.28
9.0.0   Removing Forms 3.29
10.0.0 Other Hand and Power Tools Used When Working
       with Concrete 3.29
11.0.0 Safety Precautions for Handling, Placing, and
       Finishing Concrete 3.32

Overview

On large commercial and industrial projects, carpenters are involved to some extent in the handling and placing of concrete. Concrete work is what goes on at the job site a great deal of the time. This module will provide you with knowledge of the equipment and methods used in placing and finishing concrete, as well as the use of various types of joints used to minimize or control cracking.

Objectives

When you have completed this module, you will be able to do the following:

1. Recognize the various equipment used to transport and place concrete.
2. Describe the factors that contribute to the quality of concrete placement.
3. Demonstrate the correct methods for placing and consolidating concrete into forms.
4. Demonstrate how to use a screed to strike off and level concrete to the proper grade in a form.
5. Demonstrate how to use tools for placing, floating, and finishing concrete.
6. Determine when conditions permit the concrete finishing operation to start.
7. Name the factors that affect the curing of concrete and describe the methods used to achieve proper curing.
8. Properly care for and safely use hand and power tools used when working with concrete.

Trade Terms

Bleed
Consolidating concrete
Elastomeric
Grout

Segregation
Set
Spalling

Required Trainee Materials

1. Pencil and paper
2. Appropriate personal protective equipment

Prerequisites

Before you begin this module, it is recommended that you successfully complete *Core Curriculum*, and *Construction Technology*, Modules 68101-09 and 68102-09.

This course map shows all of the modules in the *Construction Technology* curriculum. The suggested training order begins at the bottom and proceeds up. Skill levels increase as you advance on the course map. The local Training Program Sponsor may adjust the training order.

CONSTRUCTION TECHNOLOGY

68117-09
Copper Pipe and Fittings

68116-09
Plastic Pipe and Fittings

68115-09
Introduction to Drain, Waste, and Vent (DWV) Systems

68114-09
Introduction to HVAC

68113-09
Residential Electrical Services

68112-09
Electrical Safety

68111-09
Basic Stair Layout

68110-09
Exterior Finishing

68109-09
Roofing Applications

68108-09
Roof Framing

68107-09
Wall and Ceiling Framing

68106-09
Floor Systems

68105-09
Masonry Units and Installation Techniques

68104-09
Introduction to Masonry

68103-09
Handling and Placing Concrete

68102-09
Introduction to Concrete, Reinforcing Materials, and Forms

68101-09
Site Layout One: Distance Measurement and Leveling

CORE CURRICULUM:
Introductory Craft Skills

103CMAP.EPS

1.0.0 ◆ INTRODUCTION

It is important that you understand the proper methods for handling and placing concrete. There are many reasons for this. First, carpenters are often required to perform some part of this work on a specific project. Second, a considerable portion of the carpenter's time is concerned with concrete placement. This is especially true when the carpenter is working on the early part of a project when constructing the footings, foundations, walls, etc., for a structure.

This module introduces the procedures, tools, and equipment required for the handling, placement, and finishing of concrete at the job site. The information given in this module assumes that all the required concrete forms are constructed and in place. Information about the construction and use of various concrete forms is covered in the *Foundations and Slab-on-Grade, Horizontal Formwork,* and *Vertical Formwork* modules.

Advance planning is necessary in order to avoid costly delays or problems in placing or finishing concrete. To achieve the best productivity and job quality, the work must be planned to make the best use of the material, personnel, and equipment. All preparatory steps must be completed; the proper tools and personnel must be available; weather conditions must be favorable; and the concrete must be ordered and available when needed.

The basic tasks involved with the placing and finishing of concrete are:

- Moving and handling concrete
- Placing concrete in forms
- Consolidating concrete
- Finishing

2.0.0 ◆ JOINTS IN CONCRETE STRUCTURES

Before we proceed with the discussion of the equipment and methods for placing and finishing concrete, it is important to become familiar with the different kinds of concrete joints you will encounter and the specific purpose for each type of joint. This background information is presented here so that you have an appreciation for the need to make properly constructed joints as you perform related tasks during your concrete handling and finishing work. Joints are made in concrete structures to do the following:

- Control cracking of concrete due to temperature and moisture changes
- Allow movement of foundations or building components
- Increase tensile strength
- Add to the decorative element of concrete

Joints can be grouped into four classes: construction joints, expansion or isolation joints, control joints, and decorative joints.

2.1.0 Construction Joints

Construction joints are formed at the point where concrete placement is temporarily stopped. An example of a construction joint is the joint produced by a bulkhead to break up a large floor area into reasonably sized areas. Another example is the joint formed between succeeding layers of concrete placed in a wall form. *Figure 1* shows some common methods used for forming construction joints.

Ideally, the use of construction joints should be kept to a minimum because they can be a source of water leaks. When dealing with construction joints, take precautions to ensure that a good bond will exist between one layer of concrete and the next.

To achieve a good bond in a construction joint, proceed as follows:

- Clean the surface of the first layer of concrete carefully. Wire brush the concrete to expose the coarse aggregate before the first layer has become thoroughly hard.
- Dampen the surface of the first layer of concrete.
- Apply a coat of portland cement grout on the surface of the first concrete layer. Bonding agents are also available for this purpose. You can now begin placing the next layer of concrete.

THINK ABOUT IT

Concrete versus Mortar

The focus of this module is on the handling and placing of concrete. What is concrete and how does it differ from mortar?

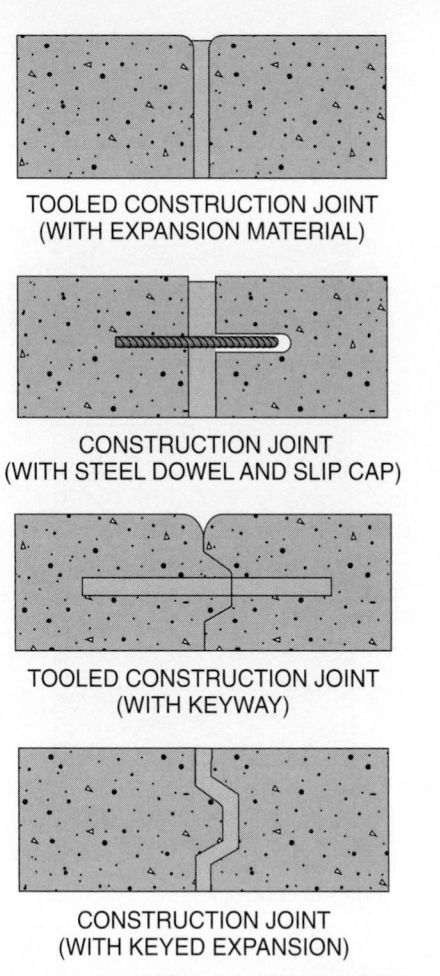

TOOLED CONSTRUCTION JOINT
(WITH EXPANSION MATERIAL)

CONSTRUCTION JOINT
(WITH STEEL DOWEL AND SLIP CAP)

TOOLED CONSTRUCTION JOINT
(WITH KEYWAY)

CONSTRUCTION JOINT
(WITH KEYED EXPANSION)

305F01.EPS

Figure 1 ◆ Types of construction joints.

To achieve a good bond between fresh concrete and previously placed concrete at a construction joint, the previously placed concrete must be properly prepared. This is especially important if the joint is to be watertight and durable. The surface of the previously placed concrete should be cleaned so that it is free of laitance, dirt, oil and grease, curing compound, soft concrete, and other debris. The surface should be slightly roughened. In addition to wire brushing, wet sandblasting and high-pressure water blasting are two methods commonly used for cleaning a construction joint. A properly prepared joint will have a clean, clear, sharp appearance similar to that of a fresh break in sound concrete.

2.2.0 Isolation Joints

Isolation joints (*Figure 2*), also called expansion joints, are used to separate different parts of a structure to permit both vertical and horizontal movement. For example, isolation joints are used to separate a new concrete slab from adjoining building materials or from older cured concrete. They are also used around the perimeter of a floating floor slab and around column piers and other isolated foundation forms. Isolation joints allow new concrete to expand and contract at its own rate, unaffected by the different curing rates of the adjoining materials.

Isolation joints are created prior to the concrete being placed by installing ¼" to ½" premolded strips, typically made from asphalt-saturated fiber or similar material, at the required locations. The strips should extend for the full thickness of the slab and can be placed so that the top edge is flush with the finished level of the slab where no safety hazard from tripping exists. If the strips are not installed flush with the finished level, the joint is typically sealed with joint sealing compound. The premolded strips remain permanently in place after the concrete has **set**. The strips can be placed at grade level to establish the screed line.

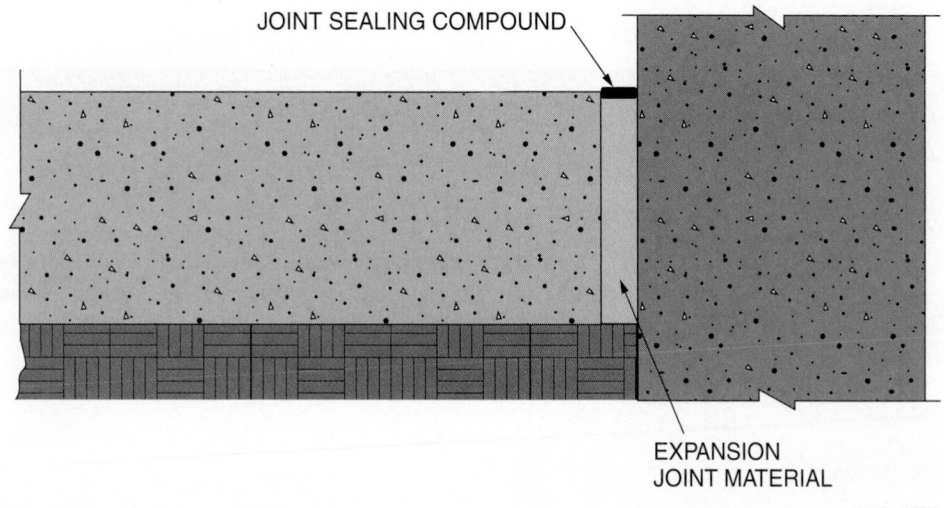

JOINT SEALING COMPOUND

EXPANSION
JOINT MATERIAL

305F02.EPS

Figure 2 ◆ Isolation joint.

2.3.0 Control Joints

Control joints, also called contraction joints, are used in a concrete structure to induce cracking at desired points rather than at random points. Control joints compensate for contraction of the concrete caused by drying shrinkage. If control joints are not used in slabs or lightly reinforced walls, random cracks will occur. *Figure 3* shows some examples of different kinds of control joints. Control joints can be made using a hand groover or they can be sawed using a power saw equipped with a concrete blade. Other methods used to make control joints include placing plastic strips, steel T-bars, or polyethylene strips in the concrete.

For slabs, the depth of control joints should be at least one-fifth to one-quarter the thickness of the slab and a minimum of 1" for all slabs. Manual tooling of control joints requires that the hand grooving tool have sufficient depth required by the thickness of the slab. If a saw cut joint is made in a slab, it must be sawed into the concrete after it is sufficiently hardened, but before the slab starts to crack. To avoid problems with cracking, some concrete finishers make the joints with a hand groover first, then cut them deeper with a saw later.

2.4.0 Decorative Joints

Decorative joints (*Figure 4*) are made to create decorative patterns and designs. For example, a hand groover can be used to make a joint that divides a concrete slab into attractive spacings. Other joints can be made by placing decorative wood (redwood, cedar, or pressure-treated) inserts into the concrete to provide attractive spacings and/or designs.

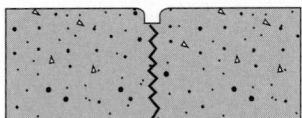

TOOLED CONTROL JOINT

SAWED CONTROL JOINT

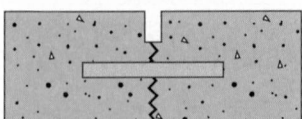

SAWED CONTROL JOINT
(WITH SMOOTH STEEL DOWEL)

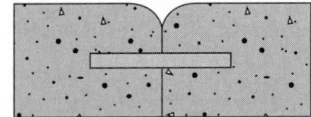

EDGED TOOLED CONTROL JOINT
(WITH SMOOTH STEEL DOWEL)

305F03.EPS

Figure 3 ◆ Types of control joints.

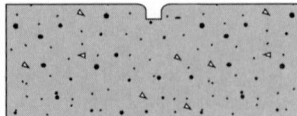 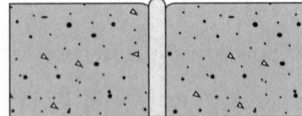

TOOLED DUMMY JOINT   DECORATIVE JOINT
(WITH WOOD SPACER)

305F04.EPS

Figure 4 ◆ Types of decorative joints.

Identifying Construction and Control Joints on Construction Drawings

Construction joints and control joints have the same abbreviation (CJ) when shown on drawings. However, construction joints are drawn as wider lines than lines shown for control joints.

3.0.0 ◆ MOVING AND HANDLING CONCRETE

The transportation and handling of concrete involves the use of both off-site and on-site equipment. Off-site equipment is used to mix and transport the concrete from the cement plant to the job site. On-site equipment is used at the job site to mix, move, and place the concrete.

3.1.0 Off-Site Equipment for Mixing and Conveying Concrete

Typically, commercial concrete used for construction is produced to specifications off-site at a batch plant or central mix plant. Components at both types of plants typically include compartmented storage bins of the different grades of aggregate, storage bins for each type of cement, weight batchers for proportioning the cement and aggregate, a means of batching water, and some means for controlling the batching process. The arrangement and size of the components is determined by the design and type of plant. Detailed information about the composition of concrete and the different types of concrete mixtures are covered in the *Properties of Concrete* module.

Batch plants (*Figure 5*) are dry-mix plants, meaning that they only proportion or batch the ingredients, but do not mix them. The actual mixing of the concrete is performed in the trucks, called truck mixers, that deliver the concrete to the job site. Because the concrete is mixed in the truck, the mixing can be controlled so that the concrete can be transported for short or long distances before it must be discharged. Truck mixers used to transport concrete can be a rear-discharge or front-discharge type (*Figure 6*). Front-discharge

mixers are popular because they are highly maneuverable and allow the driver to distribute the concrete by moving the chute via a hydraulic system without leaving the truck cab.

Central mix plants are wet-mix plants, meaning they produce fully mixed plastic concrete. Once mixed, the concrete is delivered to the job site via truck mixers. While in transit, the plastic cement is prevented from setting up prematurely via agitation by the truck's mixing mechanism. With plant-mixed concrete, it is a common practice to mix the concrete to the point where the plant slump meter indicates that the desired slump is predictable, then finish mixing the concrete on the way to the job site. This procedure is called shrink mixing. Because the concrete is plastic when it leaves the plant, the distance that it can be transported before it must be discharged is normally less than that allowed for truck-mixed concrete.

3.2.0 On-Site Equipment for Mixing, Conveying, and Placing Concrete

When the job site is in a remote area where concrete delivery is not readily available, or where sufficient quantities of concrete required for the

305F05.EPS

Figure 5 ◆ Concrete batch plant.

305F06.EPS

Figure 6 ◆ Front-discharge truck mixer.

project justify their cost, mobile concrete batching and mixing equipment located at the job site are often used. Some other reasons for using on-site mobile concrete production equipment include: the elimination of waiting times for mixer delivery trucks to arrive, the ability to customize the concrete mix on demand for different on-site pour requirements, and the ability to limit the amount of concrete produced to an as-needed basis. Mobile concrete batching plants and volumetric mixers are two types of mobile equipment commonly used to produce concrete at the job site. Mobile concrete batch plants like the one shown in *Figure 7* are mounted on one or more trailers for transportation to and from the job site. Mobile concrete batch plants all have top-loading bins for holding aggregates and cement and water tanks for holding water. Depending on the manufacturer, model, and design, mobile concrete batch plants have varying concrete production capacities.

Volumetric mixers are truck-mounted mixers (*Figure 8*) that have water tanks and top-loading bins for holding coarse aggregate, fine aggregate, and cement. Correctly proportioned dry coarse aggregates, fine aggregates, and cement supplied from the bins simultaneously drop onto and are carried by conveyors into the charging end of the mixer at the rear of the unit. There, a predetermined metered flow of water also enters the mixer. Via the action of a mixer assembly, the dry ingredients and water are rapidly and thoroughly mixed to produce a continuous discharge of concrete. The mixing and discharging of concrete can be stopped at any time and started again as determined by the vehicle operator. The capability to start and stop mixing and discharging of concrete allows the production of concrete to be controlled to the needs of the placing and finishing crews

and other job requirements. It also allows the truck to move to different concrete placing locations within the job site.

Concrete is moved around the job site and placed using a wide variety of equipment. The specific equipment used depends on the type of construction and the topography of the job site. Typically, the equipment used includes buckets handled by cranes, concrete pumps, chutes and push buggies. Concrete conveying equipment is summarized in *Table 1*.

Using improper placement methods or equipment when placing concrete can result in segregation of the concrete materials. Segregation is the tendency for the coarse aggregates to separate from the sand-cement mortar. This results in part of the batch having too little coarse aggregate and the remainder having too much. The former is likely to shrink and crack and have poor wear resistance. The latter is too harsh for full consolidation and finishing. Segregation can usually be avoided by following these placing procedures:

- The drop of the concrete should always be vertical.
- Drop chutes should be used to prevent concrete from striking the reinforcement steel or the side of the form above the level of placement.
- The freefall distance of the concrete should not exceed 4'.

3.2.1 Cranes and Buckets

Cranes equipped with concrete buckets are widely used to transport concrete to levels above or below ground level. On very large projects, this is often done using a tower-type crane similar to the one shown in *Figure 9*. Tower (hammerhead) cranes can be rotated 360 degrees, allowing them to reach all areas of a building being constructed. Tower cranes are erected at the job site. When the building is complete, they are dismantled and taken away. If the project being built is not large

305F07.EPS

Figure 7 ◆ Mobile concrete batch plant.

305F08.EPS

Figure 8 ◆ Volumetric mixer truck.

Table 1 Methods of Conveying Concrete (1 of 2)

Equipment	Type and Range of Work for Which Equipment Is Best Suited	Advantages	Points to Watch For
Truck agitator	Used to transport concrete for all uses in pavements, structures, and buildings. Haul distances must allow discharge of concrete within 1½ hours, but limit may be waived under certain circumstances.	Truck agitators usually operate from central mixing plants where quality concrete is produced under controlled conditions. Discharge from agitators is well controlled. There is uniformity of concrete on discharge.	Timing of deliveries to suit job organization. Concrete crew and equipment must be ready on site to handle concrete. Large batches.
Truck mixer	Used to mix and transport concrete to job site over short and long hauls. Hauls can be any distance.	No central mixing plant needed, only a batching plant since concrete is completely mixed in truck mixer. Discharge is same as for truck agitator.	Control of concrete quality is not as good as with central mixing. Slump tests of concrete consistency are needed on discharge. Careful preparations are needed for receiving the concrete.
Nonagitating truck	Used to transport concrete on short hauls.	Capital cost of nonagitating equipment is lower than that of truck agitators or mixers.	Concrete slump should be limited. Possibility of segregation. Height is needed for high lift of truck body upon discharge.
Mobile continuous mixer	Used for continuous production of concrete at job site.	Combination materials transporter and mobile mixing system for quick, precise proportioning of specified concrete. One-person operation.	Trouble-free operation requires good preventive maintenance program on equipment. Materials must be identical to those in original mix-design proportioning.
Crane	Used for work above and below ground level.	Can handle concrete reinforcing steel formwork and sundry items in high-rise concrete-framed buildings.	Has only one hook. Careful scheduling between trades and operations is needed to keep it busy.
Buckets	Used on cranes and cableways for construction of buildings and dams. Convey concrete directly from central discharge point to formwork or to secondary discharge point.	Enable full versatility of cranes and cableways to be exploited. Clean discharge. Wide range of capacities.	Select bucket capacity to conform with size of the concrete batch and capacity of the placing equipment. Discharge should be controllable.
Barrows and buggies	For short, flat hauls on all types of on-site concrete construction, especially where accessibility to work area is restricted.	Very versatile and therefore ideal inside and on job sites where placing conditions are constantly changing.	Slow and labor intensive.
Chutes	For conveying concrete to lower level (usually below ground level) on all types of concrete construction.	Low cost and easy to maneuver. No power required. Gravity does most of the work.	Slopes range between 1 to 2 and 1 to 3 and chutes must be adequately supported in all positions. Arrange for discharge at end (downpipe) to prevent segregation.
Belt conveyors	For conveying concrete horizontally or vertically. Usually used between main discharge point and secondary discharge point. Not suitable for conveying concrete directly into formwork.	Belt conveyors have adjustable reach traveling diverter, and variable speed both forward and reverse. Can place large volumes of concrete quickly when access is limited.	End discharge arrangements needed to prevent segregation. Leave no mortar on return belt. In adverse weather (hot or windy) long reaches of belt need cover.

Table 1 Methods of Conveying Concrete (2 of 2)

Equipment	Type and Range of Work for Which Equipment Is Best Suited	Advantages	Points to Watch For
Pneumatic guns	Used where concrete is to be placed in difficult locations and where thin sections and large areas are needed.	Ideal for placing concrete in free-form shapes, for repairing and strengthening buildings, and for protective coatings and thin linings.	Quality of work depends on skill of those using equipment. Only experienced nozzle operators should be employed.
Concrete pumps	Used to convey concrete directly from central discharge point to formwork or secondary discharge point.	Pipelines take up little space and can be readily extended. They deliver concrete in a continuous stream. Mobile boom pump can move concrete both vertically and horizontally.	Constant supply of fresh, plastic concrete is needed with average consistency and without any tendency to segregate. Care must be taken in operating pipeline to ensure an even flow and to clean out at conclusion of each operation. Pumping vertically around bends and through flexible hose will considerably reduce the maximum pumping distance.
Drop chutes	Used for placing concrete in vertical forms of all kinds. Some chutes are in one piece, while others are assembled from a number of loosely-connected segments.	Drop chutes direct concrete into formwork and carry it down to the bottom of forms without segregation. Their use avoids spillage of grout and concrete on form sides, which is harmful when off-the-form surfaces are specified. They also prevent segregation of coarse particles.	Drop chutes should have sufficiently large, splayed top openings into which concrete can be discharged without spillage. The cross section of drop chute should be chosen to permit inserting into the formwork without interfering with steel reinforcing.
Elephant trunks (tremies)	Used for placing concrete under water.	Can be used to funnel concrete down through the water into the foundation or other part of the structure being cast.	Precautions are needed to ensure that the tremie discharge end is always buried in fresh concrete so that a seal is preserved between water and concrete mass. Diameter should be 10" to 12" (200 to 300 mm) unless pressure is available. Concrete pumps can be used. Concrete mixture needs more cement—$6\frac{1}{2}$ to 8 bags per cubic yard (363 to 446 kg/m^3) and greater. Slump 6" to 9" (150 to 230 mm) because concrete must flow and consolidate without any vibration.
Screw spreaders	Used for spreading concrete over flat areas, as in pavements.	With a screw spreader, a batch of concrete discharged from bucket or truck can be spread quickly over a wide area to a uniform depth. The spread concrete has good uniformity of compaction before vibration is used for final compaction.	Screws are usually used as part of a paving train. They should be used for spreading before vibration is applied.

305T01.EPS

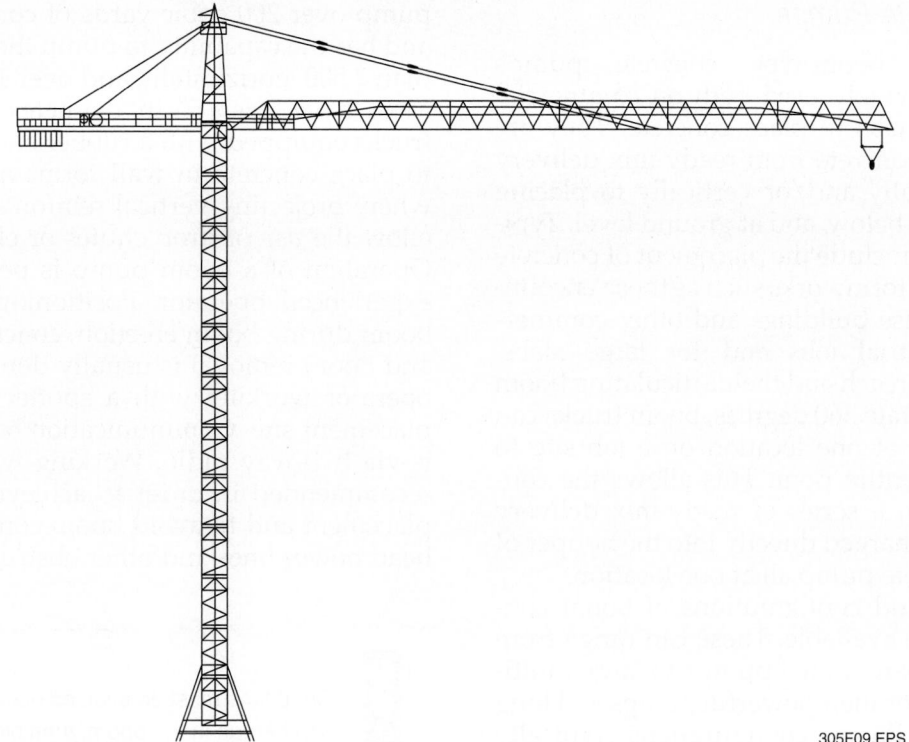

305F09.EPS

Figure 9 ◆ Tower-type (hammerhead) crane.

enough or tall enough to warrant the erection of a tower crane, then portable cranes are brought to the job site to handle concrete as well as other building materials.

A concrete bucket is used to transport concrete from one location to another when using either type of crane. *Figure 10* shows a conventional bucket. Buckets can be either round or square and are made with capacities ranging from ⅛ to 12 cubic yards. There are many bucket designs having various slopes to their sides and different sizes of gate openings. Discharge from the bucket may be either by hand-operated gates for smaller buckets or air-actuated gates for larger buckets.

305F10.EPS

Figure 10 ◆ Concrete bucket.

WARNING!

Concrete weighs approximately 4,150 pounds per cubic yard. The combined weight of the bucket and its contents must never exceed the safe capacity of the crane and its associated rigging.

INSIDE TRACK

Discharging Concrete from a Bucket

To prevent segregation, concrete being discharged from a bucket should not be allowed to drop freely in air more than 3' or 4'.

3.2.2 Concrete Pumps

Today mobile boom-type concrete pumps (*Figure 11*) are widely used both on commercial and residential jobs to place concrete. They are used to pump concrete from ready-mix delivery trucks horizontally and/or vertically to placing locations above, below, and at ground level. Typical applications include the placement of concrete in large-volume formworks such as those encountered for high-rise buildings and other commercial and industrial jobs and for large slabs. Because of their reach and their articulating boom arms that can rotate 360 degrees, boom trucks can often be placed at one location on a job site to accomplish an entire pour. This allows the concrete loads from a series of ready-mix delivery trucks to be discharged directly into the hopper of the boom concrete pump all at one location.

Many sizes and configurations of boom concrete pumps are available. These can range from single-axle truck-mounted pumps to large multi-axle units used for their powerful pumps and long reach booms. Boom configurations typically range from three to five sections. The typical concrete pump used for high-rise construction can pump over 200 cubic yards of concrete per hour and has the capability to pump the concrete more than 2,500' horizontally and over 1,000' vertically. Most have lines 5" in diameter or less. Boom trucks equipped with a rubber end hose are used to place concrete in wall forms and other areas where projecting vertical reinforcement does not allow the use of drop chutes or elephant trunks. Operation of a boom pump is performed by an experienced operator. Positioning of the pump boom during boom erection, concrete placement, and boom removal is usually done by the pump operator working with a spotter located at the placement site. Communication between the two is via two-way radio. Working with a spotter is recommended in order to achieve correct boom placement and to avoid boom contact with overhead power lines and other obstructions.

WARNING!

Dead time must be avoided because concrete can harden in the boom, then blow up the boom or shoot a high-pressure projectile.

305F11.EPS

Figure 11 ◆ Boom-type concrete pump.

Some important factors to take into consideration when using a boom concrete pump include:

- *Overhead power lines* – The American Concrete Pumping Association requires that the boom tip and other sections of the boom be at least 17 feet from power lines. This is because the boom chassis and many of its other components are all conductors of electricity. The plastic concrete being pumped can also become a conductor of electricity because of its high water content.

- *Excavations* – Proper placement of a boom-type concrete pump near an excavation or land that falls off steeply is also important. A general rule of thumb states that when pumping concrete near a vertical drop-off of an excavation, the pump should be located a minimum of one foot back from the edge of the excavation for every foot in depth of the excavation. For example, if the depth of an excavation is 20', the pump should be located a minimum of 20' from the edge of the excavation. New pump trucks have a radio remote control to position the boom.

- *Pump truck stabilization* – Proper-sized loadbearing supports called cribbing must be placed under each of the pump truck's outriggers to stabilize the pump trunk while pumping concrete. The area of cribbing under each outrigger must be big enough so that the pressure placed on each outrigger is less than the loadbearing capability of the soil. One method of determining the proper cribbing size involves laying cribbing on firm spots of ground and positioning the pump truck outriggers on them.

Following this and one at a time, the boom is extended over each outrigger and a check is made to see if the cribbing sinks into the soil. If it does, the boom must be refolded, and a larger area of cribbing placed under the outrigger. Typical cribbing materials used include steel or aluminum plates or layers of 4" × 4" or 4" × 6" boards.

- *Pump pipeline diameter* – The diameter of the pump pipeline is dependent on the largest size of the aggregate in the concrete that is to be pumped. A rule of thumb states that, in order to minimize the risk of clogging in the pump and pipeline, the pipe size should be at least three times as large as the maximum size aggregate in the concrete mix. The larger the diameter of the aggregate in the mix, the larger the pump and pipe size required to pump the concrete.

- *Pipeline length and layout* – Ideally, the boom pipeline length should be as short as possible. This is because line length directly reflects the pump line pressure. Also, to maintain the least resistance, the pipeline should contain the minimum amount of bends possible.

3.2.3 Chutes

Chutes are a widely used and simple way of transferring concrete to a lower elevation. The best example is the movable chute used to discharge concrete from the truck mixer into a bucket or other equipment, or directly into the forms. Chutes must be positioned so that they have sufficient slope to allow the concrete to readily move down them by the force of gravity.

INSIDE TRACK

Boom-Type Concrete Pump Accidents

OSHA states that more concrete pump operators die from electrocution than any other job-related cause. More than 50 percent of accidents involving boom-type concrete pumps and power lines happen when the boom is being folded, unfolded, or removed.

305SA01.EPS

3.2.4 Drop Chutes and Elephant Trunks

Drop chutes and elephant trunks (tremies) are used to place concrete at lower elevations without causing segregation. See *Figure 12*. Segregation can occur by the concrete hitting the form walls, reinforcing steel, or other obstructions. Drop chutes and/or elephant trunks are typically used when placing concrete into narrow wall forms, or at the bottom of a deep form such as a high column form. They can also be used at the end of a conveyor belt or concrete pump. Drop chutes are typically made of sheet metal or short sections of steel fastened together. Elephant trunks can be made of rubber or plastic tubing so that they are flexible and can readily be shortened. The hoppers used to distribute concrete into drop chutes or elephant trunks must be large enough and steep enough to allow for the quick discharge of the concrete without plugging.

3.2.5 Wheelbarrows, Power Buggies, and Carts

Wheelbarrows, concrete carts, and power buggies (*Figure 13*) can be used to transport concrete from the truck mixer to the location of the placement

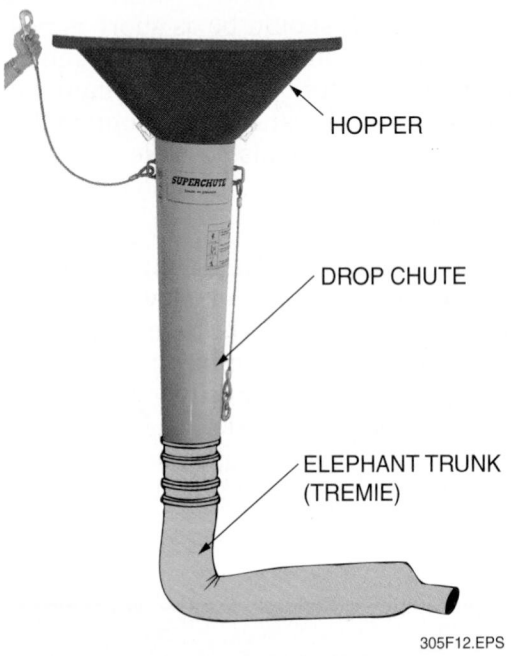

Figure 12 ◆ Hopper, drop chute, and elephant trunk.

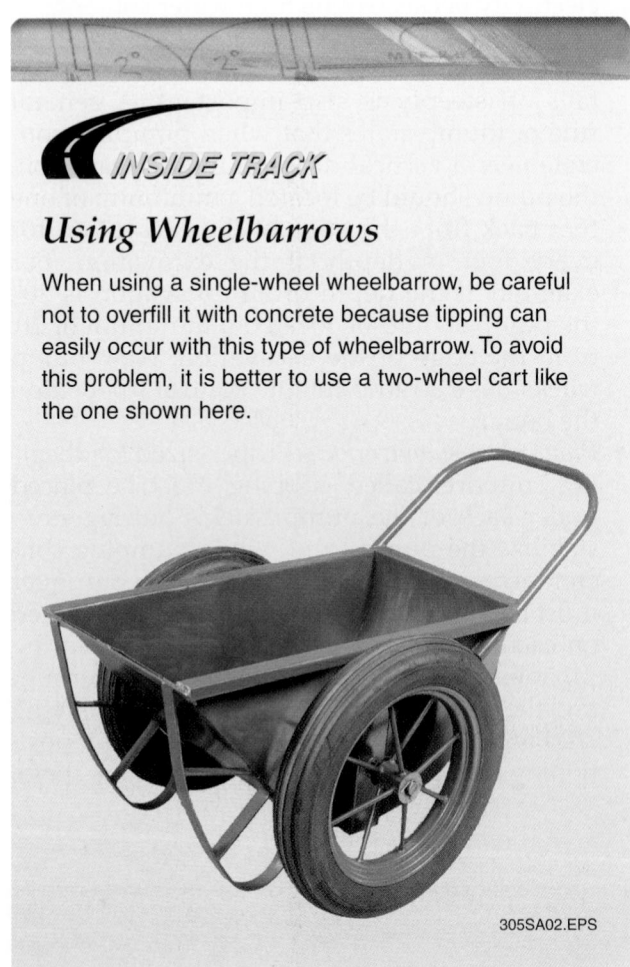

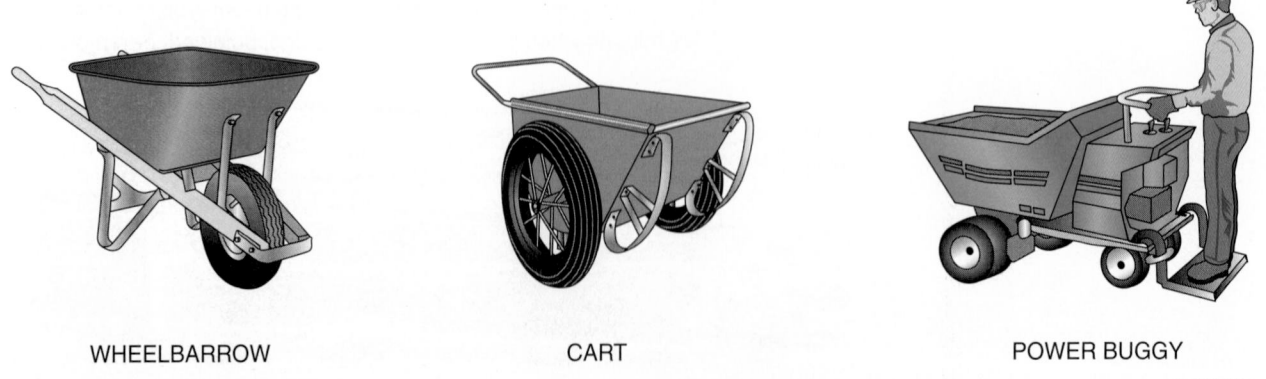

WHEELBARROW   CART   POWER BUGGY

305F13.EPS

Figure 13 ◆ Typical wheelbarrow, cart, and power buggy.

forms when other means of transport such as chutes, buckets, and belt conveyors are not used. Typically, the use of these devices is limited to small jobs, on level ground or floor surfaces, and over short distances. Whenever wheelbarrows, concrete carts, or power buggies are used, their travel should be over runways built for that purpose. The runways should be constructed of boards or plywood and made smooth to prevent bumps that tend to cause the concrete to segregate.

3.2.6 Belt Conveyors

Belt conveyors are used to transport concrete horizontally and for short vertical distances. They are typically used to transfer concrete from a truck mixer to a hopper on a concrete pump or drop chute, and into wheelbarrows, concrete carts, or power buggies. They are ideal where the ground conditions do not allow the mixer to get close to the pour.

3.2.7 Shotcrete

Another method of transferring concrete to a surface is by spraying (*Figure 14*). When cement

305F14.EPS

Figure 14 ◆ Pneumatically applied concrete (shotcrete).

mortar is sprayed on a surface under pneumatic pressure, the product is called shotcrete. Shotcrete usually requires no forms. There are dry-mix and wet-mix shotcrete processes. The original dry-mix process involves combining dry sand and cement in a mixer, then placing this mixture in a vertical double-chambered vessel. Under pneumatic pressure, the mixture flows through a rubber hose to

Cofferdam Concrete Seal Course

Cofferdams are temporary watertight structures used when building bridges, dams, and similar structures to keep water out of an excavation, such as the excavations for bridge piers. In order to remove the water from a cofferdam, the bottom of the cofferdam must be able to resist hydrostatic uplift. One method for doing this is to place a seal course of concrete at the bottom of the cofferdam structure before the water within the cofferdam structure is removed. The seal course is a concrete slab placed under water and constructed thick enough so that its weight is sufficient to counteract the force of the water trying to push its way through the bottom of the cofferdam.

The concrete for the seal course is placed through the water using a tremie pipe to supply concrete either through a hopper or by a concrete pump. Construction of the seal course is accomplished by first lowering the tremie pipe into position with a plug fitted into the pipe as a barrier between the water and concrete. Following this, concrete is charged into the tremie pipe. As the concrete is being placed, the end of the tremie pipe must always remain embedded in the concrete to a depth of at least 3 or 4 inches to prevent the water in the cofferdam from entering the concrete mix. Obviously, concrete placement and positioning of the tremie pipe must be done smoothly and deliberately. Also, the concrete mix must have good flow characteristics. Once concrete placement for the seal is started, it must continue without interruption until completion. For example, pouring of the seal course for one cofferdam during the construction of the Sidney Lanier Bridge in Georgia required 35 hours of continuous pouring.

Typically, a minimum of 5 days is allowed for the concrete to cure before attempting to remove the water from the cofferdam. For very deep cofferdams, removal of the water is sometimes done in stages. This allows for the installation of additional bracing at each stage to further support the walls of the cofferdam structure.

the nozzle, where water, applied via a second hose, is combined with the material. The mixture leaves the nozzle under high velocity. The impact of the mortar on the receiving surface causes compaction. Equipment used for the wet-mix process can involve an auger-type pump, in which plastic concrete is fed into the hose by means of a screw, then a high-velocity flow of air conveys the mixture to the nozzle where it is shot onto the receiving surface. Another type of equipment uses a pressurized tank in which rotating mixing paddles intermittently introduce air with the plastic concrete into the hose or pipe. The use of wet-mix shotcrete has an advantage over dry mix in that the amount of water can be established beforehand and maintained during the gunning operation. Shotcrete has a superior bonding ability, making it suitable for repair work on many types of structures, for earthquake-proofing historical buildings, and for the construction of swimming pools and tunnels. It is also excellent for curves and special shapes.

4.0.0 ◆ PLACING CONCRETE IN FORMS

The task of placing concrete involves depositing the concrete in the forms, and then consolidating the concrete. Before placing the concrete, it is critical to ensure that the forms are ready.

4.1.0 Checking Forms for Proper Construction and Preparation

Before concrete is placed in forms, the forms should be checked to make sure that they are accurately set, clean, tight, adequately braced, and have a form surface that will produce the desired appearance after the concrete is cured. All dirt, sawdust, shavings, tie wire, loose nails, and other debris should be removed from within the forms. The forms should be coated with a releasing agent such as form oil to prevent adhesion of the concrete and allow for easy removal. The releasing agent should be formulated for the particular usage and material to which it is to be applied. Always follow the safety precautions for form oil as specified on the product's material safety data sheet (MSDS).

All reinforcing steel should be cleaned. It should be free of mud, oil, loose rust, and mill scale. Loose, flaky, scaly rust that would affect the bond should be removed by wire brushing or other means. While following the proper safety precautions, grease and oil can be removed using a propane torch (being careful not to overheat the bars), or it can be washed off using a suitable solvent. If reinforcing steel is sticking out from a former placement of concrete, it should be cleaned of any dried mortar coatings or splashes that might have occurred during the former placement. Mortar that is so tightly bonded to the steel that it cannot be removed by vigorous wire brushing can remain.

Other points to check concerning form preparation include the following:

- Subgrade is firm and on grade and soil undisturbed or heavily tamped
- Forms properly aligned and braced
- Forms set to proper grade
- Screeds set at proper location for efficient placing and leveling of concrete
- Screeds set to proper grade
- Bulkheads properly braced and set to proper grade
- Reinforcing mesh or rods properly placed and not touching the forms
- Expansion joint material in place and on grade
- Embedded items such as anchor bolts, traps, pipe, and conduit held firmly in place using templates or by attaching to the forms or reinforcing steel

On most commercial construction jobs, a formal final inspection of the formwork is required before placement of the concrete can occur. Such inspections typically involve representatives of the contractor, an inspector representing the owner, and various municipal building inspectors.

Estimating Quantity of Concrete

On some jobs, the carpenter may be responsible for determining how much concrete is needed to fill a particular form. For example, if the form for a basement floor slab is 25' × 32' × 4", how many cubic yards of concrete are needed to fill it?

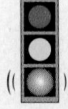

NOTE

If a formal inspection is required, never start placement of the concrete until the inspection has been performed and you have received written approval to proceed.

4.2.0 Placing the Concrete

All of the concrete handling equipment used for placement should be clean and in good working condition. Any standby equipment should also be cleaned and ready for use in case of a breakdown.

When placing concrete (*Figure 15*), it should be deposited as near as possible to its final location in order to avoid excessive movement of the concrete. The concrete must be discharged from equipment at the proper flow rate. This allows the impact force to help in its placing and consolidation. Allowing concrete to dribble slowly causes it to pile up and the coarse aggregate to separate and roll down the sides.

To minimize separation, it is best to discharge concrete vertically, rather than at an angle. A short vertical drop chute should be used at the end of sloping chutes or conveyor belts. The related hopper should be filled by dropping concrete into the center, and it should discharge the concrete vertically from a center opening. The use of hoppers with side discharge or sloping gates should be avoided. If the concrete can be placed satisfactorily without the use of drop chutes or elephant trunks, their use should be avoided. For example, if conditions allow, a crane bucket can be moved along the top of a wall form and concrete can be discharged directly into it.

When drop chutes must be used in order to prevent the concrete from striking the walls in narrow forms, they should be arranged so that they can be quickly moved and shortened, or a sufficient number should be supplied to cover the placing area without moving. The ends of elephant trunks should remain vertical. Pushing them to the side results in a sloping condition and causes separation.

305F15.EPS

Figure 15 ◆ Placing concrete from a buggy.

If required to place concrete on a sloping surface, start by placing the concrete at the bottom of the slope, then move up the slope. By using this method, the weight of the newly added concrete helps achieve proper consolidation. If using a chute for placement, a baffle should be placed at the end of the chute so separation is avoided and the concrete remains on the slope.

When placing concrete for a slab, the first discharge of the concrete should be placed along the edge of the form at one end, with each subsequent discharge placed on the heel of the fresh concrete already in place, not away from it. This method should be continued for the entire slab. Never dump concrete into separate piles or deposit it in one big pile, then move horizontally to fill a form. This practice results in segregation as the mortar flows ahead of the coarser aggregate. Concrete should be placed in a uniform thickness horizontally and must be thoroughly consolidated before the next deposit. The placement must be rapid enough so that the previous layer of concrete is still plastic when the new deposit is placed against it. This prevents flow lines and cold joints in the hardened concrete.

Placing of Concrete in Cold and Hot Temperatures

Never place the concrete for a slab on a frozen subgrade because uneven settling and cracking of the slab usually results when the subgrade thaws. Prior to placing concrete for a slab in near-freezing temperatures, the subgrade should first be thawed and the reinforcing rebar or wire mesh warmed. The form, reinforcing steel, and embedded fixtures should be free of snow and ice when the concrete is placed.

The placing of concrete for a slab in temperatures above 85°F should also be avoided because the concrete sets much faster and may result in cracked or poorly finished slabs. When working in hot weather, avoid placing the concrete in the hottest part of the day. Start early in the morning such as at daybreak. It is also better to do the placement and finishing in smaller sections, if possible, and shade the area if practical.

When placing concrete in wall forms, the concrete should be discharged from different positions along the form until an even layer about 18" thick, called a lift, has been placed in the form. This procedure is repeated until enough lifts have been placed to fill the form. For walls, beams, and girders, water should not be allowed to collect on the ends, in corners, or along the form faces. When concrete is placed in tall wall and column forms at a rapid rate, the concrete usually **bleeds;** that is, the solids settle and the water moves to the top. This is normal. Do not use concrete of a stiffer consistency (lower slump) or slow down the placement to compensate for the gain in water.

5.0.0 ◆ CONSOLIDATING CONCRETE

After being placed in forms, concrete must be consolidated into a uniform solid mass within the form and around embedded parts and reinforcement. Proper consolidation prevents defects such as rock or stone pockets, entrapped air voids, and sand or gravel streaks from occurring in the concrete. Note that entrapped air is defined as an air pocket that is 1 millimeter (mm) or larger. Do not confuse entrapped air with the microscopic entrained air incorporated in some concrete mixtures to improve their workability and resistance to cold weather conditions. Consolidation is not intended to remove the entrained air from such mixtures.

Consolidation can be performed by hand or using mechanical methods. For small concrete jobs with mixes that are rather free flowing, the concrete can be consolidated by hand using tamping rods, spades, shovels, or other suitable tools that will reach to the bottom of the lift. Thrusting up and down with these tools can accomplish the job without segregating the materials. Blows with a hammer applied to the outside of the form are also used to consolidate air pockets that can form near the form walls.

The most frequently used method of consolidation is by vibration (*Figure 16*). When concrete is vibrated, the friction between the coarse aggregate particles is temporarily destroyed and the concrete mixture flows like a liquid. The liquid concrete flows together in a compact mass via gravitational forces and any entrapped air is released to the surface. As soon as the vibration ceases, friction is reestablished. There are two basic types of vibration equipment: internal vibrators and external vibrators.

Consolidating Concrete in Column Forms

To help in the proper consolidation of concrete being placed into forms for support columns, access doors are sometimes built into the column forms at appropriate intervals along their length. These doors provide a convenient way to insert an internal vibrator into the form to consolidate the concrete as it is being placed progressively from the bottom to the top of the form.

Figure 16 ◆ Using an internal vibrator to consolidate concrete.

Figure 17 ◆ Electric internal vibrator.

MOTOR

VIBRATOR
HEAD

FLEXIBLE
SHAFT

305F16.EPS

305F17.EPS

5.1.0 Internal Vibrators

Internal vibrators, also called immersion, poker, or spud vibrators, are used to consolidate concrete in slabs, columns, beams, and walls. They consist of a flexible shaft with a vibrating head connected to a motor (*Figure 17*) that can be powered by electricity, gas, or air. However, gas motors are often preferred and are specified on many jobs. There is an unbalanced weight inside the head that rotates at a high speed, causing the head to revolve in a circular motion.

The proper use of a vibrator is very important. Hit or miss penetration with a vibrator at all angles and without sufficient depth will result in improper consolidation between layers of concrete. A systematic vibration of each new layer (lift) of concrete is necessary. The vibrator head should be lowered vertically into the concrete at regular intervals and allowed to descend by gravity to the bottom of the layer of concrete being placed and at least 6" into the preceding layer. For each insertion, the vibrator should be left in position for a period of about 5 seconds and then slowly withdrawn. The length of time for immersion depends on the slump of the concrete. Watch for changes in the surface such as the appearance of a thin film of glistening paste and the escape of large bubbles of entrapped air to the surface.

5.2.0 External Vibrators

There are a wide variety of external vibrators, including form vibrators, vibrating tables, and surface vibrators. Surface vibrators include vibrating screeds, plate vibrators, vibratory roller screeds, and vibratory hand floats or trowels. Note that for most construction applications, internal vibrators are the most common method used. For this reason, only the vibrating screed type of surface external vibrator is covered in this module. Vibrating screeds are sometimes used to both strike off and consolidate concrete when working with slab floors or other flatwork. More information about screeds, including vibrating screeds, is covered in the next section.

Use of Vibrators

To avoid segregation, internal vibrators should not be used to move concrete horizontally. The vibrator should not be allowed to come into contact with the form walls, as damage could occur to the architectural features and finishes built into the forms. Also, excessive vibration could blow out the form.

6.0.0 ◆ FINISHING CONCRETE

While the finishing of concrete is generally left to concrete finishers or masons, carpenters often get involved in this activity. This module will focus on the finishing of structural (not architectural) surfaces. Architectural-type concrete surfaces with various colors, textures, impressed patterns, and exposed aggregate are generally finished by specialists. The sequence for finishing structural concrete consists of screeding, leveling, and finishing.

Because concrete finishing involves the finishing of slab-on-grade floors and elevated floor slabs, a brief discussion about concrete floor/slab flatness and levelness is in order here. In warehouses and similar applications, good floor flatness and levelness is important in order to allow higher and safer speeds for lift trucks and other vehicles and to prevent shaking and vibration to such vehicles. Most commercial and industrial project specifications require that floors be measured for flatness and levelness in accordance with specification *ASTM E 1155*. This specification assigns F_F (flatness) and F_L (levelness) dimensionless numbers for these parameters, where the higher the F number, the greater the degree of flatness or levelness, respectively. The F_F number defines the maximum floor curvature over 24" computed on the basis of successive 12" elevation differentials. It is most affected by the finishing operation. Similarly, the F_L number defines the conformity of the floor surface to a horizontal plane as measured over a 10' distance. It is most affected by forming and the concrete screeding process.

NOTE

These specs are not applicable to elevated decks.

Specification *ASTM E 1155* defines the requirements for minimum sample size, methods of calculating F_F and F_L numbers, methods for obtaining data points, and calculations for combining the results of multiple measurements to obtain overall F_F and F_L numbers. Floor flatness and levelness must be measured within 48 hours of concrete placement by an ACI or other certified person using measurement instruments such as a rolling F-meter or dipstick floor profiler. Traffic patterns for floor slab areas are categorized as being either a defined traffic floor area or a random traffic floor area. A defined traffic floor area is one where forklifts and/or other wheeled vehicles travel in the same pattern at all times, such as in the aisles between racks in a warehouse. All areas not designated as defined traffic areas are designated as random traffic areas. Typically, the F-numbers specified for defined traffic floor areas are higher than those specified for random traffic floor areas. *Table 2* gives some examples of typical F-numbers for different floor profile categories.

6.1.0 Screeding

Screeding is the first finishing process performed after the concrete is placed in the forms. It is the process of striking off the excess concrete to bring the top surface to the proper grade or elevation. Screeding removes the humps, fills the hollows, and gives a true, even surface. Screeding is performed using a straightedge called a screed. Depending on the size and type of job, screeding can be done manually or using a powered screed.

A manual screed is a straightedge (*Figure 18*) that rests across the top of the form edges or across a pair of screed guides placed next to or within the form. Typically, two people, one on each side of the form, move the screed back and forth across the concrete in a sawing motion and advance it slightly with each movement. An excess of concrete should

Table 2 Example F-Numbers for Different Floor Flatness/Levelness Profiles

Floor Flatness/Levelness Profile	Random Traffic Floor Areas				Defined Traffic Floor Areas
	Specified Overall Value		Minimum Local Value		
	F_F	F_L	F_F	F_L	F_{min}
Good	38	26	19	13	38
Flat	50	33	25	17	50
Very Flat	75	50	38	25	75
Super Flat	100	66	50	33	100
Ultra Flat	150	100	75	50	150

305T02.EPS

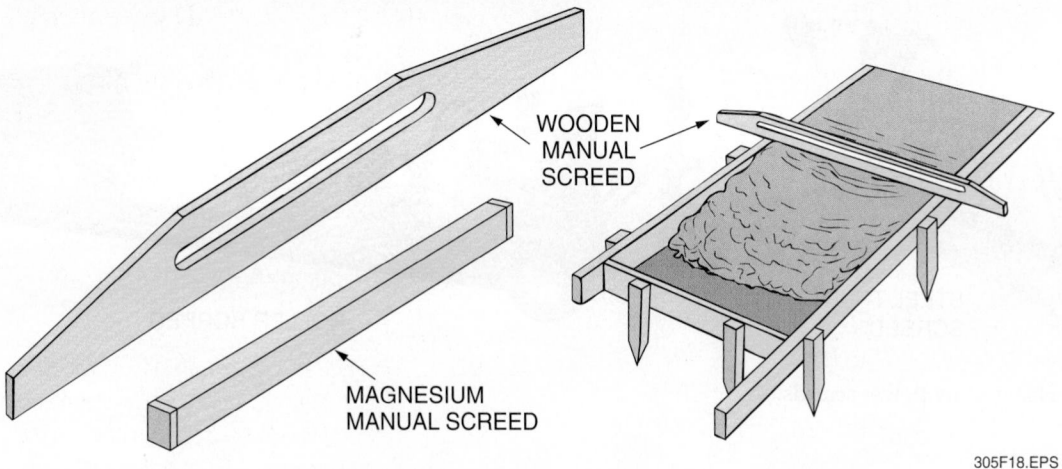

WOODEN MANUAL SCREED

MAGNESIUM MANUAL SCREED

305F18.EPS

Figure 18 ◆ Typical manual screeds (straightedge).

be carried along in front of the screed to fill low places as the screed is moved forward.

Manual screeds can be made of wood or magnesium. Common wood screeds are typically made of 1½" stock. Magnesium screeds are usually 1" thick and 4½" wide. Both types come in lengths ranging from 3' to 16', with the length used depending on the distance between the edge forms or screed guides. Both wood and magnesium screeds can be used on the same types of jobs, but magnesium screeds work better on some concrete surfaces (toppings) because they prevent drag. Magnesium screeds are also preferred as a guide when cutting joints.

Powered screeds include smaller gasoline or air-powered screeds, heavy-duty screeds, and laser screeds. Smaller gasoline and air-powered screeds (*Figure 19*) function to both consolidate and smooth the concrete. These screeds typically have interchangeable magnesium blades of varying lengths for use on different size slabs. Most are operated by either one or two men. Heavy-duty vibrating and roller-type (tube-type) units (*Figure 20*) are used to strike and consolidate wide expanses of high-volume concrete. They can be self-propelled or pulled by crank winches. Both of the types have an attached method of vibration and a metal or metal-covered straightedge, which can be adjusted to compensate for sag or designed crown or slope. Because of their heavy weight, steel truss screeds must be supported by side rails or side forms. They also require a relatively long setup time.

When working with slabs that require very flat profiles, self-propelled laser screeds similar to one shown in *Figure 21* are widely used. Laser screeds have a hydraulic powered laser-controlled screed/compacting head mounted on a telescopic boom. The self-leveling screed head consists of a plow that removes excess concrete, an auger that

305F19.EPS

Figure 19 ◆ Smaller one-man operated gasoline-driven power screed.

cuts concrete on grade, and a vibrator that consolidates the concrete material. Depending on the manufacturer and model, they have different screed head widths and boom length extensions to satisfy different job size requirements. For example, the laser screed shown in *Figure 21* has a 12' screed head width and a 20' boom reach, enabling it to level and consolidate up to 240 square feet of concrete in one pass. Laser receivers mounted at each end of the screed head receive a signal several times per second from a related transmitter located at the proper floor elevation level to provide totally automatic control of the finished floor level.

Use of a laser screed involves placing concrete in strips to match the size of the screed head at a depth of about 1" higher than final grade. Following this the laser screed is moved into position and the boom extended over the placed concrete.

STEEL TRUSS
SCREED

ROLLER SCREED

305F20.EPS

Figure 20 ◆ Heavy-duty power screeds.

305F21.EPS

Figure 21 ◆ Laser screed.

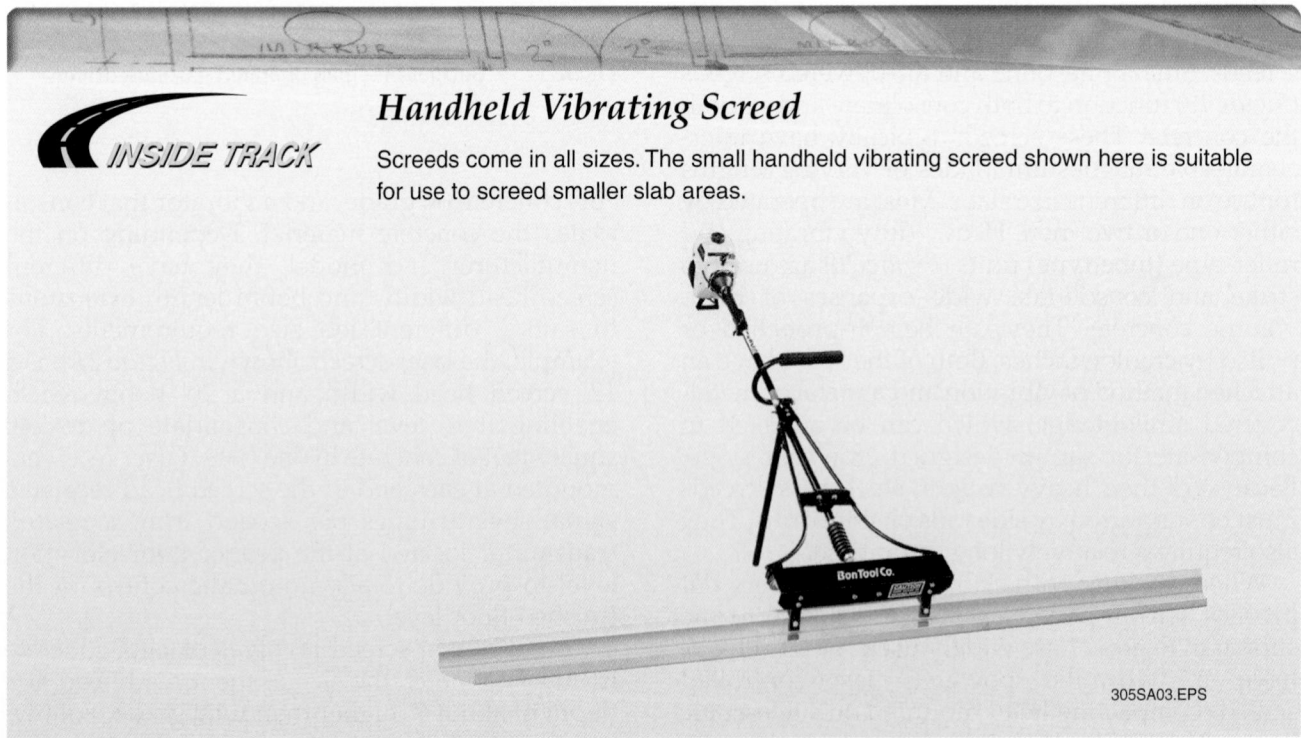

INSIDE TRACK

Handheld Vibrating Screed

Screeds come in all sizes. The small handheld vibrating screed shown here is suitable for use to screed smaller slab areas.

305SA03.EPS

The screed/compacting head is then lowered to the established grade as controlled by the laser transmitters. Retraction of the boom causes the screed head to be drawn across the fresh concrete, which is leveled and compacted in a single pass. Once a pass is completed, the machine is repositioned to the right or left of the previously screeded area and with some overlap the operation is repeated.

Before a vibrator screed is used on flatwork, an immersion-type vibrator should be used along the edges of the form. This is because surface vibration is least effective in these areas. Also, vibrating screeds should not be used on concrete with slumps in excess of 3". Surface vibration of such concrete will result in an excess accumulation of mortar and fine material on the surface and thus reduce wear resistance. For the same reason, surface vibrators should not be operated after the concrete has adequately consolidated.

6.2.0 Leveling

Immediately after screeding and before the free water can bleed to the surface, the concrete should be further leveled using a darby float or long-handled bullfloat. This process eliminates high and low spots in the concrete and embeds the large aggregate just below the surface. It also brings sufficient mortar to the surface in preparation for other finishing.

Darbies (*Figure 22*) are long, flat, rectangular pieces of wood or magnesium that have handles and come in lengths from 2½' to 6½' and widths from 3" to 4". Both can be used on the same types of jobs, but magnesium darbies are more durable, prevent drag, and are preferred on some toppings and on air-entrained concrete. When using a darby, sweep the tool in wide arcs while applying a light pressure on the blade trailing edge. Work from the center to the edges. When using a darby on wide slabs, it is necessary to support yourself on knee boards so that you can reach all areas without damaging the concrete.

Bullfloats (*Figures 23* and *24*) are more commonly used to float concrete in locations where there is enough room to accommodate their 4' to 18' long

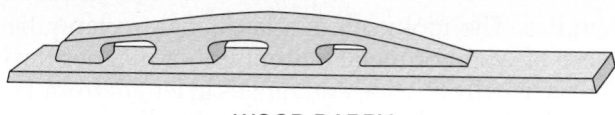

WOOD DARBY

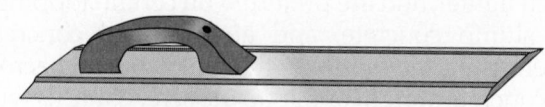

SMALL MAGNESIUM DARBY

305F22.EPS

Figure 22 ◆ Darby floats.

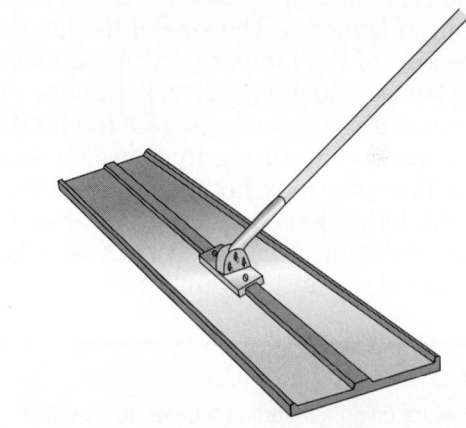

METAL (MAGNESIUM OR ALUMINUM) BULLFLOAT

305F23.EPS

Figure 23 ◆ Bullfloat.

305F24.EPS

Figure 24 ◆ Using a bullfloat.

Leveling

INSIDE TRACK

The concrete should not be overworked during the leveling process. This can cause excessive bleeding to occur, thus weakening the concrete surface.

handles. The tool itself is a large, flat, rectangular piece of wood or metal (aluminum or magnesium) that is usually 8" wide and ranges in length from 3½' to 5'. Both wood and magnesium bullfloats can be used on most floor slabs, pavement, and sidewalks. However, magnesium bullfloats prevent drag, are much lighter, and are preferred on certain toppings, low-slump concrete, and air-entrained concrete. When using the bullfloat, the tool is pushed across the concrete at right angles to the screed marks with the front edge slightly raised, then it is pulled back with the blade flat.

When dealing with low-slump concrete mixes (1" or less), it is sometimes necessary to use a tamper to consolidate the concrete and settle the larger aggregate just below the surface of the concrete. One type of tamper used for this is the 3' to 6' rollerbug (*Figure 25*). The size of the job determines the size of the tamper used. When using a rollerbug tamper, do not overwork the concrete as this will force the coarser aggregate too far down below the surface, resulting in a weaker surface condition. For some jobs, no further finishing will be required. Other jobs may require one or more additional finishing operations such as edging, jointing, troweling, and brooming.

NOTE

Before using a rollerbug tamper on a project, make sure that their use is permitted. Some job specifications prohibit the use of rollerbug tampers.

6.3.0 Finishing

Generally, finishing of the concrete surface begins after the concrete has been allowed to harden enough to support the weight of an average person, leaving only slight footprints on the surface. The bleed water and water sheen or glossy appearance should be gone. Finishing can involve the following tasks:

- Edging and jointing
- Floating
- Troweling
- Brooming

6.3.1 Edging and Jointing

Edging is required to round the upper corners of concrete at the form to prevent **spalling** of the edges and to improve the appearance. Edging compacts the concrete next to the form, where floating and troweling are least effective.

305F25.EPS

Figure 25 ◆ Rollerbug tamper.

The concrete should be cut away at the form about an inch deep using an aluminum float or a pointed trowel, as shown in *Figure 26*, and then finished with an edger (*Figures 27* and *28*). Edgers are made of steel, stainless steel, bronze, or malleable iron. They are available in many sizes, but the most common sizes are those from 6" to 10" long, from 1½" to 4" wide, with a lip that is from ⅛" to ⅝", and with a radius from ⅛" to 1½". Some edgers have a long handle so that they can be used from a standing position. When using an edger, it must be held flat with the front raised slightly to prevent digging into the surface.

Either during or immediately following the edging operation, control joints should be made in the slab, sidewalk, driveway, etc., to induce cracking at desired points rather than at random points. This can also be done in the green state using a power saw. Control joints compensate for contraction of the concrete caused by drying shrinkage.

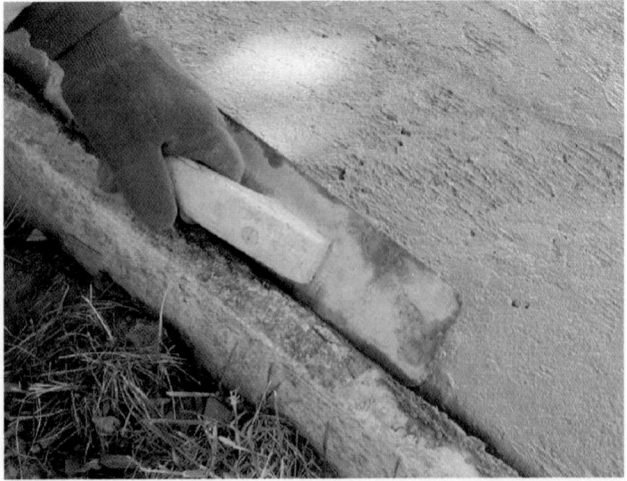

305F26.EPS

Figure 26 ◆ Cutting away a concrete edge using an aluminum float.

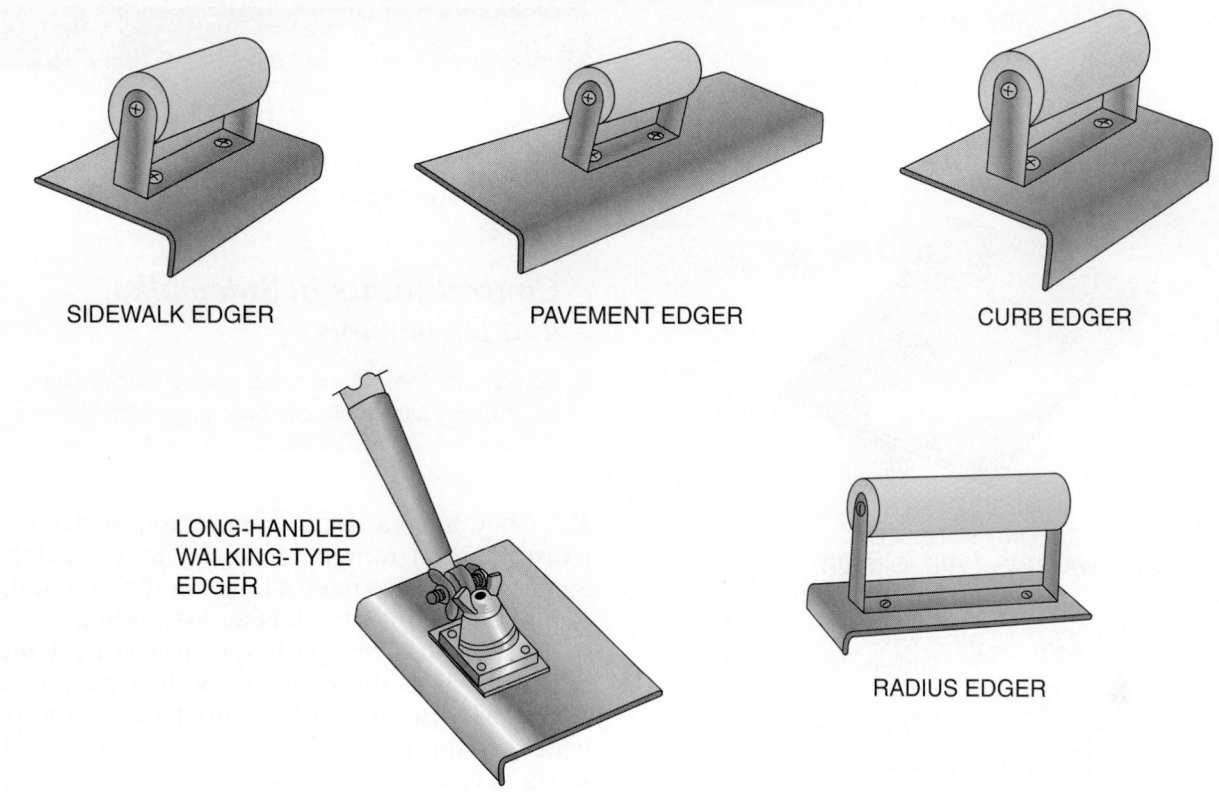

SIDEWALK EDGER   PAVEMENT EDGER   CURB EDGER

LONG-HANDLED
WALKING-TYPE
EDGER

RADIUS EDGER

305F27.EPS

Figure 27 ◆ Edgers.

305F28.EPS

Figure 28 ◆ Using an edger.

The depth and distance between control joints depend on the size, location, and depth of the slab. Typically, the depth of the control joint should be at least one-fifth to one-quarter the thickness of the slab and a minimum of 1" for all slabs. The rule of thumb for spacing control joints in a slab is to space joints, in feet, two to three times the slab thickness, in inches. Apply this rule to a 4" slab. Doubling 4 = 8, and tripling 4 = 12. Therefore, the control joints should be spaced 8' to 12' apart.

Control joints can be made using a hand jointer (groover) or power saw, or by inserting strips of plastic, hardboard, wood, metal, or preformed joint material into the plastic concrete.

Hand jointers (*Figures 29* and *30*) are made of stainless steel, bronze, or malleable iron. Common jointers are 6" long and vary in width from 2" to

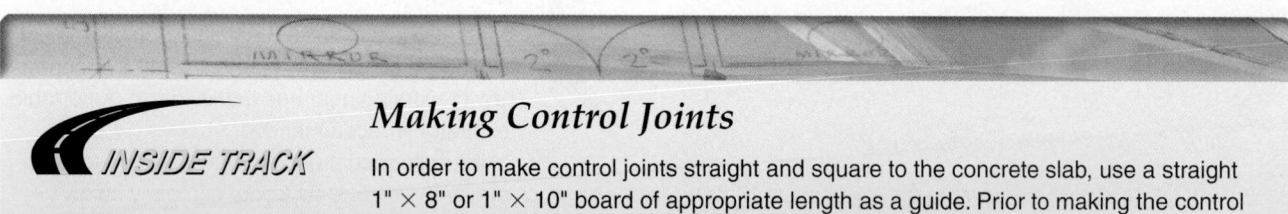

Making Control Joints

INSIDE TRACK

In order to make control joints straight and square to the concrete slab, use a straight 1" × 8" or 1" × 10" board of appropriate length as a guide. Prior to making the control joint, make sure that the board is placed square with the slab on both sides of the form.

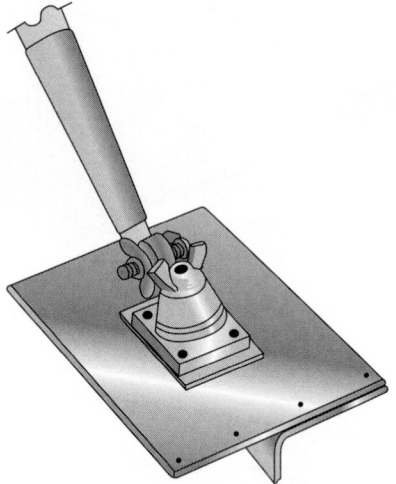

LONG-HANDLED
WALKING-TYPE JOINTER

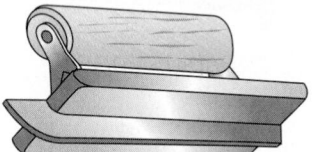

SMALL-BITE
HAND-TYPE JOINTER

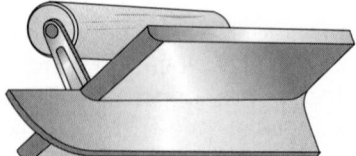

DEEP-BITE
HAND-TYPE JOINTER

305F29.EPS

Figure 29 ◆ Jointers (groovers).

305F30.EPS

Figure 30 ◆ Using a hand jointer.

4½". They have a shallow, medium, or deep bite (cutting edge) ranging from ³⁄₁₆" to 1" in depth. Some jointers also have a long handle so that they can be used while the finisher is standing.

If a saw cut joint is made in a slab, it must be sawed into the concrete with a power saw (*Figure 31*) after it is sufficiently hardened to prevent raveling of the saw cut, but before the slab starts to crack. To avoid problems with cracking, some concrete finishers make the joints with a hand jointer first, then cut them deeper with a saw later. *Figure 32* shows a typical floor-type power saw that can be used for both dry and wet sawing of concrete. It has a built-in water distribution system that supplies the water to both sides of the saw blade for wet sawing. When wet sawing, the water helps to both cool the saw blade and to keep the amount of saw dust down. Another type of widely used concrete saw, called a soft-cut saw, is shown in *Figure 33*. It is designed for use to cut control joints in concrete within the first hour or two after finishing, rather than after the final set of the concrete. The advantage of early joint cutting is that it allows the joints to be in place before significant tensile stresses that can cause random cracks develop in the concrete.

In addition to being used to cut control and decorative joints, power saws are also used in concrete work to cut out concrete sections and to aid in the demolition of structures.

 WARNING!

Concrete dust is toxic. Breathing protection must be worn at all times by workers using a concrete saw or working near one that is in use. A portable respirator is recommended.

It is also recommended that water be used to reduce the dust. Check your company safety practices manual.

Figure 31 ◆ Concrete power saw.

Figure 32 ◆ Wet and dry cut floor-type concrete saw.

Figure 33 ◆ Soft-cut concrete saw.

6.3.2 Floating

After the concrete has been edged and jointed, it should be floated. Floating further embeds the aggregate particles just beneath the surface, removes slight imperfections, compacts the mortar at the surface, and keeps the surface open so excess moisture can escape. Floating produces a relatively even (but not smooth) texture. It is often used for a final nonslip finish, especially for exterior slabs. When a nonslip finish is desired, it may be necessary to float the surface a second time. Marks left by edgers and hand jointers are normally removed during floating unless they are desired for decorative purposes. If a decorative edge is desired, those tools should be rerun after final floating. As with all finishing operations, the surface should not be overworked while it is still plastic as this will bring an excess of water and fine materials to the surface and result in surface defects.

Floating can be done using hand floats moved in a circular fashion to smooth out bullfloat or darby marks. On larger areas, support yourself on knee boards. If desired, you can also lean on a second float for additional support. Common hand floats (*Figure 34*) are made of wood or metal (steel, aluminum, or magnesium). Metal floats are usually 3½" wide and made in two lengths, 12" or 16". Wood floats are made in 12", 15", or 18" lengths and in 3½" or 4½" widths. Metal floats prevent drag, are lighter than wood, and are preferred on some types of toppings and on air-entrained concrete. Special types of hand floats made of cork, rubber, sponge, or carpet are also available to create textured finishes on rubbed concrete.

Floating can also be done mechanically using powered finishing machines (*Figure 35*). These

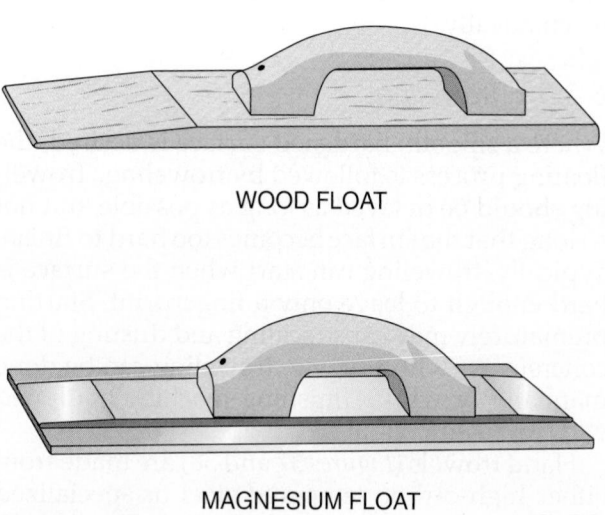

WOOD FLOAT

MAGNESIUM FLOAT

Figure 34 ◆ Hand floats.

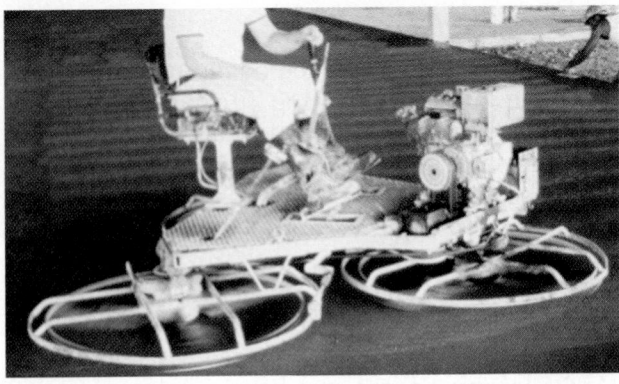

Figure 35 ◆ Powered finishing machines.

Figure 36 ◆ Using a single-rotary finishing machine.

machines can be used to both float and trowel a surface by equipping it with the applicable float or trowel blades. Single rotary machines are commonly used for smaller jobs. These typically have three blades and are controlled by an operator walking along with the machine (*Figure 36*). Operator-driven multiple rotary machines with three rotaries are used mainly for finishing jobs involving very large areas of concrete. Note that hand floating is still required around openings and other obstructions that cannot be finished mechanically.

6.3.3 Troweling

Where a smooth, hardened surface is desired, the floating process is followed by troweling. Troweling should be delayed as long as possible, but not so long that the surface becomes too hard to finish. Typically, troweling can start when the surface is hard enough to leave only a fingerprint. Starting prematurely may cause scaling and dusting of the concrete. As with floating, troweling can be done manually or with a finishing machine equipped with troweling blades.

Hand trowels (*Figures 37* and *38*) are made from either high-carbon tempered steel or specialized long-wearing stainless steel. They are available in

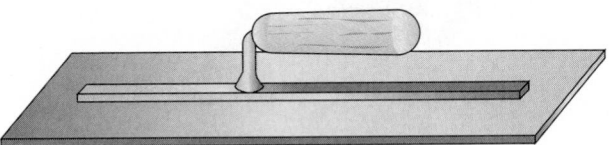

FINISHING TROWEL

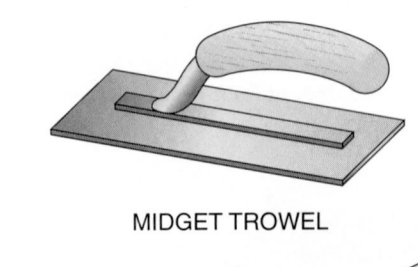

MIDGET TROWEL

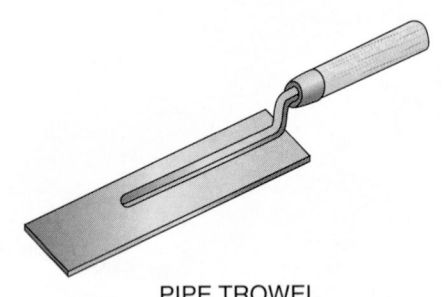

PIPE TROWEL

Figure 37 ◆ Hand trowels.

a variety of sizes ranging from 10" to 20" long and from 3" to 5" wide. The size used depends on the job. For example, a long trowel is normally used for the first troweling process, while a shorter trowel is used for final troweling.

During the first troweling, the trowel should be held flat against the surface. If it is tilted or pitched at too great an angle, an objectionable

Figure 38 ♦ Using hand trowels.

Figure 39 ♦ Brooming a concrete surface.

washboard or chatter surface will result. The first flat troweling may produce an adequate surface that is free of defects, but additional troweling with the blade tilted may be necessary to increase the smoothness and hardness. There should be a time delay between trowelings to allow the concrete to become harder. As the surface hardens, each successive troweling should be done with a smaller-sized trowel to allow enough pressure to be exerted to increase the density, strength, and hardness of the surface. The final pass should make a ringing sound as the tilted blade moves over the hardening surface. Again, when hand troweling large slab surfaces, you must support yourself on knee boards.

When the first troweling is done by machine, at least one additional troweling by hand is required to remove small irregularities. If necessary, tooled edges and joints should be rerun after troweling to maintain uniformity and true lines.

6.3.4 Brooming

A slip-resistant surface can be produced by brooming the surface of the concrete (*Figure 39*). This should be done before the concrete has thoroughly hardened, but while it is sufficiently hard to retain the scoring. Rough scoring can be achieved with a rake, steel wire broom, or stiff, coarse fiber broom. Brooming is usually done after floating. However, if a finer texture is desired, the concrete should be floated and troweled to a smooth surface, then brushed with a soft bristled broom. Best results are obtained with a broom that is made specifically for texturing concrete. Slabs are usually broomed perpendicular to the main direction of traffic.

7.0.0 ♦ CURING CONCRETE

Once concrete has been placed and finished, care must be taken to make sure it cures properly. Hydration is a chemical reaction that takes place when water is combined with cement, sand, and gravel in a concrete mix. It is this chemical reaction that causes concrete to harden. Curing concrete is the process of retaining the moisture in freshly placed concrete long enough to allow for proper hydration to take place.

Concrete must be protected against moisture loss during the early stages of hydration in order for it to cure properly. If the water contained in the concrete is allowed to evaporate too quickly, hydration of the cement will stop and there will be no further gain in the strength and durability of the concrete. As concrete dries, it shrinks, and if drying occurs when the concrete has little strength, cracks will result. In addition, since drying occurs first on the surface, the cement will not be hydrated there but will be present as a dust coating with no strength to hold the aggregate together.

The degree and speed of concrete hardening depends on many factors such as the amount of cement in the mixture, type of cement, temperature of the concrete, temperature of the surrounding air, and chemical admixtures used in the concrete mix. For standard concrete mixtures, the first three days after placing are the most critical. During this time, concrete is most exposed to damage. At seven days, it reaches about 70 percent of its strength and at two weeks, it is at about 85 percent. Under normal conditions, maximum strength is reached at about 28 days.

The greatest influence on the curing rate of concrete is the temperature. Higher temperatures speed the curing rate. Proper hydration occurs when the temperature ranges between 55°F and

73°F. Above 73°F, the hydration process should be slowed to gain the greatest strength of the concrete. Some methods that are commonly used to do this include the following:

- Adding water-reducing admixtures to the concrete mix
- Spraying the forms, reinforcing steel, and subgrade with cool water immediately before placement of the concrete
- Spraying a curing agent or compound on the concrete surface to form an impassable film that prevents or retards the escape of moisture from the concrete
- Covering the concrete surface with a curing blanket of burlap, polyethylene sheeting, or waterproof paper to prevent rapid evaporation of the moisture
- Spraying the finished surface continuously with a water mist using a system of water pipes and spray heads

Hydration takes place at a slower rate when temperatures are low. When concrete temperatures fall below freezing, no hydration takes place and the concrete can be permanently damaged. To compensate for this condition, the concrete mixture is usually heated both during and after placement. Using a concrete mix with air-entraining or accelerating admixtures is another method used to aid curing when placing concrete in cold weather.

Wind, rain, and traffic on a slab are some other factors that must be considered when curing concrete. Winds affect curing by speeding the evaporation of water. The combination of hot, dry weather and wind can dry water from the concrete surface faster than bleed water can replace it. Surprisingly, strong wind can speed evaporation in cold weather as well. In strong wind, surfaces dry faster, so scheduled wetting needs to occur more frequently. Wind shifts water from sprayers, sprinklers, and foggers away from the concrete. Wind unseals lap joints and edge seals of plastic sheeting or impervious paper. Strong wind lifts and tears unsecured plastic sheeting or paper. Windbreaks can be built out of tarpaulins to protect fresh concrete from evaporation.

Rain creates problems by adding too much moisture. Rain on freshly placed concrete can erode the surface and dilute the cement paste at and near the surface. Rain can wash away fines and cause voids in concrete being placed. Finishing of unformed concrete cannot be completed in the rain because it will be damaged. Freshly placed concrete must be sheltered completely from rain, or work stopped until the rain is over.

Temporary rain shelters can be built from tarpaulins or plastic sheeting to protect concrete surfaces. Check that the rain will not drip off the edges of the shelter onto the slab.

As concrete cures, it gains strength over time. During this time, construction traffic and loads driven over new slabs can damage the surface, can exceed the strength of the slab, and can lead to cracked or damaged slabs. The best protection is to keep traffic off the slab for the first 7 days. If activity must continue, it should start with smaller equipment, foot traffic, and light material loads. This traffic is acceptable after 3 to 7 days, the initial cure time. The loads must be light, and the concrete surface must be protected. Kraft paper, plywood, or fiber sheets can be used to provide traffic paths over a slab. Care should be taken to prevent damage to edges by routing paths away from edges and by reinforcing path material at the edges of the slab. Concrete without reinforcement is strong in compression (crushing force) but not in tension (pulling force) or flexure (bending force). Loading or heavy traffic can bend the concrete, especially at the edges of slabs, and cause cracks. Use common sense to protect new concrete from loads. Residential slabs, driveways, and walks are not designed to carry heavy loads such as ready-mix trucks or heavy materials. These types of loads should not be allowed on residential concrete.

8.0.0 ◆ JOINT SEALANTS

Sometimes the concrete finishing process requires that joint sealant be applied to the joints in the finished concrete structure. Joint sealants are flexible materials used to seal construction, isolation, and similar joints subject to movement between adjacent sections of concrete or between concrete and other construction materials.

A wide variety of sealant formulations are available for use on most interior and exterior surfaces that require sealing, including concrete, masonry, metal, glass, ceramic, and wood. Their durability, flexibility, and other features vary, depending on how they are formulated. Different applications require different types of sealant. Polyurethane and polysulfide sealants are widely used in commercial/industrial applications for sealing expansion and control joints in precast concrete panels and on joints in sidewalks, parking lots, terrace decks, etc. They are formulated for exterior use as highly elastomeric materials. This means they have excellent flexibility and elongation properties. They come in one- and two-part formulations and have excellent resistance to weather, ultraviolet light, and moisture. They cure quickly with little shrinkage and stay flexible for years in most environments. Premolded sealants such as plastic and rubber water stops and gaskets are also available.

The requirement for the use of joint sealants and the type of sealant to use are normally specified in the working drawings and/or specifications for the structure. If sealants are required but not specified, follow the sealant manufacturer's recommendations when selecting a sealant for a particular job.

Regardless of the type of sealant used, the joint must be prepared properly in order for the sealant to bond correctly. Joint preparation must be done in accordance with the manufacturer's instructions for the sealant. Generally, this involves making sure that the joint is dry and free of oil, dust, or other debris. Joints cast in new concrete can contain form oil, loose mortar, and dust. They can be cleaned by sandblasting, power and/or hand wire brushing, or by using power-driven routers. Sawed joints will be dusty and should be blown out with oil-free air. If honeycombs or faults exist on the side of the joint slot, they should be patched before application of the sealant.

9.0.0 ◆ REMOVING FORMS

Depending on strength and deflection considerations, forms for floors and similar types of slabs may have to remain in place for several days before removal. The time of removal should be determined by a structural engineer. Forms for walls, columns, beam sides, slabs, etc., where strength is not a problem are typically removed one day after concrete placement. However, these times can vary depending on the type of concrete, weather conditions, etc. On large construction jobs, the recommended times for removing various types of forms are often spelled out in the specifications.

When initially building forms, they should be constructed in a manner that allows them to be easily taken apart and removed. Form surfaces that come in contact with the concrete should be coated with form oil before the concrete is placed to prevent the concrete from adhering to the forms, making them more difficult to remove. Generally, forms are taken apart in reverse order of the way in which they were built. Removal of forms must be done carefully to avoid damaging the concrete. If the forms are to be reused, they should be carefully removed using tools and equipment that will not damage either the concrete or the forms. Wooden wedges, not steel tools, should be used to separate the forms from the concrete.

Once removed, reusable forms should be cleaned, repaired, handled, and stored in such a way that they are not damaged. Wooden forms can be cleaned using a wooden scraper and stiff fiber brush. Steel forms can be cleaned using a steel scraper and wire brush. Holes in forms can be patched with metal plates, corks, or plastic inserts. When the forms are coated with form oil, they should be allowed to dry before stacking or storing them. Reusable form hardware such as form ties, nut washers, and wedges should be sorted, cleaned, and stored for reuse. All damaged hardware should be thrown away.

10.0.0 ◆ OTHER HAND AND POWER TOOLS USED WHEN WORKING WITH CONCRETE

The basic hand and power tools used for finishing concrete have been described where applicable in the preceding sections. This section briefly describes some other tools used when working with concrete and explains when they are used.

10.1.0 Combination Tools

Combination tools such as those shown in *Figure 40* are tools designed to perform special jobs. The jobs each tool is used to perform are represented by the name of the tool. These tools can be made of metal or plastic.

10.2.0 Pointing and Margin Trowels

Pointing and margin trowels are used for patching and mixing concrete. The pointing trowel (*Figure 41*) is used to patch holes and grout machine bases. The margin trowel is used mainly to mix concrete, patch holes, and grout machine and steel column bases. It can also be used to finish patches in hard-to-reach areas.

10.3.0 Hammers

There are two types of hammers used by finishers to work with hardened concrete: the chipping hammer and the bush hammer (*Figure 42*). Chipping hammers are used to cut off projections and high places on hardened concrete. The bush hammer is used to fracture aggregate to present a rough, decorative finish.

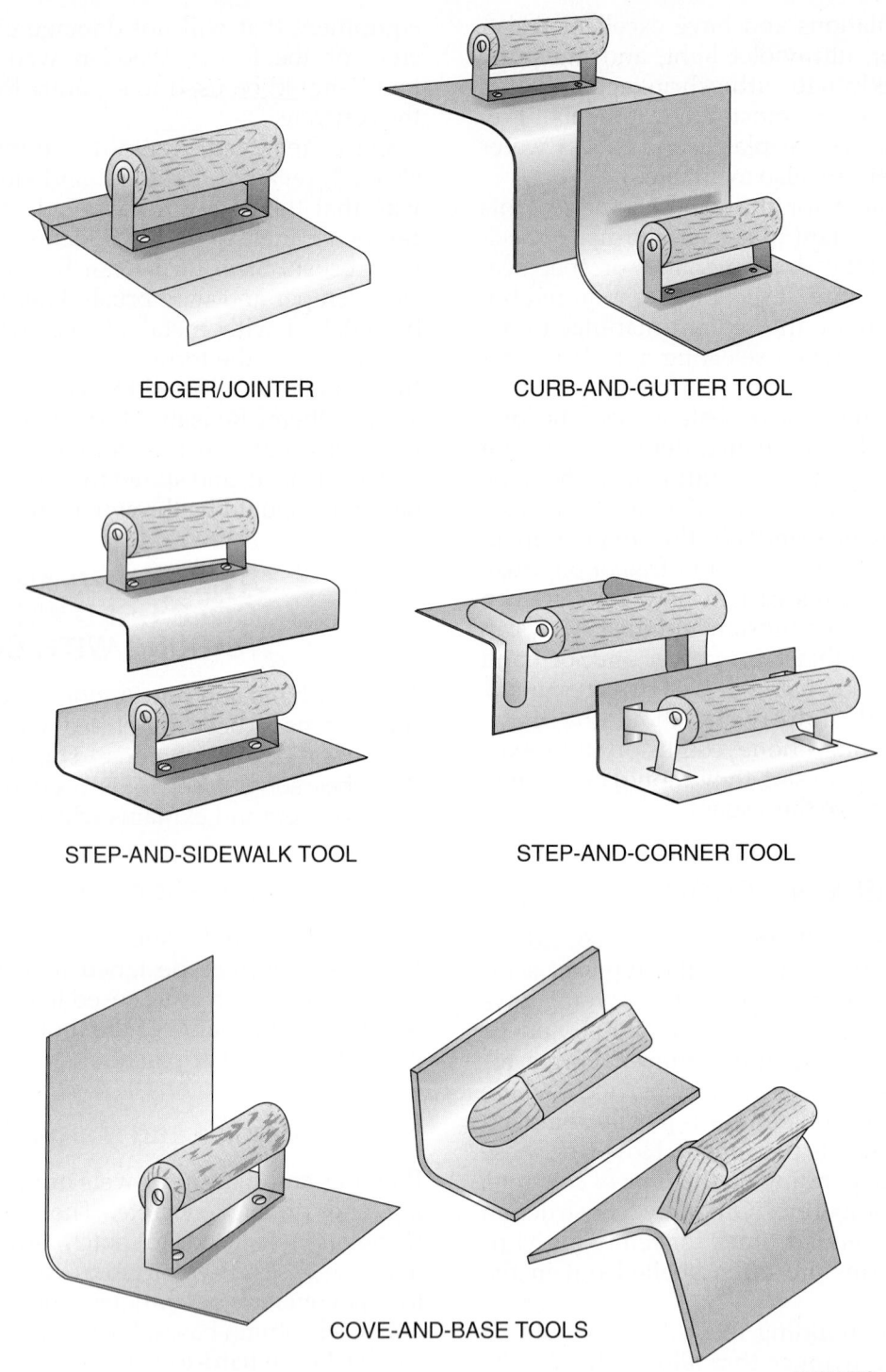

EDGER/JOINTER

CURB-AND-GUTTER TOOL

STEP-AND-SIDEWALK TOOL

STEP-AND-CORNER TOOL

COVE-AND-BASE TOOLS

305F40.EPS

Figure 40 ◆ Combination tools.

10.4.0 Carborundum Rubbing Stones

Carborundum rubbing stones (*Figure 43*) are used to smooth partly hardened concrete. Common rubbing stones are 6" to 8" long, 2" to 3½" wide, and ¾"to 2" thick. Grit sizes range from No. 24 (coarse) to No. 150 (fine).

10.5.0 Sprayers

Sprayers (*Figure 44*) are used to apply form oil to concrete forms, to spray curing compounds, and to mist or fog water over finished concrete to reduce rapid evaporation.

10.6.0 Power Grinders

Power grinders (*Figure 45*) are used to remove seams and projections, reduce high spots, and smooth surfaces. Wear appropriate personal protective equipment when using power grinders. Electric handheld grinders are used for smaller areas. Larger gasoline-powered units are used for larger areas. Single- and double-head concrete grinders are available in various sizes. Scarifiers are used for rough finishing.

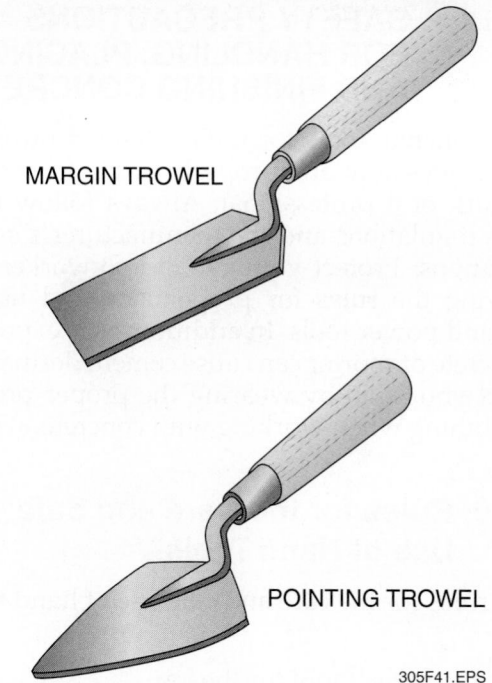

MARGIN TROWEL

POINTING TROWEL

305F41.EPS

Figure 41 ◆ Pointing and margin trowels.

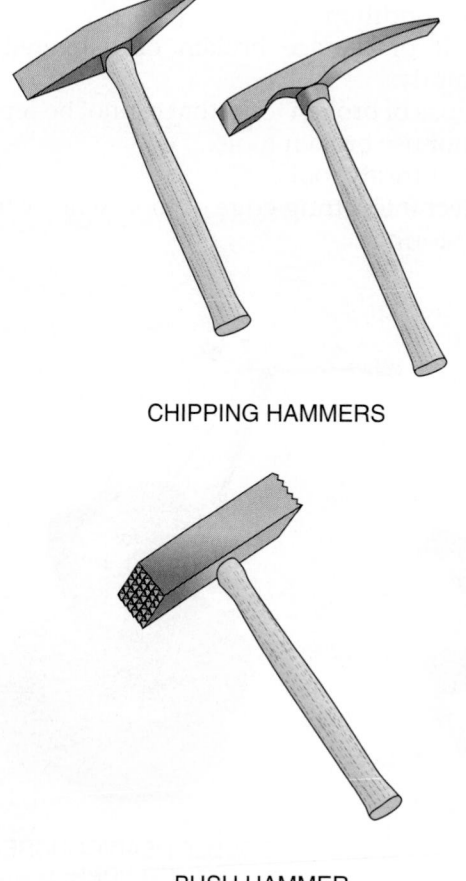

CHIPPING HAMMERS

BUSH HAMMER

305F42.EPS

Figure 42 ◆ Cement hammers.

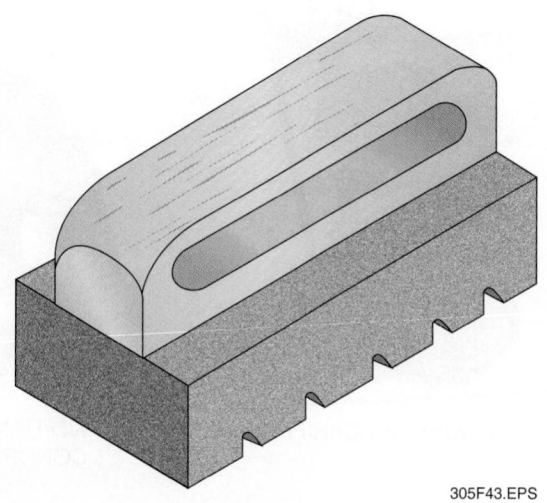

305F43.EPS

Figure 43 ◆ Typical carborundum rubbing stone.

Figure 44 ◆ Typical concrete sprayer.

305F44.EPS

11.0.0 ◆ SAFETY PRECAUTIONS FOR HANDLING, PLACING, AND FINISHING CONCRETE

Using concrete tools for their intended purpose, using them safely, and properly caring for tools is the mark of a professional. Always follow both OSHA regulations and the manufacturer's safety precautions. Protect yourself and co-workers by following the rules for proper care and use of hand and power tools. In addition, skin exposure to concrete or mortar can cause cement dermatitis. Protect your skin by wearing the proper protective clothing when working with concrete.

11.1.0 Rules for the Care and Safe Use of Hand Tools

Some rules for the care and safe use of hand tools are:

- Use the correct tool for the job.
- Keep tools clean; be sure to wash concrete from surfaces immediately after use.
- Maintain tools properly.
- Sharpen tools, when appropriate.
- Inspect tools frequently to make sure they are in good condition.
- Repair or replace broken or damaged tools promptly.
- Dispose of broken tools that cannot be repaired.
- Do not use broken tools.
- Do not throw tools.
- Protect the cutting edge of tools when carrying and storing.

CONCRETE SCARIFIER

TWO-HANDLE ELECTRIC CONCRETE GRINDER

DOUBLE HEAD CONCRETE GRINDER

305F45.EPS

Figure 45 ◆ Power grinders.

- Store tools properly when not in use. Do not carry tools in your pocket. Do not place tools where they can roll off. Lightly oil tools before storing. Store tools in a dry place.
- Wear eye, ear, and respiratory protection when appropriate.
- Stay alert when using tools. Keep fingers away from cutting edges. Work away from your body when using cutting tools. Be sure everyone is clear before you swing a sledgehammer. Use tools with insulated or wooden handles when working around electrical equipment.

11.2.0 Rules for the Care and Safe Use of Power Tools

Some rules for the care and safe use of power tools are as follows:

- Do not attempt to operate any power tool without being certified on that particular tool.
- Always wear appropriate safety equipment and protective clothing. For example, wear safety glasses and tight-fitting clothing that cannot become caught in the moving parts of the tool. Roll up long sleeves, tuck in your shirt-tail, and tie back long hair.
- Do not leave a power tool running unattended.
- Assume a safe and comfortable position before starting a power tool.
- Do not distract others or let anyone distract you while operating a power tool.
- Be sure that any electric power tool is properly grounded before using it.
- Be sure that power is not connected before performing maintenance or changing accessories.
- Do not use dull or broken accessories.
- Use a power tool only for its intended purpose.
- Do not use a power tool with guards or safety devices removed.
- Use an extension cord of sufficient length to service the particular electric tool you are using.
- Do not operate an electric power tool if your hands or feet are wet.

- Become familiar with the correct operation and adjustments of a power tool before attempting to use it.
- Keep a firm grip on the power-float or trowel-machine handle while operating the machine.
- Be sure there is proper ventilation before operating gasoline-powered equipment indoors.
- Keep a fire extinguisher nearby when filling and operating gasoline-powered equipment.
- Keep hands and feet away from cutting tools such as concrete saws, grinders, and trowel-machine blades.
- Change trowel-machine blades when they become ragged.
- Store tools properly when not in use.

11.3.0 Preventing Cement Dermatitis

Persons working with concrete need to take precautions to prevent skin irritation as a result of the skin coming in contact with concrete or mortar. When wet, the lime contained in cement causes it to heat and burn the skin. Prolonged exposure and inhalation could result in sores appearing on the skin and in the lining of the mouth and nose. This condition is called cement poisoning or cement dermatitis. Factors that contribute to its formation include excessive sweating, failure to observe precautions, and pre-existing dermatitis or an allergy. Medical care should be sought by anyone who develops dermatitis. Some actions that help reduce the risk of cement dermatitis include the following:

- Use of the proper clothing such as coveralls and long-sleeved shirts
- Use of rubber gloves and boots
- Use of goggles or face mask
- Use of a respirator or other breathing device
- Frequent washing with a lanolin-based soap or lotion
- Bathing after each shift

Summary

It is important for carpenters to understand the proper methods of handling, placing, and finishing concrete. Knowledge about the factors involved with these tasks enables the carpenter to better lay out and construct concrete forms so that the forms help rather than hinder efficient placement of the concrete. In addition, carpenters are often required to perform some or all parts of the concrete handling, placement, and finishing work. This is especially true if they are involved in residential, light commercial, or remodeling work. On large construction projects, carpenters normally work closely with the concrete trades; therefore, it is important for them to fully understand the concrete placement and finishing process so that their work is performed professionally and in an efficient manner.

Notes

1. The joint marked B in *Figure 1* represents a
   _____.
   a. construction joint with a dowel
   b. tooled expansion joint
   c. sawed control joint
   d. construction joint with a keyed expansion

2. The joint marked C in *Figure 1* represents a
   _____.
   a. tooled dummy joint
   b. tooled construction joint
   c. sawed control joint
   d. construction joint with a dowel

3. The joint marked D in *Figure 1* represents a
   _____.
   a. decorative joint with a dowel
   b. sawed control joint with a dowel
   c. tooled control joint with a dowel
   d. tooled dummy joint

4. The joint marked E in *Figure 1* represents a
   _____.
   a. sawed control joint
   b. tooled control joint
   c. tooled dummy joint
   d. tooled control joint with a dowel

5. The joint marked A in *Figure 1* represents a
   _____.
   a. decorative joint with a wood spacer
   b. tooled dummy joint
   c. construction joint with a dowel
   d. tooled construction joint

6. Concrete from a batch plant is usually
   _____.
   a. completely mixed at the batch plant
   b. completely mixed at the construction site
   c. completely mixed in a truck mixer while being delivered
   d. a wet-mix type of concrete

7. A truck-mounted, self-contained vehicle that carries all the dry materials (cement, aggregates, etc.) and water, then mixes them on-site to make concrete is called a
   _____.
   a. mobile batch plant
   b. volumetric mixer
   c. truck mixer
   d. front-discharge truck mixer

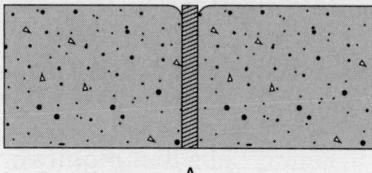

A

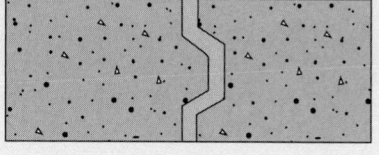

B

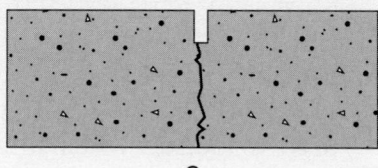

C

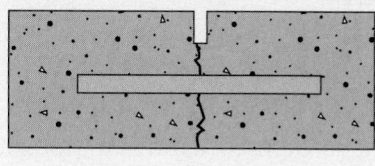

D

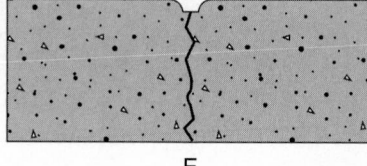

E

Figure 1

305RQ01.EPS

8. When concrete is to be placed in a difficult location, or where thin sections and large areas are needed, the delivery device commonly used is the _____.
   a. pneumatic gun
   b. concrete pump
   c. concrete buggy
   d. crane and bucket

9. The American Concrete Pumping Association requires that the boom of a boom-type concrete pump be kept a minimum of _____ feet away from overhead power lines.
   a. 5
   b. 12
   c. 17
   d. 20

10. A device used to direct concrete into narrow vertical forms and prevent separation of the concrete ingredients is a _____.
    a. belt conveyer
    b. concrete drop
    c. drop pipe
    d. drop chute

11. The process of depositing concrete in forms and consolidating the concrete is called _____.
    a. placing
    b. dropping
    c. finishing
    d. rodding

12. To minimize separation, it is best to discharge concrete _____.
    a. horizontally
    b. at a 45-degree angle
    c. vertically
    d. at a 30-degree angle

13. Entrapped air is defined as an air pocket that is _____.
    a. 1 millimeter or smaller
    b. 1 millimeter or larger
    c. 1 centimeter or smaller
    d. 1 centimeter or larger

14. The requirements for minimum sample size, methods for calculating F_F and F_L floor profile numbers, and other requirements are described in specification _____.
    a. *ASTM E 1155*
    b. *ASTM E 1221*
    c. *ASTM E 2112*
    d. *ASTM E 5511*

15. A specification requires that the flatness and levelness profile of a slab-on-grade floor be $F_F = 75$ and $F_L = 50$. This typically represents a(n) _____ flatness and levelness profile.
    a. flat
    b. very flat
    c. super flat
    d. ultra flat

16. The process of leveling concrete to an established grade is called _____.
    a. finishing
    b. plumbing
    c. placing
    d. screeding

17. Which of the following finish steps should be done first when finishing a slab?
    a. Bullfloat or darby float the slab.
    b. Trowel the surface.
    c. Edge the slab.
    d. Straightedge (screed) the slab.

18. The type of screed widely used when finishing slabs that require a very flat or finer profile is a _____ creed.
    a. steel truss
    b. gas or electric powered one- or two-man
    c. laser
    d. roller

19. Which of the following factors should be used to decide when to start finishing a concrete slab surface?
    a. Size of the slab
    b. Hardness and appearance of the concrete
    c. Weather conditions
    d. Type of concrete being used

20. A concrete finishing tool used to form a radius at the edges of a slab is called a(n) _____.
    a. float
    b. jointer
    c. edger
    d. groover

21. A tool that smoothes concrete using a series of blades rotated by a motor is called a _____.
    a. power screed
    b. power darby
    c. power grinder
    d. finishing machine

22. Where a smooth, hardened surface is desired, the floating process is followed by _____.
    a. jointing
    b. troweling
    c. grinding
    d. brooming

23. The chemical reaction of water with cement is called _____.
    a. bonding
    b. hydration
    c. curing
    d. setting

24. If you encounter a chipping hammer with a cracked handle, you should _____.
    a. tape the handle
    b. install tacks at the break
    c. replace the hammer immediately
    d. use the hammer carefully until you find a replacement

25. Cement dermatitis is an irritation caused by the reaction between lime and _____.
    a. moisture
    b. sand
    c. silica
    d. aggregates

Trade Terms
Introduced in This Module

Bleed: A condition in concrete in which the solids settle and the water moves to the top.

Consolidating concrete: Working freshly placed concrete so that each layer is compacted with the layer below and voids caused by water or air pockets are eliminated.

Elastomeric: A material having the properties of excellent flexibility and elongation.

Grout: A thin mortar.

Segregation: The separation of sand-cement ingredients from the gravel due to the improper placement of concrete.

Set: The hardening of concrete.

Spalling: The condition of concrete breakup, chipping, splitting, or crumbling.

This module is intended to be a thorough resource for task training. The following reference works are suggested for further study. These are optional materials for continued education rather than for task training.

American Concrete Institute (ACI). www.concrete.org.

Cement Association of Canada. www.cement.ca.

Portland Cement Association. www.cement.org.

NCCER makes every effort to keep these textbooks up-to-date and free of technical errors. We appreciate your help in this process. If you have an idea for improving this textbook, or if you find an error, a typographical mistake, or an inaccuracy in NCCER's Contren® textbooks, please write us, using this form or a photocopy. Be sure to include the exact module number, page number, a detailed description, and the correction, if applicable. Your input will be brought to the attention of the Technical Review Committee. Thank you for your assistance.

Instructors – If you found that additional materials were necessary in order to teach this module effectively, please let us know so that we may include them in the Equipment/Materials list in the Annotated Instructor's Guide.

Write: Product Development and Revision
              National Center for Construction Education and Research
              3600 NW 43rd St, Bldg G, Gainesville, FL 32606

Fax: 352-334-0932

E-mail: curriculum@nccer.org

Craft _____ Module Name _____

Copyright Date _____ Module Number _____ Page Number(s) _____

Description _____

_____

_____

_____

(Optional) Correction _____

_____

_____

(Optional) Your Name and Address _____

_____

_____

Module 68104-09

Introduction to Masonry

COURSE MAP

This course map shows all of the modules in the *Construction Technology* curriculum. The suggested training order begins at the bottom and proceeds up. Skill levels increase as you advance on the course map. The local Training Program Sponsor may adjust the training order.

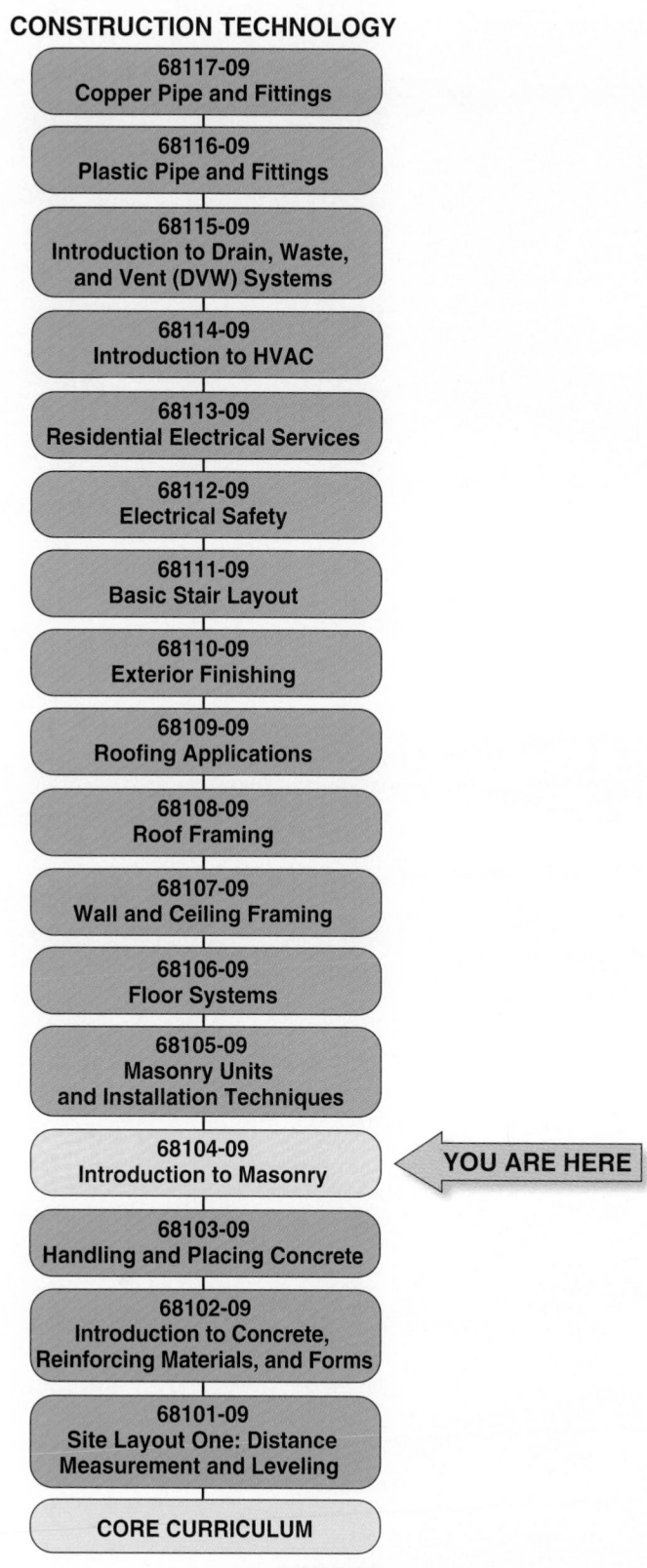

CONSTRUCTION TECHNOLOGY

68117-09
Copper Pipe and Fittings

68116-09
Plastic Pipe and Fittings

68115-09
Introduction to Drain, Waste, and Vent (DVW) Systems

68114-09
Introduction to HVAC

68113-09
Residential Electrical Services

68112-09
Electrical Safety

68111-09
Basic Stair Layout

68110-09
Exterior Finishing

68109-09
Roofing Applications

68108-09
Roof Framing

68107-09
Wall and Ceiling Framing

68106-09
Floor Systems

68105-09
Masonry Units and Installation Techniques

68104-09
Introduction to Masonry

← YOU ARE HERE

68103-09
Handling and Placing Concrete

68102-09
Introduction to Concrete, Reinforcing Materials, and Forms

68101-09
Site Layout One: Distance Measurement and Leveling

CORE CURRICULUM

104CMAP.EPS

1.0.0   **INTRODUCTION** 4.1

2.0.0   **THE HISTORY OF MASONRY** 4.2

3.0.0   **MASONRY TODAY** 4.4

    3.1.0   Clay Products 4.4

    3.1.1   *Solid Masonry Units/Brick* 4.5

    3.1.2   *Hollow Masonry Units/Tiles* 4.6

    3.1.3   *Architectural Terra-Cotta* 4.7

    3.1.4   *Brick Classifications* 4.7

    3.2.0   Brick Masonry Terms 4.7

    4.0.0   Concrete Products 4.8

    4.1.0   Block 4.11

    4.2.0   Concrete Brick 4.11

    4.3.0   Other Concrete Units 4.12

5.0.0   **STONE** 4.13

6.0.0   **MORTARS AND GROUTS** 4.13

7.0.0   **MODERN CONSTRUCTION TECHNIQUES** .4.14

    7.1.0   Wall Structures 4.14

    7.2.0   Modern Techniques 4.17

8.0.0   **MASONRY AS A CAREER** 4.18

    8.1.0   Career Stages 4.18

    8.2.0   Apprentice 4.18

    8.3.0   Journeyman 4.19

    8.4.0   Supervisors, Superintendents, and Contractors 4.20

    8.5.0   The Role of NCCER 4.20

9.0.0   **KNOWLEDGE, SKILLS, AND ABILITY** 4.20

    9.1.0   Knowledge 4.20

    9.1.1   *Job-Site Knowledge* 4.21

    9.1.2   *Learning More* 4.21

    9.2.0   Attitude and Work 4.21

    9.2.1   *Dependability* 4.21

    9.2.2   *Responsibility* 4.21

    9.2.3   *Adaptability* 4.21

    9.2.4   *Pride* 4.21

    9.3.0   Quality 4.22

10.0.0 **BASIC BRICKLAYING** 4.22

    10.1.0   Preparing Mortar 4.22

    10.2.0   Spreading Mortar 4.23

    10.3.0   Picking Up Mortar 4.23

    10.3.1   *Holding the Trowel* 4.23

10.3.2   *Picking Up Mortar from a Board* 4.24

10.3.3   *Picking Up Mortar from a Pan* 4.24

10.4.0   Spreading, Cutting, and Furrowing 4.25

10.4.1   *Spreading* 4.25

10.4.2   *Cutting or Edging* 4.25

10.4.3   *Furrowing* 4.26

10.5.0   Buttering Joints 4.26

10.6.0   General Rules 4.27

11.0.0 SAFETY PRACTICES 4.28

11.1.0   The Cost of Job Accidents 4.28

11.2.0   Wearing Safety Gear and Clothing 4.29

11.3.0   Hazards on the Job 4.30

11.4.0   Falling Objects 4.30

11.5.0   Mortar and Concrete Safety 4.31

11.6.0   Flammable Liquid Safety 4.32

11.7.0   Material Handling 4.33

11.7.1   *Materials Stockpiling and Storage* 4.33

11.7.2   *Working Stacks* 4.33

11.8.0   Gasoline-Powered Tools 4.34

11.9.0   Powder-Actuated Tools 4.34

11.10.0 Pressure Tools 4.35

11.11.0 Weather Hazards 4.36

11.11.1 *Cold Weather* 4.36

11.11.2 *Hot Weather* 4.37

12.0.0 FALL PROTECTION 4.37

12.1.0   Guardrails 4.38

12.2.0   Personal Fall-Arrest Systems 4.39

12.2.1   *Body Harnesses* 4.40

12.2.2   *Lanyards* 4.40

12.2.3   *Deceleration Devices* 4.41

12.2.4   *Lifelines* 4.41

12.2.5   *Anchoring Devices and Equipment Connectors* 4.42

12.2.6   *Selecting an Anchor Point and Tying Off* 4.42

12.2.7   *Using Personal Fall-Arrest Equipment* 4.43

12.3.0   Safety Net Systems 4.43

12.4.0   Rescue After a Fall 4.44

13.0.0 FORKLIFT SAFETY 4.44

13.1.0   Before You Operate a Forklift 4.45

13.1.1   *Training and Certification* 4.45

13.1.2   Pre-Shift Inspection 4.45

13.1.3   General Safety Precautions 4.45

13.2.0   Traveling 4.47

13.2.1   Stay Inside 4.47

13.2.2   Pedestrians 4.47

13.2.3   Passengers 4.47

13.2.4   Blind Corners and Intersections 4.47

13.2.5   Keeping the Forks Low 4.48

13.2.6   Horseplay 4.48

13.2.7   Travel Surface 4.48

13.3.0   Handling Loads 4.48

13.3.1   Picking Up Loads 4.48

13.3.2   Traveling with Loads 4.48

13.3.3   Traveling with Long Loads 4.48

13.3.4   Placing Loads 4.49

13.3.5   Placing Elevated Loads 4.49

13.3.6   Tipping 4.49

13.3.7   Using a Forklift to Rig Loads 4.50

13.3.8   Dropping Loads 4.50

13.3.9   Obstructing the View 4.50

13.4.0   Working on Ramps and Docks 4.51

13.4.1   Ramps 4.51

13.4.2   Docks 4.51

13.5.0   Fire and Explosion Hazards 4.51

13.5.1   Flammable and Combustible Liquids 4.52

13.5.2   Flammable Gases 4.52

13.5.3   Fire Fighting 4.52

13.6.0   Pedestrian Safety 4.52

SUMMARY 4.53

REVIEW QUESTIONS 4.54

PROFILE IN SUCCESS 4.55

GLOSSARY 4.57

REFERENCES 4.58

Figures

Figure 1   Notre Dame Cathedral, Paris 4.2
Figure 2   Herringbone pattern 4.2
Figure 3   Radial arch 4.3
Figure 4   Roman arch 4.4
Figure 5   Standard brick 4.5
Figure 6   Common bond patterns 4.5
Figure 7   Special brick shapes 4.6
Figure 8   Masonry units and mortar joints 4.8
Figure 9   Cavity walls 4.8
Figure 10   Common concrete block .4.10
Figure 11   Parts of a block 4.11
Figure 12   Concrete brick 4.12
Figure 13   Common pre-faced concrete units 4.12
Figure 14   Manhole and vault unit 4.13
Figure 15   Stone facing used as decorative trim 4.13
Figure 16   A block wall faced with stone 4.13
Figure 17   Types of masonry construction 4.15
Figure 18   Reinforced walls 4.16
Figure 19   Mason's trowels 4.17
Figure 20   Example of apprenticeship training recognition 4.19
Figure 21   Masonry mortar 4.22
Figure 22   Mixing mortar 4.23
Figure 23   Holding the trowel 4.23
Figure 24   Picking up mortar from a board 4.24
Figure 25   Picking up mortar from a pan 4.24
Figure 26   Spreading mortar 4.25
Figure 27   Cutting an edge 4.26
Figure 28   A furrow 4.26
Figure 29   A buttered joint 4.27
Figure 30   Placing the brick 4.27
Figure 31   Checking the level 4.27
Figure 32   Hidden costs of accidents 4.28
Figure 33   Dressed for masonry work 4.29
Figure 34   Crane hazard on the job .4.31
Figure 35   Emergency hand signals .4.31
Figure 36   Palletized brick 4.33

Figure 37   Stacking brick 4.34

Figure 38   Powder-actuated fastening tool 4.34

Figure 39   Proper and improper safety harness use 4.38

Figure 40   Guardrails 4.39

Figure 41   Full-body harnesses with sliding back D-rings 4.40

Figure 42   Harness with front chest D-ring 4.40

Figure 43   Lanyard with a shock absorber 4.41

Figure 44   Rope grab and retractable lifeline 4.41

Figure 45   Vertical lifeline 4.42

Figure 46   Horizontal lifeline 4.42

Figure 47   Eye bolt 4.42

Figure 48   Double locking snaphook 4.42

Figure 49   Forklift 4.44

Figure 50   Forklift operator's daily checklist 4.46

Figure 51   Forklift working in a storage area 4.47

Figure 52   Forklift with forks in low position 4.48

Figure 53   Center of gravity 4.49

Figure 54   Combined center of gravity .4.50

Tables

Table 1   Mortar Composition 4.14

Table 2   Powder Charge Color-Coding System 4.34

Introduction to Masonry

Objectives

When you have completed this module, you will be able to do the following:

1. Discuss the history of masonry.
2. Describe modern masonry materials and methods.
3. Explain career ladders and advancement possibilities in masonry work.
4. Describe the skills, attitudes, and abilities needed to work as a mason.
5. State the safety precautions that must be practiced at a work site, including the following:
   - Safety practices
   - Fall-protection procedures
   - Forklift-safety operations
6. Perform the following basic bricklaying procedures:
   - Mixing of mortar
   - Laying a mortar bed
   - Laying bricks
7. Put on eye protection, respiratory protection, and a safety harness.
8. Use the correct procedures for fueling and starting a gasoline-powered tool.

Recommended Prerequisites

Before you begin this module, it is recommended that you successfully complete the following: *Core Curriculum; Construction Technology,* Modules 68101-06 through 68103-06.

Required Trainee Materials

1. Pencil and paper
2. Appropriate personal protective equipment

1.0.0 ◆ INTRODUCTION

You are beginning the study of masonry, one of the world's oldest and most respected crafts. Masonry construction has been around for thousands of years. The remains of stone buildings date back 15,000 years, and the earliest manufactured bricks unearthed by archaeologists are more than 10,000 years old. These bricks were made of hand-shaped, dried mud. Among the most well-known works of **masons** are the pyramids of ancient Egypt and Notre Dame Cathedral in Paris (*Figure 1*).

Masons build structures of **masonry units**. Masonry units are blocks of brick, concrete, **ashlar**, glass, tile, **adobe**, and other materials. In the most common forms of masonry, a mason assembles walls and other structures of clay brick or **concrete masonry units (CMUs)** using **mortar** to bond the units together.

In this module, you will learn about the basic materials, tools, and techniques used by masons. With the guidance of your instructor, you will learn to mix mortar and lay brick. At first glance, building with masonry units may appear simple. It's not. The first challenge is to lay the units perfectly straight and level. The next challenge is to do it quickly. The production level will vary depending on whether the work is commercial or residential, and the type of work being done.

Laying 600 to 800 bricks a day is a common requirement. In some companies, however, a skilled bricklayer may be expected to lay 1,400 bricks a day. In a contest held by U.S. Brick in Dallas, Texas, in 1996, a bricklayer from McGee Brothers Masonry of Charlotte, North Carolina, laid more than 1,400 bricks in an hour, setting a new world record.

Figure 1 ◆ Notre Dame Cathedral, Paris.

2.0.0 ◆ THE HISTORY OF MASONRY

Brick is the oldest manufactured building material, invented thousands of years ago. The Hanging Gardens of Babylon hung down from brick towers. These hand-formed mud bricks were reinforced with straw and dried in the sun. They were stacked with wet mud between them. Sometimes they were covered with another coat of mud, which was decorated. This was a common and effective building technique for centuries. Some 8,000-year-old bricks have been recovered from the biblical city of Jericho. These sun-dried bricks have a row of thumbprints along their tops. Later, bricks had the king's name and the date stamped on the top; this practice is very useful to today's archaeologists working to date their excavations. Today's bricks are stamped with the manufacturer's name in the same place on the top of the brick. Handmade clay brick is still in use in some parts of the Near East and Africa.

Early bricks led to early brick architecture. Someone had the idea of laying brick in different patterns instead of simply stacking them. *Figure 2*

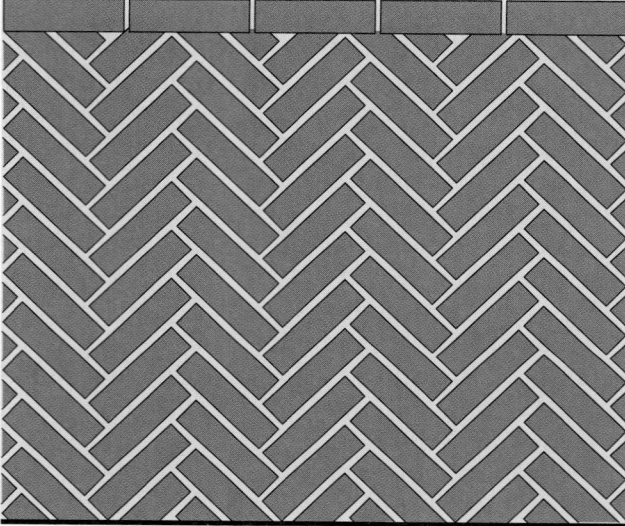

101F02.EPS

Figure 2 ◆ Herringbone pattern.

shows a herringbone pattern, seen in ancient walls still standing today. These walls used mortar to hold the bricks together. The first mortar was wet mud. Along with firing and glazing brick, the Babylonians developed two new types of mortar. These were based on mixing lime or pitch (asphalt) with the mud.

A later boost for brick architecture was the development of the dome and arch (*Figure 3*). With domes and arches, early masons could build larger and higher structures, with more open space inside.

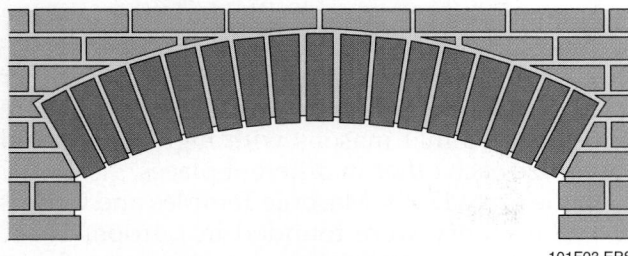

Figure 3 ◆ Radial arch.

The Romans refined arches and domes and built large-scale brickyards. They covered the Roman Empire with roads that brought Roman bricks, mortar, and Roman designs for arches (*Figure 4*) and domes, along with Roman civilization, to the known world. The Romans refined the Babylonian lime mortar by developing a form of cement that was a waterproof mortar. This mortar was useful for both brick and stone construction. It was also applied as a finish coat to the exterior of the surface as an early form of stucco.

The Romans standardized the sizes of their brick. Roman brick is a recognized standard size for bricks today. The Romans produced highly ornate brick architecture using specialized brick shapes and varied brick colors and glazes.

BUILDING BLOCKS

Brick in Construction

Brick is found in all types of construction, from tract houses to stately mansions. Brick is also used in the construction of banks, schools, churches, and office buildings. Brick not only provides an attractive appearance, it creates a sense of permanence.

Figure 4 ◆ Roman arch.

When the Normans conquered England in 1066, they built many castles in England. They imported brick as well as the masons to lay the brick. This construction boom boosted the trade economy of Europe. It also boosted the status of masons. The Norman brick is still a recognized and widely used size of brick.

As the demand for more elaborate construction grew, the need for skilled workers became greater. By the middle 1300s, masons had organized into early unions, known as guilds, across most of Europe.

Guilds controlled the practice of the craft by monitoring the skill level of the craftworker. Early masons' guilds recognized three levels of work:

• The rough mason was the equivalent of the apprentice of today. They worked where the guild would allow, under supervision of masters or journeymen, and served an apprenticeship of three to five years. They were examined by the guild before being allowed to work independently as masons.

• The journeyman, or mason, was a skilled worker allowed to work on finer jobs without supervision. Journeymen were also free to move to wherever the work was. The local guild set the time, as long as ten years, for this stage. At the end, the journeyman could take an examination before the guild and be awarded the status of master. Not all masons reached master status.

• The master mason ran the business, designed structures, employed journeymen, and trained apprentice masons. The master masons also held office, served on boards, and ran the affairs of the masons' guild.

The local guilds continued to control the practice of the masons' craft for centuries. They monitored training, judged disputes, and shared knowledge among members. They collected dues, provided some support to widows and the ill, and celebrated special masons' holidays.

Nonmembers were not allowed to know the secrets of the guilds. Guilds could choose to recognize or not recognize masons certified by other local guilds. Secret signs and passwords were developed so that masons with high skills could recognize each other in different places.

In the early 1700s, Masonic Temples and Orders of Freemasonry were founded in Europe. These organizations were political and spiritual, but based on many of the ideas of the masons' guilds. The pyramid seal on the back of the American dollar bill is an isolated legacy of the masons' guilds.

3.0.0 ◆ MASONRY TODAY

Masons are still recognized as premier craftworkers at any construction site. Their work takes advantage of twentieth-century technology. The two main types of masonry units manufactured today are made of clay or concrete. Clay products are commonly known as brick and tile; concrete products are commonly known as block or concrete masonry units (CMUs).

3.1.0 Clay Products

Brick has been developed and improved upon for centuries. The modern age in brick manufacture started with the first brickmaking machine. It was powered by a steam engine and patented in 1800. The process has not changed much. The clay is mined, pulverized, and screened. It is mixed with water, formed, and cut into shape. Some plants extrude the clay, punch holes into it, then cut it into shape. Any coating or glazing is applied before the units are air dried. After drying, the brick is fired in a kiln. Because of small variations in materials and firing temperatures, not all bricks are exactly alike. Even bricks made and fired in the same batch have variations in color and shading.

The brick is slowly cooled to prevent cracking. It is then bundled into cubes and shipped. A cube traditionally holds 500 standard bricks, or 90 blocks, although manufacturers today make cubes of varying sizes.

Today, there are over 100 commonly manufactured structural clay products. The American Society for Testing and Materials (ASTM) International has published standards for masonry design and construction. The standards cover performance specifications for manufactured masonry units. ASTM has also specified standard sizes for various kinds of brick. *Figure 5* shows the standard sizes for today's most commonly used brick. The first six types are the most widely used. The sizes shown are actual dimensions. Bricks are

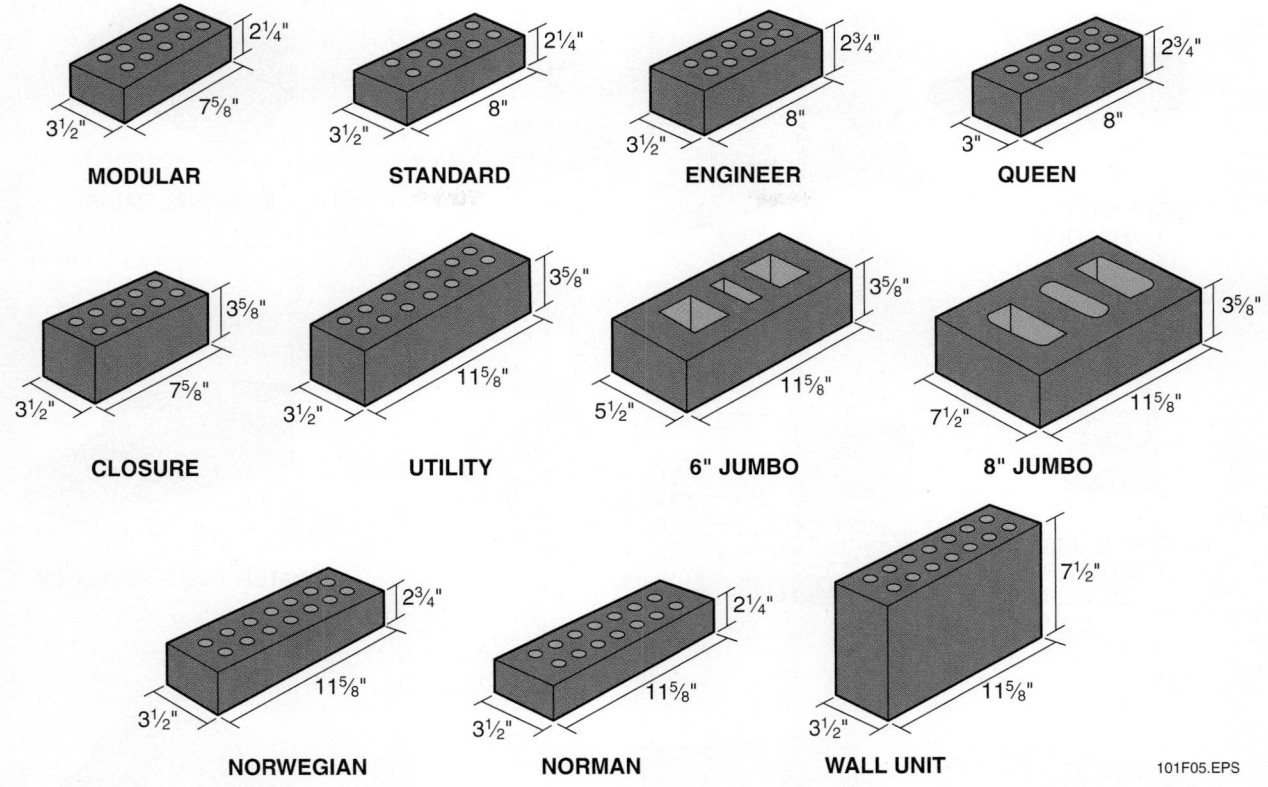

Figure 5 ◆ Standard bricks.

also identified by nominal sizes, which include the thickness of the mortar **joint**. The nominal size of the modular brick, for example, is 4" × 8" × 2⅜".

Structural clay products include the following:

- Solid masonry units, or brick
- Hollow masonry units, or tile
- Architectural terra-cotta units

The next sections provide more information about these products.

3.1.1 Solid Masonry Units/Brick

Brick is classified as solid if 25 percent or less of its surface is open (void). Brick is further divided into the following classifications: building, **facing**, hollow, paving, ceramic glazed, thin veneer, sewer, and manhole. ASTM standards exist for all of these types of brick. Fire brick has its own standard, but is not considered a major type classification.

Brick comes in modular and nonmodular sizes, in colors determined by the minerals in the clay or by additives. There is also a variety of face textures and a rainbow of glazes. The variety is dazzling. Brick can be laid in structural bonds to create patterns in the face of a wall or walkway. *Figure 6* shows several examples of commonly used bond patterns. Some bond patterns are traditional in

some parts of the country. The herringbone pattern shown earlier is still popular for walkways.

Brick is also made in special shapes to form arches, sills, copings, columns, and stair treads. Custom shapes can be made to order for architectural or artistic use. *Figure 7* shows some commonly manufactured special shapes of brick.

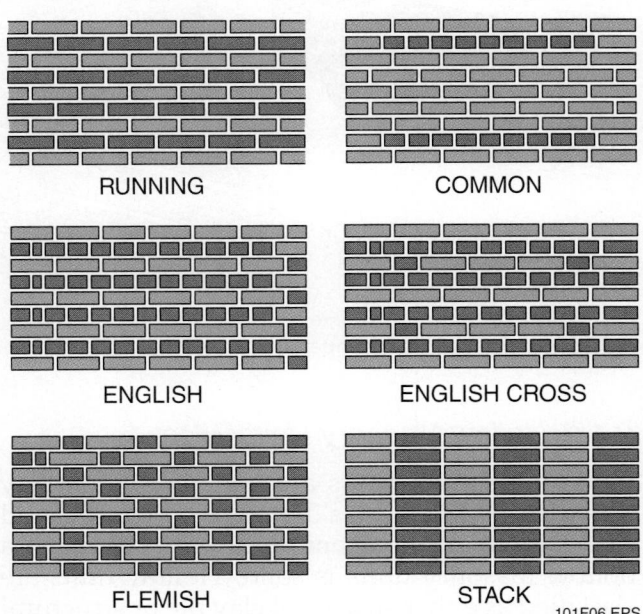

Figure 6 ◆ Common bond patterns.

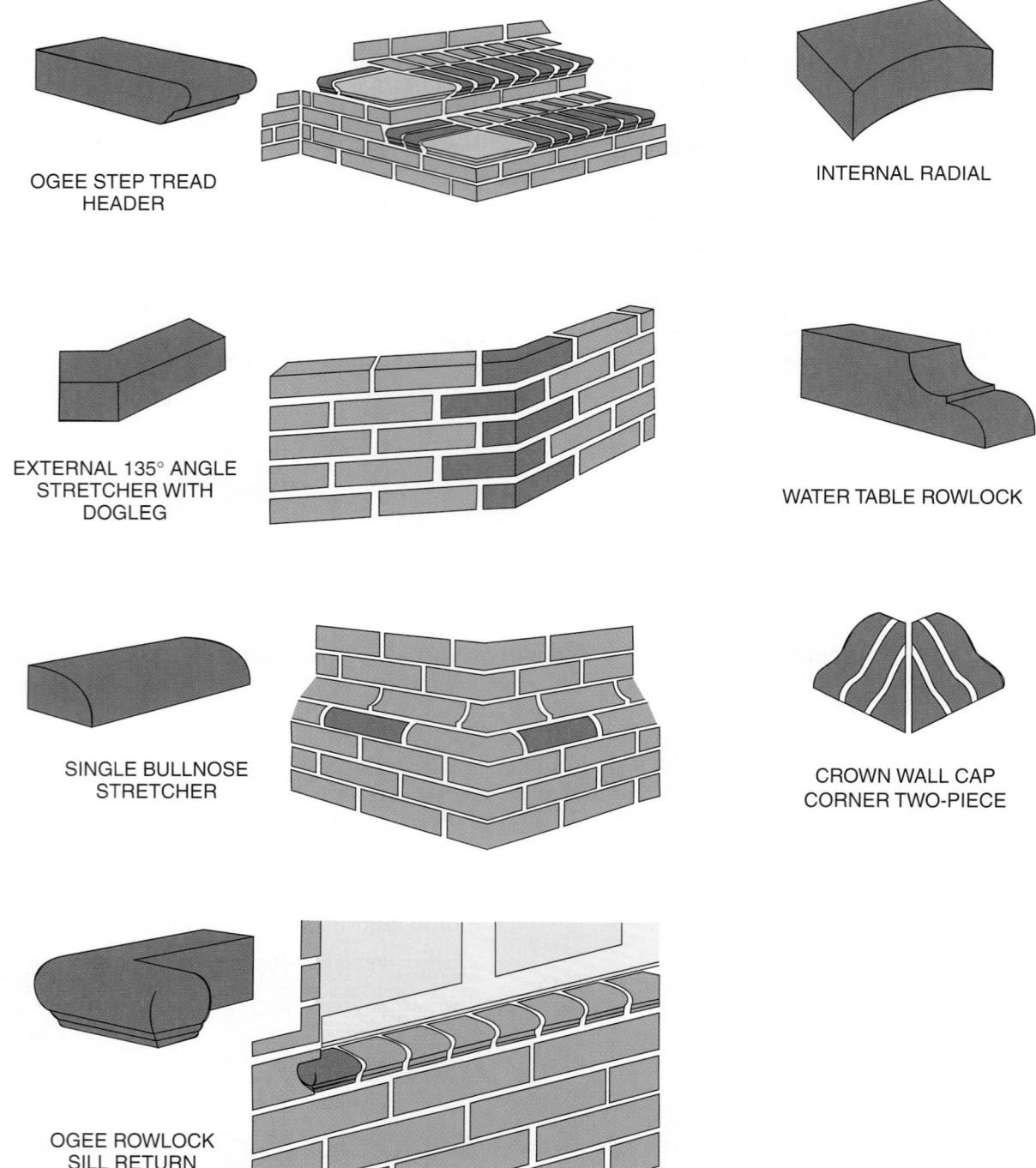

OGEE STEP TREAD
HEADER

INTERNAL RADIAL

EXTERNAL 135° ANGLE
STRETCHER WITH
DOGLEG

WATER TABLE ROWLOCK

SINGLE BULLNOSE
STRETCHER

CROWN WALL CAP
CORNER TWO-PIECE

OGEE ROWLOCK
SILL RETURN

101F07.EPS

Figure 7 ◆ Special brick shapes.

3.1.2 Hollow Masonry Units/Tiles

Hollow masonry units are machine-made clay tiles extruded through a die and cut to the desired size. Less than 75 percent of the surface area of a hollow masonry units is solid. Hollow units are classified as either structural clay tile or structural clay facing tile.

Structural clay tile comes in many shapes, sizes, and colors. It is divided into loadbearing and non-bearing types. Structural tile can be used for load-bearing on its side or on its end. In some applications, structural tile is used as a backing **wythe** behind brick. Nonbearing tile is designed for use as fireproofing, furring, or ventilating partitions.

Structural clay facing tile comes in modular sizes as either glazed or unglazed tile. It is designed for interior uses where precise tolerances are required. A special application of clay facing tile is as an acoustic barrier. The acoustic tiles have a holed face surface to absorb sound. Clay facing tile can also be patterned by shaping the surface face.

3.1.3 Architectural Terra-Cotta

Architectural terra-cotta is a made-to-order product with an unlimited color range. High-temperature fired ceramic glazes are available in an unlimited color range and unlimited arrangements of parts, shapes, and sizes.

Architectural terra-cotta is classified into anchored ceramic veneer, adhesion ceramic veneer, and ornamental or sculptured terra-cotta. Anchored ceramic veneer is thicker than 1", held in place by grout and wire anchors. Adhesion ceramic veneer is 1" or thinner, held in place by mortar. Ornamental terra-cotta is frequently used for cornices and column capitals on large buildings.

3.1.4 Brick Classifications

As previously stated, the three general types of structural brick-masonry units are solid, hollow, and architectural terra-cotta. All three can serve a structural function, a decorative function, or a combination of both. The three types differ in their formation and composition, and are specific in their use. Bricks commonly used in construction include the following:

- *Building bricks* – Also called common, hard, or kiln-run bricks, these bricks are made from ordinary clays or shales and fired in kilns. They have no special scoring, markings, surface texture, or color. Building bricks are generally used as the backing courses in either solid or cavity brick walls because the harder and more durable kinds are preferred.
- *Face bricks* – These are better quality and have better durability and appearance than building bricks because they are used in exposed wall faces. The most common face brick colors are various shades of brown, red, gray, yellow, and white.
- *Clinker bricks* – These bricks are oven-burnt in the kiln. They are usually rough, hard, durable, and sometimes irregular in shape.
- *Pressed bricks* – These bricks are made by the dry-press process rather than by kiln-firing. They have regular smooth faces, sharp edges,

and perfectly square corners. Ordinarily, they are used as face bricks.

- *Glazed bricks* – These have one surface coated with a white or other color of ceramic glazing. The glazing forms when mineral ingredients fuse together in a glass-like coating during burning. Glazed brick is particularly suited to walls or partitions in hospitals, dairies, laboratories, and other structures requiring sanitary conditions and easy cleaning.
- *Fire bricks* – These are made from a special type of fire clay to withstand the high temperatures of fireplaces, boilers, and similar constructions without cracking or decomposing. Fire brick is generally larger than other structural brick, and often is hand-molded.
- *Cored bricks* – These bricks have three, five, or ten holes extending through their beds to reduce weight. Three holes are most common. Walls built entirely from cored bricks are not much different in strength than walls built entirely from solid bricks. Both have about the same resistance to moisture penetration. Whether cored or solid, use the more easily available brick that meets building requirements.

Source: U.S. Army FM5-428

3.2.0 Brick Masonry Terms

You need to know the specific terms that describe the position of masonry units and mortar joints in a wall (*Figure 8*). These terms include the following:

- *Course* – One of several continuous, horizontal layers (or rows) of masonry units bonded together
- *Wythe* – A vertical wall section that is the width of one masonry unit
- *Stretcher* – A masonry unit laid flat on its bed along the length of a wall with its face parallel to the face of the wall
- *Header* – A masonry unit laid flat on its bed across the width of a wall with its face perpendicular to the face of the wall; generally used to bond two wythes
- *Rowlock* – A header laid on its face or edge across the width of a wall
- *Bull stretcher* – A rowlock brick laid with its bed parallel to the face of the wall
- *Bull header* – A rowlock brick laid with its bed perpendicular to the face of the wall
- *Soldier* – A brick laid on in a vertical position with its face perpendicular to the courses in the wall

Source: U.S Army FM5-428

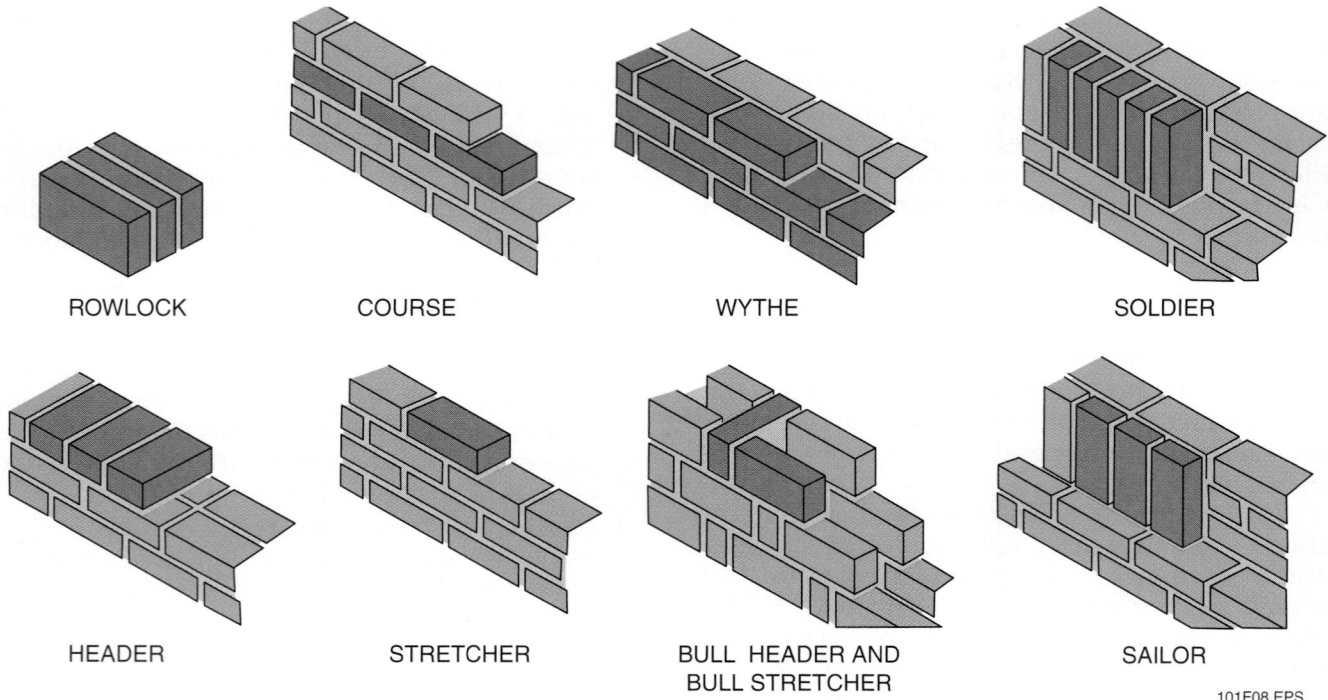

ROWLOCK COURSE WYTHE SOLDIER

HEADER STRETCHER BULL HEADER AND BULL STRETCHER SAILOR

101F08.EPS

Figure 8 ◆ Masonry units and mortar joints.

4.0.0 ◆ CONCRETE PRODUCTS

CMUs have not been around as long as brick. The first CMUs were developed in 1850, when Joseph Gibbs was trying to develop a better way to build masonry cavity walls.

Masonry is not waterproof, only water resistant. Thick walls slow down moisture so that it does not reach the inside of the wall, but thick walls are expensive to build. Cavity walls were invented to handle this problem. Over the years, many types of cavity walls have been designed to slow down or prevent moisture from reaching the inside surfaces. Today, several different designs are used based on the availability of local materials and environmental requirements. Most of them use either brick or a combination of brick and block.

Figure 9 shows examples of both the old and newer style cavity walls. Cavity walls are made of two courses of masonry units, with a 2" to 4" gap between them. Water can get through the outside wall and run down inside the cavity without wetting the inside wall.

In his search for a faster way to build a cavity wall, Gibbs developed a block with air cells in it. This idea was refined and patented by several other people. In 1882, someone took advantage of new materials developments to make a hollow block of portland cement.

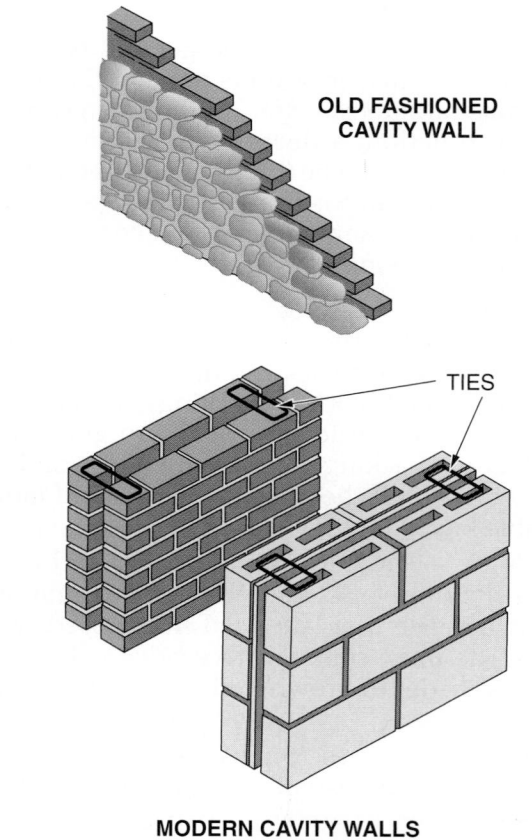

OLD FASHIONED CAVITY WALL

TIES

MODERN CAVITY WALLS

101F09.EPS

Figure 9 ◆ Cavity walls.

In 1900, Harmon Palmer patented a machine that made hollow concrete block. The blocks were 30" long, 10" high, and 8" wide. Even though these blocks were heavy and very hard to lift, cavity walls became cheaper and faster to build. Over time, other machines were developed to produce smaller blocks. These blocks were easier to handle, but were still very heavy for their size.

In 1917, Francis Straub patented a block made with the cinders left after burning coal. He used the cinders to replace the sand and small **aggregates** in the concrete mix. This new cinderblock was lighter, cheaper, and easier to handle. Straub's block made it possible to build a one-course wall with a built-in cavity very quickly and inexpensively. Faster machinery was developed to keep up with the demand for this new masonry material.

The demand for block increased with the rise of engineered masonry in the U.S. The production of concrete block surpassed that of clay brick in the 1950s. Since the 1970s, there have been more walls built of concrete block in the U.S. than those of clay brick and all other masonry materials together.

Blocks are made of water added to portland cement, aggregates (sand and gravel), and **admixtures**. The cinders have been replaced today by other lightweight aggregates. Admixtures affect the color and other properties of the cement, such as freeze resistance, weight, and speed of setting.

The block is machine-molded into shape. It is compacted in the molds and cured, typically using live steam. After curing, the blocks are dried and aged. The moisture content is checked. It must be a specified minimum amount before the blocks can be shipped for use. *Figure 10* shows commonly used sizes and shapes of concrete block.

Not all blocks are CMUs. Concrete units fall into classifications based on intended use, size, and appearance. ASTM standards exist for the following types of masonry units:

- Loadbearing and nonbearing concrete block
- Concrete brick
- Calcium silicate face brick
- Pre-faced or pre-finished facing units
- Manholes and catch basin units

Block Construction

Concrete block is often used in commercial construction. In some parts of the country, it is also used in the walls of residential construction in place of wood framing. The block can be painted on the outside or faced with brick, stucco, or other finish material.

101SA04.EPS

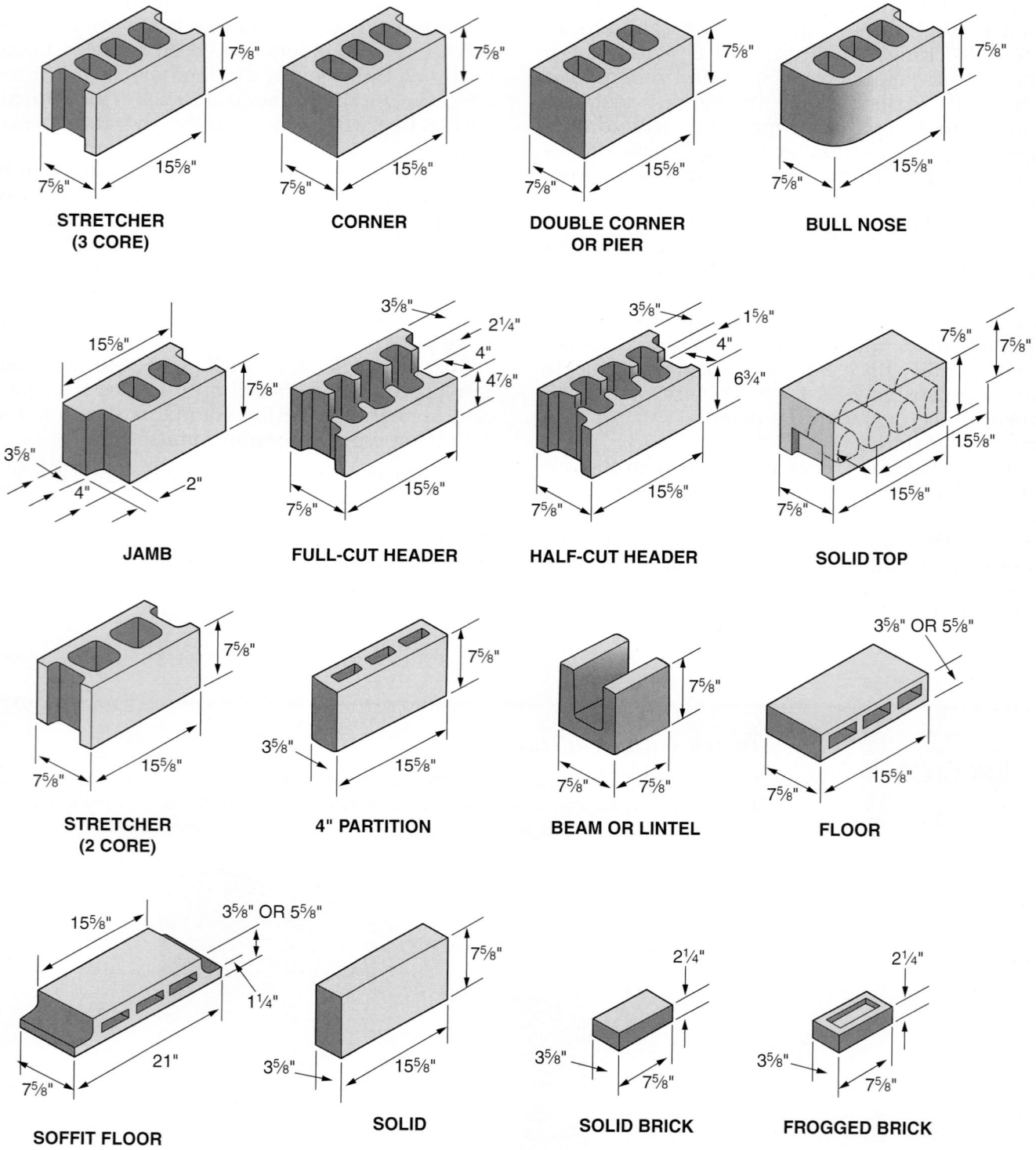

STRETCHER (3 CORE)

CORNER

DOUBLE CORNER OR PIER

BULL NOSE

JAMB

FULL-CUT HEADER

HALF-CUT HEADER

SOLID TOP

STRETCHER (2 CORE)

4" PARTITION

BEAM OR LINTEL

FLOOR

SOFFIT FLOOR

SOLID

SOLID BRICK

FROGGED BRICK

NOTE: Dimensions are actual unit sizes. A 7⅝" × 7⅝" × 15⅝" unit is an 8" × 8" × 16" nominal-size block.

101F10.EPS

Figure 10 ◆ Common concrete block.

4.1.0 Block

Concrete block is a large unit, typically 8" × 8" × 16", with a hollow core. Blocks come in modular sizes, in colors determined by the cement ingredients, the aggregates, or any admixtures. A variety of surface and mixing treatments can give block varied and attractive surfaces. Newer finishing techniques can give block the appearance of brick, rough stone, or cut stone. Like clay masonry units, block can be laid in structural pattern bonds. *Figure 11* shows the names of the parts of block.

Block takes up more space than other building units, so fewer are needed. Block bed joints usually need mortar only on the shells and webs, so there is less mortaring as well.

Concrete block comes in three weights: normal, lightweight, and aerated. Lightweight block is made with lightweight aggregates. The loadbearing and appearance qualities of the first two weights are similar; the major difference is that lightweight block is easier and faster to lay. Normal-weight block can be made of concrete with regular, high, and extra-high strengths. The last two are made with different aggregates and curing times. They are used to limit wall thickness in buildings over ten floors high. Aerated block is made with an admixture that generates gas bubbles inside the concrete for a lighter block.

Concrete blocks are classified as hollow or solid. Like clay products, less than 75 percent of the surface area of a hollow unit is solid. Common hollow units have two or three cores. The hollow cores make it easy to reinforce concrete block walls. Grout alone, or steel reinforcing rods combined with grout, can be used to fill the hollow cores. Reinforcement increases loadbearing strength, rigidity, and wind resistance. Less than 25 percent of the surface area of a solid block is hollow. Normal and lightweight solid units are intended for special needs, such as structures with unusually high loads, drainage catch basins, manholes, and firewalls. Aerated block is made in an oversize solid unit used for buildings.

Loadbearing block is used as backing for veneer walls, bearing walls, and all structural uses. Both regular and specially shaped blocks are used for paving, retaining walls, and slope protection. Nonstructural block is used for screening, partition walls, and as a veneer wall for wood, steel, or other backing. Both kinds of blocks come in a variety of shapes and modular sizes.

4.2.0 Concrete Brick

The length and height dimensions of regular concrete brick are the same as those of standard clay brick. The thickness is an additional ⅛". A popular type is slump brick, shown in *Figure 12*. Slump brick is made from very wet concrete. When the mold is removed, the brick bulges because it is not dry enough to completely hold its shape. Slump brick looks like ashlar and adds a decorative element to a wall.

Concrete brick is produced in a wide range of textures and finishes. It is available in specialized shapes for copings, sills, and stairs, just as clay brick is. Concrete brick is more popular in some areas of the country because it is less expensive.

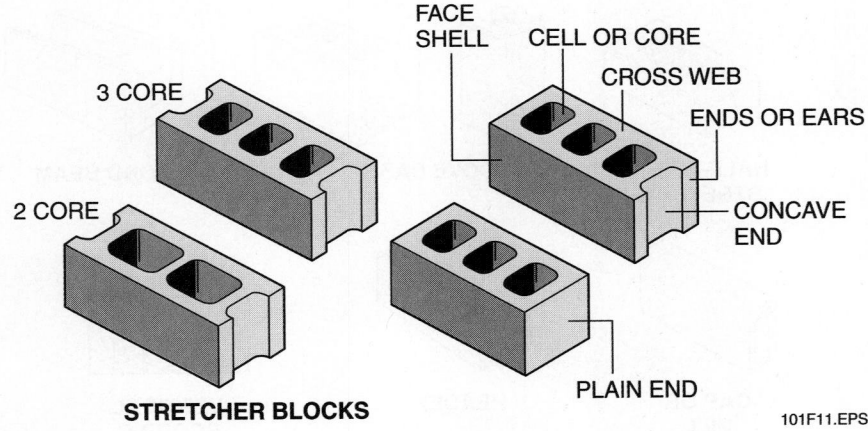

Figure 11 ◆ Parts of a block.

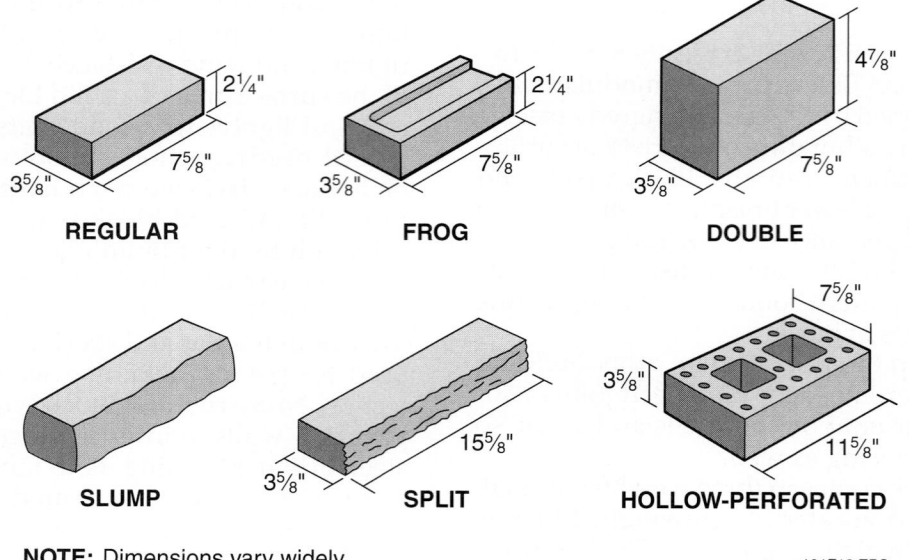

NOTE: Dimensions vary widely.

101F12.EPS

Figure 12 ◆ Concrete brick.

4.3.0 Other Concrete Units

Concrete pre-faced or pre-coated units are coated with colors, patterns, and textures on one or two face shells. The facings are made of resins, portland cement, ceramic glazes, porcelainized glazes, or mineral glazes. The slick facing is easily cleaned. These units are popular for use in gyms, hospital and school hallways, swimming pools, and food processing plants. They come in a variety of sizes and special-purpose shapes, such as

coving and bullnose corners. *Figure 13* shows commonly used concrete pre-faced units.

Concrete manhole and catch basin units are specially made with high-strength aggregates. They must be able to resist the internal pressure generated by the liquid in the completed compartment. *Figure 14* shows the shaped units manufactured for the top of a catchment vault. These blocks are engineered to fit the vault shape and are cast to specification. They are made with interlocking ends for increased strength.

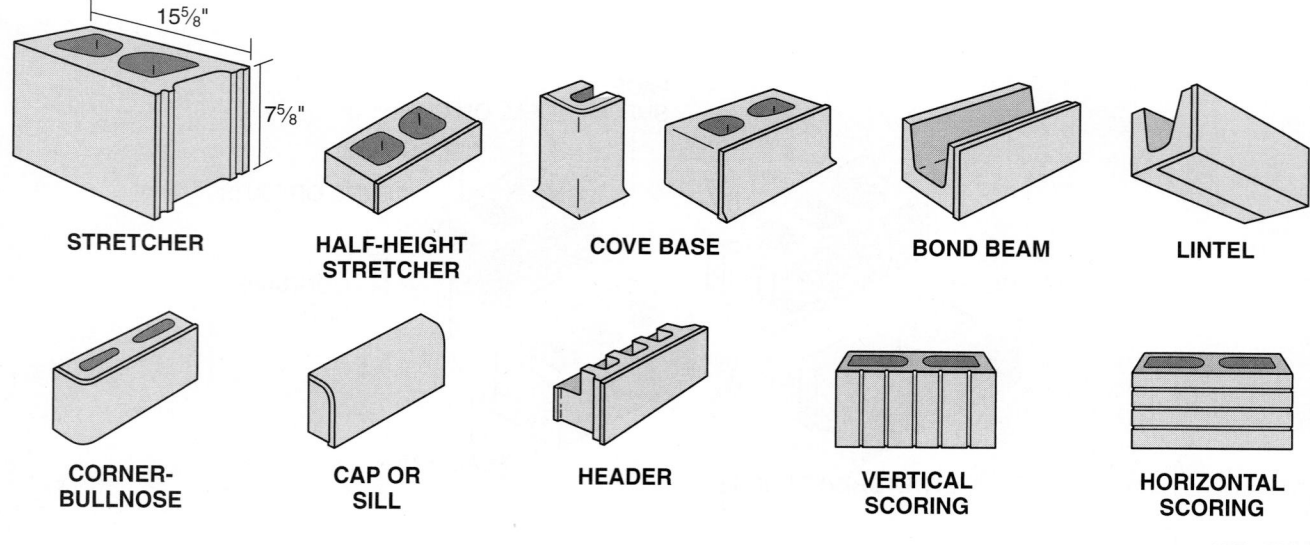

101F13.EPS

Figure 13 ◆ Common pre-faced concrete units.

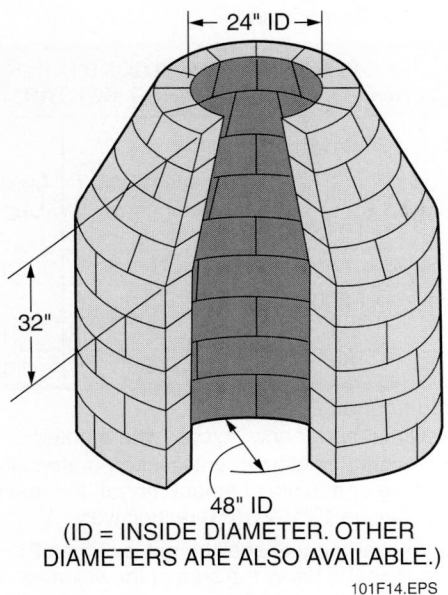

24" ID

32"

48" ID
(ID = INSIDE DIAMETER. OTHER
DIAMETERS ARE ALSO AVAILABLE.)

101F14.EPS

Figure 14 ◆ Manhole and vault unit.

5.0.0 ◆ STONE

Stone was once used in the construction of all types of buildings, especially churches, schools, and government buildings. Today, it is more commonly used as a decorative material, such as the stone trim shown on the home in *Figure 15*.

Rubble and ashlar are used for dry stone walls, mortared stone walls, retaining walls, facing walls, slope protection, paving, fireplaces, patios, and walkways.

Rubble stone is irregular in size and shape. Stones collected in a field are rubble. Rubble from quarries is left where shaped blocks have been removed. It is also irregular with sharp edges. Rubble can be roughly squared with a brick hammer to make it fit more easily.

Ashlar stone is cut at the quarry. It has smooth bedding surfaces that stack easily. Ashlar is usually granite, limestone, marble, sandstone, or slate. Other stone may be common in different parts of the country.

Flagstone is used for paving or floors. It is 2" or less thick and cut into flat slabs. Flagstone is usually quarried slate, although other stone may be popular in different areas of the country.

Stone is often used as a veneer over brick or block. The wall shown in *Figure 16* is an example. The brownstone buildings in New York and the grey stone buildings of Paris are veneer over brick. Many of the government buildings and monuments in Washington, D.C., are of stone veneer construction.

Stone, including flagstone, can be laid in a variety of decorative patterns. Concrete masonry units are made in shapes and colorings to mimic every kind of ashlar. These units are called cast stone and are more regular in shape and finish than natural stone. Cast stone has replaced natural stone in many commercial projects because it is more economical. ASTM specifications cover cast stone and natural stone.

6.0.0 ◆ MORTARS AND GROUTS

The first mortar was wet mud, and it is still in use in some parts of the world. Mortar is no longer made of mud, but sometimes it is still called mud.

101F15.EPS

Figure 15 ◆ Stone facing used as decorative trim.

101F16.EPS

Figure 16 ◆ A block wall faced with stone.

By the Roman period, sand was a common additive. Burnt limestone, or quicklime, was added as an ingredient around the first century B.C.E. Experimenting with waterproofing mortar, the Romans added volcanic ash and clay. The resulting cement made a strong, waterproof mortar. This made it possible to build aqueducts, water tanks, water channels, and baths that are still in use today. Unfortunately, some of the Roman formula was lost over time.

In 1824, portland cement was patented by Joseph Aspdin, a mason. He was trying to recreate the waterproof mortar of the Romans. By 1880, portland cement had become the major ingredient in mortar. The new, waterproof portland cement mortar began to replace the old lime and sand mixture.

Portland cement is made of ground earth and rocks burned in a kiln to make clinker. The clinker is ground to become the cement powder. Mixed with water, lime, and rocks, the cement becomes concrete. Mortar is somewhat different from concrete in consistency and use. The components and performance specifications are different also.

Modern mortar is mixed from portland cement or other cementitious material (something that has the properties of cement), along with lime, water, sand, and admixtures. The proportions of these elements determine the characteristics of the mortar.

The two main types of mortar are as follows:

- Cement-lime mortars are made of portland cement, hydrated lime, sand, and water. These ingredients are mixed at the job site by the mason.

- Masonry cement mortars are premixed with additives. The mason only adds sand and water. The additives affect flexibility, drying time, and other properties.

Mortar is mixed to meet four sets of performance specifications, as listed in *Table 1*.

- *Type M* – With high compressive strength, Type M mortar is typically used in contact with earth for foundations, sewers, and walks. This varies with geographic location.
- *Type S* – With medium strength, high bonding, and flex, Type S mortar is used for reinforced masonry and veneer walls.
- *Type N* – With heavy weather resistance, Type N mortar is used in chimneys, parapets, and exterior walls.
- *Type O* – With low strength, Type O mortar is used in nonbearing applications. It is not recommended for professional use.

Table 1 Mortar Composition

PROPERTY SPECIFICATIONS FOR LABORATORY-PREPARED MORTAR*			
Mortar Type	Minimum Compressive Strength, PSI at 28 days	Minimum Water Retention,%	Maximum Air Content,%**
M	2,500	75	12***
S	1,800	75	12***
N	750	75	14***
O	350	75	14***

* Adapted from *ASTM C270*.

** Cement-lime mortar only (except where noted).

*** When structural reinforcement is incorporated in cement-lime or masonry cement mortar, the maximum air content shall be 12% or 18%, respectively.

Note: The total aggregate shall be not less than 2¼ and not more than 3½ times the sum of the volumes of the cement and lime used.

101T01.EPS

Another type of mortar, Type K, has no cement materials but only lime, sand, and water. This type of mortar is used for the preservation or restoration of historic buildings.

Grout is a mixture of cement and water, with or without fine aggregate. Wet enough to be pumped or poured, it is used in reinforcement to bond masonry and steel together. It gives added strength to a structure when it is used to fill the cores of block walls.

7.0.0 ◆ MODERN CONSTRUCTION TECHNIQUES

This section introduces modern structures and modern construction techniques. There are several types of structures you must learn to build. These structures and the techniques used to build them are basic to the craft.

7.1.0 Wall Structures

Masonry structures today take many forms in residential, commercial, and industrial construction. Modern engineering has added loadbearing strength so masonry can carry great weight without bulk. In addition to load bearing, masonry offers these advantages:

- Durability
- Ease of maintenance
- Design flexibility
- Attractive appearance
- Weather and moisture resistance
- Competitive cost

Modern engineering and ASTM standards have been applied directly to everyday masonry work. There are six common classifications of structural wall built with masonry. *Figure 17* shows some of these walls. Masonry walls can fit into more than one classification.

Solid walls, as shown in *Figure 17A*, are built of solid masonry units with full mortar joints. Solid units have voids of less than 25 percent of their surface. These walls can have one or two loadbearing wythes tied together with mortar.

Hollow walls, as shown in *Figure 17B*, are solid walls built of masonry units with more than 25 percent of their surface hollow. These can also have one or two loadbearing wythes tied together with mortar.

Cavity walls, as shown in *Figure 17C*, have two wythes with a 2" to 4½" space between them. Sometimes insulation is put in the cavity. The wythes are tied together with metal ties. Both wythes are loadbearing.

Veneer walls, as shown in *Figure 17D*, are not loadbearing. A masonry veneer is usually built 1" to 2" away from a loadbearing stud wall or block wall. Veneer walls are used in high-rise and residential construction.

Composite walls have different materials in the facing (outer) and backing (inner) wythes. The wythes are set with a 1" air space between them and are tied together by metal ties. Unlike a veneer wall, both wythes of a composite wall are loadbearing.

Reinforced walls (*Figure 18*) have steel reinforcing embedded in the cores of block units or between two wythes. The steel is surrounded with grout to hold it in place. This very strong wall is used in high-rise construction and in areas subject to earthquake and high winds. Sometimes, grout is used alone for reinforcement. The grout is pumped into the cores of the blocks or into the cavity between the wythes.

Contemporary masonry systems are designed not as barriers to water, but as drainage walls. Penetrated moisture is collected on flashing and expelled through **weepholes**. Design, workmanship, and materials are all important to the performance of masonry drainage walls.

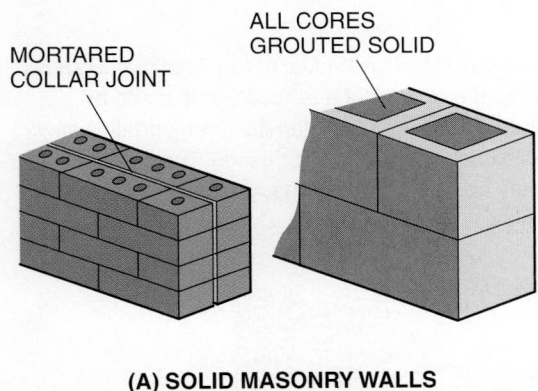

(A) SOLID MASONRY WALLS

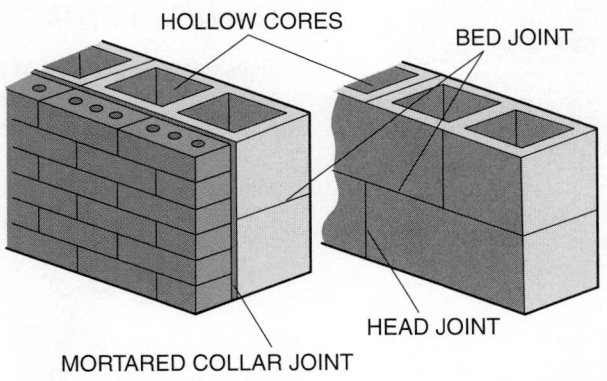

(B) SOLID WALLS OF HOLLOW UNITS

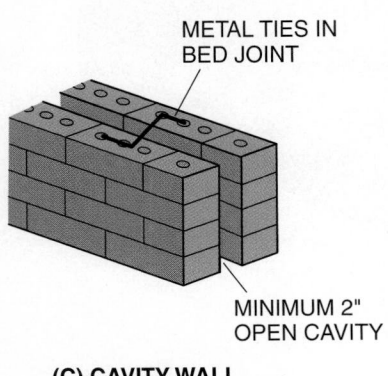

(C) CAVITY WALL

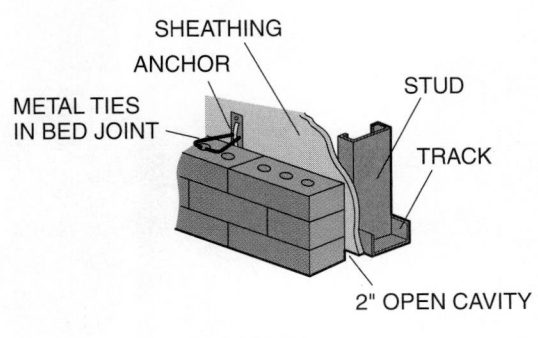

(D) VENEER WALL

101F17.EPS

Figure 17 ◆ Types of masonry construction.

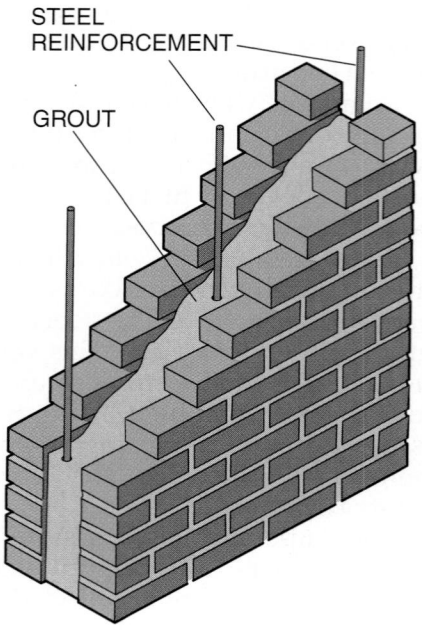

STEEL
REINFORCEMENT

GROUT

FULLY
GROUTED

101F18.EPS

Figure 18 ◆ Reinforced walls.

Curtain Walls

Skyscrapers and other tall buildings were once built with individual bricks. This method has not been used in many years, however. Today, the exterior of a tall building is made by attaching curtain wall sections to a steel or concrete structure. Manufactured curtain wall panels faced with brick are now used when a brick appearance is desired. The curtain wall panel shown here combines 2"-thick concrete brick with a heavy-gauge steel frame and insulated stainless steel anchors.

101SA05.EPS

7.2.0 Modern Techniques

Masons use a number of specialized hand and power tools. As you will learn in the module on tools and equipment, there are many kinds of special trowels (*Figure 19*) and at least six kinds each of hammers, chisels, and steel joint finishing tools. There are seven kinds of measuring and leveling tools. Power tools include several kinds each of saws, grinders, splitters, and drivers. Mortar can be mixed by hand, using special equipment, or in a power mixer. Cranes, hoists, and lifts bring the masonry units to the masons working on one of four types of steel scaffolding.

While masonry tools have changed over the centuries, on thing has not: the relation between the mason and the masonry unit. The mason uses this twentieth-century wealth of tools and equipment to perform the following tasks:

- Calculate the number and type of units needed to build a structure
- Estimate the amount of mortar needed
- Assemble the units near the work station
- Lay out the wall or other architectural structure

- Cut units to fit, as needed
- Mix the appropriate type and amount of mortar
- Place a bed of mortar on the **footing**
- **Butter** the **head joints** and place masonry units on the bed mortar
- Check that each unit is level and true
- Lay courses in the chosen bond pattern or create a new pattern
- Install ties as required for loadbearing
- Install flashing and leave weepholes as required for moisture control
- Clean excess mortar off the units as the work continues
- Finish the joints with jointing tools
- Give the structure a final cleaning
- Complete the work to specification, on time

Masonry work is still very much a craft. The relation between the mason and the masonry unit is personal. The straightness and levelness of each masonry unit in a structure—brick, block, or stone—depend on the hands and the eyes of the mason. These things have not changed in 10,000 years.

BRICK   POINTING   MARGIN   TUCKPOINTER

PARGING   DUCK BILL   BUCKET   TILESETTING

101F19.EPS

Figure 19 ◆ Mason's trowels.

The tradition of masonry calls for a bit of art, too. The mason gets trained by work to see the subtle shadings and gradations of color. He or she learns to create a pattern and to select the right unit to complete the pattern or the shading. The mason grows skilled in building something that is both enduring and attractive.

8.0.0 ◆ MASONRY AS A CAREER

Masonry offers a rewarding career for people who want to work with their hands. As masons, they will be skilled workers who understand the principles and practices of masonry construction. They will earn good pay and be rewarded for initiative. They will have opportunity for advancement.

Masons will continue to play an important part in building homes, schools, offices, and commercial structures. They can add artistic elements to their work and create beauty. They can be proud of their skills and the fact that they produce something people need.

Masons work on different projects, so each job is different and never boring. If they like to travel, masons can find good jobs all over the country. They can be independent and creative while working outdoors. Masons will be in demand as long as buildings are being constructed.

Masons can find work on large construction projects for commercial buildings, as well as projects for building homes, patios, sidewalks, or walls. They can also specialize in repair work, cleaning, and **tuckpointing** old buildings. They can specialize in restoring historic brick buildings, which is a recognized craft specialty in some parts of the country. Historically, they have been well paid.

Masonry is more than physical labor. It is a skilled occupation that calls for good hand-eye coordination, balance, and strength. It also requires good mental skills. This means ongoing study, concentration, and continued learning in an environment free from substance abuse.

Because masonry is a highly skilled craft, it takes time to learn. Your learning starts with this course, combined with, or followed by, an apprentice's job. Masons usually work as part of a team, so you will also learn to be a good team player. Masons work outdoors and do a lot of lifting and bending, so you will learn to keep yourself in good shape. Masons work on high scaffolding, so you will learn safety rules and practices. Masons bring their skills and tools wherever they go.

8.1.0 Career Stages

Masons were among the first workers to band together. During the Middle Ages, they formed influential groups that still shape trade practices. Today, as in the past, masons' organizations recognize several stages of skill:

* Helper
* Apprentice
* Journeyman
* Foreman
* Superintendent
* Contractor

The helper is a laborer, not a mason. The helper carries masonry materials, tools, and mortar and gets things for the mason. The helper mixes mortar, cleans tools, and learns by watching the mason at work. Sometimes, helpers decide they want to become masons. If they do, they may enter an apprenticeship program. Apprentices are at the beginning level of the masonry career path. Their training will lead them to full participation in the mason's trade and the opportunity for higher job levels.

8.2.0 Apprentice

An apprentice is a person who has signed an apprenticeship agreement with a local joint apprenticeship committee. The committee works with local contractors who have agreed to take apprentices. The U.S. Department of Labor regulates the apprenticeship process. In most states, the state department of labor is also involved, as state labor regulations provide guidelines on legal age requirements, pay, hours, and other aspects of apprenticeship.

The length of the apprenticeship will vary. The U.S. Department of Labor program is three years and 4,500 hours with a minimum of 432 hours of classroom instruction. Programs such as the *Contren® Learning Series* are used for classroom instruction or as part of apprenticeship training.

The apprentice is assigned to work with a contractor and to take classes. The apprentice must study and work under supervision. The apprentices agree to do the following:

* Perform the work assigned by the contractor.
* Abide by the rules and regulations of the contractor and the committee.
* Complete the hours of instruction.

- Keep records of work experience, training, and instruction.
- Learn and use safe working habits.
- Work with the assigned contractors for the entire apprenticeship period, unless reassigned by the committee.
- Conduct themselves in an ethical manner, realizing that time, money, and effort are being spent to afford them this opportunity to become a skilled worker.
- Remain free from drug and alcohol abuse.

A typical three-year apprenticeship is divided into six periods of six months each. The first six months is a trial period. The committee reviews the apprentice's performance and may end the agreement.

The apprentice attends classes and works under the supervision of a journeyman mason. As part of the supervised work, the apprentice learns to lay masonry units and perform other craftwork. The apprentice's pay increases for each six-month period as skill and performance increase. At the end of the period, the apprentice receives a certificate of completion. This type of certificate (*Figure 20*) is known and accepted everywhere in the United States. The apprentice is now a journeyman mason.

8.3.0 Journeyman

Unlike an apprentice, a journeyman is a free agent who can work for any contractor. A journeyman can work without close supervision and is skilled in most tasks. The successful journeyman knows that the end of the apprenticeship is not the end of learning.

Journeymen are people with an excellent trade. They earn good wages in a trade that is always in demand. They have the satisfaction of creating and the opportunity to grow as masonry artists. They also have the opportunity to grow as layout persons, trainers, and supervisors.

101F20.EPS

Figure 20 ◆ Example of apprenticeship training recognition.

An experienced and skilled journeyman can work as a layout person. For a pay premium, the layout person lays out the work and lays the leads. Less experienced masons and apprentices work between the leads set by the layout person. Experienced and skilled journeymen also train apprentices and supervise their work. With further experience, journeymen can supervise crews.

Journeymen can continue to learn by studying and handling more complex tasks. They can continue to develop their skills as they work. Further education in masonry innovations and techniques is available as is training in leadership and supervision.

8.4.0 Supervisors, Superintendents, and Contractors

Supervisors are responsible for managing and supervising a group of workers. This job requires a high degree of knowledge about masonry and leadership skills. Supervisors are typically responsible for training workers in safety measures and keeping work areas safe. They also train workers in new techniques and easier ways of working. They solve daily problems, keep on top of materials and supplies, and make sure workers meet job schedules. They check work to ensure it is done to standards. Supervisors may be called crew leaders or forepersons depending on the company that hires them.

Superintendents have several supervisors reporting to them. Usually, the superintendent is the lead person on a large job. For a smaller company, the superintendent may be in charge of all the work in the field for the contractor. The superintendent oversees the work of the supervisors and makes major decisions about the job under construction. The superintendent must have strong masonry, leadership, and business skills.

A masonry contractor owns the company. He or she bids on jobs, organizes the work and the workers, inspects the work, confers with the clients, and runs the business. The contractor needs to be able to plan ahead to keep up with change.

Contractors, along with journeymen, supervisors, and superintendents, need to keep up with the latest materials and methods. Like apprentices, they need to keep on studying their trade.

8.5.0 The Role of NCCER

This course is part of a curriculum produced by the National Center for Construction Education and Research (NCCER). Like every course in NCCER's curriculum, it was developed by the construction industry for the construction industry. NCCER develops and maintains a training process which is nationally recognized, standardized, and competency-based. A competency-based program requires you to show that you can perform specific job-related tasks safely to receive credit. This approach is unlike other apprenticeship programs that are based on a required number of hours in the classroom and on the job.

The construction industry knows that the future construction workforce will largely be recruited and trained in the nation's secondary and postsecondary schools.

Schools know that to prepare their students for a successful construction career they must use the curriculum that is developed and recognized by the industry. Nationwide, thousands of schools have adopted NCCER's standardized curricula.

The primary goal of NCCER is to standardize construction craft training throughout the country so that both you and your employer will benefit from the training, no matter where you or your job are located. As a trainee in a NCCER-accredited program, you will be listed in the National Registry. You will receive a certificate for each level of training you complete, which can then travel with you from job to job as you progress through your training. In addition, many technical schools and colleges use NCCER's programs.

9.0.0 ◆ KNOWLEDGE, SKILLS, AND ABILITY

Becoming a good mason takes more than the ability to lay a masonry unit and level it. A competent mason is one who can be trusted to perform the required work and meet the project specifications. This mason must have the necessary knowledge, skills, and ability, as well as good attitudes about the work itself, about safety, and about quality.

9.1.0 Knowledge

Masons need to know how to handle all aspects of masonry work. They need to know how to do all of the following:

- Read and interpret drawings and specifications
- Calculate and estimate quantities, lengths, weights, and volumes
- Select the proper materials for the job
- Lay masonry units into structural elements
- Work productively alone or as part of a team
- Assemble and disassemble scaffolding
- Keep tools and equipment in good repair and safe condition
- Follow safety precautions to protect themselves and other workers on the job

9.1.1 Job-Site Knowledge

Masons need to be skilled in applying their knowledge to the challenges they face each day on the job. The best way to do the work at a particular job site will depend on the layout of the work, what is happening around the masonry site, and the conditions surrounding the project.

Most masonry work is done outside in temperature and weather variations. You must be able to work under these conditions and not be distracted by them. You must know how to react to changing conditions around you.

Much of this knowledge can be learned as you work, if you will pay attention. Notice what others do and ask questions. Ask your supervisor questions, too. Learn to respond to conditions at the job site.

9.1.2 Learning More

Masons need to keep on learning after they finish their apprenticeships. They need to keep updating their skills all the time. The environment, tools, and expectations about masonry have evolved and will continue to change. Craftworkers and contractors alike will need to change the way they think about their work and how they do it.

National, regional, and local organizations offer continuing education for masons. Technical seminars, training sessions, publications, and classes are often free or low cost. They can bring you the latest information about tools, materials, and methods. To succeed, you must be alert to change and willing to learn new ways.

9.2.0 Attitude and Work

Attitude can build an invisible bridge, or build an invisible wall, between us and others. No one wants to hang around a grouch or count on someone who is not dependable. No one minds helping someone who can do something in return or working with a friendly, cooperative partner. On top of knowledge, skills, and ability, you need the right attitude. Your attitude comes from how you think and feel about your work and yourself.

9.2.1 Dependability

You must be dependable. Masonry work, like all construction, is a closely timed operation. Once started, it cannot stop without waste of material and money. Employers need workers who report to work on time. An undependable, absent worker will slow or stop masonry work and cost the project time and money. An undependable worker will not be able to depend on having a job for very long.

9.2.2 Responsibility

You must be responsible for doing the assigned work in a proper and safe manner, be responsible enough to work without supervision, and work until the task is complete.

Being responsible for your own work includes admitting your mistakes. It also includes learning from your mistakes. Nobody is expected to be perfect. Everyone is expected to learn and to grow more skilled.

Employers are always in need of workers who are ambitious and want to become leaders. Being responsible for what others do may be your career goal. The path to that goal starts with being responsible for what you do.

9.2.3 Adaptability

On any construction project, a large amount of work must be done in a short time. Planning and teamwork are needed in order to work efficiently and safely. Supervisors sometimes form teams of two or more workers to do specific tasks. You may work in a team to erect a scaffold, then work alone for most of the day, then team with someone else to do a cleanup.

On a job site, you may find yourself teaming with different people at different times. Being a team player becomes important. Team players accept instruction and direction. They communicate clearly, keep an eye out for potential problems, and share information. They meet problems squarely with constructive ideas, not criticism.

All team players treat each other with respect. Everyone must be willing to work together. Everyone must be willing to bring their best attitude to the team. Team members need to be able to depend on each other. Team priorities must be more important than individual priorities.

9.2.4 Pride

Pride in what you do comes from doing high-quality work in a timely manner and from knowing you are doing your best. Being proud of what you do can overflow into other areas. Proud workers take pride in their personal appearance. Their work clothes are clean, safe, neat, and suitable. Proud masons take pride in their tools. They have a complete set of well-maintained tools and other special equipment they need to do their jobs. They keep their tools safe and orderly and know how to use the right tool for the work at hand.

Proud masons work so that they can continue to be proud of what they do and how they do it. Being proud of what you do is an important part of being proud of who you are.

9.3.0 Quality

The latest ideas about quality in work are not new. Those who work in masonry construction and finishing have been concerned about quality for thousands of years. The walls unearthed at Jericho were laid true and still stand true.

The quality of masonry depends on many factors. When building a wall, you may have little control over its design or the choice of masonry units. But you do have control over the quality of the completed job. A wall out of level or with poorly finished joints is your responsibility.

The quality of the finished masonry structure depends directly on your knowledge, skill, and ability. Good work will be easily seen by all. Poor work will be seen even more easily. Given the durability of masonry, the quality of the work will be a monument to your skill for a very long time. The skilled, proud mason always strives for the highest quality that can be achieved.

10.0.0 ◆ BASIC BRICKLAYING

In this section, you will learn the basic elements of bricklaying. When you have completed the section, you should be able to set up a job, mix mortar, and lay bricks as directed by your instructor.

10.1.0 Preparing Mortar

Mortar (*Figure 21*) is a mixture of portland cement, lime, sand, and water. The first three ingredients are determined by the type of mortar being mixed. Water is added until the mix is at the proper consistency. The ability to mix mortar properly, and to produce the same consistency time after time, can only be developed through practice.

You will probably mix your first few batches by hand in a wheelbarrow, pan, or mortar box. Assuming that you will be using a wheelbarrow, proceed as follows:

Step 1 Place half the sand in the wheelbarrow and make an even spread over the bottom.

Step 2 Add the required amount of cement and lime (or masonry cement) to the wheelbarrow.

Step 3 Blend the dry ingredients with a hoe (*Figure 22*), then pull the mix to one end of the wheelbarrow.

Step 4 Add half the water to the empty end of the wheelbarrow. Begin mixing the water and dry material with short push-pull strokes of the hoe.

Step 5 Add water to obtain the required consistency. When the mix is right, the mortar will stick to a trowel when the trowel is turned upside down.

Figure 21 ◆ Masonry mortar.

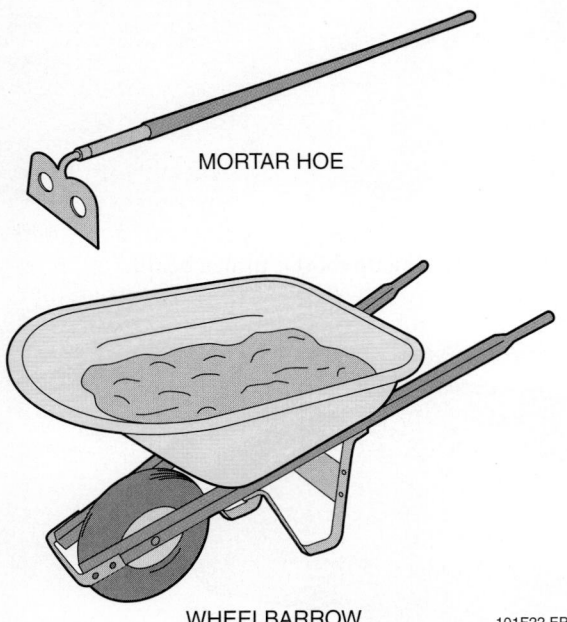

MORTAR HOE

WHEELBARROW

101F22.EPS

Figure 22 ◆ Mixing mortar.

10.2.0 Spreading Mortar

After the mortar is mixed, pick it up on your trowel and spread it. Filling and emptying the trowel is an important skill. Applying the mortar, or spreading it, is the next step.

The following sections describe holding the trowel, picking up the mortar, and laying it down. At this point, you will learn something to be experienced rather than memorized. The techniques in the next sections should be practiced until you feel comfortable using them.

10.3.0 Picking Up Mortar

There are several ways of using the trowel to pick up mortar. This section will introduce you to a general method for picking up mortar from a board and a general method for picking up mortar from a pan. There are many different ways to do these tasks. The instruction here begins with some tips on holding the trowel. All of these instructions are only approximations of the work itself. You can only learn this through watching a skilled mason and practicing until you feel comfortable with these movements.

10.3.1 Holding the Trowel

Pick up your trowel by the handle. Put your thumb along the top of the handle with the tip on the handle, not the shank, as shown in *Figure 23*.

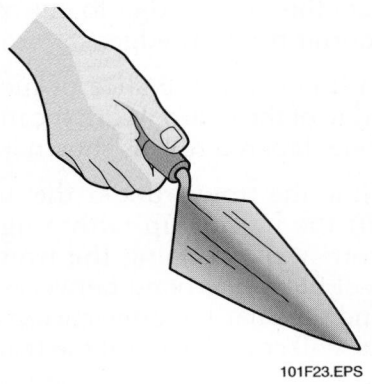

101F23.EPS

Figure 23 ◆ Holding the trowel.

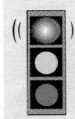

WARNING!

Keep your thumb off the shank to keep it out of the mortar. Mortar is caustic and can cause chemical burns.

Keep your second, third, and fourth fingers wrapped around the handle of the trowel. Keep the muscles of your wrist, arm, and shoulder relaxed so you can move the trowel freely.

Most of your work with the trowel will require holding the blade flat, parallel to the ground, or rotating the blade so it is perpendicular to the ground.

Rotating the blade gives you a cutting edge. The best edge for cutting is the edge on the side closest to your thumb. It is best this way because you can see what you are cutting. When you turn the trowel edge to cut, rotate your arm so your thumb moves down. This will rotate the trowel so that the bottom of the blade turns away from you.

If you rotate only your wrist, after a while you will strain it. Use the larger muscles in your arm and shoulder to rotate the trowel.

Rotating the blade also gives you a scooping motion. Turning your thumb down will give you a forehand scoop. Turning your thumb up will rotate the bottom of the blade toward you and give you a backhand scoop. Using a forehand or backhand movement will depend on the position of the material you are trying to scoop.

10.3.2 Picking Up Mortar from a Board

After putting the mortar on the board, follow these steps:

Step 1   Work the mortar into a pile in the center of the board, and smooth it off with a backhand stroke.

Step 2   Use the trowel edge to cut off a slice of mortar from the edge.

Step 3   Pull and roll the slice of mortar to the edge of the board. Work the mortar into a long, tapered roll, as shown in *Figure 24*.

Step 4   Slide the trowel under the mortar, then lift the mortar up with a light snap of your wrist. Raising the trowel quickly will break the bond between the mortar and the board. If done correctly, the mortar will completely fill the trowel blade.

10.3.3 Picking Up Mortar from a Pan

Try this method when the mortar is in a pan:

Step 1   Cut a slice of mortar, as shown in *Figure 25*.

Step 2   Without removing the trowel from the mortar, slide the trowel under the mortar so the blade becomes parallel to the floor.

Step 3   Firmly push the trowel, with the blade parallel to the floor, toward the middle of the pan. The mortar will pile up on the blade.

101F24.EPS

Figure 24 ◆ Picking up mortar from a board.

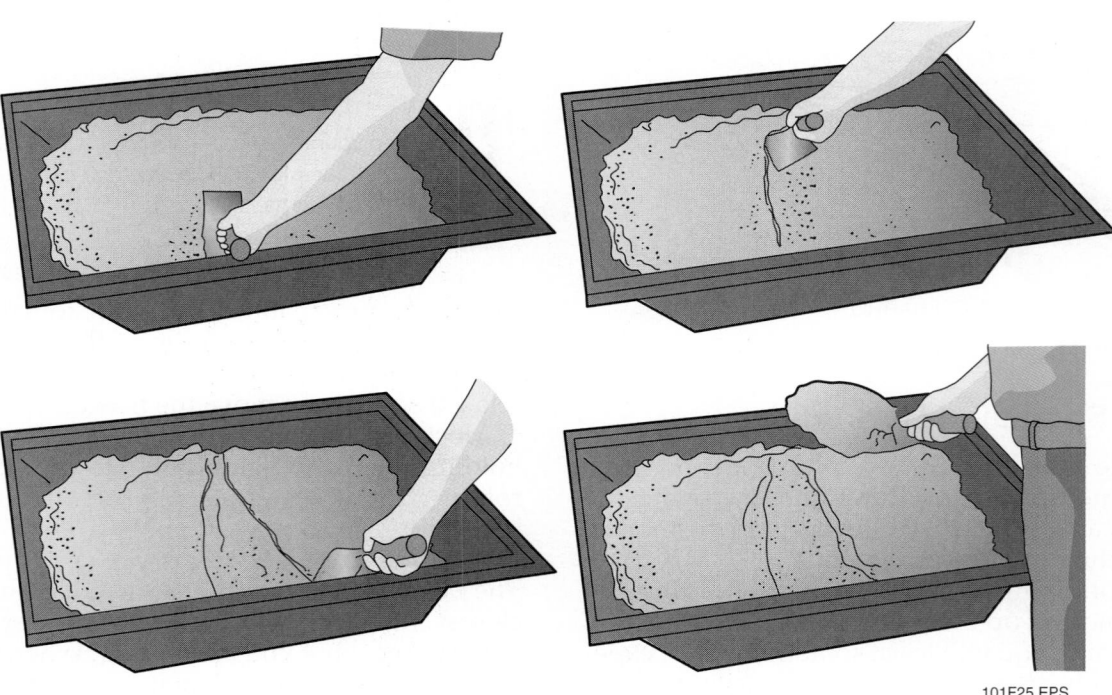

101F25.EPS

Figure 25 ◆ Picking up mortar from a pan.

Step 4 Lift the trowel from the mortar at the end of the stroke. The trowel should be fully loaded with a tapered section of mortar.

Step 5 To prevent the mortar from falling off the trowel, snap your wrist slightly to set the mortar on the trowel as you lift.

10.4.0 Spreading, Cutting, and Furrowing

The next sections describe spreading the mortar, shaping its edges, and *furrowing* it. You can practice spreading, cutting, and furrowing the mortar along a 2 × 4 board spread between two cement blocks or other props. Practice until you feel comfortable with these movements.

10.4.1 Spreading

Spreading the mortar means applying it in a desired location at a uniform thickness (see *Figure 26*). Mortar is spread for bed joints. The process of spreading the mortar for the bed joint is also called *stringing* the mortar. The spreading motion has two components to it, and they occur at the same time.

101F26.EPS

Figure 26 ◆ Spreading mortar.

Mortar application should adhere to the following guidelines:

- The joints are completely filled with no small voids for water to enter.
- The mortar is still pliable while you level and plumb the unit.
- The finished joint is the specified thickness after you level and plumb the masonry unit.
- The mortar does not smear the face of the masonry unit.

The first component of the spreading motion is a horizontal sweep from the starting point or the point where the last spread of mortar ended, back toward you. The mortar deposited is called a **spread**. Try to make the spread about two bricks long to begin with. If you are working with block, try to string the spread about one block long at first. After practice, you should be able to string the spread three to four bricks long, or two blocks long.

The second component of the motion is a vertical rotation. The trowel starts with its blade horizontal. As you move the trowel back toward you, you are also rotating it. As you rotate it, your thumb moves downward, and the back of the blade moves away from you. As the blade tilts, with the trowel traveling horizontally, the mortar is deposited along the path of the trowel.

Practice spreading until you can deposit a trail rather than a mound of mortar. Keep the trowel in the center of the wall for the length of the spread, so mortar will not get thrown on the face of the masonry. Start with a goal of 16" and work up to a spread of 24" to 32".

The joint spread should be about ¾" tall for brick and 1½" tall for block. Full joint spreads are used for all brick but not for all block. Block is usually mortared on its face shells and not its webs. However, block needs a full bed joint when it fits into any of the following categories:

- The first or starting course on a foundation, footing, or other structure
- Part of masonry columns, piers, or **pilasters** designed to carry heavy loads
- In a reinforced masonry structure, where all cores are to be grouted

Check the specifications to be sure. After the first course, the remaining block is mortared on shells, or shells and webs, according to specifications.

Whether you work with block or brick, you will need to know how to spread a full bed joint, cut it, and furrow it.

10.4.2 Cutting or Edging

After each spread, use the edge of the trowel to cut off excess mortar. To cut, hold the edge of the trowel at about a 60-degree angle, perpendicular to the edge of the mortar. Use the edge of the trowel to shave off the edge of the mortar. *Figure 27* shows the correct angle for shaving the edge of the spread.

Keep the edge of the trowel at a flat angle as shown. This will allow you to catch the mortar as you shave the edge. At this stage in your practice,

Figure 27 ◆ Cutting an edge.

101F27.EPS

learn how to catch the mortar as you cut it. The excess mortar can be returned to the mortar pan or used to fill any spaces in the bed joint.

Catching the mortar as you shave it means you do not have to go back and pick it up afterwards. On the job, having mortar stuck to the face of the masonry unit or lying in piles at the foot of a wall is unacceptable. Mortar is hard to remove when it dries, easy to clean when it is fresh. Learn to clean mortar as you lay it.

10.4.3 Furrowing

Furrowing is the act of shaping the bed joint before laying a masonry unit on it. A furrow is a shallow triangular depression, like a trough, extending the length of the bed joint. The furrow gives the mortar room to move slightly, just enough to let you adjust the masonry unit to its proper position. If the furrow is too shallow, the masonry unit will not move easily. If the furrow is too deep, it may expose the unit below and eventually cause a leak.

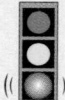
> **NOTE**
> Furrowing can be done with the trowel upright or upside down.

To make the furrow, hold the trowel blade at a 35-degree angle to the length of the spread, with the point into the spread. The point of the trowel should not go below the depth you want the finished furrow to stand. Tap the trowel point into the mortar at that angle and repeat the taps along the length of the spread. *Figure 28* shows the furrow. Notice the overlaid spacing for the trowel taps.

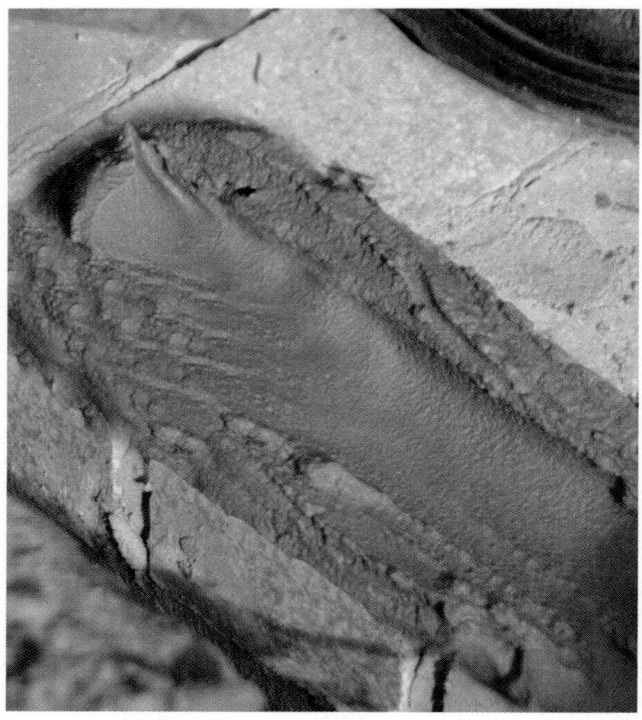

Figure 28 ◆ A furrow.

101F28.EPS

After furrowing the length of the spread, cut back the excess mortar. Use the edge of the trowel blade to shave off excess mortar hanging over the face of the wall. As you shave it, catch the excess mortar on the trowel blade. Use the excess mortar to butter the head joint on the next masonry unit to be laid.

10.5.0 Buttering Joints

Buttering the head joint is applying mortar to a header surface of a masonry unit. Buttering occurs after the bed joint is spread and the first masonry unit is laid in the bed. Buttering techniques are different for brick and block.

Buttering brick is a two-handed job. Begin by spreading the mortar on the bed joint. Keeping the trowel in your hand, pick up the first brick with your other hand. Press this brick into position in the mortar. Cut off the excess mortar on the outside face with the edge of the trowel.

Keeping the trowel in your hand, pick up a second brick in your brick hand. As you hold it, apply mortar to the header end of the brick. *Figure 29* shows a properly buttered head joint.

The buttered mortar should cover all the header surface but should not extend past the edges of the brick. Hold the trowel at an angle to the header surface to keep the mortar off the sides of the brick.

When the brick is buttered, use your brick hand to press it into position next to the first brick (*Figure 30*).

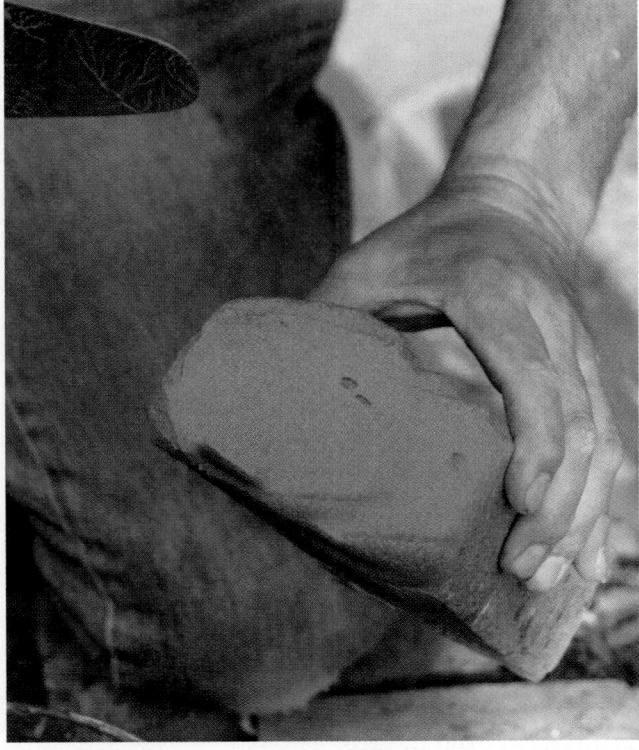

Figure 29 ◆ A buttered joint.

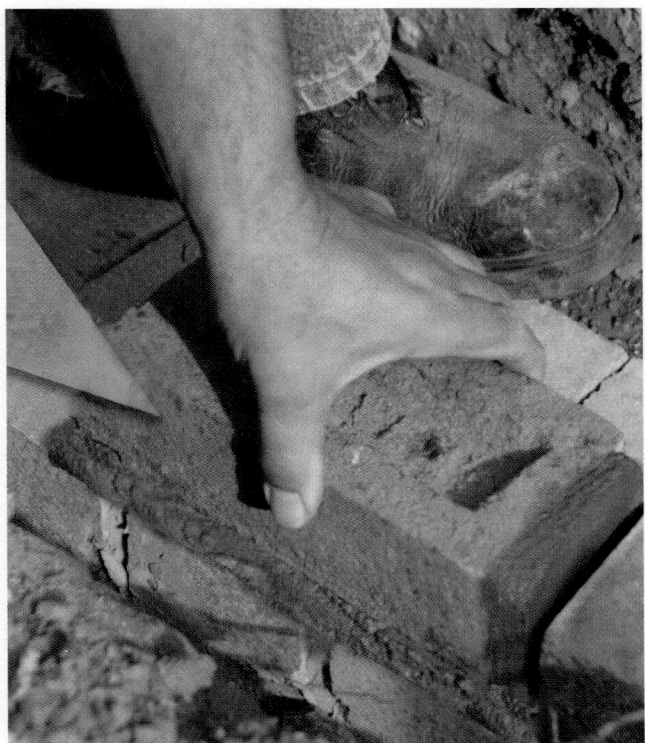

Figure 30 ◆ Placing the brick.

After placing the brick, cut off the excess mortar with the edge of your trowel.

Unlike blocks, you can easily hold a brick in one hand. Take advantage of this to use both hands for laying bricks. Try to develop a rhythmic set of movements. This will make the work faster and easier on you. Remember to use your shoulders and arms, not just your wrists.

After you have laid six bricks, check them for placement. Use your mason's level to check both plumb and level. If a brick is out of line, tap it gently with the handle of your trowel (*Figure 31*). Do not tap the level. Do not use the point or blade of your trowel or it will lose its edge.

10.6.0 General Rules

The way you work the mortar determines the quality of the joints between the masonry units. The mortar and the joints form a vital part of the structural strength and water resistance of the wall. Learning these general rules and applying them as you spread mortar will help you build good walls:

- Use mortar with the consistency of mud, so it will cling to the masonry unit, creating a good bond.
- Butter the head joints fully for brick and block; butter both ears of the head joints for block.

Figure 31 ◆ Checking the level.

- When laying a unit on the bed joint, press down slightly and sideways, so the unit goes against the one next to it.
- If mortar falls off a moving unit, replace the mortar before placing the unit.
- Put down more mortar than the size of the final joint; remember that placing the unit will compress the mortar.
- Do not string a spread that is more than 6 bricks or 3 blocks long; longer spreads will get too stiff to bond properly as water evaporates from them.
- Do not move a unit once it is placed, leveled, plumbed, and aligned.
- If a unit must be moved after it is placed, remove all the mortar on it and rebutter it.
- After placing the unit, cut away excess mortar with your trowel and put it back in the pan, or use it to butter the next joint.
- Throw away mortar after 2 to 2½ hours. At that point, it is beginning to set and will not give a good bond.

11.0.0 ◆ SAFETY PRACTICES

Masons operate in a high-risk environment. All around them are stacks of materials, trucks, and heavy equipment. Work sites have many possibilities for accidents. Workers themselves can cause accidents. They can drop masonry or tools off scaffolding and onto other workers. They can assemble scaffolding so poorly that it collapses under the weight of the load. They can fall off scaffolding. They can use damaged or poorly maintained tools that could injure themselves or others.

You must think and practice safety at all times. Your work must be planned so that it is safe as well as efficient.

All workers at a construction site must wear appropriate personal protective equipment (PPE) to protect their skin and eyes from mortar, grout, and flying masonry chips. They also need to protect themselves by being aware of what is happening around them. Workers need to keep track of the rest of their crew and of other crews. Unusual movements or noises can indicate something is moving that should not be. Masons need to have the knowledge, skill, and ability to do the following:

- Recognize an unsafe situation.
- Alert fellow crew members to the danger.
- Take evasive or corrective action.

11.1.0 The Cost of Job Accidents

Unsafe working conditions and practices can result in the following:

- Personal injury or death
- Injury or death of other workers
- Damage to equipment
- Damage to the work site

Insurance may cover some of the costs, but there are hidden (indirect or uninsured) costs as well (*Figure 32*).

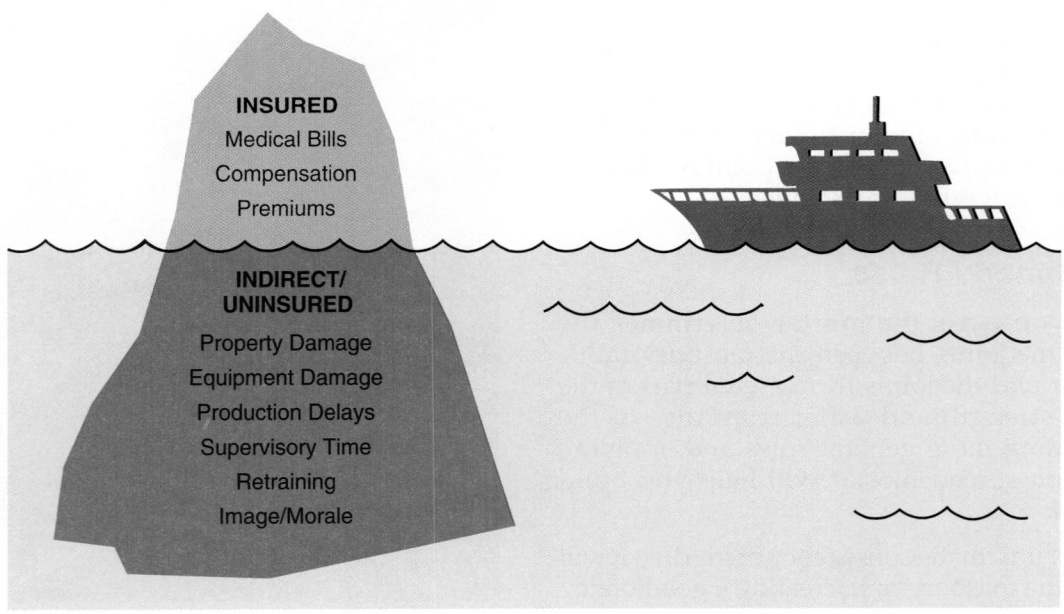

Figure 32 ◆ Hidden costs of accidents.

101F32.EPS

In addition to the pain and suffering for the individuals involved and their families, accidents can affect also others For example, accidents can slow down or stop a job, thereby putting the entire operation in jeopardy, and possibly resulting in site-wide layoffs. A high accident rate can cause an employer's insurance rates to rise, making the company less competitive with other construction companies and less likely to secure future work.

11.2.0 Wearing Safety Gear and Clothing

In general, the employer is responsible to Occupational Safety and Health Administration (OSHA) for making sure that all employees are wearing appropriate personal protective equipment whenever those employees are exposed to possible hazards to their safety. In turn, you are responsible for wearing the gear and clothing assigned to you. *Figure 33* shows a mason properly dressed and equipped for most masonry jobs.

 DID YOU KNOW?

Eye Injuries

The average cost of an eye injury is $1,463. That includes both the direct and indirect costs of accidents, not to mention the long-term effects on the health of the worker; that's priceless.

Source: The Occupational Safety and Health Administration (OSHA)

It is important to take the following safety precautions when dressing for masonry work:

- Remove all jewelry, including wedding rings, bracelets, necklaces, and earrings. Jewelry can get caught on or in equipment, which could result in a lost finger, ear, or other appendage.
- Confine long hair in a ponytail or in your hard hat. Flying hair can obscure your view or get caught in machinery.

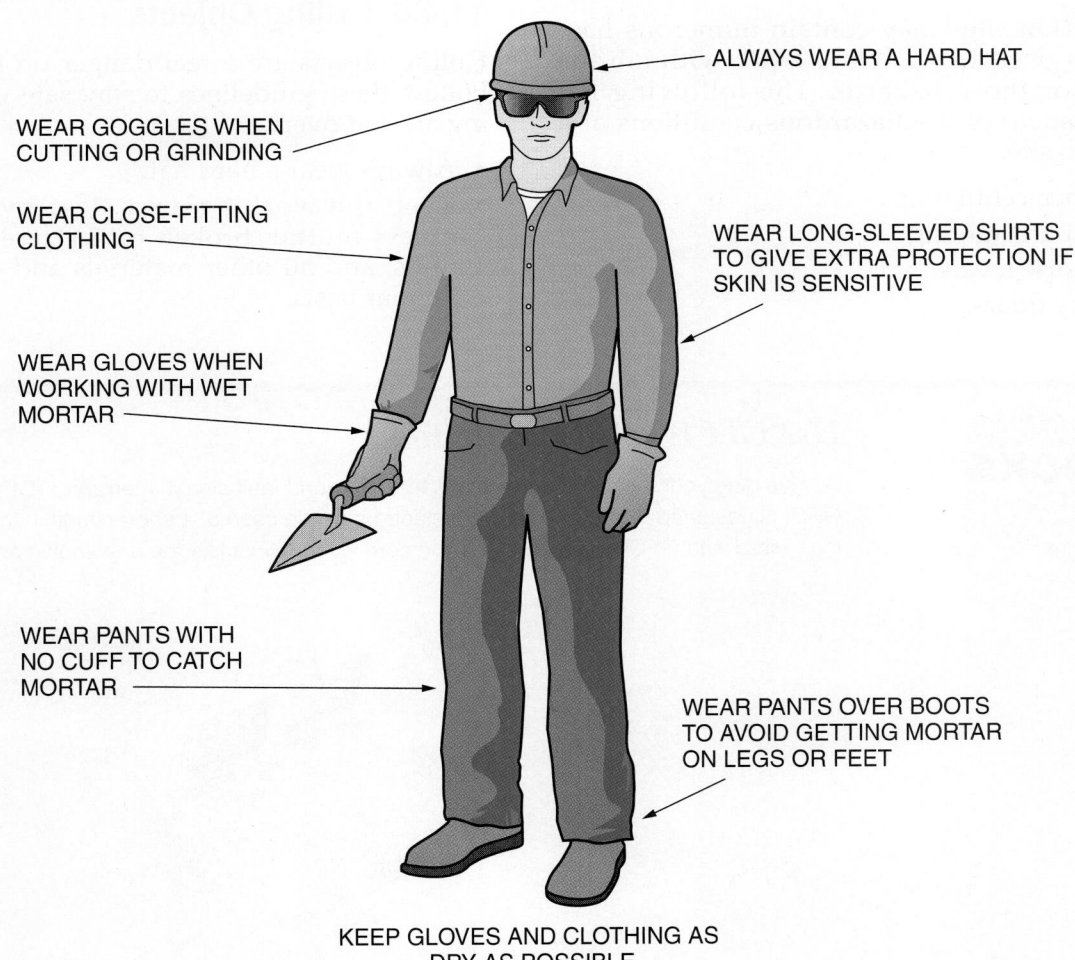

ALWAYS WEAR A HARD HAT

WEAR GOGGLES WHEN CUTTING OR GRINDING

WEAR CLOSE-FITTING CLOTHING

WEAR GLOVES WHEN WORKING WITH WET MORTAR

WEAR LONG-SLEEVED SHIRTS TO GIVE EXTRA PROTECTION IF SKIN IS SENSITIVE

WEAR PANTS WITH NO CUFF TO CATCH MORTAR

WEAR PANTS OVER BOOTS TO AVOID GETTING MORTAR ON LEGS OR FEET

KEEP GLOVES AND CLOTHING AS DRY AS POSSIBLE

101F33.EPS

Figure 33 ◆ Dressed for masonry work.

- Wear close-fitting clothing that is appropriate for the job. Clothing should be comfortable and should not interfere with the free movement of your body. Clothing or accessories that do not fit tightly, or that are too loose or torn, may get caught in tools, materials, or scaffolding.
- Wear face and eye protection as required, especially if there is a risk from flying particles, debris, or other hazards such as brick dust or chemicals.
- Wear hearing protection as required.
- Wear respiratory protection as required.
- Wear a long-sleeved shirt to provide extra protection for your skin.
- Protect any exposed skin by applying skin cream, body lotion, or petroleum jelly.
- Wear sturdy work boots or work shoes with thick soles. Never show up for work dressed in sneakers, loafers, or sport shoes.
- Wear fall protection equipment as required.

11.3.0 Hazards on the Job

Construction sites may contain numerous hazards. You need to walk and work with all due respect for those hazards. The following list includes some of the hazardous conditions at a typical job site:

- Improper ventilation
- Inadequate lighting
- High noise levels
- Slippery floors
- Unmarked low ceilings
- Excavations, holes, and open, unguarded spaces, including open, unbarricaded elevator shafts
- Poorly constructed or poorly rigged scaffolds
- Improperly stacked materials
- Live wires, loose wires, and extension cords
- Unsafe ladders
- Unsafe crane operations
- Water and mud
- Unsafe storage of hazardous or flammable materials
- Defective or unsafe tools and equipment
- Poor housekeeping

Your safety and that of your fellow workers should be a primary consideration in your work life. Some common-sense rules and ways of doing things can make the job site safer for everyone. These safety tips should be a part of your everyday thinking. Develop a positive safety attitude. It will keep you and your co-workers safe and sound.

11.4.0 Falling Objects

Falling objects are a real danger on the job site. Follow these guidelines to stay safe when working around overhead hazards:

- Always wear a hard hat.
- Keep the working area clear by removing excess mortar, broken or scattered masonry units, and all other materials and debris on a regular basis.

BUILDING BLOCKS

Use GFCIs with Power Tools

Always plug your electrical power tools into a ground fault circuit interrupter (GFCI). The GFCI is designed to protect you from electrocution in case of a short circuit. If the work site isn't wired with GFCIs, use an extension cord with a GFCI like the one shown here.

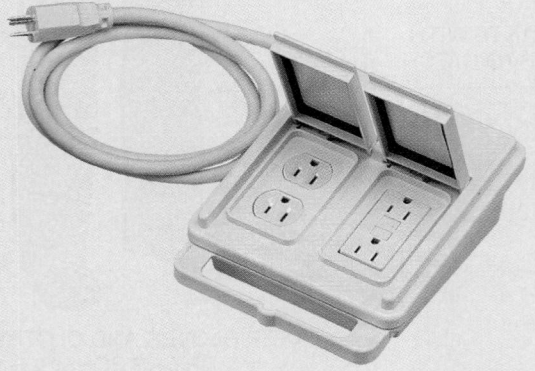

101SA06.EPS

- Keep openings in floors covered. When guardrail systems are used to prevent materials from falling from one level to another, any openings must be small enough to prevent the passage of potential falling objects.
- Do not store materials other than masonry and mortar within 4' (1.2 meters) of the working edges of a guardrail system.
- Be very careful around operating cranes. Stay clear of the crane's working area (*Figure 34*).
- Never work or walk under loads that are being hoisted by a crane.
- Learn the basic hoisting signals for cranes. Be sure to learn the stop signal and the emergency stop signal, as shown in *Figure 35*.
- Erect toeboards or guardrail systems to protect yourself from objects falling from higher levels. Toeboards should be erected along the edges of the overhead walking/working surface for a distance that is sufficient to protect the workers below.
- Erect paneling or screening from the walking/working surface or toeboard to the top of a guardrail system's top rail or mid-rail if tools, equipment, or materials are piled higher than the top edge of the toeboard.
- Raise or lower tools or materials with a rope and bucket or other lifting device. Never throw tools or materials.
- Never put tools or materials down on ladders or in other places where they can fall and injure people below. Before moving a ladder, make sure there are no tools left on it.

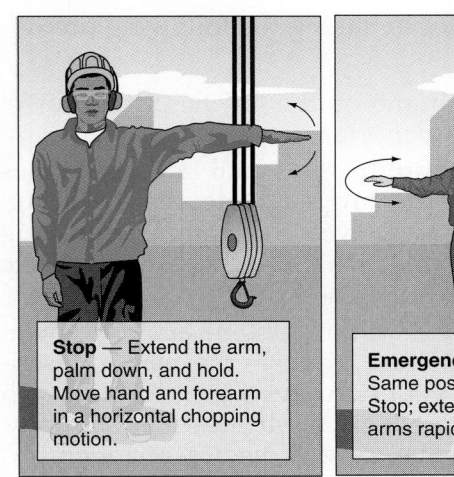

Stop — Extend the arm, palm down, and hold. Move hand and forearm in a horizontal chopping motion.

Emergency Stop — Same position as for Stop; extend and retract arms rapidly.

101F35.EPS

Figure 35 ◆ Emergency hand signals.

11.5.0 Mortar and Concrete Safety

Another hazard encountered by masonry craftworkers is exposure to mortar, grout, and concrete. These cement-based materials have ingredients that can hurt your eyes or skin. The basic ingredient of mortar, portland cement, is alkaline in nature and is therefore caustic. It is also hygroscopic, which means that it will absorb moisture from your skin. Prolonged contact between the fresh mix and skin can cause skin irritation and chemical burns to hands, feet, and exposed skin areas. It can also saturate a worker's clothes and transmit alkaline or hygroscopic effects to the skin. In addition, the sand contained in fresh mortar can cause skin abrasions through prolonged contact.

 WARNING!

Those working with dry cement or wet concrete should be aware that it is harmful. Dry cement dust can enter open wounds and cause blood poisoning. The cement dust, when it comes in contact with body fluids, can cause chemical burns to the membranes of the eyes, nose, mouth, throat, or lungs. It can also cause a fatal lung disease known as silicosis.

Wet cement or concrete can also cause chemical burns to the eyes and skin. Always wear appropriate personal protective equipment when working with dry cement or wet concrete. If wet concrete enters waterproof boots from the top, remove the boots and rinse your legs, feet, boots, and clothing with clear water as soon as possible. Repeated contact with cement or wet concrete can also cause an allergic skin reaction known as cement dermatitis.

101F34.EPS

Figure 34 ◆ Crane hazard on the job.

Avoid injuries by taking the following precautions:

- Keep your thumb on the ferrule of the trowel away from the mortar.
- Keep cement products off your skin at all times by wearing the proper protective clothing, including boots, gloves, and clothing with snug wristbands, ankle bands, and neckband. Make sure they are all in good condition.
- Prevent your skin from rubbing against cement products. Rubbing increases the chance of serious injury. If your skin does come in contact with any cement products, wash your skin promptly. If a reaction persists, seek medical attention.
- Wash thoroughly to prevent skin damage from cement dust.
- Keep cement products out of your eyes by wearing safety glasses when mixing mortar or pumping grout. If any cement or cement mixtures get in your eye, flush it immediately and repeatedly with water, and consult a physician promptly.
- Rinse off any clothing that becomes saturated from contact with fresh mortar. A prompt rinse with clean water will prevent continued contact with skin surfaces.
- Be alert! Watch for trucks backing into position and overhead equipment delivering materials. Listen for the alarms or warning bells on mixers, pavers, and ready-mix trucks.
- Never put your hands, arms, or any tools into rotating mixers.
- Be certain adequate ventilation is provide when using epoxy resins, organic solvents, brick cleaners, and other toxic substances.
- Use good work practices to reduce dust in the air when handling mortar and lime. For example, do not shake out mortar bags unnecessarily and do not use compressed air to blow mortar

dust off clothing or a work surface. Stand upwind when dumping mortar bags.
- Never use solvents to clean skin.
- Immediately remove epoxy, solvents, and other toxic substances from skin using the appropriate cleansing agents.
- Know the locations of all eye wash stations and emergency showers on your job site. Be sure you know how to use them properly.

11.6.0 Flammable Liquid Safety

Flammable liquids are particularly dangerous and require additional safety precautions. Always adhere to all relevant codes, job-site rules, and the following guidelines:

- Carefully read labels on all flammable liquids, and use flammable liquids only in open, well-ventilated areas.
- Do not inhale or ignite fumes from flammable liquids.
- Be sure that all flammable liquids are marked correctly for storage.
- Store all flammable liquids properly and only in approved safety containers, such as safety cans and safety cabinets.
- Store oily rags and flammable materials in metal containers with self-sealing lids.
- When clothing is soaked by a flammable liquid, immediately change clothing, and cleanse the body with an appropriate cleaner.
- Use flammable liquids only for their intended purposes; for example, never use gasoline as a cleaner.
- Never use flammable liquids near fire or flame.
- Always be aware of the location of an appropriate fire extinguisher.
- Learn your company's emergency response procedures for fire and explosion.

Don't Get Burned

Here are some facts about gasoline:

- One gallon of gasoline contains the same explosive force as 14 sticks of dynamite.
- Gasoline vapors are heavier than air, can travel several feet to an ignition source, and can ignite at temperatures as low as 45°F. To be safe, keep open gasoline containers well removed from all potential ignition sources.
- Gasoline has a low electrical conductivity. As a result, a charge of static electricity builds up on gasoline as it flows through a pipe or hose. Getting into and out of your vehicle during refueling can build up a static charge, especially during dry weather. That charge can cause a spark that can ignite gasoline vapors if it occurs near the fuel nozzle.

Source: U.S. Department of Energy

11.7.0 Material Handling

Approximately 25 percent of all occupational injuries occur when handling or moving construction materials. Strains, sprains, fractures, and crushing injuries can be minimized with a knowledge of safe lifting and handling procedures and proper ergonomics. General guidelines for you to keep in mind when handling or moving construction materials are as follows:

- Wear steel-toe safety shoes.
- Keep floors free of water, grease, and other slippery substances so as to prevent falls.
- Inspect materials for grease, slivers, and rough or sharp edges.
- Determine the weight of the load before applying force to move it.
- Know your own limits for how much you can lift.
- Be sure that your intended pathway is free from obstacles.
- Make sure your hands are free of oil and grease.
- Take a firm grip on the object before you move it, being careful to keep your fingers from being pinched.
- When the load is too large or heavy, get help, or, if possible, simply reduce the load and make more trips.
- Whenever possible, use mechanical means of material handling.
- When stacking materials, be sure to follow OSHA regulations as to the height, shape, and stability of the pile.

Materials in the general working area and on the stockpile should be stacked safely. Masonry units that are not stacked properly and secured in some way are very likely to fall, which could cause injury to you or a fellow worker.

11.7.1 Materials Stockpiling and Storage

OSHA has several guidelines regarding the stockpiling and handling of materials. You must adhere to these and all OSHA guidelines:

- All materials stored in tiers should be stacked, racked, blocked, interlocked, or otherwise secured to prevent sliding, falling, or collapse.
- Maximum safe load limits of floors within buildings and structures should be posted in all storage areas, and maximum safe loads should not be exceeded.
- Aisles and passageways should be kept clear to provide for the free and safe movement of material handling equipment or employees.

- Materials stored inside buildings under construction should not be placed within 6' of any hoistway or inside floor openings nor within 10' of an exterior wall that does not extend above the top of the material stored.
- Each employee required to work on stored material in silos, hoppers, tanks, and similar storage areas should be equipped with personal fall-arrest equipment.
- Noncompatible materials should be segregated in storage.
- Bagged materials should be stacked by stepping back the layers and cross-keying the bags at least every ten bags high.
- Materials should not be stored on scaffolds or runways in excess of supplies needed for immediate operations.
- Stockpiles of palletized brick should not be higher than 7'. When a loose brick stockpile reaches a height of 4', it should be tapered back 2" in every foot of height above the 4' level.
- When loose masonry blocks are stockpiled higher than 6', the stack should be tapered back one-half block per tier above the 6' level.

11.7.2 Working Stacks

Working stacks of brick are used at the wall, on the ground, or on the scaffold platform. Palletized brick (*Figure 36*) is unbundled and moved to where it will be needed. When stacking materials in working piles, keep the pile neat and vertically in line to eliminate the possibility of snagging your clothes. Keep the piles about 3' high so that you can easily get to the brick. Bricks and other materials stacked too high not only pose a safety hazard, but may reduce productivity.

101F36.EPS

Figure 36 ◆ Palletized brick.

The most common way to stack bricks is to reverse the direction of every other course so that the stack is secure. Such a stack should be no more than 3' high and no closer than 2' to the wall. Wider stacks can be made by alternating a pattern of eight bricks, as shown in *Figure 37*.

11.8.0 Gasoline-Powered Tools

In masonry work, you may need to use gasoline-powered tools. Follow these safety guidelines:

- Be sure there is proper ventilation before operating gasoline-powered equipment indoors.
- Use caution to prevent contact with hot manifolds and hoses.
- Be sure the equipment is out of gear before starting it.
- Use the recommended starting fluid.
- Always keep the appropriate fire extinguishers near when filling, starting, and operating gasoline-powered equipment. OSHA requires

that gasoline-powered equipment be turned off prior to filling.

- Do not pour gasoline into the carburetor or cylinder head when starting the engine.
- Never pour gasoline into the fuel tank when the engine is hot or when the engine is running.
- Do not operate equipment that is leaking gasoline.

WARNING!

Never operate a tool without proper training and personal protective equipment.

11.9.0 Powder-Actuated Tools

A powder-actuated fastening tool (*Figure 38*) is a low-velocity fastening system powered by gunpowder cartridges, commonly called boosters. Powder-actuated tools are used to drive specially designed fasteners into masonry and steel.

Manufacturers use color-coding schemes to identify the strength of a powder load charge. It is

Table 2 Powder Charge Color-Coding System

Power Level*	Color
1	Gray
2	Brown
3	Green
4	Yellow
5	Red
6	Purple

*From the least powerful (1) to the most powerful (6)

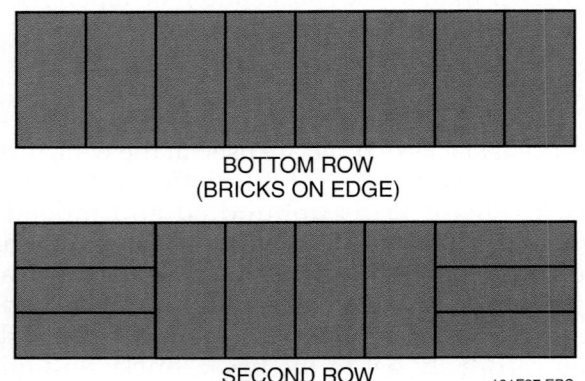

BOTTOM ROW
(BRICKS ON EDGE)

SECOND ROW

101F37.EPS

Figure 37 ◆ Stacking brick.

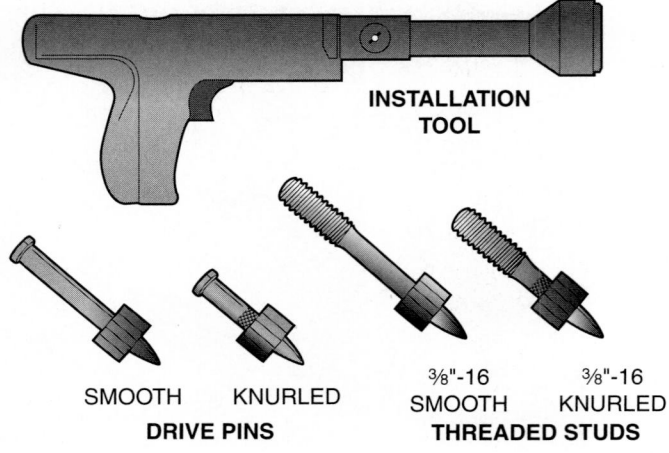

INSTALLATION
TOOL

SMOOTH   KNURLED
DRIVE PINS

⅜"-16   ⅜"-16
SMOOTH   KNURLED
THREADED STUDS

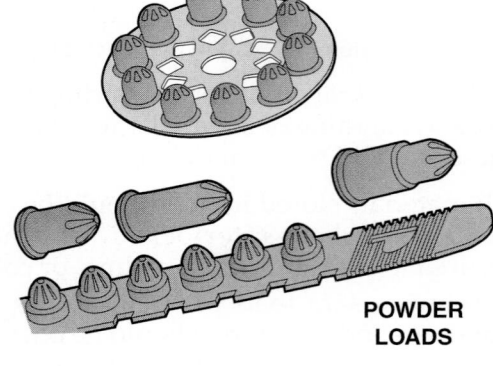

POWDER
LOADS

101F38.EPS

Figure 38 ◆ Powder-actuated fastening tool.

Fumes from Liquid-Fuel Tools

A worker at a large, enclosed construction site died of carbon monoxide poisoning after he and six other workers were exposed to high levels of the gas. He died because ventilation on the site was inadequate, and three machines were giving off carbon monoxide: a portable mixer and a trowel, both powered by gasoline, and a forklift powered by propane. This worker would have survived if the work area had been properly ventilated and if he were using the proper personal protective equipment, including a respirator.

The Bottom Line: Keep work areas well ventilated, and wear respirators when required. Hazardous air conditions can develop without warning.

Source: The Occupational Safety and Health Administration (OSHA)

WARNING!
OSHA requires that all operators of powder-actuated tools be qualified and certified by the manufacturer of the tool. You must carry a certification card whenever using the tool.

If a gun does not fire, hold it against the work surface for at least 30 seconds. Follow the manufacturer's instructions for removing the cartridge. Do not try to pry it out because some cartridges are rim-fired and could explode.

extremely important to select the right charge for the job, so learn the color-coding system that applies to the tool you are using. *Table 2* shows an example of a color-coding system.

Other rules for safely operating a powder-actuated tool are as follows:

- Do not use a powder-actuated tool unless you are certified.
- Follow all safety precautions in the manufacturer's instruction manual.
- Always wear safety goggles and a hard hat when operating a powder-actuated tool.
- Use the proper size pin for the job you are doing.

- When loading the driver, put the pin in before the charge.
- Use the correct booster (powder load) according to the manufacturer's instructions.
- Never hold your hand behind or near the material you are fastening.
- Never hold the end of the barrel against any part of your body or cock the tool against your hand.
- Do not shoot close to the edge of concrete.
- Never attempt to pry the booster out of the magazine with a sharp instrument.
- Always wear ear protection.
- Always hold the muzzle perpendicular (90 degrees) to the work.

11.10.0 Pressure Tools

You may use pressure tools on the job. These tools can be dangerous. Always use extreme caution, and never operate a pressure tool on which you have not been trained.

Pneumatic saws and grinders should have automatic overspeed controls. Runaway speeds can cause carborundum saw discs and buffers to disintegrate. Guards should also be in good condition in order to prevent flying particles.

Powder-Actuated Tool

A 22-year-old worker was killed when he was struck in the head by a nail fired from a powder-actuated tool in an adjacent room. The tool operator was attempting to anchor plywood to a hollow wall and fired the gun, causing the nail to pass through the wall, where it traveled nearly 30' before striking the victim. The tool operator had never received training in the proper use of the tool, and none of the employees in the area were wearing personal protective equipment.

The Bottom Line: The use of powder-actuated tools requires special training and certification. In addition, all personnel in the area must be aware that the tool is in use and should be wearing appropriate personal protective equipment.

All tools that produce dust need exhaust systems to collect the dust in order to prevent contamination of the air in the work areas. Automatic emergency valves should be installed at all compressed air sources to shut off the air immediately if the hose becomes disconnected or is severed. This will help prevent wild thrashing of the hose.

11.11.0 Weather Hazards

Masons usually work outdoors. Under certain environmental conditions, such as extreme hot or cold weather, work can become uncomfortable and possibly dangerous. There are specific things to be aware of when working under these adverse conditions.

11.11.1 Cold Weather

The amount of injury caused by exposure to abnormally cold temperatures depends on wind speed, length of exposure, temperature, and humidity. Freezing is increased by wind, humidity, or a combination of the two factors. Follow these guidelines to prevent injuries such as frostbite during extremely cold weather:

- Always wear the proper clothing.
- Limit your exposure as much as possible.
- Take frequent, short rest periods.
- Keep moving. Exercise fingers and toes if necessary, but do not overexert.
- Do not drink alcohol before exposure to cold. Alcohol can dull your sensitivity to cold and make you less aware of over-exposure.
- Do not expose yourself to extremely cold weather if any part of your clothing or body is wet.
- Do not smoke before exposure to cold. Breathing can be difficult in extremely cold air. Smoking can worsen the effect.

- Learn how to recognize the symptoms of over-exposure and frostbite.
- Place cold hands under dry clothing against the body, such as in the armpits.

If you live in a place with cold weather, you will most likely be exposed to it when working. Spending long periods of time in the cold can be dangerous. It's important to know the symptoms of cold weather exposure and how to treat them. Symptoms of cold exposure include the following:

- Shivering
- Numbness
- Low body temperature
- Drowsiness
- Weak muscles

Follow these steps to treat cold exposure:

Step 1   Get to a warm inside area as quickly as possible.

Step 2   Remove wet or frozen clothing and anything that is binding such as necklaces, watches, rings, and belts.

Step 3   Rewarm by adding clothing or wrapping in a blanket.

Step 4   Drink hot liquids, but do not drink alcohol.

Step 5   Check for frostbite. If you suspect frostbite, seek medical help immediately.

Frostbite is an injury resulting from exposure to cold elements. It happens when crystals form in the fluids and underlying soft tissues of the skin. The frozen area is generally small. The nose, cheeks, ears, fingers, and toes are usually affected. Affected skin may be slightly flushed just before frostbite sets in. Symptoms of frostbite include the following:

- Skin that becomes white, gray, or waxy yellow. Color indicates deep tissue damage. Victims are

Cold Weather Clothing Tips

Use the following tips to prevent injury due to cold weather:

- Dress in layers.
- Wear thermal-type woolen underwear.
- Wear outer clothing that will repel wind and moisture.
- Wear a face helmet and head and ear coverings.
- Carry an extra pair of dry socks when working in snowy or wet conditions.
- Wear warm boots, and make sure that they are not so tight that circulation becomes restricted.
- Wear wool-lined mittens or gloves covered with wind- and water-repellent material.

NOTE

In advanced cases of frostbite, mental confusion and poor judgment occur, the victim staggers, eyesight fails, the victim falls and may pass out. Shock is evident, and breathing may cease. Death, if it occurs, is usually due to heart failure.

often not aware of frostbite until someone else recognizes the pale, glossy skin.

- Skin tingles and then becomes numb.
- Pain in the affected area starts and stops.
- Blisters show up on the area.
- The area of frostbite swells and feels hard.

Use the following steps to treat frostbite.

Step 1   Protect the frozen area from refreezing.

Step 2   Warm the frostbitten part as soon as possible.

Step 3   Get medical attention immediately.

11.11.2 Hot Weather

Hot weather can be as dangerous as cold weather. When someone is exposed to excessive amounts of heat, they run the risk of overheating. Conditions associated with overheating include the following:

- Heat cramps
- Heat exhaustion
- Heat stroke

Heat cramps can occur after an attack of heat exhaustion. Cramps are characterized by abdominal pain, nausea, and dizziness. The skin becomes pale with heavy sweating, muscular twitching, and severe muscle cramps.

If you experience heat cramps, sit or lie down in a cool area, preferably indoors. Drink a half a glass of water every 15 minutes, and gently stretch and massage any cramped muscles. Do not resume work until you feel fully recovered.

Heat exhaustion is characterized by pale, clammy skin; heavy sweating with nausea and possible vomiting; a fast, weak pulse; and possible fainting.

Treat heat exhaustion by having the victim lie down with his or her feet elevated 6 to 8 inches. If nauseous, have the victim lie on his or her side, not back. Remove any heavy clothing and loosen all other clothing. Apply cool, wet cloths, and fan the victim, but stop if chills develop. If the victim is fully conscious, give him or her a half a glass of water every 15 minutes. If the condition does not improve quickly, call emergency medical services (911).

Heat stroke is an immediate, life-threatening emergency that requires urgent medical attention. It is characterized by headache, nausea, and visual problems. Body temperature can reach as high as 106°F. This will be accompanied by hot, flushed, dry skin; slow, deep breathing; possible convulsions; and loss of consciousness.

If someone experiences heat stroke, call emergency medical services (911) immediately. Then, move the victim to a cool area and have him or her lie on his back. As with heat exhaustion, if the victim is nauseous, have the victim lie on his or her side instead. Move all nearby objects, as heat stroke may cause convulsions or seizures. Apply cool, wet cloths and/or fan the victim. If the victim is not nauseous and is conscious, give small amounts of water in 15-minute intervals. Place ice packs under the armpits and in the groin area. Remain with the victim until emergency medical assistance arrives.

Follow these guidelines when working in hot weather in order to prevent heat exhaustion, cramps, or heat stroke:

- Drink plenty of water.
- Do not overexert yourself.
- Wear lightweight clothing.
- Keep your head covered and face shaded.
- Take frequent, short work breaks.
- Rest in the shade whenever possible.

12.0.0 ◆ FALL PROTECTION

Falls are the leading cause of death in the construction industry. In fact, more than one third of all deaths in the industry are the result of a fall. Fall protection is required when workers are exposed to falls from work areas with elevations that are 6' or higher. The types of work areas that put the worker at risk include the following:

- Scaffolding
- Ladders
- Leading edges
- Ramps or runways
- Wall or floor openings
- Roofs
- Excavations, pits, and wells
- Concrete forms
- Unprotected sides and edges

Falls happen because of the inappropriate use or lack of fall-protection systems (*Figure 39*). They also happen because of worker carelessness. It is your responsibility to learn how to set up, use, and maintain fall-protection equipment. Not only will this keep you alive and uninjured, it could save the lives of your co-workers.

Falls are classified into two groups: falls from an elevation and falls on the same level. Falls from an elevation can happen when you are doing work from scaffolding, work platforms, decking, concrete forms, ladders, or excavations. Falls from elevations are almost always fatal. This is not to say that falls on the same level aren't also extremely dangerous. When a worker falls on the same level, usually from tripping or slipping, head injuries often occur. Sharp edges and pointed objects such as exposed rebar could cut or stab the worker.

The following safe practices can help prevent slips and falls:

- Wear safe, strong work boots that are in good repair.
- Watch where you step. Be sure your footing is secure.
- Install cables, extension cords, and hoses so that they will not become tripping hazards.

- Do not allow yourself to get in an awkward position. Stay in control of your movements at all times.
- Maintain clean, smooth walking and working surfaces. Fill holes, ruts, and cracks. Clean up slippery material and litter.
- Do not run on scaffolding, work platforms, decking, roofs, or other elevated work areas.

The best way to survive a fall from an elevation is to use fall-protection equipment. The three most common types of fall-protection equipment are guardrails, personal fall-arrest systems, and safety nets.

12.1.0 Guardrails

Guardrails (*Figure 40*) protect workers by providing a barrier between the work area and the ground or lower work areas. They may be made of wood, pipe, steel, or wire rope and must be able to support 200 pounds of force applied to the top rail.

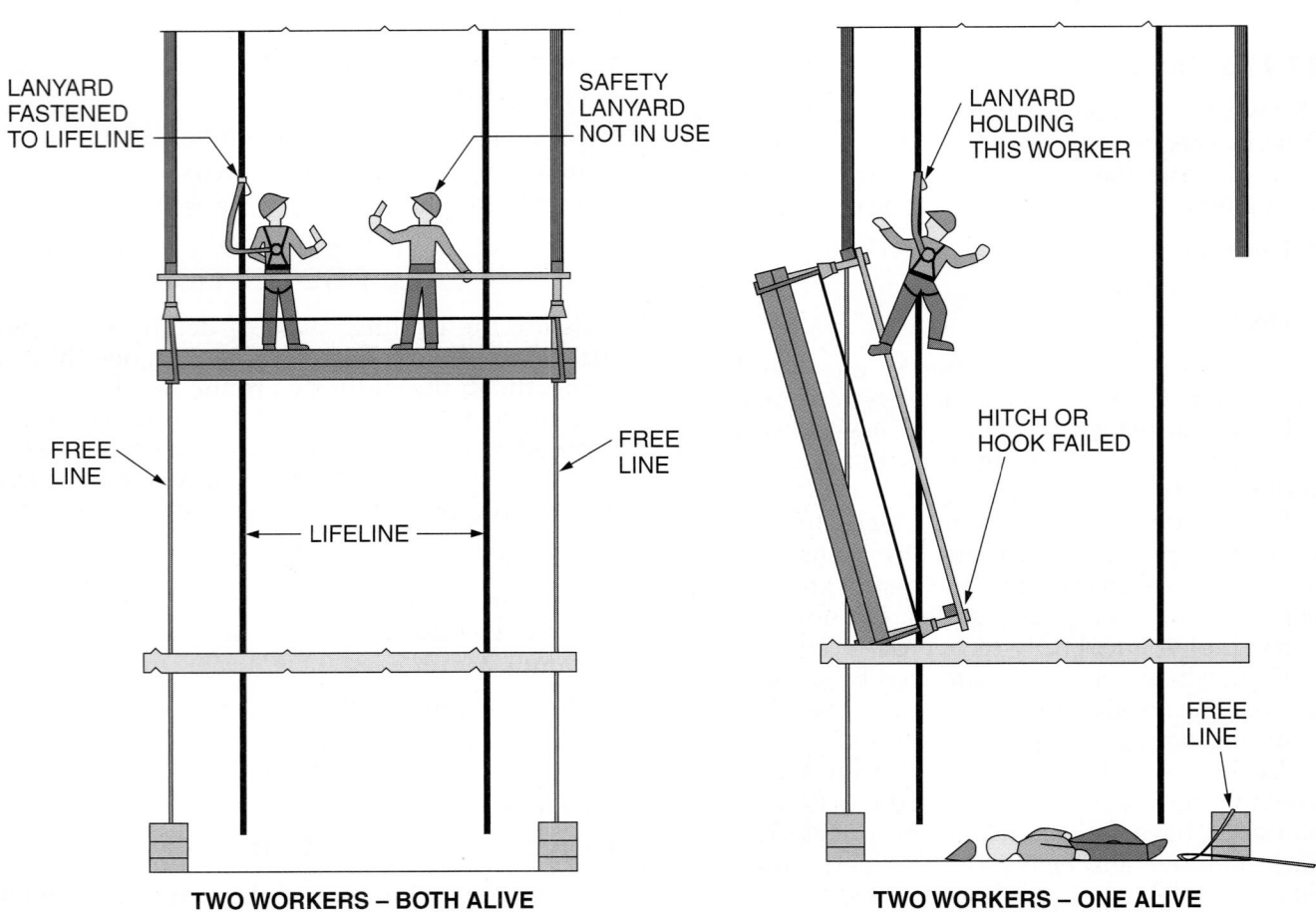

TWO WORKERS – BOTH ALIVE   **TWO WORKERS – ONE ALIVE**

A properly used body harness and lifeline will protect you if the scaffolding fails.

101F39.EPS

Figure 39 ◆ Proper and improper safety harness use.

101F40.EPS

Figure 40 ◆ Guardrails.

12.2.0 Personal Fall-Arrest Systems

Personal fall-arrest systems catch workers after they have fallen. They are designed and rigged to prevent a worker from free falling a distance of more than 6' and hitting the ground or a lower work area. When describing personal fall-arrest systems, these terms must be understood:

- *Free-fall distance* – The vertical distance a worker moves after a fall before a deceleration device is activated.
- *Deceleration device* – A device such as a shock-absorbing lanyard or self-retracting lifeline that brings a falling person to a stop without injury.
- *Deceleration distance* – The distance it takes before a person comes to a stop when falling. The required deceleration distance for a fall-arrest system is a maximum of 3½'.
- *Arresting force* – The force needed to stop a person from falling. The greater the free-fall distance, the more force is needed to stop or arrest the fall.

Personal fall-arrest systems use specialized equipment. This equipment includes the following:

- Body harnesses and belts
- Lanyards
- Deceleration devices
- Lifelines
- Anchoring devices and equipment connectors

12.2.1 Body Harnesses

Full-body harnesses with sliding back D-rings (*Figure 41*) are used in personal fall-arrest systems. They are made of straps that are worn securely around the user's body. This allows the arresting force to be distributed throughout the body, including the shoulders, legs, torso, and buttocks. This distribution decreases the chance of injury. When a fall occurs, the sliding D-ring moves to the nape of the neck. This keeps the worker in an upright position and helps to distribute the arresting force. The worker then stays in a relatively comfortable position while waiting for rescue.

Selecting the right full-body harness depends on a combination of job requirements and personal preference. Harness manufacturers normally provide selection guidelines in their product literature. Some types of full-body harnesses can be equipped with front chest D-rings, side D-rings, or shoulder D-rings. Harnesses with front chest D-rings are typically used in ladder climbing and personal-positioning systems (*Figure 42*). Those with side D-rings are also used in personal-positioning systems. Personal-positioning systems allow workers to hold themselves in place, keeping their hands free to accomplish a task.

A personal-positioning system should not allow a worker to free fall more than 2'. The anchorage that it's attached to should be able to support at least twice the impact load of a worker's fall or 3,000 pounds, whichever is greater.

12.2.2 Lanyards

Lanyards are short, flexible lines with connectors on each end. They are used to connect a body harness or body belt to a lifeline, deceleration device, or anchorage point. There are many kinds of lanyards made for different uses and climbing situations. All must have a minimum breaking strength of 5,000 pounds. They come in both fixed and adjustable lengths and are made out of steel, rope, or nylon webbing. Some have a shock absorber (*Figure 43*) that absorbs up to 80% of the arresting force when a fall is being stopped. When choosing a lanyard for a particular job, always follow the manufacturer's recommendations.

101F42.EPS

Figure 42 ◆ Harness with front chest D-ring.

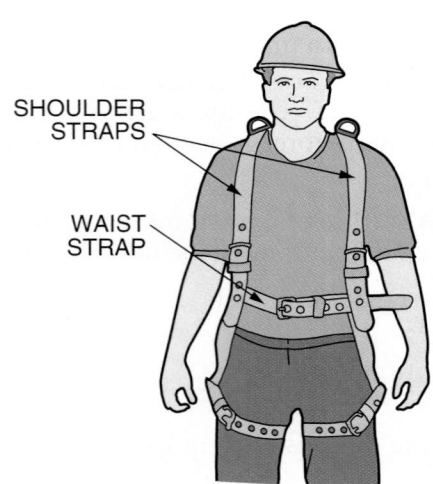

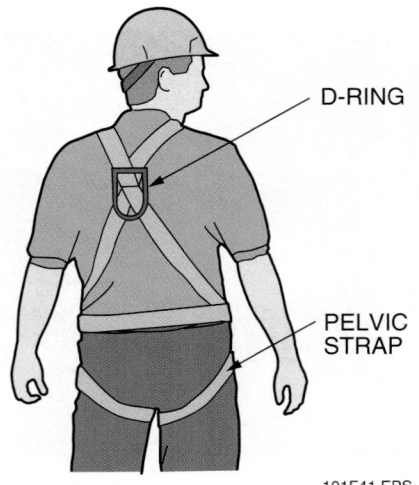

101F41.EPS

Figure 41 ◆ Full-body harnesses with sliding back D-rings.

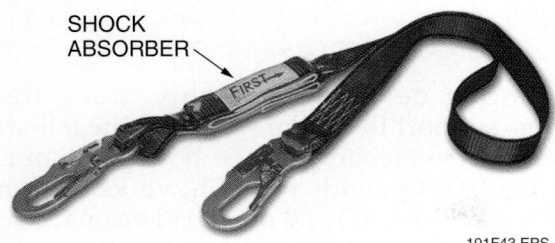

SHOCK
ABSORBER

101F43.EPS

Figure 43 ◆ Lanyard with a shock absorber.

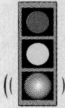

NOTE

In the past, body belts were often used instead of full-body harnesses as part of a fall-arrest system. As of January 1, 1998, however, they have been banned from such use. This is because body belts concentrate all of the arresting force in the abdominal area, which can cause significant injuries. It also causes the worker to hang in an uncomfortable and potentially dangerous position while awaiting rescue.

WARNING!

When activated during the fall-arresting process, a shock-absorbing lanyard stretches in order to reduce the arresting force. This potential increase in length must always be taken into consideration when determining the total free-fall distance from an anchor point.

12.2.3 Deceleration Devices

Deceleration devices limit the arresting force to which a worker is subjected when the fall is stopped suddenly. Rope grabs and self-retracting lifelines are two common deceleration devices (*Figure 44*). A rope grab connects to a lanyard and attaches to a lifeline. In the event of a fall, the rope grab is pulled down by the attached lanyard, causing it to grip the lifeline and lock in place. Some rope grabs have a mechanism that allows the worker to unlock the device and slowly descend down the lifeline to the ground or surface below.

Self-retracting lifelines provide unrestricted movement and fall protection while workers are climbing and descending ladders and similar equipment or when working on multiple levels. Typically, they have a 25' to 100' galvanized-steel cable that automatically takes up the slack in the attached lanyard, keeping the lanyard out of the worker's way. In the event of a fall, a centrifugal braking mechanism engages to limit the worker's free-fall distance. Self-retracting lifelines and lanyards that limit the free-fall distance to 2' or less

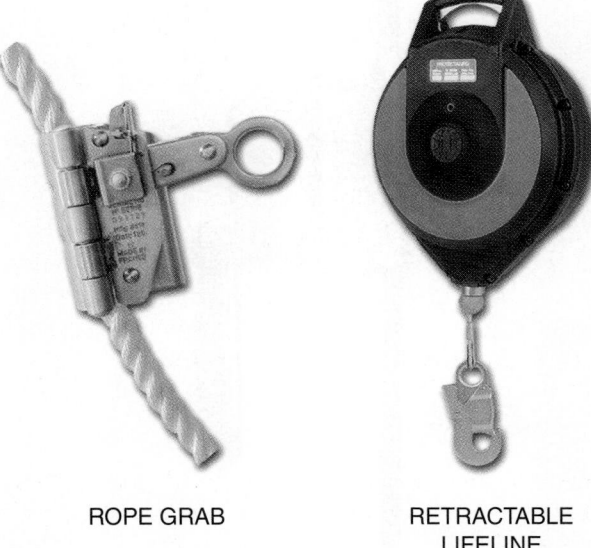

ROPE GRAB   RETRACTABLE LIFELINE

101F44.EPS

Figure 44 ◆ Rope grab and retractable lifeline.

must be able to support a minimum tensile load of 3,000 pounds. Those that do not limit the free fall to 2' or less must be able to hold a tensile load of at least 5,000 pounds.

12.2.4 Lifelines

Lifelines are ropes or flexible steel cables that are attached to an anchorage. They provide a means for tying off personal fall-protection equipment. Vertical lifelines (*Figure 45*) are suspended vertically from a fixed anchorage. A fall-arrest device such as a rope grab is attached to the lifeline. Vertical lifelines must have a minimum breaking strength of 5,000 pounds. Each worker must use his or her own line. This is because if one worker falls, the movement of the

---

lifeline during the fall arrest may also cause the other workers to fall. A vertical lifeline must be connected in a way that will keep the worker from moving past its end, or it must extend to the ground or the next lower working level.

Horizontal lifelines (*Figure 46*) are connected horizontally between two fixed anchorages. These lifelines must be designed, installed, and used under the supervision of a qualified, competent person. The more workers who are tied off to a single horizontal line, the stronger the line and anchors must be.

101F45.EPS

Figure 45 ◆ Vertical lifeline.

101F46.EPS

Figure 46 ◆ Horizontal lifeline.

12.2.5 Anchoring Devices and Equipment Connectors

Anchoring devices, commonly called tie-off points, support the entire weight of the fall-arrest system. The anchorage must be capable of supporting 5,000 pounds for each worker attached. Eye bolts (*Figure 47*) and overhead beams are considered anchorage points.

The D-rings, buckles, carabiners, and snaphooks (*Figure 48*) that fasten and/or connect the parts of a personal fall-arrest system are called connectors. There are regulations that specify how they are to be made and that require D-rings and snaphooks to have a minimum tensile strength of 5,000 pounds.

>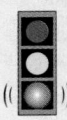
> **NOTE**
> As of January 1, 1998, only locking-type snaphooks are permitted for use in personal fall-arrest systems.

12.2.6 Selecting an Anchor Point and Tying Off

Connecting the body harness either directly or indirectly to a secure anchor point is called tying off. Tying off is always done before you get into a position from which you can fall. Follow the manufacturer's instructions on the best tie-off methods for your equipment.

In addition to the manufacturer's instructions, an anchorage point should be as follows:

101F47.EPS

Figure 47 ◆ Eye bolt.

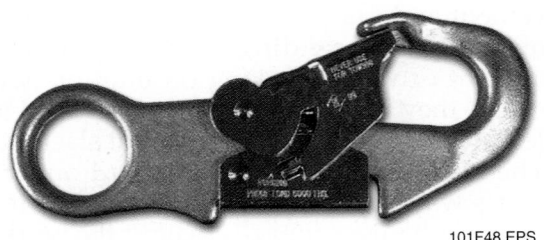

101F48.EPS

Figure 48 ◆ Double locking snaphook.

Myths and Facts About Falls in Construction

Myth 1: In the construction industry, falls are not a leading cause of death.

Fact: One third of all deaths in the construction industry are caused by falls.

Myth 2: You have to fall a long distance to kill yourself.

Fact: Half of the construction workers who die in falls fall from a height of 21' or less. If you hit your head hard enough, you can die at any height. Even if you survive a fall, you may be laid up for some time with an injury.

Myth 3: Experienced workers don't fall.

Fact: The average age of construction workers who have fallen to their death is 47. That's not exactly young and inexperienced.

> "It just happens so fast. It's when you think you're safe that you need to be more careful."
> – *Gene, Builder*

Myth 4: Working safely is costly.

Fact: Some fall-protection equipment is inexpensive, such as ladder stabilizers, guardrail holders, and fall-protection kits. Other items such as harnesses, lifelines, and safe scaffolding are more costly. Injury and death, however, are much more expensive in the end.

> "I fell three stories and was out of work for 8 weeks. I was subcontracting and didn't have comp [workman's compensation insurance]. This was a long time ago, but I probably lost around $5,000. A harness would have cost me $50 back then." – *Dan, General Contractor*

Myth 5: Fall-protection equipment is more of a hindrance than a help.

Fact: Nothing is more of a hindrance than a lifelong disability you may experience due to a fall.

Source: Electronic Library of Construction Occupational Safety and Health

---

- Directly above the worker
- Easily accessible
- Damage-free and capable of supporting 5,000 pounds per worker
- High enough so that no lower level is struck should a fall occur
- Separate from work basket tie offs

Be sure to check the manufacturer's equipment labels, and allow for any equipment stretch and deceleration distance.

12.2.7 Using Personal Fall-Arrest Equipment

Before using fall-protection equipment on the job, you must know the basics and proper usage of fall protection equipment. All equipment supplied by your employer must meet established standards for strength. Before each use, always read the instructions and warnings on any fall-protection equipment. Inspect the equipment using the following guidelines:

- Examine harnesses and lanyards for mildew, wear, damage, and deterioration.
- Ensure no straps are cut, broken, torn, or scraped.
- Check for damage from fire, chemicals, or corrosives.

- Make sure all hardware is free of cracks, sharp edges, and burrs.
- Check that snaphooks close and lock tightly and that buckles work properly.
- Check ropes for wear, broken fibers, pulled stitches, and discoloration.
- Make sure lifeline anchors and mountings are not loose or damaged.

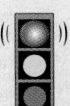 **WARNING!**
Never use fall-protection equipment that shows signs of wear or damage.

Do not mix or match equipment from different manufacturers. All substitutions must be approved by your supervisor. All damaged or defective parts must be taken out of service immediately and tagged as unusable or destroyed. If the equipment was used in a previous fall, remove it from service until it can be inspected by a qualified person.

12.3.0 Safety Net Systems

Safety nets are used for fall protection on bridges and similar projects. They must be installed as

close as possible, not more than 30', beneath the work area. There must be enough clearance under a safety net to prevent a worker who falls into it from hitting the surface below. There must also be no obstruction between the work area and the net.

Depending on the actual vertical distance between the net and the work area, the net must extend 8' to 13' beyond the edge of the work area. Mesh openings in the net must be limited to 36 square inches and 6" on the side. The border rope must have a 5,000-pound minimum breaking strength, and connections between net panels must be as strong as the nets themselves. Safety nets must be inspected at least once a week and after any event that might have damaged or weakened them. Worn or damaged nets must be removed from service.

12.4.0 Rescue After a Fall

Every elevated job site should have an established rescue and retrieval plan. Planning is especially important in remote areas where help is not readily available. Before beginning work, make sure that you know what your employer's rescue plan calls for you to do in the event of a fall. Find out what rescue equipment is available and where it is located. Learn how to use equipment for self-rescue and the rescue of others.

If a fall occurs, any employee hanging from the fall-arrest system must be rescued safely and quickly. Your employer should have previously determined the method of rescue for fall victims, which may include equipment that lets the victim rescue himself or herself, a system of rescue by co-workers, or a way to alert a trained rescue squad.

If a rescue depends on calling for outside help such as the fire department or rescue squad, all the needed phone numbers must be posted in plain view at the work site. In the event a co-worker falls, follow your employer's rescue plan. Call any special rescue service needed. Communicate with the victim, and monitor him or her constantly during the rescue.

13.0.0 ◆ FORKLIFT SAFETY

Forklifts are common on many masonry work sites. They are useful for lifting and moving heavy or awkward loads of materials, supplies, and equipment (*Figure 49*). While extremely useful and relatively easy to operate, these machines can also be very dangerous. They present several risks, including hitting other workers, dropping loads, tipping over, and causing fires and explosions.

Mechanical and hydraulic problems can cause accidents, but the most common cause of forklift

101F49.EPS

Figure 49 ◆ Forklift.

Location is Everything

A worker was placing metal bridge decking onto the stringers of a bridge deck to be welded. After the first decking was placed down on stringers, the employee stepped onto it in order to put down the next decking. The decking was not secured in place and shifted.

Although safety nets were being used under another section of the bridge, they had not been moved forward as the crew moved to another area. The worker fell approximately 80' into the river and was killed.

The Bottom Line: Make sure that all safety equipment is in place before beginning any job.

accidents is human error. That means most forklift accidents can be avoided if the operator and other workers in the area stay alert and use caution and common sense. In fact, research by Liberty Mutual Insurance Company shows that drivers with more than a year of experience operating a forklift are more likely to have an accident than someone with little experience. This is because operators tend to become too comfortable and less attentive after they gain experience on the equipment. The same study showed that the most common type of forklift accident is one in which a pedestrian is hit by the truck.

13.1.0 Before You Operate a Forklift

Before you can begin operating a forklift, you must be trained and certified on that particular piece of equipment. Once you are trained and certified, you must thoroughly inspect your forklift before you begin each shift.

13.1.1 Training and Certification

It is a common misconception that if you can drive a car, truck, or piece of heavy equipment, you can operate a forklift. However, OSHA requires forklift operators to be trained and certified on each piece of equipment before they operate it on the job site. The operator's card only applies to the specific piece of equipment on which they are trained. Powered forklift operators must have the visual, hearing, physical, and mental abilities necessary to safely operate the equipment. Personnel who have not been trained in forklift operation may only operate them for the purpose of training. The training must be conducted under the direct supervision of a qualified trainer.

13.1.2 Pre-Shift Inspection

You must perform a pre-shift inspection before operating a forklift. The more thorough you are when inspecting the forklift, the safer and more productive you will be during your shift. *Figure*

50 shows a sample checklist covering the basic items that need to be checked during a pre-shift inspection.

Your company's checklist and your supervisor can provide you with specific information about what you should check before you begin each shift on a forklift. Your training and the forklift operator's manual will help you understand what to look for when you inspect the forklift. If you find any problems during your pre-shift inspection, notify your supervisor or maintenance manager immediately. The forklift should be locked out and tagged. It cannot be used until all problems are corrected.

13.1.3 General Safety Precautions

Safe operation is the operator's responsibility. Operators must develop safe working habits and be able to recognize hazardous conditions in order to avoid equipment and property damage and to protect themselves and others from death or injury. They must always be aware of unsafe conditions, so they can protect the load and the forklift from damage. They must also understand the operation and function of all controls and instruments before operating any forklift. Operators must read and fully understand the operator's manual for each piece of equipment being used.

The following safety rules are specific to forklift operation:

- Always check the capacity chart mounted on the machine before operating any forklift.
- Never put any part of the body into the mast structure or between the mast and the forklift.
- Never put any part of the body within the reach mechanism.
- Understand the limitations of the forklift.
- Do not permit passengers to ride in the forklift unless a safe place to ride has been provided by the manufacturer.
- Never leave the forklift running unattended.
- Never carry passengers on the forks.

OPERATOR'S DAILY CHECKLIST

Check Each Item Before Start of Each Shift

Date: _____

Check One: Gas/LGP/Diesel Truck ☐ Electric Sit-Down ☐ Electric Stand-Up ☐ Electric Pallet

Truck Serial Number: _____ Operator: _____ Supervisor's OK: _____

Hour Meter Reading: _____

Check each of the following items before the start of each shift. Let your supervisor and/or maintenance department know of any problem. DO NOT OPERATE A FAULTY TRUCK. Your safety is at risk.

After checking, mark each item accordingly. Explain below as necessary.

Check boxes as follows: ☐ OK ☐ NG, needs attention or repair. Circle problem and explain below.

OK	NG	Visual Checks	OK	NG	Visual Checks
		Tires/Wheels: wear, damage, nuts tight			Steering: loose/binding, leaks, operation
		Head/Tail/Working Lights: damage, mounting, operation			Service Brake: linkage loose/binding, stops OK, grab
		Gauges/Instruments: damage, operation			Parking Brake: loose/binding, operational, adjustment
		Operator Restraint: damage, mounting, operation oily, dirty			Seat Brake (if equipped): loose/binding, operational, adjustment
		Warning Decals/Operator's Manual: missing, not readable			Horn: operation
		Data Plate: not readable, missing adjustment			Backup Alarm (if equipped): mounting, operation
		Overhead Guard: bent, cracked, loose, missing			Warning Lights (if equipped): mounting, operation
		Load Back Rest: bent, cracked, loose, missing			Lift/Lower: loose/binding, excessive drift, leaks
		Forks: bent, worn, stops OK			Tilt: loose/binding, excessive drift, "chatters," leaks
		Engine Oil: level, dirty, leaks			Attachments: mounting, damaged operation, leaks
		Hydraulic Oil: level, dirty, leaks			Battery Test (electric trucks only): indicator in green
		Radiator: level, dirty, leaks			Battery: connections loose, charge, electrolyte low while holding full forward tilt
		Fuel: level, leaks			Control Levers: loose/binding, freely return to neutral
		Covers/Sheet Metal: damaged, missing			Directional Controls: loose/binding, find neutral OK
		Brakes: linkage, reservoir fluid level, leaks			
		Engine: runs rough, noisy, leaks			

Explanation of problems marked above: _____

_____

_____

_____

_____

_____

101F50.EPS

Figure 50 ◆ Forklift operator's daily checklist.

Forklift operators must pay special attention to the safety of any pedestrians on the job site. Safeguard pedestrians at all times by observing the following rules:

- Always look in the direction of travel.
- Do not drive the forklift up to anyone standing in front of an object or load.
- Make sure that personnel stand clear of the rear swing area before turning.
- Exercise particular care at cross aisles, doorways, and other locations where pedestrians may step into the travel path.
- Always use a spotter or signal person when moving an elevated load with a telescoping-boom forklift.

13.2.0 Traveling

Traveling refers to driving your forklift both with and without a load. To move either the forklift or a load of materials, you must travel. Sometimes the job requires only short travel distances, such as from a flatbed truck or rail car to a storage or staging area (*Figure 51*). Other times you must travel longer distances on the forklift. For example, you may need to move a load of bricks from a storage area all the way across the work site to a building under construction.

13.2.1 Stay Inside

You are safest when traveling on a forklift if you keep your whole body inside the vehicle. Many experienced drivers get into the unsafe habit of hanging an elbow outside of the truck, sliding a foot off the platform, or resting one hand with the fingers hanging over the edge of the truck. This often results in crushing injuries and amputation. The operator's compartment, along with the use of seat belts, is designed to protect the operator from falling objects, impact from collisions, contact with electrical utilities, and tipping accidents. For example, if you allow your elbow to hang over the edge of the truck, and then accidentally back into a support beam, you could easily crush your arm between the forklift and the beam.

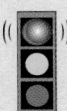 **WARNING!**
Keep all parts of your body inside the forklift cab during operation or travel.

101F51.EPS

Figure 51 ◆ Forklift working in a storage area.

13.2.2 Pedestrians

Always yield the right-of-way to pedestrians. Forklifts are heavy machines that typically require a distance equal to the length of the forklift in order to stop. Pedestrians are usually not aware of this and walk around the site expecting these large and cumbersome machines to be able to stop quickly. Because of this, it is important to look out for any pedestrians.

13.2.3 Passengers

Forklifts are designed to carry one person: the operator. No one else should ever ride in or stand on a forklift. There are a few specially designed and certified attachments that allow forklifts to be converted to personnel lifts. Other than those few situations, no one other than the operator should be on the forklift.

13.2.4 Blind Corners and Intersections

As you approach a blind corner or intersection, always assume that a pedestrian or another piece of equipment is coming the other way. Stop at the intersection. Sound your horn. Then proceed slowly through the intersection or around the corner. Be prepared to stop if necessary.

13.2.5 Keeping the Forks Low

Whether traveling with or without a load, keep the forks as low as possible (*Figure 52*). As a general rule of thumb, the forks should never be higher than 6" from the travel surface while traveling, unless you are moving over an extremely rough surface. The forks are strong and pointed. Ramming into something or someone can cause serious damage or injury. If the forks are low, the chance of critically or fatally injuring someone is greatly reduced.

101F52.EPS

Figure 52 ◆ Forklift with forks in low position.

13.2.6 Horseplay

Driving a forklift is a serious operation that requires maturity and attention. Never drive a forklift toward another person as a joke, particularly if they are in front of a solid object or another piece of equipment. Doing so could easily lead to a very serious crushing accident. Accidents caused by horseplay are the most avoidable problem on the job site. Working with heavy equipment is dangerous, and your behavior on the job should reflect that. Your safety and that of your co-workers depends on it.

13.2.7 Travel Surface

Whenever possible, take the smoothest and driest route when traveling. Rough or bumpy surfaces can cause a lot of bouncing, which may destabilize a forklift and make the forklift or its load tip. Wet or slippery surfaces can cause the tires to lose traction, resulting in loss of control of the forklift.

13.3.0 Handling Loads

A forklift's main use is to transport large, heavy, or awkward loads. If not handled correctly, loads can fall from the forks, obstruct the operator's view, or cause the forklift to tip. The most important factor to consider when using any forklift is its capacity. Each forklift is designed with an intended capacity, and this capacity must never be exceeded. Exceeding the capacity jeopardizes not only the machine and the load, but also the safety of everyone on or near the forklift. Every manufacturer supplies a capacity chart for each forklift. The operator must be aware of the capacity of the machine before being allowed to operate the forklift.

13.3.1 Picking Up Loads

Some forklifts are equipped with a sideshift device that allows the operator to shift the load sideways several inches in either direction with respect to the mast. A sideshift device enables more precise placing of loads, but it also changes the forklift's center of gravity and must be used with caution. If the forklift being used is equipped with a sideshift device, be sure to return the fork carriage to the center position before attempting to pick up a load.

13.3.2 Traveling with Loads

Always travel at a safe rate of speed with a load. Never travel with a raised load. Keep the load as low as possible, and be sure the mast is tilted rearward to cradle the load.

As you travel, stay alert, and pay attention. Watch the load and the conditions ahead of you, and alert others to your presence. Avoid sudden stops and abrupt changes in direction. Be careful when downshifting because sudden deceleration can cause the load to shift or topple. Watch the machine's rear clearance when turning.

If you are traveling with a telescoping-boom forklift, be sure the boom is fully retracted. If you have to drive on a slope, keep the load as low as possible. Do not drive across steep slopes. If you have to turn on an incline, make the turn wide and slow.

13.3.3 Traveling with Long Loads

Traveling with long loads presents special hazards, particularly if the load is flexible and subject to damage. Traveling multiplies the effect of bumps over the length of the load. A stiffener may be added to the load to give it extra rigidity.

To prevent slippage, secure long loads to the forks. A field-fabricated cradle may be used to support the load. While this is an effective method, it requires that the load be jacked up.

The forklift may be used to carry pieces of rigging equipment. This method requires the use of slings and a spreader bar.

In some cases, long loads may be snaked through openings that are narrower than the load itself. This is done by approaching the opening at an angle and carefully maneuvering one end of the load through the opening first. Avoid making quick turns because abrupt maneuvers will cause the load to shift.

13.3.4 Placing Loads

Position the forklift at the landing point so that the load can be placed where you want it. Be sure everyone is clear of the load. The area under the load must be clear of obstructions and able to support the weight of the load. If you cannot see the placement, use a signaler to guide you.

With the forklift in the unloading position, lower the load and tilt the forks to the horizontal position. When the load has been placed and the forks are clear from the underside of the load, back away carefully to disengage the forks.

13.3.5 Placing Elevated Loads

When placing elevated loads, you must be especially careful. Some forklifts are equipped with a leveling device that allows the operator to rotate the fork carriage to keep the load level during travel. When placing elevated loads, it is extremely important to level the machine before lifting the load.

One of the biggest potential safety hazards during elevated load placement is poor visibility. There may be workers in the immediate area who cannot be seen. The landing point itself may not be visible. Your depth perception decreases as the height of the lift increases. To be safe, use a signal person to help you position the load.

Use tag lines to tie off long loads to the mast of the forklift. Drive the forklift as closely as possible to the landing point with the load kept low. Set the parking brake, and then raise the load slowly and carefully while maintaining a slight rearward tilt to keep the load cradled. Under no circumstances should the load be tilted forward until the load is over the landing point and ready to be set down.

If the forks start to move, sway, or lean, stop immediately but not abruptly. Lower the load slowly. Reposition it, or break it down into smaller components if necessary. If ground conditions are poor at the unloading site, you may need to reinforce the ground with planks to provide greater stability. As the load approaches the landing point, slow the lift speed to a minimum. Continue lifting until the load is slightly higher than the landing point.

13.3.6 Tipping

There are three main causes for a forklift tipping:

- The load is too heavy.
- The load is placed too far forward on the forks.
- The operator is not driving safely.

To avoid tipping, you need to understand what the center of gravity is and how it applies to forklifts. The center of gravity is the point around which all of an object's weight is evenly distributed (*Figure 53*). Your forklift has a center of gravity, and the load you're moving will have its own center of gravity. When the forklift picks up the load, the center of gravity shifts to the combined center of gravity (*Figure 54*).

The forklift will tip if the center of gravity moves too far forward, backward, right, or left. Putting too heavy a load on the forklift or placing the load too far forward on the forks will cause the combined center of gravity to be too far forward, causing the forklift to tip forward. Turning too sharply or quickly can cause the forklift to sway or swing to the left or right, causing the combined center of gravity to veer far enough off center to tip the forklift over. These types of accidents can easily result in the operator or a bystander being crushed by the forklift.

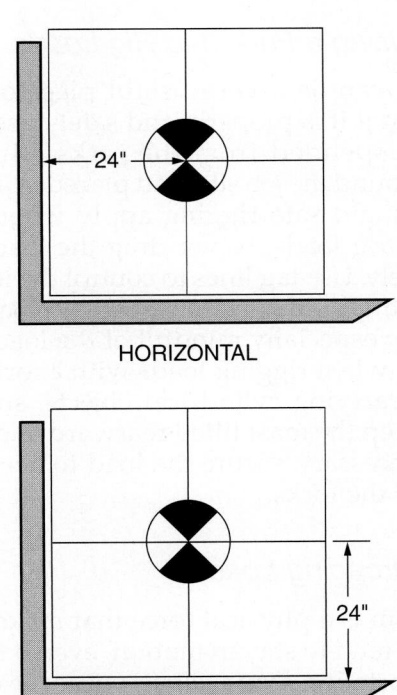

Figure 53 ◆ Center of gravity.

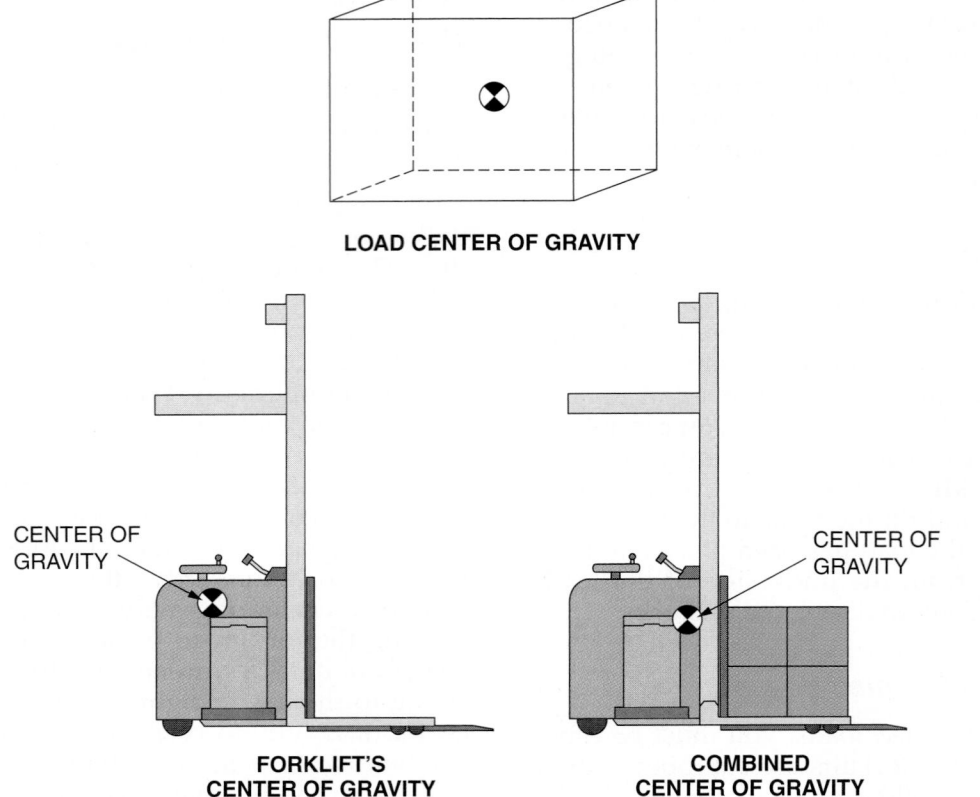

LOAD CENTER OF GRAVITY

CENTER OF
GRAVITY

CENTER OF
GRAVITY

**FORKLIFT'S
CENTER OF GRAVITY**

**COMBINED
CENTER OF GRAVITY**

101F54.EPS

Figure 54 ◆ Combined center of gravity.

13.3.7 Using a Forklift to Rig Loads

A forklift can be a very useful piece of rigging equipment if it is properly and safely used. Loads can be suspended from the forks with slings, moved around the job site, and placed. All the rules of careful and safe rigging apply when using a forklift to rig loads. Never drag the load or let it swing freely. Use tag lines to control the load.

Never attempt to rig an unstable load with a forklift. Be especially mindful of the load's center of gravity when rigging loads with a forklift.

When carrying cylindrical objects, such as oil drums, keep the mast tilted rearward to cradle the load. If necessary, secure the load to keep it from rolling off the forks.

13.3.8 Dropping Loads

Momentum is a physical force that makes objects in motion tend to stay in motion, even if the mode of transportation stops. Have you ever carried a meal on a cafeteria tray? In your experience, what happened to the items on your tray if you had to stop suddenly or if you turned too quickly? The items on your tray probably started to topple or slide. They might have fallen right off the tray, or

they may have shifted the load, so that the tray tipped, spilling everything.

The same principle applies to the load on your forklift. If you turn or stop too quickly, the load will keep going in the direction you were originally headed, causing it to slide off the forks. At the very least, this may be inconvenient, as you have to stop everything to restack your load. It can be expensive if you drop fragile materials or equipment. It can be deadly if the load falls on a co-worker or causes your truck to tip over. Avoid sudden maneuvers when operating a forklift.

13.3.9 Obstructing the View

Forklift operators commonly try to move as much as possible in the fewest number of trips in order to save time. Sometimes this causes them to stack loads too high. Doing so may cause them to exceed the safe weight limit of the forklift, and it may also block the operator's view. To operate a forklift safely, the operator must be able to clearly see what is in front of and behind the forklift without leaning outside of the operator's compartment. Leaning outside the operator's compartment can result in serious injuries.

Deadly Overload

A forklift operator was carrying a load that was stacked too high. The load obstructed his view. To make up for it, he stuck his head out the side of the operator's compartment to see around the load. Unfortunately, as he was preparing to drive forward, another forklift was backing up and sideswiped his machine, decapitating him.

The Bottom Line: Never carry a load that obstructs your view. Always keep all of your body parts inside the operator's compartment.

Source: The Occupational Safety and Health Administration (OSHA)

13.4.0 Working on Ramps and Docks

Ramps and docks have special working conditions with specific safety requirements. Ramps allow wheeled vehicles to move easily from one level to the next. However, going up and down an angled surface has an impact on the forklift's center of gravity, making it easier for the forklift to tip. Operators can be crushed by or thrown from the forklift if this happens. Docks elevate the driving surface to a convenient height for the loading and unloading of over-the-road (OTR) trucks and rail cars. However, they also create a risk that the forklift might fall off the edge of the dock.

13.4.1 Ramps

The ramp's grade increases the tipping hazard for a forklift. Follow these rules when working on ramps:

- Keep the load as low as possible.
- Always keep the load uphill.
- When working on any graded surface, make sure your load is pushed as far back onto the forks as possible.

- If possible, tilt the forks so that the load is level with the graded surface.
- Do not turn or make quick starts or stops on a ramp.

13.4.2 Docks

It is possible to drive a forklift off the edge of a dock if you are not careful and attentive. Not only can this damage the equipment, it can also kill or injure the operator and anyone near the area of the falling load.

Forklift operators must be aware of the edge of the dock. The edge is normally painted a bright color, such as yellow. Besides wheel chocks for the truck, devices called dock plates may be used to smoothly bridge the gap between a dock and the floor of a truck.

13.5.0 Fire and Explosion Hazards

Forklifts use fuels such as gasoline, liquid propane (LP) gas, and diesel fuel. All of these fuels are capable of causing a fire or explosion if not handled properly. In addition, LP gas is stored in cylinders under pressure, creating an explosion hazard if the

Chock the Wheels

As a forklift passes from a dock to an over-the-road truck and back, the force can slowly move the truck forward if the wheels are not properly chocked. In one case, a forklift operator incorrectly assumed that the truck driver had chocked the wheels of the truck. Every time the forklift passed from the dock onto the truck, the truck moved away from the dock. Because the operator was concentrating on his job, he did not notice that the gap between the dock and the truck was growing. After several trips in and out of the truck, the gap was wide enough for the forklift's front wheels to fall into the gap. This caused the forklift to tip forward into the truck and tip the truck trailer backward, crushing the operator between the truck and the forklift.

The Bottom Line: Never assume that the wheels are chocked properly. Always chock the wheels yourself, or check to make sure it was done correctly.

Source: The Occupational Safety and Health Administration (OSHA)

cylinder is exposed to extreme heat or fire. It is very important to keep these fuels away from any source of fire and to keep the areas in which the forklift is used free of any flammable materials. There are specific precautions that must be taken to avoid the possibility of a fire or explosion.

The best way to prevent a fire is to make sure that the three elements needed for fire (fuel, heat, and oxygen) are never present in the same place at the same time. Here are some basic safety guidelines for fire prevention:

- Always work in a well-ventilated area, especially when you are using flammable materials.
- Never smoke or light matches when you are working with flammable materials.
- Keep oily rags in approved, self-closing metal containers.
- Store combustible materials only in approved containers.
- Know where to find fire extinguishers, what kind of extinguisher to use for different kinds of fires, and how to use the extinguishers.
- Keep open fuel containers away from any sources of sparks, fire, or extreme heat.
- Make sure all extinguishers are fully charged. Never remove the tag from an extinguisher; it shows the date the extinguisher was last serviced and inspected.
- Don't fill a gasoline or diesel fuel container while it is resting on a truck bed liner or other ungrounded surface. The flow of fuel creates static electricity that can ignite the fuel if the container is not grounded.
- Always use approved containers, such as safety cans, for flammable liquids.

13.5.1 Flammable and Combustible Liquids

Liquids can be flammable or combustible. Flammable liquids have a flash point below 100°F. Combustible liquids have a flash point at or above 100°F. Fires can be prevented by doing the following things:

- *Removing the fuel* – Liquid does not burn. What burns are the gases (vapors) given off as the liquid evaporates. Keeping the liquid in an approved, sealed container prevents evaporation. If there is no evaporation, there is no fuel to burn.
- *Removing the heat* – If the liquid is stored or used away from a heat source, it cannot ignite.
- *Removing the oxygen* – The vapor from a liquid will not burn if oxygen is not present. Keeping safety containers tightly sealed prevents oxygen from coming into contact with the fuel.

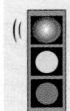 **WARNING!**
Never transfer flammable or combustible liquids between containers without proper training. Doing so can result in fire or explosion.

13.5.2 Flammable Gases

Flammable gases used on construction sites include acetylene, hydrogen, ethane, and LP gas. To save space, these gases are compressed so that a large amount can be stored in a small cylinder or bottle. As long as the gas is kept in the cylinder, oxygen cannot get to it and start a fire. The cylinders must be handled carefully and stored away from sources of heat.

Oxygen is also classified as a flammable gas. If it is allowed to escape and mix with another flammable gas, the resulting mixture can explode.

13.5.3 Fire Fighting

You are not expected to be an expert firefighter, but you may have to deal with a fire to protect your safety and the safety of others. You need to know the location of fire-fighting equipment on your job site. You also need to know which equipment to use on different types of fires. However, only qualified personnel are authorized to fight fires.

Most companies tell new employees where fire extinguishers are kept. If you have not been told, be sure to ask. Also ask how to report fires. The telephone number of the nearest fire department should be clearly posted in your work area. If your company has a company fire brigade, learn how to contact them. Learn your company's fire-safety procedures.

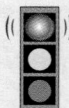 **WARNING!**
Before working with any flammable materials, be sure you know the closest escape route and the appropriate measures to take in case of emergency.

13.6.0 Pedestrian Safety

You may not be a forklift operator, but you will probably work on a site with forklifts. Here are some guidelines for working safely around forklifts.

Remember that it may be difficult for the operator to see you. The operator may be concentrating on the load and may not be paying attention to pedestrians. Therefore, always assume that the forklift operator doesn't see you. Remember, the most common type of forklift accident is hitting a pedestrian.

Forklifts are heavy and usually carry large loads. This results in a large amount of momentum, which means that it will probably take a distance equal to the length of the forklift to stop it. Never risk your safety on a forklift's ability to stop in time. Make sure that you're never positioned between a forklift and an immovable object.

Sometimes forklifts carry heavy loads high overhead. Never stand under the raised forks of a forklift. Objects may fall from the forks or, if there is a sudden loss of hydraulic pressure, the forks and load may drop suddenly, crushing any people or objects beneath them.

When working in a storage room or warehouse with racking, never work in the aisle on the other side of a racking unit where a forklift is working. Occasionally, the operator may push a load into the rack, causing it to fall off the other side. If you are unlucky enough to be on the other side, there is a good chance that you will be struck or crushed by a falling object.

Never hitch a ride on a forklift. Forklifts are designed for one operator and a load. Riding on the forks creates a high risk of falling and being run over by the forklift. Riding on the tractor of the forklift is also not permitted. It is easy to fall and/or be crushed between the forklift and other objects.

Summary

Masonry is a craft that has existed for thousands of years. Over the centuries, hand-formed dried mud brick has been replaced by molded mud brick, then by fired, molded clay brick. Today, fired clay brick is available in a variety of shapes, sizes, colors, and textures.

Modern clay products are categorized as solid and hollow brick, structural and nonstructural tile, and made-to-order architectural terra-cotta. Clay has been joined by concrete as a modern masonry material. Concrete masonry units outnumber clay units in their variety. Concrete products are block, cement brick, and special-purpose block. Stone, the oldest recovered building material, is still laid by masons. Mortars and grouts have evolved from mud to special-purpose, high-strength cements.

Science and engineering have brought masonry into the modern age. With modern construction techniques, masonry is now used in high-rise buildings as well as residential and commercial projects. This allows for widespread use of masonry construction throughout North America.

Masonry as a career dates back to the ancient kingdoms of the Middle East. Masons were among the first craftworkers to organize into guilds. The guilds protected the secrets of the craft and maintained standards for the work. The legacy of these guilds includes the apprenticeship program available today.

Masonry offers advancement through the recognized career steps of apprentice, mason, layout person, supervisor, superintendent, and masonry contractor. Your success as a mason requires the willingness to keep on learning.

Safety is a particularly important issue for you as a masonry worker. You will often work at heights where falls and falling objects are major hazards. You must also be careful when working with mortar and concrete products because they can cause skin irritations and lung ailments if not handled properly. You will often work on sites where heavy equipment is used. For that reason, you must be especially vigilant. Forklift safety is of particular concern because forklifts are commonly used to transport and lift masonry materials at a work site.

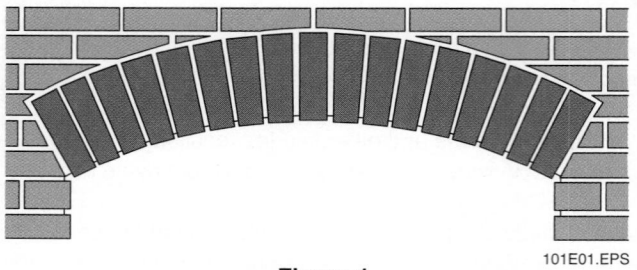

Figure 1

101E01.EPS

1. The arch shown in *Figure 1* is a _____ arch.
   a. Roman
   b. radial
   c. jack
   d. Babylonian

2. Structural clay products include _____.
   a. brick, block, and tile
   b. brick, tile, and terra-cotta
   c. tile, terra-cotta, and pre-faced units
   d. block, grout, and tile

3. A cube of bricks contains _____ standard bricks.
   a. 90
   b. 300
   c. 400
   d. 500

4. Solid masonry units have voids occupying _____.
   a. less than 25 percent of their surface
   b. more than 25 percent of their mass
   c. less than 10 percent of their mass
   d. more than 75 percent of their surface

5. Rubble stone is _____.
   a. called slump brick
   b. made of concrete
   c. cut with smooth bedding surfaces
   d. irregular in size and shape

6. Modern masonry mortar is made of _____.
   a. cement, lime, and sand
   b. cement, sand, admixtures, and water
   c. sand, lime, ceramic, and water
   d. masonry cement, sand, and water

7. Veneer walls are _____.
   a. built 6½" away from the weight-bearing wall
   b. commonly used in industrial construction
   c. not loadbearing
   d. not designed for appearance

8. The National Registry for construction craft training is maintained by _____.
   a. the U.S. Department of Labor
   b. local apprenticeship councils
   c. NCCER
   d. state labor departments

9. Using mortar with the consistency of mud will *not* enable you to create strong bonds.
   a. True
   b. False

10. The process of spreading a bed of mortar is called ____.
    a. snapping
    b. stringing
    c. leading
    d. furrowing

11. The process of carving a trough in a bed of mortar is called _____.
    a. furrowing
    b. buttering
    c. tailing
    d. tuckpointing

12. Portland cement is hygroscopic, which means that it will absorb moisture from your skin, resulting in skin irritation.
    a. True
    b. False

13. Fall protection is required when working at heights _____ or higher.
    a. 6'
    b. 8'
    c. 10'
    d. 12'

14. Short, flexible lines with connectors on each end are called _____.
    a. anchorages
    b. deceleration devices
    c. lanyards
    d. lifelines

15. The most common type of forklift accident is the _____.
    a. forklift tipping over
    b. load falling
    c. forklift falling off a dock
    d. forklift hitting a pedestrian

Arnold Shueck, Field Superintendent

Nester Brothers Masonry
Pennsburg, PA

Arnold Shueck grew up on a farm and, in doing so, learned the value of hard work that led to his success in the masonry trade. He decided early on that he wanted to work outdoors and work with his hands. This led him to enter the high school masonry program that was the beginning of a long and rewarding career.

How did you get started in the masonry trade?
I learned early on that I wanted to work outdoors. I had heard that masons make good money, so I enrolled in the masonry program at my high school. The training I received in high school helped me land a job with Nester Brothers. I started out as a laborer apprentice, but my training helped me move up pretty quickly to laying brick and block. I was promoted to foreman, and was eventually moved up to field superintendent.

What does your present job entail?
As a field superintendent I go from project to project checking on the work and helping to solve problems. It's my job to make sure all the projects flow smoothly. I estimate new projects, plan the projects, and order the materials. Nester Brothers works mostly on commercial masonry projects, so I get to work on some very challenging jobs.

What do you think it takes to become a success?
You have to be willing to work hard and stick with it. Growing up on a farm taught me about the value of hard work and gave me an appreciation for working with my hands.

What do you like most about your job?
First, I like the company I work for. I've been with the same company for my entire career and they have been very good to me. Second, I like working outdoors. I never saw myself in an indoor job. I get a lot of satisfaction from seeing the work I've accomplished over the years as I drive around the area.

What would you say to someone just entering the trade?
Be sure this is what you want to do. If it is, then give it everything you have. Don't be afraid to take advice from experienced people. Take advantage of every training and educational opportunity that comes your way. Materials and techniques are constantly changing. It's important to keep up with the industry.

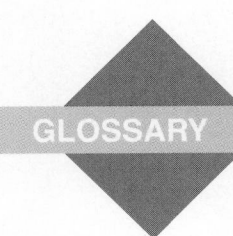

Trade Terms Introduced in This Module

Admixture: A chemical or mineral other than water, cement, or aggregate added to mortar immediately before or during mixing to change its setting time or curing time, to reduce water, or to change the overall properties of the mortar.

Aggregate: Materials such as crushed stone or gravel used as a filler in concrete and concrete block.

Adobe: Sun-dried, molded clay brick.

Alkaline: Bitter, slippery, or caustic.

American Society for Testing and Materials (ASTM) International: The publisher of masonry standards.

Ashlar: A squared or rectangular cut stone masonry unit; or, a flat-faced surface having sawed or dressed bed and joint surfaces.

Butter: Apply mortar to the end of a masonry unit.

Capital: The top part of an architectural column.

Concrete masonry unit (CMU): A hollow or solid block made from portland cement and aggregates.

Cornice: The horizontal projection crowning the wall of a building.

Course: A row or horizontal layer of masonry units.

Cube: A strapped bundle of approximately 500 standard bricks, or 90 standard blocks, usually palletized. The number of units in a cube will vary according to the manufacturer.

Facing: That part of a masonry unit or wall that shows after construction; the finished side of a masonry unit.

Footing: The base for a masonry unit wall, or concrete foundation, that distributes the weight of the structural member resting on it.

Furrowing: Making an indentation with a trowel point along the center of the mortar bed joint.

Grout: A mixture of portland cement, lime, and water, with or without fine aggregate, with a high enough water content that it can be poured into spaces between masonry units and voids in a wall.

Head joint: A vertical joint between two masonry units.

Hygroscopic: The tendency of a substance to absorb moisture.

Joints: The area between each brick or block that is filled with mortar.

Mason: A person who assembles masonry units by hand, using mortar, dry stacking, or mechanical connectors.

Masonry unit: Any building block made of brick, cement, ashlar, clay, adobe, rubble, glass, tile, or any other material, that can be assembled into a structural unit.

Mortar: A mixture of portland cement, lime, fine aggregate, and water, plastic or stiff enough to hold its shape between masonry units.

Nonstructural: Not bearing weight other than its own.

Parapet: A low wall or railing.

Pilaster: A square or rectangular pillar projecting from a wall.

Spread: A row of mortar placed into a bed joint.

Stringing: Spreading mortar with a trowel on a wall or footing for a bed joint.

Structural: Bearing weight in addition to its own.

Tuckpointing: Filling fresh mortar into cutout or defective joints in masonry.

Weephole: A small opening in mortar joints or faces to allow the escape of moisture.

Wythe: A continuous section of masonry wall, one masonry unit in thickness, or that part of a wall which is one masonry unit in thickness.

Additional Resources

This module is intended to be a thorough resource for task training. The following reference works are suggested for further study. These are optional materials for continued education rather than for task training.

Building Block Walls—A Basic Guide, 1988. Herndon, VA: National Concrete Masonry Association.

Bricklaying: Brick and Block Masonry. Reston, VA: Brick Industry Association.

Concrete Masonry Handbook. Skokie, IL: Portland Cement Association.

NCCER makes every effort to keep these textbooks up-to-date and free of technical errors. We appreciate your help in this process. If you have an idea for improving this textbook, or if you find an error, a typographical mistake, or an inaccuracy in NCCER's Contren® textbooks, please write us, using this form or a photocopy. Be sure to include the exact module number, page number, a detailed description, and the correction, if applicable. Your input will be brought to the attention of the Technical Review Committee. Thank you for your assistance.

Instructors – If you found that additional materials were necessary in order to teach this module effectively, please let us know so that we may include them in the Equipment/Materials list in the Annotated Instructor's Guide.

Write: Product Development and Revision
National Center for Construction Education and Research
3600 NW 43rd St, Bldg G, Gainesville, FL 32606

Fax: 352-334-0932

E-mail: curriculum@nccer.org

Craft _____ Module Name _____

Copyright Date _____ Module Number _____ Page Number(s) _____

Description _____

(Optional) Correction _____

(Optional) Your Name and Address _____

Masonry Units and Installation Techniques

COURSE MAP

This course map shows all of the modules in the *Construction Technology* curriculum. The suggested training order begins at the bottom and proceeds up. Skill levels increase as you advance on the course map. The local Training Program Sponsor may adjust the training order.

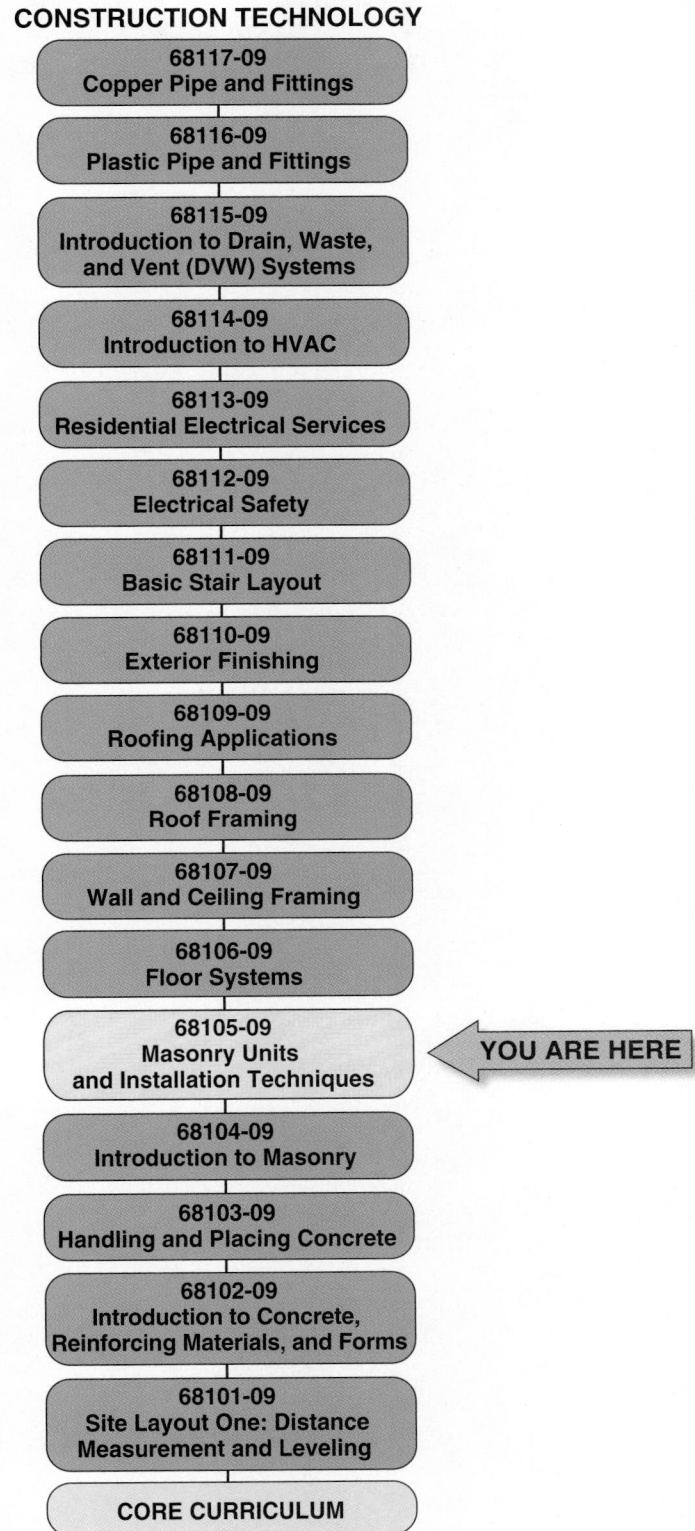

CONSTRUCTION TECHNOLOGY

68117-09
Copper Pipe and Fittings

68116-09
Plastic Pipe and Fittings

68115-09
Introduction to Drain, Waste, and Vent (DVW) Systems

68114-09
Introduction to HVAC

68113-09
Residential Electrical Services

68112-09
Electrical Safety

68111-09
Basic Stair Layout

68110-09
Exterior Finishing

68109-09
Roofing Applications

68108-09
Roof Framing

68107-09
Wall and Ceiling Framing

68106-09
Floor Systems

68105-09
Masonry Units and Installation Techniques

YOU ARE HERE

68104-09
Introduction to Masonry

68103-09
Handling and Placing Concrete

68102-09
Introduction to Concrete, Reinforcing Materials, and Forms

68101-09
Site Layout One: Distance Measurement and Leveling

CORE CURRICULUM

105CMAP.EPS

1.0.0 INTRODUCTION 5.1

2.0.0 CONCRETE MASONRY MATERIALS 5.2

2.1.0 ASTM Specifications 5.2

2.1.1 Compressive Strength 5.2

2.1.2 Moisture Absorption and Content 5.2

2.2.0 Contraction and Expansion Joints 5.2

2.3.0 Concrete Block Characteristics 5.3

2.4.0 Concrete Brick 5.8

2.5.0 Other Concrete Units 5.8

2.5.1 Pre-faced Units 5.9

2.5.2 Calcium Silicate Units .5.9

2.5.3 Catch Basins 5.10

3.0.0 CLAY AND OTHER MASONRY MATERIALS 5.10

3.1.0 Clay Masonry Units 5.10

3.2.0 Stone 5.11

3.3.0 Other Masonry Materials 5.12

3.3.1 Metal Ties 5.13

3.3.2 Veneer Ties 5.13

3.3.3 Reinforcement Bars 5.13

3.3.4 Joint Reinforcement Ties 5.13

3.3.5 Flashing 5.13

3.3.6 Joint Fillers 5.14

3.3.7 Anchors 5.14

4.0.0 SETTING UP AND LAYING OUT 5.14

4.1.0 Setting Up 5.14

4.2.0 Job Layout 5.16

4.2.1 Planning 5.16

4.2.2 Locating 5.17

4.2.3 Dry Bonding 5.18

5.0.0 BLOCK HEAD JOINTS 5.18

5.1.0 Buttering Block 5.18

5.2.0 Block Bed Joints 5.19

5.3.0 General Rules 5.20

6.0.0 BONDING MASONRY UNITS 5.20

6.1.0 Mechanical or Mortar Bond .5.20

6.2.0 Pattern Bond 5.21

6.3.0 Structural Bond and Structural Pattern Bond 5.22

6.4.0 Block Bond Patterns 5.23

7.0.0 CUTTING MASONRY UNITS 5.25

7.1.0 Brick Cuts 5.25

7.2.0 Block Cuts 5.25

7.3.0   Cutting with Hand Tools 5.26

7.3.1   Cutting with Chisels and Hammers 5.26

7.3.2   Cutting with Masonry Hammers 5.27

7.3.3   Cutting with Trowels 5.27

7.4.0   Cutting with Saws and Splitters 5.28

7.4.1   Saws and Splitters 5.28

7.4.2   Units and Cuts 5.29

8.0.0   LAYING MASONRY UNITS 5.29

8.1.0   Laying Brick in Place 5.29

8.1.1   Placing Brick 5.29

8.1.2   Checking the Height 5.30

8.1.3   Checking Level 5.30

8.1.4   Checking Plumb 5.30

8.1.5   Checking Straightness 5.31

8.1.6   Laying the Closure Unit 5.31

8.2.0   Placing Block 5.32

8.3.0   Laying to the Line 5.33

8.3.1   Setting Up the Line Using Corner Poles 5.33

8.3.2   Setting Up the Line Using Line Blocks and Stretchers 5.33

8.3.3   Setting Up the Line Using Line Pins 5.34

8.3.4   Setting Up the Line Using Line Trigs 5.34

8.3.5   Laying Brick to the Line 5.34

8.3.6   Laying Block to the Line 5.36

8.4.0   Building Corners and Leads 5.36

8.4.1   Rackback Leads 5.37

8.4.2   Brick Rackback Corners 5.37

8.4.3   Block Rackback Corners 5.39

9.0.0   MORTAR JOINTS 5.40

9.1.0   Joint Finishes 5.40

9.2.0   Striking the Joints 5.41

9.2.1   Testing the Mortar 5.41

9.2.2   Striking 5.41

9.2.3   Cleaning Up Excess Mortar 5.42

10.0.0 PATCHING MORTAR 5.42

10.1.0   Pointing 5.42

10.2.0   Tuckpointing 5.43

11.0.0 CLEANING MASONRY UNITS 5.44

11.1.0   Clean Masonry Checklist 5.44

11.2.0   Bucket Cleaning for Brick 5.44

11.3.0   Cleaning Block 5.46

SUMMARY 5.46

REVIEW QUESTIONS 5.46

PROFILE IN SUCCESS 5.48

GLOSSARY 5.49

REFERENCES 5.50

Figures

Figure 1   Locations of control joints 5.3
Figure 2   Block wall with reinforcing rods 5.3
Figure 3   Vertical control joint with plastic filler 5.4
Figure 4   Parts of blocks 5.4
Figure 5   Common concrete block shapes 5.6–5.7
Figure 6   An application of loadbearing block 5.7
Figure 7   Concrete brick 5.8
Figure 8   A wall made from slump brick 5.9
Figure 9   Concrete pre-faced units 5.9
Figure 10   Calcium silicate units 5.9
Figure 11   Manhole and vault units 5.11
Figure 12   Brick positions 5.11
Figure 13   Imitation stone made of concrete 5.12
Figure 14   Metal ties 5.13
Figure 15   Veneer ties 5.13
Figure 16   Joint reinforcement ties 5.13
Figure 17   Flashing applications 5.13
Figure 18   Joint fillers 5.14
Figure 19   Shaped anchors 5.15
Figure 20   Door and window openings 5.16
Figure 21   Foundation plan 5.17
Figure 22   Example of a dry bond 5.18
Figure 23   Block corner layouts 5.19
Figure 24   Placing block 5.19
Figure 25   Buttering block 5.20
Figure 26   Stack bond 5.21
Figure 27   Brick positions in walls 5.22
Figure 28   Running bonds 5.22
Figure 29   English and Flemish bonds 5.23
Figure 30   Common bond 5.23
Figure 31   Dutch bond 5.23

Figure 32   Garden wall bonds 5.24
Figure 33   Block bonds 5.24
Figure 34   Common brick cuts 5.25
Figure 35   Horizontal and vertical face cuts 5.25
Figure 36   End and web block cuts 5.26
Figure 37   Bond beam cut 5.26
Figure 38   Cutting with a hammer and chisel 5.27
Figure 39   Cutting with a brick hammer 5.27
Figure 40   Portable masonry saw 5.28
Figure 41   Brick splitter 5.28
Figure 42   Proper hand position 5.29
Figure 43   Mason's rules 5.30
Figure 44   Leveling the course 5.30
Figure 45   Checking plumb 5.31
Figure 46   Plumb and out-of-plumb bricks 5.31
Figure 47   Checking for straightness 5.31
Figure 48   Placing a closure unit 5.31
Figure 49   Placing a block 5.32
Figure 50   Line block 5.33
Figure 51   Using a line stretcher 5.34
Figure 52   Using a line pin 5.35
Figure 53   Using a line trig 5.35
Figure 54   Laying brick to the line 5.36
Figure 55   Building corner leads 5.37
Figure 56   Rackback lead 5.37
Figure 57   Checking alignment 5.37
Figure 58   Tailing the diagonal 5.38
Figure 59   Leveling the diagonal 5.38
Figure 60   Mortar joint finishes 5.40
Figure 61   Extruded or weeping joint 5.41
Figure 62   Using a jointer 5.41
Figure 63   Using the convex sled 5.42
Figure 64   Skate raker 5.42
Figure 65   Trimming and dressing mortar burrs 5.42
Figure 66   Preparation of joints for tuckpointing 5.43
Figure 67   Tuckpointing 5.43

Table

Table 1   Cleaning Guide for New Masonry 5.45

Masonry Units and Installation Techniques

Objectives

When you have completed this module, you will be able to do the following:

1. Describe the most common types of masonry units.
2. Describe and demonstrate how to set up a wall.
3. Lay a dry bond.
4. Spread and furrow a bed joint, and butter masonry units.
5. Describe the different types of masonry bonds.
6. Cut brick and block accurately.
7. Lay masonry units in a true course.

Recommended Prerequisites

Before you begin this module, it is recommended that you successfully complete the following: *Core Curriculum; Construction Technology,* Modules 68101-06 through 68104-06.

Required Trainee Materials

1. Pencil and paper
2. Appropriate personal protective equipment

1.0.0 ◆ INTRODUCTION

This module contains detailed information on masonry materials. It also provides instructions for building a single-wythe masonry wall. This module gives you details about the following topics:

- Cement, clay, and stone masonry units
- Setting up and laying out a wall
- Spreading mortar
- Bonding masonry units
- Cutting masonry units

- Laying masonry units
- Mortar and other joints
- Patching, pointing, and tuckpointing
- Cleaning masonry units

Safety is a continuous effort. Being a mason calls for following safe work practices and procedures. This includes performing the following work activities:

- Inspecting tools and equipment before use
- Using tools and equipment properly
- Keeping tools and equipment clean and properly maintained
- Keeping your hands out of the mortar
- Using the right tools for the job
- Assembling and using scaffolding and foot boards properly
- Using caution and common sense when working on elevated surfaces

Accidents also happen when tools and equipment are left in the way of other workers. Store tools safely, where other people cannot trip over them and where the tools cannot get damaged. Clean, well-kept tools make for safe work as well as good work.

Do not drop or temporarily store tools or masonry units in pathways or around other workers. Stack masonry units neatly so the stack will be less likely to topple over. Stack materials by reversing the direction of the units on every other layer, so they will be less prone to tip. Keep the pile neat and vertical to avoid snagging clothes. As a rule, do not stack masonry units higher than your chest. A stack that is too high is more likely to tip over.

Various materials used to build masonry structures have properties that can be harmful. Mortar and grout, for example, can be very caustic. This means that continued contact with the skin can cause burns and irritation. It is important to protect your skin and eyes when mixing and working with these materials.

Cutting masonry units can be a dangerous job. When sawing or chiseling a masonry unit, chips can fly off and hit your eyes.

The process of using a saw or chisel is dangerous. Always keep your hands away from the blade. Do not operate power saws when you are tired, sick, or otherwise unable to give the process your full attention and effort.

WARNING!
Always wear safety glasses or goggles and respiratory protection when cutting masonry units.

2.0.0 ◆ CONCRETE MASONRY MATERIALS

Most masonry materials are made of clay, concrete, or stone. The masonry unit most commonly used in the United States is the concrete block. The term CMU stands for concrete masonry unit. CMUs are classified into six types:

- Loadbearing concrete block
- Nonbearing concrete block
- Concrete brick
- Calcium silicate units
- Pre-faced or prefinished concrete facing units
- Concrete units for manholes and catch basins

2.1.0 ASTM Specifications

Each of the six types of CMUs has its own ASTM (American Society for Testing and Materials) International standards for performance characteristics. The standards describe the expected performance of the CMU in compressive strength, water absorption, loadbearing, and other characteristics. CMUs are valuable for their fire resistance, sound absorption, and insulation value.

2.1.1 Compressive Strength

The compressive strength of a CMU measures how much weight it can support without collapsing. These figures are set according to ASTM test results. The tests are performed on a specific shape, size, and weight of CMU. The CMU tested then becomes the standard for that particular compressive strength. Compressive strength is measured in pounds of pressure per square inch (psi). This is the

weight that the unit, and the structure made of the units, can support. A structure supports less imposed weight than the sum of its unit strength because the structure has to support itself as well.

The quality of a mason's work is an important part of how a masonry structure performs and whether it meets its compressive strength specification. Tests have shown that the compressive strength of a loaded wall is about 42 percent of the compressive strength of a single CMU when the mason uses a face shell mortar bedding. When the mason uses a full mortar bedding, the compressive strength increases to about 53 percent. The engineer factors these components into the equation for picking the CMU. The factoring is contained in the job specifications. This is another reason that specifications are important.

2.1.2 Moisture Absorption and Content

Moisture in the CMU has an effect on shrinking and cracking in the finished structure. Generally, the lower the moisture, the less likely the units are to shrink after they are set. Acceptable moisture content and absorption rates are set by ASTM standards and local codes. Manufacturers specify that their units meet ASTM or other standards. Unit tests are made to make sure they do. For the mason, this means keeping CMUs dry. Never wet CMUs immediately before or during the time they are to be laid. Stockpile them on planks or pallets off the ground. Use plastic or tarpaulin covers for protection against rain and snow.

When stopping work, cover the tops of masonry structures to keep rain or snow off. Be sure moisture does not get into cavities between wythes. When laying CMUs for interior use, dry them before laying. They should be dried to the average condition to which the finished wall will be exposed.

2.2.0 Contraction and Expansion Joints

CMUs have one major difference from other masonry units: like concrete slabs and sidewalks, CMU construction is prone to cracking from shrinkage. Shrinkage cracking occurs as the concrete slowly finishes drying. As the concrete shrinks, it moves slightly. The movement causes cracks in a rigid slab or wall. The shrinkage cracking in a CMU structure is controlled in the same way as for concrete slabs. This is done by using reinforcement in combination with contraction or control joints.

Figure 1 shows some typical locations for contraction joints. Walls are likely to crack at abrupt changes in wall thickness or heights, at openings,

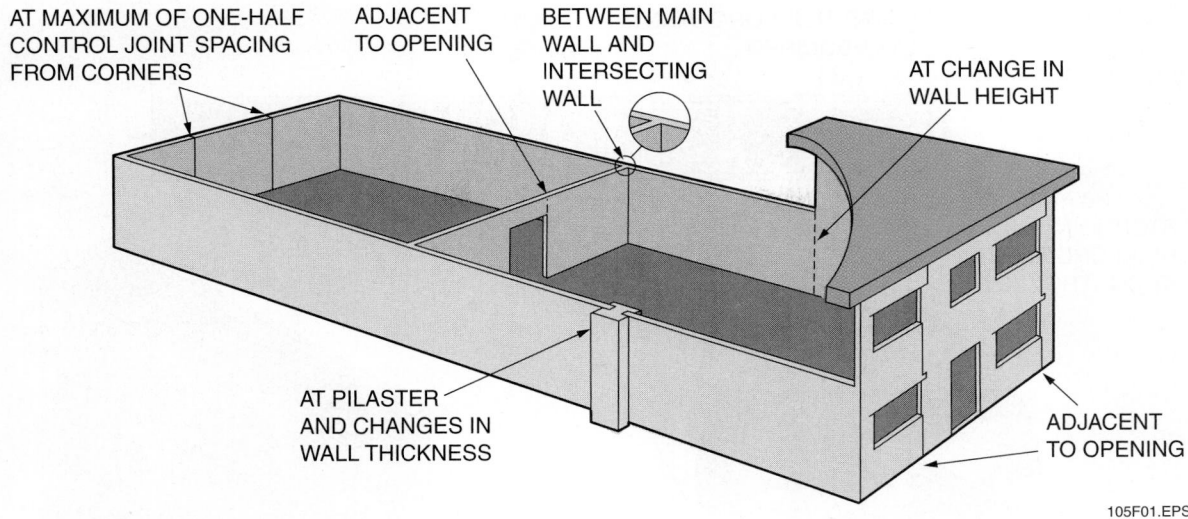

AT MAXIMUM OF ONE-HALF CONTROL JOINT SPACING FROM CORNERS

ADJACENT TO OPENING

BETWEEN MAIN WALL AND INTERSECTING WALL

AT CHANGE IN WALL HEIGHT

AT PILASTER AND CHANGES IN WALL THICKNESS

ADJACENT TO OPENING

105F01.EPS

Figure 1 ◆ Locations of control joints.

over windows, and over doors. Contraction joints are also used at intersections between loadbearing walls and partition walls.

CMU walls use two kinds of reinforcement: grout and steel with grout. Either the grout is poured into CMU cores, or steel rods are inserted in CMU cores, and grout is poured around the steel (*Figure 2*). The reinforcement gives rigidity and strength to the wall. CMU walls need reinforcement of either kind on both sides of a contraction joint.

Contraction joints control cracking by weakening the structure. Cracks occur at the weakened area instead of randomly. In a concrete slab, these control joints are made by cutting grooves. In a CMU wall, this is done by breaking the contact between two columns of units, as shown in *Figure 3*.

The control joints replace a standard mortar joint every 20' or so. Control joints must be no more than

30' apart. The control joints in CMU walls have no mortar; they are filled with silicon or another flexible material. The contraction joint filler keeps the rain out but allows slight movement. The reinforcement keeps the edges of the contraction joints aligned as they move.

Clay masonry units do not contract and shrink as they age, but they do change size very slightly with temperature and moisture changes in the air. Clay masonry structures need expansion joints to handle this type of movement. The control joints in CMU walls take care of expansion as well as contraction. Clay masonry walls must include expansion joints. These are usually soft, mortarless joints filled with foam and covered with a layer of silicon paste. As with contraction joints, the filler keeps the rain out and allows slight movement of the masonry.

2.3.0 Concrete Block Characteristics

Block is produced in four classes: solid loadbearing, solid nonbearing, hollow loadbearing, and hollow nonbearing. Block comes in two weights: normal and lightweight. Lightweight block is made with fly ash, pumice, and scoria or other lightweight aggregate. Loadbearing and appearance qualities of the two weights are similar. The major difference is that lightweight block is easier and faster to lay.

Solid blocks are for special needs, such as structures with unusually high loads, drainage catch basins, manholes, and firewalls. Like solid clay products, 25 percent or less of the surface of a solid concrete unit is hollow. Less than 75 percent of a hollow unit is solid. *Figure 4* shows the names of different parts of blocks.

105F02.EPS

Figure 2 ◆ Block wall with reinforcing rods.

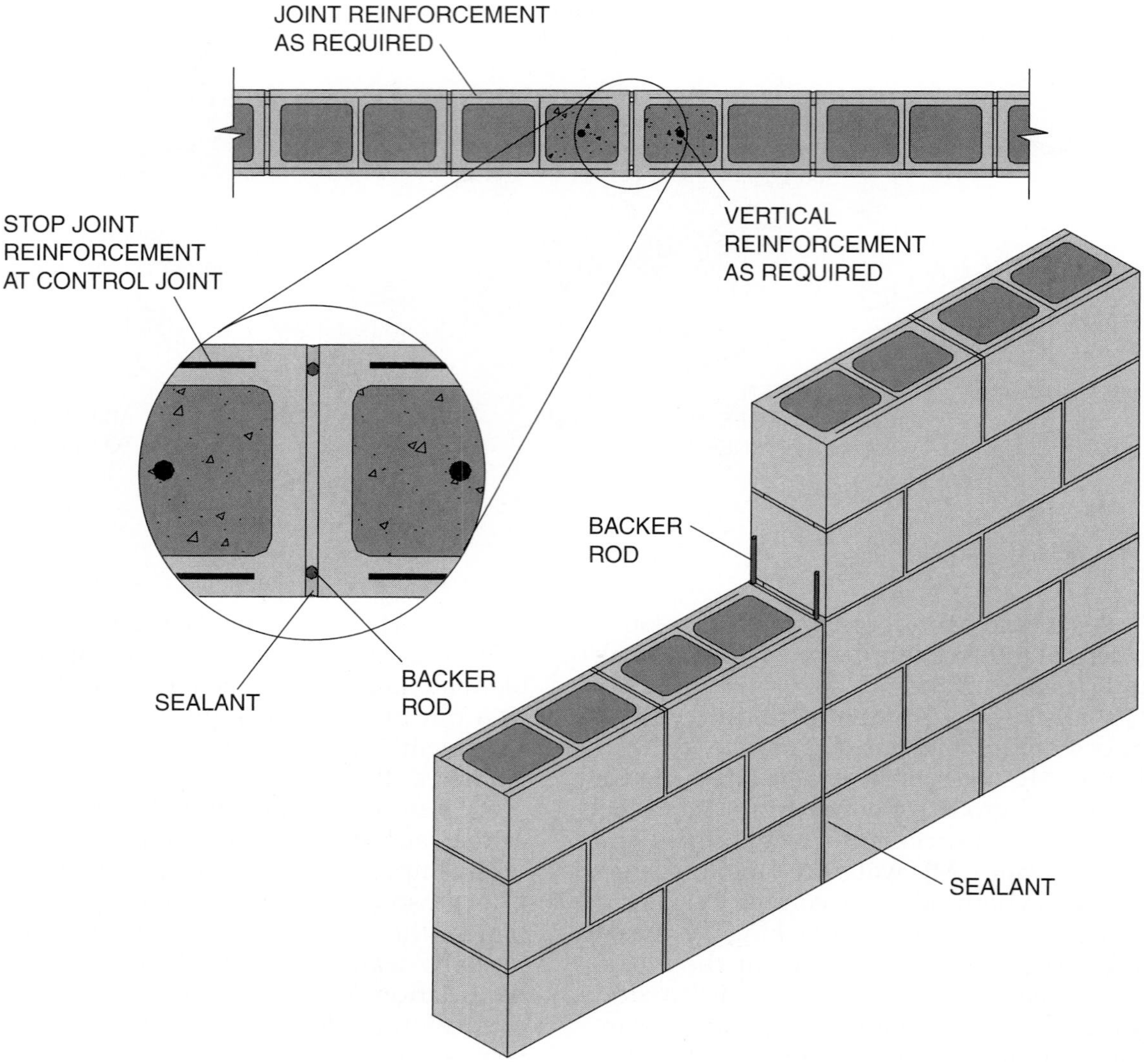

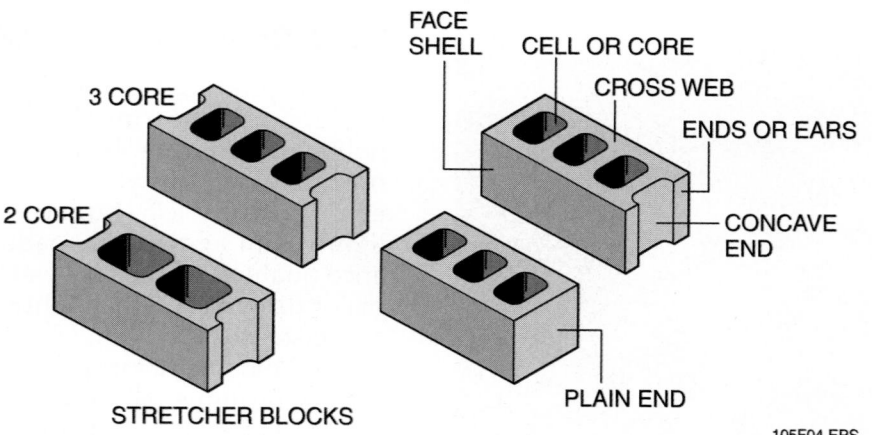

JOINT REINFORCEMENT AS REQUIRED

STOP JOINT REINFORCEMENT AT CONTROL JOINT

VERTICAL REINFORCEMENT AS REQUIRED

BACKER ROD

SEALANT

BACKER ROD

SEALANT

105F03.EPS

Figure 3 ◆ Vertical control joint with plastic filler.

3 CORE

2 CORE

STRETCHER BLOCKS

FACE SHELL

CELL OR CORE

CROSS WEB

ENDS OR EARS

CONCAVE END

PLAIN END

105F04.EPS

Figure 4 ◆ Parts of blocks.

Reinforcing Block Walls

Grout is a mixture of cementitious material and aggregate with enough liquid content to make it flow readily. Grout is often pumped into block wall cavities with special pumps.

105SA01.EPS

105SA02.EPS

Most block is governed by *ASTM C90*, which covers hollow loadbearing block. At one time, *ASTM C90* specified Grades N and S. Grade S has been discontinued for this block, and *ASTM C90* block is now ungraded. The ASTM does specify a minimum compressive strength of 800 psi, however. Solid loadbearing block falls under *ASTM C145*, which specifies Grades N and S. Grade N has a compressive strength of 1,500 psi, and the compressive strength of Grade S is 1,000 psi.

Block comes in modular sizes, with colors determined by the cement ingredients, the aggregates, and any additives. The basic block is called a stretcher. It has a nominal face size of 8" × 16" with a standard ⅜" mortar joint. The most commonly used block has a nominal width of 8", but 4", 6", 10", and 12" widths are also common. Three modular bricks have the same nominal height as one nominal 8" block. Two nominal brick lengths equal one nominal block length. So, laying one stretcher covers the same area as laying six modular bricks.

Block comes in a variety of shapes to fit common and special purposes. *Figure 5* shows a sampling of block sizes and shapes. Common hollow units can have two or three cores. Most cores are tapered slightly to provide a larger bed joint surface. Block edges may be flanged, notched, or smooth. There are local variations as well, with some shapes available only in specific parts of the country.

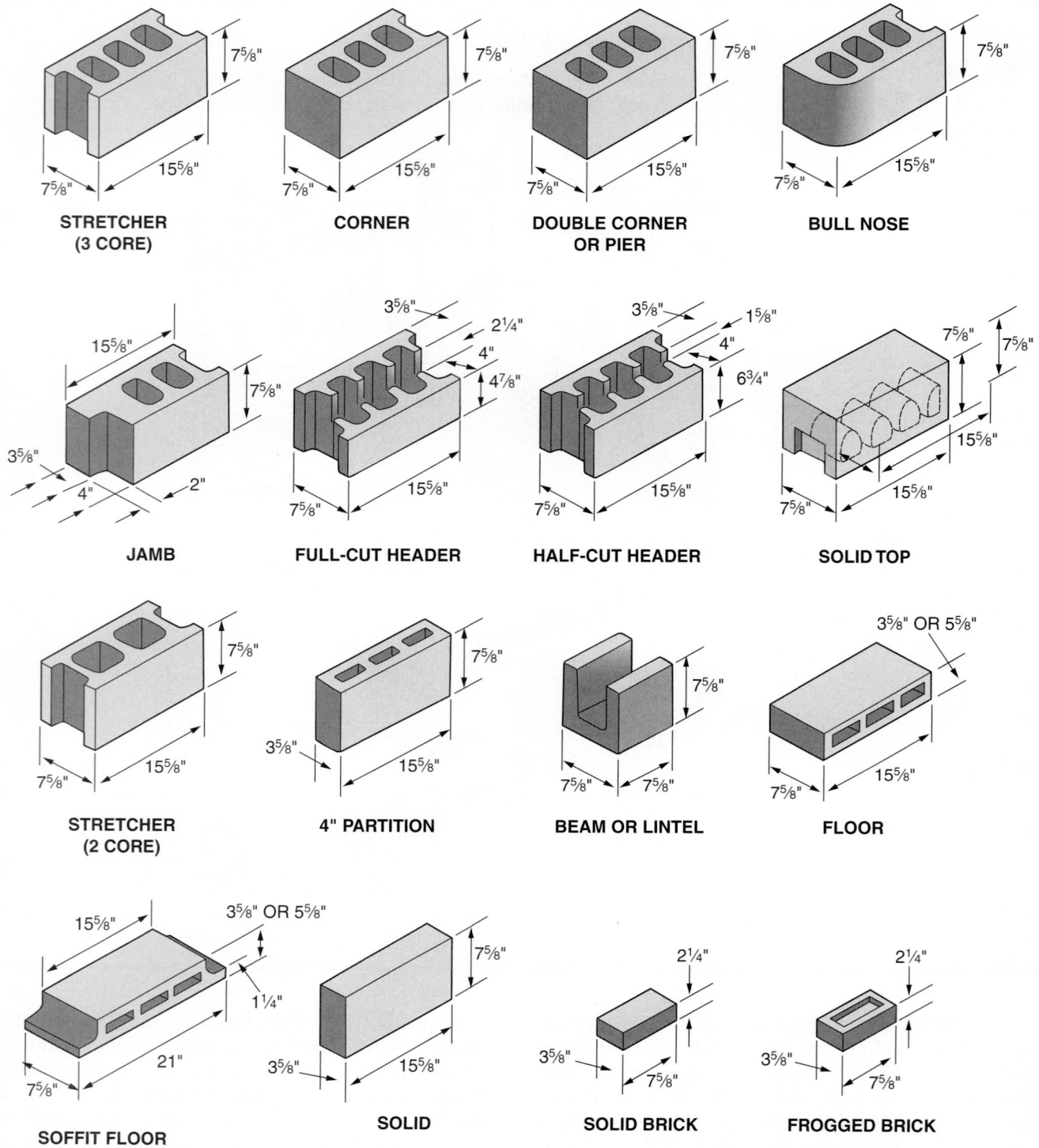

NOTE: Dimensions are actual unit sizes. A $7\frac{5}{8}$" × $7\frac{5}{8}$" × $15\frac{5}{8}$" unit is an 8" × 8" × 16" nominal-size block.

105F05A.EPS

Figure 5 ◆ Common concrete block shapes. (1 of 2)

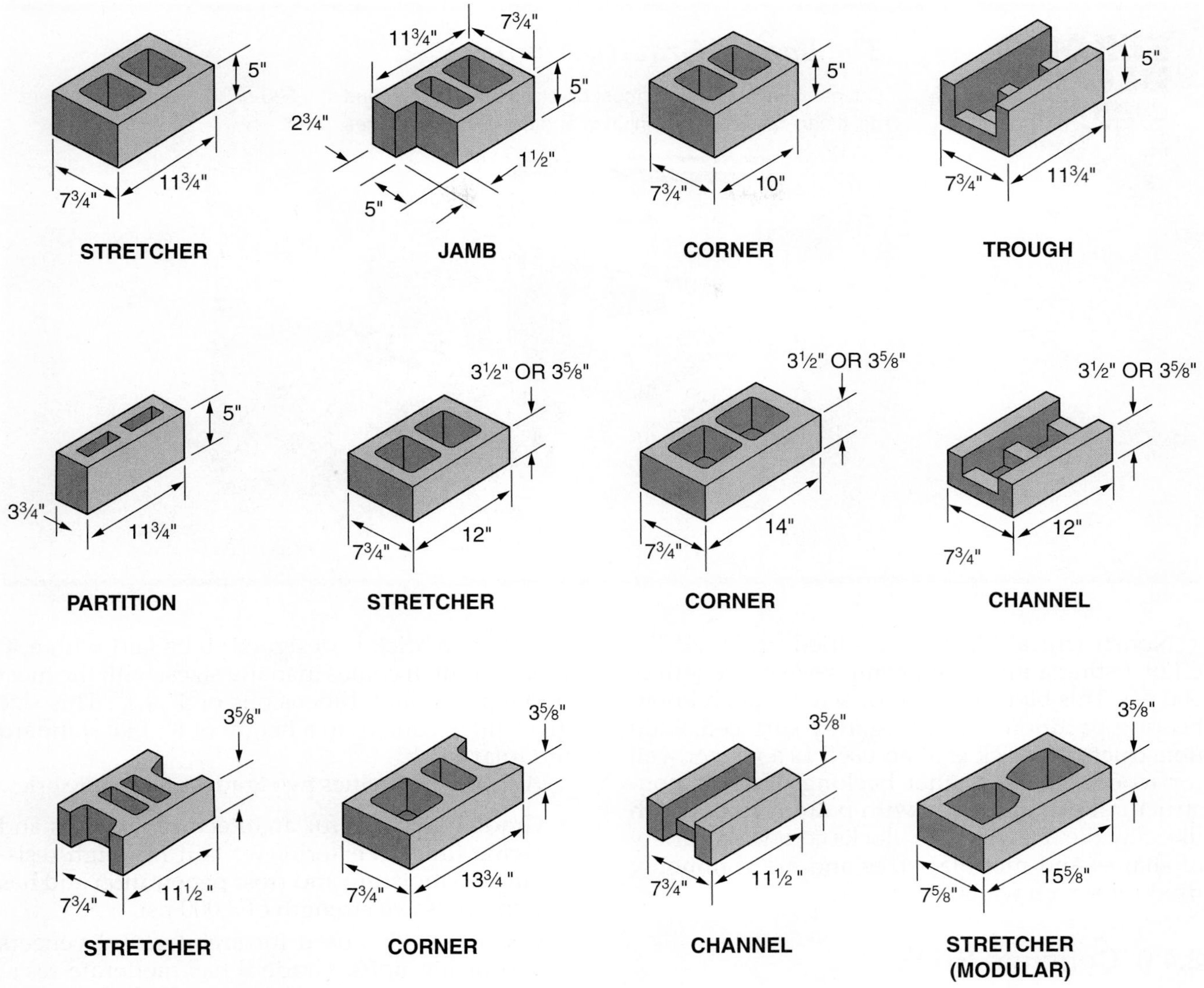

STRETCHER
5"
11¾"
7¾"

JAMB
11¾"
7¾"
2¾"
5"
5"
1½"

CORNER
5"
7¾"
10"

TROUGH
5"
7¾"
11¾"

PARTITION
5"
3¾"
11¾"

STRETCHER
3½" OR 3⅝"
7¾"
12"

CORNER
3½" OR 3⅝"
7¾"
14"

CHANNEL
3½" OR 3⅝"
7¾"
12"

STRETCHER
3⅝"
7¾"
11½"

CORNER
3⅝"
7¾"
13¾"

CHANNEL
3⅝"
7¾"
11½"

STRETCHER (MODULAR)
3⅝"
7⅝"
15⅝"

NOTE: Dimensions are actual unit sizes. A 7⅝" × 7⅝" × 15⅝" unit is an 8" × 8" × 16" nominal-size block.

105F05B.EPS

Figure 5 ◆ Common concrete block shapes. (2 of 2)

A variety of surface and mixing treatments can give block varied and attractive surfaces. Newer finishing techniques give block face the appearance of brick, stone, ribbed columns, raised patterns, or architectural fabrics. Like clay masonry units, block can be laid in structural pattern bonds.

Loadbearing block is used as backing for veneer walls, bearing walls, and all structural assemblies. Both regular and specially shaped blocks are used for paving, retaining walls, and slope protection. Landscape architects call the newer, shaped-to-interlock blocks *hardscape* and use them as part of landscape design, as shown in *Figure 6*.

105F06.EPS

Figure 6 ◆ An application of loadbearing block.

The Law of Gravity Applies

Cubes of brick and block must be placed on level ground. If a 500-pound load of bricks lets go, it can cause serious injuries and expensive damages.

105SA03.EPS

Nonstructural block is specified under *ASTM C129*, listing a minimum compressive strength of 500 psi. This block is used for screening and non-bearing partition walls. Elegantly surfaced, solid nonstructural block is often used as a veneer wall for wood, steel, or other backing. Hollow nonstructural block is made with pattern cores much like clay tile. Pattern core blocks come in a variety of shapes and modular sizes and are commonly used for screen walls.

2.4.0 Concrete Brick

Concrete brick is a solid loadbearing unit, roughly brick size, used in the same way as clay brick. Concrete brick has no voids and may be frogged as shown in *Figure 7*. A frog is a depression in the head of a brick that lightens the weight of the brick. It also makes for a better mortar joint by increasing the area of mortar contact.

Concrete brick is designed to be laid with a ⅜" mortar joint. It comes in many sizes, with the most popular nominal dimensions of 4" × 8". This size gives three courses in a height of 8", like standard modular brick.

ASTM C55 specifies two grades of concrete brick:

• Grade N is used for architectural veneers and facing units in exterior walls It has high resistance to moisture and frost penetration and has a compressive strength of 3,000 psi.

• Grade S is also used for architectural veneers and facing units. Grade S has moderate resistance to moisture and frost and is used in the southern region of the United States. Its compressive strength rating is 2,000 psi.

Slump brick is made from a wet mixture. The units sag or slump when removed from the molds. This gives an irregular face resembling stone, as shown in *Figure 8*. In other respects, slump brick meets concrete brick standards.

Solid block is also made from a slump mixture. Because of the greater surface area, the block face is very irregular. Its height, surface texture, and appearance resemble stone.

2.5.0 Other Concrete Units

CMUs include pre-faced units, calcium silicate CMUs, and catch basin units. ASTM specifications cover loadbearing, moisture retention, aggregate mix, and other characteristics of these units.

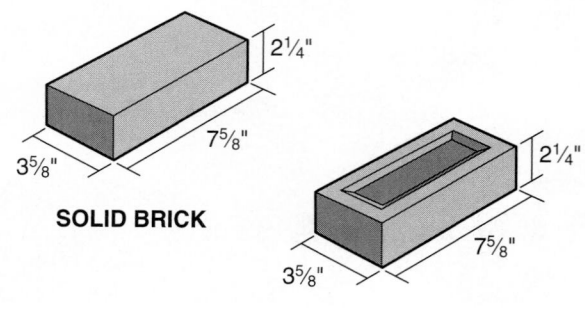

105F07.EPS

Figure 7 ◆ Concrete brick.

Figure 8 ◆ A wall made from slump brick.

The calcium silicate acts as a leavening agent and creates gas bubbles in the mix. The units are not fired or cured in a kiln but cured in an **autoclave** with pressurized live steam. In the autoclave, the lime reacts with the silica to bind the sand particles into a very lightweight, strong unit. ASTM performance specifications cover this type of brick and block with grading standards identical to those for traditional products.

The units are also called sand-lime brick or aerated block. They are used extensively in Europe, Australia, Mexico, and the Middle East. In the U.S., this brick is used mostly in flues, chimney stacks, and other high-temperature locations. They resist sulfates in soil, do not effloresce, and are not damaged by repeated freeze-thaw cycles. The block is now manufactured in the U.S. in a variety of sizes for commercial or home building.

2.5.1 Pre-Faced Units

Concrete pre-faced or pre-coated units are faced with colors, patterns, and textures on one or two face shells (*Figure 9*). The facings are made of resins, portland cement, ceramic glazes, porcelainized glazes, or mineral glazes. The slick facing is easily cleaned. These units are popular for use in gyms, hospital or school halls, swimming pools, and food processing plants. They come in a variety of sizes and special-purpose shapes, such as coving and bullnose corners.

2.5.2 Calcium Silicate Units

Calcium silicate units (*Figure 10*) are made of a mixture of sand, water, lime, and calcium silicate.

Figure 10 ◆ Calcium silicate units.

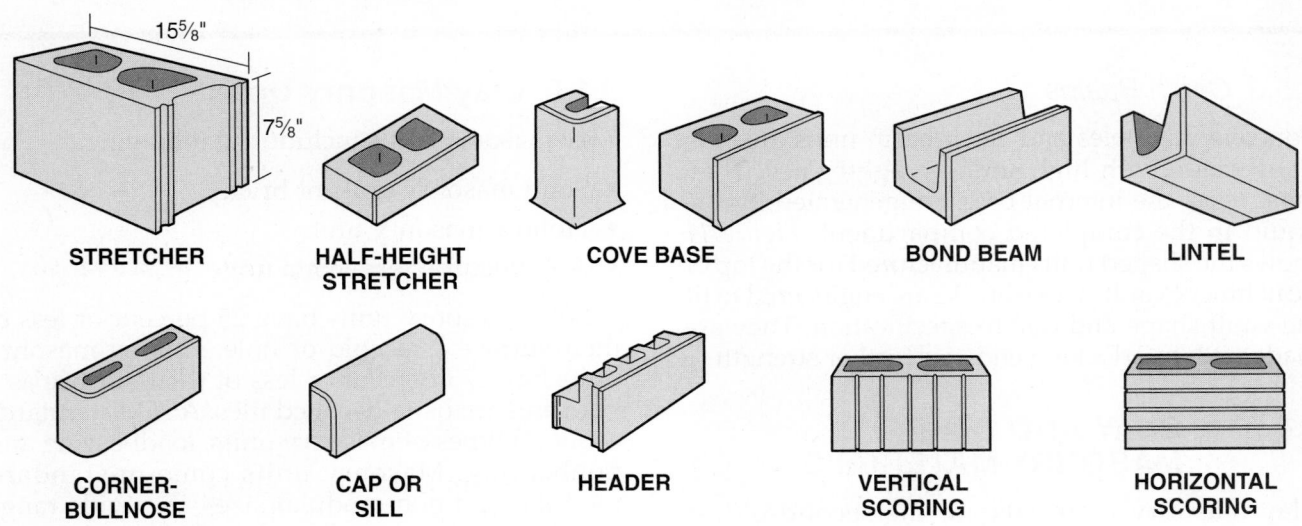

STRETCHER

HALF-HEIGHT STRETCHER

COVE BASE

BOND BEAM

LINTEL

CORNER-BULLNOSE

CAP OR SILL

HEADER

VERTICAL SCORING

HORIZONTAL SCORING

Figure 9 ◆ Concrete pre-faced units.

Concrete Pre-Faced Units

Pre-faced concrete units like those shown here are often designed to imitate stone construction.

105SA04.EPS

105SA05.EPS

2.5.3 Catch Basins

Concrete manholes and catch basin units are specially made with high strength aggregates. They must resist the internal pressure generated by the liquid in the completed compartment. *Figure 11* shows the shaped units manufactured for the top of a catchment vault. These blocks are engineered to fit the vault shape and cast to specification. They are made with interlocking ends for further strength.

3.0.0 ◆ CLAY AND OTHER MASONRY MATERIALS

Clay masonry materials are the second-oldest building material. The following sections review clay and stone masonry units and introduce metal and plastic masonry materials.

3.1.0 Clay Masonry Units

Clay masonry units include the following:

• Solid masonry units or brick
• Hollow masonry units
• Architectural terra cotta units

Solid masonry units have 25 percent or less of their surface as a void or hole. Hollow masonry units have 75 percent or less of their surface as a void and are usually called tiles. ASTM standards cover all types of masonry units, loadbearing and nonbearing. Masonry units come in standard modular and non-modular sizes, in a wide range of colors, textures, and finishes. This module will focus on brick and laying brick.

Bricks can be installed in any of six positions. *Figure 12* shows each of these six positions and

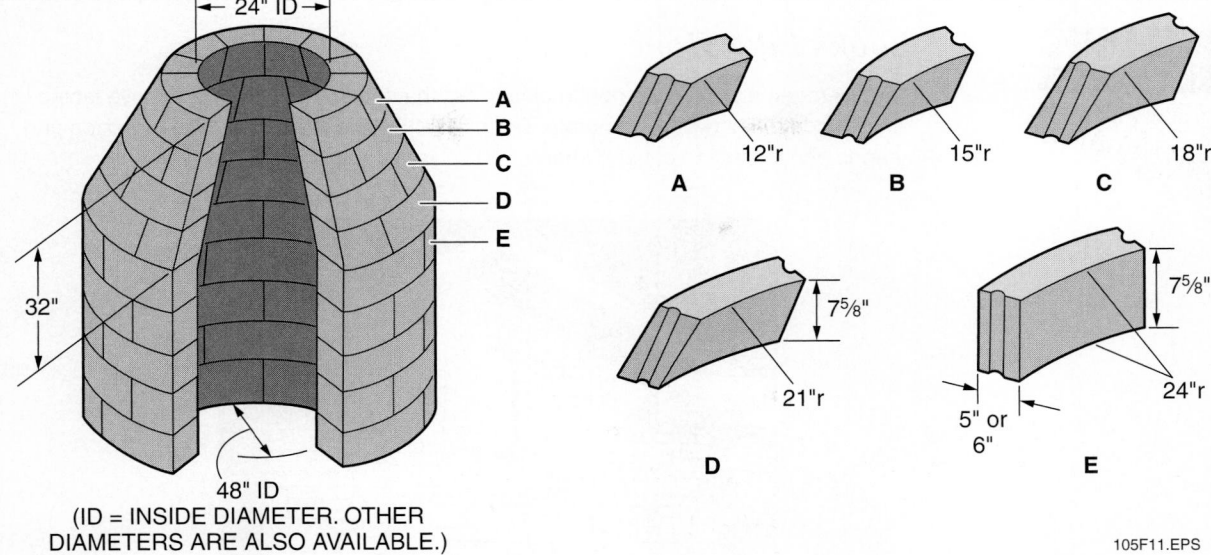

Figure 11 ◆ Manhole and vault units.

their names. The shaded part of the brick is the named part and the part that shows when the brick is laid in a pattern bond.

The most important characteristic of brick is its absorption capacity, or the amount of water it can soak up in a fixed length of time. The percentage of water present in brick affects the hardening of the mortar around the brick. If the brick contains a high percentage of moisture, the mortar will set more slowly than usual, and the bond will be poor. The brick will not absorb moisture and mortar into its microscopic irregularities. If the brick contains a low percentage of moisture, it will absorb too much moisture from the mortar. This will prevent the mortar from hardening properly because there will not be enough water left for good hydration.

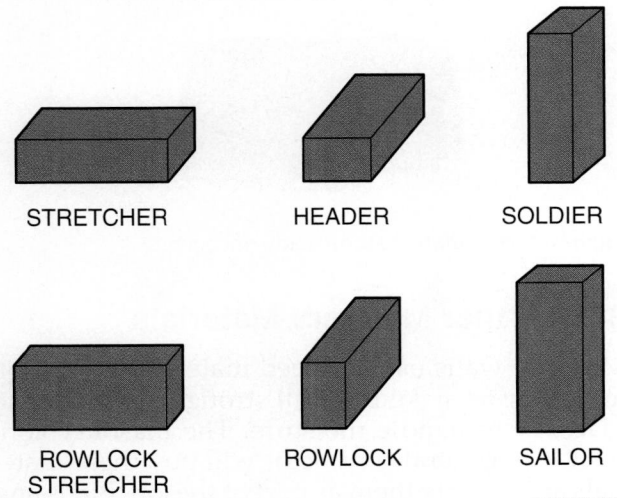

Figure 12 ◆ Brick positions.

Hard-surfaced bricks and CMUs usually need to be covered on the job site so they do not get wet. Soft-surfaced bricks are usually very absorbent and may sometimes need to be wetted down before they are used.

The mason needs to determine whether the brick is too dry for a good bond with the mortar. The following test can be used to measure the absorption rate of brick:

Step 1   Draw a circle about the size of a quarter on the surface of the brick with a crayon or wax marker.

Step 2   With a medicine dropper, place 20 drops of water inside the circle.

Step 3   Using a watch with a second hand, note the time required for the water to be absorbed.

If the time for absorption exceeds 1½ minutes, the brick does not need to be wetted. If the brick absorbs the water in less than 1½ minutes, the brick should be wetted.

Wet brick with a hose played on the brick pile until water runs from all sides. Let the surface of the bricks dry before laying them in the wall.

3.2.0 Stone

As noted in the first module, rubble stone is irregular in size and shape. Ashlar stone has been cut into a rectangular unit. Stone is expensive to assemble and time-intensive to lay. It is rarely used today except for trim and detail.

Looks Like Stone

The average person would not be able to tell that the stone in this decorative facing is really made of lightweight concrete. The material is available in a variety of sizes and shapes and is easy to cut and shape.

105SA06.EPS

Natural stone is used mostly for veneer walls, floors, and trim. Rubble and ashlar are used for dry stone walls, mortared stone walls, retaining walls, facing walls, slope protection, paving, fireplaces, patios, and walkways. Limestone ashlars still make the finest sill blocks. As they are one piece, there is no concern about water coming through into the wall underneath. They are also still used for **lintels** and as coping stones on top of brick walls.

Concrete masonry units are made in shapes and colorings to mimic every kind of ashlar. These units are called cast stone and are more regular in shape and finish than natural stone (*Figure 13*). They are lighter in weight and do not need as large a footing as natural stone. Cast stone has replaced natural stone in most commercial projects because it is less expensive. ASTM specifications cover cast and natural stone.

Masons may think stone work is a separate craft, but that is not entirely so. Masons still must know how to lay stone copings or sills or lintels. Masons must pattern, shape, and lay stone veneer for fireplaces or home walls.

105F13.EPS

Figure 13 ◆ Imitation stone made of concrete.

3.3.0 Other Masonry Materials

Masonry walls usually need material in addition to mortar to make the wall stronger, to hold it in place, or to handle moisture. The masonry contractor or general contractor will buy these materials and supply them as part of the job. Plans and specifications typically detail the locations and types of these materials to be used.

3.3.1 Metal Ties

Metal ties are used to tie cavity walls together and allow them to be loadbearing. The ties keep the walls from separating when weight is placed on them by the other parts of the structure. Metal ties are also used for composite walls. The ties equalize the loadbearing and also tie the two wythes together. Ties are made from $\frac{3}{16}$" zinc-coated steel and are placed 24" apart. *Figure 14* shows rectangular and Z-shaped ties.

3.3.2 Veneer Ties

Veneer ties (*Figure 15*) are used to tie a masonry veneer wall to a backing wall or wythe. Unlike metal ties, veneer ties do not equalize loadbearing. They do keep the veneer wall from moving away from its backing. They are made of corrugated galvanized steel and placed about 12" apart.

3.3.3 Reinforcement Bars

Steel reinforcement bars come in different thicknesses and lengths. They are inserted in block cores, and then the cores are filled with grout. They add strength and weight-bearing capacity to block walls. Sometimes they are placed in the middle of cavity walls where the cavity is to be grouted.

3.3.4 Joint Reinforcement Ties

Joint reinforcement ties are made of two 10' lengths of steel bars welded together by rectangular or triangular cross bracing. *Figure 16* shows the ladder (rectangular) and truss (triangular) versions of these ties. They are used in horizontal joints every second or third course as specified.

3.3.5 Flashing

Flashing keeps water from leaking from the top of a masonry wall into the unit below. It is placed under masonry lintels, sills, copings, and spandrels. The most common flashing is made of copper, stainless, or galvanized metal. Bituminous flashing is made of fabric saturated with asphalt. Newer types of flashing are made of plastics. They are cheaper and easier to work with. *Figure 17* shows flashing in position under a sill and a lintel.

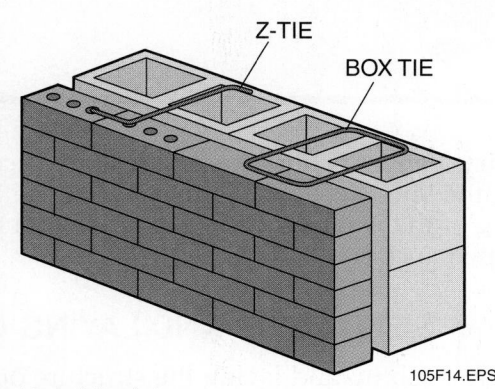

Figure 14 ◆ Metal ties.

Figure 15 ◆ Veneer ties.

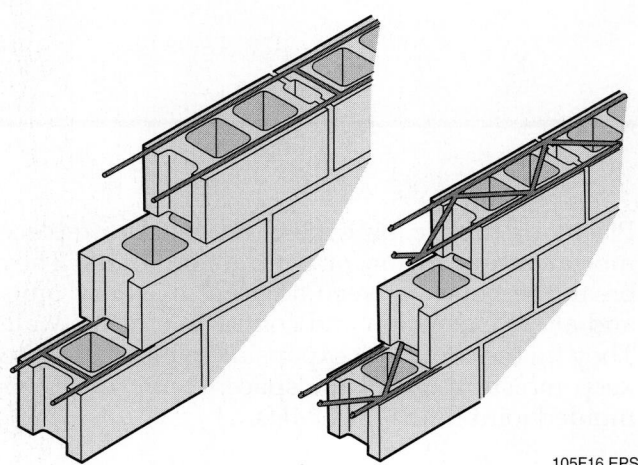

Figure 16 ◆ Joint reinforcement ties.

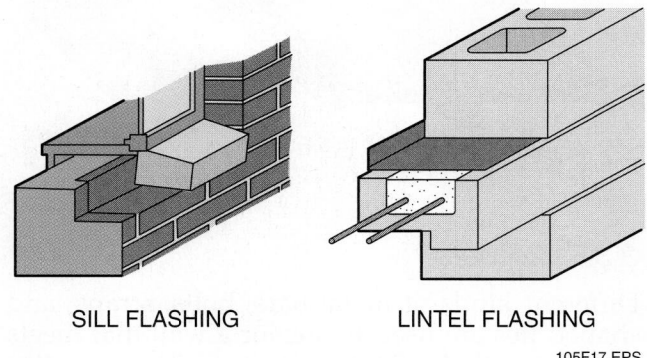

SILL FLASHING   LINTEL FLASHING

Figure 17 ◆ Flashing applications.

Chimney Flashing

Proper flashing is extremely important around chimneys, especially in areas subject to snow buildup. On a new building, roofers will apply the flashing. If existing flashing is disturbed during a chimney repair, however, it is up to the mason to make sure the flashing is secure.

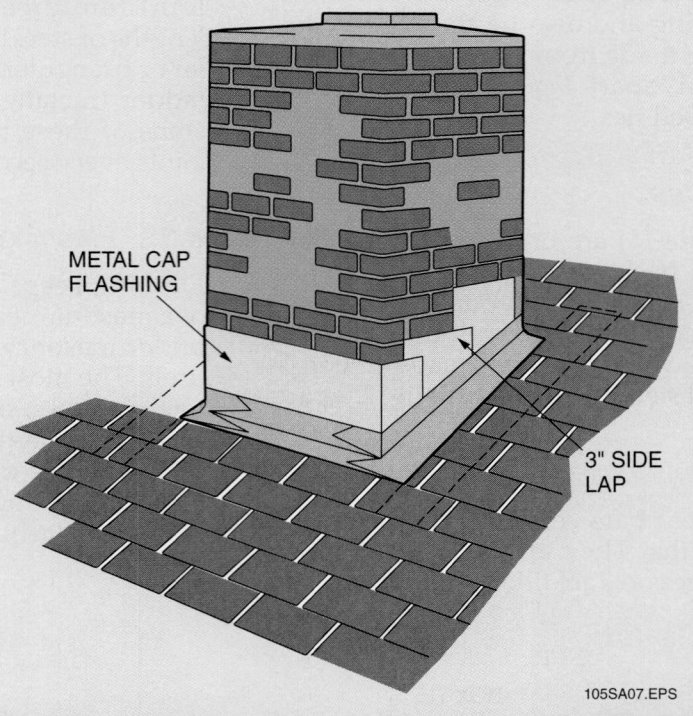

METAL CAP FLASHING

3" SIDE LAP

105SA07.EPS

3.3.6 Joint Fillers

Plastic or rubber joint fillers are used to replace mortar in expansion or contraction joints. They break the bond between adjacent masonry units and allow expansion and contraction of the wall. They fill the control or expansion joints in order to keep moisture out of the space. *Figure 18* shows molded joint fillers for CMUs.

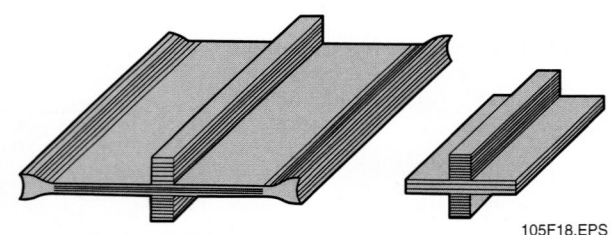

105F18.EPS

Figure 18 ◆ Joint fillers.

3.3.7 Anchors

Different kinds of metal bars, bolts, straps, and shaped ties are used to anchor a wall that meets another wall at a 90-degree angle. They are also used to tie different architectural elements to masonry walls. Anchors must be installed according to the specifications, as they affect the load-bearing of the wall. *Figure 19* shows several types of shaped anchors.

4.0.0 ◆ SETTING UP AND LAYING OUT

Setting the job up and laying the structure out are two distinct steps. Both must be complete before the mason can start to lay units. Setting up refers to materials and site preparation. Laying out refers to establishing the baseline for the masonry structure. The next sections give details for both of these tasks.

4.1.0 Setting Up

Masonry setup work starts when the contract for the job is signed. The first step for the masonry contractor is to read the contract, blueprints, and specifications. The next step is to review the schedule plus any standards and codes cited in the contract. After all of that, the masonry contractor is ready to estimate the workers, materials, and equipment needed for that job. Review the information in *Measurements, Drawings, and Specifications* to get a clearer idea of what this work entails.

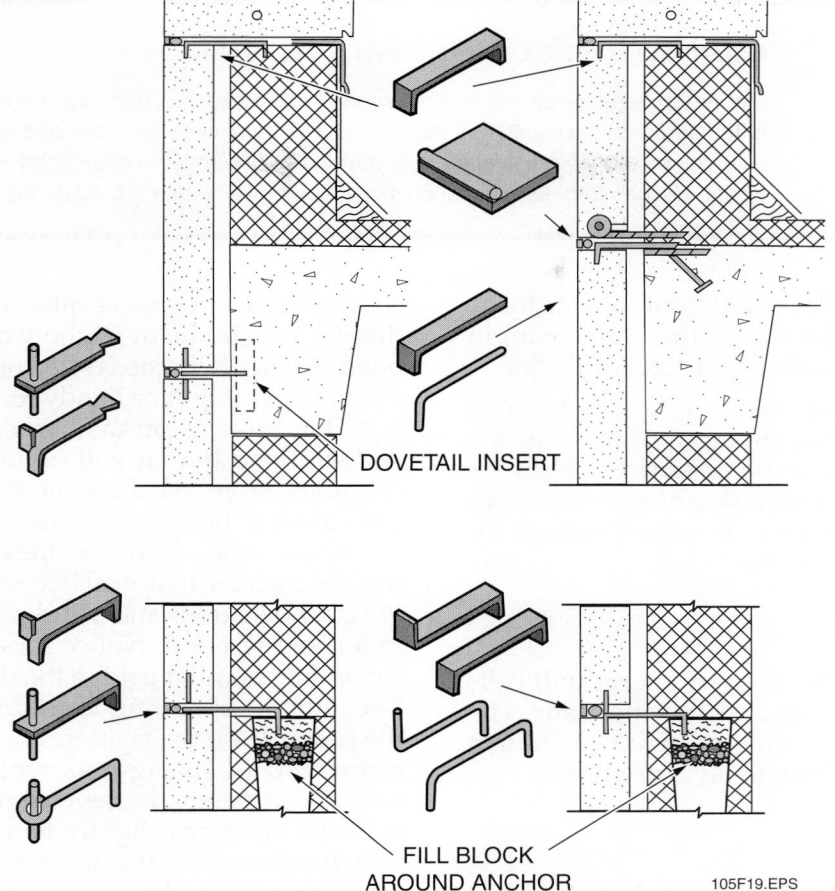

Figure 19 ◆ Shaped anchors.

DOVETAIL INSERT

FILL BLOCK
AROUND ANCHOR

105F19.EPS

The next step is to estimate again, check figures, and order the masonry equipment and materials. A visit to the job site and discussion with the engineer or construction foreman will give the masonry contractor an idea of where and how to store masonry materials. The masonry contractor must specify a delivery date and location on site. Materials must be stored close to where they will be used and protected from the weather. The crew must be hired and briefed. Then, the work is ready to begin, but there is still much to do before laying the first masonry unit. The following checklist shows some of the preliminary procedures:

• Check that all materials are stored close to work stations and protected from moisture. Masonry units must be laid dry in order to avoid shrinkage upon drying. Pile materials on pallets or planks off the ground, and cover with a tarpaulin or plastic sheet. Bagged materials can be stored in sheds or stacked on pallets and covered. Sand must be covered also, to protect it from moisture and dirt.

• Prepare mortar mixing areas within several feet of the work areas or as close as possible. Place water barrels next to them for water supply and for storage of hoes and shovels not in use. Be sure mixing equipment does not interfere with movement paths.

• Place mortar pans and boards by workstations. If you are using scaffolding, place the pans at intervals on the scaffolding near the point of final use.

• Stockpile units on each side of the mortar pans and at intervals along the wall line. If you are using scaffolding, place units along the top of the scaffold near the point of final use. Stockpiles should allow the mason to move block as little as possible once laying starts.

• Stack block in stockpiles with the bottom side down, just as they will be laid in the wall. The top of the block has a larger shell and web. Stack faced units with the faced sides in the direction they will go, just as they will be laid in the wall. Stack all units so the mason will move or turn them as little as possible.

• Check all scaffolding for proper assembly and position. Ensure that braces are attached and planks are secured at each end. Scaffolding should be level and no closer than 3" from the wall.

BUILDING BLOCKS

Cold Weather Considerations

Mortar temperature should be between 40°F and 120°F in order for proper hydration to occur. In temperatures below 40°F, it may be necessary to heat the water and/or sand in order to keep the mortar at a high enough temperature. One method often used to heat sand is to pile it over a large-diameter pipe, such as a culvert pipe, with a fire inside the pipe.

• Check all mechanical equipment, power tools, and hand tools. Make sure they are clean, in good condition, and the right size for the job.

The contractor may assign a helper to keep mortar pans full by supplying mortar from the mixer. A helper may also be assigned to keep the stockpiles refilled. The objective of all setup work is to make everything efficient and convenient for the masons once they begin laying.

4.2.0 Job Layout

Laying out the wall or other structural unit calls for a review of the plans and specifications. The first steps are to plan out the work, establish where it will go, and then to lay a **dry bond**.

4.2.1 Planning

Planning out the work means you need to check the plans for wall lengths, heights, door dimensions, and window openings. What pattern or bond is specified? What is the nominal size of the masonry unit? How are openings to be treated? Are the dimensions and the masonry units on the modular scale of 4" increments?

After answering these questions, the mason can draw a rough layout of the wall and lay out the bond pattern. If the job is sized on the modular grid, graph paper might be handy for the spacing drawing. This drawing can show where the bond pattern will start and how it will fit around the specified openings. From this drawing, the mason can count and calculate how many masonry units to cut.

The question of whether the designer did or did not use the modular grid becomes important. *Figure 20* shows door and window openings located in a running bond. Notice how the openings are set off the modular grid in the diagram on the left. The amount of cutting is enormous compared to the example on the right. Using many small units reduces wall strength as well. Sometimes the mason can persuade the designer or engineer to shift the openings slightly to avoid so much cutting. In other cases, the dimensions are critical and cannot be changed.

Drawing the bond pattern on the wall area may seem like a time-consuming exercise, but it can save a lot of time, especially with non-modular work. The starting point of the pattern determines how many masonry units will need to be cut. By adjusting the starting unit of a non-modular

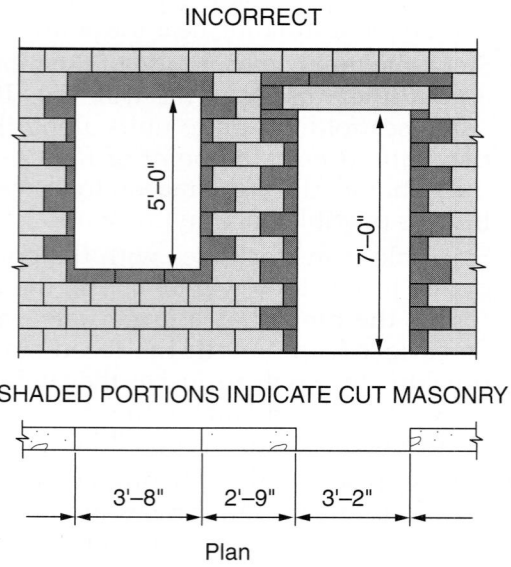

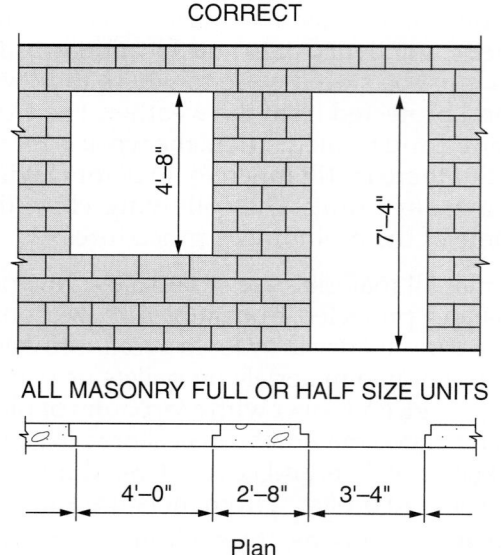

Figure 20 ◆ Door and window openings.

bond pattern, the mason can come up with a layout that calls for cutting the smallest amount of units. The mason will check these calculations by laying a dry bond before cutting any units.

4.2.2 Locating

The mason will check the location first. Masonry walls take a footing or support, usually made of concrete. The surveyor or foreman will mark the corners of the structure or slab. On some jobs, the foreman will drive nails into the wall footing to mark the building line. *Figure 21* shows a foundation layout with a footing plan for block foundation walls. At the job site, the first thing to do is to locate the footing. Next, brush it off. Remove any dried concrete particles or large aggregates to ensure a good bond between the footing and the first course.

Check that the footing is level. If the footing is not within an inch of level, it must be fixed. Do not apply a thick mortar joint to level the first course.

This can result in a joint too thick to carry the load of the wall. If the footing is out of level, notify your supervisor.

The next step is to locate the walls. Take measurements from the foundation or floor plan and transfer them to the foundation, footing, or floor slab. All measurements on the plans must be followed accurately. Be sure the door openings are placed exactly and the corners are on the footings exactly as given on the detailed drawings. Check to see that you are not confusing the measurements for the interior and exterior walls. If it appears that the wall cannot be laid out exactly because of errors in the footing, notify your supervisor.

The next task is to establish two points, corner-to-corner or corner-to-door. Then, run a chalkline between the two points and snap it on the footing or foundation. Because a chalkline is easily erased, mark key points along the chalkline with a marking pencil, nail, or screwdriver. This will allow resnapping the chalkline without refinding the points.

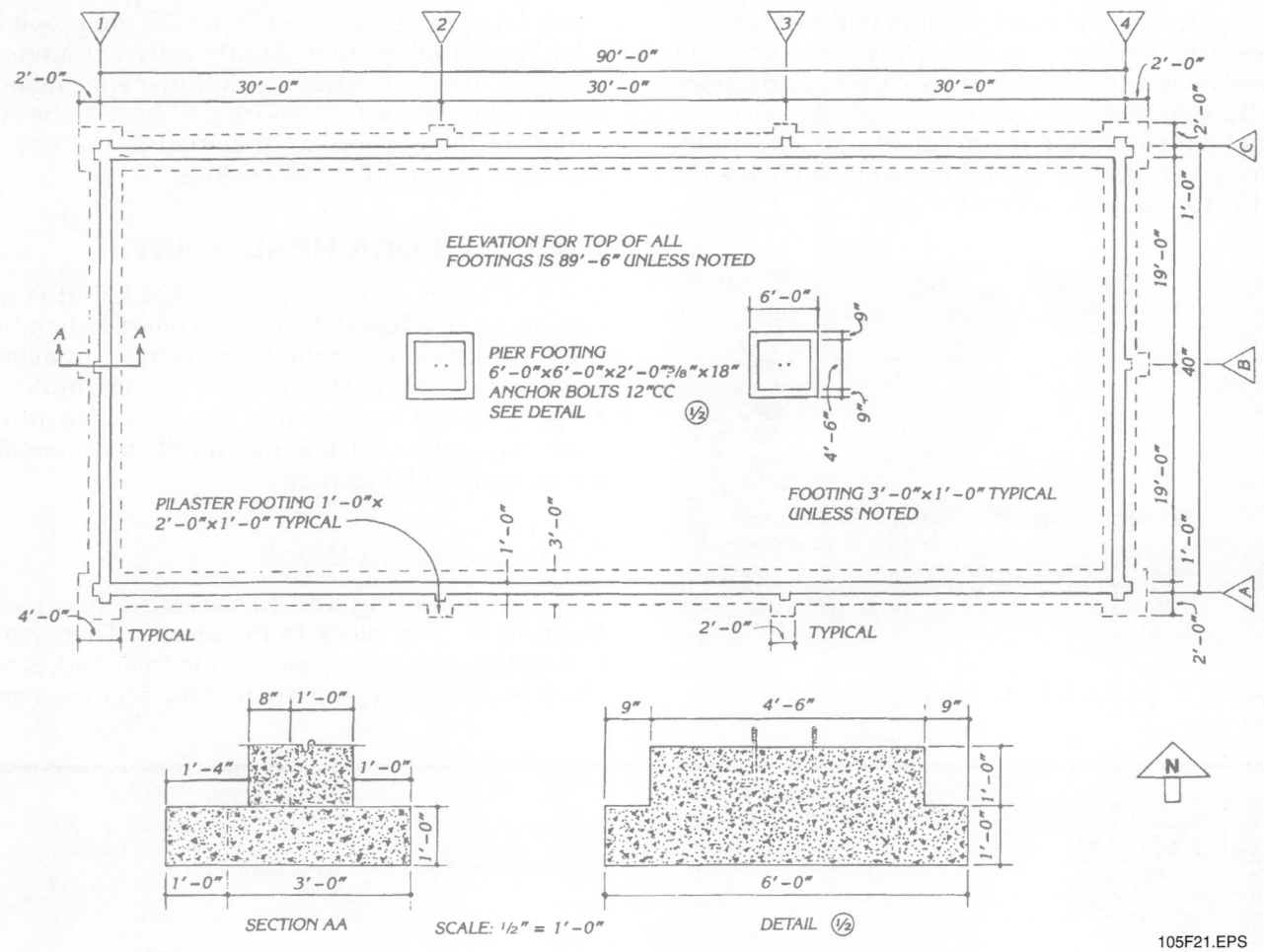

Figure 21 ◆ Foundation plan.

Mark the entire foundation for walls, openings, and control joints. After snapping the chalkline, mark over the chalk with a marker or nail. Once you have completed all markings, check the measurements of the markings against the foundation plan. Again, be sure you are reading the correct measurements. If there is to be a veneer wall, check that you are dimensioning the veneer, not the backing wythe. If everything does not fit precisely and exactly, it must be done over. It is easier to redo measurements than to redo a masonry wall.

4.2.3 Dry Bonding

Dry bonding is an alternative to measuring to establish the positioning of the masonry units. Starting with the corners, the mason can lay the first course with no mortar, or dry bond. This is a visual check of how the units will fit. It also checks the pattern bond drawing and the calculations for cut units. For CMUs, it provides a chance to check unit size and specifications.

From the corners, the mason lays units along the wall markings for the entire foundation, as in *Figure 22*. Since all mortar joints will be standard sizes, use a ⅜" or ½" piece of plywood or other material as a spacing jig. Check the specifications for the size of brick joints. Remember that all block is laid with a ⅜" joint. If you run into spacing problems, use a spacing jig, and mark any adjustments on the foundation and on the jig.

105F22.EPS

Figure 22 ◆ Example of a dry bond.

Lay the units through door openings to see how bond will be maintained above the doors. Then check spacing for openings above the first course, such as windows. Do this by taking away units from the first course and checking the spacing for the units at the higher level. These checks will show whether the joint width will work out for each course up to the top of the wall. Use the pattern bond diagram to help you. If spacing has to be adjusted slightly, mark it on the diagram and on the foundation.

After the units have been laid out correctly, mark the end of every other unit. Do this with a marking pencil directly on the foundation. This will guide you in laying mortar when the dry units are removed.

Once all of this has been done, the mason can use the steel square to mark the exact location and angle of the corners. The next step is checking the corner layout on the drawings.

The layout of the corner itself is important, especially when you are working with block and modular spacing. The architect will detail the corner layout on the working drawings. Different block layouts, as shown in *Figure 23*, are possible. Each layout takes up a slightly different amount of space. This will affect the modular spacing and determine whether any block will have to be cut. Building the corners as specified is the key to maintaining modular dimensions.

5.0.0 ◆ BLOCK HEAD JOINTS

Block is larger and heavier than brick and is not easy to lift one handed. Use two hands when lifting block to avoid strain. Block is more demanding than brick in that both blocks must be buttered to get a good head joint. Block is also more demanding than brick in that it calls for three different types of bed joints.

5.1.0 Buttering Block

Start by spreading and furrowing a bed joint. Position the first block in the mortar. Then stand two or three blocks on end next to their bed. Since block is wider at the top than at the bottom, stand

Dry Bonding

Some masons place their index finger between the bricks to account for the ⅜" mortar joint.

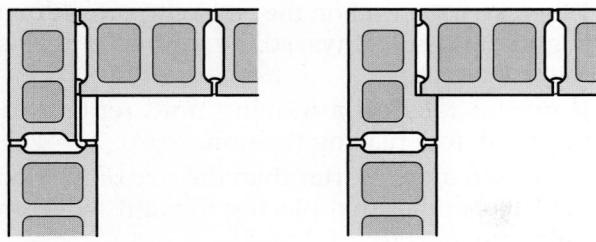
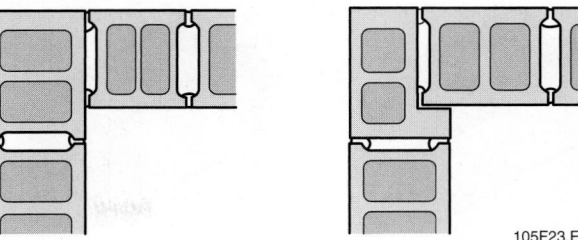

105F23.EPS

Figure 23 ◆ Block corner layouts.

the block so that the top sides will be on top when the block is placed. This makes it quick to butter several blocks at once.

You do not need to fill the trowel with mortar because one block does not take as much mortar as one brick. Butter the ear ends of the standing blocks. Wrap the mortar around the inside of each ear to help hold it in place. Then butter the ear end of the laid block. Lift the standing block by grasping the webs, or ends, with both hands, as shown in *Figure 24*. Do not jerk the block, or the mortar will fall off.

Place the block against the buttered, laid block. Tilt the block slightly toward you as you lay it into place, so that you can see the alignment of the cores and edges. Visually check that the edge of the block aligns with the block directly below.

105F24.EPS

Figure 24 ◆ Placing block.

To seat the block, gently press down and forward, so that the mortar squeezes out at the joints. Do not drop the block, but ease it into place. Continue laying the pre-buttered blocks, being sure to butter the ear end of the laid block each time.

An alternative method of buttering block is to butter one end, lift it by the webs with one hand, and butter the other end. This method is not recommended for beginners.

After you place the block, cut off the excess mortar with the edge of your trowel. Check for level and plumb with your mason's level. Use the handle of your trowel to gently tap the block into place. After you place the block, the mortar joint spacing should be the standard ⅜" for both the bed and head joints.

Do not move the block after it is pushed against its neighbor. If you must move the block, take it off and remortar the bed joint and the head joint. Unlike brick with its solid and complete mortaring, block mortaring is fragile. Because its webs are so small in area compared to its size, block mortar joints are easily disturbed by movement. Do not take the chance of a weakened mortar joint developing a leak in the wall.

5.2.0 Block Bed Joints

While bricks have one type of bed joint, the full furrowed joint, blocks can use one of three types of bed joint depending on their purpose. Check the specifications before laying a block wall to confirm which type of bed joint to use. Consider the following:

* If the block is laid as the first course on a footing, it takes a full furrowed bed joint, as does brick.
* If the block is not to be in a reinforced wall, the bed joint has mortar on the face shells only. *Figure 25* shows this type of mortaring.
* If the block is part of a reinforced wall that will have reinforcing grout in some cores, the block needs a full block bed joint. This has mortar on the face shells and on the webs, as shown in the detail in *Figure 25*. Mortaring the webs will keep the grout from oozing out of the cores.

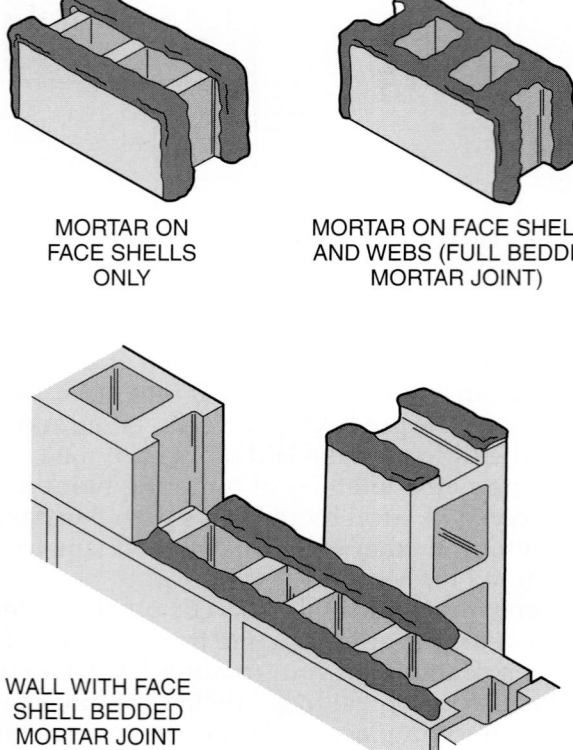

MORTAR ON FACE SHELLS ONLY

MORTAR ON FACE SHELLS AND WEBS (FULL BEDDED MORTAR JOINT)

WALL WITH FACE SHELL BEDDED MORTAR JOINT

105F25.EPS

Figure 25 ◆ Buttering block.

Sometimes, the specifications will call for an unreinforced wall to be laid with a full block bed joint. Mortaring the webs as well as the shells increases the loadbearing strength of the wall. The architect or engineer may have calculated that a full block bed joint will do the job instead of reinforcement. If you use only a shell bed joint, the wall will not have the calculated strength. This is another reason why it is important to read the specifications.

5.3.0 General Rules

These guidelines were covered in an earlier module, but they bear repeating. The way you work the mortar determines the quality of the joints between the masonry units. The mortar and the joints form a vital part of the structural strength and water resistance of the wall. Learning these general rules and applying them as you spread mortar will help you build good walls:

- Use mortar with the consistency of mud, so it will cling to the masonry unit, creating a good bond.
- Butter the head joints thoroughly for brick and block; butter both sides of the head joints for block.

- When laying a unit on the bed joint, press down slightly and sideways, so the unit goes against the one next to it.
- If mortar falls off a moving unit, replace the mortar before placing the unit.
- Put down more mortar than the size of the final joint; remember that placing the unit will compress the mortar.
- Do not string a spread more than six bricks or three blocks long; longer spreads will get too stiff to bond properly as water evaporates from them.
- Do not move a unit once it is placed, leveled, plumbed, and aligned.
- If a unit must be moved after it is placed, remove all the mortar on it and rebutter it.
- After placing the unit, cut away excess mortar with your trowel, and put it back in the pan, or use it to butter the next joint.
- Throw away mortar after 2 to 2½ hours, as it is beginning to set and will not give a good bond.

6.0.0 ◆ BONDING MASONRY UNITS

Masons deal with four types of bonds:

- A simple mechanical bond is made by the joining of mortar and a masonry unit. The strength of this bond depends on the mortar. This is also called a mortar bond.
- A pattern bond is a pattern formed by masonry units and mortar joints on the face of a surface. Unless it is the result of a structural bond, a pattern bond is purely decorative.
- A structural bond is made by interlocking or tying masonry units together so they act as a single structural unit.
- A structural pattern bond is the result of a structural bond that forms a pattern as well as a bond. Most traditional pattern bonds are structural pattern bonds.

The next sections discuss these different types of bonds. Note that the distinction between a structural bond and a pattern bond is hard to make. The act of overlapping or interlocking masonry to create a structural bond also creates a pattern. Defining a particular pattern as a structural bond or a structural pattern bond depends on local custom.

6.1.0 Mechanical or Mortar Bond

On the basic level, a mechanical bond is formed between the masonry unit and the mortar. This bond ties the masonry in a wythe into a single unit.

Brick on Block

Adding a brick veneer to a reinforced concrete block wall is a common construction method.

105SA08.EPS

For the majority of masonry construction, the most important property of mortar is bond strength. Mortar bond strength depends on the properties of the mortar and the bonding surface:

- The mortar must have the right proportions of ingredients for its use. It must stay wet enough to lay and level the masonry.
- The masonry surface should be irregular to provide mechanical bonding. It should be absorptive enough to draw the mortar into its irregularities.
- The masonry surface should not be so dry that it dries out the mortar. Slow, moist curing improves mortar bond and compressive strength.

The second most important property of mortar is bond integrity. The work of the mason defines the bond between masonry units. Bond integrity depends on the mason who does the following:

- Keeps tools and masonry units clean
- Butters every joint fully without air bubbles
- Does not move the masonry unit after it is leveled
- Levels units shortly after they are laid
- Uses mortar wet enough to dampen the masonry unit
- Keeps the mortar tempered
- Mixes fresh mortar after two hours

6.2.0 Pattern Bond

Pattern bonds add design but not strength to masonry walls. The stack bond (*Figure 26*) is only a pattern bond. It provides no structural bond as there is no overlapping of units. This pattern is more commonly used with block than brick.

If the stack bond pattern is used in a loadbearing wall, the wythe must be bonded to its backing with rigid steel ties. In loadbearing construction, this patterned wall should be reinforced with steel joint reinforcement ties.

Pattern and structural pattern bonding calls for placing brick in different positions in the wythes. *Figure 27* shows different ways of placing brick in order to make different kinds of patterns.

The stretcher is the everyday workhorse. Headers are used primarily for tying wythes together,

105F26.EPS

Figure 26 ◆ Stack bond.

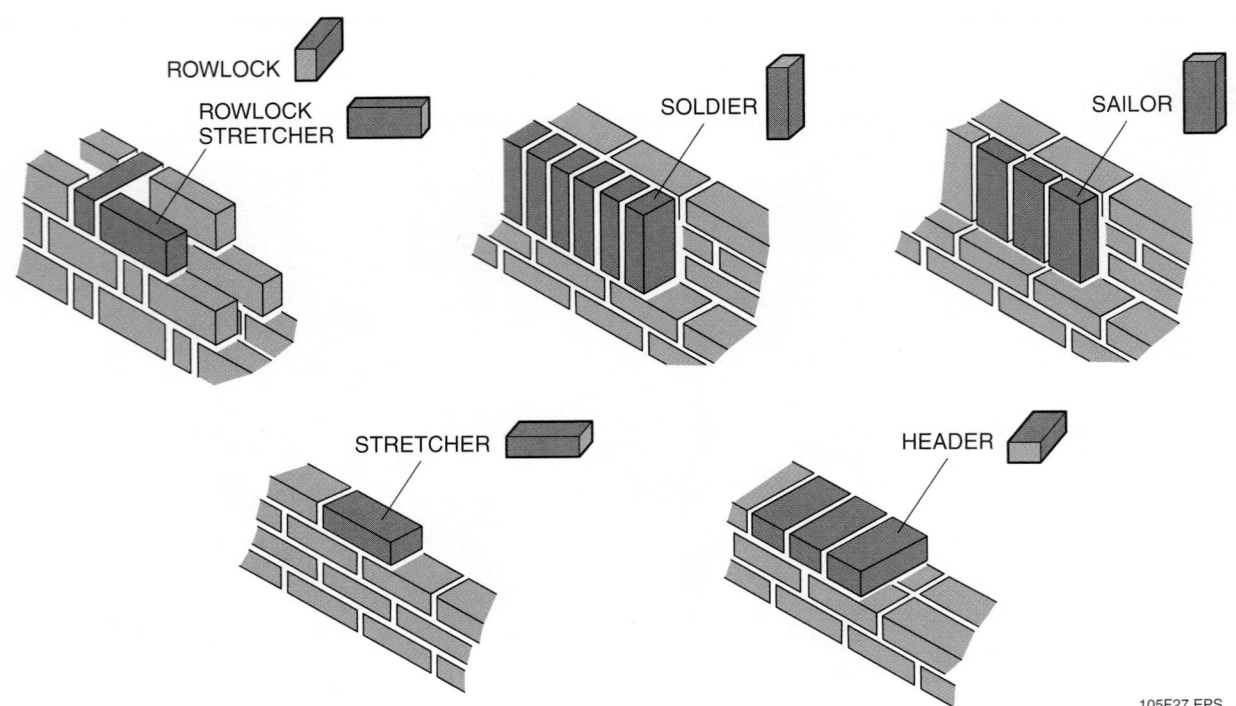

Figure 27 ◆ Brick positions in walls.

capping walls, flat windowsills, and pattern bonds. Soldiers are used over doors, windows, or other openings, and in pattern bonds. Shiners, or rowlock stretchers, are used in pattern bonds, in brick walks, and for leveling when a 4" lift is needed. Rowlocks are found in capping walls, windowsills, ornamental cornices, and pattern bonds. Sailors are rarely seen, except in pattern bonds and brick walks.

6.3.0 Structural Bond and Structural Pattern Bond

Wythes can be structurally bonded by using metal ties, joint reinforcements, anchors, grout, and steel rods. These engineering methods are used to increase strength and loadbearing by firmly tying masonry units and wythes together.

Another, older way to structurally bond a wythe is to lap masonry units. Lapping one unit halfway over the one under it provides the best distribution of weight and stress.

In a single-wythe wall, a structural bond is made by staggering the placement of the bricks. This results in the brick in one course overlapping the brick underneath. The structural pattern bond resulting from this simple overlap is the running bond, as shown in *Figure 28*. Common overlaps are the half lap and the one-third lap. Changing the proportion of the overlap changes the look of the pattern.

½-RUNNING BOND

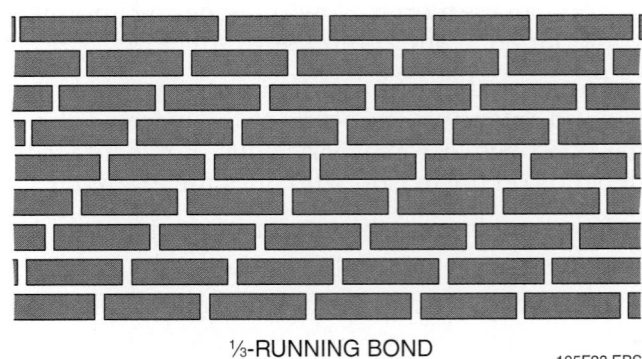

⅓-RUNNING BOND

Figure 28 ◆ Running bonds.

In two-wythe walls, a structural bond is made between the wythes. This can be made by rigid steel ties that equalize loadbearing. It can also be made by overlapping a brick from the face wythe to the backup wythe. The overlap brick is turned into the header or the rowlock position. This results in a complex structural bond that is also a structural pattern bond, with different sizes of brick facing out. The results are the traditional English bond and Flemish bond shown in *Figure 29*.

The Flemish bond consists of alternating headers and stretchers in every course. The English bond consists of alternating courses of headers and stretchers. If the headers are not needed for structural bonding, cut bricks are used. Brick can be laid to show different faces and cut in different ways.

The combination of the Flemish and English bonds with the running bond results in the common or American bond. As shown in *Figure 30*, the common bond is a running bond with headers every sixth course. The headers are in the Flemish or English pattern, according to the specifications.

The English cross, or Dutch bond, uses a structural pattern bond that repeats every four courses. The pattern courses are all stretcher, all header, and a course of three stretchers and one header. The last pattern course is all header again. *Figure 31* shows the Dutch bond.

The Dutch bond may seem complicated until you look at a traditional garden wall bond. *Figure 32* shows two variations on the garden wall struc-

Figure 30 ◆ American, or common, bond.

ENGLISH
CORNER

DUTCH
CORNER

105F31.EPS

Figure 31 ◆ English cross, or Dutch bond.

tural pattern bond. The double stretcher garden wall pattern shown in *Figure 32A* repeats every five courses. The dovetail garden wall pattern shown in *Figure 32B* repeats every 14 courses. More variations are possible. The only limit on patterning is the skill and ingenuity of the mason.

6.4.0 Block Bond Patterns

Block has its own set of commonly used bond patterns. *Figure 33* shows common block bonds, some of which, such as the herringbone, are also seen in brickwork. As with brick, the stack bonds do not provide any structural strength. With block, however, it is simple to reinforce stack bond with grout and steel or grout in the cores.

The other block bond patterns add structural strength. To get a solid face, block can only be laid in the stretcher mode. Pattern variations, such as the coursed ashlar, can be made by using different sizes of block. Modern block walls can also add visual interest through texture and surface designs.

Many designers rely on textures and surface designs, alone or in combination with bond patterns, to enhance block walls.

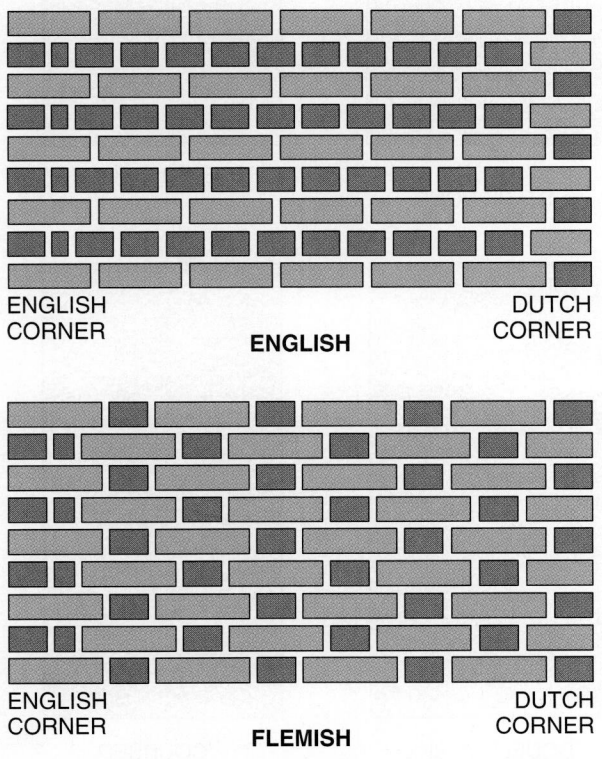

ENGLISH
CORNER

DUTCH
CORNER

ENGLISH

ENGLISH
CORNER

DUTCH
CORNER

FLEMISH

105F29.EPS

Figure 29 ◆ English and Flemish bonds.

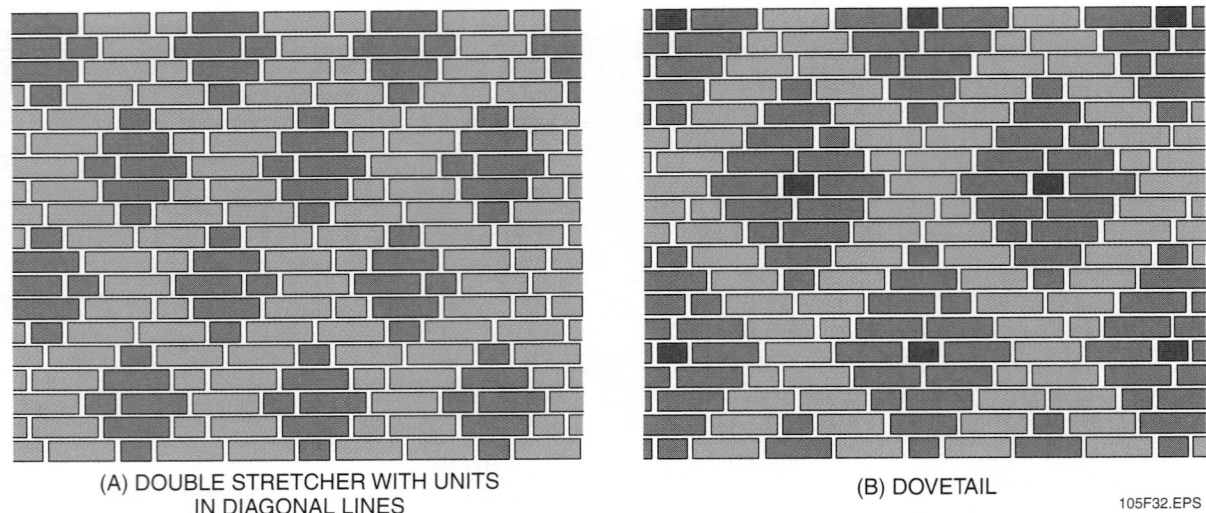

(A) DOUBLE STRETCHER WITH UNITS
IN DIAGONAL LINES

(B) DOVETAIL

105F32.EPS

Figure 32 ♦ Garden wall bonds.

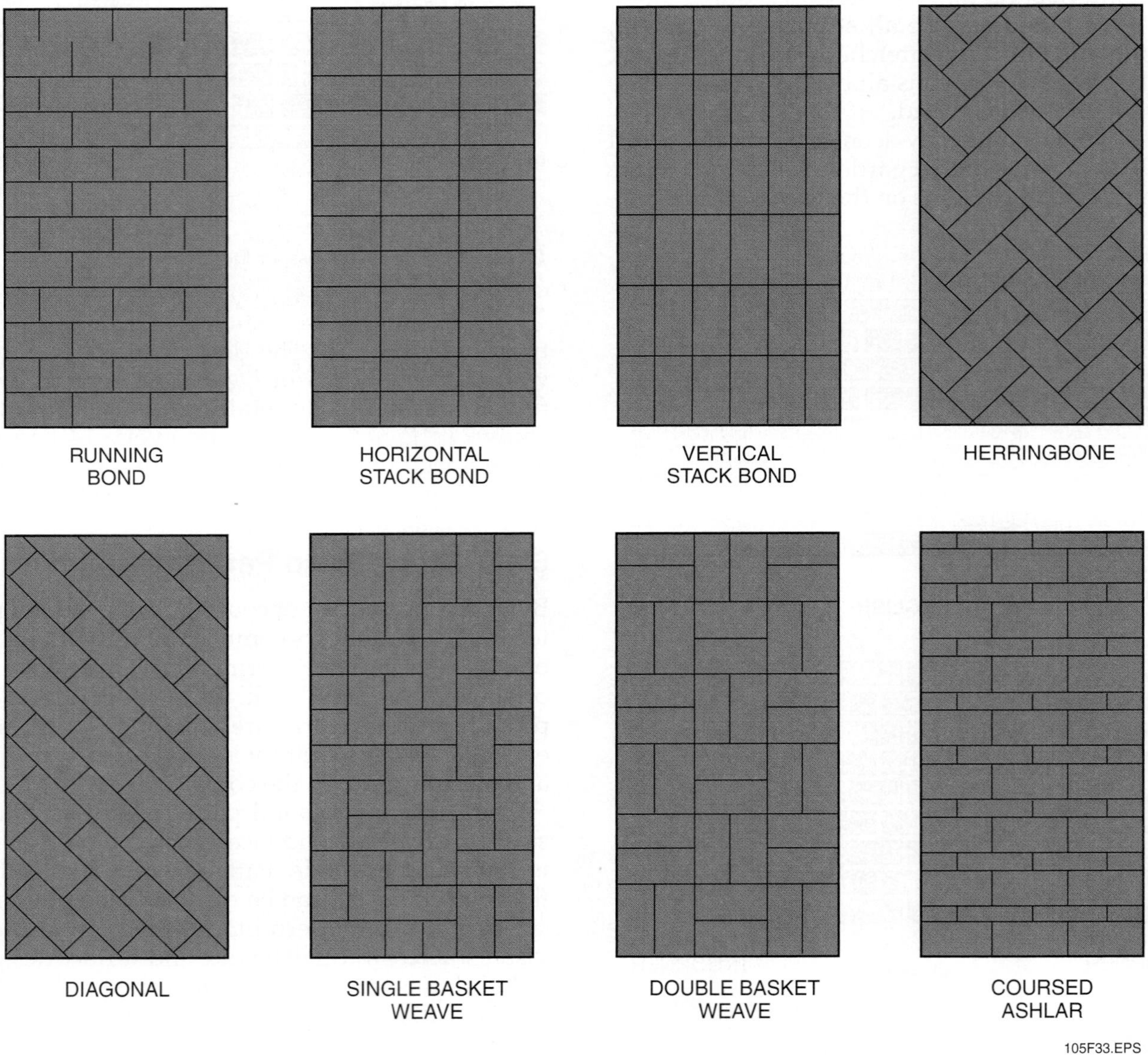

RUNNING
BOND

HORIZONTAL
STACK BOND

VERTICAL
STACK BOND

HERRINGBONE

DIAGONAL

SINGLE BASKET
WEAVE

DOUBLE BASKET
WEAVE

COURSED
ASHLAR

105F33.EPS

Figure 33 ♦ Block bonds.

7.0.0 ◆ CUTTING MASONRY UNITS

Masonry units often need to be cut to fit a specific space. Even when building on a modular grid, structural bond patterns, door and window openings, and corners usually call for some cut masonry units. English and Dutch corners specifically call for cut masonry units as part of the patterning.

On a large job, the masonry contractor or foreman will figure the pattern layouts and calculate the number of masonry units to be cut. Someone will be assigned to cut the units with a masonry saw or a splitter before they are needed. Sometimes, masons need to cut a few more units or cut to a slightly different size. This is when you need to know how to cut masonry with hand tools.

7.1.0 Brick Cuts

Brick can easily be cut by hand tool, masonry saw, or splitter. Sometimes you will need cut bricks for finishing corner patterns or for pattern bonds. *Figure 34* shows the common cut brick shapes and names. The king and queen closures are used for cornering.

7.2.0 Block Cuts

Block is usually cut in several standard ways. It can be cut across the stretcher face, both horizontally and vertically, as shown in *Figure 35*. You may easily make these cuts by hand. Blocks cut across the face horizontally are called splits or rips. If the block is cut exactly in half, it is called a

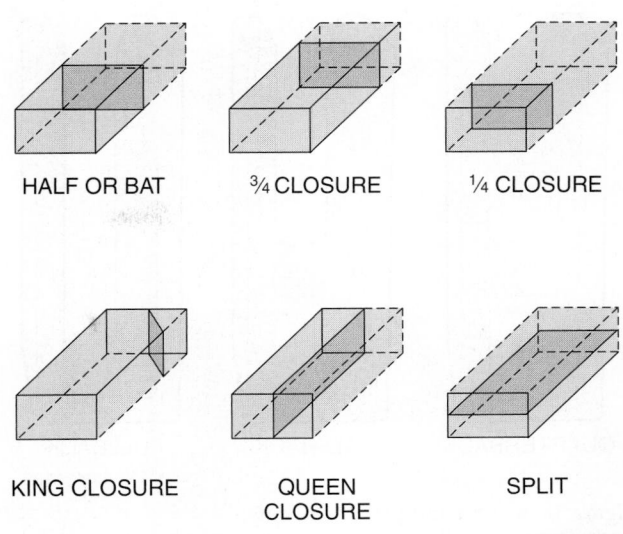

HALF OR BAT   ¾ CLOSURE   ¼ CLOSURE

KING CLOSURE   QUEEN CLOSURE   SPLIT

105F34.EPS

Figure 34 ◆ Common brick cuts.

half-high rip. Rip blocks are often used under windows. They act as a filler to reach a height of 8", so normal coursing can continue.

Blocks also get their webs cut out. Taking one end off a block makes an opening easily slipped over a pipe. Fitting the block to its location may take more cuts. You might make these cuts using a masonry saw. The cuts have their own names to save time and confusion (*Figure 36*).

The following three types of cuts are shown in *Figure 36* on a three-cell block:

- The quarterback cut has the end of one cell cut out, leaving two cells.

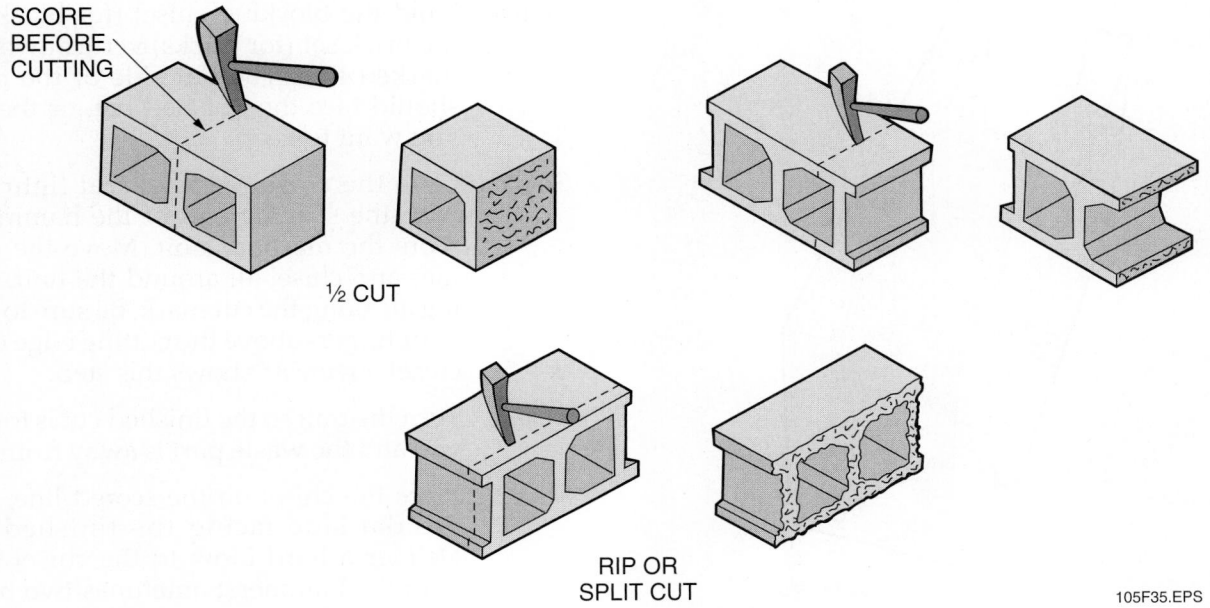

SCORE BEFORE CUTTING

½ CUT

RIP OR SPLIT CUT

105F35.EPS

Figure 35 ◆ Horizontal and vertical face cuts.

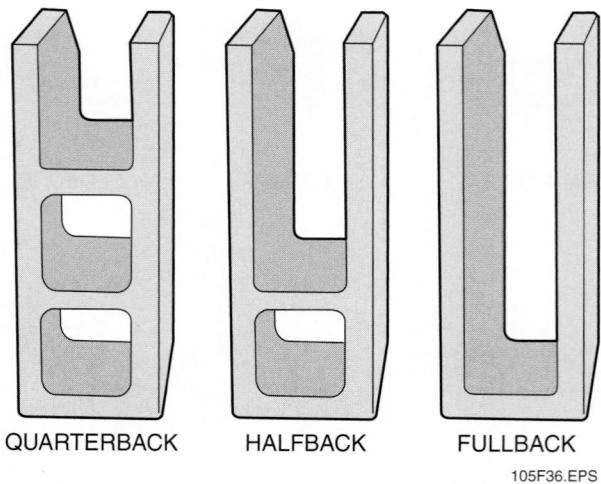

QUARTERBACK   HALFBACK   FULLBACK

105F36.EPS

Figure 36 ◆ End and web block cuts.

- The halfback cut has the end of one cell and the web of the next cell cut. This leaves one cell.
- The fullback cut has one end and both internal webs cut. This leaves only one end to hold the block together.

If a block has only two cells, the cuts are halfback and fullback; there is no quarterback cut for a two-cell block.

The bond beam block has the ends and inside webs of the block cut down about three fourths of the way. This cut (*Figure 37*) can be used for a lintel over an opening. The cuts give room for the reinforcement on top of the opening.

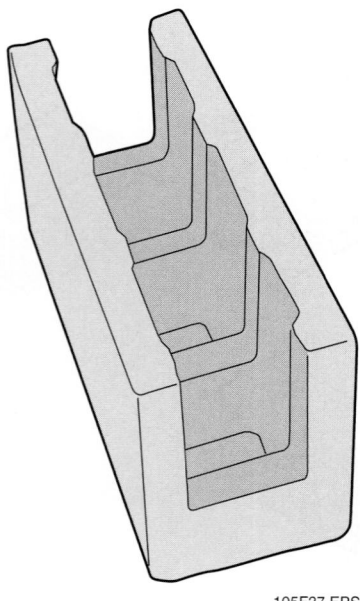

105F37.EPS

Figure 37 ◆ Bond beam cut.

7.3.0 Cutting with Hand Tools

Brick and block can be cut with chisels or a brick hammer. Brick can also be cut with the edge of a trowel. The procedures are detailed in the following sections.

 **WARNING!**
Remember to wear a hard hat and eye protection when cutting with hand tools. Never cut masonry over the mortar pan or near other workers. Chips may fly off, causing injury.

7.3.1 Cutting with Chisels and Hammers

Using the chisel and hammer can result in a smooth cut for block and brick. This procedure works well for both types of units:

Step 1   Check the tools you will use. Cutting edges should be sharp, and the hammer handle should be firmly attached.

Step 2   Put on your hard hat and safety goggles or other eye protection.

Step 3   Put the brick or block on a bag of sand, a board, or the ground to make a safe cutting surface. Make sure it is resting flat and plumb on a surface with some give to it.

Step 4   Use a steel square and a pencil to mark the cut all the way around the masonry unit.

Step 5   Hold the blocking chisel (for blocks) or the brick set (for bricks) vertically on the marked line. The flat side of the chisel should face the finished cut, or the part you want to keep.

Step 6   Give the chisel end several light taps with the striking end of the hammer to score the masonry unit. Move the hammer and chisel all around the unit, scoring all along the cut mark. Be sure to keep your fingers above the cutting edge of the chisel. *Figure 38* shows this step.

Step 7   Turn the unit so the finished cut is toward you and the waste part is away from you.

Step 8   Place the chisel on the scored line, with the flat side facing the finished cut. Deliver a hard blow to the chisel head with the hammer. Sometimes two blows are needed.

Figure 38 ◆ Cutting with a hammer and chisel.

Cutting in this way gives an accurate and clean cut. You can also make an accurate cut by using a hammer, instead of a chisel, for the final step. If you are cutting with the mason's hammer, follow Steps 1 through 7 just listed, then continue with these steps:

Step 1   Place the scored block on top of another block so that the waste part hangs free. You can hold the wanted part secure with your foot.

Step 2   Strike the waste end of the block with the striking end of the hammer. This knocks off the waste end, leaving a clean, finished cut.

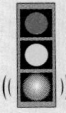 **NOTE**
Hammer and chisel cutting may not be permitted on some commercial jobs. Check the project specifications.

7.3.2 Cutting with Masonry Hammers

Cutting with the chisel end of the masonry hammer gives a rougher cut. The steps for cutting brick in this way are as follows:

Step 1   Check the tool you will use. Cutting edges should be sharp, and the hammer handle should be firmly attached.

Step 2   Put on your hard hat and safety goggles or other eye protection.

Step 3   Use a steel square and a pencil to mark the cut all the way around the masonry unit.

Step 4   Hold the brick in one hand and your hammer in the other. Hold the part of the brick you want to keep with the waste part down.

Step 5   Strike the brick lightly with the chisel end of the hammer to score it along the marks on all sides. As you turn the brick, be sure to keep your fingers and thumb off the side of the brick being scored. Figure 39 shows this step.

Step 6   Strike the face of the brick sharply with the chisel end of the hammer. Let the waste part fall to the ground.

Step 7   If necessary, use either end of the hammer to dress out any small, rough edges left by the cut.

The same procedure can be used for block, except that block is not usually held in your hand. Set the block on sand, the ground, or a board for a safe cutting surface. Follow Steps 1, 2, 3, and 5. Then tilt the block face away and prop it with another block. Hold it with your foot and apply a sharp blow with the hammer. Blocks may need to be struck on both faces. Finish by dressing out any rough edges.

7.3.3 Cutting with Trowels

Cutting with a trowel is a last resort when your hammer is not available. It is not recommended for block or very hard brick. Cutting with a trowel in cold weather can break the blade. After you have mastered cutting with the hammer and brick set, this method will be easier to learn.

Figure 39 ◆ Cutting with a brick hammer.

Mark the brick for cutting. Hold the brick in one hand by the part you want to keep. Keep your fingers well under the brick to avoid cutting them. Strike the brick using the upper edge of the trowel, close to the heel. Strike hard with a quick sharp blow and a sharp snap of your wrist. If the brick does not break, use a brick hammer and follow the steps previously outlined.

7.4.0 Cutting with Saws and Splitters

Cutting masonry units with power saws or splitters takes two kinds of awareness: you must be aware of how to operate the machinery, and also be aware of the masonry units and cuts.

7.4.1 Saws and Splitters

Masonry saws are available in freestanding and portable models (*Figure 40*). They use either diamond or carborundum blades. Diamond blades are irrigated to prevent fire. The water wets the masonry unit, which must dry out before it can be laid. Carborundum blades are not irrigated, but they make clouds of dust. The dust must be blown or vented away from the saw and nearby workers. Smaller handheld saws use dry blades and also make clouds of dust, which must be blown or vented away.

Splitters (*Figure 41*) do not use water or generate dust. They do, however, exert tremendous force through gearing and hydraulic power.

105F40.EPS

Figure 40 ◆ Portable masonry saw.

105F41.EPS

Figure 41 ◆ Brick splitter.

As with any potentially dangerous equipment, follow these general safety rules:

- Do not operate any saw or splitter until you have had specific instructions in handling that equipment.
- Check the condition of the equipment before using it.
- Follow all safety rules for using power equipment or otherwise dangerous equipment.
- Wear a hard hat, respiratory protection, goggles, gloves, and other appropriate personal protective equipment as needed.
- Never force the equipment.
- For bedded saws, use conveyor carts, pushers, or blocks to move the unit under the blade.
- For handheld saws, secure and brace the unit before cutting it.
- Do not operate equipment when you are feeling ill or are taking any medication that may slow your reaction time.

Review the safety rules as well as the operating instructions before operating any equipment.

 WARNING!

Silicosis is a serious lung disease that is caused by inhaling sand dust. Silica is a major component of sand and is therefore present in concrete products and mortar. Silica dust is released when cutting brick and cement, especially when dry-cutting with a power saw. Any time you are involved in the cutting or demolition of concrete or masonry materials, be sure to wear approved respiratory equipment.

7.4.2 Units and Cuts

After you have checked out the equipment, the safety procedures, and the operating procedures, check out the masonry units.

- Know what the finished item should look like.
- Mark all cutting lines before the blade starts running.
- Mark cutting lines in grease pencil for wet-cut saws.
- Do not cut a cracked masonry unit.

If you are not clear about the cuts to be made, ask your supervisor for more direction.

8.0.0 ◆ LAYING MASONRY UNITS

Laying masonry units is a multi-stage process. As discussed in previous sections, the first step in any masonry job is reading the specifications. This is followed by planning the layout of the job. The next tasks are to locate and lay out the wall, then do the dry bonding. Dry bonding assures that the layout will be correct and that the minimum number of cut blocks will be needed. Then, calculate the number of units to cut, and cut them. For the purpose of this module, assume all work is done in running bond on the modular grid system. Now you are ready to mix the mortar and start the actual laying.

The next tasks are to spread mortar, lay masonry units in place, and check their positioning.

An earlier module gave detailed procedures for spreading and furrowing bed joints and buttering head joints. The following sections give procedures for positioning individual masonry units, laying to the line, and building corners and leads.

8.1.0 Laying Brick in Place

Laying brick will be less of a strain if you use as few motions as possible. One way to make things easier is to have your materials close by. If you are working on a veneer wall, use both hands to pick up bricks and stack them on the completed back section of the wall. Place them along the length you will be laying before you start spreading mortar. This will eliminate the need to bend to pick bricks up off the ground as you go.

Using both hands is another efficient practice. Keep your trowel in one hand, and use the other for picking, holding, and placing bricks. This will make the work easier and faster.

Use your fingers efficiently as well. When you pick up a brick, hold it plumb. Pick it up so that your thumb is on the face of the brick. Let your fingers and thumb curl down over the top edges of the brick, slightly away from the face. In this position, your fingers will not interfere with the line as you place the brick on the wall. *Figure 42* shows the proper position of fingers and thumb for holding and laying the brick in place.

Keep your mason's level close by. After laying every six bricks, check them for position. This means checking for height, level, plumb, and straightness as you go. If you cannot adjust a unit to meet the four measures of height, level, plumb, and straightness, you must take the brick out and start over.

105F42.EPS

Figure 42 ◆ Proper hand position.

8.1.1 Placing Brick

The most important placing rule is to place gently. Do not drop the brick or block onto the bed joint; lower it down gently. Press the brick forward at the same time so that it will butt against the unit next to it. Mortar should ooze out slightly on both head and bed joints to show that there has been full contact.

Align the latest brick with the brick next to it as you place it. Line it up with the mason's line if you are using one. By standing slightly to one side, you will be able to sight down the wall. This will help maintain plumb head joints by sighting the brick below the newly laid unit.

You may need to slightly adjust the unit in its bed. First try pressing downward on the brick with the heel of your hand. Keep part of the heel of your hand on the brick next to the one you are adjusting. If this is not sufficient, you may need to tap the brick with the handle of your trowel. After adjusting the brick, cut off the extruded mortar with your trowel, and lay the next unit. Cutting mortar as you go will help you to keep the masonry clean.

8.1.2 Checking the Height

The first check is always course height. If this is off, there is no use checking anything else. Use your modular or standard course spacing rule to check the height of the brick. Follow these steps:

Step 1 After the bricks are laid on the wall, unfold the rule (*Figure 43*), and place it on the base or footing used for the mortar and brick.

Step 2 Hold the rule vertically. Check that the end of the rule is flat on the base, so the reading is accurate. If you are using standard brick, the first course should be even with number 6 on the modular rule. If you are using a different size of brick, check the appropriate scale on the modular or course rules.

Step 3 If the height of the course does not hit the right place on the rule, take the bricks out, and clear off the bed joint. Lay the bed joint again, and replace the buttered bricks. Recheck the height of the course. Then replumb and relevel.

The height, or vertical course spacing, depends on the thickness of the mortar joints. Practicing laying full bed and head joints is the fastest way to learn to make standard size joints.

If more than one course is laid, always set the modular rule on the top of the first course to measure. The base may have been irregular, and a large joint may have been used to level the first course.

8.1.3 Checking Level

After checking the height for your string of six bricks, check with your mason's level for levelness using the following steps:

Step 1 Remove any excess mortar on top of the bricks.

Step 2 Place your mason's level lengthwise on the center width of the six bricks to be checked.

Step 3 Use your trowel handle to gently tap down any bricks that are high with relation to the mason's level (*Figure 44*). Do not tap them so hard they sink too low.

Step 4 If bricks are low, pick them up. Clean and mortar the bed and the head joint again, and reposition the brick. Reposition the mason's level again, and get it level.

8.1.4 Checking Plumb

The next step after leveling is to check for plumb, or vertical straightness. Follow these steps:

Step 1 Hold the level in a vertical position against the end of the last brick laid (*Figure 45*).

Step 2 Tap the brick with the trowel handle to adjust the brick face either in or out.

Step 3 Move the level to the end of the first brick laid, and repeat the process.

Figure 46 gives profiles and names for bricks that are plumb and out of plumb. The large black dot represents the mason's line. By looking and touching, you can train your hand and eye to know bricks that are plumb and bricks that are not.

105F44.EPS

Figure 44 ◆ Leveling the course.

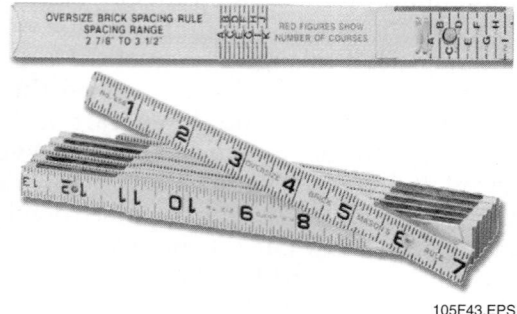

105F43.EPS

Figure 43 ◆ Mason's rules.

105F45.EPS

Figure 45 ◆ Checking plumb.

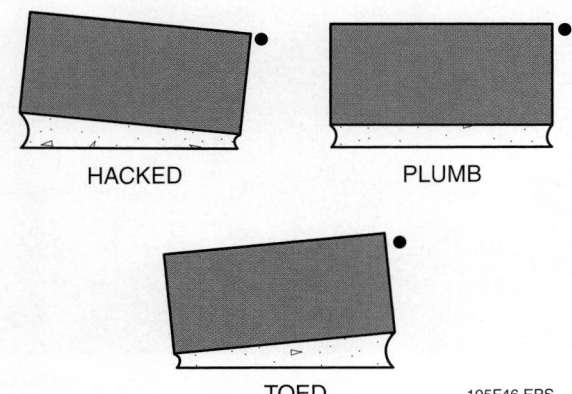

HACKED   PLUMB

TOED   105F46.EPS

Figure 46 ◆ Plumb and out-of-plumb bricks.

8.1.5 Checking Straightness

After the first and last bricks in the string have been plumbed, check the rest for straightness:

Step 1   Hold the mason's level in a horizontal position against the top of the face of the six bricks, as shown in *Figure 47*.

Step 2   Tap the bricks either forward or back until they are all aligned against the mason's level. Be careful not to move the end plumb points while you are aligning the middle four bricks.

By sighting down from above, you can train your eye to know bricks that are straight and bricks that are not.

8.1.6 Laying the Closure Unit

The last unit in a course is called the **closure unit**. Masons lay corners of a wall first then work from each corner toward the middle. The last unit, or

105F47.EPS

Figure 47 ◆ Checking for straightness.

closure unit, must fit in the gap between the masonry units that have already been laid (*Figure 48*). The closure unit should fall toward the middle of a wall. The space left for it should be large enough for the unit and its two head joints.

The process for laying the closure unit is the same for block and brick:

Step 1   Butter the closure unit on both head joints.

Step 2   Butter the adjacent units on their open head joints.

Step 3   Gently ease the unit into the space.

Step 4   If any mortar falls out of a closure unit joint, remove the unit, and reset it in fresh mortar.

If the head joints have been properly spaced, the closure unit will slide in with the specified

105F48.EPS

Figure 48 ◆ Placing a closure unit.

joint spacing. Otherwise, the closure unit will have head joints that are too large or too small. If this is the case, remove the last three or four units that were laid on either side of the closure unit. Remortar them, and relay them to correct for the closure head joint size. The objective is to avoid a sudden jump in the size of a head joint. A big change in joint size will catch the eye and can also skew the pattern bond. If you must move bricks, be sure to check them again for height, level, plumb, and straightness.

8.2.0 Placing Block

Placing block is similar to placing brick except that it is not a one-handed job (*Figure 49*). Butter the units on both sides of head joints. Use both hands to place the block. Do not drop it, but move it slowly down and forward so it butts against the adjacent unit. Slightly delaying release allows the block to absorb moisture from the mortar, which makes a good mechanical bond. If the mortar does not ooze from the joints, you are not using enough mortar. There will be voids in the joints, and the wall will eventually leak.

105F49.EPS

Figure 49 ◆ Placing a block.

Tilt the block toward you as you position it. Look down over the edge and into the cores to check alignment with the block underneath. If a block is set unevenly, check to see if a pebble or other material is wedged between the mortar and the block. If so, take the block up and clean it. Rebutter fresh mortar for both bed and head joints, and reset it.

If the block requires adjustment, lightly tap it into place. You may want to use your mason's hammer to tap the block as it may require stronger taps.

Check each block for position as described for brick. If the block cannot be adjusted, take it up, remortar it, and reset it.

8.3.0 Laying to the Line

To keep masonry courses level over a long wall, masons lay the units to a line. Working to a line allows several masons to work on the same wall without the wall moving in several directions. The line is set up between corner poles or corner lead units. The poles or leads must be carefully checked for location, plumb, level, and height. The line is placed on the outside of the course, so that will be the most precisely laid side. The mason usually works on the same side as the line but can also work from the other side depending on job conditions and experience.

Each masonry unit is placed with its outside top edge level with the line and 1/16" away from it. The distance is the same for all types of masonry units. Your eye will get trained to measure that distance automatically after some practice laying to the line.

A mason's line needs to be tied to something that will not move. It must be tied taut at a height that can be measured precisely. The mason's line is attached to corner poles or corner leads by means of line stretchers, line blocks, or line pins.

8.3.1 Setting Up the Line Using Corner Poles

Corner poles allow masons to lay to the line without laying the corners first. Attach the corner pole securely. It must not move as you pull a mason's line from it. For brick veneer walls, the corner pole can be braced against the frame or backing wall. You must check the placement of the corner poles before you string the line. If the pole has course markings, check that they are the correct distance from the footing. If the pole has no markings, transfer markings from your course rule. Make sure you start the measures from the footing.

Step 1 Attach the line to the left pole to start. If the pole has no clamps or fasteners, attach the line with a hitch or half hitch knot.

Stretch it to the right pole and gradually tighten it until it is stretched. Use a hitch or half hitch to secure it, tightening it as you tie. Check that it is at the proper height before you start laying.

Step 2 After laying each course, move the line up to the next course level. Stretch and measure it again. It is critical to make sure the line is at the proper height for each course.

Step 3 Use your modular rule or course spacing rule to check the line height at each end for every course.

8.3.2 Setting Up the Line Using Line Blocks and Stretchers

A mason's line can be set between corner leads or corners laid to mark the ends of the wall. The line can be attached by line blocks, line stretchers, or line pins.

Line blocks (*Figure 50*) have a slot cut in the center to allow the line to pass through. It takes two sets of hands to set up line blocks. The procedure is as follows:

Step 1 Pass the line through the slot of the block. Tie a knot, or tie a nail on the end of the line, to keep it from passing through the slot.

Step 2 Have one person hold the line block aligned with the top of the course to be laid. Traditionally, mason's lines start on the left side as you face the wall.

Step 3 Place the line block so that it hooks over the edge of the masonry unit, and hold it snug.

Step 4 The second person will walk the line to the right end of the wall.

105F50.EPS

Figure 50 ◆ Line block.

Step 5 The second person then pulls the line as tight as possible and wraps the line three or four turns around the middle of the line block.

Step 6 The second person hooks the tensioned line block over the edge of the corner.

Step 7 Both parties check that the line is at the correct height.

Line stretchers are put in place following the same steps. The line stretcher slips over the top of the blocks, not the edges (*Figure 51*). Line stretchers are useful when the corner lead is not higher than the course to be laid.

8.3.3 Setting Up the Line Using Line Pins

Steel line pins hold a mason's line in place. The line pin is less likely to pull out of the wall because of its shape. The peg end of the line, or the starting end, is traditionally started at the left as the mason faces the wall.

Step 1 Drive the line pin securely into the lead joint (*Figure 52*). Make sure that the top of the pin is level with the top of the course to be laid. Place the pin at a 45-degree downward angle, several units away from the corner. This will prevent the pin from coming loose as the line is pulled.

Step 2 Tie the line securely to the pin using the notches on the pin. Give the line a few very sharp, strong tugs. This tests whether the pin will come out as the line is tightened and helps to prevent injuries caused by flying line pins.

Step 3 Walk the line to the other lead. Drive the second line pin securely into the lead joint even with the top of the course. Check and

measure that the pin is secure and in the correct position before applying the line.

Step 4 Wrap the line around the pin, and start tensioning. Pull the line with your left hand, and wrap it around the line pin with your right hand. Use a clove hitch or half hitch knot to secure the taut line to the pin. Be careful not to pull the line so tight that it breaks.

Step 5 When you move the line up for another course, immediately fill the pin holes with fresh mortar. If you wait until later to fill the line pin holes, you will need to mix another batch of mortar. Taking care of the holes as you move the pins saves many steps at the end of the project.

8.3.4 Setting Up the Line Using Line Trigs

To keep a long line from sagging, masons set trigs to support the line midstring.

Step 1 Set the trig support unit in mortar in position on the wall. Be sure that this unit is set with the bond pattern of the wall, close to the middle of the wall.

Step 2 Check that the unit is level and plumb with the face of the wall. Check that the unit is at the proper height with a course rule or course pole.

Step 3 Sight down the wall to be sure that the trig unit is aligned with the wall and is set the proper distance from the line. The trig support unit is a permanent part of the wall, so place it carefully.

Step 4 After the trig support unit is in the proper place, slip a trig or clip over the taut line. Check that the line is still in position. Lay the trig on the top of the support unit with the line holder on the bottom side. Place another masonry unit on top of the trig to hold it in place. *Figure 53* shows the use of a trig.

Step 5 Check the line for accuracy once more. The line should just be level with and slightly off the corner of the trig support unit and the standard $\frac{1}{16}$" away from it.

8.3.5 Laying Brick to the Line

After you string the line and put on a trig, you can begin to lay masonry units to the line. The advantage of laying to the line is that it cuts down on the need for the mason's level.

105F51.EPS

Figure 51 ◆ Using a line stretcher.

105F52.EPS

Figure 52 ◆ Using a line pin.

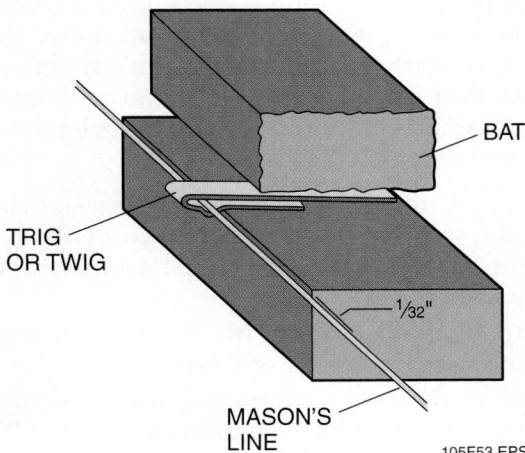

BAT

TRIG
OR TWIG

1/32"

MASON'S
LINE

105F53.EPS

Figure 53 ◆ Using a line trig.

You must lay to the line without disturbing the line. If the line is hit, other masons on the line will have to wait for the line to stop moving. Hitting the line is called **crowding the line**. Even experienced masons will crowd the line occasionally. To avoid crowding the line, hold your brick from the top, as shown in *Figure 54A*. As you release the brick, roll your fingers or thumb away from the line, then press the brick into place, as shown in *Figure 54B*.

The brick must come to sit ¹⁄₁₆" inside the line. The top of the brick must be even with the top of the line. Looking at the brick from above, you should be able to see a sliver of daylight between the line and the brick. Brick set too close is crowding the line. Brick set too far is **slack to the line**. When laid correctly, the bottom edge of the brick should be in line with the top of the course under it, and the top edge of the brick should be ¹⁄₁₆" back and even with the top edge of the line.

Adjust the brick to the line by pressing down with the heel of your hand. Check that the brick is not hacked or toed. While you are learning to lay to the line, it is a good idea to check your placement with the mason's level. After you have gained some skill in working to the line, you will find you will not need to use the mason's level so often.

With your trowel, cut off the mortar just squeezed out of the joints. Apply the mortar to the head of the brick just laid. By buttering the head joint like this, you will not have to return to the mortar pan for each individual head joint. When buttering the head, hold the trowel blade at an angle so as not to move or cut the line.

| (A) | (B) | 105F54.EPS |

Figure 54 ◆ Laying brick to the line.

Most of the mason's time is spent laying brick to the line. Practice will improve your ability to lay precisely without disturbing the line constantly. As you learn to do this, there are some additional habits you should pick up to save yourself time and energy:

- Always pick up a brick with the face out so that it is in the same position in which you will lay it. Limit turning brick in your hand as this slows and tires you.
- Pick up frogged brick with the frog down because this is the way it will be laid.
- Fill head and bed joints plump and full. This cuts down the time you will need to strike joints later. This also ensures stronger, waterproof walls.
- Stock your brick within arm's length, or approximately two feet away. When working on a veneer or cavity wall, stock your brick on the backing wythe.

8.3.6 Laying Block to the Line

The difference in laying block and brick to the line is the difference in handling the units.

Blocks should be kept dry at all times, as moisture will cause them to expand. If they are used wet, they will shrink when they dry and cause cracks in the wall joints. To cut down on handling, stack them close to the work sites with the bottom (smaller) shells and webs down.

Practicing laying block will let you discover the easiest methods for yourself. Find a way to hold the buttered block that is comfortable for you. Lift the block firmly by grabbing the web at each end

of it, and lay it on the mortar joint. Keep the trowel in your hand when laying block to save time.

As you place the block, tip it toward you a little. You can look down the face to align the block with the top of the block in the course below. Then, roll the block back slightly so that the top is in correct alignment to the line. At the same time, press the block toward the last block laid. Moving the block slowly is key to this process. Do not release the block quickly, or you will have to remortar and reposition it.

You can adjust the block by tapping. Be sure to tap in the middle of the block, away from the edges. Block face shells may chip if you tap on them. Using your trowel handle on the block is not recommended because the block roughens up the end of the handle. Use your mason's hammer.

8.4.0 Building Corners and Leads

Corners (*Figure 55*) are called leads because they lead the laying of the wall. They set the position, alignment, and elevation of the wall by serving as guides for the courses that fill the space between them. Building corners requires care as well as accurate leveling and plumbing to ensure that the corner is true.

As you learn to build corners, practice technique and good workmanship. Speed will follow. Be certain that each course is properly positioned before going on to the next. Once a corner is out of alignment, it is difficult to straighten it.

In addition to corners being leads, masons also have leads that are not corners. The next sections discuss both types of leads.

Figure 55 ◆ Building corner leads.

105F55.EPS

8.4.1 Rackback Leads

Sometimes it is necessary to build a lead or guide between corners on a long wall. This is a lead without corner angles, or **returns**. It is merely a number of brick courses laid to a given point. A **rackback** lead is **racked**, or stepped back a half brick on each end. This means that the lead is laid in a half-lap running bond with one less brick in each course.

The first course is usually six bricks long, the length of the mason's level. Each course is one brick less until the sixth course has only a single brick, pyramid fashion. *Figure 56* shows a completed racked lead.

Building a lead starts with marking the exact place in line with the corners and properly located for the bond pattern. Use a chalkline between the corners to locate the place. Lay brick in the rackback lead by following standard techniques. As each course is laid, check the course spacing on each course with the modular or spacing rule. Then check the course level, plumb, and straightness with your mason's level.

When the lead is complete, it needs to be checked for diagonal alignment. Do this by holding the mason's level at an angle along the side edges of the end bricks. As shown in *Figure 57*,

hold the rule in line with the corner of each brick. This lets you check that no bricks are protruding.

Next, **tail** the diagonal by laying your mason's level on the points of the end bricks, as shown in *Figure 58*. If the rule touches the edges of all the bricks, the head joint spacing is correct. If the head joint spacing is not correct, take out one or two units in that course and the courses above it. Clean and reset them with fresh mortar.

The rackback lead can now be used to anchor a line as detailed previously. Learning to build rackback leads will teach you three-quarters of what you need to know about building corners.

8.4.2 Brick Rackback Corners

A rackback corner is a rackback lead with a return or bend in it. The return must be a 90-degree angle unless the specifications say otherwise. Placement and alignment of the corner are crucial

Figure 56 ◆ A rackback lead.

105F56.EPS

Figure 57 ◆ Checking alignment.

105F57.EPS

Figure 58 ◆ Tailing the diagonal.

because the corner will set the location of the remainder of the wall. Laying a corner can be intricate, demanding work if there is a pattern bond to follow. The following steps are for building an unreinforced outside rackback corner in a half-lap running bond pattern.

Step 1 Check the specifications for the location of the corner. Determine how high the corner should be. The corner should reach halfway to the top of a wall built with no scaffold, or halfway to the bottom of the scaffold. Subsequent corners are built as the work progresses.

Step 2 Determine the number of courses the corner will need. Use your course rule to calculate the courses in a given height; then calculate the number of bricks in each course. The sum of the stretchers in the first course must equal the number of courses high the corner will reach. If, for instance, you need a corner 11 courses high, the first course will have 6 bricks on one leg, 5 bricks on the other.

Step 3 Locate the building line and the position of the corner on the footing or foundation. Clean off the footing, and check that it is level. Lay out the corner with a steel square. Mark the location directly on the footing. Check the plan or specifications to determine which face of the corner gets the full stretcher and which face gets the header. Some plans have detailed drawings of corners.

Step 4 Dry bond the units along the first course in each leg of the corner. Mark the spacing along the footing.

Step 5 Lay the first course in mortar, and check the height, level, plumb, and straightness. Be sure to level the corner brick and the end brick before the bricks in the center. Check height, level, plumb, and straightness for each leg of the corner.

Step 6 Check level on the diagonal, as in *Figure 59*. Lay the mason's level across each diagonal pair of bricks. This will let you make sure that the corner continues level across the angle.

Step 7 Remove excess mortar along the bed and head joints and from the leg ends. Also remove excess mortar from the inside of the corner.

Step 8 **Range** the bricks. Ranging is sighting along a string to check horizontal alignment. Ranging is done after the first course is laid. Fasten one end of the line to the edge of one corner leg, and wrap the line around the outside of the corner. Fasten the other end of the line to the outside edge of the other corner leg. Adjust any bricks in the line that are not in perfect horizontal alignment with the line. Then take the line off.

Step 9 Lay the second course, reversing the placement of the bricks at the corner. The leg that had the full stretcher before now gets the header. Use one less brick in the second course to rack the ends. Since each course is racked, stop spreading mortar half a brick from the end of each course.

105F59.EPS

Figure 59 ◆ Leveling the diagonal.

Step 10 After placing the bricks, check height, level, plumb, and straightness. Check across the diagonal as well. To train your eye, sight down the outermost point of the corner bricks from above to check plumb. Remove excess mortar from the outside and inside of the corner and from each exposed edge brick.

Step 11 Continue until you have reached the required number of courses. Use one less brick in each course. Lay and check each course. Remove excess mortar from each course, inside and outside. Be sure that aligning the bricks does not disturb the mortar bond. If the mortar bond is disturbed, take up the brick, clean it, rebutter it, and replace it.

Step 12 If the corner does not measure up at each course, take the course up, and do it again. This is easier than taking the wall up and doing that again.

Step 13 When the corner is at the required number of courses, check each leg of the corner for diagonal alignment just as you did for the rackback lead. Hold the mason's level at an angle along the sides of the end bricks as shown in *Figure 57*. This lets you check that no bricks are protruding. Also set the mason's level on top of the racked

edges as shown in *Figure 58*. This lets you check that each one touches the level and that the head joint spacing is correct.

Step 14 Check the mortar, and strike the joints. Brush the loose mortar carefully from the brick. Check the height of the corner with the modular spacing rule. Recheck the corner to make sure it is plumb, level, and straight. If it does not measure up, take the corner down, and start over.

Because the corner is so important to the wall, speed is not half as important as accuracy. Learn to be accurate, and the speed will follow.

8.4.3 Block Rackback Corners

Speed is the main advantage in building with block. It takes an experienced mason about 40 minutes to lay a block corner compared to 180 minutes to lay a brick corner. Because they are larger and heavier, blocks require some special handling. Blocks chip easily when moved or tapped down in place; therefore, they must be eased slowly into position. Their size makes them harder to keep level and plumb. Each block needs to be checked for position in all dimensions. But even with these disadvantages, they do save time and money.

The procedure for laying a block rackback outside corner follows the same steps as for brick. The main difference is that block requires different

Make Sure the Corner Is Square

You can use a framing square to make sure the corner is square. Each brick in the lead should touch the framing square.

105SA10.EPS

mortaring and more checking with the mason's level. The following steps do not repeat location material previously covered:

Step 1   Clean the foundation, and dampen it. Locate the point of the corner. Snap a chalkline from this point across the wall location to the opposite corner. Repeat the procedure on the other side of the corner. This aligns the corner with the other corners. Check the accuracy of the chalkline with a steel square before laying any block.

Step 2   Check that the footing is level. Lay the first course as a dry bond. Because actual sizes of block may vary, space out the dry bond with bits of wood for the joints. Check that the dry bond is plumb to find any irregularities in the footing. If the footing is too high in places, make an adjustment by cutting off some of the bottom of the block. If the footing is too low, add some pieces of block to bring the first course to the correct level.

Step 3   Lay the corner block first. Use a full bedded mortar joint without a furrow. Check the corner block for height, level, plumb, and straightness. Check for plumb on both sides of the block.

Step 4   Continue with the leg of the corner. Line up each block with the chalkline, checking the placement of each block for height, level, plumb, and straightness. Check for alignment as well. When both legs are finished, check them again, and check the diagonals.

Step 5   Do not remove excess mortar immediately, because this could cause the block to settle unevenly. Remove excess mortar from the first course after the second course is laid.

Step 6   For subsequent courses, apply mortar in a face shell bedding on top of the previously laid course. Check each block for height, level, plumb, and straightness.

Step 7   Check for diagonal alignment; then tail the rack ends. If the edges of all blocks do not touch the level, the head joints are not properly sized. Adjust the block if the mortar is still plastic enough, or rebuild the courses, as required.

When measuring blocks, each one must touch and be completely flush with the mason's level. They must also be completely in line with the chalk marks on the footing. Repeat all measurements often to prevent bulges or depressions in the wall and to keep the courses in line.

9.0.0 ◆ MORTAR JOINTS

Mortar joints between masonry units serve the following functions:

- Bonding units together
- Compensating for differences in the size of the units
- Bonding metal reinforcements, grids, and anchor bolts
- Making the structure weathertight
- Creating a neat, uniform appearance

Mortar joints are made by buttering masonry units with mortar and laying the units. The mason controls the amount of mortar buttered so that it fills a standard space between the units. Excess mortar oozes out between the units, and the mason trims it off. But this is not the last stage in making a mortar joint. After it dries partially, the mortar left between the masonry units must be tooled to be a proper mortar joint.

It is this last step, the tooling, that gives the mortar joints their uniform appearance and weathertight quality.

9.1.0 Joint Finishes

Mortar joints can be finished in a number of ways. *Figure 60* shows some standard joint finishes. Usually, the joint finish will be part of the detailed specifications on a project. The process of tooling

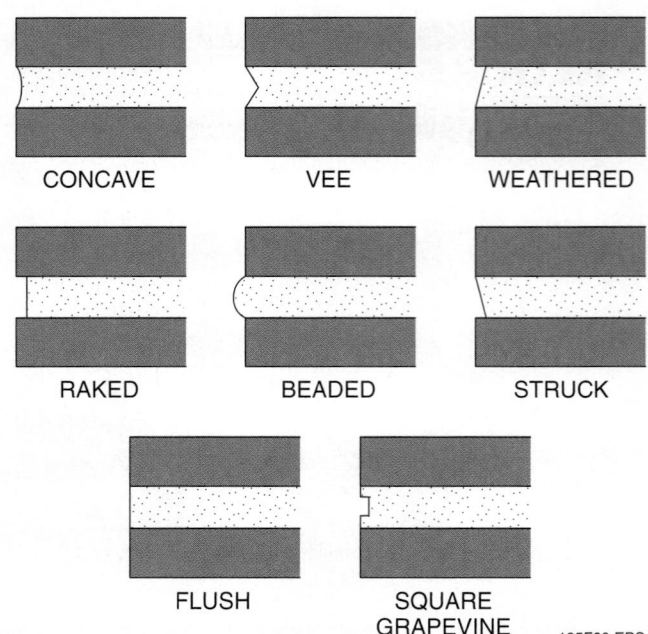

105F60.EPS

Figure 60 ◆ Mortar joint finishes.

the joint compresses the mortar and thereby increases its water resistance. Tooling also closes any hairline cracks that open as the mortar dries. Joints are tooled by shaped jointers. Raked joints are made by rakers. Struck, weathered, and flush joints are tooled by a trowel.

The raked joint is scooped out, not compressed. It does not get the extra water resistance, so it is not recommended for exterior walls in wet climates. The struck joint collects dirt and water on the ledge, so it is not recommended for exterior walls. Flush joints are not compressed, only struck off. They are recommended for walls that will be plastered or parged.

One additional joint type is the extruded or weeping joint (*Figure 61*). This joint is made when the masonry unit is laid. When the unit is placed, the excess mortar is not trimmed off. The mortar is left to harden to become the extruded joint. Since the mortar is not compressed in any way, this joint is not recommended for exterior walls.

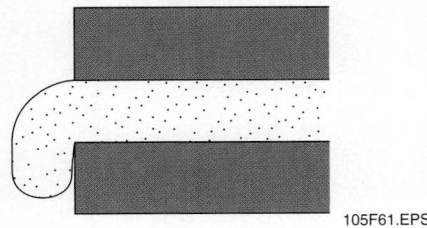

105F61.EPS

Figure 61 ◆ Extruded or weeping joint.

9.2.0 Striking the Joints

Working the joints with the jointer, raker, or trowel is called striking the joints. Whichever tool you use to strike the joint, the procedure is the same. The first step is to test the mortar.

9.2.1 Testing the Mortar

After you have laid the masonry units, the mortar must dry out before it can be tooled. The test equipment for checking proper dryness is the mason's thumb. Press your thumb firmly into the mortar joint:

- If your thumb makes an impression, but mortar does not stick to it, the joint is ready.
- If the mortar sticks to your thumb, it is too soft. The mortar is still runny, and the joint will not hold the imprint of the jointing tool.
- If your thumb does not make an impression, the mortar is too stiff. Working the steel tool will burn black marks on the joints.

The best time for striking the joints will vary because weather affects mortar drying time. Test the mortar repeatedly to find the right window of time to do the finishing work.

9.2.2 Striking

When the mortar is ready, the next step is the striking. The tool should be slightly larger than the mortar joint to get the proper impression.

- Hold the tool with your thumb on the handle, so it does not scrape on the masonry unit.
- Apply enough pressure so that the runner fits snugly against the edges of the masonry units. Keep the runner pressed against the unit edges all the way through the strike.
- Strike the head joints first (*Figure 62*). Strike head joints upward for a cleaner finish.
- Strike the bed joints last. The convex sled runner striking tool (*Figure 63*) is most commonly used. To keep the joints smooth, walk the jointer along the wall as you strike. The joints should be straight and unbroken from one end of the wall to the other. If the head joints are struck last, they will leave ridges on the bed joints.

If you are making a raked joint, follow the same order of work. Some joint rakers, or skate rakers (*Figure 64*), have adjustable set screws that set the depth of the rake-out. Do not rake out more than ½", or you will weaken the joint and possibly expose ties or reinforcements in the joint. Be sure you leave no mortar on the ledge of the raked unit.

If you are making a troweled joint, follow the same order of work. Ensure that the angle of the struck or weathered joint faces the same way on all the head joints. If you are striking flush joints with your trowel, strike up rather than down.

105F62.EPS

Figure 62 ◆ Using a jointer.

Figure 63 ◆ Using the convex sled.

Figure 64 ◆ Skate raker.

9.2.3 Cleaning Up Excess Mortar

After you strike the joints, you must clean up the excess mortar. Dried mortar sticks to masonry and is difficult to clean. Cleaning is much easier when you do it immediately after striking. Follow these steps to clean up excess mortar:

Step 1   Trim off mortar burrs by using a trowel. Hold the trowel fairly flat to the wall, as shown in *Figure 65A*. As you trim the burrs, flick them away so they do not stick to the units.

Step 2   Dress the wall after trimming off burrs. Dressing can be done with a soft brush, as shown in *Figure 65B*. In can also be done with coarse fabric, such as burlap or carpet, wrapped around a wood block. Flush joints need the fabric dressing, as a brush will not smooth them.

(A)

(B)

105F65.EPS

Figure 65 ◆ Trimming and dressing mortar burrs.

Step 3   Brush the head joints vertically first. Then brush the bed joints horizontally. If necessary, restrike the joints after brushing to get a sharp, neat joint.

Step 4   After you finish cleaning the wall, clean the floor at the foot of the wall.

10.0.0 ◆ PATCHING MORTAR

Two common methods of patching mortar are pointing and tuckpointing. Pointing is the act of putting additional mortar into a soft mortar joint. This type of patch does not require much preliminary work. Tuckpointing is the act of replacing hardened mortar with fresh mortar. This type of patch requires some preparation.

10.1.0 Pointing

Despite the best workmanship, mortar can fall out of a head joint or crack when a unit settles. A unit may get a chipped edge or a lost corner or get moved by some accident. Line pins and nails leave holes in mortar joints. Because pointing is easier than tuckpointing, it is a good idea to continuously check the surface of mortar joints. Perform the following tasks as you work:

- Fill line pin and nail holes as you move the line.
- Use mortar of the same consistency as was used for laying the units.
- Force the mortar into the holes with the tip of a pointing trowel or a slicker.
- Push all the mortar with a forward motion, in one direction for each hole.
- If the hole is deep, fill it with several thin layers of mortar, each no more than ¼" deep. This will avoid air pockets in the pointed joint.
- Clean excess mortar off masonry units with your trowel.

Inspect the condition of mortar joints after you finish a section of wall. Inspect them again before you strike them with the jointer. If there is a void, force additional mortar into the joint. If the back of the unit can be reached, use a backstop, such as the handle of a hammer, to brace the unit. This will prevent the unit from moving as you point the joint.

10.2.0 Tuckpointing

Patching or tuckpointing after the mortar has hardened is more complex. Follow these steps for proper tuckpointing:

Step 1   Mix some mortar, and let it dry out for about an hour to get partly stiff. This will reduce shrinkage after it is put into the joint. While it is stiffening, clean out the damaged joints.

Step 2   With a joint chisel or a tuckpointer's grinder, dig out the bad mortar to a depth of about ½". The damaged area may be deeper due to cracks or shrinkage. Be sure you have cleaned down to solid mortar. As a rule, the depth of mortar removed should be at least as deep as the joint is wide. *Figure 66* shows a properly excavated joint along with examples of improperly excavated joints.

Step 3   Remove all loose mortar with a stiff brush or with a jet of water from a hose.

Step 4   Thoroughly wet the surrounding masonry with water, but do not saturate it. Wetting will slow setting time and produce a better bond. However, excess moisture will bead up and prevent a good bond in the joint.

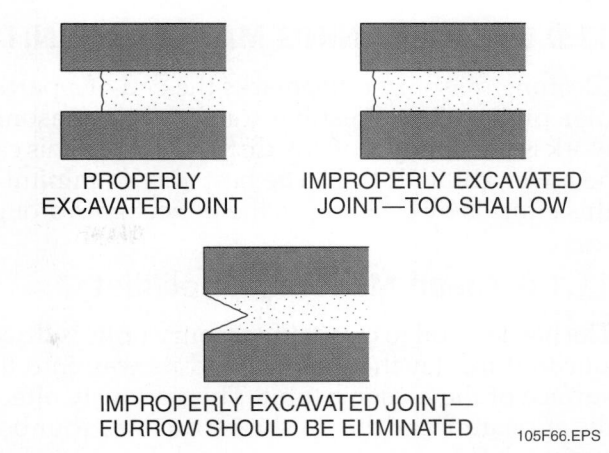

PROPERLY EXCAVATED JOINT

IMPROPERLY EXCAVATED JOINT—TOO SHALLOW

IMPROPERLY EXCAVATED JOINT— FURROW SHOULD BE ELIMINATED

105F66.EPS

Figure 66 ◆ Preparation of joints for tuckpointing.

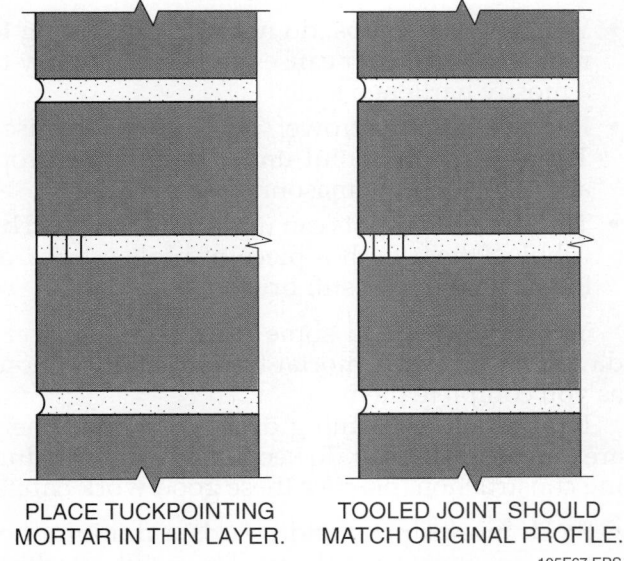

PLACE TUCKPOINTING MORTAR IN THIN LAYER.

TOOLED JOINT SHOULD MATCH ORIGINAL PROFILE.

105F67.EPS

Figure 67 ◆ Tuckpointing.

Step 5   Force fresh mortar into the damp joint. Use a trowel or slicker with a point narrower than the joint, and press the mortar hard. If the damaged area is deep, fill it with several thin layers of mortar, each no more than ¼" deep (*Figure 67*). This will prevent air pockets from forming in the pointed joint.

Step 6   Clean excess mortar off of the units with your trowel.

Step 7   Retool the joint after the mortar has set long enough.

Step 8   Clean the pointed areas after you retool the joints. Use the cleaning procedure described previously under in the section on striking the joints.

11.0.0 ◆ CLEANING MASONRY UNITS

Cleaning masonry units marks the end of a particular project. The finishing touch on all masonry work is the removal of any dirt and stains. This can be a wet, difficult task. The best way to minimize this effort is by cleaning as the project goes along.

11.1.0 Clean Masonry Checklist

The hardest soil to clean off masonry units is dried, smeared mortar that has worked its way into the surface of the masonry unit. This seriously affects the appearance of the finished structure. Your best approach is to avoid smearing and dropping mortar during construction. These guidelines will help you clean as you work:

- When mortar drops, do not rub it in. Trying to remove wet mortar causes smears. Let it dry to a mostly hardened state.

- Remove it with a trowel, putty knife, or chisel. Try to work the point under the mortar drop, and flick it off the masonry.

- The remaining spots can usually be removed by rubbing them with a piece of broken brick or block, then with a stiff brush.

You should spend some time cleaning every day, removing stray mortar from the wall sections as you complete them.

In addition to cleaning dropped mortar, there are other things to do. To keep masonry clean during construction, practice these good work habits:

- Stock mortar pans and boards a minimum of two feet away from the wall to avoid splashes.

- Temper the mortar with small amounts of water, so it will not drip or smear on the units.

- After laying units, cut off excess mortar carefully with the trowel.

- Wait until mortar hardens for striking, to avoid smearing wet mortar on masonry units.

- After tooling joints, scrape off mortar burrs with your trowel before brushing.

- Avoid any motion that rubs or presses wet mortar into the face of the masonry unit.

- Keep materials clean, covered, and stored out of the way of concrete, tar, and other staining agents. Do not store materials under the scaffolding.

- Turn scaffolding boards on edge with the clean side to the wall at the end of the day. This will prevent rain from splashing dirt and mortar on the wall.

- Always cover walls at the end of the day to keep them dry and clean.

Following these practices should reduce the amount of time you spend cleaning the masonry units after construction is complete. The clean masonry checklist is especially important when working with CMUs. Because of their rougher surface texture, mortar spilled on them is harder to clean.

11.2.0 Bucket Cleaning for Brick

The best method of cleaning any new brick masonry is the least severe method. If the daily cleaning practices listed previously are not enough, the next step is bucket and brush hand cleaning. This may include using a proprietary cleaning compound or an acid wash. Acid affects brick over time, so it should be your last resort. *Table 1* lists cleaning methods for different types of brick as developed by the Brick Institute of America.

Any chemical compound you use should first be tested on a 4' × 5' inconspicuous section of wall. Sometimes, minerals in the brick may react with some chemicals and cause stains. Read the brick manufacturer's material safety data sheet (MSDS) for recommended cleaning solutions. Read the MSDS, and follow the manufacturer's directions for mixing, using, and storing any chemical solution.

 WARNING!
Wear appropriate personal protective equipment when using chemical solutions.

When cleaning, you will need a hose, bucket, wooden scraper, chisel, and stiff brush. Follow these guidelines:

- Do not start cleaning until at least one week after the wall is finished. This gives the mortar time to cure and set. Do not wait longer than six months because the mortar will be almost impossible to remove.

- Dry scrub the wall with a wooden paddle. Go over large particles with a chisel, wood scraper, or piece of brick or block. This should remove most of the mortar.

- Before wetting, protect any metal, glass, wood, limestone, and cast stone surfaces. Mask or cover windows, doors, and fancy trim work.

- Prepare the chemical cleaning solution. Follow manufacturer's directions. Remember to pour chemicals into water, not water into chemicals.

- Presoak the wall with the hose to remove loose particles or dirt.

Table 1 Cleaning Guide for New Masonry

BRICK CATEGORY	CLEANING METHOD	REMARKS
Red and red flashed	Bucket and brush hand cleaning High-pressure water Sandblasting	Hydrochloric acid solutions, proprietary compounds, and emulsifying agents may be used. *Smooth texture:* Mortar stains and smears are generally easier to remove; less surface area is exposed; easier to pre-soak and rinse; unbroken surface, thus more likely to display poor rinsing, acid staining, and poor removal of mortar smears. *Rough texture:* Mortar and dirt tend to penetrate deep into textures; additional area for water and acid absorption; essential to use pressurized water during rinsing.
Red, heavy sand finish	Bucket and brush hand cleaning High-pressure water	Clean with plain water and scrub brush, or lightly applied high-pressure and plain water. Excessive mortar stains may require use of cleaning solutions. Sandblasting is not recommended.
Light colored units, white, tan, buff, gray, specks, pink, brown, and black	Bucket and brush hand cleaning High-pressure water Sandblasting	*Do not use muriatic acid!* Clean with plain water, detergents, emulsifying agents, or suitable proprietary compounds. Manganese colored brick units tend to react to muriatic acid solutions and stain. Light colored units are more susceptible to acid burn and stains, compared to darker units.
Same as light colored units, plus sand finish	Bucket and brush hand cleaning High-pressure water	Lightly apply either method (see notes for light colored units). Sandblasting is not recommended.
Glazed brick	Bucket and brush hand cleaning	Wipe glazed surface with soft cloth within a few minutes of laying units. Use a soft sponge or brush plus ample water supply for final washing. Use detergents where necessary and acid solutions only for very difficult mortar stain. Do not use acid on salt-glazed or metallic-glazed brick. Do not use abrasive powders.
Colored mortars	Method is generally controlled by the brick unit	Many manufacturers of colored mortars do not recommend chemical cleaning solutions. Most acids tend to bleach colored mortars. Mild detergent solutions are generally recommemded.

105T01.EPS

- Start working from the top. Keep the area immediately below the space you are scrubbing wet also to prevent the chemicals from drying into the wall.
- Scrub a small area with the chemical applied on a stiff brush. Keep the scrub area small enough so that the solution does not dry on the wall as you are working.
- To remove stubborn spots, rub a piece of brick over them. Then scrub the spot again with more chemical solution. Repeat this as needed.

- As you complete scrubbing in each area, rinse the wall thoroughly. Rinse the surrounding wall area above and below, all the way to the bottom of the wall, to keep chemicals from staining the wall.
- Flush the entire wall for ten minutes after you finish scrubbing. This will dilute any remaining chemical and prevent burns.

High-pressure water washing, steam cleaning, and sandblasting are also used to clean new and old masonry. Because these techniques can damage masonry surfaces, they require trained operators. If you have not been trained, do not use this equipment.

11.3.0 Cleaning Block

Cleaning block is difficult because the surface on standard block is very porous. It is important to follow the clean masonry checklist procedures previously described. If block is stained with mortar at the end of a job, rub it with a piece of block.

For further cleaning, it is important to check the manufacturer's MSDSs for recommended chemical cleaners and cleaning procedures. Acid is very destructive to block and cannot be used without protective countermeasures. Read the block manufacturer's MSDSs for recommended cleaning solutions. Read the MSDSs, and follow manufacturer's directions for mixing, using, and storing any chemical solution. Detergents and surfactants are often recommended for use on block. If no cleaning procedures are given, follow the bucket and brush procedures listed earlier.

Any chemical compound used should first be tested on a 4' × 5' inconspicuous section of wall. Sometimes, minerals in the block may react with some chemicals and cause stains.

Summary

Masons are skilled craft professionals whose work often stands, not only for many years, but for many decades after completion. It is up to each individual mason to learn the skills necessary to create strong, properly built structures. These skills take practice and patience to master.

In order to have a long, productive career as a mason, you must work hard toward mastering your craft. You must learn all of the relevant specifications and standards and keep yourself informed of changes in the regulations governing the masonry industry. As well, you need to be familiar with the specific challenges and special project types common to your particular region.

Learning the basic skills in this module is essential to becoming a skilled mason. However, being a successful mason is about much more than understanding how to choose masonry units, lay a strong course, and properly mix mortar. It is about having pride in your workmanship and about taking the time to do each job right. By mastering the craft, having real pride in your work, and always continuing to learn, you can ensure yourself a long, productive career in masonry.

Review Questions

1. Random cracking in CMU walls can be controlled by _____.
   a. contraction joints
   b. changes in wall thickness
   c. pilasters
   d. chases

2. Grout is used with concrete masonry units to _____.
   a. strengthen joints
   b. reinforce walls
   c. remove stains
   d. add color

3. Which of the following is a correct statement about calcium silicate brick?
   a. It is banned in the United States.
   b. It is not fired in a kiln.
   c. It cannot be used for fireplaces and chimneys.
   d. It is made without sand.

For Questions 4 through 8, refer to *Figure 1*, and match the brick placement to its name.

4. _____ Stretcher

5. _____ Sailor

6. _____ Soldier

7. _____ Rowlock

8. _____ Header

9. Solid masonry units have 75% or less of their surface as a void.
   a. True
   b. False

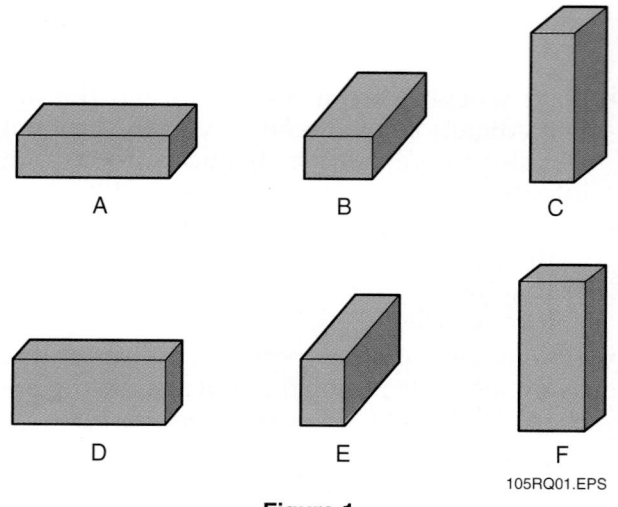

105RQ01.EPS

Figure 1

10. Cavity walls can be loadbearing when they _____.
    a. have flashing and expansion joints in the cavity
    b. are placed on a modular grid to avoid cuts
    c. are tied with metal ties to equalize bearing forces
    d. have both wythes of the same material

11. Laying the first course with no mortar is called _____.
    a. leading
    b. rackbacking
    c. hot ranging
    d. dry bonding

12. Moving a CMU after it is in place will _____.
    a. chip its edges
    b. weaken the bond
    c. set mortar firmly
    d. retemper the mortar

13. All of the following are true of pattern bonds *except* pattern bonds _____.
    a. call for cutting brick to fit
    b. add a visual design to masonry walls
    c. add structural strength to masonry walls
    d. call for placing brick in different positions

14. Flemish, common, and running are _____.
    a. stack bonds
    b. defined brick cuts
    c. corner patterns
    d. structural pattern bonds

15. Blocks cut horizontally across the face are known as _____.
    a. rowlocks
    b. rips
    c. half-highs
    d. ¾ cuts

16. Masonry is plumbed and leveled after the joints are struck.
    a. True
    b. False

17. Laying to the line calls for positioning the _____ edge of the masonry unit level with the line and _____ away from it.
    a. outside bottom; ³⁄₁₆"
    b. outside top; ¹⁄₁₆"
    c. inside top; ³⁄₃₂"
    d. inside bottom; ¹⁄₁₆"

18. Poles, blocks, pins, and trigs are used to _____.
    a. level masonry units
    b. strike mortar joints
    c. secure the mason's line
    d. shield masonry from mortar smears

19. Which of the following allows the mason to lay to the line without building corner leads?
    a. Line trig
    b. Corner block
    c. Line pin
    d. Corner pole

20. A rackback lead is _____.
    a. stepped back a half brick on each end
    b. used to check plumb
    c. slack to the line
    d. crowding the line

21. A rackback lead with a return in it is a _____.
    a. furrow
    b. rackback corner
    c. string
    d. ranged line

22. Joints should be struck when the mortar _____.
    a. brushes off the masonry units
    b. has set firm and hard
    c. will hold a thumbprint
    d. still oozes

23. When striking flush joints with a trowel, you should strike _____.
    a. upward
    b. downward

24. Laying the mason's level on the points of the end bricks laid in a rackback lead is called _____.
    a. laying the line
    b. tailing the diagonal
    c. ranging the angle
    d. leveling the corner

25. Which of the following is *not* a standard joint finish?
    a. struck
    b. convex
    c. vee
    d. concave

Zachary Reinert, Apprentice

Nester Brothers Masonry
Pennfield, PA

While still a high school student, Zack Reinert took first place in the 2004 masonry competition at the Associated Builders and Contractors Craft Olympics held in Hawaii.

How did you get started in the Masonry trade?
I enrolled in the three-year masonry program at my high school. This is a co-op program, so I work part time at Nester Brothers getting real job experience. I started as a laborer, but was given the opportunity to lay block. Fortunately, I did well at it, so I was able to continue doing it.

Describe your job.
I'm still in high school, so I attend school most of the time. In my job at Nester Brothers, I lay brick, block, and some stone.

What do you like most about your job?
I like working outdoors, and I like working with my hands. I also like that I can see the results of my work every day and take pride in what I've accomplished. I like the company I work for. They have helped and encouraged me along the way and sponsored me to compete in the Craft Olympics.

What would you say to someone just entering the trade?
Get into a training program. It is difficult to learn correct technique on the job. You will probably start out as a laborer, but when the opportunity comes to work on a wall, you have to be prepared. If you don't do well, it could be a long time before you get another chance. Taking some training, especially in high school, will help you decide if you really like the work. I like doing masonry work. I learned pretty quickly that if you enjoy what you do, it keeps you motivated and makes the time go quickly.

Trade Terms Introduced
in This Module

Autoclave: A pressurized, steam-heated tank used for sterilizing and cooking.

Closure unit: The last brick or block to fill a course.

Crowding the line: A person touching the mason's line, or a masonry unit too close to the line.

Dry bond: Laying out masonry units without mortar to establish spacing.

Lintel: The support beam over an opening such as a window or door. Also called a header.

Rackback: A lead or other structure built with each course of masonry shorter than the course below it.

Racking: Shortening each course of masonry by one unit so it is shorter than the course below it, resulting in a pyramid shape.

Ranging: Aligning a corner by using a line. Corners can be ranged around themselves or from one corner to another.

Return: A corner in a structure or lead.

Slack to the line: Masonry units set too far away from the mason's line.

Tail: To check the spacing of head joints by checking the diagonal edges of the courses on a lead or corner.

Additional Resources

This module is intended to be a thorough resource for task training. The following reference works are suggested for further study. These are optional materials for continued education rather than for task training.

Bricklaying: Brick and Block Masonry. Reston, VA: Brick Institute of America.

Building Block Walls—A Basic Guide, 1988. Herndon, VA: National Concrete Masonry Association.

Concrete Masonry Handbook. Skokie, IL: Portland Cement Association.

The ABCs of Concrete Masonry Construction, Videotape, 13:34 minutes. Skokie, IL: Portland Cement Association.

NCCER makes every effort to keep these textbooks up-to-date and free of technical errors. We appreciate your help in this process. If you have an idea for improving this textbook, or if you find an error, a typographical mistake, or an inaccuracy in NCCER's Contren® textbooks, please write us, using this form or a photocopy. Be sure to include the exact module number, page number, a detailed description, and the correction, if applicable. Your input will be brought to the attention of the Technical Review Committee. Thank you for your assistance.

Instructors – If you found that additional materials were necessary in order to teach this module effectively, please let us know so that we may include them in the Equipment/Materials list in the Annotated Instructor's Guide.

Write: Product Development and Revision
National Center for Construction Education and Research
3600 NW 43rd St, Bldg G, Gainesville, FL 32606

Fax: 352-334-0932

E-mail: curriculum@nccer.org

Craft _____ Module Name _____

Copyright Date _____ Module Number _____ Page Number(s) _____

Description _____

_____

_____

_____

_____

(Optional) Correction _____

_____

_____

(Optional) Your Name and Address _____

_____

_____

_____

An Associated Builders and Contractors 2005 National Craft Championships contestant measures a member for a metal frame. Metal framing is gaining widespread use in both commercial and residential construction.

68106-09
Floor Systems

Topics to be presented in this module include:

1.0.0   Introduction 6.2
2.0.0   Methods of Framing Houses 6.2
3.0.0   Building Working Drawings and Specifications 6.7
4.0.0   The Floor System 6.12
5.0.0   Laying Out and Constructing a Platform
        Floor Assembly 6.27
6.0.0   Installing Joists for Projections and Cantilevered
        Floors 6.38
7.0.0   Estimating the Quantity of Floor Materials 6.40
8.0.0   Guidelines for Determining Proper Girder and
        Joist Sizes 6.41

Overview

The construction of a wood-frame floor begins with placement of the sill plate on the foundation. The sill plate is typically attached to the foundation with anchor bolts that are embedded into the concrete foundation.

Once the sill plate is installed, the carpenter lays out and marks the locations of the floor joists on the sill plate, and then places and attaches the joists. In many instances, the joist span is too great for a single piece of lumber, so joists are joined over a beam or girder. Once all the joists are in place and a header board has been attached to the ends of the joists, a subfloor is attached to the joists. Bridging is attached between the joists to provide stability.

Wood I-beams and floor trusses made of wood or steel are sometimes used instead of joists. These materials are stronger than comparable lengths of dimension lumber and can therefore be placed across greater spans.

Objectives

When you have completed this module, you will be able to do the following:

1. Identify the different types of framing systems.
2. Read and interpret drawings and specifications to determine floor system requirements.
3. Identify floor and sill framing and support members.
4. Name the methods used to fasten sills to the foundation.
5. Given specific floor load and span data, select the proper girder/beam size from a list of available girders/beams.
6. List and recognize different types of floor joists.
7. Given specific floor load and span data, select the proper joist size from a list of available joists.
8. List and recognize different types of bridging.
9. List and recognize different types of flooring materials.
10. Explain the purposes of subflooring and underlayment.
11. Match selected fasteners used in floor framing to their correct uses.
12. Estimate the amount of material needed to frame a floor assembly.
13. Demonstrate the ability to:
    - Lay out and construct a floor assembly
    - Install bridging
    - Install joists for a cantilever floor
    - Install a subfloor using butt-joint plywood/OSB panels
    - Install a single floor system using tongue-and-groove plywood/OSB panels

Required Trainee Materials

1. Pencil and paper
2. Appropriate personal protective equipment

Prerequisites

Before you begin this module, it is recommended that you successfully complete *Core Curriculum*, and *Construction Technology*, Modules 68101-09 through 68105-09.

This course map shows all of the modules in *Construction Technology*. The suggested training order begins at the bottom and proceeds up. Skill levels increase as you advance on the course map. The local Training Program Sponsor may adjust the training order.

68117-09
Copper Pipe and Fittings

68116-09
Plastic Pipe and Fittings

68115-09
Introduction to Drain, Waste, and Vent (DWV) Systems

68114-09
Introduction to HVAC

68113-09
Residential Electrical Services

68112-09
Electrical Safety

68111-09
Basic Stair Layout

68110-09
Exterior Finishing

68109-09
Roofing Applications

68108-09
Roof Framing

68107-09
Wall and Ceiling Framing

68106-09
Floor Systems

68105-09
Masonry Units and Installation Techniques

68104-09
Introduction to Masonry

68103-09
Handling and Placing Concrete

68102-09
Introduction to Concrete, Reinforcing Materials, and Forms

68101-09
Site Layout One: Distance Measurement and Leveling

CORE CURRICULUM:
Introductory Craft Skills

CONSTRUCTION TECHNOLOGY

106CMAP.EPS

Trade Terms

Crown	Rafter plate
Dead load	Scab
Firestop	Scarf
Foundation	Soleplate
Header joist	Span
Joist hanger	Tail joist
Let-in	Trimmer joist
Live load	Truss
Pier	Underlayment

1.0.0 ◆ INTRODUCTION

This module briefly introduces the different methods used for framing buildings. Some of the types of framing described are rarely used today. They are seen in existing installations, however, so some knowledge about them is helpful, especially when remodeling. The remainder of the module describes floor framing with an emphasis on the platform method of framing. Included are descriptions of the materials and general methods used to construct floors. Proper construction of floors is essential because, no matter how well the foundation of a building is constructed, a structure will not stand if the floors and sills are poorly assembled.

2.0.0 ◆ METHODS OF FRAMING HOUSES

Various areas of the country have had different methods of constructing a wood frame dwelling. This variation in methods can be attributed to economic conditions, availability of material, or climate in different parts of the country. We will discuss the various types of framing and show the approved method of construction today. Structures that are framed and constructed entirely of wood above the foundation fall into several classifications:

- Platform frame (also known as western and box frame)
- Braced frame
- Balloon frame
- Post-and-beam frame

2.1.0 Platform Frame

The platform frame, sometimes called the western frame, is used in most modern residential and light commercial construction. In this type of construction, each floor of a structure is built as an individual unit (*Figure 1*). The subfloor is laid in place prior to the exterior walls being put in place. The soleplate and top plate are nailed to the studs. Window openings are constructed and put in place, and studs are notched in order to let in a 1×4 diagonal brace.

In western or platform frame construction, the walls are built lying on the subfloor. Once a section or a part of a wall is complete, it is then lifted into place and nailed to the floor system and another plate is added to the top plate. This will tie in the second floor system. This method of construction allows workers to work safely on the floor while constructing the wall systems.

Platform framing is subject to settling caused by the shrinkage of a large number of horizontal, loadbearing frame members. Settling can result in various problems such as cracked plaster, cracked wallboard joints, uneven ceilings and floors, ill-fitting doors and windows, and nail pops (nails that begin to protrude through the wallboard). Settling can be minimized by using well-dried lumber.

2.2.0 Braced Frame

In early times, the brace frame method of construction was frequently used because most bearing lumber was in the vertical position. Very little shrinkage occurred in this type of construction.

INSIDE TRACK

Platform Framing

In platform framing, it is inevitable that the structure will begin to settle. While settling is unavoidable, the degree to which the structure settles can be minimized by not cutting any corners during the framing and construction process. Compacting the foundation soil and making sure you brace the frame properly during construction will also decrease the amount of settlement.

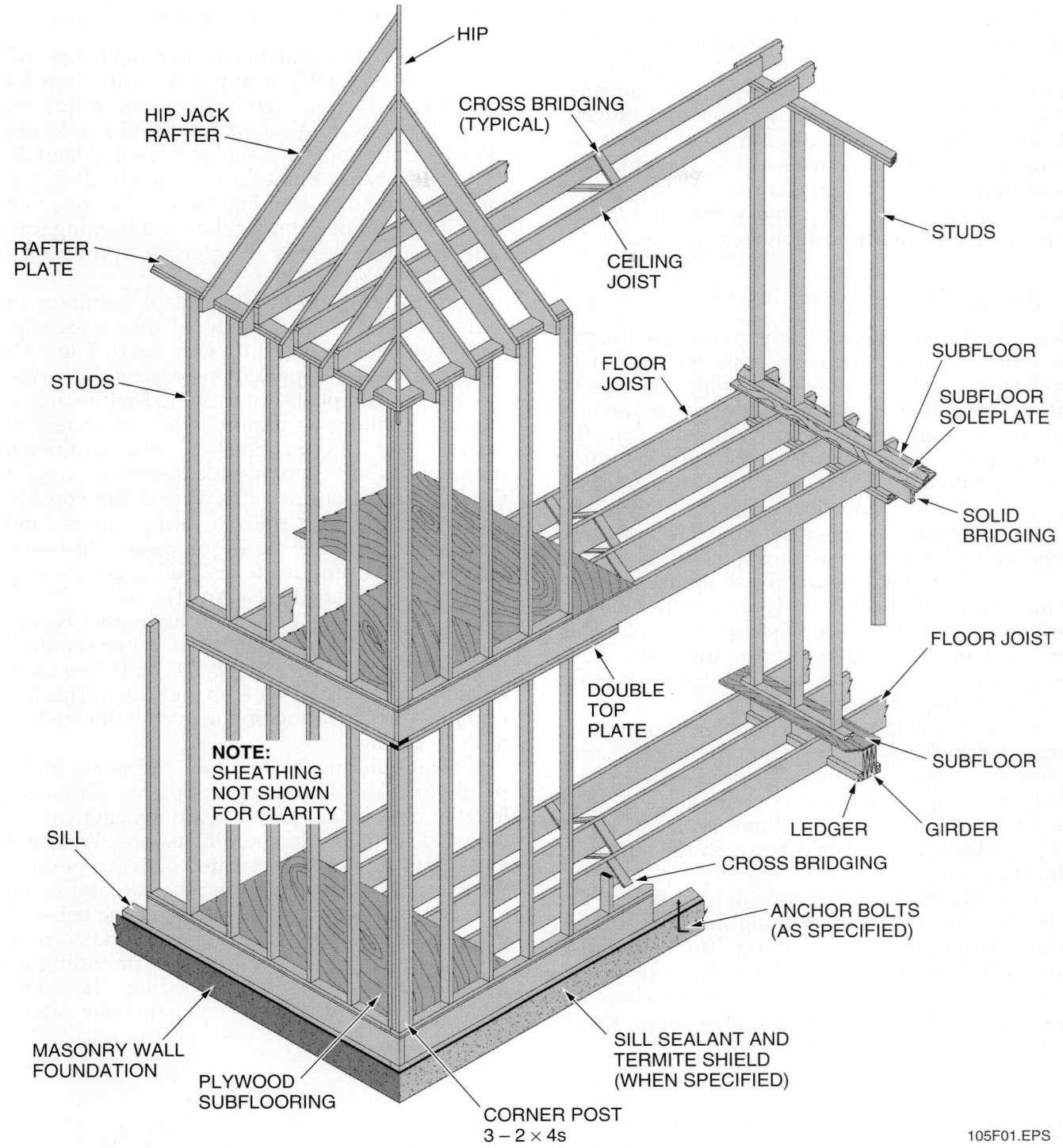

HIP

HIP JACK RAFTER

CROSS BRIDGING (TYPICAL)

RAFTER PLATE

CEILING JOIST

STUDS

STUDS

FLOOR JOIST

SUBFLOOR

SUBFLOOR SOLEPLATE

SOLID BRIDGING

FLOOR JOIST

DOUBLE TOP PLATE

NOTE:
SHEATHING
NOT SHOWN
FOR CLARITY

SUBFLOOR

SILL

LEDGER

GIRDER

CROSS BRIDGING

ANCHOR BOLTS (AS SPECIFIED)

MASONRY WALL FOUNDATION

PLYWOOD SUBFLOORING

SILL SEALANT AND TERMITE SHIELD (WHEN SPECIFIED)

CORNER POST
3 – 2 × 4s

105F01.EPS

Figure 1 ◆ Platform framing.

In this method of construction, the framework does not rely on sheathing for rigidity. The framework, which in part had posts for corners and beams or girders, was mortised and tenoned together and held in place by pins and dowels. Diagonal braces were added to give the framework rigidity, thus making it stable. Therefore, the sheathing or planking could run up the sides of the structure vertically. One common type of structure built in this fashion was the barn.

2.3.0 Balloon Frame

Balloon frame construction is rarely used today because of lumber and labor costs, but a substantial number of structures built with this type of frame are still in use. In the balloon frame method of construction, the studs are continuous from the sill to the rafter plate (*Figure 2*). The wall studs and first floor joists rest on a solid sill (usually a 2 × 6), with the floor joists being nailed to the sides of the studs. The second floor joists rest on a horizontal 1 × 4 or 1 × 6 board called a ribbon that is installed in notches (let-ins) to the faces of the studs. Braces (usually 1 × 4) are notched in on a diagonal into the outside face of the studs. This method of construction makes the frame self-supporting; therefore, it does not need to rely on the sheathing for rigidity.

In the balloon frame, firestops must be installed in the walls in several locations. A firestop is an approved material used in the space between frame members to prevent the spread of fire for a limited period of time. Typically, 2 × 4 wood blocks are installed between the studs for this purpose.

One advantage of balloon framing is that the shrinkage of the wood framing members is low, thus helping to reduce settling. This is because wood shrinks across its width, but practically no shrinkage occurs lengthwise. This provides for high vertical stability, making the balloon frame adaptable for two-story structures.

2.4.0 Post-and-Beam Frame

The post-and-beam method of framing floors and roofs has been used in heavy timber buildings for many years. It uses large, widely-spaced timbers for joists, posts, and rafters. Matched planks are often used for floors and roof sheathing (*Figure 3*).

Post-and-beam homes are normally designed for this method of framing. In other words, you cannot start a house under standard framing procedures and then alter the construction to post-and-beam framing.

In post-and-beam framing, plank subfloors or roofs are usually of 2" nominal thickness, supported on beams spaced up to 8' apart. The ends of the beams are supported on posts or piers. Wall spaces between posts are provided with supplementary framing as required for attachment of exterior and interior finishes. This additional framing also serves to provide lateral bracing for the building. Consider this versus the conventional framing that utilizes joists, rafters, and studs placed from 16" to 24" on center. Post-and-beam framing requires fewer but larger framing members spaced further apart. The most efficient use of 2" planks occurs when the lumber is continuous over more than one span. When standard lengths of lumber such as 12', 14', or 16' are used, beam spacings of 6', 7', or 8' are indicated. This factor has a direct bearing on the overall dimensions of the building.

If local building codes allow end joints in the planks to fall between supports, planks of random lengths may be used and beam spacing can be adjusted to fit the house dimensions. Windows and doors are normally located between posts in the exterior walls, eliminating the need for headers over the openings. The wide spacing between posts permits ample room for large glass areas. Consideration should be given to providing an adequate amount of solid wall siding. The siding must also provide ample and adequate lateral bracing.

Balloon Framing

Balloon framing is frequently used in hurricane-prone areas for gable ends. In fact, this type of framing may be required by local building codes. Because balloon framing has studs that span multiple stories, alternative types of studs may be required if standard wood studs aren't available. Engineered lumber or metal studs can be used for this purpose.

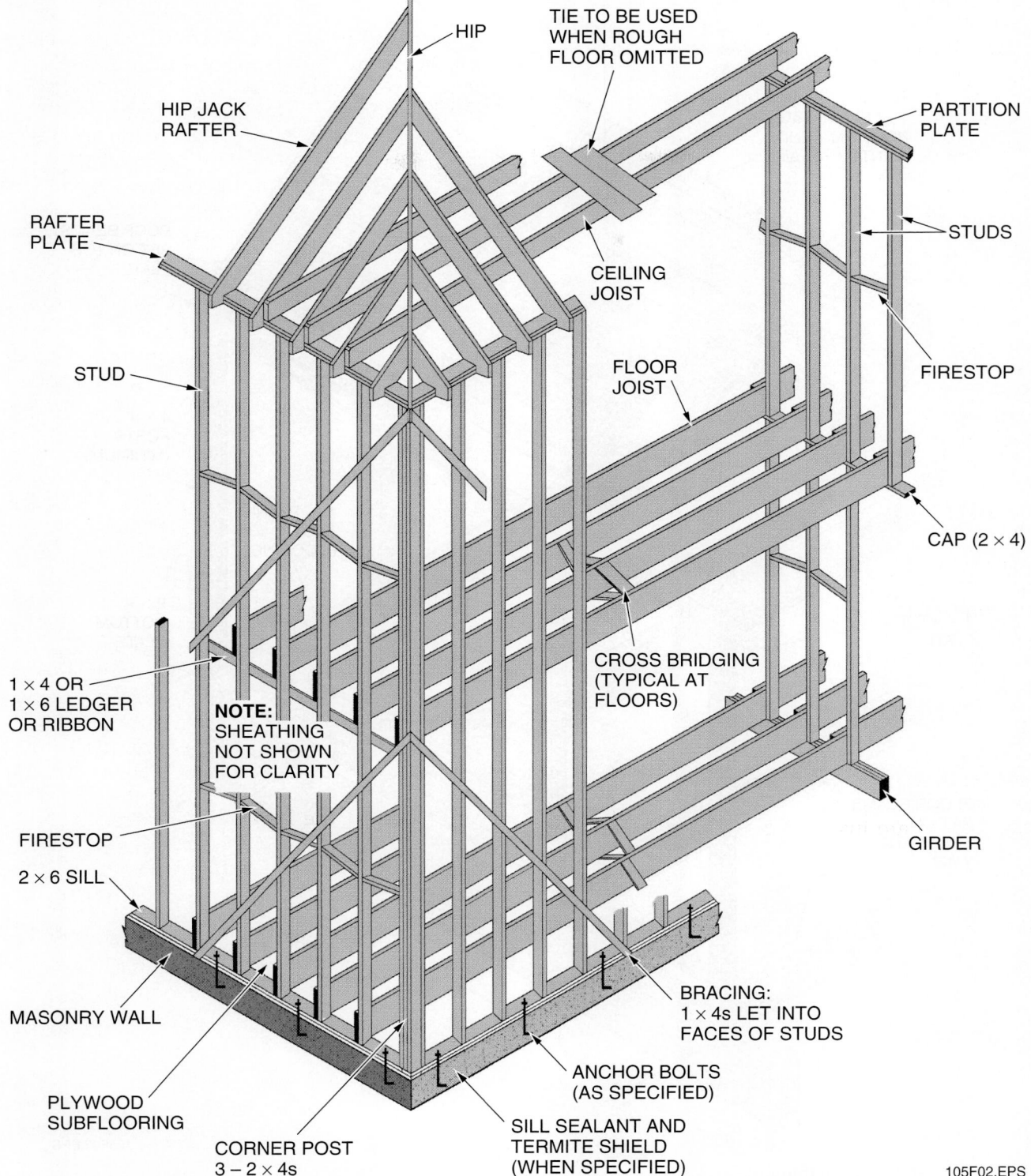

HIP

TIE TO BE USED
WHEN ROUGH
FLOOR OMITTED

HIP JACK
RAFTER

PARTITION
PLATE

RAFTER
PLATE

STUDS

CEILING
JOIST

STUD

FLOOR
JOIST

FIRESTOP

CAP (2 × 4)

1 × 4 OR
1 × 6 LEDGER
OR RIBBON

NOTE:
SHEATHING
NOT SHOWN
FOR CLARITY

CROSS BRIDGING
(TYPICAL AT
FLOORS)

FIRESTOP

2 × 6 SILL

GIRDER

MASONRY WALL

BRACING:
1 × 4s LET INTO
FACES OF STUDS

PLYWOOD
SUBFLOORING

ANCHOR BOLTS
(AS SPECIFIED)

CORNER POST
3 – 2 × 4s

SILL SEALANT AND
TERMITE SHIELD
(WHEN SPECIFIED)

105F02.EPS

Figure 2 ◆ Balloon framing.

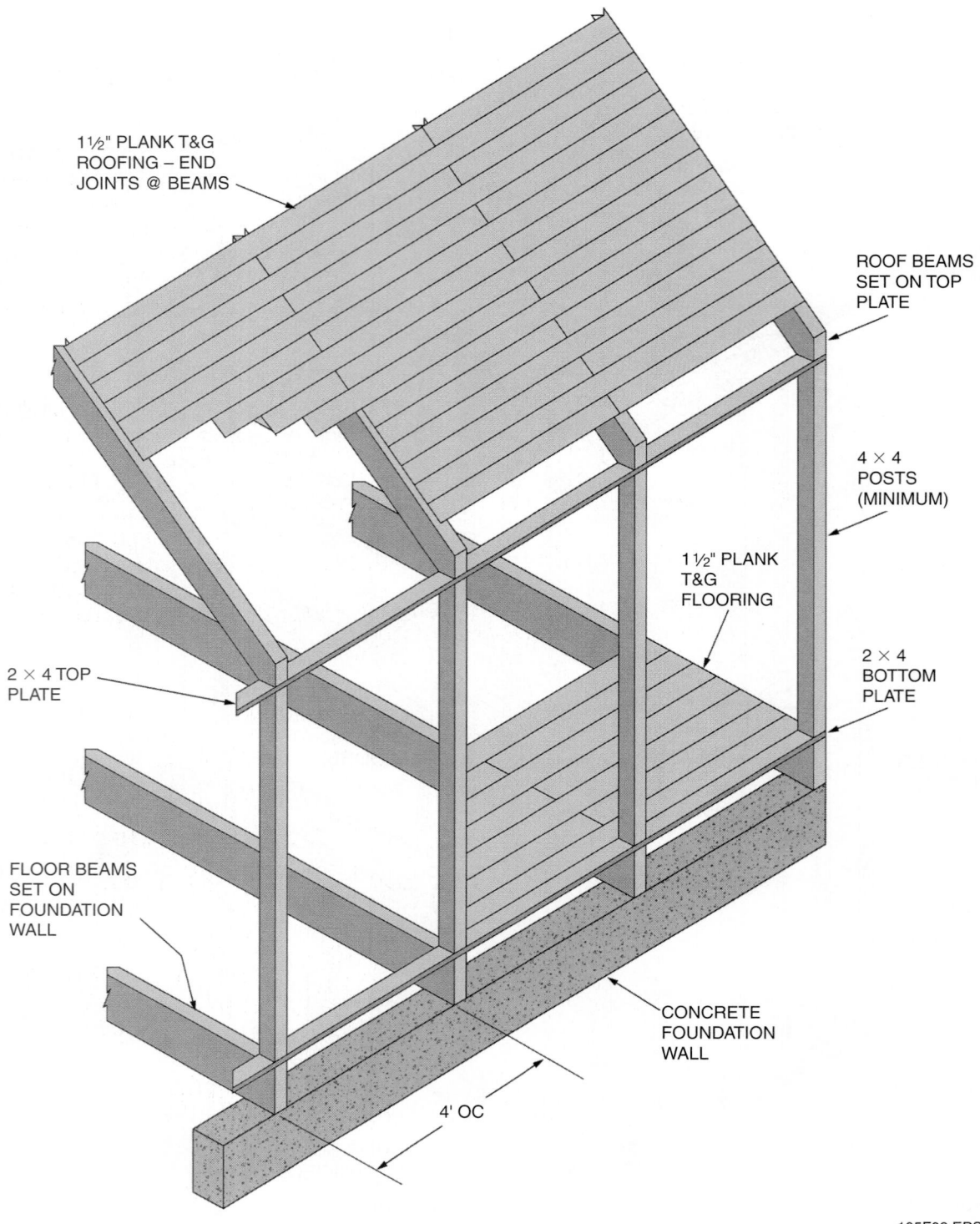

1½" PLANK T&G ROOFING – END JOINTS @ BEAMS

ROOF BEAMS SET ON TOP PLATE

4 × 4 POSTS (MINIMUM)

1½" PLANK T&G FLOORING

2 × 4 TOP PLATE

2 × 4 BOTTOM PLATE

FLOOR BEAMS SET ON FOUNDATION WALL

CONCRETE FOUNDATION WALL

4' OC

105F03.EPS

Figure 3 ◆ Post-and-beam framing.

Kobe, Japan

On January 17, 1995, a massive earthquake that measured 7.2 on the Richter scale leveled the city of Kobe, Japan. The quake caused an estimated $120 billion in initial damage and claimed the lives of some 5,500 people. Experts have attributed much of the residential destruction to the traditional Japanese post-and-beam construction methods. Many of the structures were poorly constructed, poorly fastened, and had few lower level partitions that would increase their ability to resist lateral movement. Most Japanese homes also had ceramic tile roofs that created very top-heavy structures. When the earthquake hit, these dwellings buckled under the pressure. New building codes have emerged in the wake of this earthquake, and the Japanese construction industry is now emphasizing platform framing. Architects and structural engineers from around the world have traveled to Kobe to research ways to improve the structural integrity of home construction. For example, they have discovered that increasing the size of wall paneling can strengthen the lateral resistance of a post-and-beam structure.

A combination of conventional framing with post-and-beam framing is sometimes used where the two adjoin each other. On a side-by-side basis, no particular problems will be encountered.

Where a post-and-beam floor or roof is supported on a stud wall, a post should be placed under the end of the beam to carry the conventional load. A conventional roof can be used with post-and-beam construction by installing a header between the posts to carry the load from the rafters to the posts.

3.0.0 ◆ BUILDING WORKING DRAWINGS AND SPECIFICATIONS

Construction blueprints (working drawings) and related written specifications contain all the information and dimensions needed to build or remodel a structure. The interpretation of blueprints is critically important in the construction of floor systems, walls, and roof systems. *Figure 4* shows the contents and sequence of a typical set of working drawings. The drawings that apply when building a floor system are briefly reviewed here.

3.1.0 Architectural Drawings

The architectural drawings contain most of the detailed information needed by carpenters to build a floor system. The specific categories of architectural drawings commonly used include the following:

- Foundation plan
- Floor plan
- Section and detail drawings

3.1.1 Foundation Plan

The foundation plan is a view of the entire substructure below the first floor or frame of the building. It gives the location and dimensions of footings, grade beams, foundation walls, stem walls, piers, equipment footings, and foundations. Generally, in a detail view, it also shows the location of the anchor bolts or straps in foundation walls or concrete slabs.

3.1.2 Floor Plan

The floor plan is a cutaway view (top view) of the building, showing the length and breadth of the building and the layout of the rooms on that floor. It shows the following kinds of information:

- Outside walls, including the location and dimensions of all exterior openings
- Types of construction materials
- Location of interior walls and partitions
- Location and swing of doors
- Stairways
- Location of windows
- Location of cabinets, electrical and mechanical equipment, and fixtures
- Location of cutting plane line

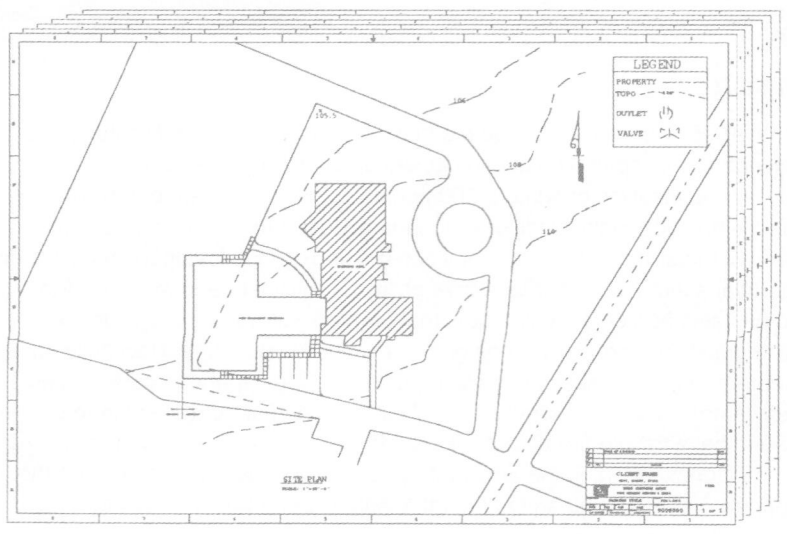

TITLE SHEET(S)
ARCHITECTURAL DRAWINGS
- SITE (PLOT) PLAN
- FOUNDATION PLAN
- FLOOR PLANS
- INTERIOR/EXTERIOR ELEVATIONS
- SECTIONS
- DETAILS
- SCHEDULES

STRUCTURAL DRAWINGS
PLUMBING PLANS
MECHANICAL PLANS
ELECTRICAL PLANS

105F04.EPS

Figure 4 ◆ Typical format of a working drawing set.

3.1.3 Section and Detail Drawings

Section drawings are cutaway vertical views through an object or wall that show its interior makeup. They are used to show the details of construction and information about walls, stairs, and other parts of construction that may not show clearly on the plan. A section view is limited to the specific portion of the building construction that the architect wishes to clarify. It may be drawn on the same sheet as an elevation or plan view or it may appear on a separate sheet. *Figure 5* shows a typical example of a section drawing. Detail views are views that are normally drawn to a larger scale. They are often used to show aspects of a design that are too small to be shown in sufficient detail on a plan or elevation drawing. Like section drawings, detail drawings may be drawn on the same sheet as an elevation or plan drawing or may appear on a separate sheet in the set of plans.

Drawings

The various drawings that are used during construction are the property of either the architect or the building owner. They cannot legally be used for any other purpose without express permission.

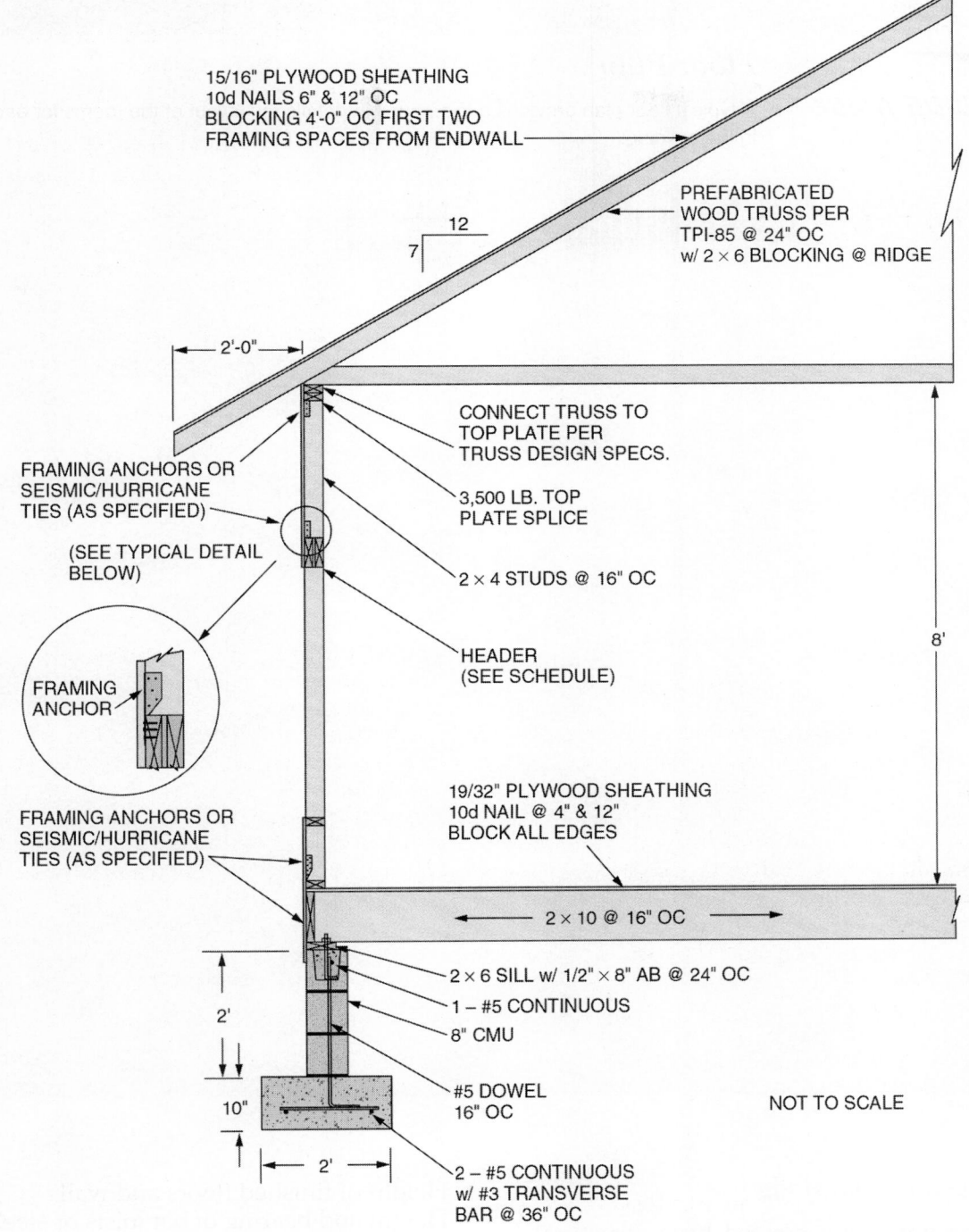

15/16" PLYWOOD SHEATHING
10d NAILS 6" & 12" OC
BLOCKING 4'-0" OC FIRST TWO
FRAMING SPACES FROM ENDWALL

PREFABRICATED
WOOD TRUSS PER
TPI-85 @ 24" OC
w/ 2 × 6 BLOCKING @ RIDGE

12
7

2'-0"

CONNECT TRUSS TO
TOP PLATE PER
TRUSS DESIGN SPECS.

3,500 LB. TOP
PLATE SPLICE

2 × 4 STUDS @ 16" OC

FRAMING ANCHORS OR
SEISMIC/HURRICANE
TIES (AS SPECIFIED)

(SEE TYPICAL DETAIL
BELOW)

FRAMING
ANCHOR

HEADER
(SEE SCHEDULE)

FRAMING ANCHORS OR
SEISMIC/HURRICANE
TIES (AS SPECIFIED)

19/32" PLYWOOD SHEATHING
10d NAIL @ 4" & 12"
BLOCK ALL EDGES

8'

2 × 10 @ 16" OC

2 × 6 SILL w/ 1/2" × 8" AB @ 24" OC

1 – #5 CONTINUOUS

8" CMU

2'

10"

#5 DOWEL
16" OC

NOT TO SCALE

2'

2 – #5 CONTINUOUS
w/ #3 TRANSVERSE
BAR @ 36" OC

105F05.EPS

Figure 5 ◆ Typical section drawing.

Floor Plan

A typical floor plan provides a top view that details the layout of the rooms for each floor in the building.

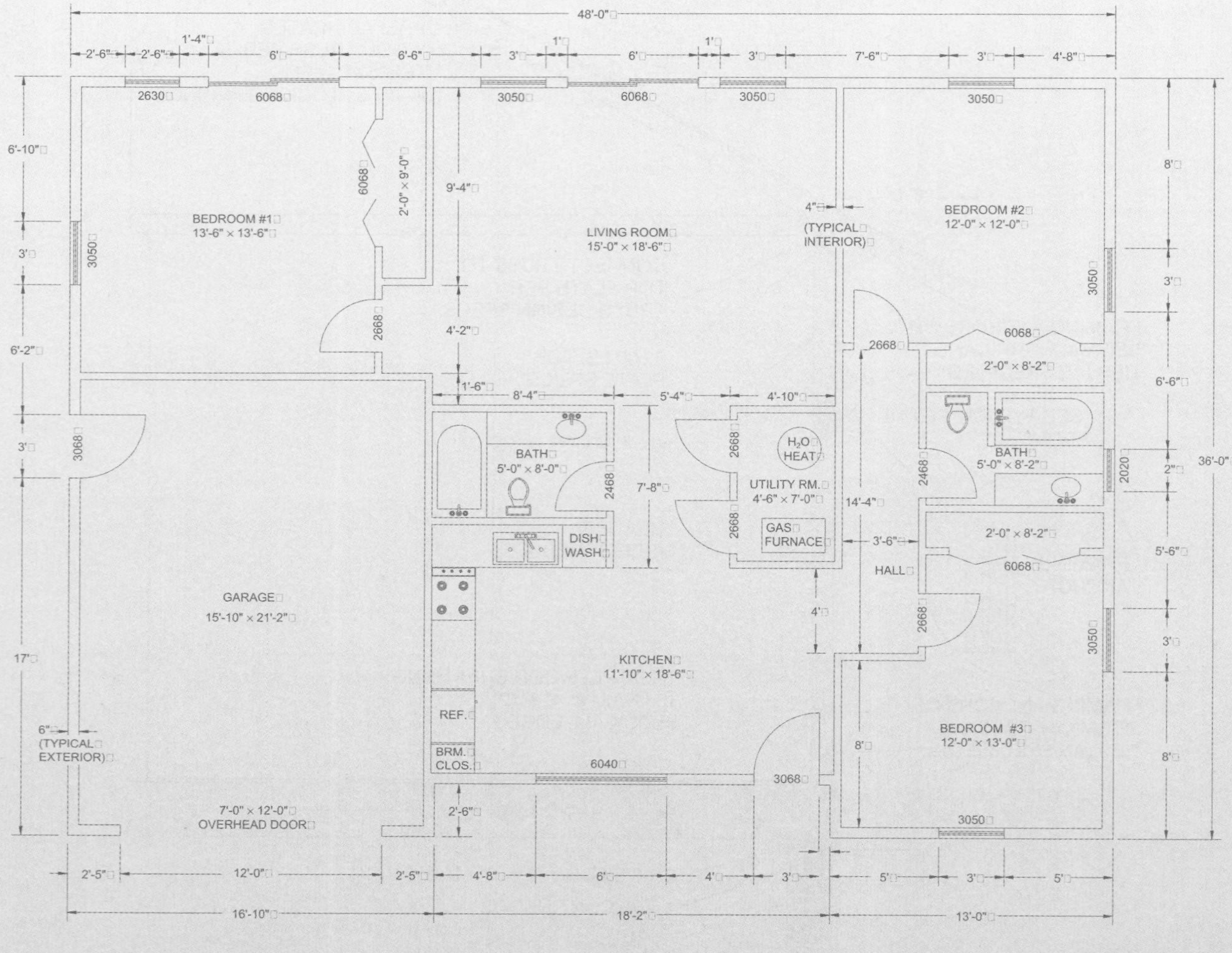

(Floor plan drawing with the following labeled rooms and dimensions:)

- BEDROOM #1 — 13'-6" × 13'-6"
- LIVING ROOM — 15'-0" × 18'-6"
- BEDROOM #2 — 12'-0" × 12'-0"
- BATH — 5'-0" × 8'-0"
- BATH — 5'-0" × 8'-2"
- H₂O HEAT
- UTILITY RM. — 4'-6" × 7'-0"
- GAS FURNACE
- HALL
- GARAGE — 15'-10" × 21'-2"
- KITCHEN — 11'-10" × 18'-6"
- BEDROOM #3 — 12'-0" × 13'-0"
- DISH WASH
- REF.
- BRM. CLOS.
- 7'-0" × 12'-0" OVERHEAD DOOR
- 4" (TYPICAL INTERIOR)
- 6" (TYPICAL EXTERIOR)

105SA01.EPS

3.1.4 Structural Drawings

Structural drawings are created by a structural engineer and accompany the architect's plans. They are usually drawn for large structures such as an office building or factory. They show requirements for structural elements of the building including columns, floor and roof systems, stairs, canopies, bearing walls, etc. They include such details as:

- Height of finished floors and walls
- Height and bearing of bar joists or steel joists
- Location of bearing steel materials
- Height of steel beams, concrete planks, concrete Ts, and poured-in-place concrete
- Bearing plate locations
- Location, size, and spacing of anchor bolts
- Stairways

Dimension vs. Scale

The scale of a drawing refers to the amount or percentage that a document has been reduced in relation to reality (full scale). Specific dimensions on documents, however, should always take precedence over the scaled graphic representation (drawing) found on the plans. The drawing itself is meant to give a general idea of the overall layout. The dimensions found on the drawing are what you should use when planning and building.

3.2.0 Plumbing, Mechanical, and Electrical Plans

Plumbing plans show the size and location of water and gas systems if they are not included in the mechanical section. Mechanical plans show temperature control and ventilation equipment including ducts, louvers, and registers. Electrical plans show all electrical equipment, lighting, outlets, etc.

It is important to note that while carpenters usually work with architectural and structural drawings, there are useful notes and views on drawings found in other sections, especially the mechanical section. For example, typical items not found on architectural drawings but that may be found on mechanical drawings are exposed heating, ventilating, and air conditioning (HVAC) ductwork; heating convectors; and fire sprinkler piping. Any one or all of these items may require the carpenter to build special framing, so make a habit of reviewing all drawings. Also, make sure to coordinate any such work with the appropriate other trades to make sure that the proper framing is done to accommodate ductwork, piping, wiring, etc.

3.3.0 Reading Blueprints

The following general procedure is suggested as a method of reading any set of blueprints for understanding:

Step 1 Read the title block. The title block tells you what the drawing is about. It contains critical information about the drawing such as the scale, date of last revision, drawing number, and architect or engineer. If you have to remove a sheet from a set of drawings, be sure to fold the sheet with the title block facing up.

Step 2 Find the north arrow. Always orient yourself to the structure. Knowing where north is enables you to more accurately describe the locations of walls and other parts of the building.

Step 3 Always be aware that blueprints work together as a team. The reason the architect or engineer draws plans, elevations, and sections is that it requires more than one type of view to communicate the whole project. Learn how to use more than one drawing, when necessary, to find the information you need.

Step 4 Check the list of blueprints in the set. Note the sequence of the various types of plans. Some blueprints have an index on the front cover. Notice that the prints are broken into several categories:
– Architectural
– Structural
– Mechanical
– Electrical
– Plumbing

Step 5 Study the site plan (plot plan) to observe the location of the building. Notice that the geographic location of the building may be indicated on the site plan.

Step 6 Check the foundation and floor plans for the orientation of the building. Observe the location and features of entries, corridors, offsets, and any special features.

Step 7 Study the features that extend for more than one floor, such as plumbing and vents, stairways, elevator shafts, heating and cooling ductwork, and piping.

Step 8 Check the floor and wall construction and other details relating to exterior and interior walls.

Step 9 Check the foundation plan for size and types of footings, reinforcing steel, and loadbearing substructures.

Step 10 Study the mechanical plans for the details of heating, cooling, and plumbing.

Step 11 Observe the electrical entrance and distribution panels, and the installation of the lighting and power supplies for special equipment.

Step 12 Check the notes on the various pages and compare the specifications against the construction details. Look for any variations.

Step 13 Thumb through the sheets of drawings until you are familiar with all the plans and structural details.

Step 14 Recognize applicable symbols and their relative locations in the plans. Note any special construction details or variations that will affect your job.

When you are building a floor system, the building plans (or specifications) should provide all the information you need to know about the floor system. Important information you should look for regarding the floor system includes:

- Type of wood or other materials used for sills, posts, girders, beams, joists, subfloors, etc.
- Size, location, and spacing of support posts or columns
- Direction of both joists and girders
- Manner in which joists connect to girders
- Location of any loadbearing interior walls that run parallel to joists
- Location of any toilet drains
- Rough opening sizes and locations of all floor openings for stairs, etc.
- Any cantilevering requirements
- Changes in floor levels
- Any special metal fasteners needed in earthquake areas
- Types of blocking or bridging
- Clearances from the ground to girder(s) and joists for floors installed over crawlspaces

3.4.0 Specifications

Written specifications are equally as important as the drawings in a set of plans. They furnish what the drawings cannot, in that they give detailed and accurate written descriptions of work to be done. They include quality and quantity of materials, methods of construction, standards of construction, and manner of conducting the work.

Specifications will be studied in detail later in your training. The basic information found in a typical specification includes:

- Contract
- Synopsis of the work
- General requirements
- Owner's name and address
- Architect's name
- Location of structure
- Completion date
- Guarantees
- Insurance requirements
- Methods of construction
- Types and quality of building materials
- Sizes

4.0.0 ♦ THE FLOOR SYSTEM

Floor systems provide a base for the remainder of the structure to rest on. They transfer the weight of people, furniture, materials, etc., from the subfloor, to the floor framing, to the foundation wall, to the footing, then finally to the earth. Floor systems are constructed over basements or crawlspaces. Single-story structures built on slabs do not have floor systems; however, multi-level structures may have both a slab and a floor system. *Figure 6* shows a typical platform floor system and identifies the various parts.

4.1.0 Sills

Sills, also called sill plates, are the lowest members of a structure's frame. They rest horizontally on the foundation and support the floor joists. The foundation is the supporting portion of a structure below the first floor construction, including the footings. Sills serve as the attachment point to the concrete or block foundation for all of the other wood framing members. The sills provide a means of leveling the top of the foundation wall and also prevent the other wood framing lumber from making contact with the concrete or masonry, which can cause the lumber to rot.

Today, sills are normally made using a single layer of 2 × 6 lumber (*Figure 7*). Local codes normally require that pressure-treated lumber and/or foundation-grade redwood lumber be used for constructing the sill whenever the sill plate comes in direct contact with any type of concrete. However, where codes allow, untreated softwood can be used.

Sills are attached to the foundation wall using either anchor bolts (*Figure 7*) or straps (*Figure 8*) embedded in the foundation. The exposed portion

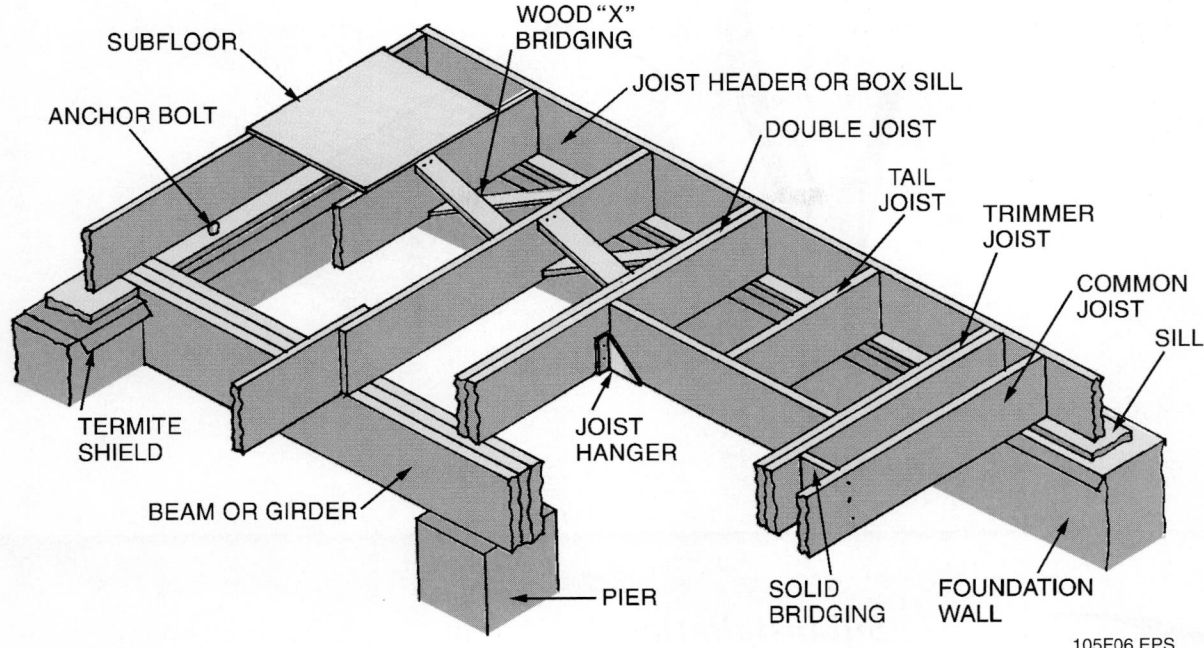

Figure 6 ◆ Typical platform frame floor system.

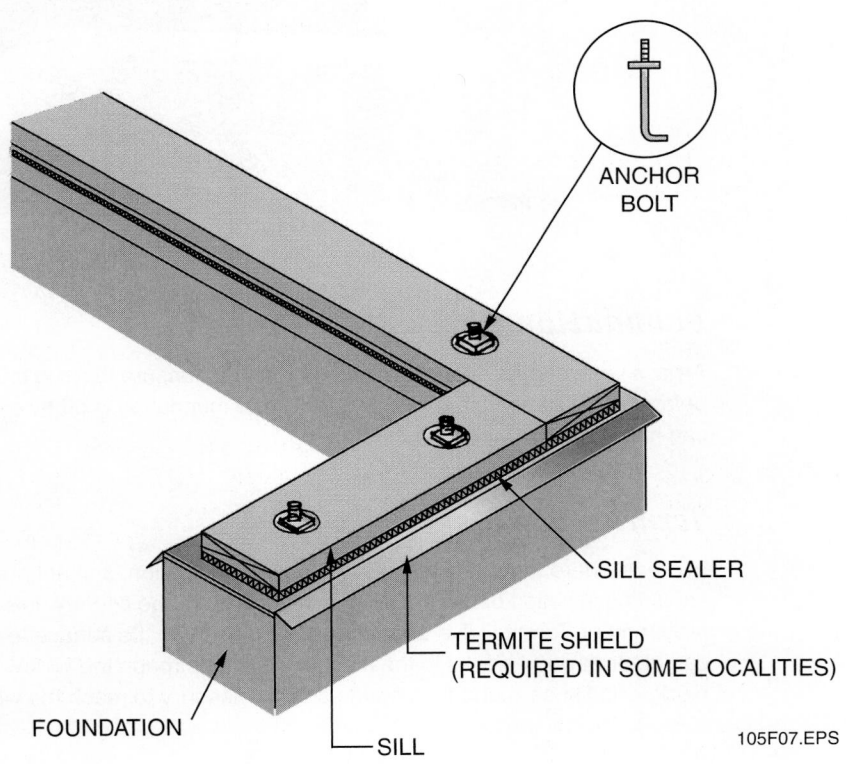

Figure 7 ◆ Typical sill installation.

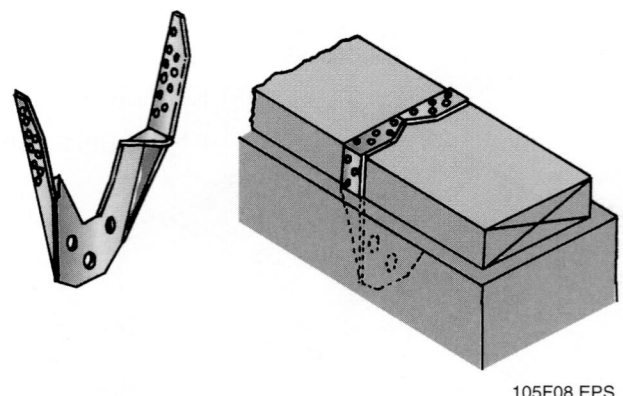

105F08.EPS

Figure 8 ◆ Typical sill anchor strap.

Sill Installation

Installing the sill plate is the first step in framing.

105SA02.EPS

Foundation Checks

Always verify the foundation measurements and ensure that the foundation is square before you start framing a floor system. If the foundation is off by even a tiny amount, it can have a major impact on the framing.

Termite Shields

In areas where there is a high risk of termite infestation, a sheet metal termite shield should be installed below the sill. In some areas of the country, this is a code requirement. Termites live underground and come to the surface to feed on wood. They can enter through cracks in the masonry, tunnel through the hollow cells of concrete block, or build earthen tubes on the side of masonry to reach the wood.

of the strap-type anchor is nailed to the sill. Some types must be bent over the top of the sill, while others are nailed to the sides. The size, type, and spacing between the anchor bolts or straps must be in compliance with local building codes. Their location and other related data are normally shown on the building blueprints.

In structures where the underfloor areas are used as part of the HVAC system, for storage, or as a basement, a glass-wool insulating material, called a sill sealer (*Figure 7*), should be installed to account for irregularities between the foundation wall and the sill. It seals against drafts, dirt, and insects. Sill sealer material is made in 6" wide, 50' rolls. Uncompressed, its thickness is 1". However, it can compress to as little as ½" when the weight of the structure is upon it. The sill sealer should be installed between the sill and the foundation wall, or between the sill and a termite shield (if used).

4.2.0 Beams/Girders and Supports

The distance between two outside walls is frequently too great to be spanned by a single joist. When two or more joists are needed to cover the span, support for the inboard joist ends must be provided by one or more beams, commonly called girders. Girders carry a very large proportion of the weight of a building. They must be well designed, rigid, and properly supported at the foundation walls and on the supporting posts or columns. They must also be installed so that they will properly support the joists. Girders may be made of solid timbers, built-up lumber, engineered lumber, or steel beams (*Figure 9*). Each type has advantages and disadvantages. Note that in some instances, precast reinforced concrete girders may also be used. A general procedure for determining how to size a girder is given later in this module.

4.2.1 Solid Lumber Girders

Solid timber girder stock used for beams is available in various sizes, with 4 × 6, 4 × 8, and 6 × 6 being typical sizes. If straight, large timbers are available, their use can save time by not having to make built-up girders. However, solid pieces of large timber stock are often badly bowed and can create a rise in the floor unless the crowns are pulled down. The crowns are the high points of the crooked edges of the framing members.

4.2.2 Built-Up Lumber Girders

Built-up girders are usually made using nominal 2" stock (2 × 8s, 2 × 10s) nailed together so that they act as one piece. Built-up girders have the advantage of not warping as easily as solid wooden girders and are less likely to have decayed wood in the center. The disadvantage is that a built-up girder is not capable of carrying the same load as an equivalent size solid timber girder. When constructing a built-up girder, the individual boards must be nailed together according to code requirements. Also, it is necessary to

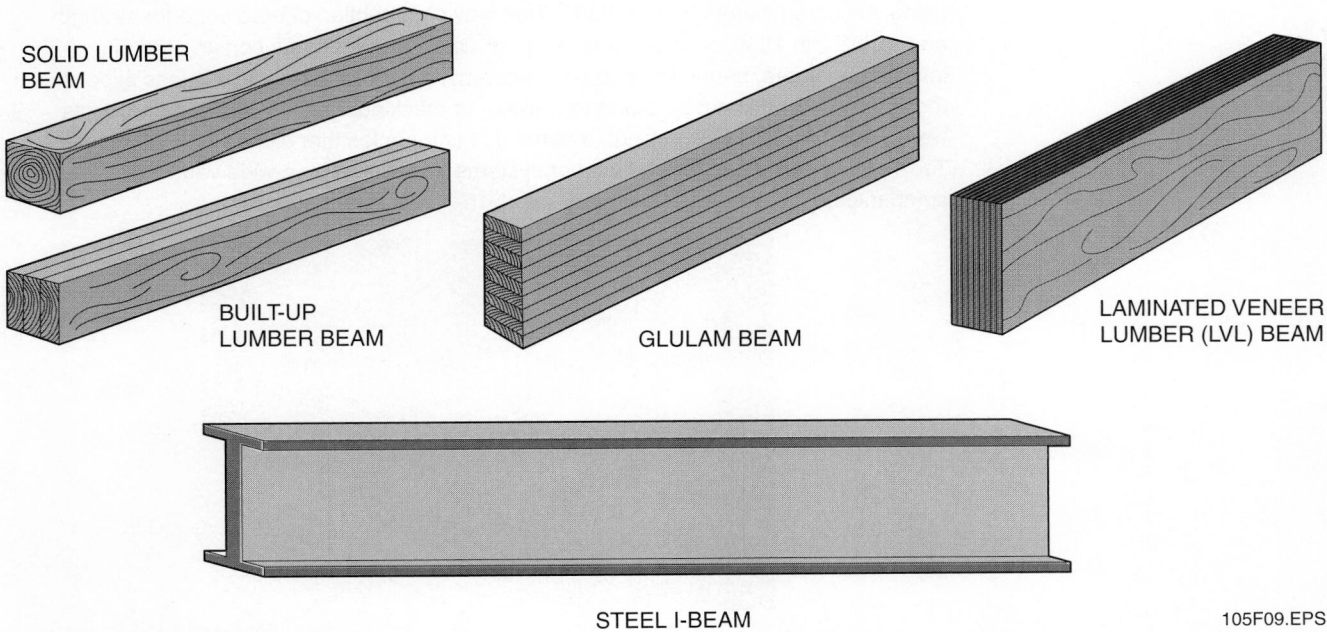

SOLID LUMBER BEAM

BUILT-UP LUMBER BEAM

GLULAM BEAM

LAMINATED VENEER LUMBER (LVL) BEAM

STEEL I-BEAM

105F09.EPS

Figure 9 ◆ Types of girders.

stagger the joints at least 4' in either direction. Construction of built-up girders is covered in more detail later in this module.

4.2.3 Engineered Lumber Girders

Laminated veneer lumber (LVL) and glue-laminated lumber (glulam) are engineered lumber products that are used for girders and other framing members. Their advantage is that they are stronger than the same size structural lumber. For a given length, the greater strength of engineered lumber products allows them to span a greater distance. Another advantage is that they are very straight with no crowns or warps.

LVL girders are made from laminated wood veneer like plywood. The veneers are laid up in a staggered pattern with the veneers overlapping to increase strength. Unlike plywood, the grain of each layer runs in the same direction as the other layers. The veneers are bonded with an exterior-grade adhesive, then pressed together and heated under pressure.

Glulam girders are made from lengths of solid, kiln-dried lumber glued together. They are commonly used where the beams are to remain exposed. Glulam girders are available in three appearance grades: industrial, architectural, and premium. For floor systems like those described in this module, industrial grade would normally be used because appearance is not a priority. Architectural grade is used where beams are exposed and appearance is important. Premium grade is used where the highest-quality appearance is needed. Glulam beams are available in various widths and depths and in lengths up to 40' long.

4.2.4 Steel I-Beam Girders

Metal beams can span the greatest distances and are often used when there are few or no piers or interior supports in a basement. Also, they can span greater distances with smaller beam sizes, thereby creating greater headroom in a basement or crawlspace. For example, a 6" high steel beam may support the same load as an 8" or 10" high wooden beam. Two types of steel beams are available: standard flange (S-beam) and wide flange (W-beam). The wide beam is generally used in residential construction. Being metal, I-beams are more expensive than wood and are harder to work with. They are normally used only when the design or building code calls for it.

I-Beams

The first plywood I-beam was created in 1969. In 1977, the first I-beam was created using laminated veneer lumber (LVL). This new construction offered superior strength and stability. In 1990, oriented strand board (OSB) web material, constructed of interlocking fibers, began to be used in I-beams, as shown here. OSB is less expensive than plywood and is not as prone to warping or cracking. Engineered wood products were once only available through a handful of companies that pioneered the industry. Today, engineered lumber and lumber systems are offered by a wide variety of companies.

105SA03.EPS

4.2.5 Beam/Girder Supports

Girders and beams must be properly supported at the foundation walls, and at the proper intervals in between, either by supporting posts, columns, or piers (*Figure 10*). Solid or built-up wood posts installed on pier blocks are commonly used to support floor girders, especially for floors built over a crawlspace. Usually, 4 × 4 or 4 × 6 posts are used. However, all posts must be as wide as the girder. Where girder stock is jointed over a post, a 4 × 6 is normally required. To secure the wood posts to their footings, pieces of ½" reinforcing rod or iron bolts are often embedded in the support footings before the concrete sets. These project into holes bored in the bottoms of the posts. The use of galvanized steel post anchors is another widely used method of fastening the bottoms of wooden posts to their footings (*Figure 11*). The

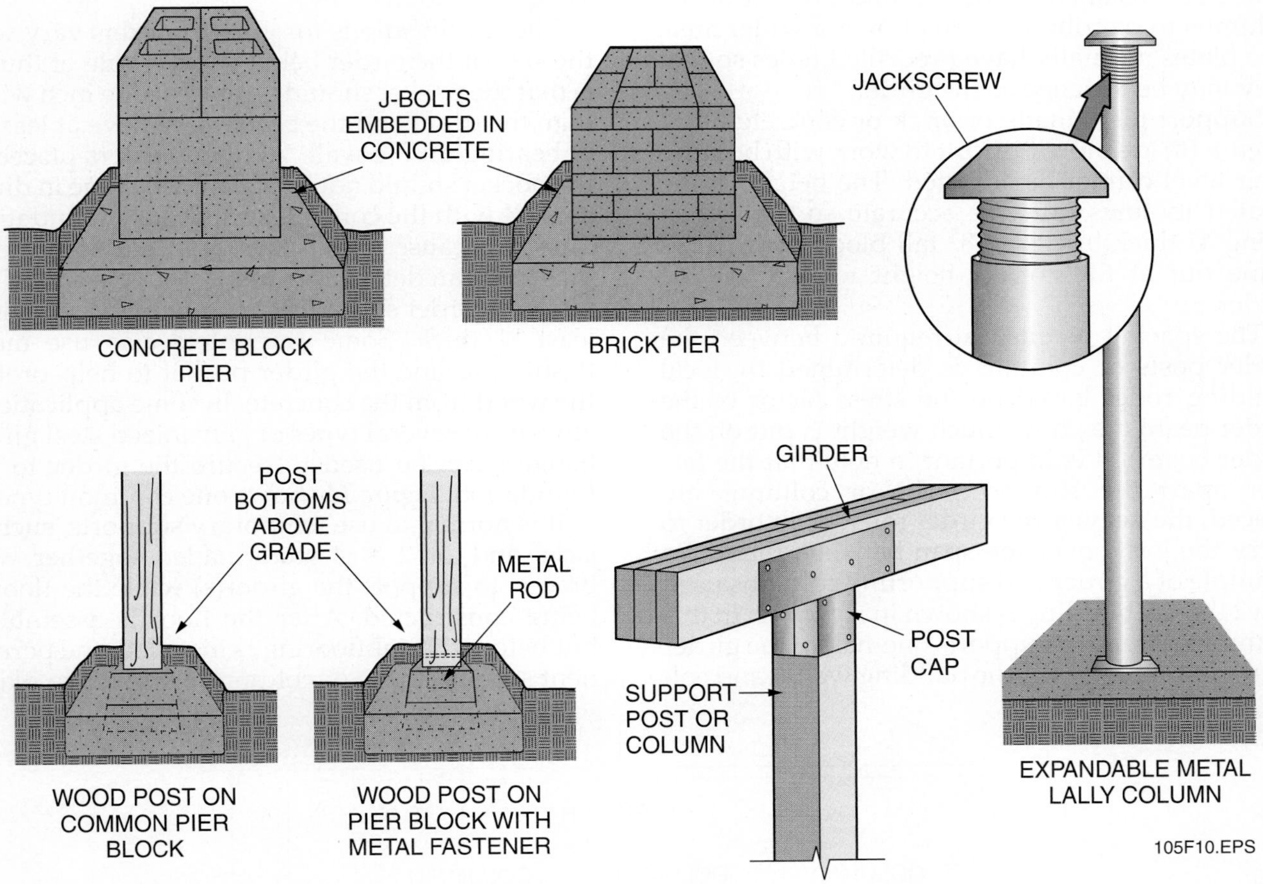

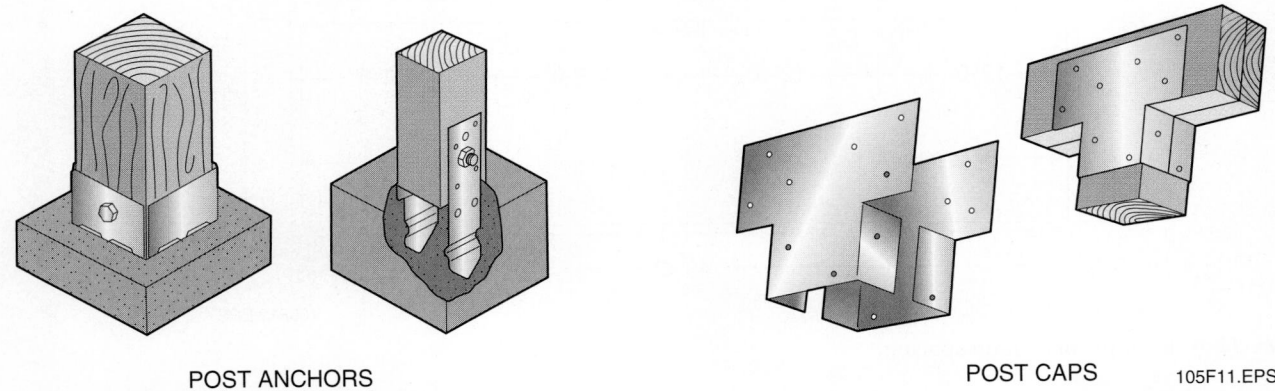

Figure 10 ◆ Typical methods of supporting girders.

Figure 11 ◆ Typical post anchors and caps.

tops of the posts are normally fastened to the girder using galvanized steel post caps. In addition to securing the post to the girder, these caps also provide for an even bearing surface.

Four-inch round steel columns filled and reinforced with concrete (*Figure 10*), called lally columns, are commonly used as support columns in floors built over basements. Some types of lally columns must be cut to the required height, while others have a built-in jackscrew that allows the column to be adjusted to the proper height. Metal plates are installed at the top and bottom of the columns to distribute the load over a wider area. The plates normally have predrilled holes so that they may be fastened to the girder.

Support piers made of brick or concrete block (*Figure 10*) are more difficult to work with because their level cannot be adjusted. The height of the related footings must be accurate so that when using 4" thick bricks or 8" tall blocks, their tops come out at the correct height to support the girder.

The spacing or interval required between the girder posts or columns is determined by local building codes based on the stress factor of the girder beam (i.e., how much weight is put on the girder beam). It is important to note that the farther apart the support posts or columns are spaced, the heavier the girder must be in order to carry the joists over the span between them. An example of a girder and supporting columns used in a 24' × 48' building is shown in *Figure 12*. In this example, column B supports one-half of the girder load existing between the building wall A and column C. Column C supports one-half of the girder load between columns B and D. Likewise, column D will share equally the girder loads with column C and the wall E.

As shown in *Figure 13*, support of girder(s) at the foundation walls can be done by constructing posts made from solid wood or piers made of concrete block or brick. Another widely used method is to construct girder (beam) pockets into the concrete or concrete block foundation walls (*Figure 14*). Provide steel reinforcement as required by the job specifications.

The specifications for girder pockets vary with the size of the girder being used. A rule of thumb is that the pocket should be at least one inch wider than the beam and the beam must have at least 4" of bearing on the wall. Wooden girders placed in the pocket should not be allowed to come in direct contact with the concrete or masonry foundation. This is because the chemicals in the concrete or masonry can deteriorate the wood. The end of the wooden girder should sit on a steel plate that is at least ¼" thick. Some carpenters also use metal flashing to line the girder pocket to help protect the wood from the concrete. In some applications, any one of several types of galvanized steel girder hangers can be used to secure the girder to the foundation. *Figure 14* shows one common type.

It is normal to use temporary supports, such as jacks and/or 2 × 4 studs nailed together with braces, to support the girder(s) while the floor is being constructed. After the floor is assembled, but before the subflooring is installed, the permanent support posts or columns are put into place.

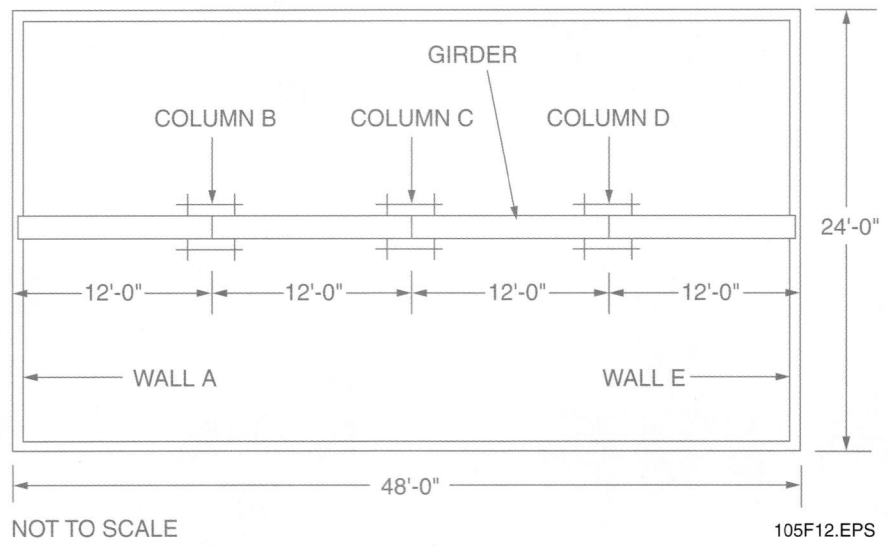

Figure 12 ◆ Example of column spacing.

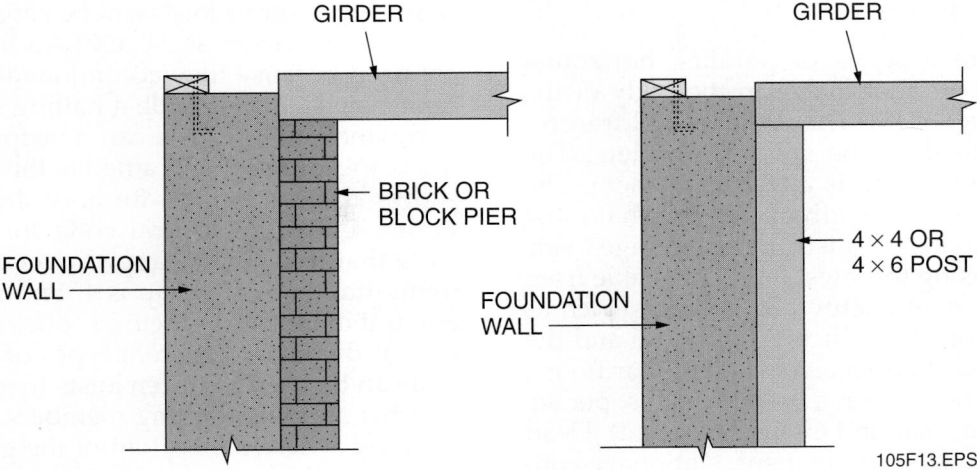

GIRDER

GIRDER

BRICK OR
BLOCK PIER

FOUNDATION
WALL

4 × 4 OR
4 × 6 POST

FOUNDATION
WALL

105F13.EPS

Figure 13 ◆ Post or pier support of a girder at the foundation wall.

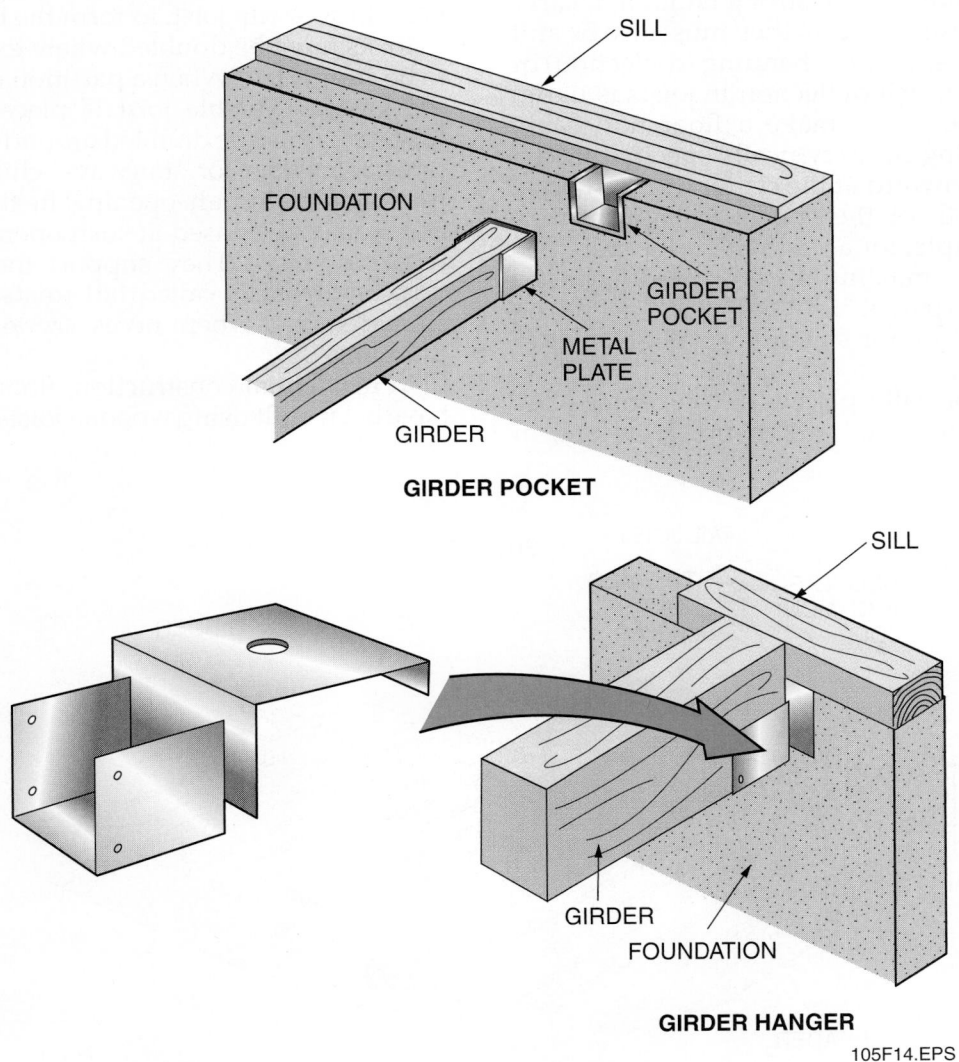

SILL

FOUNDATION

GIRDER
POCKET

METAL
PLATE

GIRDER

GIRDER POCKET

SILL

GIRDER

FOUNDATION

GIRDER HANGER

105F14.EPS

Figure 14 ◆ Girder pocket and girder hanger girder supports.

4.3.0 Floor Joists

Floor joists are a series of parallel, horizontal framing members that make up the body of the floor frame (*Figure 15*). They rest on and transfer the building load to the sills and girders. The flooring or subflooring is attached to them. The span determines the length of the joist that must be used. Safe spans for joists under average loads can be found using the latest tables available from wood product manufacturers or sources such as the National Forest Products Association and the Southern Forest Products Association. For floors, this is usually figured on a basis of 50 lbs. per sq. ft. (10 lbs. **dead load** and 40 lbs. **live load**). Dead load is the weight of permanent, stationary construction and equipment included in a building. Live load is the total of all moving and variable loads that may be placed upon a building.

Joists must not only be strong enough to carry the load that rests on them; they must also be stiff enough to prevent undue bending (deflection) or vibration. Too much deflection in joists is undesirable because it can make a floor noticeably springy. Building codes typically specify that the deflection downward at the center of a joist must not exceed ¹⁄₃₆₀th of the span with normal live load. For example, for a joist with a 15' span, this would equal a maximum of ½" of downward deflection (15' span × 12 = 180" ÷ 360 = 0.5"). A general procedure for sizing joists is given later in this module.

Joists are normally placed 16" on center (OC) and are always placed crown up. However, in some applications joists can be set as close as 12" OC or as far apart as 24" OC. All these distances are used because they accommodate 4' × 8' subfloor panels and provide a nailing surface where two panels meet. Joists can be supported by the top of the girder or be framed to the side. *Figure 16* shows three methods for joist framing at the girder. Check your local code for applicability. Note that if joists are lapped over the girder, the minimum amount of lap is 4" and the maximum amount of lap is 12". *Figure 17* shows some examples of the many different types of **joist hangers** that can be used to fasten joists to girders as well as other support framing members. Joist hangers are used where the bottom of the girder must be flush with the bottoms of the joists. At the sill end of the joist, the joist should rest on at least 1½" of wood. In platform construction, the ends of all the joists are fastened to a **header joist,** also called a band joist or rim joist, to form the box sill.

Joists must be doubled where extra loads need to be supported. When a partition runs parallel to the joists, a double joist is placed underneath. Joists must also be doubled around all openings in the floor frame for stairways, chimneys, etc., to reinforce the rough opening in the floor. These additional joists used at such openings are called **trimmer joists.** They support the headers that carry short joists called **tail joists.** Double joists should spread where necessary to accommodate plumbing.

In residential construction, floors traditionally have been built using wooden joists. However, the

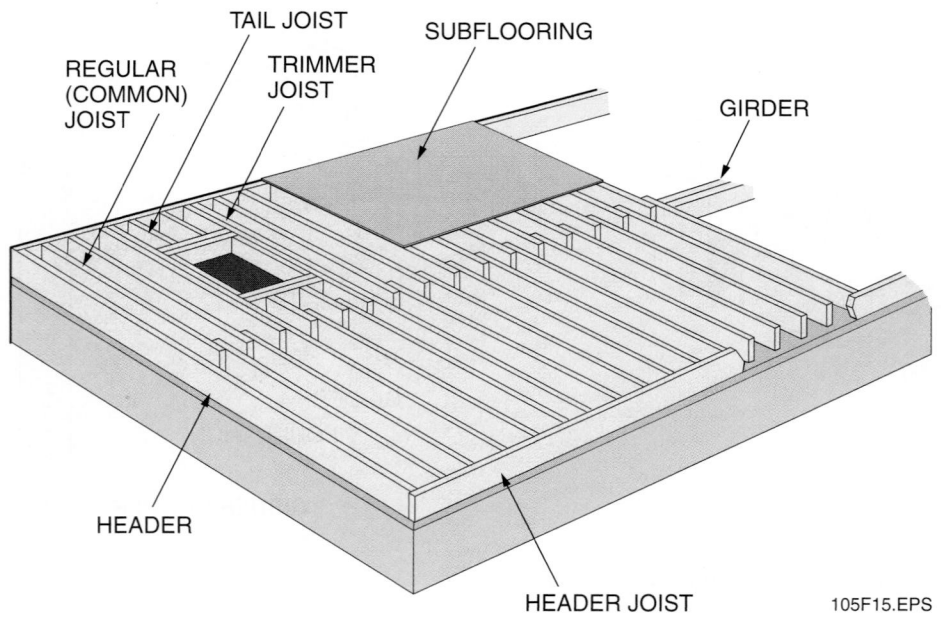

Figure 15 ◆ Floor joists.

105F15.EPS

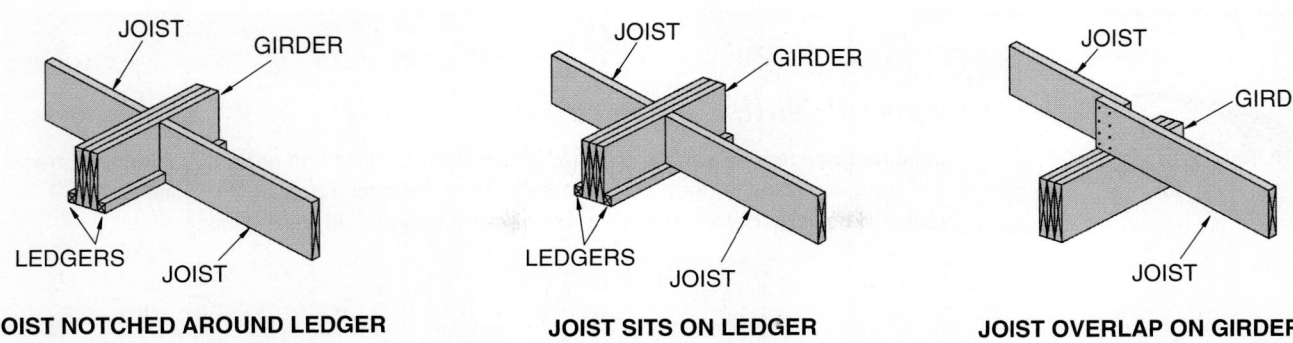

JOIST NOTCHED AROUND LEDGER　　**JOIST SITS ON LEDGER**　　**JOIST OVERLAP ON GIRDER**

105F16.EPS

Figure 16 ◆ Methods of joist framing at a girder.

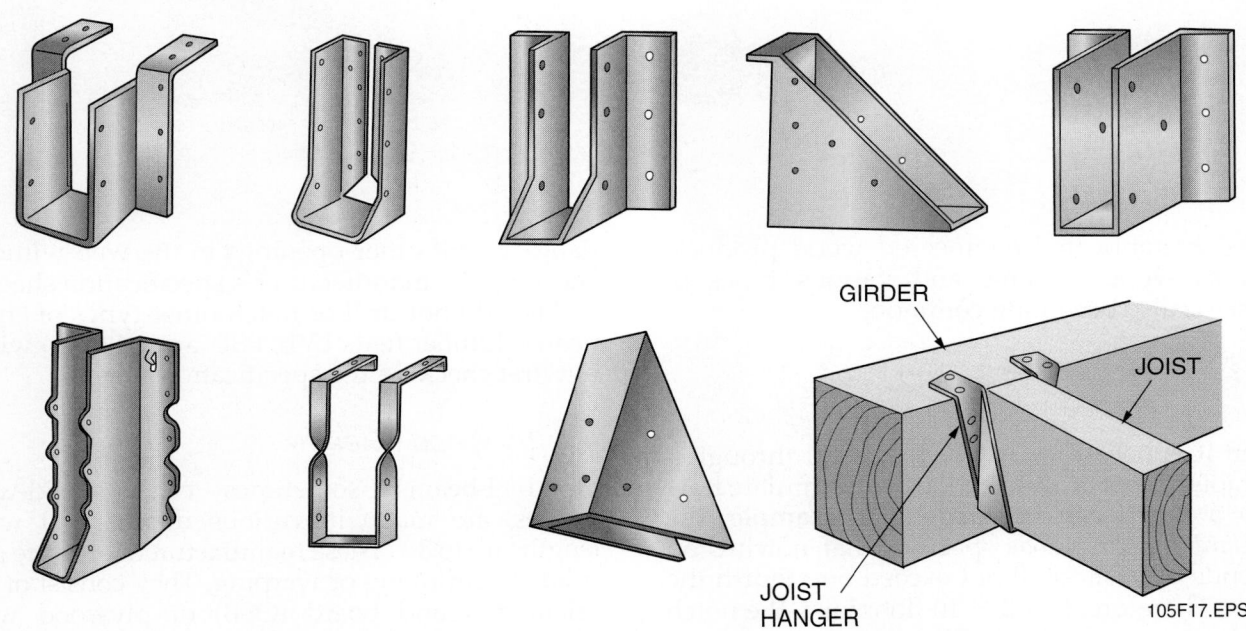

105F17.EPS

Figure 17 ◆ Typical types of joist hangers.

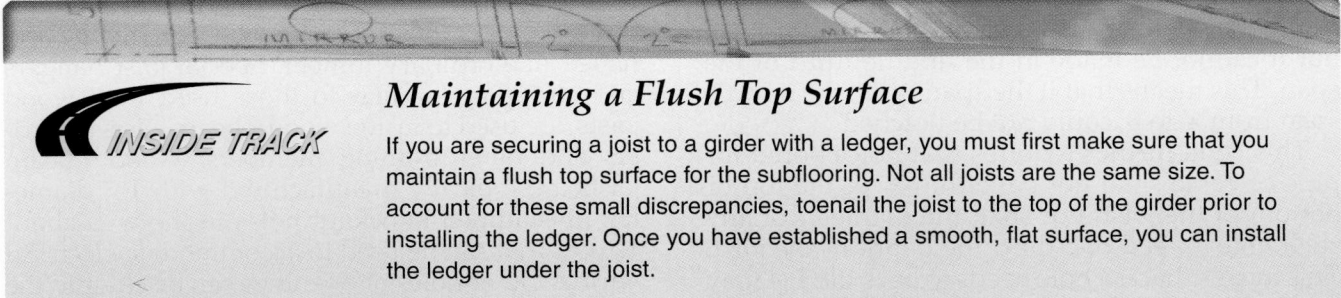

INSIDE TRACK

Maintaining a Flush Top Surface

If you are securing a joist to a girder with a ledger, you must first make sure that you maintain a flush top surface for the subflooring. Not all joists are the same size. To account for these small discrepancies, toenail the joist to the top of the girder prior to installing the ledger. Once you have established a smooth, flat surface, you can install the ledger under the joist.

Special Nails

Joist hangers require special nails to secure them to joists and girders. These nails are 1½" long and stronger than common nails. They are often referred to as joist or stub nails. The picture below shows various types of hangers as well as joist nails.

105SA04.EPS

use of prefabricated engineered wood products such as wood I-beams and various types of trusses is also becoming common.

4.3.1 Notching and Drilling of Wooden Joists

When it is necessary to notch or drill through a floor joist, most building codes will stipulate how deep a notch can be made. For example, the *Standard Building Code* specifies that notches on the ends of joists shall not exceed one-fourth the depth. Therefore, in a 2 × 10 floor joist, the notch could not exceed 2½" (see *Figure 18*).

This code also states that notches for pipes in the top or bottom shall not exceed one-sixth the depth, and shall not be located in the middle third of the span. Therefore, when using a 2 × 10 floor joist, a notch cannot be deeper than 1⅝". This notch can be made either in the top or bottom of the joist, but it cannot be made in the middle third of the span. This means that if the span is 12', the middle span from 4' to 8' could not be notched.

This code further requires that holes bored for pipe or cable shall not be within 2" of the top or bottom of the joist, nor shall the diameter of any such hole exceed one-third the depth of the joist. This means that if a hole needs to be drilled, it may not exceed 3" in diameter if a 2 × 10 floor joist is used. Always check the local codes.

Some wood I-beams are manufactured with perforated knockouts in their web, approximately 12" apart. Never notch or drill through the beam

flange or cut other openings in the web without checking the manufacturer's specification sheet.

Also, do not drill or notch other types of engineered lumber (e.g., LVL, PSL, and glulam) without first checking the specification sheets.

4.3.2 Wood I-Beams

Wood I-beams, sometimes called solid-web trusses, are made in various depths and with lengths up to 80'. These manufactured joists are not prone to shrinking or warping. They consist of an oriented strand board (OSB) or plywood web (*Figure 19*) bonded into grooves cut in the wood flanges on the top and bottom. This arrangement provides a joist that has a strength-to-weight ratio much greater than that of ordinary lumber. Because of their increased strength, wood I-beams can be used in greater spans than a comparable length of dimension lumber. They can be cut, hung, and nailed like ordinary lumber. Special joist hangers and strapping, similar to those used with wood joists, are used to fasten wood I-beam joists to girders and other framing members. Wood I-beam joists are typically manufactured with 1½" diameter, prestamped knockout holes in the web about 12" OC that can be used to accommodate electrical wiring. Other holes or openings can be cut into the web, but these can only be of a size and at the locations specified by the I-beam manufacturer. Under no circumstances should the flanges of I-beam joists be cut or notched. *Figure 20* shows a typical floor system constructed with wood I-beams.

Wood I-Beams

I-beams have specific guidelines and instructions to follow for cutting, blocking, and installation. It is important to always follow the manufacturer's instructions when installing these materials. Otherwise, you may create a very dangerous situation by compromising the structural integrity of the beam.

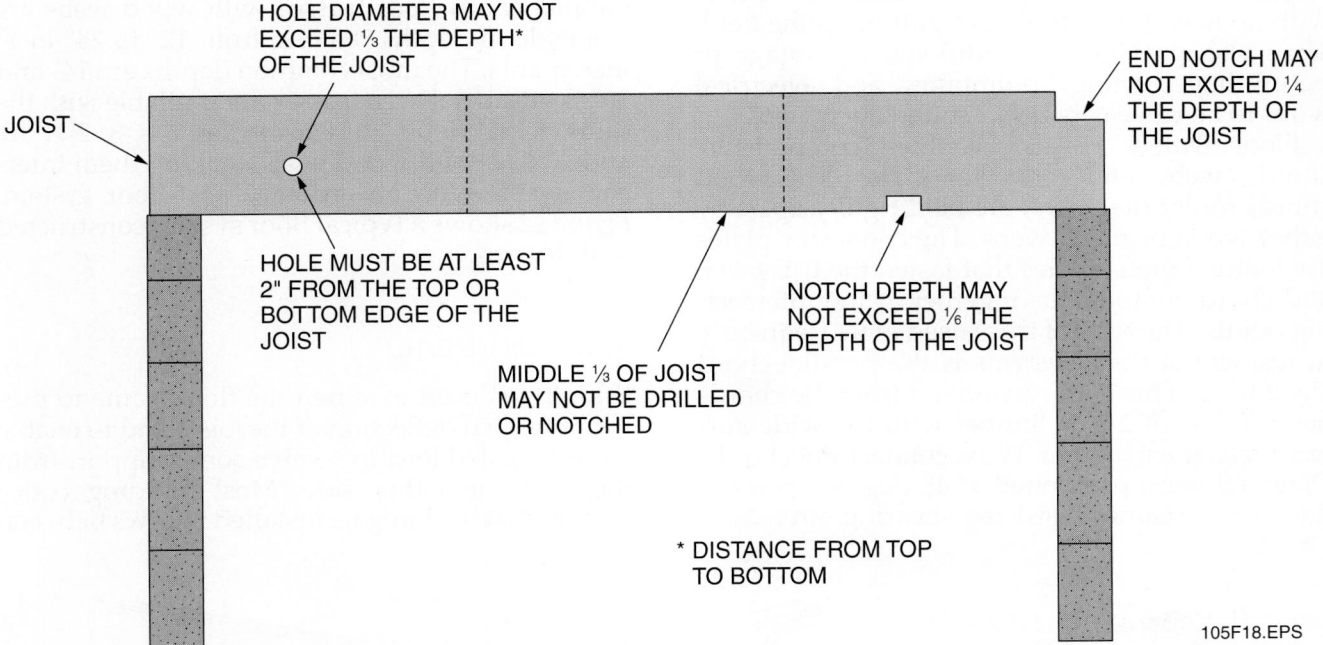

HOLE DIAMETER MAY NOT EXCEED ⅓ THE DEPTH* OF THE JOIST

END NOTCH MAY NOT EXCEED ¼ THE DEPTH OF THE JOIST

JOIST

HOLE MUST BE AT LEAST 2" FROM THE TOP OR BOTTOM EDGE OF THE JOIST

NOTCH DEPTH MAY NOT EXCEED ⅙ THE DEPTH OF THE JOIST

MIDDLE ⅓ OF JOIST MAY NOT BE DRILLED OR NOTCHED

* DISTANCE FROM TOP TO BOTTOM

105F18.EPS

Figure 18 ◆ Notching and drilling of wooden joists.

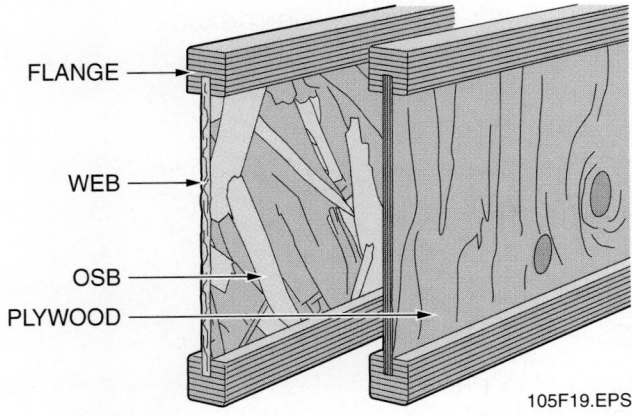

FLANGE

WEB

OSB

PLYWOOD

105F19.EPS

Figure 19 ◆ I-beams.

105F20.EPS

Figure 20 ◆ Typical floor system constructed with engineered I-beams (second floor shown).

4.3.3 Trusses

Trusses are manufactured joist assemblies made of wood or a combination of steel and wood (*Figure 21*). Solid light-gauge steel and open-web steel trusses are also made, but these are used mainly in commercial construction. Like the wood I-beams, trusses are stronger than comparable lengths of dimension lumber, allowing them to be used over longer spans. Longer spans allow more freedom in building design because interior load-bearing walls and extra footings can often be eliminated. Trusses generally erect faster and easier with no need for trimming or cutting in the field. They also provide the additional advantage of permitting ducting, plumbing, and electrical wires to be run easily between the open webs.

Floor trusses consist of three components: chords, webs, and connector plates. The wood chords (outer members) are held rigidly apart by either wood or metal webs. The connector plates are toothed metal plates that fasten the truss web and chord components together at the intersecting points. The type of truss used most frequently in residential floor systems is the parallel-chord 4×2 truss. This name is derived from the chords being made of 2×4 lumber with the wide surfaces facing each other. Webs connect the chords. Diagonal webs positioned at 45-degree angles to the chords mainly resist the shearing stresses in the truss. Vertical webs, which are placed at right angles to the chords, are used at critical load transfer points where additional strength is required. Wood is used most frequently for webs, but galvanized steel webs are also used. Trusses made with metal webs provide greater clear spans for any given truss depth than wood-web trusses. The openings in the webs are larger too, which allows more room for HVAC ducting.

Note that there are several different kinds of parallel-chord trusses. What makes each one different is the arrangement of its webs. Typically, parallel-chord floor trusses with wood webs are available in depths ranging from 12" to 24" in 1" increments. The most common depths are 14" and 16". Some metal-web trusses are available with the same actual depth dimensions as 2×8, 2×10, and 2×12 solid wood joists, making them interchangeable with an ordinary joist-floor system. *Figure 22* shows a typical floor system constructed with trusses.

4.4.0 Bridging

Bridging is used to stiffen the floor frame to prevent unequal deflection of the joists and to enable an overloaded joist to receive some support from the joists on either side. Most building codes require that bridging be installed in rows between

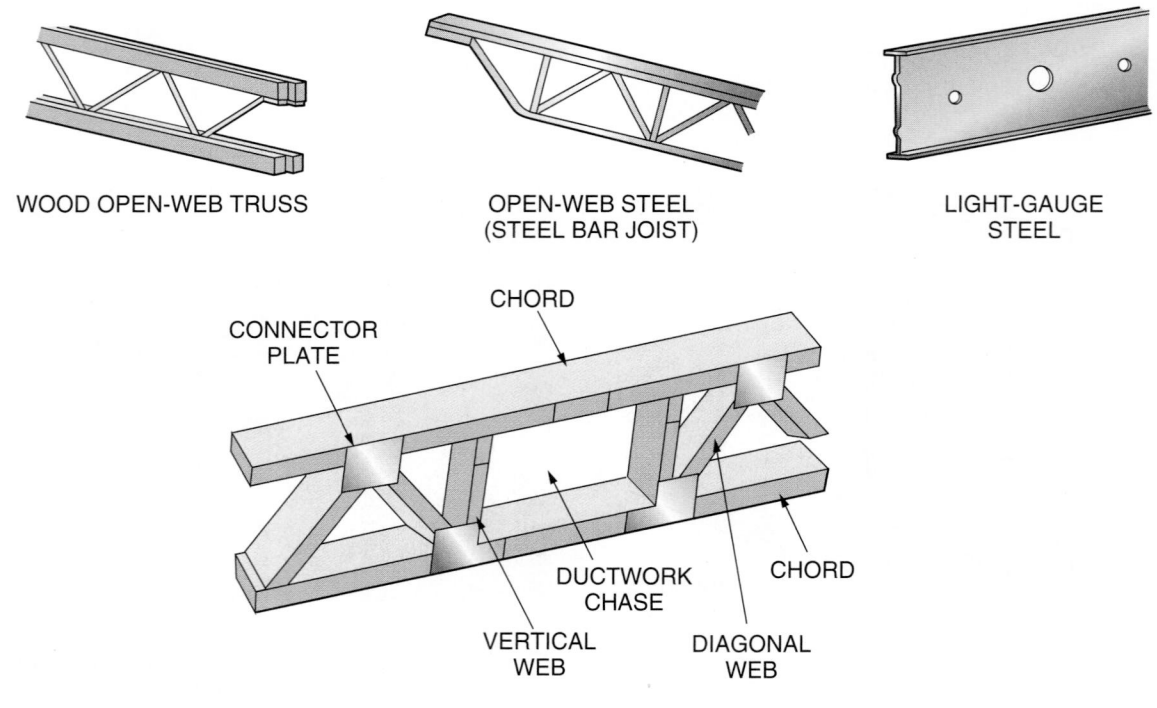

WOOD OPEN-WEB TRUSS

OPEN-WEB STEEL
(STEEL BAR JOIST)

LIGHT-GAUGE
STEEL

CONNECTOR
PLATE

CHORD

DUCTWORK
CHASE

CHORD

VERTICAL
WEB

DIAGONAL
WEB

PARALLEL-CHORD WOOD 4×2 TRUSS

105F21.EPS

Figure 21 ◆ Typical trusses.

the floor joists at intervals of not more than 8'. For example, floor joists with spans of 8' to 16' need one row of bridging in the center of the span.

Three types of bridging are commonly used (*Figure 23*): wood cross-bridging, metal cross-bridging, and solid bridging. Wood and metal cross-bridging are composed of pieces of wood or metal set diagonally between the joists to form an X. Wood cross-bridging is typically 1 × 4 lumber placed in double rows that cross each other in the joist space.

Metal cross-bridging is installed in a similar manner. Metal cross-bridging comes in a variety of styles and different lengths for use with a particular joist size and spacing. It is usually made of 18-gauge steel and is ¾" wide. When using cross-bridging, you may nail the top, but do not nail the bottom until the subfloor is installed. Solid bridging, also called blocking, consists of solid pieces of lumber, usually the same size as the floor joists, installed between the joists. It is installed in an off-set fashion to enable end nailing.

4.5.0 Subflooring

Subflooring consists of panels or boards laid directly on and fastened to floor joists (*Figure 24*) in order to provide a base for **underlayment** and/or the finish floor material. Underlayment is a material, such as particleboard or plywood, laid on top of the subfloor to provide a smoother surface for some finished flooring. This surface is normally applied after the structure is built but before the finished floor is laid. The subfloor adds rigidity to the structure and provides a surface upon which wall and other framing can be laid out and constructed. Subfloors also act as a barrier to cold and dampness, thus keeping the building

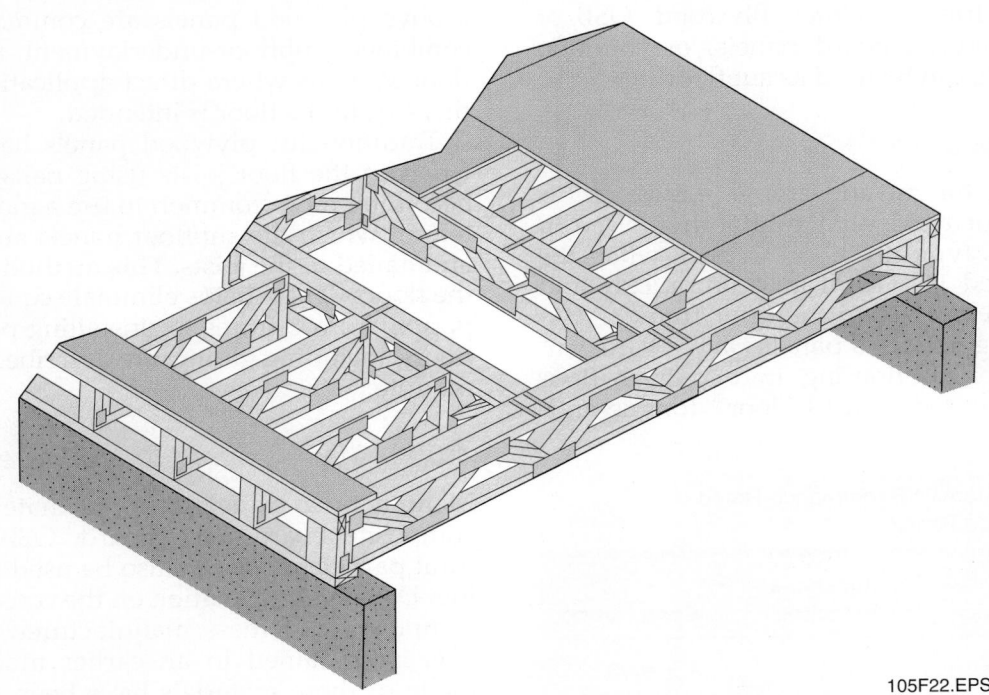

105F22.EPS

Figure 22 ◆ Typical floor system constructed with trusses.

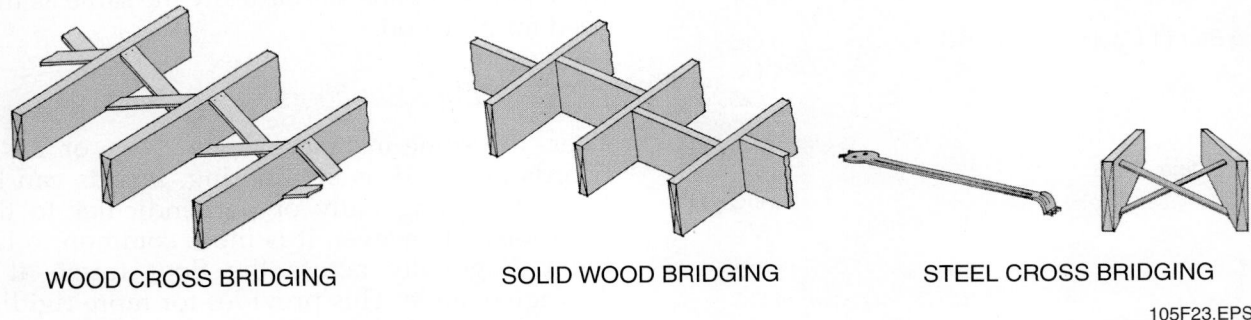

WOOD CROSS BRIDGING   SOLID WOOD BRIDGING   STEEL CROSS BRIDGING

105F23.EPS

Figure 23 ◆ Types of bridging.

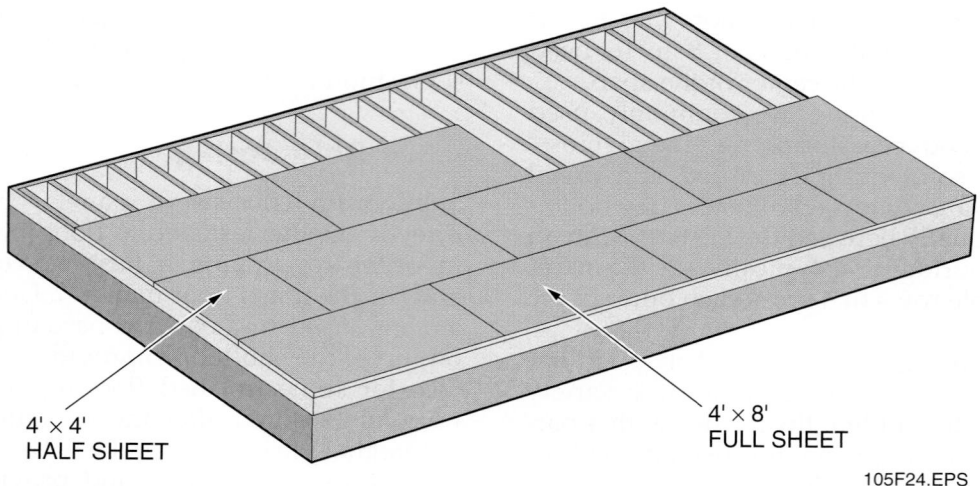

4' × 4'
HALF SHEET

4' × 8'
FULL SHEET

105F24.EPS

Figure 24 ◆ Subflooring installation.

warmer and drier in winter. Plywood, OSB or other manufactured board panels, or common wooden boards can be used as subflooring.

4.5.1 Plywood Subfloors

Butt-joint or tongue-and-groove plywood is widely used for residential subflooring. Used in 4' × 8' panels, typically ⅜" to ¾" thick when the joists are placed 16" OC, it goes on quickly and provides great rigidity to the floor frame. APA-rated sheathing plywood panels (*Table 1*) are generally used for subflooring in two-layer floor systems. APA-rated Sturd-I-Floor® tongue-and-

groove plywood panels are commonly used in combined subfloor-underlayment (single-layer) floor systems where direct application of carpet, tile, etc., to the floor is intended.

Traditionally, plywood panels have been fastened to the floor joists using nails. Today, it is becoming more common to use a glued floor system in which the subfloor panels are both glued and nailed to the joists. This method helps stiffen the floors. It also helps eliminate squeaks and nail popping. Procedures for installing plywood subfloors, including gluing, are described later in this module.

4.5.2 Manufactured Board Panel Subfloors

Manufactured panels made of materials such as composite board, waferboard, OSB, and structural particleboard can also be used for subflooring. Detailed information on the construction and composition of these manufactured wood products is contained in an earlier module. Panels made of these materials have been rated by the American Plywood Association and meet all standards for subflooring. The method for installing these kinds of panels is basically the same as that used for plywood.

4.5.3 Board Subfloors

There are some instances when 1 × 6 or 1 × 8 boards are used as subflooring. Boards can be laid either diagonally or perpendicular to the floor joists. However, it is more common to lay them diagonally across the floor frame at a 45-degree angle. This provides for more rigidity of the floor and also assists in the bracing of the

Table 1	Guide to APA Performance-Rated Plywood Panels	
Panel Grade	**Thickness in Inches**	**Span Rating**
Rated Sheathing	⅜	24/0
	⁷⁄₁₆	24/16
	¹⁵⁄₃₂	32/16
	¹⁹⁄₃₂	40/20
	²³⁄₃₂	48/24
Rated Sturd-I-Floor	¹⁹⁄₃₂	20 OC
	²³⁄₃₂	24 OC
	⅞, 1	32 OC
	1⅛	48 OC
Rated Siding	¹¹⁄₃₂	16 OC
	⁷⁄₁₆	24 OC
	¹⁵⁄₃₂	24 OC
T1-11	¹⁹⁄₃₂	16 OC
	¹⁹⁄₃₂	24 OC

OSB Subfloors

Today, many builders prefer to use OSB for subfloors. It offers acceptable structural strength at a reduced price.

105SA05.EPS

floor joists. Also, if laid perpendicular to the joist in a subfloor where oak flooring is to be laid over it, the oak flooring (instead of the subflooring) would have to be laid diagonally to the floor joist. This is necessary to prevent the shrinkage of the subfloor from affecting the joints in the finished oak floor, which would cause the oak flooring to pull apart. Board subflooring is nailed at each joist. Typically, two nails are used in each 1 × 6 board and three nails for wider boards. Note that a subfloor made of boards is normally not as rigid as one made of plywood or other manufactured panels.

5.0.0 ◆ LAYING OUT AND CONSTRUCTING A PLATFORM FLOOR ASSEMBLY

After the foundation is completed and the concrete or mortar has properly set up, assembly of the floor system can begin. Framing of the floors and sills is usually done before the foundation is backfilled. This is because the floor frame helps the foundation withstand the pressure placed on it by the soil. This section gives an overview of the procedures and methods used for laying out and constructing a basic platform floor assembly. When building any floor system, always coordinate your work with that of the other trades to ensure the framing is properly done to accommodate ductwork, piping, wiring, etc.

The construction of a platform floor assembly is normally done in the sequence shown below:

Step 1 Check the foundation for squareness.

Step 2 Lay out and install the sill plates.

Step 3 Build and/or install the girders and supports.

Step 4 Lay out the sills and girders for the floor joists.

Step 5 Lay out the joist locations for partitions and floor openings.

Step 6 Cut and attach the joist headers to the sill.

Step 7 Install the joists.

Step 8 Frame the openings in the floor.

Step 9 Install the bridging.

Step 10 Install the subflooring.

5.1.0 Checking the Foundation for Squareness

Before installing the sill plates, ensure that the foundation wall meets the dimensions specified on the blueprints and that the foundation is square. However, keep in mind that parallel and plumb take precedence over square. Checking the foundation for square is done by making measurements of the foundation with a 100' steel measuring tape. First, the lengths of each of the foundation walls are measured and recorded (*Figure 25*). The measurements must be as exact as possible. Following this, the foundation is measured diagonally from one outside corner to the opposite outside corner. A second diagonal measurement is then made between the outsides of the remaining two corners. If the measured lengths of the opposite walls are equal and the diagonals are equal, the foundation is square. For buildings where the foundation is other than a simple rectangle, a good practice is to divide the area into two or more individual square or rectangular areas and measure each area as described above.

5.2.0 Installing the Sill

For floors where the sill is installed flush with the outside of the foundation walls, installation of the sill begins by snapping chalklines on the top of the foundation walls in line with the inside edge of the sill (*Figure 26*). If the sill must be set in to

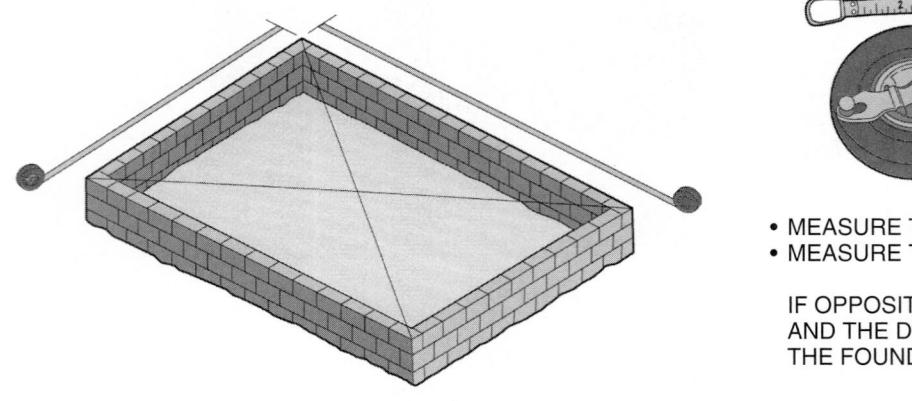

- MEASURE THE FOUR WALLS
- MEASURE THE DIAGONALS

IF OPPOSITE WALLS ARE EQUAL AND THE DIAGONALS ARE EQUAL, THE FOUNDATION IS SQUARE.

105F25.EPS

Figure 25 ◆ Checking the foundation for squareness.

Keep It Square

When it is time to install sill plates on the foundation, you may discover that the foundation wall is not exactly true and square. You shouldn't use the foundation wall as a guide. Instead, ensure that the sill plates are square with each other by using a tape measure (shown here) to measure the four plates and the diagonals. If the opposite plates and the diagonals are equal, the sill plates are square with each other. This may mean that the outside edge of the sill plates may not align exactly with the outside edge of the foundation. If necessary, some sills may overlap or underlay the wall.

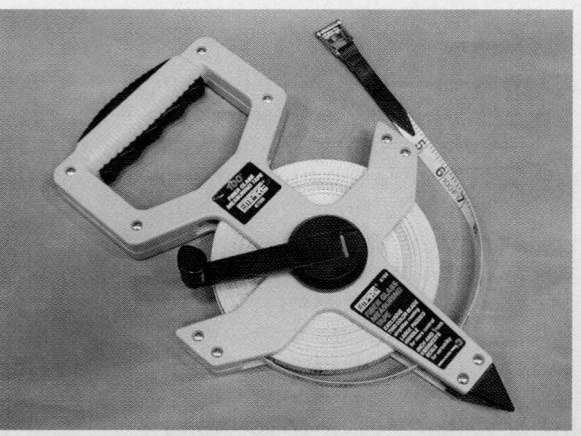

105SA06.EPS

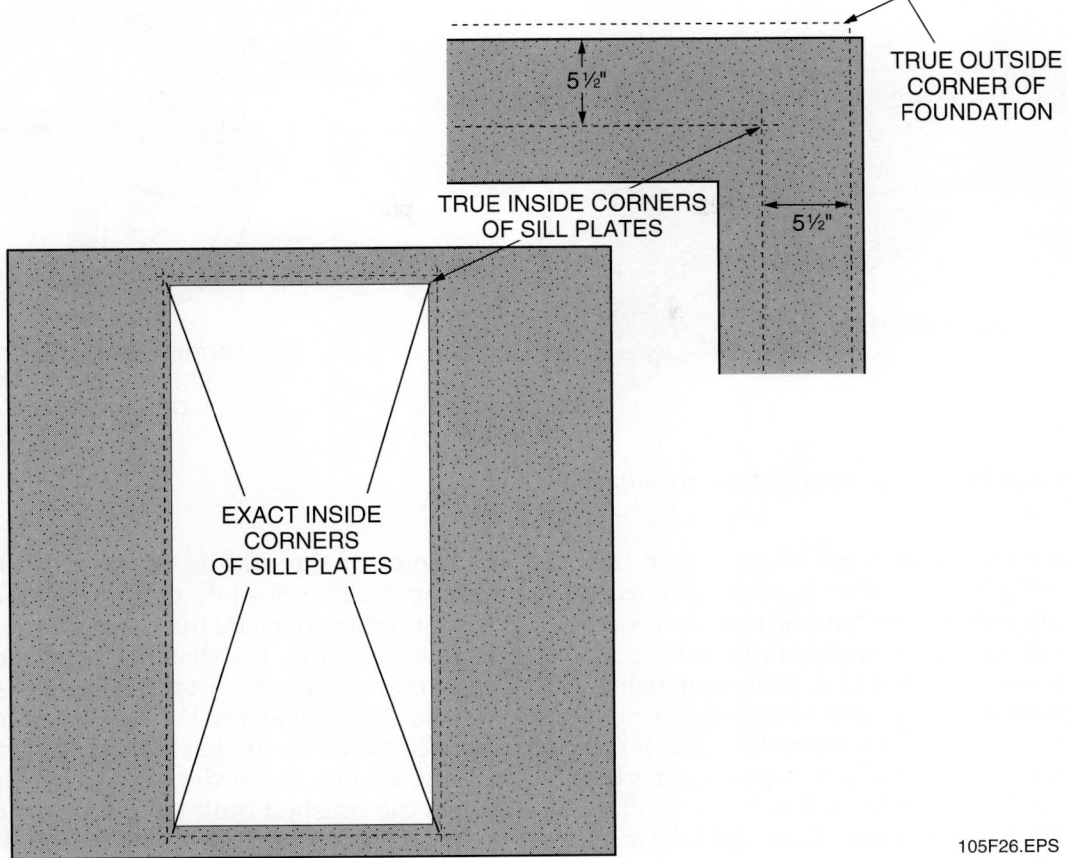

5½"

TRUE OUTSIDE
CORNER OF
FOUNDATION

TRUE INSIDE CORNERS
OF SILL PLATES

5½"

EXACT INSIDE
CORNERS
OF SILL PLATES

105F26.EPS

Figure 26 ◆ Inside edges of sill plates marked on the top of the foundation wall.

accommodate the thickness of wall sheathing, brick veneer, etc., the chalklines may be snapped for the outside edge of the sill plates or the inside edge (if the foundation size allows). At each corner, the true location of the outside corner of the sill is used as a reference point to mark the corresponding inside corner of the sill on the foundation wall. To do this, the exact width of the sill stock being used must be determined. For example, if using 2 × 6 sills, 5½" is the sill width measurement. After the exact inside corners are located and marked on the sill, chalklines are snapped between these points. This gives an outline on the top of the foundation wall of the exact inside edges of the sill plates. At this point, a good practice is to double-check the dimensions and squareness of these lines to make sure that they are accurate.

After the location of the sill is marked on the foundation, the sill pieces can be measured and cut. Take into consideration that there must be an anchor bolt within 12" of the end of any plate. Also, sill plates cannot butt together over any opening in the foundation wall. When selecting the lumber, choose boards that are as straight as possible for making the sills. Badly bowed pieces should not be used.

Holes must be drilled in the sill plates so that they can be installed over the anchor bolts embedded in the foundation wall. To lay out the location of these holes, hold the sill sections in place on top of the foundation wall against the anchor bolts (*Figure 27*). At each anchor bolt, use a combination square to scribe lines on the sill corresponding to both sides of the bolt. On the foundation, measure the distance between the center of each anchor bolt and the chalkline, then transfer this distance to the corresponding bolt location on the sill by measuring from the inside edge. After the sill hole layout is done, the holes in the sill are drilled. They should be drilled about ⅛" to ¼" larger than the diameter of the anchor bolt in order to allow for some adjustment of the sill plates, if necessary. Also, make sure all holes are drilled straight.

Before installing the sill plates, the termite shield (if used) and sill sealer are installed on the foundation. Following this, the sill sections are placed in position over the anchor bolts, making sure that the inside edges of the sill plate sections are aligned with the chalkline on top of the foundation wall

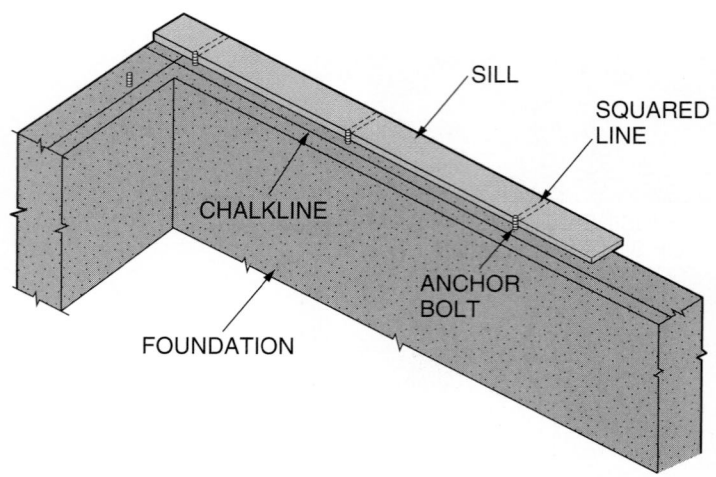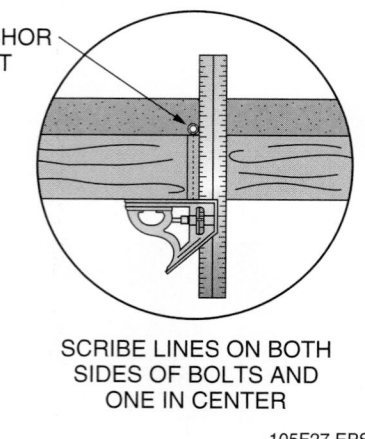

ANCHOR BOLT

SCRIBE LINES ON BOTH
SIDES OF BOLTS AND
ONE IN CENTER

105F27.EPS

Figure 27 ◆ Square lines across the sill to locate the anchor bolt hole.

and that the inside corners are aligned with their marks. The sill plates are then loosely fastened to the foundation with the anchor bolt nuts and washers and the sill checked to make sure it is level. An 8' level can be used for this task. However, using a transit or builder's level and checking the level every 3' or 4' along the sill is more accurate. It cannot be emphasized enough how important it is that the sill be level. If it is not level, it will throw off the building's floors and walls. Low spots can be shimmed with plywood wedges or filled with grout or mortar. If the sill is too high, the high areas of the concrete foundation will need to be ground or chipped away. After the sill has been made level, the anchor bolt nuts can be fully tightened. Be careful not to overtighten the nuts, especially if the concrete is not thoroughly dry and hard, because this can crack the wall.

5.3.0 Installing a Beam/Girder

Before installing a girder, use the job specifications to determine the details related to its installation. For example, assume you are working with a structure that has a foundation that is 24' wide and 48' long (*Figure 28*). The foundation is poured concrete that is 12" thick with 6"-deep and 7"-wide girder (beam) pockets centered in the short walls. The girder is to be a built-up beam containing three thicknesses of 2 × 10. The columns used to support the girder when the floor system is complete are three 4" lally columns, concrete-filled, with ¼" steel plates top and bottom. The distance between these columns is 12'-0" OC.

To lay out the distances for each of the support columns, first use a steel tape to measure from one end of the foundation to the other in the precise location where the girder will sit. Using a plumb bob, hold the line at the 24'-0" mark. This locates

the center of the middle support column. Then, measure 12'-0" to the left of center and 12'-0" to the right of center to locate the center of the other two support columns. The distance from the center of the two end columns into their girder pockets in the foundation walls is 11'-5½". This allows for ½" of space between the back of each girder pocket and the end of the girder. Given the dimensions above, the finished built-up girder for our example needs to measure 46'-11" in length. When constructing this girder, remember that the joints should fall directly over the support columns and the girder crown must face upward.

 **NOTE**

The *Uniform Building Code* specifies that in standard framing, nails should not be spaced closer than one-half their length nor closer to the edge of a framing member than one-quarter their length. Keep this requirement in mind when nailing floor systems.

When framing floors, use the fasteners as indicated below:

- *16d nails* – Used to attach the header to joists, to install solid bridging, and to construct beams and girders.
- *Special nails* – Furnished to attach the joist hangers.
- *8d nails* – Used to install the wood crossbridging and subfloor.
- *Pneumatic, ring shank, and screw nails, or etched galvanized staples* – Used to apply the subfloor and underlayment.
- *Construction adhesive* – Used to apply the subfloor and underlayment.

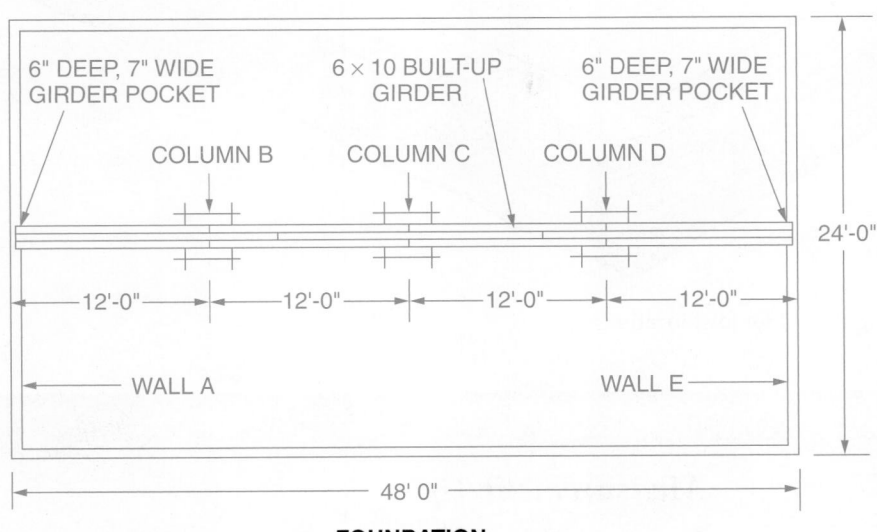

46'-11"			
11'-5½"	12'-0"	12'-0"	11'-5½"
15'-5½"	16'-0"		15'-5½"
11'-5½"	12'-0"	12'-0"	11'-5½"

BUILT-UP GIRDER

FOUNDATION

105F28.EPS

Figure 28 ◆ Example girder and support column data.

As shown in *Figure 28*, the 6 × 10 built-up girder can be constructed using eight 12' long 2 × 10s and three 16' long 2 × 10s. Four of the 12' long 2 × 10s are cut to 11'-5½" and two of the 16' long 2 × 10s are cut to 15'-5½". The 16' pieces are used in order to provide for an overlap of at least 4'-0" at the joints. To make the girder, nail the 2 × 10s together in the pattern shown using 16d nails spaced about 24" apart. Note that the nailing schedule will vary at different locations, so make sure to consult local codes for the proper schedule. Drive the nails at an angle for better holding power. Be sure to butt the joints together so they form a tight fit. Continue this process until the 46'-11" girder is finished. Once completed, the girder is put in place, supported by temporary posts or A-frames, and made level. Note that the temporary supports are removed and replaced by the permanent lally columns after the floor system joists are all installed.

5.4.0 Laying Out Sills and Girders for Floor Joists

Joists should be laid out so that the edges of standard size subfloor panels fall over the centers of the joists. There are different ways to lay out a sill plate to accomplish this. One method for laying out floor joists 16" OC is described here. Begin by using a steel tape to measure out from the end of the sill exactly 15¼" (*Figure 29*). At this point, use a speed square to square a line on the sill. To make sure of accurate spacing, drive a nail into the sill at the 15¼" line, then hook the steel tape to the nail and stretch it the length of the sill. At every point on the tape marked as a multiple of 16" (most tapes highlight these numbers), make a mark on the sill. It is important that the spacing be laid out accurately; otherwise, the subfloor panels may not fall in the center of some joists.

After the sill is marked, use your speed square to square a line at each mark. Next, mark a narrow X next to each line on the sill to show the actual position where each joist is to be placed. Note that the lines marked on the sill mark the edge of the joists, not the center. Be sure to mark the X on the proper side. If the layout has been started from the left side of the sill, as shown in *Figure 29*, the X should be placed to the right of the line. If the layout has been started from the right side, the X should be placed to the left of the line. After the locations for all common 16" OC joists have been laid out, the locations for any double joists, trimmer joists, etc., should be marked on the sill and identified with a T (or other letter) instead of an X.

After the first sill has been laid out, the process is repeated on the girder and the opposite sill. If

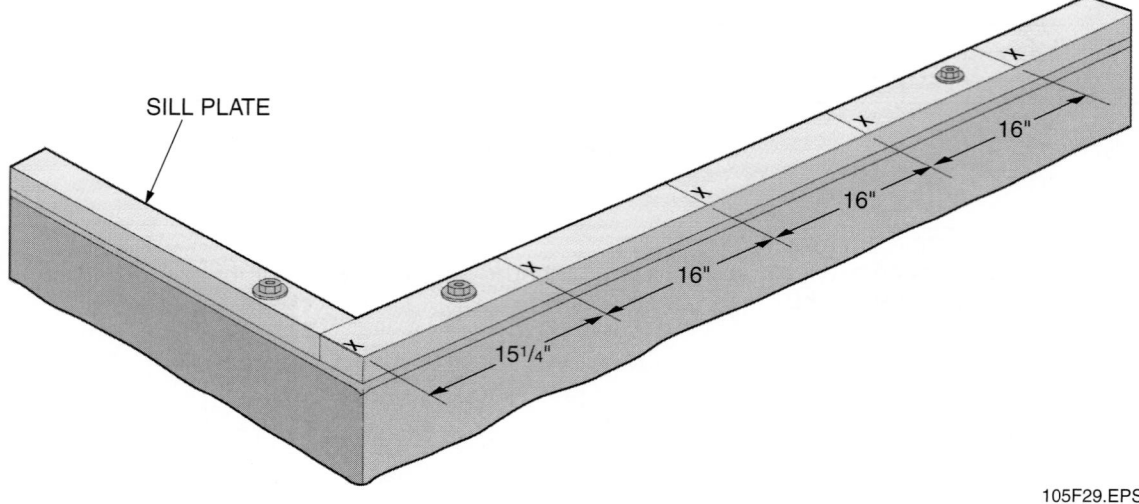

SILL PLATE

16"

16"

16"

16"

15¼"

105F29.EPS

Figure 29 ◆ Marking the sill for joist locations.

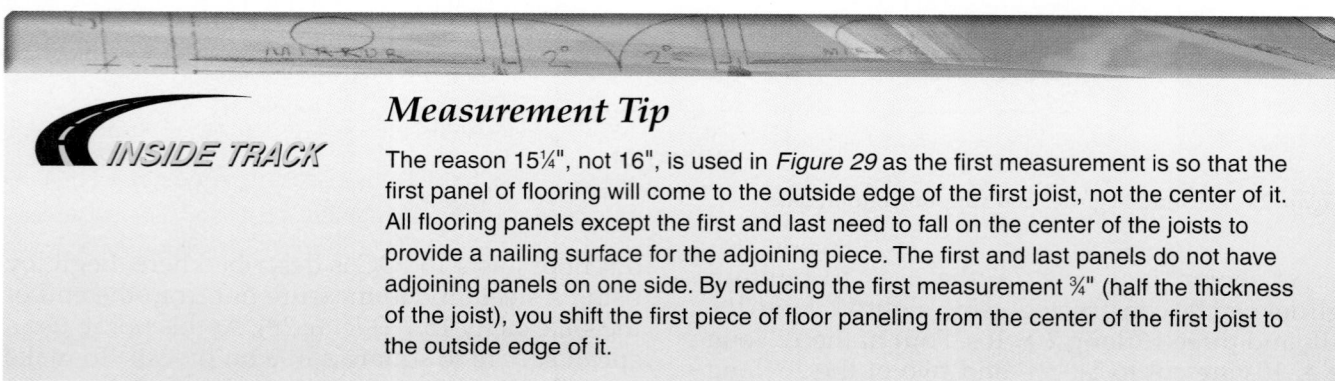

Measurement Tip

The reason 15¼", not 16", is used in *Figure 29* as the first measurement is so that the first panel of flooring will come to the outside edge of the first joist, not the center of it. All flooring panels except the first and last need to fall on the center of the joists to provide a nailing surface for the adjoining piece. The first and last panels do not have adjoining panels on one side. By reducing the first measurement ¾" (half the thickness of the joist), you shift the first piece of floor paneling from the center of the first joist to the outside edge of it.

the joists are in line, Xs should be marked on the same side of the mark on both the girder and the sill plate on the opposite wall. If the joists are lapped at the girder, an X should be placed on both sides of the mark on the girder and on the opposite side of the mark on the sill plate on the opposite wall.

The location of the floor joists can be laid out directly on the sills as described above. However, in platform construction, some carpenters prefer to lay them out on the header joists rather than the sill. If done on the header, the procedure is basically the same. Also, instead of making a series of individual measurements, some carpenters make a layout rod marked with the proper measurements and use it to lay out the sills and girders.

5.5.0 Laying Out Joist Locations for the Partition and Floor Openings

After the locations of all the common 16" OC floor joists are laid out, it is necessary to determine the

locations of additional joists needed to accommodate loadbearing partitions, floor openings, etc., as shown on the blueprints. Typically, these include:

- Double joists needed under loadbearing interior walls that run parallel with the joists (*Figure 30*). Depending on the structure, the double joists may need to be separated by 2 × 4 (or larger) blocks placed every 4' to allow for plumbing and electrical wires to pass into the wall from below. Loadbearing walls that run perpendicular to the joist system normally do not need extra joists added.
- Double joists needed for floor openings for stairs, chimneys, etc.

The sill plates and girder should be marked where the joists are doubled on each side of a large floor opening. Also, the sill and girder should be marked for the locations of the shorter tail joists at the ends of floor openings. They can be identified by marking their locations with a *T* instead of an X.

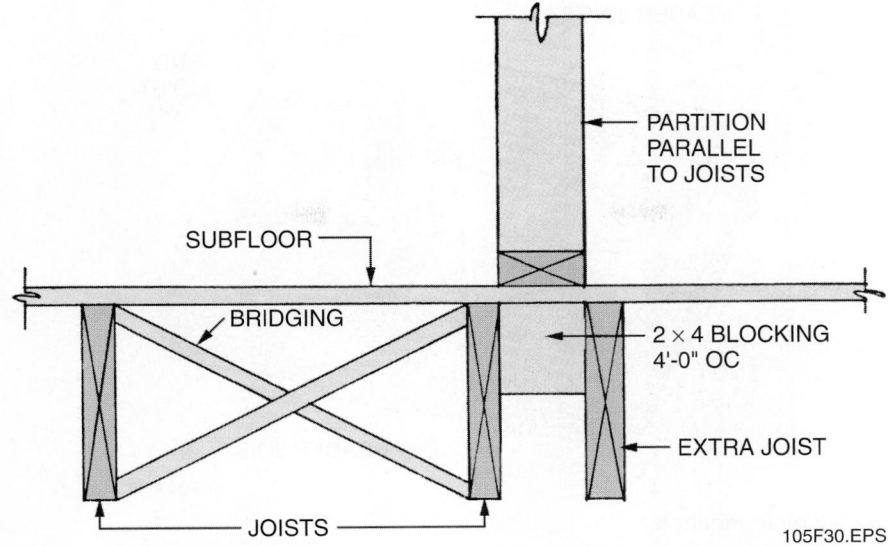

Figure 30 ◆ Double joists at a partition that runs parallel to the joists.

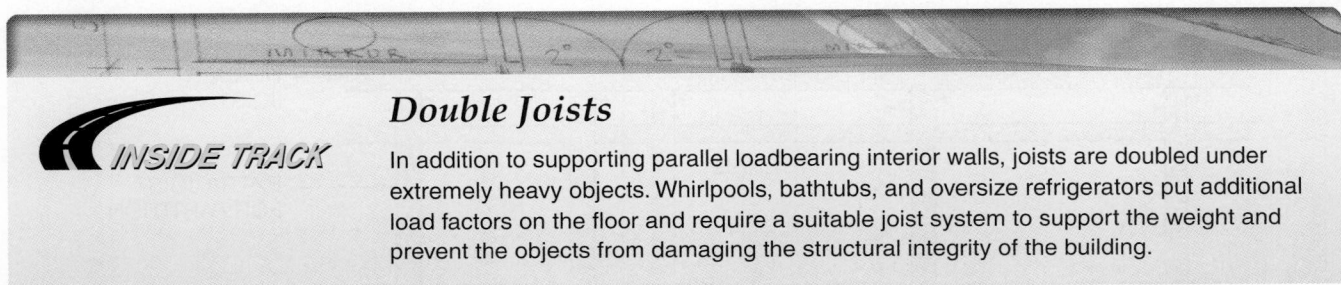

Double Joists

In addition to supporting parallel loadbearing interior walls, joists are doubled under extremely heavy objects. Whirlpools, bathtubs, and oversize refrigerators put additional load factors on the floor and require a suitable joist system to support the weight and prevent the objects from damaging the structural integrity of the building.

5.6.0 Cutting and Installing Joist Headers

After the sills and girder are laid out for the joist locations, the box sill can be built (*Figure 31*). The box sill encloses the joist system and is made of the same size stock (typically 2×10s). It consists of the two header joists that run perpendicular to the joists and the first and last joists (end joists) in the floor system. The headers are placed flush with the outside edges of the sill. Good straight stock should be used so that the headers do not rise above the sill plates. They need to sit flat on the sill so that they do not push up the wall. After the joist headers are cut, they are toenailed to the sill and face-nailed where the first and last joists meet the header joist. Any splices that need to be made in the header joist can meet either in the center of a joist or be joined by a **scab.** Note that some carpenters do not use header joists; they use blocks placed between the ends of the joists instead. Also, in areas subject to earthquakes, hurricanes, or tornadoes, codes may require that metal fasteners be used to further attach the header and end joists to the sill.

5.7.0 Installing Floor Joists

With the box sill in place, floor joists are placed at every spot marked on the sill. This includes the extra joists needed at the locations of partition walls and floor openings. When installing each joist, it is important to locate the crown and always point the crown up. With the joist in position at the header joist, hold the end tightly against the header and along the layout line so the sides are plumb, then end nail it to the header and toenail it to the sill (*Figure 32*). Repeat this procedure until all the joists are attached to their associated header. To facilitate framing of openings in the floor, a good practice is to leave the full-length joists out where the floor openings occur.

Following this, the joists are fastened at the girder. If they join end-to-end without overlapping at the girder, they should be joined with a **scarf** or metal fastener. Where the joists overlap at the girder, they should be face-nailed together and toenailed to the girder.

HEADER JOIST
END
JOIST
END
JOIST
HEADER JOIST
105F31.EPS

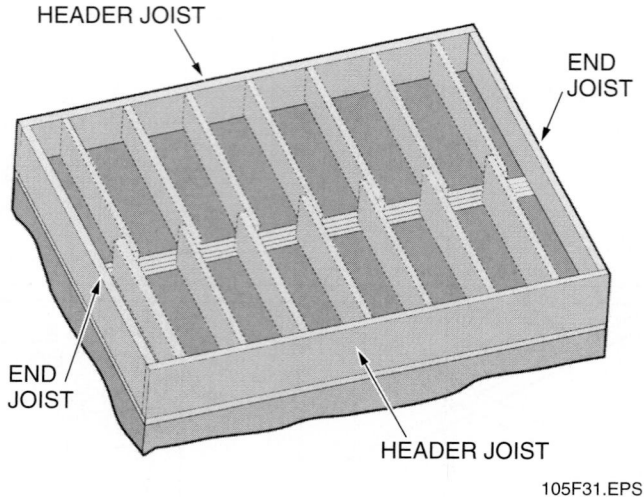

Figure 31 ◆ Box sill installed on foundation.

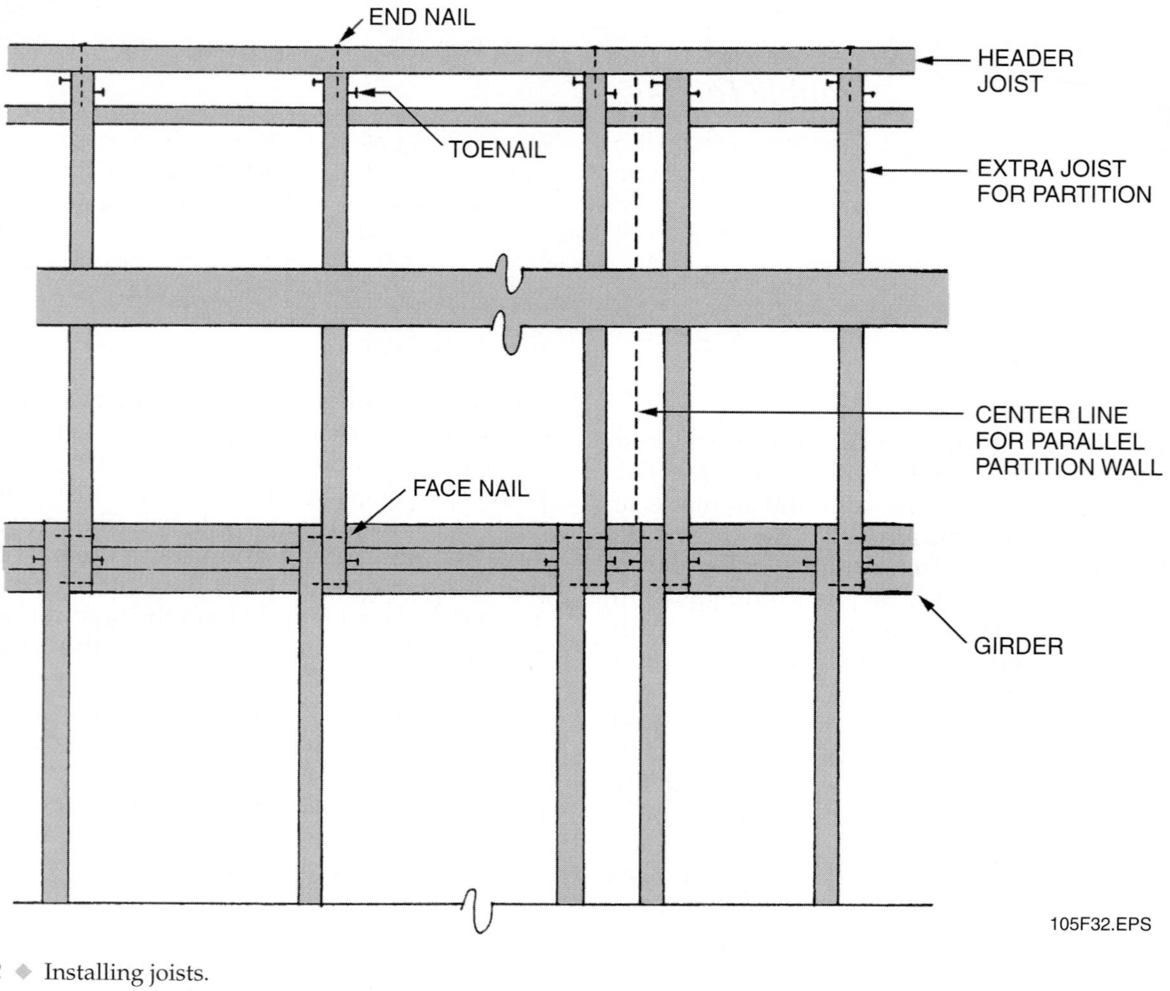

END NAIL
HEADER
JOIST
TOENAIL
EXTRA JOIST
FOR PARTITION
CENTER LINE
FOR PARALLEL
PARTITION WALL
FACE NAIL
GIRDER
105F32.EPS

Figure 32 ◆ Installing joists.

5.8.0 Framing Openings in the Floor

Floor openings are framed by a combination of headers and trimmer joists (*Figure 33*). Headers run perpendicular to the direction of the joists and are doubled. Full-length trimmer joists and short tail joists run parallel to the common joists. The blueprints show the location in the floor frame and the size of the rough opening (RO). This represents the dimensions from the inside edge of one trimmer joist to the inside edge of the other trimmer joist and from the inside edge of one header to the inside edge of the other header. The method used to frame openings can vary depending on the particular situation. Trimmers and headers at openings must be nailed together in a certain sequence so that there is never a need to nail through a double piece of stock.

A typical procedure for framing an opening like the one shown in *Figure 33* is given here:

Step 1 First install full-length trimmer joists A and C, then cut four header pieces with a length corresponding to the distance between the trimmer joists A and C.

Step 2 Nail two of these header pieces (headers No. 1 and No. 2) between trimmer joists A and C at the required distances.

Step 3 Following this, cut short tail joists X and Y and nail them to headers No. 1 and No. 2, as shown. Check the code to see if hangers are required.

Step 4 After headers No. 1 and No. 2 and tail joists X and Y are securely nailed, headers No. 3 and No. 4 can be installed and nailed to headers No. 1 and No. 2. Then, joists B and D can be placed next to and nailed to trimmer joists A and C, respectively.

5.9.0 Installing Bridging

Three types of bridging can be installed: wood cross-bridging, metal cross-bridging, and solid bridging. Wood cross-bridging is typically made of 1 × 4 pieces of wood installed diagonally between the joists to form an X. Normally, the bridging is installed every 8' or in the middle of the joist span. For example, joists with a 12' span would have a row of bridging installed at 6'.

A framing square can be used to lay out the bridging. First, determine the actual distance between the floor joists and the actual depth of the joist. For example, 2 × 10 joists 16" OC measure 14½" between them. The actual depth of the joist is 9¼". To lay out the bridging, position the framing square on a piece of bridging stock, as shown in *Figure 34*. This will give the proper length and angle to cut the bridging. Make sure to use the same side of the framing square in both places. Once the required length and angle of the bridging are determined, many carpenters build a jig to cut the numerous pieces of bridging.

Metal cross-bridging comes in a variety of styles and different lengths for use with a particular joist

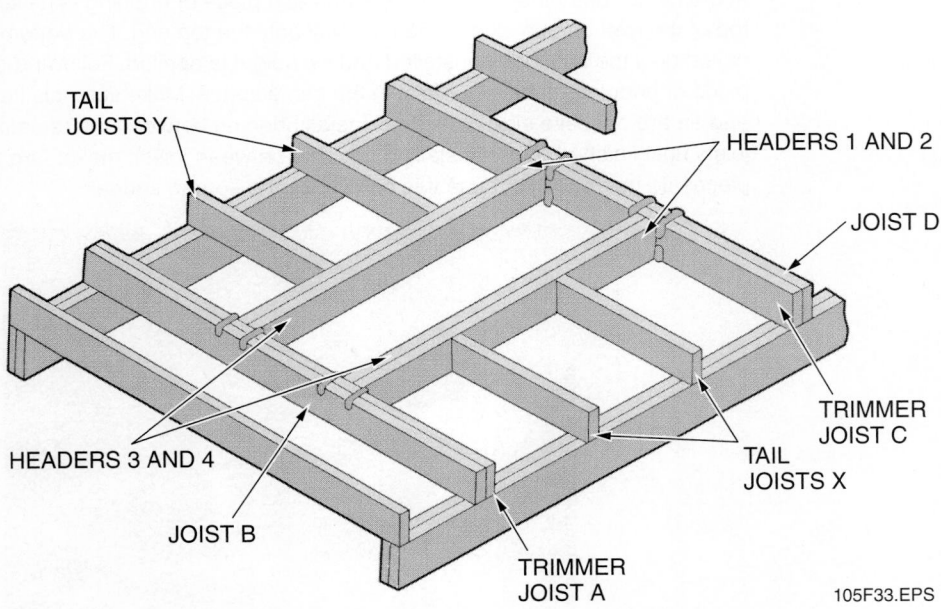

Figure 33 ◆ Floor opening construction.

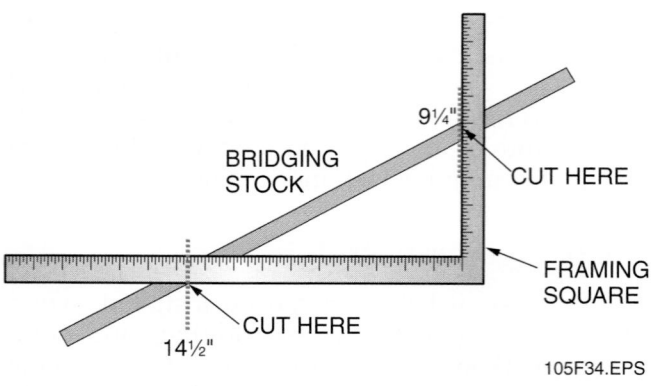

Figure 34 ◆ Framing square used to lay out cross-bridging.

size and spacing. Its installation should be done in accordance with the manufacturer's instructions. Layout of the joist system for the location of installing metal cross-bridging is done in the same way as described for wood cross-bridging.

Solid bridging consists of solid pieces of lumber joist stock installed between the joists. Layout of the joist system for the location of solid bridging is done in the same way as described for wood cross-bridging. The solid bridging is installed between pairs of joists, first on one side of the chalkline and then in the next pair of joists on the other. This staggered method of installation enables end nailing. Note that because of variations in lumber thickness and joist spacing, the length of the individual pieces of solid bridging placed between each joist pair may have to be adjusted.

5.10.0 Installing Subflooring

Installation of the subfloor begins by measuring 4' in from one side of the frame and snapping a chalkline across the tops of the floor joists from one end to the other. When installing 4' × 8' plywood, OSB, or similar floor panels, the long (8') dimension of the panels must be run across (perpendicular) to the joists (*Figure 35*). Also, the panels must be laid so that the joints are staggered in each successive course (row). Never allow an intersection of four corners. This is done by starting the first course with a full panel and continuing to lay full panels to the opposite end. Following this, start the next course using half a panel (4' × 4'), then continue to lay full panels to the opposite end. Repeat this procedure until the surface of the floor is covered. The ends of the panels that overhang the end of the building are then cut off flush with the floor frame. When butt-joint plywood panels are used,

Bridging Installation

To install the bridging, a straight chalkline must be snapped across the top of the joists at the center of their span. Then, one end of a piece of bridging is nailed flush with the top of the joist on one side of the line. Nail only the top end. The bottom ends are not nailed until the subfloor is installed and its weight is applied. Following this, nail another piece of bridging to the other joist in the same space. Make sure it is flush with the top and on the opposite side of the line. Install bridging between the remaining pairs of joists until finished. When installing bridging between joists, make sure that the two pieces do not touch because this can cause the floor to squeak.

105SA07.EPS

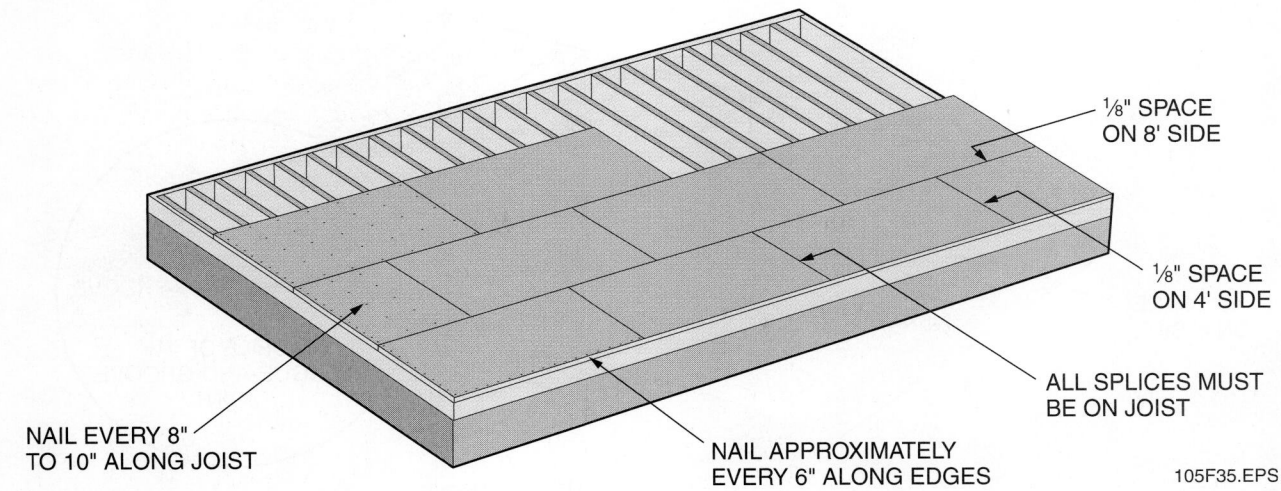

Figure 35 ◆ Installing butt-joint floor panels.

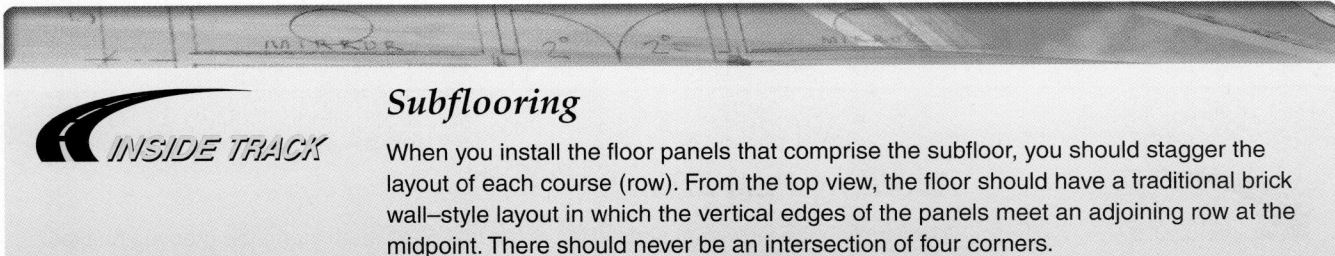

Subflooring

When you install the floor panels that comprise the subfloor, you should stagger the layout of each course (row). From the top view, the floor should have a traditional brick wall–style layout in which the vertical edges of the panels meet an adjoining row at the midpoint. There should never be an intersection of four corners.

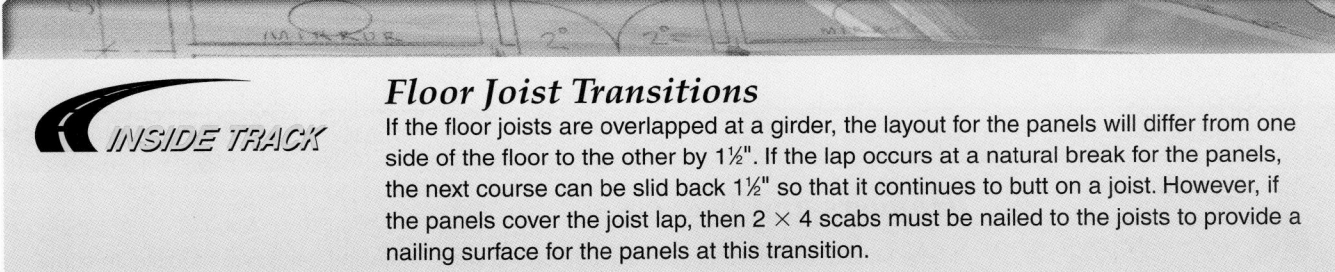

Floor Joist Transitions

If the floor joists are overlapped at a girder, the layout for the panels will differ from one side of the floor to the other by 1½". If the lap occurs at a natural break for the panels, the next course can be slid back 1½" so that it continues to butt on a joist. However, if the panels cover the joist lap, then 2 × 4 scabs must be nailed to the joists to provide a nailing surface for the panels at this transition.

at least ⅛" of space should be left between each head joint and side joint for expansion. If installing a single-layer floor, blocking is also required under the joints of the butt-edged panels. Specifications for nailing panels to floor members vary with local codes. Typically, when nailing ⅝"-thick panels to joists on 16" centers, the nailing would be done every 6" along the edges and 8" to 10" along intermediate members using ring-shank or screw nails. To avoid fatigue, pneumatic nailers are commonly used to nail the subfloors. Some carpenters use power screw guns to screw the subfloor to the joists.

Traditionally, plywood panels have been fastened to the floor joists using nails. Today, it is more common to use a glued floor system in which the subfloor panels are both glued and nailed (or screwed) to the joists, as previously discussed. Before each of the panels is placed, a ¼" bead of subflooring adhesive is applied to the joists using a caulking gun. Two beads are applied on joists where panel ends butt together. Following this, the panel should immediately be nailed to the joists before the adhesive sets. Be sure all nails hit the floor joists.

Building a subfloor with tongue-and-groove panels is done in basically the same way as described for butt-joint panels, with the following exceptions. Begin the first course of paneling with the tongue (*Figure 36*) of the panels facing the outside of the house, not the inside. This leaves the groove as the leading edge. The next course of

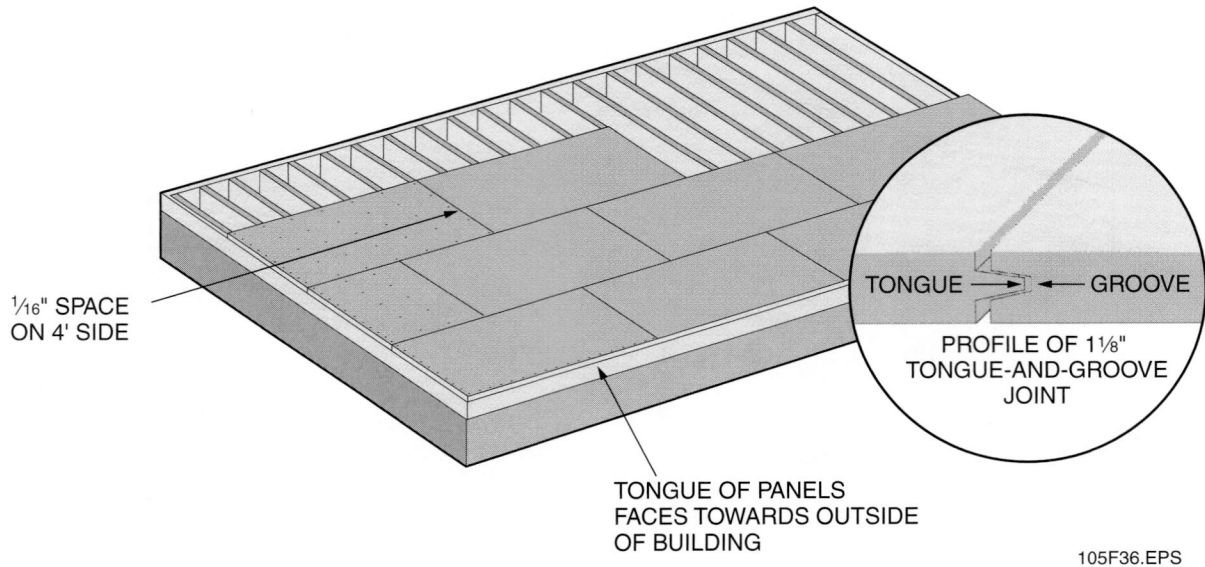

¹⁄₁₆" SPACE
ON 4' SIDE

TONGUE → ← GROOVE

PROFILE OF 1⅛"
TONGUE-AND-GROOVE
JOINT

TONGUE OF PANELS
FACES TOWARDS OUTSIDE
OF BUILDING

105F36.EPS

Figure 36 ◆ Installing tongue-and-groove floor panels.

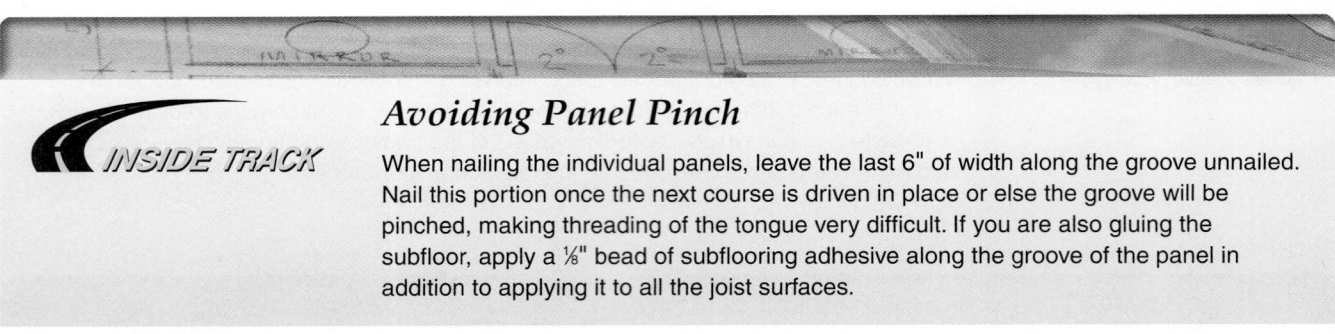

Avoiding Panel Pinch

When nailing the individual panels, leave the last 6" of width along the groove unnailed. Nail this portion once the next course is driven in place or else the groove will be pinched, making threading of the tongue very difficult. If you are also gluing the subfloor, apply a ⅛" bead of subflooring adhesive along the groove of the panel in addition to applying it to all the joist surfaces.

Hangers and Ledgers

Make sure to check local construction codes when installing cantilevered joists. In some cases, the hangers are installed upside down (*Figure 37*) and ledgers are used to support the floor load.

panels has to be interlocked into the previous course by driving the new sheets with a 2 × 4 block and a sledgehammer. The grooved edge of the panels can take this abuse; the tongued edge cannot.

6.0.0 ◆ INSTALLING JOISTS FOR PROJECTIONS AND CANTILEVERED FLOORS

Porches, decks, and other projections from a building present some special floor-framing situations. Projections overhang the foundation wall

and are suspended without any vertical support posts. When constructing a projection, it makes a difference whether the joists run parallel or perpendicular to the common joists in the floor system. If they are parallel, longer joists are simply run out past the foundation wall (*Figure 37*). If they must be run perpendicular to the common joists, the projection must be framed with cantilevered joists. This means that you have to double up a common joist, then tie the cantilevered joists into this double joist. Regardless of whether you are using parallel or cantilevered joists, there should always be at least a ⅓ to ⅔ relationship

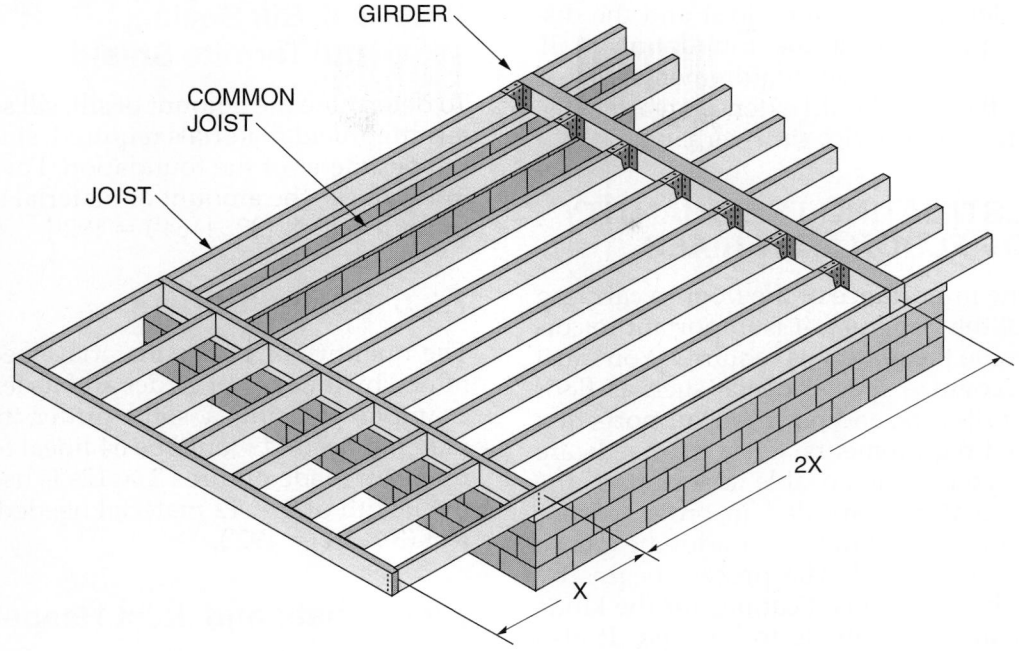

GIRDER

COMMON
JOIST

JOIST

2X

X

EXAMPLE OF PROJECTION JOIST RUNNING IN SAME DIRECTION AS COMMON JOISTS

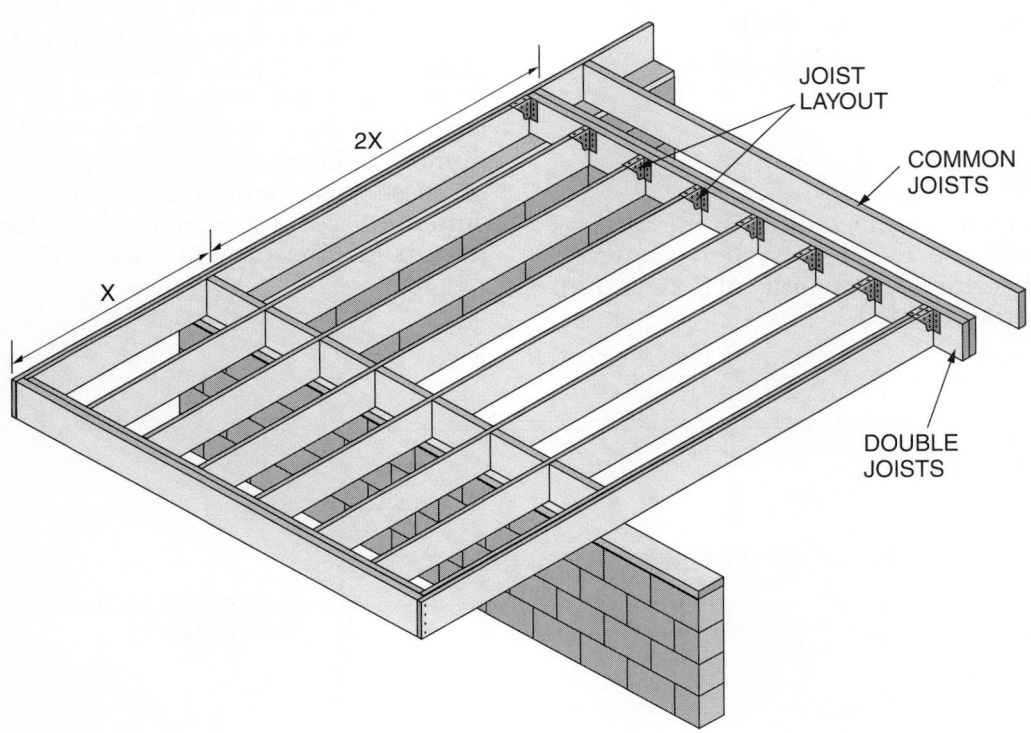

JOIST
LAYOUT

COMMON
JOISTS

2X

X

DOUBLE
JOISTS

EXAMPLE OF CANTILEVERING WHEN JOISTS RUN PERPENDICULAR TO COMMON JOISTS

105F37.EPS

Figure 37 ◆ Cantilevered joists.

between the length of the total joist and the distance it can project past the foundation wall. Check local codes for exact requirements. Stated another way, the joist should extend inward a distance equal to at least twice the overhang.

7.0.0 ◆ ESTIMATING THE QUANTITY OF FLOOR MATERIALS

Because of the importance of the floor in carrying the weight of the structure, it is important to correctly determine the materials required. You must be able to recognize special needs such as floor openings, cantilevers, and partition supports that affect material requirements. Once the needs are determined, you must be able to estimate the quantities of materials needed in order to construct the floor without delays or added expense from too much material. The process begins by checking the building specifications for the kinds and dimensions of materials to be used. It also requires that the blueprints be checked or scaled to determine the dimensions of the various components needed. These include:

- Sill sealer, termite shield, and sill
- Girders or beams
- Joists
- Joist headers
- Bridging
- Subflooring

For the purpose of an example in the sections below, we will use the floor system shown in *Figure 38* to determine the quantity of floor and sill framing materials needed.

7.1.0 Sill, Sill Sealer, and Termite Shield

To determine the amount of sill, sill sealer, and/or termite shield materials required, simply measure the perimeter of the foundation. For our example in *Figure 38*, the amount of material needed is 192 lineal feet [2 × (32' + 64') = 192'].

7.2.0 Beams/Girders

The quantity of girder material needed is determined by the type of girder and its length. For our example, if using a solid girder, the length of material needed would be 64 lineal feet. If a built-up beam made of three 2 × 12s is used as shown, the length of 2 × 12 material needed is 192 lineal feet (3 × 64' = 192').

7.3.0 Joists and Joist Headers

To determine the number of floor joists in a frame, divide the length of the building by the joist spacing and add one joist for the end and one joist for each partition that runs parallel to the joists. For our example, there are no partitions; therefore, the number of 2 × 8 joists is 49 [(64' × 12") ÷ 16" OC = 48 + 1 = 49]. Because there are two rows of joists (one on each side of the girder), the total number of joists needed is 98 (2 × 49 = 98). Each of these joists would be about 18' long. The amount of 2 × 8 material needed for the header joists is 128 lineal feet (2 × 64' = 128').

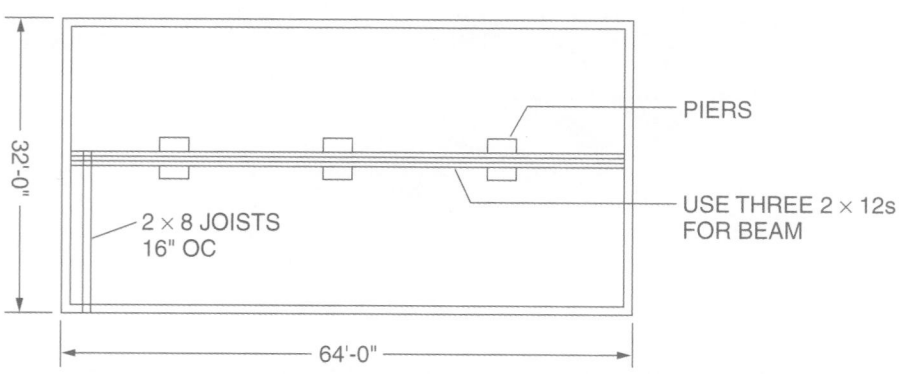

Figure 38 ◆ Determining floor system materials.

7.4.0 Bridging

Codes require one row of bridging in spans over 8' and less than 16' in length. Two rows of bridging are required in spans over 16'.

To find the amount of wood cross-bridging needed, determine the number of rows of bridging needed and the length of each row of bridging to find the total lineal footage for the bridging rows. For our example, this is 128 lineal feet (2 rows × 64' = 128'). Next, multiply the total lineal footage of bridging rows by the appropriate factor given in *Table 2* to get the total lineal feet of bridging needed. For our example, we are using 2 × 8 joists; therefore, the total amount of bridging needed is 256 lineal feet (2 × 128' = 256').

To find the amount of solid bridging needed, determine the number of rows of bridging needed and the length of each row of bridging. Then, multiply the number of rows by the length of each row to determine the total lineal feet needed. For our example, we need 128 lineal feet (2 × 64' = 128').

To determine the amount of steel bridging needed, multiply the number of rows of bridging needed by the length of each row of bridging to determine the total lineal footage of bridging rows. For our example, this is 128 lineal feet (2 × 64' = 128'). Then, multiply the total lineal footage of bridging rows by 0.75 (¾) to find the number of spaces between joists that are 16" OC. For our example, there are 96 spaces (0.75 × 128 = 96). Then, multiply the number of spaces by 2 to determine the total number of steel bridging pieces needed. For our example, we need 192 pieces (2 × 96 = 192).

7.5.0 Flooring

To determine the number of 4' × 8' plywood/OSB sheets needed to cover a floor, divide the total floor area by 32 (the area in square feet of one panel). For our example, the total floor area is 2,048 square feet (64 × 32 = 2,048). Therefore, we need 64 panels (2,048 ÷ 32 = 64). For any fractional sheets, round up to the next whole sheet.

Table 2	Wood Cross Bridging Multiplication Factor	
Joist Size	Spacing (Inches OC)	Lineal Feet of Material (per Foot of Bridging Row)
2 × 6, 2 × 8, 2 × 10	16	2
2 × 12	16	2.25
2 × 14	16	2.5

If using lumber boards for flooring, calculate the total floor area to be covered. To this amount, add a quantity of material to allow for waste. When using 1 × 6 lumber, a rule of thumb is to add ⅙ to the total area for waste; for 1 × 8 lumber, add ⅛ for waste.

8.0.0 ◆ GUIDELINES FOR DETERMINING PROPER GIRDER AND JOIST SIZES

Normally, the sizes of the girder(s) and joists used in a building are specified by the architect or structural engineer who designs the building. However, a carpenter should be familiar with the procedures used to determine girder and joist sizes.

8.1.0 Sizing Girders

The following discussion on how to size a girder is keyed to the example first-floor plan shown in *Figure 39*. Assume that a built-up girder will be used and the ceilings are drywall (not plaster).

Step 1 Determine the distance (span) between girder supports. For our example floor, the span is 8'-0".

Step 2 Find the girder load width. The girder must be able to carry the weight of the floor on each side to the midpoint of the joist that rests upon it. Therefore, the girder load is half the length of the joist span on each side of the girder multiplied by 2. For our example floor, the girder load width is 12'-0" (6'-0" on each side of the girder).

Step 3 Find the total floor load per square foot carried by the joists and bearing partitions. This is the sum of the loads per square foot as shown in *Figure 40*, with the exception of the roof load. Roof loads are not included because these are carried on the outside walls unless braces or partitions are placed under the rafters. With a drywall ceiling and no partitions, the total load per square foot for our example floor is 50 pounds per square foot (40 pounds live load + 10 pounds dead load).

Step 4 Find the total load on the girder. This is the product of the girder span multiplied by the girder width multiplied by the total floor load. For our example floor, the total load on the girder is 4,800 pounds (8 × 12 × 50 = 4,800).

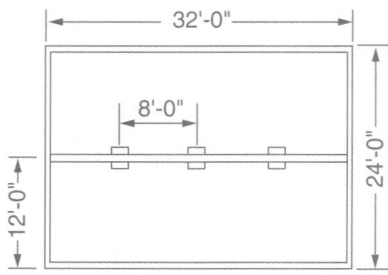

- DETERMINE LENGTH OF JOIST SPAN AND GIRDER WIDTH
- FIND TOTAL FLOOR LOAD PER SQUARE FOOT CARRIED BY JOIST AND BEARING PARTITIONS
- CALCULATE TOTAL LOAD ON GIRDER
- SELECT PROPER SIZE OF GIRDER IN ACCORDANCE WITH LOCAL CODES

105F39.EPS

Figure 39 ◆ Sizing girders.

GIRDER LOADS (POUNDS PER SQUARE FOOT)

	LIVE LOAD*	DEAD LOAD
ROOF	20	10
ATTIC FLOOR	20	20 (FLOORED)
		10 (NOT FLOORED)
SECOND FLOOR	40	20
PARTITIONS		20
FIRST FLOOR	40	20 (CEILING PLASTERED)
		10 (CEILING NOT PLASTERED)
PARTITIONS		20

*USUAL LOCAL REQUIREMENTS

EXAMPLE:

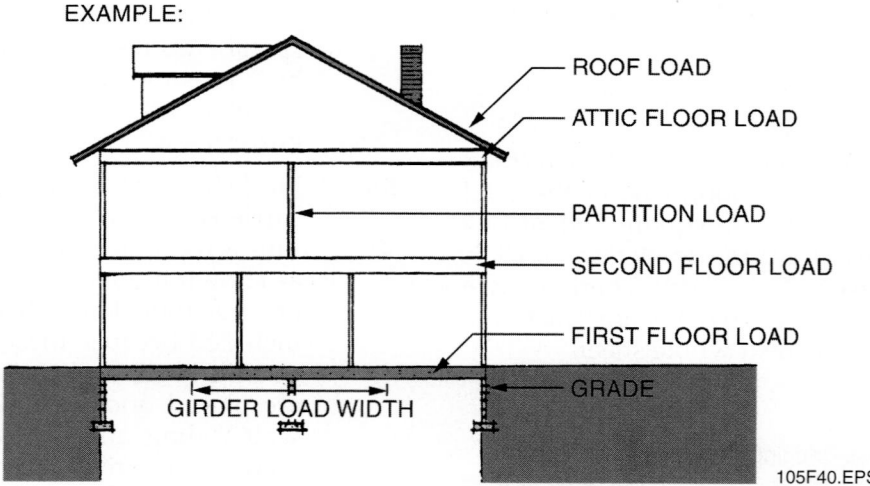

105F40.EPS

Figure 40 ◆ Floor loads.

Step 5 Select the proper size of girder according to local codes. *Table 3* is typical. It indicates safe loads on standard-size girders for spans from 6' to 10'. For our example floor, a 6 × 8 built-up girder is needed to carry a 4,800-pound load at an 8' span. Note that shortening the span is the most economical way to increase the load that a girder will carry.

8.2.0 Sizing Joists

The following discussion on how to size joists is keyed to the example first-floor plan shown in *Figure 41*. Assume that the plan falls into the 40-pound live load category and will be built with joists on 16" centers and drywall ceilings.

Step 1 Determine the length of the joist span. For our example, the span is 16'.

Step 2 Determine if there is a dead load on the ceiling. For our example, the ceiling is drywall; therefore, there is a dead load of 10 pounds per square foot.

Step 3 Select the proper size of joists according to local codes or by using the latest tables available from wood product manufacturers or sources such as the National Forest Products Association and the Southern Forest Products Association. *Table 4* indicates maximum safe spans for various sizes of wood joists under ordinary load conditions. For floors, this is usually figured on a basis of 50 pounds per square foot (10 pounds dead load and 40 pounds live load). For our example, *Table 4* shows that for a 40-pound live load, a 2 × 10 joist 16" OC will carry the load up to a span of 19'-2". Note that a 2 × 8 joist 16" OC will carry the required load up to 15'-3", which is not long enough for the required span.

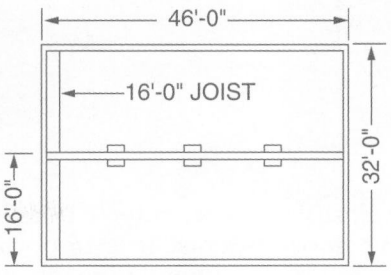

- DETERMINE LENGTH OF JOIST SPAN
- DETERMINE LIVE LOAD PER SQUARE FEET
- DETERMINE IF THERE IS A DEAD LOAD ON CEILING
- SELECT PROPER SIZE JOISTS IN ACCORDANCE WITH LOCAL CODES

105F41.EPS

Figure 41 ◆ Sizing joists.

Table 3 Typical Safe Girder Loads

Nominal Girder Size	Safe Load in Pounds for Spans Shown				
	6 ft	7 ft	8 ft	9 ft	10 ft
6 × 8 solid	8,306	7,118	6,220	5,539	4,583
6 × 8 built-up	7,359	6,306	5,511	4,908	4,062
6 × 10 solid	11,357	10,804	9,980	8,887	7,997
6 × 10 built-up	10,068	9,576	8,844	7,878	7,086
8 × 8 solid	11,326	9,706	8,482	7,553	6,250
8 × 8 built-up	9,812	8,408	7,348	6,554	5,416
8 × 10 solid	15,487	14,732	13,608	12,116	10,902
8 × 10 built-up	13,424	12,968	11,792	10,504	9,448

Table 4 Safe Joist Spans

Nominal Joist Size	Spacing	30# Live Load	40# Live Load	50# Live Load	60# Live Load
2 × 6	12"	14'-10"	13'-2"	12'-0"	11'-1"
	16"	12'-11"	11'-6"	10'-5"	9'-8"
	24"	10'-8"	9'-6"	8'-7"	7'-10"
2 × 8	12"	19'-7"	17'-5"	15'-10"	14'-8"
	16"	17'-1"	15'-3"	13'-10"	12'-9"
	24"	14'-2"	13'-6"	11'-4"	10'-6"
2 × 10	12"	24'-6"	21'-10"	19'-11"	18'-5"
	16"	21'-6"	19'-2"	17'-5"	16'-1"
	24"	17'-10"	15'-10"	14'-4"	13'-3"
2 × 12	12"	29'-4"	26'-3"	24'-0"	22'-2"
	16"	25'-10"	23'-0"	21'-0"	19'-5"
	24"	21'-5"	19'-1"	17'-4"	16'-9"
3 × 8	12"	24'-3"	21'-8"	19'-10"	18'-4"
	16"	21'-4"	19'-1"	17'-4"	16'-0"
	24"	17'-9"	15'-9"	14'-4"	13'-3"
3 × 10	12"	30'-2"	27'-1"	34'-10"	23'-0"
	16"	26'-8"	23'-10"	21'-9"	20'-2"
	24"	22'-3"	19'-10"	18'-1"	16'-8"

Summary

A great majority of a carpenter's time is devoted to building floor systems. It is important that a carpenter not only be knowledgeable about both traditional and modern floor framing techniques, but, more importantly, be able to construct modern flooring systems.

The construction of a platform floor assembly involves the tasks listed below and is normally done in the sequence listed:

- Check the foundation for squareness.
- Lay out and install the sill plates.
- Build and/or install the girders and supports.
- Lay out the joist locations for partitions and floor openings.
- Cut and attach the joist headers to the sill.
- Install the joists.
- Frame the openings in the floor.
- Install the bridging.
- Install the subflooring.

Notes

1. In braced frame construction, the corner posts and beams are _____.
   a. nailed
   b. mortised and tenoned
   c. pinned
   d. glued

2. In balloon framing, the second floor joists sit on a _____.
   a. plate nailed to the top plate of the wall assembly below
   b. 1 × 4 or 1 × 6 ribbon let into the wall studs
   c. sill attached to the top of the wall assembly below
   d. 2 × 4 let into the wall studs

3. Shrinkage in wood framing members occurs mainly _____.
   a. lengthwise
   b. both lengthwise and on the wide width
   c. on the wide width
   d. in the middle

4. The most common framing method used in modern residential and light commercial construction is _____ framing.
   a. brace
   b. balloon
   c. platform
   d. post-and-beam

5. The method of construction that experiences a relatively large amount of settling as a result of shrinkage is _____ framing.
   a. platform
   b. brace
   c. balloon
   d. post-and-beam

6. The method of construction that features widely spaced, heavy framing members is _____ framing.
   a. brace
   b. balloon
   c. platform
   d. post-and-beam

7. In a set of working drawings, the details about the floor used in a building most likely will be defined in the _____.
   a. architectural drawings
   b. structural drawings
   c. mechanical plans
   d. site plans

8. The letter *H* in *Figure 1* is pointing to the _____.
   a. sill
   b. termite shield
   c. bearing plate
   d. sill sealer

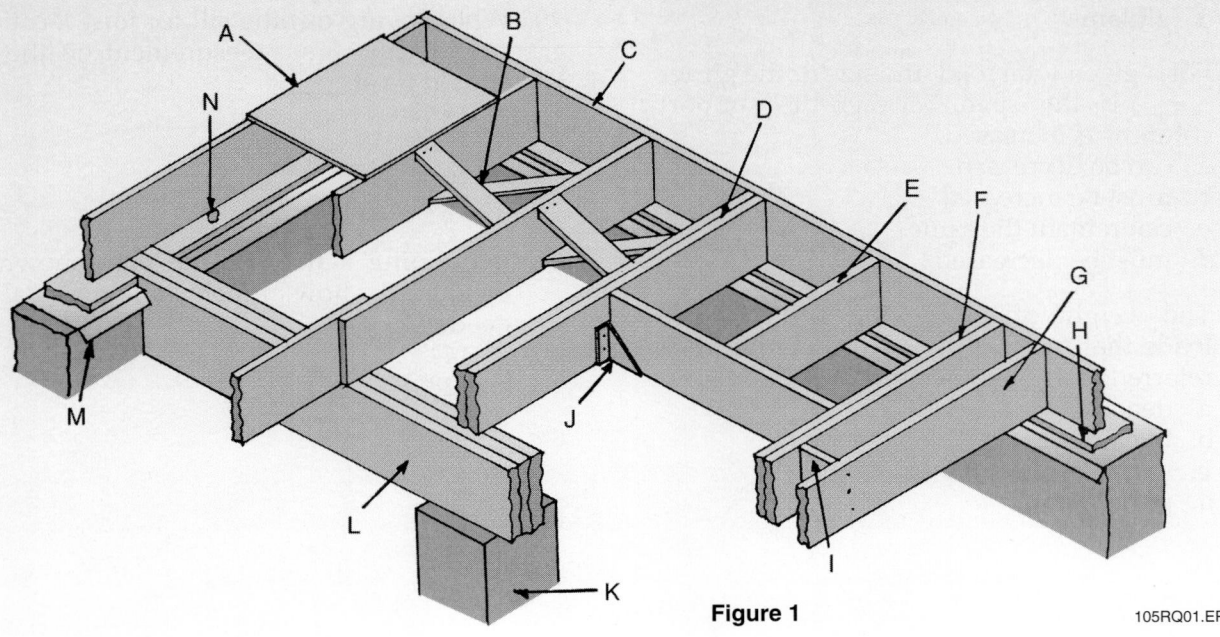

Figure 1

105RQ01.EPS

9. The letter *C* in *Figure 1* is pointing to the
   _____.
   a. tail joist
   b. trimmer joist
   c. joist header
   d. common joist

10. The letter *F* in *Figure 1* is pointing to the
    _____.
    a. tail joist
    b. trimmer joist
    c. joist header
    d. common joist

11. The letter *G* in *Figure 1* is pointing to the
    _____.
    a. tail joist
    b. trimmer joist
    c. joist header
    d. common joist

12. The letter *L* in *Figure 1* is pointing to the
    _____.
    a. joist header
    b. beam or girder
    c. column
    d. triple joist

13. For a given size, the type of girder with the
    least strength is the _____ girder.
    a. built-up
    b. solid lumber
    c. LVL
    d. glulam

14. For a given total load, the size of the girder
    _____ if the span between its support
    columns is increased.
    a. can be decreased
    b. must be increased
    c. can remain the same
    d. must be decreased

15. The weight of all moving and variable
    loads that may be placed on a building is
    referred to as _____ weight.
    a. dead
    b. live
    c. variable
    d. permanent

16. According to the *Standard Building Code,*
    notches for pipes made in the top or bottom
    of floor joists shall not exceed _____ of the
    depth.
    a. one-eighth
    b. one-sixth
    c. one-fourth
    d. one-half

17. Which of the following panels creates the
    least rigid subflooring?
    a. Plywood
    b. Oriented strand board (OSB)
    c. Tongue-and-groove plywood
    d. 1 × 6 boards

18. The first task that should be done when
    constructing a floor is to _____.
    a. lay out and install sill plates
    b. build and install girders and supports
    c. check the foundation for squareness
    d. lay out the sill plates and girders for joist
       locations

19. When nailing floor systems, nails should
    not be spaced closer to one another than
    _____ their length *nor* closer to the edge of
    a framing member than _____ their length.
    a. one-half; one-half
    b. one-half; one-quarter
    c. one-quarter; one-quarter
    d. one-quarter; one-half

20. When laying out the sill for joist locations
    16" OC, the first measurement on the sill
    should be at _____.
    a. 14¼"
    b. 15⅜"
    c. 16"
    d. 15¼"

21. If building a floor to the plan shown in
    Figure 28, how much sill material is
    needed?
    a. 144'
    b. 128'
    c. 96'
    d. 72'

22. For the same structure, how many feet of 2 × 10s are needed for a built-up girder?
  a. 48'
  b. 96'
  c. 144'
  d. 192'

23. For the same structure, what is the total number of 2 × 10 joists needed if the joists are spaced 16" OC?
  a. 36
  b. 37
  c. 72
  d. 74

24. For the same structure, how many feet of wood cross-bridging are needed?
  a. 96'
  b. 192'
  c. 48'
  d. 216'

25. For the same structure, how many 4' × 8' panels of subflooring material are needed?
  a. 96
  b. 72
  c. 36
  d. 48

Trade Terms Quiz

1. The furniture placed in a building would be considered part of the _____.

2. The HVAC equipment in a building would be considered part of the _____.

3. Before installing hardwood flooring, you would most likely install _____ over the subfloor.

4. The roof of a structure is supported by a(n) _____ assembly.

5. A joist can be secured to the structure using a(n) _____.

6. The distance between girders is known as the _____.

7. The rafters of a building are attached to a horizontal piece of lumber known as the _____.

8. A(n) _____ can be used to support the girders of a structure.

9. In balloon framing, the studs must be blocked with approved _____.

10. Lumber must be installed _____ side up.

11. In platform construction, the ends of the joists are attached to a(n) _____ to form the box sill.

12. A(n) _____ can be applied over a joint to strengthen it.

13. A(n) _____ can be used to hide a joint.

14. A notch in a joist used to support another piece is known as a(n) _____.

15. A(n) _____ runs from an opening to a bearing.

16. A rough opening in a floor such as that for a stairway would be reinforced by a(n) _____.

17. Every building is supported by a(n) _____.

18. The bottom horizontal member of a wall frame is known as the _____.

Trade Terms

Crown	Joist hanger	Scab	Trimmer joist
Dead load	Let-in	Scarf	Truss
Firestop	Live load	Soleplate	Underlayment
Foundation	Pier	Span	
Header joist	Rafter plate	Tail joist	

Adam Deeds

SkillsUSA Carpentry Silver Medal Winner—Two Years in a Row

Adam grew up in Kansas. He participated in the SkillsUSA competition in his junior and senior years in high school. In his junior year, he won a gold medal at the state level and a silver at the nationals. The following year he also won the gold at the state competition and silver at the nationals. He is currently working for a residential home builder and is a freshman at Pittsburgh State University majoring in construction management with a minor in business.

How did you become interested in carpentry?
My father is a carpenter and I have three siblings. When we were younger, my dad used to tell us to go out to the garage and pound nails if we were bored. We weren't building anything in particular, I would just hammer nails into a block of wood. So I pretty much had a hammer in my hand for as long as I can remember. My father encouraged all of us to be carpenters, but I am the only one that decided to pursue it.

My father has been a carpenter for 30 years. He owned his own business and now teaches carpentry. He was a general contractor and taught me the trade as I was growing up. I built on my skills in high school. During our sophomore, junior, and senior years our class built a house. I have always enjoyed everything about construction and carpentry.

What do you think it takes to be a success in your trade?
To be successful, you have to have a strong ambition and a will to do your best. Always be willing to learn new things. Be willing to try anything because you don't know what you are good at until you try it. You can never know everything about construction—there is always more to learn. I think I have been successful because I have always enjoyed working and I am not afraid to ask questions.

What positions have you held and how did they help you get to where you are now?
I helped my father build several houses. We also built a shop in the backyard and remodeled both of our houses. It taught me a lot about residential construction. I worked for a few summers building modular homes. I learned different aspects of home construction there. Now I am working with a general contractor while I am going to school.

What do you do in your job?
Anything my boss asks me to do. I do many different tasks depending on what is needed. I have done framing, painting, fascia and soffit work, and siding, and I have only had this job for a few months.

What do you like about the work you do?
I enjoy it all. I enjoy working with my hands and building things. When you learn something you can put it to good use. You learn something and then go out and do it.

What would you say to someone entering the trade today?
Always be willing to learn a new skill. I learn best by watching people. You can learn when you are standing around watching someone. Later, you can repeat it and put those skills to work.

Trade Terms Introduced in This Module

Crown: The high point of the crooked edge of a framing member.

Dead load: The weight of permanent, stationary construction and equipment included in a building.

Firestop: An approved material used to fill air passages in a frame to retard the spread of fire.

Foundation: The supporting portion of a structure, including the footings.

Header joist: A framing member used in platform framing into which the common joists are fitted, forming the box sill. Header joists are also used to support the free ends of joists when framing openings in a floor.

Joist hanger: A metal stirrup secured to the face of a structural member, such as a girder, to support and align the ends of joists flush with the member.

Let-in: Any type of notch in a stud, joist, etc., which holds another piece. The item that is supported by the notch is said to be let in.

Live load: The total of all moving and variable loads that may be placed upon a building.

Pier: A column of masonry used to support other structural members, typically girders or beams.

Rafter plate: The top or bottom horizontal member at the top of a wall.

Scab: A length of lumber applied over a joint to strengthen it.

Scarf: To join the ends of stock together with a sloping lap joint so there appears to be a single piece.

Soleplate: The bottom horizontal member of a wall frame.

Span: The distance between structural supports such as walls, columns, piers, beams, or girders.

Tail joist: Short joists that run from an opening to a bearing.

Trimmer joist: A full-length joist that reinforces a rough opening in the floor.

Truss: An engineered assembly made of wood, or wood and metal members, that is used to support floors and roofs.

Underlayment: A material, such as particleboard or plywood, laid on top of the subfloor to provide a smoother surface for the finished flooring.

This module is intended to present thorough resources for task training. The following reference works are suggested for further study. These are optional materials for continued education rather than for task training.

Builder Tips: Steps to Construct a Solid, Squeak-Free Floor System. Tacoma, WA: APA – The Engineered Wood Association.

Building with Floor Trusses. Madison, WI: Wood Truss Council of America (11-minute DVD or video).

Field Guide for Prevention and Repair of Floor Squeaks. Boise, ID: Trus Joist, a Weyerhauser business.

I-Joist Construction Details: Performance-Rated I-Joists in Floor and Roof Framing. Tacoma, WA: APA – The Engineered Wood Association.

Quality Floor Construction. Tacoma, WA: APA – The Engineered Wood Association (15-minute video).

Storage, Handling, Installation & Bracing of Wood Trusses. Madison, WI: Wood Truss Council of America (69-minute DVD or video).

American Wood Council. A trade association that develops design tools and guidelines for wood construction. www.awc.org.

Western Wood Products Association. A trade association representing softwood lumber manufacturers in 12 western states and Alaska. www.wwpa.org

Wood I-Joist Manufacturers Association. An organization representing manufacturers of prefabricated wood I-joist and structural composite lumber. www.i-joist.org

Wood Truss Council of America. An international trade association representing structural wood component manufacturers. www.woodtruss.com

NCCER makes every effort to keep these textbooks up-to-date and free of technical errors. We appreciate your help in this process. If you have an idea for improving this textbook, or if you find an error, a typographical mistake, or an inaccuracy in NCCER's Contren® textbooks, please write us, using this form or a photocopy. Be sure to include the exact module number, page number, a detailed description, and the correction, if applicable. Your input will be brought to the attention of the Technical Review Committee. Thank you for your assistance.

Instructors – If you found that additional materials were necessary in order to teach this module effectively, please let us know so that we may include them in the Equipment/Materials list in the Annotated Instructor's Guide.

Write:	Product Development and Revision
	National Center for Construction Education and Research
	3600 NW 43rd St, Bldg G, Gainesville, FL 32606
Fax:	352-334-0932
E-mail:	curriculum@nccer.org

Craft _____ Module Name _____

Copyright Date _____ Module Number _____ Page Number(s) _____

Description _____

_____

_____

_____

(Optional) Correction _____

_____

_____

(Optional) Your Name and Address _____

_____

_____

Luis Gomez, a contestant at the Associated Builders and Contractors 2005 National Craft Championships, assembles a metal frame. Metal studs are used for exterior and interior loadbearing and non-loadbearing walls.

68107-09
Wall and Ceiling Framing

Topics to be presented in this module include:

1.0.0	Introduction	7.2
2.0.0	Components of a Wall	7.2
3.0.0	Laying Out a Wall	7.8
4.0.0	Measuring and Cutting Studs	7.12
5.0.0	Assembling the Wall	7.14
6.0.0	Erecting the Wall	7.16
7.0.0	Ceiling Layout and Framing	7.19
8.0.0	Estimating Materials	7.23
9.0.0	Wall Framing in Masonry	7.26
10.0.0	Steel Studs in Framing	7.28

Overview

The walls of most single-family dwellings are framed with 2 × 4 or 2 × 6 lumber. Exterior sheathing and siding, along with interior finishes such as drywall are then attached to the framing. There are two critical steps in the framing process. The first is accurate measuring and layout. The second is accurate leveling and plumbing of the walls. Any deviation in level or plumb will result in serious problems later in the construction process.

Once the walls are erected, the ceiling joists are placed on the walls to serve as the supporting frame for the next level. Again, proper measuring, layout, and placement of these framing members is critical.

The use of steel studs for wall framing is common in commercial construction and is becoming increasingly popular in residential construction as the cost of lumber rises. While the layout process is essentially the same as that of lumber framing, different tools and fastening methods are used.

Objectives

When you have completed this module, you will be able to do the following:

1. Identify the components of a wall and ceiling layout.
2. Describe the procedure for laying out a wood frame wall, including plates, corner posts, door and window openings, partition Ts, bracing, and firestops.
3. Describe the correct procedure for assembling and erecting an exterior wall.
4. Identify the common materials and methods used for installing sheathing on walls.
5. Lay out, assemble, erect, and brace exterior walls for a frame building.
6. Describe wall framing techniques used in masonry construction.
7. Explain the use of metal studs in wall framing.
8. Describe the correct procedure for laying out ceiling joists.
9. Cut and install ceiling joists on a wood frame building.
10. Estimate the materials required to frame walls and ceilings.

Trade Terms

Blocking	Hip roof
Cripple stud	Jamb
Double top plate	Ribband
Drying-in	Sill
Furring strip	Strongback
Gable roof	Top plate
Header	Trimmer stud

Required Trainee Materials

1. Pencil and paper
2. Appropriate personal protective equipment

Prerequisites

Before you begin this module, it is recommended that you successfully complete *Core Curriculum*, and *Construction Technology*, Modules 68101-09 through 68106-09.

This course map shows all of the modules in *Construction Technology*. The suggested training order begins at the bottom and proceeds up. Skill levels increase as you advance on the course map. The local Training Program Sponsor may adjust the training order.

CONSTRUCTION TECHNOLOGY

68117-09
Copper Pipe and Fittings

68116-09
Plastic Pipe and Fittings

68115-09
Introduction to Drain, Waste, and Vent (DWV) Systems

68114-09
Introduction to HVAC

68113-09
Residential Electrical Services

68112-09
Electrical Safety

68111-09
Basic Stair Layout

68110-09
Exterior Finishing

68109-09
Roofing Applications

68108-09
Roof Framing

68107-09
Wall and Ceiling Framing

68106-09
Floor Systems

68105-09
Masonry Units and Installation Techniques

68104-09
Introduction to Masonry

68103-09
Handling and Placing Concrete

68102-09
Introduction to Concrete, Reinforcing Materials, and Forms

68101-09
Site Layout One: Distance Measurement and Leveling

CORE CURRICULUM:
Introductory Craft Skills

107CMAP.EPS

1.0.0 ◆ INTRODUCTION

In this module, you will learn about laying out and erecting walls with openings for windows and doors. Also covered in this module are instructions for laying out and installing ceiling joists.

Precise layout of framing members is extremely important. Finish material such as sheathing, drywall, paneling, etc., is sold in 4 × 8 sheets. If the studs, rafters, and joists are not straight and evenly spaced for their entire length, you will not be able to fasten the sheet material to them. A tiny error on one end becomes a large error as you progress toward the other end. Spacings of 16" and 24" on center are used because they will divide evenly into 48".

2.0.0 ◆ COMPONENTS OF A WALL

Figure 1 identifies the structural members of a wood frame wall. Each of the members shown on the illustration is then described. You will need to know these terms as you proceed through this module.

- *Blocking* (*spacer*) – A wood block used as a filler piece and support between framing members.
- *Cripple stud* – In wall framing, a short framing stud that fills the space between a **header** and a **top plate** or between the **sill** and the soleplate.
- *Double top plate* – A plate made of two members to provide better stiffening of a wall. It is also used for connecting splices, corners, and partitions that are at right angles (perpendicular) to the wall.
- *Header* – A horizontal structural member that supports the load over an opening such as a door or window.
- *King stud* – The full-length stud next to the trimmer stud in a wall opening.
- *Partition* – A wall that subdivides space within a building. A bearing partition or wall is one that supports the floors and roof directly above in addition to its own weight.
- *Rough opening* – An opening in the framing formed by framing members, usually for a window or door.
- *Rough sill* – The lower framing member attached to the top of the lower cripple studs to form the base of a rough opening for a window.
- *Soleplate* – The lowest horizontal member of a wall or partition to which the studs are nailed. It rests on the rough floor.
- *Stud* – The main vertical framing member in a wall or partition.
- *Top plate* – The upper horizontal framing member of a wall used to carry the roof trusses or rafters.
- *Trimmer stud* – The vertical framing member that forms the sides of rough openings for doors and windows. It provides stiffening for the frame and supports the weight of the header.

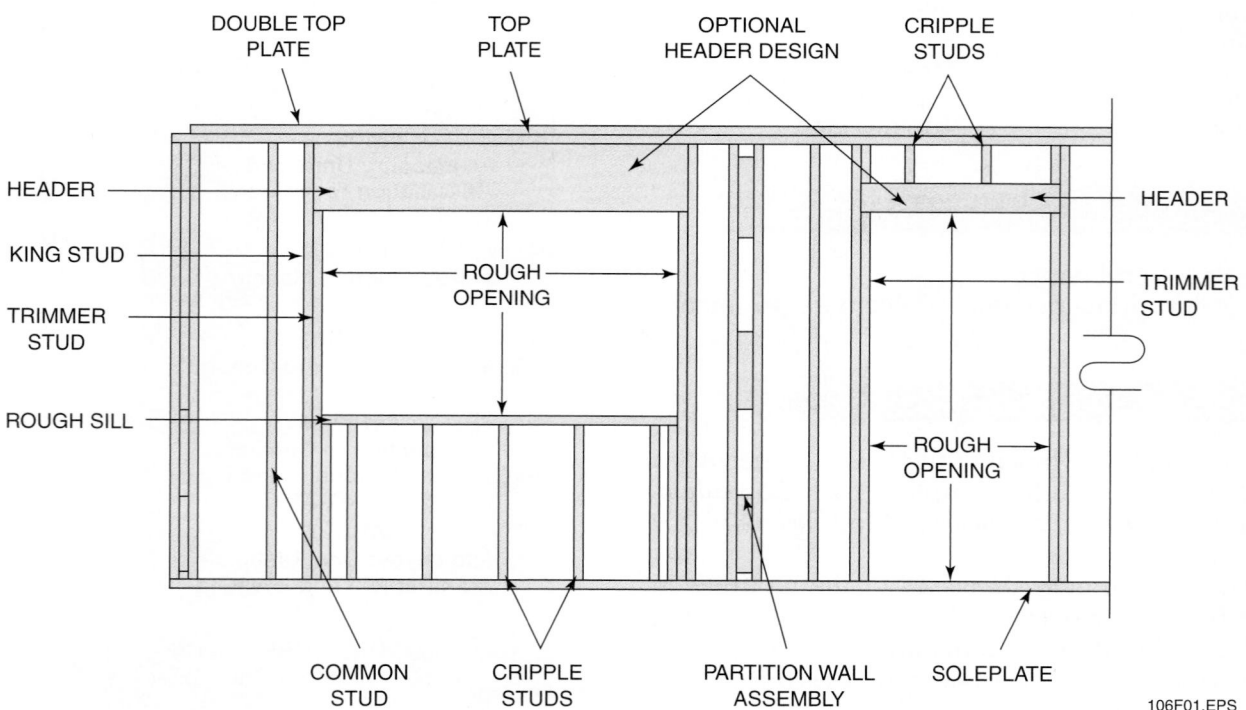

Figure 1 ◆ Wall and partition framing members.

This section contains an overview of the layout and assembly requirements and procedures for walls.

2.1.0 Corners

When framing a wall, you must have solid corners that can take the weight of the structure. In addition to contributing to the strength of the structure, corners must provide a good nailing surface for sheathing and interior finish materials. Carpenters generally select the straightest, least defective studs for corner framing.

Figure 2 shows the method typically used in western platform framing. In one wall assembly, there are two common studs with blocking between them. This provides a nailing surface for the first stud in the adjoining wall. Notice the use of a double top plate at the top of the wall to provide greater strength.

Figure 3 shows a different way to construct a corner. It has several advantages:

- It doesn't require blocking, which saves time and materials.
- It results in fewer voids in the insulation.
- It promotes better coordination among trades. For example, an electrician running wiring through the corner shown in *Figure 2* would need to bore holes through two or three studs and, possibly, a piece of blocking. However, an electrician wiring through the corner shown in *Figure 3* would need to bore through only two studs.

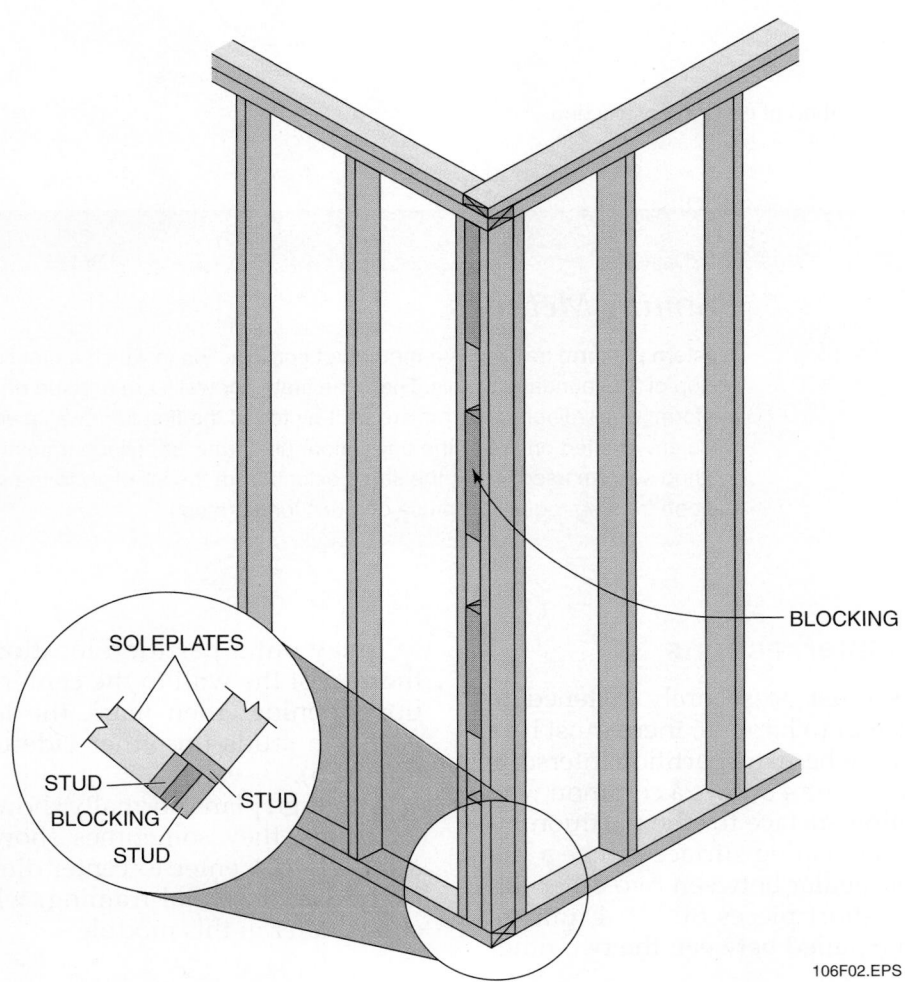

BLOCKING

SOLEPLATES

STUD

STUD

BLOCKING

STUD

106F02.EPS

Figure 2 ◆ Corner construction typical of western platform framing.

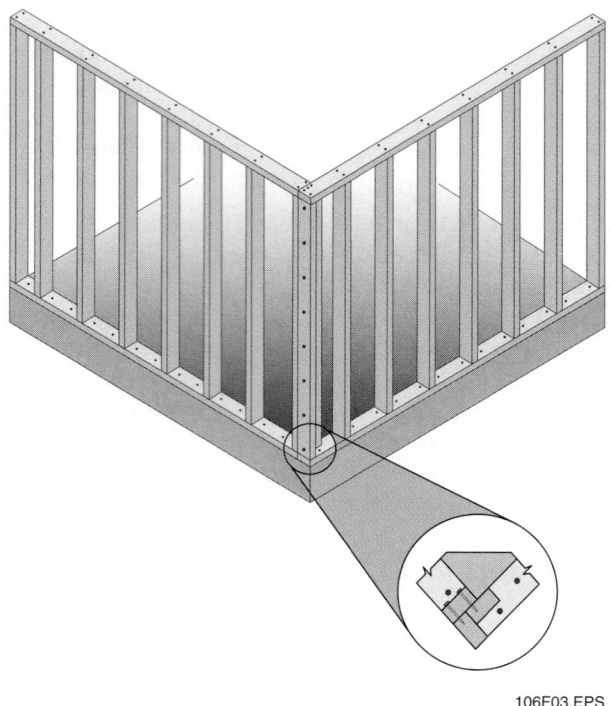

106F03.EPS

Figure 3 ◆ Alternative method of corner construction.

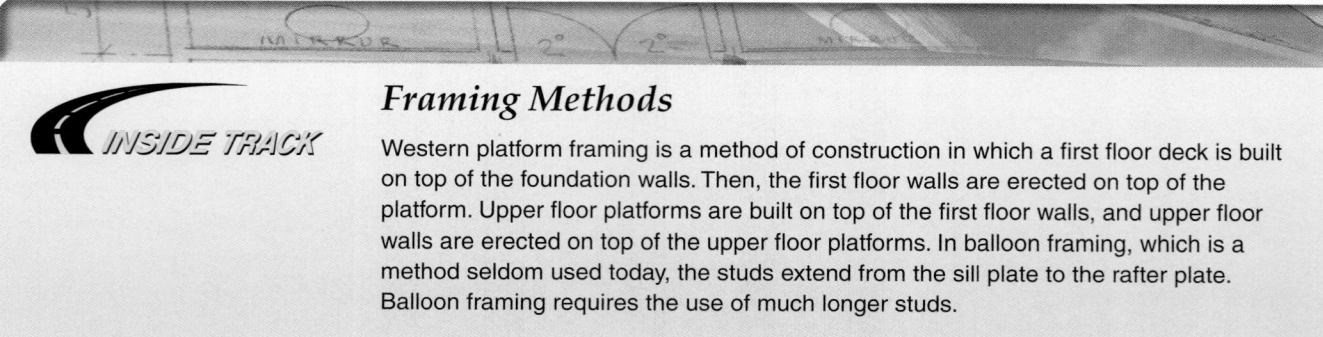

Framing Methods

Western platform framing is a method of construction in which a first floor deck is built on top of the foundation walls. Then, the first floor walls are erected on top of the platform. Upper floor platforms are built on top of the first floor walls, and upper floor walls are erected on top of the upper floor platforms. In balloon framing, which is a method seldom used today, the studs extend from the sill plate to the rafter plate. Balloon framing requires the use of much longer studs.

2.2.0 Partition Intersections

Interior partitions must be securely fastened to outside walls. For that to happen, there must be a solid nailing surface where the partition intersects the exterior frame. *Figure 4* shows a common way to construct a nailing surface for the partition intersections or Ts. The nailing surface can be a full stud nailed perpendicular between two other full studs, or it can be short pieces of 2 × 4 lumber, known as blocking, nailed between the two other full studs.

Figure 5 shows two other ways to prepare a nailing surface. Compare *Figures 4* and *5*. Notice that in *Figure 4* the spacing between studs differs, but in *Figure 5* the spacing between studs remains the same.

To lay out a partition location, measure from the end of the wall to the centerline of the partition opening, then mark the locations for the partition studs on either side of the centerline (*Figure 6*).

Although plans normally show stud-to-stud dimensions, they sometimes show finish-to-finish dimensions. Center-to-center dimensions are typically used in metal framing, which will be discussed later in this module.

2.3.0 Headers

When wall framing is interrupted by an opening such as a window or door, a method is needed to distribute the weight of the structure around the

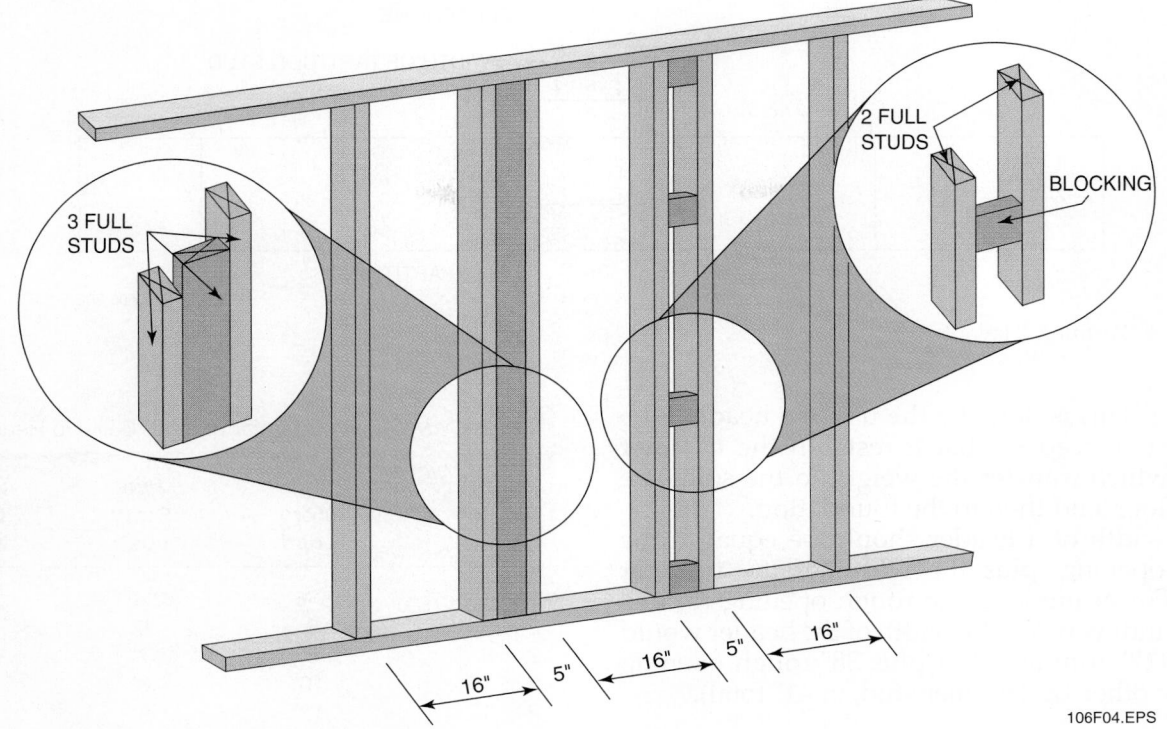

Figure 4 ◆ Constructing nailing surfaces for partitions.

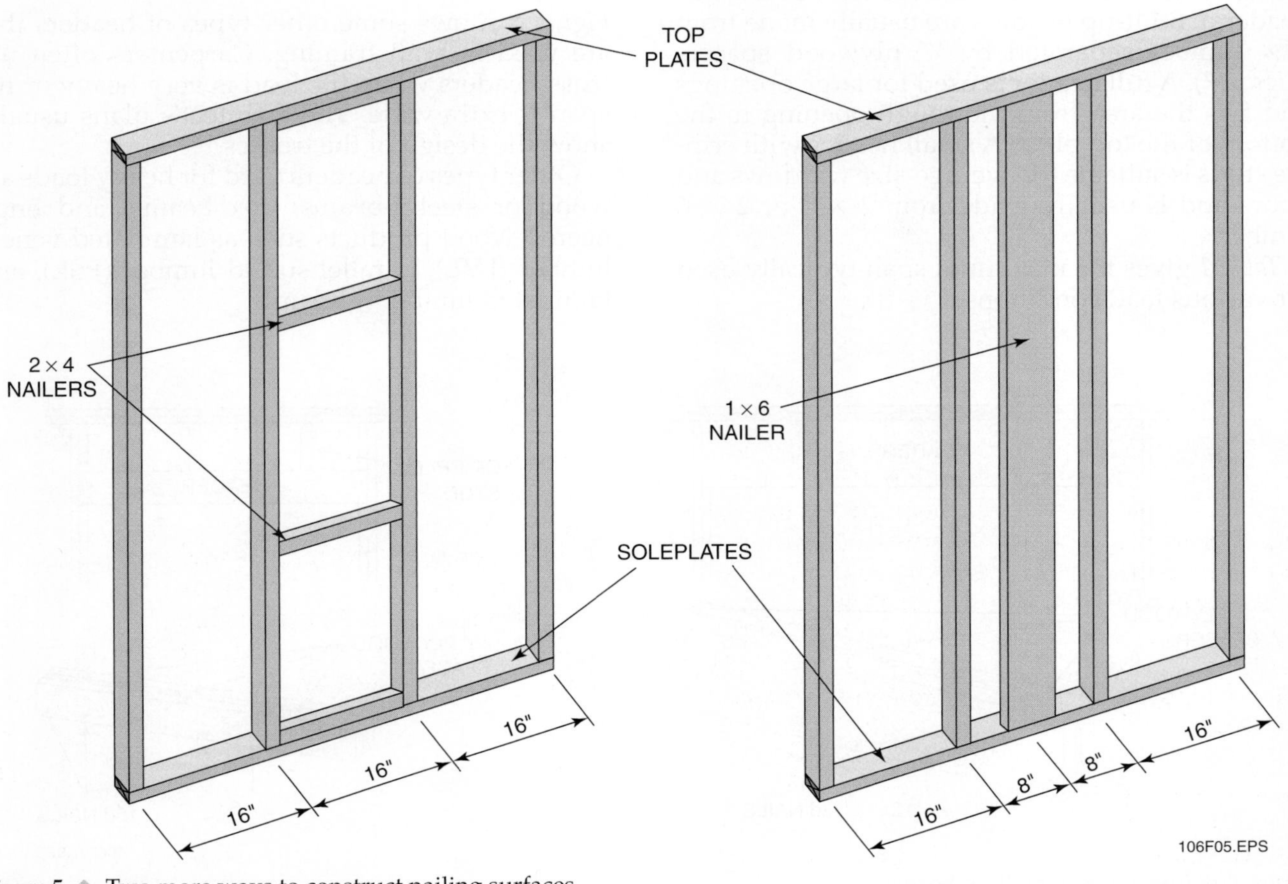

3 FULL STUDS

2 FULL STUDS

BLOCKING

16"   5"   16"   5"   16"

106F04.EPS

TOP PLATES

2 × 4 NAILERS

1 × 6 NAILER

SOLEPLATES

16"   16"   16"

16"   8"   8"   16"

106F05.EPS

Figure 5 ◆ Two more ways to construct nailing surfaces.

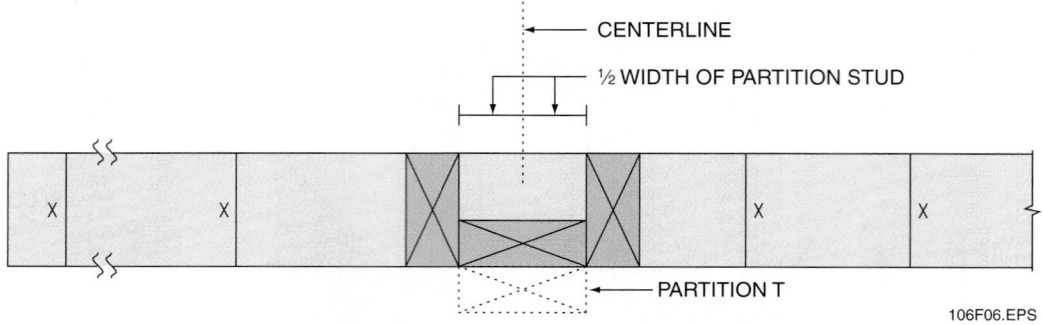

Figure 6 ◆ Partition T layout.

opening. This is done by the use of a header. The header is placed so that it rests on the trimmer studs, which transfer the weight to the soleplate or subfloor, and then to the foundation.

The width of a header should be equal to the rough opening, plus the width of the trimmer studs. For example, if the rough opening for a 3' wide window is 38", the width of the header would be 41" (1½" trimmer stud plus 38" rough opening plus the other 1½" trimmer stud, or 41" total).

2.3.1 Built-Up Headers

Headers are usually made of built-up lumber (although solid wood beams are sometimes used as headers). Built-up headers are usually made from 2 × lumber separated by ½" plywood spacers (*Figure 7*). A full header is used for large openings and fills the area from the rough opening to the bottom of the top plate. A small header with cripple studs is suitable for average-size windows and doors and is usually made from 2 × 4 or 2 × 6 lumber.

Table 1 gives the maximum span typically used for various load conditions.

Table 1	Maximum Span for Exterior Built-Up Headers		
Built-Up Header Size	Single-Story Load	Two-Story Load	Three-Story Load
2 × 4	3'-6"	2'-6"	2'
2 × 6	6'	5'	4'
2 × 8	8'	7'	6'
2 × 10	10'	8'	7'
2 × 12	12'	9'	8'

2.3.2 Other Types of Headers

Figure 8 shows some other types of headers that are used in wall framing. Carpenters often use truss headers when the load is very heavy or the span is extra wide. The architect's plans usually show the design of the trusses.

Other types of headers used for heavy loads are wood or steel I-beams, box beams, and engineered wood products such as laminated veneer lumber (LVL), parallel strand lumber (PSL), and laminated lumber (glulam).

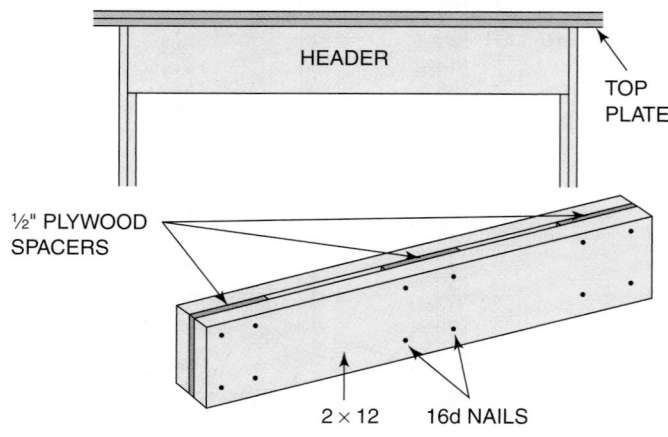
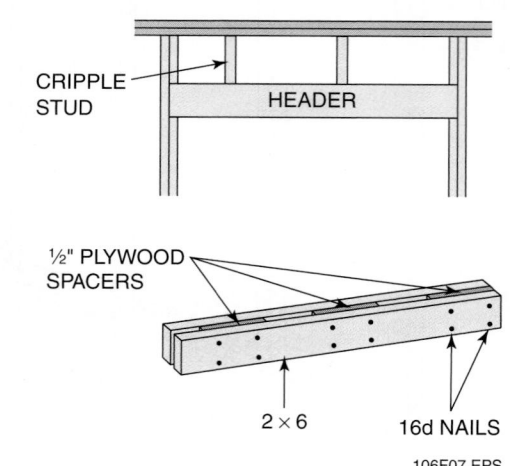

Figure 7 ◆ Two types of built-up headers.

Headers

Headers can be constructed in many ways, some of which are shown here.

HEADER WITH CRIPPLES

106SA01.EPS

SOLID HEADER

106SA02.EPS

GARAGE DOOR HEADER
USING ENGINEERED LUMBER

106SA03.EPS

TRUSS HEADERS

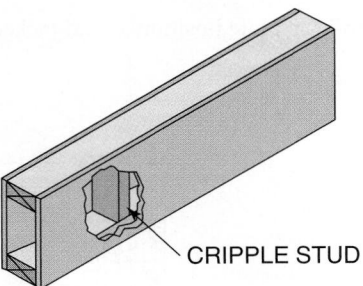

CRIPPLE STUD

BOX BEAM HEADER

106F08.EPS

Figure 8 ◆ Other types of headers.

3.0.0 ◆ LAYING OUT A WALL

This section covers the basic procedures for laying out wood frame walls with correctly sized window and door openings and partition Ts. Later in this module, you will be introduced to methods for framing with metal studs, and framing window and door openings in masonry walls.

Wall framing is generally done with 2 × 4 studs spaced 16" on center. In a one-story building, 2 × 4 spacing can be 24" on center. If 24" spacing is used in a two-story building, the lower floor must be framed with 2 × 6 lumber. The following provides an overview of the procedure for laying out a wall.

Step 1 Mark the locations of the soleplates by measuring in the width of the soleplate (e.g., 3½") from the outside edge of the sill on each corner. Snap a chalkline to mark the soleplate location, then repeat this for each wall.

Step 2 The top plate and soleplate are laid out together. Start by placing the soleplate as indicated by the chalkline and tacking it in place (*Figure 9*). Lay the top plate against the soleplate so that the location of framing members can be transferred from the soleplate to the top plate. Also tack the top plate. Tacking prevents the plates from moving, which would make the critical layout lines inaccurate.

Step 3 Lay out the common stud positions. To begin, measure and square a line 15¼" from one end. Subtracting this ¾" ensures that sheathing and other panels will fall at the center of the studs because the first sheet goes to the edge rather than the center of the corner stud. Drive a nail at that point and use a continuous tape to measure and mark the stud locations every 16" (*Figure 10*). Align your framing square at each mark. Scribe a line along each side of the framing square tongue across both the soleplate and top plate. These lines will show the outside edges of each stud, centered on 16" intervals.

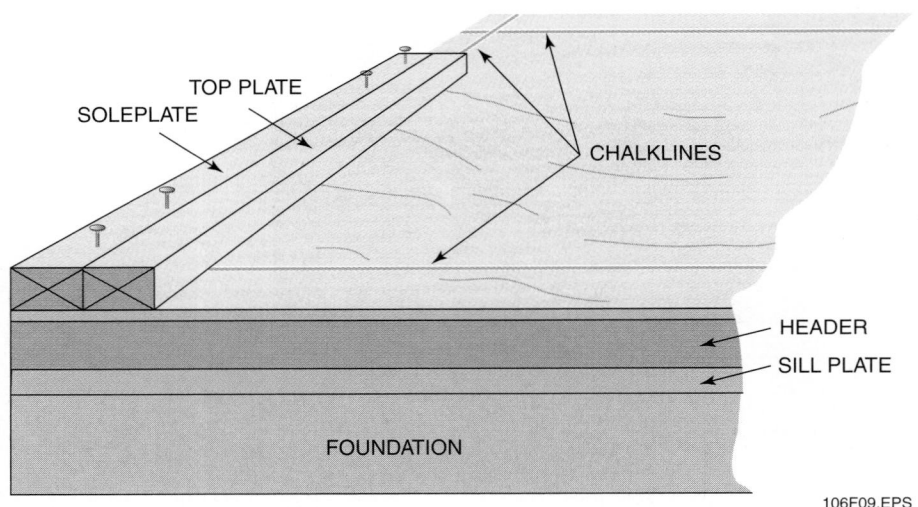

Figure 9 ◆ Soleplate and top plate positioned and tacked in place.

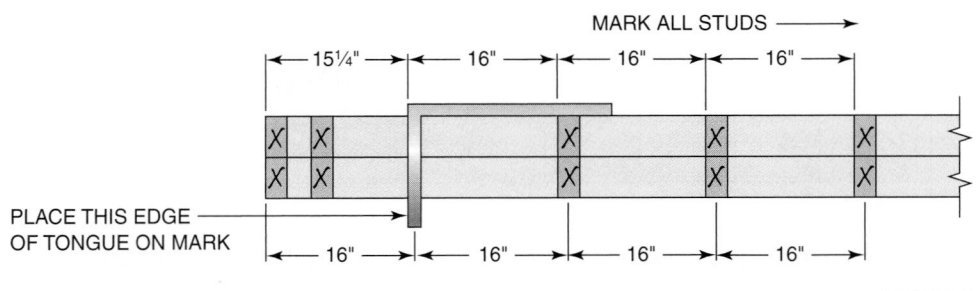

Figure 10 ◆ Marking stud locations.

Chalkline Protection

If the chalkline will be exposed to the weather, spray it with a clear protective coating to keep it from being washed away.

3.1.0 Laying Out Wall Openings

The floor plan drawings for a building (*Figure 11*) show the locations of windows and doors. Notice that each window and door on the floor plan is identified by size. In this case, the widths are shown in feet and inches, but the complete dimension information is coded. Look at the window in bedroom #1 coded 2630. This means the window is 2'-6" wide (the width is always given first) and 3'-0" high. Similar codes give widths and heights for the doors.

The window and door schedules (*Figure 12*) provided in the architect's drawings will list the dimensions of windows and doors, along with types, manufacturers, and other information.

Placement of windows is important, and is normally dealt with on the architectural drawings. A good rule of thumb is to avoid placing horizontal framework at eye level. This means you have to consider whether people will generally be sitting or standing when they look out the window. Unless they are architectural specialty

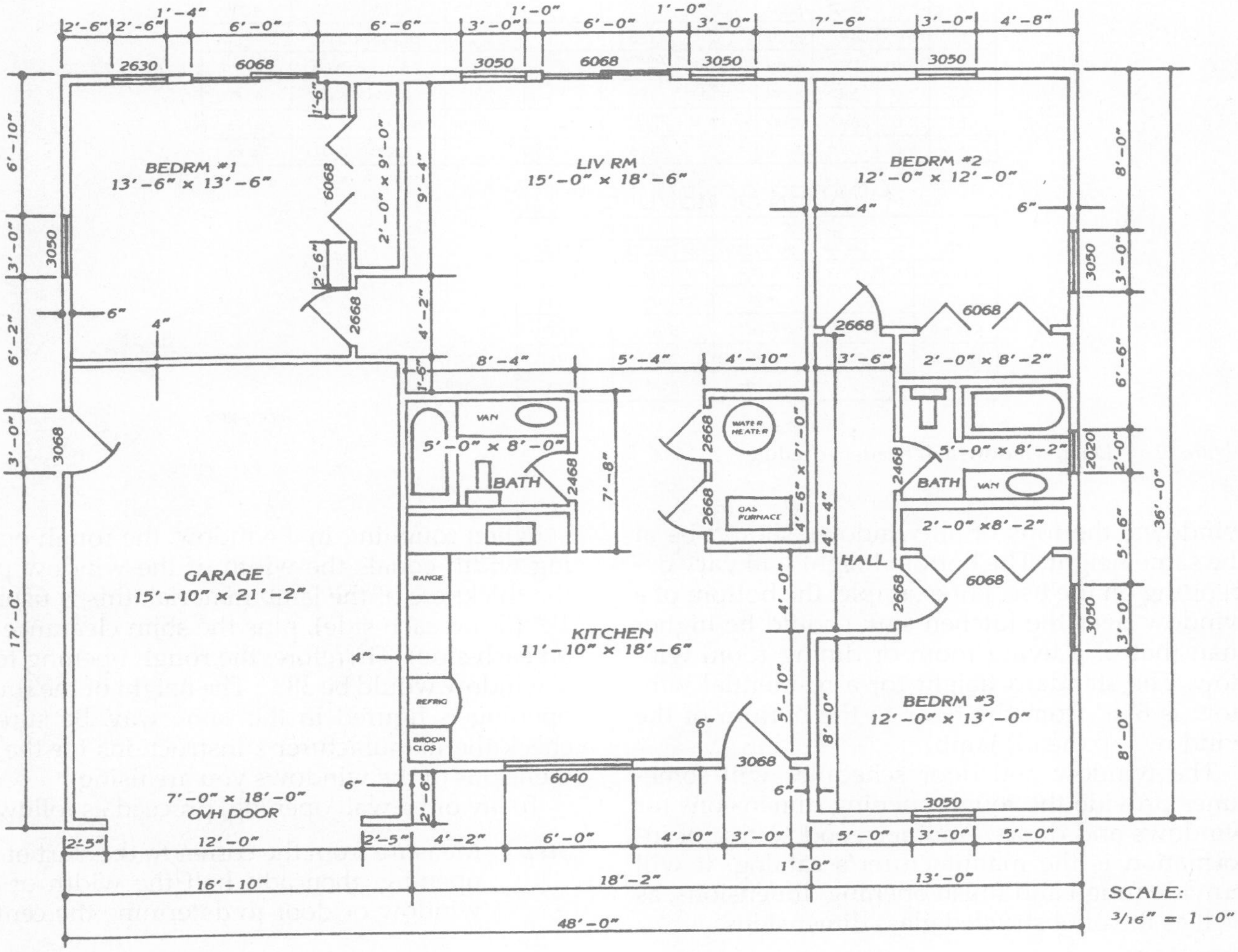

Figure 11 ◆ Sample floor plan.

106F11.EPS

DOOR SCHEDULE

NO.	DOOR SIZE	MAT'L	TYPE	H.W.#	FRAME MAT'L	TYPE	HEAD	JAMB	SILL	REMARKS
104A	PR. 3'-0" x 7-8⁵/₁₆" x 1-³/4"	W.D.	IV	11	W.D.	—	8/16	4/16	3/16	
104B	PR. 3'-0" x 8'-0" x 1-³/4"	GLASS	VI	10	—	—	3/7	15/15, 16/15	4/7	W/ FULL GLASS SIDELITE
105A	3'-0" x 7-2" x 1-³/4"	W.D.	I	1	H.M.	II	5/15	6/7, 7/7	—	
105B	↓		I	1		I	SIM. 3/15, 10	SIM. 4/15, 10	—	
106			I	1		I	1/15	2/15	—	
107	↓ ↓ ↓		I	1		I	1/15	2/15	—	
108	PR. 3'-0" x 7-2" x 1-³/4"		II	4		II	8/5	9/15	—	"C" LABEL
109	3'-0" x 7-2" x 1-³/4"		I	7		I	3/15	4/15	—	
110			I	7		I	3/15	4/15	—	
112			I	13		I	SIM. 3/15, 10	SIM. 4/15, 10	—	
113A			I	13		III	5/15	6/7	—	"C" LABEL
113B	↓ ↓ ↓		I	13		I	SIM. 1/15, 10	SIM. 2/15, 10	—	
114	2'-6" x 7-2" x 1-³/4"	↓	I	13	↓	I	SIM. 1/15, 10	SIM. 2/15, 10	—	
115A	PR. 3'-0" x 7'-0" x 1-³/4"	ALUM.	V	14	ALUM.	—	28/15	33/15	38/15	
115B	PR. 3'-0" x 7-2" x 1-³/4"	W.D.	III	5	H.W.	I	10/15	11/15, 12/15	—	"C" LABEL NOTE 1
121D	↓	↓	↓	17	↓	I	4/15	35/10	—	

WINDOW SCHEDULE

SYMBOL	WIDTH	HEIGHT	MAT'L	TYPE	SCREEN & DOOR	QUANTITY	REMARKS	MANUFACTURER	CATALOG NUMBER
A	3'-8"	3'-0"	ALUM.	DOUBLE HUNG	YES	2	4 LIGHTS, 4 HIGH	LBJ WINDOW CO.	141 PW
B	3'-8"	5'-0"	ALUM.	DOUBLE HUNG	YES	1	4 LIGHTS, 4 HIGH	LBJ WINDOW CO.	145 PW
C	3'-0"	5'-0"	ALUM.	STATIONARY	STORM ONLY	2	SINGLE LIGHTS	H & J GLASS CO.	59 PY
D	2'-0"	3'-0"	ALUM.	DOUBLE HUNG	YES	1	4 LIGHTS, 4 HIGH	LBJ WINDOW CO.	142 PW
E	2'-0"	6'-0"	ALUM.	STATIONARY	STORM ONLY	2	20 LIGHTS	H & J GLASS CO.	37 TS
F	3'-6"	5'-0"	ALUM.	DOUBLE HUNG	YES	1	16 LIGHTS, 4 HIGH	LBJ WINDOW CO.	141 PW

HEADER SCHEDULE

HEADER SIZE	EXTERIOR 26' + UNDER	26' TO 32'	INTERIOR 26' + UNDER	26' TO 32'
(2) 2 x 4	3'-6"	3'-0"	USE (2) 2 x 6	
(2) 2 x 6	6'-6"	6'-0"	4'-0"	3'-0"
(2) 2 x 8	8'-6"	8'-0"	5'-6"	5'-0"
(2) 2 x 10	11'-0"	10'-0"	7'-0"	6'-6"
(2) 2 x 12	13'-6"	12'-0"	8'-6"	8'-0"

106F12.EPS

Figure 12 ◆ Door, window, and header schedules.

windows, the tops of all windows should be at the same height. The bottom height will vary depending on the use. For example, the bottom of a window over the kitchen sink should be higher than that of a living room or dining room window. The standard height for a residential window is 6'-8" from the floor to the bottom of the window top (head) jamb.

The window and door schedules will sometimes provide the rough opening dimensions for windows and doors. Another good source of information is the manufacturer's catalog. It will provide rough and finish opening dimensions, as well as the unobstructed glass dimensions.

When roughing-in a window, the rough opening width equals the width of the window plus the thickness of the jamb material; this is usually 1½" (¾" on each side), plus the shim clearance (½" on each side). Therefore, the rough opening for a 3' window would be 38½". The height of the rough opening is figured in the same way. Be sure to check the manufacturer's instructions for the dimensions of the windows you are using.

To lay out a wall opening, proceed as follows:

Step 1 Measure from the corner to the start of the opening, then add half the width of the window or door to determine the center-

Rough Opening Dimensions

The residential plan shown in *Figure 11* shows dimensions to the sides of rough openings—that is, from the corner of the building to the near side of the first rough opening, then from the far side of the first rough opening to the near side of the second rough opening. However, it is more common for plans to show centerline dimensions— that is, from the corner of the building to the center of the first rough opening, then from the center of the first rough opening to the center of the second rough opening.

line (*Figure 13*). Mark the locations of the full studs and trimmer studs by measuring in each direction from the centerline. Mark the cripples 16" on center starting with one trimmer.

Each window opening requires a common stud (king stud) on each side, plus a header, cripple studs, and a sill (*Figure 14*). If the window is more than 4' wide, local codes may require a double sill. Door openings also require trimmer studs, king

studs, and a header. Cripple studs will be needed unless the door is double-wide. In that case, a full header may be called for.

Step 2 Mark the location of each common stud and king stud (X), trimmer (T), and cripple (C), as shown in *Figure 15*. (This is a suggested marking method. The only important thing is to mark the locations with codes that you and other members of your crew will recognize.)

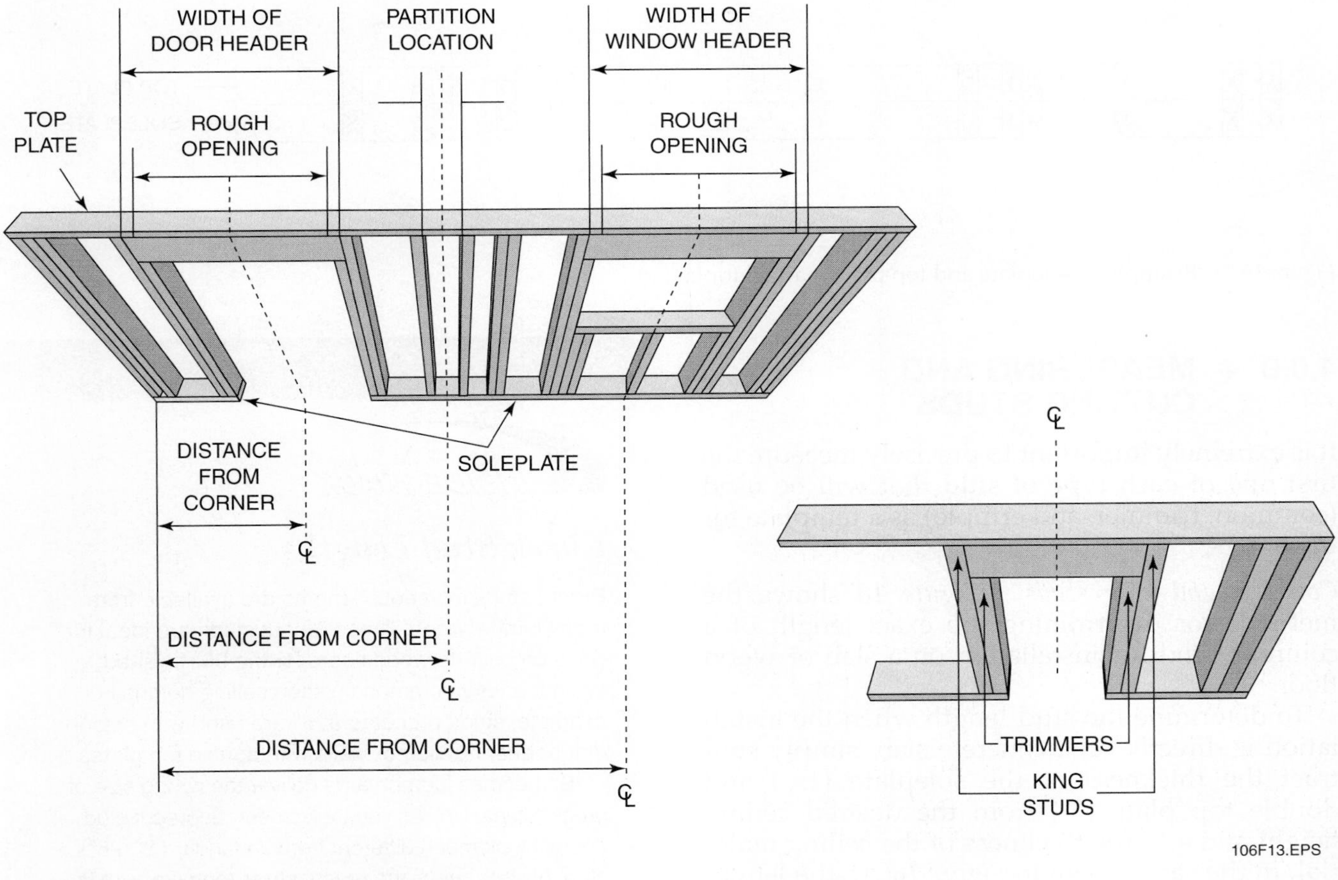

Figure 13 ◆ Laying out a wall opening.

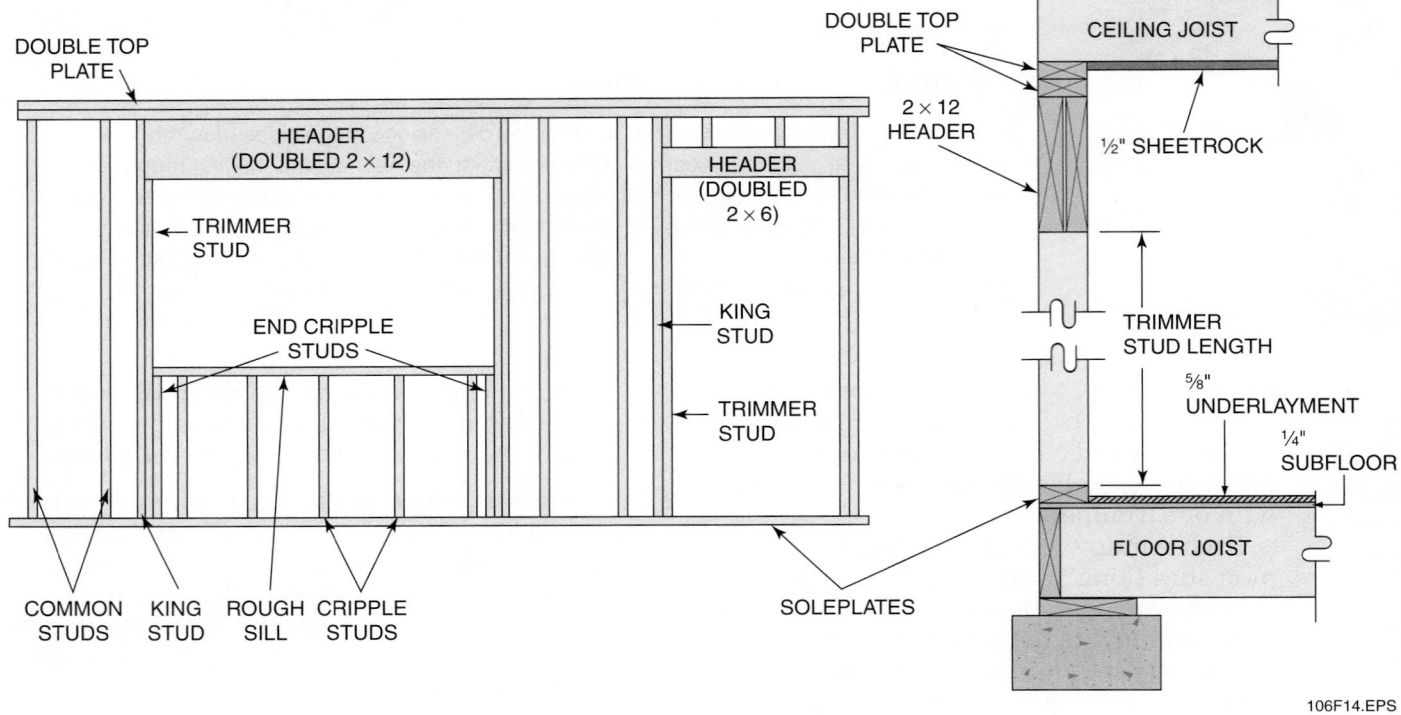

Figure 14 ◆ Window and door framing.

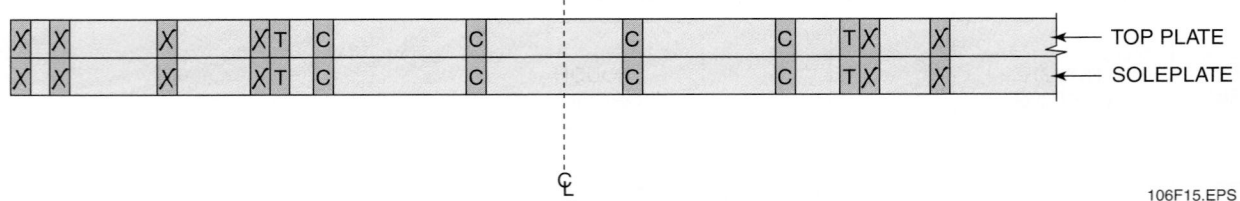

Figure 15 ◆ Example of soleplate and top plate marked for layout.

4.0.0 ◆ MEASURING AND CUTTING STUDS

It is extremely important to precisely measure the first one of each type of stud that will be used (common, trimmer, and cripple) as a template for the others.

Common and king studs – Figure 16 shows the methods for determining the exact length of a common stud for installation on a slab or wood floor.

To determine the stud length when the installation is directly on a concrete slab, simply subtract the thickness of the soleplate (1½") and double top plate (3") from the desired ceiling height and add the thickness of the ceiling material. In the case shown in *Figure 16(A)*, the length of the stud is based on the ceiling height, which is 96", plus the ½" thickness of the ceiling material, less the combined plate thicknesses of 4½", or 92".

INSIDE TRACK

Check Stud Lengths

Precut studs in various lengths are available from many lumberyards. These precut studs are ideal for walls on built-up wood floors for the 96" finished ceiling, a very common finished ceiling height. For example, studs precut to 92⅝" are used with exterior or loadbearing interior walls with double top plates.

Sometimes lumberyards deliver the wrong size of precut studs. Unless you look closely, a precut stud doesn't look much different from a standard 8' (96") stud. Always check the precut studs to make sure they are the right length. Taking a few seconds to measure before you start might save hours of rebuilding later on.

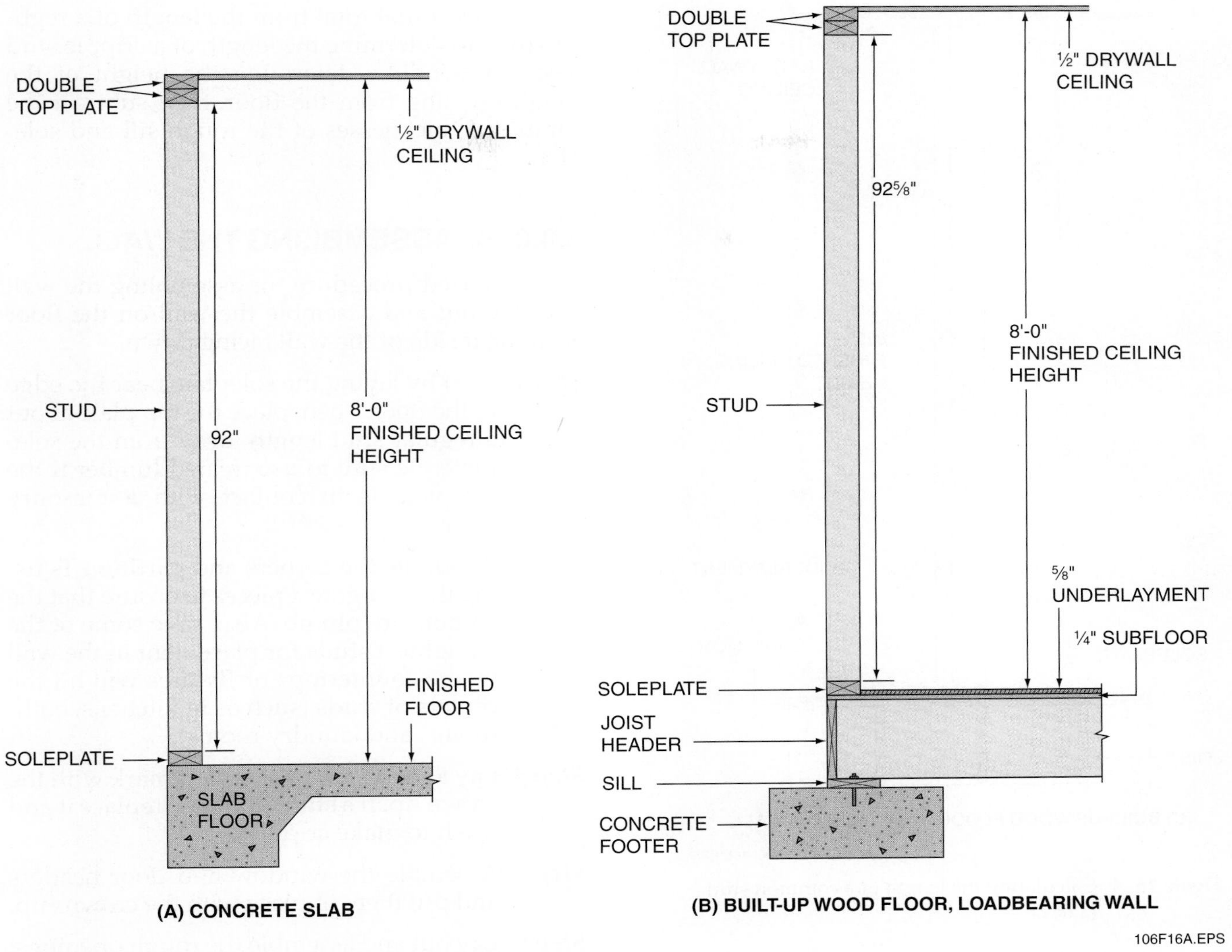

DOUBLE
TOP PLATE

½" DRYWALL
CEILING

STUD

92"

8'-0"
FINISHED CEILING
HEIGHT

SOLEPLATE

FINISHED
FLOOR

SLAB
FLOOR

(A) CONCRETE SLAB

DOUBLE
TOP PLATE

½" DRYWALL
CEILING

92⅝"

8'-0"
FINISHED CEILING
HEIGHT

STUD

⅝"
UNDERLAYMENT

¼" SUBFLOOR

SOLEPLATE

JOIST
HEADER

SILL

CONCRETE
FOOTER

(B) BUILT-UP WOOD FLOOR, LOADBEARING WALL

106F16A.EPS

Figure 16 ◆ Calculating the length of a common stud. (1 of 2)

This example assumes that the flooring material has no appreciable thickness.

In the example shown in *Figure 16(B)*, the thickness of the underlayment must also be considered. Therefore, the length of the stud should be 92⅝"; i.e., ceiling height plus the combined thicknesses of the ceiling material and underlayment

(½" + ⅝"), less the combined thicknesses of the plates (4½"). Again, this example assumes a flooring material of no appreciable thickness.)

Figure 16(C) shows an interior, nonbearing wall that does not require the use of a double top plate. Therefore, the calculated stud length is 1½" longer (94⅛").

Measure Twice, Cut Once

INSIDE TRACK

Before cutting all the studs, double-check your measurements, or have someone else do it. You do not want to find out there was an error in calculation after you have cut 200 pieces of 2 × 4. When you are satisfied with the measurements, cut the required amounts for each type of stud and stack them neatly near the wall locations. Be sure to allow ample room to assemble the walls.

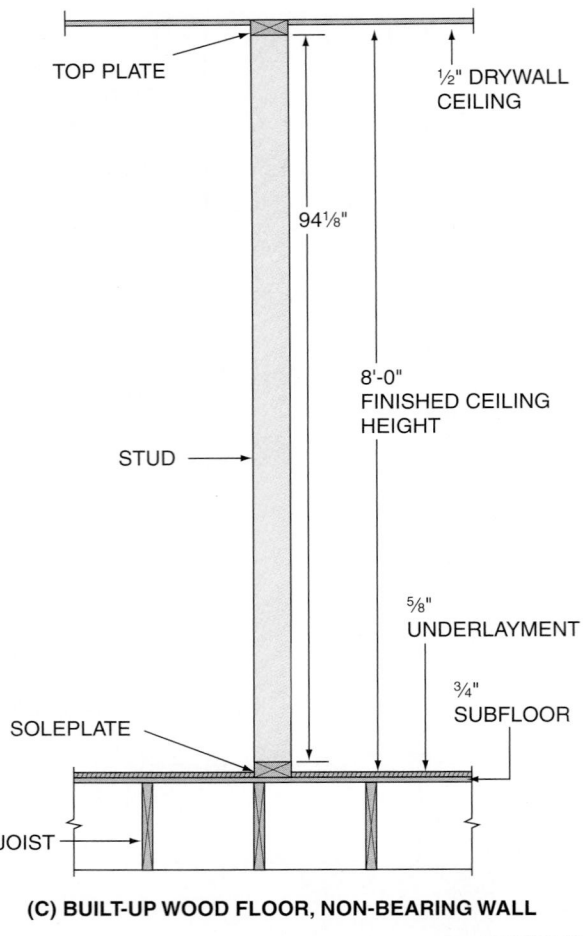

(C) BUILT-UP WOOD FLOOR, NON-BEARING WALL

106F16B.EPS

Figure 16 ◆ Calculating the length of a common stud. (2 of 2)

Trimmers – The length of a window or door trimmer stud is determined by subtracting the thickness of the soleplate from the height of the header. If the installation is on a wood floor, the thickness of the underlayment must also be subtracted.

Cripples – To determine the length of a cripple stud above a door or window, combine the height of the trimmer and the thickness of the header,

then subtract that total from the length of a regular stud. To determine the length of a cripple stud below a window, determine the height of the rough opening from the floor, then subtract the combined thicknesses of the rough sill and soleplate.

5.0.0 ◆ ASSEMBLING THE WALL

The preferred procedure for assembling the wall is to lay out and assemble the wall on the floor with the inside of the wall facing down.

Step 1 Start by laying the soleplate near the edge of the floor. Then, place the top plate about a regular stud length away from the soleplate. Be sure to use treated lumber if the soleplate is in contact with a masonry floor.

Step 2 Assemble the corners and partition Ts using the straightest pieces to ensure that the corners are plumb. Also, save some of the straightest studs for placement in the wall where countertops or fixtures will hit the centers of studs (such as in kitchens, bathrooms, and laundry rooms).

Step 3 Lay a regular stud at each X mark with the crown up. If a stud is bowed, replace it and use it to make cripples.

Step 4 Assemble the window and door headers and put them in place with the crowns up.

Step 5 Lay out and assemble the rough openings, making sure that each opening is the correct size and that it is square.

Step 6 Nail the framework together. For 2 × 4 framing, use two 16d nails through the plate into the end of each stud. For 2 × 6 framing, use three nails. The use of a nail gun is recommended for this purpose; however, do not use this tool if you have not received proper training.

Cripple Studs

It's good to know how to calculate the proper lengths as the text describes; however, in practice, lumber dimensions and assembly tolerances can vary. Therefore, it's better to hold off until after the headers, trimmers, and rough sills are assembled, and then actually measure and cut the cripples to the lengths needed.

Determining Stud Length

1. If the finish ceiling height of a building with built-up wood floors is supposed to be 9'-6", how long are the studs in exterior walls (assume ½" underlayment and ⅝" drywall)?
2. If the top of the rough sill of a window is supposed to be 32" above the floor, how long are the cripple studs below the sill?

INSIDE TRACK

Coping with Natural Defects

Because wood is a natural material, it will have variations that can be considered defects. Lumber is almost never perfectly straight. Even when a piece of lumber is sawn straight, it will most likely curve, twist, or split as it dries.

In the old days, wood cut at a sawmill was just stacked and left to air dry. Normal changes in daily air temperature, humidity, and other conditions would result in lumber with very interesting shapes. Nowadays, lumber is dried in a kiln (a large, low-temperature oven), which will reduce the amount of twisting and curving. But even modern lumber is still somewhat distorted.

Virtually all lumber is slightly curved along its narrow side (the 1½" dimension in 2 × 4s). This is called a crown. It's normal, and, unless it is extreme, the crown doesn't prevent the lumber from being used.

When assembling a floor, you should sight down each piece of lumber to determine which side has the crown. Mark the crowned side. When using the lumber in a floor, position all the lumber crown side up. Weight on the floor will cause the crown to flatten out.

In a wall, there's no force that will flatten the crown. But, for a more uniform nailing surface for the sheathing, you should position all crowns in the same direction. In the example to follow, the crowns are placed up, so when the wall is erected the crown side will be toward the exterior of the building.

A curve along the wide side of lumber (the 3½" dimension in 2 × 4s) is called a bow. If a noticeably bowed stud was used in a wall, sheathing could not be nailed to it in a straight line. Either the nail would miss the bowed stud, or you would have to take extra time to lay out a curved line to nail through. That's not good use of your time, so discard the bowed stud and use a straight one instead.

5.1.0 Firestops

In some areas, local building codes may require firestops. Firestops are short pieces of 2 × 4 blocking (or 2 × 6 pieces if the wall is framed with 2 × 6 lumber) that are nailed between studs. See *Figure 17*.

Without firestops, the space between the studs will act like a flue in a chimney. Any holes drilled through the soleplate and top plate create a draft, and air will rush through the space. In a fire, air, smoke, gases, and flames can race through the chimney-like space.

The installation of firestops has two purposes. First, it slows the flow of air, which feeds a fire through the cavity. Second, it can actually block flames (temporarily, at least) from traveling up through the cavity.

If the local code requires firestops, it may also require that holes through the soleplate and top plate (for plumbing or electrical runs) be plugged with a firestopping material to prevent airflow.

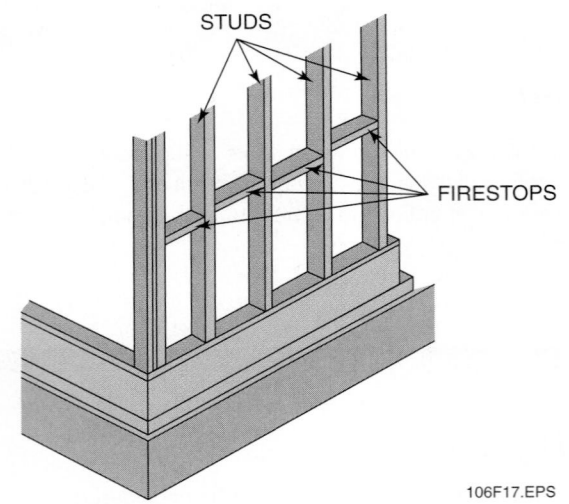

STUDS

FIRESTOPS

106F17.EPS

Figure 17 ◆ Firestops.

6.0.0 ◆ ERECTING THE WALL

There are four primary steps in erecting a wall:

Step 1 If the sheathing was installed with the wall lying down, or if the wall is very long, it will probably be too heavy to be lifted into place by the framing crew. In that case, use a crane or the special lifting jacks made for that purpose (*Figure 18*). Use cleats to prevent the wall from sliding.

Step 2 Raise the wall section and nail it in place using 16d nails on every other floor joist. On a concrete slab, use preset anchor bolts or powder-actuated pins. Do not use these tools if you have not received proper training and certification.

Step 3 Plumb the corners and apply temporary exterior bracing. Then erect, plumb, and brace the remaining walls. The bracing helps keep the structure square and will prevent the walls from being blown over by the wind. Generally, the braces remain in place until the roof is complete.

Step 4 As the walls are erected, straighten the walls and nail temporary interior bracing in place.

6.1.0 Plumbing and Aligning Walls

Accurate plumbing of the corners is possible only after all the walls are up. Always use a straightedge along with a hand level (*Figure 19*). The straightedge can be a piece of 2 × 4 lumber. Blocks ¾" thick are nailed to each end of the 2 × 4. The blocks make it possible to accurately plumb the wall from the bottom plate to the top plate. (If you just placed the level directly against the wall, any bow or crown in the end stud would give a false reading.)

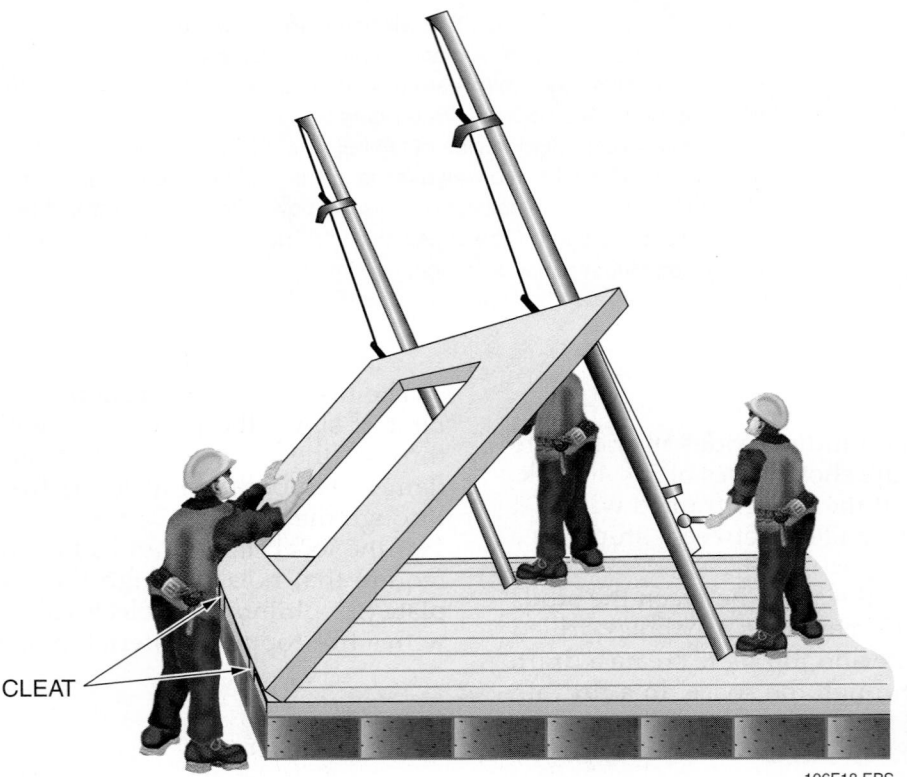

CLEAT

106F18.EPS

Figure 18 ◆ Wall lifting jack.

Raising a Wall

When a wall is being erected, either by hand or with a lifting jack, the bottom of the wall can slide forward. If the wall slides off the floor platform, the wall or objects on the ground below it can be damaged and workers can be injured.

Use wood blocks or cleats securely nailed to the outside of the rim joist to catch the wall as it slides forward, preventing damage or injury.

Some carpenters use metal banding (used to bundle loads of wood from the mill) to achieve the same effect. One end of a short length of banding is nailed to the floor platform. The other end is bent up 90° and nailed to the bottom of the soleplate. When the wall is raised, the flexible band straightens out horizontally, much like a hinge, but the wall cannot slide forward.

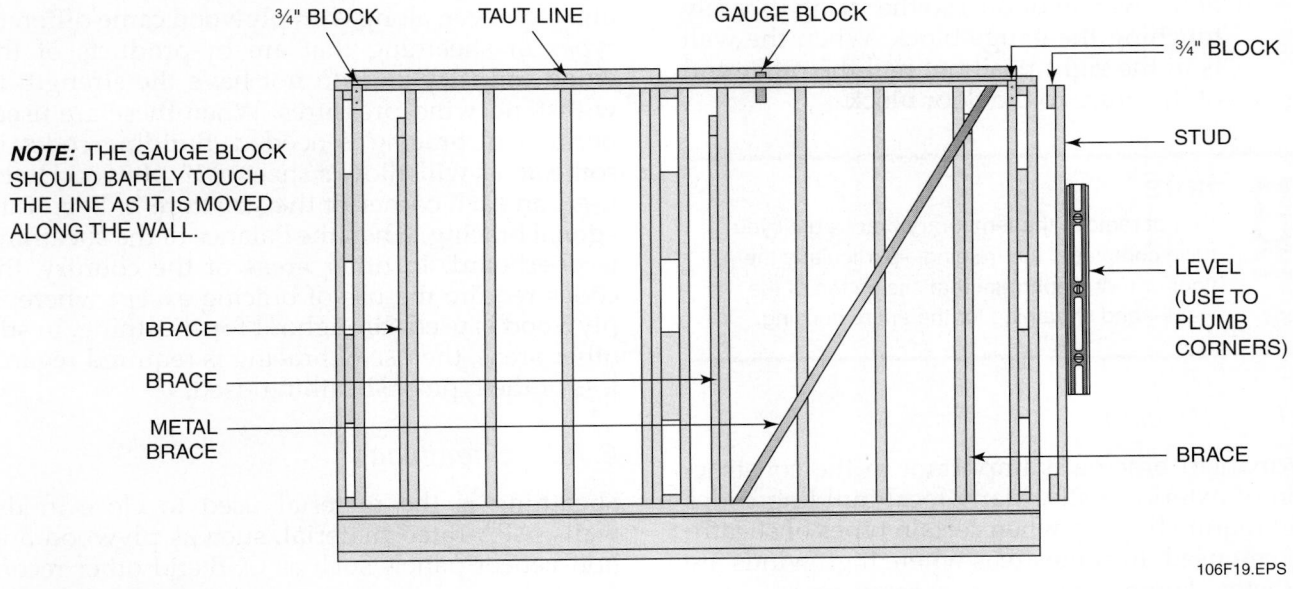

Figure 19 ◆ Plumbing and aligning a wall.

The plumbing of corners requires two people working together. One carpenter releases the nails at the bottom end of the corner brace so that the top of the wall can be moved in or out. At the same time, the second carpenter watches the level. The bottom end of the brace is renailed when the level shows a plumb wall.

Install the second plate of all the double top plates (*Figure 20*). In addition to adding strength in bearing walls, the second plate helps to straighten a bowed or curved wall. If bows are turned opposite each other, intersections of walls should be double plated after walls are erected. Overlap the corners and partition Ts. Drive two 16d nails at each end, then drive one 16d nail at each stud location.

After you have plumbed all the corners, line up the tops of the walls. This must be done before you nail the ceiling to the tops of the walls. To line up the walls, proceed as follows (refer to *Figure 19*):

Step 1 Start at the top plate at one corner of the wall. Fasten a string at that corner. Stretch the string to the top plate at the corner at the opposite end of the wall, and fasten the string.

Step 2 Cut three small blocks from 1 × 2 lumber. Place one block under each end of the string so that the line is clear of the wall. Use the third block as a gauge to check the wall at 6' or 8' intervals.

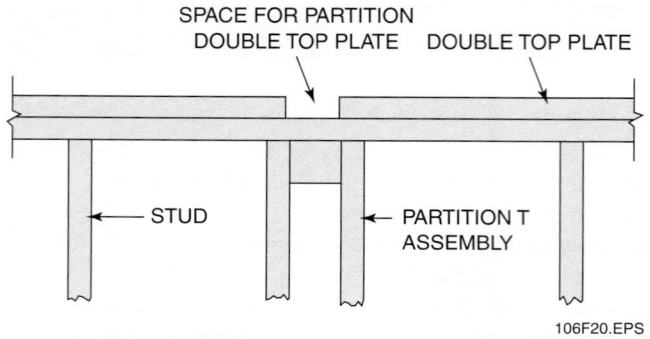

Figure 20 ◆ Double top plate layout.

Step 3 At each checkpoint, nail one end of a temporary brace near the top of a wall stud. Also attach a short 2 × 4 block to the subfloor. Adjust the wall (by moving the top of the wall in or out) so the string is barely touching the gauge block. When the wall is in the right position, nail the other end of the brace to the floor block.

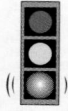

> **NOTE**
>
> Do not remove the temporary braces until you have completed the framing—particularly the floor or roof diaphragm that sits on top of the walls—and sheathing for the entire building.

6.1.1 Bracing

Permanent bracing is important in the construction of exterior walls. Many local building codes will require bracing when certain types of sheathing are used. In some areas where high winds are a factor, lateral bracing is a requirement even when ½" plywood is used as the sheathing.

Several methods of bracing have been used since the early days of construction. One method is to cut a notch or let-in for a 1 × 4 or 1 × 6 at a 45-degree angle on each corner of the exterior walls. Another method is to cut 2 × 4 braces at a 45-degree angle for each corner. Still another type of bracing used (where permitted by the local code) is metal wall bracing (*Figure 21*). This product is made of galvanized steel.

Metal strap bracing is easier to use than let-in wood bracing. Instead of notching out the studs for a 1 × 4 or 2 × 4, you simply use a circular saw to make a diagonal groove in the studs, top plate, and soleplate for the rib of the bracing strap and nail the strap to the framing.

With the introduction of plywood, some areas of the country have done away with corner brac-

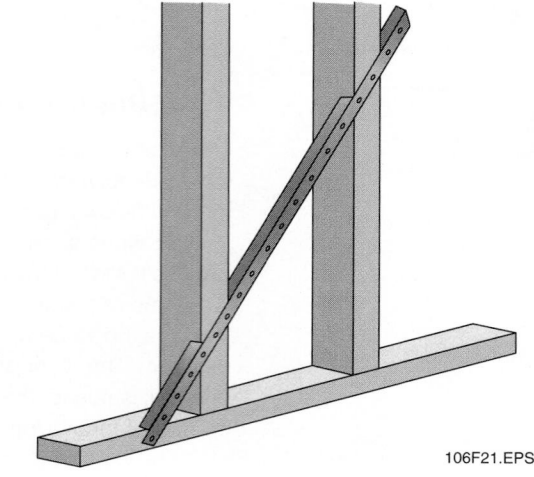

Figure 21 ◆ Use of metal bracing.

ing. However, along with plywood came different types of sheathing that are by-products of the wood industry and do not have the strength to withstand wind pressures. When these are used, permanent bracing is needed. Building codes in some areas will allow a sheet of ½" plywood to be used on each corner of the structure in lieu of diagonal bracing, when the balance of the sheathing is fiberboard. In other areas of the country, the codes require the use of bracing except where ½" plywood is used throughout for sheathing. In still other areas, the use of bracing is required regardless of the type of sheathing used.

6.1.2 Sheathing

Sheathing is the material used to close in the walls. APA-rated material, such as plywood and non-veneer panels such as OSB and other reconstituted wood products, are generally used for sheathing.

Some carpenters prefer to apply the sheathing to a squared wall while the wall frame is still lying on the subfloor. Although this helps to ensure that the wall is square, it has two drawbacks:

- It may make the wall too heavy for the framing crew to lift.
- If the floor is not perfectly straight and level, it will be a lot more difficult to square and plumb the walls once they are erected.

When plywood is used, the panels will range from ⁵⁄₁₆" to ¾" thick. A minimum thickness of ⅜" is recommended when siding is to be applied. The higher end of the range is recommended when the sheathing acts as the exterior finish surface. The panels may be placed with the grain running horizontally or vertically. If they are placed horizontally, local building codes may require that blocking be used along the top edges.

Typical nailing requirements call for 6d nails for panels ½" thick or less and 8d nails for thicker panels. Nails are spaced 6" apart at the panel edges and 12" apart at intermediate studs.

Other materials that are sometimes used as sheathing are fiberboard (insulation board), exterior-rated gypsum wallboard, and rigid foam sheathing. A major disadvantage of these materials is that siding cannot be nailed to them. It must either be nailed to the studs or special fasteners must be used. If you are installing any of these materials, keep in mind that the nailing pattern is different from that of rated panels. In addition, they take roofing nails instead of common nails. Check the manufacturer's literature for more information.

When material other than rated panels is used as sheathing, rated panels can be installed vertically at the corners to eliminate the need for corner bracing in some applications. Check the plans and local codes.

6.1.3 Panelized Walls

Instead of building walls on the job site, they can be prefabricated in a shop and trucked to the job site. The walls, or panels, are either set with a small crane or by hand. The wall sections or panels will vary in length from 4' to 16'.

When working with a pre-engineered structure, the **drying-in** time is much quicker than with a field-built structure. A 1,200-square-foot residence, for example, can be dried-in within two working days. The siding would be applied at the factory and the walls erected the first day. The soffit and fascia would be installed on the morning of the second day, and the roof dried-in and ready to shingle by the morning of the third day. The residence would be ready for rough-in plumbing, electrical, heating, and cooling by the third day.

7.0.0 ◆ CEILING LAYOUT AND FRAMING

After you assemble and erect the walls, what is next? Traditionally, in a one-story building, carpenters would install ceiling joists on top of the structure, spanning the narrow dimension of the building from top plate to top plate. Then, if the carpenters intended to build a common **gable roof,** they would install rafters that extend from the ends of the ceiling joists to the peak, forming a triangle.

Nowadays, it is more likely that carpenters will install roof trusses instead of joists in a one-story building. The lowest member (or bottom chord) of the truss serves the same purpose as the ceiling joist. Alternatively, if the building is taller than one story, carpenters will install floor joists or floor trusses and build a platform and erect the walls for the second story. Again, the floor joist or truss, when viewed from below, serves the same purpose as the ceiling joist.

You will learn more about roof trusses and modern roof framing methods in later modules. For now, let's focus on the older, simpler way of framing with ceiling joists.

Ceiling joists have two important purposes:

- The joists are the top of the six-sided box structure of a building. They keep opposite walls from spreading apart.
- The joists provide a nailing surface for ceiling material, such as drywall, which is attached to the underside of joists.

As noted previously, joists extend all the way across the structure, from the outside edge of the double top plate on one wall to the outside edge of the double plate of the opposite wall. Ordinarily, carpenters lay the ceiling joists across the narrow width of a building, usually at the same positions as the wall studs.

If the spacing of the ceiling joists is the same as that of the wall studs, lay out the joists directly above the studs. This makes it easier to run ductwork, piping, flues, and wiring above the ceiling.

Laying out ceiling joists for a gable roof is very similar to laying out floor joists and wall studs. Measure along the double top plate to a point 15¼" in from the end of the building. Mark it, square the line, then use a long steel tape to mark every 16" (or 24", depending on the architect's plans). To the right of each line, mark an X for the joist location. Then mark an R for each rafter on the left side of each mark (*Figure 22*). Repeat this procedure on the opposite wall and on any bearing partitions.

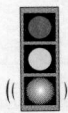

NOTE

If you are installing a **hip roof,** it is also necessary to mark the end double top plates for the locations of hip rafters. Start by marking the center of the end double top plate. Then measure and place marks for joists and rafters by measuring from each corner toward the center mark.

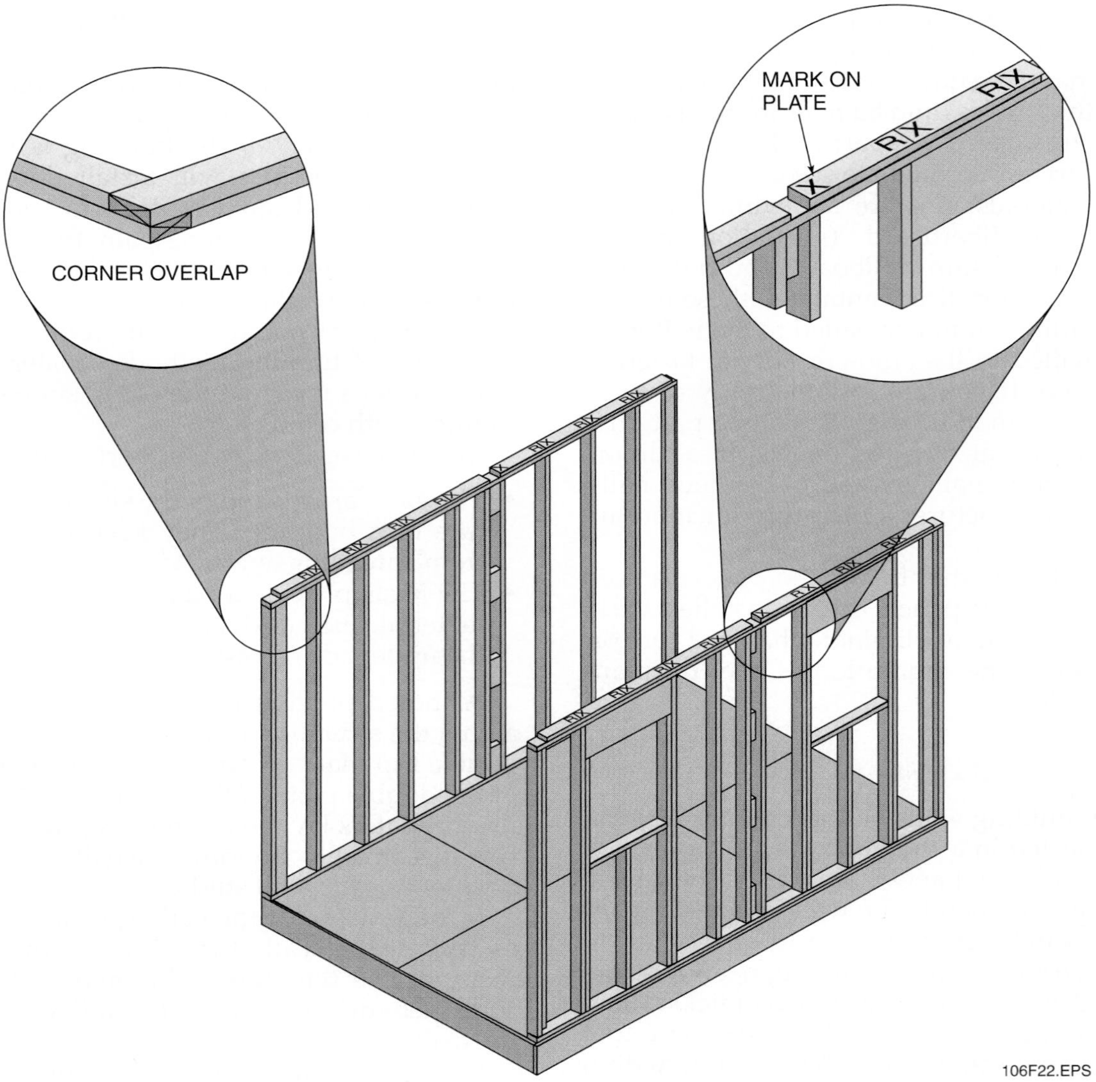

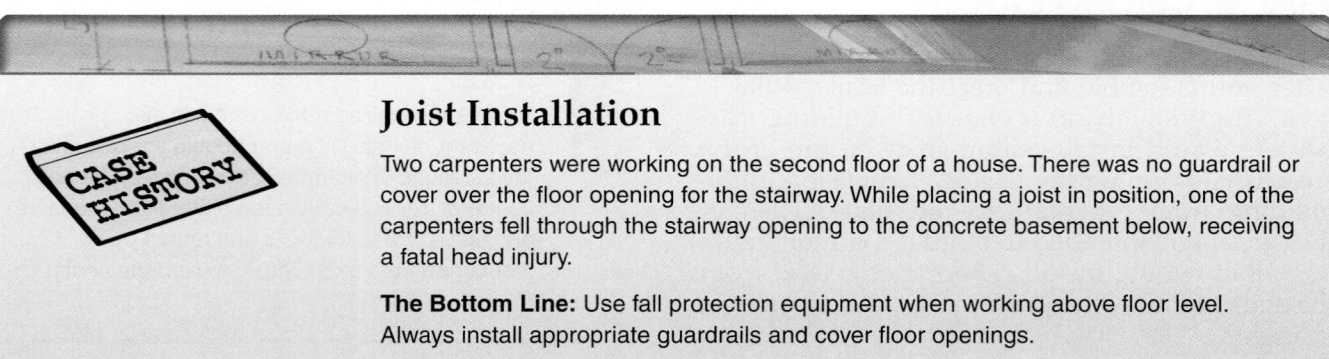

Figure 22 ◆ Marking joist and rafter locations on the double top plate.

As you learned in the module on floor layout, the actual allowable span for joists depends on the species, size, and grade of lumber, as well as the spacing and the load to be carried. If the joist exceeds the allowable span, two pieces of joist material must be spliced over a bearing wall or partition. *Figure 23* shows two ways to splice joists. In the first method, as shown in *Figure 23(A)*, two joists are overlapped above the center of the bearing partition. The overlap should be no less than 6".

Joist Installation

Two carpenters were working on the second floor of a house. There was no guardrail or cover over the floor opening for the stairway. While placing a joist in position, one of the carpenters fell through the stairway opening to the concrete basement below, receiving a fatal head injury.

The Bottom Line: Use fall protection equipment when working above floor level. Always install appropriate guardrails and cover floor openings.

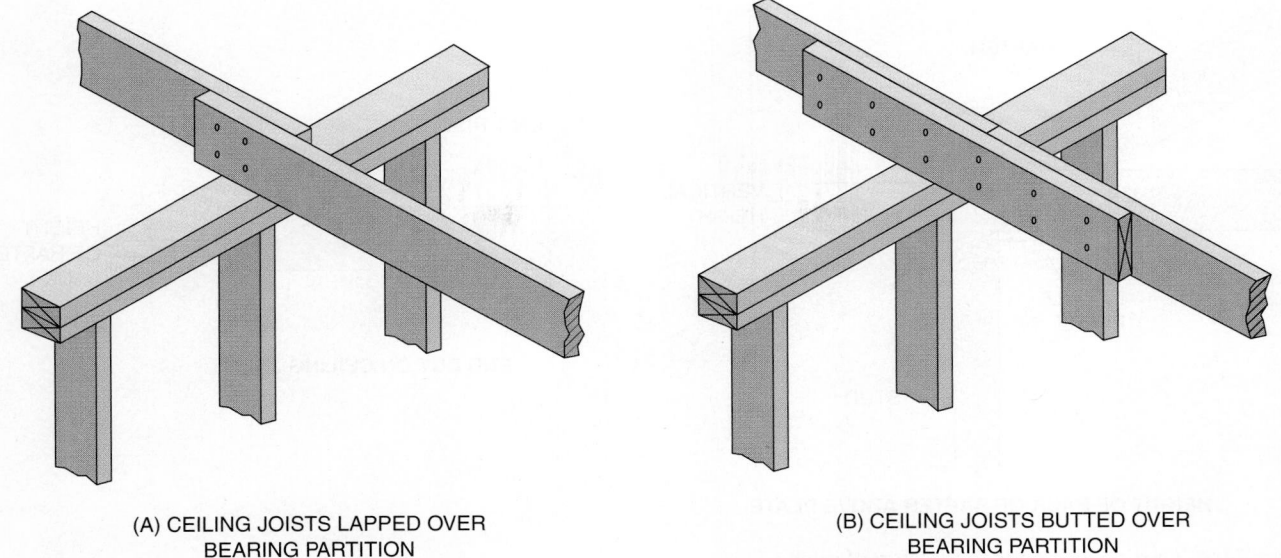

(A) CEILING JOISTS LAPPED OVER
BEARING PARTITION

(B) CEILING JOISTS BUTTED OVER
BEARING PARTITION

106F23.EPS

Figure 23 ◆ Splicing ceiling joists.

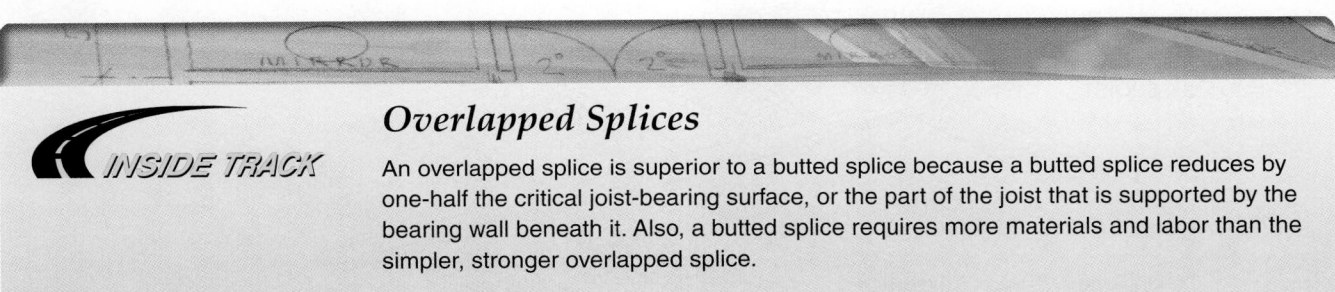

Overlapped Splices

An overlapped splice is superior to a butted splice because a butted splice reduces by one-half the critical joist-bearing surface, or the part of the joist that is supported by the bearing wall beneath it. Also, a butted splice requires more materials and labor than the simpler, stronger overlapped splice.

Figure 23(B) shows another way to splice joists. Instead of overlapping, the two joists butt together directly over the center of the bearing partition. A shorter piece of joist material is nailed to both joists to make a strong joint. Other materials used to reinforce the joist splice include plywood, 1 × lumber, steel strapping, and special anchors. The architectural plans will specify the proper method for splicing the joists.

7.1.0 Cutting and Installing Ceiling Joists

The joists must be cut to the proper length, so the ends of the joists will be flush with the outside edge of the double top plate. As you learned in the previous section, you must allow enough extra length for any overlap of the joists above bearing partitions.

You must also cut the ends of the joists at an angle matching the rafter pitch, as shown in *Figure 24*. This is so the roof sheathing will lie flush on the roof framing. You will learn more about this in the module about roof framing.

After you have cut the joists to the right length and have positioned them in the right places, toenail them into the double top plate. The architectural plans might also call for the installation of metal anchors.

After you install the joists, you must nail a rib-band or strongback across the joists to prevent twisting or bowing (*Figure 25*). The strongback is used for longer spans. In addition to holding the joists in line, the strongback provides support for the joists at the center of the span.

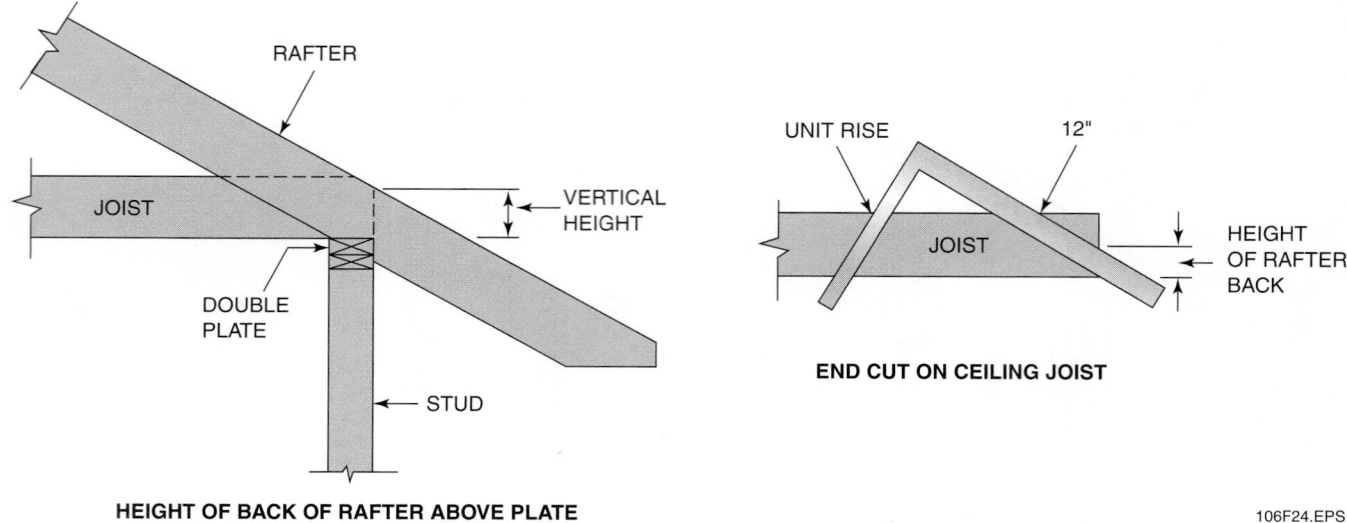

VERTICAL HEIGHT

JOIST

RAFTER

DOUBLE PLATE

STUD

HEIGHT OF BACK OF RAFTER ABOVE PLATE

UNIT RISE

12"

JOIST

HEIGHT OF RAFTER BACK

END CUT ON CEILING JOIST

106F24.EPS

Figure 24 ◆ Cutting joist ends to match roof pitch.

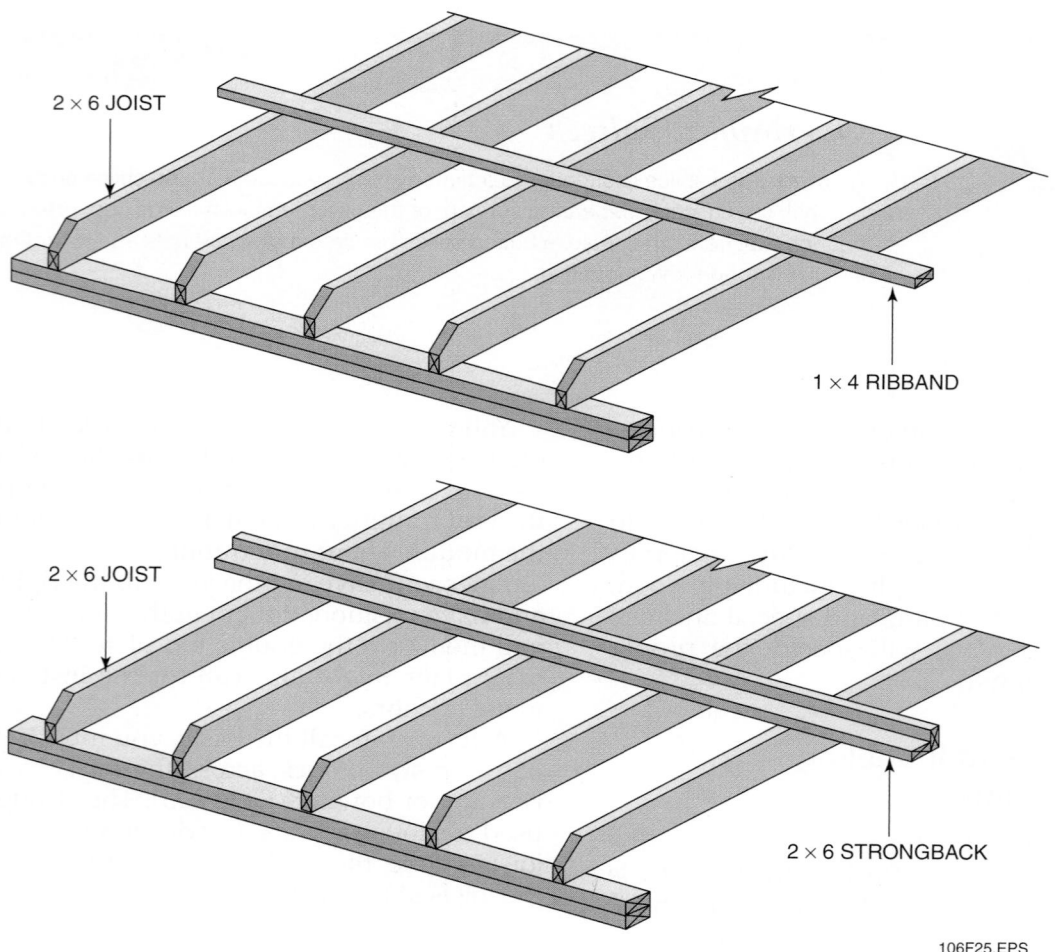

2 × 6 JOIST

1 × 4 RIBBAND

2 × 6 JOIST

2 × 6 STRONGBACK

106F25.EPS

Figure 25 ◆ Reinforcing ceiling joists.

8.0.0 ◆ ESTIMATING MATERIALS

In this section, you will follow the basic steps to estimate the amount of lumber you will need to frame the walls and ceilings of a building. Here's some important information that you need:

- For this example, the structure is 24' × 30'. (See *Figure 26*.)
- The walls are 8' tall.
- You will construct the building with 2 × 4s with 16" on center spacing.
- You will construct the ceiling from 2 × 6s.

- The building has two 24" wide windows on each narrow side, and two 48" wide windows and one 36" wide door on each wide side.
- Inside, the building is divided into three rooms: one is 12' × 30', two are each 11'-1½" × 12', divided by a 4'-10" wide hallway.
- The long wall that separates the large room from the two smaller rooms and hallway is a loadbearing partition.
- Each of the smaller rooms has a 36" door leading to the hallway.
- There is a 36" opening, without a door, from the large room to the hallway.

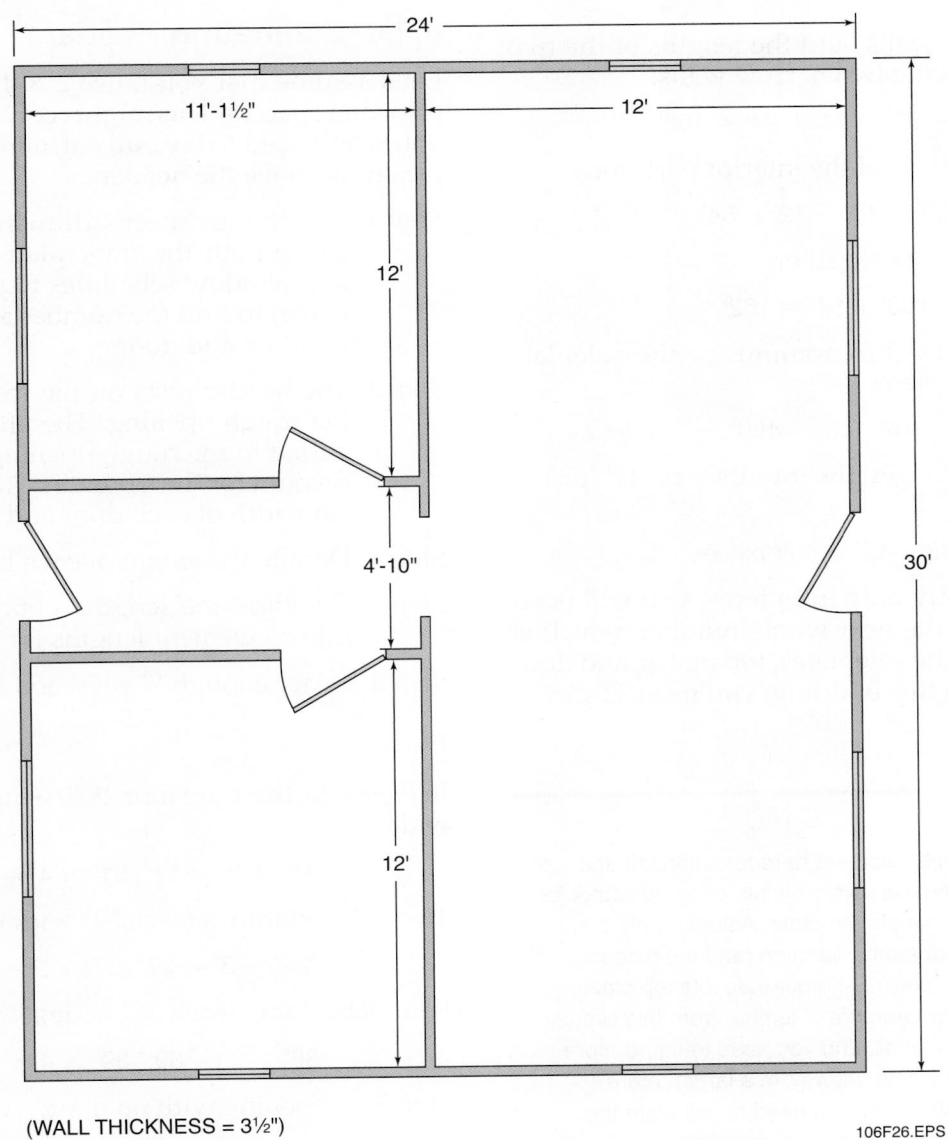

(WALL THICKNESS = 3½")

106F26.EPS

Figure 26 ◆ Sample floor plan.

8.1.0 Estimating Soleplates and Top Plates

Assume that the plate material is ordered in 12' lengths.

Step 1 Determine the length of the walls in feet. For loadbearing walls, multiply this number by 3 to account for the soleplates and double top plates (remember that a double top plate is made of two 2 × 4s stacked together).

Step 2 Divide the result by 12 to get the number of 12' pieces. Round up to the next full number, and allow for waste.

Example:

For the exterior walls, add the lengths of the two long walls, plus the two narrow walls:

$$30' + 30' + 24' + 24' = 108'$$

Also add the length of the interior partitions:

$$30' + 12' + 12' = 54'$$

Add those numbers together:

$$108' + 54' = 162'$$

Then multiply by 3 to account for the soleplate and double top plate:

$$162' \times 3' = 486'$$

Divide by 12 to get the number of 12' pieces needed:

$$486' \div 12' = 40\frac{1}{2} \text{ pieces}$$

Since you can buy only full pieces, you will need to round up to the next whole number, which is 41. So, for all of the soleplates, top plates, and double top plates in this building, you need 41 pieces of 12' 2 × 4s.

NOTE

The example estimate here for soleplate and top plate material is a very simple one and works for this small, simple structure. Actually, only the center loadbearing partition (and the exterior walls, of course) will require double top plates. But the extra amount of lumber from this simpler estimate is small, and you need to figure more lumber for waste anyway. In a larger, more complex structure, you need to calculate the different quantities of lumber for loadbearing and nonbearing partitions to get a more precise estimate.

8.2.0 Estimating Studs

Step 1 Determine the length in feet of all the walls.

Step 2 The general industry standard is to allow one stud for each foot of wall length, even when you are framing 16" on center. This should cover any additional studs that are needed for openings, corners, partition Ts, and blocking.

When you were estimating for soleplate and top plate material, you found that there were 162' of exterior walls and interior partitions. You will need one stud for every foot of wall, or 162 studs.

8.3.0 Estimating Headers

Let's assume that you'll use 2 × 12 headers with plywood spacers. This might cost a little more, but you won't need to lay out, cut, and assemble cripple studs above the headers.

Step 1 Use the architect's drawings (to be safe, check both the floor plans and the door and window schedules to make sure they agree) to find the number and size of each window and door.

Step 2 The header rests on the trimmer studs in the rough opening. Therefore, 3" must be added to the rough opening dimension to account for the trimmers. Add 3" to the finish width of each door and window.

Step 3 Double the length of each header.

Step 4 Combine the lengths obtained in Step 3 into convenient lengths for ordering.

Step 5 Order enough ½" plywood for spacers.

Example:

In *Figure 26*, there are four 4830 windows (each 48" wide):

$$48" + 3" = 51"; 51" \times 4 = 204"$$

Two 2430 windows (each 24" wide):

$$24" + 3" = 27"; 27" \times 2 = 54"$$

Four 3680 doors (each 36" wide):

$$36" + 3" = 39;" 39" \times 4 = 156"$$

One 3'-0" opening with no door:

$$36" + 3" = 39"$$

Add those numbers together:

$$204" + 54" + 156" + 39" = 453"$$

How Accurate Is an Estimate?

How close does the estimate come to actual needs? Sometimes, it does not come close enough. In our example building, we estimated 162 studs based on one stud per lineal foot of wall space. However, if you calculate the number based on the actual framing spaced 16" on center, it takes:

- 17 studs to span each of the two 24' walls
- 21 studs to span each of the two 30' walls
- 21 studs for the 28'-10" interior partition (excludes the width of the two interior and two exterior stud walls, which is 1'-2" total)
- 9 studs to span each 11'-1½" interior partition

So, that subtotal is:

$$17 + 17 + 21 + 21 + 21 + 9 + 9 = 115$$

But you need to add an extra full stud and trimmer on each side of each opening. There are thirteen openings, so …

$$13 \times 4 = 52$$

…add that to the previous subtotal …

$$115 + 52 = 167$$

However, about 15 full studs will be eliminated where the openings will be, so subtract that from the subtotal …

$$167 - 15 = 152$$

Then add two extra studs for each of the six partition Ts …

$$152 + 12 = 164$$

…and two extra studs for each exterior corner …

$$164 + 8 = 172$$

Of course, you'll need studs for cripples and for extra blocking here and there. Then, some of the delivered lumber will be unusable because it is twisted or warped. Plus, measuring and cutting mistakes will be made. So, you're well over the 162 pieces calculated by allowing one stud every 12". Be smart and add a safety factor. It might vary depending on the size and complexity of the job, but for our example here, 15% (for a total of about 186 studs) would have been about right. On bigger jobs, the safety factor might be smaller or even nil. That's because the total amount of lumber is greater, and the safety margin already built into the one-stud-per-foot will cover all needs.

Then double that total:

$$453" \times 2 = 906"$$

Divide by 12 to get feet:

$$906" \div 12 = 75'\text{-}6"$$

Round up to 76'.

8.4.0 Estimating Diagonal Bracing

You need to install diagonal bracing at each end of all exterior walls. You can use metal strips or 1 × 4 lumber. Braces run from the top plate to the soleplate at a 45-degree angle. The walls for this building are 8' high, so they would need a 12' brace. (Review the *Core* module, *Introduction to Construction Math,* to determine how to find the length of the brace using the formula for the hypotenuse of a right triangle.)

Step 1 Determine the number of outside corners.

Step 2 Figure the length of each brace based on the height of the walls, then multiply by the number of corners.

Example:

This part is pretty easy. *Figure 26* shows that this building is a rectangle with four corners. Each corner has two legs, so there are eight places where you would install diagonal braces. Since you need 12' of bracing at each location, 8 × 12' = 96' of bracing altogether.

Remember that let-in or diagonal bracing is typically applied to the interior side of the exterior walls.

8.5.0 Estimating Ceiling Joists

As mentioned previously, it is more likely that carpenters will install roof trusses instead of joists in a one-story building. Roof trusses and modern roof framing methods are covered later in this book. Here the focus is on framing with ceiling joists.

Step 1 Determine the span of the building.

Step 2 Figure the number of joists based on the spacing (remember that you'll place joists 16" on center), then add one for the end joist.

Step 3 Multiply the span by the number of joists to get total length. Add 6" per joist per splice where joists will be spliced at bearing walls, partitions, and girders.

Example:

The building in *Figure 26* is 30' by 24'. Let's take the short dimension (24'). Fortunately, there is a partition down the long dimension of the building. That breaks the 24' span into two 12' spans.

Divide the long dimension of the building (30') by 16" to determine the number of joist locations. First, convert 30' to inches:

$$30' \times 12" = 360"$$

Divide that by 16":

$$360" \div 16" = 22\frac{1}{2} \text{ (round up to 23)}$$

Then add one for the end joist:

$$23 + 1 = 24$$

The 24' width is too big to span with one long joist, so you'll span it with two 12' joists. Multiply the last number by two:

$$24 \times 2 = 48 \text{ pieces}$$

Now calculate the total length, adding 6" for each joist at each splice.

$$48 \times 12'\text{-}6" = 600'$$

So how much lumber do you order? Twelve-foot lengths of 2 × 6 would be 6" short, so 14-foot lengths would be a better choice. You'd end up with 1'-6" of waste from each piece of lumber, but you can use the waste for bridging and blocking.

9.0.0 ◆ WALL FRAMING IN MASONRY

You must be aware of the methods used in furring masonry walls in order to install the interior finish. As a general rule, furring of masonry walls should be done on 16" centers. Some contractors will apply 1 × 2 **furring strips** 24" OC. This may save material, but it does not provide the same quality as a wall constructed on 16" centers. Carpenters have little control over the quality of the masonry structure. It is important that the walls are put up square and plumb. If a structure starts off level and plumb, little difficulty will be encountered with measuring, cutting, and fitting. If the walls or floor are not plumb and level, problems can be expected throughout the structure.

Furring is applied to a masonry wall with masonry nails measuring 1¼" to 1¾" long. In addition to the furring strips, 1 × 4 and 1 × 6 stock is used. Remember, all material that comes in contact with concrete or masonry must be pressure-treated.

Backing for partitions against a masonry block wall is done using one of the following methods. In the first method, locate where the partition is going to be, then nail a 1 × 6 board, centered on the center of the partition (i.e., locate the center of the partition at the bottom corner of the wall, move back 2¾", and mark the wall by plumbing with a straightedge and level). Attach the 1 × 6 to the wall with one edge on the mark. This will allow an even space on either side of the partition to receive the drywall, as shown in *Figure 27*.

In the second method, secure the partition to the block wall and install a furring strip on each side of the partition. This will cause a problem when the drywall is installed; i.e., the partition stud is 1½", and by adding a ¾" furring strip to the wall we have only ¾" of stud left to work with. Nailing the ½" drywall to the furring strip on the wall will leave only ¼" of nailing on the stud. This does not allow enough room to nail the ½" drywall, so it may become necessary to install a 16" 2 × 4 block to the bottom, center, and top of the partition, as shown in *Figure 28*.

The proper nailing surface for drywall is ¾" to 1". In preparing the corners of the block wall to receive the furring strips, enough space should be allowed for the drywall to slip by the furring strips. Come away from the corner ⅝" in either direction with the strips (*Figure 29*).

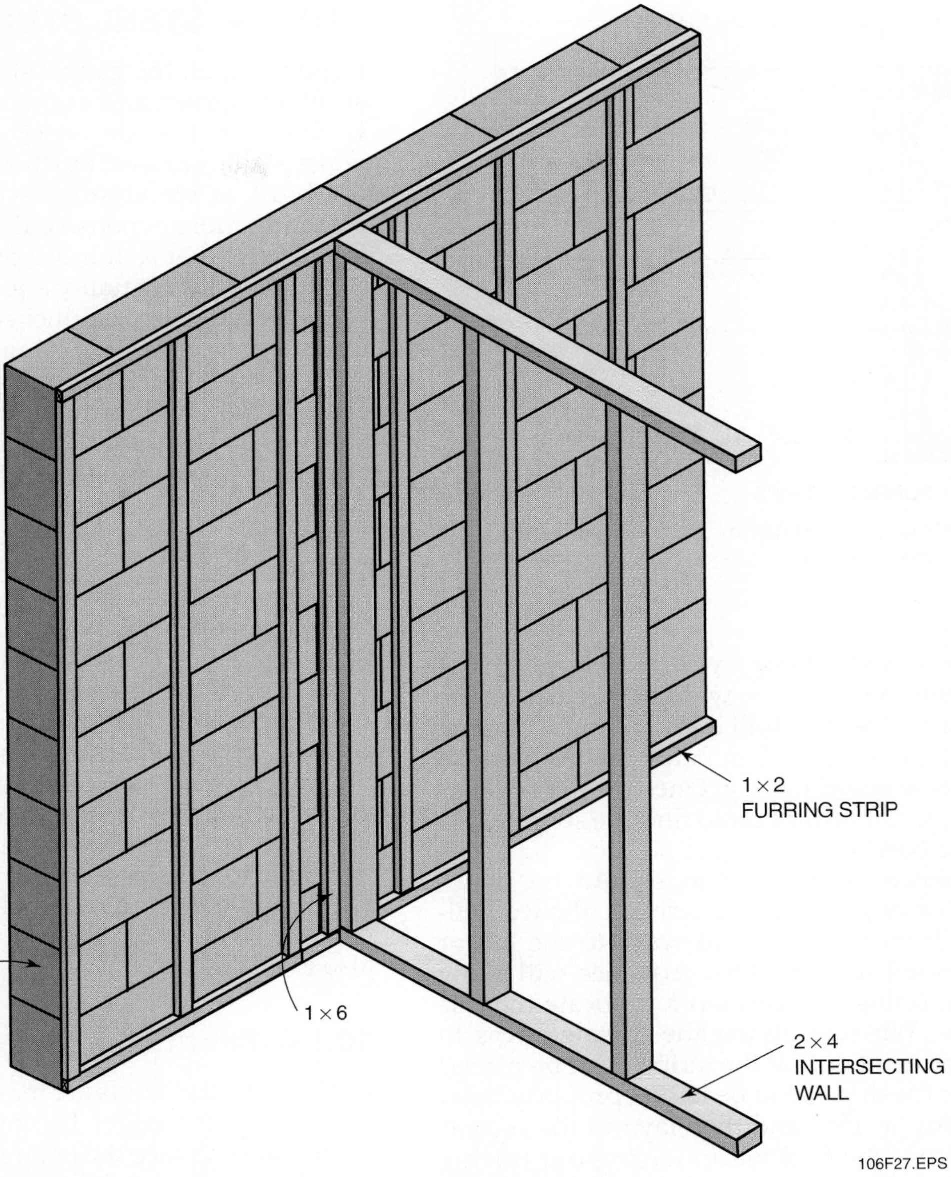

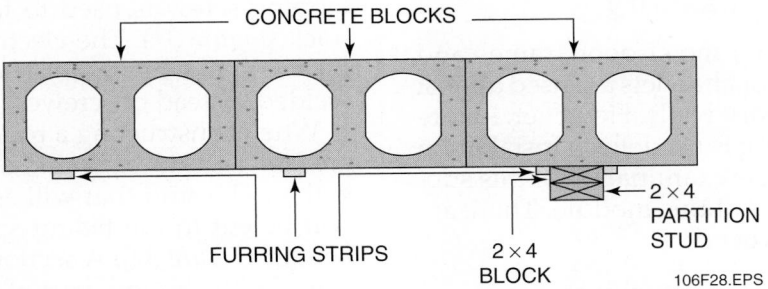

Figure 27 ◆ Partition backing using a 1 × 6.

1 × 2 FURRING STRIP

1 × 6

MASONRY WALL

2 × 4 INTERSECTING WALL

106F27.EPS

CONCRETE BLOCKS

FURRING STRIPS

2 × 4 BLOCK

2 × 4 PARTITION STUD

106F28.EPS

Figure 28 ◆ Partition backing using 2 × 4 blocks.

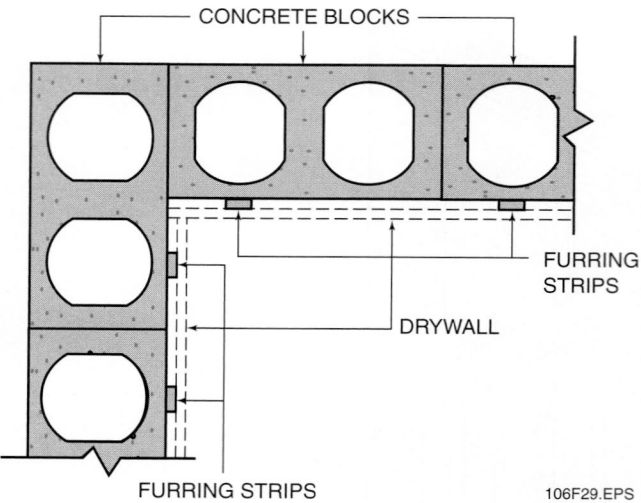

Figure 29 ◆ Placement of furring strips at corners of masonry walls.

106F29.EPS

A 1 × 4 is used at floor level to receive the baseboard. Either a narrow or wide baseboard can be used. Some carpenters will install a simple furring strip at floor level and depend on the vertical strips for baseboard nailing. Once the drywall has been installed, it is difficult to find the strips when nailing the baseboard.

A sequence of installation should be established for nailing the furring strips to the wall. Either start from the right and work to the left or work from left to right. This sequence will allow the person doing the trim work to locate the furring strips. When applying the furring strips to the wall, remember that the strips must be placed 16" OC for the drywall to be nailed properly. Start the first strip at 15¼" and then lay out the second strip 16" from the first mark. Laying out furring strips is done in the same way as laying out wall studs.

9.1.0 Framing Door and Window Openings in Masonry

In modern construction, metal door frames and metal window frames or channels are used almost exclusively with masonry walls. However, in rare instances, wood framing is used. Each installation is unique and a complete examination of this subject is beyond the scope of this module. There are two general rules, however:

• Use treated wood in contact with masonry.
• Use cut nails, expansion bolts, anchor bolts, or other fasteners to attach the wood to the masonry securely.

10.0.0 ◆ STEEL STUDS IN FRAMING

Depending on the gauge, steel studs are typically stronger, lighter, and easier to handle than wood. Unlike wood studs, steel studs will not split, warp, swell, or twist. Furthermore, steel studs will not burn as wood studs would. Steel studs are currently more expensive than wood studs, but it is expected that as lumber prices continue to rise, the costs will eventually equalize.

Steel studs are prepunched to permit quick installation of piping, wiring, and bracing members.

Steel studs have become popular in residential, commercial, and industrial construction. Steel studs may be spaced 16" or 24" OC. On a nonbearing wall, spacing is determined by drywall type and thickness. Unlike wood (which has defects), steel studs are consistent in material composition.

There are three types of steel studs. The first is used for nonbearing walls that have facings to accept drywall. The second will accept lath and plaster on both interior and exterior walls. The third type is a wide-flange steel stud, which is used for both loadbearing and nonbearing walls.

A wide variety of accessories are available for steel studs. There are tracks for floors and ceilings to which the studs are fastened. Tracks are also available for sills, fascia, and joint-end enclosures. Other accessories include channels, angles, and clips. For residential construction, steel trusses are also available.

10.1.0 Fabrication

For layout, steel studs are marked to the centerline rather than the edge. The open side of the stud should always face the beginning point of the layout. The bottom channel is fastened to the concrete floor with small powder-actuated fasteners (*Figure 30*). Note that the T-channel is held back slightly to allow room for the drywall to slide between the two channel sections. A metal self-tapping screw is used to fasten the studs to the track (*Figure 31*). The electric screwgun takes the place of the hammer. The studs may also be welded instead of screwed.

When constructing a rough opening, two studs are put back to back and screwed or welded together. The stud that will act as the trimmer stud will be cut to the height specified to receive the header (*Figure 32*). A section of floor track can be used for the bottom part of the header, with short pieces of studs put in place over the header and secured in place. Blocking may be required to fasten millwork.

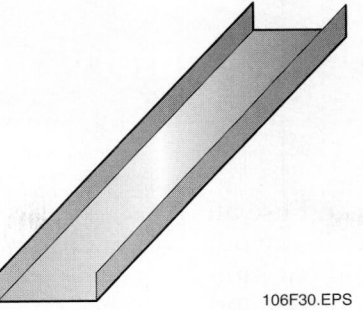

106F30.EPS

Figure 30 ◆ Steel stud channel.

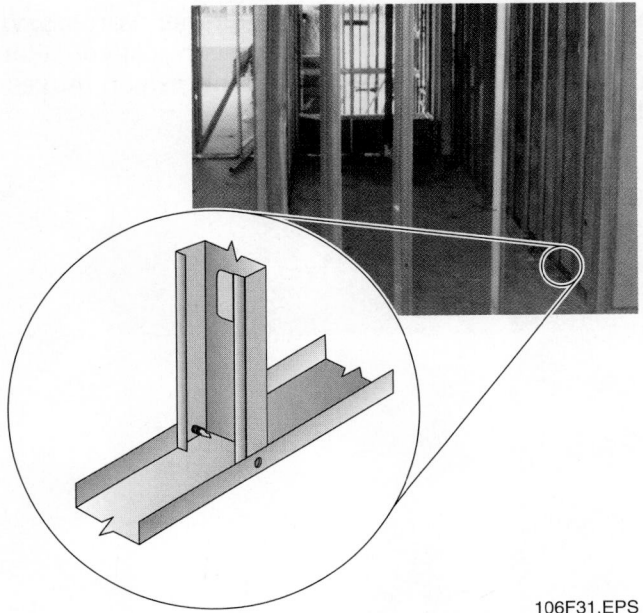

106F31.EPS

Figure 31 ◆ Stud wall section.

106F32.EPS

Figure 32 ◆ Window opening framed with steel studs.

Summary

This module covered how to identify and use all the wall and ceiling components; how to lay out walls and ceilings; how to measure and cut lumber; how to assemble, erect, plumb, brace, and sheath walls; how to lay out and build doors, windows, and other wall openings; how to install ceilings; how to estimate materials; how to work with masonry walls; how to construct walls with steel studs; and much more.

The most important skill in wall and ceiling framing is accuracy. Walls must be straight, plumb, and square. Precise layout and measuring of studs and headers is also critical.

Even a very small error can cause big problems. For example, over a span of 20', an error of only $\frac{1}{16}$" at one end of the span could become $1\frac{1}{4}$" at the other end. The closer you get to the end of the wall, the more patching and fitting you will have to do to get the sheathing, windows, and exterior doors to fit.

In other words, saving a little time at the beginning can cost you a lot of time at the end. Remember this as you move forward in your training.

Notes

1. A short framing member that fills the space between the rough sill and the soleplate is a _____.
   a. spacer
   b. cripple stud
   c. trimmer stud
   d. top plate

2. The framing member that forms the side of a rough opening for a window or door is the _____.
   a. trimmer stud
   b. header
   c. soleplate
   d. cripple

3. The straightest, least defective studs are normally used for _____.
   a. window and door trimmers
   b. common studs
   c. corners
   d. partition Ts

4. The framing member that distributes the weight of the structure around a door or window opening is the _____.
   a. trimmer stud
   b. cripple stud
   c. top plate
   d. header

5. The width of a header is equal to the _____.
   a. width of the rough opening
   b. width of the opening plus the width of two trimmer studs plus the width of two common studs
   c. width of the opening plus the width of two trimmer studs
   d. width of the trimmer plus the width of two cripple studs

6. Fabricated headers normally use _____ as spacers.
   a. furring strips
   b. ½" plywood
   c. ¾" plywood
   d. 2 × 4 blocks

7. The length of a cripple stud above a window opening equals the length of a common stud less the height of the trimmer, combined with the thickness of the _____.
   a. soleplate
   b. rough sill
   c. top plate
   d. header

8. When calculating the length of a common stud for a wood frame floor, the thickness of the _____ is not considered.
   a. soleplate
   b. subfloor
   c. underlayment
   d. double top plate

9. Pieces of 2 × 4 placed horizontally between each pair of studs are used to _____.
   a. brace the frame
   b. retard the spread of fire
   c. provide nailers for siding
   d. provide a place for carpenters to stand when installing ceiling joists

10. A long 2 × 4 with a ¾" standoff block nailed to each end is used as a _____.
    a. wall brace
    b. straightedge
    c. gauge block
    d. trimmer stud

11. When the sheathing acts as the finish surface, you should use _____ sheets.
    a. ⅜"
    b. ½"
    c. ¾"
    d. 1"

12. When you are installing APA-rated sheathing material, you should space the nails _____ apart at the panel edges.
    a. 6"
    b. 12"
    c. 16"
    d. 24"

13. A ribband is a(n) _____.
    a. brace placed between studs
    b. L-shaped joist brace
    c. type of joist hanger
    d. 1 × 4 strip used to prevent joists from twisting or bowing

14. The purpose of prepunched holes in steel studs is to _____.
    a. make them lighter
    b. save material
    c. provide runs for piping and wiring
    d. make it easier to attach drywall

15. Steel studs are attached to the bottom channel using _____.
    a. wood screws
    b. powder-actuated fasteners
    c. construction adhesive
    d. self-tapping metal screws

Trade Terms Quiz

1. The part of a door frame that comes in contact with the door is the _____.

2. A roof with four sides running toward the center is a(n) _____.

3. A roof with two slopes that meet at a center ridge is a(n) _____.

4. A(n) _____ is an L-shaped support member used to strengthen and align the ceiling joists.

5. _____ is used to provide filler and support between framing members.

6. The framing at a wall corner is stiffened using a(n) _____.

7. A(n) _____ is a short framing member that fills the space between longer framing members.

8. A(n) _____ is a framing member that forms the sides of the rough opening for a door.

9. The top support member of a rough door or window opening is known as the _____.

10. The bottom framing member of a rough door or window opening is known as the _____.

11. A(n) _____ provides a nailing surface when a wall is attached to masonry.

12. The roof trusses or rafters are supported by the _____.

13. A(n) _____ is installed between ceiling joists to prevent twisting or bowing.

14. The installation of sheathing, doors, and windows makes up the _____ stage of building construction.

Trade Terms

Blocking	Furring strip	Jamb	Top plate
Cripple stud	Gable roof	Ribband	Trimmer stud
Double top plate	Header	Sill	
Drying-in	Hip roof	Strongback	

Barry Caldwell

Silver Medal Winner Associated Builders and Contractors National Championships

Barry Caldwell won the silver medal in carpentry at the 2005 ABC Craft Olympics. He remembers the tough competition and the testing areas. In a mock building, contestants had to build a concrete form, construct sub-flooring, a metal wall, and a wooden wall. The competition was timed over six hours, and Barry finished all task areas with just five minutes remaining. The judge's criteria were safety, workmanship, job-site etiquette, cleanliness, efficient use of materials, and of course—accuracy. There was an eighth-of-an-inch tolerance, Barry says.

How did you become interested in carpentry?
I've been working around the trade basically my whole life. I started helping my father (who's also a carpenter) on job sites when I was five years old. After high school, I went to a vocational school for two years, and then I entered an apprenticeship program with Cleveland Construction. I've been there for the past three years.

What are some of the things you do in your job?
I work for the Interior Division of Cleveland Construction, which specializes in interior and exterior framing, and drywall and ceiling tile installation. The Interior Division is where people start out. As they progress, they can move on to other areas in the company, or they can stay with that division. Eventually, I'd like to move into a management position with the company.

What do you like most about carpentry?
I like it because there's a real sense of accomplishment. Fifty years later, you can go back and see what you've done. I love being able to go down the road and say "I did that" or "I helped with that." I love that new technologies, new tools, are always coming out. For example, right now there's a move toward using metal studs in framing instead of wood studs. I also like working outside. And there's always something new—a new project, a new site.

You're still in your apprenticeship now. What separates a good apprentice from the rest?
Successful apprentices want to stay busy—they want to work. A good apprentice is motivated. The trade is not for everybody. If you don't want to do it, then don't do it. The best apprentices are those people who want to do the work. I love my job. I have fun at work every day.

Trade Terms
Introduced in This Module

Blocking: A wood block used as a filler piece and a support between framing members.

Cripple stud: In wall framing, a short framing stud that fills the space between a header and a top plate or between the sill and the soleplate.

Double top plate: A plate made of two members to provide better stiffening of a wall. It is also used for connecting splices, corners, and partitions that are at right angles (perpendicular) to the wall.

Drying-in: Applying sheathing, windows, and exterior doors to a framed building.

Furring strip: Narrow wood strips nailed to a wall or ceiling as a nailing base for finish material.

Gable roof: A roof with two slopes that meet at a center ridge.

Header: A horizontal structural member that supports the load over a window or door.

Hip roof: A roof with four sides or slopes running toward the center.

Jamb: The top (head jamb) and side members of a door or window frame that come into contact with the door or window.

Ribband: A 1 × 4 nailed to the ceiling joists at the center of the span to prevent twisting and bowing of the joists.

Sill: The lower framing member attached to the top of the lower cripple studs to form the base of a rough opening for a window.

Strongback: An L-shaped arrangement of lumber used to support ceiling joists and keep them in alignment.

Top plate: The upper horizontal framing member of a wall used to carry the roof trusses or rafters.

Trimmer stud: A vertical framing member that forms the sides of rough openings for doors and windows. It provides stiffening for the frame and supports the weight of the header.

This module is intended to present thorough resources for task training. The following reference works are suggested for further study. These are optional materials for continuing education rather than for task training.

Builder's Essentials: Advanced Framing Methods. Kingston, MA: R.S. Means Company.

Builder's Essentials: Framing & Rough Carpentry. Kingston, MA: R.S. Means Company.

Framing Floors, Walls and Ceilings. Newton, CT: Taunton Press.

Framing Walls (DVD). Newton, CT: Taunton Press.

Graphic Guide to Frame Construction. Newton, CT: Taunton Press.

Precision Framing for Pros by Pros. Newton, CT: Taunton Press.

The Proper Construction and Inspection of Ceiling Joists and Rafters (DVD and workbook). Falls Church, VA: International Code Council.

Residential Steel Framing Handbook. New York, NY: McGraw-Hill.

International Code Council. A membership organization dedicated to building safety and fire prevention through development of building codes. www.iccsafe.org

National Association of Home Builders. A trade association whose mission is to enhance the climate for housing and the building industry. www.nahb.org

NCCER makes every effort to keep these textbooks up-to-date and free of technical errors. We appreciate your help in this process. If you have an idea for improving this textbook, or if you find an error, a typographical mistake, or an inaccuracy in NCCER's Contren® textbooks, please write us, using this form or a photocopy. Be sure to include the exact module number, page number, a detailed description, and the correction, if applicable. Your input will be brought to the attention of the Technical Review Committee. Thank you for your assistance.

Instructors – If you found that additional materials were necessary in order to teach this module effectively, please let us know so that we may include them in the Equipment/Materials list in the Annotated Instructor's Guide.

Write: Product Development and Revision
National Center for Construction Education and Research
3600 NW 43rd St, Bldg G, Gainesville, FL 32606

Fax: 352-334-0932

E-mail: curriculum@nccer.org

Craft _____   Module Name _____

Copyright Date _____   Module Number _____   Page Number(s) _____

Description _____
_____
_____
_____
_____

(Optional) Correction _____
_____
_____

(Optional) Your Name and Address _____
_____
_____

A Carpentry contestant in the SkillsUSA 2005 National Championships inspects his miniature building. The miniature building consisted of a wooden frame, a metal frame, a stair stringer, and gypsum wallboard. The wooden frame component tested a contestant's ability to lay out and construct studs, jamb studs, sills, headers, joists, rafters, a double top plate, and a ridge beam. Carpentry contestants had 6½ hours to complete the project, which often left little time to review their work.

68108-09
Roof Framing

Topics to be presented in this module include:

1.0.0   Introduction 8.2
2.0.0   Types of Roofs 8.2
3.0.0   Basic Roof Layout 8.3
4.0.0   Installing Sheathing 8.15
5.0.0   Rafter Layout Using a Speed Square 8.18
6.0.0   Truss Construction 8.20
7.0.0   Determining Quantities of Materials 8.25
8.0.0   Dormers 8.26
9.0.0   Plank-and-Beam Framing 8.27
10.0.0   Metal Roof Framing 8.28

Overview

Before the invention of truss systems, all the rafters of a framed roof had to be measured and cut by carpenters. This was time-consuming work involving precise calculations and measurement, and often resulted in a lot of rework.

Today, roof frame members are generally integrated into truss systems that are designed and fabricated off site by companies that specialize in manufacturing trusses. However, there are occasions when a carpenter will have to stick-frame all or part of a roof. In such cases, it is essential to be able to perform the length and angle calculations needed to correctly size and cut each type of rafter.

Objectives

When you have completed this module, you will be able to do the following:

1. Understand the terms associated with roof framing.
2. Identify the roof framing members used in gable and hip roofs.
3. Identify the methods used to calculate the length of a rafter.
4. Identify the various types of trusses used in roof framing.
5. Use a rafter framing square, speed square, and calculator in laying out a roof.
6. Identify various types of sheathing used in roof construction.
7. Frame a gable roof with vent openings.
8. Frame a roof opening.
9. Erect a gable roof using trusses.
10. Estimate the materials used in framing and sheathing a roof.

Trade Terms

Barge rafter
False fascia
Gable

Lookout
Purlin

Required Trainee Materials

1. Pencil and paper
2. Appropriate personal protective equipment
3. Framing square

Prerequisites

Before you begin this module, it is recommended that you successfully complete *Core Curriculum*, and *Construction Technology*, Modules 68101-09 through 68107-09.

This course map shows all of the modules in *Construction Technology*. The suggested training order begins at the bottom and proceeds up. Skill levels increase as you advance on the course map. The local Training Program Sponsor may adjust the training order.

68117-09
Copper Pipe and Fittings

68116-09
Plastic Pipe and Fittings

68115-09
Introduction to Drain, Waste, and Vent (DWV) Systems

68114-09
Introduction to HVAC

68113-09
Residential Electrical Services

68112-09
Electrical Safety

68111-09
Basic Stair Layout

68110-09
Exterior Finishing

68109-09
Roofing Applications

68108-09
Roof Framing

68107-09
Wall and Ceiling Framing

68106-09
Floor Systems

68105-09
Masonry Units and Installation Techniques

68104-09
Introduction to Masonry

68103-09
Handling and Placing Concrete

68102-09
Introduction to Concrete, Reinforcing Materials, and Forms

68101-09
Site Layout One: Distance Measurement and Leveling

CORE CURRICULUM:
Introductory Craft Skills

CONSTRUCTION TECHNOLOGY

103CMAP.EPS

1.0.0 ◆ INTRODUCTION

Roof framing is the most demanding of the framing tasks. Floor and wall framing generally involves working with straight lines. Residential roofs are usually sloped in order to shed water from rain or melting snow. In areas where there is heavy snowfall, the roof must be constructed to bear extra weight. Because a roof is sloped, laying out a roof involves working with precise angles in addition to straight lines.

In this module, you will learn about the different types of roofs used in residential construction. You will also learn how to lay out and frame a roof.

2.0.0 ◆ TYPES OF ROOFS

The most common types of roofs used in residential construction are shown in *Figure 1* and described below.

- *Gable roof* – A **gable** roof has two slopes that meet at the center (ridge) to form a gable at each end of the building. It is the most common type of roof because it is simple, economical, and can be used on any type of structure.
- *Hip roof* – A hip roof has four sides or slopes running toward the center of the building.

Rafters at the corners extend diagonally to meet at the ridge. Additional rafters are framed into these rafters.

- *Mansard roof* – The mansard roof has four sloping sides, each of which has a double slope. As compared with a gable roof, this design provides more available space in the upper level of the building. The upper slope is typically not visible from the ground.
- *Gable and valley roof* – This roof consists of two intersecting gable roofs. The part where the two roofs meet is called a valley.
- *Hip and valley roof* – This roof consists of two intersecting hip roofs.
- *Gambrel roof* – The gambrel roof is a variation on the gable roof in which each side has a break, usually near the ridge. The gambrel roof provides more available space in the upper level.
- *Shed roof* – Also known as a lean-to roof, the shed roof is a flat, sloped construction. It is common on high-ceiling contemporary construction, and is often used on additions.

There are two basic roof framing systems. In stick-built framing, ceiling joists and rafters are laid out and cut by carpenters on site and the frame is constructed one stick at a time.

In truss-built construction, the roof framework is prefabricated off site. The truss contains both the rafters and the ceiling joist.

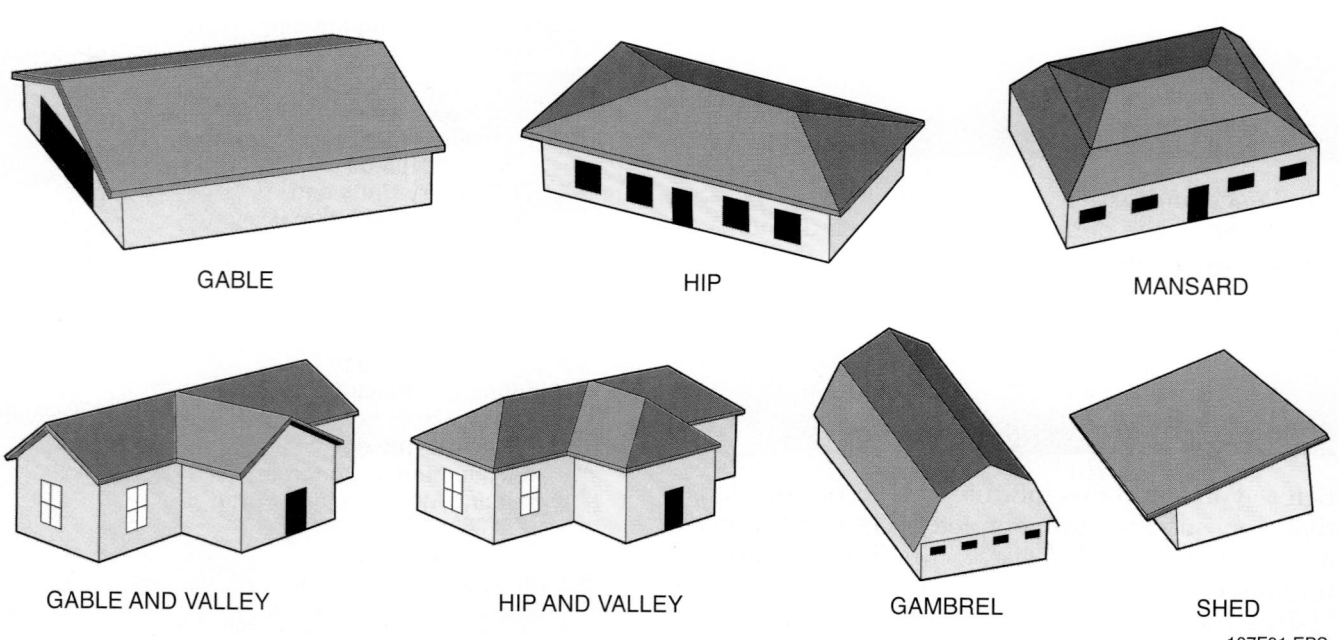

GABLE   HIP   MANSARD

GABLE AND VALLEY   HIP AND VALLEY   GAMBREL   SHED

107F01.EPS

Figure 1 ◆ Types of roofs.

Mansard Roof

Mansard roofs are popular in the design of many fast-food restaurants and small office buildings.

107SA01.EPS

3.0.0 ◆ BASIC ROOF LAYOUT

Rafters and ceiling joists provide the framework for all roofs. The main components of a roof are shown in *Figure 2* and described below.

- *Ridge (ridgeboard)* – The highest horizontal roof member. It helps to align the rafters and tie them together at the upper end. The ridgeboard is one size larger than the rafters.
- *Common rafter* – A structural member that extends from the top plate to the ridge in a direction perpendicular to the wall plate and ridge. Rafters often extend beyond the roof plate to form the overhang (eaves) that protect the side of the building.
- *Hip rafter* – A roof member that extends diagonally from the outside corner of the plate to the ridge.
- *Valley rafter* – A roof member that extends from the inside corner of the top plate to the ridge along the lines where two roofs intersect.
- *Jack rafter* – A roof member that does not extend the entire distance from the ridge to the top plate of a wall. The hip jack and valley jack are shown in *Figure 2*. A rafter fitted between a hip rafter and a valley rafter is called a cripple jack. It touches neither the ridge nor the plate.
- *Plate* – The wall framing member that rests on top of the wall studs. It is sometimes called the rafter plate because the rafters rest on it. It is also referred to as the top plate.

As you can see in *Figure 3*, on any pitched roof, rafters rise at an angle to the ridgeboard. Therefore, the length of the rafter is greater than the horizontal distance from the plate to the ridge. In order to calculate the correct rafter length, the carpenter must factor in the slope of the roof. Here are some additional terms you will need to know in order to lay out rafters:

- *Span* – The horizontal distance from the outside of one exterior wall to the outside of the other exterior wall.
- *Run* – The horizontal distance from the outside of the top plate to the center line of the ridgeboard (usually equal to half of the span).
- *Rise* – The total height of the rafter from the top plate to the ridge. This is stated in inches per foot of run.
- *Pitch* – The angle or degree of slope of the roof in relation to the span. Pitch is expressed as a fraction; e.g., if the total rise is 6' and the span is 24', the pitch would be ¼ (6 over 24).
- *Slope* – The inclination of the roof surface expressed as the relationship of rise to run. It is stated as a unit of rise to so many horizontal units; e.g., a roof that has a rise of 5" for each foot of run is said to have a 5 in 12 slope (*Figure 3*). The roof slope is sometimes referred to as the roof cut.

The first step in determining the correct length of a rafter is to find the unit rise, which is usually

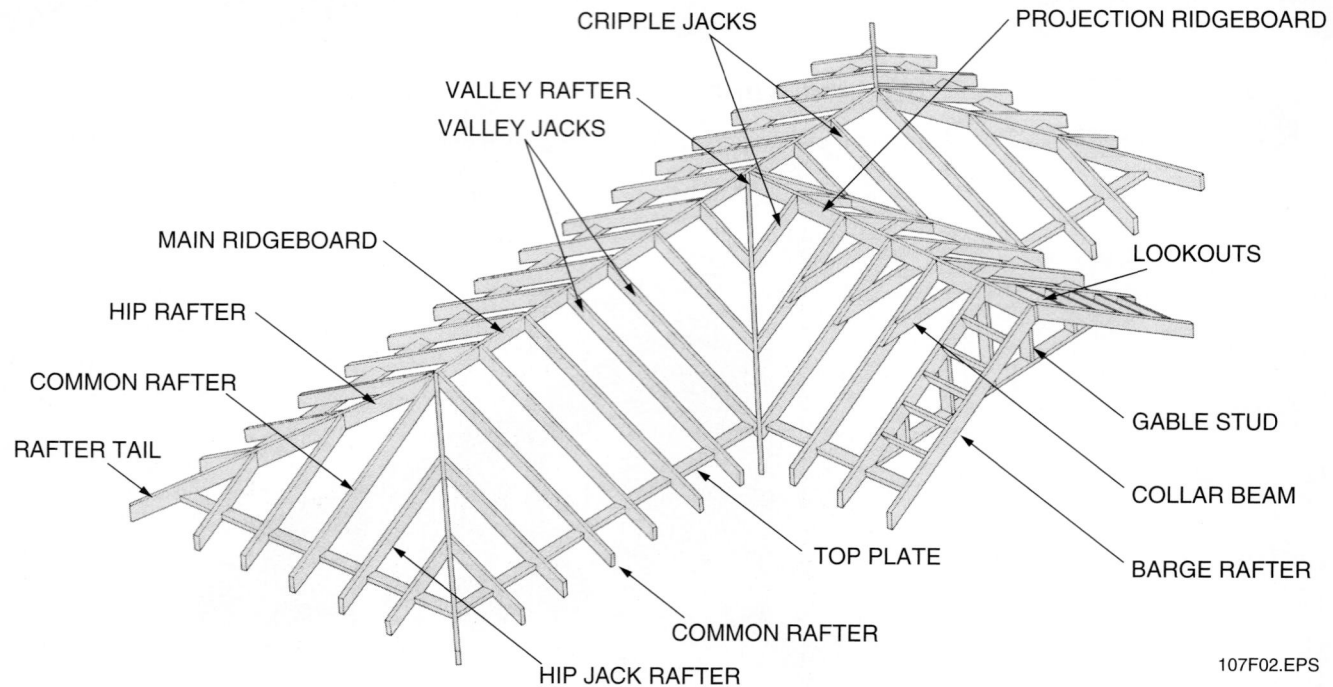

Figure 2 ◆ Roof framing members.

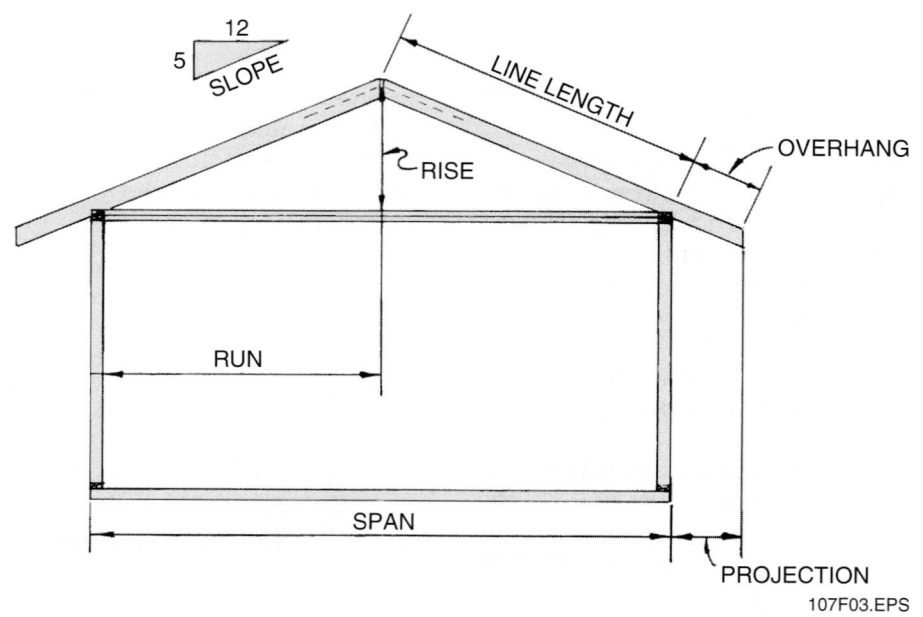

Figure 3 ◆ Roof layout factors.

shown on the building's elevation drawing. The unit rise is the number of inches the rafter rises vertically for each foot of run. The greater the rise per foot of run, the greater the slope of the roof.

There are several ways to calculate the length of a rafter. It can be done with a framing (rafter) square or speed square, or it can be done using a calculator. We will show the framing square method now, then discuss the speed square.

3.1.0 Rafter Framing Square

The rafter framing square is a special carpenter's square that is calibrated to show the length per foot of run for each type of rafter (*Figure 4*). Note that the tongue is the short (16") section of the square and the blade (or body) is the long (24") section. The corner is known as the heel. Rafter tables are normally provided on the back of the

Pitch and Slope

Carpenters may use the terms pitch and slope interchangeably on the job site, but the two terms actually refer to two different concepts. Slope is the amount of rise per foot of run and is always referred to as a number in 12. For example, a roof that rises 6" for every foot of run has a 6 in 12 slope (the 12 simply refers to the number of inches in a foot). Pitch, on the other hand, is the ratio of rise to the span of the roof and is expressed as a fraction. For example, a roof that rises 8' over a 32' span is said to have a pitch of ¼ (⁸⁄₃₂ = ¼).

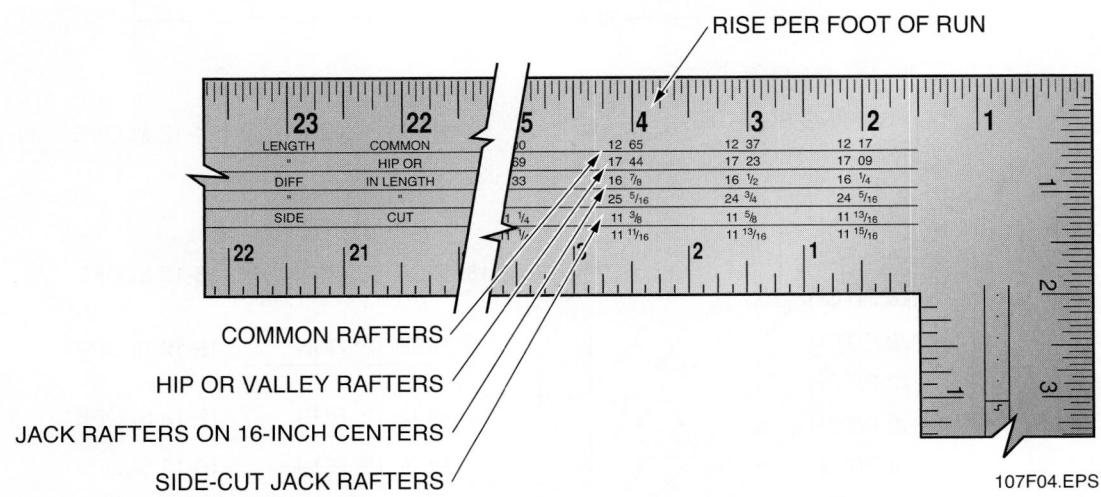

RISE PER FOOT OF RUN

COMMON RAFTERS
HIP OR VALLEY RAFTERS
JACK RAFTERS ON 16-INCH CENTERS
SIDE-CUT JACK RAFTERS

107F04.EPS

Figure 4 ◆ Rafter tables on a framing square.

square. The rafter tables usually give the rafter dimensions in length per foot of run, but some give length per given run.

The framing square is used to determine the rafter length and to measure and mark the cuts that must be made in the rafter (*Figure 5*). As you can see, you can relate the pitch and slope to the rise per foot of run. The rise per foot of run is always the same for a given pitch or slope. For example, a pitch of ½, which is the same as a 12 in 12 slope, equals 12" of rise per foot of run.

3.2.0 Basic Rafter Layout

Laying out the framing for a roof involves four tasks:

• Marking off the rafter locations on the top plate
• Determining the length of each rafter
• Making the plumb cuts at the ridge end and tail end of each rafter
• Making the bird's mouth cut in each rafter

Rafters must be laid out and cut so that the ridge end will fit squarely on the ridge and the tail end

Cutting Sequence

Some carpenters prefer to make the tail end plumb cut while laying out the rafter. Others prefer to tail cut all the rafters at once after the rafters are installed. One method isn't necessarily better than the other and is simply a matter of preference.

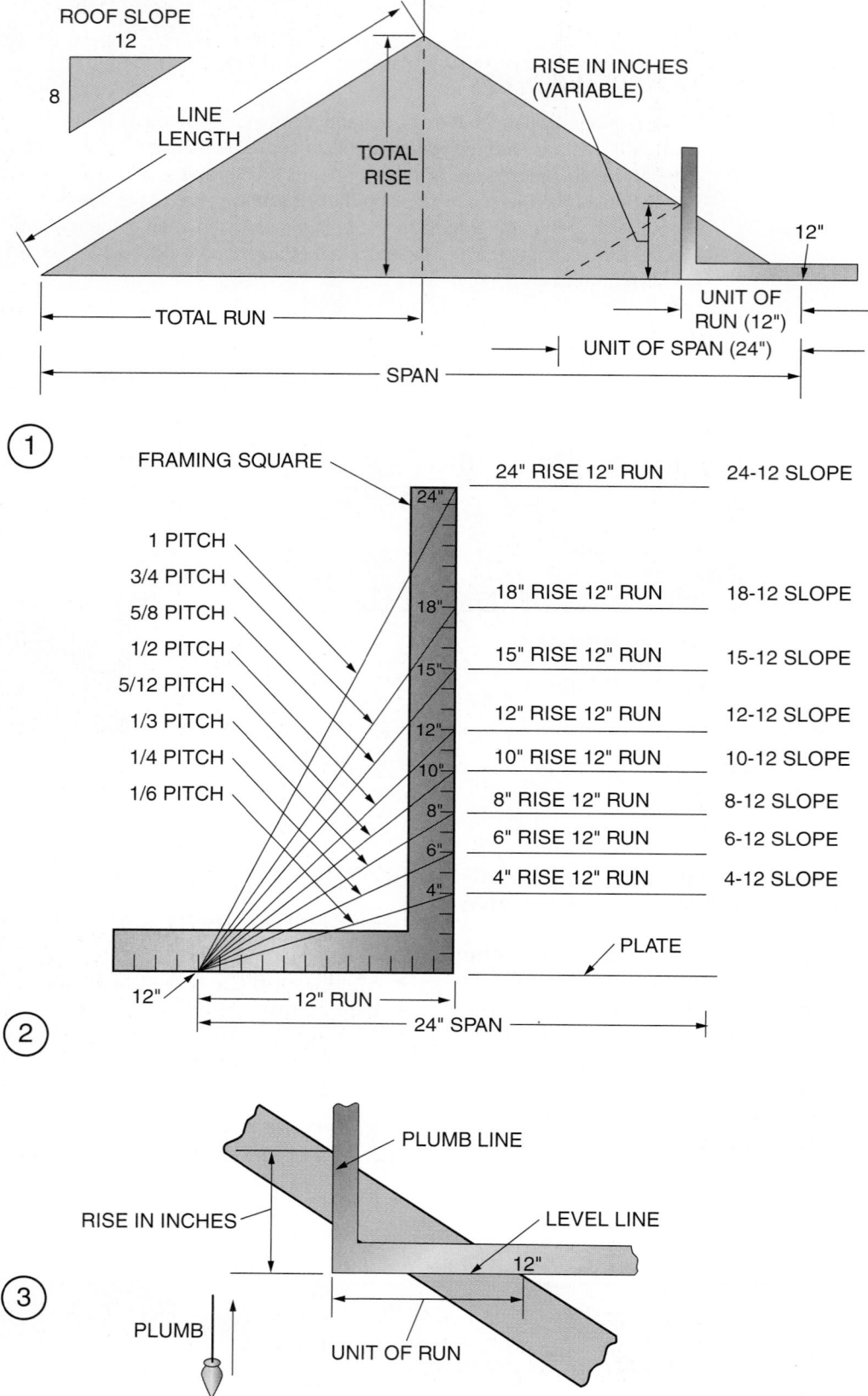

ROOF SLOPE
12
8

LINE LENGTH

TOTAL RISE

RISE IN INCHES (VARIABLE)

12"

TOTAL RUN

UNIT OF RUN (12")

UNIT OF SPAN (24")

SPAN

①

FRAMING SQUARE

24" RISE 12" RUN   24-12 SLOPE

1 PITCH
3/4 PITCH
5/8 PITCH
1/2 PITCH
5/12 PITCH
1/3 PITCH
1/4 PITCH
1/6 PITCH

24"

18"

15"

12"

10"

8"

6"

4"

18" RISE 12" RUN   18-12 SLOPE

15" RISE 12" RUN   15-12 SLOPE

12" RISE 12" RUN   12-12 SLOPE

10" RISE 12" RUN   10-12 SLOPE

8" RISE 12" RUN   8-12 SLOPE

6" RISE 12" RUN   6-12 SLOPE

4" RISE 12" RUN   4-12 SLOPE

PLATE

12"

12" RUN

24" SPAN

②

PLUMB LINE

LEVEL LINE

RISE IN INCHES

12"

PLUMB

UNIT OF RUN

③

107F05.EPS

Figure 5 ◆ Application of the rafter square.

will present a square surface for the **false fascia** board. In addition, a bird's mouth cut (*Figure 6*) must be made at the correct location and angle for the rafter to rest squarely on the plate.

3.2.1 Laying Out Rafter Locations

The following is a basic procedure for marking the locations of the rafters on the top plate (*Figure 7*) for 24" on center (OC) construction. Keep in mind

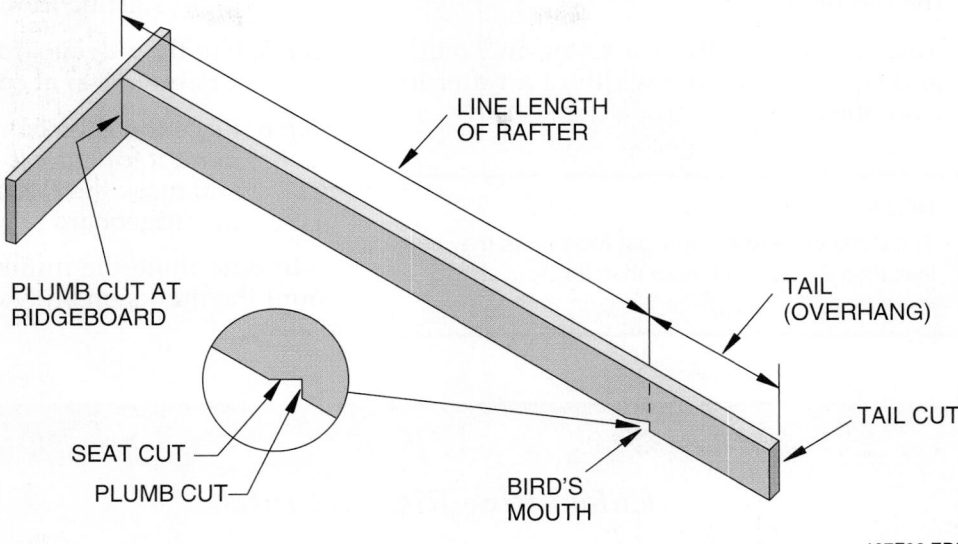

LINE LENGTH
OF RAFTER

PLUMB CUT AT
RIDGEBOARD

TAIL
(OVERHANG)

TAIL CUT

SEAT CUT

PLUMB CUT

BIRD'S
MOUTH

Figure 6 ◆ Parts of a rafter.

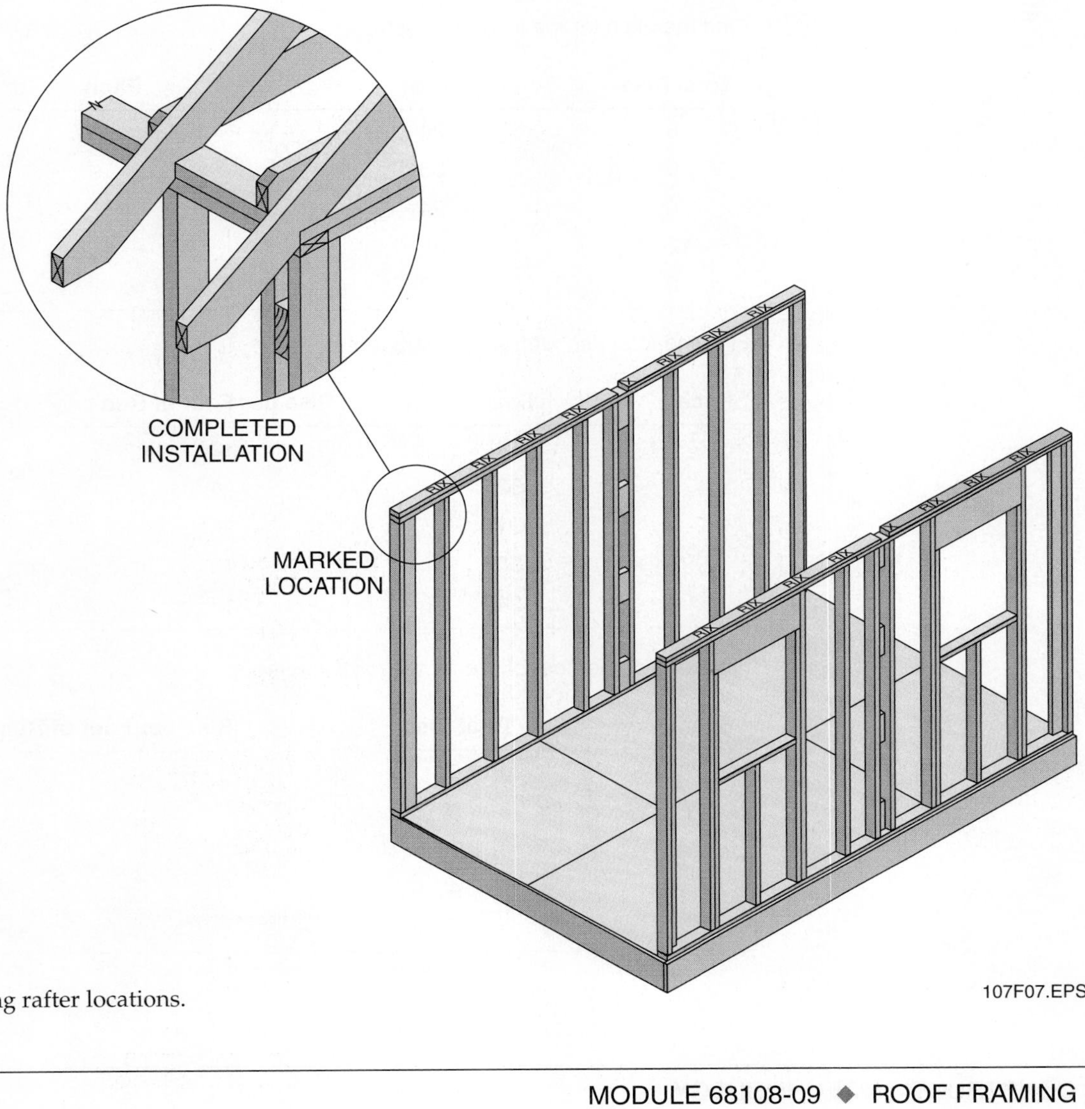

107F06.EPS

COMPLETED
INSTALLATION

MARKED
LOCATION

Figure 7 ◆ Marking rafter locations.

107F07.EPS

that in most cases, the ceiling joists would be in place and the rafter locations would have already been marked.

Step 1 Locate the first rafter flush with one end of the top plate.

Step 2 Measure the width of a rafter and mark and square a line the width of a rafter in from the end.

NOTE
The distance between the last two rafters may be less than 24", but not more than 24".

Step 3 Use a measuring tape to space off and mark the rafter locations every 24" all the way to the end. Square the lines.

Step 4 Repeat the process on the opposite top plate, starting from the same end as before.

Step 5 Cut the ridgeboard to length, allowing for a **barge rafter** at each end, if required.

Step 6 Place the ridgeboard on the top plate and mark it for correct position. Then measure and mark the rafter locations with an R on the ridgeboard.

To determine the number of rafters you need, count the marks on both sides of the ridge or top plate.

Calculating Rise and Pitch

Now that you are familiar with the concepts of pitch, rise per foot of run, and total rise, practice your skills by completing the following example problems.

Find the pitch for the following roofs:

Total Rise	Span	Pitch
8'	24'	
9'	36'	
6'	24'	
12'	36'	
8'	32'	

Find the total rise for the following roofs:

Pitch	Span	Rise per Foot of Run	Total Rise
½	24'	12"	
⅙	32'	4"	
⅓	24'	8"	
⁵⁄₁₂	30'	10"	
¾	24'	18"	

Find the rise per foot of run for the following roofs:

Span	Total Rise	Rise per Foot of Run
18'	6'	
24'	8'	
28'	7'	
16'	4'	
32'	12'	

3.2.2 Determining the Length of a Common Rafter

The following is an overview of the procedure for laying out the rafters and joists of a gable roof.

Here is an easy way of determining the required length of a rafter:

Step 1 Start by measuring the building span, then divide that in half to determine the run.

Step 2 Determine the rise. This can be done in either of the following ways:
- Calculate the total rise by multiplying the span by the pitch (e.g., 40' span × ¼ pitch = 10' rise).
- Look for it on the slope diagram on the roof plan as discussed previously.

Step 3 Divide the total rise (in inches) by the run (in feet) to obtain the rise per foot of run.

Step 4 Look up the required length on the rafter tables on the framing square.

For example, if the roof has a span of 20', the run would be 10'. Then, assuming that the blueprint shows the rise per foot of run to be 8", the correct rafter length would be 14.42" (14⁵⁄₁₂") per foot of run. Since you have 10' of run, the rafter length would be 144⅛" (*Figure 8*). If an overhang is used, the overhang length must be added to the rafter length.

Another way to calculate rafter length is to measure the distance between 8 on the framing square blade and 12 on the tongue, as shown in *Figure 5*.

Yet another method of determining the approximate length of a rafter is the framing square step-off method shown in *Figure 9*. In this procedure,

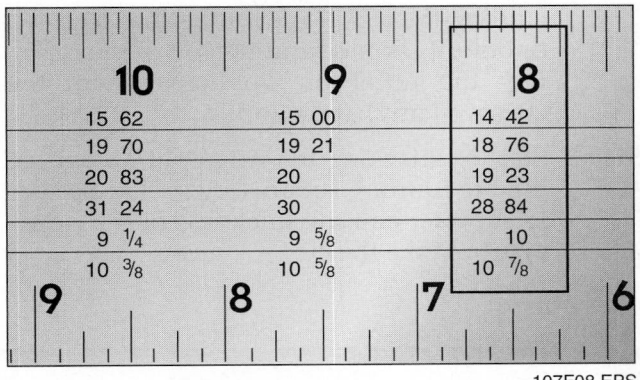

Figure 8 ◆ Determining rafter length with the framing square.

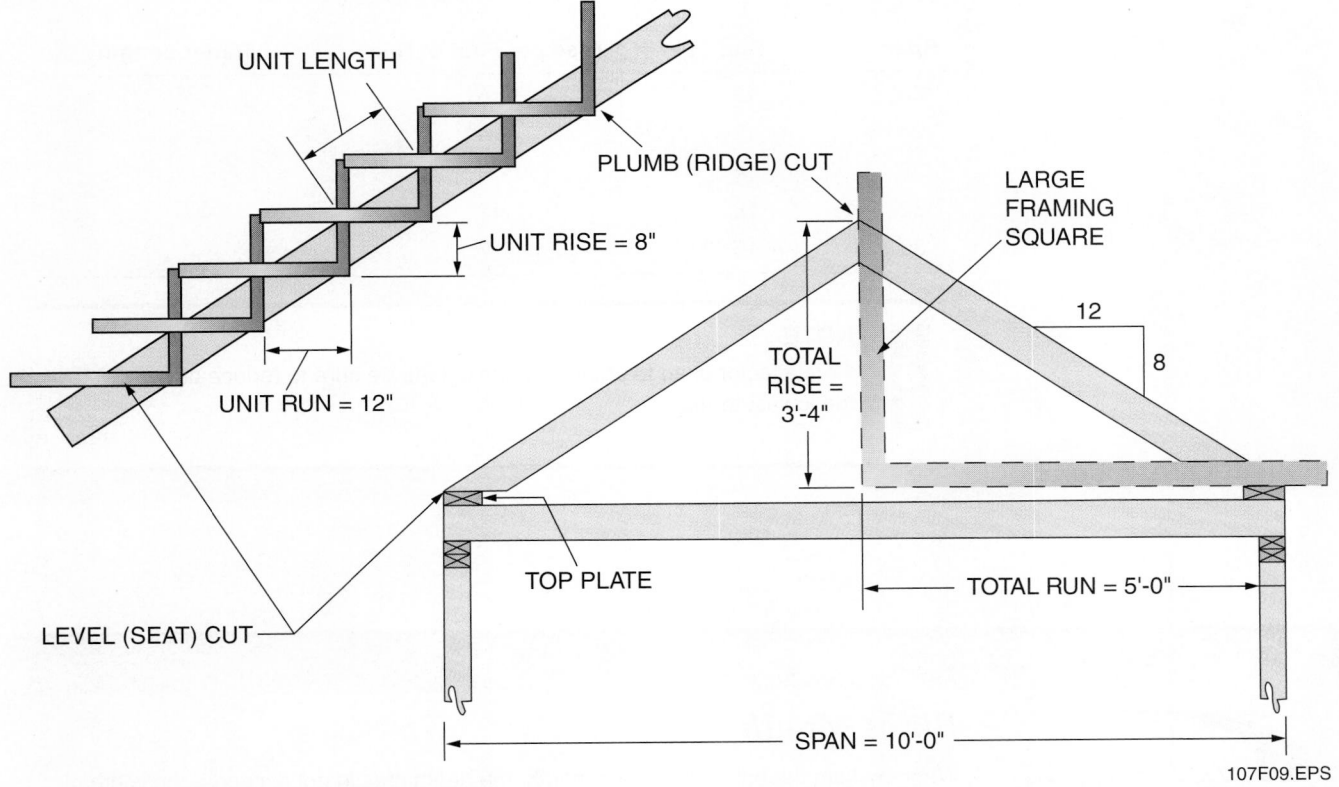

Figure 9 ◆ Framing square step-off method.

the ridge plumb cut is determined, then the framing square is stepped once for every foot of run. The final step marks the plumb cut for the bird's mouth. This is not the preferred method because it is not as precise as other methods.

3.2.3 Laying Out a Common Rafter

The following is an overview of the procedure for laying out and cutting a common rafter:

Step 1 Start with a piece of lumber a little longer than the required length of the rafter, including the tail. If the lumber has a crown or bow, it should be at the top of the rafter. Lay the rafter on sawhorses with the crown (if any) at the top.

Step 2 Start by marking the ridge plumb cut using the framing square (*Figure 10*). Be sure to subtract half the thickness of the ridgeboard. Make the cut.

Step 3 Measure the length of the rafter from the plumb cut mark to the end (excluding the tail) and mark another plumb cut for the bird's mouth. Reposition the framing square and mark the bird's mouth seat.

Step 4 Make the end plumb cut, then cut out the bird's mouth. Cut the bird's mouth partway with a circular saw, then use a hand saw to finish the cuts.

Step 5 Use the first rafter as a template for marking the remaining rafters. As the rafters are cut, stand them against the building at the joist locations.

Calculating Common Rafter Lengths

Now that you are familiar with the method of arriving at the length of a rafter using the square, find the lengths of common rafters for the following spans:

Span	Run	Rise per Foot of Run	Rafter Length
26'	13'	6"	
24'	12'	4"	
28'	14'	8"	
32'	16'	12"	
30'	15'	7"	

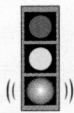

NOTE

12 is a factor used to obtain a value in feet. Be sure to reduce or convert to the lowest terms.

Bird's Mouth

When making the cut for the bird's mouth, the depth should not exceed ⅓ the width of the rafter.

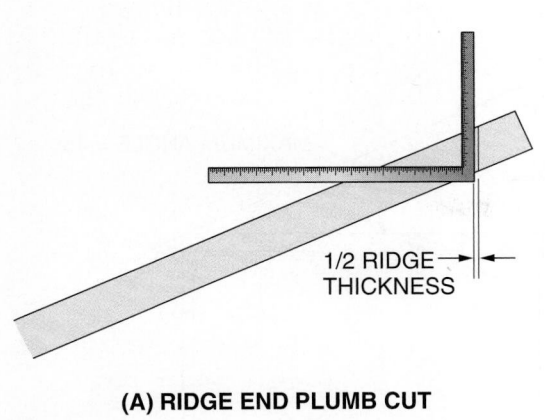

(A) RIDGE END PLUMB CUT

1/2 RIDGE THICKNESS

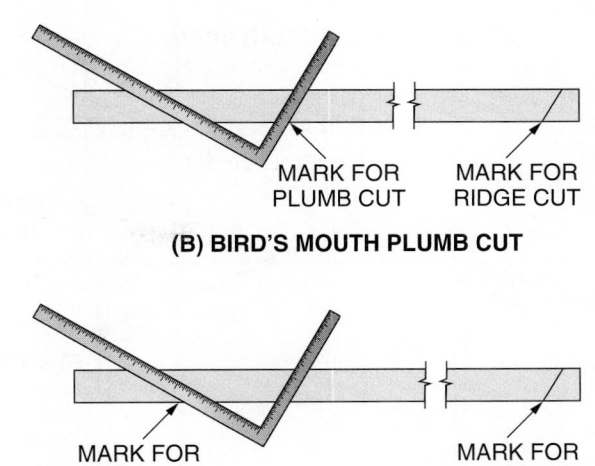

MARK FOR PLUMB CUT   MARK FOR RIDGE CUT

(B) BIRD'S MOUTH PLUMB CUT

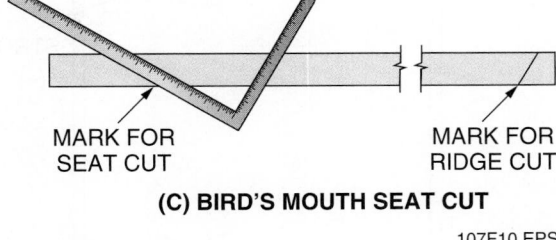

MARK FOR SEAT CUT   MARK FOR RIDGE CUT

(C) BIRD'S MOUTH SEAT CUT

107F10.EPS

Figure 10 ◆ Marking the rafter cuts.

3.3.0 Erecting a Gable Roof

This section contains an overview of the procedure for erecting a gable roof. The layout and construction of hips and valleys is covered in the *Appendix*.

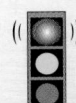

WARNING!

Be sure to follow applicable fall protection procedures.

3.3.1 Installing Rafters

Rafters are installed using the following procedure.

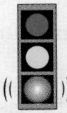

NOTE

It is a good idea to mark a 2 × 4 in advance with the total rise and use it as a guide for the height of the ridgeboard.

Step 1 Start by placing boards over the joists to walk on. Nail a rafter at each end of the ridgeboard, then lift the ridgeboard to a temporary position, secure it, and nail the bird's mouth of each rafter to the joists (*Figure 11*). Nail the rafters in pairs.

Step 2 On the opposite side, start by nailing the bird's mouth to the joist, then toenail the plumb cut into the ridgeboard. Once this is done, use a temporary brace to hold the ridgeboard in place while installing the remaining rafters. Remember to keep the ridgeboard straight and the rafters plumb.

Step 3 Run a line and trim the rafter tails.

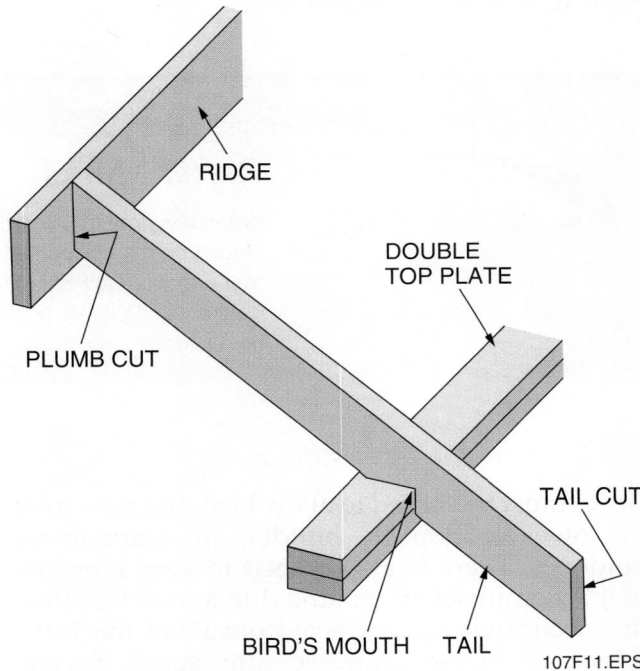

RIDGE

DOUBLE TOP PLATE

PLUMB CUT

TAIL CUT

BIRD'S MOUTH   TAIL

107F11.EPS

Figure 11 ◆ Rafter installation.

If the rafter span is long, additional support will be required. *Figure 12* shows the use of strongbacks, **purlins**, braces, and collar beams (collar ties) for this purpose. Two by six collar ties are installed at every second rafter. Two by four diagonal braces are notched into the purlins. Strongbacks are L-shaped members that run the length of the roof. They are used to straighten and strengthen the ceiling joists.

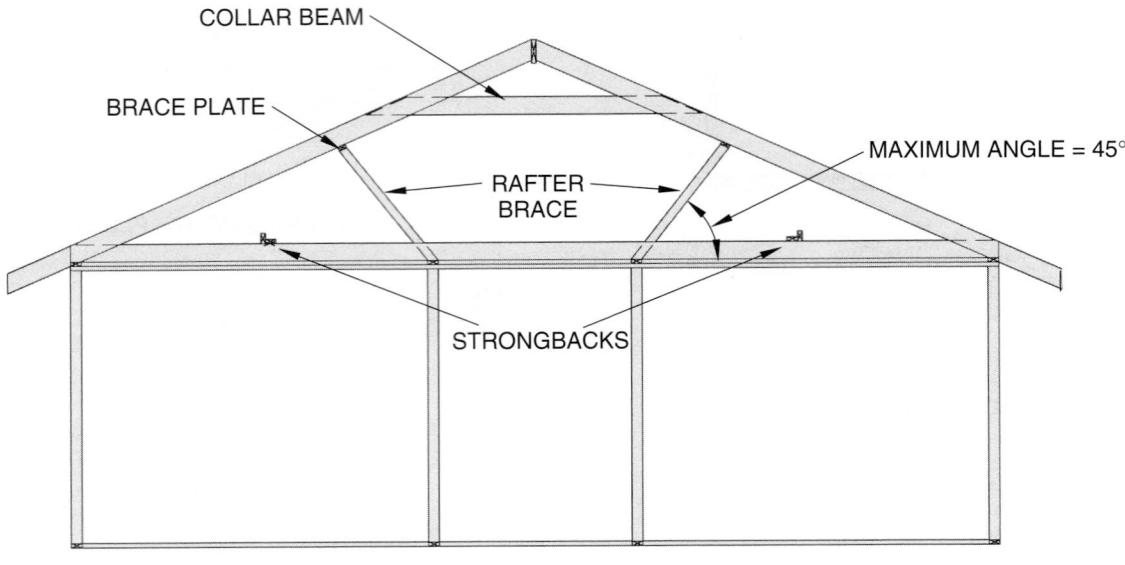

COLLAR BEAM

BRACE PLATE

MAXIMUM ANGLE = 45°

RAFTER
BRACE

STRONGBACKS

107F12.EPS

Figure 12 ◆ Bracing a long roof span.

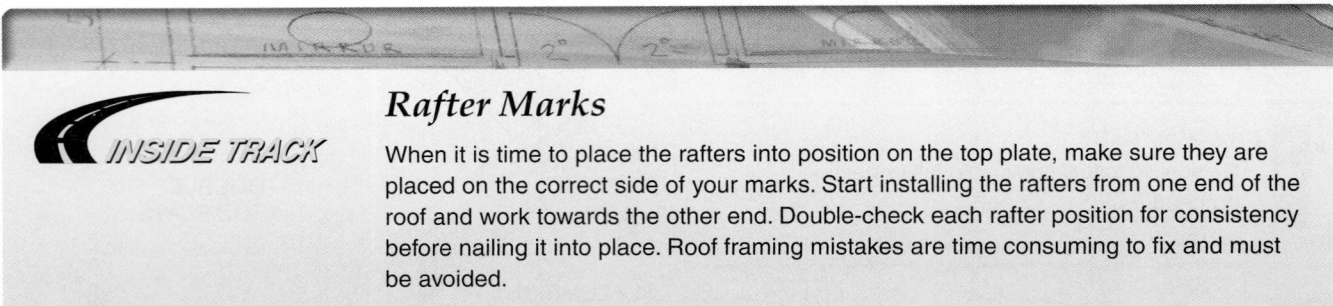

Rafter Marks

INSIDE TRACK

When it is time to place the rafters into position on the top plate, make sure they are placed on the correct side of your marks. Start installing the rafters from one end of the roof and work towards the other end. Double-check each rafter position for consistency before nailing it into place. Roof framing mistakes are time consuming to fix and must be avoided.

3.3.2 Framing the Gable Ends

Attics must be vented to allow heat that rises from the lower floors of the building to escape to the outdoors. There is also a need to vent moisture that accumulates in an attic due to condensation that occurs when rising heat from below meets the cooler air in the unheated attic space. Several methods are used to vent roofs. *Figure 13* shows two types of gable end vents.

Notice that the lengths of the studs decrease as they approach the sides. Each pair of studs must be measured and cut to fit.

Step 1 Start by plumbing down from the center of the rafter to the top plate, then mark and square a line on the top plate at this point.

Step 2 Lay out the header and sill for the vent opening by measuring 8" (in this case) on either side of the plumb line.

Step 3 Mark the stud locations above the wall studs, then stand a stud upright at the first position, plumb it, and mark the diagonal cut for the top of the stud.

Step 4 Measure and mark the next stud in the sequence. The difference in length between the first and second studs is the common difference, which can be applied to all remaining studs.

Step 5 Cut and install the studs as shown in *Figure 14*. Toenail or straight nail the gable stud to the double top plate. The notch should be as deep as the rafter.

Step 6 When studs are in place, cut and install the header and sill for the vent opening. Then lay out, cut, and install cripple studs above and below the vent using the same method as before.

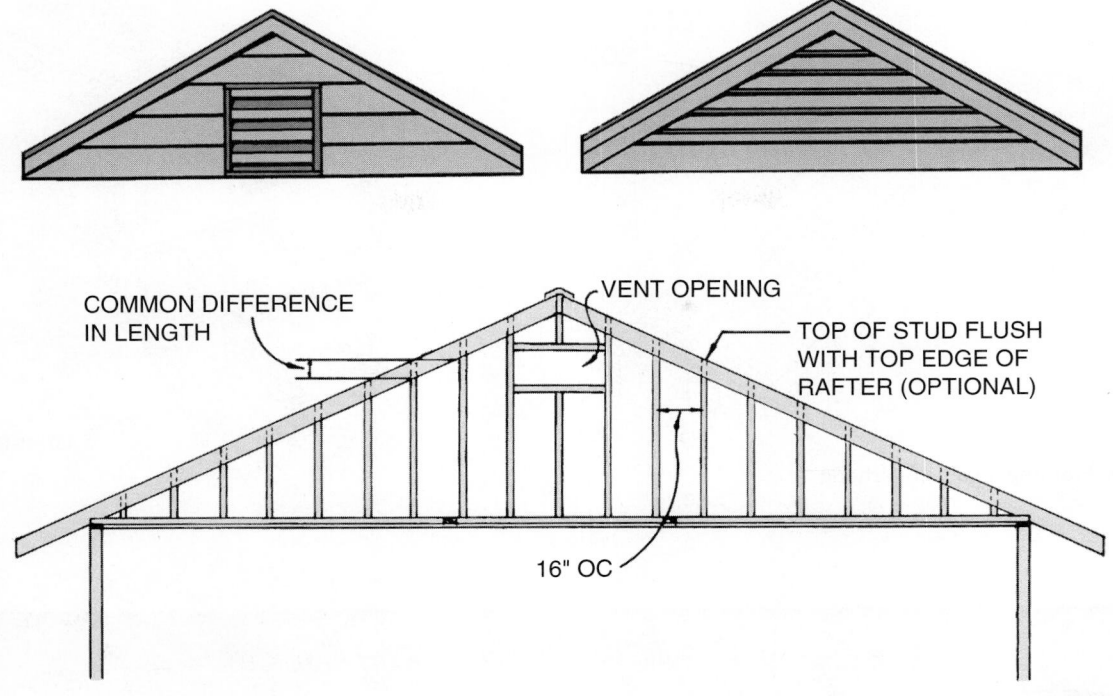

Figure 13 ◆ Gable end vents and frame.

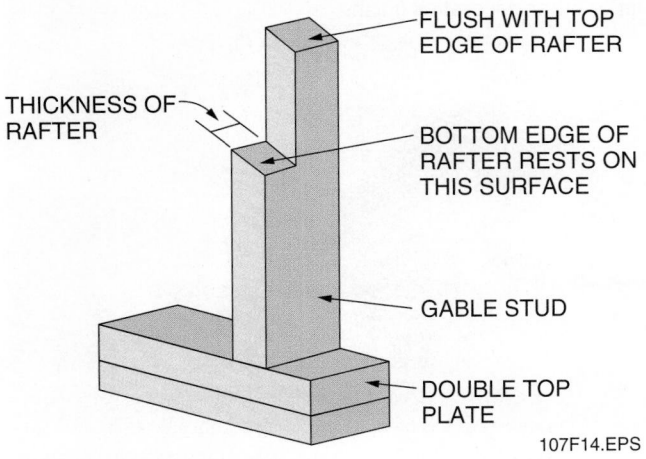

Figure 14 ◆ Gable stud.

3.3.3 Framing a Gable Overhang

If an extended gable overhang (rake) is required, the framing must be done before the roof framing is complete. *Figure 15* shows two methods of framing overhangs. In the view on the right, 2 × 4 **lookouts** are laid into notched rafters. In the view on the left, which can be used for a small overhang, a 2 × 4 barge rafter and short lookouts are used.

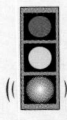

NOTE

Install a brace across the rafters to be cut. It will temporarily hold the framing in place until the headers are installed.

3.3.4 Framing an Opening in the Roof

It is sometimes necessary to make an opening in a roof for a chimney, skylight, or roof window (*Figure 16*). The following is a general procedure for framing such an opening.

Step 1 Lay out the opening on the floor beneath the opening, then use a plumb bob to transfer the layout to the roof. If you are framing a chimney, be sure to leave adequate clearance. If the opening is large, allow for double headers.

Step 2 Cut the rafters per the layout. Install the headers, then install a double trimmer rafter on either side of the opening, as shown in *Figure 16*.

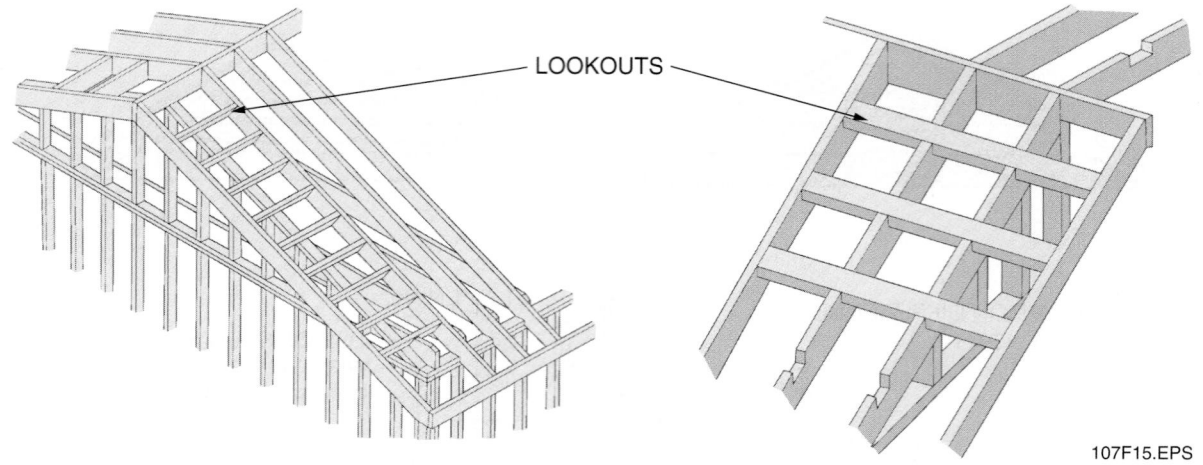

LOOKOUTS

107F15.EPS

Figure 15 ◆ Framing a gable overhang.

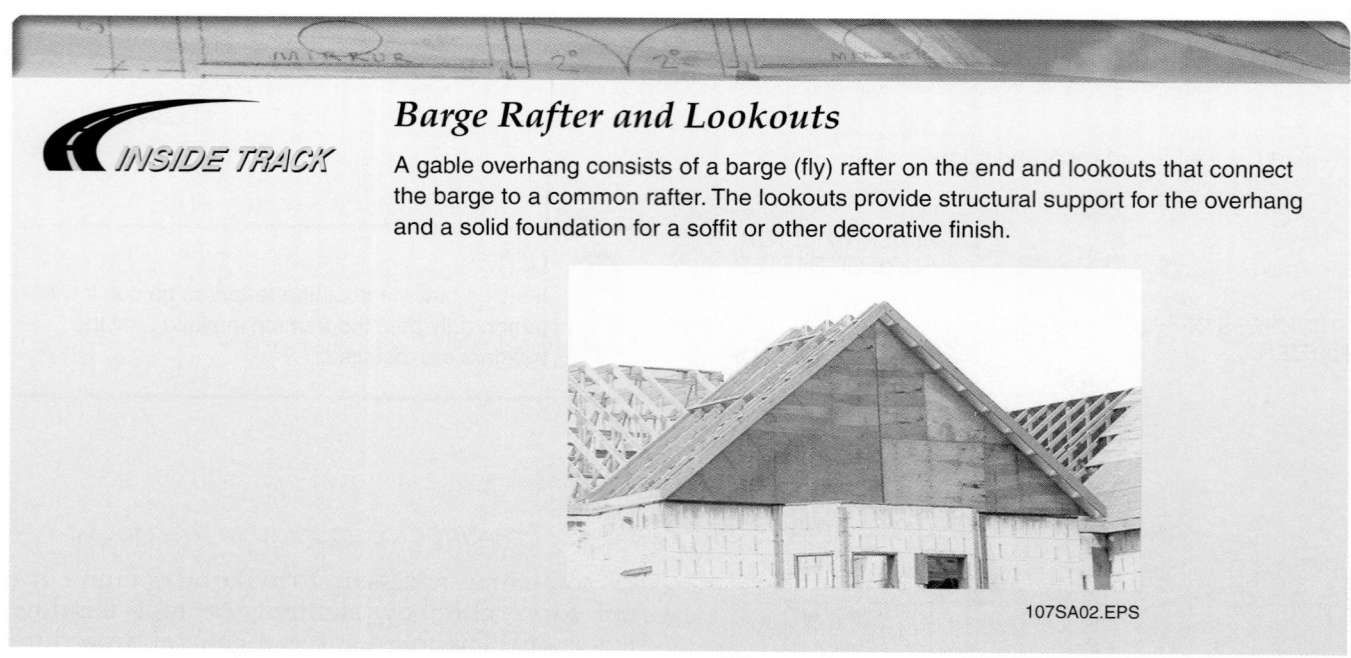

Barge Rafter and Lookouts

A gable overhang consists of a barge (fly) rafter on the end and lookouts that connect the barge to a common rafter. The lookouts provide structural support for the overhang and a solid foundation for a soffit or other decorative finish.

107SA02.EPS

Framing Roof Openings

Before you cut an opening in a roof, check with the site engineer to make sure that you are proceeding according to specifications. If you don't double-check with the engineer, you may cut an incorrect opening and new rafters or trusses may be required. This type of error is very time-consuming and expensive to fix. Also, you should always be sure to check local fire codes for the proper clearance around a chimney and other roof openings. If the framing for the opening is too close to a chimney, it could cause a fire.

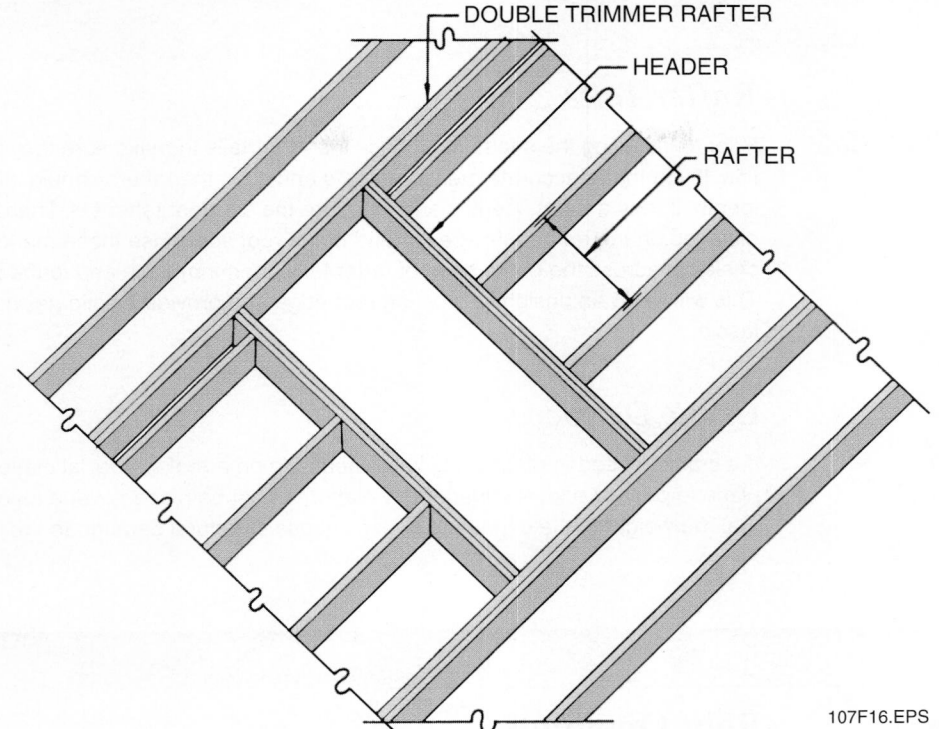

Figure 16 ◆ Framing an opening in a roof.

Labels on figure: DOUBLE TRIMMER RAFTER, HEADER, RAFTER, 107F16.EPS

4.0.0 ◆ INSTALLING SHEATHING

The sheathing should be applied as soon as the roof framing is finished. The sheathing provides additional strength to the structure, and provides a base for the roofing material.

Some of the materials commonly used for sheathing are plywood, OSB, waferboard, shiplap, and common boards. When composition shingles are used, the sheathing must be solid. If wood shakes are used, the sheathing boards may be spaced. When solid sheathing is used, leave a ⅛" space between panels to allow for expansion.

The following is an overview of roof sheathing requirements using plywood or other 4 × 8 sheet material.

Step 1 Start by measuring up 48¼" from where the finish fascia will be installed. Chalk a line at that point, then lay the first sheet down and nail it. Install H-clips midway between the rafters or trusses before starting the next course (*Figure 17*). These clips eliminate the need for tongue-in-groove panels.

Step 2 Apply the remaining sheets. Stagger the panels by starting the next course with a half sheet. Let the edges extend over the hip, ridge, and gable end. Cut the extra sheathing off with a circular saw.

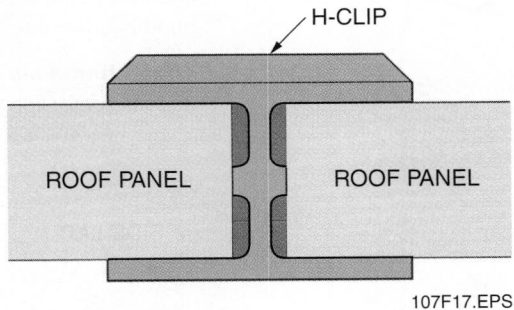

Figure 17 ◆ Use of H-clips.

Labels on figure: H-CLIP, ROOF PANEL, ROOF PANEL, 107F17.EPS

Once the sheathing has been installed, an underlayment of asphalt-saturated felt or other specified material must be installed to keep moisture out until the shingles are laid. For roofs with a slope of 4" or more, 15-pound roofer's felt is commonly used.

Material such as coated sheets or heavy felt that could act as a vapor barrier should not be used. They can allow moisture to accumulate between the sheathing and the underlayment. The underlayment is applied horizontally with a 2" top lap and a 4" side lap, as shown in *Figure 18(A)*. A 6" lap should be used on each side of the centerline of hips and valleys. A metal drip edge is installed along the rakes and eaves to keep out wind-driven moisture.

Rafter Tails

Prior to installing the sheathing, check the rafter tails to make sure they form a straight line. If you made accurate measurements and cuts, the rafters should all be the same length. If they are not, identify and measure the shortest rafter tail. Then mark the same distance on the rafter tails at each end of the roof span. Use these marks to snap a chalkline across the entire span of rafter tails, then trim each end to the same length. This will avoid an unsightly, crooked roof edge and provide a solid nailing base for the fascia.

Crane Delivery

If a crane is used to place a stack of sheathing on a roof, a special platform must be in place to provide a level surface. The platform must be placed over a loadbearing wall so that the weight of the sheathing doesn't cause structural damage to the framing.

Roof Openings

A 21-year-old apprentice was installing roofing on a building with six unguarded skylights. During a break, he sat down on one of the skylights. The plastic dome shattered under his weight and he fell to a concrete floor 16 feet below, suffering fatal head injuries.

The Bottom Line: Never sit or lean on a skylight. Always provide appropriate guarding and fall protection for work around skylights and other roof openings.

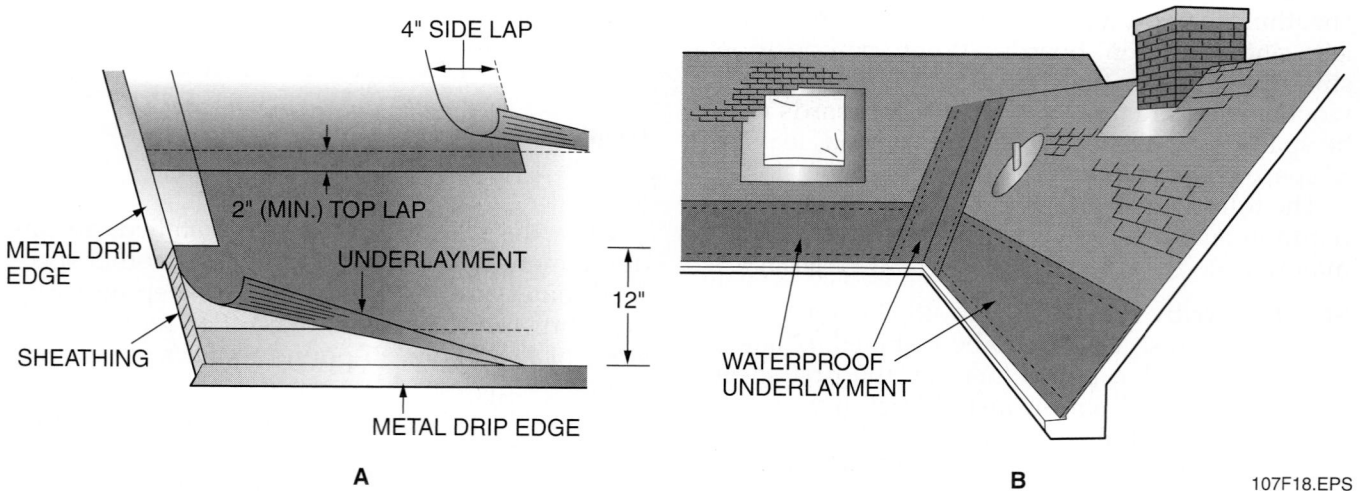

Figure 18 ◆ Underlayment installation.

In climates where snow accumulates, a waterproof underlayment, as shown in *Figure 18(B)*, should be used at roof edges and around chimneys, skylights, and vents. This underlayment has an adhesive backing that adheres to the sheathing. It protects against water damage that can result from melting ice and snow that backs up under the shingles.

Sheathing Safety

When installing sheathing on an inclined roof, always be sure to use a toeboard and safety harness. You can gain additional footing by wearing skid-resistant shoes. Falling from a roof can result in serious injury or death. Taking the time to follow the proper precautions will prevent you from slipping or falling. The picture shows a safety harness anchor and toeboard.

107SA03.EPS

Sheathing

If the trusses or rafters are on 24" centers, use ⅝" sheathing. With 16" centers, ½" sheathing may be used. This may vary depending on the local codes. Before starting construction, be sure to check the sheathing and fastener code requirements.

Felt Installation

The felt underlayment should be applied to the installed sheathing as soon as possible. It is important that both the sheathing and felt be dry and smooth at the time of installation. If the roof is damp or wet, wait a couple of days for it to dry completely before installing the felt. Moisture will cause long-term damage to both the sheathing and underlayment.

5.0.0 ◆ RAFTER LAYOUT USING A SPEED SQUARE

The speed square, also known as a super square or quick square, is a combination tool consisting of a protractor, try miter, and framing square. A standard speed square is a 6" triangular tool with a large outer triangle and a smaller inner triangle (*Figure 19*). The large triangle has a 6" scale on one edge, a full 90-degree scale on another edge, and a T-bar on the third edge. The inner triangle has a 2" square on one side. The speed square is the same on both sides. A 12" speed square is used for stair layout.

To use a square, you need to know the pitch of the roof. When you buy a speed square, it usually comes with an instruction booklet. This booklet normally contains (among other information) tables that show the required rafter length for every pitch.

5.1.0 Procedure for Laying Out Common Rafters

Step 1 Choose a piece of lumber that is slightly longer than that needed for the rafter. Remember, the eaves or overhang are not included in the measurements found in the information booklet. The length of the overhang is added after you have determined the length of the rafter. Place the lumber on a pair of sawhorses with the top edge of the rafter stock facing away from you. Be sure the crown is facing away from you (the crown, if any, will be the top edge of the rafter). The rafter will be cut for a building that is 24' wide with a 4" rise per foot of run. Therefore, the length of the rafter from the center of the ridge to the seat cut (bird's mouth) will be 12'-7¾". Any overhang must be added to this.

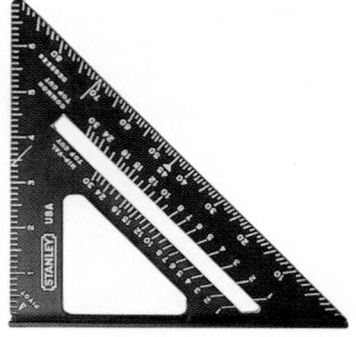

107F19.EPS

Figure 19 ◆ Speed square.

NOTE

When working with the speed square, there is no measuring line. The length is measured on the top edge of the rafter (this will be shown later in this module). Assume that the right side of the lumber will be the top of the rafter. Place the speed square against the top edge of the lumber and set the square with the 4 on the common scale.

Step 2 Draw a line along the edge of the speed square, as shown in *Figure 20(A)*. From the mark just made, measure with a steel tape along the top edge of the rafter the length required for the total length of the rafter (12'-7¾") and make a mark, as shown in *Figure 20(B)*.

Step 3 Place the speed square with the pivot point against that mark, and move the square so the 4 on the common scale is even with the edge. Draw a line along the edge of the square, as shown in *Figure 20(C)*. This will establish the vertical seat cut (bird's mouth) or plumb cut.

Step 4 To lay out the horizontal cut, reverse the speed square so that the short line at the edge of the square is even with the line previously drawn. Place the line so that the edge of the square is even with the lower edge of the rafter, as shown in *Figure 20(D)*.

Step 5 Draw a second line at a right angle to the plumb line. This line will establish the completed seat cut (bird's mouth), as shown in *Figure 20(E)*.

Step 6 The top and bottom cuts have been established for the total length of the rafter. If a ridgeboard is being used, be sure to deduct half of the thickness of the ridge from the top plumb cut. If any overhang is needed, it should be measured at right angles to the plumb cut of the seat cut (bird's mouth), as shown in *Figure 20(F)*.

Step 7 Once the measurement has been established, place the square against the top edge of the rafter, lining up the 4 on the common scale and marking it, as shown in *Figure 20(G)*.

Step 8 If a bottom or vertical cut is required, follow the procedure for cutting the horizontal cut of the seat cut (bird's mouth). In following the procedure described above,

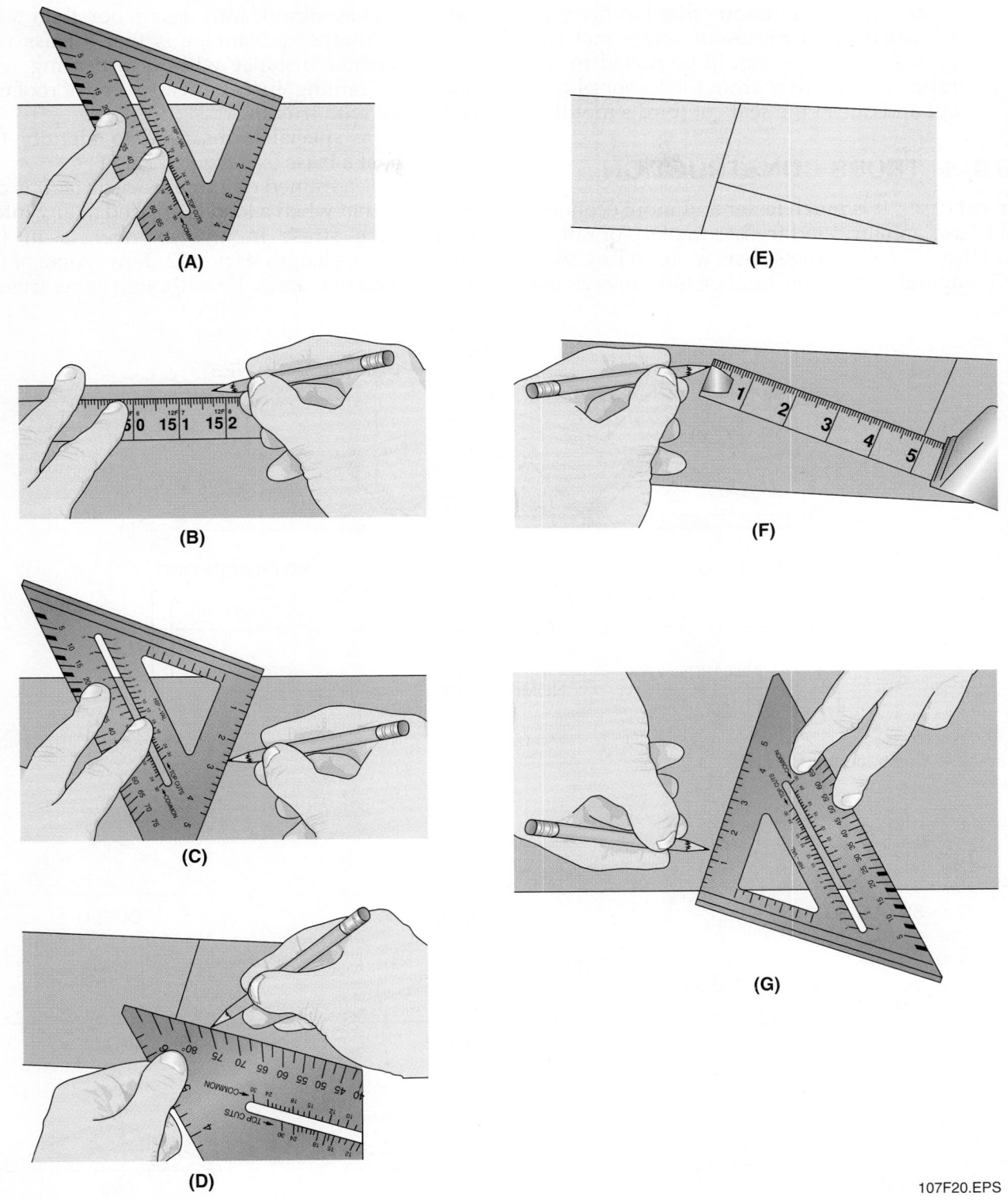

(A)

(B)

(C)

(D)

(E)

(F)

(G)

107F20.EPS

Figure 20 ◆ Laying out a common rafter.

a pattern for a common rafter has been established. Two pieces of scrap material (1 × 2 or 1 × 4) should be nailed to the rafter, one about 6" from the top or plumb cut and one at the seat cut (bird's mouth).

6.0.0 ◆ TRUSS CONSTRUCTION

In most cases, it is much faster and more economical to use prefabricated trusses in place of rafters and joists. Even if a truss costs more to buy than the comparable framing lumber (not always the case), it takes significantly less labor than stick framing. Another advantage is that a truss will span a greater distance without a bearing wall than stick framing. Just about any type of roof can be framed with trusses.

There are special terms used to identify the members of a truss (see *Figure 21*).

A truss is a framed or jointed structure. It is designed so that when a load is applied at any intersection, the stress in any member is in the direction of its length. *Figure 22* shows some of the many kinds of trusses. Even though some trusses

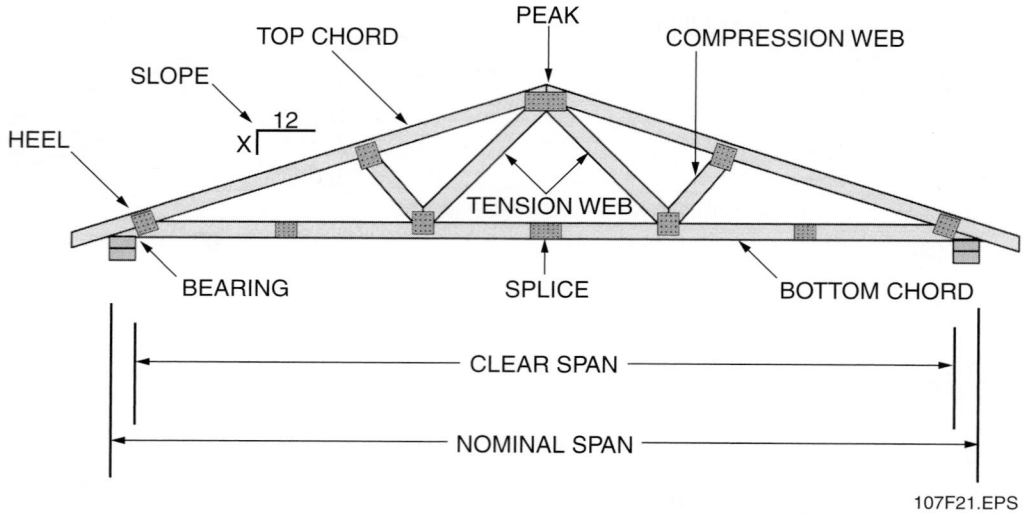

107F21.EPS

Figure 21 ◆ Components of a truss.

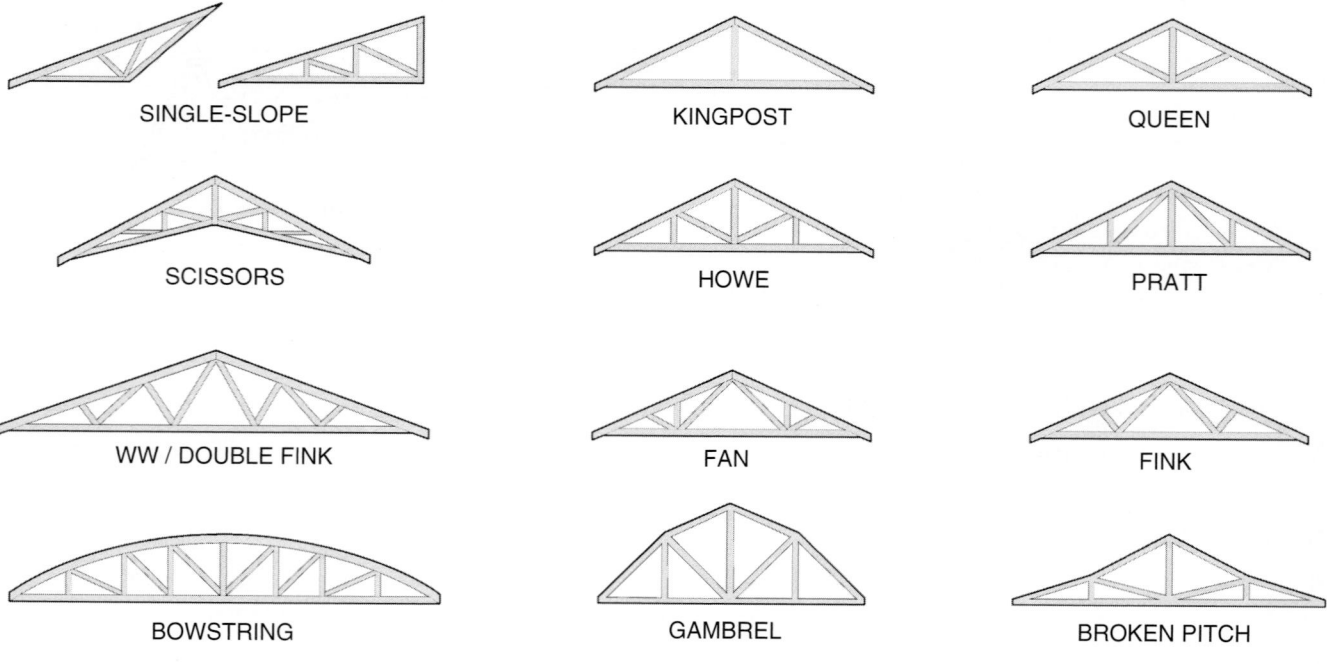

107F22.EPS

Figure 22 ◆ Types of trusses.

look nearly identical, there is some variation in the interior (web) pattern. Each web pattern distributes weight and stress a little differently, so different web patterns are used to deal with different loads and spans. The decision of which truss to use for a particular application will be made by the architect or engineer and will be shown on the blueprints. Do not substitute or modify trusses on site, because it could affect their weight and stress-bearing capabilities. Also, be careful how you handle trusses. They are more delicate than stick lumber and cannot be thrown around or stored in a way that applies uneven stress to them.

Trusses are stored or carried on trucks either lying on their sides or upright in cradles and protected from water. It is a good idea to use a crane to lift the trusses from the truck, and unless the trusses are small and light, it is usually recommended that a crane be used to lift them into place on the building frame. *Figure 23* shows examples of erecting methods used for trusses. Note the tag lines. These lines are held by someone on the ground to stabilize the truss while it is being lifted.

A single lift line can be used with trusses with a span of less than 20'. From 20' to 40', two chokers are needed. If the span exceeds 40', a spreader bar is required.

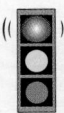
WARNING!
Installing trusses can be extremely dangerous. It is very important to follow the manufacturer's instructions for bracing and to follow all applicable safety procedures.

6.1.0 Truss Installation

Before installing roof trusses, refer to the framing plans for the proper locations. *Figure 24* is an example of a truss placement diagram. If a truss is damaged before erection, obtain a replacement or instructions from a qualified individual to repair the truss. Remember, repairs made on the ground are usually a lot better and are always easier. Never alter any part of a truss without consulting the job superintendent, the architect/engineer, or the manufacturer of the roof truss. Cutting, drilling, or notching any member without the proper approval could destroy the structural integrity of the truss and void any warranties given by the truss manufacturer.

Girders are trusses that carry other trusses or a relatively large area of roof framing. A common truss or even a double common truss will rarely serve as a girder. If there is a question about

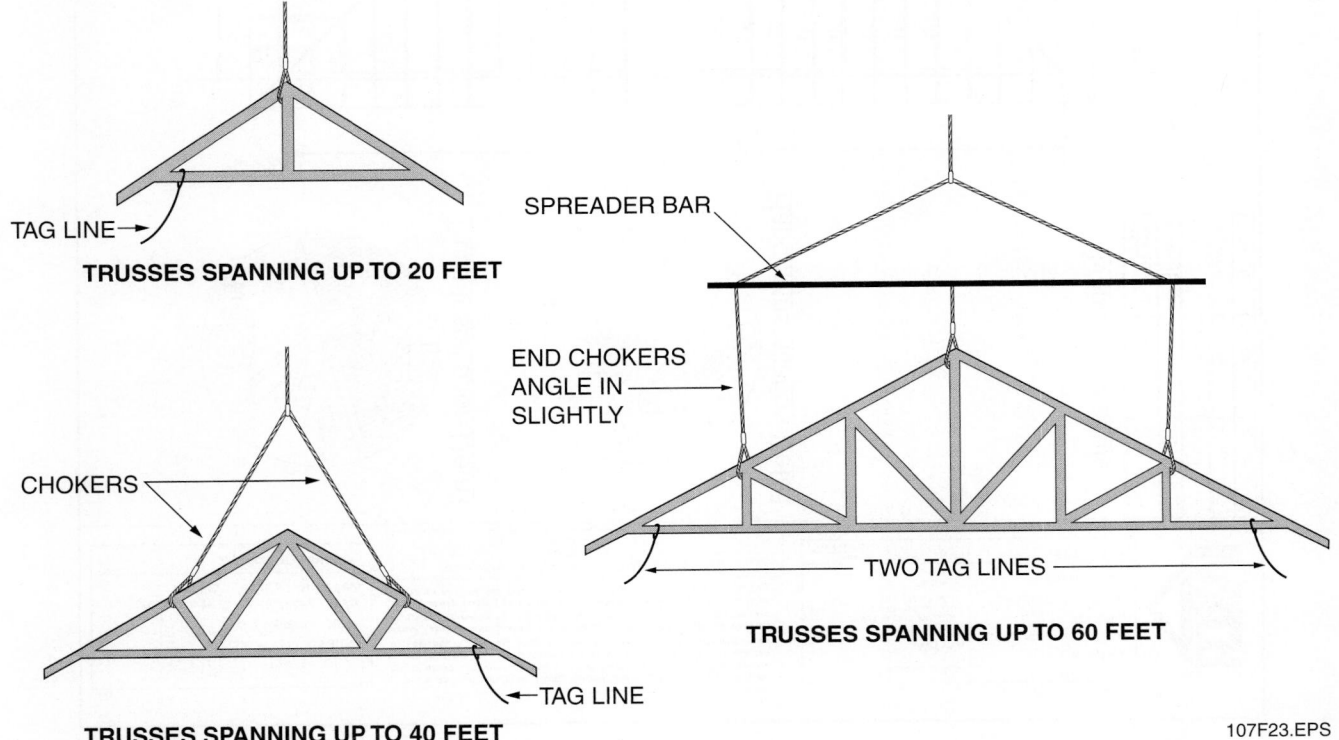

TAG LINE →

TRUSSES SPANNING UP TO 20 FEET

CHOKERS

← TAG LINE

TRUSSES SPANNING UP TO 40 FEET

SPREADER BAR

END CHOKERS ANGLE IN SLIGHTLY →

TWO TAG LINES

TRUSSES SPANNING UP TO 60 FEET

107F23.EPS

Figure 23 ◆ Erecting trusses.

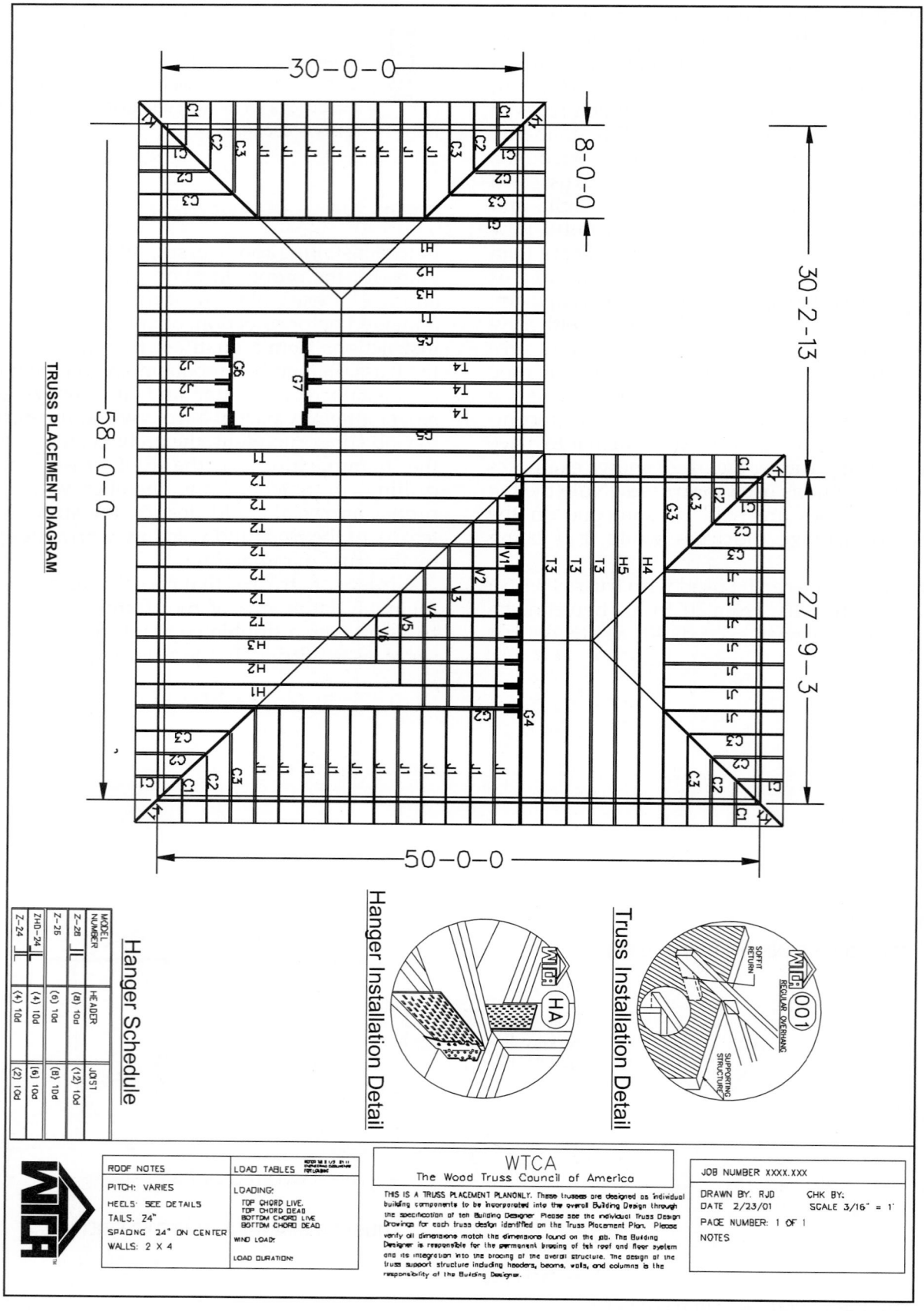

Figure 24 ◆ Example of a truss placement diagram.

107F24.EPS

whether a girder is needed, your job superintendent should confer with the truss manufacturer. Double girders are commonplace. Triple girders are sometimes required to ensure the proper load-carrying capacity. Always be sure that multiple-member girders are properly laminated together. Spacing the trusses should be done in accordance with the truss design. In some cases, very small deviations from the proper spacing can create a big problem. Always seek the advice of the job superintendent whenever you need to alter any spacing. Make certain that the proper temporary bracing is installed as the trusses are being set.

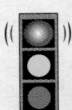 **WARNING!**
Never leave a job at night until all appropriate bracing is in place and secured well.

Light trusses (under 30' wide, for example) can be installed by having them lifted up and anchored to the top plate, then pushed into place with Y-shaped poles by crew members on the ground. Larger trusses require a crane.

When the bottom chord is in place, the truss is secured to the top plate. This can be done by toe-nailing with 10d nails. In some cases, however, metal tiedowns are required (*Figure 25*). An example is a location where high winds occur.

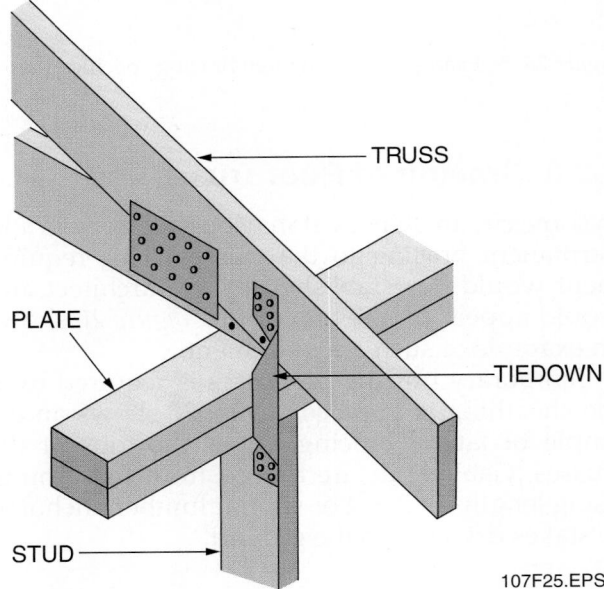

107F25.EPS

Figure 25 ◆ Use of a tiedown to secure a truss.

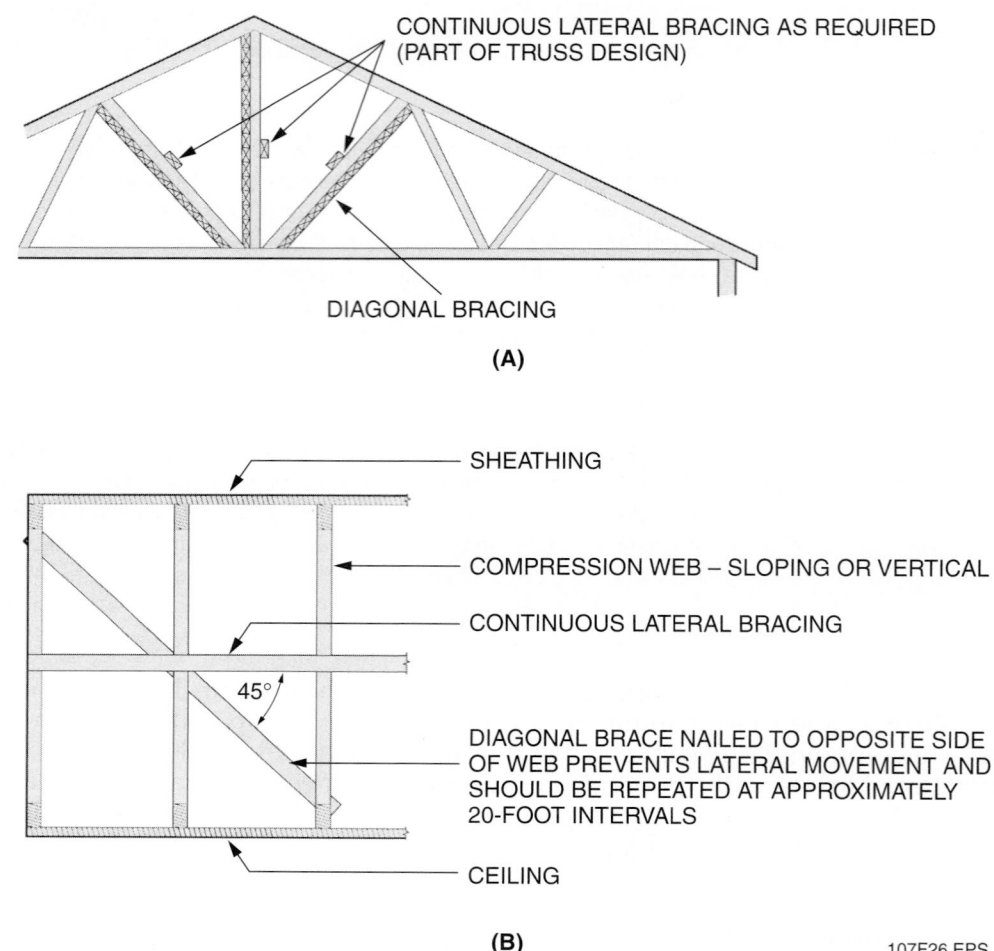

CONTINUOUS LATERAL BRACING AS REQUIRED
(PART OF TRUSS DESIGN)

DIAGONAL BRACING

(A)

SHEATHING

COMPRESSION WEB – SLOPING OR VERTICAL

CONTINUOUS LATERAL BRACING

45°

DIAGONAL BRACE NAILED TO OPPOSITE SIDE
OF WEB PREVENTS LATERAL MOVEMENT AND
SHOULD BE REPEATED AT APPROXIMATELY
20-FOOT INTERVALS

CEILING

(B)

107F26.EPS

Figure 26 ◆ Example of permanent bracing specification.

6.2.0 Bracing of Roof Trusses

In some circumstances, it may be necessary to add permanent bracing to the trusses. This requirement would be established by the architect and would appear on the blueprints. *Figure 26* shows an example of such a requirement.

Temporary bracing of trusses is required until the sheathing is in place. *Figure 27* shows an example of lateral bracing across the tops of the trusses. Gable ends are braced from the ground using lengths of 2 × 4 or similar lumber anchored to stakes driven into the ground.

107F27.EPS

Figure 27 ◆ Example of temporary bracing of trusses.

Temporary Truss Bracing

The leading cause of truss collapses during construction is insufficient temporary bracing. One crucial aspect of bracing that is often overlooked is diagonal bracing (shown in red in the figure below). The truss industry provides guides for safe and efficient handling, installation, and bracing of metal plate connected wood trusses called the Building Component Safety Information (BCSI) series. These documents cover recommendations for correct hoisting and bracing procedures based on the span of the truss, installation tolerances, and limits on construction loading. Following BCSI recommendations will minimize truss damage during construction, lead to better long-term performance of the truss system, and create a safer work site.

Text courtesy of the Wood Truss Council of America (WTCA). For more information, visit www.woodtruss.com.

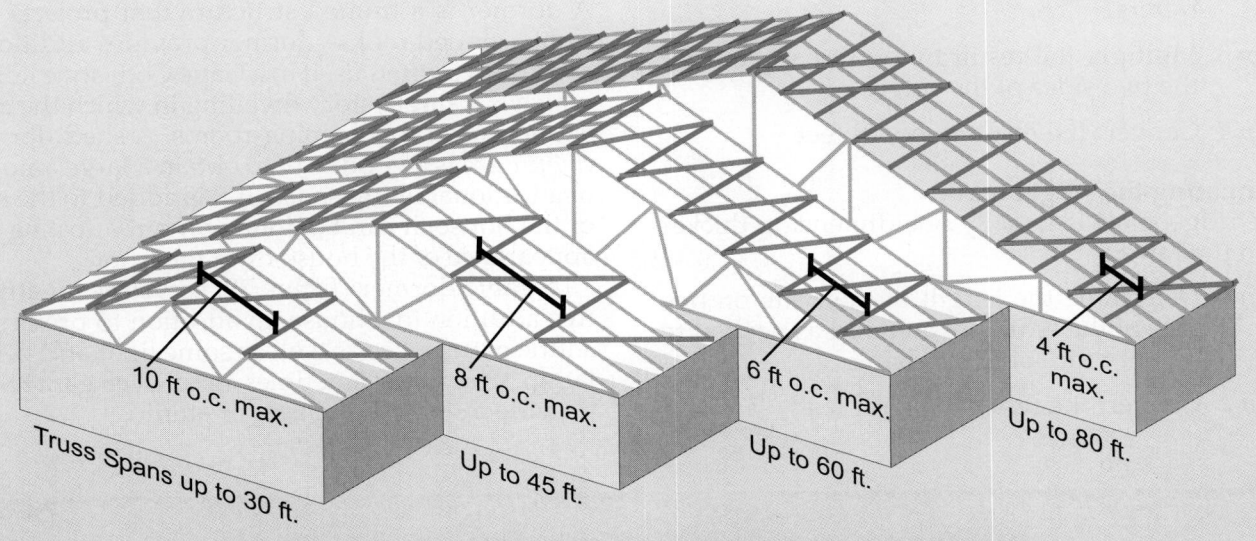

Spacing of Top Chord Temporary Lateral Bracing on Trusses

107SA04.EPS

7.0.0 ◆ DETERMINING QUANTITIES OF MATERIALS

Estimating the material you will need for a roof depends first of all on the type of roof you are planning to construct, the size of the roof, the spacing of the framing members, and the load characteristics. Local building codes will usually dictate these factors and they will be disclosed on the building plans. Lumber for conventional framing may be from 2 × 4 to 2 × 10, depending on the span and the load. For example, 2 × 4 framing on 16" centers might support a 9' span. By comparison, 2 × 10 framing on 16" centers would support a span of more than 25'.

7.1.0 Determine Materials Needed for a Gable Roof

Determining the rafter material:

Step 1 To determine how much lumber you will need for rafters on a gable roof, first determine the length of each common rafter (including the overhang) using the framing square or another method.

Step 2 Figure out the number of rafters based on the spacing (16", 24", etc.). Remember that you will need one rafter for each gable end. You will also need barge rafters in the gable overhang. Note that these are usually one size smaller than the common rafters.

Step 3 Multiply the result by two to account for the two sides of the ridge.

Step 4 Convert the result to board feet.

Estimating the ridgeboard:

The ridgeboard is usually one dimension thicker than the rafters.

Step 1 Determine the length of the plate on one side of the structure and add as needed to account for gable overhang.

Step 2 Convert the result to board feet.

Estimating the sheathing:

Step 1 Multiply the length of the roof including overhangs by the length of a common rafter. This yields half the area of the roof.

Step 2 Divide the roof area by 32 (the number of square feet in a 4 × 8 sheet) to get the approximate number of sheets of sheathing you will need. Round up if you get a fractional number.

Step 3 Multiply by 2 to obtain the number of sheets needed for the full roof area.

8.0.0 ◆ DORMERS

A dormer is a framed structure that projects out from a sloped roof. A dormer provides additional space and is often used in a Cape Cod–style home, which is a single-story dwelling in which the attic is often used for sleeping rooms. A shed dormer (*Figure 28*) is a good way to obtain a large amount of additional living space. If it is added to the rear of the house, it can be done without affecting the appearance of the house from the front.

A gable dormer (*Figure 29*) serves as an attractive addition to a house, in addition to providing a little extra space as well as some light and ventilation. They are sometimes used over garages to provide a small living area or studio.

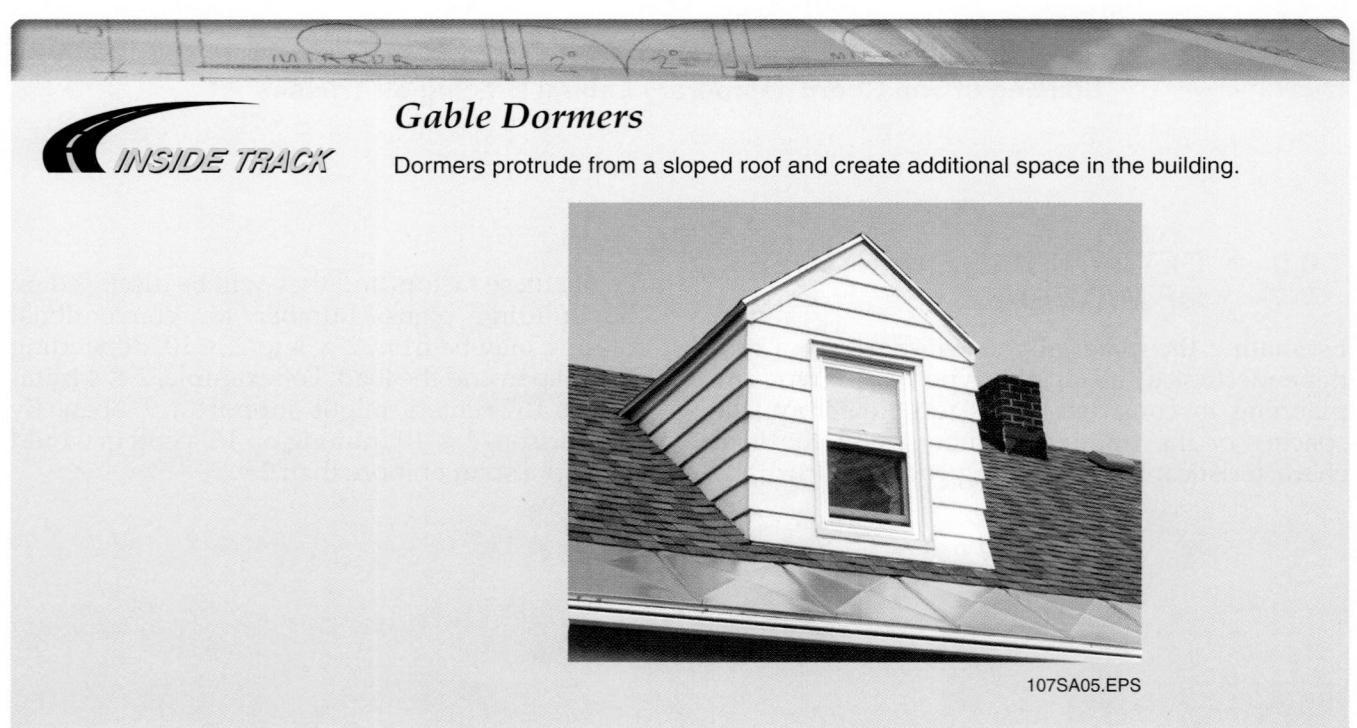

INSIDE TRACK

Gable Dormers
Dormers protrude from a sloped roof and create additional space in the building.

107SA05.EPS

Figure 28 ◆ Shed dormer.

9.0.0 ◆ PLANK-AND-BEAM FRAMING

Plank-and-beam framing, also known as post-and-beam framing (*Figure 30*), employs much sturdier framing members than common framing. It is often used in framing roofs for luxury residences, churches, and lodges, as well as other public buildings where a striking architectural effect is desired.

Because the beams used in this type of construction are very sturdy, wider spacing may be used. Vertical supports are typically spaced 48" OC, as compared with 16" OC used in conventional framing. When plank-and-beam framing is used for a roof, the beams and planking can be finished and left exposed. The underside of the planking takes the place of an installed ceiling.

In light construction, solid posts or beams such as 4 × 4s are used. In heavy construction, laminated beams made of glulam, LVL, and PSL are used.

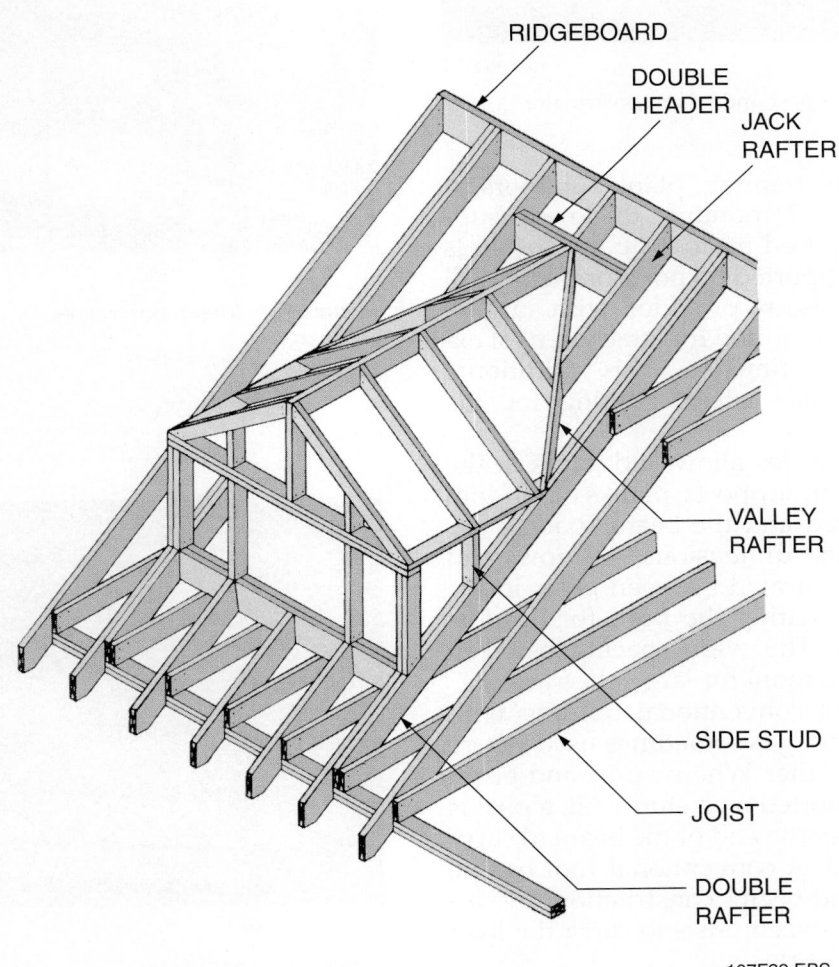

RIDGEBOARD

DOUBLE HEADER

JACK RAFTER

VALLEY RAFTER

SIDE STUD

JOIST

DOUBLE RAFTER

Figure 29 ◆ Gable dormer framing.

107F30.EPS

Figure 30 ◆ Example of post-and-beam construction.

In post-and-beam framing, plank subfloors or roofs are usually of 2" nominal thickness, supported on beams spaced up to 8' apart. The ends of the beams are supported on posts or piers. Wall spaces between posts are provided with supplementary framing as required for attachment of exterior and interior finishes. This additional framing also provides lateral bracing for the building.

If local building codes allow end joints in the planks to fall between supports, planks of random lengths may be used and the beam spacing adjusted to fit the house dimensions. Windows and doors are normally located between posts in the exterior walls, eliminating the need for headers over the openings. The wide spacing between posts permits ample room for large glass areas.

A combination of conventional framing with post-and-beam framing is sometimes used where the two adjoin each other. Where a post-and-beam floor or roof is supported on a stud wall, a post is usually placed under the end of the beam to carry a conventional load. A conventional roof can be used with post-and-beam construction by installing a header between posts to carry the load from the rafters to the posts.

10.0.0 ◆ METAL ROOF FRAMING

When steel framing is used, the roof framing is often done with prefabricated metal trusses. If the trusses are fabricated on site, the chords and webs are cut with a portable electric saw and placed into a jig. They are then welded and/or connected with self-tapping screws.

Trusses that look like wood are used in some applications (*Figure 31*). Commercial projects may use open web joists, known as bar joists, to support the roof (*Figure 32*).

107F31.EPS

Figure 31 ◆ Metal roof trusses.

107F32.EPS

Figure 32 ◆ Bar joists.

Summary

The correct layout and framing of a roof requires patience and skill. If the measurement, cutting, and installation work are not done carefully and precisely, the end result will never look right. Fortunately, there are many tools and reference tables that help to simplify the process. The important thing is to be careful and precise with the layout and cutting of the first rafter of each type. This is the rafter that is used as a pattern for the others.

Notes

1. The type of roof that has four sides running toward the center of the building is the _____ roof.
   a. gable
   b. hip
   c. shed
   d. gable and valley

2. The letter A in *Figure 1* is pointing to a _____.
   a. hip jack
   b. valley jack
   c. hip rafter
   d. cripple jack

3. The letter B in *Figure 1* is pointing to the _____.
   a. main ridgeboard
   b. projection ridgeboard
   c. top plate
   d. collar beam

4. The letter D in *Figure 1* is pointing to a _____.
   a. valley cripple
   b. hip jack
   c. valley rafter
   d. valley jack

5. The letter E in *Figure 1* is pointing to the _____.
   a. top plate
   b. projection ridgeboard
   c. main ridgeboard
   d. gable stud

6. The letter G in *Figure 1* is pointing to a _____.
   a. gable end
   b. common rafter
   c. collar beam
   d. gable stud

7. The letter M in *Figure 1* is pointing to a _____ rafter.
   a. hip jack
   b. valley jack
   c. cripple jack
   d. hip

8. The roof run measurement is usually equal to _____.
   a. twice the span
   b. half the span
   c. the distance from the top plate to the ridgeboard
   d. the length of a common rafter

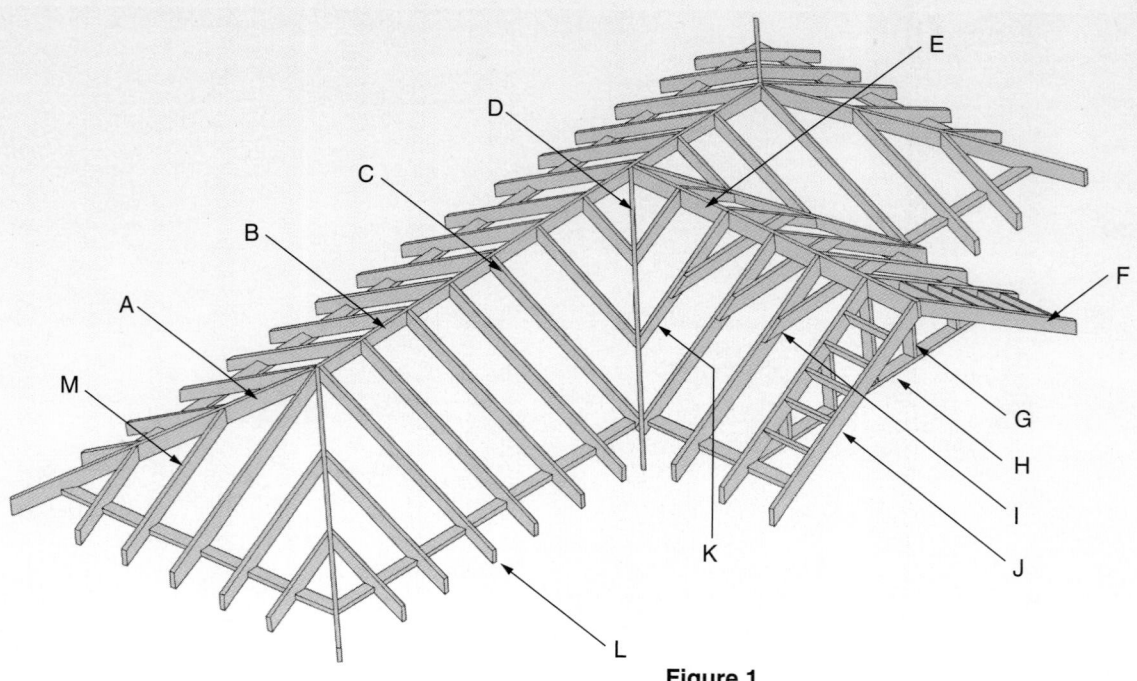

Figure 1

107RQ01.EPS

Review Questions

9. The horizontal distance from the outside of the top plate to the center of the ridgeboard is called the _____.
   a. slope
   b. span
   c. pitch
   d. run

Refer to *Figure 2* to answer Questions 10 and 11.

10. The length of a common rafter for a 20' wide building with a 10" rise per foot of run is _____.
    a. 19.7'
    b. 20'
    c. 13'
    d. 14'-8"

11. The rise per foot of run of a roof with a pitch of ½ is _____.
    a. 6"
    b. 12"
    c. 18"
    d. 24"

12. A roof with a pitch of ½ has a(n) _____ rise per foot of run.
    a. 6"
    b. 8"
    c. 10"
    d. 12"

13. When laying out a common rafter, it is necessary to deduct _____ in order to arrive at the final measurement.
    a. half the thickness of the ridgeboard
    b. half the thickness of the rafter
    c. half the thickness of the top plate
    d. nothing

14. An L-shaped brace used to strengthen and straighten ceiling joists is known as a _____.
    a. purlin
    b. strongback
    c. collar tie
    d. ridgeboard

15. Double trimmers are used when framing _____.
    a. roof openings
    b. gable ends
    c. overhangs
    d. valleys

16. When placing underlayment over roof sheathing, it is best to use the heaviest felt you can find.
    a. True
    b. False

17. Immediately after nailing the plywood sheathing on a roof frame, you should install _____.
    a. a vapor barrier
    b. a felt underlayment
    c. a drip edge
    d. plywood clips

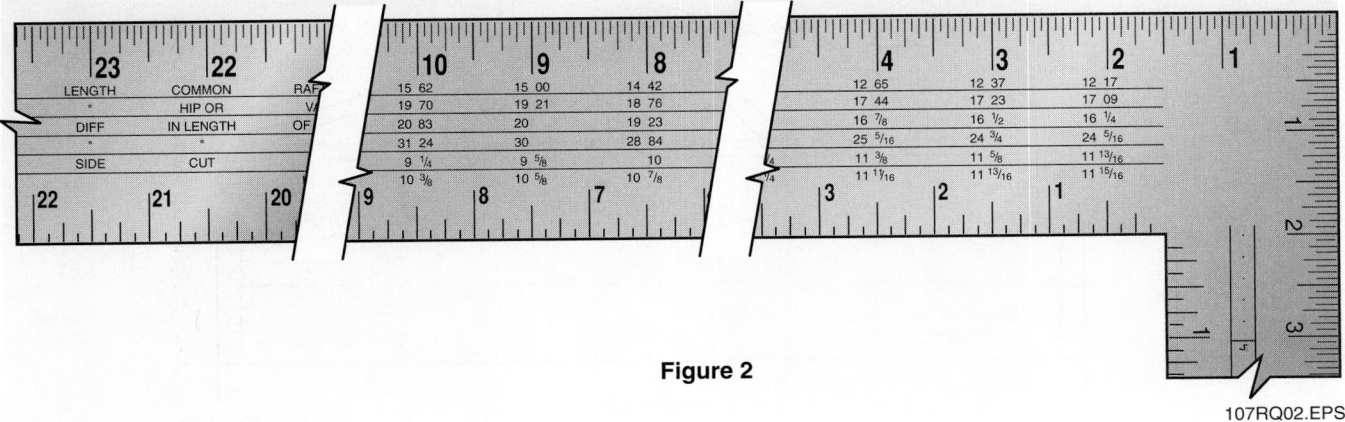

Figure 2

107RQ02.EPS

18. When you are using a speed square, the length of the rafter is determined by_____.
    a. stepping the square
    b. reading it directly from the square
    c. using rafter tables in the instruction manual
    d. multiplying the run by the rise

19. When you are laying out rafters with a speed square, you don't have to add the length of the overhang because it is included in the rafter tables provided with the square.
    a. True
    b. False

Refer to the following illustration to answer Questions 21 and 22.

20. Item 6 in *Figure 3* is pointing to the _____.
    a. center chord
    b. slope
    c. tension web
    d. compression web

21. Item 4 in *Figure 3* is pointing to the _____.
    a. top chord
    b. clear span
    c. peak
    d. slope

22. When rigging a truss with a span of 45', you should use a _____.
    a. spreader bar
    b. single choker
    c. double choker
    d. triple choker

23. The number of common rafters needed to frame a 20' long gable roof for a house framed on 16" centers with no overhang is _____.
    a. 15
    b. 32
    c. 30
    d. 16

24. The number of sheets of plywood sheathing needed for a house 30' long with a span of 24' and ¼ pitch is _____. (Assume no overhang.)
    a. 14
    b. 18
    c. 26
    d. 28

25. Vertical supports for plank-and-beam framing are typically spaced _____ on center.
    a. 16"
    b. 24"
    c. 36"
    d. 48"

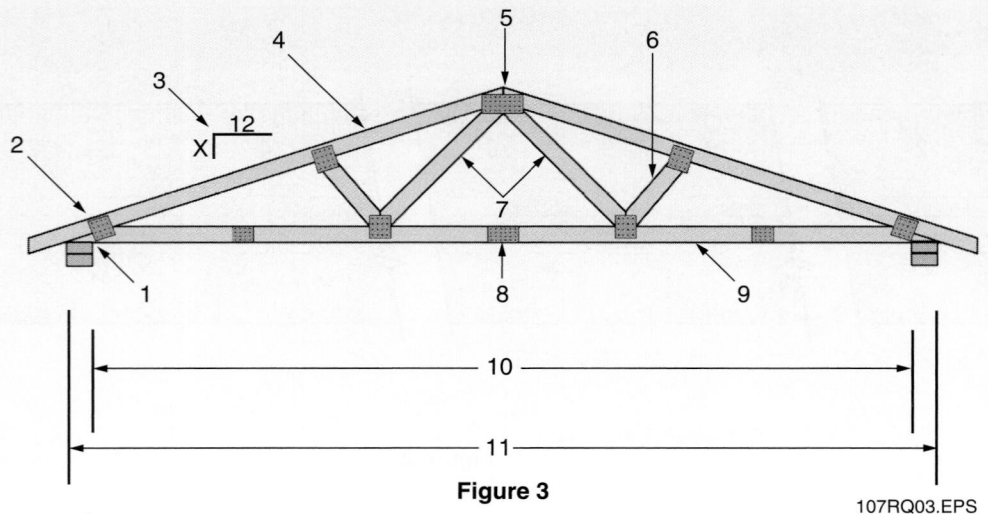

Figure 3

107RQ03.EPS

Trade Terms Quiz

1. A _____ roof has two slopes that meet at the center to form a point.

2. The board nailed to the rafter tails in order to space the rafters is the _____.

3. A gable with an overhang uses a _____ to support the extension and provide a nailing surface for a decorative edge.

4. A _____ is a horizontal support member used to strengthen a long rafter span.

5. An overhang is framed using short lengths of boards known as a _____.

Trade Terms

Barge rafter
False fascia
Gable
Lookout
Purlin

Alan Van Holten

Gold Medal Winner, Associated Builders and Contractors National Championships

Before becoming a carpenter, Alan was a welder for 12 years. Although he enjoyed welding, he believes that carpentry offers a wider range of job opportunities. He won a gold medal at both the regional and national ABC championships. He is currently working as a foreman for a building company that specializes in tilt-up construction.

How did you become interested in carpentry?

I learned to weld in high school when I was 13. I enjoyed it and worked as a welder for 12 years. But I found that I was often doing the same types of work and I wanted a change. I went to work as a laborer for a building company. They sent me to school for carpentry. I really enjoyed carpentry because you are always doing different kinds of jobs. Each day has new challenges and new things to figure out.

What positions have you held and how did they help you get to where you are now?

In carpentry I started as a laborer, then I was promoted to lead man. Then I became a journeyman and a foreman. Each position was a stepping-stone to achieving my goals. Eventually, I want to become a project manager and superintendent, but I have a lot to learn before I can achieve that goal.

Did you enjoy the Associated Builders and Contractors National Championships?

I think that Associated Builders and Contractors is a good organization. They have been very good to me. The competition itself was pretty intense but it was also really gratifying. The first day is a written test. On the second day, you have six hours to complete a project. At first I was pretty nervous. When you start, you are in a big room and have no idea what you will be doing. Then they give you the drawings. Once I had the drawings, I relaxed and did what I had to do. It was like going to work. Just figure out what to do with the materials you have on hand and build the project.

Later on in my career, I was a judge for the SkillsUSA Championship. I have to say that it was pretty nice being on the other side of the competition.

What are some of the things you do in your job?

As a foreman I do a lot of paperwork. We do tilt-up construction for commercial buildings, retail stores, and office buildings. Everything is done on site. We build the forms for the concrete to make the panels. Then we assemble the building.

What do you think it takes to be a success in your trade?

Frankly, it takes lots of hard work. You need to have a desire to achieve success and grow as a person. You have to want to become the best at what you do. That's what drives me. I always want to be the best at whatever I am doing.

What do you like about the work you do?

I like that I get to do many different things. It's not the same tasks every day. There is much more variety in carpentry than there was in welding.

What advice would you give someone just starting out?

Stay focused on where you want to be in your life. There will be rough spots and you will stumble. But don't give up in the hard times. Keep striving for what you want.

Work hard for your company. The harder you work, the more a company will appreciate you and the more opportunities you will have. Foremen notice the people who work really hard every day.

Trade Terms
Introduced in This Module

Barge rafter: A gable end roof member that extends beyond the gable to support a decorative end piece. Also known as a fly rafter.

False fascia: The board that is attached to the tails of the rafters to straighten and space the rafters and provide a nailer for the fascia. Also called sub fascia and rough fascia.

Gable: The triangular wall enclosed by the sloping ends of a ridged roof.

Lookout: A structural member used to frame an overhang.

Purlin: A horizontal roof support member parallel to the plate and installed between the plate and the ridgeboard.

Laying Out and Erecting Hips and Valleys

An intersecting roof contains two or more sections sloping in different directions. Examples are the connection of two gable sections or a gable and hip combination such as that shown in *Figure A-1*. *Figure A-2* shows an overhead view of the same layout.

A valley occurs wherever two gable or hip roof sections intersect. Valley rafters run at a 45-degree angle to the outside walls of the building.

The material that follows provides an overview of the procedures for laying out hip and valley sections and the various types of rafters used in framing these sections. The first step, as always, is to lay out the rafter locations on the top plate (*Figure A-3*). The layout of the common rafters for a hip roof is the same as that for a gable roof. The next step would be to lay out, cut, and install the common rafters and the main and projection ridgeboards, as described earlier in this module. At that point, you are ready to lay out the hip or valley section.

HIP RAFTERS

A hip rafter is the diagonal of a square formed by the walls and two common rafters (*Figure A-2*). Because they travel on a diagonal to reach the ridge, the hip rafters are longer than the common rafters. The unit run is 17" (16.97" rounded up), which is the length of the diagonal of a 12" square. You can see this on the top line of the rafter tables on the framing square. There are two hip rafters in every hip section.

For every hip roof of equal pitch, for every foot of run of common rafter, the hip rafter has a run of 17". This is a very important fact to remember. It can be said that the run of a hip rafter is 17 divided by 12 times the run of a common rafter. The total rise of the hip rafter is the same as that of a common rafter. For example, if a common rafter has an 8" rise per foot of run, the hip rafter would also have a rise of 8" per 17" run. Therefore, the rise of a hip rafter would be the same as a common rafter at any given corresponding point.

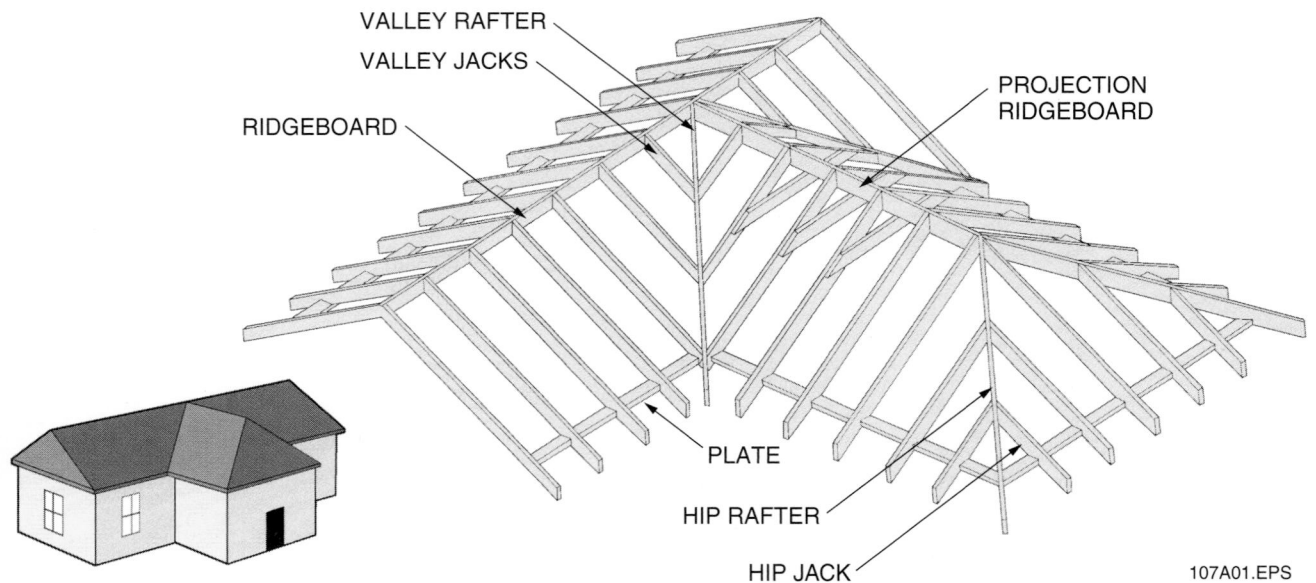

VALLEY RAFTER

VALLEY JACKS

RIDGEBOARD

PROJECTION RIDGEBOARD

PLATE

HIP RAFTER

HIP JACK

107A01.EPS

Figure A-1 ◆ Example of post-and-beam construction.

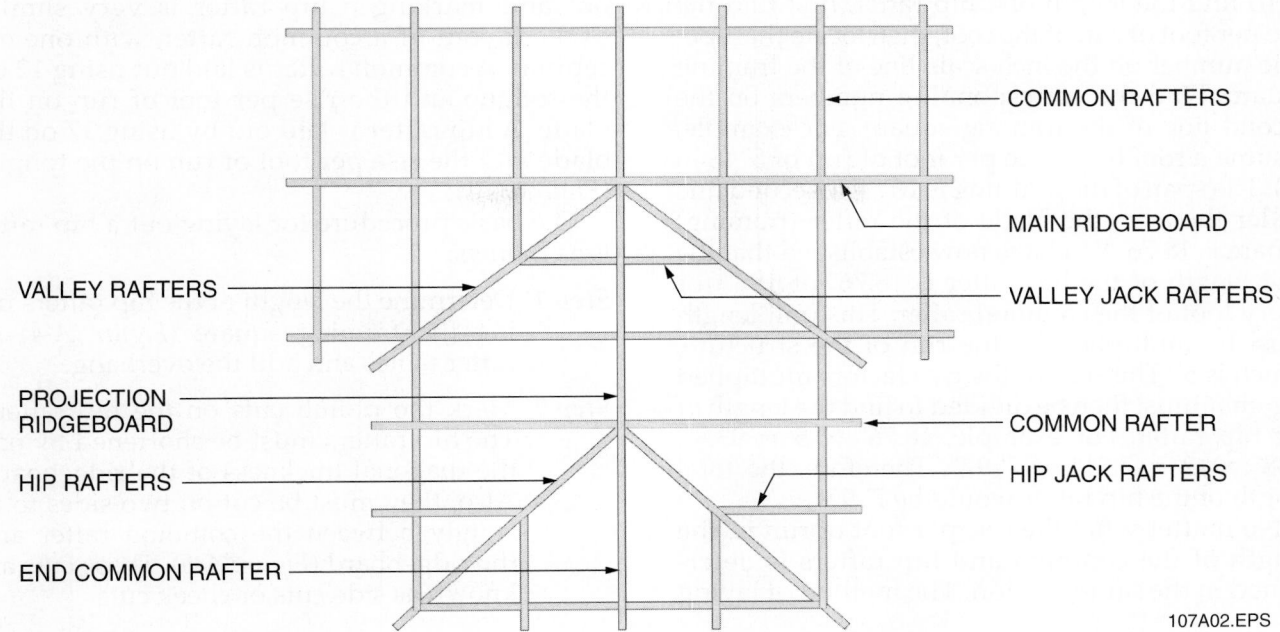

COMMON RAFTERS

MAIN RIDGEBOARD

VALLEY JACK RAFTERS

VALLEY RAFTERS

PROJECTION RIDGEBOARD

HIP RAFTERS

END COMMON RAFTER

COMMON RAFTER

HIP JACK RAFTERS

107A02.EPS

Figure A-2 ◆ Overhead view of a roof layout.

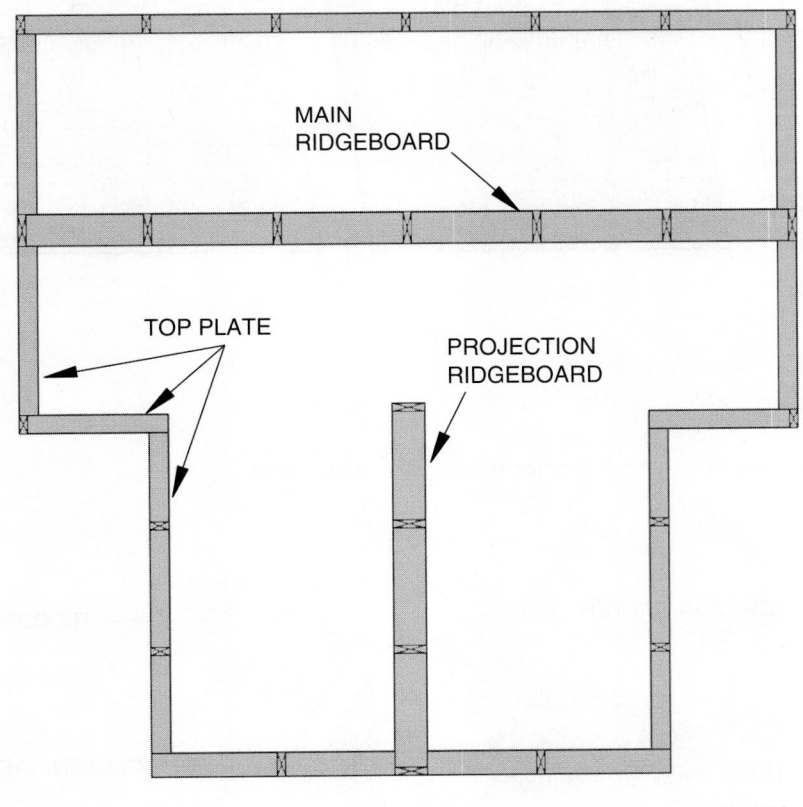

MAIN RIDGEBOARD

TOP PLATE

PROJECTION RIDGEBOARD

107A03.EPS

Figure A-3 ◆ Layout of rafter locations on the top plate.

To find the length of a hip rafter, first find the rise per foot of run of the roof, then locate that specific number on the inch scale line of the framing square. Find the corresponding numbers on the second line of the framing square. For example, assume a roof has a rise per foot of run of 8" (8 in 12). The span of the building is 10'. The second line under the 8 on the blade of the rafter (framing) square is 18.76. You have now established that the unit length of the hip rafter is 18.76" or 18¾" for every foot of the common rafter. This unit length must be multiplied by the run of the structure, which is 5'. The sum of the two factors multiplied together must then be divided to find the length of the hip rafter. For example, 18.76" × 5 = 93.8". 93.8" ÷ 12 = 7.81" or 7'-9¾". Therefore, the total length of the hip rafter would be 7'-9¾".

No matter what the rise per foot of run is, the length of the common and hip rafters is determined in the same fashion. The method of laying out and marking a hip rafter is very similar to the layout of a common rafter, with one exception. A common rafter is laid out using 12 on the tongue and the rise per foot of run on the blade. A hip rafter is laid out by using 17 on the blade and the rise per foot of run on the tongue (*Figure A-4*).

The basic procedure for laying out a hip rafter is as follows:

Step 1 Determine the length of the hip rafters using the framing square (*Figure A-4*) or rafter tables and add the overhang.

Step 2 Mark the plumb cuts on the hip rafters. The hip rafters must be shortened by half the diagonal thickness of the ridgeboard. Also, they must be cut on two sides to fit snugly between the common rafter and the ridgeboard (*Figure A-5*). These cuts are known as side cuts or cheek cuts.

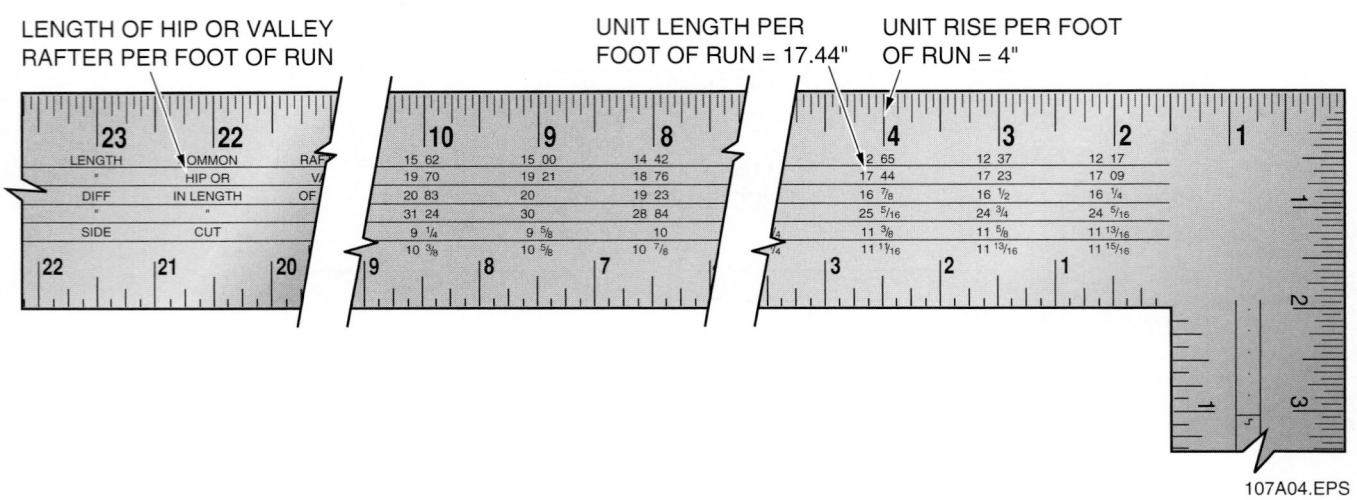

107A04.EPS

Figure A-4 ◆ Using the framing square to determine the length of a hip rafter.

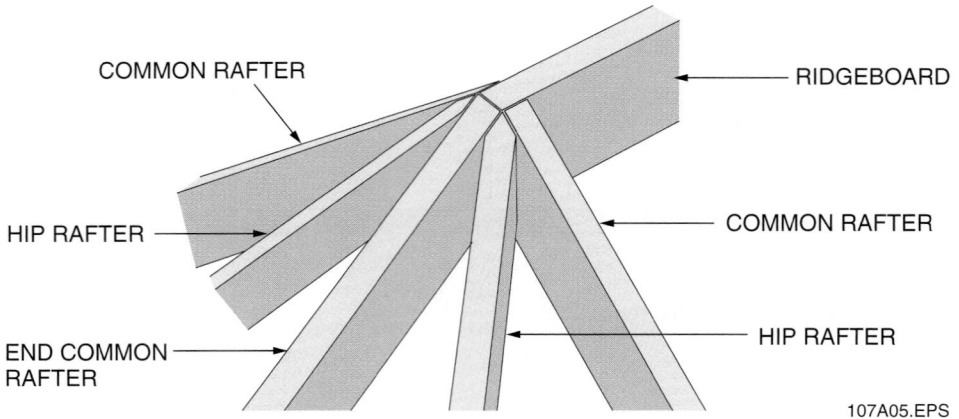

107A05.EPS

Figure A-5 ◆ Hip rafter position.

Step 3 The bird's mouth plumb and seat cuts are determined in the same way as for a common rafter. However, the seat cut is dropped half the thickness of the rafter to align it with the top plane of the common rafter (distance A, *Figure A-6*). This drop is necessary because the corners of the rafters would otherwise be higher than the plane on which the roof surface is laid. Another way to accomplish this is to chamfer the top edges of the rafter. This procedure is known as backing. Most carpenters use the dropping method because it is faster.

VALLEY LAYOUT

Each valley requires a valley rafter and some number of valley jack rafters (*Figure A-7*). The layout of a valley rafter is basically the same as that of the hip rafter, with 17" used for the unit run. The only difference in layout between the hip and valley rafters is in the seat and tail cuts.

For a valley rafter, the bird's mouth plumb cut must be angled to allow the rafter to drop down into the inside corner of the building (*Figure A-8*). In addition, the tail end cuts must be made so that corner made by the valley will line up with the rest of the roof overhang. Like the hip rafter, the valley rafter must be aligned with the plane of the common rafter. Cheek cuts are also required on valley rafters to allow them to fit between the two ridgeboards.

The layout of valley jacks is the same as that for hip jacks, with the exception that the valley jacks usually run in the opposite direction; i.e., toward the ridge.

JACK RAFTER LAYOUT

Hip jack rafters run from the top plate to the hip rafters. Valley jacks run from the valley rafter to the ridge (see *Figure A-1*). Notice that layout of valley jack rafters usually starts at the building line and moves toward the ridge. As with gable

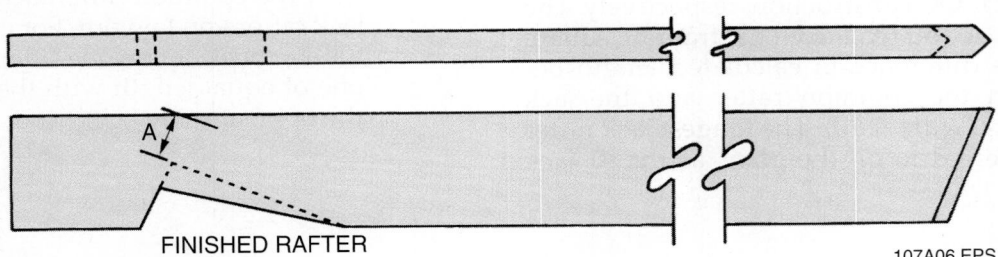

FINISHED RAFTER

107A06.EPS

Figure A-6 ◆ Dropped bird's mouth cut.

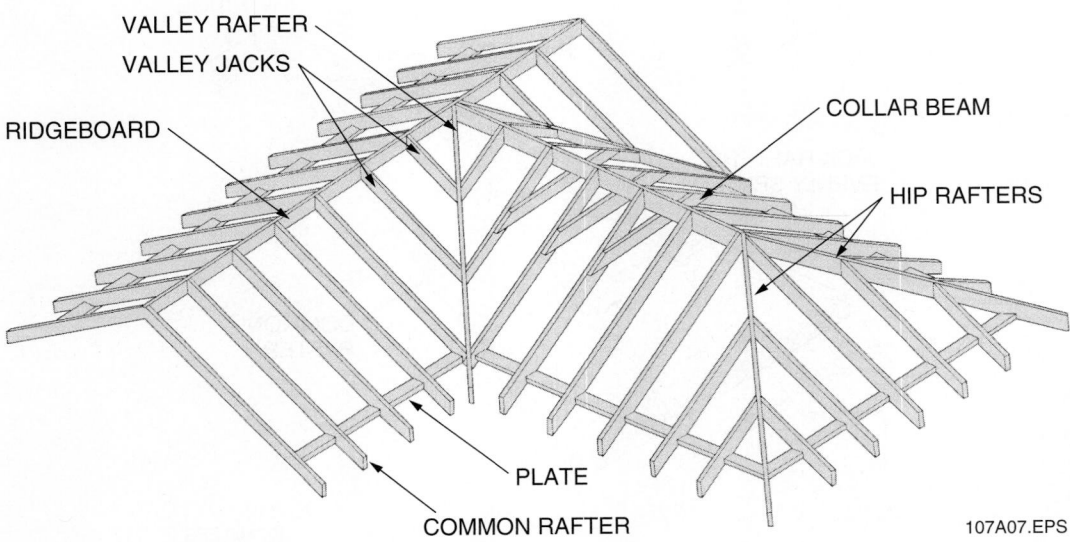

Figure A-7 ◆ Valley layout.

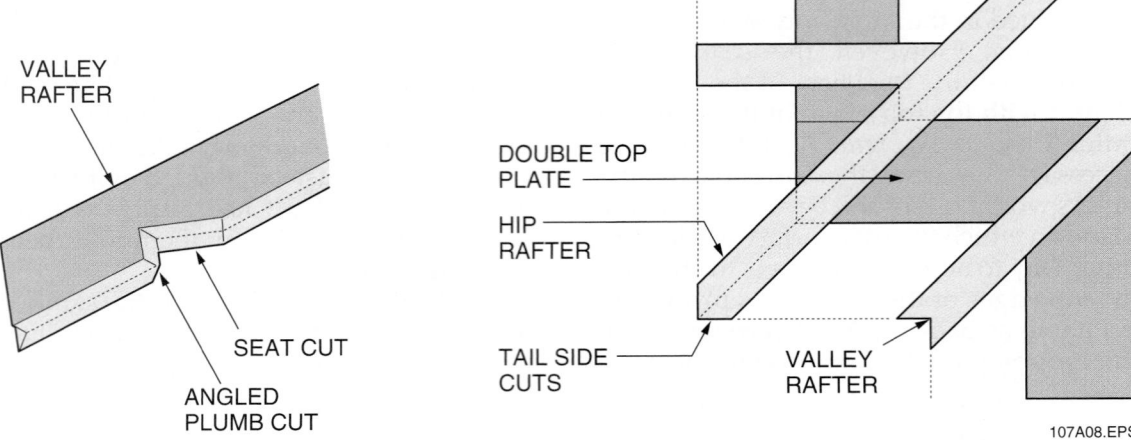

Figure A-8 ◆ Valley rafter layout.

end studs, there is a common difference from one jack rafter to the next (*Figure A-9*).

Here is an overview of the jack rafter layout process.

Step 1 The third and fourth lines of the framing square have the information you need to determine the lengths of jack rafters for 16" and 24" OC construction, respectively. The number you read from the framing square is the difference in calculated length between the common rafter and the jack rafter (*Figure A-10*). The longest jack rafter is referred to on the plans as the #1 jack rafter. Jack rafters must be cut and installed in pairs to prevent the hip or valley from bowing.

Step 2 Use the common rafter to mark the bird's mouth.

Step 3 To lay out the additional jack rafters, subtract the common difference from the last jack rafter you laid out. For each jack rafter with a cheek cut on one side, there must be one of equal length with the cheek cut on the opposite side.

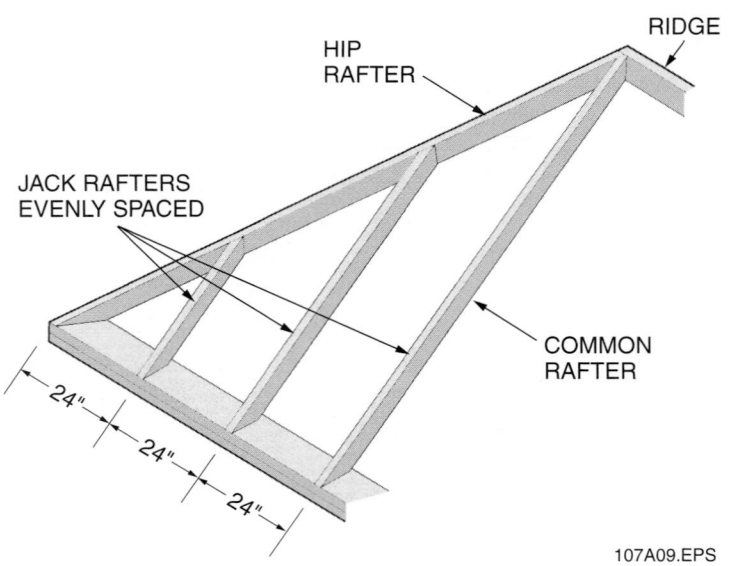

Figure A-9 ◆ Jack rafter locations.

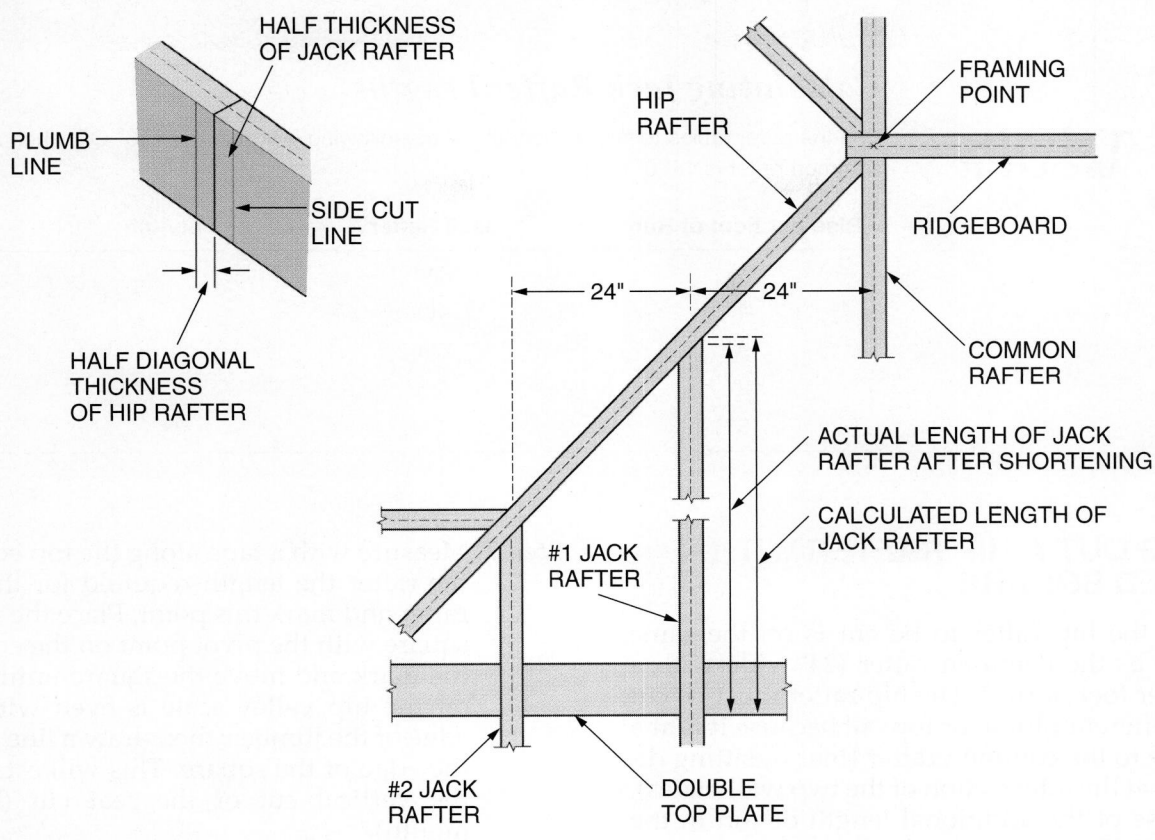

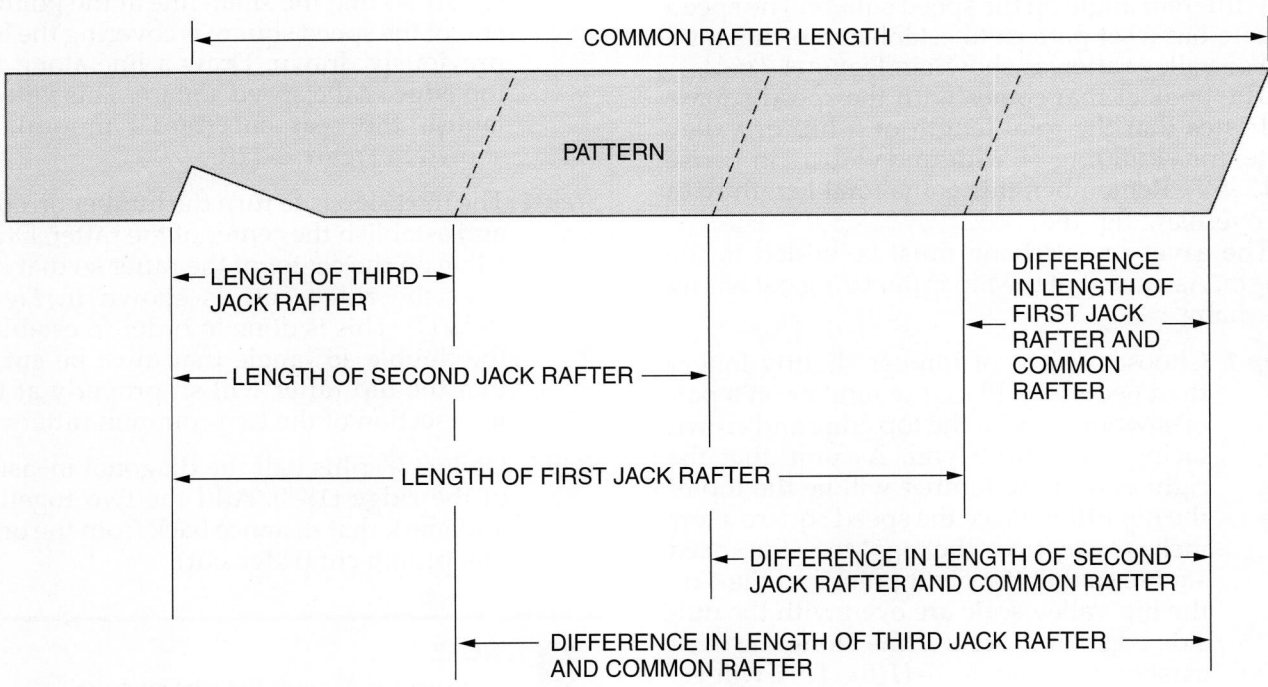

107A10.EPS

Figure A-10 ◆ Hip and jack rafter layout.

Calculating Jack Rafter Lengths

Use the rafter tables to find the lengths of the following jack rafters at 16" OC when a common rafter is 14'-6" long.

Rise per Foot of Run	Jack Rafter	Length
6"	2nd	
4"	4th	
9"	5th	
12"	3rd	
5"	4th	

LAYING OUT A HIP RAFTER WITH A SPEED SQUARE

Assume the hip rafter to be cut is on the same building as the common rafter (24' wide with a 4" rise per foot of run). The hip rafter must be cut with a different plumb or top cut because it is at a 45° angle to the common rafter (that is, sitting diagonally at the intersection of the two wall plates).

Because of the additional length of run in the hip rafter (17" of the run to 12" of run for a common rafter), the plumb or top cut must be figured at a different angle on the speed square. The speed square has a set pattern to establish the cuts for a hip or valley rafter, as shown in *Figure A-11(A)*.

The booklet that comes with the speed square indicates that the total length of a hip or valley rafter for a building 24' wide and with a 4 in 12 rise is 17'-5¼". Remember, this is the total length with no overhang figured.

The eaves or overhang must be added to the length. Assume that the hip rafter being cut has no overhang.

Step 1 Choose a piece of lumber slightly longer than necessary. Place the lumber on a pair of sawhorses with the top edge and crown facing away from you. Assume that the right end of the lumber will be the top of the hip rafter. Place the speed square a few inches away from the top. Move the speed square so that the pivot point and the 4 on the hip valley scale are even with the outside edge, and draw a line along the edge as shown in *Figure A-11(B)*. This will establish the top or plumb cut of the hip rafter.

Step 2 Measure with a tape along the top edge of the rafter the length required for the hip rafter and mark this point. Place the speed square with the pivot point on the edge of the mark and move the square until the 4 on the hip valley scale is even with the edge of the lumber; then draw a line along the edge of the square. This will establish the vertical cut of the seat cut (bird's mouth).

Step 3 To obtain the horizontal cut, reverse the square so that the small line at the pointed end of the speed square is covering the line previously drawn. Draw a line along the top edge of the speed square. This will establish the seat cut (bird's mouth), as shown in *Figure A-11(C)*.

Step 4 The next step is to turn the lumber on edge and establish the center of the rafter. Draw a line on the center of the rafter so that it is over the ridge cut, as shown in *Figure A-11(D)*. This is done in order to establish the double 45° angle that must be cut so that the hip rafter will sit properly at the intersection of the two common rafters.

Step 5 Deduct ¾" plus half the diagonal measure of the ridge (1⁄16"). Add the two together and mark that distance back from the original plumb cut (ridge cut).

NOTE

If the ridge is 2" stock, the total measure deducted is 1¹³⁄₁₆".

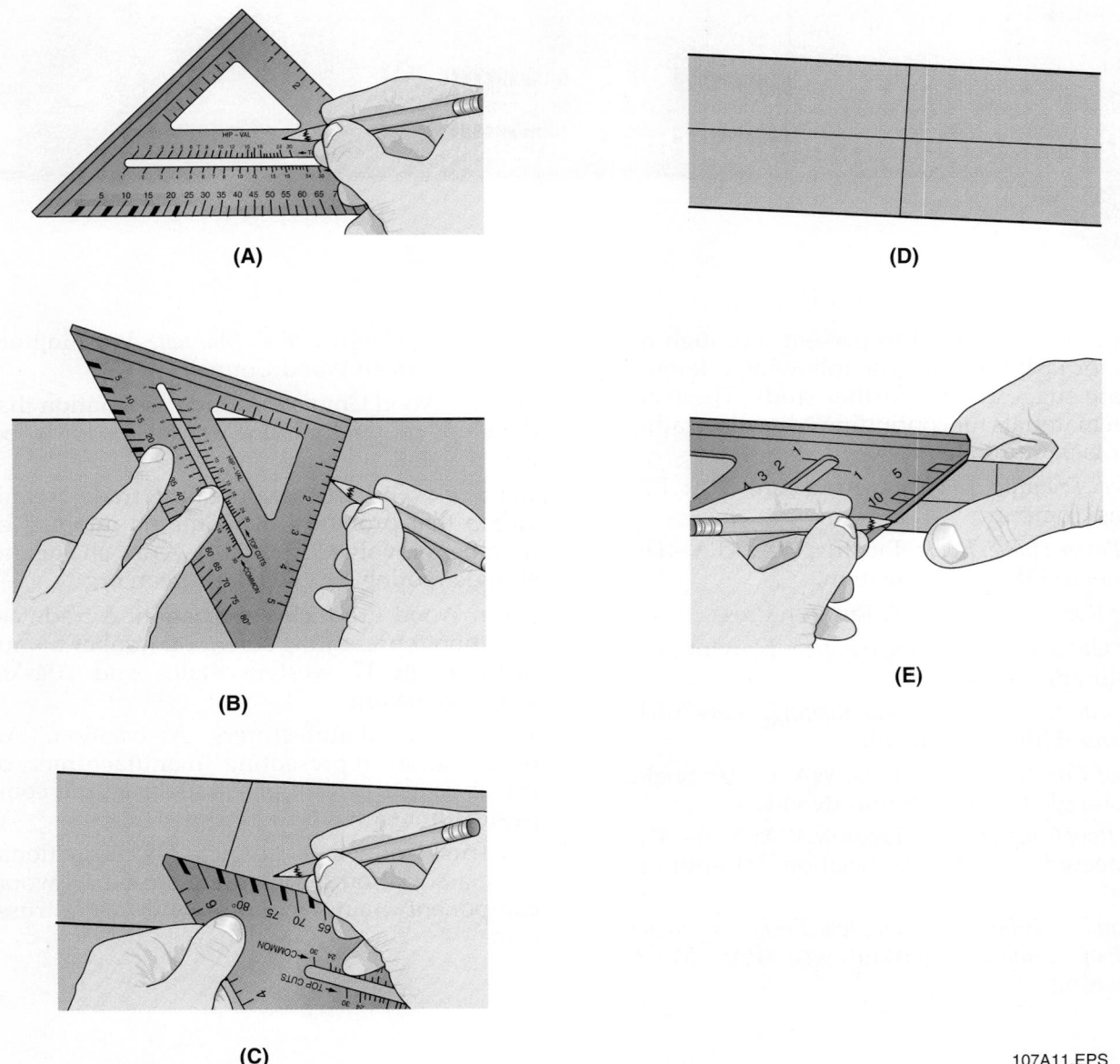

(A)

(B)

(C)

(D)

(E)

107A11.EPS

Figure A-11 ◆ Hip rafter layout.

Step 6 Using the speed square, draw a 45° line from the edge of the second line drawn to the center line and do the same on the opposite side of the center line, as shown in *Figure A-11(E)*.

Step 7 Make the ridge cuts and the seat cut. The hip rafter is now ready to be put in place. Be sure to check that the hip rafter is not up beyond the common rafter. The common rafter may need to be dropped before nailing the hip rafter in place. Use the hip rafter as a pattern to cut the other hip rafters.

The procedure outlined for making a hip rafter is the same procedure used to cut a valley rafter. There are two alternative methods of cutting the seat cut (bird's mouth) of the hip rafter. The first is to cut the rafter plate at the intersection of the two plates. The other alternative is to cut the seat cut at a double 45° angle.

The method of cutting a valley rafter is to cut a reverse 45° cut to fit into the intersecting corners.

Additional Resources and References

This module is intended to present thorough resources for task training. The following reference works are suggested for further study. These are optional materials for continuing education rather than for task training.

Advanced Framing Methods. Kingston, MA: R.S. Means Company.

Build a Better Home: Roofs. Tacoma, WA: APA – The Engineered Wood Association.

Framing Roofs. Newton, CT: Taunton Press.

Graphic Guide to Frame Construction. Newton, CT: Taunton Press.

Miller's Guide to Framing and Roofing. New York: McGraw-Hill Professional.

New Roof Construction. Sumas, WA: Cedar Shake and Shingle Bureau (15-minute video).

Quality Roof Construction. Tacoma, WA: APA – The Engineered Wood Association (15-minute video).

Roof Framer's Bible: The Complete Pocket Reference to Roof Framing. Jenkintown, PA: M.E.I. Publishing.

Wood Frame Construction Manual. Washington, D.C.: American Wood Council.

American Wood Council. A trade association that develops design tools and guidelines for wood construction. www.awc.org

Cedar Shake and Shingle Bureau. A trade organization that promotes the common interests of members involved in quality cedar shake and shingle roofing. www.cedarbureau.org

Western Wood Products Association. A trade association representing softwood lumber manufacturers in 12 western states and Alaska. www.wwpa.org

Wood I-Joist Manufacturers Association. An organization representing manufacturers of prefabricated wood I-joist and structural composite lumber. www.i-joist.org

Wood Truss Council of America. An international trade association representing structural wood component manufacturers. www.woodtruss.com

NCCER makes every effort to keep these textbooks up-to-date and free of technical errors. We appreciate your help in this process. If you have an idea for improving this textbook, or if you find an error, a typographical mistake, or an inaccuracy in NCCER's Contren® textbooks, please write us, using this form or a photocopy. Be sure to include the exact module number, page number, a detailed description, and the correction, if applicable. Your input will be brought to the attention of the Technical Review Committee. Thank you for your assistance.

Instructors – If you found that additional materials were necessary in order to teach this module effectively, please let us know so that we may include them in the Equipment/Materials list in the Annotated Instructor's Guide.

Write: Product Development and Revision
National Center for Construction Education and Research
3600 NW 43rd St, Bldg G, Gainesville, FL 32606

Fax: 352-334-0932

E-mail: curriculum@nccer.org

Craft _____ Module Name _____

Copyright Date _____ Module Number _____ Page Number(s) _____

Description _____

_____

_____

_____

_____

(Optional) Correction _____

_____

_____

(Optional) Your Name and Address _____

_____

_____

Roofing Applications

68109-09
Roofing Applications

Topics to be presented in this module include:

1.0.0   Introduction 9.2
2.0.0   Typical Roofing Materials 9.2
3.0.0   Tools 9.14
4.0.0   Safety 9.17
5.0.0   Preparation for Roofing Application 9.27
6.0.0   Composition Shingle Installation 9.30
7.0.0   Roll Roofing Installation 9.52
8.0.0   Wood Shingles and Shakes 9.57
9.0.0   Common Metal Roofing 9.68
10.0.0 Slate and Tile Roofing 9.73
11.0.0 Single-Ply Roofing Application 9.83
12.0.0 Torch-Down Roofing Application 9.85
13.0.0 Roof Ventilation and Ice Edging 9.87

Overview

As you travel around, take note of the different types of residential and commercial structures. Note how many different kinds of roof construction there are, and note how many different types of roofing materials are used. Part of your work as a carpenter will involve preparing roof decks to receive the finish roofing material, and you may even find yourself installing roofing material.

The roof is the most vulnerable part of a building. If it is not properly installed, it will leak. In some situations, it could even collapse. Safety is always a major consideration when working on a roof.

Objectives

When you have completed this module, you will be able to do the following:

1. Identify the materials and methods used in roofing.
2. Explain the safety requirements for roof jobs.
3. Install fiberglass shingles on gable and hip roofs.
4. Close up a valley using fiberglass shingles.
5. Explain how to make various roof projections watertight when using fiberglass shingles.
6. Complete the proper cuts and install the main and hip ridge caps using fiberglass shingles.
7. Lay out, cut, and install a cricket or saddle.
8. Install wood shingles and shakes on roofs.
9. Describe how to close up a valley using wood shingles and shakes.
10. Explain how to make roof projections watertight when using wood shakes and shingles.
11. Complete the cuts and install the main and hip ridge caps using wood shakes/shingles.
12. Demonstrate the techniques for installing other selected types of roofing materials.

Trade Terms

Asphalt roofing cement	Scrim
Base flashing	Selvage
Bundle	Side lap
Cap flashing	Slope
Exposure	Square
Head lap	Top lap
Overhang	Underlayment
Pitch	Valley
Ridge	Valley flashing
Roof sheathing	Vent stack flashing
Saddle	Wall flashing

Required Trainee Materials

1. Pencil and paper
2. Appropriate personal protective equipment

Prerequisites

Before you begin this module, it is recommended that you successfully complete *Core Curriculum,* and *Construction Technology,* Module 68101-09 through 68108-09.

This course map shows all of the modules in *Construction Technology.* The suggested training order begins at the bottom and proceeds up. Skill levels increase as you advance on the course map.

CONSTRUCTION TECHNOLOGY

68117-09
Copper Pipe and Fittings

68116-09
Plastic Pipe and Fittings

68115-09
Introduction to Drain, Waste, and Vent (DWV) Systems

68114-09
Introduction to HVAC

68113-09
Residential Electrical Services

68112-09
Electrical Safety

68111-09
Basic Stair Layout

68110-09
Exterior Finishing

68109-09
Roofing Applications

68108-09
Roof Framing

68107-09
Wall and Ceiling Framing

68106-09
Floor Systems

68105-09
Masonry Units and Installation Techniques

68104-09
Introduction to Masonry

68103-09
Handling and Placing Concrete

68102-09
Introduction to Concrete, Reinforcing Materials, and Forms

68101-09
Site Layout One: Distance Measurement and Leveling

CORE CURRICULUM:
Introductory Craft Skills

103CMAP.EPS

The local Training Program Sponsor may adjust the training order.

1.0.0 ◆ INTRODUCTION

Roofing materials are used to protect a structure and its contents from the elements. Besides rain protection, some materials are especially suitable for use in areas where fire, high wind, or extreme heat problems exist or in areas where cold weather, snow, and ice are problems. Materials can contribute to the attractiveness of the structure due to the careful selection of texture, color, and pattern. However, the design of the structure as well as local building codes may limit the choice of materials because of the pitch of the roof or because of other considerations at a particular location. In any case, the project specifications must be checked to determine the type of roofing materials to be used.

2.0.0 ◆ TYPICAL ROOFING MATERIALS

Many types of roofing materials as well as commercial roofing systems are available today. This section will cover the most common materials used on residential and small commercial structures. Commercial roofing systems used on larger or more expensive structures will be described in more detail in a later level.

2.1.0 Composition Shingle Roofing

Composition shingles (*Figure 1*) are the most common roofing material in North America. They are available in a wide variety of colors, textures, types, and weights (thicknesses). The standard shingle is made of a fiber or fiber-mat material, coated or impregnated with asphalt, and then coated with various mineral granules to provide color, fire resistance, and ultraviolet protection.

In the past, composition shingles were made using asbestos fiber or organic fiber and were commonly referred to as asphalt shingles. The manufacture of asbestos-fiber shingles has been prohibited because the asbestos poses a cancer risk and environmental disposal hazard.

Organic fiber shingles, which have a lifespan of 15 to 20 years, have been largely replaced by asphalt-coated, fiberglass mat shingles, simply called fiberglass shingles, which have a lifespan of 20 to 25 years.

More expensive architectural shingles with lifespans of 25 to 40 years or more are also available. Architectural shingles are constructed of multiple layers of fiberglass that are laminated

INSIDE TRACK

Metric Shingles

Most shingles manufactured today are available in foot-pound (English) and metric dimensions. The standard foot-pound, three-tab shingle or laminated architectural shingle is 36" long. The metric versions are 39⅜" long.

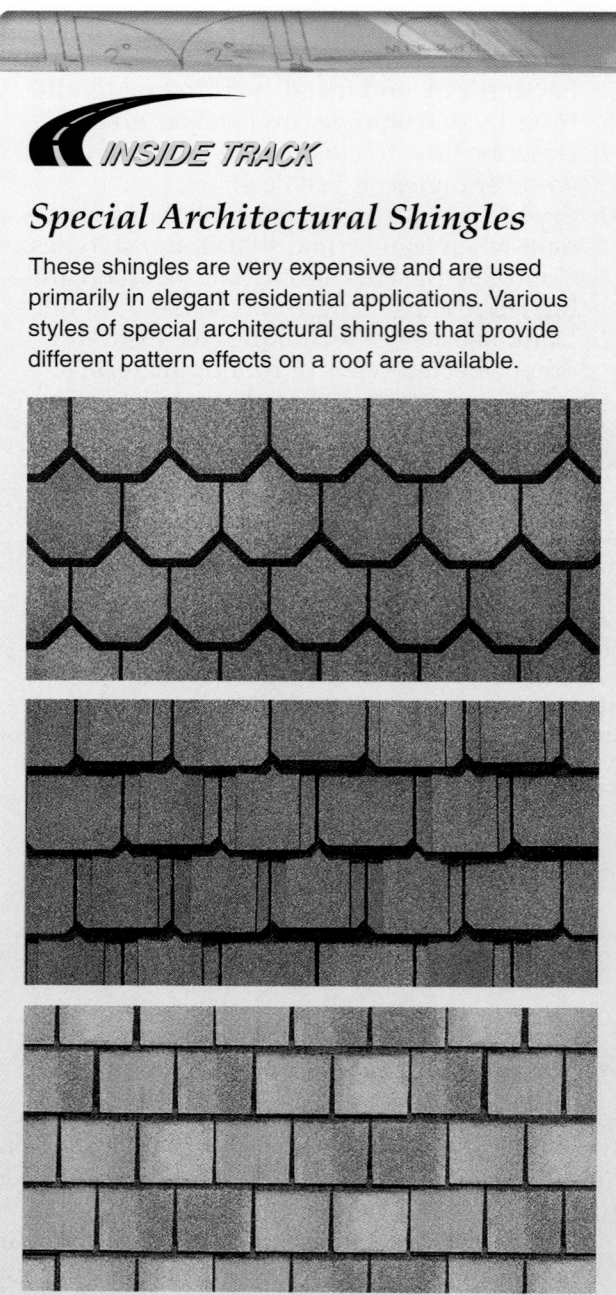

INSIDE TRACK

Special Architectural Shingles

These shingles are very expensive and are used primarily in elegant residential applications. Various styles of special architectural shingles that provide different pattern effects on a roof are available.

202SA01.EPS

together when manufactured or job-applied in layers to create a heavy shadow effect (*Figure 2*). Manufacturers may also provide their shingles in fungus/algae-resistant versions for use in damp locations where unsightly fungus growth or black streaks caused by algae tend to be a problem. The fungus/algae resistance is provided by copper granules incorporated with the mineral aggregates that are bonded to the surface of the shingle.

Composition shingles generally suit every climate in North America and can normally be applied to any roof with a **slope** of 4 in 12 (*Figure 3*) up to 21 in 12, provided they have factory-applied, seal-down adhesive strips. By applying double-lap **underlayment**, they can be

ARCHITECTURAL SHINGLE

THREE-TAB SHINGLE

202F01.EPS

Figure 1 ◆ Typical composition shingles.

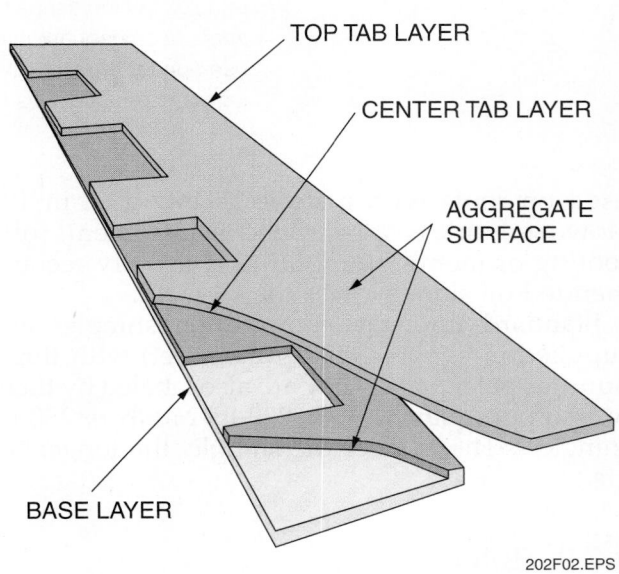

TOP TAB LAYER

CENTER TAB LAYER

AGGREGATE SURFACE

BASE LAYER

202F02.EPS

Figure 2 ◆ Typical architectural shingle.

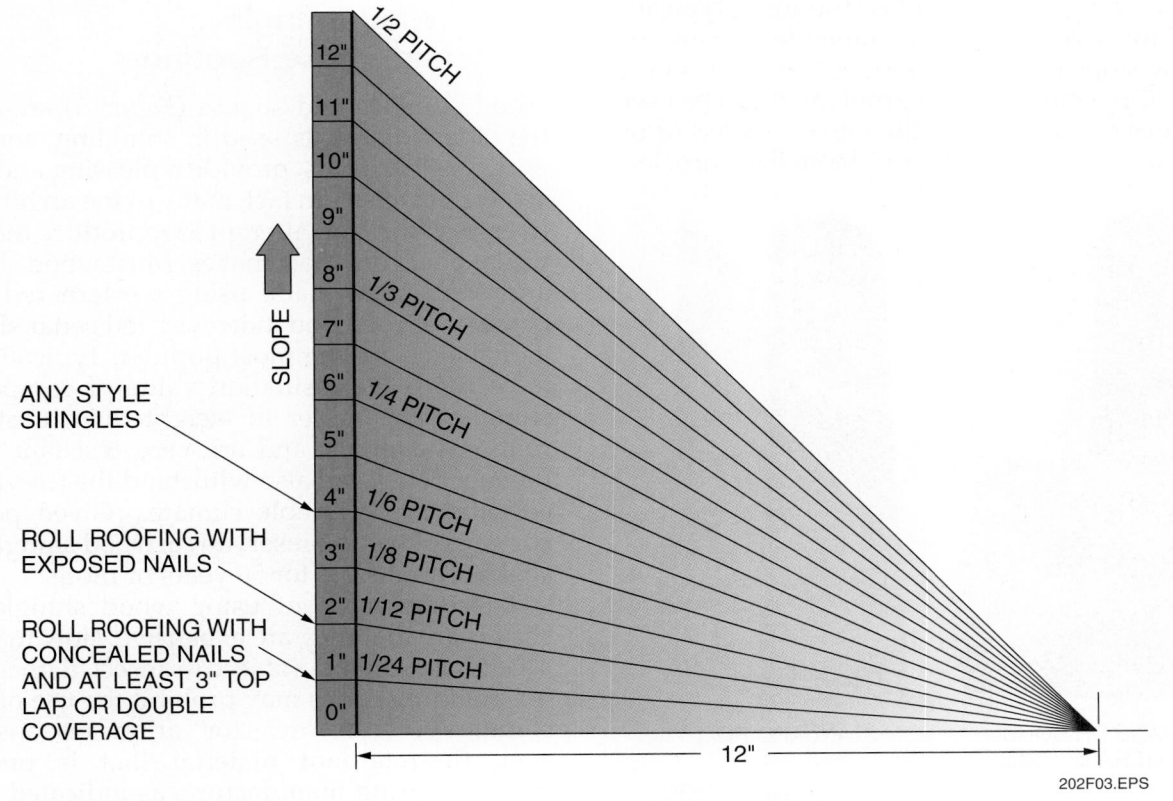

12"
1/2 PITCH
11"
10"
9"
8"
1/3 PITCH
7"
6"
1/4 PITCH
5"
4"
1/6 PITCH
3"
1/8 PITCH
2"
1/12 PITCH
1"
1/24 PITCH
0"

SLOPE

ANY STYLE SHINGLES

ROLL ROOFING WITH EXPOSED NAILS

ROLL ROOFING WITH CONCEALED NAILS AND AT LEAST 3" TOP LAP OR DOUBLE COVERAGE

12"

202F03.EPS

Figure 3 ◆ Typical shingle or roll roofing applications for various roof slopes/pitches.

Steep Slope Applications

On steep-slope roofs greater than 21 in 12, such as a mansard or simulated mansard roof, the built-in sealant strips on shingles do not generally adhere to the underside of the overlaying shingles. Because of this, each shingle must be sealed down with spots of quick-setting asphalt cement at installation. On standard three-tab shingles, two spots of cement are used under each tab. On strip shingles, three or four spots are used under the strip. On laminated architectural shingles, extra nails and four spots of cement are usually used for the shingle.

used on roofs with a slope as low as 2 in 12. However, unless appearance is a problem, roll-roofing or membrane roofing is usually recommended on slopes lower than 4 in 12.

Standard three-tab composition shingles are supplied in **squares** (100 square feet) with three **bundles** per square. They are also labeled by their weight per square (210-lb, 230-lb, 240-lb, or 250-lb shingles). The heavier the shingle, the longer its life.

2.2.0 Roll Roofing

Roll roofing (*Figure 4*) is available in 50-lb to 90-lb weights with the same materials and colors as composition shingles. While it is inexpensive and quick to apply, the life of rolled roofing is typically only 5 to 12 years. It is recommended for use on shallow-slope roofs with slopes of less than 4 in 12 where appearance is not a problem. It can be used on slopes of 2 in 12 with the nails revealed or on slopes as low as 1 in 12 with the nails concealed.

SELVAGE OR DOUBLE COVERAGE ROLL   STANDARD ROLL

202F04.EPS

Figure 4 ◆ Typical roll roofing.

Roll roofing can also be used as **valley flashing** to match the color of the roof shingles. It is supplied in three types. One type is a smooth surface roofing not covered with any granules. It is used on roofs that are subsequently covered with a hot or cold asphalt and separately applied aggregate. The other two types are for finish roofs. One is completely covered with granules and the other type, called **selvage** or double-coverage roll roofing, is only half covered with granules and is designed to be applied with a cemented-down half lap, with concealed fasteners used on very low slopes. The rolls are normally 36" wide. However, granule-covered, half-width, starter-strip rolls are also available.

2.3.0 Wood Shingle and Shake Roofing

Wood shingles and shakes (*Figure 5*) are among the oldest materials used in shingling, and both, particularly shakes, provide a pleasing and desirable visual effect. In fact, many of the architectural fiberglass shingles attempt to reproduce the rustic visual effect of wood shakes. Most wood shingles and shakes are made using western red cedar, cypress, or redwood; however, red cedar shingles and shakes are the most popular. Typically, they have twice the insulation value of composition shingles, are lighter in weight than most other roofing materials, and are very resistant to hail damage. They can also withstand the freeze-thaw conditions of variable climates. Given periodic coatings of wood preservative, wood shingles and shakes should last for 50 years or more.

The drawbacks of using wood shingles and shakes are that they are expensive, slow to install, a fire hazard, and subject to insect damage and rot. Building codes may prohibit the use of wood shingles in certain areas for various reasons; however, fire-retardant material that is pressure-injected during manufacture, as indicated by the industry designation Certi-Guard™, is supposed

Figure 5 ◆ Typical wood shake and shingle.

to conform to all state and local building codes for use in fire hazard regions.

Wood shingles are machine-cut and smoothed on both sides. They are uniform in thickness and length but vary in width. Hand-split wood shakes, which are more expensive, are either hand-split on one side and machine smoothed on the other side, or hand-split on both sides. The hand splitting produces a rough, rustic surface. Hand-split shakes vary in thickness and width, but are uniform in length. Because of the varying thickness, as well as the rough surface of at least one face of the shakes, an underlayment is used between each course of the shakes to prevent wind-driven rain from being forced back under the shakes. Machine sawn and grooved shakes that simulate hand-split shingles are also available.

Wood shingles and shakes are also available as 4' to 8' pre-bonded panels that are in two-ply and three-ply sections with **exposures** of 5½" to 9" (*Figure 6*). These panelized shingles and shakes, with course-separating underlayment pre-installed, are more expensive than traditional shakes and shingles. However, they are claimed to be up to two times faster to install than composition shingles and up to four times faster to install than traditional wood shakes and shingles. A simulated wood shingle made of wood fiber composition hardboard is also available in panel form. The panels are 12" × 48" and are applied lengthwise across the roof. They are embossed with deep shadow lines and random-cut grooves that mimic the look of shakes. The panels overlap with a shiplap joint between courses and between panels in the same course. After exposure, they weather to a gray that is similar to cedar shingles. Simulated shingle and shake metal roofing panels that are 4' long and completely fireproof are also available. These metal panels are very much like metal siding. They are applied horizontally and have an interlocking joint at the top and bottom edges.

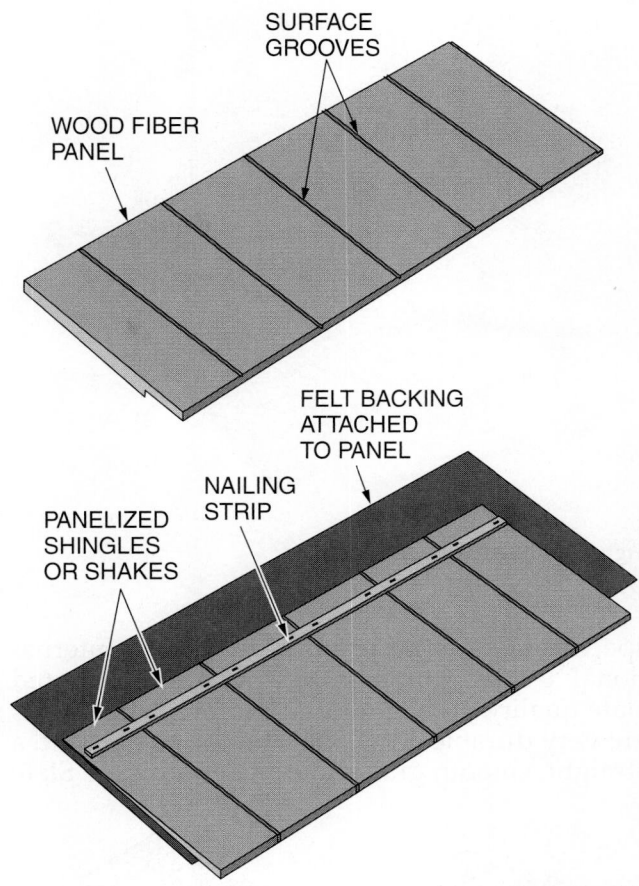

Figure 6 ◆ Panelized shakes or shingles and wood fiber hardboard panel.

2.4.0 Slate Roofing

Today, real slate roofing (*Figure 7*) is probably the most expensive roofing option in terms of roof framing materials, roofing materials, and increased installation time. Even though it has a very long service life of 60 to 100 years or more, it is usually not financially practical to use slate unless it is required or desired strictly for architectural purposes. No other material can match the high-quality look of slate. Unfortunately, besides being expensive, slate is heavy—about 7 to 10 lbs per square foot. As a result, the roof framing must be engineered to be substantially stronger to support the slate load, as well as any anticipated snow and ice loads.

Slate is completely fireproof and rot-proof. It is available in a wide variety of grades, thicknesses, and colors ranging from gray or black to shades of green, purple, and red. The colors are qualified as unfading or, if subject to some change, weathering. The industry often uses the old federal grading system of A (best) through C to indicate the quality of slate; however, architectural

Figure 7 ◆ Slate roofing.

202F07.EPS

types with streaks of color usually contain intrusions of sand or other impurities and are not as durable. Premium quality slate in various colors can be obtained from quarries from Maine to Georgia. Copper slater's nails and flashing are normally used with slate roofs.

Synthetic slate is made of fiber mat (usually fiberglass) that has been impregnated and coated with cement. It looks like real slate, but is lighter in weight. Even though it is lighter than slate, it still requires strong roof framing for support. Synthetic slate is fireproof and can last 40 years or more. Synthetic slate is still a relatively expensive material and is typically used on structures where historically appropriate materials are required. The synthetic slate material is difficult to cut and must be carefully fastened to avoid cracking the shingles.

specifications normally use the ASTM International testing numbers to specify the desired slate quality. New York and Vermont slate types are very durable and have a uniform color and a straight, smooth grain running lengthwise. Slate

2.5.0 Tile Roofing

Like slate, glazed and unglazed clay and ceramic tile roofing products (*Figure 8*) are fireproof and rot-proof, and last from 50 to 100 years. Tile roofing is expensive and heavy (7 to 10 pounds per

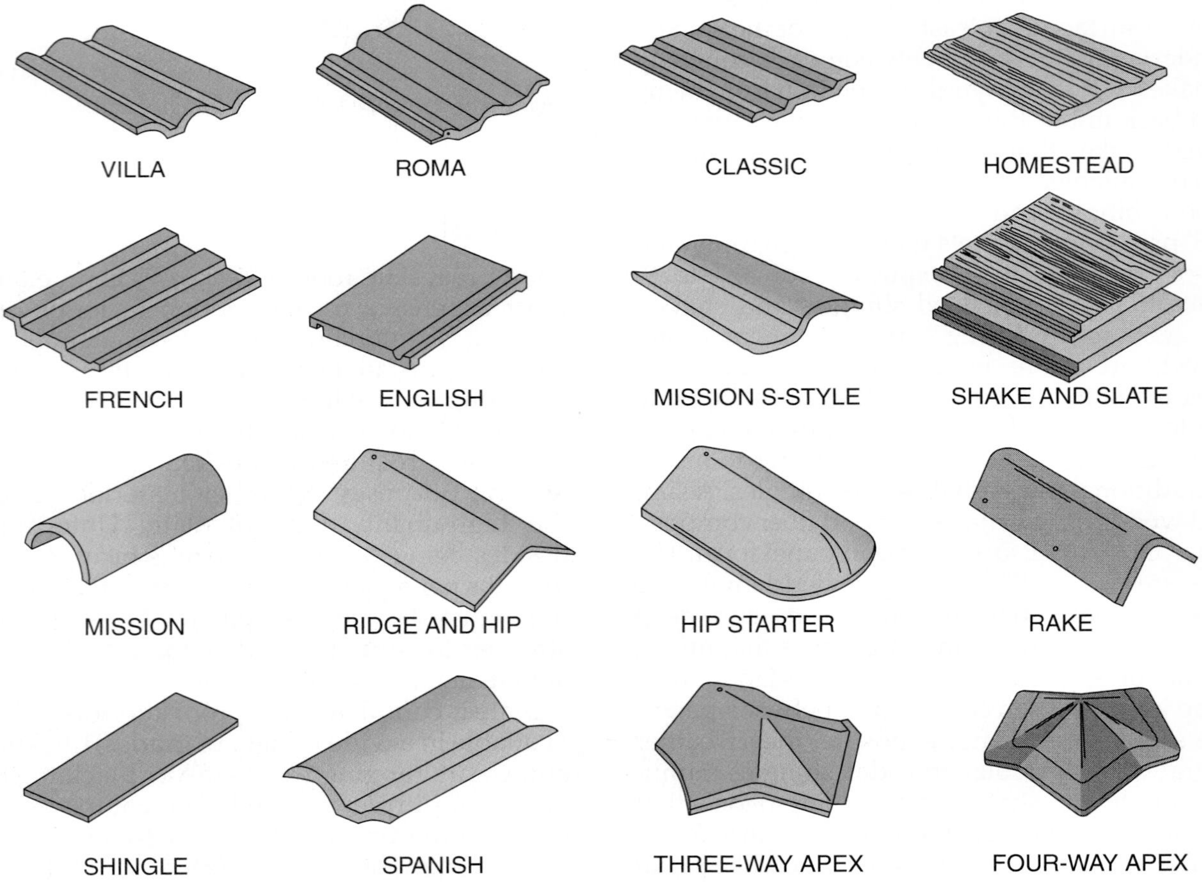

VILLA

ROMA

CLASSIC

HOMESTEAD

FRENCH

ENGLISH

MISSION S-STYLE

SHAKE AND SLATE

MISSION

RIDGE AND HIP

HIP STARTER

RAKE

SHINGLE

SPANISH

THREE-WAY APEX

FOUR-WAY APEX

202F08.EPS

Figure 8 ◆ Typical tile roofing styles.

Synthetic Tiles, Shakes, and Shingles

One alternative to clay and ceramic tiles is synthetic tile made of fiber-reinforced concrete or shredded tire material. These tiles are very durable and lighter than clay or ceramic tiles. In some cases, they may be light enough to install on roofs intended to support composition shingles. Synthetic tiles are available in many colors and styles. Like clay and ceramic tiles, synthetic tiles are fireproof and rot-proof. They last almost as long as clay and ceramic tiles. Other alternatives to clay, ceramic, or concrete tiles are interlocking baked-enamel steel and vinyl-coated aluminum tiles and shingles. These types of tiles and shingles have a life expectancy exceeded only by real slate or tile. They are available in styles similar to shakes, slate, and tile roofing and are light as well as less expensive.

SLATE

SHAKE

202SA02.EPS

square foot) and requires appropriate roof framing to support the tile weight and any other anticipated loads. It is used in the South and to a great extent in the southwest areas of the country because it is fireproof and impervious to damage caused by intense sunlight. It is available in Spanish, Mission (barrel or S-type), and other styles known by various names such as French, English, Roma, and Villa. Other styles may also include flat tiles that may resemble slate or wood shakes. Matching hip and **ridge** tiles, rake/barge tiles, and hip starter tiles are also available. Every tile for a particular style furnished by a manufacturer is a uniform size in width, thickness, and length.

Tiles are fastened with non-corrosive nails (copper, galvanized, or stainless steel). Copper nails have the advantage of being soft enough to allow expansion and contraction without causing the tiles to crack if they are fastened too tight. They also allow easier replacement of any broken tiles. Because of its durability and resistance to any alkaline corrosion, copper is generally used as flashing.

2.6.0 Metal Roofing

Metal roofing is available in a great variety of materials and styles. These materials can be purchased with a baked enamel, ceramic, or plastic coating. They can also be purchased without a coating so that a separate roof coating can be applied after installation. The following are some common metal roofing materials:

- Aluminum (plain or coated)
- Galvanized steel (plain or coated)
- Terne metal (heat-treated, copper-bearing steel hot-dipped in terne metal comprised of 80 percent lead and 20 percent tin)
- Aluminum/zinc-coated steel (plain or coated)
- Stainless steel

Besides the common residential panel roofing styles (*Figure 9*), many new engineered/pre-formed, architectural metal fascia/roofing systems (*Figure 10*) are available for commercial or residential use in a wide selection of colors and in styles that include shingles, panels, and tiles.

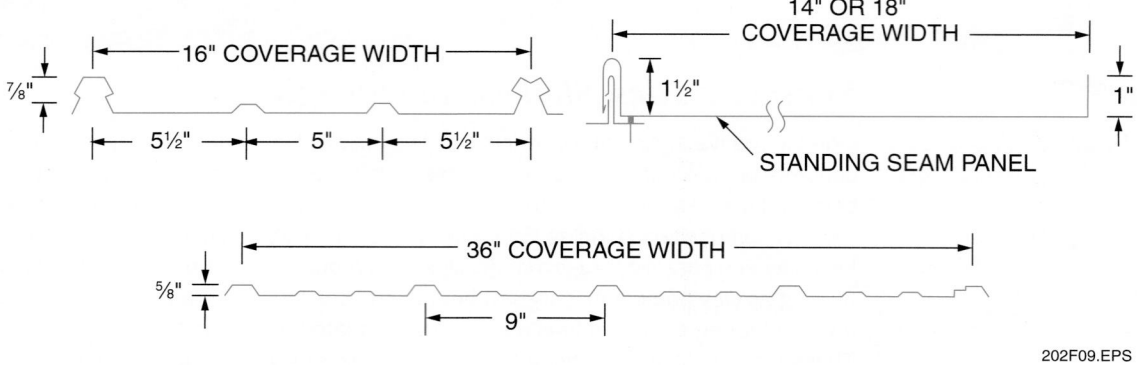

Figure 9 ◆ Common metal roofing styles.

202F09.EPS

202F10.EPS

Figure 10 ◆ Example of architectural metal fascia/roofing system.

2.7.0 Built-Up Roofing Membrane

Conventional built-up roofing (BUR) membrane (*Figure 11*) has been used for over 100 years on very low-slope roofs of residential and commercial structures. While it is still a viable form of roofing, it is gradually being replaced by premanufactured membrane roofing systems.

Built-up roofing, which is field-fabricated, consists of three to five layers of heavy, asphalt-coated polyester/fiberglass felt embedded in alternate layers of hot-applied or cold-applied bitumens (coal tar or asphalt based) that are the waterproofing material. Most commercial applications use hot asphalt as the waterproofing material. The top surface layer of bitumen is sometimes left smooth but is most often covered with either a mineral-coated cap sheet or embedded with a loose mineral surface consisting of small aggregates such as washed gravel, pea rock, or crushed stone.

Hot asphalt applications require special heating equipment along with some method of transporting the hot asphalt to the roof unless the heating equipment is lifted and positioned onto the roof. Installing this type of roof is labor-intensive and requires experienced roofers. Because the membrane is created in the field, its quality is subject to many variables, including the weather, application techniques, and the experience of the roofers. Most BUR is placed over one or more insulation boards that are bonded or fastened to the roof substrate.

The lifespan of a correctly applied, built-up roof is about 10 to 20 years, depending on the number of layers. Generally, specific damage to this type of roof can be easily repaired. However, the normal, gradual deterioration of the roof, which may result in deep splits over much of the surface or delamination of the layers, will generally require that the roof be completely removed and a new roof applied.

2.8.0 Premanufactured Membrane Roofing

There are two general categories of premanufactured membrane roofing systems: modified bitumen systems or single-ply systems. The single-ply membrane systems are wholly synthetic roofing materials that exhibit elastomeric (rubber-like) properties to various degrees. There are a number of types within these two general systems. Today, there are many manufacturers of both systems, and it is important to note that each manufacturer requires specific, compatible materials and accessories for their versions of each type within each system. These materials and accessories include flashing, fasteners, drain and vent boots, inside/outside corners, cant strips (coving), cleaners, solvents, adhesives, caulking, sealers, and tapes. To obtain the maximum performance from the system

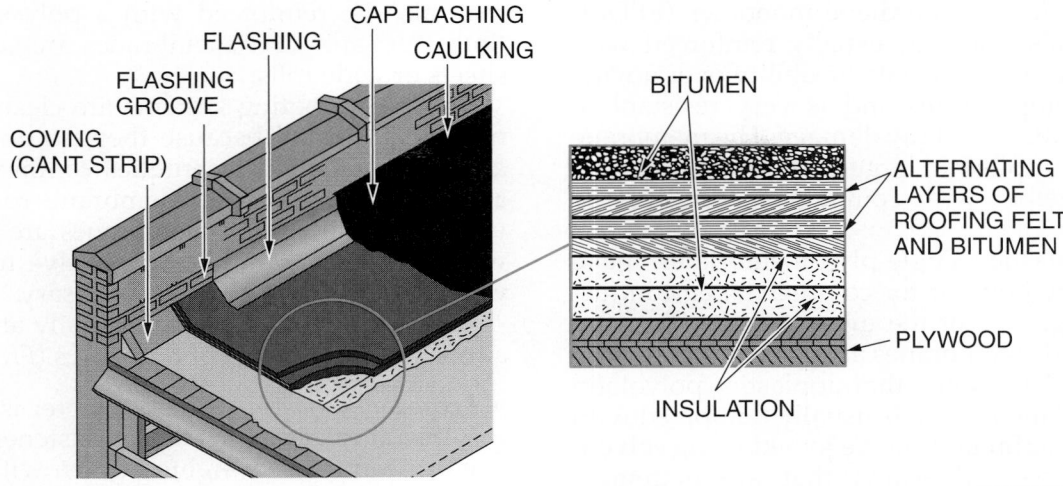

COVING (CANT STRIP)
FLASHING GROOVE
FLASHING
CAP FLASHING
CAULKING
BITUMEN
ALTERNATING LAYERS OF ROOFING FELT AND BITUMEN
PLYWOOD
INSULATION

202F11.EPS

Figure 11 ◆ Conventional built-up roofing membrane.

and for warranty purposes, compatible materials and accessories must be used and manufacturer-specified application procedures must be strictly followed. Normally, these systems are installed over insulation boards by roofers experienced with a specific manufacturer's product.

2.8.1 Modified Bitumen Membrane Roofing Systems

Modified bitumen roofing systems can be classified as either styrene butadiene styrene (SBS) or atactic polypropylene (APP) modified bitumen products. The SBS products are usually a composite of polyester or glass fiber and modified asphalts coated with an elastomeric blend of asphalt and SBS rubber. The APP products are coated with an elastomeric blend of asphalt and atactic polypropylene. Both products are the weatherproofing medium and are used in a hybrid BUR as a cap sheet over one or more base/felt plies that have been secured with hot asphalt or cold adhesive to a deck board covering the insulation or directly to the insulation.

Depending on the manufacturer, SBS products are either supplied with a preapplied adhesive and the product is rolled for adhesion, or the product is secured with hot asphalt or a cold adhesive applied separately. APP products, called torch-down roofing, are usually secured to the layers below by heat welding. Using flame heating equipment, the back of the product roll is heated as the product is unrolled and pressed down. The back coating of the product is heated to the point where the bitumen coating acts as an adhesive, bonding the product to the layer below and the overlapped edges of the adjacent sheet.

Flame heating equipment (*Figure 12*) can be used for BUR and torch-down roofing systems. Both types of systems require a mineral covering or protective coating to prevent ultraviolet destruction of the modified bitumen materials. In some cases, the product is available with various colors of ceramic roofing granules preapplied to the exposed surface.

2.8.2 Single-Ply Membrane Roofing Systems

Single-ply roofing systems can be classified as either thermoplastic (plastic polymer) or thermoset (rubber polymer) systems. Within these classifications, a number of types exist. One of the most common thermoset polymer membranes is

202F12.EPS

Figure 12 ◆ Flame heating equipment for BUR and torch-down roofing.

an ethylene propylene diene monomer (EPDM) product. This polymer, usually reinforced with polyester scrim, retains its flexibility over a wide range of temperatures and is very resistant to ozone and ultraviolet ray damage. The membrane is usually supplied in large sheets or rolls and is spliced together with compatible adhesives or tapes.

Thermoplastic single-ply membranes have become very popular for commercial and industrial roofing. Two of the most common types of thermoplastic membranes available are polyvinyl chloride (PVC) and thermoplastic polyolefin (TPO). PVC membrane is usually reinforced with a polyester scrim and can be joined using solvent or hot-air welded seams that are extremely durable (*Figure 13*). PVC membrane is lightweight and aesthetically pleasing. It is very resistant to ozone and ultraviolet ray damage. It is also puncture- and tear-resistant. TPO membrane combines the advantages of the solvent or hot-air seam welding capability of PVC with the greater weatherability and flexibility benefits of the more traditional EPDM membranes. TPO membranes

may also be reinforced with a polyester scrim. Both PVC and TPO membranes are supplied in sheets or wide rolls.

Single-ply roofing systems are clean and economical to install because they do not use hot-asphalt installation techniques common to BUR and modified bitumen membrane roofing systems. Most single-ply membranes are underlaid with insulation boards or protective mats along with a fireproof slipsheet, if necessary.

Single-ply membranes are usually anchored to a roof structure in one of four ways (*Figure 14*).

- *Loose laid/ballasted* – The perimeter is anchored with adhesives or mechanical fasteners, and the entire surface is weighted down with a round stone ballast or walking pavers (thin concrete blocks). This method is used only on very low-slope roofs capable of supporting the ballast load. It is fast and economical.
- *Partially adhered* – The entire area of the membrane is spot-adhered to the roof with mechanical fasteners and/or adhesives. This method produces a membrane with a dimpled, wind-resistant surface.
- *Mechanically adhered* – The entire area of the membrane is spot-adhered to the roof with mechanical fasteners. This method produces a dimpled, wind-resistant installation and allows easy removal of the membrane during future roof replacement.
- *Fully adhered* – The entire area of the membrane is completely cemented down with an adhesive. This method produces a smooth, windproof surface.

2.9.0 Drip Edge and Flashing

Figure 15 shows various types of drip edges that are available. Some are used in reroofing applications only. Drip edges and any other flashing must be made of materials that are compatible with the roofing and any items that are being flashed. They must last as long as the finish roof

202F13.EPS

Figure 13 ◆ Single-ply membrane installation.

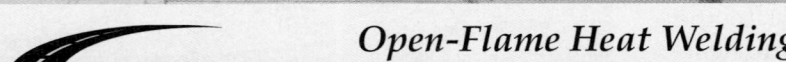

INSIDE TRACK

Open-Flame Heat Welding

Today, most seam-sealing methods and equipment recommended by manufacturers of single-ply membrane roofing systems make use of hot-air welding methods that are quite safe. However, some contractors still use open-flame heating equipment for seam welding. Open-flame seam sealing can be hazardous and may cause damage to the membrane. For these reasons, open-flame seam sealing of polymer membrane roofing should be avoided.

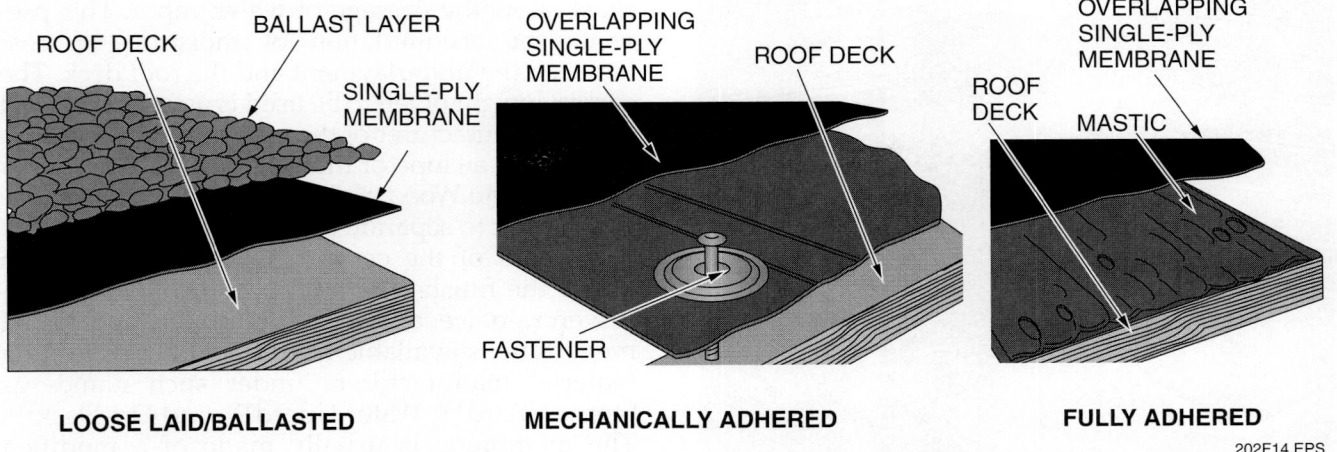

LOOSE LAID/BALLASTED   MECHANICALLY ADHERED   FULLY ADHERED

202F14.EPS

Figure 14 ◆ Typical methods of anchoring single-ply membrane to a roof.

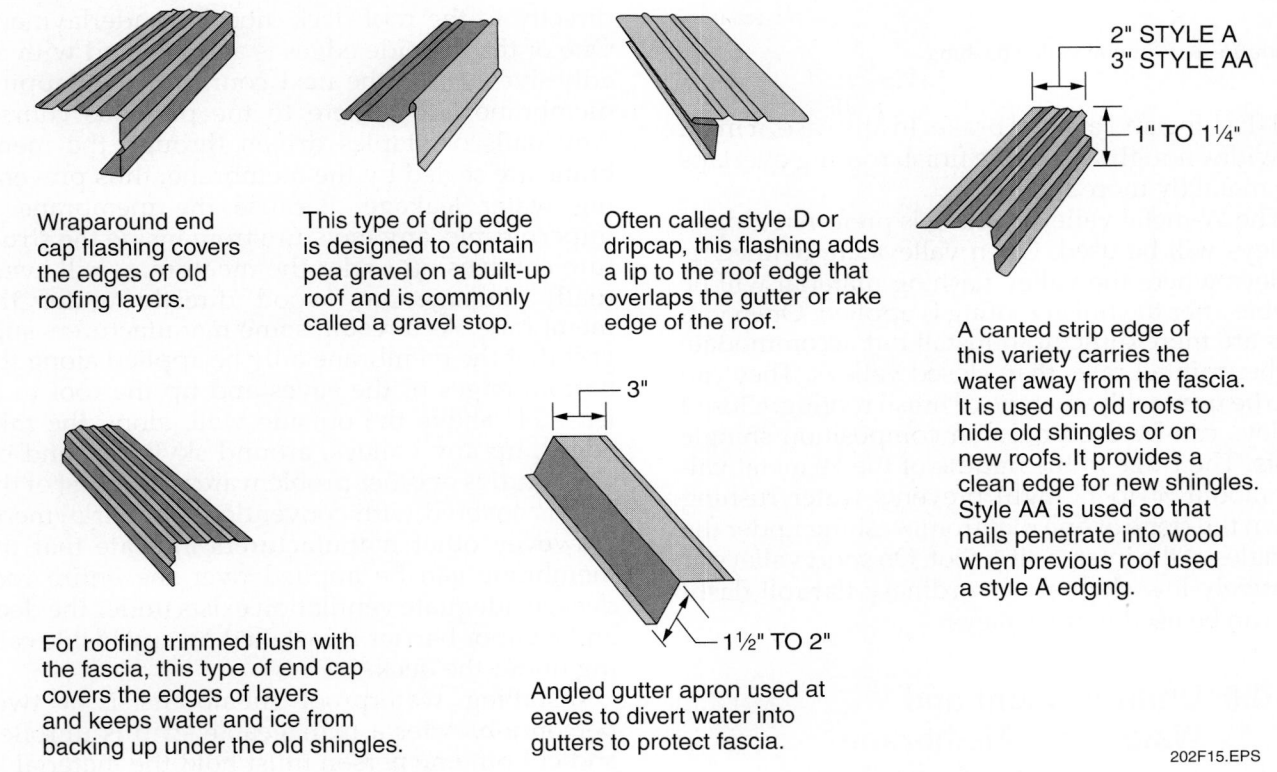

Wraparound end cap flashing covers the edges of old roofing layers.

This type of drip edge is designed to contain pea gravel on a built-up roof and is commonly called a gravel stop.

Often called style D or dripcap, this flashing adds a lip to the roof edge that overlaps the gutter or rake edge of the roof.

2" STYLE A
3" STYLE AA
1" TO 1¼"

A canted strip edge of this variety carries the water away from the fascia. It is used on old roofs to hide old shingles or on new roofs. It provides a clean edge for new shingles. Style AA is used so that nails penetrate into wood when previous roof used a style A edging.

For roofing trimmed flush with the fascia, this type of end cap covers the edges of layers and keeps water and ice from backing up under the old shingles.

3"
1½" TO 2"

Angled gutter apron used at eaves to divert water into gutters to protect fascia.

202F15.EPS

Figure 15 ◆ Various types of drip edges.

material. Today, aluminum, galvanized steel, copper, vinyl, and stainless steel are the most common materials used for drip edges and flashing. Galvanized steel, copper, or stainless steel must be used with any cement-based roofing material or against masonry materials due to the corrosive nature of cement. Copper is normally used with slate roofing because of its long life and resistance to corrosion.

For best results, any galvanized drip edge or flashing should be coated with an appropriate primer before installation. All drip edges and flashing must be fastened with nails or staples made of a compatible material to prevent electrolytic corrosion between the fasteners and the flashing. The roofing material manufacturer's recommendations for flashing must be followed.

Figure 16 shows a W-metal (so named because its end profile looks like a letter W) or standing-seam **valley** flashing. This type of valley flashing is available in 8' to 10' lengths in widths of 20" to 24". If desired, it can be field-fabricated using flat

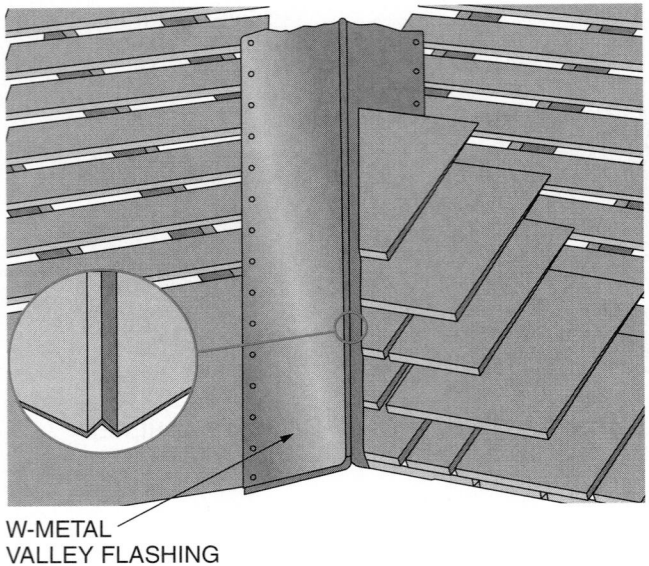

W-METAL
VALLEY FLASHING

202F16.EPS

Figure 16 ◆ W-metal valley flashing.

roll flashing and a metal brake. In any case, it must be wide enough so that the finish roofing overlaps the metal by more than 6".

The W-metal valley flashing is preferred if open valleys will be used. Open valleys are defined as valleys where the valley flashing material will be visible after the finish roofing is applied. Open valleys are more difficult to install but accommodate higher rainfall rates than closed valleys. They can also be used with any type of finish roofing. Closed valleys can be used only on composition shingle roofs. The ridge in the middle of the W-metal valley (about ¾" to 1" high) prevents water rushing down the slope of one roof from washing under the shingles of the intersecting roof. On short valleys or relatively low-slope roofs, ordinary flat-roll flashing can be used in the valleys.

2.10.0 Underlayment and Waterproof Membrane

Underlayment is available as 60-lb rolls of nonperforated, 15-lb, 30-lb, and 60-lb asphalt-saturated felt (*Table 1*). The felt used under roofing materials must allow the passage of water vapor. This prevents the accumulation of moisture or frost between the underlayment and the roof deck. The correct weight of felt to be used is usually specified by the manufacturer of the finish roofing material. Roofs with a slope of more than 4 in 12 normally use 15-lb felt. Wood shakes usually require the use of 30-lb felt to separate the courses.

In areas of the country where water backup under the finish roof is a problem due to wind-driven rain, ice, and snow buildup, a waterproof membrane is available from a number of roofing material manufacturers under such names as Storm-Guard™, Water-Guard™, and Dri-Deck™. The membrane is usually made of a modified asphalt-impregnated fiberglass mat, coated on the bottom side with an elastic-polymer sealer that is also an adhesive. The membrane must be applied directly to the roof deck, not the underlayment. One of the top side edges is also covered with an adhesive so that the next course of overlapping membrane will adhere to the previous course. Any nails or staples driven through the membrane are sealed by the membrane, thus preventing water leakage. Because the membrane is impermeable, any moisture from inside the structure condensing under the membrane will eventually damage any wood directly under the membrane. As a result, some manufacturers suggest that the membrane only be applied along the bottom edges of the eaves and up the roof to at least 24" above the outside wall, along the rake edges, up any valleys, around skylights, and on any **saddles** or other problem areas. The rest of the roof is covered with conventional underlayment. However, other manufacturers indicate that the membrane can be applied over the entire roof deck, if adequate ventilation exists under the deck and a vapor barrier is installed on any inside ceiling under the deck.

Installing waterproof membrane is a two-person job. After a manageable strip is unrolled and cut off, one person must hold the material in place while another peels away a protective sheet covering the bottom adhesive. The membrane adhesive remains tacky during installation and

Table 1	Sizes, Weights, and Coverage of Asphalt-Saturated Felt						
Approximate Weight per Roll	Approximate Weight per Square	Squares per Roll	Roll Length	Roll Width	Side or End Laps	Top Lap	Exposure
60#	15#	4	144'	36"	4" to 6"	2"	34"
60#	30#	2	72'	36"	4" to 6"	2"	34"
60#	60#	1	36'	36"	4" to 6"	2"	34"

Waterproof Membranes

Waterproof membranes used on roof edges and valleys are self-healing. This means that if the membrane is intentionally or accidentally penetrated by a screw or nail, a modified asphalt coating will flow to seal the penetration when the roof and membrane are heated by the sun. Because they have an adhesive that secures them to the roof deck, waterproof membranes are virtually impossible to remove once the adhesive is heat-set by the sun. This makes reroofing of a structure very expensive if the membrane must be removed. This is because the roof deck must be replaced if it is wood or a wood product. Newer versions of membranes are available with a granular surface that allows the overlying shingles to be removed without damaging the membrane.

the membrane can be lifted off the roof deck and repositioned, if necessary. However, after a short time it will set up and cannot be removed without damaging the membrane.

2.11.0 Roofing Nails

Common roofing materials must be fastened with nails of the proper length and made of a material that is compatible or the same as the drip edges and flashing. *Figures 17* and *18* show the most common nails used for composition shingles and wood shakes or shingles. These nails are usually available in galvanized steel; some other types are aluminum, stainless steel, or copper. Normally, copper slater's nails are used for slate roofs, and stainless steel nails are used for tile roofs. About 1 pound of nails per square (100 square feet) is required to fasten composition shingles. For wood shakes and shingles, 2 to 4 pounds of nails will be required. For other types of nailed roofing, follow the

manufacturer's recommendations. Check your local code for nail penetration requirements.

For composition roofs or underlayment, nails can be installed using pneumatic-powered or electric-powered nailing guns.

2.12.0 Cold Asphalt Roofing Cement

Cold **asphalt roofing cements**, consisting of modified asphalt and/or coal tar products, are used in the installation of composition shingles, underlayment, roll roofing, and cold asphalt BUR. They are available in liquid non-fibered form or in plastic fibered form. The non-fibered forms are usually used in the lapped installation of underlayment and roll roofing and are spread over large areas with a spreader or mop. The plastic fibered cement is used for spot repairs or cementing nail heads, shingle tabs, valley overlaps, and saddle materials, and as an exposed sealer for gaps or flashing. Asphalt cements are generally available in one-gallon and five-gallon pails.

NAILING APPLICATION	⅜" PLYWOOD OR WAFER BOARD	1" SHEATHING
STRIP OR SINGLE (NEW CONSTRUCTION)	⅞"	1¼"
OVER OLD ASPHALT LAYER	1"	1½"
REROOFING OVER WOOD SHINGLES	–	1¾"

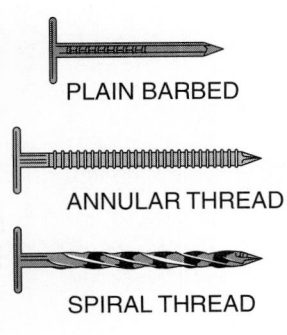

202F17.EPS

Figure 17 ◆ Typical composition shingle nails.

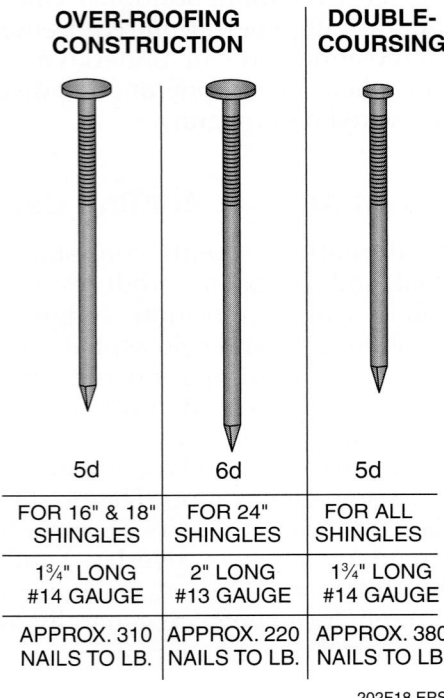

NEW ROOF CONSTRUCTION

3d	3d	4d
FOR 16" AND 18" SHINGLES		FOR 24" SHINGLES
1¼" LONG	1¼" LONG #14½ GAUGE	1½" LONG #14 GAUGE
APPROX. 376 NAILS TO LB.	APPROX. 515 NAILS TO LB.	APPROX. 382 NAILS TO LB.

OVER-ROOFING CONSTRUCTION		**DOUBLE-COURSING**
5d	6d	5d
FOR 16" & 18" SHINGLES	FOR 24" SHINGLES	FOR ALL SHINGLES
1¾" LONG #14 GAUGE	2" LONG #13 GAUGE	1¾" LONG #14 GAUGE
APPROX. 310 NAILS TO LB.	APPROX. 220 NAILS TO LB.	APPROX. 380 NAILS TO LB.

202F18.EPS

Figure 18 ◆ Typical wood shake or shingle nails.

3.0.0 ◆ TOOLS

Many of the tools used for roofing are common to other trades. Some of these common tools are:

- Backsaw
- Power circular saw
- Crowbar
- Hand saw
- Carpenter's level
- Nail apron
- Sliding T-bevel
- Keyhole saw
- Pop riveter
- Chalkline
- Tape measure
- Power saber saw
- Angle square
- Power drill
- Caulking gun
- Tin snips
- Prybar
- Scribing compass
- Utility knife
- Drill bit sets (regular and masonry)
- Framing square
- Claw hammer
- Pneumatic nailers
- Flat spade or spud bar (for roofing material removal)

Other tools that are specific to the installation of certain types of roofing are also used. Some of these tools are shown in *Figure 19*.

The roofing hammer, also referred to as a shingle hatchet, is used primarily for wood shingle and shake installation. The hatchet end is used to split shingles or shakes and the top edge is marked or equipped with a sliding gauge to set a dimension for the amount of weather exposure for the shingle or shake.

A composition shingle knife is used to trim or cut all types of composition shingles, including architectural shingles, during installation. It is also used to cut underlayment, cap shingles, roll roofing, and membrane roofing.

Slate roofing installation usually requires three specialized tools: a slater's hammer, a nail ripper, and a slate cutter. The slater's hammer is equipped with a sharp edge for cutting slate and a point for poking nail holes through the slate. The nail ripper has sharp-edged barbs on one end that are used to shear off nails under a piece of slate. To shear nails, the ripper is struck on the face of the anvil with a hammer. The slate cutter aids in the trimming of slate and the punching of nail holes.

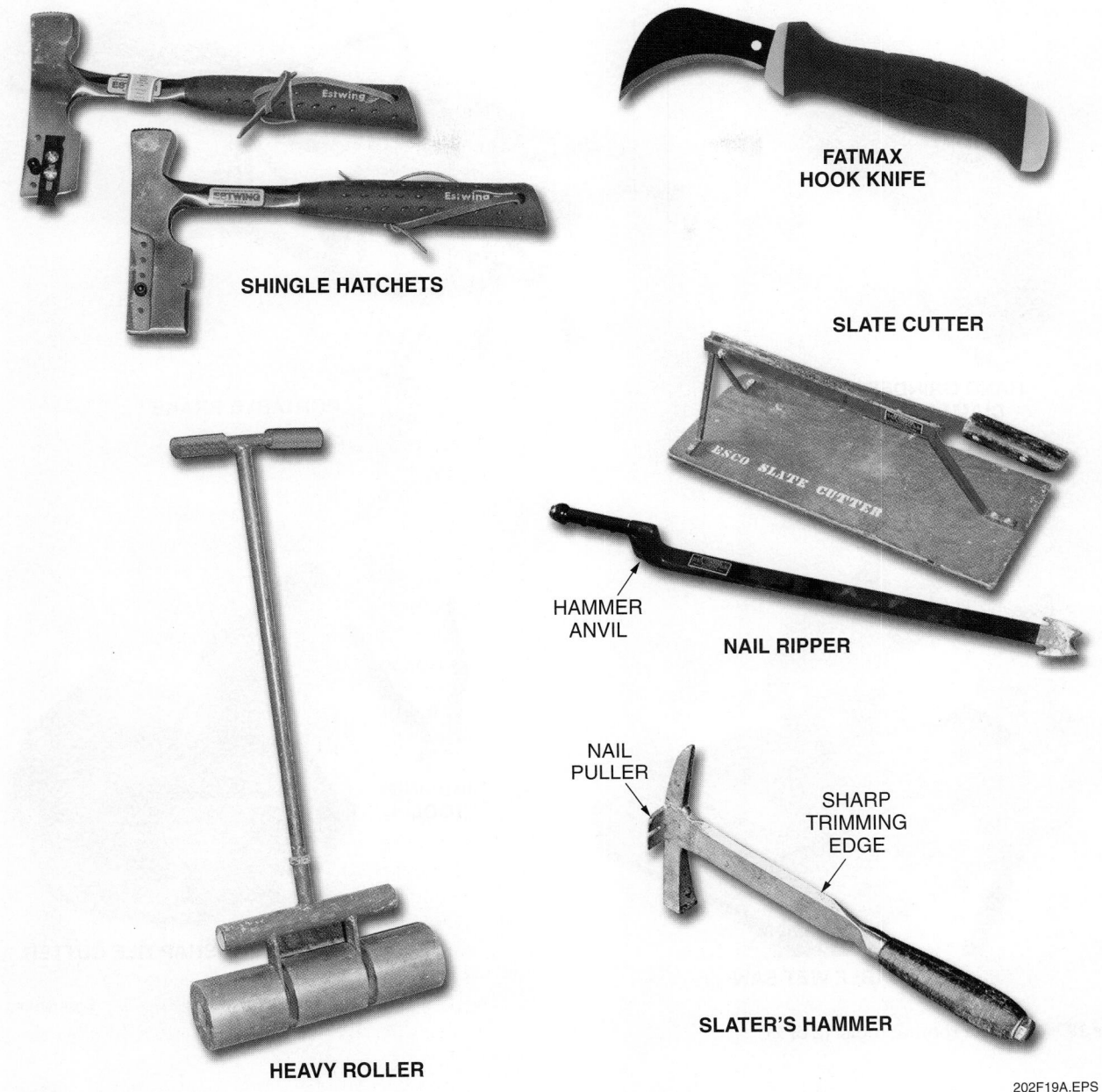

FATMAX
HOOK KNIFE

SHINGLE HATCHETS

SLATE CUTTER

HAMMER
ANVIL

NAIL RIPPER

NAIL
PULLER

SHARP
TRIMMING
EDGE

HEAVY ROLLER

SLATER'S HAMMER

202F19A.EPS

Figure 19 ◆ Special roofing tools. (1 of 2)

Various types of tile cutters and nibblers are used in the installation of tile roofs for splitting or trimming tiles. A nail ripper, like the one used for slate roofing, can be used to shear off nails under a tile. A wet saw with a diamond wheel can be used on large projects for flat or shaped tile cutting.

Portable brakes are used for the custom bending of flashing material for any type of roof installation. Some roofers use heavy rollers to flatten underlayment to eliminate buckling under the finish roof and for the application of cold-cement,

fully-adhered roll roofs, BURs, or single-ply membrane roofing. These rollers are sold as vinyl flooring rollers in weights ranging from 75 to 150 pounds.

Hand grinders with diamond wheels are used to cut slots in masonry for flashing installation.

In addition to tools, other equipment such as scaffolding, material movement equipment, ladders, and ladder jacks or pump jacks may be required. All roofing installation jobs will require some type of fall protection system.

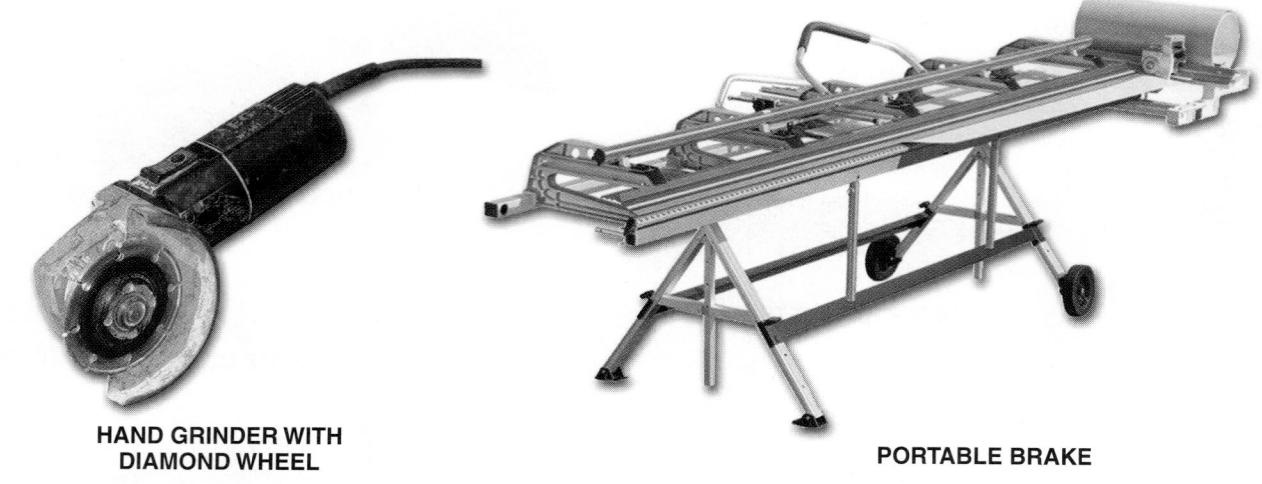

**HAND GRINDER WITH
DIAMOND WHEEL**

PORTABLE BRAKE

PORTABLE WET SAW

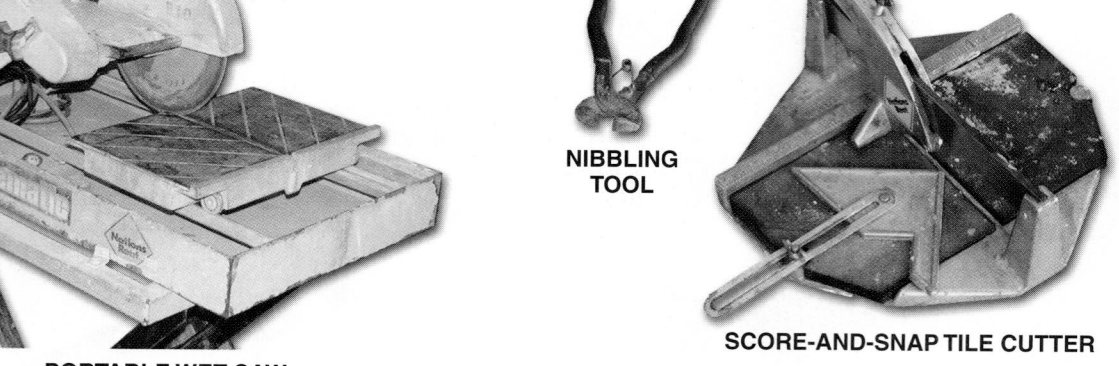

**NIBBLING
TOOL**

SCORE-AND-SNAP TILE CUTTER

202F19B.EPS

Figure 19 ◆ Special roofing tools. (2 of 2)

Power Nailers

Make sure that the nailer is equipped with a flush-mount attachment or that the impact pressure of the tool can be regulated at the tool to prevent overdriving the nails and cutting through the roofing material.

PNEUMATIC ROOFING NAILER WITH PLASTIC WASHER ATTACHMENT FOR UNDERLAYMENT APPLICATION

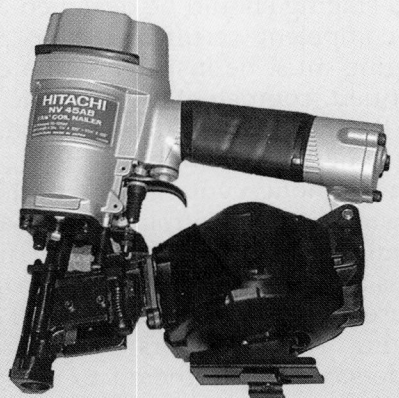

TYPICAL PNEUMATIC ROOFING NAILER

202SA03.EPS

4.0.0 ◆ SAFETY

Worker safety is important on a construction site. Every work site must have a fall protection plan for working on roofs or at certain heights off the ground.

4.1.0 Safety Summary

The following guidelines must be observed to ensure your safety and the safety of others:

- Wear boots or shoes with rubber or crepe soles that are in good condition.
- Always wear fall protection devices, even on shallow-pitch roofs.
- Rain, frost, and snow are all dangerous because they make a roof slippery. If possible, wait until the roof is dry; otherwise, wear special roof shoes with skid-resistant cleats in addition to fall protection.
- Brush or sweep the roof periodically to remove any accumulated dirt or debris.
- Install any required underlayment as soon as possible. Underlayment usually reduces the danger of slipping. On sloped roofs, do not step on underlayment until it is properly fastened.
- On pitched roofs, install necessary roof brackets as soon as possible. They can be removed and repositioned as shingle-type roofing is installed.

- Remove any unused tools, cords, and other loose items from the roof. They can be a serious hazard.
- Check and comply with any federal, local, and state code requirements when working on roofs.
- Be alert to any other potential hazards such as live power lines.
- Use common sense. Taking chances can lead to injury or death.

When working outdoors or in high-heat conditions for extended periods of time, take precautions to avoid heat exhaustion and exposure to the sun's ultraviolet rays. Preventive measures include the following:

- Wear a hard hat.
- Wear light clothing that is made of natural fibers.
- If possible, wear tinted glasses or goggles.
- Use a sun protection factor (SPF) 30 or higher sunblock on exposed skin.
- Drink adequate amounts of water to prevent dehydration, especially in arid parts of the country.

4.2.0 Scaffolding and Staging

Scaffolding and staging have been the causes of many minor and serious accidents due to faulty or incomplete construction or inexperience on the part of the designer or craftsperson constructing them. Therefore, to avoid hazards caused by faulty or incompetent construction, all scaffolding and staging should be designed and constructed by competent, certified persons. Scaffolding and staging must be inspected on a daily basis by a certified, competent person. The inspector must tag the scaffolding/staging for safety every morning or at every change of shift.

Even though you have no part in the design or construction of scaffolding, for safety's sake you should be familiar with safety rules and regulations that govern its construction. If you are going to use it, you should know how it is built. The following safety factors should be thoroughly understood and adhered to by everyone on the job site:

- Any type of scaffold used should have a minimum safety factor ratio of four to one; that is, it should be constructed so that it will carry at least four times the load for which it is intended. Roofing material weight adds up quickly when placed in one location.
- All staging or platform planks must have end bearings on scaffold edges with adequate support throughout their lengths to ensure the minimum safety factor ratio of four to one.
- Scaffolding timbers, if used, must be carefully selected and maximum nailing used for added strength.
- Because scaffolds are built for work that cannot be done safely from the ground, makeshift scaffolds using unstable objects for support such as boxes, barrels, or piles of bricks are prohibited.

When the scaffolding is placed on a solid, firm base and erected correctly, the roofing applicator should be able to work in confidence.

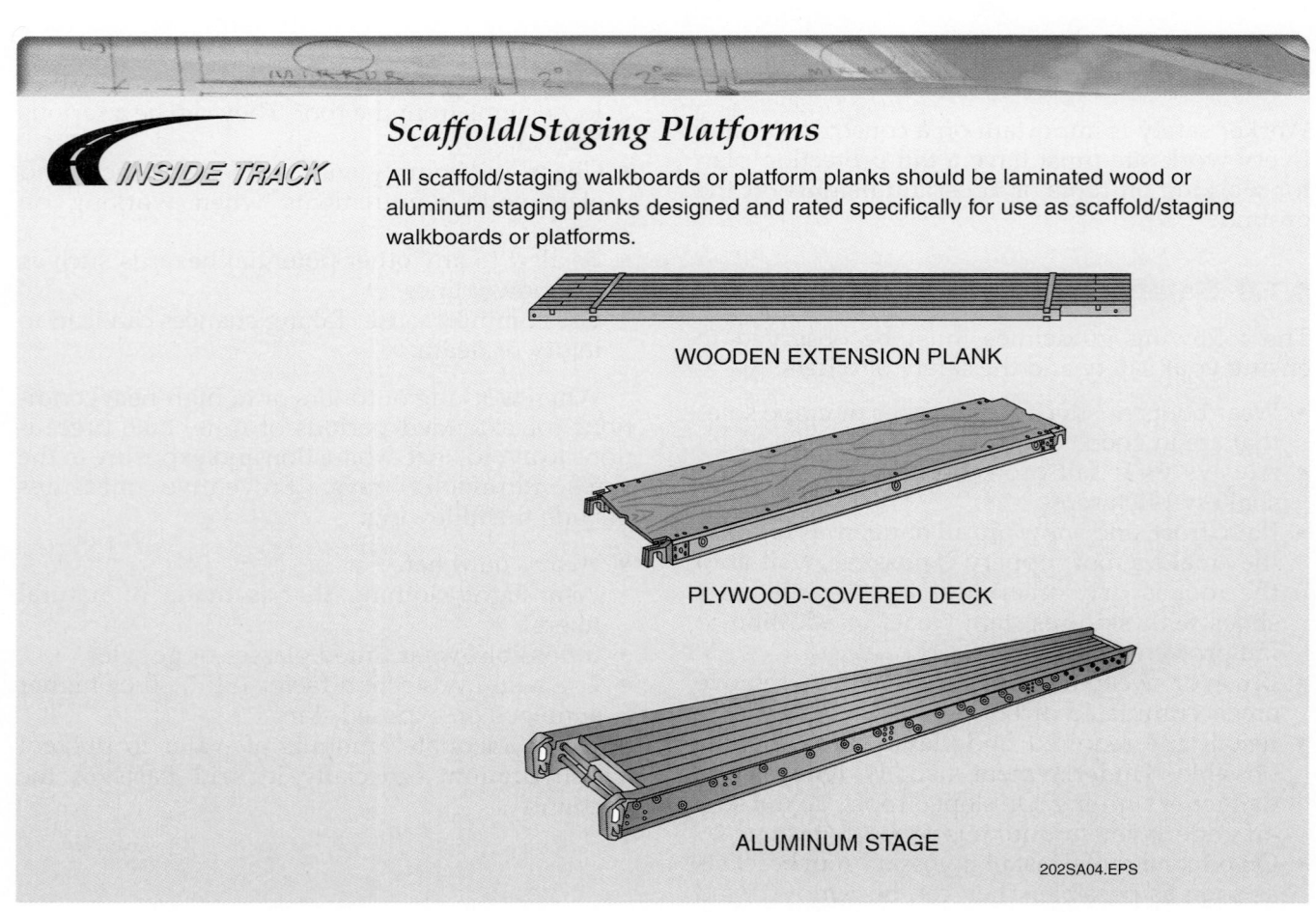

INSIDE TRACK

Scaffold/Staging Platforms

All scaffold/staging walkboards or platform planks should be laminated wood or aluminum staging planks designed and rated specifically for use as scaffold/staging walkboards or platforms.

WOODEN EXTENSION PLANK

PLYWOOD-COVERED DECK

ALUMINUM STAGE

202SA04.EPS

Any scaffolding assembled for use should be tagged. Three tag colors are used:

- *Green* – A green tag identifies a scaffold that is safe for use. It meets all OSHA standards.
- *Yellow* – A yellow tag means the scaffolding does not meet all applicable standards. An example is a scaffold where a railing cannot be installed because of equipment interference. A yellow-tagged scaffold may be used; however, a safety harness and lanyard are mandatory. Other precautions may also apply.
- *Red* – A red tag means a scaffold is being erected or taken down. You should never use a red-tagged scaffold.

Other more common types of scaffolding include ladder jacks (*Figure 20*) and pump jacks (*Figure 21*). Pump jacks and, to a lesser extent, ladder jacks are useful for applying the starter strip and lower courses of roofing. Ladder jacks can usually support a 2'-wide adjustable-length platform up to 10' or 18' long. Ladder jacks, which must not be used for heights over 20', can be attached to either side of a ladder and must be separated by not more than 8' intervals along the length of the platform. Pump jacks using aluminum posts for heights under 50' are movable platform supports that are raised or lowered vertically. They are operated by a foot lever and can raise a person plus a rated load.

4.3.0 Ladders

Ladders are useful and necessary pieces of equipment for roofing application and do not cause accidents if properly used and maintained. Ladder accidents are caused by:

- Improper use of ladders
- Ladders not secured properly
- Structural failure of ladders
- Improper handling of objects while on a ladder

Figure 22 shows a ladder erected correctly in relation to the roof eaves. If the ladder is to be left standing for a long period of time, it should be securely fastened at both the top and bottom.

202F20.EPS

Figure 20 ◆ Ladder jack and adjustable platform.

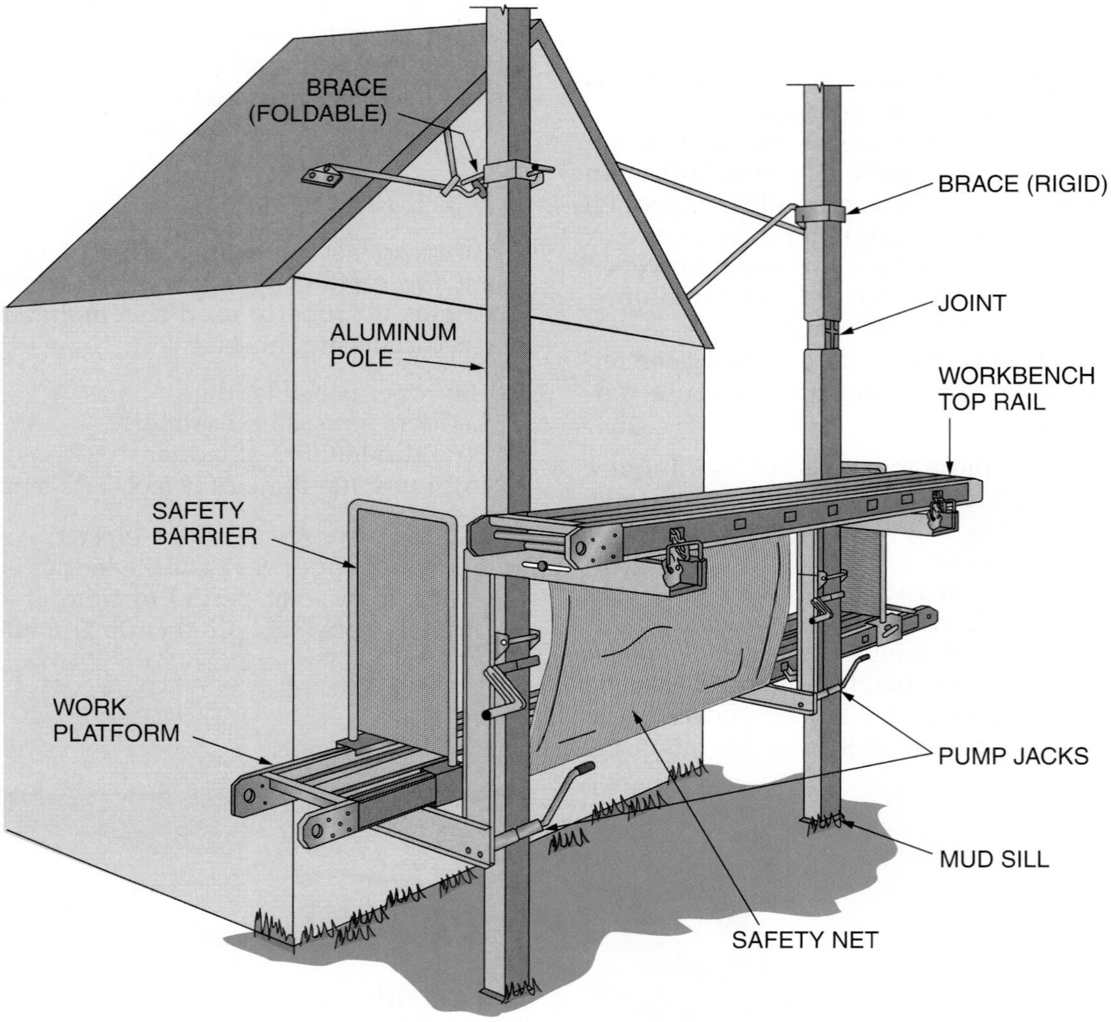

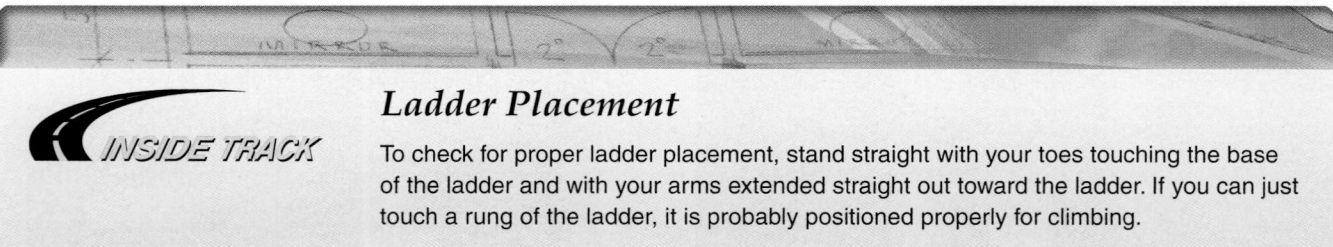

BRACE (FOLDABLE)

BRACE (RIGID)

JOINT

ALUMINUM POLE

WORKBENCH TOP RAIL

SAFETY BARRIER

WORK PLATFORM

PUMP JACKS

MUD SILL

SAFETY NET

202F21.EPS

Figure 21 ◆ Pump jacks.

Ladder Placement

INSIDE TRACK

To check for proper ladder placement, stand straight with your toes touching the base of the ladder and with your arms extended straight out toward the ladder. If you can just touch a rung of the ladder, it is probably positioned properly for climbing.

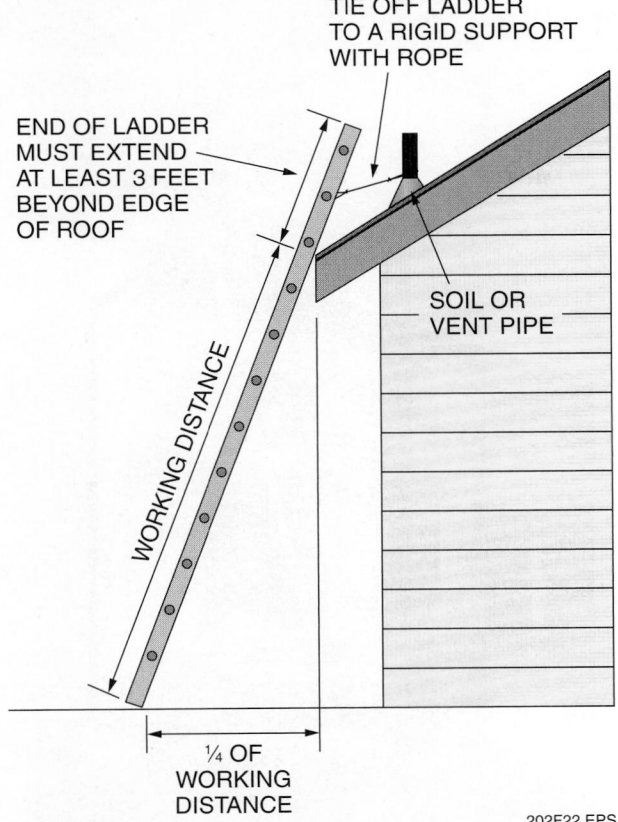

TIE OFF LADDER
TO A RIGID SUPPORT
WITH ROPE

END OF LADDER
MUST EXTEND
AT LEAST 3 FEET
BEYOND EDGE
OF ROOF

WORKING DISTANCE

SOIL OR
VENT PIPE

¼ OF
WORKING
DISTANCE

202F22.EPS

Figure 22 ◆ Safe ladder placement.

The normal purpose of a ladder used in a roofing application is solely to gain access to the roof itself. For ladder safety, follow these precautions:

- Always use a Type 1 (250 lbs/rung) OSHA-rated ladder that is 15" or more in width with rungs that are 12" apart.
- Fiberglass ladders should always be used to reduce the possibility of accidental electrocution resulting from contact with power lines. Avoid using aluminum ladders. Aluminum ladders, if used, should never be raised or placed in situations where they can fall or accidentally come into contact with power lines.
- For longer ladders, two people should carry, position, and erect the ladder.
- Always face a ladder and grasp the side rails or rungs with both hands when going up or down.
- Take one step at a time.

- Remember that an ordinary straight ladder is built to support only one person at a time.
- Before using a ladder, be sure there is no oil, grease, or sand on the soles of your shoes. Due to the tread composition, some shoe types easily attract foreign objects that can cause you to slip on a ladder.
- Never carry tools or materials up or down a ladder. A rope or other device should be used to raise or lower everything so that you can always have both hands free when climbing the ladder.
- Make sure that the base of the ladder is level and has adequate support. Shim the legs or use levelers, if necessary.
- Make sure the ladder is at the correct angle. Always tie off the ladder prior to use (see *Figure 22*). The ladder needs to be secured at its top to a rigid support.
- Never overreach.
- Fixed ladders must be provided with cages, wells, ladder safety devices, or self-retracting lifelines where the length of climb is less than 24', but the top of the ladder is at a distance greater than 24' above lower levels.

4.4.0 Material Movement

Many mechanical devices are used to move materials on a job site. These devices include conveyor belts, power ladder conveyers (attached to a ladder), forklift trucks, and truck-mounted hydraulic lifts. Make sure all safety devices are in place before starting.

Exercise extreme caution to ensure that the lift does not make contact with the roof surface or the person unloading the bundles. Immediately distribute the bundles around the roof area. Do not pile them in any one spot. Roof structures have been designed to carry a specific dead load. Placing unnecessary strain on the roof structure by concentrating the shingles in one place on the roof can jeopardize the safety of the workers. The end result might prove to be disastrous with the collapse of the roof itself. An immediate dispersal of the bundles will prevent any problems from occurring and also make the installation more efficient because you only want to move and place the load once.

202SA06.EPS

4.5.0 Roofing Brackets

Roofing brackets provide firm footing and material storage points on steep-slope roofs. If any type of roofing bracket is to be used, the decision will be based on the slope of the roof. Most roofers feel comfortable on a roof with a 4 in 12 or a 5 in 12 slope. When the slope increases to 6 in 12, more strain is placed on the feet and the body. Therefore, the roofer has to be very conscious of the height off the ground and be careful with each and every movement. *Figure 23* shows two types of roofing brackets that can be used with a 10' to 14', defect-free, 2" thick plank.

Both types of brackets can be nailed firmly to the roof, but the adjustable bracket, which is installed to correspond to the slope of the roof, makes standing and moving around more comfortable. Never get overconfident due to bracket usage. Be aware of the height at which you are working and be cautious.

When installing roof brackets, make sure that they are nailed to the rafters, not just to the roof sheathing.

Roof brackets and toe boards alone are not sufficient to meet OSHA fall standards. A proper rail or safety harness is required above six feet.

4.6.0 Power Nailers

There are many types of power nailers. In the application of asphalt or fiberglass shingles, the use of one of these tools can cut the labor time in half. Portable units are used by carpenters on the construction site. Portable units are electric or pneumatic (air or carbon dioxide gas powered).

Some of the listed safety practices for power nailers are as follows:

- Always wear safety glasses. They must be an OSHA-approved type and are usually recommended by the manufacturer of the unit being used.
- Because operating principles vary, study the manufacturer's operating manual.
- Be certain to use the type of fastener required by the manufacturer.

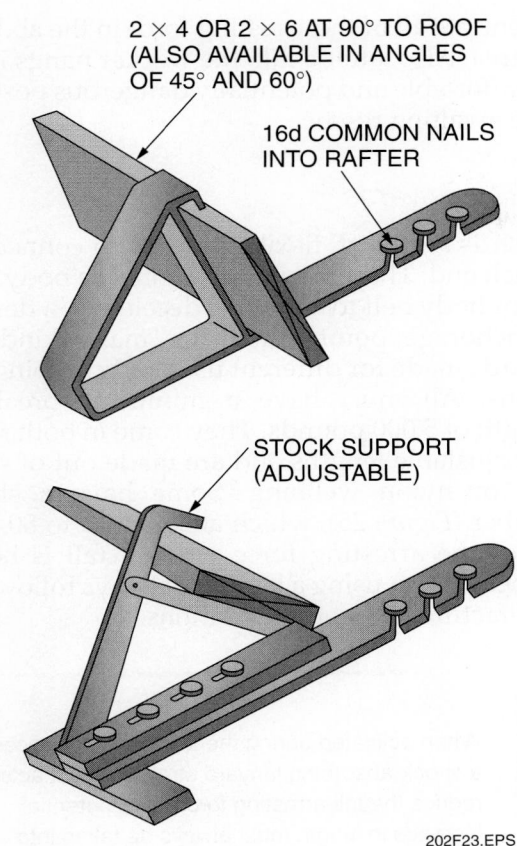

2 × 4 OR 2 × 6 AT 90° TO ROOF
(ALSO AVAILABLE IN ANGLES
OF 45° AND 60°)

16d COMMON NAILS
INTO RAFTER

STOCK SUPPORT
(ADJUSTABLE)

202F23.EPS

Figure 23 ◆ Roofing brackets.

- If pneumatic, make sure that air pressure can be adjusted at the nailer.
- Treat the machine as you would a gun. Do not point it at yourself or others.
- Always keep the unit tight against the surface to drive the fastener correctly.
- When not in use, disconnect the unit from the power source to prevent accidental release of fasteners.
- Keep air lines untangled on the roof to prevent tripping.

4.7.0 Fall Protection Equipment

Roofers spend a major part of their time working on sloped roofs. Most construction injuries and deaths are caused by falls. Falls from high places can cause serious injury or death when the wrong type of fall protection equipment is used, or when the right equipment is used improperly. A fall protection plan must be prepared for any project where workers will be more than 6' off the ground.

There are three common types of fall protection equipment: guardrails, personal fall arrest systems, and safety nets. This section will focus on personal fall arrest systems. These devices and their use are governed by *OSHA Safety and Health Standards for the Construction Industry, Part 1926, Subpart M*. The rules covering guardrails on scaffolds are contained in Subpart L. Basically, OSHA requires that all workers use guardrail systems, safety net systems, or personal fall arrest systems to protect themselves from falling more than 6' (1.8 meters) and hitting the ground or a lower work level.

When describing personal fall arrest systems, the following terms must be understood:

- *Free-fall distance* – The vertical distance a worker moves after a fall before a deceleration device is activated.
- *Deceleration device* – A device such as a shock-absorbing lanyard, rope grab, or self-retracting lifeline that brings a falling person to a stop without injury.
- *Deceleration distance* – The distance it takes before a person comes to a stop. The required deceleration distance for a fall arrest system is a maximum of 3½'.
- *Arresting force* – The force needed to stop a person from falling. The greater the free-fall distance, the more force is needed to stop, or arrest, the fall.

The following sections discuss equipment used in personal fall arrest systems:

- Body harnesses and belts
- Lanyards
- Deceleration devices
- Lifelines
- Anchoring devices and equipment connectors

4.7.1 Body Harnesses

Full body harnesses (*Figure 24*) with sliding back D-rings are used in personal fall arrest systems. They are made of straps that are designed to be worn securely around the user's body. This allows the arresting force to be distributed via the harness straps throughout the body, including the shoulders, legs, torso, and buttocks. This distribution decreases the chance of injury. When a fall occurs, the sliding D-ring moves to the nape of the neck, keeping the worker in an upright position and helping to distribute the arresting force. This keeps the worker in a relatively comfortable position while awaiting rescue.

Selecting the right full body harness depends on a combination of job requirements and personal

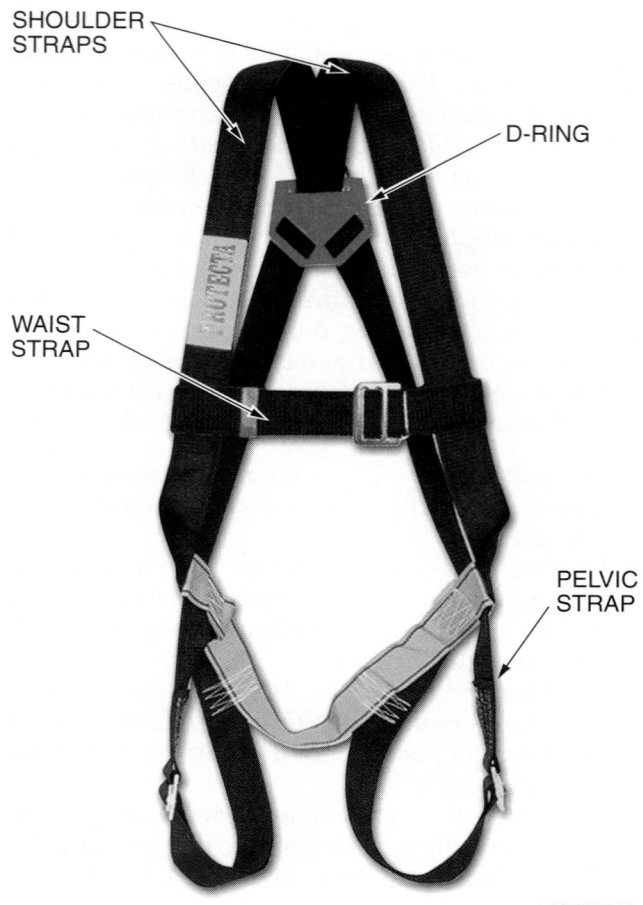

SHOULDER STRAPS

D-RING

WAIST STRAP

PELVIC STRAP

202F24.EPS

Figure 24 ◆ Full body harness.

preference. Harness manufacturers normally provide selection guidelines in their product literature. Other types of full body harnesses can be equipped with front chest D-rings, side D-rings, or shoulder D-rings. Harnesses with front chest D-rings are typically used in ladder climbing and personal positioning systems. Those with side D-rings are also used in personal positioning systems. Personal positioning systems are systems that allow workers to hold themselves in place, keeping their hands free to accomplish a task. Per OSHA regulations, a personal positioning system should not allow a worker to free-fall more than 2', and the anchorage to which it is attached should be able to support at least twice the impact load of a worker's fall or 3,000 pounds, whichever is greater. Harnesses equipped with shoulder D-rings are typically used with a spreader bar or rope yoke for entry into and retrieval from confined spaces.

Note that in the past, body belts were frequently used instead of a full body harness as part of a fall arrest system. As of January 1, 1998, OSHA banned them from such use. This is because body belts

concentrate all of the arresting force in the abdominal area. Also, after a fall, the worker hangs in an uncomfortable and potentially dangerous position while awaiting rescue.

4.7.2 Lanyards

Lanyards are short, flexible lines with connectors on each end. They are used to connect a body harness or body belt to a lifeline, deceleration device, or anchorage point. There are many kinds of lanyards made for different uses and climbing situations. All must have a minimum breaking strength of 5,000 pounds. They come in both fixed and adjustable lengths and are made out of steel, rope, or nylon webbing. Some have a shock absorber (*Figure 25*), which absorbs up to 80 percent of the arresting force when a fall is being stopped. When using a lanyard, always follow the manufacturer's recommendations.

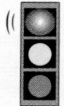 **WARNING!**

When activated during the fall arresting process, a shock-absorbing lanyard stretches as it acts to reduce the fall-arresting force. This potential increase in length must always be taken into consideration when determining the total free-fall distance from an anchor point.

4.7.3 Deceleration Devices

Deceleration devices limit the arresting force that a worker is subjected to when the fall is stopped suddenly. Rope grabs with shock-absorbing lanyards and self-retracting lifelines are two common deceleration devices. A rope grab (*Figure 26*) connects to a shock-absorbing lanyard and attaches to a lifeline. In the event of a fall, the rope grab is pulled down by the attached lanyard, causing it to grip the lifeline and lock in place. Some rope grabs have a mechanism that allows the worker to unlock the device and slowly descend the lifeline to the ground or surface below.

SHOCK ABSORBER

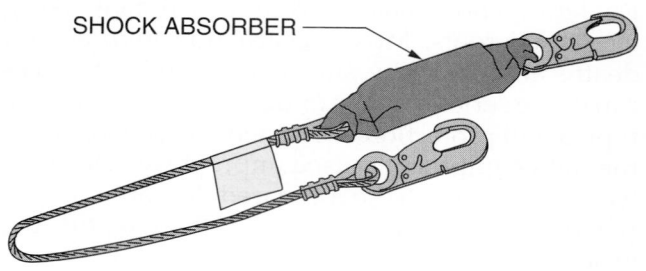

202F25.EPS

Figure 25 ◆ Typical shock-absorbing lanyard.

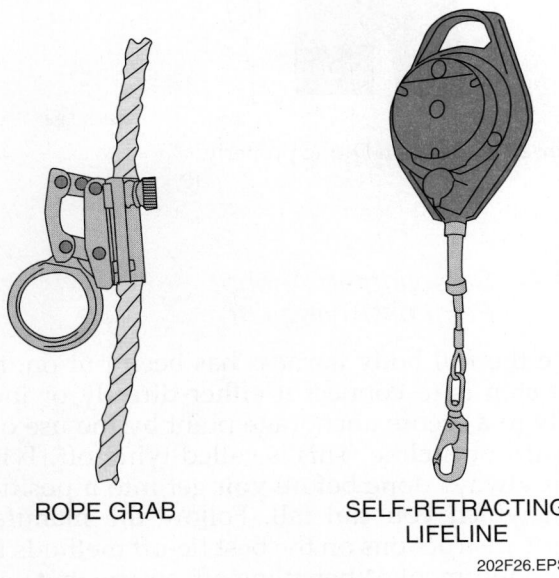

ROPE GRAB   SELF-RETRACTING LIFELINE

202F26.EPS

Figure 26 ◆ Deceleration devices.

Self-retracting lifelines (*Figure 26*) allow unrestricted movement and fall protection while climbing and descending ladders or when working on multiple levels. Typically, they have a 25' to 100' galvanized steel cable that automatically takes up the slack in the attached lanyard, keeping the lanyard out of the worker's way. In the event of a fall, a centrifugal braking mechanism engages to limit the worker's fall. Per OSHA requirements, self-retracting lifelines and lanyards that limit the free-fall distance to 2' or less must be able to support a minimum tensile load of 3,000 pounds. Those that do not limit the free-fall distance to 2' or less must be able to hold a tensile load of at least 5,000 pounds.

4.7.4 Lifelines

Lifelines are ropes or flexible steel cables that are attached to an anchorage. They provide a means for tying off personal fall protection equipment.

Vertical lifelines are suspended vertically from a fixed anchorage at the upper end to which a fall arrest device such as a rope grab is attached. Vertical lifelines must have a minimum breaking strength of 5,000 pounds. Each worker must use his or her own line. This is because if one worker falls, the movement of the lifeline during the fall arrest may also cause the other workers to fall. Vertical lifelines must be terminated in a way that will keep the worker from moving past its end, or they must extend to the ground or the next lower working level.

Horizontal lifelines are connected horizontally between two fixed anchorage points to which a fall arrest device is attached. Horizontal lifelines must be designed, installed, and used under the supervision of a qualified and competent person. The required strength of a horizontal line and its anchors increases substantially for each worker attached to it.

4.7.5 Anchoring Devices and Equipment Connectors

Anchoring devices, commonly called tie-off points, support the entire weight of the fall arrest system. The anchorage must be capable of supporting 5,000 pounds for each worker attached. Eye bolts, overhead beams, and integral parts of building structures are all types of anchorage points.

The D-rings, buckles, and snaphooks that fasten and/or connect the parts of a personal fall arrest system are called connectors. OSHA regulations specify how they are to be made, and require D-rings and snaphooks to have a minimum tensile strength of 5,000 pounds. All such components should be designed for use with the attached hardware. As of January 1, 1998, only locking-type snaphooks are permitted for use in personal fall arrest systems.

4.8.0 Procedures for Safely Using Personal Fall Arrest Equipment

Before using fall protection equipment on the job, your employer should provide you with training in the basics of fall protection and the proper use of the equipment. In addition, a job-specific fall protection plan must be available for the project. All equipment supplied by your employer must meet OSHA standards for strength. Before each use, always read the instructions and warnings on any fall protection equipment. Inspect the equipment using the following guidelines:

- Examine harnesses and lanyards for mildew, wear, damage, and deterioration.
- Make sure no straps are cut, broken, torn, or scraped.
- Check for damage due to fire, chemicals, or corrosives.
- Check that hardware is free of cracks, sharp edges, or burrs.
- Check that snaphooks close and lock tightly and that buckles work properly.
- Check ropes for wear, broken fibers, pulled stitches, and discoloration.
- Make sure lifeline anchors and mountings are not loose or damaged.

4.8.1 Wearing a Full Body Harness

The general procedure for using a sliding back D-ring, full body harness is as follows:

Step 1 Hold the harness by the back D-ring, then shake the harness, allowing all the straps to fall into place.

Step 2 Unbuckle and release the waist and/or leg straps.

Step 3 Slip the straps over your shoulders so that the D-ring is located in the middle of your back (*Figure 27*).

Step 4 Fasten the waist strap. It should be tight, but not binding.

Step 5 Pull the straps between each leg and buckle the straps.

Step 6 After all the straps have been buckled, tighten all friction buckles so that the harness fits snugly but allows a full range of movement.

Step 7 Pull the chest strap around the shoulder straps and fasten it in the mid-chest area. Tighten it enough to pull the shoulder straps taut.

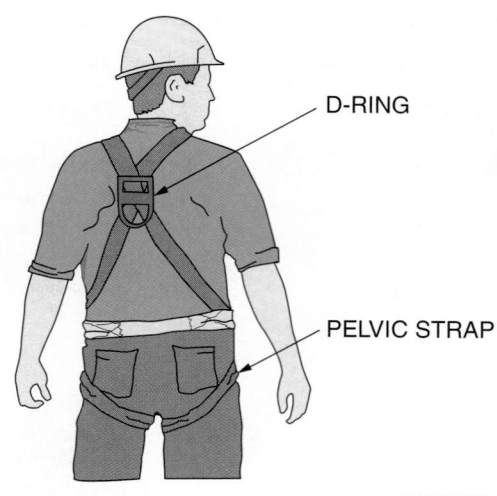

202F27.EPS

Figure 27 ◆ Position D-ring properly.

4.8.2 Selecting an Anchor Point and Tying Off

Once the full body harness has been put on, the next step is to connect it either directly or indirectly to a secure anchorage point by the use of a lanyard or lifeline. This is called tying off. Tying off is always done before you get into a position from which you can fall. Follow the manufacturer's instructions on the best tie-off methods for your equipment. When tying off, ensure that your anchorage point has the following characteristics:

- Directly above you
- Easily accessible
- Damage-free and capable of supporting 5,000 pounds per worker
- Never on the same point as a workbasket tie-off

Be sure to check the manufacturer's equipment labels and allow for any equipment stretch and deceleration distance.

When tying off, consider the following:

- Tie-offs that use knots are weaker than other methods of attachment. Knots can reduce the lifeline or lanyard strength by 50 percent or more. A stronger lifeline or lanyard should be used to compensate for this effect.
- To protect equipment from cuts, do not tie off around rough or sharp surfaces. Tying off around H-beams or I-beams can weaken the line because of the cutting action of the beam's edge. This can be prevented by using a webbing-type lanyard or wire-core lifeline.
- Never tie off in a way that would allow you to fall more than 6'.

- A shorter fall can reduce your chances of falling into obstacles, being injured by the arresting force, and damaging your equipment. To limit your fall, a shorter lanyard can be used between the lifeline and your harness. Also, the amount of slack in your lanyard can be reduced by raising your tie-off point on the lifeline. The tie-off point to the lifeline or anchor must always be higher than the connection to your harness.

4.8.3 Rescue after a Fall

Every elevated job site should have a rescue and retrieval plan in case it is necessary to rescue a fallen worker. Planning is especially important in remote areas that are not readily accessible to a telephone. Before there is a risk of a fall, make sure that you know what your employer's rescue plan calls for you to do. Find out what rescue equipment is available and where it is located. Learn how to use the equipment for self-rescue and the rescue of others.

If a fall occurs, any employee hanging from the fall arrest system must be rescued safely and quickly. Your employer should have previously determined the method of rescue for fall victims, which may include equipment that lets the victim rescue himself or herself, a system of rescue by co-workers, or a way to alert a trained rescue squad. If a fall rescue depends on calling for outside help, such as the fire department or rescue squad, all the needed phone numbers must be posted in plain view at the work site. In the event a co-worker falls, follow your employer's rescue plan. Call any special rescue service needed. Communicate with the victim and monitor him or her constantly during the rescue.

4.9.0 Testing Fall Protection Systems and Equipment

The testing of fall arrest equipment should be performed regularly to make sure it complies with OSHA requirements. Depending on the prevailing state and local laws, the tests may be either voluntary or mandatory. Guidelines for testing personal fall arrest equipment and systems are given in *OSHA Safety and Health Standards for the Construction Industry, Part 1926, Appendices C and D to Subpart M*. A good practice is to tag or label all items of fall protection equipment with the date when the equipment was last tested and the date it is due for the next test.

Safety nets should be drop-tested at the job site after the initial installation, whenever relocated, after a repair, and at least every six months if left in one place. The drop test consists of dropping a 400-pound bag of sand into the net from at least 42" above the highest walking/working surface at which workers are exposed to fall hazards.

5.0.0 ◆ PREPARATION FOR ROOFING APPLICATION

Regardless of the type of roofing to be installed, the amount of material required must first be estimated. After the amount of material has been determined and obtained, the roof deck must be prepared before the finish roofing material is raised to the roof deck and installed.

5.1.0 Estimating Roofing Materials

If the building plans for the structure are available, the roof dimensions can be determined directly from the plans. If the plans are not available, the length and width of each section of the roof can be measured. Then, the area of each section of the roof is calculated and a percentage is added for waste. The result is converted into the number of squares (100 sq ft) of material required.

Step 1   Measure the length and width of each triangular and rectangular roof section of the structure, including any **overhangs** (*Figure 28*).

Step 2   Calculate the area for one half of each roof section and add the areas together. Then, multiply by two to obtain the total roof area and subtract any triangular areas covered by roof intersections.

To calculate the total area of the roof shown in *Figure 28*, proceed as follows:

$$30' \times 35' \qquad = \quad 1,050.0 \text{ sq ft (Area A)}$$
$$60' \times 35' \qquad = + 2,100.0 \text{ sq ft (Area B)}$$
$$15' \times 35' \times \tfrac{1}{2} = + \quad 262.5 \text{ sq ft (Area C)}$$
$$3,412.5 \text{ sq ft } (\tfrac{1}{2} \text{ roof area})$$
$$3,412.5 \times 2 \quad = \quad 6,825.0 \text{ sq ft}$$
$$35' \times 20' \times \tfrac{1}{2} = - \quad 350.0 \text{ sq ft (Area D*)}$$
$$= \quad 6,475.0 \text{ sq ft (roof area)}$$

*Roof projection area

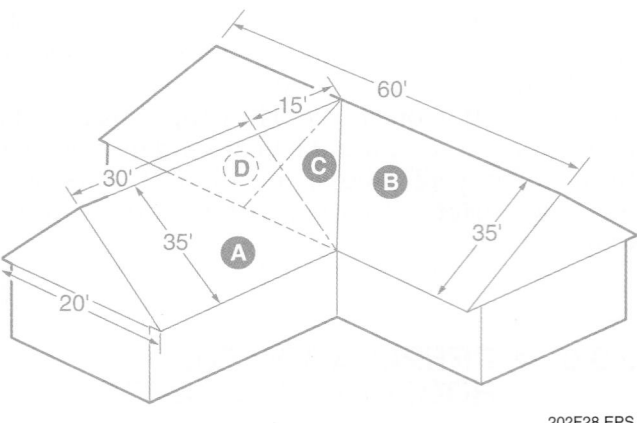

Figure 28 ◆ Roof example (including overhangs).

202F28.EPS

Step 3 Add an average of 10 percent for hips, valleys, and waste. For a complicated roof with a number of valleys or hips, more than 10 percent may be required. For a plain, straight gable roof, less is usually required. In addition, for wood or slate roofs, an additional 100 square feet is usually required for each 100 lineal feet of hips and valleys. Assuming that our example is a simple roof with standard three-tab composition shingles, we will use only the extra 10 percent figure.

$$\text{Total material} = (10\% \times 6{,}475) + 6{,}475$$
$$= 648 + 6{,}475$$
$$= 7{,}123 \text{ sq ft}$$

Step 4 Convert the total square feet of material required to squares by dividing by 100 sq ft (if using wood roofing, round up or down to the nearest bundle or square). For standard three-tab composition shingles, the rounding would be to the nearest $\frac{1}{3}$ or $\frac{2}{3}$ of a square (one or two bundles) or whole square.

$$\text{Squares} = 7{,}123 \div 100 = 71.23 = 71\tfrac{1}{3}$$

Step 5 The number of rolls of underlayment required is determined from the same total material requirement of 7,123 square feet. Starter strips, eave flashing, valley flashing, and ridge shingles must be added to complete the estimate. All of these are determined with linear measurements.

5.2.0 Roof Deck Preparation

A typical roof installation is shown in *Figure 29*. Before the finish roofing is applied, the roof deck must be flashed with a drip edge along the eaves and any valleys must be flashed. Then, an underlayment and/or a waterproofing membrane is usually installed and capped at the rake edges with metal drip edge flashing. On bare wood roof decks, the underlayment/membrane must be applied on dry wood as soon as possible. If the wood is moist due to rain or morning dew, allow it to dry before applying the underlayment/membrane. If the roof deck is damp, the membrane may not adhere to the roof deck, or the underlayment will buckle and cause the final roof to appear wavy. The underlayment/membrane prevents the finish roof materials from having direct contact with any damaging resinous or corrosive areas of the roof deck and helps resist or eliminate any water penetration into the roof deck. *Figure 30* and *31* show the recommended underlayment/waterproof membrane placement and drip edge installation.

Normally, the drip edge is installed along the length of the eaves first, followed by any valley flashing. The drip edge should be held against the fascia and nailed to the roof deck every 8" to 10". When installing valley flashing, it should overhang the valley at the upper and lower ends. The flashing is nailed every 6" to 8" on both sides, ½" from the edges.

After the flashing is secured, both ends are carefully trimmed flush with the roof deck. After the eave drip edge is in place, the exposed nail heads are covered with asphalt. Starting at the bottom of the roof, the underlayment and/or a waterproof membrane is rolled out and flattened with a roof roller before being tacked to the roof. In valleys, the waterproof membrane should extend over the flashing nails. The membrane will adhere and seal to the valley flashing; however, the underlayment should be trimmed to cover the flashing nails and should be cemented to the valley. In some cases on lower sloped roofs, the underlayment is half-lapped and cemented with asphalt to provide more of a water barrier. After the underlayment/waterproof membrane is in place, the rake edges of the roof are capped with a drip edge that is nailed every 8" to 10" to the roof deck. The bottom end of the rake drip edge overlaps the eave drip edge, and the fascia flange is cut to interlock behind the fascia flange of the eave drip edge. The nail heads should be covered with asphalt cement.

After the roof preparation is complete, the finish roof materials can be lifted and distributed equally over the roof deck.

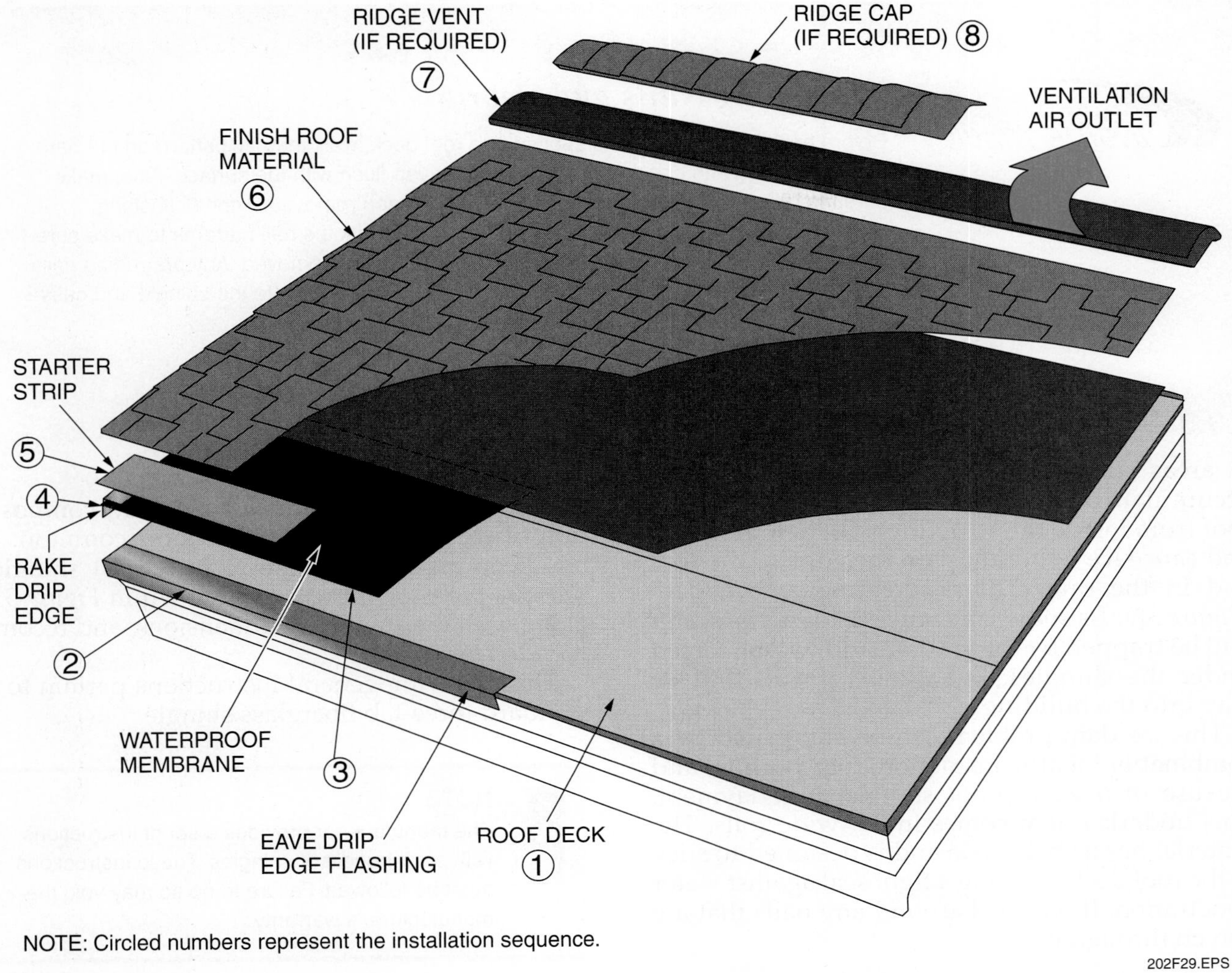

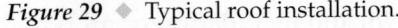

RIDGE VENT
(IF REQUIRED) ⑦

RIDGE CAP
(IF REQUIRED) ⑧

VENTILATION
AIR OUTLET

FINISH ROOF
MATERIAL
⑥

STARTER
STRIP
⑤
④
RAKE
DRIP
EDGE
②

WATERPROOF
MEMBRANE
③

EAVE DRIP
EDGE FLASHING

ROOF DECK
①

NOTE: Circled numbers represent the installation sequence.

202F29.EPS

Figure 29 ◆ Typical roof installation.

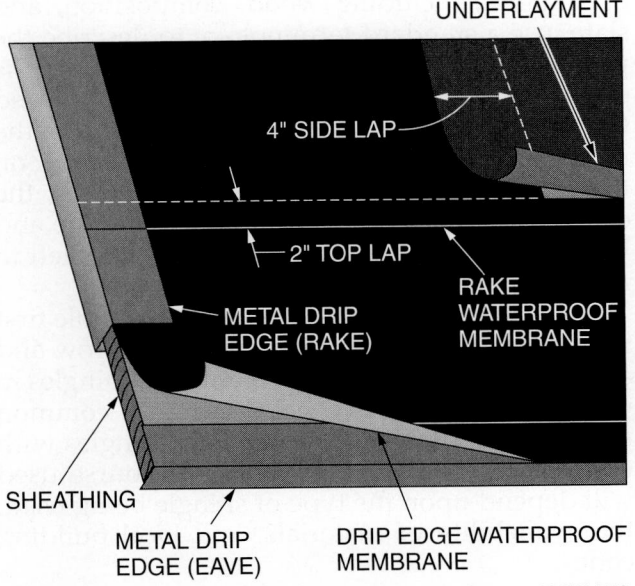

UNDERLAYMENT

4" SIDE LAP

2" TOP LAP

RAKE
WATERPROOF
MEMBRANE

METAL DRIP
EDGE (RAKE)

SHEATHING

METAL DRIP
EDGE (EAVE)

DRIP EDGE WATERPROOF
MEMBRANE

202F30.EPS

Figure 30 ◆ Drip edge and waterproof membrane
placement.

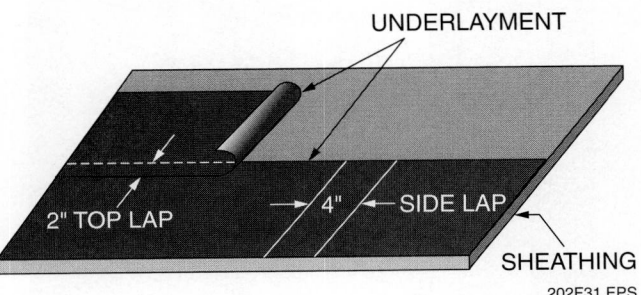

UNDERLAYMENT

2" TOP LAP

4"

SIDE LAP

SHEATHING

202F31.EPS

Figure 31 ◆ Underlayment or waterproof membrane
placement over roof deck.

5.3.0 Protection Against Ice Dams

In areas subject to heavy snow, the snow will accumulate on the roof. Heat rising through the roof from inside the structure will melt the snow and cause ice to build up on the edge of the roof and in the rain gutters, creating an ice dam (*Figure 32*). As snow continues to melt, the water will be trapped by the ice dam and will be forced under the shingles. Eventually, it will find its way into the building.

This ice dam problem can be eliminated by a combination of attic insulation, roof venting, and the use of a waterproof shingle underlayment. This underlayment comes in 36" wide rolls. The material has a sticky side and is designed to stick to the roof deck, forming a tight seal against water penetration. It will seal around any nails that are driven through it.

202F32.EPS

Figure 32 ◆ Ice dam.

6.0.0 ◆ COMPOSITION SHINGLE INSTALLATION

At one time, the three-tab, square-butt composition fiberglass shingle was the most common. It has been replaced by the architectural shingle. Various types of shingles are shown in *Figure 33*, along with their weights, dimensions, and recommended exposures.

The following general instructions pertain to a standard three-tab fiberglass shingle.

> **NOTE**
>
> The manufacturer provides a set of instructions with each bundle of shingles. These instructions must be followed. Failure to do so may void the manufacturer's warranty.

The instructions for the installation of all types of shingles, including wood, composition, and slate, use a standard terminology to describe the placing of the shingles. This terminology is explained in *Figure 34*. Roof shingles can be placed from the left side of the roof to the right or the right side of the roof to the left, depending on the preference of the roofer. On wide roofs, the courses are sometimes started in the middle and laid toward both ends. In this module, the left to right convention is used.

All strip shingles are started with a double first row, which may be made up of a starter row and a row of shingles or a double course of shingles in which two joints have been offset. A common practice is to place a starter row of shingles with the tabs cut off. The type of starter course used will depend upon the type of shingle being used, the availability of materials, and local building codes.

Unopened bundles of shingles are usually placed at various points on the roof for the roofing applicator.

	PRODUCT*	CONFIGURATION	PER SQUARE			SIZE		EXPOSURE, INCHES
			APPROXIMATE SHIPPING WEIGHT	SHINGLES	BUNDLES	WIDTH, INCHES	LENGTH, INCHES (NOMINAL)	
ARCHITECTURAL	WOOD APPEARANCE STRIP SHINGLE; MORE THAN ONE THICKNESS PER STRIP 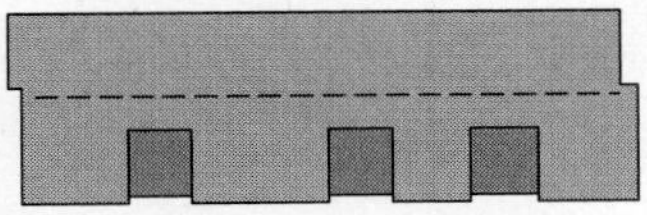 PRELAMINATED OR JOB-APPLIED LAYERS	VARIOUS EDGE SURFACE TEXTURE AND APPLICATION TREATMENTS	285# TO 390#	67 TO 90	4 OR 5	11½ TO 15	36 OR 40	4 TO 6
ARCHITECTURAL	WOOD APPEARANCE STRIP SHINGLE; SINGLE THICKNESS PER STRIP 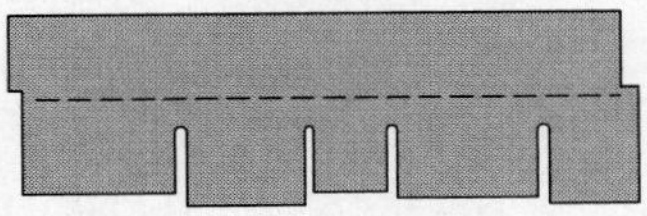	VARIOUS EDGE SURFACE TEXTURE AND APPLICATION TREATMENTS	VARIOUS, 250# TO 350#	78 TO 90	3 OR 4	12 OR 12¼	36 OR 40	4 TO 5⅛
STANDARD	3-TAB SELF-SEALING STRIP SHINGLE 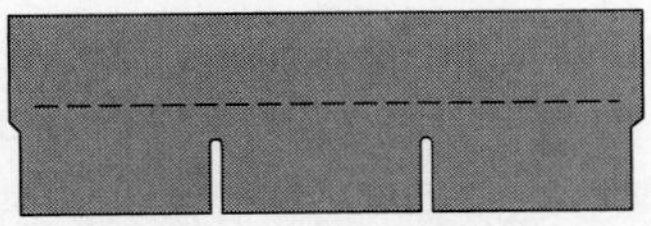	CONVENTIONAL 3-TAB	205# TO 240#	78 OR 80	3	12 OR 12¼	36	5 OR 5⅛
STANDARD	2-TAB (OR 4-TAB) VERSION 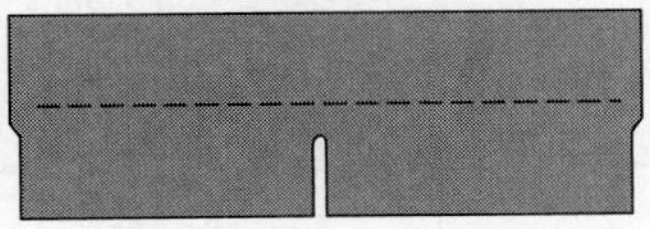	2- OR 4-TAB	VARIOUS, 215# TO 325#	78 OR 80	3 OR 4	12 OR 12¼	36	5 OR 5⅛
STANDARD	SELF-SEALING STRIP SHINGLE 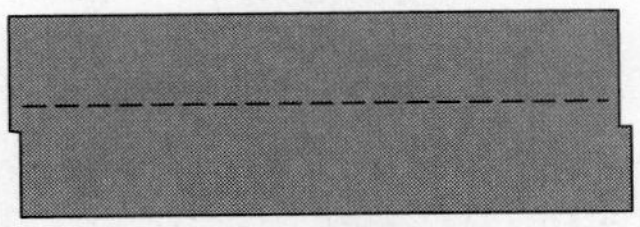 NO CUTOUT	VARIOUS EDGE AND TEXTURE TREATMENTS	VARIOUS, 215# TO 290#	78 TO 81	3 OR 4	12 OR 12¼	36 OR 40	5

*Other types available from some manufacturers in certain areas of the country.
Consult your regional asphalt roofing manufacturers' association.

202F33.EPS

Figure 33 ◆ Typical composition shingle characteristics.

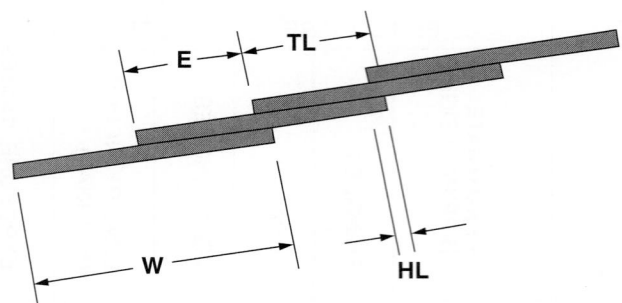

W = Width: The total width of strip shingles or the length of an individual shingle.

E = Exposure: The distance between the exposed edges of overlapping shingles.

TL = Top Lap: The distance that a shingle overlaps the shingle in the course below.

HL = Head Lap: The distance from the lower edge of an overlapping shingle to the upper edge of the shingle in the second course below.

202F34.EPS

Figure 34 ◆ Roofing terminology used in instructions.

CAUTION

Failure to scatter unopened shingle bundles can result in a broken rafter or collapsed roof, due to the weight of the bundles.

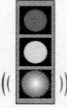

NOTE

If you are going to stand on a scaffold, it should be erected in such a manner that you will be working approximately waist high with the eave line. This places you in a safe, comfortable position to install the critical double course of shingles along the eaves.

Under normal circumstances, the shingle can be fastened with four aluminum or galvanized roofing nails positioned at a nailing line from the bottom of the shingle, one at each end and the other two above and adjacent to each cutout. See *Figure 35*. Depending upon the area of the country in which you live, this cutout may be referred to as a notch or gusset.

In areas with very high winds, the number of nails above the cutout can be increased to two, positioned about 3" apart and forming a triangle with the top of the cutout as a low point. In high wind areas, staples are not normally used.

When laying a shingle, butt the shingle to the previous shingle in the course and align the shingle. Then fasten the butted end to the roof. Keep the shingle aligned and fasten across the shingle to the other end. Alignment can be done by using the guide built into a roofing hammer or by chalking lines.

6.1.0 Gable Roofs

This section describes the installation of long and short runs of standard shingles on gable roofs. It does not include instructions for metric or architectural shingles.

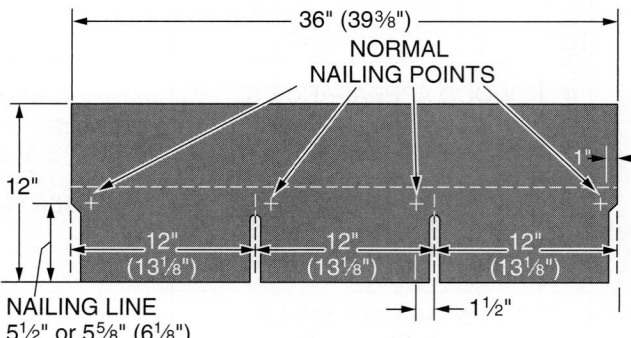

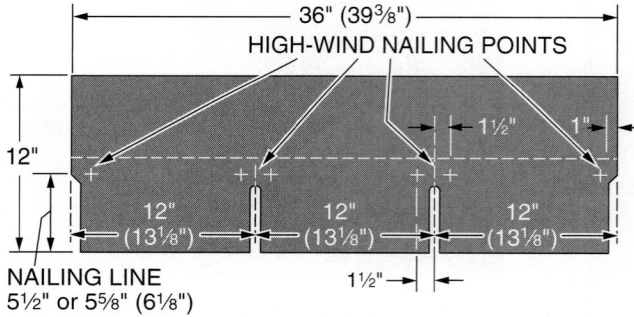

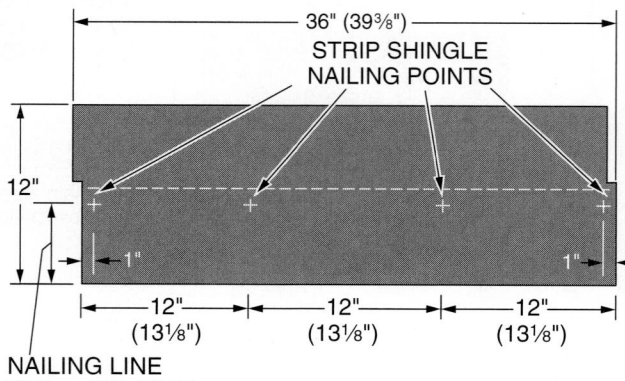

NOTE: Numbers in parentheses represent metric size shingle dimensions in English units.

202F35.EPS

Figure 35 ◆ Nailing points.

6.1.1 Gable Roofs—Long Runs

On large roofs, start applying the shingles at the center of the long run. The following procedure explains how to lay out and mark the roof. By beginning in the center, there is less chance of misalignment as you proceed in both directions. Shingle manufacturers suggest that you run shingles horizontally first rather than stacking them one row above another. The color of the shingles will blend better that way.

To install a long run on a gable roof, proceed as follows:

Step 1 Measure along the length of the roof and find the halfway point. Do this along the ridge and along the eaves. Mark both of these places. Snap a chalkline vertically up the roof at these two marks.

Step 2 Measure 6" to the right and left of this center line at the ridge and at the eaves. Snap a line vertically for these marks as well. This will provide three lines to start the rows on.

Step 3 Starting at the eaves at both ends of the roof, measure 6" up from the eave drip edge and place a mark. Then go up to 11" and make a mark there. Proceed up the tape and mark every 5" interval. Once both sides are done, snap lines horizontally across the roof to align the shingles.

Step 4 For the starter shingles, cut the tabs off a tabbed shingle. Place the remaining part of the shingle so that the tar strip is nearest the edge of the roof. With standard shingles, you should have a starter row that is 7" wide by 36" long. If the starter strip is not done this way, then the first full row of shingles will not be sealed down. The starter strip can be started on one of the vertical lines in the center of the roof. Make sure that the starter shingle stays on the first line snapped horizontally across the roof. If drip edges have been installed, position the starter strip with a ½" overhang on the eave and rake drip edges; oth-

erwise, position the strip with a ¾" overhang. Place and fasten the starter strip both ways from center.

Step 5 Take a full shingle and place it directly over the starter strip and 6" to the right or left of the vertical line that was used for the starter strip. This will ensure that the cutouts will be spaced 6" away from the joints of the starter strip. This is considered the first full row of shingles.

Step 6 The second row is in the center, 6" to the right or left of the starting place where the first row was started. Proceed right and left from that point.

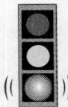

> **NOTE**
> Starting 6" over helps to minimize waste. Most of the time the scrap you have left over on the right end will work on the left and vice versa.

6.1.2 Gable Roofs—Short Runs

To install a short run on a gable roof, proceed as follows:

Step 1 Cut the tabs off of the number of strip shingles it will take to go across the roof. Mark a spot 6" up on both eaves. Snap a line across the roof at this location. This line will keep the starter strip running straight across the roof. With the first left starter strip shortened by 6", place the starter strip shingles so that the tar strip is nearest the eave. If drip edges have been installed, position the starter strip with a ½" overhang on the eave and rake drip edge; otherwise, position the strip with a ¾" overhang. Nail the starter close to the top in four locations. After the starter strip is laid as far to the right as possible, return to the rake on your left and start to double up this first course by placing a full shingle directly over the top of the first upside down shingle. See *Figure 36*. Continue this

Starter Strips

Pre-cut starter strips are available from many manufacturers. In many areas of the country, they are available in two sizes: 5" wide for roofing over existing shingles and 7" wide for new and tear-off installations.

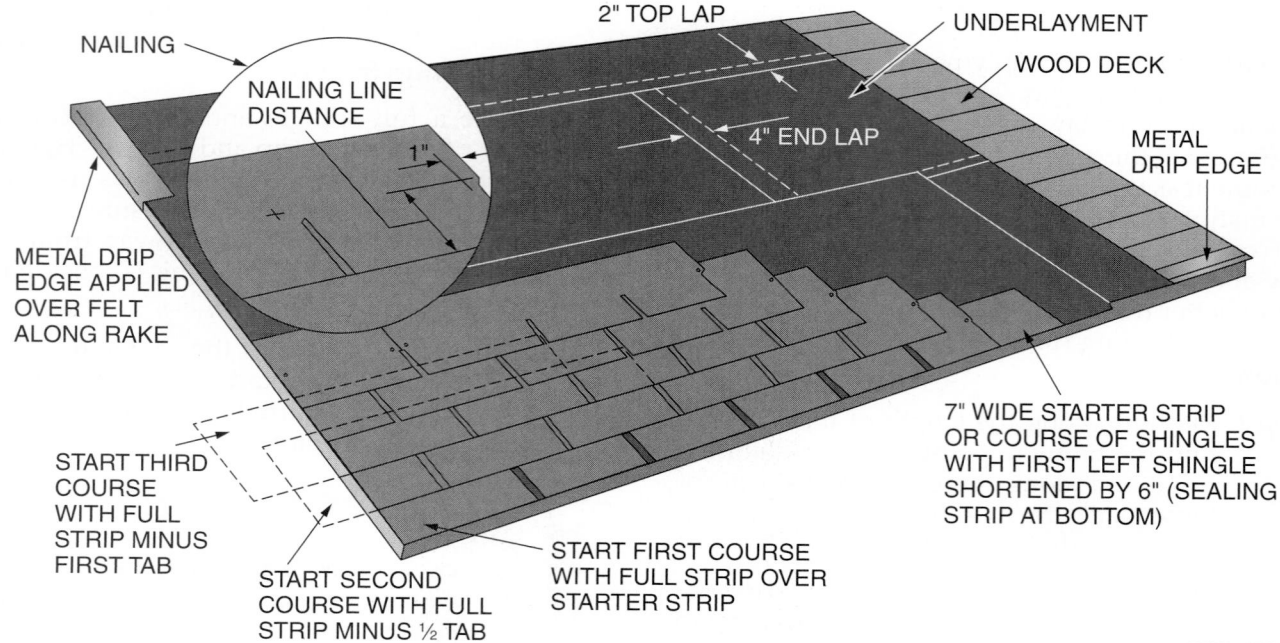

Figure 36 ◆ Shingle layout – 6" pattern.

202F36.EPS

process with full shingles as you move to the right and end before reaching the last starter strip shingle. You will observe that all cutouts and joints are covered with the full 12" tab. Make sure to nail the shingles correctly, as shown in *Figure 37*.

Step 2 Start the second course with a full strip minus 6". This layout is called a 6", half-tab pattern, or a 6-up/6-off layout, which means that half a tab is deliberately cut off and produces vertical aligned cutouts (refer to *Figure 36*). Overhang the cut edge at the rake by the same ¼" margin as the first course. Nail this shingle in place and proceed to the right with full shingles. The gusset openings can be used as a checking procedure to obtain the proper 5" exposure. End the second course before reaching the end of the first course. This is called stair stepping and allows the maximum number of shingles to be placed before moving ladders or scaffolding.

Step 3 Start the third course with a full strip minus a full tab (12"). Follow the same procedure with the same rake overhang. Again, nail the shingles in place moving to the right and gauging your exposure by the 5" gusset.

Step 4 Returning now to the fourth course, start this row with half a strip (for example, with 18" removed). This will show a 6" tab. Nail this in place with the same overhang margin on the rake. Proceed with full shingles, nailing to the right and ending short of the previous course while constantly checking the alignment.

Step 5 Start the fifth course. This row starts with a full tab only (12"). Use the same overhang margin on the rake and full shingles as you move to the right. End before reaching the last shingle on the previous course.

Step 6 The sixth course starts with a 6" tab, the same overhang margin at the rake, and then full shingles. It continues to the right as the nailing proceeds. Again, end before reaching the last shingle on the previous course.

Step 7 As the process is repeated, the seventh course starts with a full shingle. Each successive course of shingles is shortened by an additional 6". This continues as previously described until the twelfth course.

Depending upon the individual or working team, two or more courses may be carried or nailed at a time as the shingling proceeds across

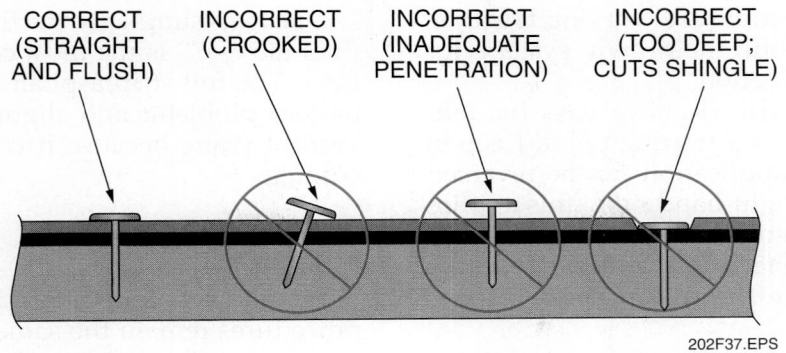

CORRECT (STRAIGHT AND FLUSH) INCORRECT (CROOKED) INCORRECT (INADEQUATE PENETRATION) INCORRECT (TOO DEEP; CUTS SHINGLE)

202F37.EPS

Figure 37 ◆ Correct and incorrect nailing.

Correct Nailing

Improper setting of nails and crooked nails can prevent the shingles from tabbing correctly and allow the wind to lift or tear the tabs of the shingles. Practice your fastening procedures so you drive the nail straight. The head should be flush with the surface of the shingle. Since your goal is to make the roof watertight, no pinholes or breaks are acceptable. If an accident should happen, a dab of asphalt cement spread with a putty knife will remedy the problem.

Other Shingle Alignment Methods

Several other alignment methods can be used for exposure. One popular method begins by snapping a chalkline along the top edge of the shingle of the first course, or 12" above and parallel with the eave line. Snap several other chalklines parallel with this first line. If you make the lines 10" apart, they can be used to check every other course by aligning the top of the shingle.

High Wind Areas

In high wind areas, the fifth and sixth courses of the 6" pattern are usually eliminated because of the possibility of the small starter tabs being torn off the roof. The fifth course tab (12" long) should be saved for use as a ridge cap. The sixth course tab (6" long) would be discarded. The shingles would also be secured using double fastening at each cutout.

the roof. When two people are working together, they usually work out their own system for speedy, accurate installation.

The procedure described above uses the left-hand rake of the roof as a starting point. Keep in mind that the entire application can be reversed by starting from the right-hand rake (this applies to gable roof construction). On small roofs, strip shingles may be laid starting at either end with a successful result since the roof measurement is usually symmetrical.

To obtain different variations of roof patterns using tabbed shingles, only a change of starting measurement is required. *Figure 38* shows one possibility using a course with a full shingle (36") followed by a second course using a reduced size shingle (32"). The third course would be reduced again to a shorter measurement (28"). Repeat these three measurements starting with the fourth course and continuing up the rake. This is called a 4" pattern and produces a diagonal cutout pattern.

Ribbon courses (*Figure 39*) are a way to add interest to a standard 6" pattern. After six courses have been applied, cut a 4"-wide strip lengthwise off the upper section of a full course of shingles. Fasten the strips as the seventh course correctly aligned with the cutouts of the sixth course. Then reverse the 8"-wide leftover pieces of shingle and align them directly over the 4" strip. Fasten the 8" pieces to the roof deck at the top of the tabs. Cover both with a full-width seventh course of shingles. This creates a three-ply edge known as the ribbon. Repeat the pattern every seventh course.

Due to its simplicity, the first pattern mentioned (half tab = 6") is the most commonly used in the field. The full strip asphalt shingle eliminates all pattern problems and alignment concerns on the vertical plane because it contains no gussets or cutouts.

6.2.0 Hip Roofs

When you encounter a hip roof, the basic nailing procedures remain the same, but the shingle lay-out starting point has to be at the center of the roof, as described for long gable roofs.

To begin, the starter strip is applied as previously described for long gable roofs. Return to the vertical line and use a full shingle for the doubling of this starter course. Offset this shingle 6" on either side of this vertical line. This will automatically cover the seam and gussets underneath and seal the roof against water penetration. You have the option of continuing this dual shingle starting course in either direction until it terminates at the hip rafter. See *Figure 40*.

At this point, the shingles should be cut to match the angle of the hip rafter and covered with a hip cap (the same as a ridge cap), completing the installation of the shingles. The hip cap is centered on the hip rafter and usually consists of a 12" tab showing a 5" exposure.

Exposed nails in the last caps should be covered with roof sealant. The ridge cap should cover the hip cap to prevent leaks.

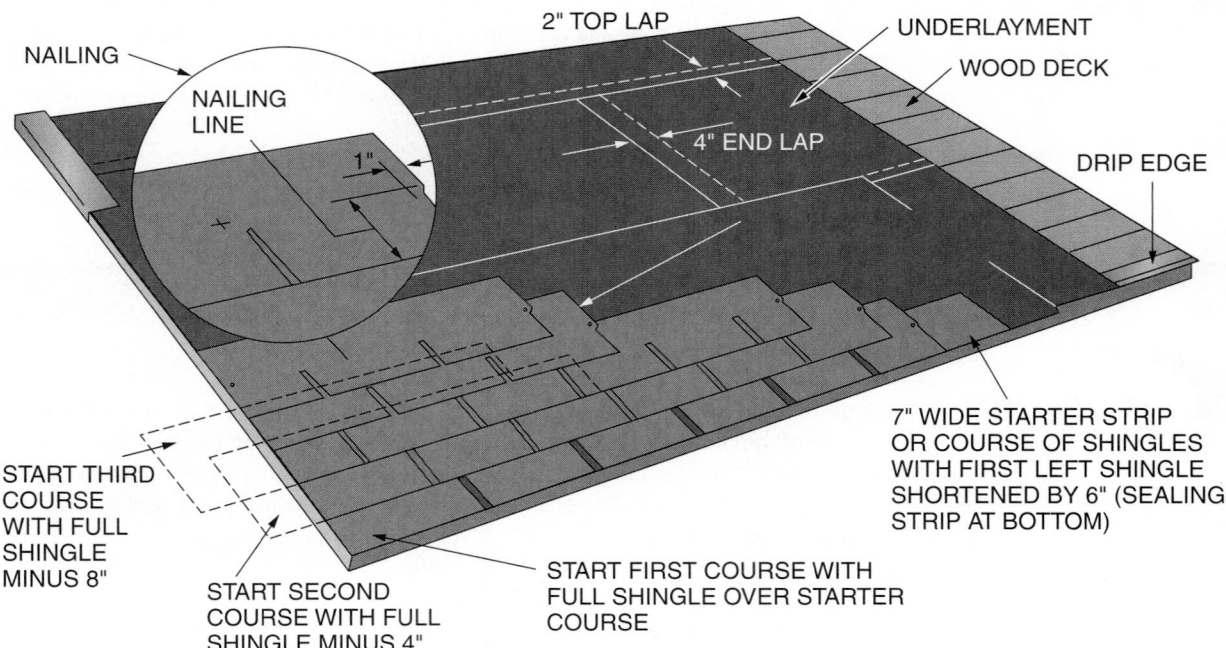

202F38.EPS

Figure 38 ◆ Shingle layout—4" pattern.

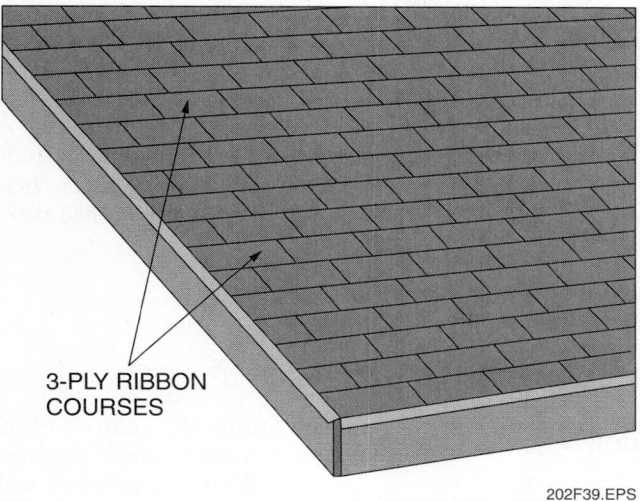

Figure 39 ◆ Ribbon courses.

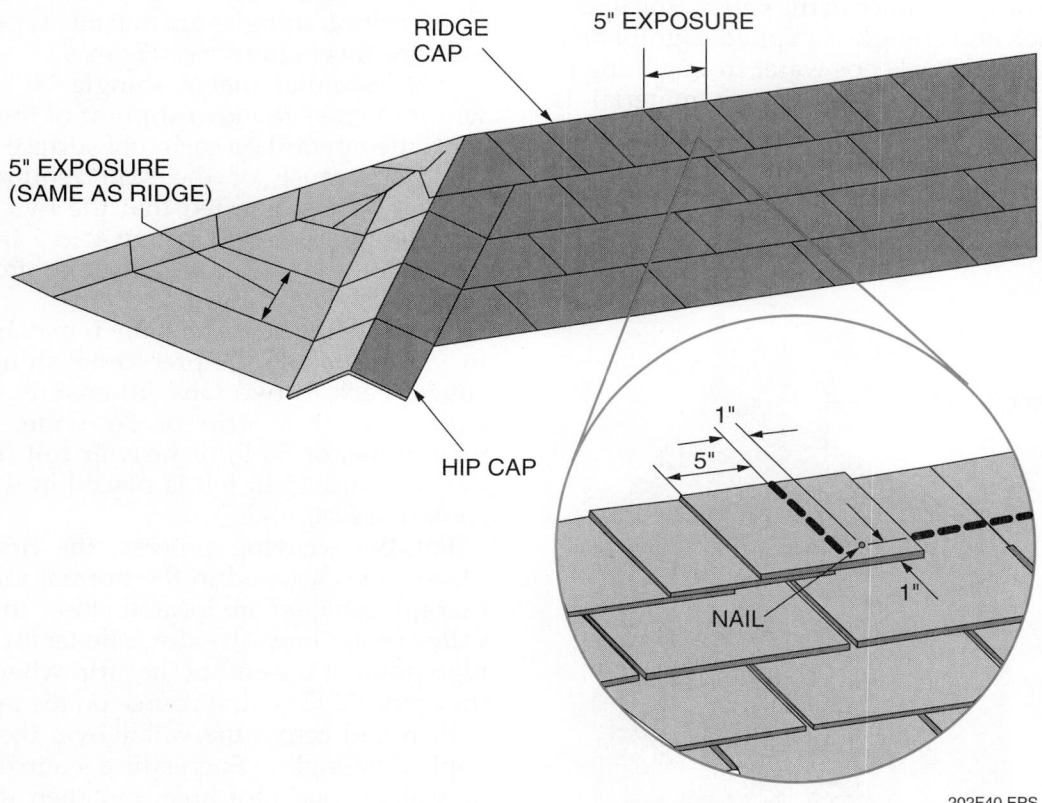

Figure 40 ◆ Hip and ridge layout.

6.3.0 Valleys

If the building you are constructing is not a perfect rectangle, you may encounter an L or T shape, which calls for another variation of shingling procedures. Where two sloping roofs meet, this intersecting valley has to be able to carry a high concentration of water drainage. Shingling becomes very critical, and the application must be done with extreme care.

6.3.1 Open Valley

With the valley flashing installed as previously described, snap two chalklines the full length of the valley. They should be 6" apart at the ridge or uppermost point. This means they should measure 3" apart when measured from the center of the valley. The marks diverge at the rate of ⅛" per foot as they approach the eaves. For example, a valley 8' in length will be 7" wide at the eaves; one

Unequally Pitched Roof Intersections

If the roof valley is formed by an intersection of two unequally pitched roofs, the woven valley will creep up one side, making it nearly impossible to maintain the correct overlap of shingles. The open valley or the closed-cut valley should be used with unequally pitched roof intersections.

16' long will be 8" wide at the eaves. The enlarged spacing provides adequate flow as the amount of water increases, passing down the valley. See *Figure 41*.

The chalkline you have snapped serves as a guide in trimming and cutting the last shingle to fit in the valley. This ensures a clean, sharp edge and a uniform appearance in the valley. The upper corner of each end shingle is clipped slightly on a 45-degree angle. This keeps water from getting in between the courses. The roofing material is cemented to the valley lining and to itself where an overlap occurs. Use plastic asphalt cement and spread a 6" to 8" bed. Do not overdo it, and clean up all excess cement so no tar shows.

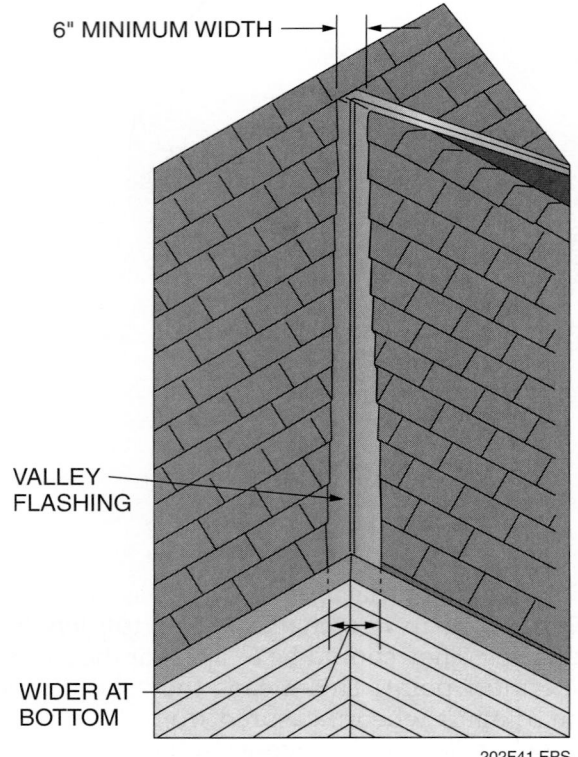

Figure 41 ◆ Open valley flashing (steep pitch).

6.3.2 Closed-Woven Valley

Some applicators of asphalt shingles prefer to use a closed-woven valley design, sometimes called a full-weave or laced valley. It is faster to install, and some feel it gives a tighter bond. Others believe closed valleys are inferior to open valleys because they do not shed high volumes of water very well. Composition shingles are the only type that can be used for this pattern. See *Figure 42*.

It is essential that a shingle be of sufficient width to cross the lowest point of the valley and continue upward on each roof surface a minimum of 12". Because of the skill required for this process, it is suggested that the two converging roofs be completed to a point 4' to 5' from the center of the valley. Then the weaving process can be accomplished carefully.

To create the 12" extension, it may be necessary to cut some of the preceding shingles in the course back to two tabs. To ensure a watertight valley, either a strip of 36"-wide, waterproof membrane, or 50-lb or heavier roll roofing over the standard 15-lb felt is placed in the valley, as shown in *Figure 42*.

For the weaving process, the first course is placed and fastened in the normal manner. Note that no fasteners are located closer than 6" to the valley center line. An extra fastener is placed at the high point at the end of the strip where it extends the extra 12". The first course on the opposite side is then laid across the valley over the previously applied shingles. Succeeding courses alternate, first along one roof area and then the other, as shown in *Figure 42*. Extreme care must be taken to maintain the proper exposure and alignment. As the shingles are woven over each other, they must be pressed tightly into the valley to provide a smooth surface where the roof surfaces join.

6.3.3 Closed-Cut Valley

In a closed-cut valley, sometimes called a half-weave or half-laced valley, the underlayment and valley flashing materials are the same as for the woven application.

36" ROLL ROOFING
(50-LB OR HEAVIER)
OR WATERPROOF MEMBRANE

EACH STRIP TO EXTEND
AT LEAST 12" BEYOND
CENTER OF VALLEY

6"

EXTRA NAIL IN END OF STRIP

NO NAILS WITHIN 6"
OF CENTER OF VALLEY

202F42.EPS

Figure 42 ◆ Closed-woven valley.

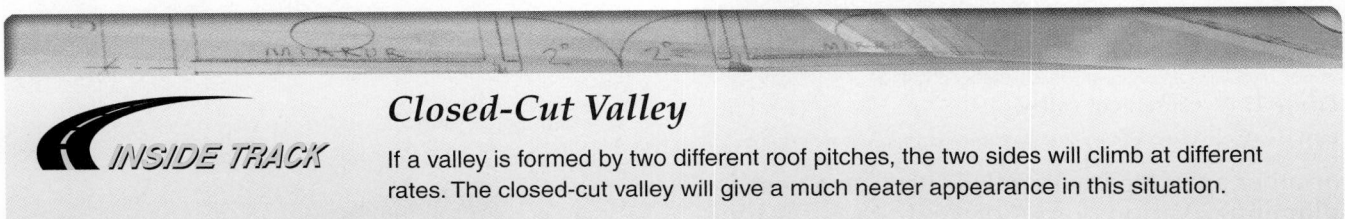

Closed-Cut Valley

INSIDE TRACK

If a valley is formed by two different roof pitches, the two sides will climb at different rates. The closed-cut valley will give a much neater appearance in this situation.

To create a closed-cut valley, proceed as follows:

Step 1 Lay the first double-course of shingles along the eaves of one roof area up to and over the valley. See *Figure 43*. Extend it up along the adjoining roof section. The distance of this extension should be at least 12" or one full tab. Follow the same procedure when applying the next course of shingles. Make sure that the shingles are pressed tightly into the valley.

Step 2 Fasten in the normal manner, except that a fastener is to be located at the end of the terminal strip. This procedure is followed up the entire length of one side of the valley. If there is a high and a low slope, the first application should always be done on the low slope side.

Step 3 When this roof surface is complete, you are ready to proceed with the intersecting roof surface, which will overlap the preceding application. Measure over 2" from the center line of the valley in the direction of the intersecting roof. Carefully snap a chalkline from top to bottom. This will be your guideline for the trim cut on the shingles.

Step 4 Now apply the first course of shingles on the intersecting roof. Use extreme care to match your chalkline angle exactly. Also, trim off the upper corner of the shingle to prevent water from running back along the top edge. Embed the end of the shingle in a 2" to 3" strip of plastic asphalt cement, being careful to allow no tar to show on the original shingle opposite. Succeeding courses are applied and completed as shown in *Figure 43*, making the valley watertight.

6.4.0 Roof Projections and Flashing

The following sections describe various types of roof projections and flashing.

6.4.1 Soil Stacks

Another roofing task is waterproofing around soil or vent stacks. Most building roofs have pipes or vents emerging from them. Most are circular. They call for special flashing methods. Asphalt products combined with metals may be used for this purpose. A soil pipe made of cast iron, copper,

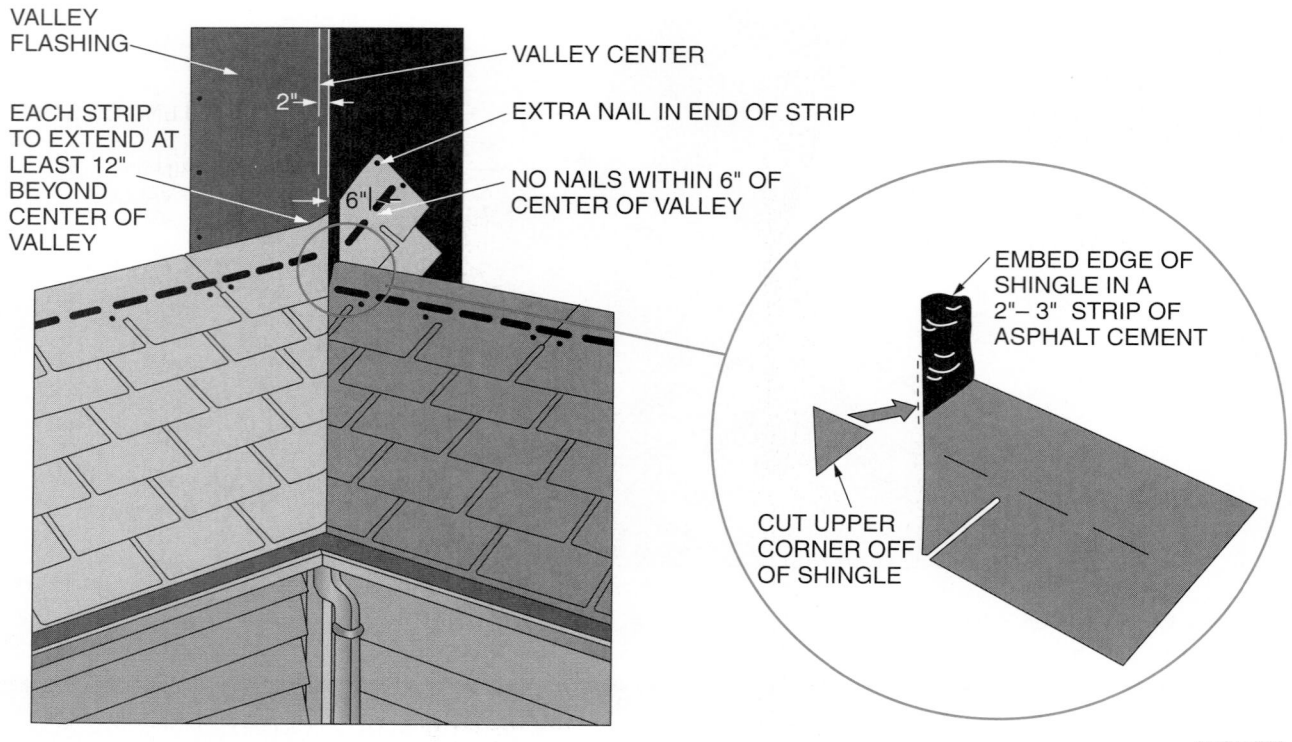

VALLEY FLASHING

VALLEY CENTER

EACH STRIP TO EXTEND AT LEAST 12" BEYOND CENTER OF VALLEY

2"

EXTRA NAIL IN END OF STRIP

6"

NO NAILS WITHIN 6" OF CENTER OF VALLEY

EMBED EDGE OF SHINGLE IN A 2"– 3" STRIP OF ASPHALT CEMENT

CUT UPPER CORNER OFF OF SHINGLE

202F43.EPS

Figure 43 ◆ Closed-cut valley.

or other approved materials is used as a vent for plumbing. Various types of **vent stack flashing** are available for this purpose. See *Figure 44*.

To apply vent stack flashing, proceed as follows:

Step 1 First, apply the roofing up to where the stack projects. Use extreme care when cutting and fitting the shingles around the stack. See *Figure 45*.

Step 2 Slip the flange over the stack and place it down into a bed of asphalt cement that has been carefully spread to the same size as the flange. The flange, sometimes called a boot or collar, is usually made of metal, but can also be made of plastic or rubber. See *Figure 46*. Prefabricated boots that slip over the stack come in different pitches.

Step 3 Mold the flange boot to the soil stack to ensure a snug fit. Use the manufacturer's recommended sealant to close up any opening. When the next course of shingles is laid, it covers the upper portion of the flange. Prior to this course, a bed of cement can be spread on the top of the flange. The end result seals and waterproofs the vented stack opening. See *Figure 47*.

Step 4 After the installation is completed, install the remainder of the shingles as previously described.

202F44.EPS

Figure 44 ◆ Vent stack flashing.

6.4.2 Vertical Wall Flashing

In the process of roof construction, there are times when the roof abuts a vertical wall horizontally or at an angle. This is a very critical spot to make watertight. Extreme care must be taken to follow correct procedures. An example of a sloped or angled abutment is shown in *Figure 48*. The initial step is to let the underlayment turn up on the vertical wall a minimum of 3" to 4". As an alternative, a strip of waterproof membrane may be applied to the roof deck and turned up on the wall.

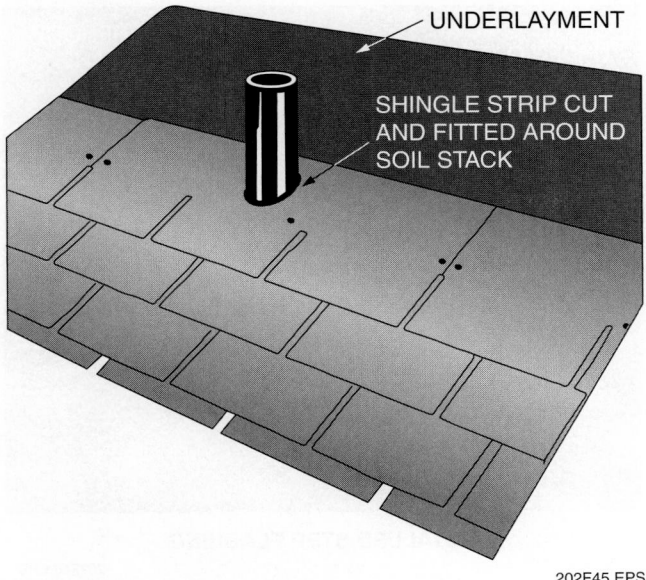

Figure 45 ◆ Layout around stack.

202F45.EPS

202F47.EPS

Figure 47 ◆ Covering flashing.

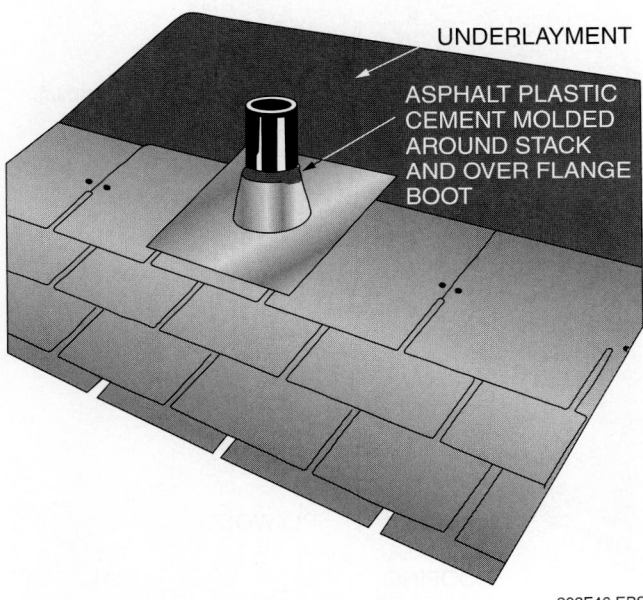

202F46.EPS

Figure 46 ◆ Placement of flashing.

This turn-up bend has to be done very carefully to maintain the seal and overlap of material without creasing or tearing the felt or membrane. Regular shingling procedures are used as each course is brought close to the vertical wall. **Wall flashing** (step flashing) is used when the rake of the roof abuts the vertical wall.

Metal step flashing shingles are applied over the end of each course of shingles and covered by the next succeeding course. The flashing shingles are usually rectangular. They are approximately 6" to 7" long and from 5" to 6" wide. When used with shingles laid 5" to the weather, they are bent so half of the flashing piece is over the roof deck with the remaining half turned up on the wall. The 7" length enables one to completely seal under the 5" exposure with asphalt cement and provide a 2" overlap up the entire length of the rake.

Nailing Step Flashing

INSIDE TRACK

Step flashing should only be nailed to the roof deck, never to the wall sheathing. This will allow settling or shifting of the structure without tearing the flashing and roofing away from the roof deck.

NAILING STEP FLASHING TO ROOF DECK

SINGLE NAIL AT UPPER OUTER EDGE

WATERPROOF MEMBRANE TURNED UP SIDE WALL

INSTALLED STEP FLASHING

202F48.EPS

Figure 48 ◆ Wall (step) flashing.

A careful study of *Figure 48* shows that each flashing shingle is placed just up the roof from the exposed edge of the shingle that overlaps it. It is secured to the deck sheathing with one nail in the top corner.

When the finished siding or clapboards are brought down over this flashing, they serve as a **cap flashing**, sometimes referred to as counter flashing. Usually, a 1" reveal margin is used and the ends of the boards are fully painted or stained to exclude dampness and prevent rot. With proper application of flashing and shingles, the joint between a sloping roof and a vertical wall should be watertight.

On a horizontal abutment (*Figure 49*), continuous flashing must be applied horizontally across the entire top of the abutting roof and against the vertical wall under the siding.

Continuous flashing can be formed with a metal brake or by hand, as shown in *Figure 50*. The flashing should be at least 9" wide and bent to match the angle of the joint to be flashed. Position the bend so that there will be at least 4" of flashing on the roof and 5" on the wall.

Before applying the flashing, adjust the last two courses of shingles so that the last course, which will be trimmed to butt against the wall, is at least 8" wide. After this abutting course is installed, place roofing cement on top of the last course of shingles. Place the flashing against the wall (slipping it under any siding, if necessary) and press it into the cement. Do not nail the flashing to the wall or roof deck. If desired, apply several beads of roofing cement to the top of the flashing. Press the tabs cut from shingles into the cement to cover

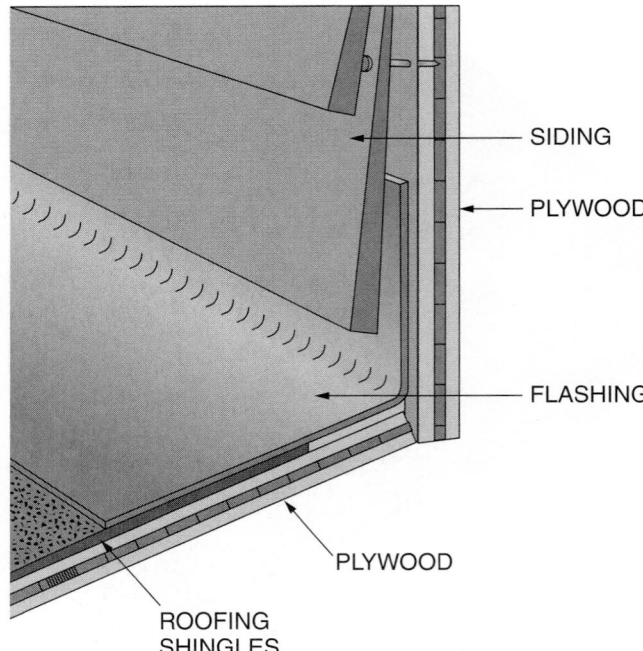

SIDING

PLYWOOD

FLASHING

PLYWOOD

ROOFING SHINGLES

202F49.EPS

Figure 49 ◆ Continuous flashing.

and hide the flashing (*Figure 51*). Position the tabs the same distance apart as the cutouts on the shingles and stagger them to match the pattern on the roof deck.

6.4.3 Dormer Roof Valley

The installation of a dormer roof valley will require you to combine some of the procedures previously covered. *Figure 52* shows an open

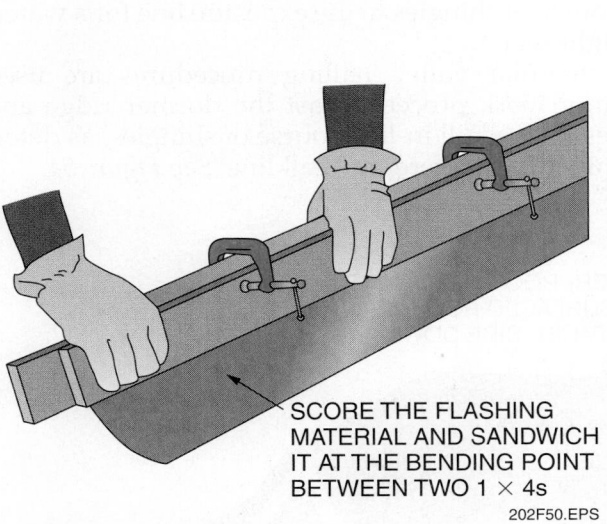

SCORE THE FLASHING
MATERIAL AND SANDWICH
IT AT THE BENDING POINT
BETWEEN TWO 1 × 4s

202F50.EPS

Figure 50 ◆ Bending continuous flashing.

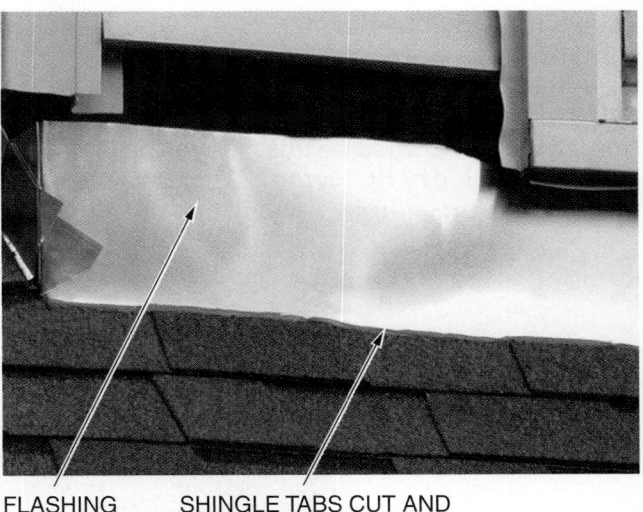

FLASHING   SHINGLE TABS CUT AND
CEMENTED TO FLASHING

202F51.EPS

Figure 51 ◆ Covering flashing.

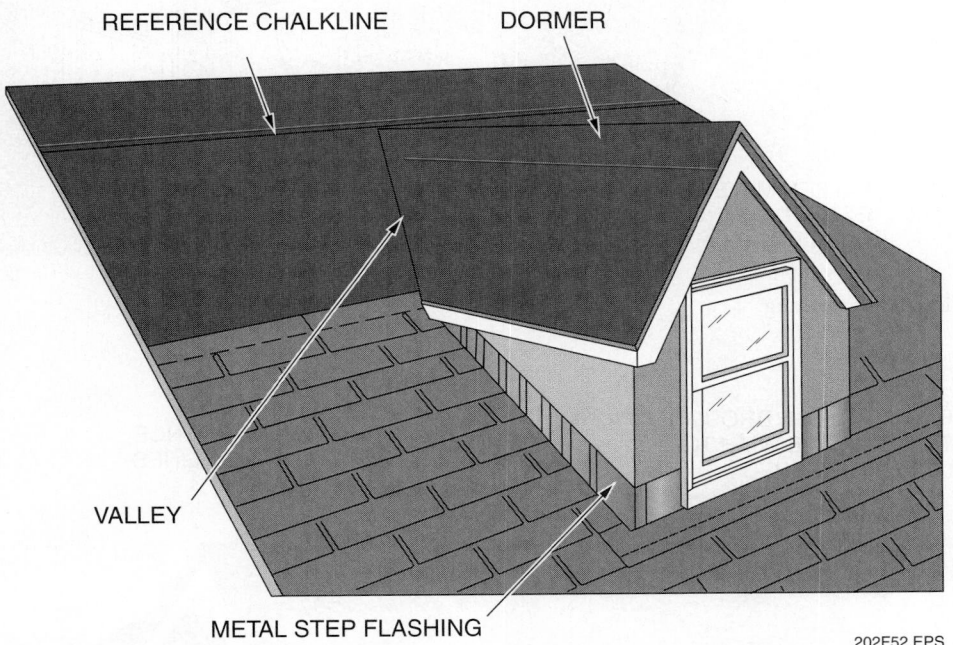

REFERENCE CHALKLINE   DORMER

VALLEY

METAL STEP FLASHING

202F52.EPS

Figure 52 ◆ Dormer flashing.

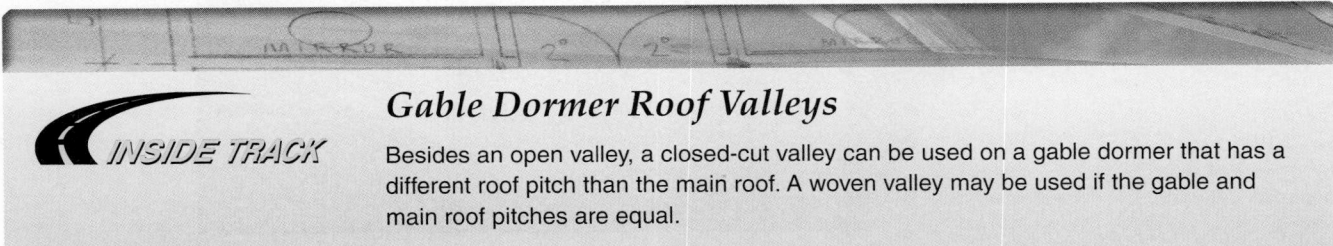

Gable Dormer Roof Valleys

INSIDE TRACK

Besides an open valley, a closed-cut valley can be used on a gable dormer that has a different roof pitch than the main roof. A woven valley may be used if the gable and main roof pitches are equal.

valley for a gable dormer roof. Note that the shingles have been laid on the main roof up to the lower end of the valley.

Extreme care must be used during the installation of the last course against the vertical wall to ensure a tight, dry fit. *Figure 53* displays the standard valley procedures for dormer flashing and shows how the valley material overlaps the course of shingles to the exposure line for a watertight seal.

Regular valley nailing procedures are used until work proceeds past the dormer ridge and resumes a full in-line course of shingles, as determined by a reference chalkline. See *Figure 54*.

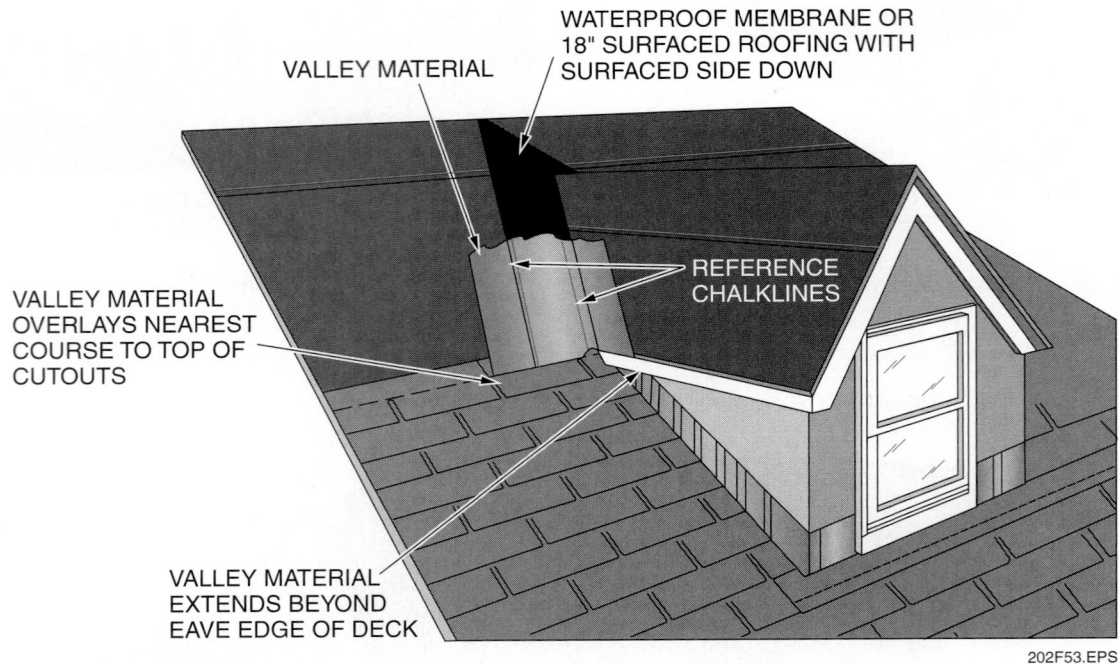

Figure 53 ◆ Dormer valley flashing.

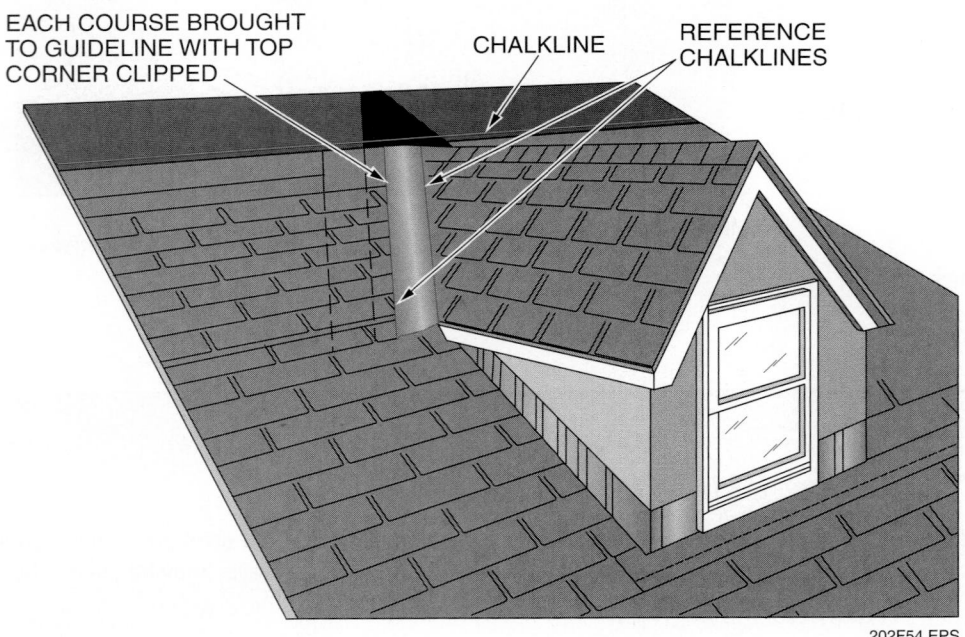

Figure 54 ◆ Dormer valley coverings.

6.4.4 Chimneys

Chimneys are subject to varying loads and certain opposing structural movements due to winds, temperature changes, and settling. Therefore, roofing materials and **base flashing** should not be attached or cemented to both the chimney and roof deck. The process of shingling around chimneys must be approached with extreme care. Due to the size of the opening in the roof and of the chimney itself, additional work must be done on the roof deck around chimneys prior to shingling. A cricket, also called a saddle, must be made. See *Figure 55*.

A cricket placed behind the chimney keeps rainwater or melting snow/ice from building up in back of the chimney. It steers flowing water around the chimney. The cricket is usually supported by a horizontal ridge piece and a vertical piece at the back of the chimney, as shown in *Figure 55*. The height of the cricket is typically half the width of the chimney, although these requirements will vary. Check local codes. The ridge, which is level, extends back to the roof slope. On wide chimneys, it may be necessary to frame the cricket, as shown in *Figure 56*. Either type of cricket may be covered with two triangular pieces of ¾" exterior plywood cut to fit from the ridge to the edge of the chimney and the roof slope. Heavy-gauge metal can also be used to form the cricket. The covering is then nailed to the support and roof deck.

Flashing at the point where the chimney comes through the roof requires something that will allow movement without damage to the water seal. It is necessary to use base flashing. The counter or cap flashing is secured to the masonry. Metal is used for base flashing and cap flashing.

To apply the chimney flashing, proceed as follows:

Step 1 Apply shingles over the roofing felt up to the front face of the chimney (*Figure 57*). Cut the base flashing for the front cut according to the pattern shown in *Figure 58*.

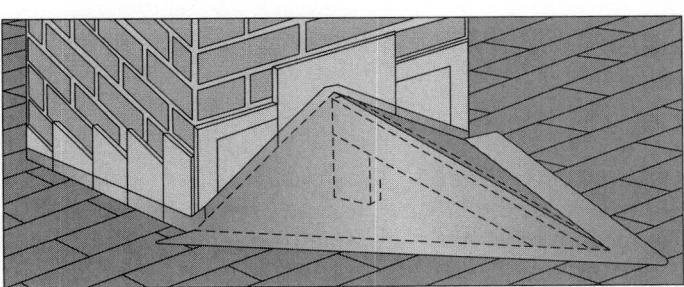

202F55.EPS

Figure 55 ◆ Simple chimney cricket.

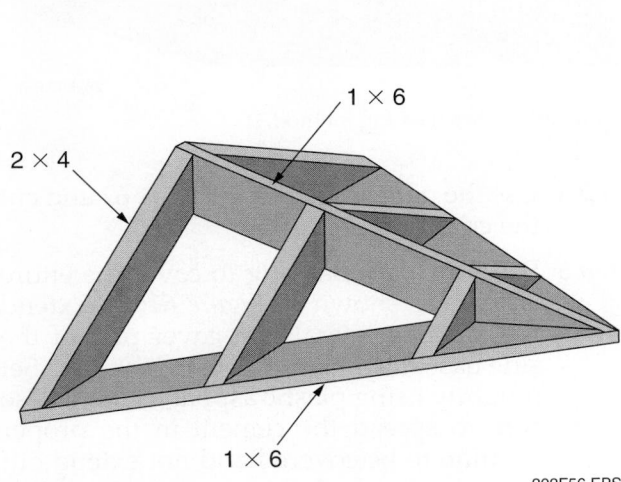

202F56.EPS

Figure 56 ◆ Cricket frame.

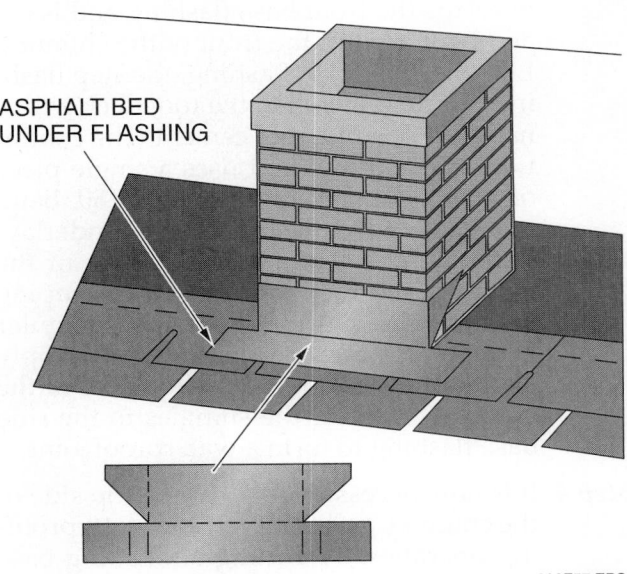

ASPHALT BED UNDER FLASHING

202F57.EPS

Figure 57 ◆ Front base flashing.

Combustible Material Spacing Requirements for a Cricket

Some building codes require that the wood framing and sheathing for a cricket must be spaced up to 1" from the chimney masonry. Always check your local codes for spacing requirements pertaining to combustible materials near chimneys.

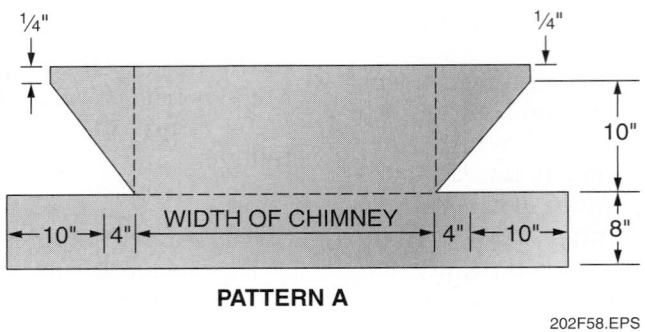

PATTERN A

202F58.EPS

Figure 58 ◆ Pattern for front base flashing.

Step 2 The base flashing for the front cut is applied first. The lower section is laid over the shingles in a bed of plastic asphalt cement. Bend the triangular ends of the upper section around the corners of the chimney.

Step 3 The sides of the chimney are base-flashed next. Either step flashing or continuous flashing can be used. Step flashing is applied as the shingles are applied up to the top side of the chimney (*Figure 59*). Note that the first piece of step flashing overlaps the front base flashing and is cut and bent around the front of the chimney. Like the front base flashing, the step flashing is fastened only to the roof deck with a nail or, if desired, roof cement. The continuous flashing method uses a single piece of metal, cut as shown in *Figure 60*. Bend and cement the flashing to the underlayment along the slope at the sides of the chimney, with the lower end overlapping the front base flashing. Bend the triangular end pieces around the chimney. Apply shingles up the roof to the top side of the chimney. Cement the shingles to the side base flashing to form a waterproof joint.

Step 4 It is now necessary to go to the top side of the chimney and complete the waterproofing operation by cutting and fitting base flashings over the cricket, known as cricket flashing, as shown in *Figure 61*.

202F59.EPS

Figure 59 ◆ Step flashing method.

Step 5 Use the pattern shown in *Figure 62* and cut the cricket base flashing.

Step 6 Bend the base flashing to cover the entire cricket, as shown in *Figure 61(C)*. Extend the flashing laterally to cover part of the side base flashing previously installed. Set it tightly using plastic asphalt cement. Use care to spread the cement in the proper location to be covered and not extend out onto the finished shingles. Neatness is a must. Bend the ends around the chimney.

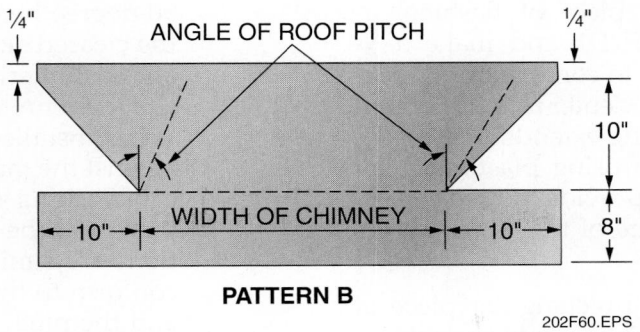

Figure 60 ◆ Pattern for continuous side flashing.

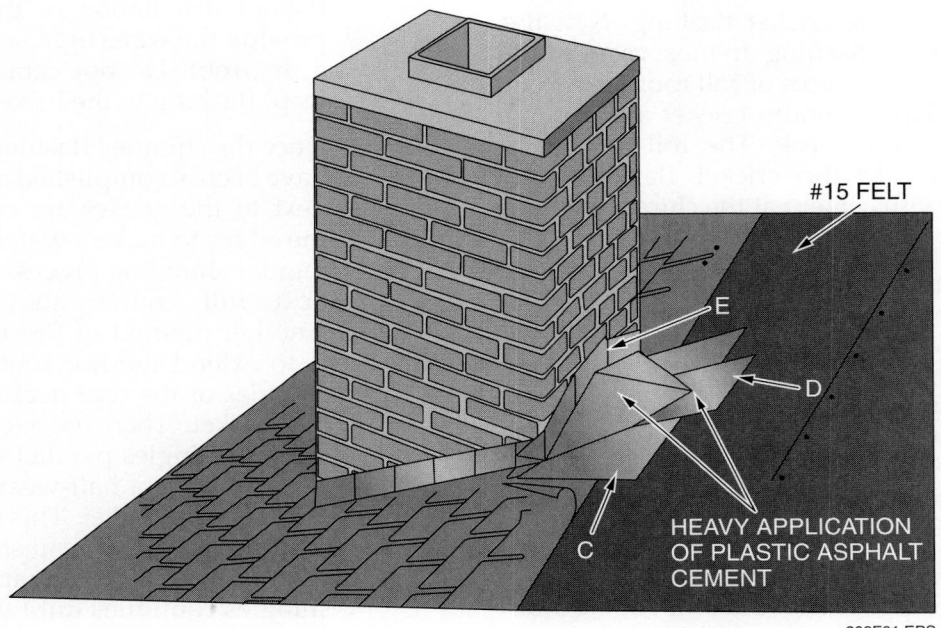

Figure 61 ◆ Cricket flashing.

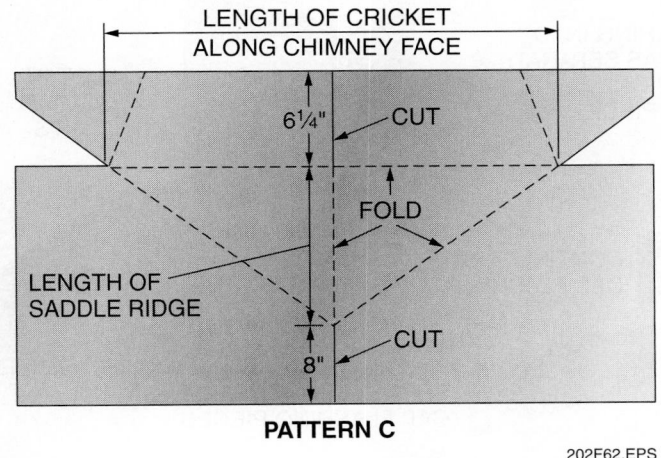

Figure 62 ◆ Cricket base flashing pattern.

Step 7 Cut a rectangular piece of flashing, as shown in *Figure 61(D)*, and make a V-cutout on one side to conform to the rear angle of the cricket. Center it over that part of the cricket flashing extending up to the deck. Set it tightly using plastic asphalt cement. This piece provides added protection where the ridge of the cricket meets the deck.

Step 8 Cut a second small rectangular piece of flashing. Cut a V on one side to conform to the pitch of the cricket, as shown in *Figure 61(E)*. Place it over the cricket ridge and against the flashing that extends up the chimney. Embed it in plastic asphalt cement to the cricket flashing. Nail the edges of the flashing. In most cases, similarly colored pieces of roll roofing are cut to overlap the entire cricket and extend onto the roof deck. The roll roofing is cemented to the cricket flashing and sealed with cement at the chimney edge.

Step 9 For completion, cap flashing (also called counter flashing) is installed. It is usually made of sheet copper 16 ounces or heavier. It can also be made of 24-gauge galvanized steel. If steel is used, it should be painted on both sides. *Figures 63* and *64* show metal cap flashing on the face of the chimney and on the sides. Cap flashing is secured to the brickwork, as shown in *Figure 65*.

Step 10 *Figure 65* shows a good method of securing the cap flashing. Cut a slot in the mortar joint to a depth of ¼" to ½". Insert a

90-degree bent edge of the flashing into the cleared slot between the bricks using an elastomeric sealant or mortar in the slot to secure the flashing to the masonry. When installed, the cap should lie snugly against the masonry. The front unit of the cap flashing should be one continuous piece. On the sides and the rear, the sections are similar in size. They are cut to conform to the locations of mortar joints and the pitch of the roof. If the sides are lapped, they must lap each other by at least 3". The slots are refilled with the brick mortar mix or elastomeric sealant and conform to the original brickwork. Patient installation of the flashing will provide the watertight seal necessary for a dry roof. Do not cement the counter (cap) flashing to the base flashing.

Step 11 Once the chimney flashing and shingling have been accomplished and the shingles next to the cricket are cemented under the edges to make a waterproof joint, the regular shingling process resumes on the next full course above the cricket. Another method of finishing the cricket is to extend the horizontal composition shingles of the roof deck up the pitch of the cricket. Then use step flashing and cement shingles parallel with the cricket ridge to form a half-weave valley at the edges of the cricket. This second method requires cementing cap shingles over the ridge of the cricket. The application of the shingles continues until the roof ridge is reached.

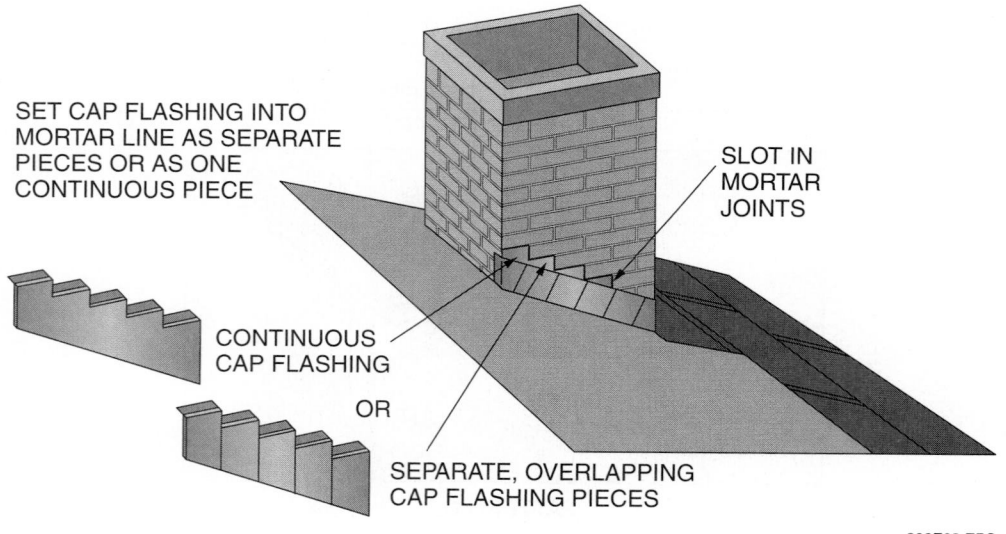

SET CAP FLASHING INTO MORTAR LINE AS SEPARATE PIECES OR AS ONE CONTINUOUS PIECE

SLOT IN MORTAR JOINTS

CONTINUOUS CAP FLASHING

OR

SEPARATE, OVERLAPPING CAP FLASHING PIECES

202F63.EPS

Figure 63 ◆ Cap flashing methods at sides.

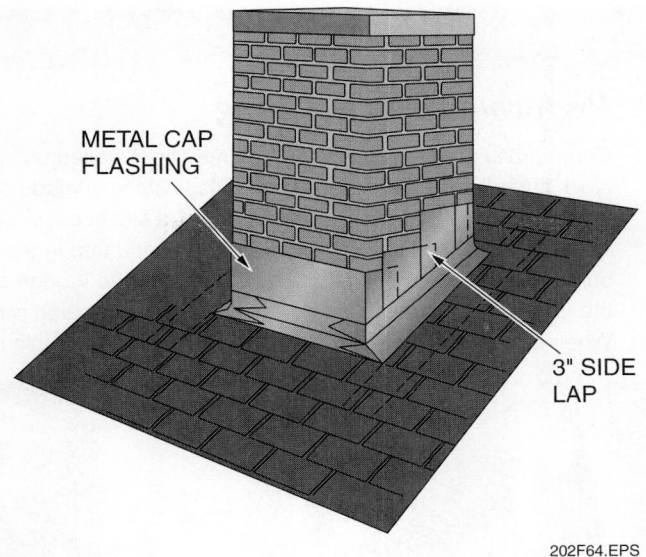

Figure 64 ◆ Flashing cap and lap.

METAL CAP
FLASHING

3" SIDE
LAP

202F64.EPS

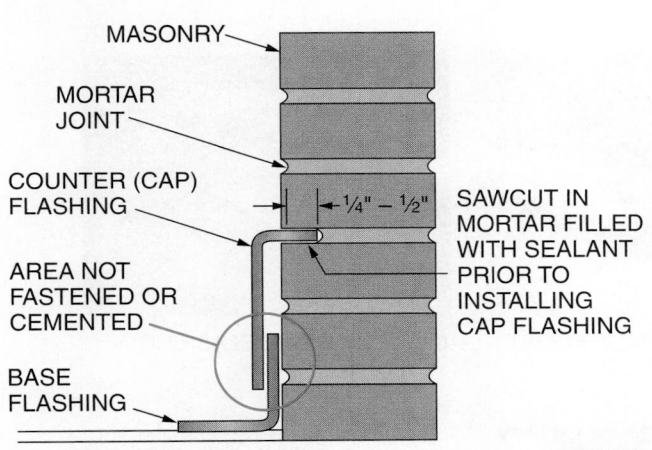

MASONRY

MORTAR
JOINT

COUNTER (CAP)
FLASHING

AREA NOT
FASTENED OR
CEMENTED

BASE
FLASHING

¼" – ½"

SAWCUT IN
MORTAR FILLED
WITH SEALANT
PRIOR TO
INSTALLING
CAP FLASHING

202F65.EPS

Figure 65 ◆ Counter (cap) flashing installation.

202F66.EPS

Figure 66 ◆ Architectural cap shingle.

CUT TABS FROM WHOLE SHINGLES AND
TAPER EACH TAB AS SHOWN

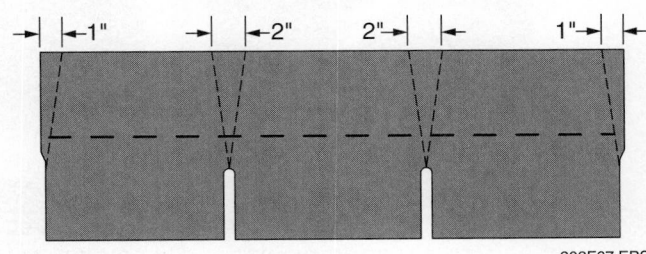

1"   2"   2"   1"

202F67.EPS

Figure 67 ◆ Cutting cap shingles.

6.4.5 Hip or Ridge Row (Cap Row)

Special shingles are required to complete the hip or ridge rows. In some cases, ridge caps and hip caps are premanufactured.

Architectural shingles have a matching cap row shingle (*Figure 66*) that must be used. Cap row shingles cannot be cut from architectural shingles. Most of the time if the roof was shingled with standard three-tab shingles, cap rows are cut from the shingles and the 12 × 12 tab is used (*Figure 67*). The tab can be reduced to 9 × 12, but nothing less.

Since the hip or ridge is a potential spot for water leaks, precautions must be taken. If ridge venting will be used, do not apply the ridge caps.

Preformed Cap Flashing

Commercial preformed cap flashing may be installed on vertical brick, concrete, or block surfaces including chimneys. This flashing is made of an aluminum-coated steel. As shown in the sequenced photographs, a slot is cut in all sides of a chimney using a ¼"-thick diamond impregnated steel wheel mounted in a small, high-speed electric grinder. The flashing is trimmed to shape and the V-edge of the flashing is pressed into the groove. The flashing is formed to shape and sealed with an elastomeric sealant. When set, the sealant and flashing may be painted to blend with the roof.

1. PREFORMED CAP FLASHING

2. CUTTING ¼" GROOVE

3. COMPLETED GROOVES

4. TRIMMING FRONT FLASHING IN GROOVE TO SIZE

5. SEATING V-EDGE OF FRONT FLASHING IN GROOVE

6. FITTING SIDE FLASHING

7. SEATING V-EDGE OF SIDE FLASHING IN GROOVE

8. SIDE FLASHING INSTALLED AND FORMED TO FRONT FLASHING

9. SEALANT APPLIED TO GROOVE AND FLASHING

10. FLASHING PAINTED TO MATCH ROOFING

202SA07.EPS

To install a ridge or hip row, proceed as follows:

Step 1 Butt and nail shingle roofing as they come up on either side of a hip or ridge. On a ridge, lay the last course and trim the shingles, as shown in *Figure 68*. On a hip, trim the shingles at an angle on the hip line.

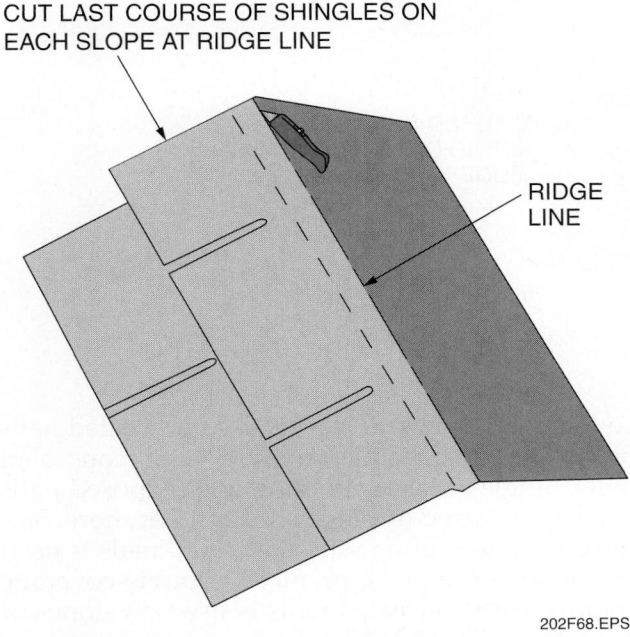

CUT LAST COURSE OF SHINGLES ON EACH SLOPE AT RIDGE LINE

RIDGE LINE

202F68.EPS

Figure 68 ◆ Applying the last course of ridge shingles.

Step 2 After the cap shingles are cut, bend them lengthwise in the center line. In cold weather, warm the shingles before bending to prevent cracks. Begin at the bottom of any hips. Cut the first tab to conform to the dual angle at the eaves.

Step 3 Lap the units to provide a 5" exposure of the granular surface. See *Figure 69*. Secure with one nail on each side, 6" back from the exposed end and 1" from the edge. As each succeeding tab is nailed going up the hip, the nail penetrates and secures two tabs. This tight bond prevents the wind from getting underneath and lifting the tab.

Nailing of the ridge row is similar to that described for the overall hip. Nailing takes place from both ends of the ridge. A final cap piece joins the ridge together in the center, and the exposed nails are covered with roof cement. An exception to this may occur in a very windy area. In that case, the ridge cap would be started at the point on the roof opposite the wind direction. As each ridge shingle is placed, it automatically allows the wind to pass over it, and there is no possibility of shingles blowing off. The junction of the roof ridge and any hip ridges can be capped with a special molded cap or by a fabricated end cap, as shown in *Figure 70*. Bed the final ridge cap or hip/ridge caps in asphalt and secure with nails as shown in the figure. Cover the nail heads with roof cement or sealant.

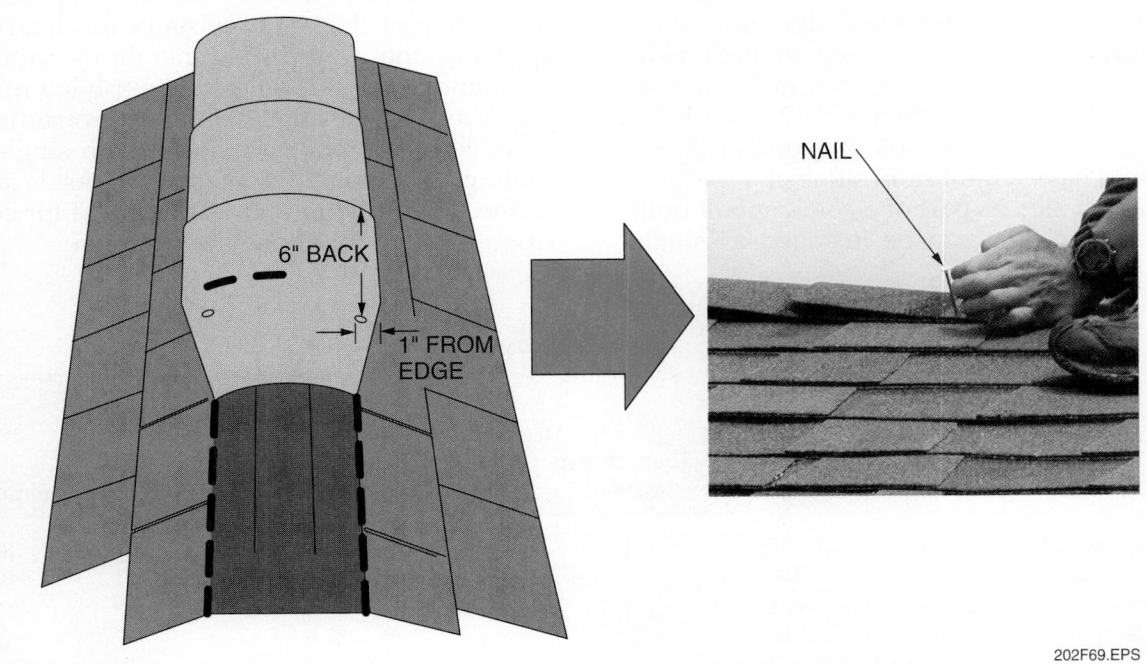

6" BACK

1" FROM EDGE

NAIL

202F69.EPS

Figure 69 ◆ Installing a ridge cap.

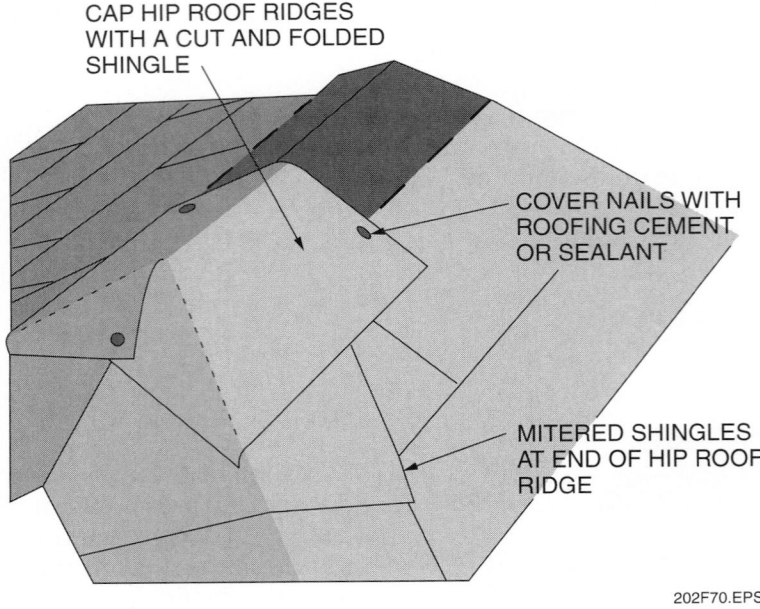

CAP HIP ROOF RIDGES WITH A CUT AND FOLDED SHINGLE

COVER NAILS WITH ROOFING CEMENT OR SEALANT

MITERED SHINGLES AT END OF HIP ROOF RIDGE

202F70.EPS

Figure 70 ◆ Hip and ridge end cap.

7.0.0 ◆ ROLL ROOFING INSTALLATION

Nearly flat roofs can be roofed with a hot-asphalt BUR, a single-ply membrane system, or roll roofing. Roll roofing can be installed on underlayment by itself, as part of a cold asphalt built-up roof, or on a waterproof membrane. The weights, characteristics, side lap, top lap, and recommended exposures for double-coverage, single-coverage, and uncoated roll roofing are shown in *Table 2*.

All flat roofs must have some pitch, either to an edge or roof drains, so that water does not collect. Composition shingles are not used on roofs with a slope of less than 2 in 12; wood shingles are not used with a slope of less than 3 in 12; and shakes are not used with a slope of less than 4 in 12. Flat roofs should have a minimum slope of ¼" per foot. When not installed as part of a cold-asphalt built-up roof, roll roofing can be installed as single-coverage roofing with exposed or concealed nails or as double-coverage roofing with concealed nails. Single-coverage roofing with exposed nails is generally used on slopes of 2 in 12 or more. Single-coverage roofing with concealed nails is used on slopes of 1 in 12 or more. Double-coverage roofing with concealed nails is used on slopes of more than ¼ in 12.

Apply a drip edge and waterproof membrane, or, as a minimum, apply a 15-lb underlayment to the roof deck. Make sure all debris is removed from the roof deck and all nails are flush before applying the waterproof membrane/underlayment and that it is clean before applying roll roofing. Even a very small pebble or protruding nail will eventually poke a hole through single-layer roofing. Flashing of roof projections is accomplished in the same way as described for composition shingles in the previous section.

Table 2   Typical Weights, Characteristics, and Recommended Exposures for Roll Roofing

| Product | Approximate Shipping Weight | | Squares per Package | Length | Width | Side or End Laps | Top Lap | Exposure |
	Per Roll	Per Square						
Mineral surface roll, double-coverage	75# to 90#	75# to 90#	1	36' 38'	36" 38"	6"	2" 4"	34" 32"
Mineral surface roll, single-coverage	55# to 70#	55# to 70#	½	36'	36"	6"	19"	17"
Uncoated roll	50# to 65#	50# to 65#	1	36'	36"	6"	2"	34"

Removing the Curl from Roll Roofing

INSIDE TRACK

Before installing roll roofing in cool to cold ambient temperatures, cut the roofing into 12' to 18' sections and stack it for a sufficient length of time to remove any curl. The length of time required will depend on the ambient temperature.

7.1.0 Single-Coverage Roll Roofing Installation

This section describes the installation of single-coverage roll roofing using the exposed and concealed nail methods.

7.1.1 Exposed Nail Method

Single-coverage roll roofing with exposed nails is generally applied horizontally over underlayment, as shown in *Figure 71*. To install roll roofing using the exposed nail method, proceed as follows:

Step 1 Protect each valley with 18"-wide metal flashing.

Step 2 For horizontal application (*Figure 72*), snap a chalkline 35½" above the eaves. Apply a 2" band of roof cement to the eaves and rake edges.

Step 3 Using the chalkline as a reference, run the first course so that it overhangs the eaves by ½" and the rake edges by 1". Cement and overlap any vertical seams by 6". Nail all seams and the bottom and rake edges of the first course every 3" with galvanized or aluminum roofing nails. Use a utility knife to trim the roofing edges to the drip edges.

Step 4 Snap another chalkline 3" down from the top edge of the first course and apply a 2" band of roof cement within the band and up the rake edges. Lay the second course to the chalkline and nail every 3" along all seams, the bottom edge, and the rake edges, as shown in *Figure 73*.

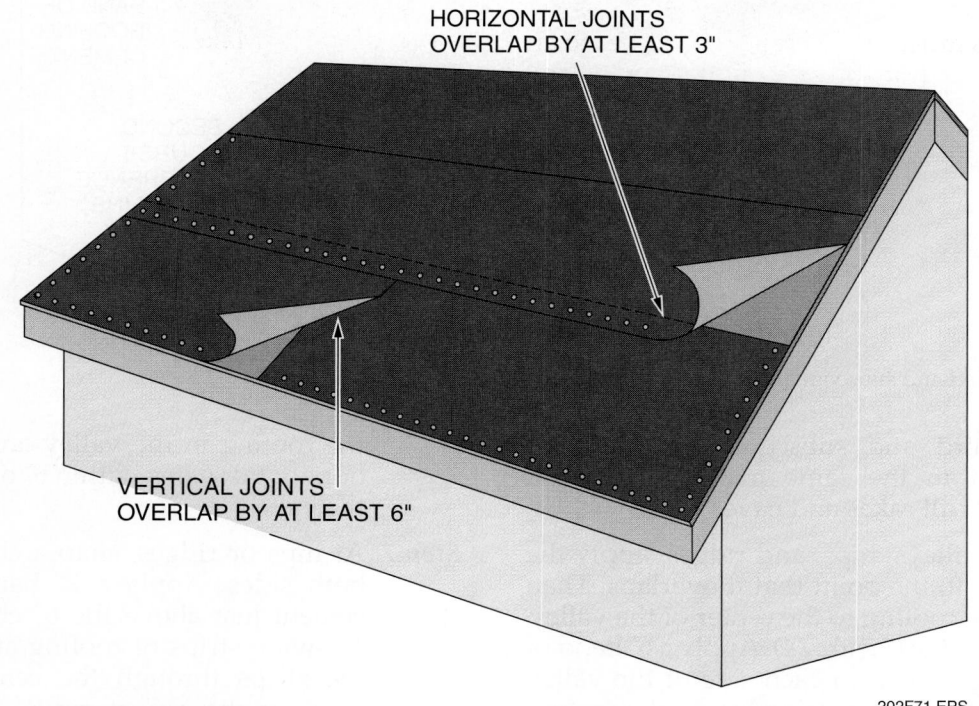

HORIZONTAL JOINTS
OVERLAP BY AT LEAST 3"

VERTICAL JOINTS
OVERLAP BY AT LEAST 6"

202F71.EPS

Figure 71 ◆ Typical roll roofing installation.

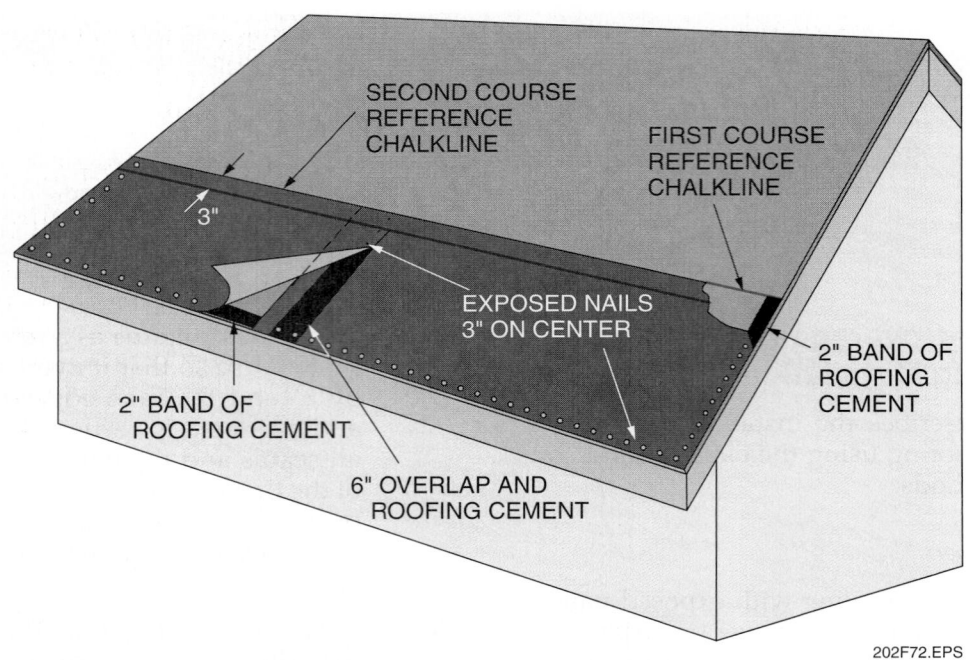

SECOND COURSE
REFERENCE
CHALKLINE

FIRST COURSE
REFERENCE
CHALKLINE

3"

EXPOSED NAILS
3" ON CENTER

2" BAND OF
ROOFING
CEMENT

2" BAND OF
ROOFING CEMENT

6" OVERLAP AND
ROOFING CEMENT

202F72.EPS

Figure 72 ◆ First course of exposed nail roll roofing.

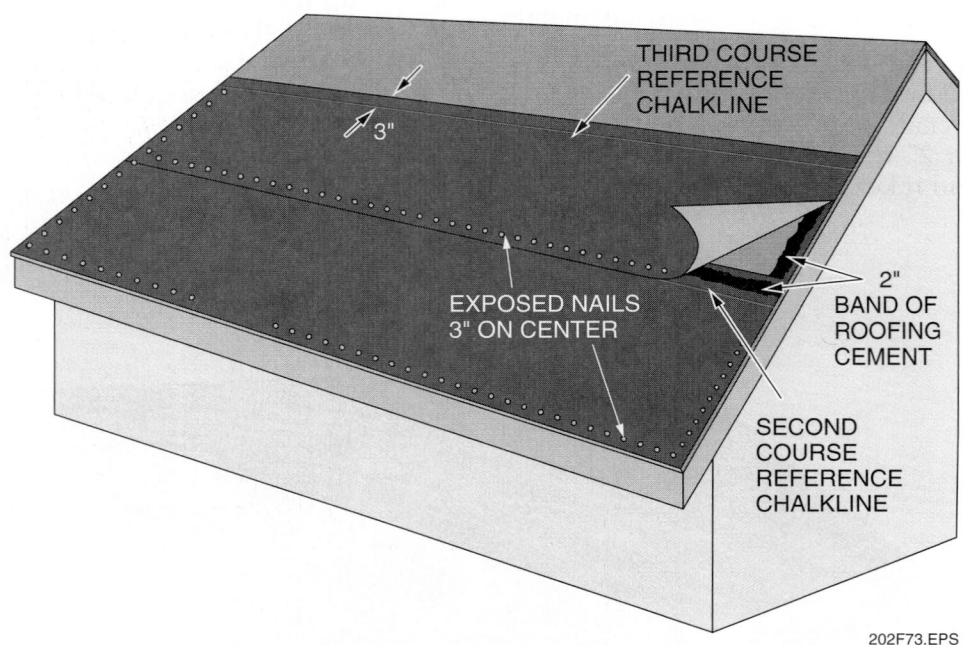

THIRD COURSE
REFERENCE
CHALKLINE

3"

EXPOSED NAILS
3" ON CENTER

2"
BAND OF
ROOFING
CEMENT

SECOND
COURSE
REFERENCE
CHALKLINE

202F73.EPS

Figure 73 ◆ Second and subsequent courses of exposed nail roll roofing.

Step 5 The third and subsequent courses are applied in the same manner. Trim the edges at all rakes and eaves.

Step 6 At all valleys, hips, and ridges, apply the roofing to the point that it overlaps. Then trim the roofing to the center of the valley, hip, or ridge (*Figure* 74). Apply a 6" band of roofing cement to each side of the valley flashing and press the edges of the roofing into the cement. Do not nail the edges of the roofing in the valley and do not nail horizontal seams within 6" of the center of the valley.

Step 7 At hips or ridges, snap a chalkline 6" on both sides. Apply a 2" band of roofing cement just above the 6" chalklines. Cut 12"-wide strips of roofing and nail down the strips through the cement on both sides of the hip or ridge. As necessary, overlap the strips by 6" and cement.

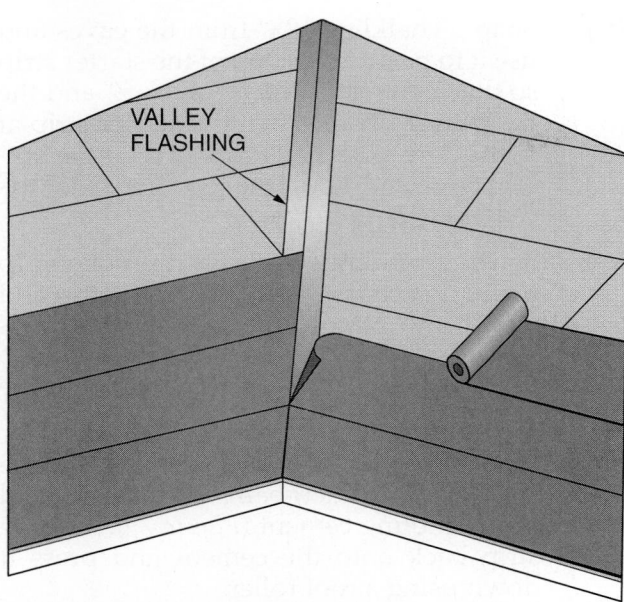

Figure 74 ◆ Roll roofing in a valley.

202F74.EPS

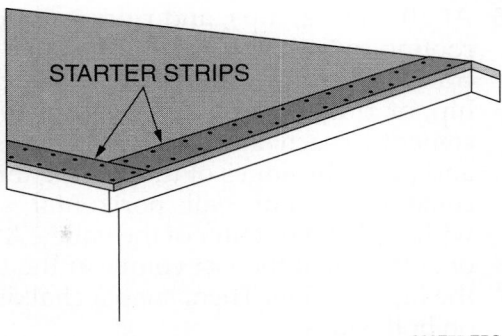

Figure 75 ◆ Roll roofing starter strips.

202F75.EPS

7.1.2 Concealed Nail Method

To install roll roofing using the concealed nail method, use the following procedure.

Step 1 Cut and install 9"-wide roofing material starter strips along the eaves and rakes (*Figure 75*). Nail the strips to the roof deck on both edges with galvanized or aluminum nails spaced ¾" from the edges and 4" apart. These strips provide a surface for adherence of roofing cement. Protect each valley with 18"-wide metal flashing.

Step 2 Snap a chalkline 35½" above the eaves. Using the chalkline as a reference, run the first course so that it overhangs the eaves by ½" and the rake edges by 1". Nail along the top edge, ¾" from the edge, every 4". Then apply a 6" band of roof cement to the eaves and rake edges (*Figure 76*). Press the roofing down into the cement with a roof roller. Cement and overlap any vertical seams by 6". Use a utility knife to trim the roofing edges to the drip edges.

Step 3 Snap another chalkline 6" down from the top edge of the first course. Lay the second course to the chalkline and nail along the top edge, ¾" from the edge, every 4". Apply a 6" band of roof cement under the bottom of the second course and up the rake edges. Press the roofing down into the cement with a roof roller.

Step 4 The third and subsequent courses are applied in the same manner (*Figure 77*). Trim the edges at all rakes and eaves.

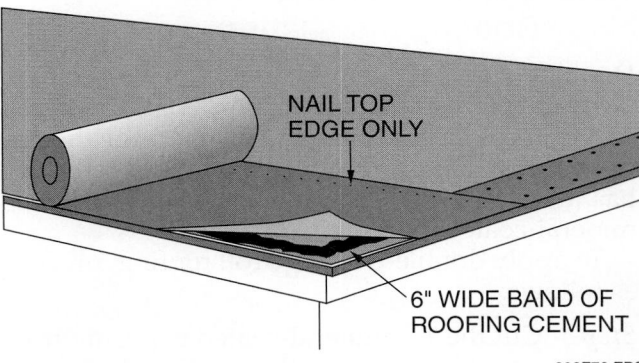

Figure 76 ◆ First course of concealed nail roll roofing.

202F76.EPS

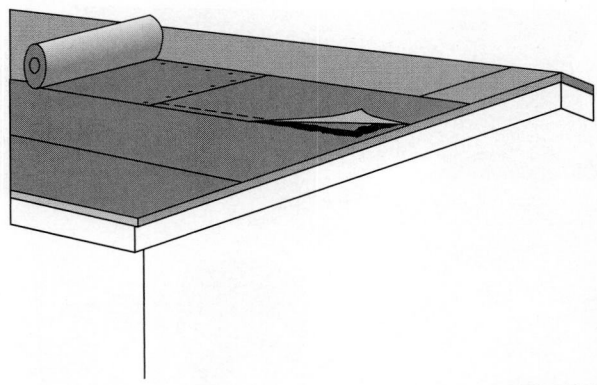

Figure 77 ◆ Third and subsequent courses of concealed nail roll roofing.

202F77.EPS

Step 5 At all valleys, hips, and ridges, apply the roofing to the point that it overlaps. Then trim the roofing to the center of the valley, hip, or ridge. Apply a 6" band of roofing cement to each side of the valley flashing and press the edges of the roofing into the cement. Do not nail horizontal seams within 6" of the center of the valley. At hips or ridges, nail the last course at the top of the hip or ridge. Then, snap a chalkline 6" on both sides.

Step 6 Apply a 6" band of roofing cement just above the 6" chalklines (*Figure 78*). Using 12"-wide strips cut from the roofing material, press the strips down into the cement on both sides of the hip or ridge with a roof roller. As necessary, overlap the strips by 6" and cement.

7.2.0 Double-Coverage Roll Roofing Installation

Double-coverage roll roofing is available for both hot and cold asphalt application. Make sure that only the cold asphalt version is used for the following installation. Double-coverage roll roofing has a 19" overlap (called the selvage) and a 17" mineral-coated exposure.

To apply double-coverage roll roofing, proceed as follows:

Step 1 Cut the 17" mineral-coated exposure from enough strips of the roofing to extend across all eaves. The mineral-coated pieces will be used as starter strips. Save the 19" selvage strips for use when the ridge caps are installed.

Step 2 Snap a chalkline 18½" from the eaves and use it to position the top of the starter strip so that it overlaps the eaves by ½" and the rakes by 1". Nail down the starter strip at both edges with nails spaced 3" at the bottom edge and 12" at the top edge. Nail the center at 12" intervals.

Step 3 Starting from one end, position one strip of the first course over the starter strip and nail it into place using two rows of nails spaced 4½" and 13" from the top of the strip in the selvage area. Space the nails in the rows about 12" apart.

Step 4 Roll back the strip and thickly coat the starter strip underneath with non-fibered liquid roofing cement (*Figure 79*). Roll the strip back onto the cement and press it down using a roof roller.

Step 5 Overlap the next strip by 6" and repeat the nailing and cementing procedure. Vertical seams are cemented, as shown in *Figure 80*, and are not nailed except in the selvage area.

Step 6 Continue applying strips and courses until the roof deck is covered. Trim the edges at all rakes and eaves.

Step 7 At all valleys, hips, and ridges, apply the roofing to the point that it overlaps. Then trim the roofing to the center of the valley, hip, or ridge.

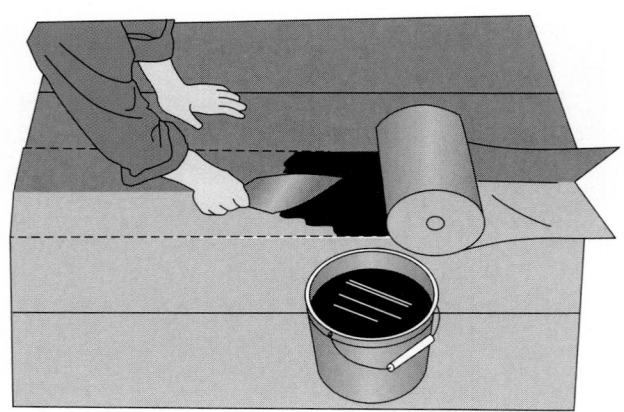

202F78.EPS

Figure 78 ◆ Covering hip or ridge joints.

202F79.EPS

Figure 79 ◆ Coating the starter strip.

Figure 80 ◆ Cementing a vertical seam.

Step 8 Apply a 6" band of roofing cement to each side of the valley and press the edges of the roofing into the cement. Do not nail horizontally to within 6" of the center of the valley.

Step 9 At hips or ridges, nail the last course at the top of the hip or ridge. Snap a chalkline 6" on both sides.

Step 10 Cut 12"-wide pieces from a roll of double-coverage roofing; include both the selvage and the mineral-coated exposure. These pieces are treated like the shingle tabs used to make a ridge cap, but are applied in double-coverage just like the roof.

Step 11 To begin, cut the selvage from one shingle. Then, starting at one end of the ridge or the bottom of a hip, place the selvage piece over the ridge and nail it down, spacing the nails 1" from the edges and at 4" intervals. Coat the selvage piece with cement.

Step 12 Next, place a full shingle over the cement with the mineral side up and press it into the cement. Nail the selvage of this shingle like the starting piece of selvage. Coat the selvage of that shingle with cement and apply another shingle.

Step 13 Repeat the process until the hip or ridge is completed. The junction of two hips and a ridge can be end capped in the same way as for a shingle roof.

8.0.0 ◆ WOOD SHINGLES AND SHAKES

Wood shingles and shakes for finish roofing are available in Grades 1 through 3.

- *Grade 1* – Grade 1 is the best and is the grade normally used. It is cut from heartwood and is clear (knot-free) and straight-grained. It is generally more rot resistant and also more expensive than the other grades.
- *Grade 2* – Grade 2 has a limited amount of sapwood, which is less rot resistant. It also has some knots and is flat grained. Grade 2 is acceptable for residential roofing.
- *Grade 3* – Grade 3 should only be used on outbuildings.

Shingles (not shakes) are also available as Grade 4 (utility). They are suitable only for use as starter courses or for shim stock because they have big knots.

Wood shingles are not recommended for roofs with a slope of less than 3 in 12, and shakes are not recommended for roofs with a slope of less than 4 in 12. The reason is that voids between the courses are not protected from wind-blown snow or water.

Exposure also must be limited for slight pitches. For example, with a 3 in 12 slope, 16" shingles must have a maximum of 3¾" exposure (5" on a 4 in 12). A 24" shingle can have a 5¾" exposure on a 3 in 12 and 7½" on a 4 in 12.

Table 3 shows the grades, characteristics, sizes, and coverage for 16", 18", and 24" wood shakes at different exposures.

Table 4 shows the grades, characteristics, sizes, and coverage for 16", 18", and 24" wood shingles at different exposures.

Table 3 Summary of Wood Shakes

Grade	Length and Thickness	Courses per Bundle	Bundles per Square	Description
		18' Pack*		
No.1–Hand split and Resawn	16" starter-finish	9/9	5	These shakes have split faces and sawn backs. Cedar logs are first cut into desired lengths. Blanks or boards of the proper thickness are split and then run diagonally through a band saw to produce two tapered shakes from each blank.
	18" × ½" mediums	9/9	5	
	18" × ¾" heavies	9/9	5	
	24" × ⅜" mediums	9/9	5	
	24" × ½" mediums	9/9	5	
	24" × ¾" mediums	9/9	5	
No. 1–Tapersawn	24" × ⅝"	9/9	5	These shakes are sawn on both sides.
	18" × ⅝"	9/9	5	
No. 1–Tapersplit	24" × ½ "	9/9	5	Produced largely by hand using a sharp-bladed steel froe (a cleaving tool) and a wooden mallet. The natural shingle-like taper is achieved by reversing the block, end-for-end, with each split.
		20' Pack*		
No. 1–Straight split	18" × ⅜" True-Edge**	14 straight	5	Produced in the same way as taper split shakes, except that by splitting from the same end of the block, the shakes acquire the same thickness throughout.
	18" × ⅜"	19 straight	5	
	24" × ⅜"	16 straight	5	

*Pack used for majority of shakes.

Table 4 Summary of Wood Shingles

Grade	Length	Thickness at Butt	Courses per Bundle	Bundles/Cartons per Square	Description
No. 1–Blue Label	16"	.40"	20/20	4 bundles	The premium grade of shingles for roofs and side walls. These top-grade shingles are 100% heartwood, 100% clear, and 100% edge grain.
	18"	.45"	18/18	4 bundles	
	24"	.50"	13/14	4 bundles	
No. 2–Red Label	16"	.40"	20/20	4 bundles	A good grade for many applications. Not less than 10" clear on 16" shingles, 11" clear on 18" shingles, and 16" clear on 24" shingles. Flat grain and limited sapwood are permitted in this grade.
	18"	.45"	18/18	4 bundles	
	24"	.50"	13/14	4 bundles	
No. 3–Black Label	16"	.40"	20/20	4 bundles	A utility grade for economy applications and secondary buildings. Not less than 5" clear on 16" and 18" shingles; 10" clear on 24" shingles.
	18"	.45"	18/18	4 bundles	
	24"	.50"	13/14	4 bundles	
No. 4–Undercoursing	16"	.40"	14/14 or	2 bundles	A utility grade for undercoursing on double-coursed side wall applications or for interior accent walls.
	18"	.45"	20/20	2 bundles	
			14/14 or	2 bundles	
			18/18	2 bundles	

In most cases, shingles or shakes are installed on open or spaced roof sheathing (*Figure 81*). This type of decking costs less to install and allows the shingles/shakes to dry out, preventing rot. In high-moisture or snowy areas of the country, a waterproofing membrane is applied up the roof to extend 24" beyond the inside wall. In areas of high humidity, shingles or shakes applied over completely sheathed roofs (closed sheathing) that are covered with underlayment or rigid insulation are usually spaced off the roof with horizontal pressure-treated 1 × 3s nailed to the roof deck and spaced at the exposure of the product. For additional ventilation or when applied over rigid insulation, vertical furring strips are applied and centered over the rafters prior to the horizontal strips. In addition, the roof is ringed with pressure-treated 1 × 3s nailed parallel to the eaves, ridge, and rakes. These ventilation strips, sometimes called skip sheathing, allow the wood shingles or shakes to dry out. Roof projections are flashed in the same way as composition shingle roofs except that flashing projections under and up the wood shingles and shakes must be wider and longer.

8.1.0 Roof Exposure

The exposure represents the area of the shingle or shake that contacts the weather. See *Figure 82*. It depends upon the pitch of the roof. A good shingle or shake installation is never less than three layers thick. There are three lengths of shingles: 16", 18", and 24".

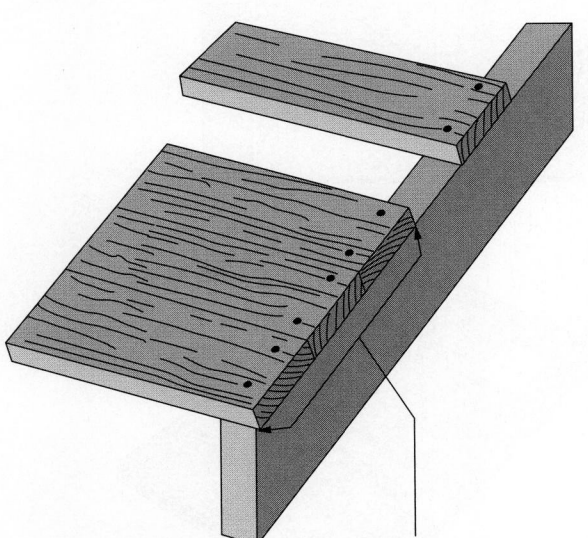

Eave edge solid sheathed for support and, if exposed, for underside appearance. Solid sheathing may extend up the roof as necessary for application of waterproofing membrane.

202F81.EPS

Figure 81 ◆ Open or spaced roof sheathing.

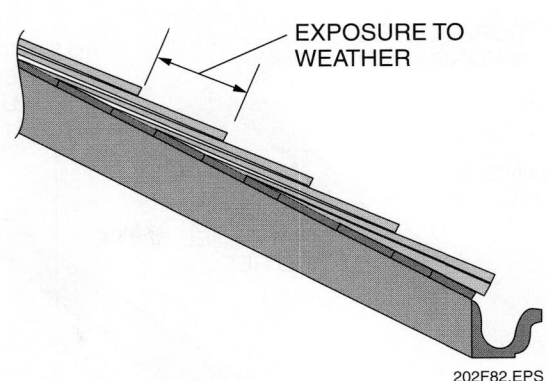

EXPOSURE TO WEATHER

202F82.EPS

Figure 82 ◆ Roof exposure.

If the roof slope is 4 in 12 or steeper (three-ply roof), use the following exposure guidelines:

- For 16" shingles, allow a 5" exposure.
- For 18" shingles, allow a 5½" exposure.
- For 24" shingles, allow a 7½" exposure.
- For 18" shakes, allow a 7½" exposure.
- For 24" shakes, allow a 10" exposure (⅜" thick shake is limited to a 5" exposure).

If the roof slope is less than 4 in 12, but not less than 3 in 12 (four-ply roof), use the following exposure guidelines:

- For 16" shingles, allow a 3¾" exposure.
- For 18" shingles, allow a 4¼" exposure.
- For 24" shingles, allow a 5¾" exposure.

Recommended exposures on low slopes are for No. 1 grade shingles. If applying a different grade, such as No. 3 shingles, make sure you check with the manufacturer. Normally, if the roof slope is less than 3 in 12, shingles are not recommended.

8.2.0 Application of Wood Shingles

The open or spaced sheathing begins at the eave line with a double snug fit or triple nailing of roof boards close together to provide a solid surface for the initial doubling of the first course of wood shingles. A 36"-wide strip of 30 lb felt or waterproof membrane should be applied at the eaves.

When the roof sheathing is completed, the roof application is ready to begin. The tools of the trade for wood shingling are minimal. A straightedge and a shingle hatchet are normally all that are necessary. The shingle hatchet should be lightweight and have a sharp blade and heel and a gauge for checking shingle exposure. See *Figure 83*.

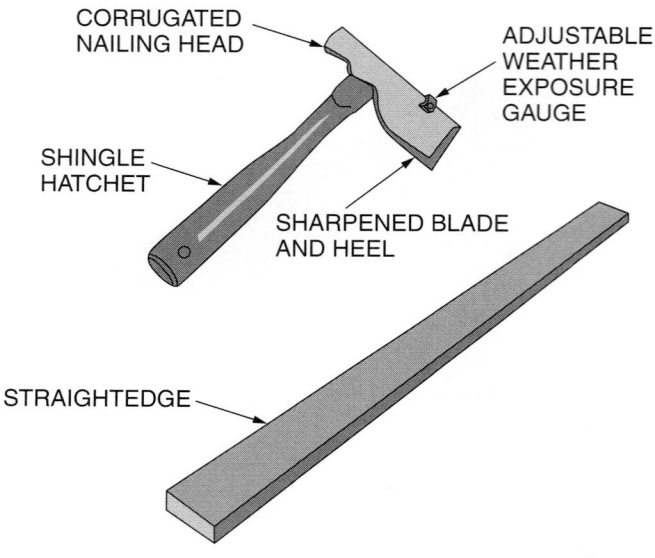

Figure 83 ◆ Shingle hatchet and straightedge.

When using scaffolding, it should be waist-high to the eaves to allow you to do a proper job of doubling up the first course and comfortably reaching successive courses. *Figure 84* shows the recommended shingle spacing on building roofs.

When applying shingles, only rust-resistant nails should be used. Hot-dipped, zinc-coated nails, which have the strength of steel and the corrosive resistance of zinc, are recommended. Sizes of nails for various jobs were discussed earlier in this module.

Most carpenters prefer to use a shingle hatchet to lay wood shingles. The sharpened blade and heel enable them to split and trim the shingles. The adjustable gauge for weather exposure makes spacing easier and quicker.

The first course of shingles at the eaves should be doubled or tripled. The starter course can be a double layer of 15" starter shingles overlayed with a third layer of regular shingles. All shingles should be spaced ¼" apart to provide space for expansion when they become rain-soaked. Let the shingles protrude over the edge to ensure proper spillage into the eave trough or gutter. See *Figure 85*.

Use only two nails to attach each shingle. Use care when nailing wood shingles. Drive the nail just flush with the surface. The corrugated nail head enables you to accomplish this due to its raised surface. The wood in shingles is soft and can be easily crushed and damaged under the nail heads.

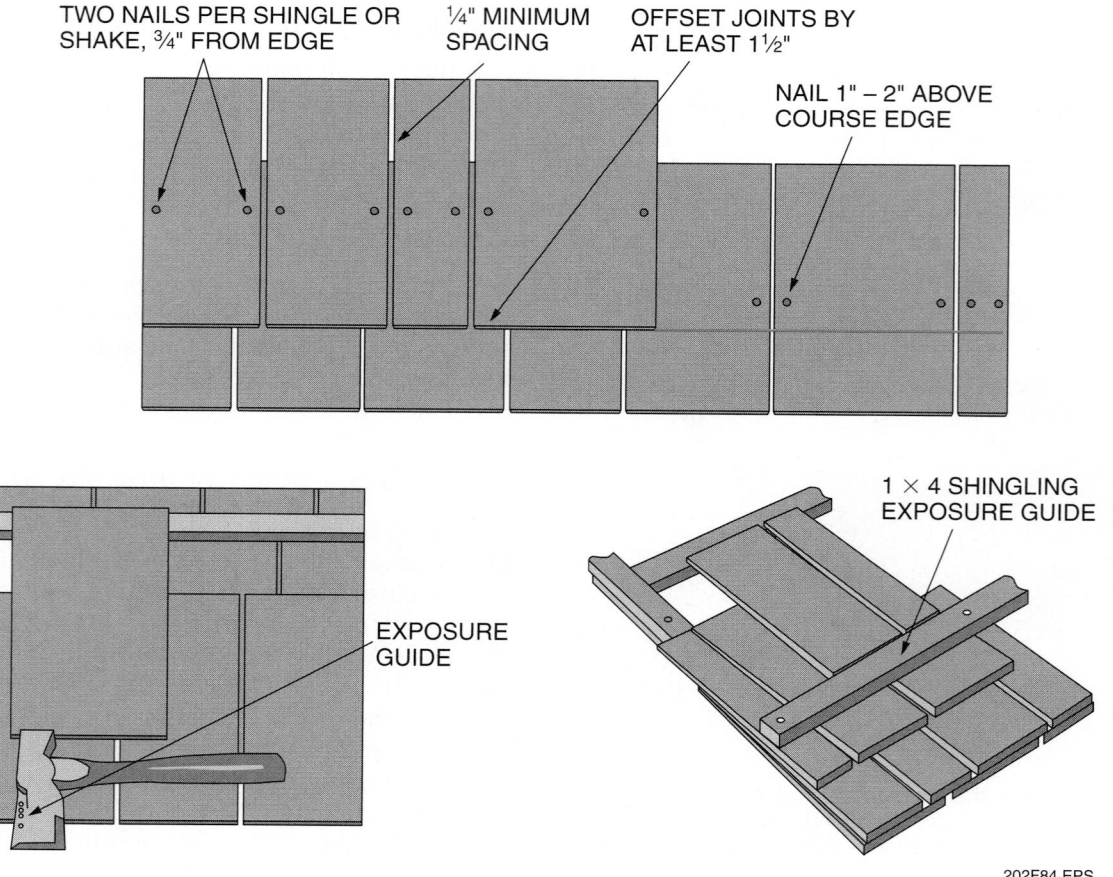

Figure 84 ◆ Shingle/shake nailing and spacing guide.

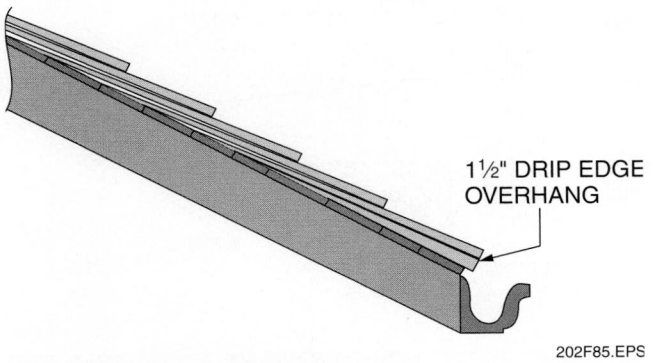

1½" DRIP EDGE OVERHANG

202F85.EPS

Figure 85 ◆ Eave overhang.

Proper nail placement is very important. The nails should be near the butt line of the shingles in the next course that is to be applied over the course being nailed, but should never be driven below this line so they will be exposed to the weather. Driving the nails 1" to 1½" above the butt line is good practice, with 2" as an allowable maximum. Each nail should be placed not more than ¾" from each side of the edge of the shingle. When nailed in this manner, the shingles will lie flat and provide good service.

The second layer of shingles in the first course should be nailed over the first layer so the joints in each course are offset by not less than 1½". See *Figure 86*.

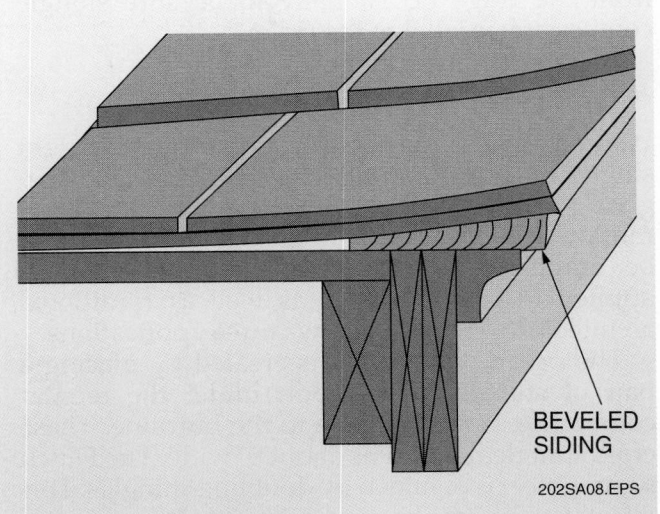

BEVELED SIDING

202SA08.EPS

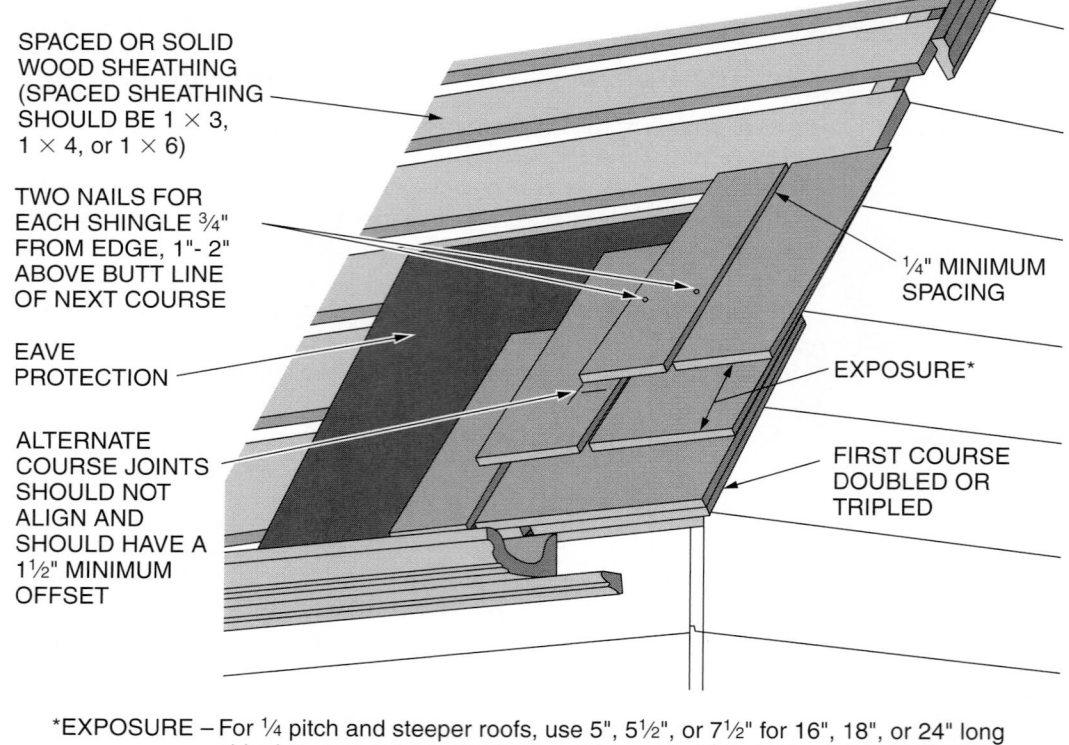

SPACED OR SOLID
WOOD SHEATHING
(SPACED SHEATHING
SHOULD BE 1 × 3,
1 × 4, OR 1 × 6)

TWO NAILS FOR
EACH SHINGLE ¾"
FROM EDGE, 1"- 2"
ABOVE BUTT LINE
OF NEXT COURSE

EAVE
PROTECTION

ALTERNATE
COURSE JOINTS
SHOULD NOT
ALIGN AND
SHOULD HAVE A
1½" MINIMUM
OFFSET

¼" MINIMUM
SPACING

EXPOSURE*

FIRST COURSE
DOUBLED OR
TRIPLED

*EXPOSURE – For ¼ pitch and steeper roofs, use 5", 5½", or 7½" for 16", 18", or 24" long
shingles, respectively. For flatter pitches, use 3½", 4½", or 5½" for the
same shingles, respectively.

202F86.EPS

Figure 86 ◆ Application of wood shingles.

Although it is not required, some contractors prefer to separate each course with underlayment strips that extend 4" above the top of a shingle down to and over the nails.

Two people often work together; one distributes and lays the shingles along the straightedge while another nails them in place. As the shingling progresses, check the alignment every five or six courses with a chalkline. Measure down from the ridge occasionally to be sure shingle courses are parallel to the ridge.

8.2.1 Special Effects

Wood shingles are normally applied in a straight single course, but variations are possible. By staggering or building up wood shingles, usually in random patterns, shadow lines and textures can be emphasized. This is a feature sometimes applied to contemporary as well as traditional architecture. *Figure 87* shows three applications.

The ocean wave effect is created by placing a pair of shingles butt-to-butt under the regular course and at right angles to the butt line. These cross shingles should be about 6" wide. The Dutch weave effect is obtained by doubling shingles. This effect can be emphasized by using two shingles

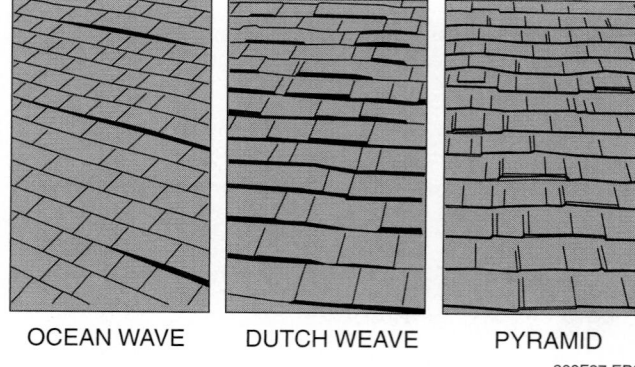

OCEAN WAVE   DUTCH WEAVE   PYRAMID

202F87.EPS

Figure 87 ◆ Shingle effects.

instead of one and is generally referred to as a pyramid pattern. The joints are always broken by at least 1½".

8.2.2 Hips and Valleys

Procedures for nailing wood shingles on hip roofs are similar to the procedures for asphalt shingles. Start at the center and work to the hip rafter. Use the largest shingles to finish off at the hip and carefully miter cut on the center line of the hip rafter.

Staggered Shingles/Shakes

Another special effect that creates a very rustic appearance is the random staggering of the bottom edge of the shingles or shakes. Some of the architectural composition shingles are designed to create this effect. With wood shingles or shakes, it can be achieved by extending the exposure of every other shingle/shake by random distances of between ¾" and 1" instead of using a fixed exposure distance. Another method is to use different lengths of shakes in a random pattern. This method, shown in the illustration, was achieved using synthetic wood shingles.

SIMULATED WOOD SHINGLE ROOF

202SA09.EPS

Valley shingling works in the opposite direction. Start from the valley and work outward. Select wide shingles for this application. Carefully cut the proper angle at the butts, following the guidelines snapped on the flashing. Valley flashing should be heavy galvanized steel or copper W-metal. Valley construction and proper cuts are shown in *Figure 88*.

8.2.3 Ridge and Hip Caps

Tight ridges and hips are required to avoid roof leakage. In good ridge and hip construction, nails are not exposed to the weather. Shingles of approximately the same width as the exposed shingle surface are sorted out to be used as caps. Two lines are then marked on the shingles on the roof the correct distance back from the center line of the ridge or hip. Waterproof membrane or flashing is applied over the ridge or hip.

Figure 89 shows the standard application of hips and ridges. Be sure to use longer nails, which will penetrate into the sheathing. The alternate overlap pattern ensures a dry seal. Factory-assembled ridge and hip units (*Figure 90*) are available.

ON ROOFS FLATTER THAN ½ PITCH, VALLEY SHEETS SHOULD EXTEND AT LEAST 10" FROM VALLEY CENTER

ON ½ PITCH AND STEEPER ROOFS, VALLEY SHEETS SHOULD EXTEND AT LEAST 7" FROM VALLEY CENTER

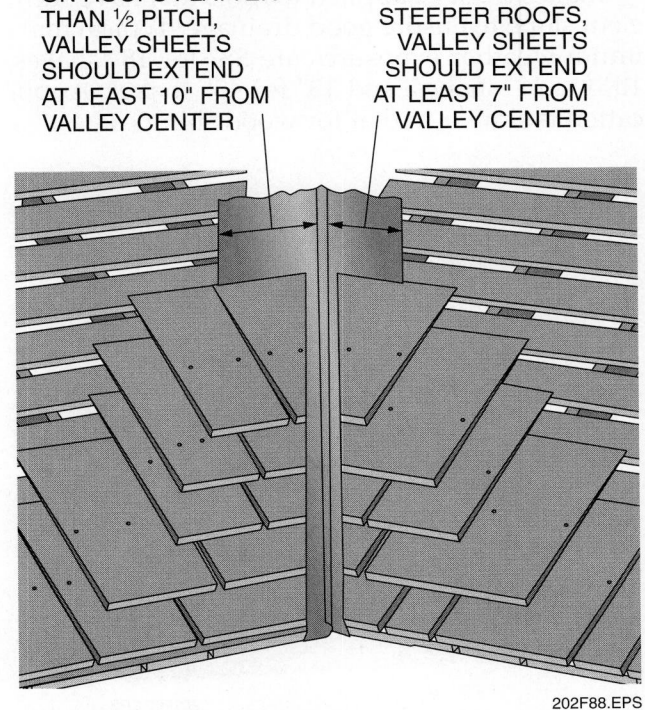

202F88.EPS

Figure 88 ◆ Valley layout—wood shingles.

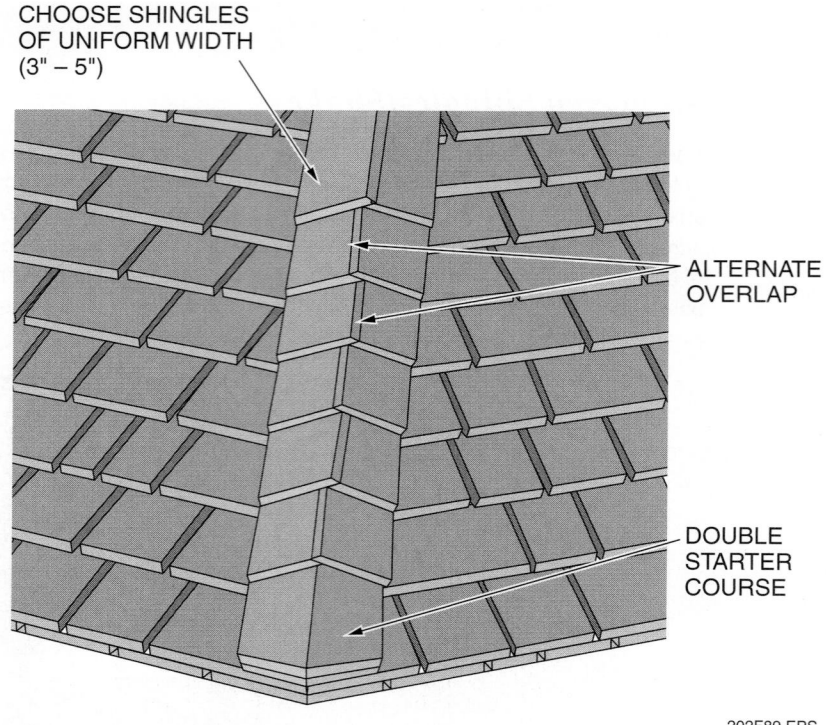

CHOOSE SHINGLES
OF UNIFORM WIDTH
(3" – 5")

ALTERNATE
OVERLAP

DOUBLE
STARTER
COURSE

202F89.EPS

Figure 89 ◆ Hip layout—wood shingles.

8.3.0 Application of Wood Shakes

Shakes are made in several lengths, thicknesses, and widths. Because of their length and thickness, they have greater exposure than shingles. *Figure 91* shows various types of wood shakes.

Shakes must be applied to roofs that have sufficient slope to ensure good drainage. Typical maximum weather exposures are 8½" for 18" shakes, 10" for 24" shakes, and 13" for 43" shakes. Application is similar to that for wood shingles.

Start the application by placing a 36" strip of 30-lb roofing felt or waterproof membrane along the eaves. The beginning or starter course is doubled, just as it is for regular shingles. After each course is applied, an 18" (or wider) strip of 30-lb felt is applied over the top portion of the shakes. The 18" strips are cut from the full 36" roll using a center line. The strips extend onto the sheathing.

The bottom edge of the felt is placed above the butt a distance equal to twice the exposure. For example, if 24" shakes are being laid at a 10" exposure, place the roofing felt 20" above the cuts of the shake. As you can see in *Figures 92* and *93*, the felt will then cover 4" of the top end of the shake and the remaining 14" will rest on the sheathing.

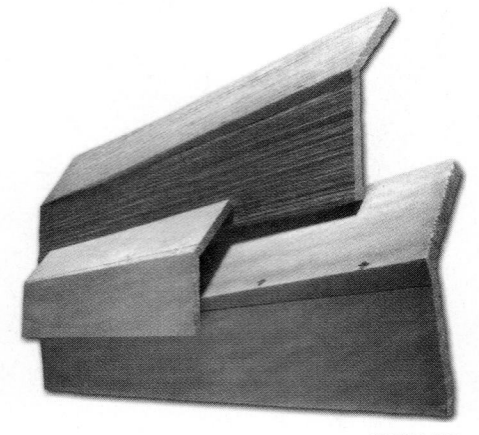

202F90.EPS

Figure 90 ◆ Prefabricated hip and ridge units.

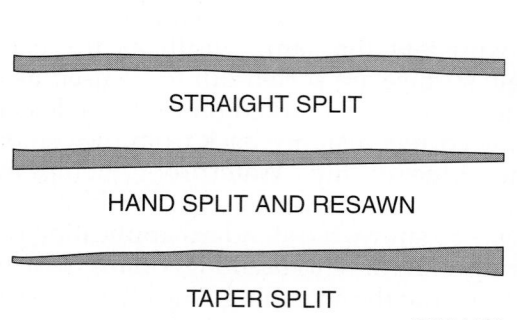

STRAIGHT SPLIT

HAND SPLIT AND RESAWN

TAPER SPLIT

202F91.EPS

Figure 91 ◆ Types of wood shakes.

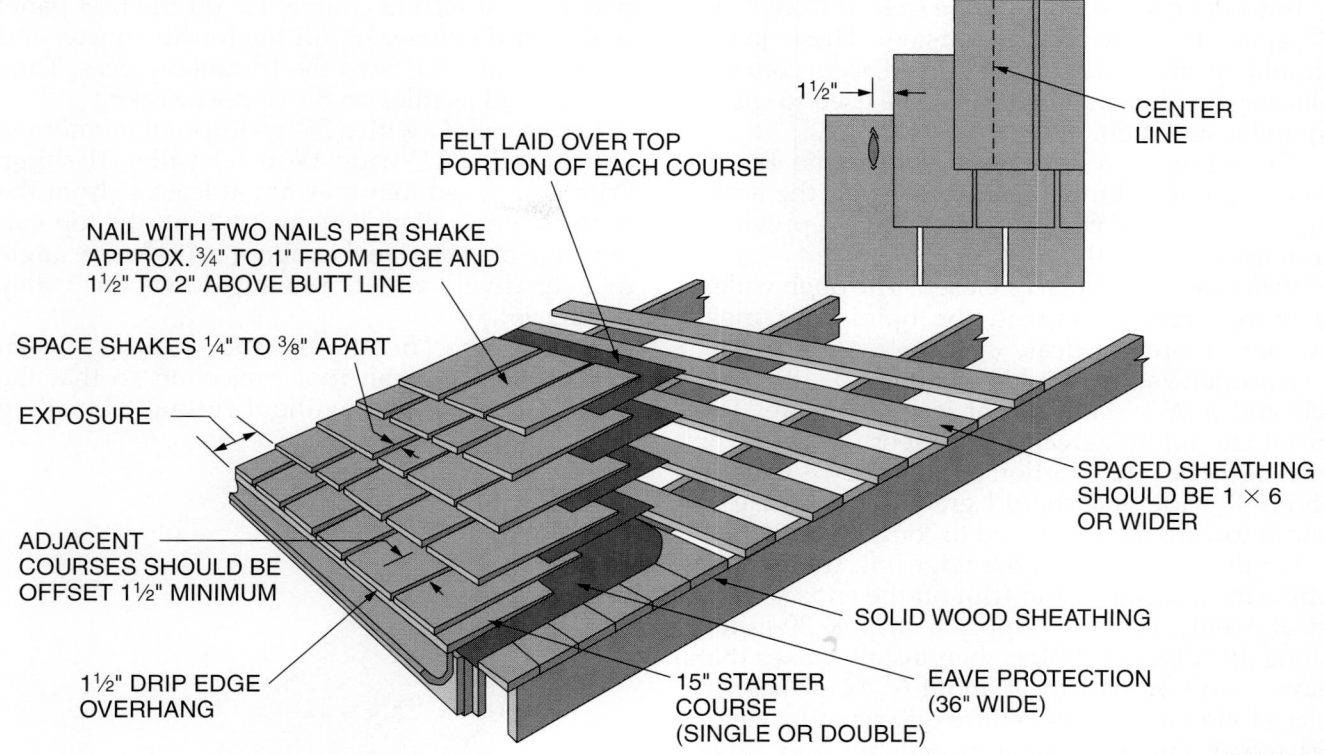

Figure 92 ◆ Shake application.

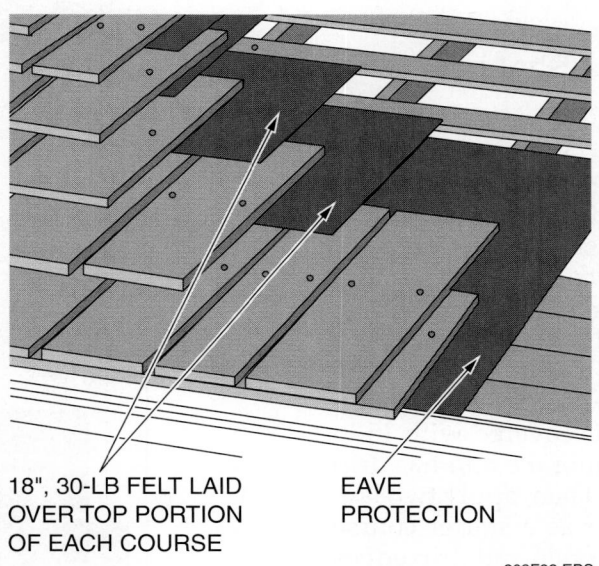

Figure 93 ◆ Shake felt underlayment.

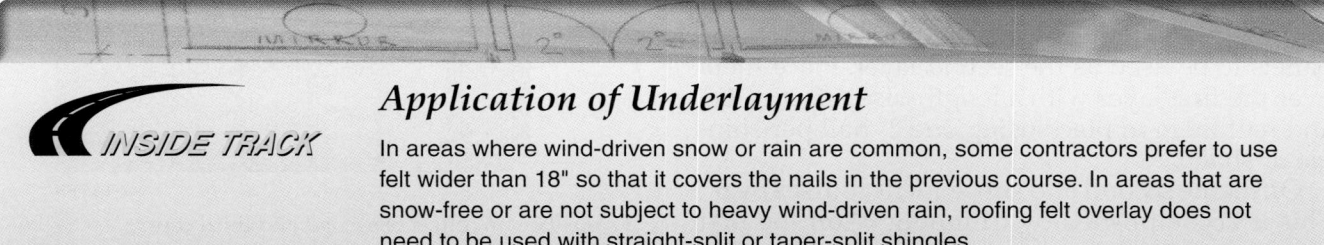

Application of Underlayment

In areas where wind-driven snow or rain are common, some contractors prefer to use felt wider than 18" so that it covers the nails in the previous course. In areas that are snow-free or are not subject to heavy wind-driven rain, roofing felt overlay does not need to be used with straight-split or taper-split shingles.

Individual shakes should be spaced from ¼" to ⅜" apart to allow for expansion. These joints should be offset by at least 1½" in adjacent courses. Once again, the straightedge can be used to speed up the installation.

The nailing procedures are the same as for wood shingles. Emphasis is placed on the accuracy of nailing to make sure the nails in previous courses are correctly covered.

Valleys may be open or closed. The open valley is more common and, in the opinion of most builders, more practical.

The open valley should be underlaid with 30-lb felt and a W-metal sheet at least 20" wide. The metal should be galvanized copper (26 gauge or heavier). The open portion of the valley is usually about 4" wide and should gradually increase in width toward the lower end to control water flow.

For the final course at the ridge line, try to select uniform size shakes and trim off the ends so they meet evenly. Carefully apply a strip of 30-lb felt along all ridges and hips, then install shakes that have a uniform width of about 6". Nail these in place following the procedure described for regular wood shingles. Prefabricated hip and ridge units are available. The use of these will save time and provide uniformity.

8.4.0 Panelized Shingle/Shake Application

Panelized shingles or shakes are much faster to apply than individual shingles or shakes. Each panel has a felt backing and a score line that is used to align courses. One fastener per shingle or shake holds each panel in place. If the felt backing extends to the right of the panel, the panels must be laid left to right. If the felt extends to the left, the panels must be laid right to left.

If the roof will be applied directly over sheathing, make sure that the roof is covered with 15-lb felt and/or a waterproof membrane and has drip edges at the eaves and rakes. Then, apply two layers of panels over each other as a starter course (*Figure 94*). If the roof deck is open and drip edges are not used, make sure that both panels overhang the eaves by 1½" and the rakes by 2". Nail the first layer of the starter course with two nails halfway up both ends. Remove the felt backing from the panels to be used as the second layer. Place them over the first layer with a lengthwise offset of 1½" and nail them in place using one 2" nail per shingle or shake.

Offset the second and subsequent courses by 6". This is accomplished by trimming 6" off the first panel of the second course, 12" off the first panel of the third course, 18" off the fourth course, and so on (*Figure 95*). Save the trimmed pieces. They can be used as filler on the opposite rake.

Cover valleys with a 36" waterproof membrane and install a 24"-wide W-metal valley flashing. Trim courses so that they are at least 4" from the center of the valley. Trim about 2" off the top valley edge of each panel at about a 45-degree angle to help divert water running down the valley (*Figure 96*).

Adjust the depth of the courses when approaching a small roof projection so that the panel can be notched without cutting the nailing bar (*Figure 97*).

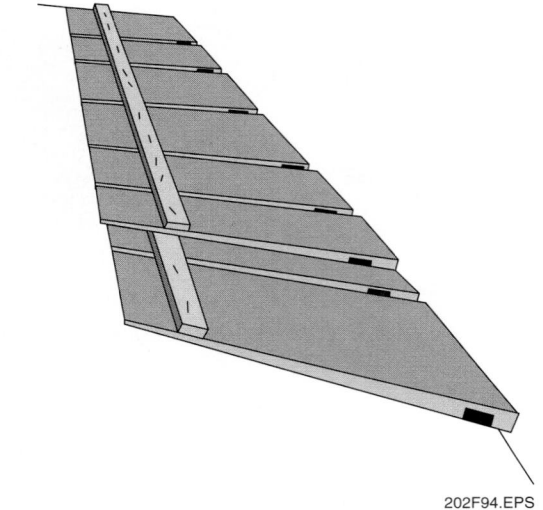

202F94.EPS

Figure 94 ◆ Panel starter course.

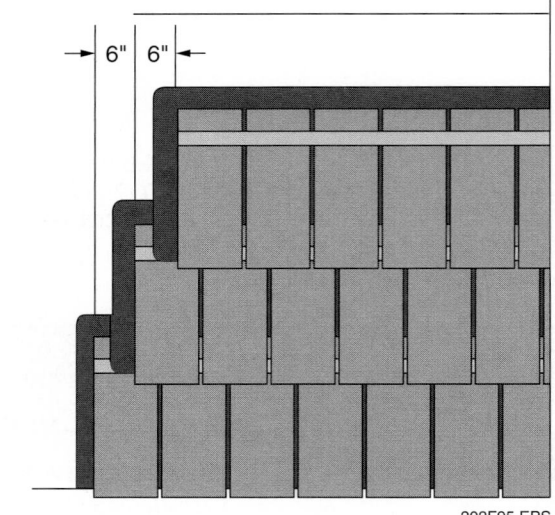

202F95.EPS

Figure 95 ◆ 6" offset of second and third course.

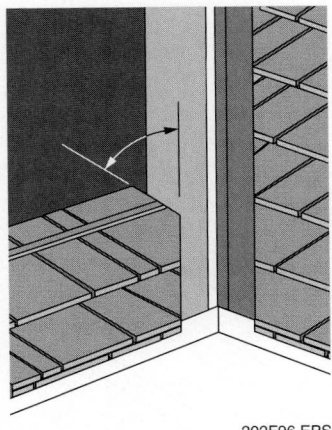

Figure 96 ◆ Trimming shingle/shake panels for an open valley.

202F96.EPS

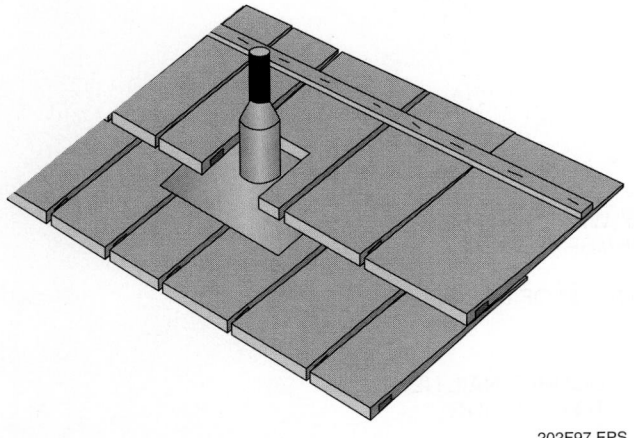

202F97.EPS

Figure 97 ◆ Notching panels for a roof projection.

8.5.0 Hardboard Shingle/Shake Panel Application

These wood fiber panels, which mimic wood shingles/shakes, are another product that is easy to apply. The panels are 4' long and are laid horizontally only on a closed roof deck. The bottom edge forms a lap joint over the top of the panel below, and the side edges form a shiplap joint with the adjacent panels. They must only be used on roofs with a slope of 4 in 12 or more.

First, make sure that the roof is covered with two layers of 30-lb felt or a waterproof membrane and has drip edges at the eaves and rakes. Cut 2½"-wide starter strips lengthwise from panels and nail them to the eaves so that they overlap the rakes by 2" and the eaves by 1" (*Figure 98*). Apply

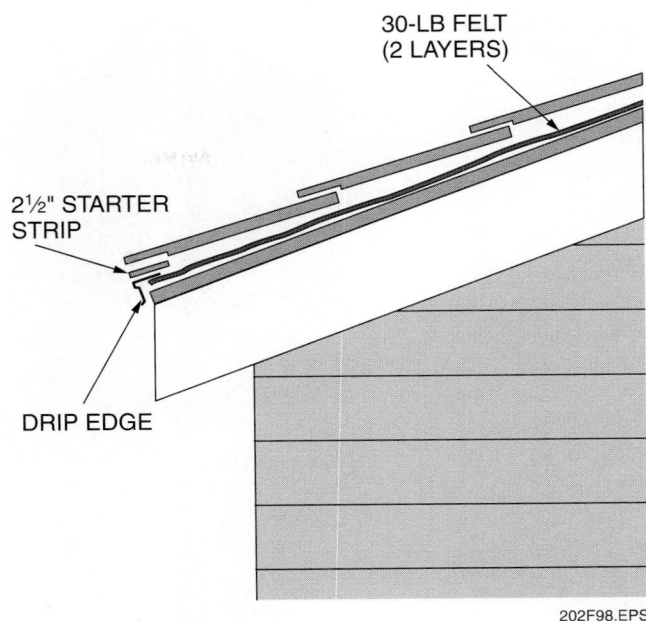

202F98.EPS

Figure 98 ◆ Starter strip and panel overlap.

the first course of panels starting at the 2" overhang at the left side of the roof (*Figure 99*). Fasten each panel with eight nails or staples along the nailing guideline at the top of the panel. Start fastening at the top center of the shiplap joint on the right end and stop 3" from the left end.

After the first course is laid, shorten the first panel of the second and subsequent courses from the right end of the panel (see *Figure 99*) to stagger the joints of the panels. Save the trimmed left end of the panels for use on the right-hand rake edge. When laying the second and subsequent course panels, make sure that the lap joint at the bottom edge of the panel is fully seated over the panel below before nailing the top edge of the panel.

Cover valleys with a 36" waterproof membrane and install a 24"-wide W-metal valley flashing (*Figure 100*). Trim courses so that they are at least 4" from the center of the valley. Trim about 2" off the top valley edge of each panel at about a 45-degree angle to help divert water running down the valley. Adjust the depth of the courses when approaching a small roof projection so that the panel can be notched less than half the width of the panel.

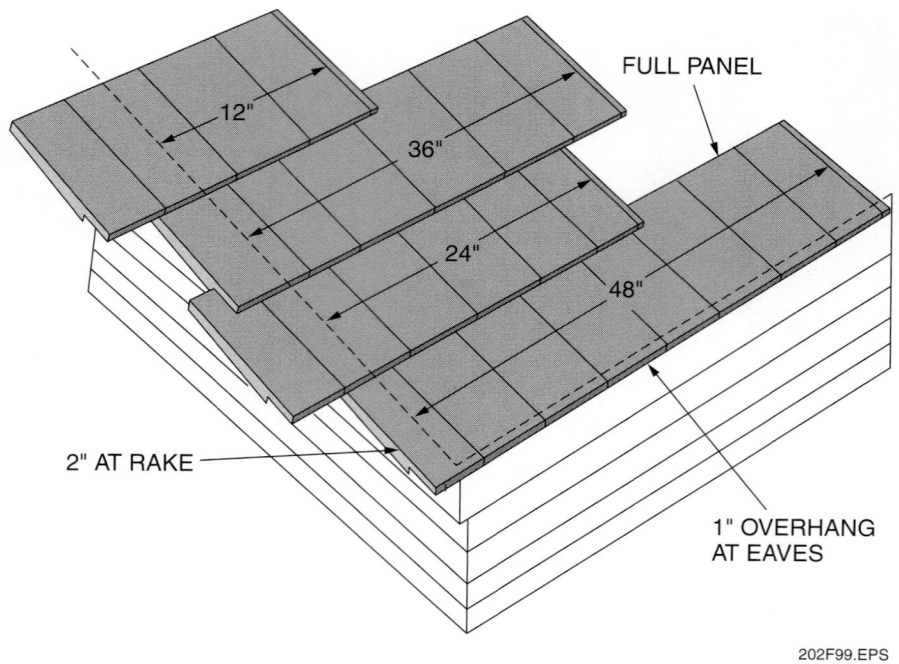

FULL PANEL

12"

36"

24"

48"

2" AT RAKE

1" OVERHANG
AT EAVES

202F99.EPS

Figure 99 ◆ Staggering of hardboard panels.

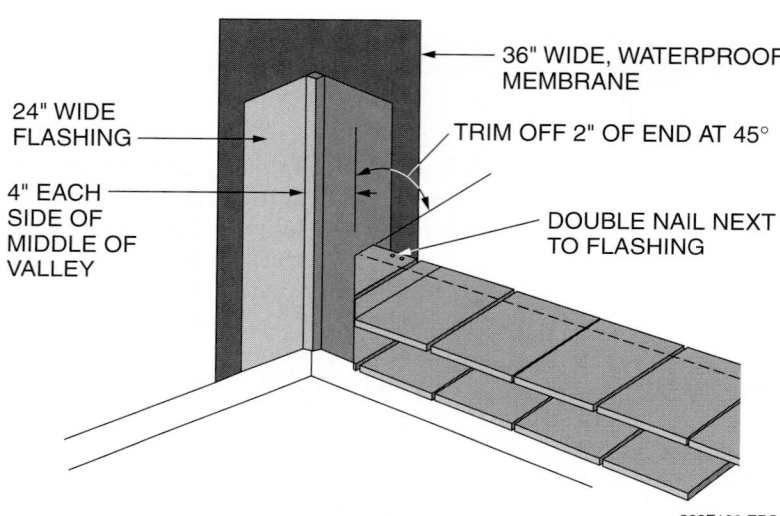

36" WIDE, WATERPROOF
MEMBRANE

24" WIDE
FLASHING

TRIM OFF 2" OF END AT 45°

4" EACH
SIDE OF
MIDDLE OF
VALLEY

DOUBLE NAIL NEXT
TO FLASHING

202F100.EPS

Figure 100 ◆ Trimming hardboard panels for an open valley.

9.0.0 ◆ COMMON METAL ROOFING

This section covers various types of metal roofing systems.

9.1.0 Corrugated Metal Roofing

Another type of roofing material is corrugated metal roofing, or galvanized metal roofing. Only galvanized sheets that are heavily coated with zinc (2.0 oz. per square foot) are recommended for permanent construction.

Galvanized sheets may be laid on slopes as low as a shallow 3" rise to the foot (⅛ pitch). If more than one sheet is required to reach the top of the roof, the ends should overlap by at least 8". When the roof has a pitch of ¼ or more, 4" end laps are usually satisfactory. To make a tight roof, sheets should be overlapped by 1½ corrugations at either side. See *Figure 101(A)*.

When using roofing that is 27½" wide with 2½" corrugations and a corrugation lap of 1½, each sheet covers a net width of 24" on the roof. If 26-gauge galvanized sheets are used, supports may be 24" apart. If 28-gauge galvanized sheets are used, supports should not be more than 12" apart. The heavier gauge has no particular advantage

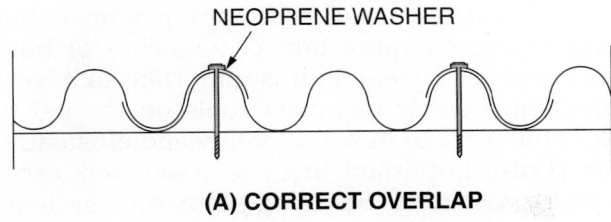

(A) CORRECT OVERLAP

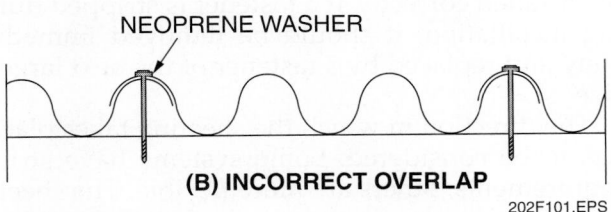

(B) INCORRECT OVERLAP

202F101.EPS

Figure 101 ◆ Corrugated roofing.

except its added strength, because the zinc coating is what gives this type of roofing its durability.

When 27½" roofing is not available, sheets of 26" width may be used. When laying the narrower sheets, every other one should be turned upside down so that each alternate sheet overlaps the two intermediate sheets, as shown in *Figure 101(A)*.

For best results, galvanized sheets should be fastened with neoprene-headed nails, galvanized nails and neoprene washers, or screws with neoprene washers. Fasteners are used only in the tops of the corrugations to prevent leakage. To avoid unnecessary corrosion, use the fasteners specified by the roofing manufacturer.

Corrugated metal roofing panels (*Figure 102*) are used on garages, storage buildings, and farm buildings. They are available in widths up to 4'-0" and lengths up to 24'-0". Normally, these panels are used on roofs with slopes of 4 in 12 or steeper. They can be used on 2 in 12 roofs if a single panel reaches from the ridge to the eave.

The panels are fastened to purlins. A purlin is a structural member running perpendicular to the rafters. See *Figure 102*. Usually, 2 × 4 wood stock is used. The spacing should follow the directions specified by the manufacturer. Filler strips are sold with the panels. They are set at the eave and the ridge. Normally, the panel is cut so it overhangs 2" to 3" at the eave. The installation of metal roofs should be done in accordance with the manufacturer's specifications.

9.2.0 Simulated Standing-Seam Metal Roofing

The character of each standing-seam roof system dictates the amount and type of planning. Each system has different components and slightly different requirements for tools and equipment.

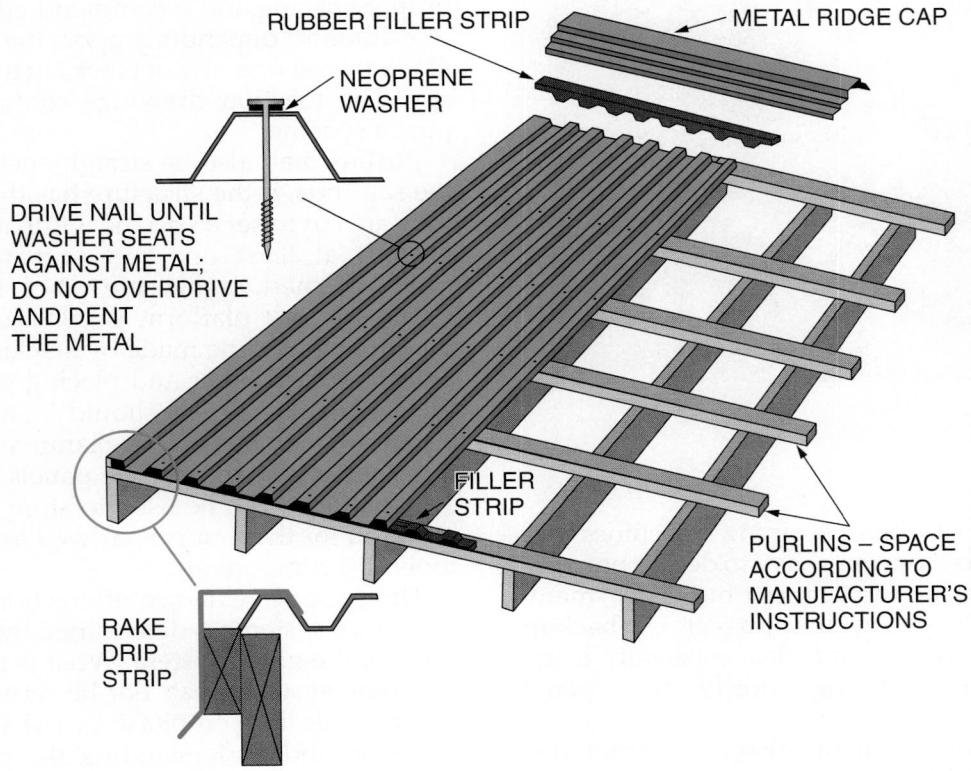

Figure 102 ◆ Corrugated roof layout.

INSIDE TRACK

Corrugated Roofing Light Panels

Corrugated translucent fiberglass light panels can be used to allow daylight into the interior of a structure. They can also be used as the entire roof, if desired. These panels are the same size and are installed in the same manner as corrugated metal roofing.

FIBERGLASS PANEL

202SA10.EPS

In some cases, where seaming machines are required, supervisors may have to decide not only when to lease the machines, but how many machines to have on the project. A backup machine is always a good idea, especially if the system requires seaming shortly after panel installation.

Power tool requirements also differ from system to system. Most systems use screw guns to install self-drilling and self-tapping screws, but some systems require impact wrenches or bulb rivet guns for fastening. It is important to have a sufficient number of power tools on the job to allow the work to move smoothly and efficiently.

It is also important that the assemblers carry oversize fasteners. Standing-seam roof systems minimize through-the-panel fasteners by up to 90 percent. Therefore, it is crucial that the fasteners be installed correctly. If a fastener is stripped during installation, it should be removed immediately and replaced by a fastener of the next larger size.

The direction in which the sheeting takes place has to be considered. Some systems have strict requirements; others are more flexible. The sheeting direction can be found on the construction drawings.

Perhaps the biggest demand each standing-seam roof system makes is to be installed by competent assemblers. Improper installation techniques cause the majority of standing-seam roof failures. This underscores the need for special training in the particular system being used.

Before roofing can begin, the structure must be plumb and level. The purlins must also be straight. Z-purlins, in particular, have a tendency to roll. If there are no purlin braces, use wood blocking. Most manufacturers suggest that wood blocks be driven tightly between purlins to ensure proper spacing and recommend either 2 × 6 or 2 × 8 lumber, depending upon the purlin depth. Place at least one row of blocks in the center of the bay. The erection drawings contain the proper purlin spacing.

Purlins may also be straightened by adjusting the sag rods, if the structure has them. Many sag rods are cut to set a specific width automatically.

Until at least one run of panels has been installed, there is no safe place to work from unless a work platform is constructed. A work platform should be made by stacking two panels on top of each other and placing walk boards in the center. The panels should be attached to the structure with locking C-clamp pliers or some other means to prevent the panels from moving. This platform can be used to store the insulation required for the first run, as well as the necessary tools and components.

The specific sequence of erection of standing-seam roof systems is determined by the manufacturer of the given system. What is recommended for some systems may not be recommended for others. This fact emphasizes the importance of knowing and understanding the particular system before installing it.

Simulated Standing-Seam Roofing Systems

A number of different methods of joining metal roofing panels are employed for these types of systems. Older systems used galvanized steel or copper standing seams that were soldered together to form a weather seal. Other systems require crimping machines or use snap-type seals.

MACHINE-CRIMPED PANELS

SNAP-TYPE SEAL

202SA11.EPS

Usually, a blanket of insulation is stretched across the structural members prior to sheeting. The blanket width depends on the panel. Keep the stapling edge ahead of the panel, but no more than a foot ahead of it.

As previously mentioned, installing metal roofing requires special tools and experience. The following is an overview of the installation procedure for one type of metal roofing that is designed to be laid over a closed, fully-sheathed roof deck. It is applied in panels that are 12" to 16½" wide and that are precut to run from the eaves to the ridge of the structure. The joints between panels are weatherproofed by means of a C-clip with a neoprene seal.

Cover the roof with a 30-lb felt or waterproof membrane. The membrane or felt must overlap the edges of the eaves and rakes.

The eave trim (*Figure 103*) is then screwed in place before panels are applied. The panels are placed one at a time and secured to the roof deck with a T-clip on one side of the panel.

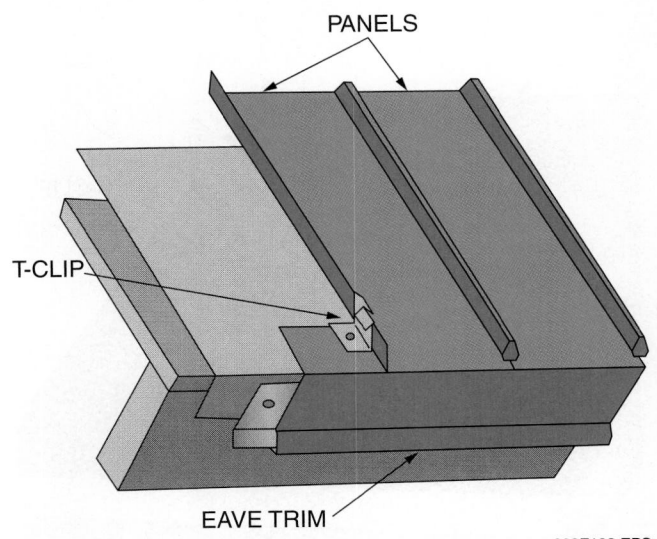

Figure 103 ◆ Eave trim.

202F103.EPS

The next panel is then inserted under the T-clip that is holding the first panel and secured on the opposite side with another T-clip (*Figures 104* and *105*). At the joint of any two panels, the joint is weather-sealed with a C-clip running the full length of the joints, as shown in *Figures 104* and *105*. The T-clips are used every 12" in high-wind areas and every 18" elsewhere. After panels are placed under each side of the T-clips, the T-clip wings are bent down, and the C-clip is forced down on the seam. The neoprene flaps inside the C-clip seal the seam against water penetration, and the bent-down wings of the T-clip keep the C-clip from being dislodged from the seam.

After the first panel is placed at one of the rakes, a rake edge is applied, as shown in *Figure 106*. If necessary at the opposite rake, the panel is trimmed off and Z-strips are sealed and secured to the panel before the rake edge is installed.

At valleys, the panels are trimmed to the angle of the valley. Channel strips running parallel in the valley are sealed and screwed to the valley and hold the edges of the panels (*Figure 107*).

At the ridge, Z-strips are fastened and sealed to the panels between the seams. They are also sealed to the seams. The cover flashing is then sealed and secured to the Z-strips (*Figure 108*).

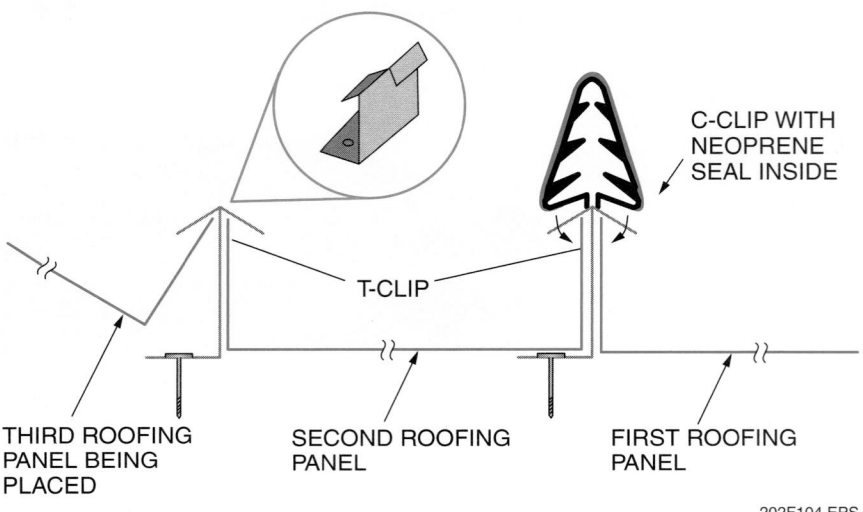

202F104.EPS

Figure 104 ◆ Placing, securing, and sealing panels.

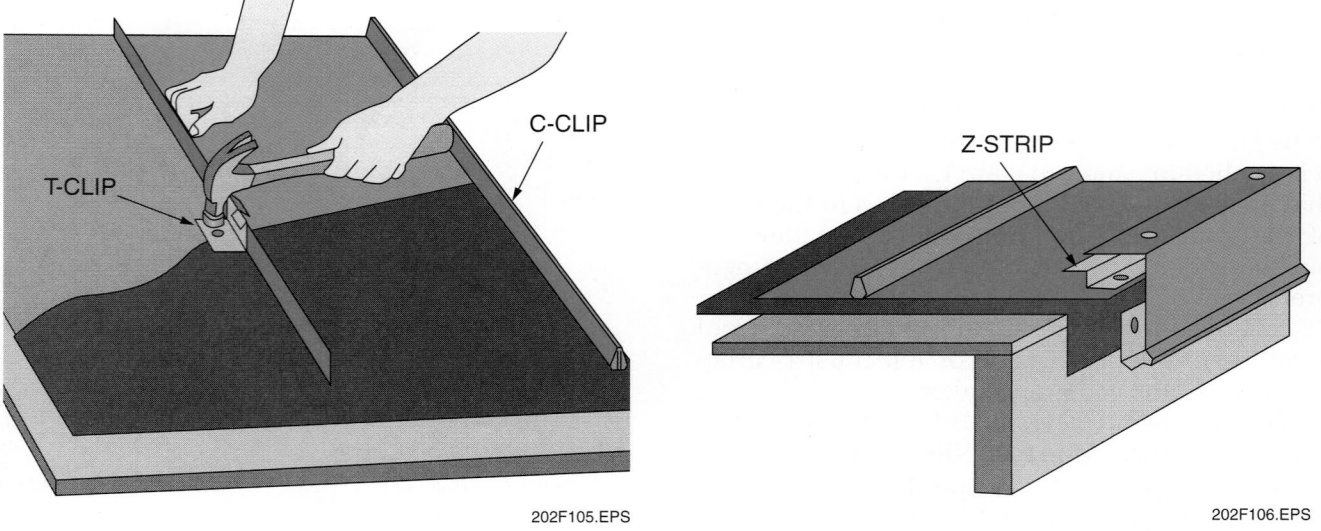

202F105.EPS

Figure 105 ◆ Nailing a T-clip to a roof deck.

202F106.EPS

Figure 106 ◆ Rake edge.

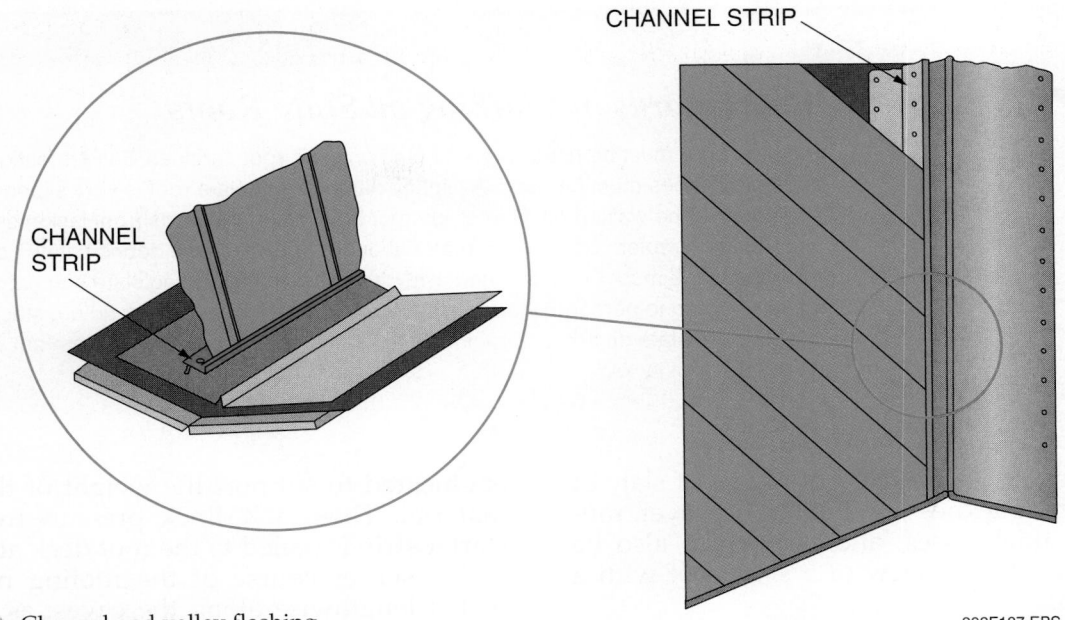

Figure 107 ◆ Channel and valley flashing.

202F107.EPS

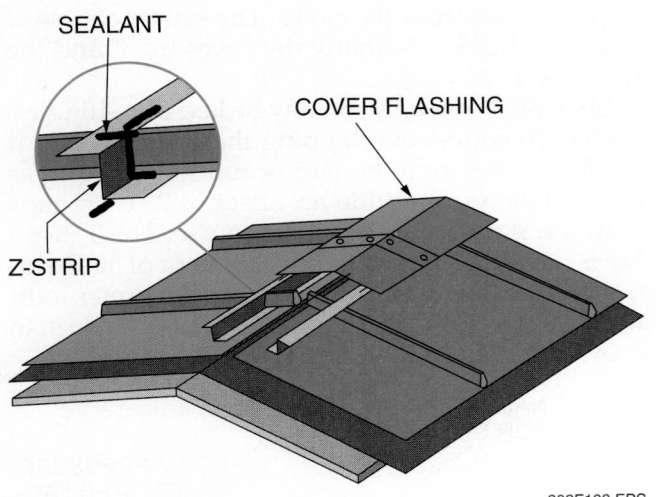

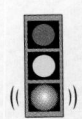

Figure 108 ◆ Ridge flashing.

202F108.EPS

9.3.0 Snug-Rib System

Another type of roofing sheet is called the snug-rib system. It utilizes a concealed fastener, which is leak resistant and eliminates through fasteners. This combination of a V-beam industrial sheet and the snug seam joint makes for a greater beam strength and a deeper corrugation than other roofing profiles. Because of the greater strength of this type of roofing panel, the purlin spacing can be increased.

The snug-rib joint is a highly efficient weather-tight joining system of a simple nature. The joint is created by engaging the hooked edges of two panels into a Y-shaped extruded spline, previously measured and anchored to the purlins with a self-templating clip. A neoprene gasket is then rolled into the extrusion between the panel edges, where it holds the panel edges securely in place and creates a watertight seal.

> **NOTE**
>
> No primary fasteners penetrate the weatherproofing membrane.

There is a 19½" covering width. The material has a V-shaped corrugation that has a 4⅞" pitch and a 1¾" depth. Lengths vary from 77" to 163", depending upon the gauge of the material. The end laps must be a minimum of 12", located over roof purlins, and staggered with the end laps in adjacent panels. Install this system according to the manufacturer's specifications.

10.0.0 ◆ SLATE AND TILE ROOFING

This section covers the installation of various types of slate and tile roofing.

10.1.0 Slate or Synthetic Slate Roofing Installation

Roofing slate is hard, tough, and usually has a bright metallic luster when freshly split. Slate has a great variety of colors such as gray, green, dark blue, purple, and red. It is available in any size from 6" to 14" wide, 12" to 24" long, and ⅛" to 2" thick. The most common sizes are 12" × 16" and 14" × 20", and either ³⁄₁₆" or ¼" thick.

The Hazards of Working on Slate Roofs

Extreme care must be used when walking on slate roof surfaces. Fall protection and soft-soled shoes must be used. A slightly damp or wet slate roof is very slippery and should never be walked on. Numerous roof brackets and boards (roof jacks) should be used to keep roofers and roofing material on the roof. An eave debris safety net and a roped-off zone must be maintained away from the eaves. Falling slate can cause severe injury or death to persons under the eaves. Because of the weight, do not stack large quantities of slate on any one roof jack. Spread smaller quantities of slate out over numerous roof jacks.

Roofs may be constructed of pieces of slate of uniform size, thickness, and color; however, random sizes, thicknesses, and colors can also be used. *Figure 109* is a view of a slate roof with a mitered hip.

An alternative to natural slate is a fiber-cement material that looks and installs the same as slate. This synthetic slate is somewhat lighter and will last 40 to 50 years.

To begin the installation, 30-lb felt and/or a waterproof membrane is laid over a roof deck engineered to support the weight of the roofing material. Then, a ¼"-thick pressure-treated lath starter strip is nailed to the roof deck at the eaves and a starter course of the roofing material is nailed lengthwise along the eaves, as shown in *Figure 110*. The upward tilt or extra pitch caused by the wood starter strip is referred to as providing extra canting at the eaves. The starter course of material should overhang the eaves by ½" and the rakes by 1".

Slate or synthetic slate may be laid like shingles, with each course overlapping the course below it (*Table 5*). They may be laid at random as long as care is taken to provide an offset of 2" over gaps between slates in the previous course. The 3" **head lap** is usually laid on an underlayment of 30-lb saturated felt; however, some contractors prefer individual felt strips under each course (like shakes) to provide extra cushioning and weatherproofing.

202F109.EPS

Figure 109 ◆ Slate roof with mitered hip.

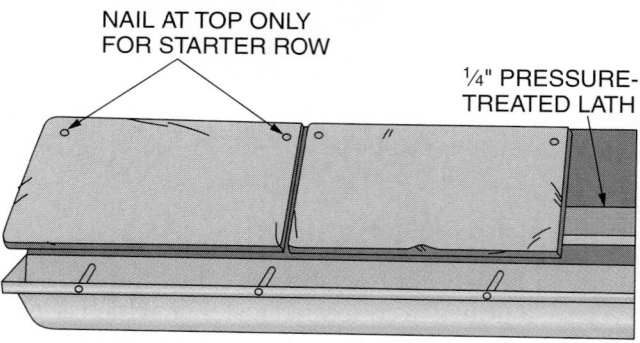

NAIL AT TOP ONLY FOR STARTER ROW

¼" PRESSURE-TREATED LATH

202F110.EPS

Figure 110 ◆ Shimming and applying starter course.

Table 5	Slate Shingle Head Lap	
Roof Slope	**Minimum Lap**
4 in 12 to 8 in 12	4" (102 mm)
8 in 12 to 20 in 12	3" (76 mm)
20 in 12 or more	2" (51 mm)

Slate or synthetic slate shingles are nailed through two prepunched or predrilled holes into wood sheathing or solid plywood decking (*Figure 111*). Drive nails flush with the shingle, but not tight. Copper or slater's nails are preferred for fastening, although redipped galvanized nails and copper-coated nails are often used where permitted by local building codes. Make sure gaps in a covering course are offset by at least 2" from the gaps in the course below. Also, make sure that the gap between shingles is at least $\frac{1}{16}$".

Slate is joined at the ridges by a saddle ridge or a combed ridge, both requiring extreme care in installation. Slate is brittle and must be cut carefully to avoid waste.

Typical capping of a ridge using a saddle ridge is shown in *Figure 112*. The slates are all the same size and the flush overlap at the peak is alternated from one side to the other. The slate is positioned horizontally like the starter course and is sometimes referred to as combing slate. The slate is fastened to the roof with two nails at one end. Holes must be drilled with a $\frac{1}{8}$" masonry bit through the underlying slate to accommodate the nails.

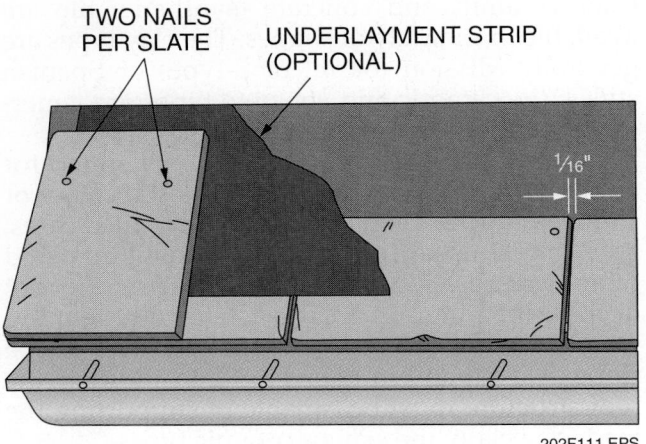

Figure 111 ◆ Installing the slates.

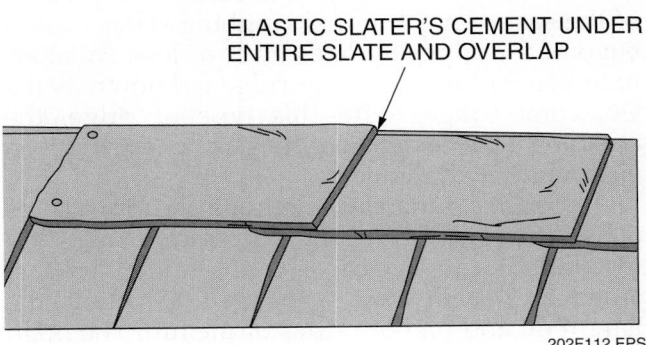

Figure 112 ◆ Capping a ridge.

Cutting and Punching Slate

Slate is usually cut and trimmed with a slate cutter. The slate is inserted in the cutter with the beveled side down, and the cutter bar is forced downward in a series of short chopping strokes to cut through the slate.

The cutter forms a properly beveled edge at the cut line. Nail holes can be punched in the slate using the spike near the handle end of the cutter.

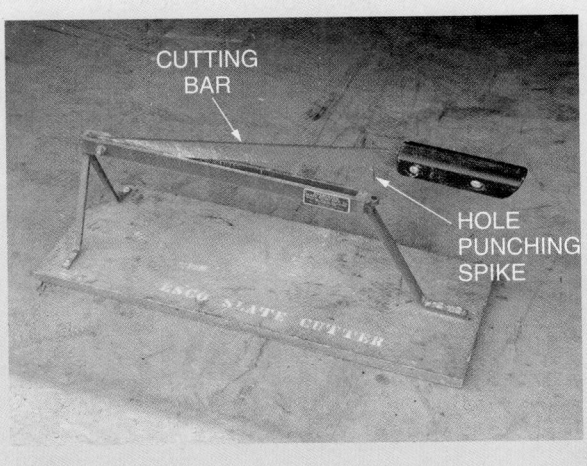

Before placing the caps, snap a chalkline the width of the cap shingles on both sides of the ridge. Then, either waterproof membrane or three layers of generously cemented felt must be placed across the open ridge between the chalklines. Elastic slater's cement should be used. After the ridge underlayment is placed, cover with elastic slater's cement and set the slate. Make sure the overlap is also covered with elastic cement.

Other versions of ridge caps include the strip-saddle ridge and combing ridge. A strip-saddle ridge is laid in a similar manner to the saddle ridge except that the slate is butted end-to-end and does not overlap. A combing ridge is laid like a saddle ridge except that the top edges of the slates do not alternate. In this case, the top edges of the east or north facing combing slates extend beyond the opposite slates by 1" or less. An alternate version of a combing ridge is known as the Cox-comb ridge. With this type of ridge, the combing slate extensions project alternately on either side of the ridge.

There are a number of methods of capping hips on slate roofs. Saddle or strip-saddle hips are installed like a ridge cap except that they are fastened to 3"-wide cant strips installed along the length of, and on both sides of the hip. The main roof slates are trimmed and installed up to the cant strips. Another type of hip is the mitered hip. This type of hip cap is formed by mitering the slates with a saw so that they fit tightly together. It is installed on a cant strip like strip-saddle or saddle hips. Sometimes, slip-sheet underlayment is used under each course of slate on mitered hips. A variation of the mitered hip is known as the fantail hip. It is installed like a mitered hip except that the pointed lower ends of the slates are clipped off at a 90-degree angle to the slope of the hip.

Another type of hip cap is known as the Boston hip. This type of cap is like a saddle hip, but it is woven in with the regular slate courses and does not require cant strips.

In dry regions, slate or synthetic slate shingles may be nailed to properly spaced wood strips without a felt underlayment. These roofs should not be used on slopes of less than 4 in 12.

10.2.0 Clay, Ceramic, and Concrete Tile Installation

Clay, ceramic, and concrete (synthetic) tile are available in a variety of shapes. The clay types are generally Mission (barrel or S-type) or Spanish styles; the ceramic and concrete types are generally flat tiles.

Conventional tiles are not generally suited for application in areas where large amounts of ice or snow buildup are expected. In some limited cases, they can be used if a waterproof membrane and other water-sealing methods are used on the roof deck. Depending on local or state code requirements, tiles should not be applied to roofs with a slope of less than 3 in 12 or 4 in 12. In addition, the roof structure must be capable of supporting the load, especially the clay or ceramic types.

Clay, ceramic, and concrete tiles are normally fastened with copper nails if allowed by local or state building codes; otherwise, galvanized or stainless steel nails should be used. All nails should be driven to a nickel's thickness above the tile to prevent cracking the tile. In certain high-wind areas of the country, special hurricane fastening systems that use stainless steel clips, brackets, and nails, along with silicon mastic sealing of each tile, are required by local or state codes in order to secure the tiles (*Figures 113* and *114*).

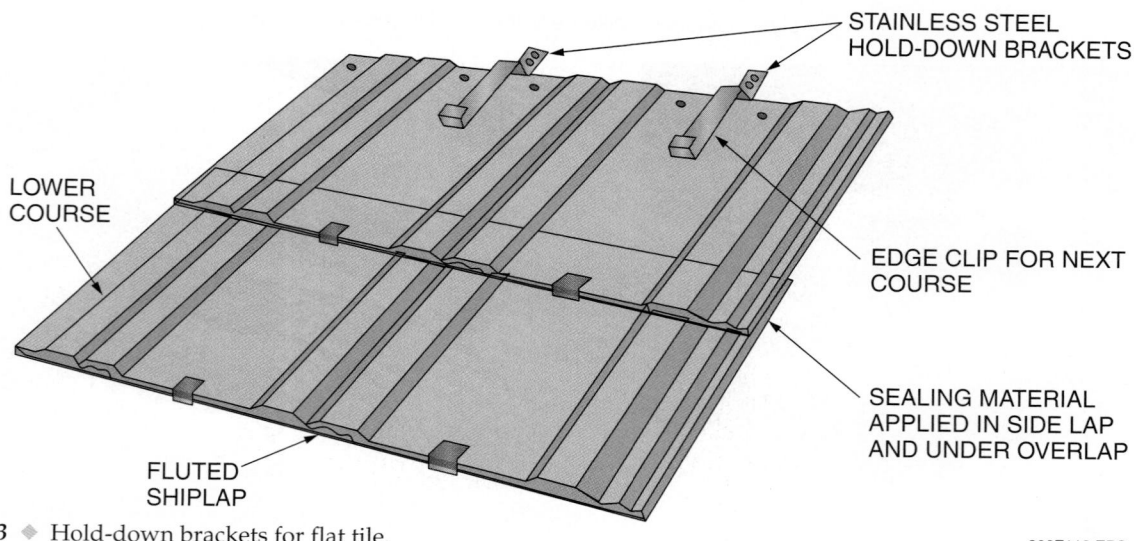

Figure 113 ◆ Hold-down brackets for flat tile.

STAINLESS STEEL HOLD-DOWN BRACKETS

LOWER COURSE

EDGE CLIP FOR NEXT COURSE

SEALING MATERIAL APPLIED IN SIDE LAP AND UNDER OVERLAP

FLUTED SHIPLAP

202F113.EPS

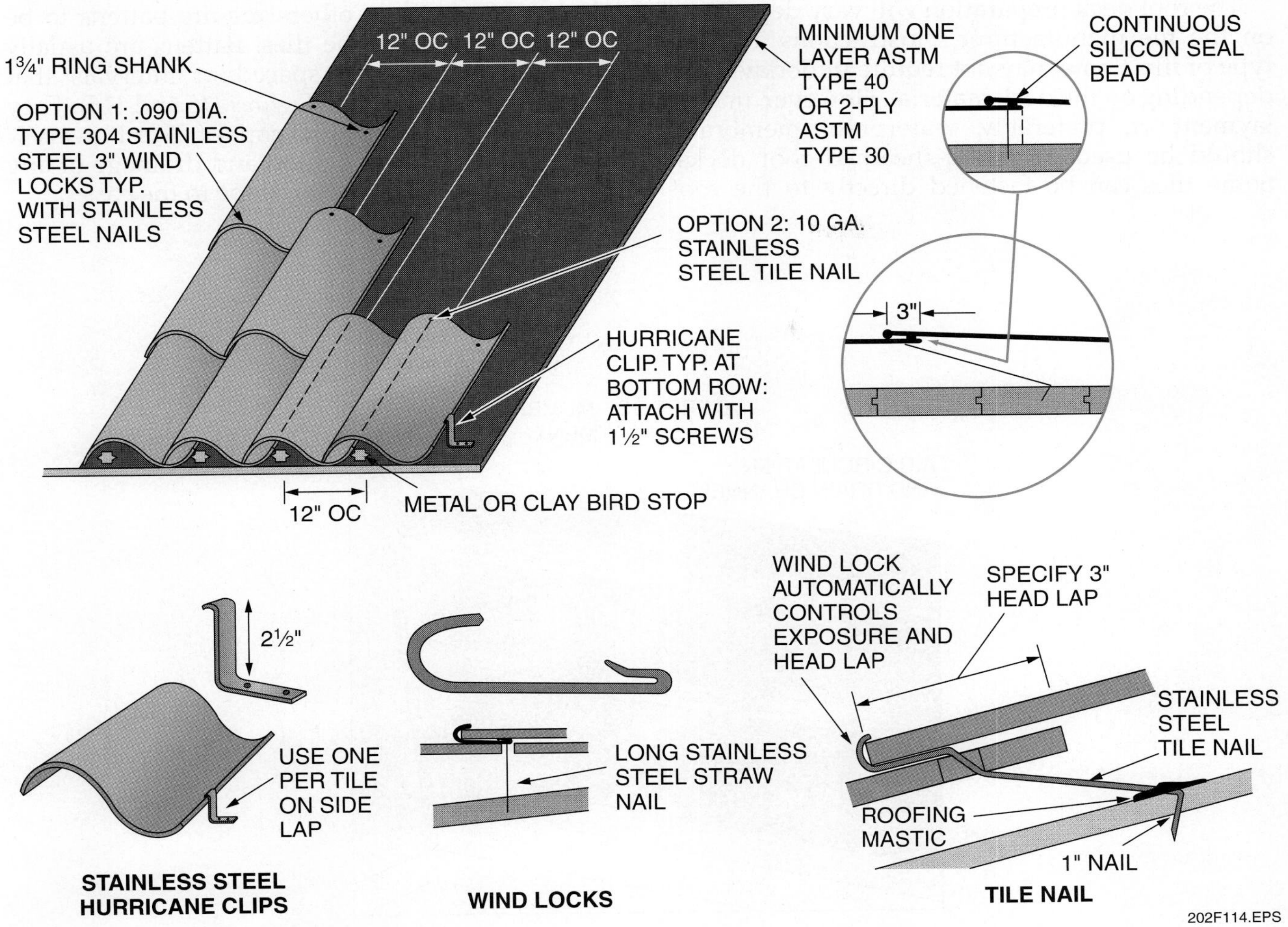

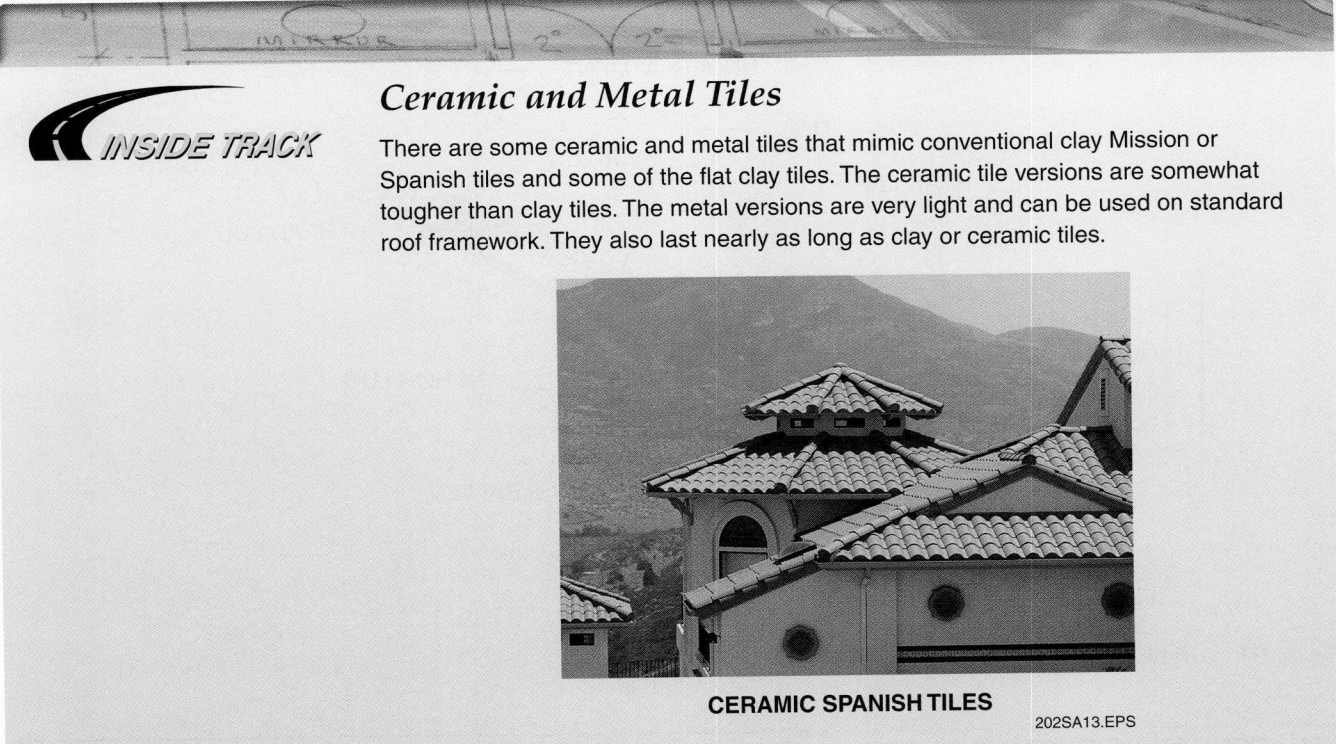

1¾" RING SHANK

OPTION 1: .090 DIA. TYPE 304 STAINLESS STEEL 3" WIND LOCKS TYP. WITH STAINLESS STEEL NAILS

12" OC   12" OC   12" OC

MINIMUM ONE LAYER ASTM TYPE 40 OR 2-PLY ASTM TYPE 30

CONTINUOUS SILICON SEAL BEAD

OPTION 2: 10 GA. STAINLESS STEEL TILE NAIL

3"

HURRICANE CLIP. TYP. AT BOTTOM ROW: ATTACH WITH 1½" SCREWS

12" OC

METAL OR CLAY BIRD STOP

2½"

USE ONE PER TILE ON SIDE LAP

STAINLESS STEEL HURRICANE CLIPS

LONG STAINLESS STEEL STRAW NAIL

WIND LOCKS

WIND LOCK AUTOMATICALLY CONTROLS EXPOSURE AND HEAD LAP

SPECIFY 3" HEAD LAP

STAINLESS STEEL TILE NAIL

ROOFING MASTIC

1" NAIL

TILE NAIL

202F114.EPS

Figure 114 ◆ Examples of a high-wind, hold-down system for barrel-style tiles.

Ceramic and Metal Tiles

INSIDE TRACK

There are some ceramic and metal tiles that mimic conventional clay Mission or Spanish tiles and some of the flat clay tiles. The ceramic tile versions are somewhat tougher than clay tiles. The metal versions are very light and can be used on standard roof framework. They also last nearly as long as clay or ceramic tiles.

CERAMIC SPANISH TILES

202SA13.EPS

The roof deck preparation will vary depending on the tile manufacturer's instructions and the type of tile. Some may not require underlayment, depending on the deck material. However, underlayment or, preferably, waterproof membrane should be used on wood-sheathed roof decks. Some tiles can be fastened directly to the roof deck surface, while others require battens to be laid first to anchor the tiles. Battens are usually pressure-treated pine spaced at intervals that match the tile exposure (*Figures 115* and *116*). They are placed with horizontal gaps of ½" to 1" every 3' to 4' to allow air circulation and drainage of any incidental water down the slope to the eaves.

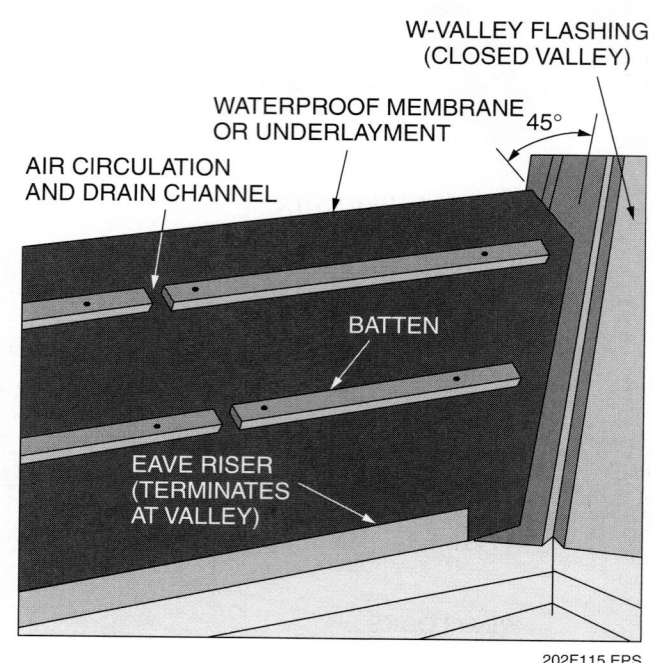

202F115.EPS

Figure 115 ◆ Typical flat tile deck preparation.

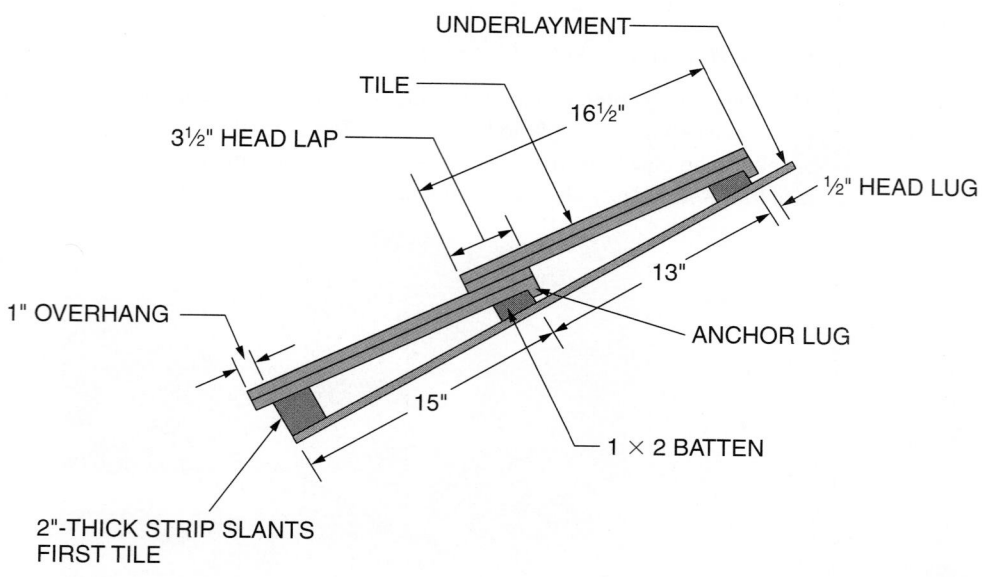

202F116.EPS

Figure 116 ◆ Typical batten spacing for flat tile.

Valleys should be underlaid with 36"-wide waterproof membrane and 24"-wide W-metal flashing made of copper, stainless steel, or galvanized steel. Further preparations may also be required, including 2 × 2s along all ridges and hips, starter strips, and flashing, or 1 × 3s nailed to rake rafters to allow the tiles to extend further sideways. Always check the manufacturer's instructions for specific requirements.

Flat styles of tile have a plain or fluted shiplap along the sides, and sometimes a fluted overlap along the top and bottom as well (*Figure 117*). The fluted shiplaps and overlaps help to prevent water leakage and wind-driven rain penetration between the tiles.

Flat tiles that use battens have a lug on the bottom of the tile that hooks over the batten to help hold the tile in place after it is fastened down (*Figures 116* and *117*). Flat tiles usually require a starter strip and some type of flashing at the eaves, as shown in *Figure 117*.

The roof projection flashing for flat tile roofs is accomplished in essentially the same way as for shingle roofs (*Figure 118*). Some high-profile tiles will require a very soft copper flashing that can be molded and cemented to the surface of the tile.

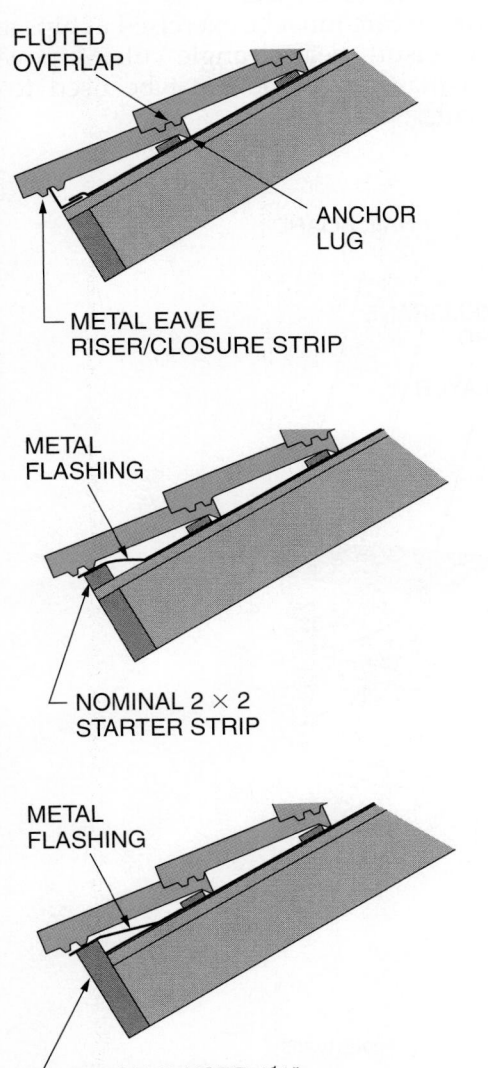

202F117.EPS

Figure 117 ◆ Typical board application.

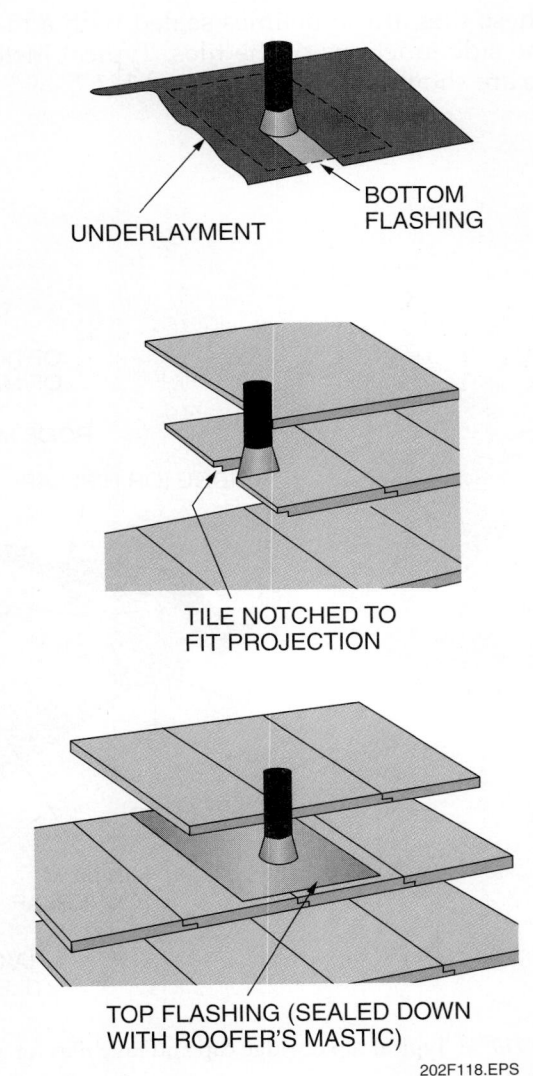

202F118.EPS

Figure 118 ◆ Typical roof projection flashing for flat tiles.

Eliminating Leakage Points

Hips, valleys, and ridges are the most likely areas for leaks in tile roofs. The manufacturer's instructions must be followed exactly to prevent these kinds of leaks.

Figure 119 shows typical rake tiles and a hip or ridge cap treatment used for flat tile roofs.

Barrel-style tiles (Mission or Spanish) are applied as a two-piece pan and cover (*Figure 120*) or as a one-piece pan combined with a cover (*Figure 121*).

These tiles are sometimes sealed with a mastic at the side junctions of the tiles. Typical installations are shown in *Figures 120* and *121*.

Common types of barrel-style flashing are shown in *Figure 122*. These types of flashing may be constructed of copper, galvanized steel, or stainless steel and are used to protect roof projections, valleys, and walls.

Extreme care must be exercised at hips and valleys to ensure perfect angle cuts. A power saw with a masonry blade must be used to obtain these cuts.

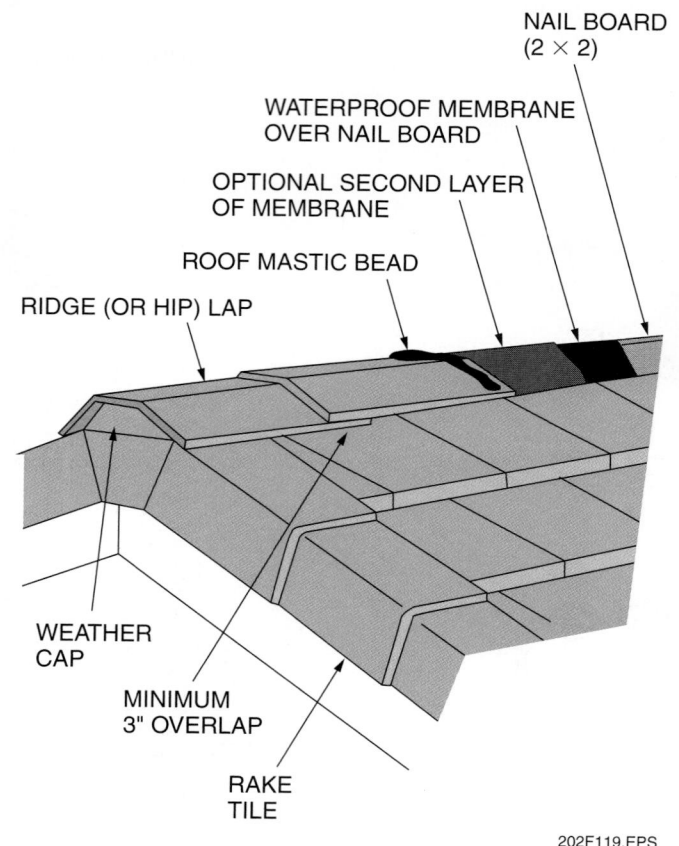

202F119.EPS

Figure 119 ◆ Typical hip or ridge cap and rake tiles for a flat tile roof.

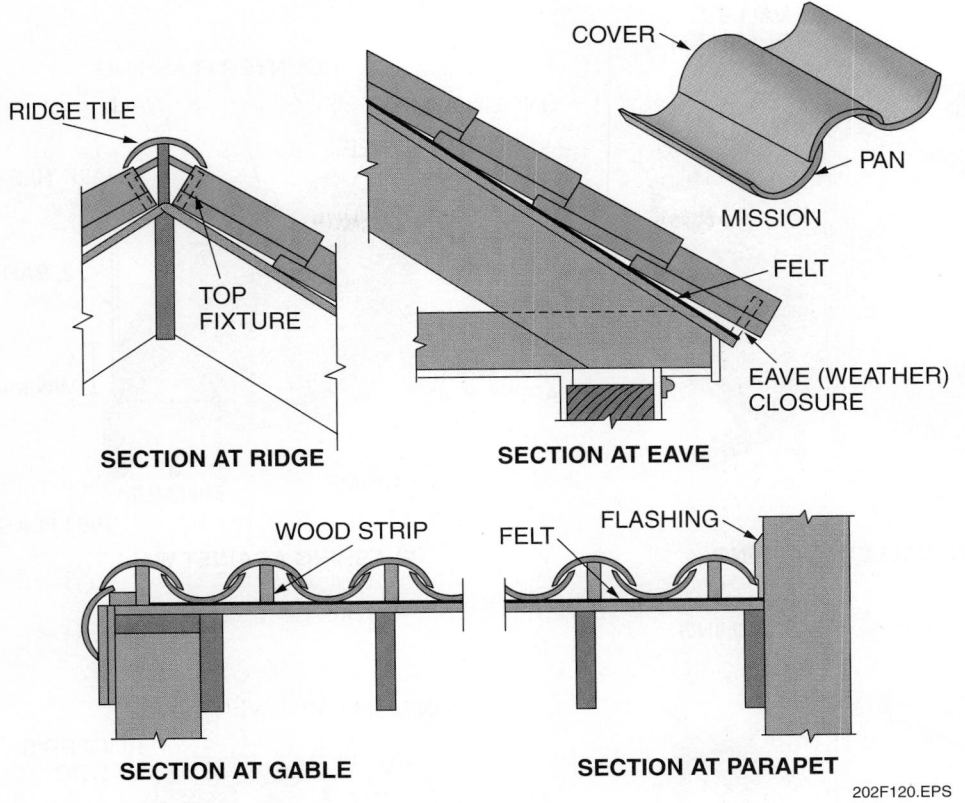

Figure 120 ◆ Mission barrel-style tile.

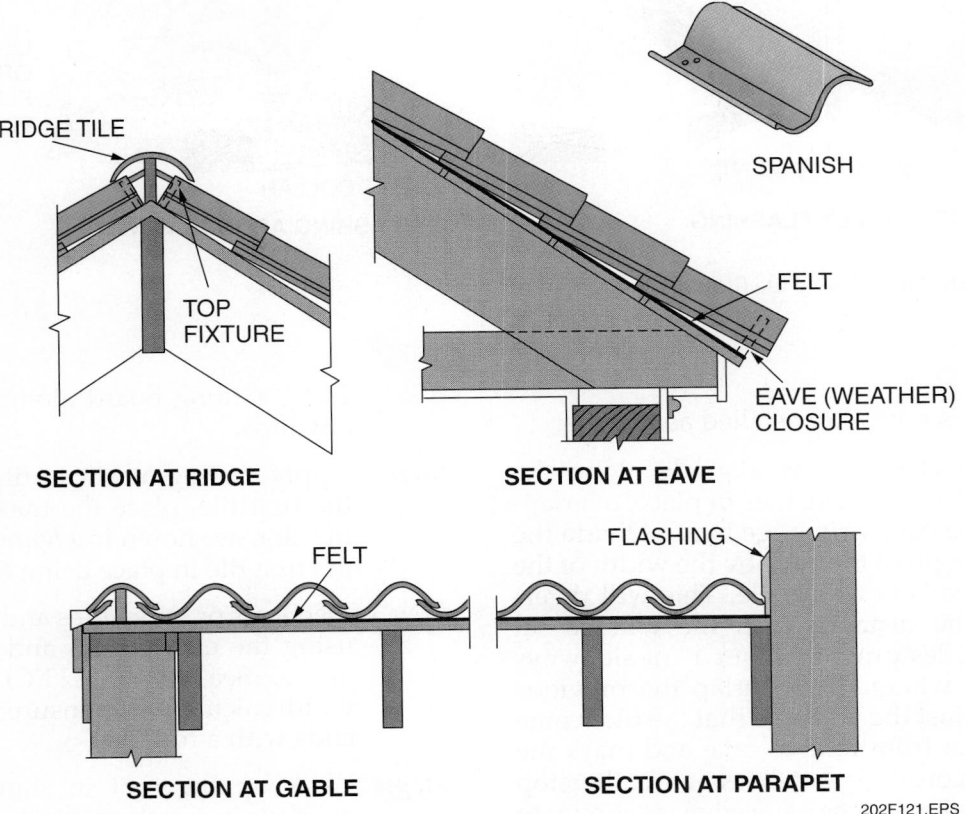

Figure 121 ◆ Spanish barrel-style tile.

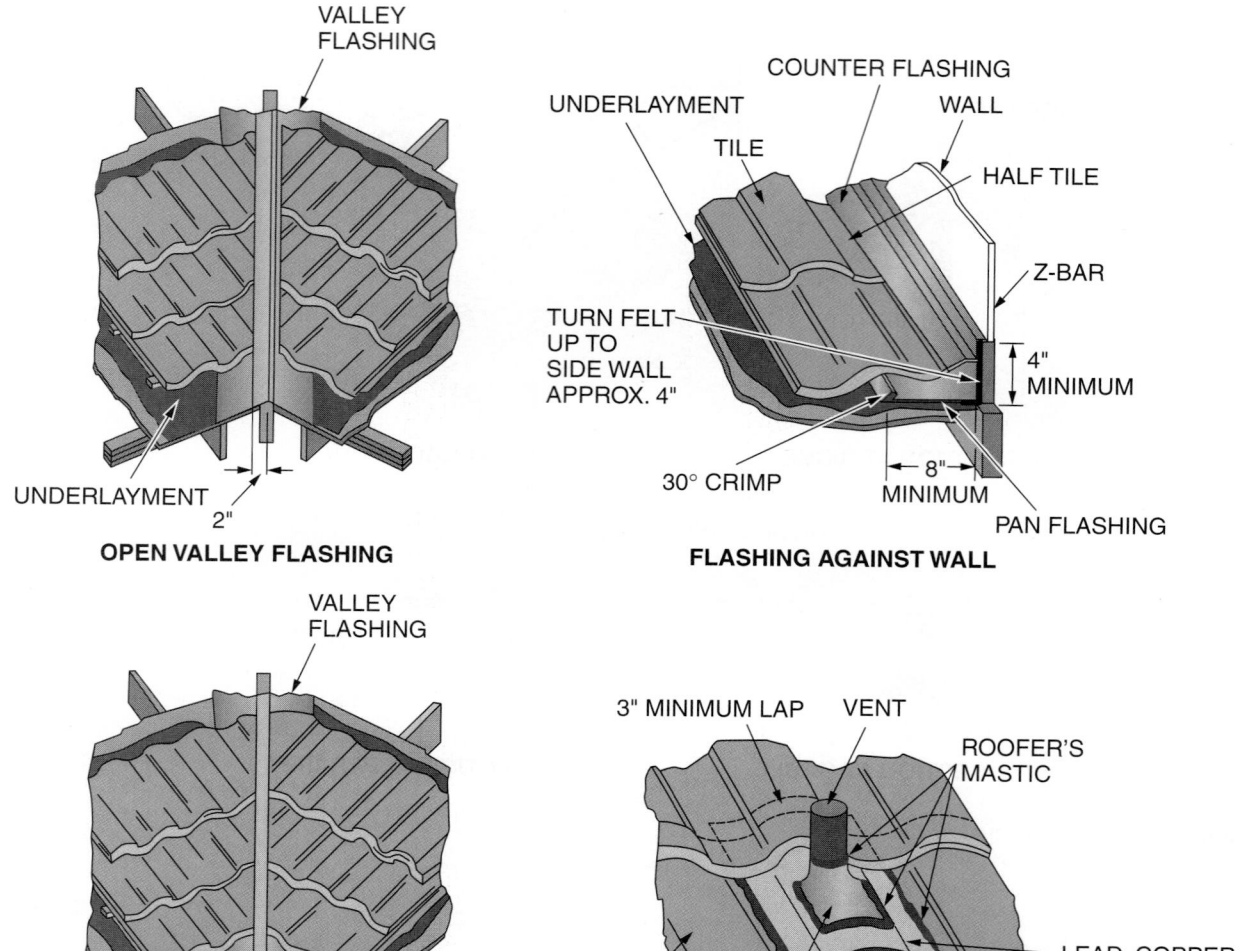

OPEN VALLEY FLASHING

VALLEY FLASHING

UNDERLAYMENT

2"

FLASHING AGAINST WALL

COUNTER FLASHING

UNDERLAYMENT

TILE

WALL

HALF TILE

Z-BAR

4" MINIMUM

TURN FELT UP TO SIDE WALL APPROX. 4"

30° CRIMP

8" MINIMUM

PAN FLASHING

CLOSED VALLEY FLASHING

VALLEY FLASHING

SHEATHING

UNDERLAYMENT

FLASHING AROUND VENTS

3" MINIMUM LAP

VENT

ROOFER'S MASTIC

LEAD, COPPER, OR OTHER FLEXIBLE FLASHING

TILE

VENT COLLAR

202F122.EPS

Figure 122 ◆ Typical barrel-style tile flashing.

Spanish tile is typically installed as follows:

Step 1 Snap a chalkline as a guide along the eaves. Nail the rake tiles in place, overlapping the eave drip edge by 2". Divide the total length of the eave by the width of the cover portion of the tile. This will determine the distance each tile will be set apart. Tiles provide about 1" of sideways leeway where they overlap the previous tile. Adjust the width so that the tiles come out even from rake to rake and mark the eave accordingly. Set the eave weatherstop tiles flush with the eave edge, as shown in *Figure 123(A)*, and nail it into place. Nail a

2 × 2 nailing board along the ridge and any hips.

Step 2 Apply roofing mastic at tile overlaps. For the first tile, place the mastic in the rake tile slot, as shown in *Figure 123(B)*. Secure the first tile in place using 6d nails.

Step 3 Install six or seven tiles and weather-stops using the eave marks and chalkline as a guide. See *Figure 123(C)*. Check your width calculation to ensure that the course ends with a full tile.

Step 4 Work up the roof in stair-step fashion, overlapping each course by at least 3", as shown in *Figure 123(D)*.

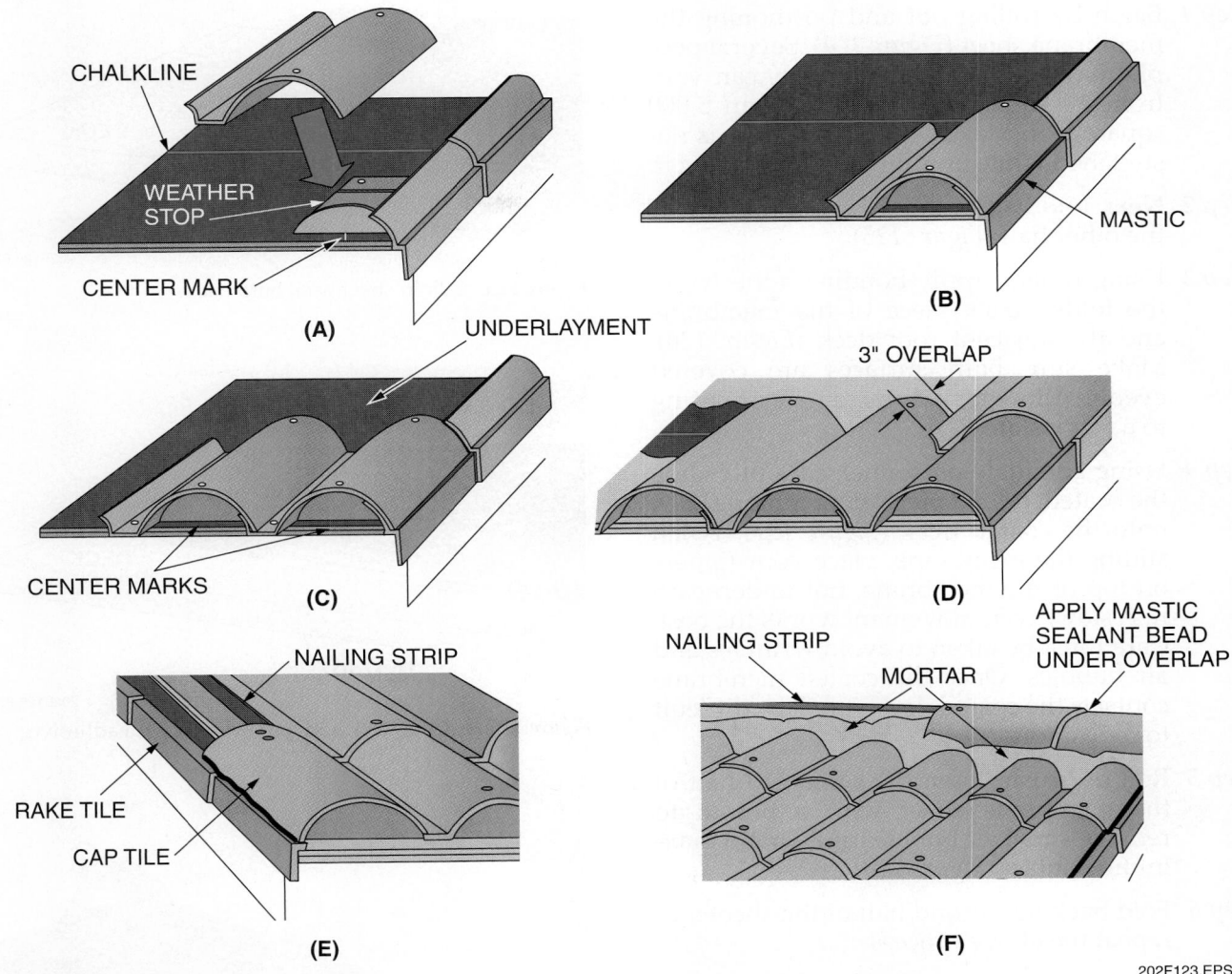

Figure 123 ◆ Typical Spanish tile installation.

Step 5 When the opposite rake is reached, stop at the next to the last tile and nail the rake tiles in place. Then, nail a 1 × 3 nailing strip from the eaves to the ridge. Position it so that it will line up with the holes at the top of the last tile. Apply mastic to the rake tile slot and place a ridge (cap) tile with mastic on the opposite edge into the rake slot and the pan of the previous tile, as shown in *Figure 123(E)*. Nail the tile into place through the nailing strip. Repeat for each course of tile.

Step 6 When the roof is complete to the ridge 2 × 2 nailing board, lay a bed of tinted mortar and set the ridge starter tiles. Then, lay mortar along both sides of the nailer and over the roof tiles. See *Figure 123(F)*. Position the cap tiles, setting them into the mortar and nailing them to the ridge nailing board. When installing the cap tiles, make sure that they overlap by at least 3".

11.0.0 ◆ SINGLE-PLY ROOFING APPLICATION

Among the membrane-type roofing systems, non-heat bonded single-ply roofing is usually the easiest to install. In most cases, it is accomplished by roofing contractors experienced with the product and its application.

Normally, single-ply membrane roofing is applied over several layers of insulation board on flat roofs. In the following description, the methods of applying only a few sections of fully-adhered EPDM-type membrane are covered without consideration for flashing of roof projections, walls, or eaves and with the assumption that the insulation has been installed.

Most manufacturers of membrane roofing systems market specific accessories such as flashing and vent boots for use with their product. The installation of these accessories varies considerably among manufacturers, so the manufacturer's specific instructions must be followed exactly. Single-ply roofing is installed as follows:

Step 1 Begin by rolling out and positioning the membrane sheet (*Figure 124*). Several people may be needed. The sheets can vary from 10 square feet to more than 5,000 square feet. Make sure the sheeting is not stretched while unrolling or positioning it.

Step 2 Next, fold half of the sheet back on top of the other half (*Figure 125*).

Step 3 Using rollers, apply bonding adhesive to the folded back piece of the membrane and the adjacent roof deck (*Figure 126*). Make sure both surfaces are covered evenly. Allow the adhesive to start curing to a tacky state.

Step 4 Using adequate personnel, carefully slide the coated, folded piece of membrane back onto the coated deck (*Figure 127*). When sliding the membrane, place your fingers on top of the membrane, not underneath it. A slow, even movement works the best. Care must be taken to avoid wrinkles and air bubbles. Once the coated membrane contacts the coated deck, it will be difficult to reposition.

Step 5 Roll or brush down the cemented half of the membrane sheet with a broom to remove small air bubbles and obtain maximum contact (*Figure 128*).

Step 6 Fold back the second half of the sheet and repeat the above process.

Step 7 Position adjoining sheets, as shown in *Figure 129*, allowing for the specified overlap splice. Before splicing, bond the sheet to the deck. Avoid getting bonding adhesive on the splice area of both sheets. Apply splicing cement to the splice area and overlap the sheets. After the splice is complete, protect the splice with lap sealant.

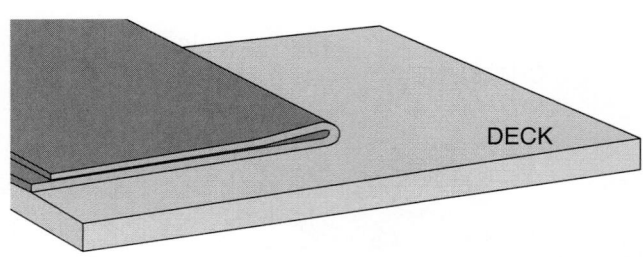
202F125.EPS

Figure 125 ◆ Fold sheet back onto itself.

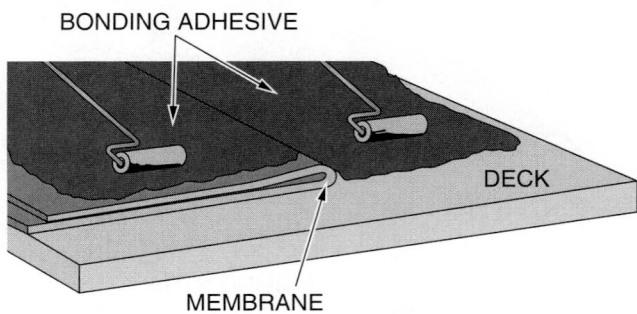
202F126.EPS

Figure 126 ◆ Coat deck and half of sheet with adhesive.

202F127.EPS

Figure 127 ◆ Slide half of sheet onto coated deck.

202F124.EPS

Figure 124 ◆ Place and position membrane sheet.

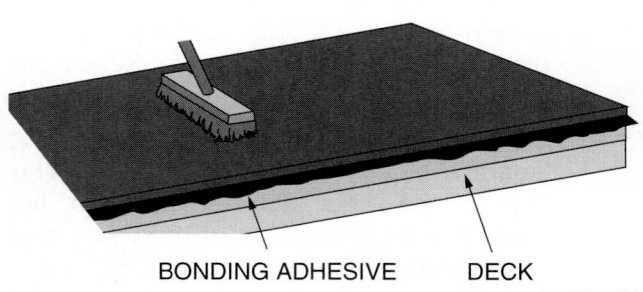
202F128.EPS

Figure 128 ◆ Brush cemented half of sheet to obtain contact.

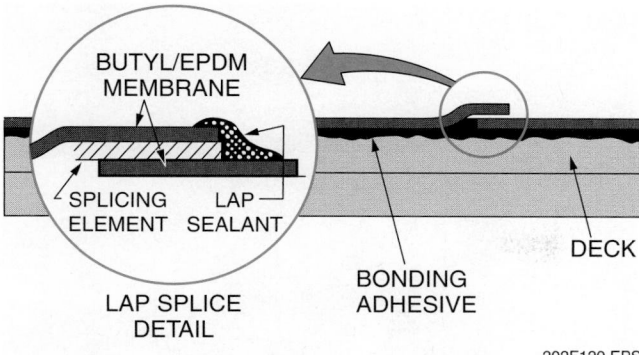

BUTYL/EPDM
MEMBRANE

SPLICING   LAP
ELEMENT   SEALANT

LAP SPLICE
DETAIL

BONDING
ADHESIVE

DECK

202F129.EPS

Figure 129 ◆ Apply and splice adjacent sheets.

Step 8 If desired, a fully-adhered EPDM roof can be covered with round ballast stone about ¾" to 1½" in diameter; however, it is not necessary.

12.0.0 ◆ TORCH-DOWN ROOFING APPLICATION

One of the most popular types of membrane roofing material for flat-roof residential applications is an APP product (torch-down roofing). This type of material is installed using the following guidelines:

Step 1 Install a manufacturer-specified underlayment starting at the lowest edge of the roof deck. Overlap the remaining courses of underlayment as recommended by the manufacturer. Fasten the underlayment to the roof deck with washer-headed nails at the manufacturer's recommended spacing (*Figure 130*).

Step 2 After igniting a propane torch (*Figure 131*), start applying the roofing material at the lowest edge of the roof deck (*Figure 132*). Heat the exposed bottom-side asphaltic surface of the roofing roll until it is melted. Be careful not to overheat the material so that the surface begins to run or catches fire. While the surface section is melted, unroll that section on to the underlayment and press it down to bond it to the underlayment. Progressively heat and unroll until the roof edge is covered or the roll is exhausted. If a new roll must be spliced, heat the end of the new roll and overlap the end of the previous roll by about 6" or as recommended by the manufacturer. Once the first course is completed, continue applying roofing material up the roof, overlapping each lower course by about 6" or as recommended by the manufacturer.

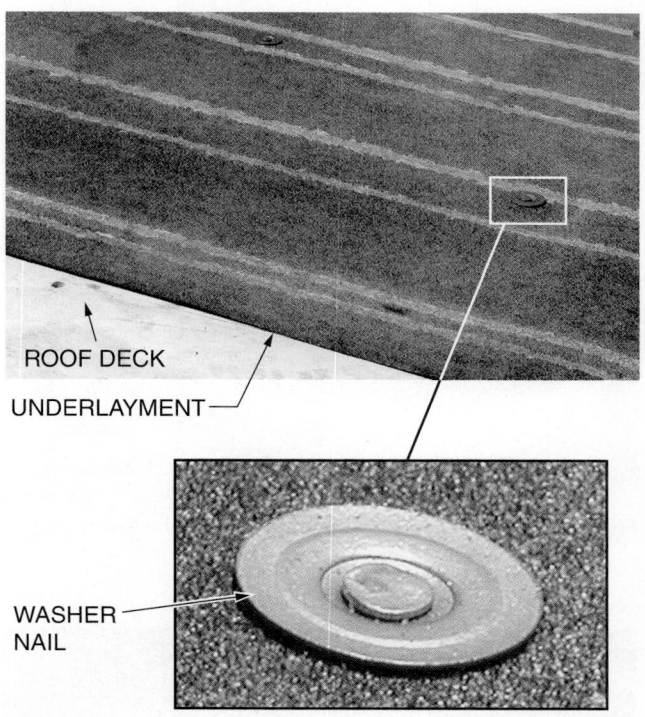

ROOF DECK

UNDERLAYMENT

WASHER
NAIL

202F130.EPS

Figure 130 ◆ Underlayment installation.

WARNING!

Be careful when using the torch to avoid setting fire to the roofing material or underlayment. Have a fire extinguisher available in case of fire.

Step 3 Once the roof deck is covered, seal all overlapping seams by melting the surface of the roofing material adjacent to the edge of each seam and troweling the melted material over the seam edge (*Figure 133*).

202F131.EPS

Figure 131 ◆ Torch-down roofing propane torch.

Figure 132 ◆ Applying roofing material to the lower edge of the roof.

Figure 134 ◆ Drip edge installation.

Figure 133 ◆ Sealing roofing-course seams.

Figure 135 ◆ Applying drip-edge strips.

Step 4 Apply manufacturer-recommended drip edge to the roof deck (*Figure 134*).

Step 5 Cut roofing material into 8"- to 9"-wide strips, then heat-bond the strips over the drip edges and main roofing material (*Figure 135*). Once the strips are applied, melt the surface of the strips and adjacent roofing material and seal the edges with a trowel (*Figure 136*).

CAUTION

Torch-down roofing material usually must be covered with a coating to shield it from the UV rays of the sun. Make sure to follow any manufacturer's instructions for applying protective coatings.

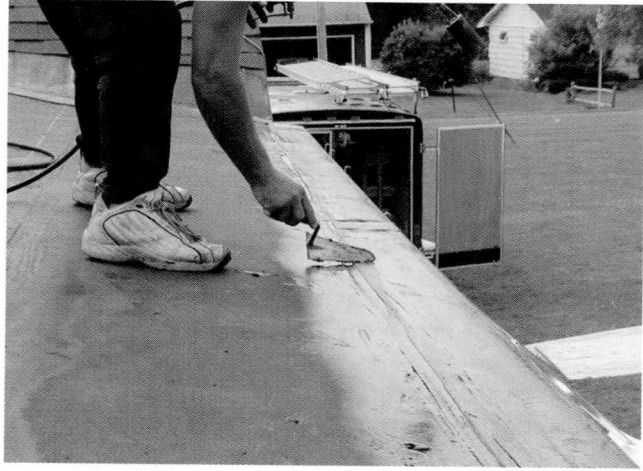

Figure 136 ◆ Sealing drip-edge strips.

302F137.EPS

Figure 137 ◆ Applying a protective coating.

Step 6 After the roof has cured for a day or so, coat the surface with a manufacturer-recommended roof coating or material such as a fibered aluminum coating (*Figure 137*).

13.0.0 ◆ ROOF VENTILATION AND ICE EDGING

This section covers the installation of various types of roof venting and ice edging systems.

13.1.0 Roof Ventilator Installation

Proper attic ventilation is necessary to allow heat and moisture to escape so that damage to the roofing and roof deck does not occur. In the winter, ventilation keeps the roof deck cold and reduces the buildup of ice on the eaves. This helps prevent water penetration through the roof and subsequent water damage to the structure. It also carries away moisture so that it does not condense on the roof deck, which can cause rotting. In the summer, excess heat can cause overheating of composition shingles, resulting in early failure of the roof.

Residential attics are generally ventilated by convection vents in the form of gable vents or roof-mounted ridge or box vents. Sometimes, electric-powered or wind-powered turbine vents or fans are used (*Figure 138*). The air used for ventilation is provided by soffit vents at the eaves of the roof. The amount of soffit ventilation (in square feet or inches) must be equal to or greater than the amount of roof ventilation.

In residential structures, the proper amount of convection-type ventilation for an unheated attic space is usually defined as 1 sq ft of ventilation for every 300 sq ft of attic area with 50 percent in the roof for exhaust and 50 percent in the eaves for intake. The attic area is calculated using the exterior foundation dimensions of the structure. For example, the amount of ventilation for a residence with an exterior foundation measurement of 40' × 60' would be calculated as follows:

40' × 60' = 2,400 sq ft (attic area)

2,400 sq ft ÷ 300 sq ft = 8 sq ft
(total ventilation required)

8 sq ft × 144 sq in per sq ft = 1,152 sq in

1,152 sq in ÷ 2 = 576 sq in for the ridge
and 576 sq in for the soffits

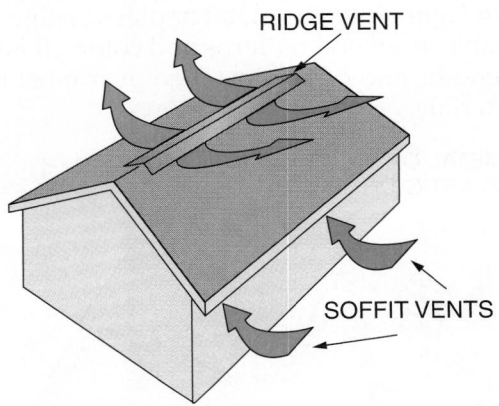

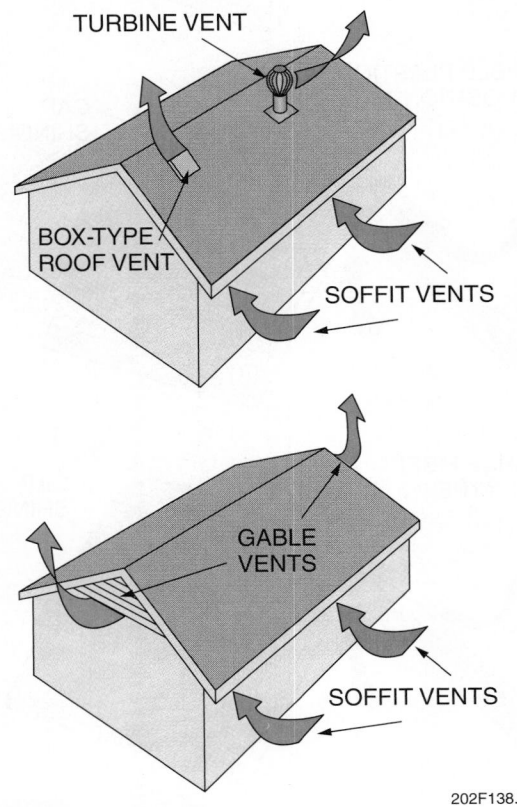

202F138.EPS

Figure 138 ◆ Residential roof vents.

Ventilation Requirements

Always check local codes for the proper ventilation requirements of a structure. In some areas, the amount of air exchange required may dictate fan-assisted ventilation if the capacity of free-air ventilation devices is not adequate.

Based on the calculated ventilation requirement, appropriately sized roof and soffit convection ventilation devices would be selected and installed.

Ridge vents, box vents, and turbine vents are available in a variety of styles and sizes for residential use. Different types of ridge vents are illustrated in *Figure 139*. The metal or plastic ridge vent is available in several patterns and colors. It is sold in 10' lengths and can be installed over most roofing materials.

The flexible, plastic composition vent is available in 4' lengths. An inert, coarse fiber vent is available in rolls. Flexible plastic and rolled coarse fiber vents can only be used over composition shingles because both are designed to be covered with cap shingles that match the roof. Most residential customers prefer the shingle-covered vents because they tend to blend into the overall roof.

The manufacturer's specifications for a ventilation device must be consulted to determine the amount of free-air ventilation (in square feet or inches) that the device will provide. Ridge ventilators are probably the most efficient and are usually rated in square inches of free-air ventilation per linear foot of the product.

13.1.1 Box Vent Installation

Box vents (*Figure 140*) are easily installed on most roofing materials. On new roofs, the proper size hole is cut in the roof and the hole is surrounded with at least a 24"-wide waterproof membrane. When the roofing courses reach the area of the ventilator, the ventilator is installed with the lower edge of its flashing overlapping the course below it. The flashing is fastened to the roof at the top and sides. Then, roofing courses are applied over and cemented to the flashing at the top and sides of the ventilator.

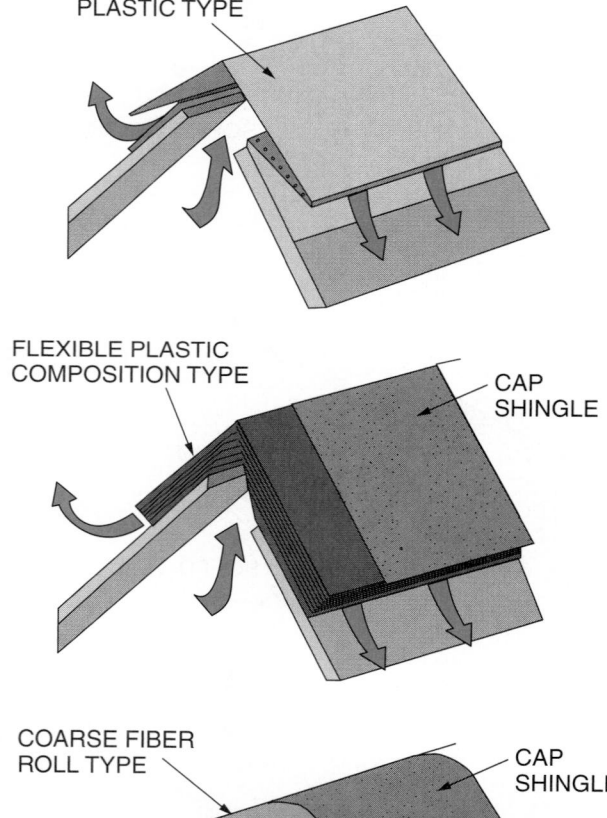

METAL OR PLASTIC TYPE

FLEXIBLE PLASTIC COMPOSITION TYPE

CAP SHINGLE

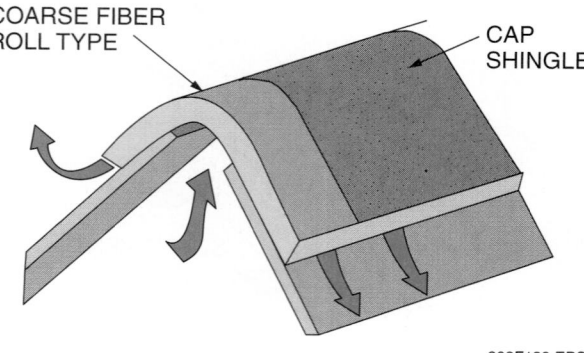

COARSE FIBER ROLL TYPE

CAP SHINGLE

202F139.EPS

Figure 139 ◆ Typical ridge vents.

202F140.EPS

Figure 140 ◆ Typical box vent.

On existing roofs, the hole is cut, and the upper flashing of the ventilator is slid under the courses above the hole and fastened to the roof. The side and bottom flashing is not fastened; however, the side flashing of the ventilator is completely cemented down to the roofing under the sides. The bottom flashing is usually not cemented.

13.1.2 Ridge Vent Installation

Use the following procedure to install ridge vents:

Step 1 Determine where the roof vent slots will be cut on the ridges and any hips (*Figure 141*). Slots cut along a ridge should start and stop approximately 12" from the gable (terminal) ends, any vertical walls, any higher intersecting roof, any roof projections at the ridge, any valleys, and any hip joints. Slots cut in hips should end 24" above the eaves and should be in 24" sections separated by 12" to maintain maximum roof strength. For appearance, the roof ridges and hips should be covered completely with the roof vent material to maintain a continuous roof line, as shown in *Figure 142*.

Step 2 Next, refer to the manufacturer's specifications and note the width of the slot required for the vent being used. Determine the roof construction (ridge board or no ridge board). If a ridge board is used (*Figure 143*), add 1½" to the required slot width and divide the result by two to find the total slot width to be cut on each side of the peak. If no ridge board is used, only divide the required slot width by two to find the slot width to be cut on each side of the peak.

Step 3 At the peak, measure down the roof on both sides for half of the required slot width as determined above and snap a chalkline along any ridges or hips at both sides of the peak or hip (*Figure 144*). Mark the start and stop point for each slot, as previously specified.

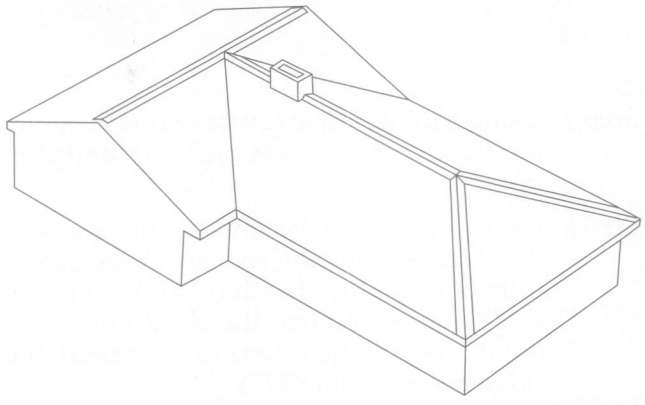

202F142.EPS

Figure 142 ◆ Example of a continuous roof line.

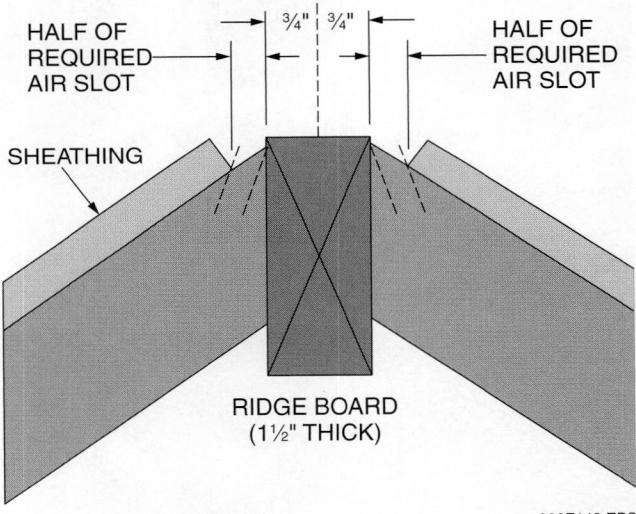

HALF OF REQUIRED AIR SLOT

¾" ¾"

HALF OF REQUIRED AIR SLOT

SHEATHING

RIDGE BOARD (1½" THICK)

202F143.EPS

Figure 143 ◆ Determining total slot width.

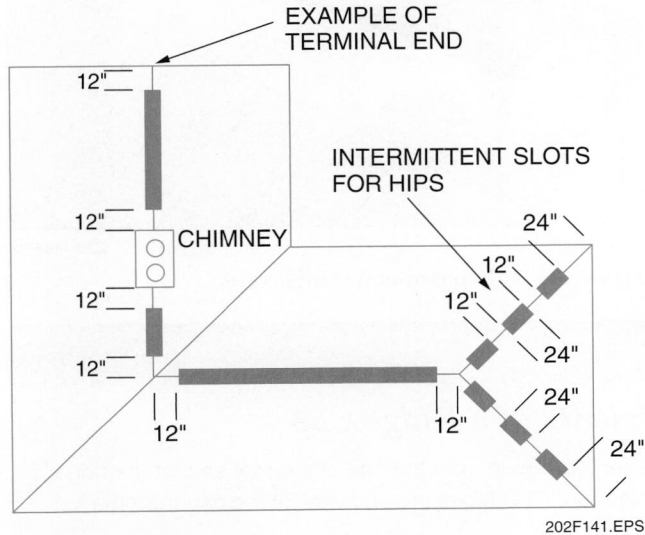

EXAMPLE OF TERMINAL END

12"

12"   CHIMNEY

12"

12"

12"

INTERMITTENT SLOTS FOR HIPS

24"

12"

12"

24"

24"

12"

24"

202F141.EPS

Figure 141 ◆ Example of slot cutout placement.

202F144.EPS

Figure 144 ◆ Snapping a chalkline for slot width.

Step 4 Using a knife, cut away the shingles along the chalklines between each start and stop point.

Step 5 Using a power saw set to the thickness of the shingles plus the roof sheathing, carefully cut into the sheathing and along the chalklines to remove the sheathing without damaging the rafters or, if present, the ridge board (*Figure 145*).

Step 6 Place a shingle ridge cap at each gable end, before and after each roof projection, at each vertical wall, at each higher intersecting roof joint, and at each eave end for any hips (*Figure 146*). This prevents any water intrusion at or under the exposed ends of the vent from penetrating into or through the sheathing.

Step 7 If packaged in rolls, unroll the vent material and temporarily tack it in place over the ridges and hips or secure the flexible plastic composition sections to the roof over the ridges and hips. Make sure that the vent covers the entire length of all ridges and hips (*Figure 147*). If rigid vent sections (*Figure 148*) are being used, make sure that vents are secured through the prepunched holes.

Step 8 Cut and taper the tabs from a number of three-tab standard or architectural cap shingles for use as cap shingles.

202F145.EPS

Figure 145 ◆ Cutting slots.

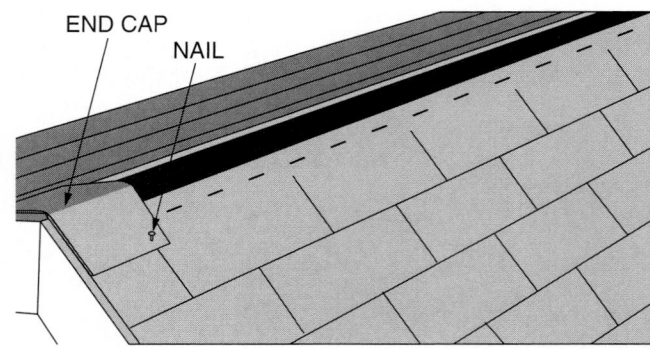

END CAP

NAIL

202F146.EPS

Figure 146 ◆ Exposed-end shingle caps.

Plastic or Metal Ridge Vent

If using a rigid plastic or metal vent not intended to be capped with shingles, fasten the vent to the roof through the flanges using a non-corrosive fastener with a neoprene seal washer; then insert an end cap in the exposed ends of the vents, if required.

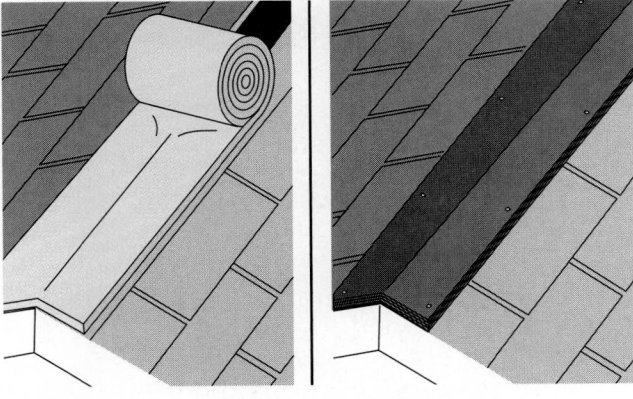

ROLLED VENT MATERIAL   RIGID VENT MATERIAL

202F147.EPS

Figure 147 ◆ Positioning vent material.

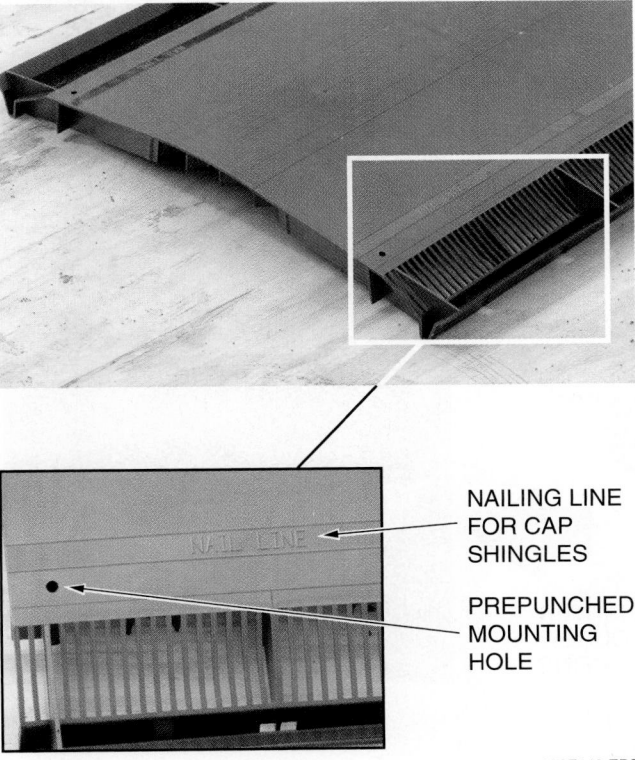

NAILING LINE FOR CAP SHINGLES

PREPUNCHED MOUNTING HOLE

202F148.EPS

Figure 148 ◆ Typical rigid vent section.

Step 9 Starting from the exposed ends of a ridge or hip and using sufficiently long nails, fasten the cap shingles to the roof through the vent material nailing line (*Figure 148*). The cap shingles are applied and mated over the ridges and hips in the same manner as described previously. See *Figure 149*. However, make sure that the shingles are nailed snugly without compressing the vent material. It is advisable to place a bead of roofing cement under the overlapping sections of the shingles to help secure them.

13.2.0 Ice Edging Installation

Many sloped-roof residences in the north have ice damming problems on the roof, usually at the eaves. This is especially true for those residences with finished attics where insulation and venting under the roof deck is limited to the rafter space. In most cases, when a new roof is installed, the application of a waterproof membrane from the eaves up the slope of the roof to a point that is at least 24" beyond the inside wall will prevent water backed up behind any ice dams from penetrating into the structure. However, on a problem residence with an existing roof that lacks a waterproof membrane underlayment, it may be desirable to install an ice edge at the eaves. On most roofing materials, ice dams cannot be easily removed. In addition, ice dams can damage some roofing materials, including slate.

An ice edge is an exposed metal sheeting mounted from the eaves up the slope of the roof from 18" to 36". It provides a shield against water penetration and allows any ice dams that form to be quickly shed from the roof during any brief thaws, thus reducing the chances of a large ice dam buildup. The width of the ice edging used is a factor of how steep the roof is and how thick the ice usually becomes, which determines how much water backs up on the roof. Normally, continuous sheets of plain or tinted/painted aluminum flashing or special, standing-seam aluminum panels are used as the ice edge (*Figure 150*); however, copper flashing can be used where corrosion is a

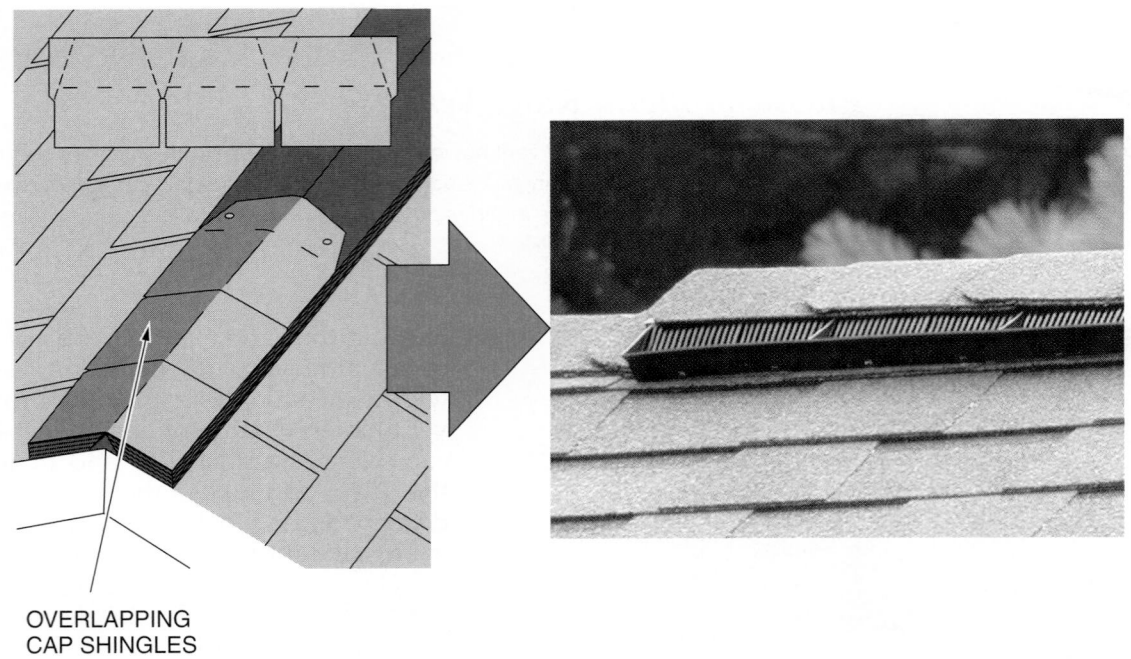

OVERLAPPING
CAP SHINGLES

202F149.EPS

Figure 149 ◆ Applying cap shingles over a vent.

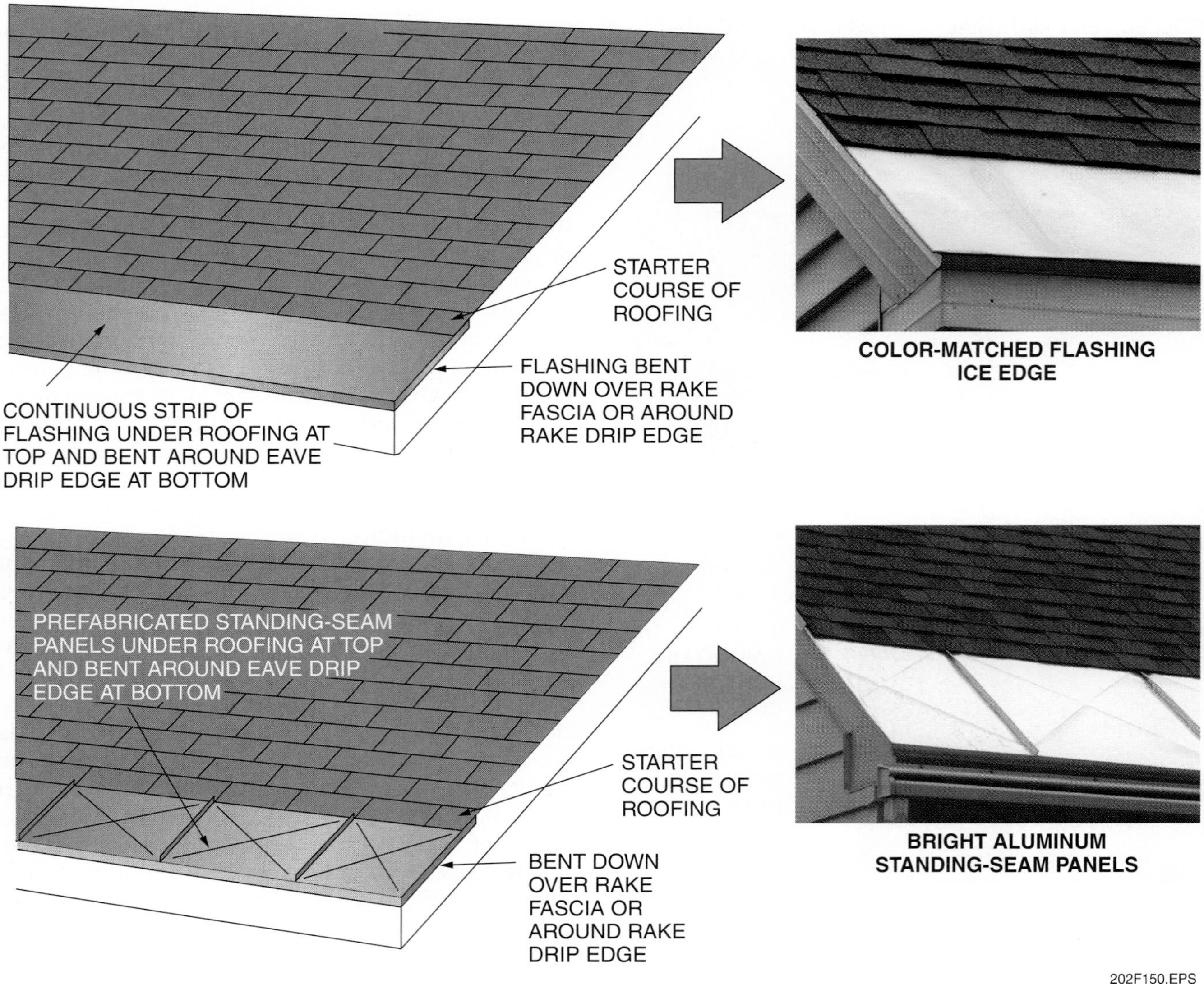

STARTER
COURSE OF
ROOFING

FLASHING BENT
DOWN OVER RAKE
FASCIA OR AROUND
RAKE DRIP EDGE

CONTINUOUS STRIP OF
FLASHING UNDER ROOFING AT
TOP AND BENT AROUND EAVE
DRIP EDGE AT BOTTOM

**COLOR-MATCHED FLASHING
ICE EDGE**

PREFABRICATED STANDING-SEAM
PANELS UNDER ROOFING AT TOP
AND BENT AROUND EAVE DRIP
EDGE AT BOTTOM

STARTER
COURSE OF
ROOFING

BENT DOWN
OVER RAKE
FASCIA OR
AROUND RAKE
DRIP EDGE

**BRIGHT ALUMINUM
STANDING-SEAM PANELS**

202F150.EPS

Figure 150 ◆ Types of ice edging.

Residential Secondary Roof Systems

An alternative for roofs that cannot be insulated properly from inside the residence is an insulated, secondary roof system installed over the original roof deck. These systems have spacing between a rigid insulation layer and a second roof deck to allow free airflow over the insulation and under the second roof deck. This greatly reduces the melting of snow and the resulting ice dam on the second roof deck. These secondary roof systems are available from several manufacturers. They can also be used in hot climates to reduce the heat load in a residence with similar insulation problems. The airflow of these systems can also reduce the heat load on the roofing materials used for the second deck.

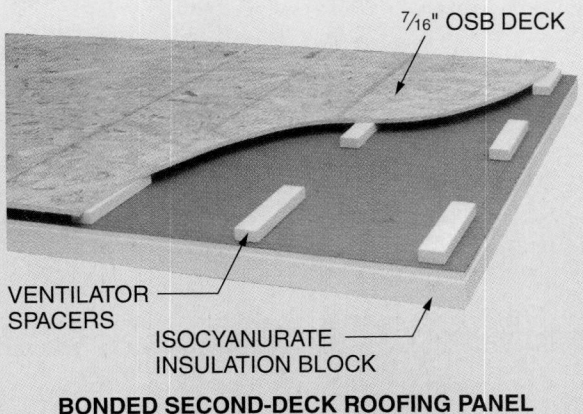

7/16" OSB DECK

VENTILATOR SPACERS

ISOCYANURATE INSULATION BLOCK

BONDED SECOND-DECK ROOFING PANEL

202SA14.EPS

problem. The standing-seam panels are more expensive and are not subject to buckling due to temperature extremes. Their surface appearance is more uniform and does not appear wavy.

There are disadvantages in the use of ice edges. One is that they are generally not considered attractive unless colored to match the roof. The possible exception is plain copper used with slate roofing. Another disadvantage is that for ice edges to be effective in shedding ice, eave troughs cannot be used on the building. This can lead to several severe hazards such as injury or death to people or damage to property or foliage, including damage to the siding of the residence caused by falling ice. In addition, water draining from the roof can collect and penetrate under the foundation and/or into a basement.

Ice Edge Hazards

Before ice edges are considered for use on the eaves of a building, the hazard of falling ice causing injury, death, or property damage under each eave must be carefully evaluated. With ice edging, a long, heavy ice dam (1,000 lbs or more) along an entire roof eave can be released from the eave without warning. In addition, attempts to remove icicles may cause an ice dam along an entire eave to be released.

202SA15.EPS

Summary

Many types of roofing materials and commercial roofing systems are available. They are used to protect a structure and its contents from the elements and vary depending on geographic location. This module covered the most common materials used on residential and small commercial structures, including composition shingles, roll roofing, wood shingles/shakes, slate, tile, metal, and membrane roofing.

While carpenters may not usually be required to install roofing, they can be involved in preparing the roof surface before the installation of roofing materials. Therefore, carpenters should be familiar with the basic preparation and installation of common roofing materials.

Notes

1. Organic asphalt composition shingles have been largely replaced by _____ shingles.
   a. asbestos
   b. wood
   c. fiberglass
   d. synthetic tile

2. Architectural shingles have a typical lifespan of _____ years.
   a. 5 to 12
   b. 15 to 20
   c. 20 to 25
   d. 25 to 40

3. Roll roofing is available in weights of _____ lbs.
   a. 20 to 40
   b. 30 to 70
   c. 50 to 90
   d. 70 to 120

4. The most popular wood shingles and shakes are made of _____ .
   a. yellow pine
   b. red cedar
   c. cypress
   d. redwood

5. Slate roofing weighs about _____ pounds per square foot.
   a. 3 to 7
   b. 5 to 9
   c. 7 to 10
   d. 9 to 12

6. Tile roofing weighs about _____ pounds per square foot.
   a. 3 to 7
   b. 5 to 9
   c. 7 to 10
   d. 9 to 12

7. Terne metal is metal coating comprised of _____ .
   a. 50 percent brass and 50 percent lead
   b. 80 percent lead and 20 percent tin
   c. 60 percent tin and 40 percent copper
   d. 20 percent copper and 80 percent zinc

8. There are _____ general categories of premanufactured membrane roofing systems.
   a. two
   b. three
   c. four
   d. five

9. The most common thermoset single-ply membrane is known as _____ .
   a. APP
   b. SBS
   c. PVC
   d. EPDM

10. Two common thermoplastic single-ply membranes are _____ .
    a. APP and BUR
    b. SBS and PVC
    c. PVC and TPO
    d. EPDM and TPO

11. The standard roofing hammer is also called a _____ .
    a. shingle hatchet
    b. slater's hammer
    c. claw hammer
    d. cricket hammer

12. A slate cutter can be used to _____ .
    a. remove nails
    b. bend metal flashing
    c. punch nail holes
    d. cut tile

13. The edge of a ladder must extend _____ above the edge of a roof.
    a. 1'
    b. 2'
    c. 3'
    d. 4'

14. What is the normal percentage that is added to a roof estimate for waste?
    a. 5 percent
    b. 10 percent
    c. 15 percent
    d. 20 percent

15. The most common composition shingle applied to residential structures is the _____ shingle.
    a. two-tab
    b. three-tab
    c. architectural
    d. strip

16. The abbreviation that defines the distance that a shingle overlaps the shingle below is _____ .
    a. E
    b. TL
    c. HL
    d. W

17. The 6" pattern of laying three-tab shingles means that 6" are _____ .
    a. added to each course
    b. subtracted from each course
    c. subtracted from every other course
    d. subtracted from the second course, 12" from the third course, 18" from the fourth, and so on

18. There are _____ types of closed valleys.
    a. two
    b. three
    c. four
    d. five

19. Step flashing is used on _____ .
    a. valleys
    b. slopes against walls
    c. hips
    d. vent pipes

20. Double-coverage roll roofing with concealed nails can be used on slopes as low as _____ .
    a. ¼ in 12
    b. 1 in 12
    c. 2 in 12
    d. 3 in 12

21. If permitted by local codes, clay, concrete, or ceramic tiles should be installed with _____ .
    a. ungalvanized roofing nails
    b. drywall screws
    c. copper nails
    d. construction adhesive

22. A lug under the top edge of a flat tile is used to _____ .
    a. anchor the tile
    b. space the tile
    c. prevent breakage of the tile
    d. keep the edges of the tile even

23. The flutes in the overlap at the top and bottom of some flat tiles are primarily used to _____ .
    a. hold the tile in place
    b. space the vertical position of the tile
    c. prevent water penetration under the tile
    d. hold mastic for sealing the tile

24. The attic ventilation area is normally divided between exhaust and inlet vents in the proportion of _____ .
    a. 50 percent exhaust, 50 percent inlet
    b. 60 percent exhaust, 40 percent inlet
    c. 70 percent exhaust, 30 percent inlet
    d. 80 percent exhaust, 20 percent inlet

25. One of the hazards of ice edging is _____.
    a. poor appearance
    b. moisture accumulation
    c. falling ice
    d. water penetration

Trade Terms Introduced in This Module

Asphalt roofing cement: An adhesive that is used to seal down the free tabs of strip shingles. This plastic asphalt cement is mainly used in open valley construction and other flashing areas where necessary for protection against the weather.

Base flashing: The protective sealing material placed next to areas vulnerable to leaks, such as chimneys.

Bundle: A package containing a specified number of shingles or shakes. The number is related to square foot coverage and varies with the product.

Cap flashing: The protective sealing material that overlaps the base and is embedded in the mortar joints of vulnerable areas of a roof, such as a chimney.

Exposure: The distance (in inches) between the exposed edges of overlapping shingles.

Head lap: The distance between the top of the bottom shingle and the bottom edge of the one covering it.

Overhang: The part that extends beyond the building line. The amount of overhang is always given as a projection from the building line on a horizontal plane.

Pitch: The ratio of the rise to the span indicated as a fraction. For example, a roof with a 6' rise and a 24' span will have a ¼ pitch.

Ridge: The horizontal line formed by the two rafters of a sloping roof that have been nailed together. The ridge is the highest point at the top of the roof where the roof slopes meet.

Roof sheathing: Usually 4 × 8 sheets of plywood, but can also be 1 × 8 or 1 × 12 roof boards, or other new products approved by local building codes. Also referred to as decking.

Saddle: An auxiliary roof deck that is built above the chimney to divert water to either side. It is a structure with a ridge sloping in two directions that is placed between the back side of a chimney and the roof sloping toward it. Also referred to as a cricket.

Scrim: A loosely knit fabric.

Selvage: The section of a composition roofing roll or shingle that is not covered with an aggregate.

Side lap: The distance between adjacent shingles that overlap, measured in inches.

Slope: The ratio of rise to run. The rise in inches is indicated for every foot of run.

Square: The amount of shingles needed to cover 100 square feet of roof surface. For example, square means 10' square or 10' × 10'.

Top lap: The distance, measured in inches, between the lower edge of an overlapping shingle and the upper edge of the lapping shingle.

Underlayment: Asphalt-saturated felt protection for sheathing; 15-lb roofer's felt is commonly used. The roll size is 3' × 144' or a little over four squares.

Valley: The internal part of the angle formed by the meeting of two roofs.

Valley flashing: Watertight protection at a roof intersection. Various metals and asphalt products are used; however, materials vary based on local building codes.

Vent stack flashing: Flanges that are used to tightly seal pipe projections through the roof. They are usually prefabricated.

Wall flashing: A form of metal shingle that can be shaped into a protective seal interlacing where the roof line joins an exterior wall. Also referred to as step flashing.

Additional Resources and References

This module is intended to present thorough resources for task training. The following reference works are suggested for further study. These are optional materials for continued education rather than for task training.

Asphalt Manufacturers Association website, *www.asphaltroofing.org*.

National Roofing Contractors Association website, *www.nrca.net*.

Roof Coating Manufacturers Association website, *www.roofcoatings.org*.

NCCER makes every effort to keep these textbooks up-to-date and free of technical errors. We appreciate your help in this process. If you have an idea for improving this textbook, or if you find an error, a typographical mistake, or an inaccuracy in NCCER's Contren® textbooks, please write us, using this form or a photocopy. Be sure to include the exact module number, page number, a detailed description, and the correction, if applicable. Your input will be brought to the attention of the Technical Review Committee. Thank you for your assistance.

Instructors – If you found that additional materials were necessary in order to teach this module effectively, please let us know so that we may include them in the Equipment/Materials list in the Annotated Instructor's Guide.

Write: Product Development and Revision
National Center for Construction Education and Research
3600 NW 43rd St, Bldg G, Gainesville, FL 32606

Fax: 352-334-0932

E-mail: curriculum@nccer.org

Craft

Module Name

Copyright Date

Module Number

Page Number(s)

Description

(Optional) Correction

(Optional) Your Name and Address

68110-09
Exterior Finishing

Topics to be presented in this module include:

1.0.0 Introduction 10.2
2.0.0 Safety 10.2
3.0.0 Insulation and Flashing 10.4
4.0.0 Cornices 10.6
5.0.0 Wood Siding 10.17
6.0.0 Fiber-Cement Siding 10.44
7.0.0 Vinyl and Metal Siding 10.49
8.0.0 Stucco (Cement) Finishes 10.64
9.0.0 Brick and Stone Veneer 10.65
10.0.0 DEFS and EIFS 10.66

Overview

Like roofing materials, there is a wide variety of siding materials used to finish the exteriors of homes and some commercial buildings. They include wood, brick, vinyl, metal, and fiber cement board. Each type of siding has its own preparation requirements and installation practices. The siding is not installed just to make the building attractive. It serves to protect the building from the elements. Improperly installed siding can allow water to penetrate the exterior and damage the interior. Siding material that is not installed per the manufacturer's instructions can also result in voiding the warranty. Another element of exterior finish is the closing of roof overhangs with soffit and fascia. This is work that requires skill and attention to detail.

Objectives

When you have completed this module, you will be able to do the following:

1. Describe the purpose of wall insulation and flashing.
2. Install selected common cornices.
3. Demonstrate lap and panel siding estimating methods.
4. Describe the types and applications of common wood siding.
5. Describe fiber-cement siding and its uses.
6. Describe the types and styles of vinyl and metal siding.
7. Describe the types and applications of stucco and masonry veneer finishes.
8. Describe the types and applications of special exterior finish systems.
9. Install three types of siding commonly used in your area.

Trade Terms

Board-and-batten
Brown coat
Building paper
Cornice
Course
Eave
Fascia
Felt paper
Finish coat
Frieze board
Ledger

Lookout
Louver
Plancier
Rabbet
Rake
R-value
Scratch coat
Shakes
Soffit
Veneer
Vent

Required Trainee Materials

1. Pencil and paper
2. Appropriate personal protective equipment

Prerequisites

Before you begin this module, it is recommended that you successfully complete *Core Curriculum*, and *Construction Technology*, Modules 68101-09 through 68109-09.

This course map shows all of the modules in *Construction Technology*. The suggested training order begins at the bottom and proceeds up. Skill levels increase as you advance on the course map. The local Training Program Sponsor may adjust the training order.

CONSTRUCTION TECHNOLOGY

68117-09
Copper Pipe and Fittings

68116-09
Plastic Pipe and Fittings

68115-09
Introduction to Drain, Waste, and Vent (DWV) Systems

68114-09
Introduction to HVAC

68113-09
Residential Electrical Services

68112-09
Electrical Safety

68111-09
Basic Stair Layout

68110-09
Exterior Finishing

68109-09
Roofing Applications

68108-09
Roof Framing

68107-09
Wall and Ceiling Framing

68106-09
Floor Systems

68105-09
Masonry Units and Installation Techniques

68104-09
Introduction to Masonry

68103-09
Handling and Placing Concrete

68102-09
Introduction to Concrete, Reinforcing Materials, and Forms

68101-09
Site Layout One: Distance Measurement and Leveling

CORE CURRICULUM:
Introductory Craft Skills

103CMAP.EPS

1.0.0 ◆ INTRODUCTION

The primary purpose of any exterior finish is to provide protection from the elements. Some of the most common siding materials are wood, stone, brick, stucco, metal, fiber-cement, and vinyl. Wood, because of its availability and workability, is the most widely used.

Before the installation of any siding, the material to which the siding will be fastened must be made weather resistant. Building paper, house wrap, aluminum paper, or insulating boards are usually installed over lumber, plywood, or chipboard exterior sheathing. For a building wrap, a 4" lap at every horizontal and vertical joint is required. At corners, an 8" lap from both sides is usually recommended. All window and door openings must be wrapped with the material to prevent water penetration.

After the sheathing and water barrier are installed, all boxed cornices, rake sections, windows, and exterior door frames are installed. Exterior window and door trim, if not part of the assembly, should be installed in the same way as interior trim, which is described in another module. For unsheathed structures, jambs and sills of doors and windows should be flashed with either a 6"-wide strip of metal, 3 oz. copper-coated paper, or 6-mil plastic film, as well as other house wrap materials. After any cornices are installed and finished, the siding is applied, and the roof drainage system gutters and downspouts are selected and installed.

Most of the common cornices and wall finishes, along with their installation methods, are covered in this module. Be sure to check the manufacturers' instructions and local building codes, which may require different or more specific construction/installation methods.

2.0.0 ◆ SAFETY

A good carpenter always remembers to put safety first in all situations. Follow all applicable OSHA standards, as well as local and national building codes.

The tools you will be using will most often be of the cutting and hammering type. Power staplers, power hand saws, power or hand sheet metal tin snips, utility knives, and hammers are the most common tools used, and each has its own potential dangers. In most cases, work will be done from the surface of scaffolding or the rung of a ladder, so greater caution is required.

All scaffolding must be placed on a firm footing and leveled. As per OSHA requirements, a ladder must also be placed on a firm footing at the proper

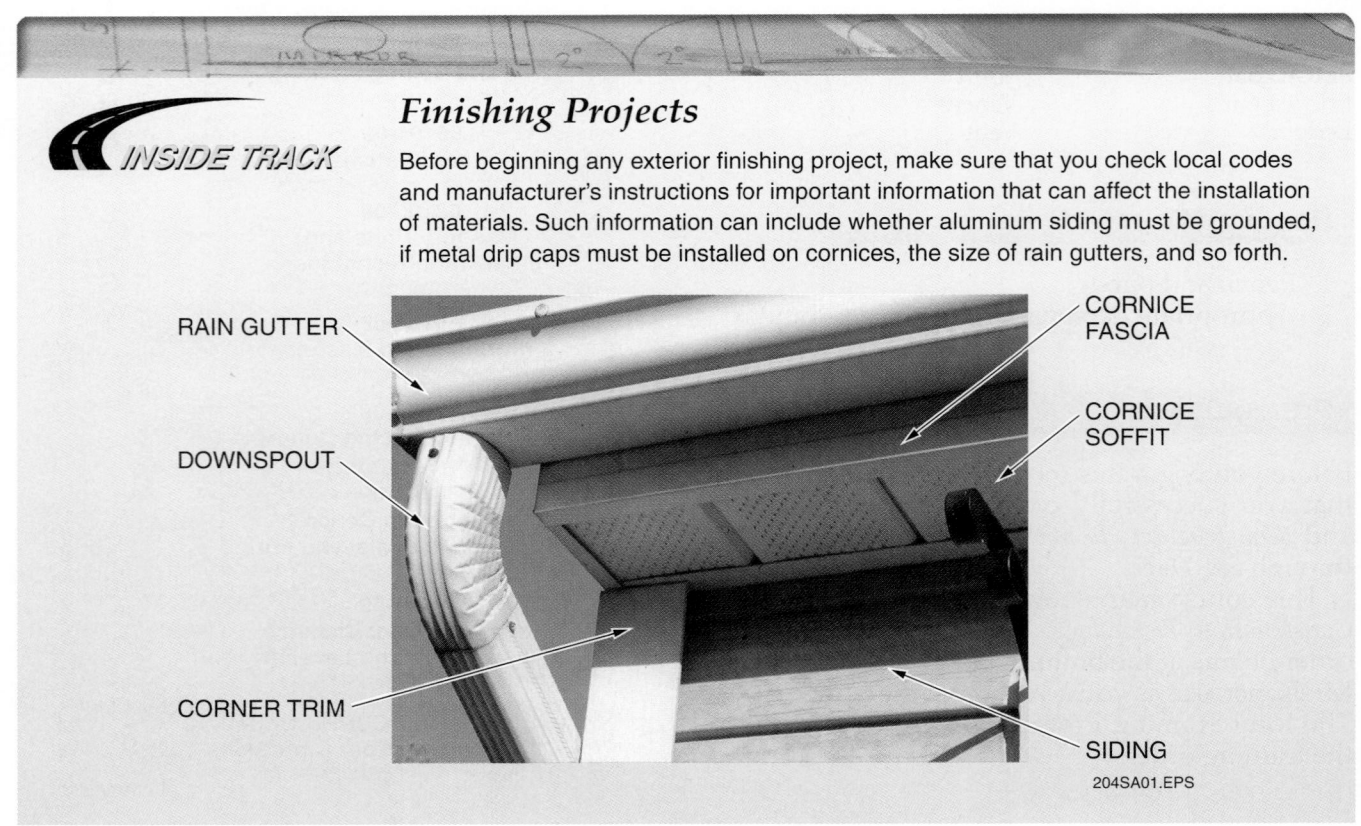

INSIDE TRACK

Finishing Projects

Before beginning any exterior finishing project, make sure that you check local codes and manufacturer's instructions for important information that can affect the installation of materials. Such information can include whether aluminum siding must be grounded, if metal drip caps must be installed on cornices, the size of rain gutters, and so forth.

RAIN GUTTER

DOWNSPOUT

CORNER TRIM

CORNICE FASCIA

CORNICE SOFFIT

SIDING

204SA01.EPS

Building Wrap

One of the more popular building wraps is a spun-bonded olefin material such as ProWrap® or Tyvek®. This material is airtight but breathes to allow water vapor to pass from inside a structure to outside. It is very tough and resists tearing and liquid water penetration. Many local codes do not require building wrap, but it is necessary for a wind and weather-resistant seal. All horizontal and vertical joints must be sealed with a tape approved for that purpose by the manufacturer. Proper installation and taping is especially important to prevent the wrap from being damaged by high winds before the siding is completed.

204SA02.EPS

Scaffolding Systems

Besides sectional, free-standing, manufactured scaffolds that can be assembled in any number of configurations, continuously adjustable aluminum scaffolding systems are available for cornice and siding working heights up to 50'. Typically, an OSHA-recognized system of this type consists of aluminum poles and a standing platform assembly with a safety railing/workbench and safety net. The standing platform can be raised and lowered as an assembly on the supporting poles to obtain the optimum working height. Most systems may be joined both vertically and horizontally for security as well as portability.

204SA03.EPS

angle and be tied off. The usual work clothing may be worn, keeping in mind that loose shirt and trouser cuffs are a hazard. No jewelry should be worn. Work shoes are probably the most important item of clothing. The shoes should be of the safety type, provide good support, and have a thick, non-skid sole for standing on scaffolds and ladders. Always look up for overhead hazards before climbing a ladder. Climbing near power lines must be avoided.

Last, but not least, always consider the weather. The weather conditions are often overlooked, and this can be fatal. When working above the ground, surface wind can be the carpenter's worst enemy. When carrying large pieces of exterior siding, soffit, or fascia boards, make sure you always point the edge into the wind. If wind strikes the flat surface of these items, it could carry you over the edge of the work area and cause serious injuries. Strong winds can also carry away unsecured ladders and scaffolding, causing severe injury to persons working below. Be aware of the danger of snow, ice, and rain. These conditions create slippery surfaces.

 WARNING!

Health precautions must be observed when cutting certain siding materials, including western red cedar, stucco, masonry coatings, pressure-treated lumber, and fiber-cement siding. The dust resulting from cutting or mixing such products can be hazardous to inhale or may be an allergen. Material safety data sheets (MSDS) furnished by the manufacturers of siding materials must be consulted for any applicable hazards before cutting siding products.

3.0.0 ◆ INSULATION AND FLASHING

This section covers insulation and flashing. Insulation materials and installation procedures are fully covered in the module entitled *Thermal and Moisture Protection*.

3.1.0 Insulation

A conventionally insulated house can suffer a 55 percent heat loss either by losing heat through frame walls or by way of air infiltration. Because of the need for energy conservation, the building industry has been working to devise new products and methods of installation to prevent heat loss in buildings.

Heat loss can be significantly reduced by using insulation sheathing. There are several brands and types of insulation sheathing. These include Dow Styrofoam™, Celotex Tuff-R™ and Celotex Thermax™ sheathing, expanded polystyrene, fiberboard sheathing, Monsanto Fome-Cor™, and thermo-ply foil-faced paper board. When selecting a particular insulating sheathing, the **R-value** and local building codes should be considered.

Installation procedures for insulation sheathing may vary, depending on the type of sheathing selected for use. Insulation sheathing is nonstructural. Adequate corner bracing, such as diagonal 1×4 let-in wood bracing, flat or profiled steel bracing, or plywood at corners overlaid with foam sheathing, should be used to comply with local building codes. Dow Styrofoam™ residential sheathing is laminated with a durable plastic film for added resistance to damage and abuse. This sheathing comes with tongue-and-groove edges on ¾" to 1" thicknesses. Celotex foam sheathings (Tuff-R™ and Thermax™) can be easily cut with a utility knife to any shape needed to conform to irregular wall angles or to fit snugly around window or door openings and other projections. Tuff-R™ insulation sheathing is semirigid and can bend around corners to reduce air infiltration.

Siding may be applied directly over the insulating sheathing. Brick, wood, hardboard, aluminum, and vinyl siding are fastened to the wood frame construction by nailing through the sheathing. Care must be taken when driving nails so that the sheathing is not crushed.

Shakes and shingles can also be applied by installing furring strips or a plywood nailer base over the insulating sheathing. A stucco finish can also be applied over an acceptable lath fastened over the sheathing.

Many other factors must be considered when insulating an exterior wall where exterior finish is to be applied. Windows are available with insulating glass. The insulating glass consists of two or three pieces of glass with an air space between them. The edges are sealed and the air space is then turned into a partial vacuum. All materials in a wall system are rated for various insulation and sound values, and different combinations can achieve large differences in insulation and acoustic characteristics.

Aluminum and plastic siding are available with an insulation board backing. This aids in insulating an exterior wall, but is not effective by itself.

3.2.0 Flashing

Before the siding is nailed in place or before the masonry veneer is laid up, the flashing must be installed around all openings. Flashing usually consists of galvanized sheet metal, aluminum, or a synthetic material; however, on rare occasions, copper and stainless steel have been used. Normally, aluminum is not used for flashing masonry because of corrosion problems. The primary purpose of flashing is to prevent water that may penetrate the siding from eventually entering the exterior walls and causing rot, water damage of interior surfaces, or mildew. If water does penetrate the siding, flashing is designed to channel it back out again, thus avoiding any water damage. See *Figure 1*.

When brick or stone veneer is used for a frame building, flashing is installed at the base of the sheathing and above door and window openings to channel the water to the outside through weep holes. Frame construction at the water table also requires that flashing be used, as illustrated in *Figure 1*.

Some metal-covered or vinyl-covered windows and doors are manufactured with a flashing flange at the tops and sides and do not require separately installed flashing; however, it is usually a good practice to install flashing as a precaution.

Box, Closed, and Open Cornices

Of the three types of cornices, box cornices are the most commonly used. A closed cornice is the least desirable type of cornice because it does not allow roof ventilation and provides little protection to the side of the building. Open cornice construction is not commonly used except on porches with directly exposed roof beams. If a ceiling is installed, the roof area above any ceiling cannot be ventilated unless vertical vents through the frieze board are provided. Depending on the rafter tail cut, fascia is sometimes applied to the rafter ends of open cornices to provide better weather protection to the ends of the rafters.

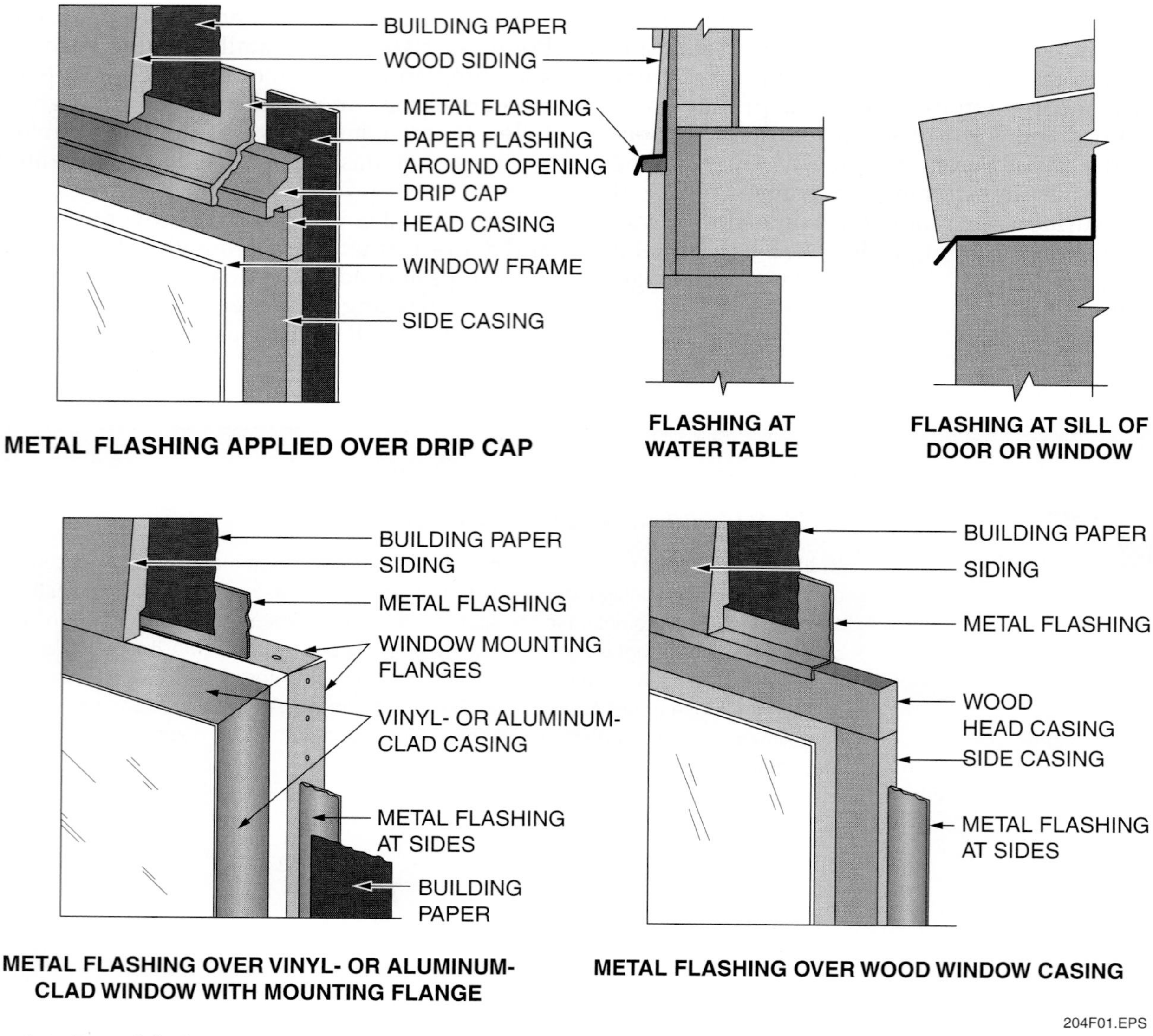

BUILDING PAPER

WOOD SIDING

METAL FLASHING

PAPER FLASHING
AROUND OPENING

DRIP CAP

HEAD CASING

WINDOW FRAME

SIDE CASING

METAL FLASHING APPLIED OVER DRIP CAP

**FLASHING AT
WATER TABLE**

**FLASHING AT SILL OF
DOOR OR WINDOW**

BUILDING PAPER
SIDING

METAL FLASHING

WINDOW MOUNTING
FLANGES

VINYL- OR ALUMINUM-
CLAD CASING

METAL FLASHING
AT SIDES

BUILDING
PAPER

**METAL FLASHING OVER VINYL- OR ALUMINUM-
CLAD WINDOW WITH MOUNTING FLANGE**

BUILDING PAPER

SIDING

METAL FLASHING

WOOD
HEAD CASING

SIDE CASING

METAL FLASHING
AT SIDES

METAL FLASHING OVER WOOD WINDOW CASING

204F01.EPS

Figure 1 ◆ Typical flashing installation.

4.0.0 ◆ CORNICES

Cornices are constructed of lumber, as well as aluminum, vinyl, and other man-made products. The type of cornice required for a particular structure is shown on the wall sections of the construction drawings. The three general types of cornices are the closed cornice, open cornice, and box cornice.

A roof with no rafter overhang normally has a closed cornice. See *Figure 2*. This cornice consists of a single strip called a frieze board. The frieze board is beveled on its upper edge to fit close under the overhang of the eaves and rabbeted on its lower edge to overlap the upper edge of the top siding course. A strip of wood shingles is used to provide a roof overhang several inches beyond the molding. In this instance, the strip can serve as both a drip edge and starter strip for a wood shingle roof or as a support for the starter strip of an asphalt shingle roof. Some codes may require the

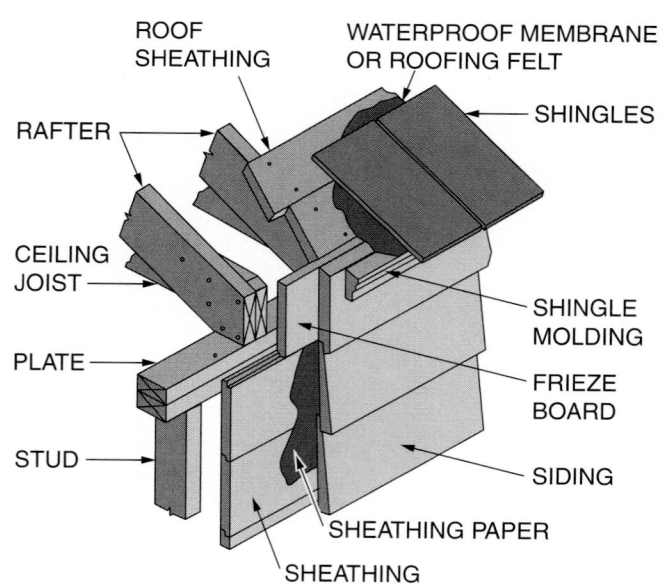

ROOF
SHEATHING

WATERPROOF MEMBRANE
OR ROOFING FELT

SHINGLES

RAFTER

CEILING
JOIST

PLATE

STUD

SHINGLE
MOLDING

FRIEZE
BOARD

SIDING

SHEATHING PAPER

SHEATHING

Figure 2 ◆ Closed cornice.

204F02.EPS

installation of a metal drip cap on the edge of the strip of shingles. If trim is used, it usually consists of molding installed as shown in *Figure 2*.

A roof with a rafter overhang may have either an open cornice or a box cornice. The simplest type of open cornice consists of only a frieze board cut to fit between the rafters, as shown in *Figure 3*. If trim is used, it usually consists of molding cut to fit between the rafters.

Another type of open cornice consists of a frieze board and a fascia (*Figure 3*). A fascia is a strip nailed to the tail cuts of the rafters. Shingle molding can be attached to the top of the fascia, but it is seldom used. *Figure 4* shows five types of tail rafter cuts.

With a box cornice, the rafter overhang is entirely boxed in by the roof covering, the fascia, and a bottom strip called a **plancier** or soffit. *Figure 5* shows examples of various types of box cornices.

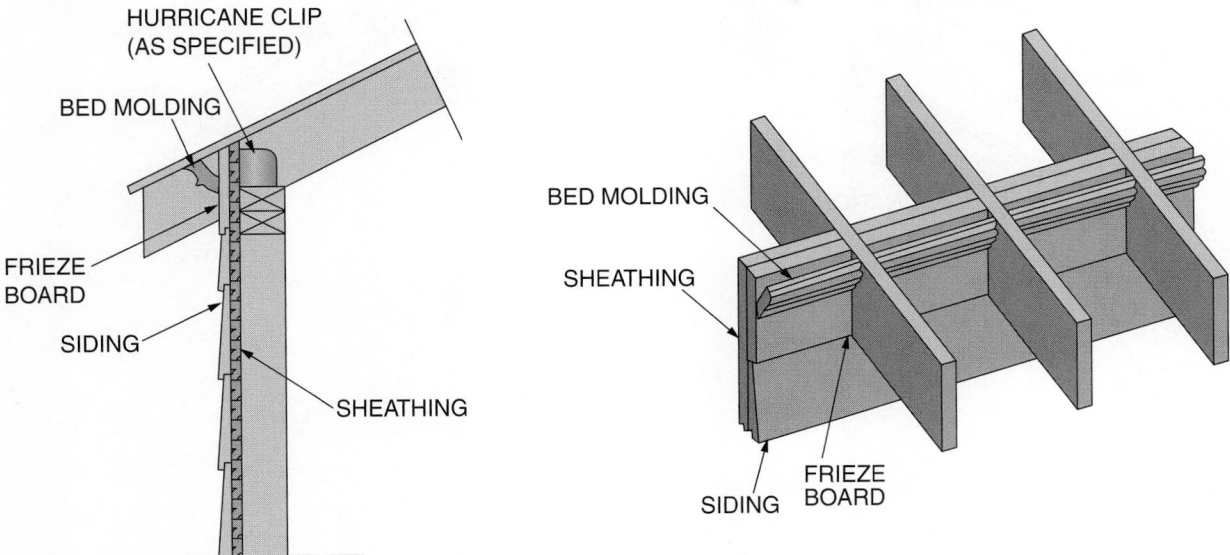

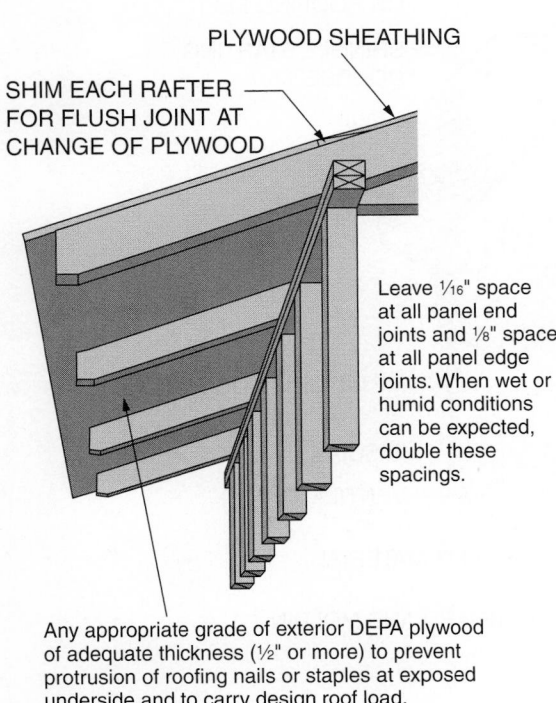

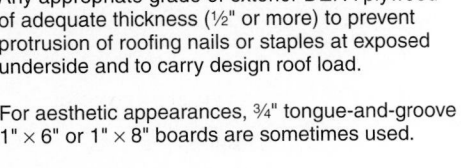

Figure 3 ◆ Open cornices.

204F03.EPS

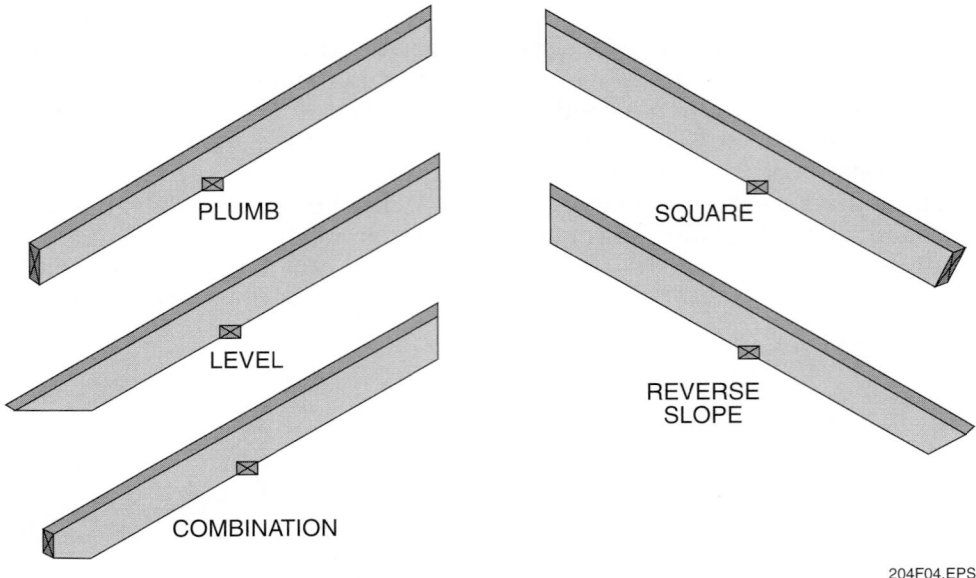

Figure 4 ◆ Types of tail rafter cuts.

204F04.EPS

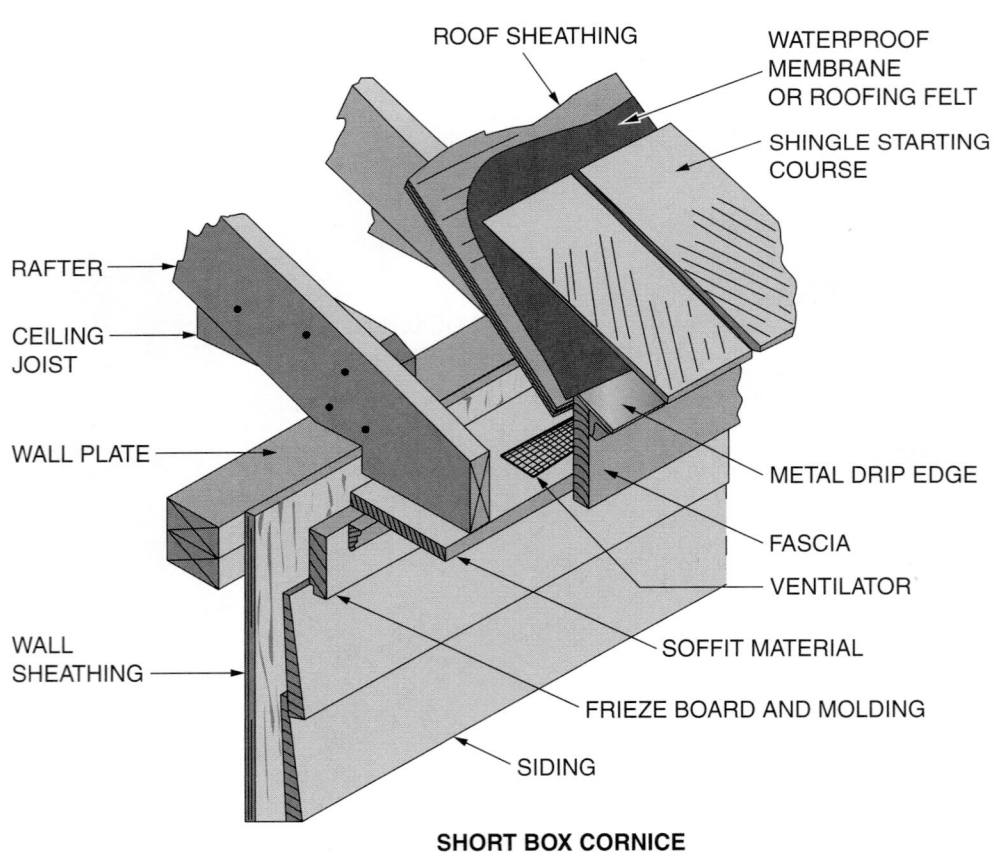

SHORT BOX CORNICE

204F05A.EPS

Figure 5 ◆ Box cornices (1 of 5).

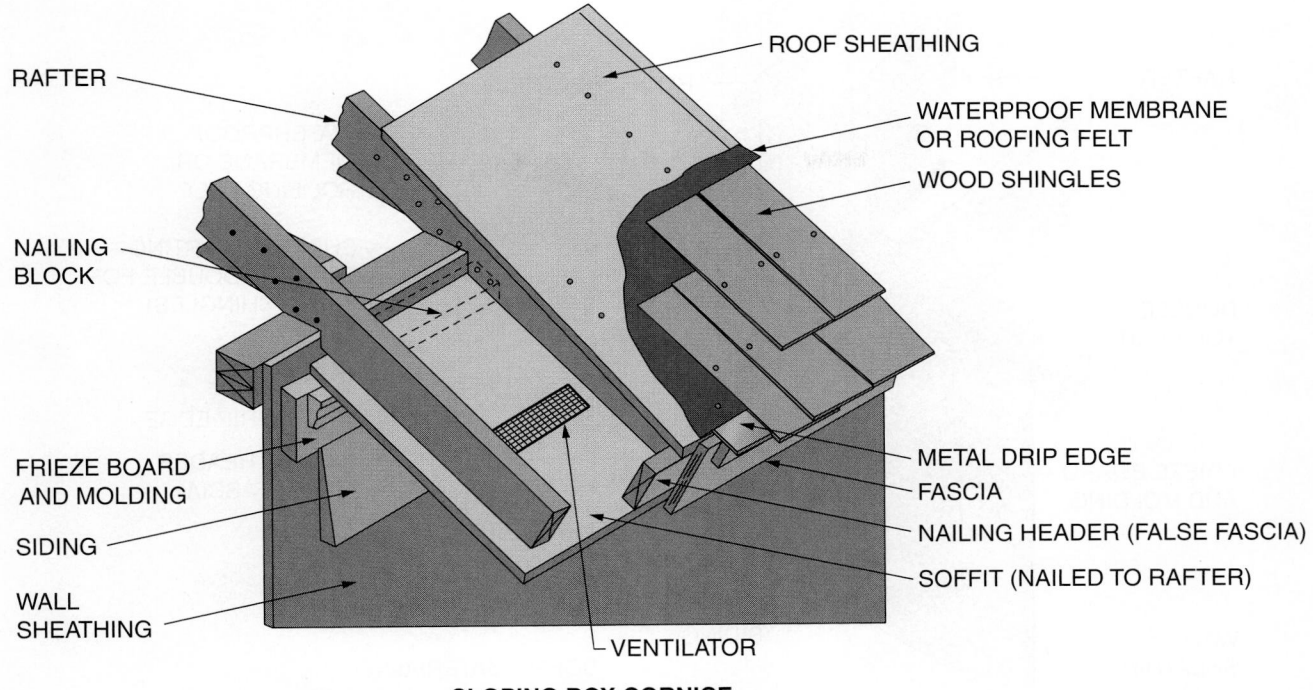

RAFTER

ROOF SHEATHING

WATERPROOF MEMBRANE
OR ROOFING FELT

WOOD SHINGLES

NAILING
BLOCK

FRIEZE BOARD
AND MOLDING

SIDING

WALL
SHEATHING

METAL DRIP EDGE

FASCIA

NAILING HEADER (FALSE FASCIA)

SOFFIT (NAILED TO RAFTER)

VENTILATOR

SLOPING BOX CORNICE

204F05B.EPS

Figure 5 ◆ Box cornices (2 of 5).

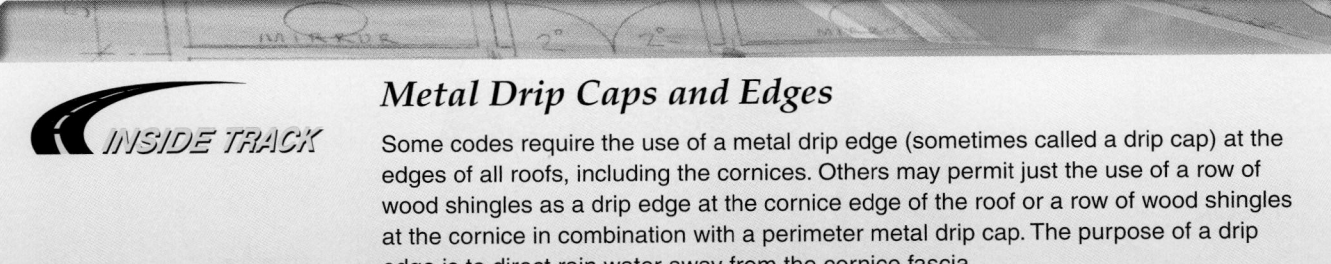

INSIDE TRACK

Metal Drip Caps and Edges

Some codes require the use of a metal drip edge (sometimes called a drip cap) at the edges of all roofs, including the cornices. Others may permit just the use of a row of wood shingles as a drip edge at the cornice edge of the roof or a row of wood shingles at the cornice in combination with a perimeter metal drip cap. The purpose of a drip edge is to direct rain water away from the cornice fascia.

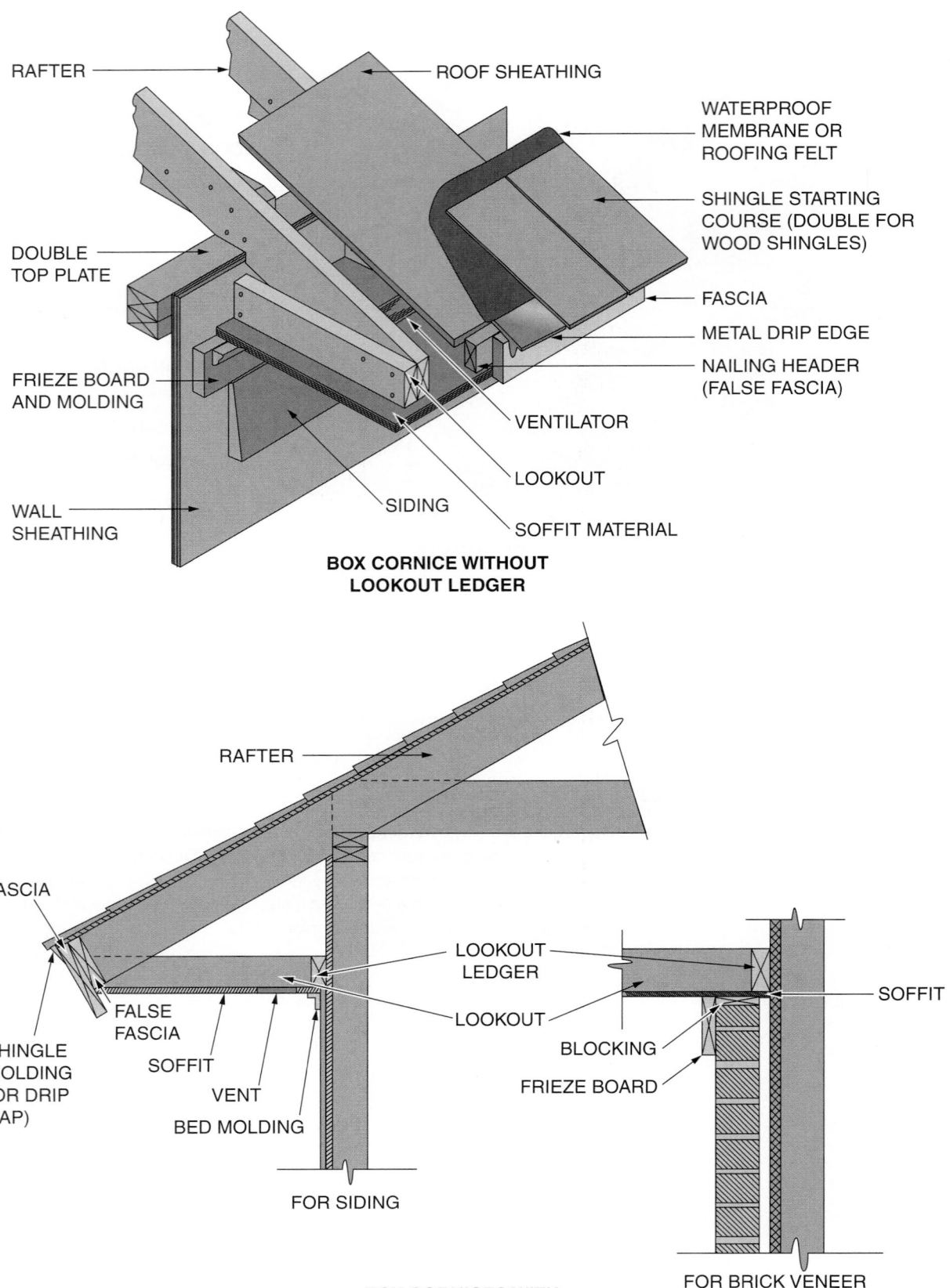

RAFTER

ROOF SHEATHING

WATERPROOF MEMBRANE OR ROOFING FELT

SHINGLE STARTING COURSE (DOUBLE FOR WOOD SHINGLES)

DOUBLE TOP PLATE

FASCIA

METAL DRIP EDGE

NAILING HEADER (FALSE FASCIA)

FRIEZE BOARD AND MOLDING

VENTILATOR

LOOKOUT

WALL SHEATHING

SIDING

SOFFIT MATERIAL

BOX CORNICE WITHOUT LOOKOUT LEDGER

RAFTER

FASCIA

LOOKOUT LEDGER

SOFFIT

LOOKOUT

SHINGLE MOLDING (OR DRIP CAP)

FALSE FASCIA

SOFFIT

BLOCKING

FRIEZE BOARD

VENT

BED MOLDING

FOR SIDING

FOR BRICK VENEER

BOX CORNICES WITH LOOKOUT LEDGER

204F05C.EPS

Figure 5 ◆ Box cornices (3 of 5).

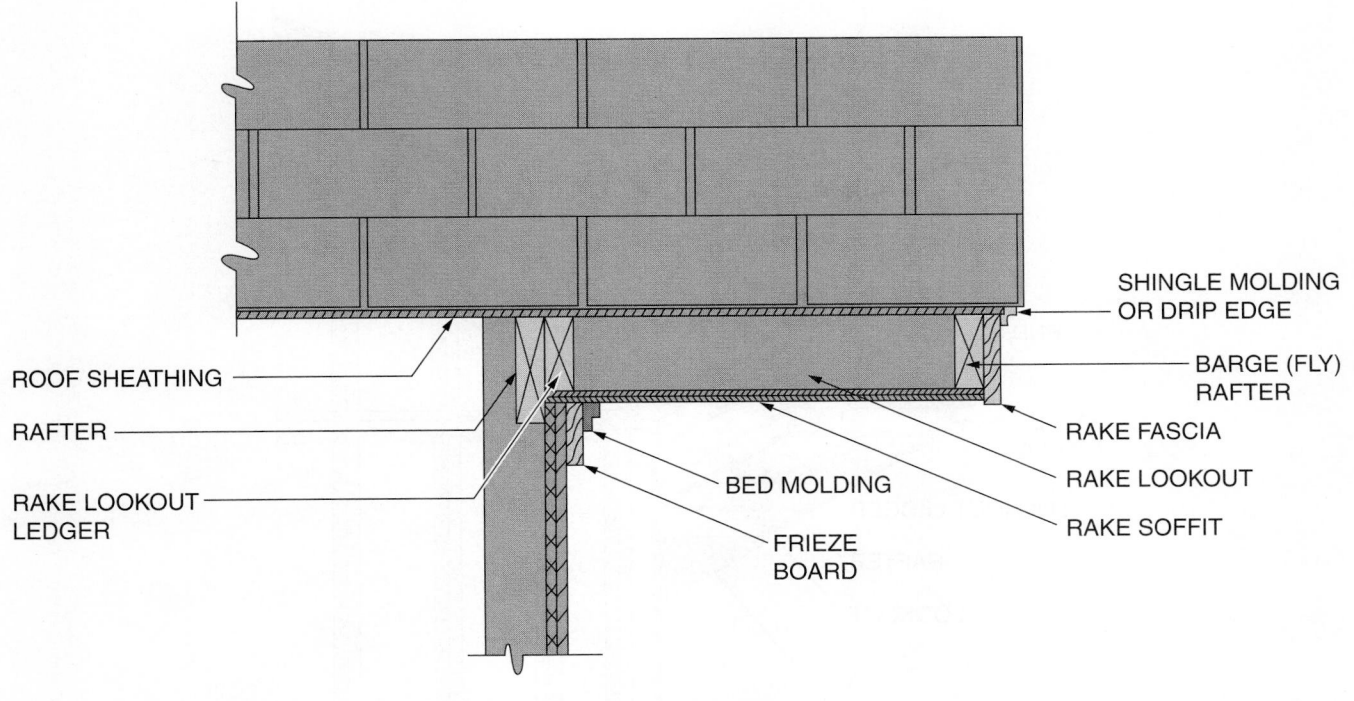

ROOF SHEATHING

RAFTER

RAKE LOOKOUT
LEDGER

BED MOLDING

FRIEZE
BOARD

SHINGLE MOLDING
OR DRIP EDGE

BARGE (FLY)
RAFTER

RAKE FASCIA

RAKE LOOKOUT

RAKE SOFFIT

RAKE SECTION FOR SIDING

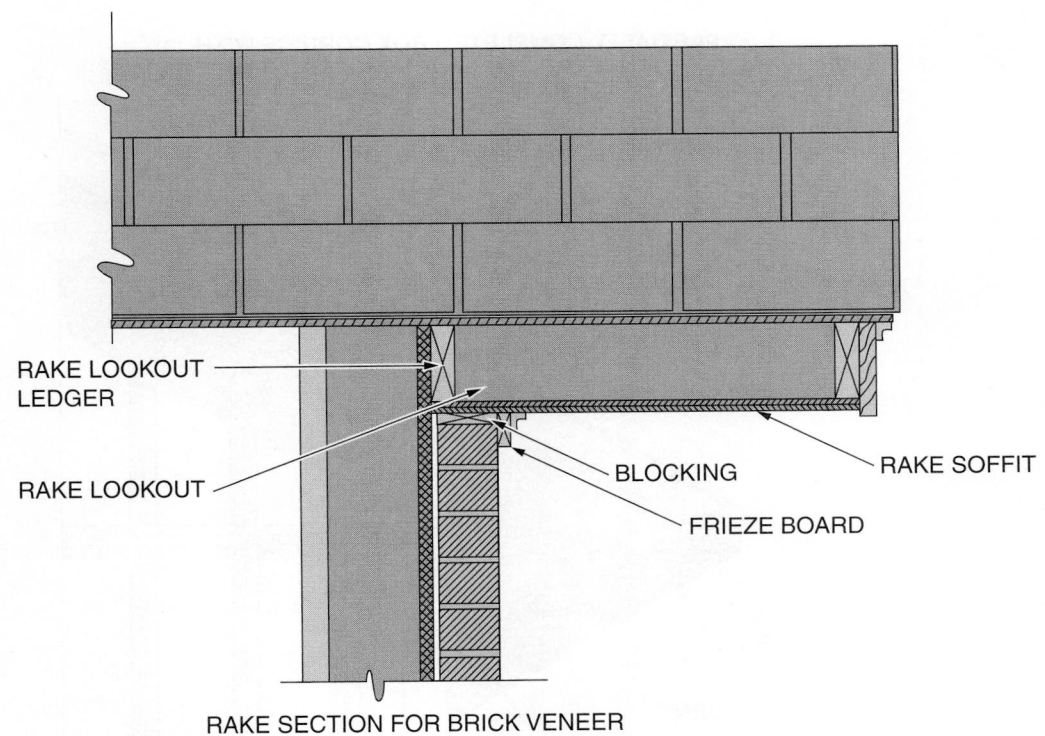

RAKE LOOKOUT
LEDGER

RAKE LOOKOUT

BLOCKING

FRIEZE BOARD

RAKE SOFFIT

RAKE SECTION FOR BRICK VENEER

BOX CORNICE RAKE SECTIONS

204F05D.EPS

Figure 5 ◆ Box cornices (4 of 5).

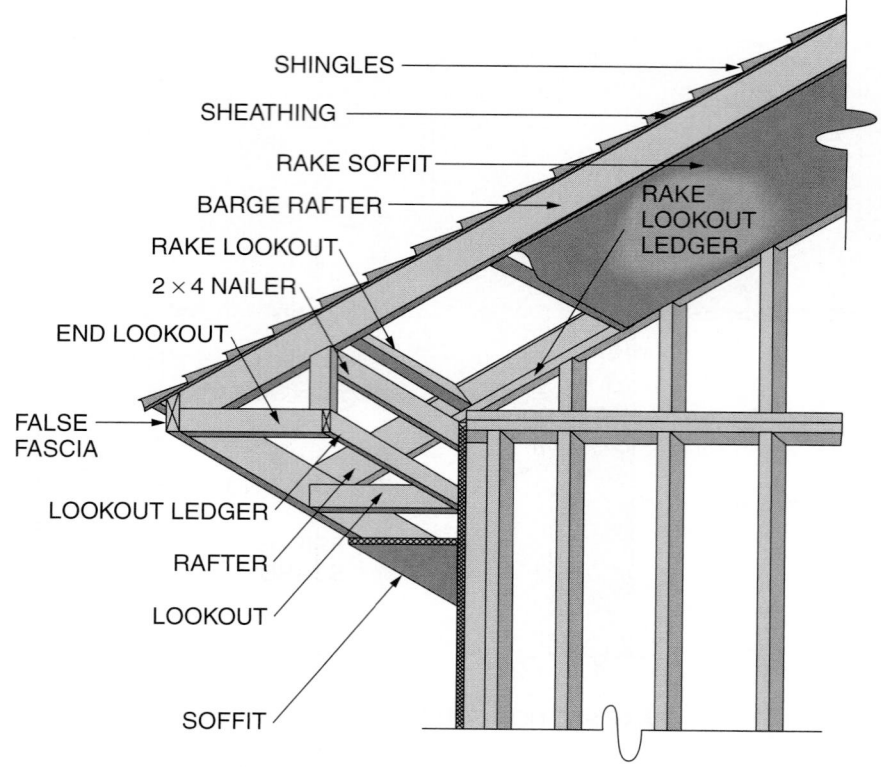

SHINGLES

SHEATHING

RAKE SOFFIT

BARGE RAFTER

RAKE LOOKOUT

2 × 4 NAILER

END LOOKOUT

FALSE FASCIA

LOOKOUT LEDGER

RAFTER

LOOKOUT

SOFFIT

RAKE LOOKOUT LEDGER

**PARTIALLY COMPLETED BOX CORNICE WITH
CORNICE RETURN AND RAKE SECTION**

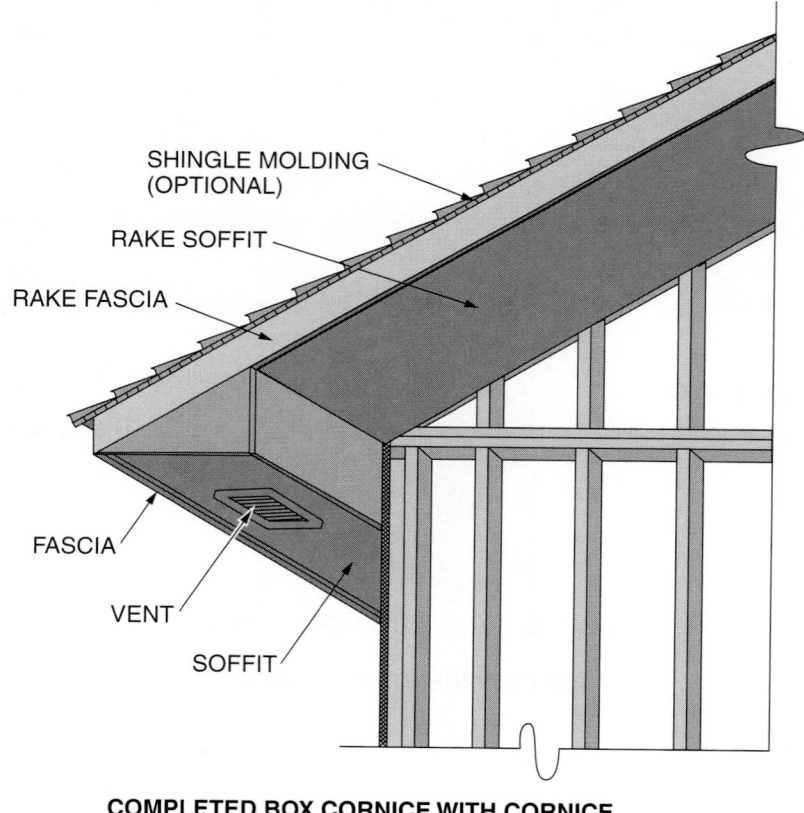

SHINGLE MOLDING
(OPTIONAL)

RAKE SOFFIT

RAKE FASCIA

FASCIA

VENT

SOFFIT

**COMPLETED BOX CORNICE WITH CORNICE
RETURN AND RAKE SECTION**

204F05E.EPS

Figure 5 ◆ Box cornices (5 of 5).

Box cornices use wood, aluminum, vinyl, or exterior gypsum materials for soffits and sometimes for fascia. *Figure 6* shows various types of trim molding used on cornices.

4.1.0 Cornice Installation

The fasteners used in wood cornice work normally consist of various sizes and types of nails. Common steel nails and cement-coated box nails are used in areas where they will be completely enclosed and protected from the weather.

Hot-dipped galvanized nails are commonly used to fasten any cornice member exposed to the weather. These nails should be of sufficient length to hold the material in place and may be any of the types commonly manufactured.

Aluminum and stainless steel nails are available for use on exterior trim to eliminate rust streaks. Stainless steel nails are used on the highest-quality work where the added cost of the nails is incidental.

Figure 7 is a nail spacing chart for plywood used as a closed soffit. Many times, the failure of plywood to maintain a level surface after installation is not due to a product failure, but is caused by improper installation.

The tools used in the installation of the cornice and related finish are the same as those used for rough and finish carpentry.

The following are guidelines for building a box cornice:

Step 1 Start by marking the location of the **ledger** (*Figure 8*), then snap a chalkline to ensure the ledger will be level. Secure the ledger to the wall. The ledger makes it easier to install the **lookouts**.

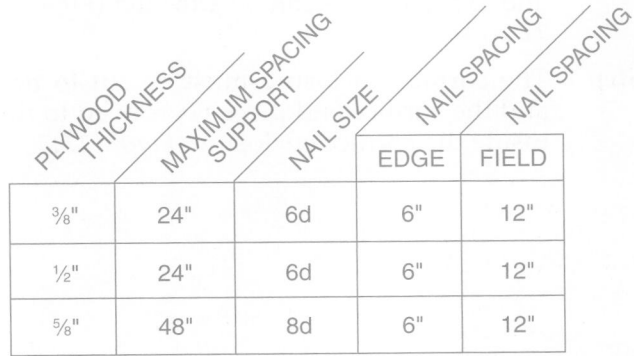

PLYWOOD THICKNESS	MAXIMUM SPACING SUPPORT	NAIL SIZE	NAIL SPACING EDGE	NAIL SPACING FIELD
³⁄₈"	24"	6d	6"	12"
½"	24"	6d	6"	12"
⁵⁄₈"	48"	8d	6"	12"

204F07.EPS

Figure 7 ◆ Nail spacing chart.

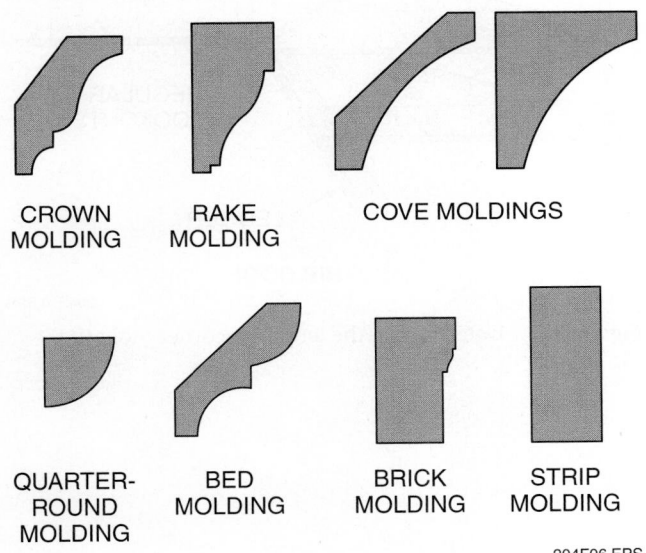

CROWN MOLDING

RAKE MOLDING

COVE MOLDINGS

QUARTER-ROUND MOLDING

BED MOLDING

BRICK MOLDING

STRIP MOLDING

204F06.EPS

Figure 6 ◆ Types of wood cornice trim molding.

MARK HERE

204F08.EPS

Figure 8 ◆ Determining the ledger position.

Stainless Steel Nails

While aluminum nails are rustproof, they may be corroded by certain chemicals in industrial or high vehicular traffic environments. Quality stainless steel nails are not affected by any environment.

Step 2 Use a straight board to mark the lookout positions on the ledger (*Figure 9*).

Step 3 Measure the distance from the face of the ledger to the ends of the rafters to determine the length of the lookouts.

Step 4 Trim the corner lookouts to match the slope of the rafters (*Figure 10*).

Step 5 Attach the lookouts to the ledger, then install the assembly, nailing through the sheathing into studs where possible.

Step 6 Once the lookouts are in place, the soffit and fascia boards can be attached (*Figures 11* and *12*).

Step 7 The cornice enclosure must be cut to fit, and the cornice end piece is grooved to fit inside the cornice enclosure (*Figure 13*).

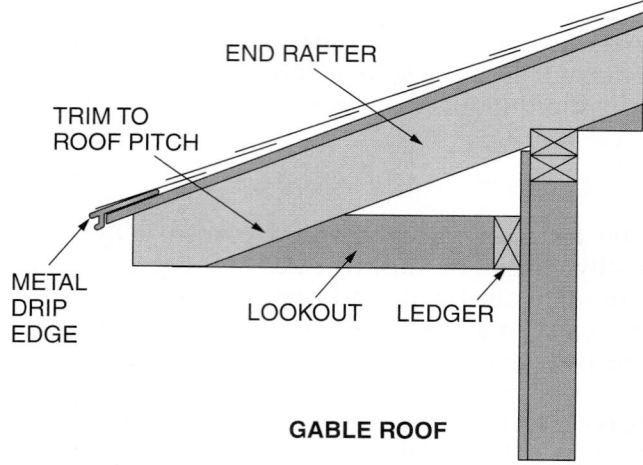

GABLE ROOF

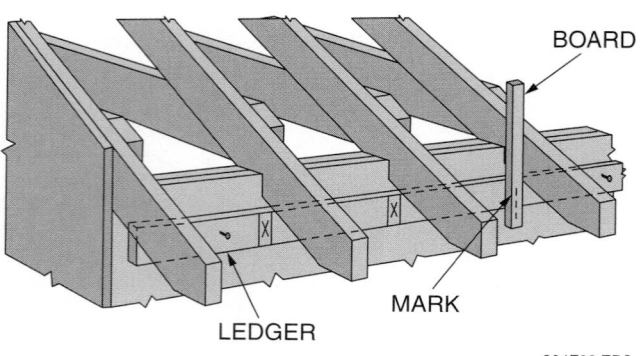

204F09.EPS

Figure 9 ◆ Marking lookout locations on a ledger.

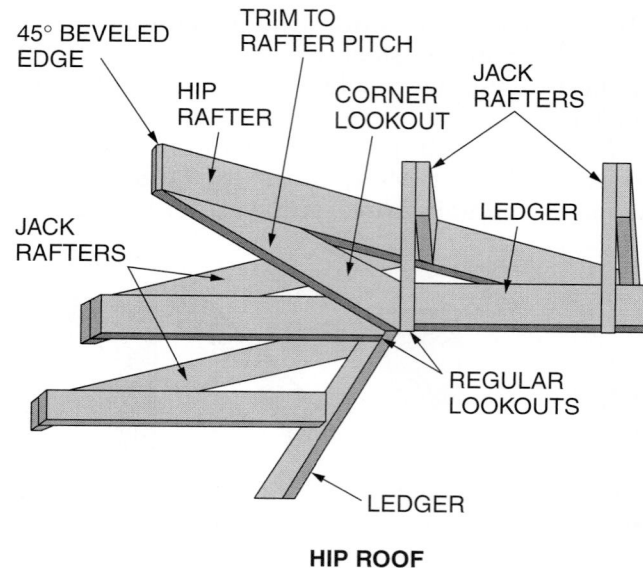

HIP ROOF

204F10.EPS

Figure 10 ◆ Determining the length of corner lookouts.

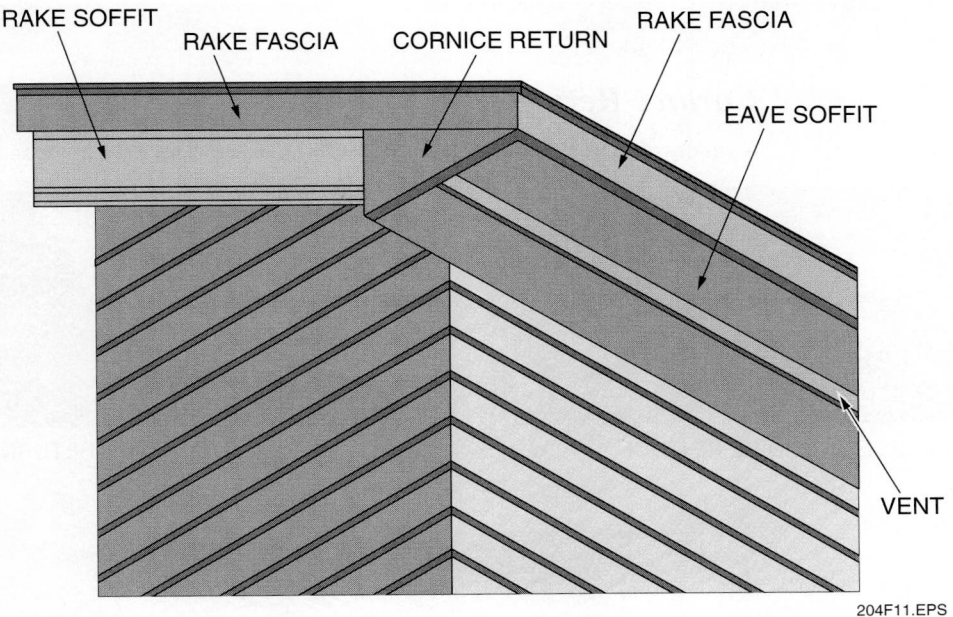

Figure 11 ◆ Cornice return.

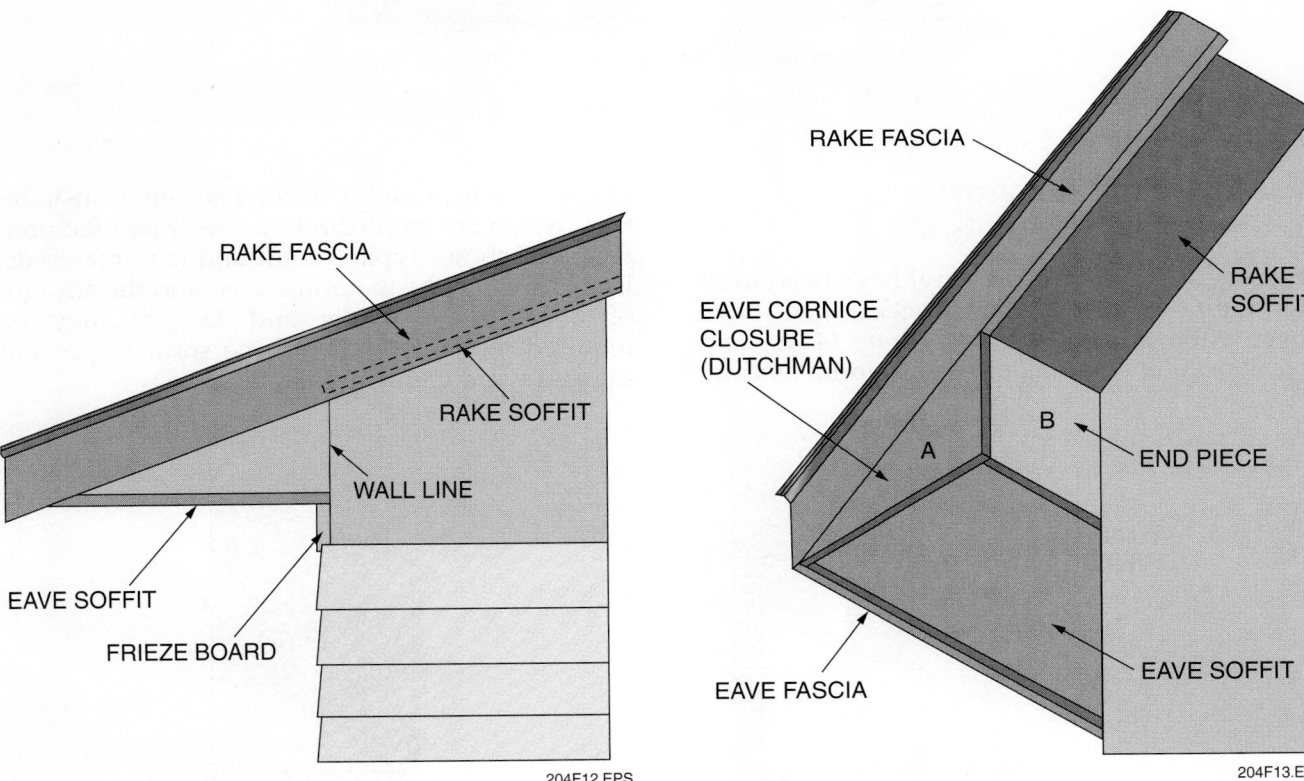

Figure 12 ◆ Rake soffit cut to wall line.

Figure 13 ◆ Boxing in the cornice return.

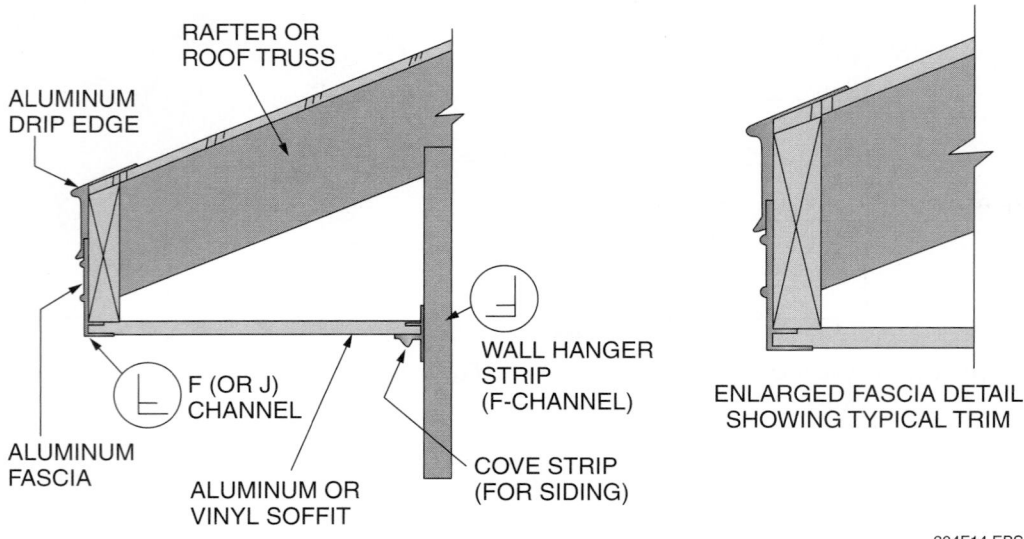

Cornice Returns

INSIDE TRACK

A variety of more complicated cornice returns exists, as shown in these illustrations.

REVERSED CORNICE RETURN

ANGLED CORNICE RETURNS

COLONIAL MOLDING CORNICE RETURN

204SA05.EPS

4.2.0 Aluminum or Vinyl Fascia and Soffits

Prefinished aluminum and vinyl have been used extensively in cornice construction and have proven to be very satisfactory. *Figure 14* shows a typical installation detail of an aluminum or vinyl fascia eave trim and a soffit. The actual installation will vary according to the manufacturer. *Figure 15* shows typical soffit and trim materials. Depending on the materials used and the amount of overhang, lookouts and ledgers may be required for the support of the soffit to prevent sag and wind damage.

RAFTER OR ROOF TRUSS

ALUMINUM DRIP EDGE

F (OR J) CHANNEL

WALL HANGER STRIP (F-CHANNEL)

ALUMINUM FASCIA

ALUMINUM OR VINYL SOFFIT

COVE STRIP (FOR SIDING)

ENLARGED FASCIA DETAIL SHOWING TYPICAL TRIM

204F14.EPS

Figure 14 ◆ Typical aluminum/vinyl soffit and trim installation.

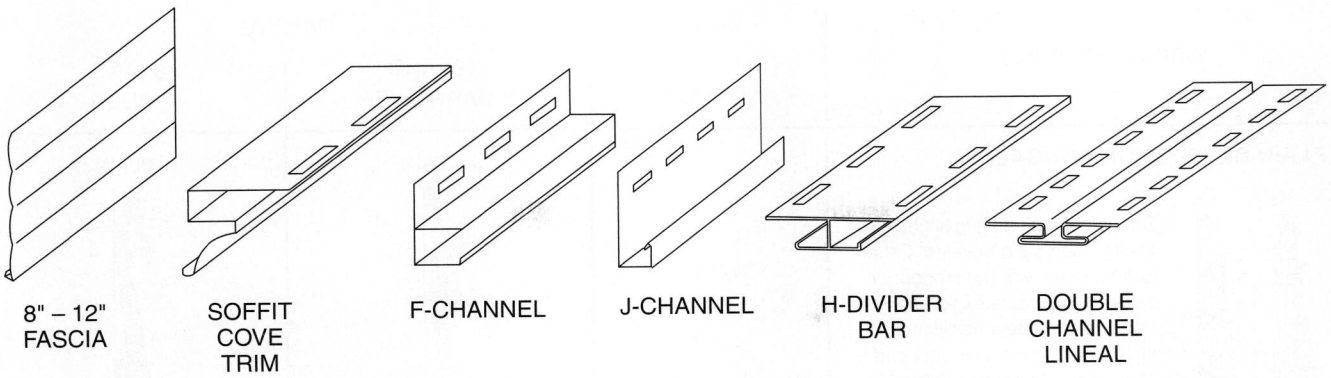

8" – 12"
FASCIA

SOFFIT
COVE
TRIM

F-CHANNEL

J-CHANNEL

H-DIVIDER
BAR

DOUBLE
CHANNEL
LINEAL

EACH SOFFIT PANEL PIECE IS CUT TO THE DESIRED WIDTH AND THEN SLID INTO PLACE FROM THE END
OF THE CORNICE. EACH PIECE INTERLOCKS WITH THE ADJOINING PIECES.

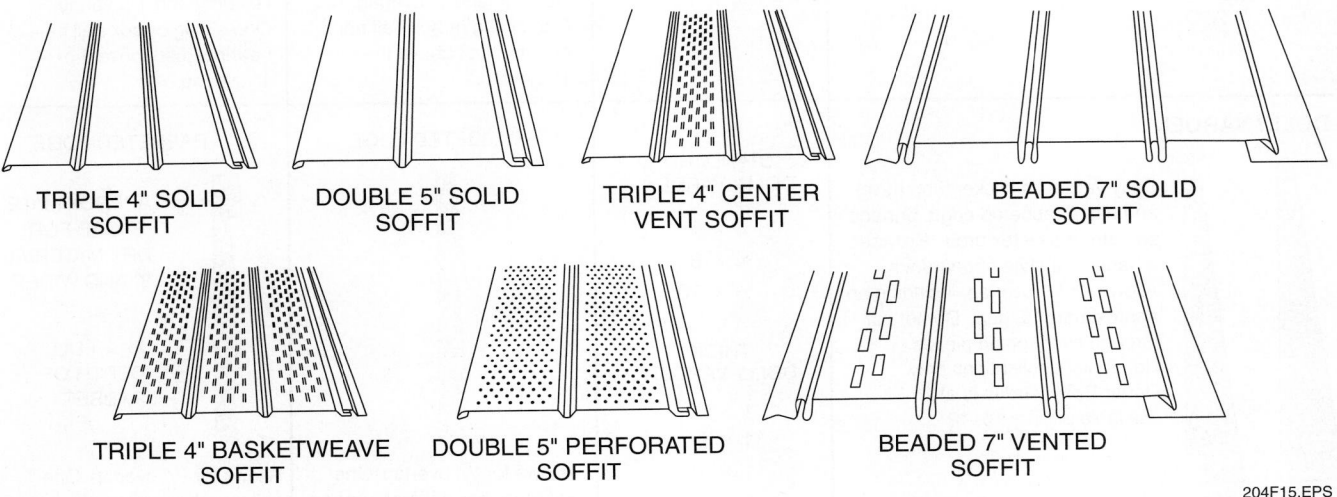

TRIPLE 4" SOLID
SOFFIT

DOUBLE 5" SOLID
SOFFIT

TRIPLE 4" CENTER
VENT SOFFIT

BEADED 7" SOLID
SOFFIT

TRIPLE 4" BASKETWEAVE
SOFFIT

DOUBLE 5" PERFORATED
SOFFIT

BEADED 7" VENTED
SOFFIT

204F15.EPS

Figure 15 ◆ Typical aluminum/vinyl soffit and trim materials.

5.0.0 ◆ WOOD SIDING

The woods used most often for siding are western red cedar (WRC), bald cypress, douglas fir, western hemlock, western larch, ponderosa pine, red pine, southern white pine, sugar pine, and redwood. These woods are shaped into many different siding styles. *Figure 16* is a summary of the most common styles, sizes, and nailing patterns.

Siding, casing, box, finish, ring, or spiral-shanked stainless steel or steel nails with hot-dipped galvanized or noncorrosive coatings (see *Figure 17*) are commonly used to apply wood siding. The size will vary from 6d to 10d. The siding

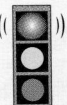

WARNING!

The dust from western red cedar is an allergen and can cause respiratory ailments including asthma and rhinitis. It can also cause eye irritation and skin disorders, including dermatitis, itching, and rashes. Avoid inhaling the dust or getting it on your skin or in your eyes.

nail is considered the best nail for wood siding except in high-wind areas. Then, a wider-headed face nail is required.

SIDING PATTERNS	SIZES (THICKNESS AND WIDTH)	NAILING	
		6" AND NARROWER	8" AND WIDER

PLAIN BEVELED OR BUNGALOW

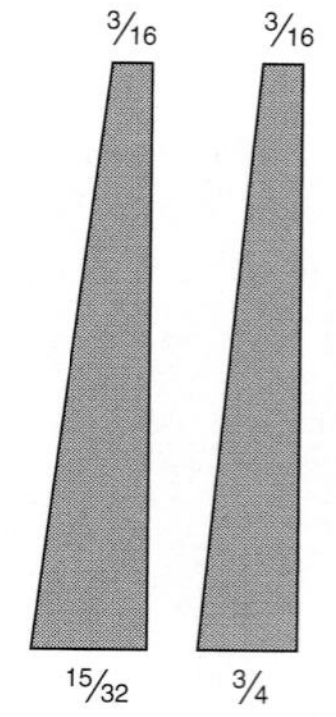

3/16   3/16

15/32   3/4

Bungalow (colonial) is slightly thicker than plain beveled. Either can be used with the smooth or saw-faced surface exposed. Patterns provide a traditional-style appearance. Recommend a 1" overlap. Do not nail through overlapping pieces. Horizontal applications only. Cedar bevel is also available in 7/8" × 10, 12.

Sizes:
½ × 4
½ × 5
½ × 6

⅝ × 8
⅝ × 10

¾ × 6
¾ × 8
¾ × 10

PLAIN

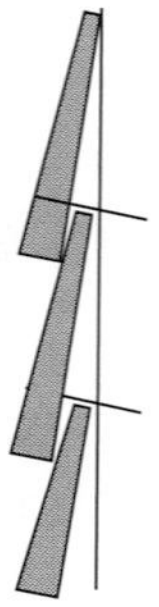

Recommend 1" overlap. One siding or box nail per bearing, just above the 1" overlap.

PLAIN

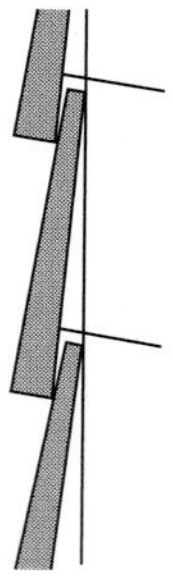

Recommend 1" overlap. One siding or box nail per bearing, just above the 1" overlap.

DOLLY VARDEN

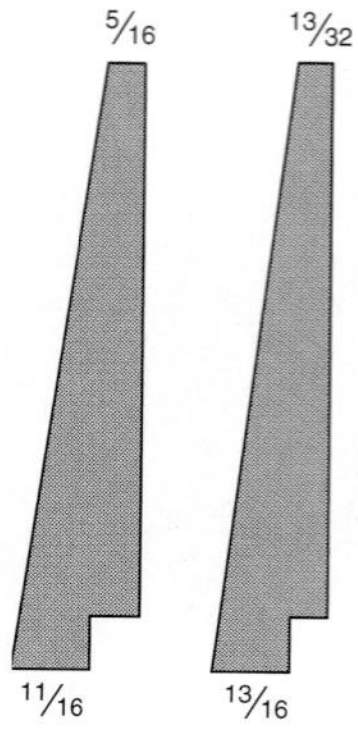

5/16   13/32

11/16   13/16

Dolly Varden is thicker than bevel and has a rabbeted edge. Surface smooth or saw textured. Provides a traditional-style appearance. Allows for ½" overlap, including an approximate ⅛" gap. Do not nail through overlapping pieces. Horizontal applications only. Cedar Dolly Varden is also available in 7/8" × 10, 12.

STANDARD DOLLY VARDEN
¾ × 6
¾ × 8
¾ × 10

THICK DOLLY VARDEN
1 × 6
1 × 8
1 × 10
1 × 12

RABBETED EDGE

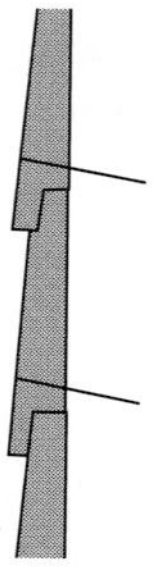

Allows for ½" overlap. One siding or box nail per bearing, 1" up from bottom edge.

RABBETED EDGE

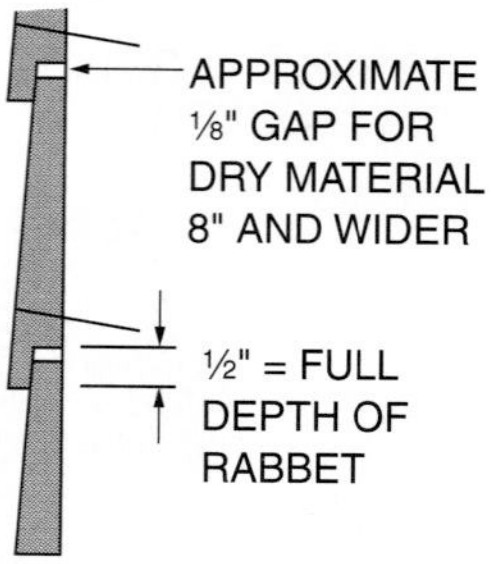

APPROXIMATE ⅛" GAP FOR DRY MATERIAL 8" AND WIDER

½" = FULL DEPTH OF RABBET

Allows for ½" overlap. One siding or box nail per bearing, 1" up from bottom edge.

TONGUE-AND-GROOVE (T&G)

T&G siding is available in a variety of patterns. T&G lends itself to different effects aesthetically. Refer to WWPA "standard patterns" (G-16) for pattern profiles. Sizes given are for plain T&G. Do not nail through overlapping pieces. Vertical, diagonal, or horizontal applications.

1 × 4
1 × 6
1 × 8
1 × 10

NOTE: T&G PATTERNS MAY BE ORDERED WITH ¼", ⅜", OR 7/8" TONGUES. FOR WIDER WIDTHS, SPECIFY THE LONGER TONGUE AND PATTERN.

PLAIN

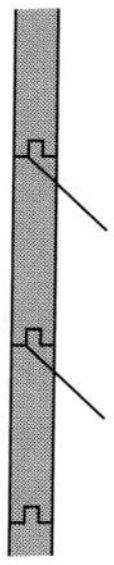

Use one casing nail per bearing to blind nail.

PLAIN

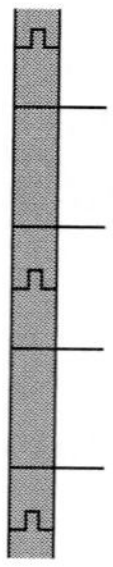

Use two siding or box nails 3" – 4" apart to face nail.

204F16A.EPS

Figure 16 ◆ Common wood siding styles (1 of 2).

SIDING PATTERNS		SIZES (THICKNESS AND WIDTH)	NAILING	
			6" AND NARROWER	8" AND WIDER

DROP

Drop siding is available in 13 patterns, in smooth, rough, and saw-textured surfaces. Some are T&G (as shown), others are shiplapped. Refer to WWPA "standard patterns" (G-16) for pattern profiles with dimensions. A variety of looks can be achieved with different patterns. Do not nail through overlapping pieces. Horizontal or vertical applications.

¾ × 6
¾ × 8
¾ × 10

T&G PATTERN   SHIPLAP PATTERN

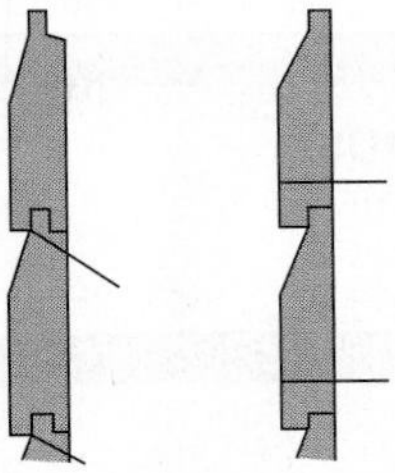

Use casing nails to blind nail T&G patterns, one nail per bearing. Use siding or box nails to face nail shiplap patterns 1" up from bottom edge.

T&G PATTERN   SHIPLAP PATTERN

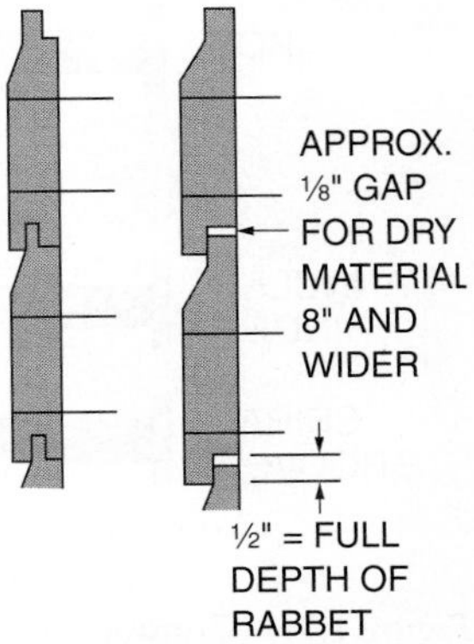

APPROX. ⅛" GAP FOR DRY MATERIAL 8" AND WIDER

½" = FULL DEPTH OF RABBET

Use two siding or box nails 3" – 4" apart to face nail starting 1" up from bottom edge.

CHANNEL RUSTIC

Channel rustic has a ½" overlap (including an approximate ⅛" gap) and a 1" to 1¼" channel when installed. The profile allows for maximum dimensional change without adversely affecting appearance in climates of highly variable moisture levels between seasons. Available smooth, rough, or saw-textured. Do not nail through overlapping pieces. Vertical, diagonal, or horizontal applications.

¾ × 6
¾ × 8
¾ × 10

Use one siding or box nail to face nail once per bearing, 1" up from bottom edge.

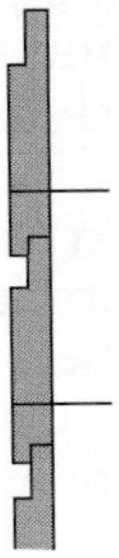

APPROXIMATE ⅛" GAP FOR DRY MATERIAL 8" AND WIDER

½" = FULL DEPTH OF RABBET

Use two siding or box nails 3" – 4" apart to face nail starting 1" up from bottom edge.

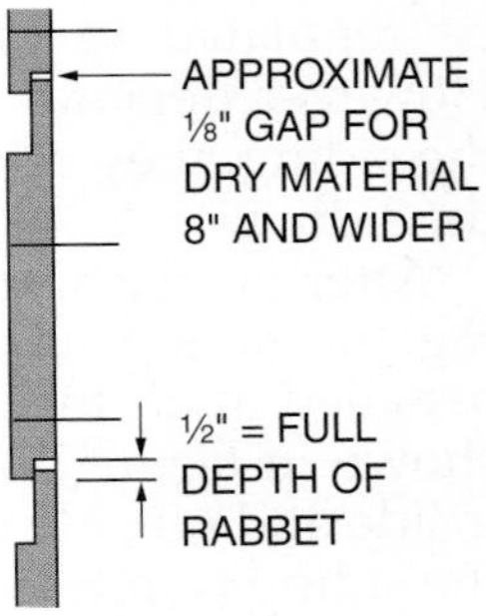

LOG CABIN

Log cabin siding is 1½" thick at the thickest point. Ideally suited to informal buildings in rustic settings. The pattern may be milled from appearance grades (commons) or dimensional grades (2× material). Allows for ½" overlap, including an approximate ⅛" gap. Do not nail through overlapping pieces. Vertical or horizontal applications.

1½ × 6
1½ × 8
1½ × 10
1½ × 12

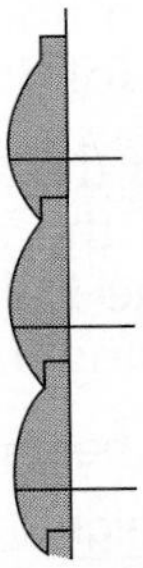

Use one siding or box nail to face nail once per bearing, 1½" up from bottom edge.

APPROXIMATE ⅛" GAP FOR DRY MATERIAL 8" AND WIDER

½" = FULL DEPTH OF RABBET

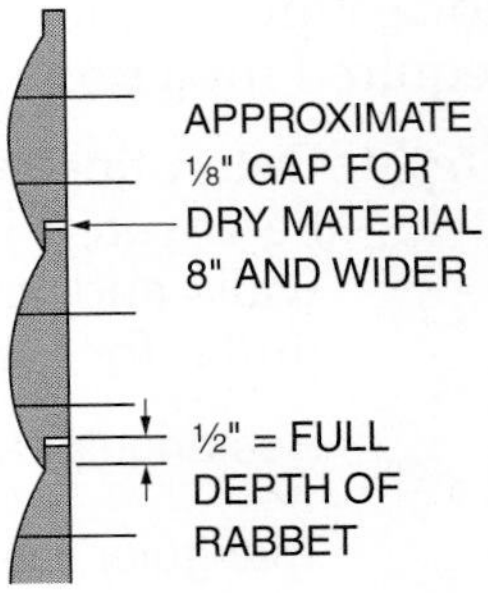

Use two siding or box nails 3" – 4" apart, per bearing to face nail starting 1½" up from bottom edge.

204F16B.EPS

Figure 16 ◆ Common wood siding styles (2 of 2).

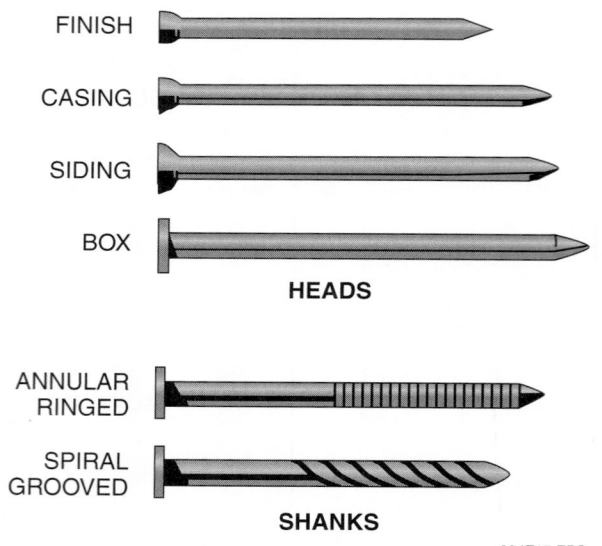

FINISH

CASING

SIDING

BOX

HEADS

ANNULAR
RINGED

SPIRAL
GROOVED

SHANKS

204F17.EPS

Figure 17 ◆ Commonly used nails.

Pneumatic nailers can be used with a flush-mount attachment or if the pressure of the tool can be regulated to prevent overdriving the nails. However, depending on the type and thickness of the siding, they often cause excessive splitting of the siding.

After the flashing is installed and before applying the siding, inside corner strips are usually installed at all inside corners of the structure, as shown in *Figure 18*. In some more costly projects, inside and outside corner pieces are not used and the siding is mitered to fit.

5.1.0 Estimating Procedure for Panel and Board Siding

To estimate the amount of siding material required for a project, proceed as follows:

Step 1 Determine the total area of the structure to be covered by adding up the areas of all walls and gables using the following formulas for the area of a rectangle or triangle:

Rectangular area = width × height

Triangular area = $\dfrac{\text{width} \times \text{height}}{2}$

Step 2 Subtract the total area of all openings.

Step 3 For wood board-type siding, add the waste percentages for the size and lap, as listed in *Table 1*. For metal or vinyl siding, add 10 percent.

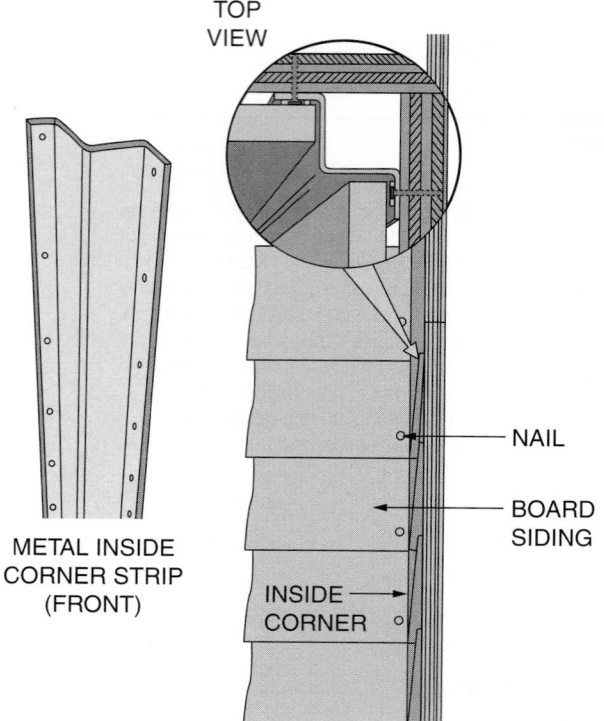

TOP
VIEW

METAL INSIDE
CORNER STRIP
(FRONT)

NAIL

BOARD
SIDING

INSIDE
CORNER

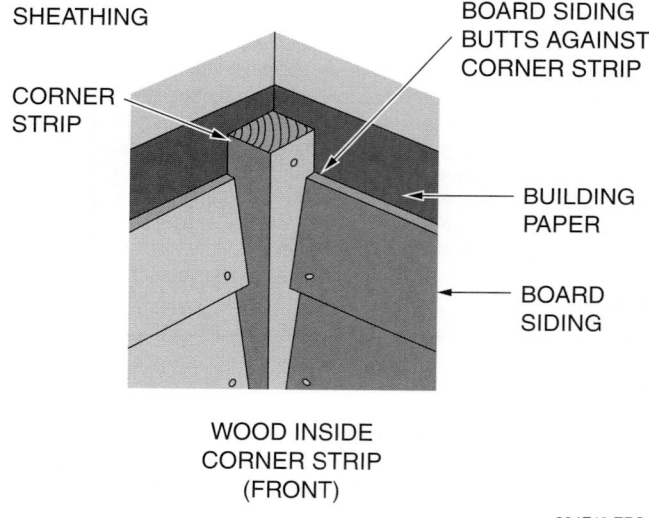

SHEATHING

BOARD SIDING
BUTTS AGAINST
CORNER STRIP

CORNER
STRIP

BUILDING
PAPER

BOARD
SIDING

WOOD INSIDE
CORNER STRIP
(FRONT)

204F18.EPS

Figure 18 ◆ Wood or metal inside corner strips.

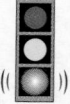 **NOTE**

Except for large openings such as a garage or sliding doors, omit Step 2 for 2 × 8 or 4 × 8 sheet material.

Table 1 Waste Allowances

Siding	Size and Lap	Percent to Add
Beveled	1 × 4 − ¾"	45 percent
	1 × 5 − ⅞"	38 percent
	1 × 6 − 1"	33 percent
	1 × 8 − 1¼"	33 percent
	1 × 10 − 1½"	29 percent
	1 × 12 − 1½"	23 percent
Drop siding and rustic (shiplapped)	1 × 4	28 percent
	1 × 5	21 percent
	1 × 6	19 percent
	1 × 8	16 percent
Drop siding and rustic (dressed and matched)	1 × 4	23 percent
	1 × 5	18 percent
	1 × 6	16 percent
	1 × 8	14 percent
Triangular areas or diagonal installation*		10 percent

*The 10 percent is in addition to other allowances.

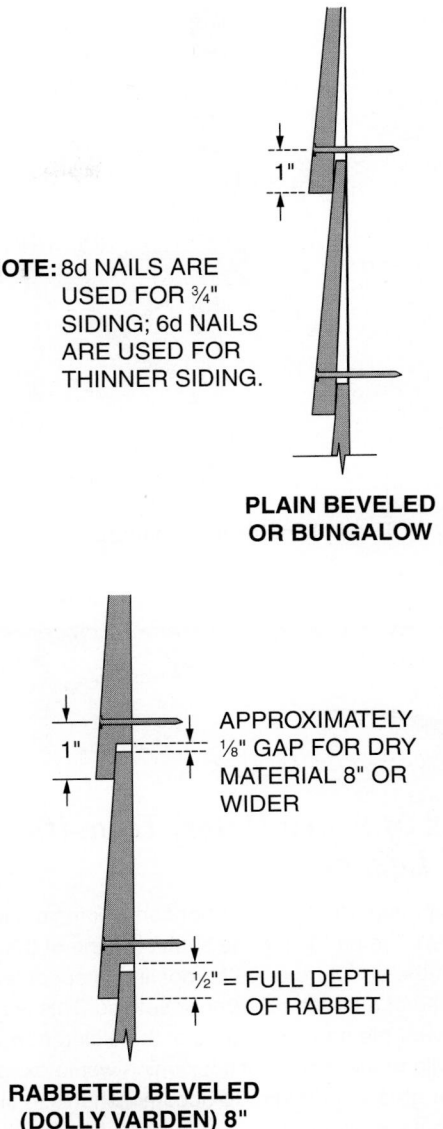

NOTE: 8d NAILS ARE USED FOR ¾" SIDING; 6d NAILS ARE USED FOR THINNER SIDING.

PLAIN BEVELED OR BUNGALOW

APPROXIMATELY ⅛" GAP FOR DRY MATERIAL 8" OR WIDER

½" = FULL DEPTH OF RABBET

RABBETED BEVELED (DOLLY VARDEN) 8" AND WIDER

204F19.EPS

Figure 19 ◆ Beveled siding.

Step 4 To determine the number of squares of board-type siding required, divide the total area plus the waste percentage by 100 sq ft. Round up to the next whole square.

$$\text{Number of squares} = \frac{\text{Total area} + \text{waste percentage}}{100}$$

Step 5 For 2 × 8 or 4 × 8 panel siding, divide the total area by the area of a panel to determine the number of panels required. Round up to the next whole panel.

$$\text{Number of } 2 \times 8 \text{ panels} = \frac{\text{total area}}{16}$$

$$\text{Number of } 4 \times 8 \text{ panels} = \frac{\text{total area}}{32}$$

5.2.0 Beveled Siding

Beveled siding is a pattern most often associated with traditional architecture, but it can also be used with success in contemporary structures. Beveled siding comes in plain, bungalow (colonial), and rabbeted (Dolly Varden) styles (see *Figure 19*). Plain beveled and bungalow styles each produce a strong shadow line. Rabbeted beveled siding provides a somewhat snugger lap and lays up faster, with a greater coverage than beveled siding. The surfaced side is normally used for painted finishes and the rough side for natural finishes and a more informal look.

Beveled siding is face nailed with one siding nail per bearing (8d for ¾" siding and 6d for thinner siding), so that the shank of the nail clears the tip of the under course. Allow ⅛" above the tip for expansion. Lap beveled siding by at least 1" and rabbeted beveled siding by ½". Use aluminum nails, hot-dipped galvanized nails, or stainless steel nails. You will need about 1 lb of 6d nails per 100 square feet of siding or 1½ lbs of 8d nails per 100 square feet of siding. Do not nail through the overlap (*Figure 20*).

204F20.EPS

Figure 20 ◆ Example of incorrect nailing.

As mentioned previously, plain or bungalow beveled siding must have a minimum lap of at least 1". However, it may be larger to allow for spacing adjustment.

5.2.1 Beveled Siding Installation

Before installing beveled siding, make a story pole that is as long as the height of the wall. Draw a line about 1" below the top of the foundation, but at least 6" above grade level. Follow these guidelines to install the siding:

• Use a divider set at the siding exposure to mark the story pole, then transfer the marks to the wall at the corners, windows, and doors. Make sure the story pole is plumb before marking the wall.

◤ *INSIDE TRACK*

Uses of Story Poles, Transits, and Lasers

It is very important that the horizontal lines of siding on a structure are parallel to the soffit/roof line of the structure even if the structure or soffit/roof line is out of level because of a structural error or settling. This is necessary to prevent the horizontal lines of the siding from being at an angle to the soffit/roof line, which would exaggerate the out-of-level problem. To accomplish this, a siding story pole, measured from the soffits, is typically used to establish vertical location marks for the siding at all corners and at all door and window openings of a structure. The marking is done after application of the external building insulation/wrap and before the application of any siding.

If the soffit/roof lines of the structure are level, a transit can be used as an alternative to set the lower edge of the siding at the corners and intermediate points. In this case, a story pole only needs to be used to place the siding marks up the structure.

On long walls or gable ends of a structure, a transit can be used to set the lower edge of the siding at intermediate points even if the soffit/roof lines are out-of-level. This is accomplished by setting the transit to an out-of-level position (in the plane parallel to the wall) that allows the transit cross hairs to sweep across the same story pole location mark at both outside corners of the wall. However, the transit must be level in the plane perpendicular to the wall.

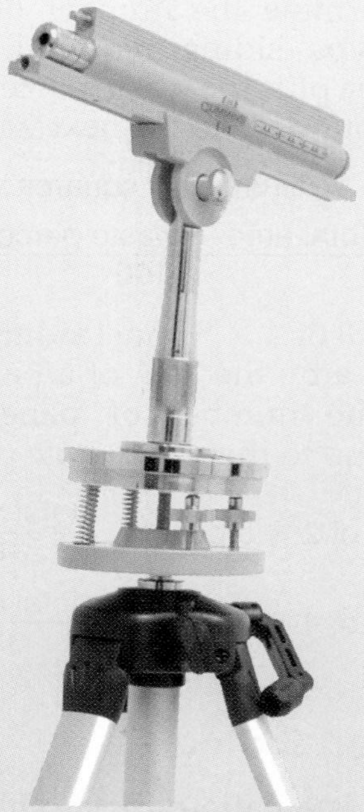

204SA06.EPS

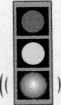

NOTE

The exposure may be adjusted, as required, so that the spacing to the top comes out even, as long as a minimum overlap of 1" is observed. If possible, also adjust spacing so single pieces of siding will run continuously above and/or below the majority of windows or other wall openings without requiring notching or small slivers of siding above or below the opening.

- It is a good idea to snap a chalkline through the siding location marks on long sides of the building, so that each piece of siding may be nailed to a mark. This will ensure that the siding will have a straight, neat appearance, devoid of waves and sags. If desired, a spacing gauge can be used to place siding, as shown in *Figure 21*. However, a chalkline should be used as a reference.

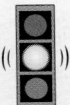

CAUTION

Before installing siding, make sure that the sheathing and siding are dry. If excessive moisture is present during the application, later drying and shrinking may cause end gaps and stress warping of the siding.

- Start by placing the bottom course of siding around the perimeter of the building.

INSIDE TRACK

Siding Reference Marks

It may be desirable to set nails at the siding reference marks so that a line can be attached to them for siding alignment. They can also be used for a chalkline if that is the method used to mark a reference line for siding alignment.

NOTE

When using plain beveled siding, place a furring strip behind the bottom edge of the starting course. The strip should be approximately the thickness of the top 1" of the siding. This will allow the siding to project the same distance that the next course of siding will project above it (refer to *Figure 21*). Also, it is important that the lowest edge of the siding be at least 6" above the ground level. The high humidity and free water often present at the base of a foundation because of landscaping can cause finish difficulties and structural problems. It is also very important that the end grain of the siding butt joints and the bottom edge of the first course of siding be given a water-repellent treatment.

- When placing the remainder of the siding, do not allow the butt joints to align vertically. Tight-fitting butt joints can be obtained by cutting the siding about $\frac{1}{16}$" longer than the measurement. Bow the piece slightly to position the ends, then snap into place.
- If the gable ends will use a different type of finish siding, provide a siding juncture as shown in *Figure 22*.

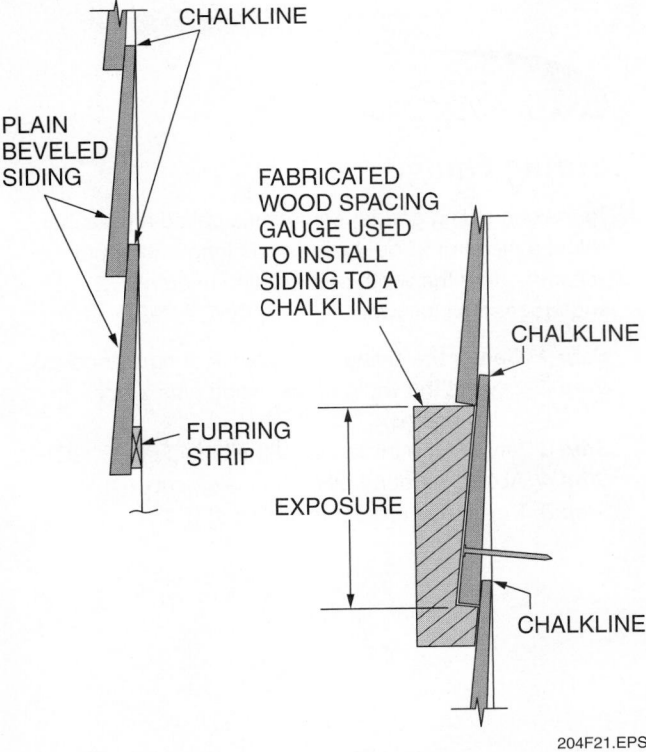

Figure 21 ◆ Installation of plain beveled siding.

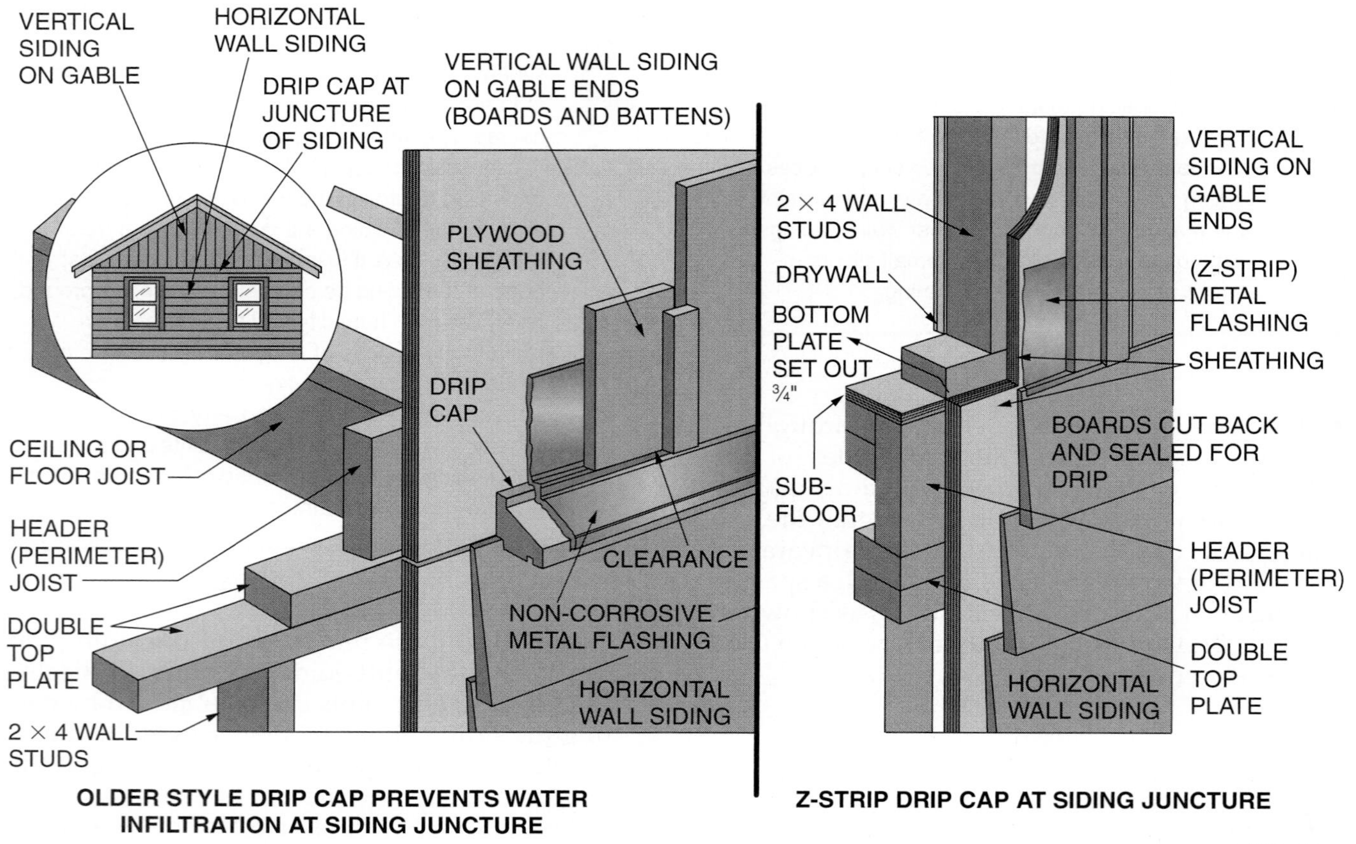

OLDER STYLE DRIP CAP PREVENTS WATER INFILTRATION AT SIDING JUNCTURE

Z-STRIP DRIP CAP AT SIDING JUNCTURE

204F22.EPS

Figure 22 ◆ Siding junctures at gable ends.

INSIDE TRACK

Siding Gauge

To make a siding gauge, sometimes called a preacher, select a piece of ⅜" or ½" hardwood long enough to accommodate the width of the siding used plus 2¼" and proceed as follows:

Step 1 Center the siding on the block of hardwood (A).

Step 2 Lay out the width of the siding plus ¼" for clearance (B).

Step 3 Lay out the thickness of the siding plus ⅝" (C).

Step 4 Allow 1" around all of the inside cuts (D).

Step 5 Bevel the corners as shown at (E).

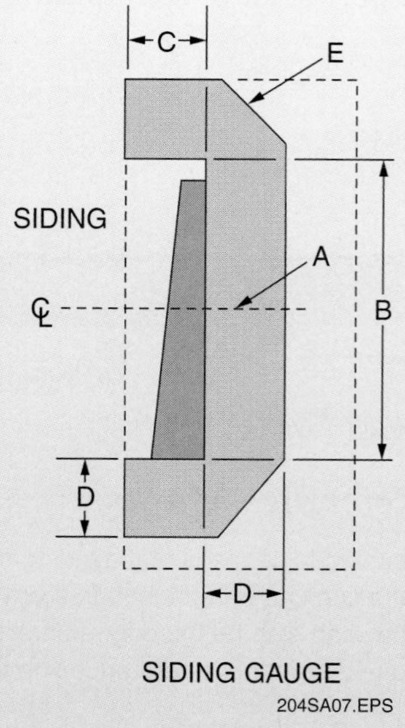

SIDING

SIDING GAUGE

204SA07.EPS

- Use a siding gauge when installing siding to fit between two window casings or a door and window casing (*Figure 23*).
- When fitting siding around windows, it may be necessary to notch the siding to fit (*Figure 24*). Flashing must be installed at the top of window and door casings to keep out water.
- The three ways to finish corners for beveled siding are metal corner caps (*Figures 25 and 26*), mitered corners (*Figure 27 and 28*), and corner boards (*Figure 29*).

Mitered corners (*Figure 27*) provide an attractive way to finish corners, but because of the additional labor involved, this method is normally used only on more expensive structures. The angle shown in *Figure 27* is 47 degrees instead of the conventional 45 degrees used for most miter cuts. This prevents gapping at the corners when the siding dries. When applying the siding, force the mitered corners together firmly and nail the mitered ends to the sheathing, not to each other.

Corner boards (*Figure 28)* are thicker than the siding projection and are nailed at all outside corners of the building. The beveled siding simply butts snugly to these boards.

5.3.0 Board-and-Batten Siding

Board-and-batten is an attractive, versatile, squared-edge siding that is widely used and accepted by architects and contractors throughout the building industry (see *Figure 29*).

The amount of nails needed to apply 100 square feet of siding will depend on their size. For 8d nails, 2½ to 3 lbs will be needed.

Board-and-batten is easy to apply and is weathertight. Because it is surfaced on four sides (S4S), it does not require expensive millwork. All of these factors contribute to making it an economical and practical vertical siding.

There are many variations of board-and-batten siding, but the most widely used is the vertical placement of wide boards, with the joints covered by narrow battens. There are a number of different sizes and textures of lumber used.

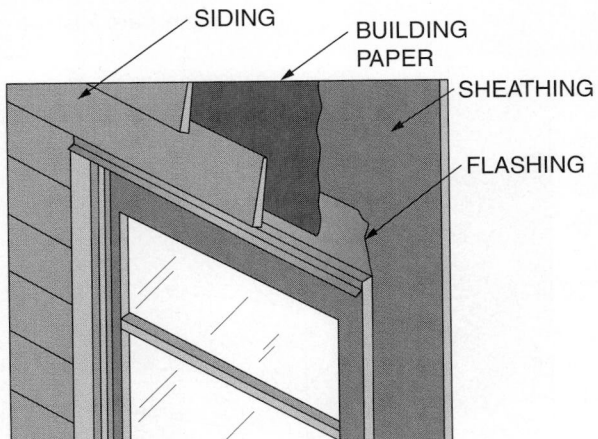

NEW WOOD WINDOW WITHOUT DRIP CAP

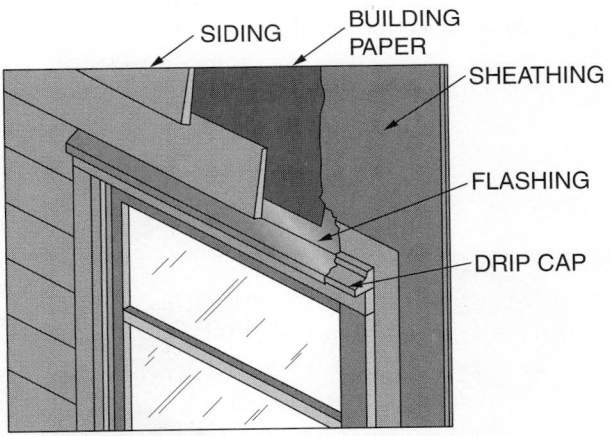

WINDOW WITH OLDER STYLE DRIP CAP

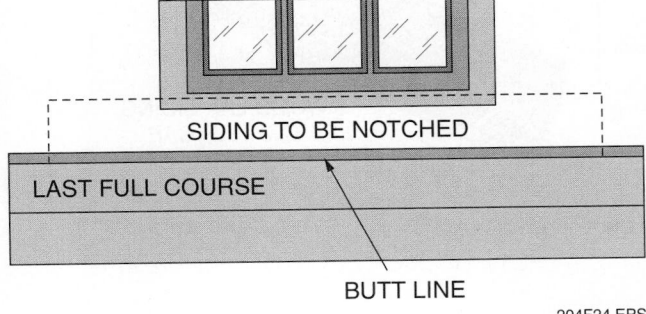

204F24.EPS

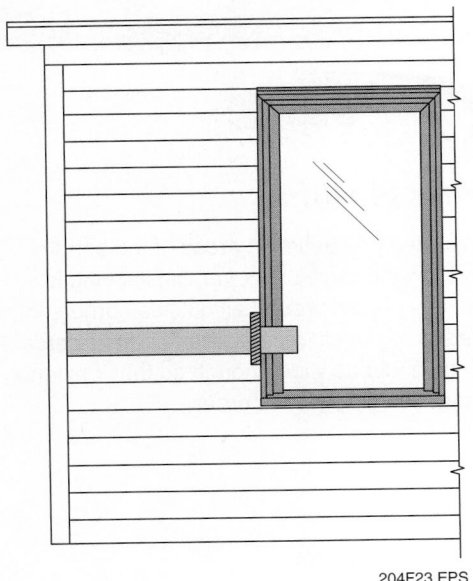

204F23.EPS

Figure 23 ◆ Using a siding gauge.

Figure 24 ◆ Installing siding around windows.

Installing Corner Caps

Corner caps are usually used to cover the outside corners of wood siding. They are available in various lengths for different siding widths and are usually installed as each course of siding is applied. Some caps can be installed after all courses are applied. Each course of siding is cut off flush or slightly back from the sheathing on the adjoining wall at each corner and nailed to the sheathing. The lips of the corner cap are tapped up under the course of siding and the cap is nailed to the sheathing above the lower edge of the next course of siding. Always make sure that the cap and siding panels are flush before nailing the cap.

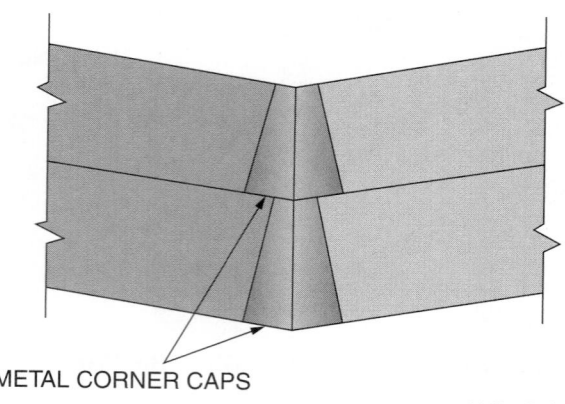

METAL CORNER CAPS

204F25.EPS

Figure 25 ◆ Metal corner caps.

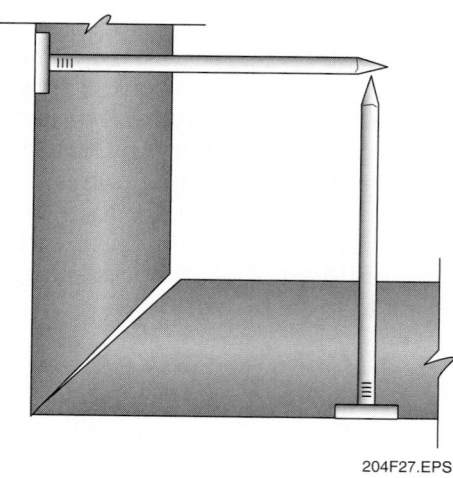

204F27.EPS

Figure 27 ◆ Mitered corner.

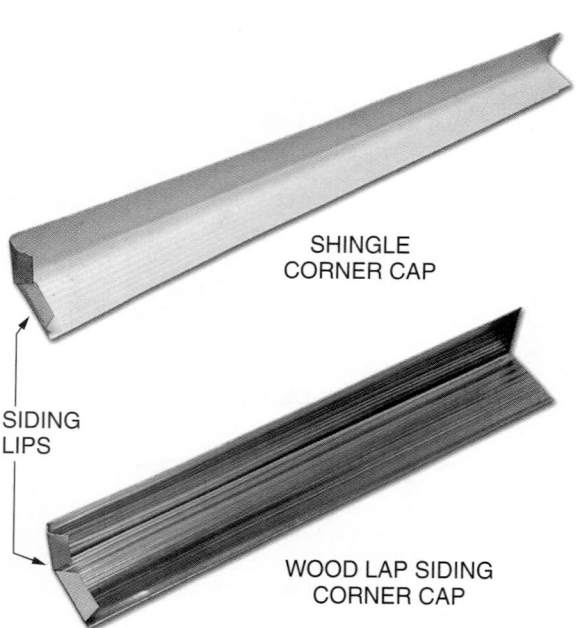

SHINGLE
CORNER CAP

SIDING
LIPS

WOOD LAP SIDING
CORNER CAP

204F26.EPS

Figure 26 ◆ Corner caps.

Corner Flashing

As an added precaution in areas of the country subject to wind-driven rain, vertical flashing is sometimes placed around all outside corners for mitered siding or siding that uses corner boards. The flashing should be wide enough so that it extends 3" to 4" on each side of the corner.

Board-and-Batten Siding as an Architectural Accent

INSIDE TRACK

Board-and-batten siding can be used as an architectural accent on a portion of a structure, such as a front-facing gable or part of a wall.

BOARD-AND-BATTEN SIDING

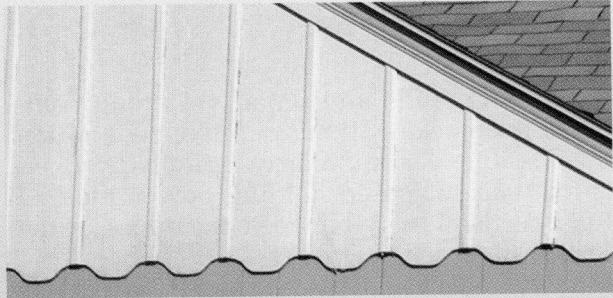

SCALLOPED BOARD-AND-BATTEN ACCENT

204SA08.EPS

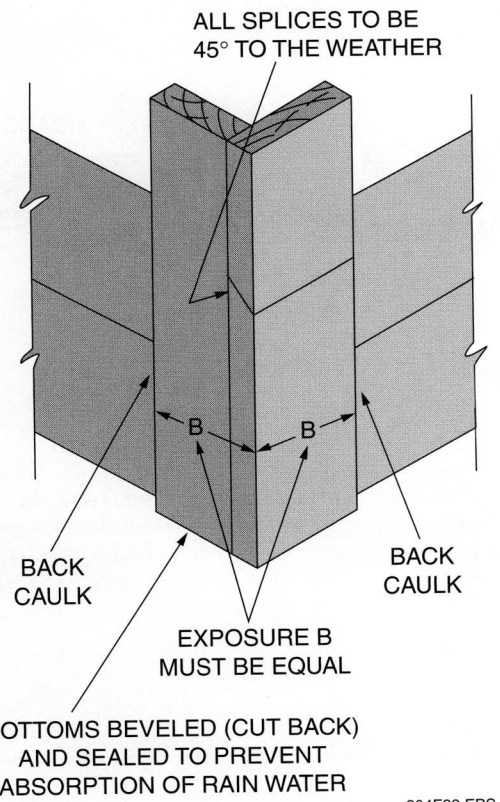

ALL SPLICES TO BE 45° TO THE WEATHER

B   B

BACK CAULK

BACK CAULK

EXPOSURE B MUST BE EQUAL

BOTTOMS BEVELED (CUT BACK) AND SEALED TO PREVENT ABSORPTION OF RAIN WATER

204F28.EPS

Figure 28 ◆ Outside corner using corner boards.

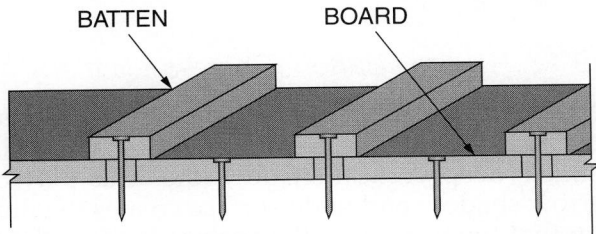

TO AID THE ESTIMATOR, THE LUMBER REQUIREMENTS TO COVER ONE SQUARE (100 SQUARE FEET) ARE INDICATED IN THE FOLLOWING CHART.

WIDTH OF BOARDS	BOARD FEET
6	147
8	139
10	132

204F29.EPS

Figure 29 ◆ Board-and-batten siding.

When framing an exterior wall that is to receive vertical siding, it is necessary to install horizontal blocking between the studs, from top to bottom at 24" OC.

When applying the siding, space the underboards ½" apart and drive the nails midway between the edges at each bearing. A major advantage of board-and-batten construction is that with proper nailing, the boards are free to move slightly with changes in moisture content, but are held snugly in place by the battens. To allow for this movement capability, only one nail should be used through the center of a board at each bearing. The nails should penetrate 1½" into

When the lumber arrives on the job site, it should be placed on blocking off the ground and covered so it will not pick up too much moisture.

Furring Strips

INSIDE TRACK

If Styrofoam™ or wood fiber insulation (⅝" or more) is used on a wall, some plans require the use of horizontal furring strips to provide an adequate surface for nailing the vertical boards. The furring strips can be nailed on 16" or 24" centers.

the studs, the studs and wood sheathing combined, or the blocking. If this depth of penetration is not possible, use annular or spirally grooved nails for their increased holding power. One 8d nail is nailed midway between the edges of the underboard at each bearing.

With the boards in place, fasten the battens using 10d nails. These should overlap each edge of the board underneath by at least 1". The nails should be driven directly through the center of the batten so that the shank passes between the underboards.

5.3.1 Reverse Batten (Board-on-Batten)

Reverse batten *(Figure 30)* is also an attractive vertical siding, giving the building a very sharp, well-defined, deep vertical shadow line. This play of narrow shadow and wide surface creates the illusion that the boards on the surface are free floating. This method can be especially attractive when using rough-sawn boards.

5.3.2 Board-on-Board Siding

Board-on-board is another type of vertical siding. Not only does this method create a vertical shadow line, but it allows the architect or builder to maintain a uniformity in the width of the material used (see *Figure 31*).

For board-on-board siding, apply the underboards first, spacing them to allow a 1½" overlap by the outer boards at both edges. Use standard nailing for underboards, with one 8d nail per bearing. The outer boards must be nailed twice per bearing to ensure proper fastening. Nails having some free length do not hold the outer boards so rigidly as to cause splitting if there is movement from humidity changes. Drive 10d siding nails so that the shanks clear the edges of the underboards by approximately ¼". This provides sufficient bearing for nailing, while allowing clearance for the underboards to expand slightly.

To aid the estimator, the lumber requirements to cover one square (100 square feet) are indicated in *Table 2*.

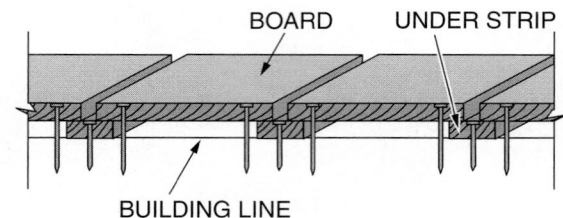

REVERSE BATTEN:
Drive one 8d nail per bearing through the center of the under strip and two 10d nails per bearing through the outer boards.

204F30.EPS

Figure 30 ◆ Reverse batten.

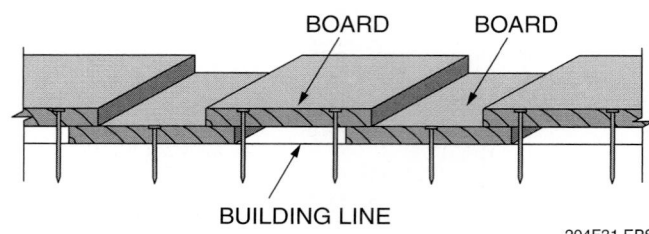

204F31.EPS

Figure 31 ◆ Board-on-board.

Table 2	Board-on-Board Lumber Requirements
Width of Board	**Board Feet/100 ft.²**
6"	150
8"	139
10"	129
12"	123

You will need approximately 25 lbs. of 8d nails per 1,000 board feet (more if using 10d nails).

When applying flat grain boards, orient each board as indicated in *Figure 32*. When the crown surface is exposed to the weather, cupping and grain raising will be prevented.

Simulated Board-and-Batten Siding

A board-and-batten or reverse-batten effect can be obtained on an exterior plywood sheathing by covering the vertical seams with a board or batten and then spacing additional boards or battens between the seams for the desired effect. Sometimes a complete board-and-batten system is applied over a wood sheathing.

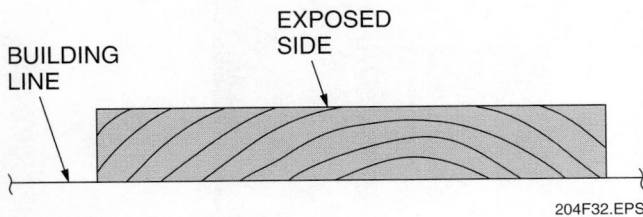

BUILDING LINE

EXPOSED SIDE

204F32.EPS

Figure 32 ◆ Applying flat grain board.

5.4.0 Tongue-and-Groove Siding

Tongue-and-groove (T&G) siding (*Figure 33*) can be applied vertically, horizontally, or at an angle and provides a perfectly weathertight wall. It is often installed diagonally. The diagonal application of siding creates an interesting exterior pattern and is pleasing to the eye. As shown, T&G siding is generally available in several styles.

On the exterior elevations of the building plans, the architect will indicate the location and the application angle. The most common slope is 45 degrees, but make sure of this by checking the plans very carefully. In vertically applied siding, do not use boards larger than 1 × 4 or 1 × 6, because using wider boards will cause problems.

Tongue-and-groove drop siding (refer to *Figure 33*) is normally applied only horizontally or vertically. Horizontal T&G drop siding is more water resistant than plain T&G because the top joint is protected by the overhang of the board above, making water penetration of the joint improbable. Like plain T&G siding, T&G drop siding can be blind-nailed.

The advantage that T&G siding has over shiplap siding is that 6" or narrower boards can be blind-nailed, while shiplap siding cannot. Both types of T&G siding are self-aligning, so they take practically no effort to apply after the first piece is set in the correct position.

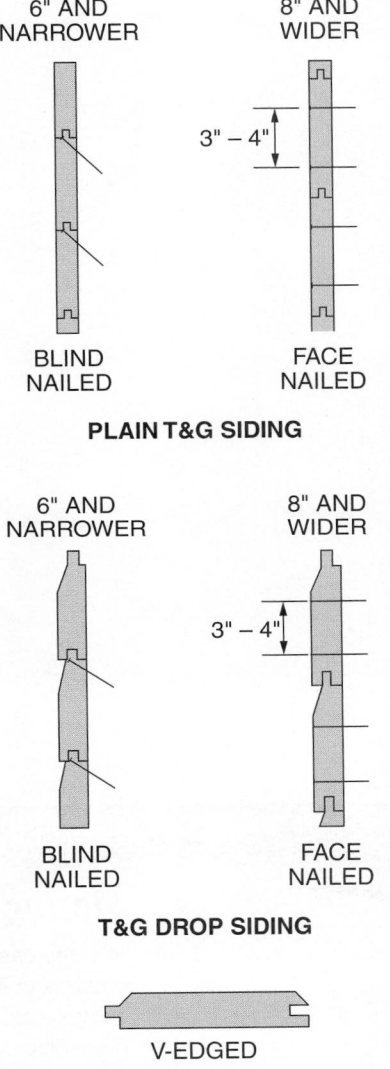

6" AND NARROWER

8" AND WIDER

3" – 4"

BLIND NAILED

FACE NAILED

PLAIN T&G SIDING

6" AND NARROWER

8" AND WIDER

3" – 4"

BLIND NAILED

FACE NAILED

T&G DROP SIDING

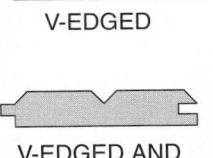

V-EDGED

V-EDGED AND CENTER V-GROOVED

OTHER T&G STYLES

204F33.EPS

Figure 33 ◆ Tongue-and-groove siding.

Architectural Accents

Vertical or diagonal applications of T&G V-edged or V-grooved and shiplap-style siding are often used as architectural accents on interior or external wall surfaces.

CHANNEL RUSTIC STYLE SHIPLAP SIDING

V-EDGED T&G SIDING

204SA09.EPS

Mating T&G Siding

In some cases, the tongue and groove of T&G siding may have to be forced together to obtain a uniform mating of each course of siding. If necessary, use a hammer and a scrap block of siding on the course being installed to force it onto the tongue of the preceding course. If a board is slightly warped and does not mate evenly along its entire length, secure the board with nails at one or both ends up to the point that the warp begins. Set additional nails in the siding beyond the point of warp. Then, drive a flat, broad chisel into the underlayment-nailing surface with its beveled edge against the siding. Use the chisel as a lever to force the siding into position and then nail the siding into place. Repeat along the length of the board as necessary until the board is seated and secured. Make sure to maintain any inside-groove spacing that may be required.

5.5.0 Shiplap Siding

Plain shiplap siding can be installed vertically, horizontally, or diagonally. In addition to the plain shiplap patterns, it is generally available in four other patterns, with the most common being the V-edged. Plain shiplap lays up with a flush edge. This tends to minimize the direction of the courses and instead accentuates the texture and grain of the wood. On the other hand, V-edged shiplap creates a definite shadow line and indicates the direction of the courses. See *Figure 34*. Other styles include drop, channel rustic, and log cabin. They can be used either horizontally or vertically.

Any style shiplap siding that is 6" or narrower can be face-nailed 1" from the bottom with one nail per bearing. Siding that is wider than 8" should be face-nailed with two nails per bearing. The general rule is that the nails should be long enough to penetrate at least 1½" into the studs, or the studs and wood sheathing combined.

Use 8d nails (25 lbs for 1,000 board feet) for 1" siding and 6d nails (15 lbs per 1,000 board feet) for thinner stock. Nails should be spaced 1½" from the edge of the overlap and 2" from the edge of the underlap for 8" boards. Nail other widths proportionately.

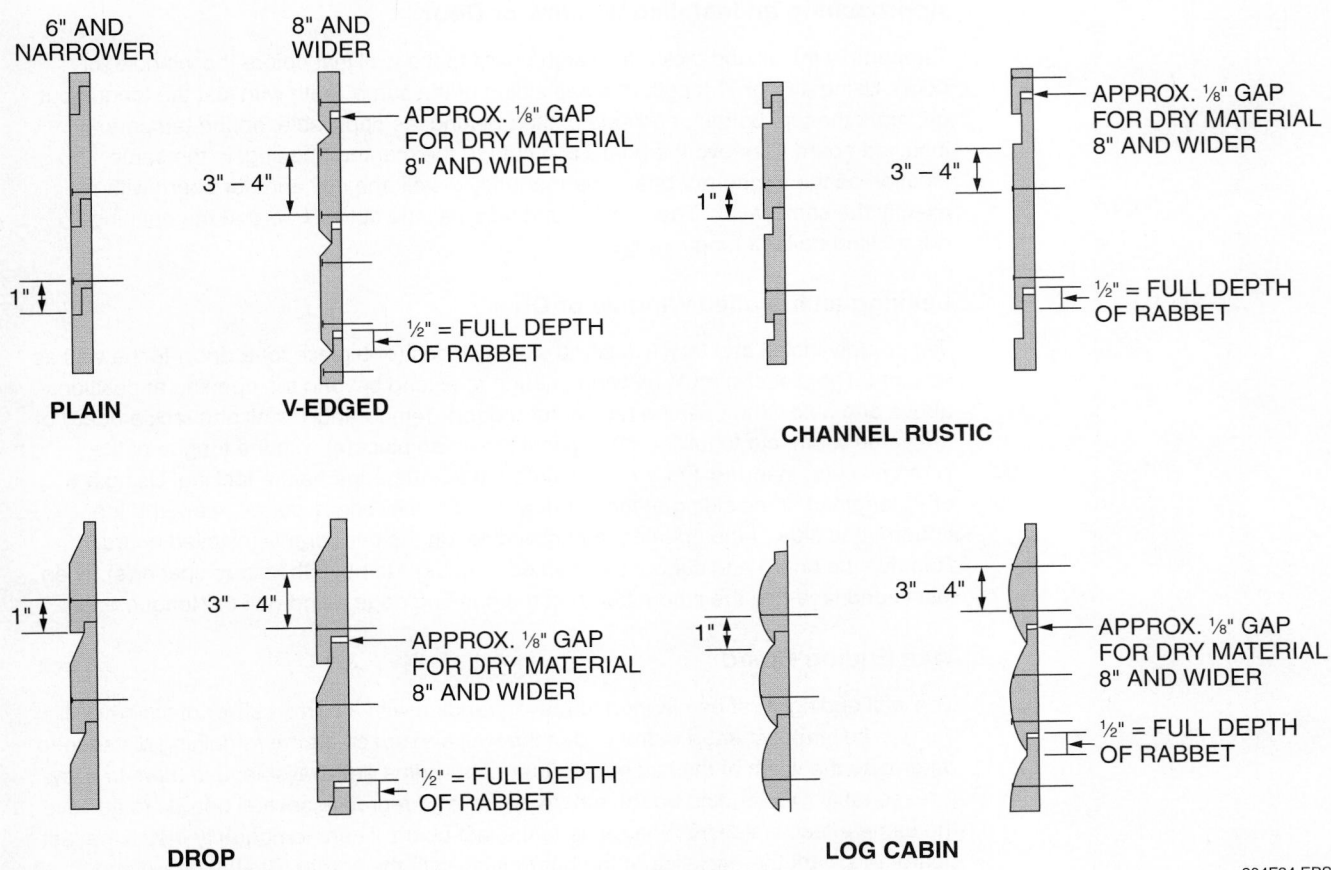

Figure 34 ◆ Styles of shiplap siding.

204F34.EPS

Vertical T&G and Shiplap Application Hints

Here are some application hints for installing 6" to 8" (or less) vertical siding:

Wall Starting Board

Plumb the tongue edge of the board with the grooved edge beyond an outside corner or against an inside corner. Temporarily tack the board in place. Mark the board along the length of the corner. For an outside corner, mark along the backside of the board flush with the corner. For an inside corner, mark along the face of the board using a spacer 6" or 7" long and wide enough to extend beyond the groove depth of the board. Rip off the grooved edge to the mark and slightly back-bevel the edge. Position the board at the corner with ripped edge flush with the corner. Install the board and recheck the plumb of the tongue edge, then face-nail the board at the corner. Blind-nail the tongue edge.

Approaching an Installed Window or Door

Temporarily install and tack a full-width board to the wall just before the window (or door). Using a 6" or 7" length of scrap siding of the same width with just the tongue cut off, mark the top, bottom, and side of the opening, as applicable, on the temporarily installed board. Remove the board and cut out the marked opening. In the same location as the temporary board, permanently install and nail another board with exactly the same width. Then, install and face-nail the cutout board at the opening edge. Blind-nail the tongue edge.

Leaving an Installed Window or Door

Temporarily install and tack two scrap siding pieces (one piece for a door) to the wall as spacers. The piece(s) must be wide enough to extend beyond the opening at positions above and below the opening (above for a door). Temporarily install and tack a board of the same width, cut to full length, against the scrap piece(s) with the tongue of the scrap piece(s) inserted. For a door, plumb the tongue edge before tacking. Using a 6" or 7" length of scrap siding of the same width with the tongue cut off, mark the top, bottom, and side of the opening, as applicable, on the temporarily installed board. Remove the board and cut out the marked opening. Remove the scrap spacer(s). Then, install and face-nail the cutout board at the opening edge. Blind-nail the tongue edge.

Wall Ending Board

The wall ending must be planned to prevent ending with a narrow sliver of siding.

Stop several feet short of the end of the wall and space off the remaining distance to determine the width of the last board. If random widths are available, use them to allow a reasonably wide ending board; otherwise, rip and regroove several boards to achieve the same effect. Install the boards up to the last board. Then, temporarily install the last board and mark the backside of the board flush with the corner. Rip the board and permanently install it by face-nailing at the corner.

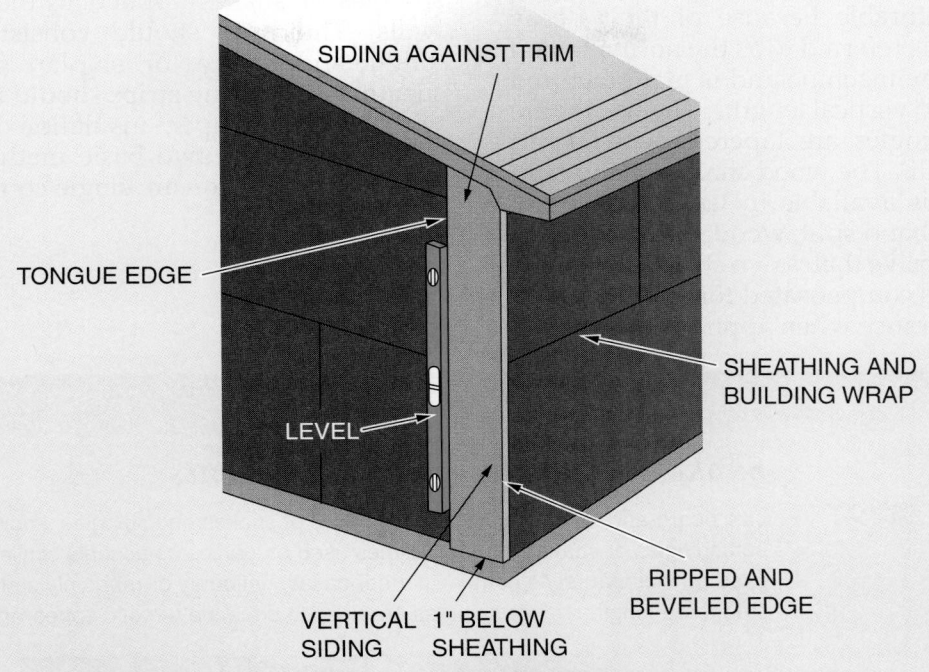

SIDING AGAINST TRIM

TONGUE EDGE

LEVEL

SHEATHING AND
BUILDING WRAP

RIPPED AND
BEVELED EDGE

VERTICAL
SIDING

1" BELOW
SHEATHING

WALL STARTING BOARD

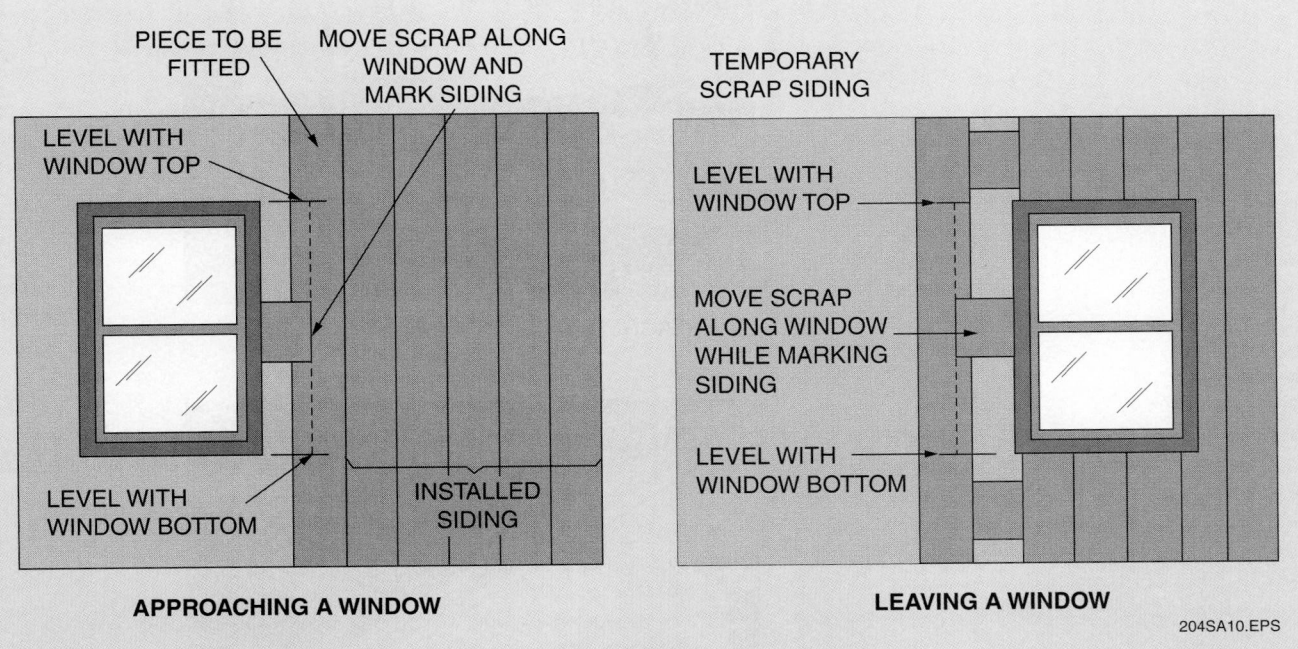

PIECE TO BE
FITTED

MOVE SCRAP ALONG
WINDOW AND
MARK SIDING

LEVEL WITH
WINDOW TOP

LEVEL WITH
WINDOW BOTTOM

INSTALLED
SIDING

APPROACHING A WINDOW

TEMPORARY
SCRAP SIDING

LEVEL WITH
WINDOW TOP

MOVE SCRAP
ALONG WINDOW
WHILE MARKING
SIDING

LEVEL WITH
WINDOW BOTTOM

LEAVING A WINDOW

204SA10.EPS

5.6.0 Shingle Siding or Shakes

The use of shingle siding or shakes as a sidewall finish results in a very attractive, rustic, architecturally interesting siding. Red cedar or cypress is normally used to make the shingles or shakes. They are very durable because of their decay resistance. What is referred to as the normal wood shingle is sawn by machine and is manufactured in 16", 18", and 24" vertical lengths. The widths are random. The shingles are tapered, with a butt thickness of ⅜"to ¾". The wood shake is hand split from a log and is available in taper split form. Because they are hand split, wood shakes are generally more expensive than sawn shingles, but this price difference is compensated for by their beautiful rustic appearance when applied.

Before applying wood shingles or shakes, check the local building codes or with the building inspector in your area and make sure that the fire codes will permit their use.

Solid nailing or stapling is a must for wood shingles or shakes applied to the exterior sidewalls. The base should consist of plywood, tongue-and-groove, or shiplap sheathing. The sheathing or furring strips should first be covered with building paper, insulation board, and/or house wrap. The two basic methods of shingle sidewall application are single course and double course.

Shake/Shingle Architectural Accents

Like board-and-batten siding, special styles of shakes and shingles, known as fancy-butt shingles or shakes, are sometimes used on gables as accents similar to those used on Victorian-style homes. In other cases, uniformly or randomly spaced staggered-length shakes/shingles are used to enhance a rustic appearance.

STAGGERED-LENGTH SHAKES/SHINGLES

FANCY-BUTT SHINGLE ACCENT

204SA11.EPS

In single-course application, the shingles are applied as in roof construction, but greater weather exposures are permitted. The maximum recommended weather exposures with single-course wall construction are 8½" for 18" lengths and 11½" for 24" lengths. Shingle walls will have two plies of shingles at every point, whereas shingle roofs will have three-ply construction.

The first course of shingles may be doubled at the bottom (*Figure 35*). After the first course is applied, lay out the story pole with all of the courses indicated on it. Make sure you have the courses arranged to line up as closely as possible to the top of all door and window openings. Sometimes it may be necessary to change the exposure slightly on the course at the door and window heads and at the window sills so that they will line up. If such an adjustment is necessary, make sure it is slight so that it is not noticeable when viewing the other exposures.

Using a story pole, transfer those markings to the ends of the building and to all door and window openings. Snap a chalkline at long runs. A furring strip or 1 × 2 straightedge may be temporarily tacked to the building at each course so that the shingle butts may be placed on it for alignment before they are nailed. To form closed joints on outside corners of the sidewalls, shingles in adjoining courses may be alternately overlapped and edge shaved to a close fit.

Another method is to miter the two adjoining shakes in each course, but because of the time consumed in doing this, it is usually too expensive. Inside corners may be woven in alternate overlaps or may be closed by nailing a 1 × 1 square molding or corner board in the corner before the shingles are applied.

The nailing for single coursing is accomplished by using 3d, 1¼" corrosion-resistant nails, such as hot-dipped zinc, aluminum, or stainless steel. Only two nails are used per shingle, and each is placed approximately ¾" from the side edge of the shingle and approximately 1" above the butt line of the next course. Drive the nails flush, but not so hard that the head crushes the wood.

The double-course method of wood shingle sidewall application is much the same as the single-course method, with a few exceptions. Double coursing allows for the application of extended weather exposure shingles over coursing-grade shingles. Double coursing also provides deep, intense, bold shadow lines. When double-coursed, a shingle wall should be tripled at the foundation line by using a double undercourse (see *Figure 36*).

The double-course nailing requires that the outer-course shingle be secured with two 5d (1¾") small head, corrosion-resistant nails, driven 1" to 2" above the butts, approximately ¾" in from each side. Additional nails are driven about 4" apart across the face of the shingle in a straight line.

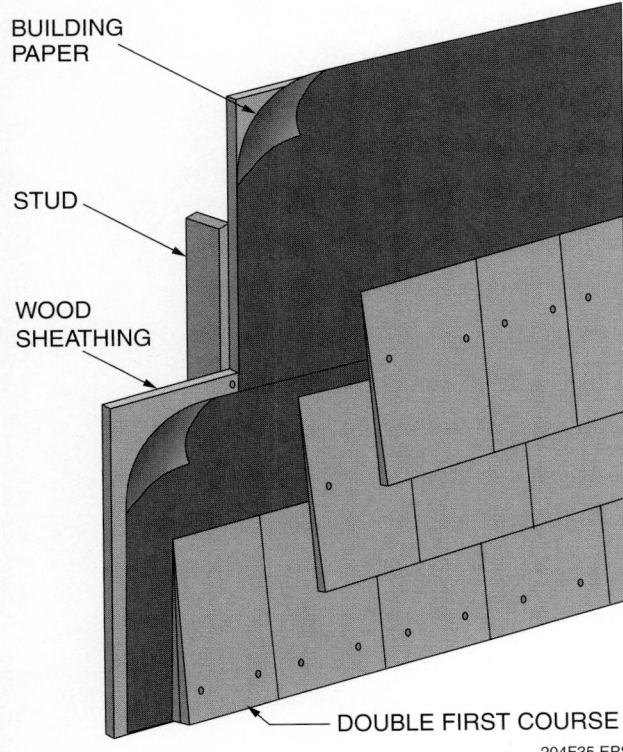

BUILDING PAPER

STUD

WOOD SHEATHING

DOUBLE FIRST COURSE

204F35.EPS

Figure 35 ◆ Installing wood shingles.

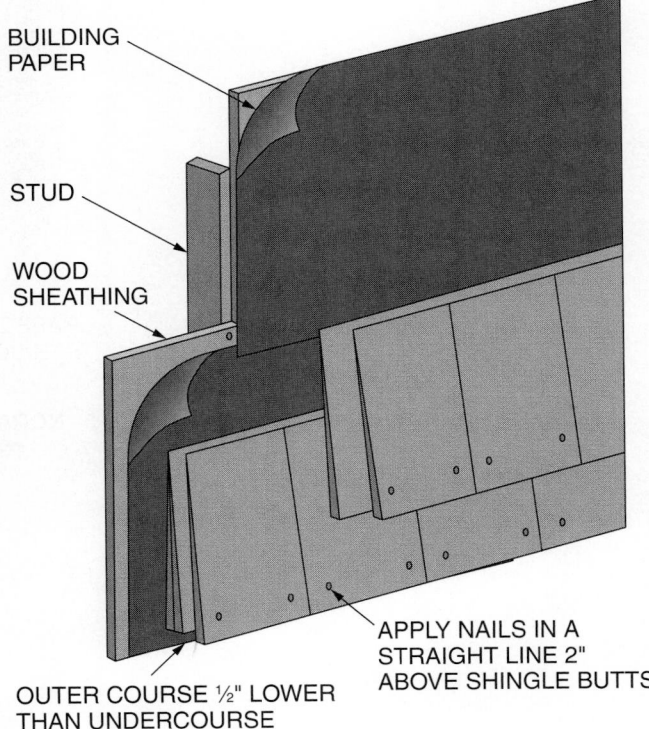

BUILDING PAPER

STUD

WOOD SHEATHING

OUTER COURSE ½" LOWER THAN UNDERCOURSE

APPLY NAILS IN A STRAIGHT LINE 2" ABOVE SHINGLE BUTTS

204F36.EPS

Figure 36 ◆ Double coursing a shingle wall.

Straightedge Shake/Shingle Application

For single coursing, use straightedges tacked at the butt line to rest shingles on for spacing selection. For double coursing, use straightedges with a rabbeted edge so that the outer course is about ¼" below the inner course. Sort the shakes/shingles for proper seam overlap and lay them butt down on the straightedge. Then, nail the shakes/shingles to the wall. For a ribbon-style double coursing, a reversed straightedge with a deeper rabbet can be used to shift the outer shake/shingle up so that about 1" to 1½" of the lower part of the inner shake/shingle is exposed. With any method, use a shingling hatchet to trim and fit the edges if necessary. Butt ends are not trimmed. If rebutted and rejoined shakes/shingles are used, no trimming should be necessary.

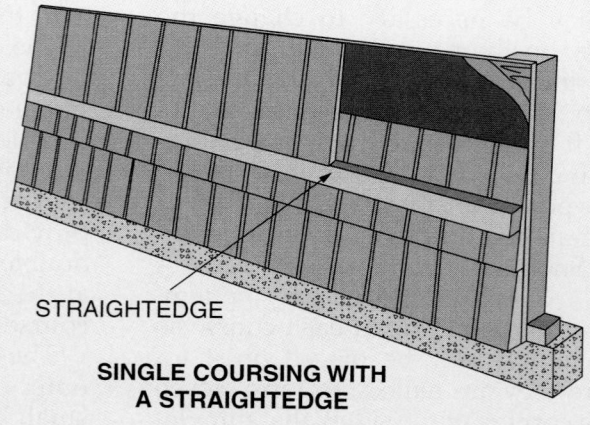

STRAIGHTEDGE

**SINGLE COURSING WITH
A STRAIGHTEDGE**

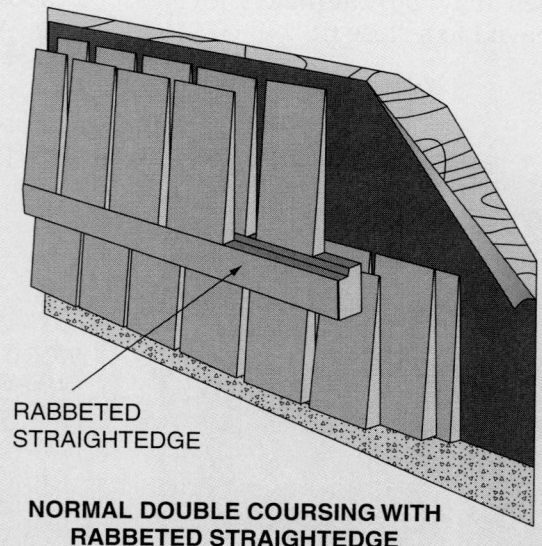

RABBETED
STRAIGHTEDGE

**NORMAL DOUBLE COURSING WITH
RABBETED STRAIGHTEDGE**

204SA12.EPS

The outside corners should be constructed with an alternate overlap of shingles between successive courses. Inside courses may be mitered or woven over a metal flashing, or they may be made by nailing an S4S 1½" or 2" square strip in the corner, after which the shingles of each course are fastened to the strip.

5.6.1 Panelized Shakes/Shingles

In most cases, panelized shakes/shingles are simply shakes glued or stapled to a backer board of plywood. They are available in widths of 4' and 8'. The panels are available in natural wood and are pre-finished in several basic colors of stain. The panels can be applied rapidly, using threaded nails colored to match the panels. The manufacturer's installation instructions should be followed.

5.7.0 Plywood Siding

The use of plywood as an exterior finish siding has been rapidly growing among architects and builders because of the speed of installation and the reliability of the waterproof glues being used. The beauty and diversity of the available surfaces have also added to its growing popularity. Because of its strength, plywood can be nailed directly to the studs, eliminating the need for sheathing. This is another plus in favor of plywood siding, because it saves not only the cost of the plywood sheathing, but also the cost of its installation.

Plywood siding is available in thicknesses of ⅜", ½", ⅝", and ¾"; however, ⅝" is the most commonly used. Some of the common textures and designs of plywood siding are shown in *Figure 37*.

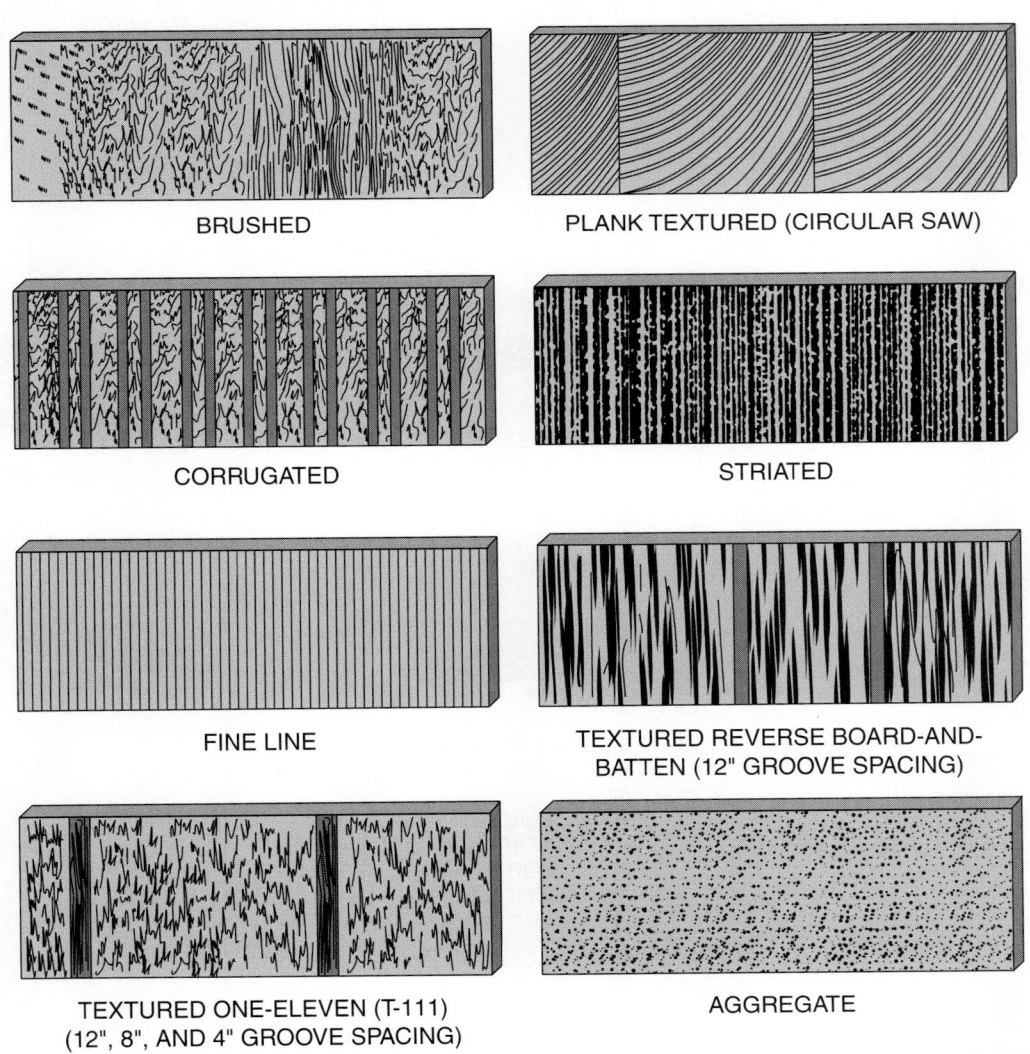

BRUSHED

PLANK TEXTURED (CIRCULAR SAW)

CORRUGATED

STRIATED

FINE LINE

TEXTURED REVERSE BOARD-AND-BATTEN (12" GROOVE SPACING)

TEXTURED ONE-ELEVEN (T-111)
(12", 8", AND 4" GROOVE SPACING)

AGGREGATE

204F37.EPS

Figure 37 ◆ Surface textures and designs of common plywood siding.

Panelized Shakes/Shingles

INSIDE TRACK

Panelized shakes/shingles can be used to cover as much as 14 square feet at a time.

204SA13.EPS

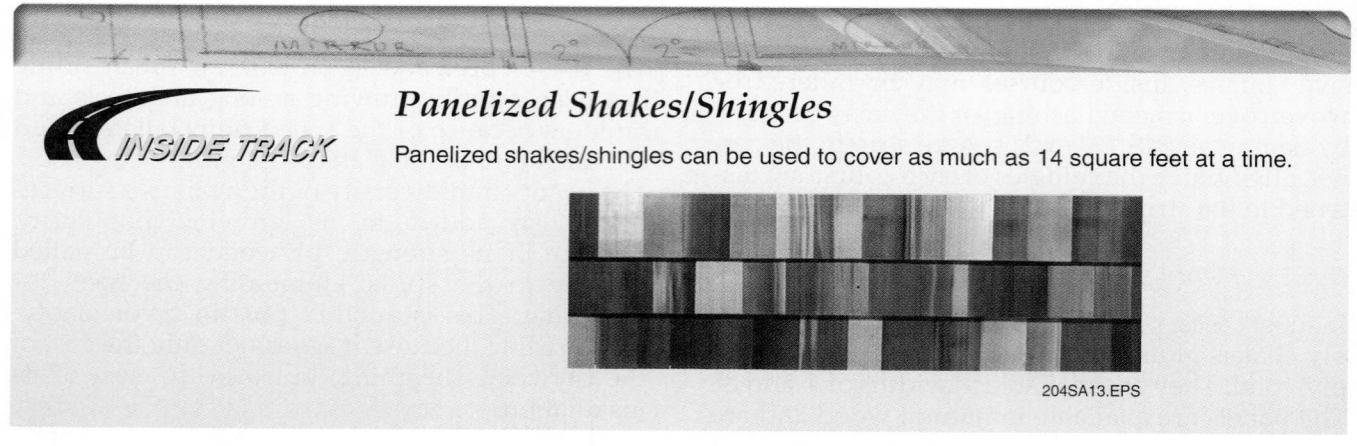

Manufacturers generally recommend 7d or 9d siding nails or box nails made of hot-dipped galvanized, aluminum, or stainless steel. Ring shank and box nails are the types usually specified. The nailing pattern is 6" OC on the edges and 12" OC in the field. Staples can also be used as long as they are the appropriate size and type.

The spacing of wall studs or nailing supports for the 3/8" plywood is a maximum of 16" OC, but thicker plywood (1/2", 5/8", and 3/4") permits a spacing of 24" OC. Blocking is usually required at all horizontal joints (subject to local building codes). Some of the joint suggestions are indicated in *Figure 38*.

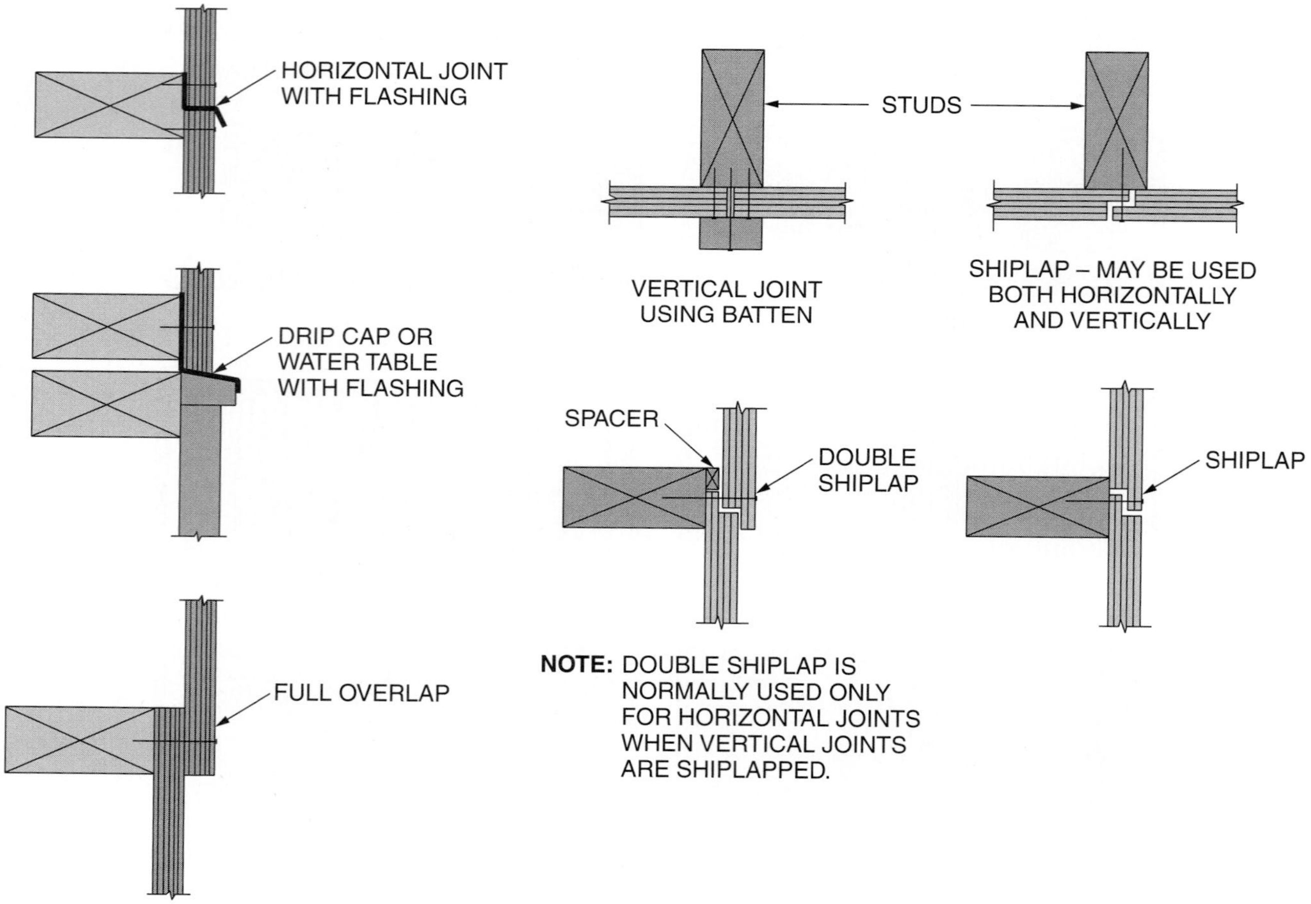

HORIZONTAL JOINT WITH FLASHING
DRIP CAP OR WATER TABLE WITH FLASHING
FULL OVERLAP
STUDS
VERTICAL JOINT USING BATTEN
SHIPLAP – MAY BE USED BOTH HORIZONTALLY AND VERTICALLY
SPACER
DOUBLE SHIPLAP
SHIPLAP
NOTE: DOUBLE SHIPLAP IS NORMALLY USED ONLY FOR HORIZONTAL JOINTS WHEN VERTICAL JOINTS ARE SHIPLAPPED.

204F38.EPS

Figure 38 ◆ Plywood joint suggestions.

Be sure to refer to the manufacturer's technical specifications prior to installing the siding.

To apply lap plywood siding, follow these suggestions:

- If sheathing is not used, place horizontal blocking at 4'−0" centers.
- Install building paper between the siding and the studs.
- Use a starter strip that is the same thickness as the siding for the first course.
- Coat the edges of the siding with a primer or a water-repellent finish before application.
- Vertical joints should be staggered. These joints must be centered over studs with a tapered wedge at least 1⅝" wide behind the joint.
- Use 8d noncorrosive siding or box nails. Insert one nail at each stud on the bottom edge of the siding.

At all vertical joints, nail 4" OC for siding 12" wide or less. Nail 8" OC for siding 16" wide or more. All nails should be placed ¼" back from the edge of the plywood. Set and putty all casing nails. Box nails are driven flush.

5.8.0 Hardboard and Particleboard Siding

Hardboard siding is fabricated for use as lap siding or as large panels up to 16' in length by 4' in width. It is impregnated with a baked-on tempering compound. This process produces a tough, dense siding that will not split or splinter and is highly resistant to denting. Various pattern grains and profiles are available.

 Hardboard siding reinforces wall construction, goes up quickly, and can be easily worked with both power tools and regular woodworking tools. Hardboards take 100 percent acrylic latex paints and stains, as well as gloss or satin oil/alkyd paints. Flat oil/alkyd paints or stains and vinyl acetate or vinyl acrylic copolymer paints and stains are not recommended. This type of siding is available unprimed or factory-primed.

Particleboard can also be used as siding material. It is available in many of the same shapes and surfaces as hardboard or plywood and is applied in the same manner.

Hardboard and particleboard siding is very sensitive to water damage. To achieve satisfactory performance and for warranty purposes, the siding must be installed and finished as specified by the manufacturer. In addition, the siding should never be cleaned using high-pressure or low-pressure washing methods.

The following are guidelines that may apply to hardboard and particleboard sidings, subject to the manufacturer's instructions:

- Hardboard siding may be applied over walls sheathed with wood or insulation and with studs spaced not more than 24" OC or over unsheathed walls with studs spaced not more than 16" OC. The lowest edge of the siding should be at least 6" to 8" above the finished grade level. When cutting, be sure to sand and prime all cut ends.
- In accordance with the manufacturer's instructions, when hardboard siding is applied directly to studs or over wood sheathing, moisture-resistant building paper or **felt paper** (non-vapor barrier) should always be laid directly under the siding. In most cases, a vapor barrier is required on the inside heated wall.
- When applying hardboard siding, use rust-proof siding nails. Nail only at stud locations and on special members around doors and windows. Use 6d nails for nailing directly into studs and 8d nails when nailing over sheathing. Nails must be kept back ½" from the ends and edges of the siding pieces.
- At inside corners, siding should be butted (with approximately ¹⁄₁₆" space) against a 1⅛" × 1⅛" wood, metal, or vinyl corner member. Outside corners may be 1⅛" wood corner boards, or metal/vinyl corners may be used. Caulking should be applied wherever the siding butts against wood corner boards, windows, and door casings. *Figures 39* and *40* show hardboard siding details.

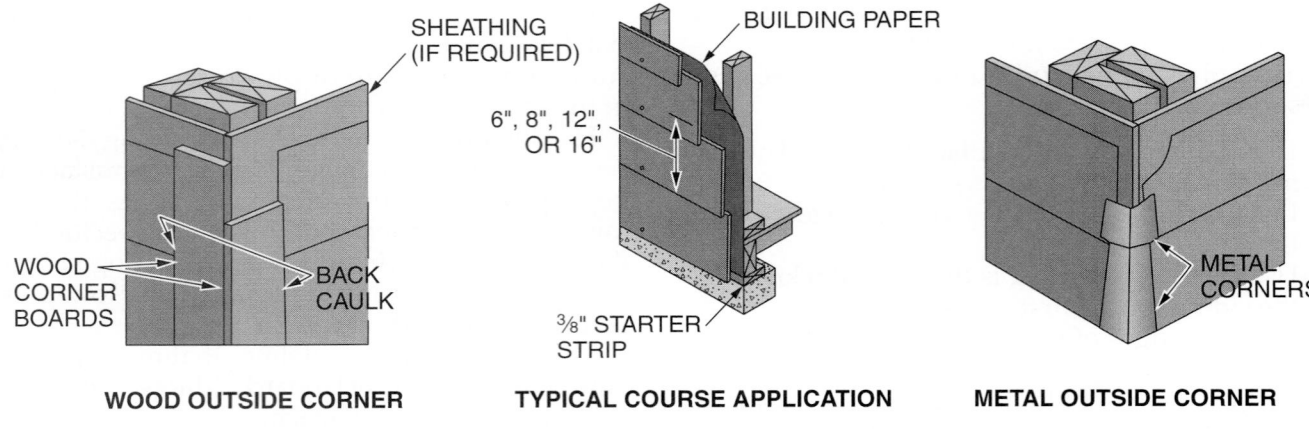

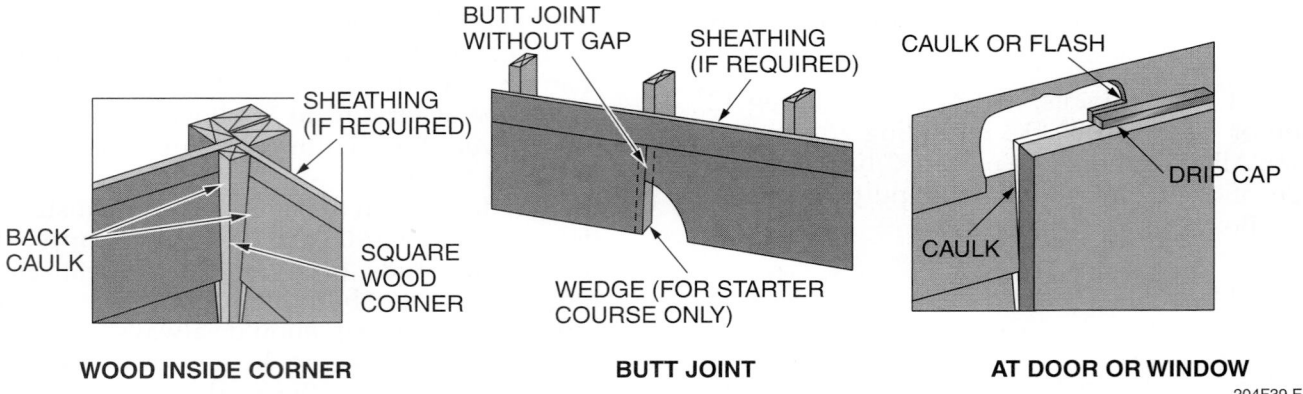

Figure 39 ◆ Typical application of lap siding.

5.8.1 Installing Flat Panels

Flat panels are installed vertically. All joints and panel edges should fall on the center of the framing members. If it is necessary to make a joint with a panel that has been field cut and the shiplap joint removed, use a butt joint. Butter the edges with caulk and gently attach them. Do not force or spring the panels into place. Leave a slight space where the siding butts against the window or door trim, and apply caulk to the space *(Figure 41)*. Horizontal joints should be spaced ⅛" apart, flashed, and sealed.

5.8.2 Installing Lap Siding

Lap siding is installed horizontally. Start the application by fastening a ⅜" × 1⅜" wood starter strip along the bottom edge of the sill. Level and install the first course of siding with the bottom edge at least ⅛" below the starter strip. Fasten the first course by nailing 1½" from the drip edge of the siding and ½" from the butt end.

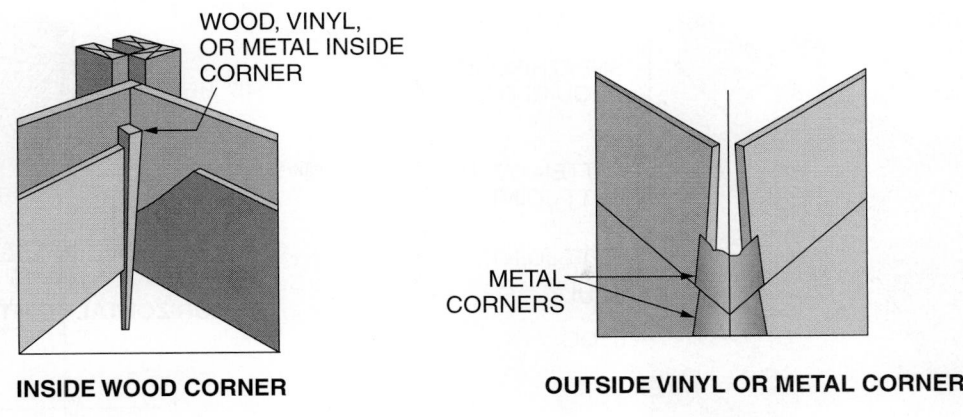

INSIDE WOOD CORNER

WOOD, VINYL, OR METAL INSIDE CORNER

OUTSIDE VINYL OR METAL CORNER

METAL CORNERS

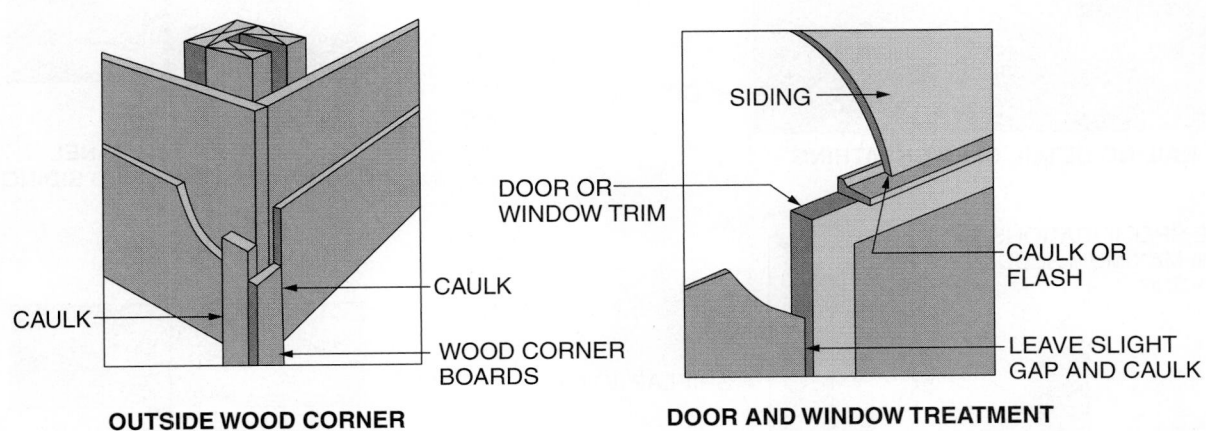

OUTSIDE WOOD CORNER

CAULK

CAULK

WOOD CORNER BOARDS

DOOR AND WINDOW TREATMENT

SIDING

DOOR OR WINDOW TRIM

CAULK OR FLASH

LEAVE SLIGHT GAP AND CAULK

204F40.EPS

Figure 40 ◆ Hardboard siding corner details.

Install subsequent siding courses using a minimum overlap of 1". Butt joints should occur only at stud locations. Metal or vinyl butt joints may be used, if desired (*Figures 42* and *43*). Factory-primed ends should be used for all vertical butt joints that will not be covered. Adjacent siding pieces should just touch at butt joints; if they do not, then a 3/16" space may be left and filled with a butyl caulk. Never force or spring the siding into place.

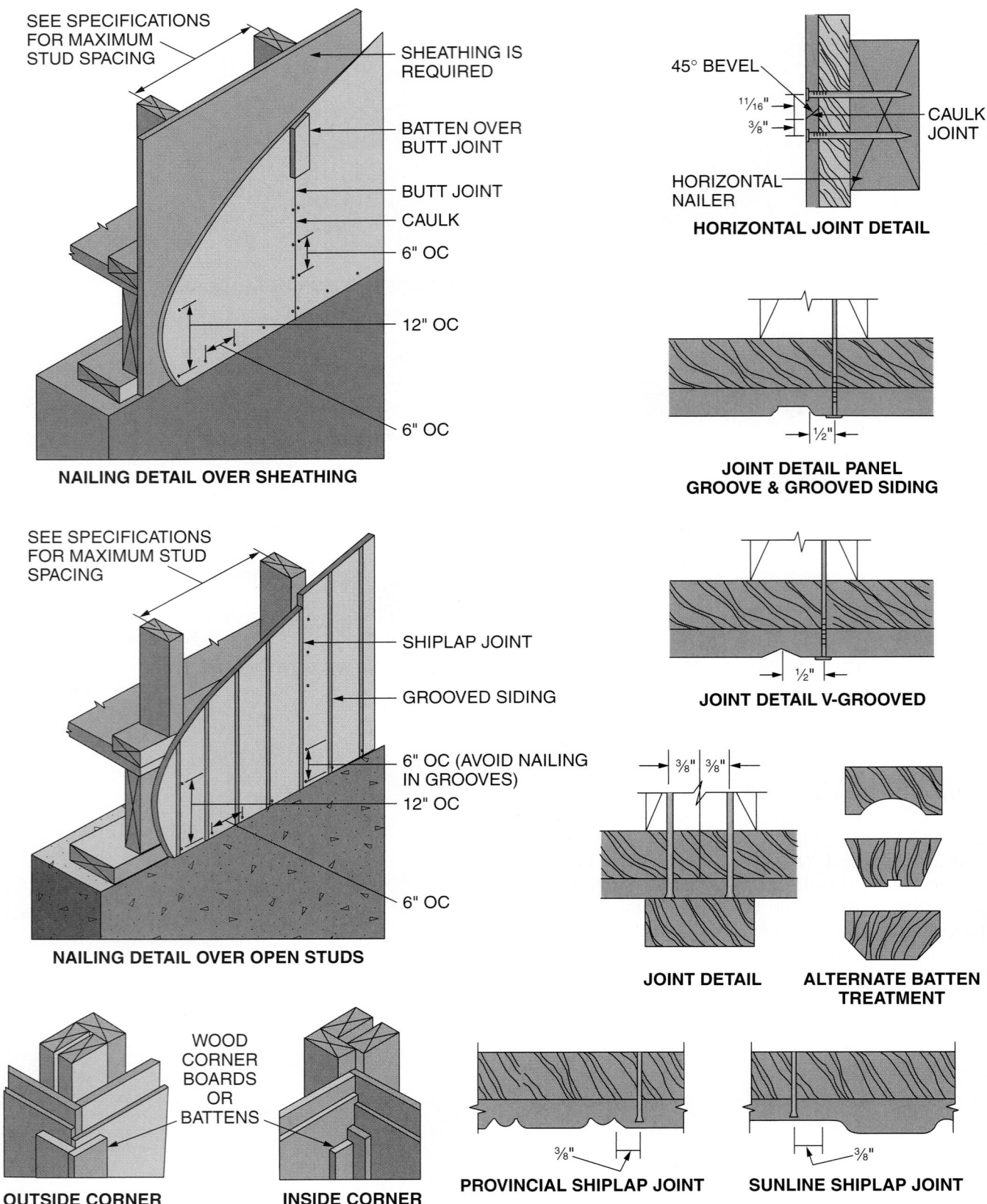

Figure 41 ◆ Installing lapped hardboard panels.

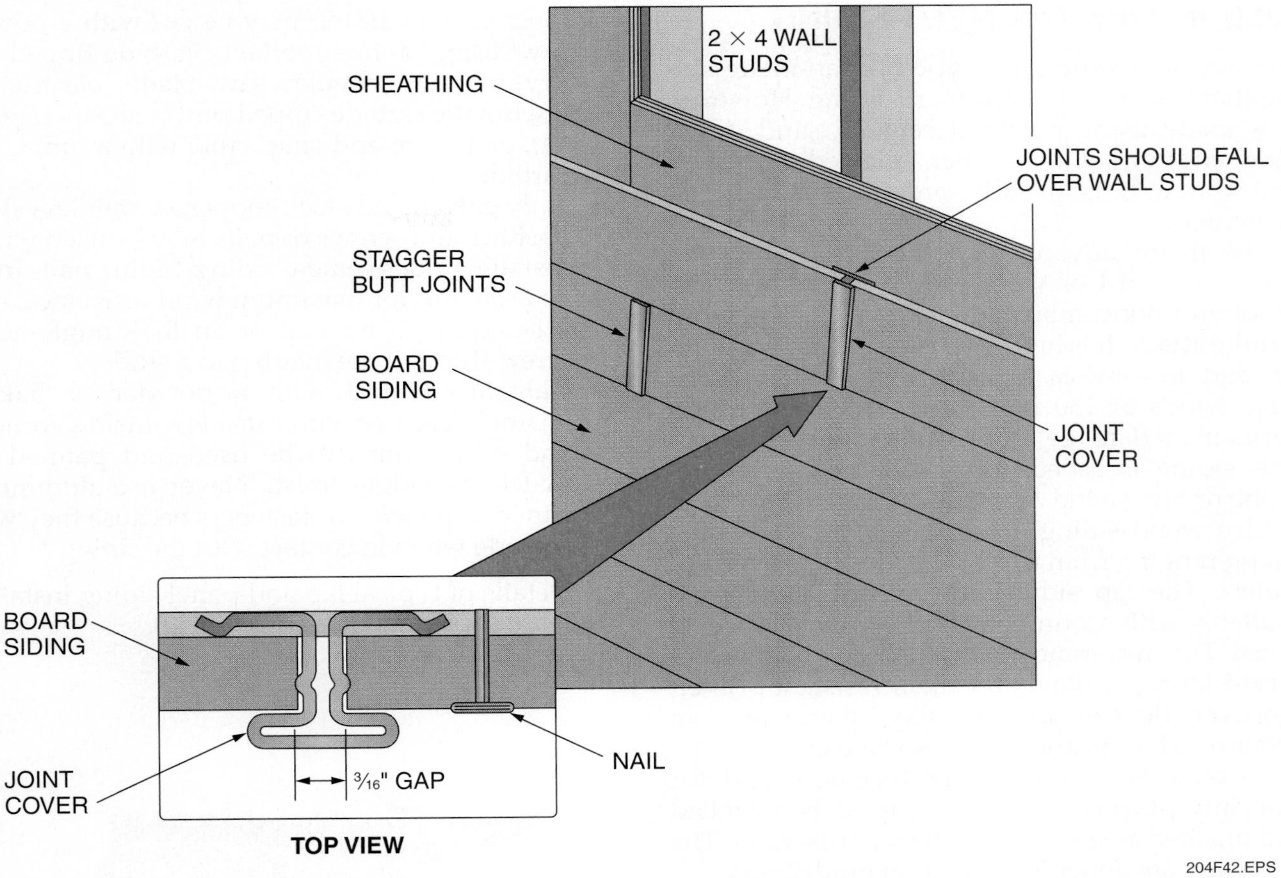

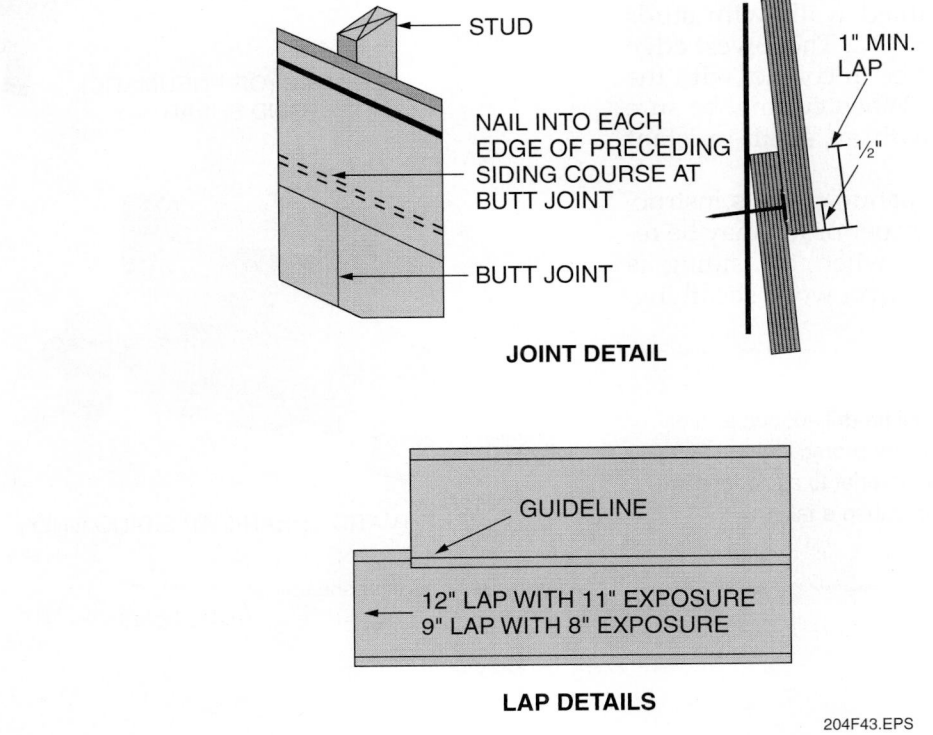

TOP VIEW

SHEATHING

2 × 4 WALL STUDS

JOINTS SHOULD FALL OVER WALL STUDS

STAGGER BUTT JOINTS

BOARD SIDING

JOINT COVER

BOARD SIDING

JOINT COVER

³⁄₁₆" GAP

NAIL

204F42.EPS

Figure 42 ◆ Metal or plastic butt joints.

STUD

NAIL INTO EACH EDGE OF PRECEDING SIDING COURSE AT BUTT JOINT

BUTT JOINT

1" MIN. LAP

¹⁄₂"

JOINT DETAIL

GUIDELINE

12" LAP WITH 11" EXPOSURE
9" LAP WITH 8" EXPOSURE

LAP DETAILS

204F43.EPS

Figure 43 ◆ Installing lapped hardboard siding.

6.0.0 ◆ FIBER-CEMENT SIDING

Fiber-cement siding is a type of man-made siding that is similar to hardboard siding. However, it is made using portland cement, sand, fiberglass and/or cellulose fiber, selected additives, and water. It is usually pressure-formed and heatcured.

The major advantage of fiber-cement siding over hardboard or wood siding is that it is rotproof and noncombustible, and can withstand a termite attack. It is highly resistant to impact damage and, in some cases, can withstand hurricaneforce winds of 130 mph or more. It also resists permanent damage from water and salt spray. This siding is especially suited for use in fireprone or high-wind areas.

Like wood siding, it is available in single-lap siding ranging from 6" to 12" wide and as vertical panels. The lap siding and vertical panels are available with a number of different surface patterns. The recommended finish is 100 percent acrylic latex paint over an alkali-resistant primer; however, gloss or satin oil/alkyd paints over an alkali-resistant primer may also be used.

To achieve satisfactory performance and for warranty purposes, the siding must be installed and finished as specified by the manufacturer. The following are general installation guidelines:

- Fiber-cement siding may be applied over walls sheathed with wood or insulation board up to 1" thick and with studs spaced not more than 24" OC or over unsheathed walls with studs spaced not more than 16" OC. The lowest edge of the siding should not be in contact with the earth or standing water. When cutting, be sure to prime all cut ends with an alkali-resistant primer.
- In accordance with the manufacturer's instructions, moisture-resistant paper or felt may be required under the siding when the siding is applied directly to studs or over wood sheathing.

 **WARNING!**

Because dry material will be drilled, cut, and/or abraded, proper respiratory protection must be used when cutting this material to avoid inhaling toxic silica dust that can cause a fatal lung disease called silicosis.

- Fiber-cement siding may be cut with a power saw using a fine-toothed, carbide-tipped or dry-diamond circular saw blade, electric or pneumatic carbide-tipped power shears (*Figure 44*), or a score-and-snap knife with a tungsten-carbide tip.
- Only galvanized steel, copper, or stainless steel flashing and screws or nails may be used when installing fiber-cement siding. Siding nails may be used, but for maximum wind resistance, use a standard 2" 6d nail or an 8-18 bugle-head screw through the overlap to a stud.
- Galvanized steel with a powder or baked enamel finish or vinyl inside/outside corners and other trim can be used and painted to match the siding finish. Never use aluminum trim components or fasteners because they will corrode when in contact with the siding.

Details of typical lap and panel siding installation are shown in *Figures 45* and *46*.

ELECTRIC (OR PNEUMATIC)
HAND SHEAR

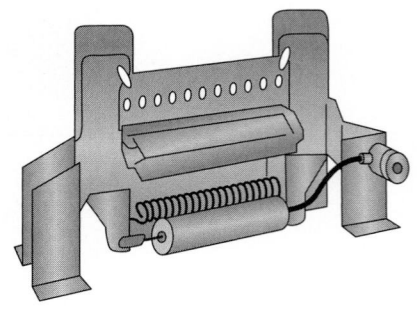

PNEUMATIC SHEAR (LAP SIDING ONLY)

204F44.EPS

Figure 44 ◆ Power shears.

Fiber-Cement Siding Styles

A number of architectural styles of fiber-cement siding are available and can be used to obtain different effects. Planks are available as smooth or wood grained, and panels are available as plain, stucco, and vertical wood grained.

HARDIPANEL® SMOOTH VERTICAL SIDING

HARDIPANEL® STUCCO VERTICAL SIDING

HARDIPANEL® SIERRA-8 VERTICAL SIDING

HARDIPANEL® SIERRA-4 VERTICAL SIDING

204SA14.EPS

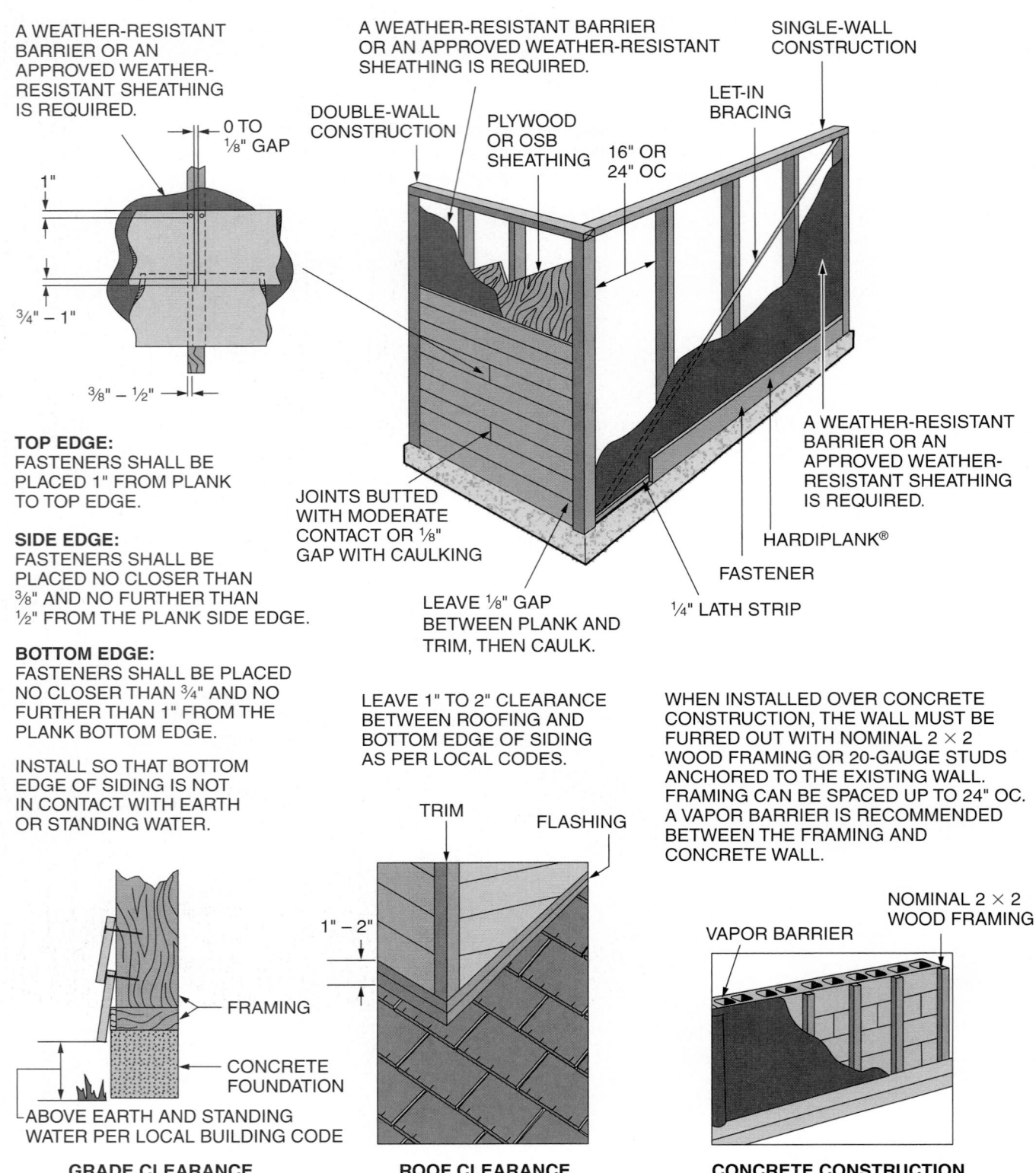

A WEATHER-RESISTANT BARRIER OR AN APPROVED WEATHER-RESISTANT SHEATHING IS REQUIRED.

0 TO 1/8" GAP

1"

3/4" – 1"

3/8" – 1/2"

A WEATHER-RESISTANT BARRIER OR AN APPROVED WEATHER-RESISTANT SHEATHING IS REQUIRED.

DOUBLE-WALL CONSTRUCTION

PLYWOOD OR OSB SHEATHING

16" OR 24" OC

LET-IN BRACING

SINGLE-WALL CONSTRUCTION

A WEATHER-RESISTANT BARRIER OR AN APPROVED WEATHER-RESISTANT SHEATHING IS REQUIRED.

HARDIPLANK®

FASTENER

1/4" LATH STRIP

JOINTS BUTTED WITH MODERATE CONTACT OR 1/8" GAP WITH CAULKING

LEAVE 1/8" GAP BETWEEN PLANK AND TRIM, THEN CAULK.

TOP EDGE:
FASTENERS SHALL BE PLACED 1" FROM PLANK TO TOP EDGE.

SIDE EDGE:
FASTENERS SHALL BE PLACED NO CLOSER THAN 3/8" AND NO FURTHER THAN 1/2" FROM THE PLANK SIDE EDGE.

BOTTOM EDGE:
FASTENERS SHALL BE PLACED NO CLOSER THAN 3/4" AND NO FURTHER THAN 1" FROM THE PLANK BOTTOM EDGE.

INSTALL SO THAT BOTTOM EDGE OF SIDING IS NOT IN CONTACT WITH EARTH OR STANDING WATER.

LEAVE 1" TO 2" CLEARANCE BETWEEN ROOFING AND BOTTOM EDGE OF SIDING AS PER LOCAL CODES.

WHEN INSTALLED OVER CONCRETE CONSTRUCTION, THE WALL MUST BE FURRED OUT WITH NOMINAL 2 × 2 WOOD FRAMING OR 20-GAUGE STUDS ANCHORED TO THE EXISTING WALL. FRAMING CAN BE SPACED UP TO 24" OC. A VAPOR BARRIER IS RECOMMENDED BETWEEN THE FRAMING AND CONCRETE WALL.

FRAMING

CONCRETE FOUNDATION

ABOVE EARTH AND STANDING WATER PER LOCAL BUILDING CODE

GRADE CLEARANCE

TRIM

FLASHING

1" – 2"

ROOF CLEARANCE

VAPOR BARRIER

NOMINAL 2 × 2 WOOD FRAMING

CONCRETE CONSTRUCTION

204F45A.EPS

Figure 45 ◆ Typical fiber-cement lap siding installation details (1 of 2).

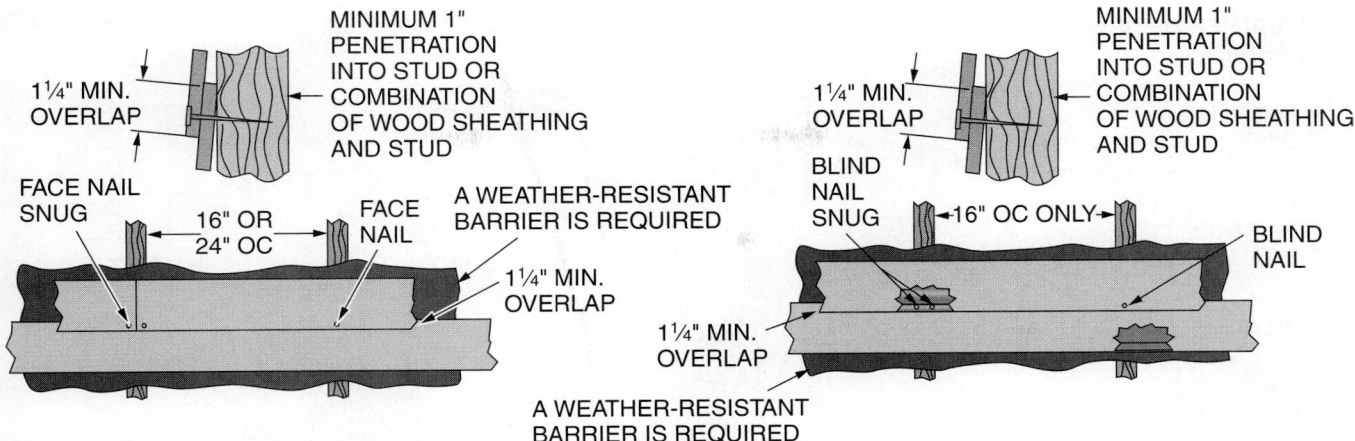

CORROSION RESISTANT NAILS
(GALVANIZED‡ OR STAINLESS STEEL)
• 6d (0.118" shank × 0.267" HD × 2" long)
• Siding nail (0.089" shank × 0.221" HD × 2" long)†
• Siding nail (0.091" shank × 0.221" HD × 1½" long)*
• ET & F pin (0.100" shank × 0.25" HD × 1½" long)†

CORROSION RESISTANT SCREWS
• Ribbed bugle-head or equivalent (No. 8-18 × 0.323" HD × 1⅝" long) Screw must penetrate ¼" or 3 threads into metal framing.

FACE NAILED

CORROSION RESISTANT NAILS
(GALVANIZED‡ OR STAINLESS STEEL)
• Siding nail (0.089" shank × 0.221" HD × 2" long)†
• 11 gauge roofing nail (0.121" shank × 0.371" HD × 1¼" long)
• ET & F Panelfast™ (0.100" shank × 0.25" HD × 1½" long)†

CORROSION RESISTANT SCREWS
• Ribbed bugle-head or equivalent (No. 8–18 × 0.375" HD × 1¼" long) Screws must penetrate ¼" or 3 threads into metal framing.

BLIND NAILED
(NOT APPLICABLE FOR 12" WIDE SIDING)

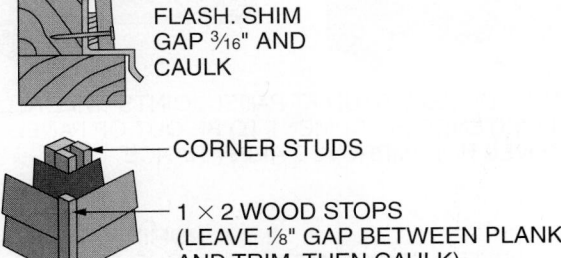

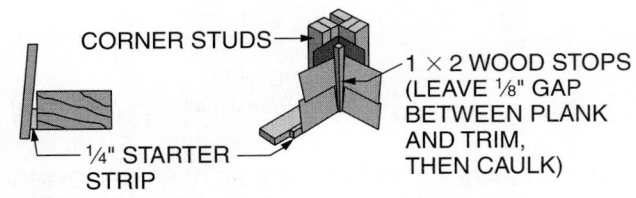

TRIM DETAILS

NOTES:
* For face nail application of 9½" wide or less siding to OSB, fasteners are spaced a maximum of 12" OC.
† The use of a siding nail or roofing nail may not be applicable to all installations where greater windloads or higher exposure categories of wind resistance are required by the Local Building Code. Consult the applicable Building Code Compliance Report.
‡ Hot dipped galvanized nails are recommended.

FASTENING REQUIREMENTS:
• Drive fasteners perpendicular to siding and framing.
• Fastener heads should fit snug against siding (no air space). (Examples 1 & 2)
• Do not underdrive nail heads or drive nails at an angle. (Example 4)
• If nail is countersunk, caulk nail hole and add a nail. (Example 3)

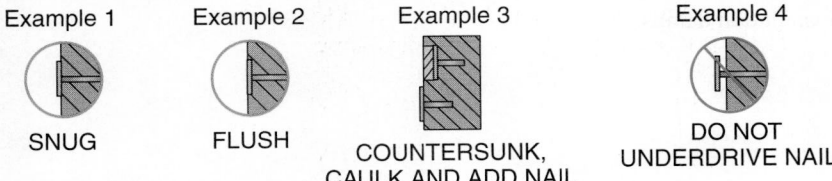

Example 1   Example 2   Example 3   Example 4

SNUG   FLUSH   COUNTERSUNK,   DO NOT
                             CAULK AND ADD NAIL   UNDERDRIVE NAILS

204F45B.EPS

Figure 45 ◆ Typical fiber-cement lap siding installation details (2 of 2).

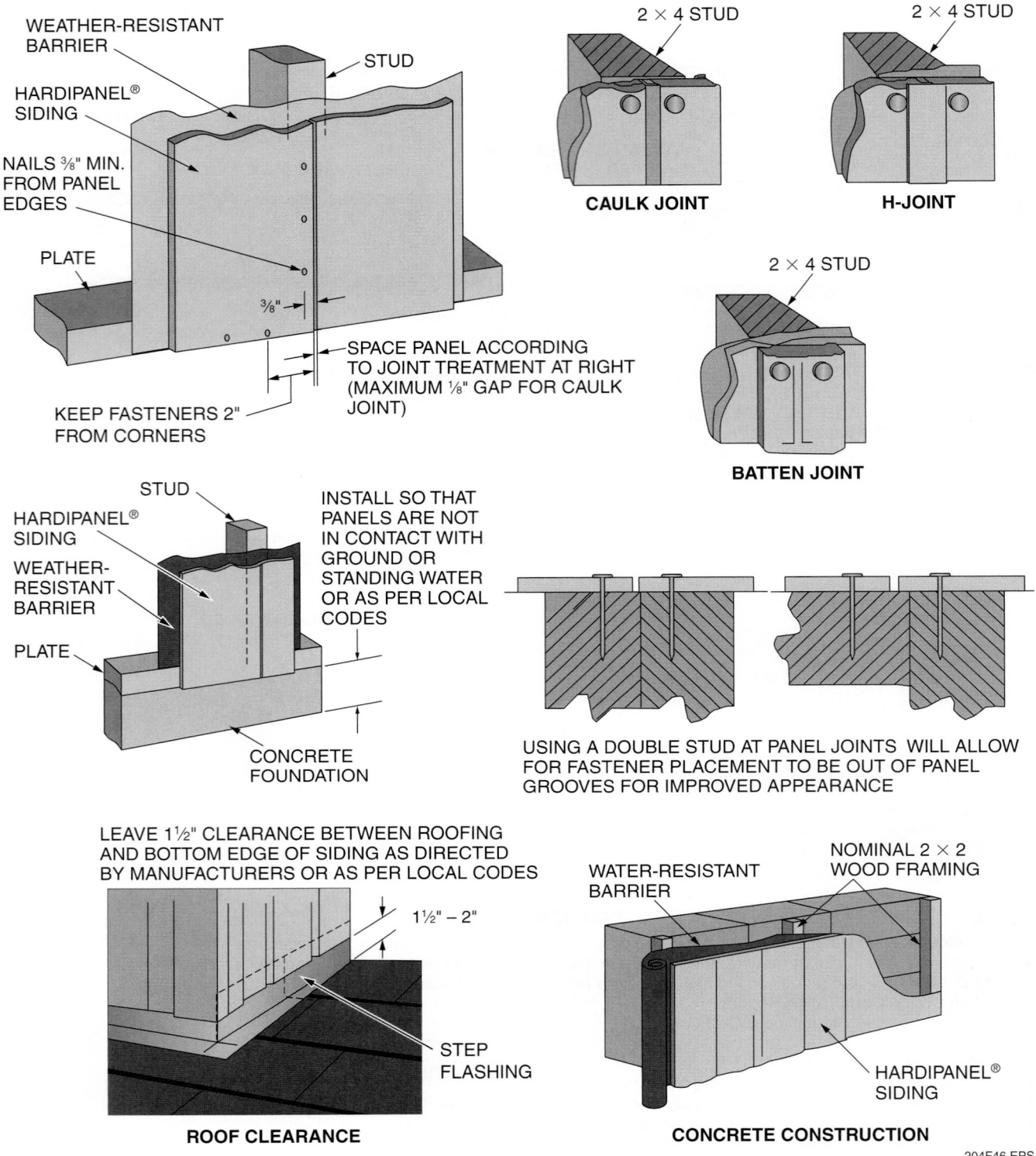

WEATHER-RESISTANT BARRIER

STUD

HARDIPANEL® SIDING

NAILS ⅜" MIN. FROM PANEL EDGES

PLATE

⅜"

SPACE PANEL ACCORDING TO JOINT TREATMENT AT RIGHT (MAXIMUM ⅛" GAP FOR CAULK JOINT)

KEEP FASTENERS 2" FROM CORNERS

2 × 4 STUD

CAULK JOINT

2 × 4 STUD

H-JOINT

2 × 4 STUD

BATTEN JOINT

STUD

HARDIPANEL® SIDING

WEATHER-RESISTANT BARRIER

PLATE

INSTALL SO THAT PANELS ARE NOT IN CONTACT WITH GROUND OR STANDING WATER OR AS PER LOCAL CODES

CONCRETE FOUNDATION

USING A DOUBLE STUD AT PANEL JOINTS WILL ALLOW FOR FASTENER PLACEMENT TO BE OUT OF PANEL GROOVES FOR IMPROVED APPEARANCE

LEAVE 1½" CLEARANCE BETWEEN ROOFING AND BOTTOM EDGE OF SIDING AS DIRECTED BY MANUFACTURERS OR AS PER LOCAL CODES

1½" – 2"

STEP FLASHING

ROOF CLEARANCE

WATER-RESISTANT BARRIER

NOMINAL 2 × 2 WOOD FRAMING

HARDIPANEL® SIDING

CONCRETE CONSTRUCTION

204F46.EPS

Figure 46 ◆ Typical fiber-cement panel siding installation details.

7.0.0 ◆ VINYL AND METAL SIDING

Vinyl, aluminum, and steel siding are applied in new construction as well as over existing finishes for remodeling work. Both vinyl and metal siding are manufactured to look like beveled siding and are available in many colors and finishes. Inside corner posts, door and window trim, individual corner pieces, starter strips, and butt supports are also available. Metal siding and trim are usually supplied with a baked-on or plastic finish. Many manufacturers offer siding with a rigid insulating backing board. This makes the siding less susceptible to exterior damage and also increases its rigidity.

7.1.0 Vinyl and Metal Siding Materials and Components

Figure 47 shows some of the styles of horizontal siding materials currently available. *Figure 48*

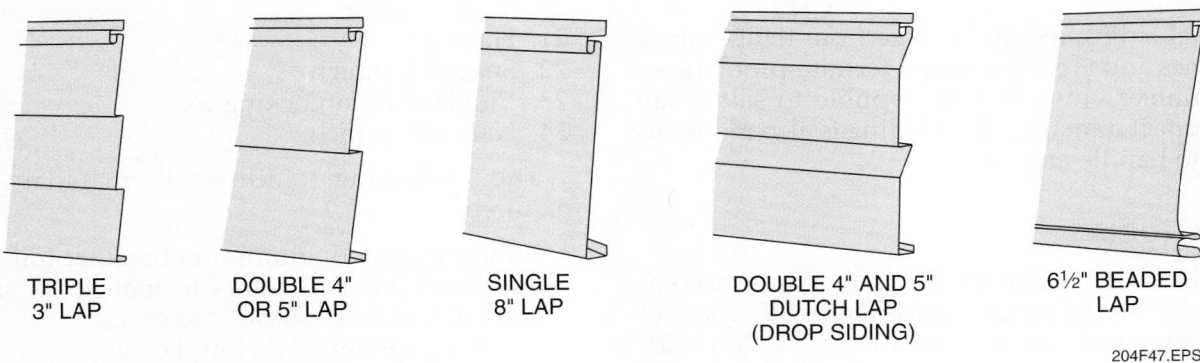

| TRIPLE 3" LAP | DOUBLE 4" OR 5" LAP | SINGLE 8" LAP | DOUBLE 4" AND 5" DUTCH LAP (DROP SIDING) | 6½" BEADED LAP |

204F47.EPS

Figure 47 ◆ Typical vinyl and metal siding materials.

HORIZONTAL STARTER STRIP   STURDY VINYL STARTER STRIP   J-CHANNEL (VARIOUS SIZES)   WIDE FLANGE J   WIDE CROWN MOLDING

UNDERSILL TRIM   H-MOLDING   NARROW COVE MOLDING   DRIP CAP

TEXTURED OR SMOOTH F-TRIM   WINDOW/ DOOR TRIM   OUTSIDE CORNER POST (VARIOUS SIZES)   SNAP-ON OUTSIDE CORNER POST (USE WITH J-CHANNEL)

FLUTED CORNER POST   INSIDE CORNER POST (VARIOUS SIZES)   8" FASCIA   FLEXIBLE J-CHANNEL   ALUMINUM TRIM COIL

TRIM COIL   ALUMINUM STARTER STRIP   J-VENT™   J-BLOCK®   MINI J-BLOCK®

204F48.EPS

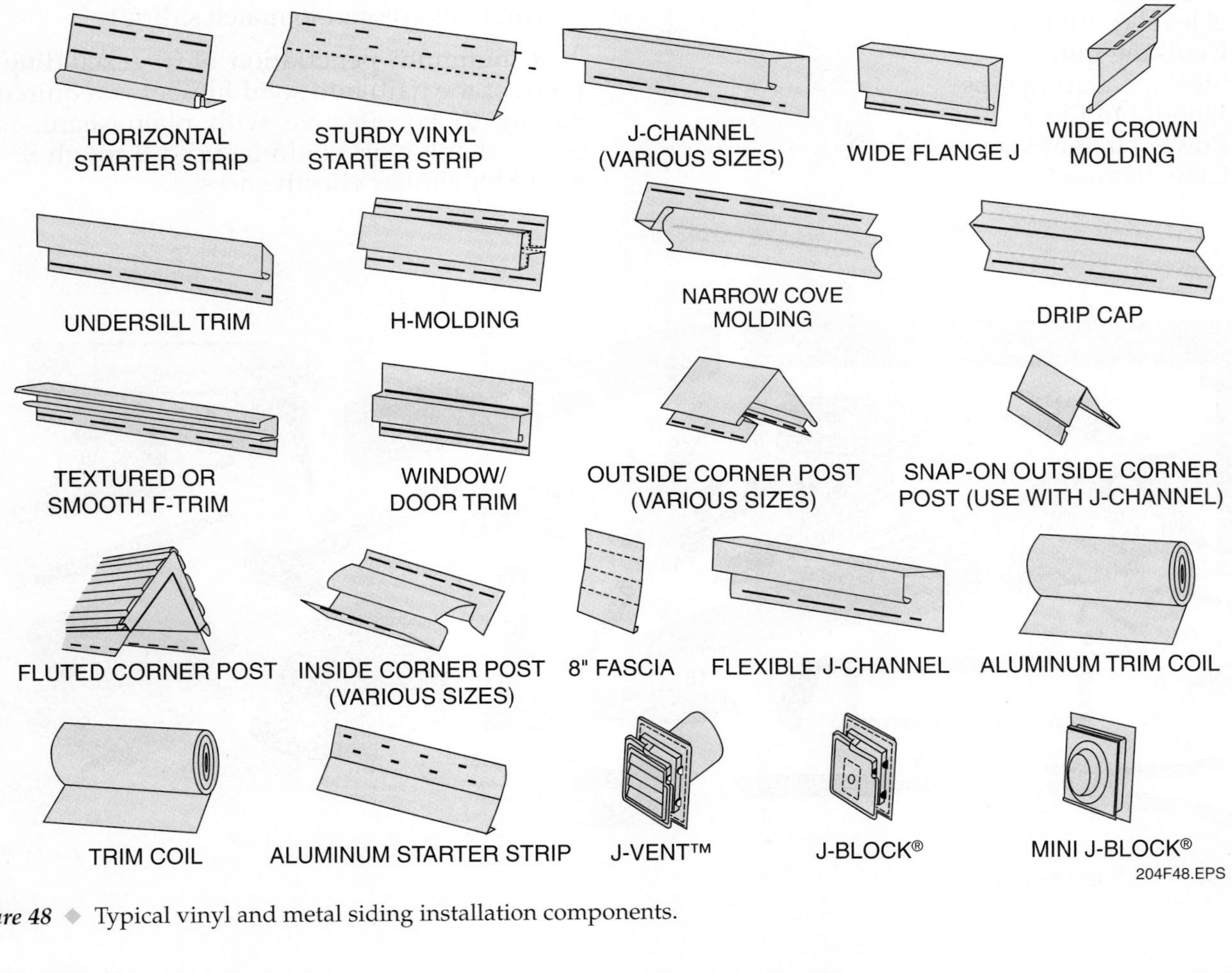

Figure 48 ◆ Typical vinyl and metal siding installation components.

shows a variety of installation components used with vinyl or metal siding. Horizontal metal siding is usually limited to single- or double-lap styles.

The major advantages of vinyl siding are its low cost, ease of handling and installation, and resistance to denting. It is also colorfast in sunlight, waterproof, rot-proof, and termite-proof. Its major disadvantage is that its resistance to impact damage is very low in cold temperatures. If installed improperly, it will break during expansion or contraction over wide temperature variations.

Metal siding resists damage from temperature extremes and is rot-proof and termite-proof; however, unlike vinyl, it is susceptible to salt spray and impact damage. Metal siding is also more difficult to handle and install.

7.2.0 Tools

The installation of metal siding should be accomplished with the proper tools *(Figure 49)*. Most of these same tools are also used to install vinyl siding.

1. Carpenter's metal square
2. Carpenter's folding rule
3. 2' level (minimum)
4. Caulking gun
5. Steel measuring tape
6. Fine-tooth file
7. Power circular saw (optional)
8. Claw hammer

9. Chalkline
10. Screwdriver
11. Pliers
12. Tin snips (duckbill type) or power hand shears
13. Aviation shears (double acting)
14. Carpenter's saw (crosscut)
15. Safety goggles
16. Steel awl
17. Fine-tooth hacksaw (24 teeth per inch)
18. Utility knife
19. Line level or water level
20. 3" putty knife
21. Hard hat
22. Snaplock punch
23. Vinyl siding unlocking tool
24. Nail hole punch

The following additional materials are required:

- Building wrap or aluminum breather foil
- Touch-up paint in colors to match the siding (for kitchen fans, service cables, etc.)
- Caulking (preferably a butyl caulk)
- Aluminum, plain, or screw-shank nails (1½" for general use; 2" for re-siding; 2½" or more to nail insulated siding into soffit sheathing; 1" to 1½" trim nails colored to match siding)

A minimum penetration of ¾" (excluding the point of the nail) into solid lumber is required for nailing to be effective with plain shank nails. Screw shank nails could be used through ½" plywood for similar effectiveness.

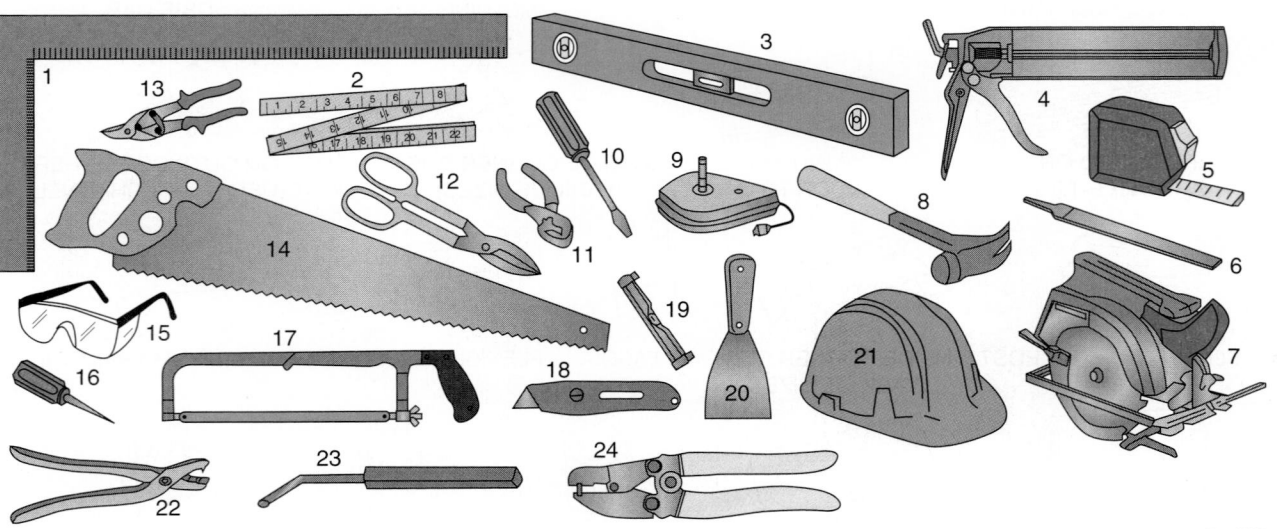

Figure 49 ◆ Siding application tools.

204F49.EPS

7.3.0 Equipment

The following equipment is required when applying metal or vinyl siding:

- *Ladders and scaffolds* – Proper ladders and scaffolds are necessary. The aluminum adjustable scaffolding systems are widely used to provide a working platform. With these systems, the distance from the building facade remains the same from the bottom to the top. Exact specifications on spacing dimensions, planking, permissible heights and loads, and other details are contained in *Occupational Safety and Health Standards for the Construction Industry (CFR 29, Part 1926)*.
- *Portable brake* – For job-site bending of custom trim sections such as fascia trim, window casing, and sill trim, a portable metal-bending machine (brake) is extremely useful (*Figure 50*). Utilizing white or colored coil stock, precision bending, including multiple bends, can be accomplished. These machines are lightweight and can be carried to the job site and set in place. Various sizes and brake styles are available. As shown in *Figure 50*, some are equipped with a lengthwise rolling cutter to allow sizing the trim stock to the desired width.
- *Cutting table* – A cutting table allows a standard portable circular power saw to be mounted in a carrier and held away from the work to avoid damaging the siding. This table can be used for measuring and crosscutting, as well as for making miters and bevels. The table is constructed of lightweight aluminum and can be easily set up on the job site by one worker.

7.4.0 Surface Preparation

The quality of the finished job depends on good preparation of the work surface, especially for remodeling work. Keep the following points in mind when preparing the surface:

- Check for low places in the plane of the wall and build out (shim out) if required.
- Prepare the entire building a few courses at a time. Securely nail all loose boards and loose wood trim. Replace any rotted boards.
- Scrape away old paint buildup, old caulking, and hardened putty, especially around windows and doors where it might interfere with the positioning of new trim. New caulk should be applied to prevent air infiltration.
- Remove downspouts and other items that will interfere with the installation of new siding.
- Tie shrubbery and trees back from the base of the building to avoid damage.
- Window sill extensions may be cut off so that J-trim can be installed flush with the window casing (*Figure 51*). However, if the building owner wishes to maintain the original window design, coil stock can be custom-formed around the sill instead of cutting away the sill extensions.

7.5.0 Furring and Insulation Techniques

Furring strips may be required in order to provide a smooth, even base for nailing on the new siding. Normally, ⅜" thick wood lath strips are used over

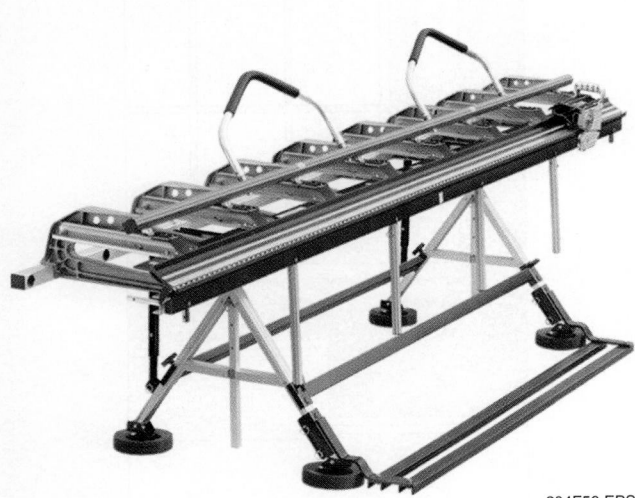

Figure 50 ◆ Portable brake.

204F50.EPS

Figure 51 ◆ Cutting sill extensions.

204F51.EPS

wood construction and 1 × 3 strips are used over brick and masonry. Furring is not usually necessary in new construction, but older homes often have uneven walls, and furring out low spots or shimming can help to prevent the siding from appearing wavy. If at all possible, it is preferable to remove any old exterior siding down to the sheathing to avoid furring. If extensive furring is used under vinyl siding, a backer board may be required for reinforcement of the siding.

For horizontal siding, the furring should be installed vertically at 16" OC. The air space at the base of the siding should be closed off with strips applied horizontally. Window, door, gable, and eave trim may have to be built out to match the thickness of the wall furring.

The furring for vertical siding is essentially the same as for horizontal siding, except the wood strips are securely nailed horizontally into structural lumber on 16" to 24" centers. When using 1 × 3 furring, be sure to check what effect, if any, the additional thickness will have on the use of trim (see *Figure 52*).

7.5.1 Aluminum Foil Underlayment

Aluminum reflector foil, if used, is a good insulator and can be used advantageously as an underlayment to siding. It may be stapled directly to the existing wall or over ¾" furring strips to provide an additional air space and better insulation. Reflector foil for remodeling must be of the perforated or breather type to allow for the passage of water vapor.

Foil should be installed with the shiny side facing the air space (outward with no furring; inward if applied over furring). Foil is generally available in 36"- and 48"-wide rolls. Nail or staple the foil just before applying the siding.

When applying foil over furring, be careful not to let the foil collapse into the air space. Place the foil as close as possible to openings and around corners where air leaks are likely to occur. Overlap the side and end joints by 1" to 2".

7.5.2 Window and Door Build-Out

Some trim build-out at windows and doors may be required to maintain the original appearance of the house when using furring strips or underlayment board. This is particularly true when the strips or underlayment board are more than ½" thick. Thicker furring and underlayment generally provide added insulation value and are usually a good investment for the homeowner, particularly if the home is uninsulated.

7.5.3 Undersill Furring

Building out below the window sill is often required in order to maintain the correct slope angle if a siding panel needs to be cut to less than full height. The exact thickness required will be apparent when the siding courses have progressed up the wall and reached this point (see *Figure 53*).

7.5.4 Under-Eave Furring

For the same reason, furring is usually required to maintain the correct slope angle if the last panel needs to be cut to less than full height. The exact thickness required will be apparent when the siding courses have progressed up the wall and reached this point (see *Figure 54*).

204F52.EPS

Figure 52 ◆ Furring for vertical siding.

204F53.EPS

Figure 53 ◆ Undersill furring.

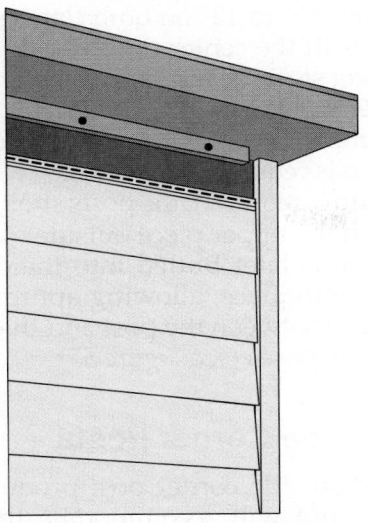

Figure 54 ◆ Under-eave furring.

7.6.0 Establishing a Straight Reference Line

The key element in a successful siding installation is establishing a straight reference line upon which to start the first course of siding. The suggested procedure is to measure equal distances downward from the eaves. This ensures that the siding appears parallel with the eaves, soffits, and windows, regardless of any actual settling of the house from true level. See *Figure 55*.

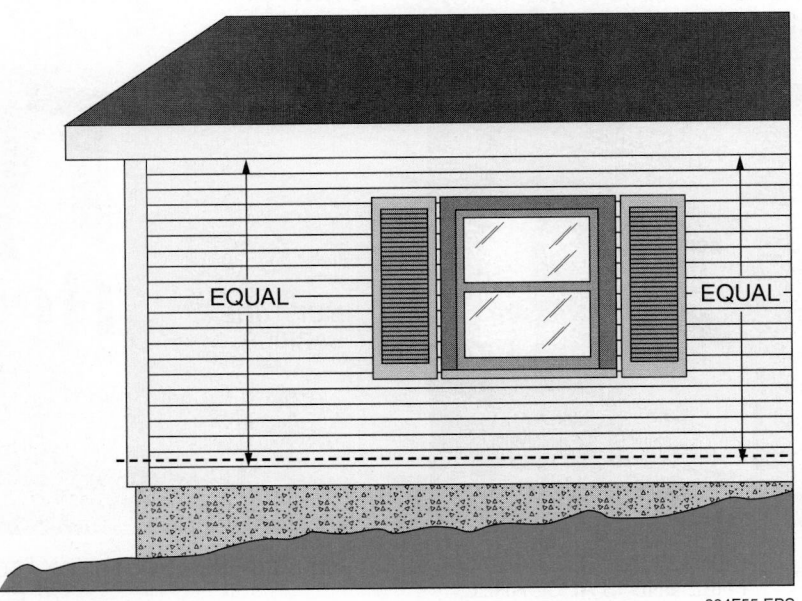

Figure 55 ◆ Straight reference line.

To establish a reference line, find the lowest corner of the house. Partly drive a nail about 10" above the lowest corner, or high enough to clear the height of a full siding panel. Stretch a taut chalkline from this corner to a similar nail installed at another corner. Reset this line based upon measuring down equal distances from points on the eaves. Repeat this procedure on all sides of the house until the chalklines meet at all corners.

Before snapping the chalklines, check for straightness. Be alert to sag in the middle, particularly if a line is more than 20' long. If preferred, lines may be left in place while installing the starter strip, as long as they are checked periodically for excess sag.

If the house is level, an alternative is to use a water level (*Figure 56*) or a transit to set the chalkline approximately 2" (or the width of the starter strip) from the lowest point of the old siding and locate the top of the starter strip at that line. Take the level reading at the corners and centers of the chalkline for best results. The water level can be used for measurements up to 100', and is accurate to ±$\frac{1}{16}$" at 50'.

7.7.0 Inside Corner Posts

The inside corner posts are installed before the siding is hung. Depending on the type of siding (insulated or non-insulated), deeper or narrower posts may be required. The post is set in the corner full length, reaching from $\frac{1}{4}$" below the bottom of the starter strip up to $\frac{1}{4}$" from the eave or gable trim. Nail the upper slot at the top of the slot, then nail

approximately 8" to 12" on both flanges with aluminum nails in the center of the slots. Make sure the post is set straight and true. The flange should be nailed securely to the adjoining wall, but do not overdrive the nails so as to cause distortion. If a short section is required, use a hacksaw to cut it. If a long section is required, the posts should be overlapped, with the upper piece outside.

The siding is later butted into the corner and then nailed into place, allowing approximately a $\frac{1}{16}$" to $\frac{1}{4}$" space between the post and the siding for expansion purposes (see *Figure 57*).

7.8.0 Outside Corner Posts

If used, the outside corner post produces a trim appearance and will accommodate the greatest variety of siding types. Most outside corner posts are designed to be installed before the siding is hung, in a manner similar to the inside corner post. If desired, old corner posts may sometimes be removed. Set a full-length piece over the existing corner, running from $\frac{1}{4}$" below the bottom of the starter strip to $\frac{1}{4}$" from the underside of the eave. If a long corner post is needed, overlap the corner post sections, with the upper piece outside.

Nail the uppermost slot at the top of the slot, then nail approximately every 8" to 12" with aluminum nails on both flanges in the center of the slots. Make sure the flanges are securely nailed (*Figure 58*), but avoid distortion caused by overdriving nails. Use a hacksaw to cut short sections, if required. If insulated siding is being used, wider corner posts are needed.

ELECTRONIC LEVEL SENSOR UNIT WITH BEEPER PLACED AT DESIRED LEVEL

BEEPER SOUNDS WHEN FREE END IS AT CORRECT LEVEL

204F56.EPS

Figure 56 ◆ Electronic water level.

MOISTURE BARRIER

INSIDE CORNER POST

$\frac{1}{16}$" TO $\frac{1}{4}$" EXPANSION GAP AT END OF SIDING INSERTED IN SLOT OF INSIDE CORNER POST

204F57.EPS

Figure 57 ◆ Inside corner post.

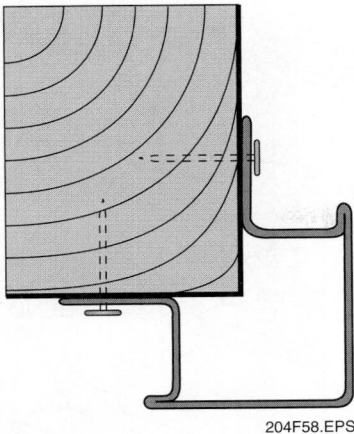

Figure 58 ◆ Correct nailing of flanges.

Figure 59 ◆ Installing a starter or J-channel strip.

7.9.0 Starter Strip

Using the chalkline previously established as a guide, take equal distance measurements, as shown in *Figure 59*, and install the starter strip or J-channel strip all the way around the bottom of the building depending on the material at the base of the building. If insulated siding is used, the starter strip should be furred out to a distance equal to the thickness of the backer. It is very important that the starter strip be straight and meet accurately at all corners because it will determine the line of all siding panels installed. Where hollows occur in the old wall surface, shim out behind the starter strip to prevent a wavy appearance in the finished siding.

When using individual corner caps, install the starter strip up to the edge of the house corner. Use aluminum nails spaced not more than 8" apart to fasten the starter strip. Nail the starter strip as low as possible. Be careful not to bend or distort the strip by overdriving the nails. The strip should not be nailed tight. Cutting lengths of starter strip is best accomplished with tin snips. Butt the sections together.

Starter strips may not work in all situations. For example, other accessory items, such as J-channels or all-purpose trim, may work better in starting the siding course over garage doors and porches or above brick. These situations must be handled on an individual basis as they occur.

7.10.0 Window and Door Trim

For a superior job in remodeling work, old window sills and casings can be covered with aluminum coil stock that is bent to fit on the job site. The advantage is freedom from maintenance.

Sometimes, window and door casings need to be built out to retain the original appearance of the house or to improve the appearance. To do this, use appropriate lengths and thicknesses of good-quality lumber and nail them securely to the existing window casings. Remove any storm windows before covering the casings with aluminum coil stock sections custom-formed on the job site.

Forming aluminum sections to fit window casings is done using a portable brake. Door casings are handled in a similar manner.

Figure 60 shows the installation of aluminum window trim.

If there is a step in the wood sill, it can best be covered by bending two separate sill cover pieces with interlocking flanges, as shown in *Figure 61*. By using tin snips and bending flanges on the job, the old sill ends can be boxed in to provide a neat appearance and to prevent water penetration.

J-channel is used around windows and doors to receive the siding. Side J-channel members are cut longer than the height of the window or door and notched at the top. Notch the top J-channel member at a 45-degree angle and bend the tab down to provide flashing over the side members. Caulking should be used behind J-channel members to prevent water infiltration between the window and the channel. See *Figures 62* and *63*.

To provide protection against water infiltration, a flashing piece, which is cut from coil stock or a precut piece of step flashing, is slipped under the base of the side J-channel members. It should be positioned so that it overlaps the top lock of the panel below, as shown in *Figure 64*.

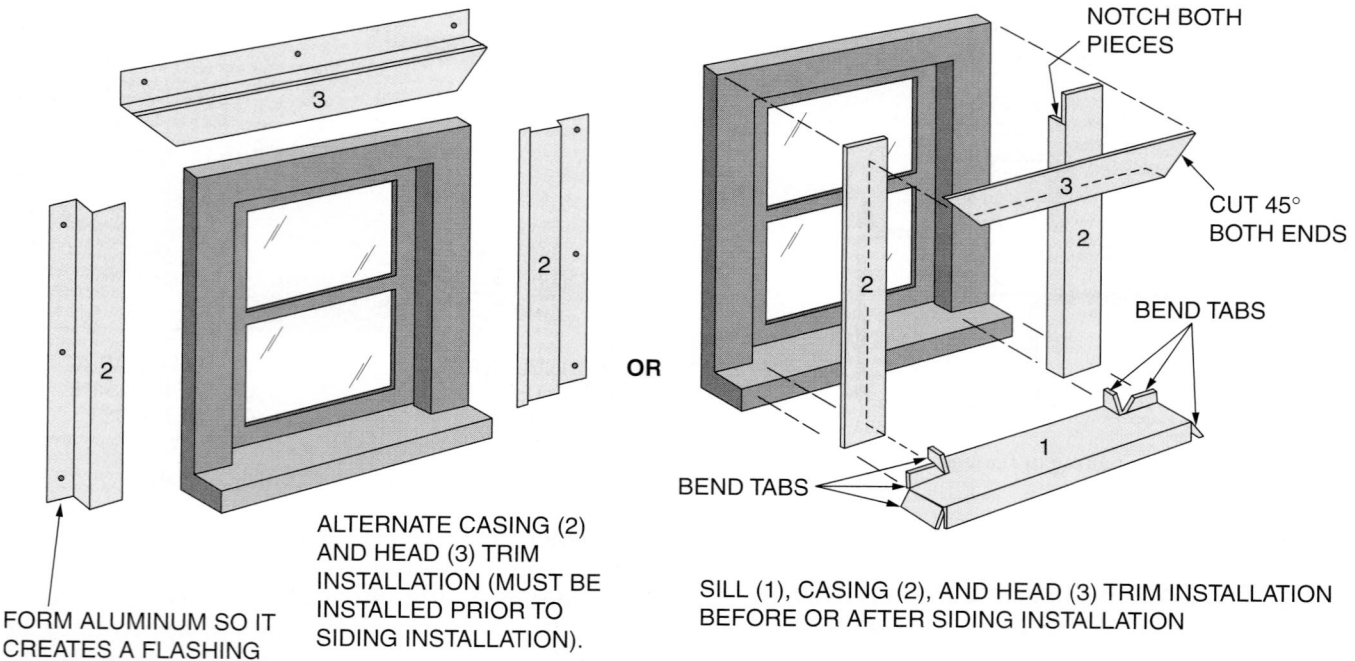

FORM ALUMINUM SO IT CREATES A FLASHING STRIP AGAINST SHEATHING (ALSO CAN BLIND NAIL FOR BETTER APPEARANCE).

ALTERNATE CASING (2) AND HEAD (3) TRIM INSTALLATION (MUST BE INSTALLED PRIOR TO SIDING INSTALLATION).

NOTCH BOTH PIECES

CUT 45° BOTH ENDS

BEND TABS

BEND TABS

SILL (1), CASING (2), AND HEAD (3) TRIM INSTALLATION BEFORE OR AFTER SIDING INSTALLATION

204F60.EPS

Figure 60 ◆ Installing aluminum window trim.

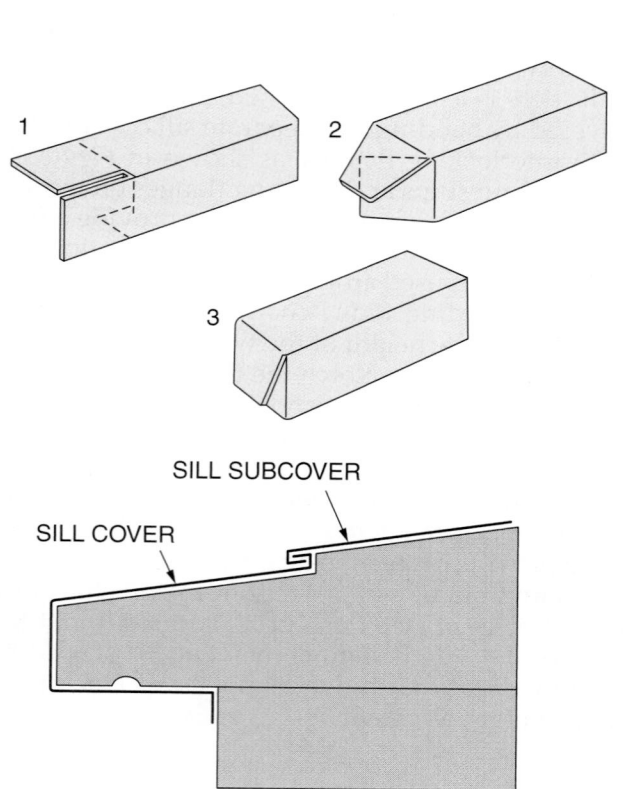

SILL SUBCOVER

SILL COVER

204F61.EPS

Figure 61 ◆ Boxing in sill ends.

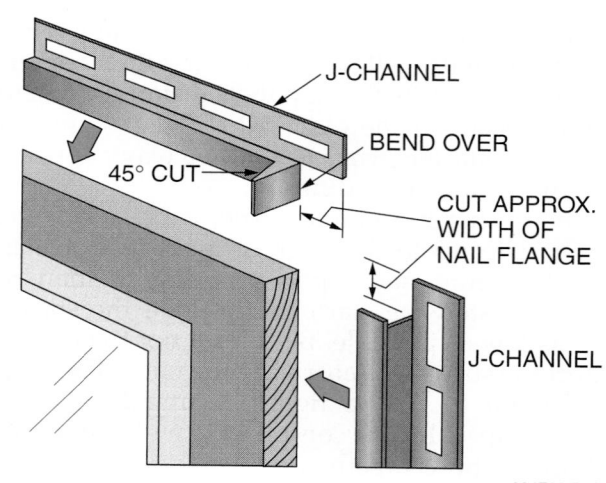

J-CHANNEL

BEND OVER

45° CUT

CUT APPROX. WIDTH OF NAIL FLANGE

J-CHANNEL

204F62.EPS

Figure 62 ◆ Cutting J-channel.

7.11.0 Gable End Trim

Before applying the siding, J-channels should be installed to receive the siding at the gable ends, as shown in *Figure 65*. Where the left and right sections meet at the gable peak, allow one of the sections to butt into the peak, with the other section overlapping it. A miter cut is made on the face flange of this piece to provide a better appearance. All old paint buildup should be removed before installing J-channels. Nail the J-channels every 12" using aluminum nails.

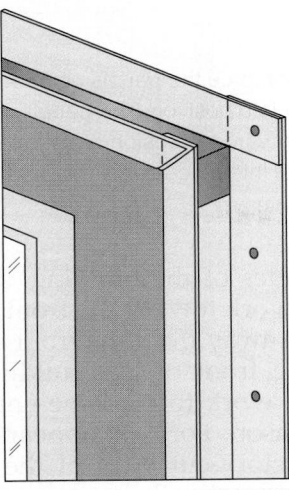

Figure 63 ◆ J-channel.

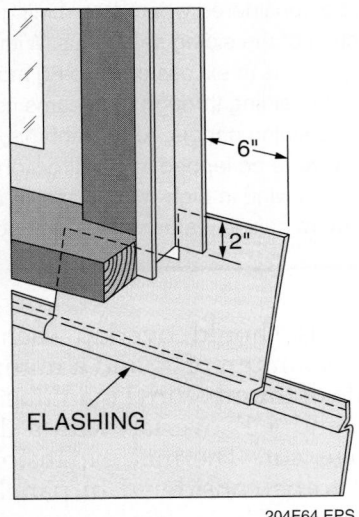

FLASHING

Figure 64 ◆ Installing a piece of flashing.

7.12.0 Cutting Procedures

For precision cutting, a power saw is the most convenient tool to use. Cutting one panel at a time is recommended. A special jig that will keep the saw base clear of the work is preferred in order to prevent damaging panels. For vinyl, reverse a fine-tooth blade to produce a smoother cut. For aluminum or steel, use a minimum 10-tooth aluminum cutting blade or an abrasive cutting blade. A bar of soap may be rubbed on the blade to produce a smoother cut on the siding panel and prolong blade life. Feed the saw through the work slowly to prevent flutter against the blade. Safety goggles must be worn at all times while operating a power saw.

Individual panels can be cut with tin snips. Start by drawing a line across the panel using a square. Begin cutting at the top lock first *(Figure 66)* and continue toward the bottom of the panel. For metal panels, break the panel across the butt edge and snip through the bottom lock. For metal siding, use a screwdriver to reopen the lock, which may become flattened by the tin snips *(Figure 67)*.

Aviation shears are sometimes used to cut the top and bottom locks, and a utility knife is used to score and break the face of the panel and to cut vinyl panels. For straight cuts, the best choice is duckbill snips. For aluminum, a heavy score is made on the panel and the piece is bent back and forth until it snaps cleanly along the score line.

On window cutouts, a utility knife and tin snips may be used for both vinyl and aluminum. Use duckbill tin snips to cut accessories such as all-purpose trim, J-channel, and starter strips. Use a hacksaw to cut accessories such as corner posts.

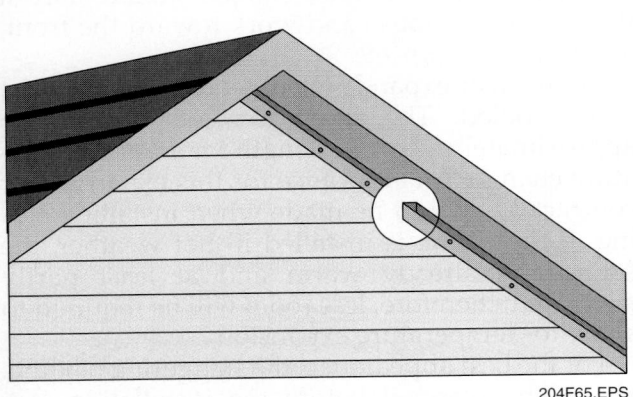

Figure 65 ◆ Gable end trim.

Figure 66 ◆ Using tin snips.

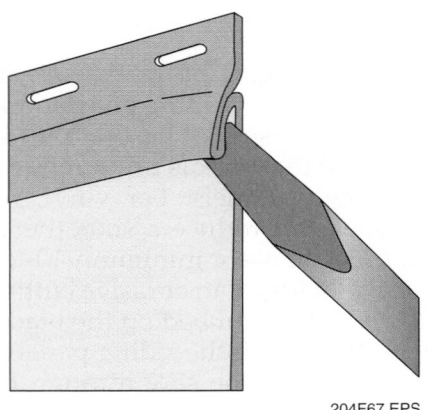

Figure 67 ◆ Reopening a lock after cutting.

204F67.EPS

7.13.0 Siding and Corner Cap Installation

This section covers siding and corner cap installation.

7.13.1 Siding Installation

For metal siding, check with local building codes to see if they require that the first course of siding be grounded to reduce the danger from lightning.

Extra care must be taken when applying the first course of siding because it establishes the base for all other courses. Apply a panel by hooking the bottom lock of the panel into the interlock bead of the starter strip. Make sure the lock is engaged. Do not force it, which might cause distortion of the panel and result in a warped shadow line. Double check for continuous locking along the panel before proceeding further. Also, check carefully for proper alignment at the corners.

At the corner posts, slide the panel into the recess first, then exert upward pressure to lock the panel into place along its entire length. Allow clearance for expansion, as necessary. If individual corner caps are being used, keep the panels back from the corner edges (¾" for non-insulated siding and ¼" for insulated siding) to allow for later fitting of the individual corners. Panels must be hung with aluminum nails through the center of the factory-slotted holes every 16" to 24" along their entire lengths or as specified by the manufacturer. Nails must be driven into sound lumber, such as ¾" penetration into house framing with plain shank nails or through ½" plywood with screw shank nails. Nails or screws should be set about ¹⁄₁₆" to ⅛" away from the panel. If they are set tight, the panels may warp, bend, or crack due to expansion and contraction. On low spots, fasten the panel on both sides of the low spot and allow the panel to float over the low spot.

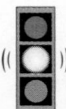

CAUTION

Do not force the panels up or down when nailing them into position. The panels should not be under vertical tension or compression after being nailed into place.

On the sides of the building, start at the rear corner and work toward the front so that the lapping will be away from the front and less noticeable. On the front of the building, start at the corners and work toward the entrance door for the same reason. For best appearance when lapping, the factory-cut ends of the panels should cover the field-cut ends.

NOTE

In some cases, prevailing wind direction must also be considered when determining the direction of the siding seam laps. With some vinyl siding, winds in excess of 50 to 60 mph entering under the siding through the seams can tear off multiple siding panels. To prevent this, the seams may have to be lapped in the direction of the prevailing wind in high-wind areas. In addition, the maximum nailing distance must not be exceeded.

Metal panels should overlap each other by about ½". A maximum of ⅝" and a minimum of ⅜" is a good rule of thumb. Vinyl manufacturers usually recommend a 1" overlap with a double-size nailing flange cut. Thermal expansion requirements need to be considered in panel overlaps. Cut away the top lock strip on the overlapped panel by twice the amount of the intended overlap (see *Figure 68*).

Avoid short panel lengths of under 24" and make sure that the factory-cut ends are always on top of the field-cut ends. The job should start at the rear of the house and work toward the front, as shown in *Figure 69*.

Siding will expand when heated and contract when cooled. The expansion will amount to approximately ⅛" per 10' length for a 100° temperature change. An allowance for this expansion or contraction should be made when installing siding. If the siding is installed in hot weather, the product is already warm and at least partly expanded; therefore, less room will be required to allow for temperature expansion.

For the best appearance, the staggering of joints should be planned before the installation (see *Figure 70*). Avoid installing siding in a set pattern.

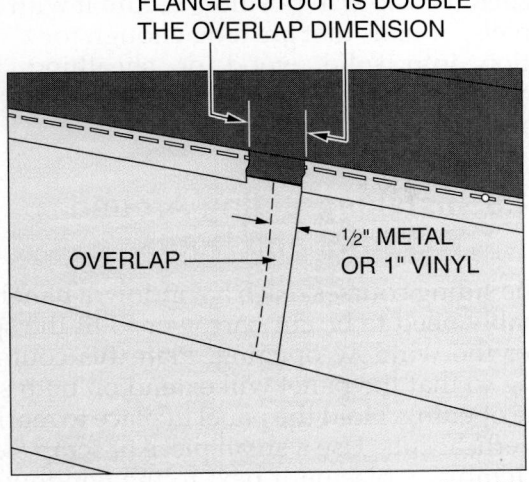

FLANGE CUTOUT IS DOUBLE THE OVERLAP DIMENSION

OVERLAP

½" METAL OR 1" VINYL

204F68.EPS

Figure 68 ◆ Overlapping panels.

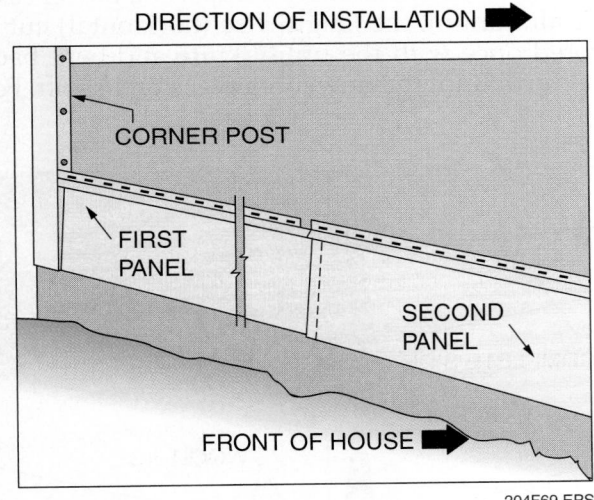

DIRECTION OF INSTALLATION ▶

CORNER POST

FIRST PANEL

SECOND PANEL

FRONT OF HOUSE ▶

204F69.EPS

Figure 69 ◆ Sequence of installation.

204F70.EPS

Figure 70 ◆ Proper staggering of joints.

INSIDE TRACK

A High-Windload Vinyl Siding

One manufacturer offers a flexible-hem vinyl siding that is designed to be installed under vertical tension and nailed tight. This allows the panel to float over low spots and move to accommodate expansion and contraction. When fastened with staples, its specifications indicate that it will withstand winds of up to 235 mph.

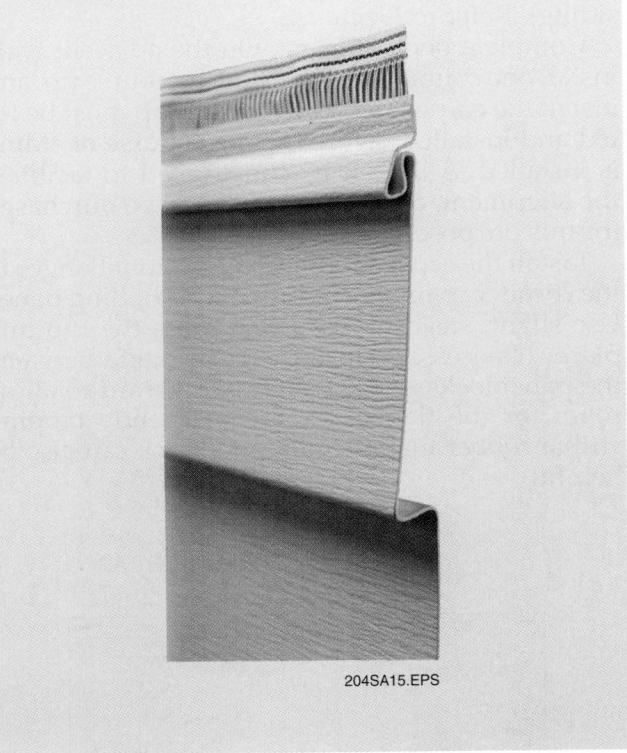

204SA15.EPS

A set pattern may be more labor- and cost-effective, but results in a poor overall appearance. It is best to plan the job so that any two joints in line vertically will be separated by at least two courses. At a bare minimum, separate panel overlaps on the next course by at least two feet. Joints should be avoided on panels directly above and below windows. Shorter pieces that develop as work proceeds can be used for smaller areas around windows and doors.

Backer tabs are used with 8" horizontal noninsulated aluminum siding only. They ensure rigidity, evenness of installation, and tight endlaps. They are used at all panel overlaps and behind panels entering corners. After the panel has been locked into place, slip the backer tab behind the panel with the flat side facing out *(Figure 71)*. The backer tab should be directly behind and even with the edge of the first panel of the overlap. Nail the backer into place.

7.13.2 *Corner Cap Installation for Aluminum Siding*

Individual corner caps, if used, may be used for 8" horizontal aluminum lap siding instead of outside corner posts. The siding courses on adjoining walls must meet evenly at the corners. To allow room for the cap, install the siding with ¾" clearance from the corner (¼" clearance for insulated siding). Refer to *Figure 72*.

Complete one wall first. On the adjacent wall, install one course of siding, line the course up, and install the corner cap. Each corner cap must be fitted and installed before the next course of siding is installed. A jig can be constructed to facilitate the alignment, or a special tool may be purchased for this purpose.

Install the cap by slipping the bottom flanges of the corner cap under the butt of each siding panel. Use slight, steady pressure to press the cap into place. If necessary, insert a putty knife between the panel locks, prying slightly outward to allow room for the flanges to slip in. Gentle tapping with a rubber mallet and wood block can also be helpful.

When the cap is in position, secure it with 2" or 2½" nails, or nails that are long enough for ¾" penetration into solid wood or sheathing. Nail through at least one of the prepunched nail holes in the top of the corner cap.

7.14.0 Installing Siding Around Windows and Doors

As the siding courses reach a window, a panel will probably need to be cut narrower to fit the space under the window opening. Plan this course of siding so that the panel will extend on both sides of the opening. Hold the panel in place to mark for the vertical cuts. Use a small piece of scrap siding as a template, placing it next to the window and locking it into the panel below *(Figure 73)*. Make a mark on this piece ¼" below the sill height to allow clearance for all-purpose trim. Do the same on the other side of the window, because windows are not always absolutely level.

The vertical cuts are made from the top edge of the panel with duckbills, tin snips, or a power saw. For aluminum, the lengthwise (horizontal) cut is scored once with the utility knife and bent back and forth until the unwanted piece breaks off. For

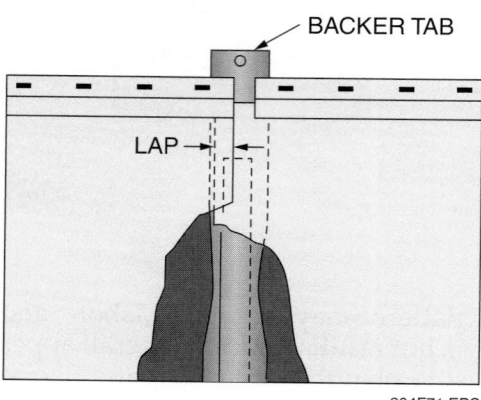

204F71.EPS

Figure 71 ◆ Inserting a backer tab for metal siding.

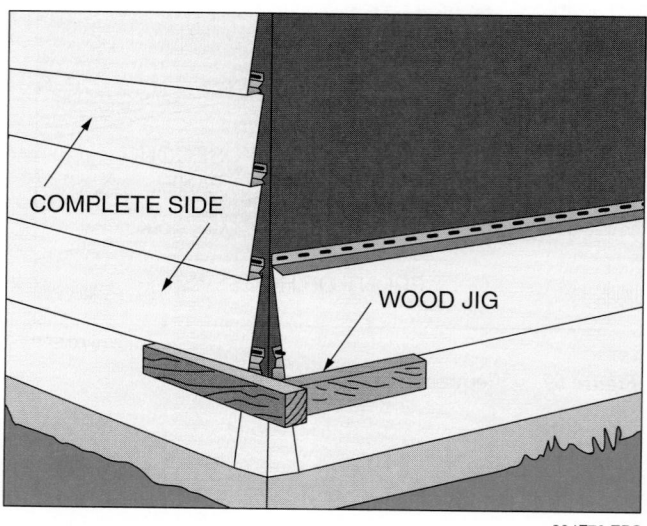

204F72.EPS

Figure 72 ◆ Leaving room for corner caps.

Aluminum Siding Corner Caps

Aluminum siding corner caps are similar to those used for wood lap siding corners. Before nailing the corner caps, always make sure that the cap and siding panels are flush.

vinyl, the horizontal cut is made with a utility knife. For steel, the horizontal cut is made with tin snips.

The raw edge of the panel should be trimmed with all-purpose trim for the exact width of the sill. First, determine if furring is required behind the cut edge to maintain the slope angle with adjacent panels. Nail the correct thickness of furring under the sill and install all-purpose trim over it with aluminum nails, close up under the sill for a tight fit (see *Figure 74*). On vinyl panels, use a snaplock punch to place raised ears on the raw edge of the panel. Slide the panel upward so as to engage the undersill or J-trim, the J-channels on the window sides, and the lock of the panels below.

Fitting panels over door and window openings is almost the same as making undersill cutouts, except that the clearances for fitting the panel are different. The cut panel on top of the opening needs more room to move down to engage the interlock of the siding panel below on either side of the window. Mark a scrap piece template without allowing clearance and then make saw cuts ¼" to ⅜" deeper than the mark *(Figure 75)*. This will provide the necessary interlock clearance.

Check the need for furring over the top of the window or door in order to maintain the slope angle and install it, if required *(Figure 76)*. Make sure the furring is pressure-treated and is spaced off the bottom of the J-channel.

Cut a piece of all-purpose trim the same width as the raw edge of the cut panel and slip it over this edge of the panel before installing it. Drop the panel into position, engaging the interlocks on the siding panels below. The all-purpose trim can now be pushed downward to close any gap at the juncture with the J-channel. Refer to *Figure 77*.

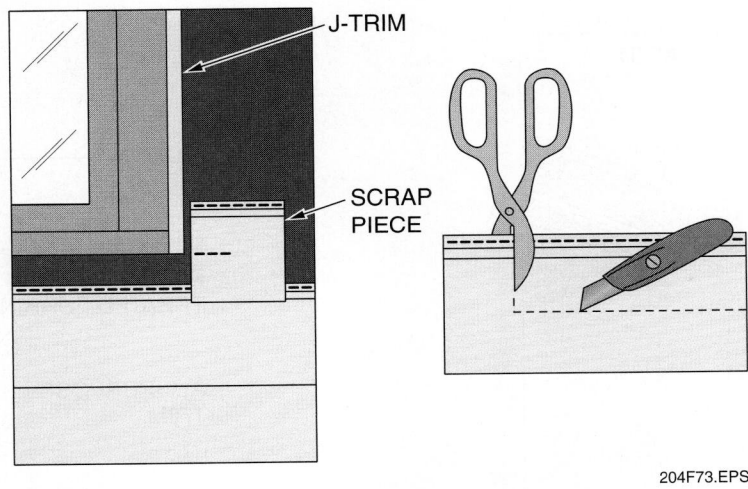

204F73.EPS

Figure 73 ◆ Cutting a panel around a window.

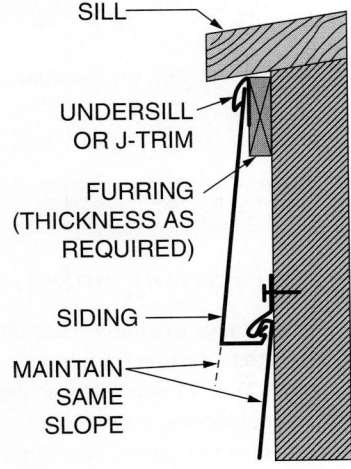
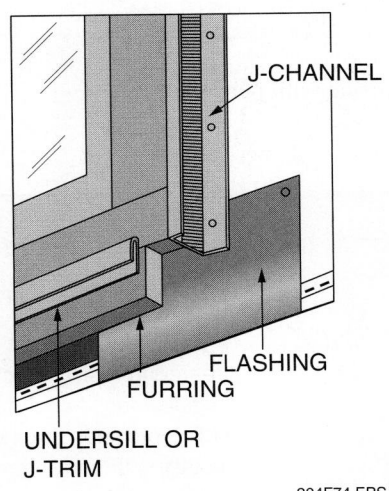

204F74.EPS

Figure 74 ◆ Fitting panels at door and window sills.

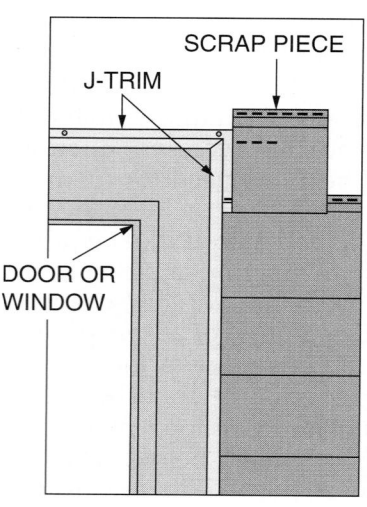
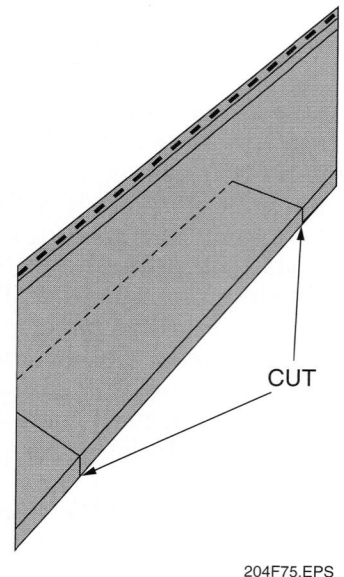

204F75.EPS

Figure 75 ◆ Using a scrap piece to measure clearance.

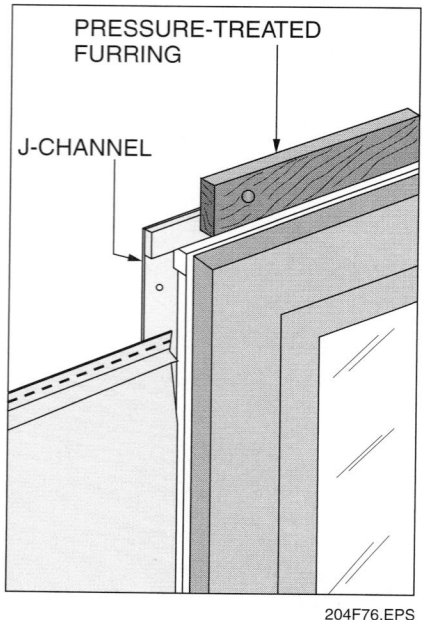

204F76.EPS

Figure 76 ◆ Using furring with J-channel.

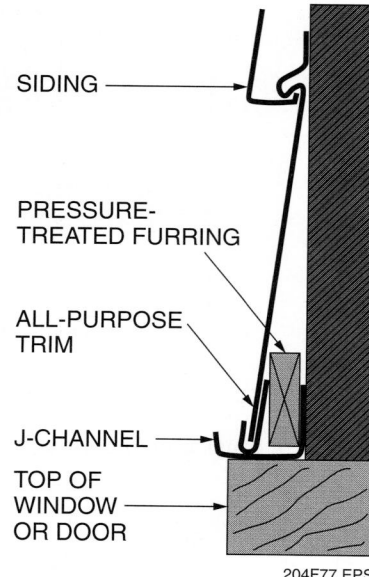

204F77.EPS

Figure 77 ◆ Completing the installation at a window top.

Pressure-Treated Furring and Furring Substitutes

If used, always make sure that any wood furring for vinyl or metal siding is pressure-treated to resist insects and rot. In place of pressure-treated furring in some installations, J-channel is used under the sill. Then undersill trim installed under the siding provides the furring spacing and a panel locking point. At the top of the opening, installing undersill trim under the siding provides the furring spacing and panel locking point.

7.15.0 Installing Siding at Gable Ends

When installing siding on gables, diagonal cuts will have to be made on some of the panels. To make a pattern of cutting panels to fit the gable slope, use two short pieces of siding as templates (*Figure 78*). Interlock one of these pieces into the panel below. Hold the second piece against the J-channel trim on the gable slope. Along the edge of this second piece, scribe a line diagonally across the interlock end panel and cut along this line with tin snips or a power saw.

This cut panel is a pattern that can be used to transfer cutting marks to each successive course along the gable slope. All roof slopes can be handled in the same manner as gable end slopes.

Slip the angled end of the panel into the J-trim previously installed along the gable end. Lock the butt into the interlock of the panel below. Remember to allow for expansion or contraction where required. If necessary, face nail with 1¼" (or longer) painted-head aluminum nails in the apex of the last panel at the gable peak. Touch-up enamel in matching siding colors can also be used for exposed nail heads.

Do not cover existing **louvers** or **vents**. Attic ventilation is necessary in summer to reduce temperatures and in winter to prevent the accumulation of moisture.

7.16.0 Installing Siding Under Eaves

The last panel course under the eaves will almost always have to be cut lengthwise to fit in the remaining space. Usually, furring will be needed under this last panel to maintain the correct slope angle. Determine the proper furring thickness and install it. Nail all-purpose trim or J-channel to the furring with aluminum nails. The trim should be cut long enough to extend the length of the wall (*Figure 79*).

To determine the width of the cut required, measure from the bottom of the top lock to the eave, subtract ¼", and mark the panel for cutting. Take measurements at several points along the eaves to ensure accuracy. Score the panel with the utility knife and bend it until it snaps. For vinyl panels, use a snaplock punch to place raised ears (16" or 24" apart) along the top cut edge so that it will lock into the J-channel.

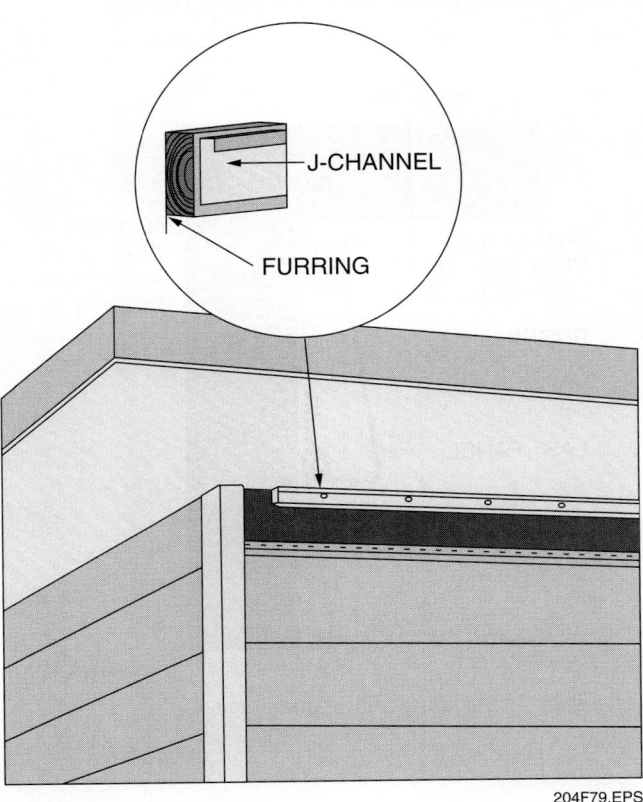

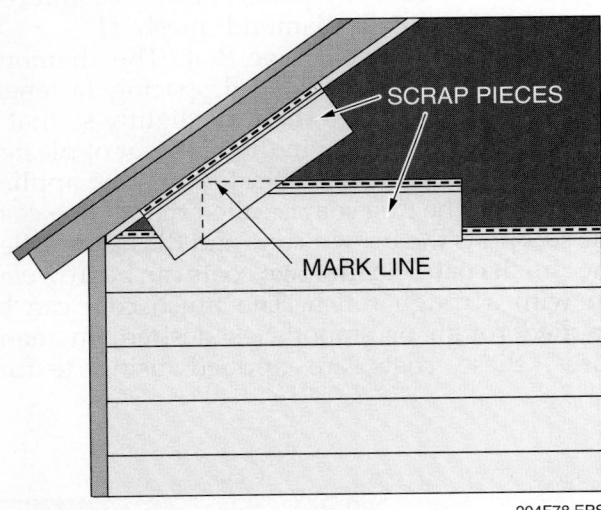

Figure 78 ◆ Installing siding at gable ends.

Figure 79 ◆ Trim extends the length of the wall.

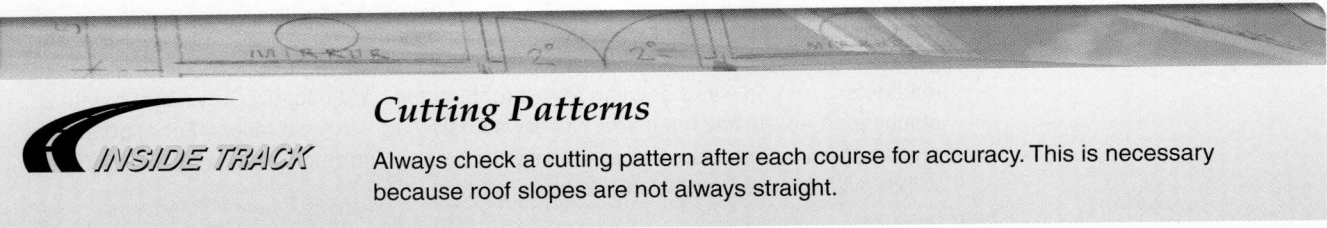

Cutting Patterns

INSIDE TRACK

Always check a cutting pattern after each course for accuracy. This is necessary because roof slopes are not always straight.

For aluminum siding, apply gutter seal to the nail flange of the all-purpose trim. Slide the final panel into the trim. Engage the interlock of the panel below.

On metal panels, the lock may be flattened slightly using a hammer and a 2' or 3' piece of lumber before the final panel is installed so it will grip more securely. Press the panel into the gutter seal adhesive. With this technique, face nails will not be required. Refer to *Figure 80*.

7.17.0 Caulking and Cleanup

In general, caulking is done around doors, windows, and gables where the siding meets wood or metal, except where accessories are used to make caulking unnecessary. Caulking is also needed where siding or siding accessories meet brick or stone around chimneys and walls. Surface caulking required around faucets, meter boxes, and other panel cutouts must be done neatly.

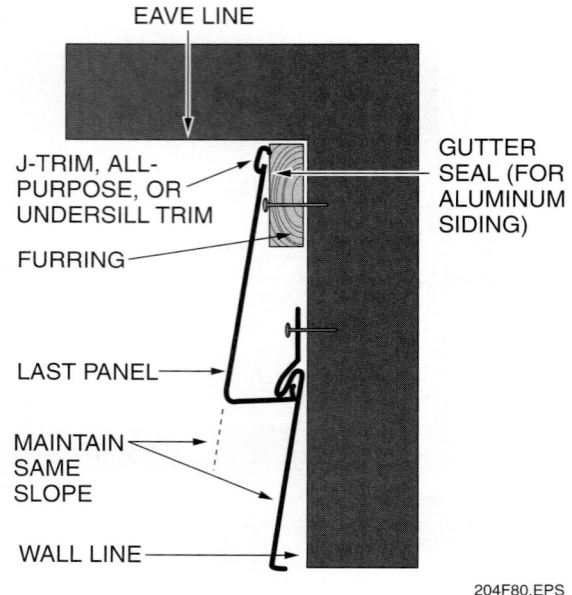

EAVE LINE

J-TRIM, ALL-PURPOSE, OR UNDERSILL TRIM

FURRING

GUTTER SEAL (FOR ALUMINUM SIDING)

LAST PANEL

MAINTAIN SAME SLOPE

WALL LINE

204F80.EPS

Figure 80 ◆ Installing siding under eaves.

It is important to get a deep caulking bead that is ¼" minimum in depth, not just a wide bead. To achieve this, cut the plastic caulking cartridge tip straight across rather than at an angle. Move the gun evenly and apply steady, even pressure on the trigger. The butyl type of caulking is preferred as it has greater flexibility. Most producers supply caulking in colors to match siding and accessories. Do not depend on caulking to fill gaps more than ⅛" wide, as the expansion or contraction of the siding may cause the caulking to crack.

Reinstall all fixtures, brackets, downspouts, etc. that were removed. Accessories that were not replaced, such as kitchen fan outlets or service cables, may be painted to match the new siding color. Most manufacturers have touch-up paint or matching paint formulas, which can be purchased at a local paint store. All scrap pieces, cartons, nails, and other materials should be removed and the job site left neat and clean each day.

8.0.0 ◆ STUCCO (CEMENT) FINISHES

Stucco is a durable cement-based coating for exterior walls that is normally applied by painters or masons.

When applying stucco over frame walls, use wood sheathing, exterior gypsum, or cement board. Building paper is placed over the sheathing and 2×8 or 4×8 panels of metal reinforcement mesh, called diamond mesh (1⅜" × ⅜" openings), are placed over that. The diamond mesh is applied with special spacing fasteners that hold it away from the wall slightly so that it will become embedded in the first coat of plaster.

Three coats of cement plaster must be applied (*Figure 81*). The first coat is called the scratch coat, the second is the brown coat, and the last is called the finish coat. The first two coats can be troweled on with a rough finish. The finish coat can be applied rough or smooth, as desired. In many cases, these coats are applied using texture

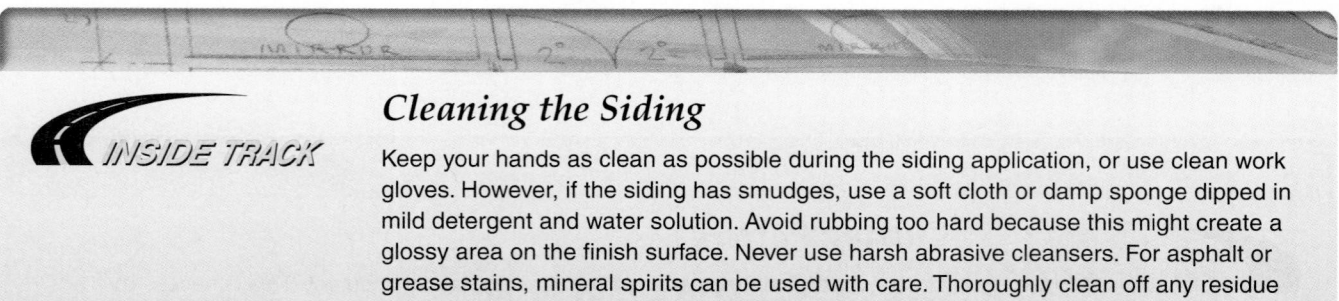

Cleaning the Siding

Keep your hands as clean as possible during the siding application, or use clean work gloves. However, if the siding has smudges, use a soft cloth or damp sponge dipped in mild detergent and water solution. Avoid rubbing too hard because this might create a glossy area on the finish surface. Never use harsh abrasive cleansers. For asphalt or grease stains, mineral spirits can be used with care. Thoroughly clean off any residue with a mild detergent and water solution.

INSIDE TRACK

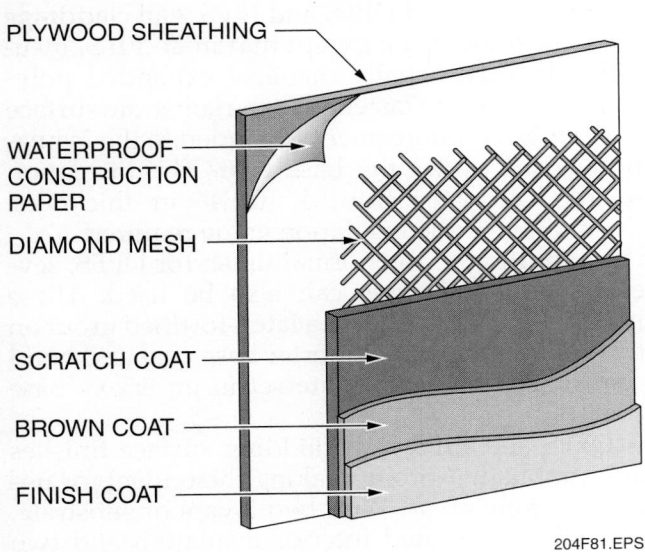

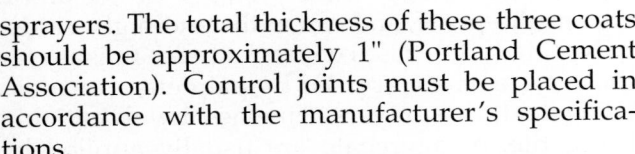

Figure 81 ◆ Stucco section.

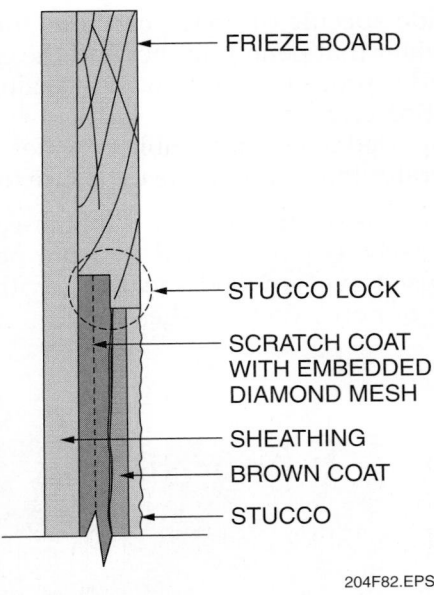

Figure 82 ◆ Stucco lock.

sprayers. The total thickness of these three coats should be approximately 1" (Portland Cement Association). Control joints must be placed in accordance with the manufacturer's specifications.

When applying wood trim next to stucco, such as the frieze board or half timbers used to simulate Tudor architecture, the trim should have a rabbet along the back joining edge, called a stucco lock (*Figure 82*). The stucco lock prevents water penetration around the stucco at the juncture of the stucco and the wood trim.

When applying stucco to concrete block walls, the diamond mesh is not required, but all other requirements apply.

9.0.0 ◆ BRICK AND STONE VENEER

The most durable materials that can be used for exterior wall finishes are brick, synthetic stone, and natural stone masonry. Brick and synthetic stone are man-made products, while stone is a natural material. All three materials are attractive and require little or no maintenance. Brick is made from a clay mixture and is hardened by baking it in a kiln. The nominal size of a brick is $2\frac{1}{4}" \times 3\frac{3}{4}" \times 8"$. This size may vary slightly due to the hardening processes in the kiln. Synthetic stone is made from a cement mixture. Natural stone is either found on the surface or is mined from an open pit quarry.

10.0.0 ◆ DEFS AND EIFS

Direct-applied exterior finish systems (DEFS) and exterior insulation and finish systems (EIFS) are designed as water-managed systems (*Figure 83*). In appearance, they are similar to traditional stucco or masonry finishes, but they employ different types and applications of material.

Water-managed systems are usually defined as wall cladding systems that:

- Provide specific drainage methods for intruding water that penetrates beyond the cladding
- Provide protection for water-sensitive construction elements
- Are applied to water-durable or water-resistant substrates that can tolerate exposure to water

In most cases, the substrate is a fiber-cement or fiberglass-coated and treated gypsum panel that can be used over a wood sheathing with underlayment or non-structural sheathing.

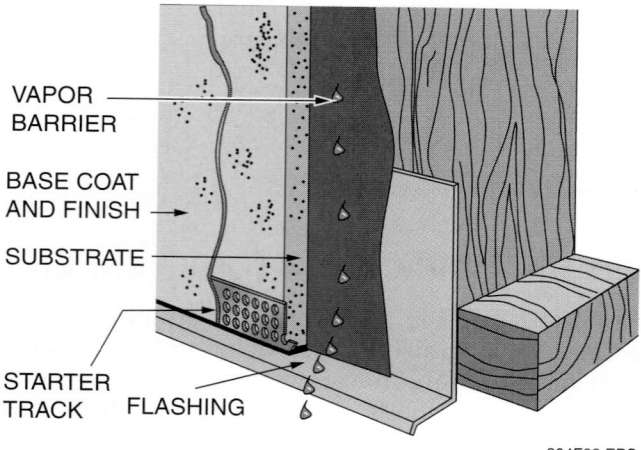

Figure 83 ◆ Typical DEFS water management.

VAPOR BARRIER

BASE COAT AND FINISH

SUBSTRATE

STARTER TRACK   FLASHING

204F83.EPS

All water-managed systems for framed construction or masonry construction have very specific means and methods for flashing and directing incidental water that enters around or through windows, doors, and other openings. For this reason, and to achieve satisfactory performance as well as warranty protection, the manufacturer's instructions must be rigorously followed when installing the components of these systems. In addition, a vapor barrier is usually required on the interior side of exterior walls.

Water-managed DEFS and EIFS wall claddings are almost identical, except that in an EIFS, insulating boards, usually made of expanded polystyrene (EPS), are fastened to the substrate surface and a mesh reinforcement is bonded to the insulation board under the base coat. The insulation boards can vary from 1" to 4" in thickness, depending on the insulation value required.

In addition to the normal finish for DEFS, several masonry facings can also be used. These include ceramic tile set in a latex-fortified grout on top of a latex-fortified mortar base coat and bond coat or exposed aggregate set in an epoxy base coat.

DEFS and EIFS wall cladding surface finishes are combustion-proof, making them ideal for use in fire-prone areas. With two layers of substrate, as well as fire-rated interior insulation and two layers of fire-rated interior drywall, these systems are rated for up to two hours as a firestop. Like fiber-cement siding, these systems are rot-proof, are termite-proof, and can withstand high winds.

Normally, the base coat and surface finishes for these systems are applied by painters or masons using texture sprayers. Finishes involving thin brick, tile, or aggregate are usually applied by masons. Carpenters normally install the flashing, water barrier, substrate(s), seam tape, insulation, and/or fiber mesh.

The following are typical general application characteristics/guidelines for these systems:

- When applied to standard wood studs or 20-gauge metal studs at 16" OC, the walls can withstand wind loads of 40 pounds per square foot (psf). Greater wind loads are allowable with the use of larger studs or closer stud spacing.
- These systems can be used for ceilings, soffits, curtain walls, bearing walls, panelization elements, and privacy fences.
- The final surface finishes can be tinted any color.
- These systems cannot be used as sill finishes.

- The substrate may not be used as structural sheathing. Racking resistance must be accomplished by separate bracing.
- All windows, doors, and other openings must be properly flashed.
- DEFS on steel framing must be laterally braced. To enhance crack resistance over steel framing, mesh reinforcement is required over the entire substrate to reinforce the base coat.
- Multiple layering of the substrate can be used to achieve various architectural effects such as banding on various levels of a building.
- The drawings and manufacturer's specific instructions must be rigidly followed.

DEFS/EIFS Codes

Always check state and local building codes before using DEFS or EIFS cladding. Some states and/or localities have prohibited its use on residential structures due to deficiencies in earlier versions of the water-management systems, compounded by faulty installation that resulted in severe damage. Some manufacturers offer improved versions of DEFS and EIFS cladding systems that are designed to alleviate these problems. Make sure that any system used is the latest version offered by the manufacturer. Also, make sure that the manufacturer's instructions for installation are rigorously followed.

Summary

You must be aware of the materials and the general methods of installation for a variety of exterior finishes and roof drainage systems. This module covered the installation methods for flashing and insulation, types of cornices and their fabrication/installation, along with descriptions and installation methods for a variety of wood, metal, and vinyl siding. It also discussed other exterior finishes, including stucco, masonry, and various special exterior finish systems. When installing any type of exterior finish or related accessory, always refer to the manufacturer's installation instructions.

Notes

1. The primary purpose of an exterior finish is to _____.
   a. provide a base for the final finish
   b. prevent entry of water at all openings
   c. provide protection from the elements
   d. allow the use of the most cost-effective interior construction

2. When building wrap is applied to the exterior of a structure, the wrap should be overlapped at the corners by _____.
   a. 2"
   b. 4"
   c. 6"
   d. 8"

3. The cutting of some wood and manufactured siding materials can produce a hazardous dust.
   a. True
   b. False

4. A conventionally insulated house can suffer a heat loss through frame walls or by air infiltration of _____.
   a. 55 percent
   b. 65 percent
   c. 75 percent
   d. 85 percent

5. The most common flashing material is _____.
   a. stainless steel
   b. asphalt building paper
   c. galvanized sheet metal
   d. copper

6. Plancier is another name for a _____.
   a. soffit
   b. ledger board
   c. frieze board
   d. lookout

7. A ledger board is used to simplify the installation of _____.
   a. fascia boards
   b. rafters
   c. frieze boards
   d. lookouts

8. The horizontal framing member that the soffit is attached to is called a _____.
   a. jack rafter
   b. hip rafter
   c. lookout
   d. cornice return

9. To the nearest square, how many squares of 1 × 6 wood siding would be required for a hip roof structure that is 50' long and 35' wide with 8' walls, twenty 3' × 5' windows, two 8' × 7' garage doors, and two 3' × 6½' entrance doors?
   a. 12 squares
   b. 14 squares
   c. 15 squares
   d. 16 squares

10. To the nearest panel, how many 4 × 8 panels of vertical siding would be required for the same structure described in Question 9?
   a. 37 panels
   b. 39 panels
   c. 41 panels
   d. 43 panels

11. Two common styles of beveled siding are _____.
   a. colonial and log cabin
   b. log cabin and rustic
   c. colonial and Dolly Varden
   d. rabbeted and tongue-and-groove

12. A vertical siding where boards overlap boards underneath that are the same size as the overlapping boards is called _____ siding.
   a. board-and-batten
   b. reverse batten
   c. board-on-board
   d. shiplap

13. Of the following, _____ is a style of tongue-and-groove siding.
   a. colonial
   b. Dolly Varden
   c. V-edged
   d. log cabin

14. The type of shiplap siding shown below is known as _____ siding.
    a. channel rustic
    b. drop
    c. V-edged
    d. plain

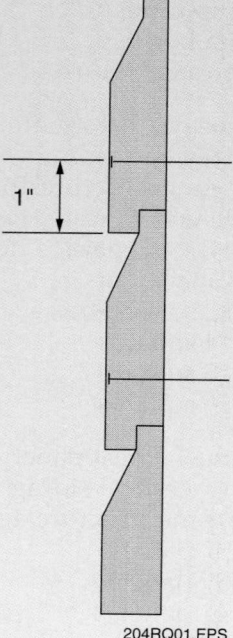

1"

204RQ01.EPS

15. The two fasteners used to apply wood shingles are _____.
    a. nails and staples
    b. nails and drywall screws
    c. staples and construction adhesive
    d. sheet metal screws and roofing nails

16. A plywood siding panel called textured one-eleven (T-111) has a surface pattern that _____.
    a. is striated
    b. is brushed
    c. has widely spaced grooves
    d. has a plank texture

17. When plywood is used as exterior siding, it is _____.
    a. nailed to the sheathing
    b. nailed to the studs
    c. glued to the sheathing
    d. glued to the studs

18. All of the following finishes are recommended for hardboard siding except _____.
    a. 100 percent acrylic latex paint
    b. flat alkyd stain
    c. gloss alkyd paint
    d. satin alkyd paint

19. Which of these types of siding should *not* be cleaned using pressure-washing equipment?
    a. Fiber-cement
    b. Vinyl
    c. Hardboard
    d. Aluminum

20. When installing lapped hardboard siding, the minimum overlap is _____.
    a. ½"
    b. ¾"
    c. 1"
    d. 2"

21. The maximum stud spacing for installation of fiber-cement board applied over sheathed walls is _____ OC.
    a. 12"
    b. 16"
    c. 20"
    d. 24"

22. Which of the following metals will corrode when placed in contact with fiber-cement board siding?
    a. Aluminum
    b. Galvanized steel
    c. Copper
    d. Stainless steel

23. When cutting vinyl siding with a power saw, the blade used should be a _____ fine-toothed blade for a smooth finish.
    a. carbide-tipped
    b. reversed
    c. dry diamond
    d. high-speed

24. A stucco lock overlaps the _____ of a stucco finish.
    a. sheathing
    b. scratch coat
    c. brown coat
    d. finish coat

25. In a water-managed system, a substrate is _____.
    a. plywood sheathing
    b. brick veneer
    c. asphalt-impregnated felt
    d. fiber-cement panels

Trade Terms
Introduced in This Module

Board-and-batten: A type of vertical siding consisting of wide boards with the joint covered by narrow strips known as battens.

Brown coat: A coat of plaster with a rough face on which a finish coat will be placed.

Building paper: A heavy paper used for construction work. It assists in weatherproofing the walls and prevents wind infiltration. Building paper is made of various materials and is not a vapor barrier.

Cornice: The construction under the eaves where the roof and side walls meet.

Course: One row of brick, block, or siding as it is placed in the wall.

Eave: The lower part of a roof, which projects over the side wall.

Fascia: The exterior finish member of a cornice on which the rain gutter is usually hung.

Felt paper: An asphalt-impregnated paper.

Finish coat: The final coat of plaster or paint.

Frieze board: A horizontal finish member connecting the top of the sidewall, usually abutting the soffit. Its bottom edge usually serves as a termination point for various types of siding materials.

Ledger: A board to which the lookouts are attached and which is placed against the out-side wall of the structure. It is also used as a nailing edge for the soffit material.

Lookout: A member used to support the overhanging portion of a roof.

Louver: A slatted opening used for ventilation, usually in a gable end or a soffit.

Plancier: The same as a soffit, but the member is usually fastened to the underside of a rafter rather than the lookout.

Rabbet: A groove cut in the edge of a board so as to receive another board.

Rake: The slope or pitch of the cornice that parallels the roof rafters on the gable end.

R-value: A numerical designation given to a material based on its insulating ability, such as R-19.

Scratch coat: The first coat of cement plaster consisting of a fine aggregate that is applied through a diamond mesh reinforcement or on a masonry surface.

Shakes: Hand- or machine-split wood shingles.

Soffit: The underside of a roof overhang.

Veneer: A brick face applied to the surface of a frame structure.

Vent: A small opening to allow the passage of air.

Additional Resources and References

This module is intended to present thorough resources for task training. The following reference works are suggested for further study. These are optional materials for continued education rather than for task training.

The Vinyl Siding Institute website, *www. vinylsiding.org.*

Cedar Shake & Shingle Bureau website, *www. cedarbureau.org.*

NCCER makes every effort to keep these textbooks up-to-date and free of technical errors. We appreciate your help in this process. If you have an idea for improving this textbook, or if you find an error, a typographical mistake, or an inaccuracy in NCCER's Contren® textbooks, please write us, using this form or a photocopy. Be sure to include the exact module number, page number, a detailed description, and the correction, if applicable. Your input will be brought to the attention of the Technical Review Committee. Thank you for your assistance.

Instructors – If you found that additional materials were necessary in order to teach this module effectively, please let us know so that we may include them in the Equipment/Materials list in the Annotated Instructor's Guide.

Write: Product Development and Revision
National Center for Construction Education and Research
3600 NW 43rd St, Bldg G, Gainesville, FL 32606

Fax: 352-334-0932

E-mail: curriculum@nccer.org

Craft

Module Name

Copyright Date   Module Number   Page Number(s)

Description

(Optional) Correction

(Optional) Your Name and Address

Basic Stair Layout
68111-09

Two Carpentry contestants in the 2005 SkillsUSA National Championships work on miniature building framing. In addition to building a wooden frame and common rafter, contestants were also required to construct a metal frame, stair stringer, and install gypsum wallboard. Contestants were judged based on safety, project quality, and the proper use of tools.

68111-09
Basic Stair Layout

Topics to be presented in this module include:

1.0.0   Introduction 11.2
2.0.0   Types of Stairs 11.2
3.0.0   Stairway Components and Typical Code Requirements . . .11.5
4.0.0   Stair Framing 11.8
5.0.0   Stairway and Stairwell Design and Layout 11.10
6.0.0   Forms for Concrete Stairs 11.21

Overview

Although prefabricated stairways are available in a variety of designs, a carpenter will sometimes have to lay out and build stairways, or build a form for concrete stairs. Laying out and cutting stair stringers is an art requiring precise measuring and the ability to perform math calculations. In addition, stairway construction is more code-driven than most other construction tasks because of the potential tripping and falling hazards. When you have mastered stair layout, you will be well on your way to becoming a professional carpenter.

Objectives

When you have completed this module, you will be able to do the following:

1. Identify the various types of stairs.
2. Identify the various parts of stairs.
3. Identify the materials used in the construction of stairs.
4. Interpret construction drawings of stairs.
5. Calculate the total rise, number and size of risers, and number and size of treads required for a stairway.
6. Lay out and cut stringers, risers, and treads.
7. Build a small stair unit with a temporary handrail.

Trade Terms

Baluster	Pitch board
Balustrade	Rise
Closed stairway	Rise and run
Geometrical stair	Riser
Guardrail	Run
Handrail	Skirtboard
Headroom	Stairwell
Housed stringer	Stringer
Landing	Tread
Newel post	Unit rise
Nosing	Unit run
Open stairway	Winding stairway

Required Trainee Materials

1. Pencil and paper
2. Appropriate personal protective equipment

Prerequisites

Before you begin this module, it is recommended that you successfully complete *Core Curriculum,* and *Construction Technology,* Modules 69101-09 through 68110-09.

This course map shows all of the modules in *Construction Technology.* The suggested training order begins at the bottom and proceeds up. Skill levels increase as you advance on the course map. The local Training Program Sponsor may adjust the training order.

CONSTRUCTION TECHNOLOGY

68117-09
Copper Pipe and Fittings

68116-09
Plastic Pipe and Fittings

68115-09
Introduction to Drain, Waste, and Vent (DWV) Systems

68114-09
Introduction to HVAC

68113-09
Residential Electrical Services

68112-09
Electrical Safety

68111-09
Basic Stair Layout

68110-09
Exterior Finishing

68109-09
Roofing Applications

68108-09
Roof Framing

68107-09
Wall and Ceiling Framing

68106-09
Floor Systems

68105-09
Masonry Units and Installation Techniques

68104-09
Introduction to Masonry

68103-09
Handling and Placing Concrete

68102-09
Introduction to Concrete, Reinforcing Materials, and Forms

68101-09
Site Layout One: Distance Measurement and Leveling

CORE CURRICULUM:
Introductory Craft Skills

103CMAP.EPS

1.0.0 ◆ INTRODUCTION

This module deals with the construction of **stairwells**, stair framing, **risers**, and **treads**. The module begins with definitions of stair terminology and a brief introduction to stair construction. Various types of stairs are discussed and procedures for designing, laying out, cutting, and installing the stair framing are detailed.

Stairways result in roughly 4,000 deaths and a million injuries requiring hospital treatment each year; therefore, stairway design and construction is strictly controlled by building codes and regulations.

Many states and localities specify minimum and maximum requirements that may differ from the national codes. Before starting any stair construction, you must be aware of any national code and any superseding state or local codes that govern the construction of the stairs. Furthermore, if construction drawings pertain to the project, they must be followed for the construction of the stairs. If the drawings are incomplete or conflict with national, state, or local codes, a supervisor should be notified to obtain or clarify the needed information.

Normally, as construction of a structure proceeds, the stairs, with temporary stair treads, are put in as soon as possible to make it easier for workers to move themselves and materials from one level to the next with minimum delay and maximum safety. Stairways are not finished until all danger of damage from workers and materials is eliminated.

DID YOU KNOW?

The World's Longest Stairway

At 5,476 feet, the service stairway for the Niesenbahn funicular railway near Spiez, Switzerland is the world's longest. This stairway, which is used only by employees, has 11,674 steps.

Source: *Guinness World Records*

2.0.0 ◆ TYPES OF STAIRS

There are many ways to classify stairs. One way depends on whether or not they are open, closed, or a combination of open and closed. *Figure 1* shows an example of an **open stairway**, which is one with a single wall in the middle. *Figure 2* illustrates an example of a **closed stairway**, and *Figure 3* shows an example of a combined open/

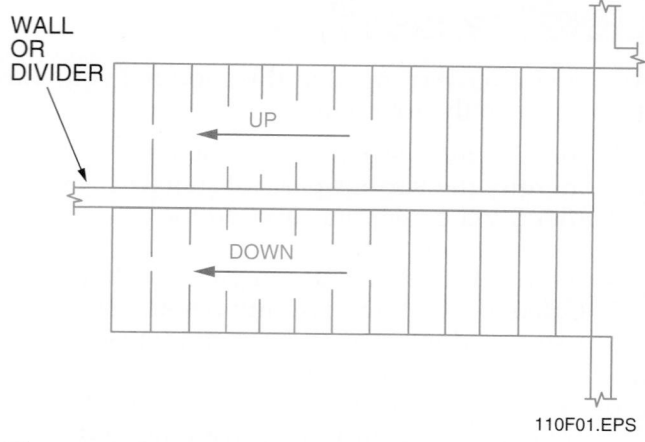

Figure 1 ◆ Open straight-run stairs.

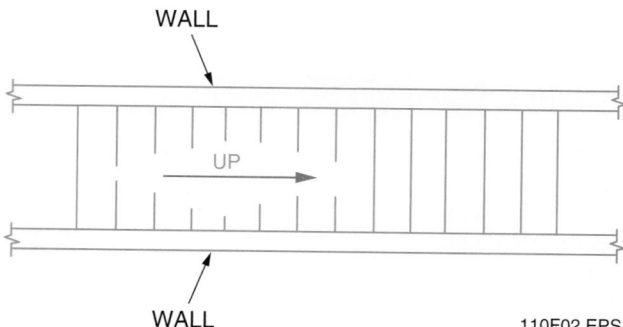

Figure 2 ◆ Closed straight-run stairs.

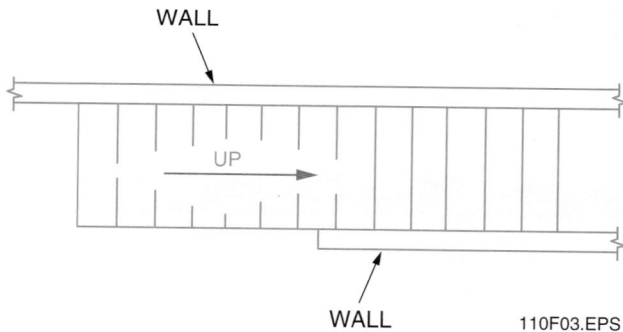

Figure 3 ◆ Combination open/closed straight-run stairs.

closed stairway. Building codes usually have special requirements for closed stairs. Refer to your local codes before construction.

Another classification relates to the shape of the stairway. A few examples are shown in *Figures 4* through *9*.

When a series of steps are straight and continuous without breaks formed by a **landing (platform)** or other construction, they are referred to as a flight of stairs, staircase, or stairway. The various classifications shown in *Figures 1* through *9* can be categorized under one of three headings: straight-run or

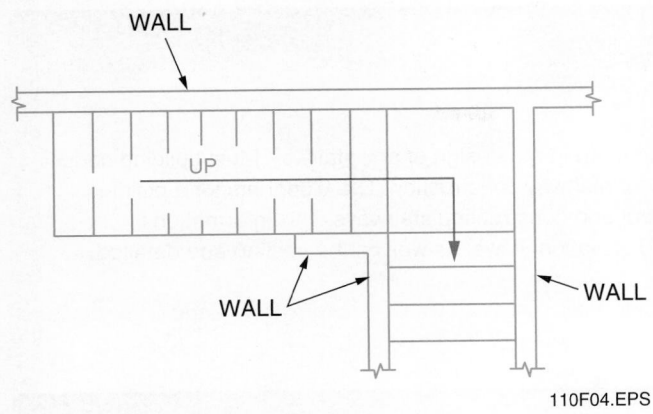

110F04.EPS

Figure 4 ◆ Long-L stairway.

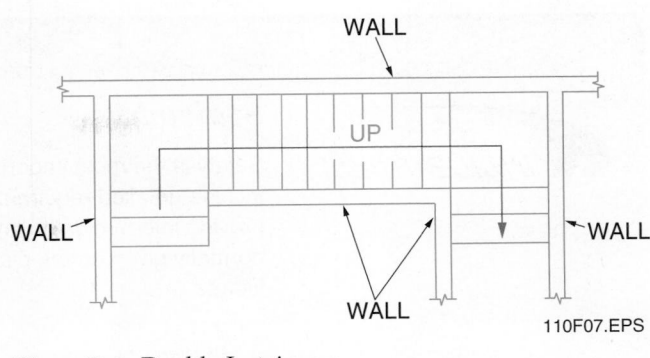

110F07.EPS

Figure 7 ◆ Double-L stairway.

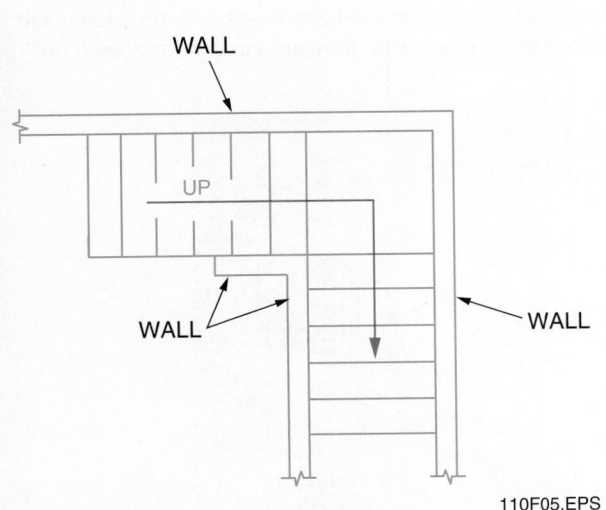

110F05.EPS

Figure 5 ◆ Wide-L stairway.

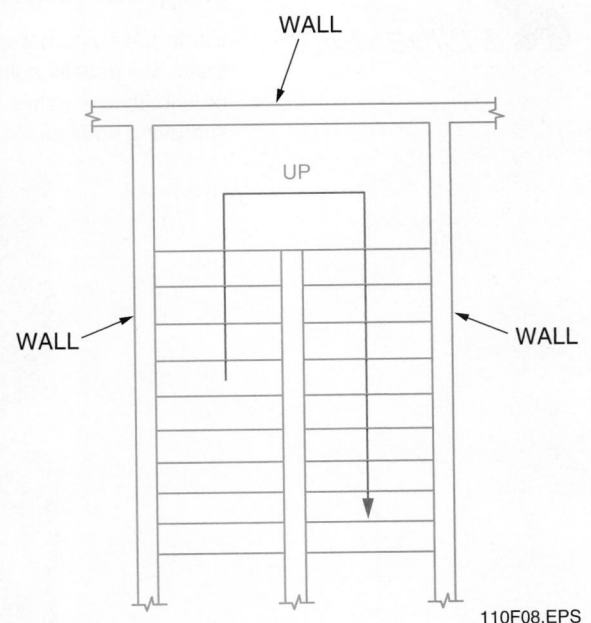

110F08.EPS

Figure 8 ◆ Narrow-U stairway.

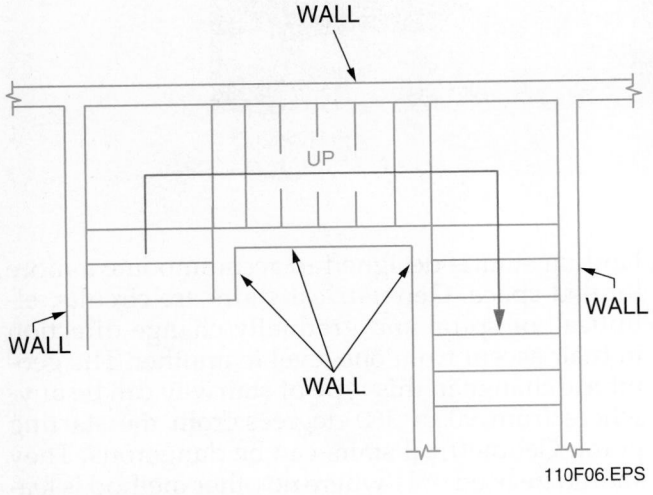

110F06.EPS

Figure 6 ◆ Wide-U stairway.

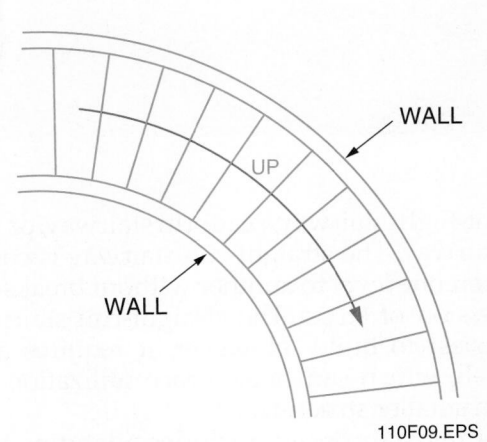

110F09.EPS

Figure 9 ◆ Winding stairway.

Stairways

Safety is the most important concern in the design of any stairway. Most building codes include detailed requirements for stairway construction. The blueprints for a building provide information on laying out and constructing stairways. This information is normally given on the plan and elevation views, as well as the section and detailed views.

Elegant Stairs

Elaborate custom-built stairs are normally installed in expensive residences. These stairs are usually built to order in a shop from expensive hardwoods and are assembled on site. In rare cases, including some restoration work, they are custom-fabricated on site using specialized tools.

110SA01.EPS

straight-flight stairway, platform stairway, or **winding stairway**. The straight-run stairway is continuous from one level to another without breaks in the progression of steps. The straight-run stairway is the easiest to build. However, it requires a long stairwell, which can cause space utilization problems in smaller structures.

Landing stair design includes a landing where the direction of the stair run usually changes. The landing stair is designed to accommodate a more limited space. Geometrical stairs are circular, elliptical, or spiral and gradually change direction in their ascent from one level to another. The geometric change in this type of stairway can be anywhere from 90 to 360 degrees from the starting point. Geometrical stairs can be dangerous. They should be used only where no other method is feasible.

3.0.0 ◆ STAIRWAY COMPONENTS AND TYPICAL CODE REQUIREMENTS

The main components of a flight of stairs are explained in the following paragraphs and shown in *Figures 10* and *11*.

- *Tread* – The horizontal surface of a step.
- *Riser* – The piece forming the vertical face of the step. Commercial stair codes sometimes require a riser piece. Other stair codes may allow the omission of the riser piece (open risers).
- *Cutout stringer* – A cutout stringer provides the main support for the stairway. Center cutout stringers may be required by codes when a wooden stairway width exceeds 30", or multiples

of 30", or if tread material thinner than 1½" is used on stairs that are 36" wide. Other types of stringers used as side supports for stairs are the cleated, dadoed, or housed stringers (see *Figure 11*). A cleated stringer is a stringer with cleats fastened to it to support the treads. A similar design is a dadoed stringer, where the treads fit into slots cut in the stringer. Both of these designs are normally used for open-riser stairs. A housed stringer is similar to a dadoed stringer, except that tapered slots are provided for the treads and risers. The treads and risers are secured with wedges and glue. Housed stringers are usually prefabricated units that are custom-made in a shop.

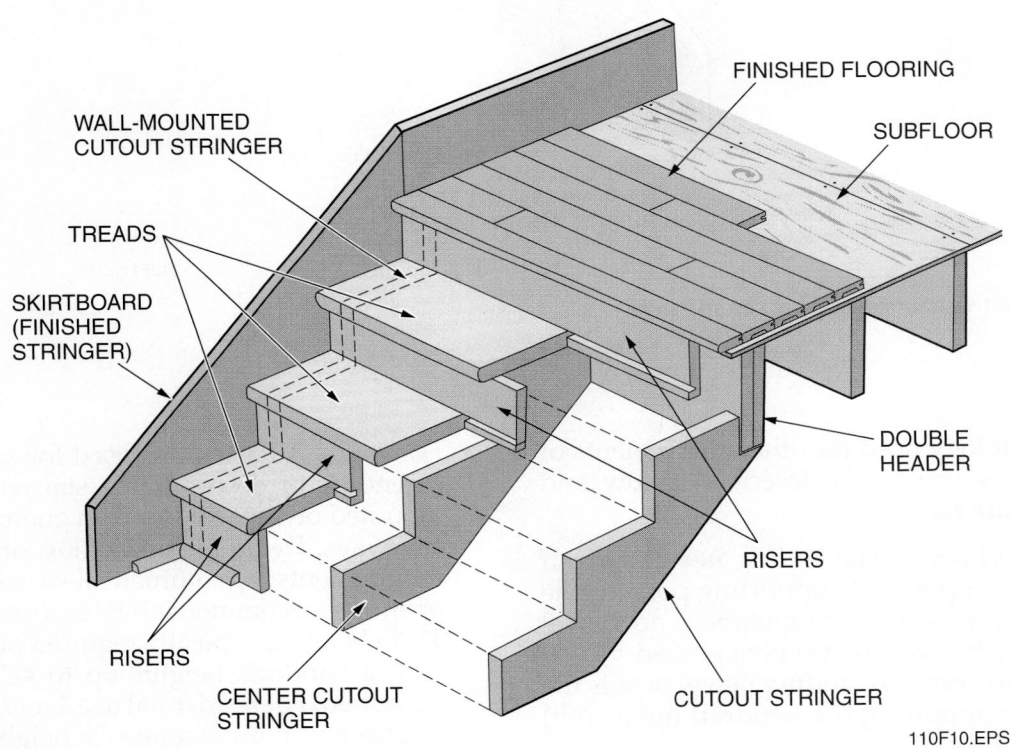

110F10.EPS

Figure 10 ◆ Main stair components.

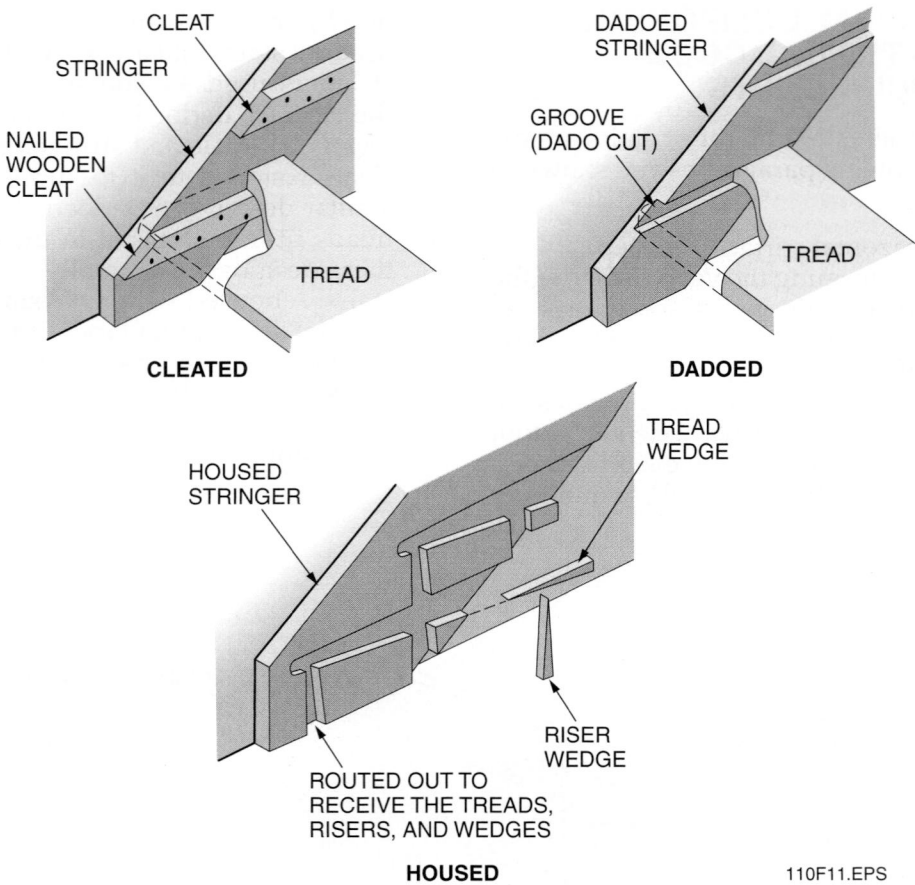

Figure 11 ◆ Cleated, dadoed, and housed stringers.

110F11.EPS

The terminology used for other components of most interior stairways is described below and shown in *Figure 12*.

- *Ending newel post* – The ending newel post, if used, is the uppermost supporting post for the handrail or guardrail. Sometimes, no newel post or only half of a newel post is used.
- *Landing newel post* – A landing newel post is the main post supporting the handrail (guardrail) at the landing.
- *Gooseneck* – A gooseneck is a bent or curved section in the handrail or guardrail.
- *Spindle* – A spindle or baluster is the upright piece that runs between the handrail or guardrail and the treads or a closed stringer. The balusters, handrail or guardrail, and newel posts make up the balustrade.

- *Handrail* – A handrail is used for support when ascending or descending a stairway. It may be mounted on the walls or on a guardrail for open stairways. *Figure 13* shows most of the handrail requirements for commercial structures. On the open side of commercial stairs, a guardrail that is 42" in height is typically required or if combined with a handrail, heights up to 42" are usually permitted. For residential use, handrails without extensions at lower ranges of height are usually permitted. In some cases, additional lower handrails are permitted for children. Guardrails used in residential applications are usually permitted at heights of 36".
- *Starting newel post* – A starting newel post is the main post supporting a handrail or guardrail at the bottom of the stairway.

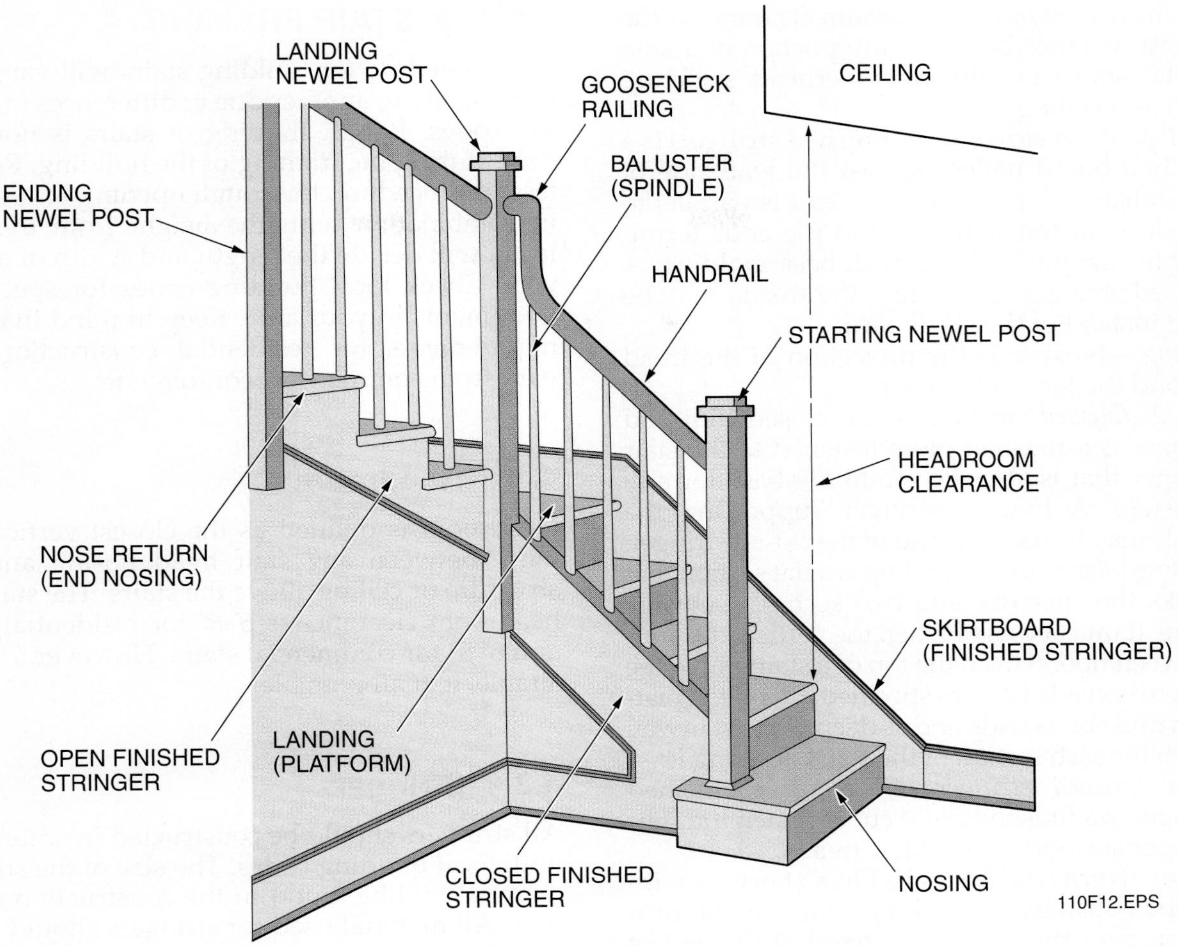

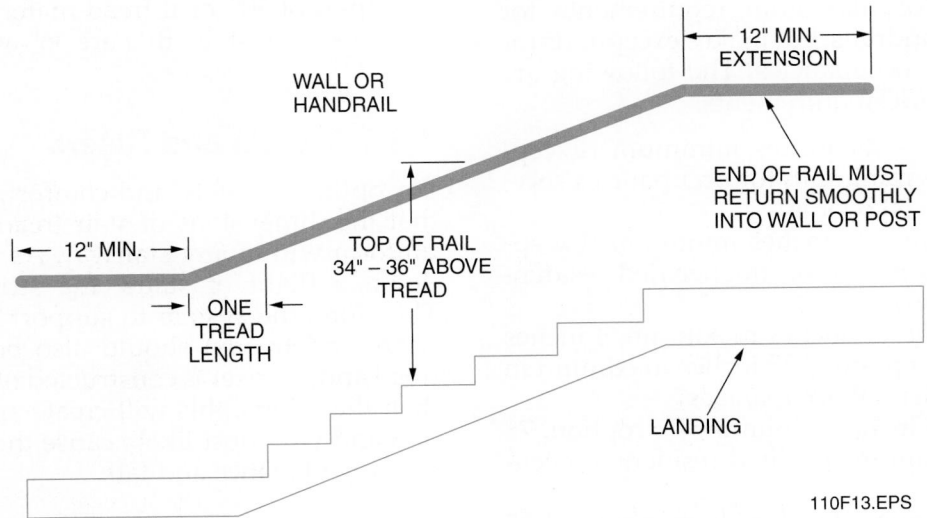

LANDING
NEWEL POST

GOOSENECK
RAILING

CEILING

ENDING
NEWEL POST

BALUSTER
(SPINDLE)

HANDRAIL

STARTING NEWEL POST

HEADROOM
CLEARANCE

NOSE RETURN
(END NOSING)

SKIRTBOARD
(FINISHED STRINGER)

OPEN FINISHED
STRINGER

LANDING
(PLATFORM)

CLOSED FINISHED
STRINGER

NOSING

110F12.EPS

Figure 12 ◆ Components of an interior stairway.

12" MIN.
EXTENSION

WALL OR
HANDRAIL

END OF RAIL MUST
RETURN SMOOTHLY
INTO WALL OR POST

12" MIN.

TOP OF RAIL
34" – 36" ABOVE
TREAD

ONE
TREAD
LENGTH

LANDING

110F13.EPS

Figure 13 ◆ Commercial handrail requirements.

- *Headroom clearance* – **Headroom** clearance is the closest distance between any portion of a step of the stairway and any overhead structure such as a ceiling.
- *Skirtboard* – A **skirtboard** (finished stringer) is a finished board nailed against the wall side of the stairway. The top of the board is parallel to the slope of the stairway, and the ends terminate horizontal with the wall base molding. A housed stringer supporting the treads may be used instead of the skirtboard.
- *Nosing* – **Nosing** is the projection of the tread beyond the face of the riser.
- *Closed finished stringer* – A closed finished stringer is a finished piece fastened to the stair stringer that is also the mounting base for any balusters. A housed stringer supporting the treads may be used instead of the closed stringer.
- *Landing (platform)* – A landing is a flat section that breaks the stairway into two sections between floors. It must be used when the vertical distance (between floors) from the top of a stairway to the bottom exceeds 12' or as specified by code. A platform must be as wide and as deep as the stairway width for each section of the stairs abutting it.
- *Open finished stringer* – An open finished stringer is a finished piece cut to match the stair stringer supporting the stair treads.
- *Nosing return (end nosing)* – On a stairway with an open stringer, the nosing return is the projection over the face of the tread at the end of the tread.

The *International Building Code* (IBC) specifies minimum and/or maximum requirements for stairways and handrails, but makes exceptions for some residential occupancies. The following are examples of the *IBC* requirements:

- *Stairway width* – 44 inches minimum (Exception: 36 inches minimum for occupancies serving fewer than 50 people)
- *Stair tread depth* – 11 inches minimum (Exception: 10 inches minimum in specified residential occupancies)
- *Stair riser height* – 7 inches maximum, 4 inches minimum (Exception: 7.75 inches maximum in specified residential occupancies)
- *Headroom* – 80 inches minimum (Exception: 78 inches minimum in specified residential occupancies)
- *Nosing* – Minimum of 0.75 inch, maximum 1.25 inches (applies only to specified residential occupancies)
- *Vertical rise* – 12 feet maximum between floor levels or landings
- *Handrail height* – 34 inches minimum, 38 inches maximum

4.0.0 ◆ STAIR FRAMING

The procedure for building stairs will vary from one locality to another due to differences in building codes. Rough framing of stairs is normally done during the framing of the building. Regardless of procedure, the rough opening in the floor, in combination with the height from the floor level, will decide the length and width of a stairway. Check local building codes for special requirements in your area. Keep in mind that code requirements for residential construction may vary from commercial requirements.

4.1.0 Headroom

Headroom is defined as the closest vertical distance between any stair tread nosing and any structure or ceiling above the stairs. The standard headroom clearance is 6'-6" for residential stairs and 6'-8" for commercial stairs. However, 7' is desirable, if at all possible.

4.2.0 Stringers

All stringers should be constructed in accordance with local building codes. The size of the stringer material will be found in the construction drawings. All material used for stringers should be selected at one time and the crowns should be matched. Normally, cutout center stringers are required on wooden stairways wider than 30", in multiples of 30", or if tread material thinner than 1½" is used on stairs that are 36" wide.

4.3.0 Treads and Risers

For optimum safety and comfort, it is important that the dimensions of stair treads and risers be uniform within any stairway. As a person walks down a flight of stairs, the stride is uniform. Therefore, the system to support the person (the risers and treads) should also be uniform. If a tread and/or riser is constructed of a different size than the others, this will create an unsafe condition and will most likely cause the person to lose his or her balance and fall.

4.4.0 Width Requirement

The minimum width required is also specified in building codes. The standard is a minimum stair width of 36" to 44", depending on the type and occupancy of the building. See *Figure 14*.

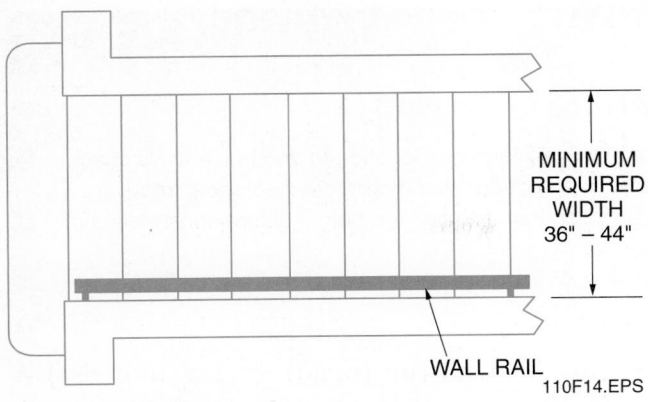

Figure 14 ◆ Minimum stairway width.

4.5.0 Handrails

A handrail is used on stairways to assist people when ascending or descending a stairway by furnishing a continuous rail for support along the side. This differs from a guardrail in that a guardrail is erected on the exposed sides and ends of stairs and platforms and incorporates a handrail. Open stairways have a low partition or banister. The handrail in a closed stairway is called a wall rail because it is attached to the wall with special brackets. As a rule, standard stairs will have a rail only on one side; however, stairs wider than 44" usually require handrails on both sides. Always refer to the local codes for specific requirements.

4.6.0 Stairwells

Stairwells must be constructed so that they are wide enough and long enough to provide the code-required width and headroom clearance for the stairs below them. If the architectural drawings do not specifically detail the stairs, carpenters can be confronted with several situations at a job site.

In one situation, the stairwells may not have been constructed and the carpenter must lay out and frame any stairwell openings as well as the stairs. In another circumstance, someone else may have already framed the openings and the task will be to frame the stairs to provide at least the minimum headroom clearance with an optimum rise and run for the stairs.

In a one-story structure, the stairway would be a set of service stairs to the basement. The stairwell opening should be constructed to allow code-required headroom clearance. When framing any stairwell, the construction must include double trimmers and headers to support weight placed on or around the stairwell opening. See *Figure 15*.

The stairwell framing opening will vary with the type of stairway being installed. As stated earlier, building codes have restrictions on headroom clearance, riser height, and tread width. Always check local codes.

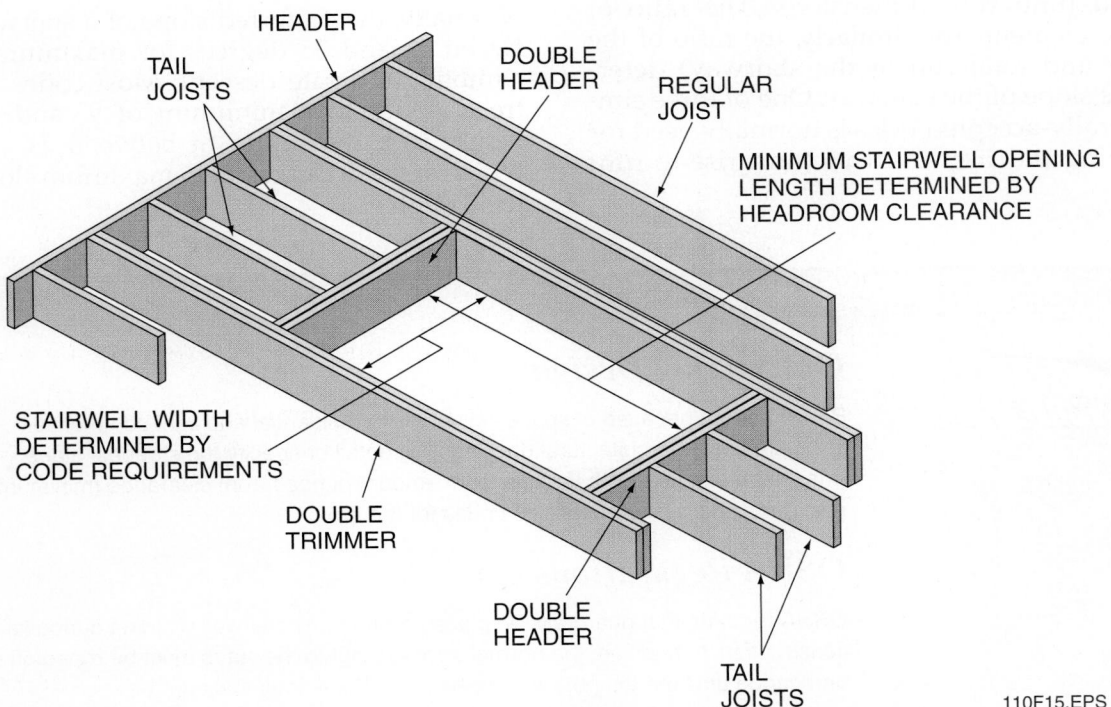

Figure 15 ◆ Typical framed stairwell opening.

5.0.0 ◆ STAIRWAY AND STAIRWELL DESIGN AND LAYOUT

If architectural building plans are being used for a project, the rise and run of the stairways in the building have usually been designed by the architect, and the unit rise (riser height), unit run (tread width), and number of risers and treads for each stairway have already been determined, along with the stairwell opening. In these cases, the stairwell openings are framed first, in accordance with the plans. Then, the carpenter lays out and constructs the rough stairways using the architect's plan. If there is no architect's plan, the carpenter may be required to design and construct any stairways and stairwells.

5.1.0 Stairway Design

An important element in stair design is the mathematical relationship between the riser (unit rise) and tread (unit run) dimensions. The ratio of these two elements (or similarly, the ratio of the total rise and total run of the stairway) determines the slope of the stairway. One of three simple, generally-accepted rules is normally used for determining the riser-to-tread (unit rise-to-run) ratio:

- *Rule 1* – Unit run (tread) + (2 × unit rise) = 24" to 25"
- *Rule 2* – Unit run + unit rise = 17" to 18"
- *Rule 3* – Unit run × unit rise = 70" to 75"

You must check the code to determine which of the above is the governing rule. The most common is Rule 1, which allows a maximum rise of 8" for the usual code-minimum tread depth of 9". Rules 2 and 3 allow a maximum rise of 8¼" for the usual code-minimum tread depth of 9".

Given the unit rise of the stairway, the applicable rule can be used to determine the longest and shortest recommended unit run (tread) length. Using any recommended tread length equal to or greater than the code minimum (usually 9"), the total run length of the stairway can then be determined. The stairway riser height (unit rise) is normally determined first.

5.1.1 Determining Riser Height

Normally, the preferred slope of a stairway is between 30 and 35 degrees for maximum ease in climbing and safe descent. Most codes limit the tread depth to a minimum of 9" and the riser height to a maximum of between 7¾" and 8¼", which, if used, can result in maximum slopes of 40 to 42 degrees.

Whenever possible, a unit rise between 7" and 7½" and a unit run between 10½" and 12" is recommended; this will result in slopes of 30 to 35 degrees. The maximum unit rise and minimum unit run permitted by code are often used in building plans or by contractors to minimize stairway space.

The riser height (unit rise) can be determined as outlined in the following procedure:

Step 1 From the building drawings, determine the final thickness, in inches, of the floor assembly above the stairs. Include the anticipated or actual finished floor, subfloor, floor joists, any furring, and the anticipated or actual finished ceiling. As shown in *Figure 16*, the floor assembly thickness for our example stairway is 8½".

Step 2 Add the floor thickness (Step 1) to the vertical distance, in inches, from the anticipated or actual finished ceiling to the anticipated or actual finished floor used at the base of the stairs. This sum is the total rise of the stairway. For our example:

$$\text{Total rise} = 8.5" + 7'\text{-}6"$$
$$= 8.5" + (7' \times 12") + 6$$
$$= 98.5"$$

NOTE

The total rise of a straight stairway normally must never exceed 12' (144") or as specified by applicable codes. If the limit is exceeded, an intermediate landing must be used.

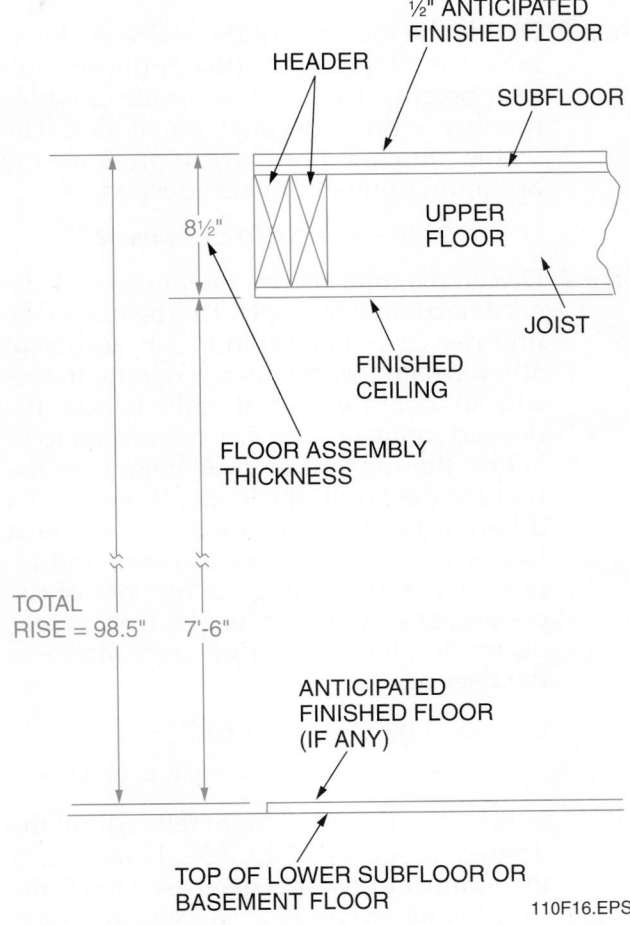

Figure 16 ◆ Method of determining total stair rise and unit rise.

Determining Riser Height

A simple summary of the procedure for determining riser height is to take the total unit rise and divide it by 7". This gives the number of risers for the stairs. By dividing the number of risers into the total unit rise, the exact height of each riser can be determined.

Step 3 Divide the total rise of the stairway determined in Step 2 by 7" (the optimum unit riser height). The result is usually a whole number with a decimal remainder. The whole number normally represents the optimum number of risers desired.

$$98.5" \div 7" = 14.07 \text{ or } 14 \text{ risers}$$

Step 4 Divide the total rise by the number of risers determined in Step 3. This results in the unit rise (riser height) in inches and, usually, a decimal remainder. Examine the result to determine if it falls within the desired range of 7" to 7.5" or is equal to or below the maximum riser height limitation for the applicable code. If not, or if a different height is required, use the next higher or lower number of risers and repeat Step 4. Increasing the number of risers decreases the riser height. Conversely, decreasing the number of risers increases the riser height.

$$\text{Unit rise} = 98.5" \div 14 = 7.036"$$
$$= 7" + \text{decimal remainder of } 0.036"$$

In this case, the riser height falls within the desired range of 7" to 7.5". However, if the number of risers is decreased to 13, the riser height would be 7.58", which is only a small fraction of an inch more than the desired range of 7" to 7.5". Although 13 risers is within code, 14 risers would yield a rise closer to 7", which would be more comfortable for use by the occupants.

Step 5 Convert the decimal remainder to fractions of an inch (32nds) by multiplying the decimal remainder by 32 and rounding up or down to the nearest whole number. This whole number represents the number of 32nds of an inch equal to the decimal remainder. Reduce the fraction to 16ths or 8ths if possible and combine with the

whole number of inches of the unit rise determined in Step 4.

$$\text{Unit rise} = 7" + 0.036"$$
$$= 7" + (0.036" \times 32 = 1.15 = \tfrac{1}{32}")$$
$$= 7\tfrac{1}{32}"$$

5.1.2 Determining Tread Run and Total Stair Run

The tread run (unit run) or depth is measured from the face of one riser to the face of the next and does not include any nosing (*Figure 17*).

To find the unit run, use the unit rise determined by either Method 1 or 2 and apply one of the appropriate rules (Rule 1, 2, or 3) as defined by the applicable code. Normally, the minimum unit run allowed is used to conserve the space required for the total run of the stairway.

To find the total run if the stringer uses the stairwell header as the last riser, subtract one from the number of risers (one less tread required) and multiply the result by the unit run.

If the stringer is mounted flush with the top of the floor as shown in the alternate configuration on the figure, multiply the number of risers by the unit run (same number of treads as risers). For our example, we will use *Figure 17*.

Step 1 Using the unit rise of 7½2" determined from the previous example and the relationship defined in Rule 1, solve for the unit run as follows:

$$\text{Maximum unit run} = 25" - (2 \times \text{unit rise or } 7\tfrac{1}{32}")$$

$$\text{Minimum unit run} = 24" - (2 \times \text{unit rise or } 7\tfrac{1}{32}")$$

The maximum and minimum allowable unit runs are:

$$\text{Maximum unit run} = 25" - 14\tfrac{1}{16}" = 10\tfrac{15}{16}"$$
$$\text{Minimum unit run} = 24" - 14\tfrac{1}{16}" = 9\tfrac{15}{16}"$$

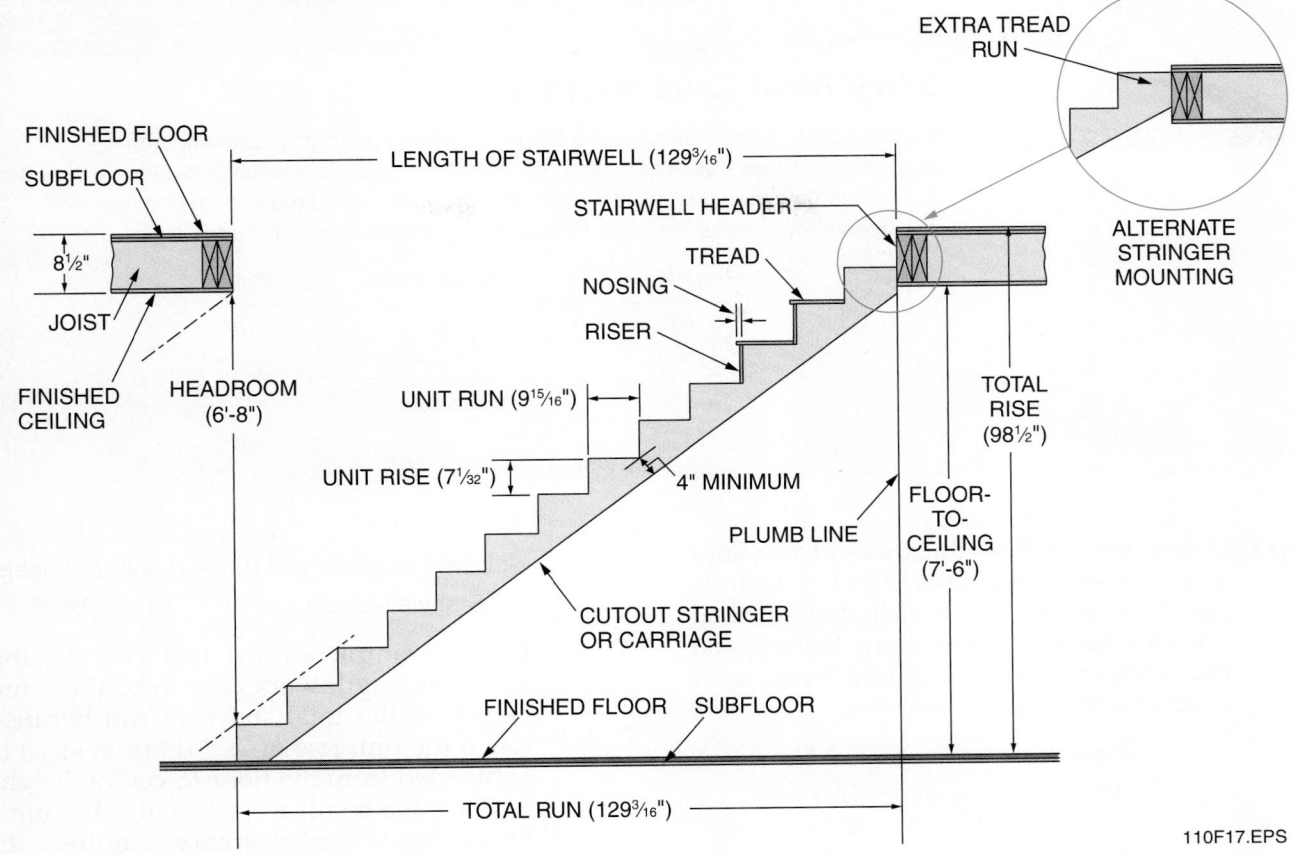

FINISHED FLOOR
SUBFLOOR
8½"
JOIST
FINISHED CEILING
HEADROOM (6'-8")
LENGTH OF STAIRWELL (129³⁄₁₆")
STAIRWELL HEADER
TREAD
NOSING
RISER
UNIT RUN (9¹⁵⁄₁₆")
UNIT RISE (7¹⁄₃₂")
4" MINIMUM
PLUMB LINE
CUTOUT STRINGER OR CARRIAGE
FINISHED FLOOR   SUBFLOOR
TOTAL RISE (98½")
FLOOR-TO-CEILING (7'-6")
TOTAL RUN (129³⁄₁₆")
EXTRA TREAD RUN
ALTERNATE STRINGER MOUNTING
110F17.EPS

Figure 17 ◆ Example stairway and stairwell with terminology and dimensions.

Step 2 For our example, a minimum unit run of 9¹⁵⁄₁₆" will be used. If the stairwell header is used as the last riser, we will require one less tread than the number of risers and the total run will be:

Total run = (number of risers − 1) × unit run
Total run = (14 − 1) × 9¹⁵⁄₁₆" = 129³⁄₁₆"

If the top end of the stringer will be flush with the top of the floor as depicted in the alternate mounting shown in *Figure 17*, then the same number of treads as risers are required, and the total run would be:

Total run = number of risers × unit run
Total run = 14 × 9¹⁵⁄₁₆" = 139⅛"

5.2.0 Stairwells

If a stairwell has been framed but the unit rise and run of the stairway are not specified, they must be determined. If the stairwell hasn't been framed, the stairwell opening should be determined as follows for the desired headroom clearance.

5.2.1 Determining Stairwell Openings

The stairwell opening width for a straight stairway is determined by the desired width of the stairway plus the thickness of any skirtboard(s). The stairwell width for an open U-shaped stairway could be as wide as both stairways, plus the skirtboards, plus the amount of space required to turn the U-shaped handrails.

The length of the stairwell depends on the slope and total rise of the stairway. Stairs with a low angle and/or low total rise require a longer stairwell to provide adequate headroom clearance. To find the required length of a stairwell, use the previously determined unit rise and run for the stairway along with the desired headroom clearance and the following procedure.

Riser/Tread Combinations

To construct a staircase having the proper angle for comfortable climbing and descending, a riser height of between 7" and 7½" and a tread width of between 10" and 11" is recommended. In addition, a rule of thumb states that the sum of one riser and one tread should equal between 17" and 18". Some examples include:

Riser Height	Tread Width	Total
7"	11"	18"
7¼"	10"	17¼"
7⅜"	10½"	17⅞"
7½"	10½"	18"

Step 1 Determine the final thickness of the floor assembly above the stairs in inches. Include the thickness of the anticipated finished floor, subfloor, floor joists, any furring, and the anticipated finished ceiling (from *Figure 17* and the previous examples).

$$\text{Floor thickness} = 8\tfrac{1}{2}" = 8.5"$$
$$\text{Headroom required} = 6'\text{-}8"$$
$$= (6' \times 12") + 8" = 80"$$
$$\text{Unit rise} = 7\tfrac{1}{32}" = 7.03"$$
$$\text{Unit run} = 9\tfrac{15}{16}"$$

Step 2 Add the floor thickness (Step 1) to the code-required or desired headroom clearance in inches and divide the sum by the riser height (unit rise) in inches. Round the answer up to the next whole number. This number represents the number of risers from the top of the stairs down to, or slightly below, the headroom clearance point.

$$\text{Number of risers required} =$$
$$(8.5" + 80") \div 7.03"$$
$$= 12.59 \text{ (round up to 13)}$$

Step 3 Multiply the tread depth (unit run) by the whole number obtained in Step 2 to obtain the stairwell length in inches. This length will be correct if the header at the top of the stairway will be the top riser. If the stringers will be framed flush with the top of the header, add one additional tread width (a unit run) to the overall length of the stairwell. In this example, assume that the stairwell header is used as the last riser and the addition of one unit run to the length will not be required.

$$\text{Stairwell length} = \text{unit run} \times \text{number of risers}$$
$$\text{Stairwell length} = 9\tfrac{15}{16}" \times 13 = 129\tfrac{3}{16}"$$

In the example shown in *Figure 17*, the stairwell length works out to be the same length as the total stairway run because when the unit rise of 7½" (bottom step) is subtracted from the floor-to-ceiling height of 7'-6", the result is essentially the minimum headroom clearance required. In stairways involving higher floor-to-ceiling heights, the stairwell will usually be shorter than the total stairway run.

5.2.2 Determining Stairway Unit Rise and Run Used for a Pre-Framed Stairwell

In some instances, you may arrive at a job site and find that someone has framed the stairwell openings but the stairway unit rise, run, and headroom clearance used are not available. In these rare cases, you must determine the stairway unit rise and run, along with the total run, based on the length of the stairwell opening, the total stairway rise, and, initially, the minimum headroom clearance allowed by the applicable code. This can be accomplished using the following process:

Step 1 Determine the optimum unit rise (riser height) in inches, as previously described. Using the values shown in *Figure 17* as an example, the total rise of the stairway is determined to be 98½" and the optimum unit rise is 7½".

Step 2 Determine the final thickness of the floor assembly above the stairs in inches. Include the thickness of the anticipated finished floor, subfloor, floor joists, any furring, and the anticipated finished ceiling.

Apparent Run Length Errors

If the unit run appears to be too large and the resultant total run appears to be too long for the space allowed, the stairwell may have been calculated using more than the minimum required headroom or it may have been lengthened by an extra tread width to accommodate for flush-mounting the stringers with the floor above. In that case, repeat the required calculations using one extra riser or greater headroom clearance (7' or more, if possible) to determine if either method provides the correct total run.

Referring to *Figure 17*, the floor thickness is 8½" and the minimum headroom is 6'-8".

Step 3 Add the floor thickness (Step 2) to the code-required headroom clearance in inches and divide the sum by the unit rise determined in Step 1. Round the answer up to the next whole number. This number represents the number of risers of optimum height from the top of the stairs down to, or slightly below, the headroom clearance point.

Number of risers required =
(8.5" + 6'-8") ÷ 7¹⁄₃₂" = 88.5" ÷ 7.03"

= 12.59 (round up to 13)

Step 4 Divide the stairwell length by the number of risers calculated in Step 3 to obtain the unit run (tread width) that may have been used to determine the length of the stairwell.

Unit run = 129³⁄₁₆" ÷ 13 = 9.93" or 9¹⁵⁄₁₆"

Step 5 Check the unit run-to-rise ratio using the applicable rule to determine if the calculated tread width meets the minimum requirement. If it does not, the maximum allowable unit rise may have been used in the original calculations for the stairs. In that case, repeat all the steps using the maximum allowable unit rise. In this example, the unit run (and consequently, the unit rise) appears to be valid because it will satisfy Rule 1 for the minimum tread width.

5.3.0 Laying Out and Cutting a Stringer

Marking a cutout stringer is a simple task. It can be done with either a framing square or a pitch board. When using a framing square, the blade will represent the unit run and the tongue will represent the unit rise. See *Figure 18*. To make the stair layout go faster, a set of stair gauges (*Figure 19*) can be used to set the unit rise and run measurements on the framing square.

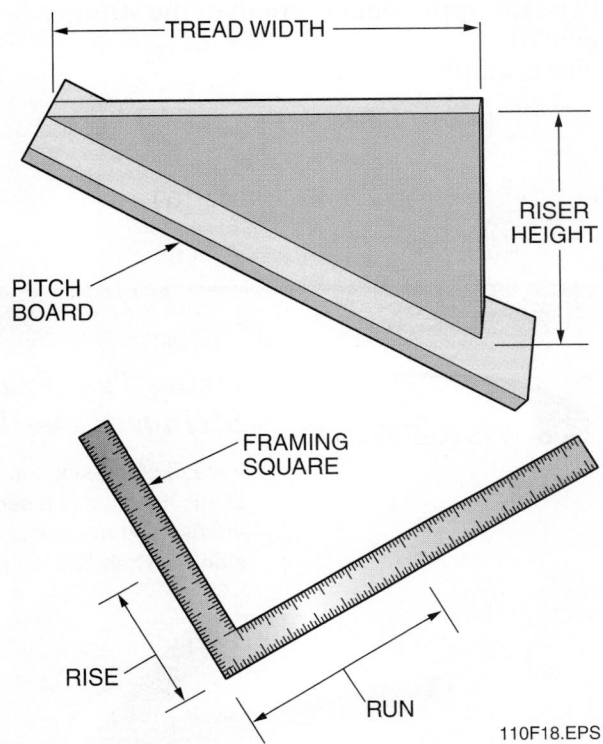

110F18.EPS

Figure 18 ◆ Pitch board and framing square.

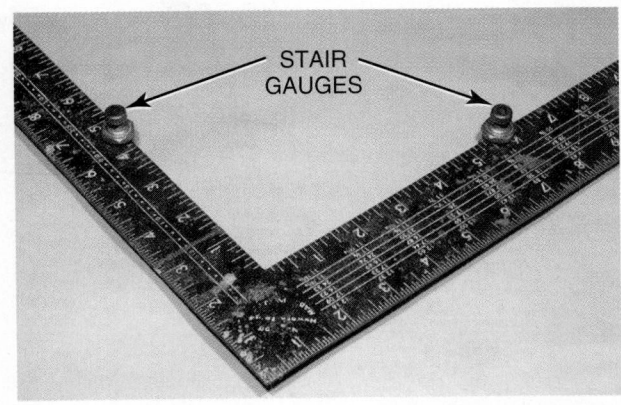

110F19.EPS

Figure 19 ◆ Stair gauges.

To obtain the approximate length of the stringer, the principles of the right triangle can be used. The rise and run are known; therefore, they form two sides of a right triangle. The stringer is the hypotenuse or third side of the right triangle. Using the outside back of the framing square, locate the total run on the blade and the rise on the tongue. Each inch increment on the outside back of the framing square represents 1' and each small increment represents 1". Mark the points on paper. Measure the hypotenuse (diagonal) between the two points along the same side of the blade. This will be the approximate length of the stringer. See *Figure 20*.

For example:

Rise = 9'-3"

Run = 12'-6"

Stringer = 15'-7" (use 16')

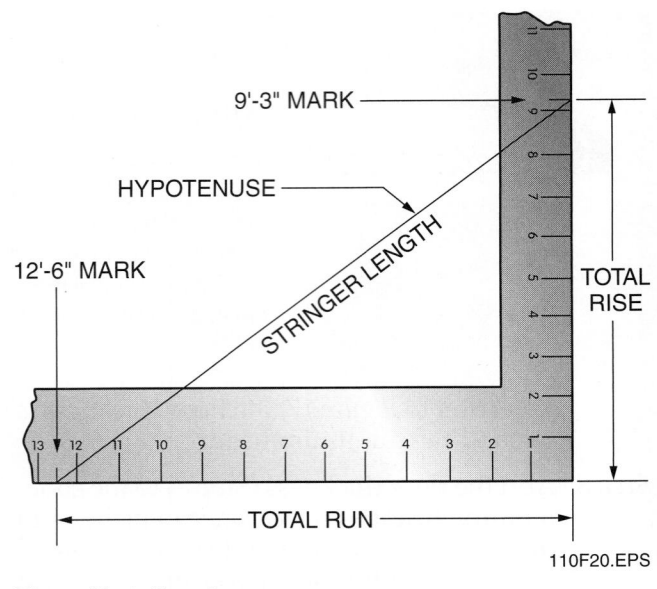

110F20.EPS

Figure 20 ◆ Framing square.

Using Two Framing Squares to Determine Stringer Length

INSIDE TRACK

Instead of marking the rise and run on a piece of paper and measuring between the points, the side of a second framing square, marked in twelfths, can be positioned over the rise and run points on the first square to form the hypotenuse representing the stringer. Then, the length of the stringer can be read directly on the second square.

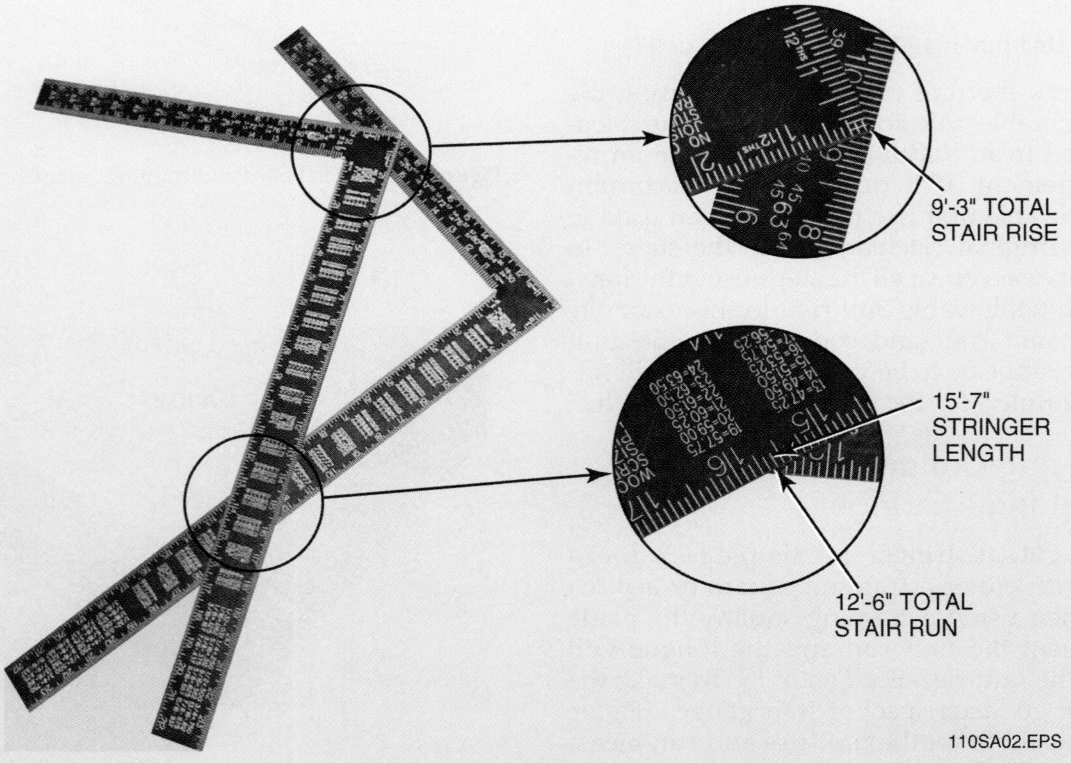

9'-3" TOTAL STAIR RISE

15'-7" STRINGER LENGTH

12'-6" TOTAL STAIR RUN

110SA02.EPS

This measurement can also be calculated mathematically by applying the Pythagorean theorem: $(a^2 + b^2 = c^2)$. In this case: $(9'\text{-}3'')^2 + (12'\text{-}6'')^2 = (15'\text{-}7'')^2$ or $(16')^2$.

The cutting and installation of stringers must be done with caution so that the carpenter finishing the stairs will not encounter difficulty in fitting the risers and treads. When laying out a stringer, all lines drawn must be thin and accurate. Cutting the risers and treads from the stringer can be done with a hand saw or power saber saw, or with a circular saw, provided you do not allow the circular saw to cut farther than the riser line. Then, a hand or saber saw can be used to finish the cuts. A stringer must be cut accurately. If the stringer is too long, the treads will slant backwards; if the stringer is too short, the treads will slant forward. Laying out a stringer can be accomplished using one of the following two methods.

Method One:

Step 1 The stock should be 2 × 10, 2 × 12, or some other structurally sound size and of a species and grade of wood approved by the local code. The stock must be as straight as possible.

Step 2 Lay the stock on sawhorses or a work table. Place the top edge (good edge) away from you. All layout work is done from the top edge. See *Figure 21*.

Step 3 Place the framing square at the end of the stringer, as shown in *Figure 22(A)*. Mark the unit rise and run on the stringer using the inside edge of the square. Remove the square and check that there is a minimum of 4" between the junction of the rise/run and the bottom of the stringer, as shown in *Figure 22(B)*. If not, a larger stringer must be used. Place the square back on the stringer, as shown in *Figure 22(C)*, and extend the unit run mark to the bottom of the stringer. This will be the portion of the stringer that rests on the floor. Identify the extended unit run mark as a cut line.

WARNING!
Overcutting at the junctions of the rise/run can significantly weaken the stringer.

Step 4 Move the framing square to the next tread and mark the rise and run. Proceed to mark the rise and run for all the other steps. If the riser is to be installed as shown in *Figure 23*, then the stringer will be cut as it was laid out because the header will be the top riser (one more riser than the number of treads). If the stringer is to be placed so that the top tread is at the level of the finished floor, the stringer will be cut with one more tread. See *Figure 24*.

Step 5 Starting at the left side, cut the first line that goes from the top to the bottom. All cuts must be done cautiously. Be sure not to overcut. See *Figure 25*.

Step 6 Cut the rest of the treads and risers.

Step 7 If the stringer is to sit on top of a finished floor, the stringer must be dropped (cut) the thickness of a tread so that the top and bottom step of the stairs will be at the same height. See *Figure 26*. The stringer may need to be scribed. If the floor that the stringers will rest on will have some type of finished floor added later, the bottom of the stringer must be cut to compensate. The amount to be cut would be the difference between the finished flooring thickness and the thickness of the tread. See *Figure 27*.

Step 8 Use the completed stringer as a pattern to lay out the remaining stringer(s). Lay out the opposite wall stringer so that any bow faces in toward the stairway.

Step 9 If a stairway is 30" or less in width, only two stringers are required. For added strength and stability of the stairway, a third stringer may be centered between the two stringers. A stairway that is 30" or wider usually requires three stringers. Check local codes for requirements.

Step 10 Install the skirtboards, if required. Then install the stringers, one on each side of the stairway. Nail the stringers to the stud wall with 16d common nails. Complete any other related construction activities as shown on the drawings.

Step 11 Install rough treads of 2" nominal stock. These treads will stay in place until the finished risers and treads are ready to be installed. The rough treads can be reused, if applicable.

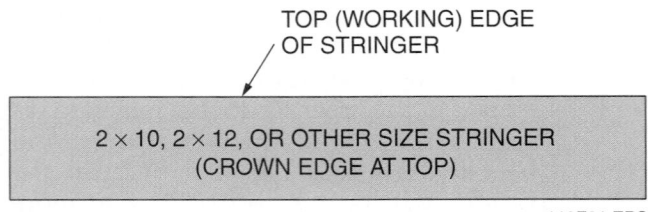

TOP (WORKING) EDGE
OF STRINGER

2 × 10, 2 × 12, OR OTHER SIZE STRINGER
(CROWN EDGE AT TOP)

110F21.EPS

Figure 21 ◆ Positioning of a stringer for layout.

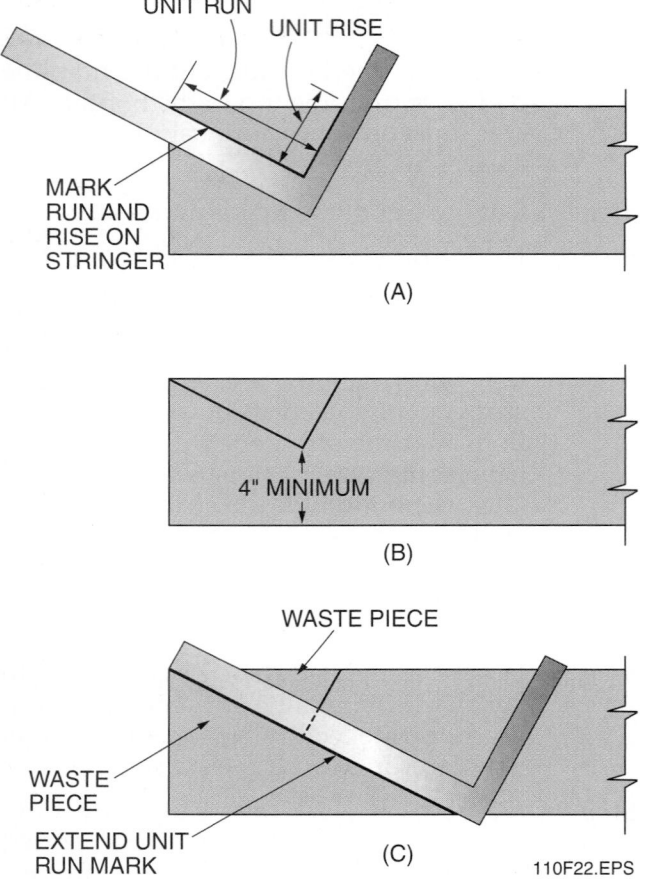

UNIT RUN

UNIT RISE

MARK
RUN AND
RISE ON
STRINGER

(A)

4" MINIMUM

(B)

WASTE PIECE

WASTE
PIECE

EXTEND UNIT
RUN MARK

(C)

110F22.EPS

Figure 22 ◆ Marking a stringer for the floor level cut line.

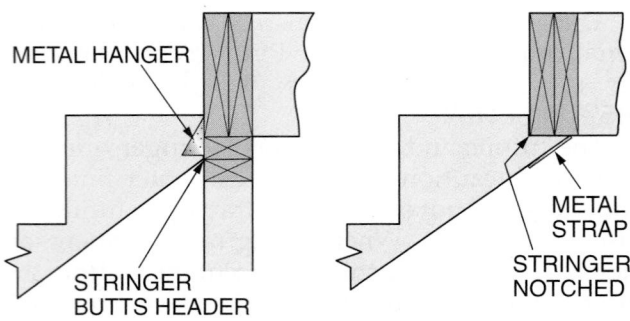

METAL HANGER

STRINGER
BUTTS HEADER

METAL
STRAP
STRINGER
NOTCHED

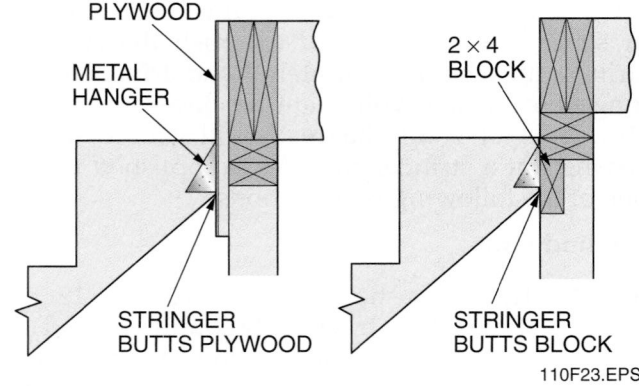

PLYWOOD

METAL
HANGER

STRINGER
BUTTS PLYWOOD

2 × 4
BLOCK

STRINGER
BUTTS BLOCK

110F23.EPS

Figure 23 ◆ Stringer mounting attachments.

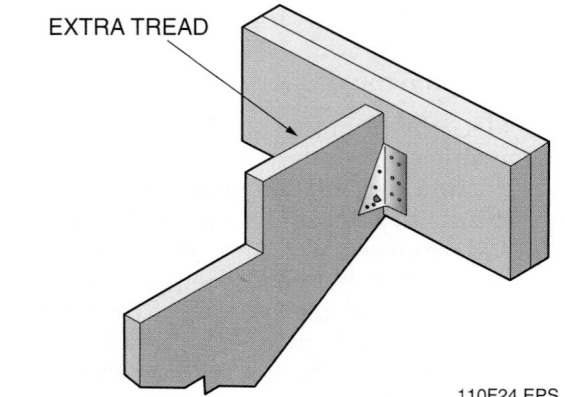

EXTRA TREAD

110F24.EPS

Figure 24 ◆ Flush stringer attachment.

Compensating for Bottom Riser Height of Stringers

INSIDE TRACK

Be sure to allow for the thickness of the finished floor and/or treads when fitting the stringers. One of the most common mistakes is not adjusting the bottom riser height of the stringers so that the top and bottom steps will be the same height as the rest of the steps.

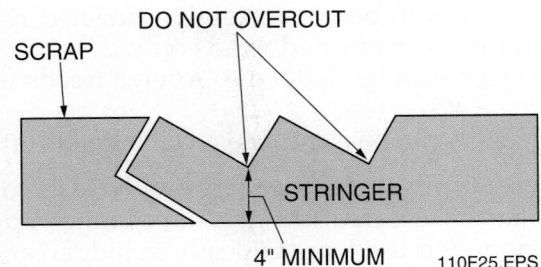

Figure 25 ◆ Stringer cuts.

SCRAP

DO NOT OVERCUT

STRINGER

4" MINIMUM

110F25.EPS

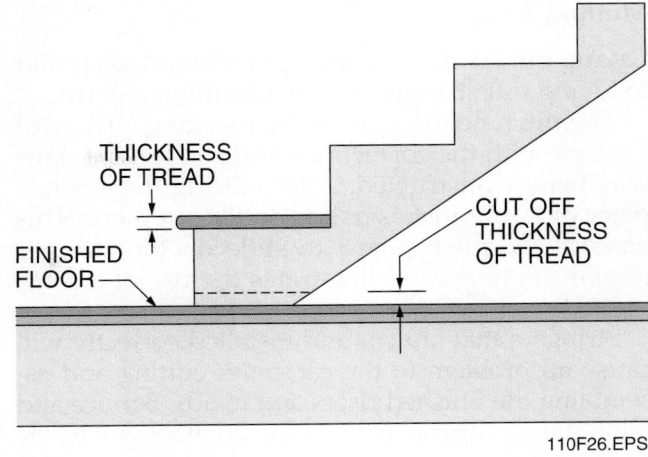

THICKNESS OF TREAD

FINISHED FLOOR

CUT OFF THICKNESS OF TREAD

110F26.EPS

Figure 26 ◆ Dropping the stringer for tread thickness.

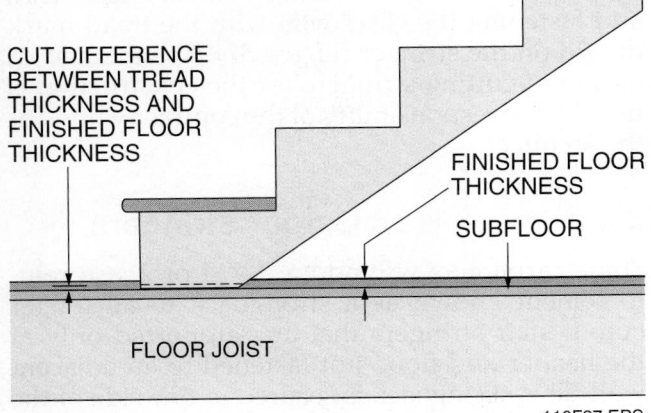

CUT DIFFERENCE BETWEEN TREAD THICKNESS AND FINISHED FLOOR THICKNESS

FINISHED FLOOR THICKNESS

SUBFLOOR

FLOOR JOIST

110F27.EPS

Figure 27 ◆ Dropping the stringer for tread thickness minus the finished floor thickness.

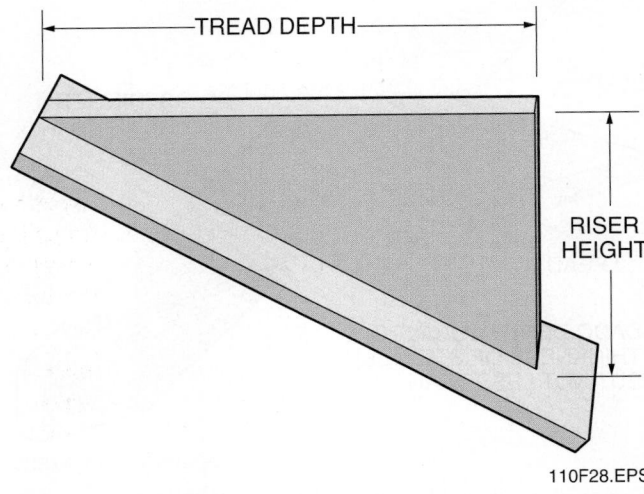

TREAD DEPTH

RISER HEIGHT

110F28.EPS

Figure 28 ◆ Pitch board.

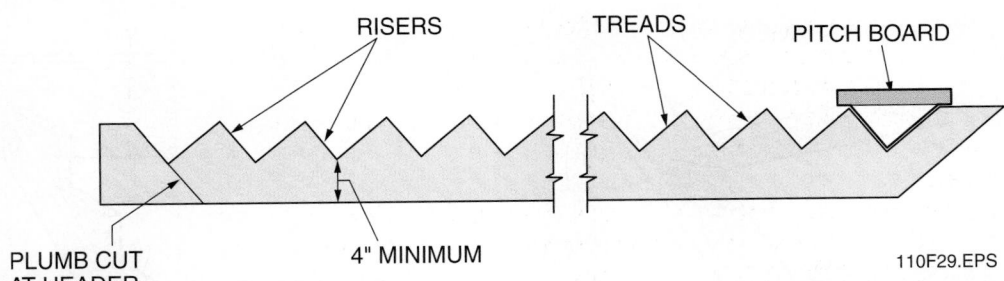

RISERS

TREADS

PITCH BOARD

PLUMB CUT AT HEADER

4" MINIMUM

110F29.EPS

Figure 29 ◆ Using a pitch board.

Field-Fabricated Housed Stringers

Field mortising a housed stringer can be done with the use of a specialized European-made jig and router. The layout of a mortised stringer starts with the same procedures described previously, including the procedure with a framing square or pitch board.

Method Two:

Laying out stringers with a pitch board is similar to laying out stringers using a framing square.

The pitch board is made by marking a piece of 1" stock with the correct rise and run for the stairway being constructed. After cutting the stock, a piece of 1 × 2 stock is nailed on the cut piece. This enables the pitch board to slide on the stringer (*Figure 28*). *Figure 29* illustrates the use of a pitch board.

Stringers that are cut and installed correctly will cause no problem to the carpenter cutting and assembling the finished risers and treads. Service and main stair stringers can be prepared in several ways:

- Stringers can be cut to receive finished risers and treads or finished treads only.
- Stringers can be dadoed to receive treads only (*Figure 30*).
- Stringers can be cleated to receive treads only.

Figure 31 illustrates a stringer with cleats to receive treads. Cleated stringers have limited applications and, if used, are typically in hidden areas, such as utility stairs for basements. Installing cleats on a stringer is done by first laying out the stringer as previously described, cutting 1" stock to the proper length or using a metal angle cleat, and fastening the cleat even with the tread mark drawn on the stringer (*Figure 31*). The method or option of cutting stringers for the construction is usually the responsibility of the contractor and/or the architect.

5.4.0 Reinforced Cutout Stringers

Some carpenters will add a 2 × 4 or 2 × 6 reinforcement known as a strongback to all longer cutout stair stringers that are supported only at the header and floor, not fastened to an adjacent wall. This strongback is secured to one side of the cut stringer below the cut notches to add strength and rigidity. See *Figure 32*.

It is also a common practice to secure a 2 × 4 ledger to the floor and/or header to add strength to the stair assembly. The stair stringers are notched to fit over the 2 × 4 ledger. See *Figure 33*.

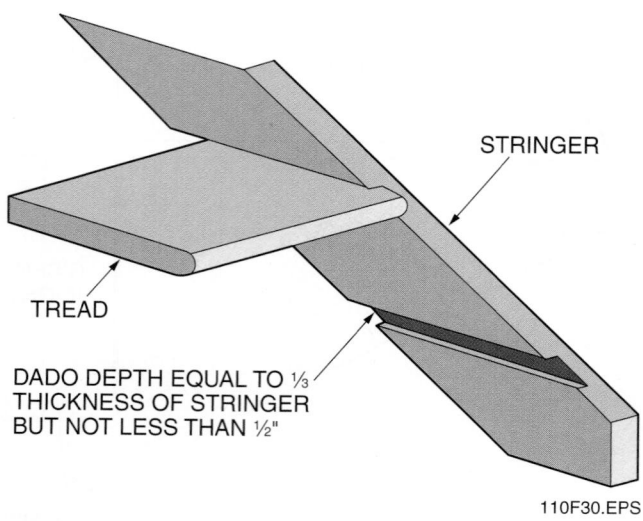

DADO DEPTH EQUAL TO ⅓ THICKNESS OF STRINGER BUT NOT LESS THAN ½"

110F30.EPS

Figure 30 ◆ Dadoed stringer.

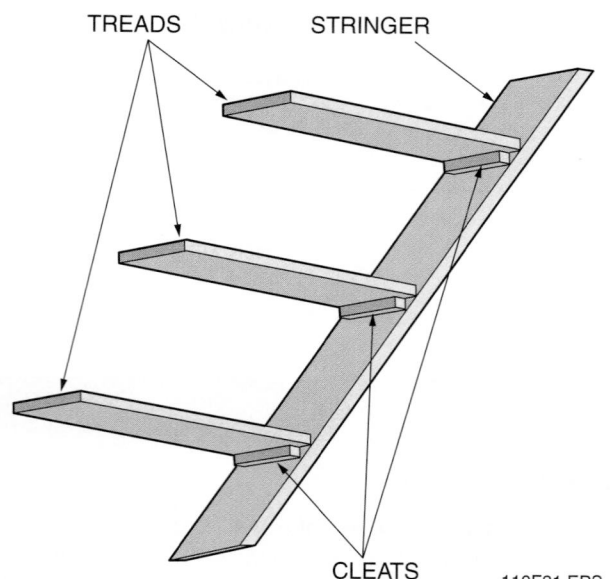

110F31.EPS

Figure 31 ◆ Cleated stringer.

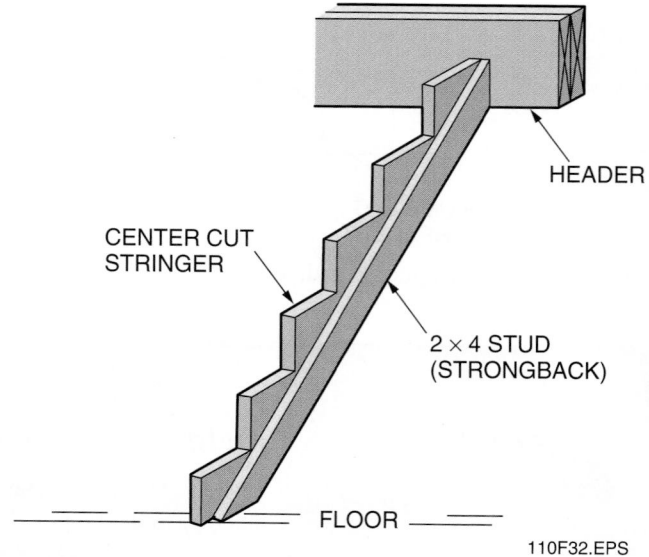

110F32.EPS

Figure 32 ◆ Reinforced cutout stringer.

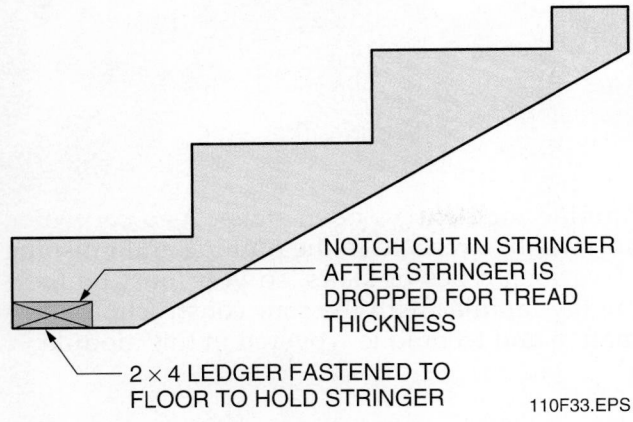

NOTCH CUT IN STRINGER AFTER STRINGER IS DROPPED FOR TREAD THICKNESS

2 × 4 LEDGER FASTENED TO FLOOR TO HOLD STRINGER

110F33.EPS

Figure 33 ◆ Reinforced stringer mounting.

NOTE

If the ledger is fastened to a concrete floor, treated lumber must be used.

6.0.0 ◆ FORMS FOR CONCRETE STAIRS

Figure 34 shows a basic concrete stair form. Note that the size and position of the riser boards establish the height, depth, and spacing of the stairs. When the stairs are wide, a center brace is needed for additional support.

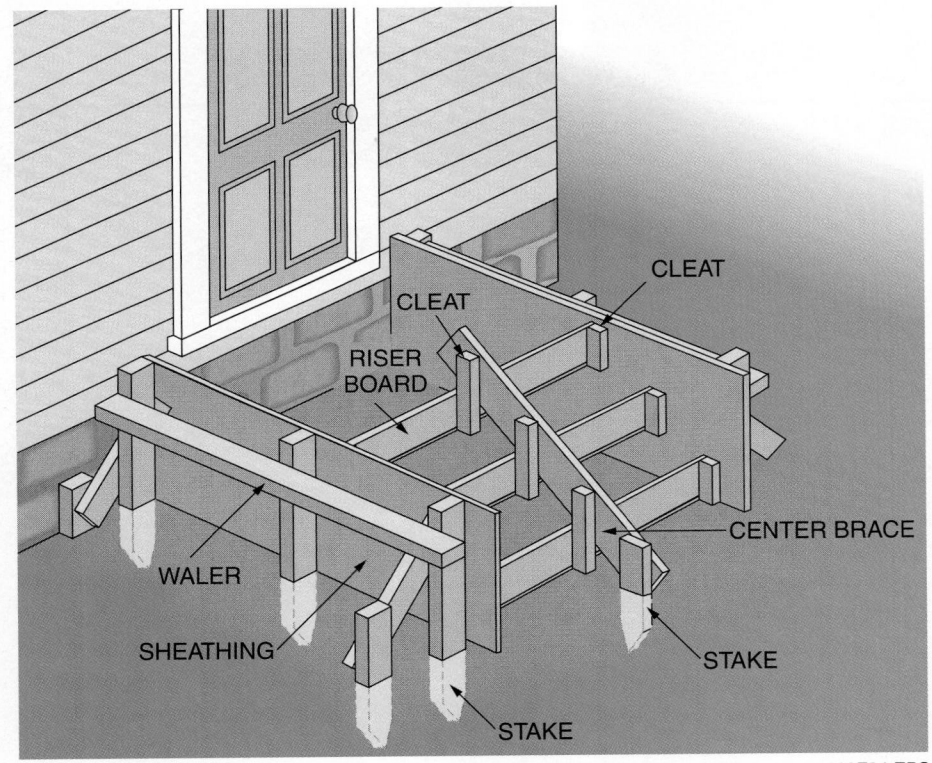

CLEAT

CLEAT

RISER BOARD

CENTER BRACE

WALER

SHEATHING

STAKE

STAKE

110F34.EPS

Figure 34 ◆ Basic stair form.

Summary

A variety of stairs are used in residential and commercial construction. This module covered stair terminology and general building code requirements. It also described the methods used when designing, laying out, cutting, and installing the framing for basic wooden stairs. As a carpenter, you will be involved in the framing and finishing of various types of stairs, so you must be thoroughly familiar with the stair construction information and techniques covered in this module.

Notes

1. The easiest type of stairway to build is a _____ stairway.
   a. straight-run
   b. platform
   c. wide-U
   d. winding

2. The stairway in *Figure 1* would be classified as a(n) _____.
   a. open stairway
   b. closed stairway
   c. combination open/closed stairway
   d. landing

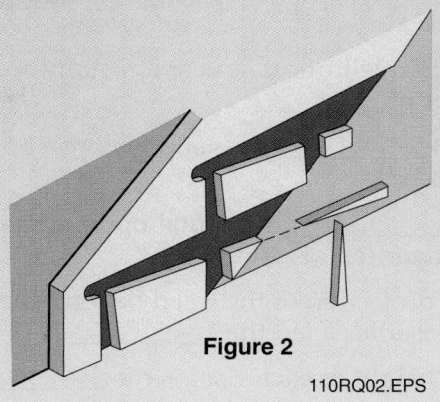

Figure 2

110RQ02.EPS

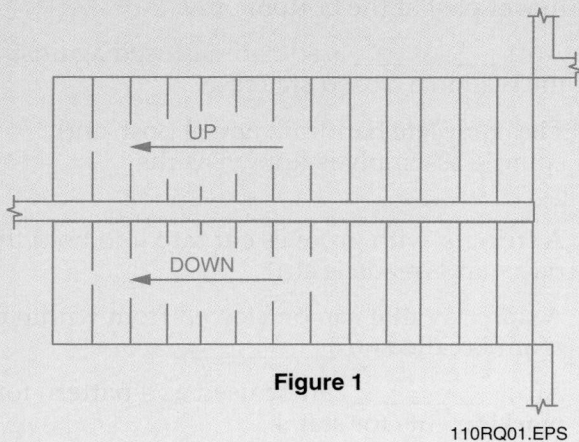

Figure 1

110RQ01.EPS

3. Which of the following is a geometrical stairway?
   a. Long-L
   b. Wide-U
   c. Elliptical
   d. Double-L

4. Center cutout stringers may be required on stairs that are wider than _____.
   a. 30"
   b. 36"
   c. 40"
   d. 45"

5. The stringer in *Figure 2* is known as a _____ stringer.
   a. dadoed
   b. housed
   c. common
   d. cleated

6. The main support for a handrail at the bottom of a stairway is called a(n) _____.
   a. baluster
   b. main newel post
   c. ending newel post
   d. starting newel post

7. A typical minimum headroom specified by the *IBC* for residential use is _____.
   a. 6'-4"
   b. 6'-6"
   c. 6'-8"
   d. 6'-10"

8. Stairways wider than _____ usually require handrails on both sides.
   a. 32"
   b. 36"
   c. 44"
   d. 88"

9. Which of the following is used to establish the rise and run of a staircase?
   a. Unit run + (2 × unit rise) = 17" to 18"
   b. Unit run × unit rise = 24" to 25"
   c. Unit run + (2 × unit rise) = 24" to 25"
   d. Unit run + (2 × unit rise) = 70" to 75"

10. The total rise of a straight stairway should *not* exceed _____.
    a. 10'
    b. 12'
    c. 14'
    d. 16'

1. Circular, elliptical, and spiral stairs are all types of _____.

2. A(n) _____ has solid walls on both sides.

3. The _____ is the main post supporting the handrail of a staircase.

4. The projection of the tread beyond the face of the riser is called the _____.

5. The height of each riser is the _____.

6. The depth of each tread is the _____.

7. When you walk up several flights of stairs, you are likely to pause on the _____ between flights.

8. An open fire escape stairway is likely to be enclosed by a(n) _____.

9. An enclosed stairway will have a(n) _____ along the wall to serve as a handhold.

10. The spindles of a staircase are also called _____.

11. The distance between any portion of a step and the ceiling is the _____.

12. A(n) _____ provides the main support for a stairway.

13. The horizontal part of a step is known as the _____.

14. The vertical part of a step is known as the _____.

15. A(n) _____ is a vertical compartment in a building into which stairs are placed.

16. The _____ is the horizontal distance from the face of the first riser to the face of the last riser.

17. The _____ is the vertical dimension of a set of stairs.

18. A(n) _____ is a curved stairway with a newel post at the bottom only.

19. A(n) _____ is sometimes used against the wall in a closed staircase.

20. The complete handrail, newel post, and spindle assembly is known as the _____.

21. A stringer with grooves cut into it to hold the risers and treads is a(n) _____.

22. A stairway that can be viewed from within a room is called a(n) _____.

23. A(n) _____ can be used as a pattern for marking cuts for stairs.

24. The degree of incline of a set of stairs is the _____.

Trade Terms

Baluster
Balustrade
Closed stairway
Geometrical stair
Guardrail
Handrail

Headroom
Housed stringer
Landing
Newel post
Nosing
Open stairway

Pitch board
Rise
Rise and run
Riser
Run
Skirtboard

Stairwell
Stringer
Tread
Unit rise
Unit run
Winding stairway

Curtis McLawhorn

2004 SkillsUSA Carpentry Championship Gold Medal Winner

Curt has participated in many carpentry competitions. He competed in three regional SkillsUSA competitions and won several medals. Winning at the regional level allowed him to go to the national competition where he won the gold medal in carpentry. He is now studying civil engineering in college and plans to continue working in the construction industry.

How did you become interested in carpentry?

My father is a carpenter and general contractor and I have been working with him for as long as I can remember. I helped him on many smaller projects over the years. When I was 12 or 13, I started working for his company during the summer. I liked the work and took some classes to improve my skills.

What kind of formal training have you had?

I took some carpentry classes in high school and that is how I got involved with the SkillsUSA competitions. When I was working for my father, I also enrolled in an apprenticeship program through the North Carolina Department of Labor. They also sponsor competitions. I won a second place in the state competition. Now I am studying civil engineering at North Carolina State University.

What do you think it takes to be a success?

I would say that it takes a good background in math to be a successful contractor. You have to be able to read plans and prepare estimates. If your estimates are off, you won't be able to make a profit.

What do you do in your job?

As a laborer working for my father, I do it all. We build residential houses, so I have done framing, roofing, plumbing, electrical work, and everything else involved in building houses.

What do you like about carpentry?

I really enjoy framing houses. I like the satisfaction of building something. When you start there is nothing there, just an empty lot. When you finish, there is the basic framework of a house. Now, when I drive by, it feels pretty good, knowing that I helped build those houses.

I also enjoyed the carpentry competitions. I was never really good at sports but I do have some trade skills. It was pretty cool to be able to show off my skills in the competitions. I won some nice prizes. I won some tools and a trip. On the trip I had the chance to meet Matt Kenseth, a NASCAR driver. It is great that SkillsUSA has competitions where you can win medals that are not sports related.

What advice would you give someone just starting out?

Try hard and don't be afraid to ask other people with more experience for help. If you get into it, get all the knowledge and experience you can.

Trade Terms
Introduced in This Module

Baluster: A supporting column or member; a support for a railing, particularly one of the upright columns of a balustrade.

Balustrade: A stair rail assembly consisting of a handrail, balusters, and posts.

Closed stairway: A stairway that has solid walls on each side.

Geometrical stair: A winding stairway built around a well. Examples include circular, elliptical, and spiral stairs.

Guardrail: A rail secured to uprights and erected along the exposed sides and ends of platform stairs, etc.

Handrail: A member supported on brackets from a wall or partition to furnish a handhold.

Headroom: The vertical and clear space in height between a stair tread and the ceiling or stairs above.

Housed stringer: A stair stringer with horizontal and vertical grooves cut (mortised) on the inside to receive the ends of the risers and treads. Wedges covered with glue are often used to hold the risers and treads in place in the grooves.

Landing: A horizontal area at the end of a flight of stairs or between two flights of stairs.

Newel post: An upright post supporting the handrail at the top and bottom of a stairway or at the turn of a landing; also, the main post about which a circular staircase winds or a stone column carrying the inner ends of the treads of a spiral stone staircase.

Nosing: The portion of the stair tread that extends beyond the face of the riser.

Open stairway: A stairway that is open on at least one side.

Pitch board: A board that serves as a pattern for marking cuts for stairs. The shortest side is the height of the riser cut, and the next longer side is the width of the tread. This is used mainly when there is great repetition such as in production housing.

Rise: The vertical dimension of a set of stairs. Also called the total rise.

Rise and run: A term used to indicate the degree of incline.

Riser: A vertical board under the tread of a stair step; in other words, a board set on edge for connecting the treads of a stairway.

Run: The horizontal distance from the face of the first or upper riser to the face of the last or lower riser. Also called the total run.

Skirtboard: A baseboard or finishing board at the junction of the interior wall and floor. Also called a finished stringer.

Stairwell: A compartment extending vertically through a building into which stairs are placed.

Stringer: The inclined member that supports the treads and risers of a stairway.

Tread: The horizontal member of a step.

Unit rise: The vertical distance from the top of one stair tread to the top of the next one above it; also called the stair rise.

Unit run: The horizontal distance from the face of one riser to the face of the next riser.

Winding stairway: A type of geometrical staircase that changes direction by means of winders or a landing and winders. The stair opening is relatively wide, and the balustrade follows the curve with only a newel post at the bottom.

This module is intended to present thorough resources for task training. The following reference works are suggested for further study. These are optional materials for continued education rather than for task training.

Basic Stairbuilding. Newton, CT: Taunton Press, Inc. (Book with companion video or DVD.)

Constructing Staircases, Balustrades & Landings. New York, NY: Sterling Publishing Co., Inc.

For Pros By Pros: Building Stairs. Newton, CT: Taunton Press, Inc.

Framing Floors and Stairs. Berkeley, CA: Publishers Group West. (Book with companion video or DVD.)

A Simplified Guide to Custom Stairbuilding and Tangent Handrailing. Fresno, CA: Linden Publishing.

Stair Builders Handbook. Carlsbad, CA: Craftsman Book Company.

Staircases. New York, NY: Watson-Guptill Publications.

Stair Layout. Homewood, IL: American Technical Publishers.

Stairs: Design and Construction. New York, NY: Birkhauser.

Arcways, Inc. Builders of custom stairways. www.arcways.com.

Classic Stairworks, Ltd. Builders of classic custom staircases. www.classicstairworks.com.

Coffman Stairs, LLC. Hardwood stair parts manufacturer. www.coffmanstairs.com.

L.J. Smith Stair Systems. Manufacturer of stair products. www.ljsmith.net.

NCCER makes every effort to keep these textbooks up-to-date and free of technical errors. We appreciate your help in this process. If you have an idea for improving this textbook, or if you find an error, a typographical mistake, or an inaccuracy in NCCER's Contren® textbooks, please write us, using this form or a photocopy. Be sure to include the exact module number, page number, a detailed description, and the correction, if applicable. Your input will be brought to the attention of the Technical Review Committee. Thank you for your assistance.

Instructors – If you found that additional materials were necessary in order to teach this module effectively, please let us know so that we may include them in the Equipment/Materials list in the Annotated Instructor's Guide.

Write: Product Development and Revision
National Center for Construction Education and Research
3600 NW 43rd St, Bldg G, Gainesville, FL 32606

Fax: 352-334-0932

E-mail: curriculum@nccer.org

Craft _____   Module Name _____

Copyright Date _____   Module Number _____   Page Number(s) _____

Description _____

_____

_____

_____

(Optional) Correction _____

_____

_____

(Optional) Your Name and Address _____

_____

_____

Electrical Safety

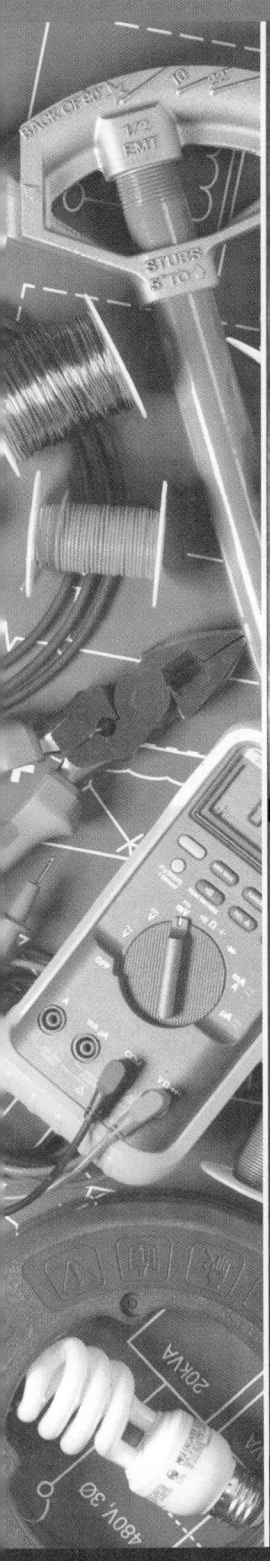

Ponnequin Wind Farm

The Ponnequin Wind Farm generates electrical power from the wind and is located on the plains of eastern Colorado just south of the Wyoming border. It consists of 44 wind turbines and can generate up to 30 megawatts of electricity. Each turbine weighs nearly 100 tons and stands 181 feet tall.

68112-09

68112-09
Electrical Safety

Topics to be presented in this module include:

1.0.0   Introduction 12.2
2.0.0   Electrical Shock 12.2
3.0.0   Reducing Your Risk 12.5
4.0.0   OSHA 12.12
5.0.0   *NFPA 70E* 12.19
6.0.0   Ladders and Scaffolds 12.20
7.0.0   Lifts, Hoists, and Cranes 12.23
8.0.0   Lifting 12.24
9.0.0   Basic Tool Safety 12.26
10.0.0   Confined Space Entry Procedures 12.29
11.0.0   First Aid 12.30
12.0.0   Solvents and Toxic Vapors 12.30
13.0.0   Asbestos 12.32
14.0.0   Batteries 12.33
15.0.0   PCBs and Vapor Lamps 12.33
16.0.0   Fall Protection 12.34

Overview

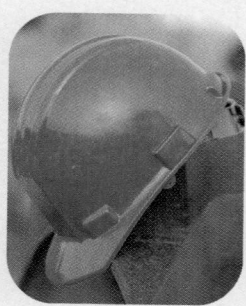

Electricians work in all areas of a job site. They are exposed to safety hazards that other workers encounter, including hazards that can cause fall-related injuries, crushing injuries in excavations, electrical shock, injuries from being struck by falling objects, cuts, burns, punctures, chemical exposure, and other injuries. Electricians are exposed to the risk of electrical shock more often than other workers, which puts them at a higher risk for electrical burns and arc burns.

Safety regulations and company policies are designed to protect those working in the electrical field, but these regulations are only effective if the worker recognizes and understands the hazards that may be present and takes the proper precautions to avoid them. For that reason, the proper use of personal protective equipment and other safety gear is a critical element of the electrician's job.

In order to protect yourself and those around you from injury and possible death, you must become familiar with the various hazards on the job site, follow established safety procedures, and always keep safe work practices foremost in your mind.

Objectives

When you have completed this module, you will be able to do the following:

1. Recognize safe working practices in the construction environment.
2. Explain the purpose of OSHA and how it promotes safety on the job.
3. Identify electrical hazards and how to avoid or minimize them in the workplace.
4. Explain safety issues concerning lockout/tagout procedures, confined space entry, respiratory protection, and fall protection systems.
5. Develop a task plan and a hazard assessment for a given task and select the appropriate PPE and work methods to safely perform the task.

Trade Terms

Double-insulated/ungrounded tool
Fibrillation
Grounded tool
Ground fault circuit interrupter (GFCI)
Polychlorinated biphenyls (PCBs)

Required Trainee Materials

1. Paper and pencil
2. Copy of the latest edition of the *National Electrical Code®*
3. Appropriate personal protective equipment

Prerequisites

Before you begin this module, it is recommended that you successfully complete *Core Curriculum*, and *Construction Technology*, Modules 68101-09 through 68111-09.

This course map shows all of the modules in *Construction Technology*. The suggested training order begins at the bottom and proceeds up. Skill levels increase as you advance on the course map. The local Training Program Sponsor may adjust the training order.

CONSTRUCTION TECHNOLOGY

68117-09
Copper Pipe and Fittings

68116-09
Plastic Pipe and Fittings

68115-09
Introduction to Drain, Waste, and Vent (DWV) Systems

68114-09
Introduction to HVAC

68113-09
Residential Electrical Services

68112-09
Electrical Safety

68111-09
Basic Stair Layout

68110-09
Exterior Finishing

68109-09
Roofing Applications

68108-09
Roof Framing

68107-09
Wall and Ceiling Framing

68106-09
Floor Systems

68105-09
Masonry Units and Installation Techniques

68104-09
Introduction to Masonry

68103-09
Handling and Placing Concrete

68102-09
Introduction to Concrete, Reinforcing Materials, and Forms

68101-09
Site Layout One: Distance Measurement and Leveling

CORE CURRICULUM:
Introductory Craft Skills

103CMAP.EPS

1.0.0 ◆ INTRODUCTION

In order to be safe, you need to be aware of potential hazards and stay constantly alert to these hazards. You must take the proper precautions and practice the basic rules of safety. You must be safety-conscious at all times and report any unsafe conditions to your supervisor and co-workers. Safety should become a habit. Keeping a safe attitude on the job will go a long way in reducing the number and severity of accidents. Remember that your safety is up to you.

As an apprentice electrician, you need to be especially careful. You should only work under the direction of experienced personnel who are familiar with the various job site hazards and the means of avoiding them.

The most life-threatening hazards on a construction site are:

- Falls when you are working in high places
- The possibility of being crushed by falling materials or equipment
- Electric shock and arc-related burns caused by coming into contact with live electrical circuits
- The possibility of being struck by flying objects or moving equipment/vehicles such as trucks, forklifts, and construction equipment

Other hazards include cuts, burns, back sprains, and getting chemicals or objects in your eyes. Most injuries, both those that are life-threatening and those that are less severe, are preventable if the proper precautions are taken.

2.0.0 ◆ ELECTRICAL SHOCK

Electricity can be described as a potential that results in the movement of electrons in a conductor. This movement of electrons is called electrical current. Some substances, such as silver, copper, steel, and aluminum, are excellent conductors. The human body is also a conductor. The conductivity of the human body greatly increases when the skin is wet or moistened with perspiration.

Electrical current flows along any path in which the voltage can overcome the resistance. If the human body contacts an electrically energized point and is also in contact with the ground or another point in the circuit, the human body becomes a path for the current. *Table 1* shows the effects of current passing through the human body. One mA is one milliamp, or one one-thousandth of an ampere.

What's wrong with this picture?

102SA01.EPS

Table 1	Current Level Effects on the Human Body
Current Value	**Typical Effects**
1mA	Perception level. Slight tingling sensation.
5mA	Slight shock. Involuntary reactions can result in serious injuries such as falls from elevations.
6 to 30mA	Painful shock, loss of muscular control.
50 to 150mA	Extreme pain, respiratory arrest, severe muscular contractions. Death possible.
1000mA to 4300mA	Ventricular fibrillation, severe muscular contractions, nerve damage. Typically results in death.

Source: Occupational Safety and Health Administration

102T01.EPS

A primary cause of death from electrical shock is when the heart's rhythm is overcome by an electrical current. Normally, the heart's operation uses a very low-level electrical signal to cause the heart to contract and pump blood. When an abnormal electrical signal, such as current from an electrical shock, reaches the heart, the low-level heartbeat signals are overcome. The heart begins twitching in an irregular manner and goes out of rhythm with the pulse. This twitching is called **fibrillation**. The use of cardiopulmonary resuscitation (CPR) can keep oxygen flowing to the body, but unless the normal heartbeat is restored using special defibrillation equipment (paddles), the individual will die. Other effects of electrical shock may include immediate heart stoppage and burns. In addition, the body's reaction to the shock can cause a fall or other accident. Delayed internal problems can also result. This is why it is critical for you to have a medical exam if you receive even a minor shock.

2.1.0 The Effect of Current

The amount of current measured in amperes that passes through a body determines the outcome of an electrical shock. The higher the voltage, the greater the chance for a fatal shock. In a one-year study in California, the following results were observed by the State Division of Industry Safety:

- Thirty percent of all electrical accidents were caused by contact with conductors. Of these accidents, 66% involved low-voltage conductors (those carrying 600 volts [V] or less).

NOTE

Electric shocks or burns are a major cause of accidents in the construction industry. According to the National Institute for Occupational Safety and Health, workers in the construction industry are four times more likely to be electrocuted at work than all other industries combined.

- Portable, electrically operated hand tools made up the second largest number of injuries (15%). Almost 70% of these injuries happened when the frame or case of the tool became energized. These injuries could have been prevented by following proper safety practices, using properly maintained grounded or **double-insulated/ungrounded tools**, and using **ground fault circuit interrupter (GFCI)** protection.

In one ten-year study, investigators found 9,765 electrical injuries in the U.S. A little more than 13% of the high-voltage injuries (over 600V) resulted in death. These high-voltage totals included limited-amperage contacts, which are often found on electronic equipment. When tools or equipment touch high-voltage overhead lines, the chance that a resulting injury will be fatal climbs to 28%. Of the low-voltage injuries, 1.4% were fatal.

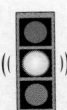

CAUTION

High voltage, defined as 600V or more, is almost ten times as likely to kill as low voltage. However, on the job you spend most of your time working on or near lower voltages. Due to the frequency of contact, most electrocution deaths actually occur at low voltages. Attitude about the harmlessness of lower voltages undoubtedly contributes to this statistic.

These statistics have been included to help you gain respect for the environment where you work and to stress how important safe working habits really are.

2.1.1 Body Resistance

Electricity travels in closed circuits, and its normal route is through a conductor. Shock occurs when the body becomes part of the electric circuit (*Figure 1*). The current must enter the body at one point and leave at another. Shock normally occurs in one of three ways: the person must come in contact with both wires of the electric circuit; one wire of the electric circuit and the ground; or a metallic part that has become live by being in contact with an energized wire while the person is also in contact with the ground.

To fully understand the harm done by electrical shock, we need to understand something about the physiology of certain body parts: the skin, the heart, and muscles.

Skin covers the body and is made up of three layers. The most important layer, as far as electric shock is concerned, is the outer layer of dead cells referred to as the horny layer. This layer is composed mostly of a protein called keratin, and it is the keratin that provides the largest percentage of the body's electrical resistance. When it is dry, the outer layer of skin may have a resistance of several thousand ohms, but when it is moist, there is a radical drop in resistance, as is also the case if there is a cut or abrasion that pierces the horny layer. The amount of resistance provided by the skin will vary widely from individual to individual. A worker with a thick horny layer will have a much higher resistance than a child. The resistance will also vary widely at different parts of the body. For instance, the worker with high-resistance hands may have low-resistance skin on the back of his calf.

The heart is the pump that sends life-sustaining blood to all parts of the body. The blood flow is caused by the contractions of the heart muscle, which is controlled by electrical impulses. The electrical impulses are delivered by an intricate system of nerve tissue with built-in timing mechanisms, which make the chambers of the heart contract at exactly the right time. An outside electric current of as little as 75 milliamperes can upset the rhythmic, coordinated beating of the heart by disturbing the nerve impulses. When this happens, the heart is said to be in fibrillation, and the pumping action stops. Death will occur quickly if the normal beat is not restored. Remarkable as it may seem, what is needed to defibrillate the heart is a shock of an even higher intensity.

The other muscles of the body are also controlled by electrical impulses delivered by nerves. Electric shock can cause loss of muscular control, resulting in the inability to let go of an electrical conductor. Electric shock can also cause injuries of

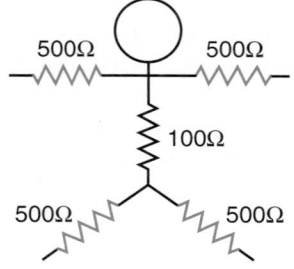

- HAND TO HAND 1000Ω
- 120 VOLT
- FORMULA: $I = E/R$
- 120/1000 = 0.120 AMPS OR 120 MILLIAMPS

102F01.EPS

Figure 1 ◆ Body resistance.

an indirect nature in which involuntary muscle reaction from the electric shock can cause bruises, fractures, and even death resulting from collisions or falls.

The severity of shock received when a person becomes a part of an electric circuit is affected by three primary factors: the amount of current flowing through the body (measured in amperes), the path of the current through the body, and the length of time the body is in the circuit. Other factors that may affect the severity of the shock are the frequency of the current, the phase of the heart cycle when shock occurs, and the general health of the person prior to the shock. Effects can range from a barely perceptible tingle to immediate cardiac arrest. Although there are no absolute limits, or even known values that show the exact injury at any given amperage range, *Table 1* lists the general effects of electric current on the body for different current levels. As this table illustrates, a difference of only 100 milliamperes exists between a current that is barely perceptible and one that is likely to kill you.

A severe shock can cause considerably more damage to the body than is visible. For example, a person may suffer internal hemorrhages and destruction of tissues, nerves, and muscle. In addition, shock is often only the beginning in a chain of events. The final injury may well be from a fall, cuts, burns, or broken bones.

2.1.2 Burns

The most common shock-related injury is a burn. Burns suffered in electrical accidents may be of three types: electrical burns, arc burns, and thermal contact burns.

Electrical burns are the result of electric current flowing through the tissues or bones. Tissue damage is caused by the heat generated by the current flow through the body. An electrical burn is one of the most serious injuries you can receive, and should be given immediate attention. Since the most severe burning is likely to be internal, what may appear at first to be a small surface wound could, in fact, be an indication of severe internal burns.

Arc burns make up a substantial portion of the injuries from electrical malfunctions. The electric arc between metals can be up to 35,000°F, which is about four times hotter than the surface of the sun. Workers several feet from the source of the arc can receive severe or fatal burns. Since most electrical safety guidelines recommend safe working distances based on shock considerations, workers can be following these guidelines and still be at risk from arc. Electric arcs can occur due to poor

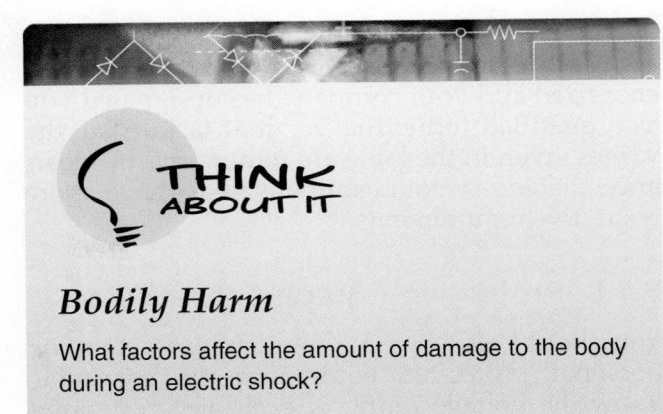

THINK ABOUT IT

Bodily Harm

What factors affect the amount of damage to the body during an electric shock?

electrical contact or failed insulation. Electrical arcing is caused by the passage of substantial amounts of current through the vaporized terminal material (usually metal or carbon).

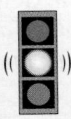

CAUTION

Since the heat of the arc is dependent on the short circuit current available at the arcing point, arcs generated by low-voltage systems can be just as dangerous as those generated at 13,000V.

The third type of burn is a thermal contact burn. It is caused by contact with objects thrown during the blast associated with an electric arc. This blast comes from the pressure developed by the near-instantaneous heating of the air surrounding the arc, and from the expansion of the metal as it is vaporized. (Copper expands by a factor in excess of 65,000 times in boiling.) The pressure wave can be great enough to hurl people, switchgear, and cabinets considerable distances. It can stop your heart, impale you with shrapnel, blow off limbs, cause deafness, and cause you to inhale vaporized metal. Another hazard associated with the blast is the hurling of molten metal droplets, which can also cause thermal contact burns and associated damage.

3.0.0 ◆ REDUCING YOUR RISK

There are many things that can be done to greatly reduce the chance of receiving an electrical shock. Always comply with your company's safety policy and all applicable rules and regulations, including job site rules. In addition, the Occupational Safety and Health Administration (OSHA) publishes the *Code of Federal Regulations (CFR)*. *CFR Part 1910* covers the OSHA standards for general industry and *CFR Part 1926* covers the OSHA standards for the construction industry.

Do not approach any electrical conductors closer than indicated in *Table 2* unless they are de-energized and your company has designated you as a qualified individual for that task. Also, the values given in the table are minimum safe clearance distances; your company may have more restrictive requirements.

3.1.0 Protective Equipment

You should also become familiar with common personal protective equipment. In particular, know the voltage rating of each piece of equipment. Rubber gloves are used to prevent the skin from coming into contact with energized circuits. A separate leather cover protects the rubber glove from punctures and other damage (see *Figure 2*). The leather protectors also provide the wearer with a certain level of protection against arc flash. OSHA addresses the use of protective equipment, apparel, and tools in *CFR 1910.335(a)*. This article is divided into two sections: *Personal Protective Equipment* and *General Protective Equipment and Tools*.

CASE HISTORY

Arc Blast

An electrician in Louisville, KY was cleaning a high-voltage switch cabinet. He removed the padlock securing the switch enclosure and opened the door. He used a voltage meter to verify absence of voltage on the three load phases at the rear of the cabinet. However, he did not test all potentially energized parts within the cabinet, and some components were still energized. He was wearing standard work boots and safety glasses, but was not wearing protective rubber gloves or an arc suit. As he used a paintbrush to clean the switch, an arc blast occurred, which lasted approximately one-sixth of a second. The electrician was knocked down by the blast. He suffered third-degree burns and required skin grafts on his arms and hands. The investigation determined that the blast was caused by debris, such as a cobweb, falling across the open switch.

The Bottom Line: Always wear the appropriate personal protective equipment and test all components for voltage before working in or around electrical devices.

Table 2 Limited Approach Boundaries to Live Parts

Nominal System Voltage Range (Phase-to-Phase)	Limited Approach Boundary
50 to 300	3 ft 6 in
301 to 750	3 ft 6 in
751 to 15kV	5 ft 0 in
15.1kV to 36kV	6 ft 0 in
36.1kV to 46kV	8 ft 0 in
46.1kV to 72.5kV	8 ft 0 in
72.6kV to 121kV	8 ft 0 in
138kV to 145kV	10 ft 0 in
161kV to 169kV	11 ft 8 in
230kV to 242kV	13 ft 0 in
345kV to 362kV	15 ft 4 in
500kV to 550kV	19 ft 0 in
765kV to 800kV	23 ft 9 in

102T02.EPS

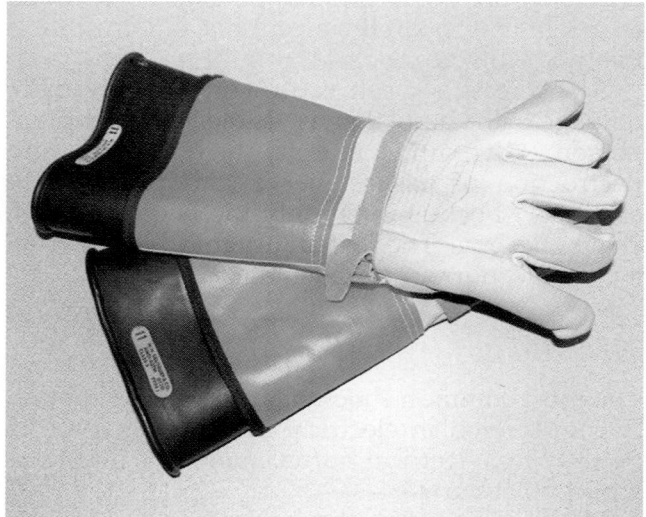

102F02.EPS

Figure 2 ◆ Rubber gloves and leather protectors.

The first section, *Personal Protective Equipment*, includes the following requirements:

- Employees working in areas where there are potential electrical hazards shall be provided with, and shall use, electrical protective equipment that is appropriate for the specific parts of the body to be protected and for the work to be performed.
- Protective equipment shall be maintained in a safe, reliable condition and shall be periodically inspected or tested, as required by *CFR 1910.137/1926.95*.
- If the insulating capability of protective equipment may be subject to damage during use, the insulating material shall be protected.

- Employees shall wear nonconductive head protection wherever there is a danger of head injury from electric shock or burns due to contact with exposed energized parts.
- Employees shall wear protective equipment for the eyes and face wherever there is danger of injury to the eyes or face from electric arcs or flashes or from flying objects resulting from an electrical explosion.

The second section, *General Protective Equipment and Tools*, includes the following requirements:

- When working near exposed energized conductors or circuit parts, each employee shall use insulated tools or handling equipment if the tools or handling equipment might make contact with such conductors or parts. If the insulating capability of insulated tools or handling equipment is subject to damage, the insulating material shall be protected.
- Fuse pullers, insulated for the circuit voltage, shall be used to remove or install fuses.

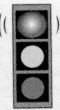

WARNING!

Fuses should never be installed or removed when energized. This could cause an electrical arc, which can result in death or a serious injury.

- Ropes and handlines used near exposed energized parts shall be nonconductive.
- Protective shields, protective barriers, or insulating materials shall be used to protect each employee from shock, burns, or other electrically related injuries while that employee is working near exposed energized parts that might be accidentally contacted or where dangerous electric heating or arcing might occur. When normally enclosed live parts are exposed for maintenance or repair, they shall be guarded to protect unqualified persons from contact with the live parts.

The types of electrical safety equipment, protective apparel, and protective tools available for use are quite varied. This module will discuss the most common types of safety equipment. These include the following:

- Currently tested rubber protective equipment, including gloves and blankets
- Protective apparel
- Natural fiber clothing
- Hot sticks
- Fuse pullers
- Shorting probes
- Safety glasses
- Face shields

3.1.1 Rubber Protective Equipment

All electrical workers may be exposed to energized circuits or equipment. Two of the most important articles of protection for electrical workers are insulated rubber gloves and rubber blankets, which must be matched to the voltage rating for the circuit or equipment. Rubber protective equipment is designed for the protection of the user. If it fails during use, a serious injury could occur.

Rubber protective equipment is available in two types. Type 1 designates rubber protective equipment that is manufactured of natural or synthetic rubber that is properly vulcanized, and Type 2 designates equipment that is ozone resistant, made from any elastomer or combination of elastomeric compounds. Ozone is a form of oxygen that is produced from electricity and is present in the air surrounding a conductor under high voltages. Normally, ozone is found at voltages of 10kV and higher, such as those found in electric utility transmission and distribution systems. Type 1 protective equipment can be damaged by corona cutting, which is the cutting action of ozone on natural rubber when it is under mechanical stress. Type 1 rubber protective equipment can also be damaged by ultraviolet rays. However, it is very important that the rubber protective equipment in use today be made of natural rubber or Type 1 equipment. Type 2 rubber protective equipment is very stiff and is not as easily worn as Type 1 equipment.

Various classes – The American National Standards Institute (ANSI) and the American Society for Testing and Materials International (ASTM) have designated a specific classification system for rubber protective equipment. The maximum AC use voltage and equipment tag colors are as follows:

- Class 00 (beige tag)   500V
- Class 0 (red tag)   1,000V
- Class 1 (white tag)   7,500V
- Class 2 (yellow tag)   17,000V
- Class 3 (green tag)   26,500V
- Class 4 (orange tag)   36,000V

Referring back to *Figure 2*, note that the gloves shown in the picture have a yellow tag and are therefore Class 2 equipment.

Inspection of protective equipment – Before rubber protective equipment can be worn by personnel in the field, all equipment must have a current test date stenciled on the equipment. Insulating gloves must be inspected each day by the user before they can be used. They must also be electrically tested every six months and any time the insulating value is in question. Because rubber protective equipment is used for personal protection and serious injury could result from its misuse or failure, it is important that an adequate safety factor be provided between the voltage on which it is to be used and the voltage at which it was tested.

All rubber protective equipment must be marked with the appropriate voltage rating and last inspection date. The markings that are required to be on rubber protective equipment must be applied in a manner that will not interfere with the protection that is provided by the equipment.

 WARNING!
Never work on anything energized without direct instruction from your employer.

Gloves – Both high- and low-voltage rubber gloves are of the gauntlet type and are available in various sizes. To get the best possible protection and service life, here are a few general rules that apply whenever they are used in electrical work:

- Always wear leather protectors over your gloves as they provide burn protection not provided by the gloves themselves. Any direct contact with sharp or pointed objects may cut, snag, or puncture the gloves and take away the protection you are depending on.
- Always wear rubber gloves right side out (serial number and size to the outside).
- Always keep the gauntlets up. Rolling them down sacrifices a valuable area of protection. Tuck the sleeves of your shirt or protective suit under the glove cuffs to prevent an arc blast from coming inside your clothing.
- Always inspect and field check gloves before using them. Always check the inside for any debris.
- Use light amounts of manufacturer-approved glove dust or cotton liners with the rubber gloves. This gives the user more comfort, and it also helps to absorb some of the perspiration that can damage the gloves over years of use.
- Wash the rubber gloves in lukewarm, clean, fresh water after each use. Dry the gloves inside and out prior to returning to storage. Never use any type of cleaning solution on the gloves.
- Once the gloves have been properly cleaned, inspected, and tested, they must be properly stored. Store them in a cool, dry, dark place that is free from ozone, chemicals, oils, solvents, or other materials that could damage the gloves.

Do not store gloves near hot pipes or in direct sunlight. Store both gloves and sleeves in their natural shape in a bag or box inside their leather protectors. They should be undistorted, right side out, and unfolded.

- Gloves can be damaged by many different chemicals, especially petroleum-based products such as oils, gasoline, hydraulic fluid inhibitors, hand creams, pastes, and salves. If contact is made with these or other petroleum-based products, the contaminant should be wiped off immediately. If any signs of physical damage or chemical deterioration are found (e.g., swelling, softness, hardening, stickiness, ozone deterioration, or sun checking), the protective equipment must not be used.
- Never wear watches or rings while wearing rubber gloves; this can cause damage from the inside out and defeats the purpose of using rubber gloves. Never wear anything conductive.
- Rubber gloves must be electrically tested every six months by a certified testing laboratory. Always check the inspection date before using gloves.
- Use rubber gloves only for their intended purpose, not for handling chemicals or other work. This also applies to the leather protectors.

Before rubber gloves are used, a visual inspection and an air test should be made. This should be done prior to use and as many times during the day as you feel necessary. To perform a visual inspection, stretch a small area of the glove, checking to see that no defects exist, such as:

- Embedded foreign material
- Deep scratches
- Pinholes or punctures
- Snags or cuts

Gloves and sleeves can be inspected by rolling the outside and inside of the protective equipment between the hands. This can be done by squeezing together the inside of the gloves or sleeves to bend the outside area and create enough stress to the inside surface to expose any cracks, cuts, or other defects. When the entire surface has been checked in this manner, the equipment is then turned inside out, and the procedure is repeated. It is very important not to leave the rubber protective equipment inside out as that places stress on the preformed rubber.

Remember, any damage at all reduces the insulating ability of the rubber glove. Look for signs of deterioration from age, such as hardening and slight cracking. Also, if the glove has been exposed to petroleum products, it should be considered suspect because deterioration can be caused by such exposure. If the gloves are suspect, turn them in for evaluation. If the gloves are defective, turn them in for disposal. Never leave a damaged glove lying around; someone may think it is a good glove and not perform an inspection prior to using it.

After visually inspecting the glove, other defects may be observed by applying the air test (*Figure 3*).

Step 1 Stretch the glove and look for any defects.

Step 2 Twirl the glove around quickly or roll it down from the glove gauntlet to trap air inside.

Step 3 Trap the air by squeezing the gauntlet with one hand. Use the other hand to squeeze the palm, fingers, and thumb to check for weaknesses and defects.

Step 4 Hold the glove up to your ear to try to detect any escaping air.

Step 5 If the glove does not pass this inspection, it must be turned in for disposal.

CAUTION

Never blow the gloves up like a balloon or use compressed gas for the air test as this can damage the glove.

Insulating blankets – An insulating blanket is a versatile cover-up device best suited for the protection of maintenance technicians against accidental contact with energized electrical equipment.

These blankets are designed and manufactured to provide insulating quality and flexibility for use in covering. Insulating blankets are designed only for covering equipment and should not be used on the floor. Special rubber floor mats, called switchboard matting, are available for floor use. Use caution when installing these over sharp edges or when covering pointed objects.

Blankets must be tested yearly and inspected before each use. To check rubber blankets, place the blanket on a flat surface and roll the blanket from one corner to the opposite corner. If there are any irregularities in the rubber, this method will expose them. After the blanket has been rolled from each corner, it should then be turned over and the procedure repeated.

Insulating blankets are cleaned in the same manner as rubber gloves. Once the protective equipment has been properly cleaned, inspected, and tested, it must be properly stored. It should be stored in a cool, dry, dark place that is free from ozone, chemicals, oils, solvents, or other materials that could damage the equipment. Such storage should not be in the vicinity of hot pipes or direct sunlight. Blankets may be stored rolled in containers that are designed for this use; the inside diameter of the roll should be at least two inches.

3.1.2 Protective Apparel

Besides rubber gloves, there are other types of special application protective apparel, such as fire suits, face shields, and rubber sleeves.

Manufacturing plants should have other types of special application protective equipment available for use, such as high-voltage sleeves, high-voltage boots, nonconductive protective helmets,

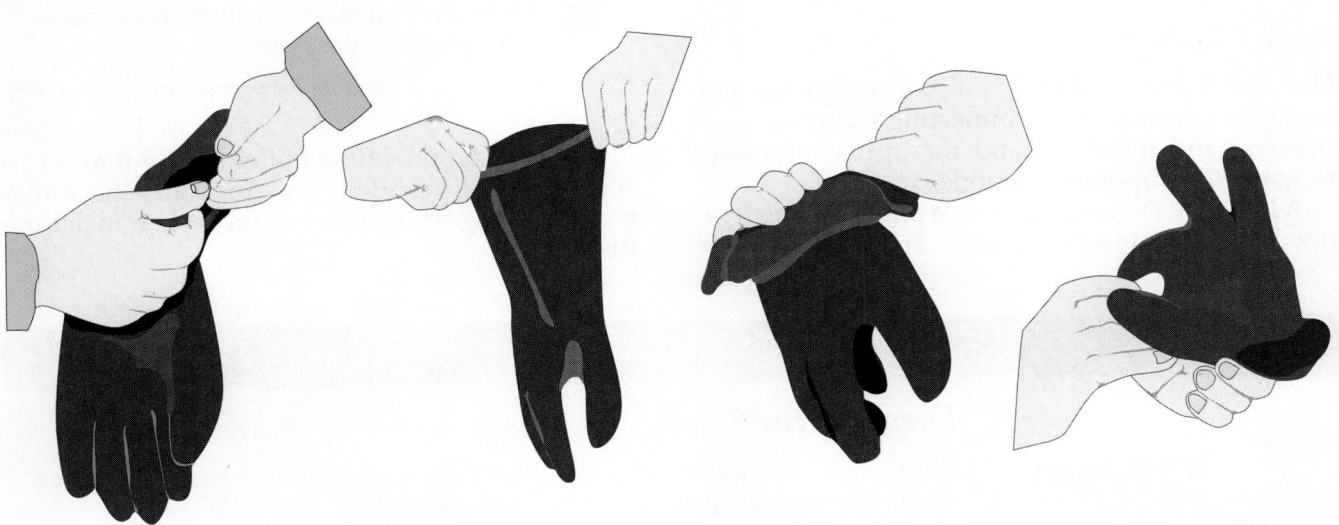

102F03.EPS

Figure 3 ◆ Glove inspection.

Field-Marked Flash Protection Signs

In other than dwelling units, *NEC Section 110.16* requires that all switchboards, panelboards, industrial control panels, meter socket enclosures, and motor control centers be clearly marked to warn qualified persons of potential electric arc flash hazards.

nonconductive eyewear and face protection, and switchboard blankets.

All equipment must be inspected before use and during use, as necessary. The equipment used and the extent of the precautions taken depend on each individual situation; however, it is better to be overprotected than underprotected when you are trying to prevent electric shock, arc blast, and burns.

When working with energized equipment, flash suits may be required in some applications.

Face shields must also be worn during all switching operations where arcs are a possibility. Always use a rated face shield.

3.1.3 Personal Clothing

Any individual who will perform work in an electrical environment or in plant substations should dress accordingly. Never wear synthetic-fiber clothing on a job site; these types of materials will melt when exposed to high temperatures and because they burn hotter, will actually increase the severity of a burn. Wear cotton clothing, fiberglass-toe boots or shoes, safety glasses, and hard hats. Use hearing protection where needed.

3.1.4 Hot Sticks

Hot sticks are insulated tools designed for the manual operation of disconnecting switches, fuse removal and insertion, and the application and removal of temporary grounds.

A hot stick is made up of two parts, the head or hood and the insulating rod. The head can be made of metal or hardened plastic, while the insulating section may be wood, plastic, laminated wood, or other effective insulating materials. There are also telescoping sticks available.

Most plants have hot sticks available for different purposes. Select a stick of the correct type and size for the application and inspect it before use. Look for signs of obvious damage, deep scratches, dust, or surface contaminants. Never use a damaged hot stick.

Storage of hot sticks is important. They should be hung up vertically on a wall to prevent any damage. They should also be stored away from direct sunlight and prevented from being exposed to petroleum products.

3.1.5 Fuse Pullers

Use the plastic or fiberglass style of fuse puller for removing and installing low-voltage cartridge fuses. All fuse pulling and replacement operations must be done using fuse pullers.

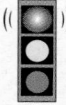 **WARNING!**
Fuses should only be pulled when they are safely de-energized. Failure to do so can result in a serious injury or death.

The best type of fuse puller is one that has a spread guard installed. This prevents the puller from opening if resistance is met when installing fuses.

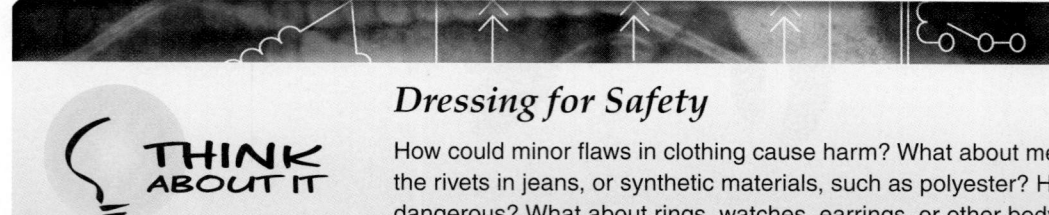

Dressing for Safety

How could minor flaws in clothing cause harm? What about metal components, such as the rivets in jeans, or synthetic materials, such as polyester? How are these dangerous? What about rings, watches, earrings, or other body piercings? How does protective apparel prevent accidents?

3.1.6 Shorting Probes

Before working on de-energized circuits that have capacitors installed, you must discharge the capacitors using a safety shorting probe. This is a procedure that requires special training and may only be performed by qualified individuals.

3.1.7 Eye and Face Protection

NFPA 70E and OSHA require that you wear safety glasses at all times whenever you are working on or near energized circuits. Face protection is worn over your safety glasses, and is required whenever there is danger of electrical arcs or flashes, or from flying or falling objects resulting from an electrical explosion. *NFPA 70E* is discussed in detail later in this module.

3.2.0 Verify That Circuits Are De-energized

You should always assume that all the circuits are energized until you have verified that the circuit is de-energized. This is called a live-dead-live test. Follow these steps to verify that a circuit is de-energized:

Step 1 Ensure that the circuit is properly tagged and locked out *(CFR 1910.333/1926.417)*.

Step 2 Verify the test instrument operation on a known source using the appropriately rated tester.

Step 3 Using the test instrument, check the circuit to be de-energized. The voltage should be zero.

Step 4 Verify the test instrument operation, once again on a known power source.

3.3.0 Other Precautions

There are several other precautions you can take to help make your job safer. For example:

- Always remove all jewelry (e.g., rings, watches, body piercings, bracelets, and necklaces) before working on electrical equipment. Most jewelry is made of conductive material and wearing it can result in a shock, as well as other injuries if the jewelry gets caught in moving components.
- When working on energized equipment, it is safer to work in pairs. In doing so, if one of the workers experiences a harmful electrical shock, the other worker can call for help.
- Plan each job before you begin it. Make sure you understand exactly what it is you are going to do. If you are not sure, ask your supervisor.

 **WARNING!**
The first time you perform a specific task, it cannot be done energized.

Using a Phasing Tester

Using a phasing tester or other voltage detection equipment can expose you to a considerable arc flash hazard. Always take the time to use the appropriate personal protective equipment. Remember, you only have one chance to protect yourself.

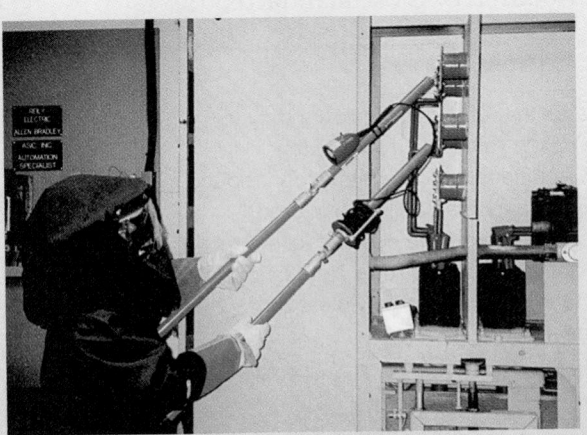

102SA02.EPS

- You will need to look over the appropriate prints and drawings to locate isolation devices and potential hazards. Never defeat safety interlocks. Remember to plan your escape route before starting work. Know where the nearest phone is and the emergency number to dial for assistance.
- If you realize that the work will go beyond the scope of what was planned, stop and get instructions from your supervisor before continuing. Do not attempt to plan as you go.
- It is critical that you stay alert. Workplaces are dynamic, and situations relative to safety are always changing. If you leave the work area to pick up material, take a break, or have lunch, reevaluate your surroundings when you return. Remember, plan ahead.

4.0.0 ◆ OSHA

The purpose of OSHA is "to ensure safe and healthful working conditions for working men and women." OSHA is authorized to enforce standards and assist and encourage the states in their efforts to ensure safe and healthful working conditions. OSHA assists states by providing for research, information, education, and training in the field of occupational safety and health.

The law that established OSHA specifies the duties of both the employer and employee with respect to safety. Some of the key requirements are outlined below. This list does not include everything, nor does it override the procedures called for by your employer.

- Employers shall provide a place of employment free from recognized hazards likely to cause death or serious injury.
- Employers shall comply with the standards of the act.
- Employers shall be subject to fines and other penalties for violation of those standards.

 WARNING!

OSHA states that employees have a duty to follow the safety rules laid down by the employer. Additionally, some states can reduce the amount of benefits paid to an injured employee if that employee was not following known, established safety rules. Your company may also terminate you if you violate an established safety rule.

 *INSIDE TRACK*

Safety on the Job Site

Uncovered openings present several hazards:

- Workers may trip over them.
- If they are large enough, workers may fall through them.
- If there is a work area below, tools or other objects may fall through them, causing serious injury to workers below.

Any hole deeper than 6' and more than 2" in any direction must be protected with a cover or with guardrails, as shown here. A cover must be able to handle twice the anticipated load, be secured in place, and have the word "HOLE" or "COVER" on it.

102SA03.EPS

4.1.0 Safety Standards

The OSHA standards are split into several sections. As discussed earlier, the two that affect you the most are *CFR 1926*, construction specific, and *CFR 1910*, which is the standard for general industry. Either or both may apply depending on where you are working and what you are doing. Your company may also have its own policies and procedures. In addition, you will be required to follow the safety procedures of any plant or facility where you are working.

4.2.0 Safety Philosophy and General Safety Precautions

The most important piece of safety equipment required when performing work in an electrical environment is common sense. All areas of electrical safety precautions and practices draw upon common sense and attention to detail. One of the most dangerous conditions in an electrical work area is a poor attitude toward safety.

As stated in *CFR 1910.333(a)/1926.403,* safety-related work practices shall be employed to prevent electric shock or other injuries resulting from either direct or indirect electrical contact when work is performed near or on equipment or circuits that are or may be energized. The specific safety-related work practices shall be consistent with the nature and extent of the associated electrical hazards. The following are considered some of the basic and necessary attitudes and electrical safety precautions that lay the groundwork for a proper safety program. Before going on any electrical work assignment, these safety precautions should be reviewed and adhered to.

- *All work on electrical equipment should be done with circuits de-energized and cleared or grounded –* It is obvious that working on energized equipment is much more dangerous than working on equipment that is de-energized. Work on energized electrical equipment should be avoided if at all possible. *CFR 1910.333(a)(1)/1926.403* states that live parts to which an employee may be exposed shall be de-energized before the employee works on or near them, unless the employer can demonstrate that de-energizing introduces additional or increased hazards or is not possible because of equipment design or operational limitations. Live parts that operate at less than 50 volts to ground need not be de-energized if there will be no increased exposure to electrical burns or to explosion due to electric arcs.

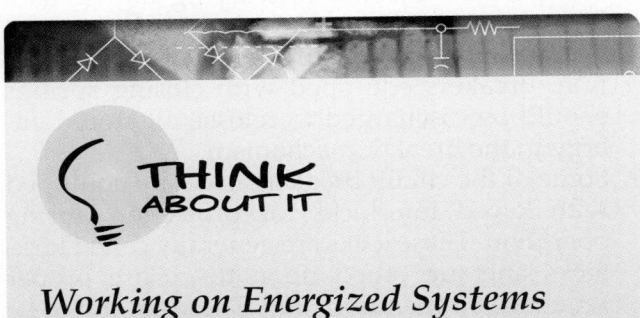

- *All conductors, buses, and connections should be considered energized until proven otherwise* – As stated in *1910.333(b)(1)/1926.417,* conductors and parts of electrical equipment that have not been locked out or tagged out in accordance with this section should be considered energized. Routine operation of the circuit breakers and disconnect switches contained in a power distribution system can be hazardous if not approached in the right manner. Several basic precautions that can be observed in switchgear operations are:
  - Wear proper clothing made of natural fiber or fire-resistant fabric.
  - Wear eye, face, and head protection.
  - Whenever operating circuit breakers in low-voltage or medium-voltage systems, always stand off to the side of the unit.
  - Always try to operate disconnect switches and circuit breakers under a no-load condition.
  - Never intentionally force an interlock on a system or circuit breaker.

Often, a circuit breaker or disconnect switch is used for providing lockout on an electrical system. To ensure that a lockout is not violated, perform the following procedures when using the device as a lockout point:

- Breakers must always be locked out and tagged as discussed previously whenever you are working on a circuit that is tied to an energized breaker. Breakers capable of being opened and racked out to the disconnected position should have this done. Afterward, approved safety locks must be installed. The breaker may be removed from its cubicle completely to prevent unexpected mishaps. Always follow the standard rack-out and removal procedures that were supplied with the switchgear. Once removed, a sign must be hung on the breaker identifying its use as a lockout point, and approved safety locks must be installed when the breaker is used for isolation. Breakers equipped with closing springs should be discharged to release all stored energy in the breaker mechanism.
- Some of the circuit breakers used are equipped with keyed interlocks for protection during operation. These locks are generally called kirklocks and are relied upon to ensure proper sequence of operation only. These are not to be used for the purpose of locking out a circuit or system. When opening or closing a disconnect manually, it should be done quickly with a positive force. Lockouts should be used when the disconnects are open.

- Whenever performing switching or fuse replacements, always use the protective equipment necessary to ensure personnel safety. Never make the assumption that because things have gone fine the last 999 times, they will not go wrong this time. Always prepare yourself for the worst case accident when performing switching.
- Whenever re-energizing circuits following maintenance or removal of a faulted component, use extreme care. Always verify that the equipment is in a condition to be re-energized safely. All connections should be insulated and all covers should be installed. Have all personnel stand clear of the area for the initial re-energization. Never assume everything is in perfect condition. Verify the conditions.

The following procedure is provided as a guideline for ensuring that equipment and systems will not be damaged by reclosing low-voltage circuit breakers into faults. If a low-voltage circuit breaker has opened for no apparent reason, perform the following:

Step 1 Verify that the equipment being supplied is not physically damaged and shows no obvious signs of overheating or fire.

Step 2 Make all appropriate tests to locate any faults.

Step 3 Reclose the feeder breaker. Stand off to the side when closing the breaker.

Step 4 If the circuit breaker trips again, do not attempt to reclose the breaker. In a plant environment, Electrical Engineering should be notified, and the cause of the trip must be isolated and repaired.

The same general procedure should be followed for fuse replacement, with the exception of transformer fuses. If a transformer fuse blows, the transformer and feeder cabling should be inspected and tested before re-energizing. A blown fuse to a transformer is very significant because it normally indicates an internal fault. Transformer failures are catastrophic in nature and can be extremely dangerous. If applicable, contact the in-plant Electrical Engineering Department prior to commencing any effort to re-energize a transformer.

Power must always be removed from a circuit when removing and installing fuses. The air break disconnects (or quick disconnects) provided on the upstream side of a large transformer must be opened prior to removing the transformer's fuses. Otherwise, severe arcing will occur as the fuse is removed. This arcing can result in personnel injury and equipment damage.

To replace fuses servicing circuits below 600 volts:

- Turn off the power to the disconnect.
- Verify that the fuses are de-energized.
- Remove the blown fuse.
- Install the new fuse. Push it in firmly and verify that it is seated properly.
- Turn the power back on.

When replacing fuses servicing systems above 600 volts:

- Open and lock out the disconnect switches.
- Unlock the fuse compartment.
- Verify that the fuses are de-energized.
- Attach the fuse removal hot stick to the fuse and remove it.

4.3.0 Electrical Regulations

OSHA has certain regulations that apply to job site electrical safety. These regulations include:

- All electrical work shall be in compliance with the latest *NEC®* and OSHA standards.
- The noncurrent-carrying metal parts of fixed, portable, and plug-connected equipment shall be grounded. Choose either grounded tools or double-insulated tools. *Figure 4* shows an example of a double-insulated tool.

- Extension cords shall be the three-wire type, shall be protected from damage, and if hung overhead, shall not be fastened with staples or bare wire or hung in a manner that could cause damage to the outer jacket or insulation. Never run an extension cord through a doorway or window that can pinch the cord. Also, never allow vehicles or equipment to drive over cords.
- Exposed lamps in temporary lights shall be guarded to prevent accidental contact, except where lamps are deeply recessed in the reflector. Temporary lights shall not be suspended, except in accordance with their listed labeling.
- Receptacles for attachment plugs shall be of an approved type and properly installed. Installation of the receptacle will be in accordance with the listing and labeling for each receptacle.
- *NEC Section 590.6(A)* states that all 125V, single-phase, 15A, 20A, and 30A receptacle outlets shall be ground fault protected.
- Each disconnecting means for motors and appliances and each service feeder or branch circuit at the point where it originates shall be legibly marked to indicate its purpose and voltage.
- Flexible cords shall be used in continuous lengths (no splices) and shall be of a type listed in *NEC Table 400.4*.
- Receptacles other than 125V, single-phase, 15A, 20A, and 30A should be ground fault protected if possible. If that is not possible, a written assured equipment grounding conductor program must be continuously enforced at the site for all cord sets per *NEC Section 590.6(B)*. *Figure 5* shows a typical ground fault circuit interrupter (GFCI).

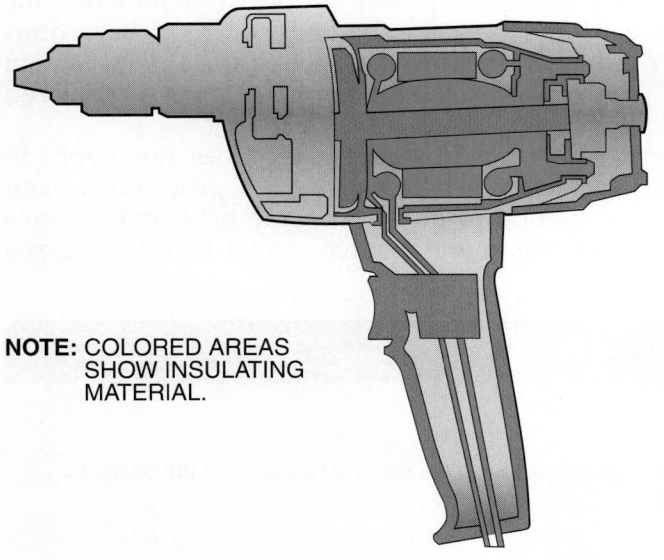

NOTE: COLORED AREAS SHOW INSULATING MATERIAL.

102F04.EPS

Figure 4 ◆ Double-insulated electric drill.

Potential Hazards

A self-employed builder was using a metal cutting tool on a metal carport roof and was not using GFCI protection. The male and female plugs of his extension cord partially separated, and the active pin touched the metal roofing. When the builder grounded himself on the gutter of an adjacent roof, he received a fatal shock.

The Bottom Line: Always use GFCI protection and be on the lookout for potential hazards.

Figure 5 ◆ Typical GFCI receptacle.

102F05.EPS

4.3.1 OSHA Lockout/Tagout Rule

OSHA released the *29 CFR 1926* lockout/tagout rule in December 1991. This rule covers the specific procedure to be followed for the "servicing and maintenance of machines and equipment in which the unexpected energization or startup of the machines or equipment, or releases of stored energy, could cause injury to employees." This standard establishes minimum performance requirements for the control of such hazardous energy.

The first step to be completed before working on a circuit is to ensure that equipment is isolated from all potentially hazardous energy (for example, electrical, mechanical, hydraulic, chemical, or thermal), and tagged and locked out before employees perform any servicing or maintenance activities in which the unexpected energization, startup, or release of stored energy could cause injury. All employees shall be instructed in the lockout/tagout procedure.

CAUTION

Although 99% of your work may be electrical, be aware that you may also need to lock out mechanical and other types of energy equipment.

The following is an example of a lockout/tagout procedure. Make sure to use the procedure that is specific to your employer or job site.

WARNING!

This procedure is provided for your information only. The OSHA procedure provides only the minimum requirements for lockouts/tagouts. Consult the lockout/tagout procedure for your company and the plant or job site at which you are working. Remember that your life could depend on the lockout/tagout procedure. It is critical that you use the correct procedure for your site. The *NEC®* requires that remote-mounted motor disconnects be permanently equipped with a lockout feature.

I. *Introduction*
   A. This lockout/tagout procedure has been established for the protection of personnel from potential exposure to hazardous energy sources during construction, installation, service, and maintenance of electrical energy systems.
   B. This procedure applies to and must be followed by all personnel who may be potentially exposed to the unexpected startup or release of hazardous energy (e.g., electrical, mechanical, pneumatic, hydraulic, chemical, or thermal).

Exception: This procedure does not apply to process and/or utility equipment or systems with cord and plug power supply systems when the cord and plug are the only source of hazardous energy, are removed from the source, and remain under the exclusive control of the authorized employee.

Exception: This procedure does not apply to troubleshooting (diagnostic) procedures and installation of electrical equipment and systems when the energy source cannot be de-energized

THINK ABOUT IT

GFCIs

Explain how GFCIs protect people. Where should a GFCI be installed in the circuit to be most effective?

because continuity of service is essential or shutdown of the system is impractical. Additional personal protective equipment for such work is required and the safe work practices identified for this work must be followed.

II. *Definitions*
- *Affected employee* – Any person working on or near equipment or machinery when maintenance or installation tasks are being performed by others during lockout/tagout conditions.
- *Appointed authorized employee* – Any person appointed by the job site supervisor to coordinate and maintain the security of a group lockout/tagout condition.
- *Authorized employee* – Any person authorized by the job site supervisor to use lockout/tagout procedures while working on electrical equipment.
- *Authorized supervisor* – The assigned job site supervisor who is in charge of coordination of procedures and maintenance of security of all lockout/tagout operations at the job site.
- *Energy isolation device* – An approved electrical disconnect switch capable of accepting approved lockout/tagout hardware for the purpose of isolating and securing a hazardous electrical source in an open or safe position.
- *Lockout/tagout hardware* – A combination of padlocks, danger tags, and other devices designed to attach to and secure electrical isolation devices.

III. *Training*
A. Each authorized supervisor, authorized employee, and appointed authorized employee shall receive initial and as-needed user-level training in lockout/tagout procedures.
B. Training is to include recognition of hazardous energy sources, the type and magnitude of energy sources in the workplace, and the procedures for energy isolation and control.
C. Retraining will be conducted on an as-needed basis whenever lockout/tagout procedures are changed or there is evidence that procedures are not being followed properly.

IV. *Protective Equipment and Hardware*
A. Lockout/tagout devices shall be used exclusively for controlling hazardous energy sources.

B. All padlocks must be numbered and assigned to one employee only.
C. No duplicate or master keys will be made available to anyone except the site supervisor.
D. A current list with the lock number and authorized employee's name must be maintained by the site supervisor.
E. Danger tags must be of the standard white, red, and black *DANGER—DO NOT OPERATE* design and shall include the authorized employee's name, the date, and the appropriate network company (use permanent markers).
F. Danger tags must be used in conjunction with padlocks, as shown in *Figure 6*.

V. *Procedures*
A. Preparation for lockout/tagout:
1. Check the procedures to ensure that no changes have been made since you last used a lockout/tagout.
2. Identify all authorized and affected employees involved with the pending lockout/tagout.
B. Sequence for lockout/tagout:
1. Notify all authorized and affected personnel that a lockout/tagout is to be used and explain the reason why.
2. Shut down the equipment or system using the normal OFF or STOP procedures.
3. Lock out energy sources and test disconnects to be sure they cannot be moved to the ON position and open the control cutout switch. If there is no cutout switch, block the magnet in the switch open position before working on electrically operated equipment/apparatus such as motors, relays, etc. Remove the control wire.
4. Lock and tag the required switches in the open position. Each authorized employee must affix a separate lock and tag. An example is shown in *Figure 7*.
5. Dissipate any stored energy by attaching the equipment or system to ground.
6. Verify that the test equipment is functional via a known power source.
7. Confirm that all switches are in the open position and use test equipment to verify that all parts are de-energized.
8. If it is necessary to temporarily leave the area, upon returning, retest to ensure that the equipment or system is still de-energized.

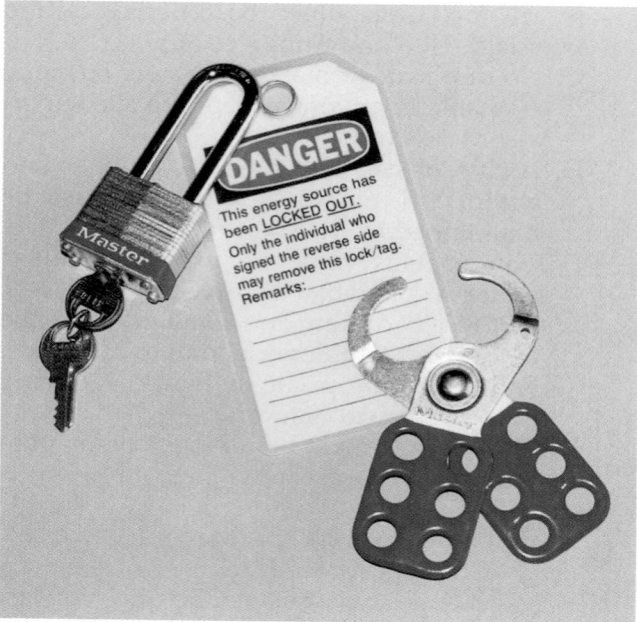

ELECTRICAL LOCKOUT

102F06A.EPS

PNEUMATIC LOCKOUT

102F06B.EPS

Figure 6 ◆ Lockout/tagout devices.

C. Restoration of energy:
  1. Confirm that all personnel and tools, including shorting probes, are accounted for and removed from the equipment or system.
  2. Completely reassemble and secure the equipment or system.
  3. Replace and/or reactivate all safety controls.

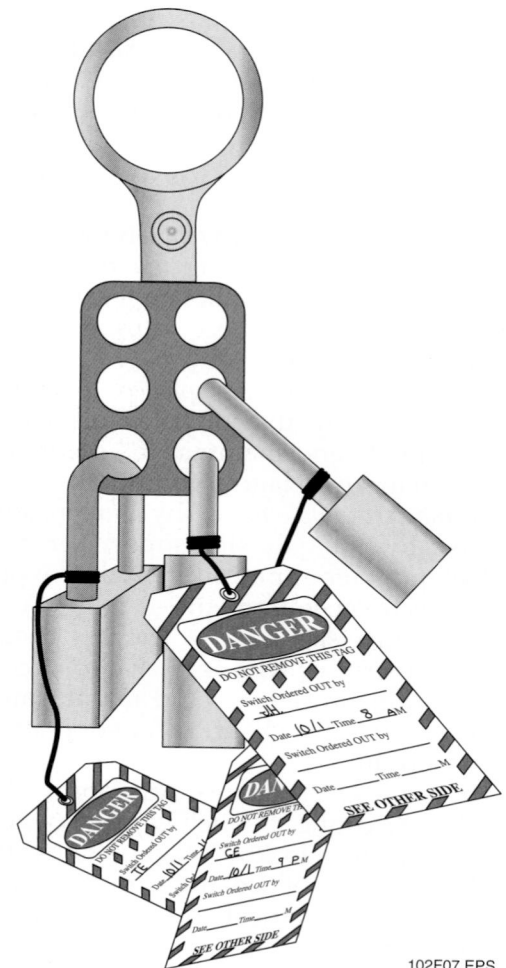

102F07.EPS

Figure 7 ◆ Multiple lockout/tagout device.

  4. Remove locks and tags from isolation switches. Authorized employees must remove their own locks and tags.
  5. Notify all affected personnel that the lockout/tagout has ended and the equipment or system is energized.
  6. Operate or close isolation switches to restore energy.

VI. *Emergency Removal Authorization*
  A. In the event a lockout/tagout device is left secured, and the authorized employee is absent, or the key is lost, the authorized supervisor can remove the lockout/tagout device.
  B. The authorized employee must be informed that the lockout/tagout device has been removed.
  C. Written verification of the action taken, including informing the authorized employee of the removal, must be recorded in the job journal.

What's wrong with this picture?

102SA05.EPS

4.4.0 Other OSHA Regulations

There are other OSHA regulations that you need to be aware of on the job site. For example:

- OSHA requires the posting of hard hat areas. Be alert to those areas and always wear your hard hat properly, with the bill in front. Hard hats should be worn whenever overhead hazards exist, or there is the risk of exposure to electric shock or burns.
- You should wear safety shoes on all job sites. Keep them in good condition.

CASE HISTORY

Lockout/Tagout Dilemma

In Georgia, electricians found energized switches after the lockout of a circuit panel in an older system that had been upgraded several times. The existing wiring did not match the current site drawings. A subsequent investigation found many such situations in older facilities.

The Bottom Line: Never rely solely on drawings. It is mandatory that the circuit be tested after lockout to verify that it is de-energized.

THINK ABOUT IT

Lockout/Tagout – Who Does It and When?

What situations are likely to require lockout/tagout? Who is responsible for performing the lockout/tagout? When would more than one person be responsible?

- Do not wear clothing with exposed metal zippers, buttons, or other metal fasteners. Avoid wearing loose-fitting or torn clothing.
- Protect your eyes. Your eyesight is threatened by many activities on the job site. Always wear safety glasses with full side shields. In addition, the job may also require protective equipment such as face shields or goggles.

5.0.0 ◆ *NFPA 70E*

In addition to the *NEC*®, the National Fire Protection Association also publishes the *Standard for Electrical Safety in the Workplace (NFPA 70E)*. The *NEC*® specifies the minimum installation provisions necessary to safeguard persons and property from electrical hazards, while *NFPA 70E* addresses the hazards that arise during the installation, operation, and maintenance of electrical equipment. In other words, the *NEC®* applies to installations, while *NFPA 70E* applies to workplaces. *NFPA 70E* is a national consensus standard, which means that it is a standard developed by the same people it affects and is then adopted by a nationally recognized institution. Other national consensus standards include those developed by ASTM International and the American National Standards Institute (ANSI).

The electrical safety requirements in *OSHA 1910.331* to *1910.335* are to use appropriate safe work practices, including de-energization of equipment, and the use of PPE appropriate to the hazard. *NFPA 70E* is really the first consensus standard to provide practical guidance in hazard assessment, training of qualified persons, and management of electrical hazards. For many, *NFPA 70E* provided the first functional understanding of the electrical arc flash hazard, direction for establishing flash protection boundaries, and levels of protection when qualified persons must work within the flash protection boundary. Like the *NEC®*, *NFPA 70E* is arranged by chapter, article, and section; for example, *NFPA 70E Chapter 4, Article 420, Section 420.1(A)*. *NFPA 70E* is organized as follows:

- *Chapter 1, Safety-Related Work Practices* – This chapter covers the basic safety requirements for working on or near electrical equipment. It covers safety training programs, explains the difference between qualified and unqualified individuals, and includes requirements for analysis of both the electrical shock hazard and the electrical arc flash/blast hazard for a given task. The analysis includes assessment of the presence and extent of hazardous voltage and the available fault current for arc flash. The end result of each hazard analysis is also selection of safe work practices and appropriate PPE to be used in performing the task.
- *Chapter 2, Safety-Related Maintenance Requirements* – This chapter covers the maintenance requirements for various types of electrical equipment. This includes premises wiring, various types of electric equipment and devices, and personal safety and protective equipment.
- *Chapter 3, Safety Requirements for Special Equipment* – This chapter covers electrolytic cells, batteries, lasers, and various types of power electronic equipment, such as radio and television transmitters, UPS systems, arc welding equipment, and lighting controllers.

- *Chapter 4, Installation Safety Requirements* – This chapter provides an abbreviated version of the requirements in the *NEC®*.
- *Annexes A through M* – The annexes provide various technical resources. These include safe approach distances, determination of flash protection boundaries and incident energy exposure from arc flash, a sample electrical lockout/tagout procedure, and other useful information.

6.0.0 ◆ LADDERS AND SCAFFOLDS

Ladders and scaffolds account for about half of the injuries from workplace electrocutions. The involuntary recoil that can occur when a person is shocked can cause the person to be thrown from a ladder or high place.

6.1.0 Ladders

Many job site accidents involve the misuse of ladders. Make sure to follow these general rules every time you use any ladder. Following these rules can prevent serious injury or even death.

- Before using any ladder, inspect it. Look for loose or missing rungs, cleats, bolts, or screws. Also check for cracked, bent, broken, or badly worn rungs, cleats, or side rails. See *Figure 8*.
- Before you climb a ladder, make sure you clear any debris from the base of the ladder so you do not trip over it when you descend.
- If you find a ladder in poor condition, do not use it. Report it and tag it for repair or disposal.
- Never modify a ladder by cutting it or weakening its parts.
- Do not set up ladders where they may be run into by others, such as in doorways or walkways. If it is absolutely necessary to set up a ladder in such a location, protect the ladder with barriers.
- Do not increase a ladder's reach by putting it in a mechanical lift or standing it on boxes, barrels, or anything other than a flat, solid surface.
- Check your shoes for grease, oil, or mud before climbing a ladder. These materials could make you slip.
- Always face the ladder and maintain three-point contact with the ladder (either have both feet and one hand on the ladder or both hands and one foot as you climb).
- Never lean out from the ladder. Keep your belt buckle centered between the rails. If something is out of reach, get down and move the ladder.

(A) CRUMBLING RAIL

(B) CRACKED STEPLADDER

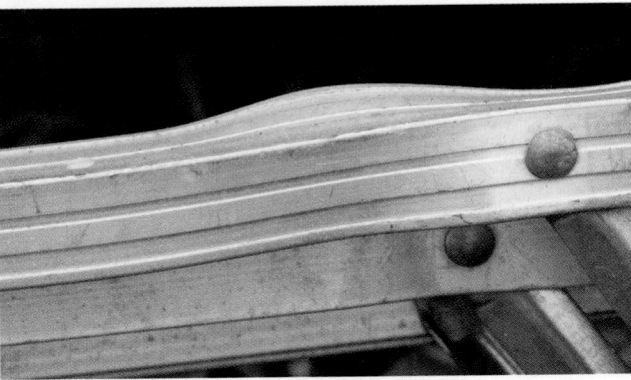

(C) BENT BACK BRACE

102F08.EPS

Figure 8 ◆ Types of ladder damage.

 **WARNING!**
When performing electrical work, always use ladders made of nonconductive material.

6.1.1 Straight and Extension Ladders

There are some specific rules to follow when working with straight and extension ladders:

- Always place a straight ladder at the proper angle. The horizontal distance from the ladder feet to the base of the wall or support should be about one-fourth the working height of the ladder. See *Figure 9*.
- Secure straight ladders to prevent slipping. Use ladder shoes or hooks at the top and bottom. Another method is to secure a board to the floor against the ladder feet. For brief jobs, someone can hold the straight ladder.
- Side rails should extend above the top support point by at least 36 inches.
- It takes two people to safely extend and raise an extension ladder. Extend the ladder only after it has been raised to an upright position.
- Never carry an extended ladder.
- Never use two ladders spliced together.
- Ladders should not be painted because paint can hide defects.

6.1.2 Step Ladders

There are also a few specific rules to use with a step ladder:

- Always open the step ladder all the way and lock the spreaders to avoid collapsing the ladder accidentally.

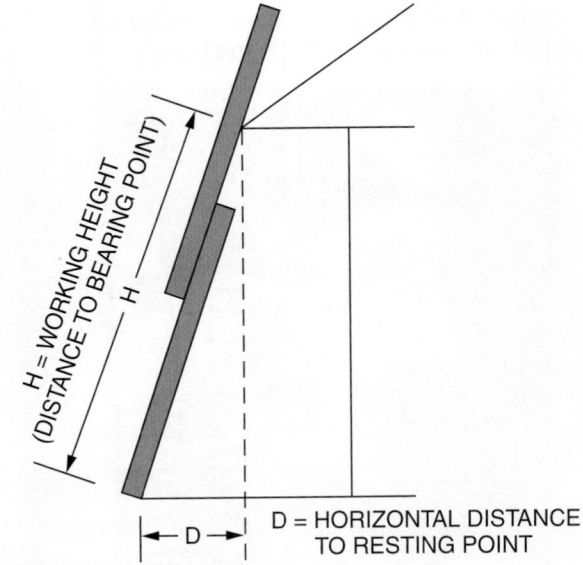

THE RATIO OF H TO D SHOULD BE 4 TO 1.

102F09.EPS

Figure 9 ◆ Straight ladder positioning.

- Use a step ladder that is high enough for the job so that you do not have to reach. Get someone to hold the ladder if it is more than 10 feet high.
- Never use a step ladder as a straight ladder.
- Never stand on or straddle the top two rungs of a step ladder.
- Ladders are not shelves.

 WARNING!
Do not leave tools or materials on a step ladder.

Sometimes you will need to move or remove protective equipment, guards, or guardrails to complete a task using a ladder. Remember, always replace what you moved or removed before leaving the area.

Fireman's Rule

To set up a straight ladder, place your feet against the side rails, stand straight up, and put your hands at right angles to your body directly in front of you. If you can grab the side rails, the ladder is at the correct angle. This is called the Fireman's Rule.

102SA06.EPS

What's wrong with this picture?

102SA07.EPS

6.2.0 Scaffolds

Working on scaffolds (*Figure 10*) also involves being safe and alert to hazards. In general, keep scaffold platforms clear of unnecessary material or scrap. These can become deadly tripping hazards or falling objects. Carefully inspect each part of the scaffold as it is erected. Your life may depend on it! Makeshift scaffolds have caused many injuries and deaths on job sites. Use only scaffolding and planking materials designed and marked for their specific use. When working on a scaffold, follow the established specific requirements set by OSHA for the use of fall protection. When appropriate, wear an approved harness with a lanyard properly anchored to the structure.

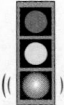 **NOTE**
The following requirements represent a compilation of the more stringent requirements of both *CFR 1910* and *CFR 1926*.

The following are some of the basic OSHA rules for working safely on scaffolds:

- Scaffolds must be erected on sound, rigid footing (referred to as a mud sill) that can carry the

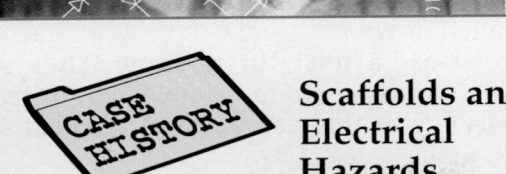

Scaffolds and Electrical Hazards

Remember that scaffolds are excellent conductors of electricity. Recently, a maintenance crew needed to move a scaffold and although time was allocated in the work order to dismantle and rebuild the scaffold, the crew decided to push it instead. They did not follow OSHA recommendations for scaffold clearance and did not perform a job site survey. During the move, the five-tier scaffold contacted a 12,000V overhead power line. All four members of the crew were killed and the crew chief received serious injuries.

The Bottom Line: Never take shortcuts when it comes to your safety and the safety of others. Trained safety personnel should survey each job site prior to the start of work to assess potential hazards. Safe working distances should be maintained between scaffolding and power lines.

maximum intended load. If the scaffolding is not erected on concrete or another firm surface, you can use 2 × 10 or 2 × 12 scaffold grade lumber to support the base plates of the scaffolding.

- Scaffolds must be erected straight and plumb with no bent or deformed pieces. Correctly erected scaffolding will be symmetrical with the same parts on both sides (unless it has an outrigger attachment).

Figure 10 ◆ Scaffolding.

102F10.EPS

- Guardrails and toe boards must be installed on the open sides and ends of platforms higher than ten feet above the ground or floor.
- There must be a screen of ½-inch maximum openings between the toe board and the midrail where persons are required to work or pass under the scaffold.
- Scaffold planks must extend over their end supports not less than 6 inches nor more than 12 inches and must be properly blocked.
- If the scaffold does not have built-in ladders that meet the standard, then it must have an attached ladder access.
- All employees must be trained to erect, dismantle, and use scaffold(s).
- Unless it is impossible, fall protection must be worn while building or dismantling all scaffolding.
- Work platforms must be completely decked for use by employees.
- Use trash containers or other similar means to keep debris from falling and never throw or sweep material from above.

7.0.0 ◆ LIFTS, HOISTS, AND CRANES

On the job, you may be working in the operating area of lifts, hoists, or cranes. The following safety rules are for those who are working in the area with overhead equipment but are not directly involved in its operation.

- Stay alert and pay attention to the warning signals from operators.
- Never stand or walk under a load, regardless of whether it is moving or stationary.

What's wrong with this picture?

102SA08.EPS

Vertical Towers

This electrician is installing a light fixture using a vertical tower or Genie® lift. This lift is designed to fold up small enough so that it can be maneuvered through house doorways.

102SA09.EPS

- Always warn others of moving or approaching overhead loads.
- Never attempt to distract signal persons or operators of overhead equipment.
- Obey warning signs.
- Do not use equipment that you are not qualified to operate.
- Cranes that are operated in areas with places in which a person can become trapped or pinched must have barricades placed around them to warn away workers.

- Hoists rigged in shafts or the outside of buildings must be secured to prevent them from being pulled down.
- Never overload a hoist, lift, or crane. Always follow lift ratings—there is no such thing as a safety factor that can be used to cheat a load.
- Cranes, lifts, and hoists must be inspected daily.
- Never lift a person using a crane, hoist, or material lift.
- Only people who have been trained in rigging should ever do rigging for a lift. It is extremely easy for a load to fall out of inappropriately rigged lifts.
- Personnel hoists require gates on every landing which can only be opened from the hoist side. This ensures that they are not opened onto a fall hazard.

8.0.0 ◆ LIFTING

Back injuries cause many lost working hours every year. That is in addition to the misery felt by the person with the hurt back! Learn how to lift properly and size up the load. To lift, first stand close to the load. Then, squat down and keep your back straight. Get a firm grip on the load and keep the load close to your body. Lift by straightening your legs. Make sure that you lift with your legs and not your back. Do not be afraid to ask for help if you feel the load is too heavy. See *Figure 11* for an example of proper lifting.

Keep the following precautions in mind when lifting:

- Make the lift smoothly and under control.
- Move your feet to pivot; do not twist or you may injure yourself.
- Constantly scan the path ahead for obstructions. If you cannot see your path over or around the object being carried, then you must have help to transport the object.
- Avoid lifting objects over your head.
- Never lift over the side or tailgate of a pickup truck.
- Don't twist your body when lifting up or setting an object down.
- Never reach over an obstacle to lift a load.
- Don't step over objects in your way.

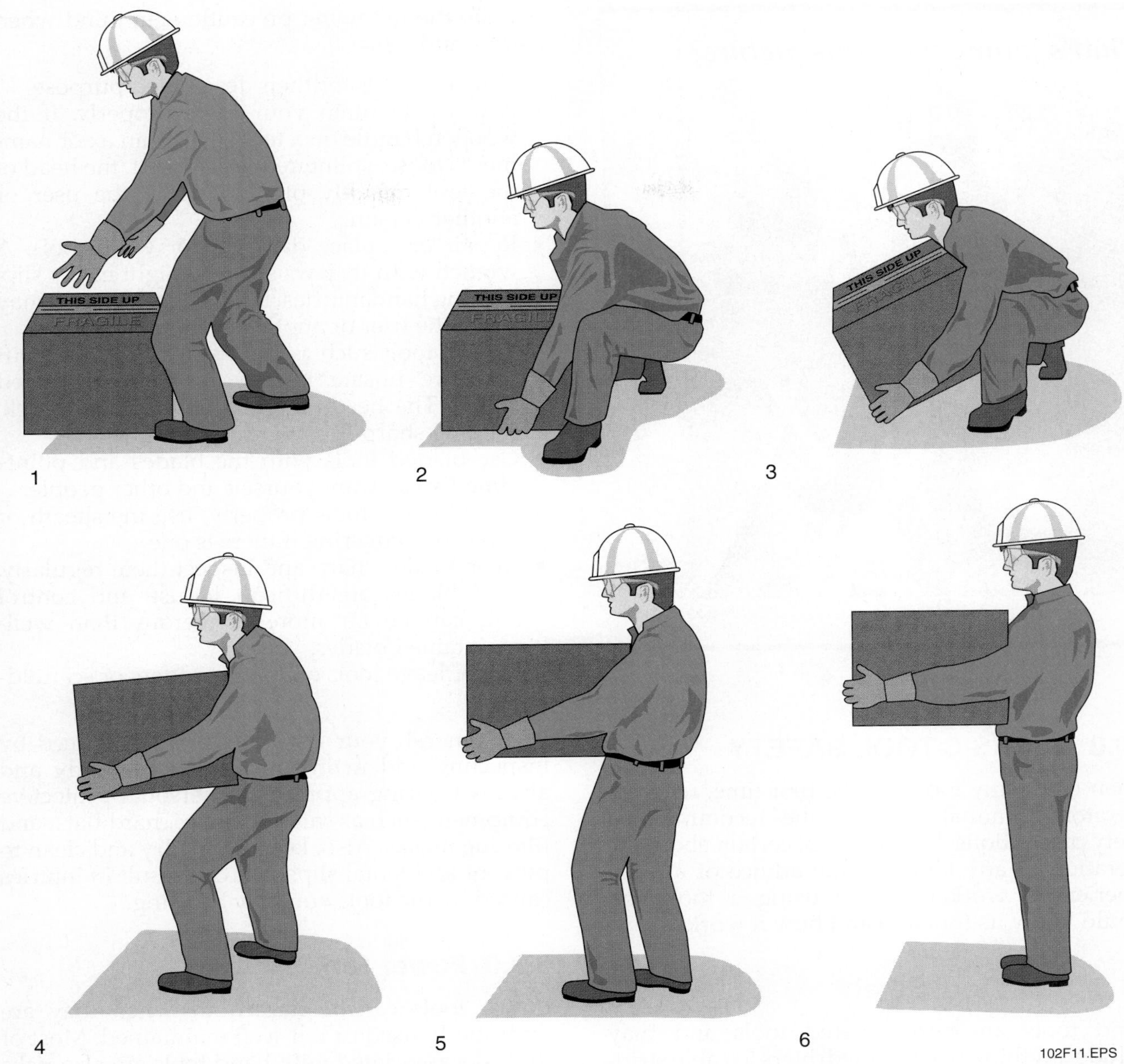

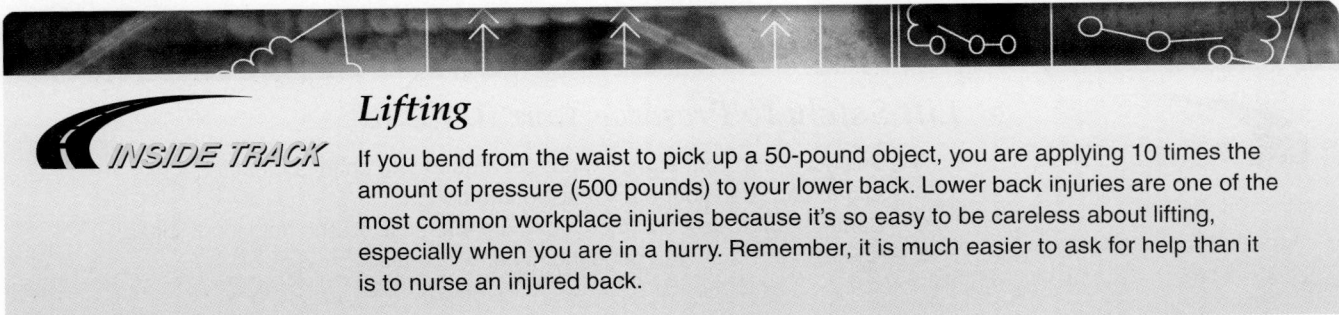

Figure 11 ◆ Proper lifting.

Lifting

INSIDE TRACK

If you bend from the waist to pick up a 50-pound object, you are applying 10 times the amount of pressure (500 pounds) to your lower back. Lower back injuries are one of the most common workplace injuries because it's so easy to be careless about lifting, especially when you are in a hurry. Remember, it is much easier to ask for help than it is to nurse an injured back.

What's wrong with this picture?

102SA10.EPS

9.0.0 ◆ BASIC TOOL SAFETY

When using any tools for the first time, read the operator's manual to learn the recommended safety precautions. If you are not certain about the operation of any tool, ask the advice of a more experienced worker. Before using a tool, you should know its function and how it works.

9.1.0 Hand Tool Safety

Hand tools are non-powered tools and may include anything from screwdrivers to cable strippers (*Figure 12*). Hand tools are dangerous if they are misused or improperly maintained.

Keep the following precautions in mind when using hand tools:

- Only use tools for their designated purpose.
- Always maintain your tools properly. If the wooden handle on a tool such as an ax or hammer is loose, splintered, or cracked, the head of the tool may fly off and strike the user or another person.
- Repair or replace damaged or worn tools. A wrench with its jaws sprung might easily slip, causing hand injuries. If the wrench flies, it may strike the user or another person.
- Impact tools such as chisels, wedges, and drift pins are unsafe if they have mushroomed heads. The heads might shatter when struck, sending sharp fragments flying.
- Use bladed tools with the blades and points aimed away from yourself and other people.
- Store bladed tools properly; use the sheath or protective covering if there is one.
- Keep blades sharp and inspect them regularly. Dull blades are difficult to use and control and can be far more dangerous than well-maintained blades.
- Never leave tools on top of ladders or scaffolding.

In general, your risks are greatly reduced by inspecting and maintaining tools regularly and always wearing appropriate personal protective equipment such as safety goggles, hard hats, and filtering masks. Also, keep floors dry and clean to prevent accidental slips that can result in injuries caused by the tools you may be using.

9.2.0 Power Tool Safety

Power tools can be hazardous when they are improperly used or not well maintained. Most of the risks associated with hand tools are also risks when using power tools. When you add a power source to a tool, however, the risk factors increase.

Lift Safely to Preserve Your B.A.C.K.

B – Balance: keep your stance wide, get a good grip on the object.
A – Alignment: keep your back relaxed and upright.
C – Contract and close: contract your stomach muscles and hold loads close.
K – Knees: make sure you bend them, not your waist.

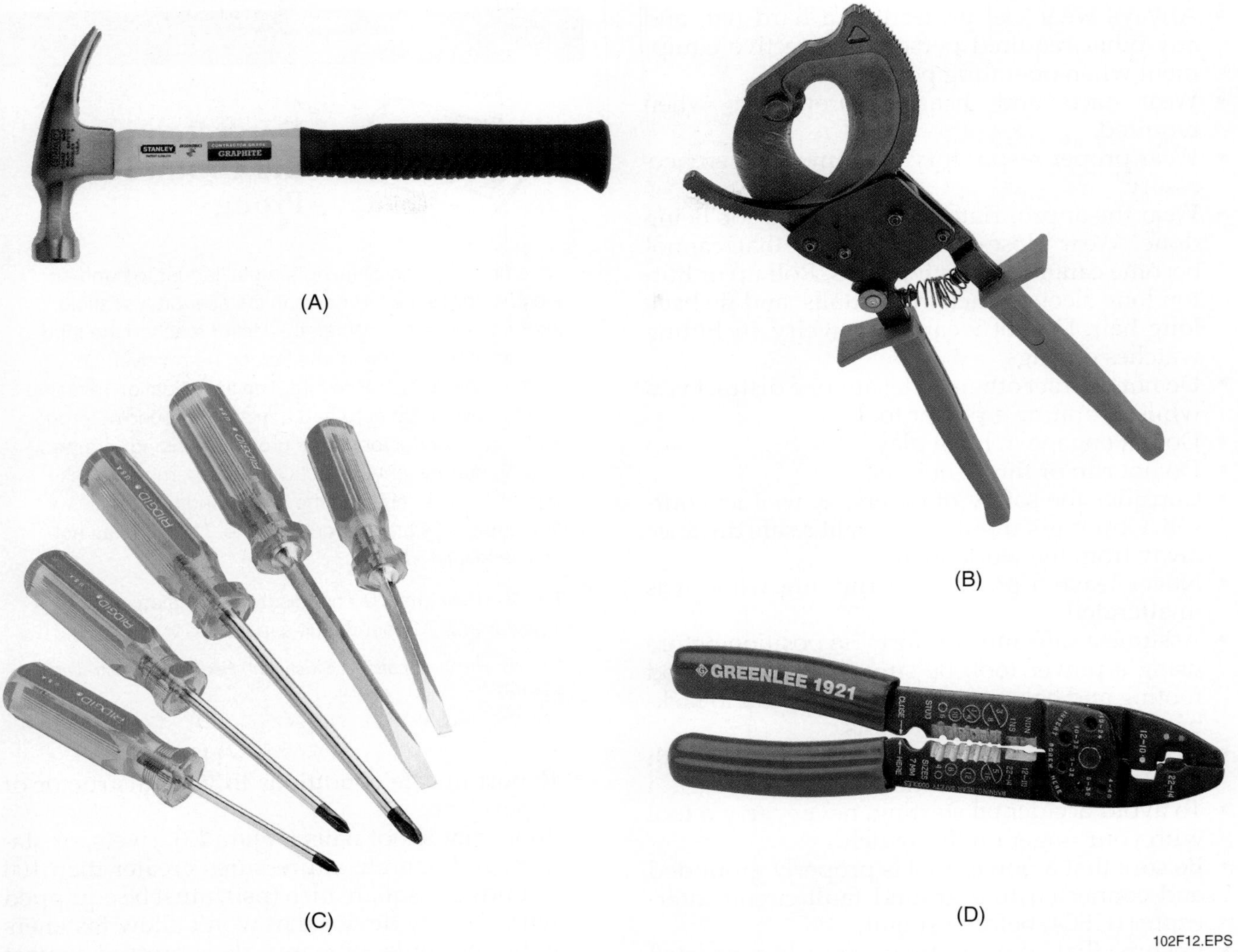

(A)

(B)

(C)

(D)

102F12.EPS

Figure 12 ◆ Hand tools.

Power tools are powered by different sources. Some examples of power sources for power tools include:

- Electricity
- Pneumatics (air pressure)
- Liquid fuel (gasoline or propane)
- Hydraulics (fluid pressure)

You must know the safety rules and proper operating procedures for each tool you use. Specific operating procedures and safety rules for using a tool are provided in the operator's/user's manual supplied by the manufacturer. Before operating any power tool for the first time, always read the manual to familiarize yourself with the tool. If the manual is missing, contact the manufacturer for a replacement.

WARNING!

Never use a tool if you are unsure of how to use it correctly.

Follow these general guidelines in order to prevent accidents and injury:

- Inspect all tools for damage before use. Remove damaged tools from use and tag DO NOT USE.
- Never carry or lower a tool by the cord or hose.
- Keep cords and hoses away from heat, oil, and sharp edges.
- Do not attempt to operate any power tool before being checked out by your instructor on that particular tool.

- Always wear eye protection, a hard hat, and any other required personal protective equipment when operating power tools.
- Wear face and hearing protection when required.
- Wear proper respiratory equipment when necessary.
- Wear the appropriate clothing for the job being done. Wear close-fitting clothing that cannot become caught in moving tools. Roll up or button long sleeves, tuck in shirttails, and tie back long hair. Do not wear any jewelry, including watches or rings.
- Do not distract others or let anyone distract you while operating a power tool.
- Do not engage in horseplay.
- Do not run or throw objects.
- Consider the safety of others, as well as yourself. Observers should be kept at a safe distance away from the work area.
- Never leave a power tool running while it is unattended.
- Assume a safe and comfortable position before using a power tool. Be sure to maintain good footing and balance in order to respond to kickbacks, jumps, or sudden shifts.
- Secure work with clamps or a vise, freeing both hands to safely operate the tool.
- To avoid accidental starting, never carry a tool with your finger on the switch.
- Be sure that a power tool is properly grounded and connected to a ground fault circuit interrupter (GFCI) before using it.
- Ensure that power tools are disconnected before performing maintenance or changing accessories.
- Use a power tool only for its intended use.
- Keep your feet, fingers, and hair away from the blade and/or other moving parts of a power tool.
- Never use a power tool with guards or safety devices removed or disabled.
- Never operate a power tool if your hands or feet are wet.
- Keep the work area clean at all times.
- Become familiar with the correct operation and adjustments of a power tool before attempting to use it.
- Keep tools sharp and clean for the best performance.
- Follow the instructions in the user's manual for lubricating and changing accessories.
- Keep a firm grip on the power tool at all times.
- Use electric extension cords of sufficient size to service the particular power tool you are using.
- Do not run extension cords across walkways where they will pose a tripping hazard.

Don't Remove the Grounding Prong

CASE HISTORY

An employee was climbing a metal ladder to hand an electric drill to the journeyman installer on a scaffold about 5' above him. When the victim reached the third rung from the bottom of the ladder, he received an electric shock that killed him. The investigation revealed that the extension cord had a missing grounding prong and that a conductor on the green grounding wire was making intermittent contact with the energized black wire, thereby energizing the entire length of the grounding wire and the drill's frame. The drill was not double insulated.

The Bottom Line: Do not disable any safety device on a power tool. A ground fault can be deadly.

Source: The Occupational Safety and Health Administration (OSHA)

- Report unsafe conditions to your instructor or supervisor.
- Tools that shoot nails (*Figure 13*), rivets, or staples, and operate at pressures greater than 100 pounds per square inch (psi), must be equipped with a safety device that won't allow fasteners to be shot unless the muzzle is pressed against a work surface.
- Compressed-air guns should never be pointed toward anyone and the muzzle should never be pressed against a person.

102F13.EPS

Figure 13 ◆ Pneumatic nail gun.

- The use of powder-actuated tools requires special training and certification.
- Never use explosive or flammable materials around powder-actuated tools.
- Never point powder-actuated tools at anybody.
- Never pick up an unattended powder-actuated tool. Instead, tell your supervisor that a powder-actuated tool has been left unattended.
- Never play with powder-actuated tools. These tools are as dangerous as a loaded gun.

10.0.0 ◆ CONFINED SPACE ENTRY PROCEDURES

Occasionally, you may be required to do your work in a manhole or vault. If this is the case, there are some special safety considerations that you need to be aware of. For details on the subject of working in manholes and vaults, refer to *CFR 1910.146/1926.21(a)(6)(i) and (ii)*. The general precautions are listed in the following paragraphs.

10.1.0 General Guidelines

A confined space includes (but is not limited to) any of the following: a manhole (*Figure 14*), boiler, tank, trench (four feet or deeper), tunnel, hopper, bin, sewer, vat, pipeline, vault, pit, air duct, or vessel. A confined space is identified as follows:

- It has limited entry and exit.
- It is not intended for continued human occupancy.
- It has poor ventilation.
- It has the potential for entrapment/engulfment.
- It has the potential for accumulating a dangerous atmosphere.

102F14.EPS

Figure 14 ◆ Manhole.

Entry into a confined space occurs when any part of the body crosses the plane of entry. No employee shall enter a confined space unless the employee has been trained in confined space entry procedures. Other requirements for confined spaces include the following:

- All hazards must be eliminated or controlled before a confined space entry is made.
- The air quality in the confined space must be continually monitored.
- All appropriate personal protective equipment shall be worn at all times during confined space entry and work. The minimum required equipment includes a hard hat, safety glasses, full body harness, and life line.
- Ladders used for entry must be secured.
- A rescue retrieval system must be in use when entering confined spaces and while working in permit-required confined spaces (discussed later). Each employee must be capable of being rescued by the retrieval system.
- Only no-entry rescues will be performed by company personnel. Entry rescues will be performed by trained rescue personnel identified on the entry permit.
- The area outside the confined space must be properly barricaded, and appropriate warning signs must be posted.
- Entry permits can only be issued and signed by a qualified person such as the job site supervisor. Permits must be kept at the confined space while work is being conducted. At the end of the shift, the entry permits must be made part of the job journal and retained for one year.

10.2.0 Confined Space Hazard Review

Before determining the proper procedure for confined space entry, a hazard review shall be performed. The hazard review shall include, but not be limited to, the following conditions:

- The past and current uses of the confined space
- The physical characteristics of the space including size, shape, air circulation, etc.
- Proximity of the space to other hazards
- Existing or potential hazards in the confined space, such as:
  - Atmospheric conditions (oxygen levels, flammable/explosive levels, and/or toxic levels)
  - Presence/potential for liquids
  - Presence/potential for particulates
- Potential for mechanical/electrical hazards in the confined space (including work to be done)

Once the hazard review is completed, the supervisor, in consultation with the project managers and/or safety manager, shall classify the confined space as one of the following:

- A nonpermit confined space
- A permit-required confined space controlled by ventilation
- A permit-required confined space

Once the confined space has been properly classified, the appropriate entry and work procedures must be followed.

 WARNING!
Only qualified and trained individuals may enter a confined space.

11.0.0 ◆ FIRST AID

You should be prepared in case an accident does occur on the job site or anywhere else. First aid training that includes certification classes in CPR and artificial respiration could be the best insurance you and your fellow workers ever receive. Make sure that you know where first aid is available at your job site. Also, make sure you know the accident reporting procedure. Each job site should also have a first aid manual or booklet giving easy-to-find emergency treatment procedures for various types of injuries. Emergency first aid telephone numbers should be readily available to everyone on the job site. Refer to *CFR 1910.151/ 1926.23* and *1926.50* for specific requirements.

12.0.0 ◆ SOLVENTS AND TOXIC VAPORS

The solvents that are used by electricians may give off vapors that are toxic enough to make people temporarily ill or even cause permanent injury. Many solvents are skin and eye irritants. Solvents can also be systemic poisons when they are swallowed or absorbed through the skin.

Solvents in spray or aerosol form are dangerous in another way. Small aerosol particles or solvent vapors mix with air to form a combustible mixture with oxygen. The slightest spark could cause an explosion in a confined area because the mix is perfect for fast ignition. There are procedures and methods for using, storing, and disposing of most solvents and chemicals. These procedures are normally found in the material safety data sheets (MSDSs) available at your facility.

An MSDS is required for all materials that could be hazardous to personnel or equipment.

These sheets contain information on the material, such as the manufacturer and chemical makeup. As much information as possible is kept on the hazardous material to prevent a dangerous situation; or, in the event of a dangerous situation, the information is used to rectify the problem in as safe a manner as possible. See *Figure 15* for an example of MSDS information you may find on the job.

12.1.0 Precautions When Using Solvents

It is always best to use a nonflammable, nontoxic solvent whenever possible. However, any time solvents are used, it is essential that your work area be adequately ventilated and that you wear the appropriate personal protective equipment:

- Wear a chemical face shield with chemical goggles to protect the eyes and skin from sprays and splashes.
- Wear a chemical apron to protect your body from sprays and splashes. Remember that some solvents are acid-based. If they come into contact with your clothes, solvents can eat through your clothes to your skin.
- A paper filter mask does not stop vapors; it is used only for nuisance dust. In situations where a paper mask does not supply adequate protection, chemical cartridge respirators might be needed. These respirators can stop many vapors if the correct cartridge is selected. In areas where ventilation is a serious problem, a self-contained breathing apparatus (SCBA) must be used.
- Make sure that you have been given a full medical evaluation and that you are properly trained in using respirators at your site.

12.2.0 Respiratory Protection

The best respiratory protection is to avoid the hazard entirely. Off-shift work, ventilation, and/or rescheduled work schedules should always be used to eliminate the need for working in areas with poor air quality. For example, in an area where hazardous solvents are used, the electrical work can be done off schedule when the solvents are not being used. When this cannot be done, protection against high concentrations of dust, mist, fumes, vapors, and gases is provided by appropriate respirators.

Appropriate respiratory protective devices should be used for the hazardous material involved and the extent and nature of the work performed.

Section VI – Spill and Leak Procedures

Steps to Be Taken in Case Material is Released or Spilled
Isolate from oxidizers, heat, sparks, electric equipment, and open flames.

Waste Disposal Method
Recycle or incinerate observing local, state and federal health, safety

and pollution laws.

Precautions to Be Taken in Handling and Storing
Store in a cool dry area. Observe label cautions and instructions.

Other Precautions
SEE ATTACHMENT PARA #3

Section VII – Personal Protection Information

Respiratory Protection (*Specify Type*)
Suitable for use with organic solvents

Ventilation	Local Exhaust	preferable	Special	none
	Mechanical (*General*)	acceptable	Other	none
Protective Gloves	recommended (must not dissolve in solvents)		Eye Protection	goggles

Other Protective Clothing or Equipment
none

Work/Hygenic Practices
Use with adequate ventilation. Observe label cautions.

102F15.EPS

Figure 15 ◆ Portion of an MSDS.

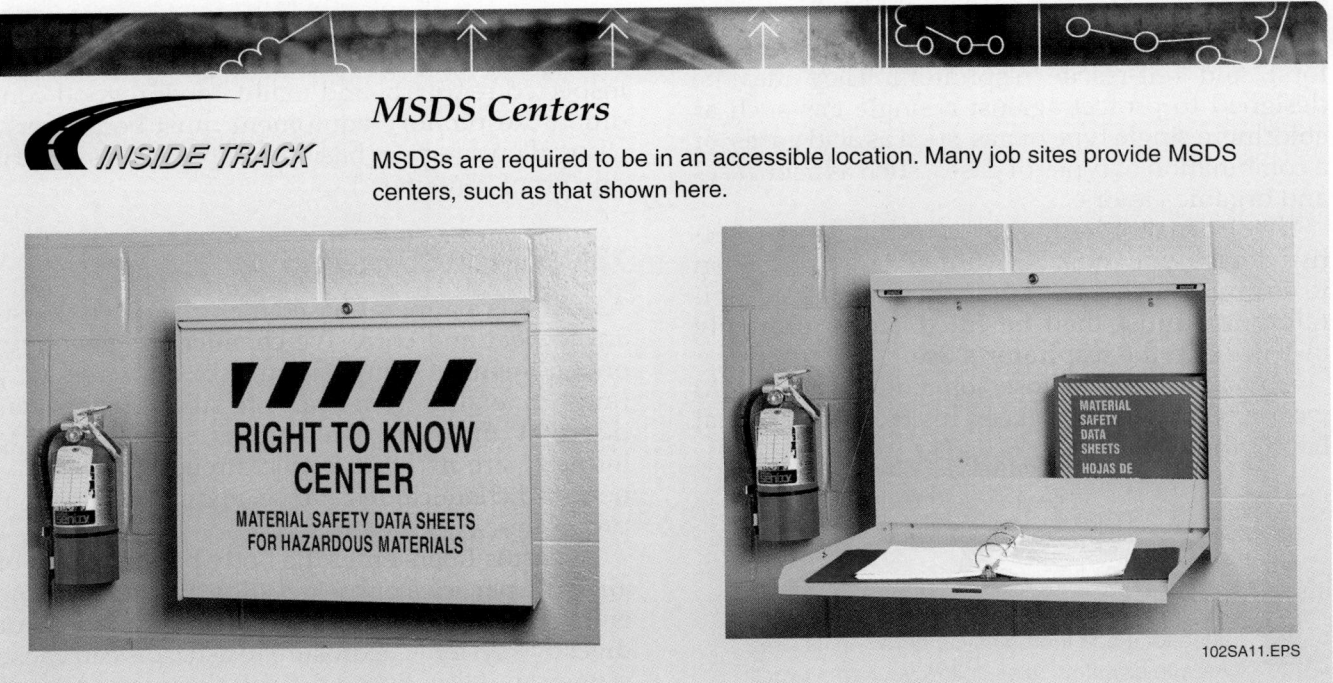

MSDS Centers

INSIDE TRACK

MSDSs are required to be in an accessible location. Many job sites provide MSDS centers, such as that shown here.

102SA11.EPS

Chemical Safety

INSIDE TRACK

The first line of defense with chemicals is to read and follow the directions found on the container. If you follow these instructions, you should be safe from chemical exposure. Be aware that everyone reacts differently to chemicals and you may be hypersensitive to a particular chemical that does not bother your co-workers. Leave the area at the first sign of an allergic reaction and seek medical attention.

An air-purifying respirator is, as its name implies, a respirator that removes contaminants from air inhaled by the wearer. The respirators may be divided into the following types: particulate-removing (mechanical filter), gas- and vapor-removing (chemical filter), and a combination of particulate-removing and gas- and vapor-removing.

Particulate-removing respirators are designed to protect the wearer against the inhalation of particulate matter in the ambient atmosphere. They may be designed to protect against a single type of particulate, such as pneumoconiosis-producing and nuisance dust, toxic dust, metal fumes or mist, or against various combinations of these types.

Gas- and vapor-removing respirators are designed to protect the wearer against the inhalation of gases or vapors in the ambient atmosphere. They are designated as gas masks, chemical cartridge respirators (nonemergency gas respirators), and self-rescue respirators. They may be designed to protect against a single gas such as chlorine; a single type of gas, such as acid gases; or a combination of types of gases, such as acid gases and organic vapors.

If you are required to use a respiratory protective device, you must be evaluated by a physician to ensure that you are physically fit to use a respirator. You must then be fitted and thoroughly instructed in the respirator's use.

Any employee whose job entails having to wear a respirator must keep his face free of facial hair in the seal area.

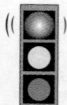 **WARNING!**
Do not use any respirator unless you have been fitted for it and thoroughly understand its use. As with all safety rules, follow your employer's respiratory program and policies.

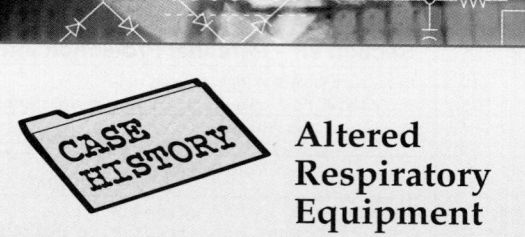

CASE HISTORY — Altered Respiratory Equipment

A self-employed man applied a solvent-based coating to the inside of a tank. Instead of wearing the proper respirator, he used nonstandard air supply hoses and altered the face mask. All joints and the exhalation slots were sealed with tape. He collapsed and was not discovered for several hours.

The Bottom Line: Never alter or improvise safety equipment.

Respiratory protective equipment must be inspected regularly and maintained in good condition. Respiratory equipment must be properly cleaned on a regular basis and stored in a sanitary, dustproof container.

13.0.0 ◆ ASBESTOS

Asbestos is a mineral-based material that is resistant to heat and corrosive chemicals. Depending on the chemical composition, asbestos fibers may range in texture from coarse to silky. The properties that make asbestos fibers so valuable to industry are its high tensile strength, flexibility, heat and chemical resistance, and good frictional properties.

Asbestos fibers enter the body by inhalation of airborne particles or by ingestion and can become embedded in the tissues of the respiratory or digestive systems. Exposure to asbestos can cause numerous disabling or fatal diseases. Among these diseases are asbestosis, an emphysema-like condition; lung cancer; mesothelioma, a cancerous tumor that spreads rapidly in the cells of membranes covering the lungs and body organs; and gastrointestinal cancer. The use of asbestos was banned in 1978.

Because asbestos was still in the manufacturing pipeline for a while after it was banned, you need to assume that any facility constructed before 1980 has asbestos in it. The owner must have a survey with any work rules needed to work safely around the asbestos. Common products that contain asbestos include thermal pipe insulation, mastic for ducts and insulation, spray-on fireproofing, floor tiles, ceiling tiles, roof insulation, exterior building sheathing, old wire insulation, and even pipe. As an electrician, you must not drill through or otherwise work with asbestos—you can only be trained to work around it when it can be done safely. Asbestos work is a trade that requires special training and protective equipment.

The following signs must be placed in areas containing asbestos.

> ## DANGER
> ### ASBESTOS
> ### CANCER AND LUNG DISEASE HAZARD
> ### AUTHORIZED PERSONNEL ONLY
> ### RESPIRATORS AND PROTECTIVE CLOTHING ARE REQUIRED IN THIS AREA

> ## DANGER
> ### CONTAINS ASBESTOS FIBERS
> ### AVOID CREATING DUST
> ### CANCER AND LUNG DISEASE HAZARD

14.0.0 ◆ BATTERIES

Working around wet cell batteries can be dangerous if the proper precautions are not taken. Batteries often give off hydrogen gas as a byproduct. When hydrogen mixes with air, the mixture can be explosive in the proper concentration. For this reason, smoking is strictly prohibited in battery rooms, and only insulated tools should be used. Proper ventilation also reduces the chance of explosion in battery areas. Follow your company's procedures for working near batteries. Also, ensure that your company's procedures are followed for lifting heavy batteries.

WARNING!
Battery-powered scissor lifts may have unsealed batteries that require water levels to be checked and topped off. Never use a flame to look inside a battery—it can cause an explosion. Charging batteries with low water levels can damage them.

14.1.0 Acids

Batteries also contain acid, which will eat away human skin and many other materials. Personal protective equipment for battery work typically includes chemical aprons, sleeves, gloves, face shields, and goggles to prevent acid from contacting skin and eyes. Follow your site procedures for dealing with spills of these materials. Also, know the location of first aid when working with these chemicals.

14.2.0 Wash Stations

Because of the chance that battery acid may contact someone's eyes or skin, wash stations are located near battery rooms. Do not connect or disconnect batteries without proper supervision. Everyone who works in the area should know where the nearest wash station is and how to use it. Battery acid should be flushed from the skin and eyes with large amounts of water or with a neutralizing solution.

CAUTION
If you come in contact with battery acid, flush the affected area with water and report it immediately to your supervisor.

15.0.0 ◆ PCBs AND VAPOR LAMPS

Polychlorinated biphenyls (PCBs) are chemicals that were marketed under various trade names as a liquid insulator/cooler in older transformers. In addition to being used in older transformers, PCBs are also found in some large capacitors and in the small ballast transformers used in street lighting and ordinary fluorescent light fixtures. Disposal of these materials is regulated by the Environmental Protection Agency (EPA) and must be done through a regulated disposal company; use extreme caution and follow your facility procedures.

In addition, any vapor lamps, such as fluorescent, halide, or mercury vapor lamps, must be recycled. The tubes must be packaged and handled carefully to avoid breakage.

16.0.0 ◆ FALL PROTECTION

All employees must receive documented training before working in any area where there is the possibility of exposure to a fall of 6' or more. This training must be renewed annually and must include ladder safety. The 6' rule does not apply to ladders, scaffolds, and mechanical lifts, which have their own standards.

16.1.0 Fall Protection Procedures

Fall protection must be used when employees are on a walking or working surface that is 6' or more above a lower level and has an unprotected edge or side. The areas covered include, but are not limited to the following:

- Finished and unfinished floors or mezzanines
- Temporary or permanent walkways/ramps
- Finished or unfinished roof areas
- Elevator shafts and hoistways
- Floor, roof, or walkway holes
- Working 6' or more above dangerous equipment

Exception: If the dangerous equipment is unguarded, fall protection must be used at all heights regardless of the fall distance.

Figure 16 shows a simple temporary guardrail installed on the stairway of a residence under construction. *Figure 17* shows a complex guardrail system used during the construction of a department store. The large central opening will eventually house the building escalators.

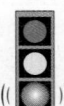

NOTE

Walking/working surfaces do not include ladders, scaffolds, vehicles, or trailers. Also, an unprotected edge or side is an edge/side where there is no guardrail system at least 39" high.

102F16.EPS

Figure 16 ◆ Temporary guardrail on stairs.

102F17.EPS

Figure 17 ◆ Complex guardrail system.

According to OSHA, an employee can never be exposed to a fall of more than 6'. This is called 100% fall protection. The employee must be protected by one of the following in this order of preference:

1. Guardrail systems
2. Personal fall arrest systems (PFAS)
3. Controlled access zone or other administrative system

16.1.1 Guardrail Systems

Guardrail systems must be constructed as follows:

- The top rails must be 42" (±3") and must be capable of withstanding 200 lbs.
- The mid-rails must be at 21" (±3") and must be capable of withstanding 150 lbs.

What's wrong with this picture?

102SA12.EPS

- The toe board must be 4" tall and no more than ¼" above the floor for drainage.
- Guardrails can be made of 2 × 4s, pipes, chains, or cables.
- Chains and cables require flags every 6'. If cables are used, they must be secured to avoid deflection greater than 3" in, out, or down from the 42" requirement.
- Banding material is not allowed for guardrail construction. Cable guardrails require the use of cable clamps on the cable. Clamps must be forged and not malleable. They must be torqued and installed properly. Ensure that the clamp does not damage the loadbearing line.

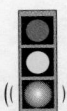

NOTE

These ratings apply to all portions of the rail system such as anchors, anchorage material, and clamps. Guardrails cannot be used to secure a PFAS unless they are designed to support 5,000 lbs per person.

Warning lines, signs, or barricades must be installed back from the edge at least 6'. That way, even if someone falls over the barrier, they will not fall to the level below.

16.1.2 Personal Fall Arrest Systems (PFAS)

PFAS provide fall arrest after an employee falls. This equipment must be selected, inspected, donned, anchored, and maintained to be effective. The complete system usually consists of a full-body harness, lanyard, and anchorage device.

Full-body harnesses – Full-body harnesses are the only acceptable equipment to wear for PFAS. Select the appropriate harness based on size and gender. Inspect the equipment before use. Harnesses must be worn snug (but not tight) with all required straps attached. When properly applied, you should be able to slide two fingers under the straps with little difficulty. The D-ring in the back of the harness must be centered between the shoulder blades. After donning the harness, have a co-worker pull sharply up on the D-ring. You

What's wrong with this picture?

102SA13.EPS

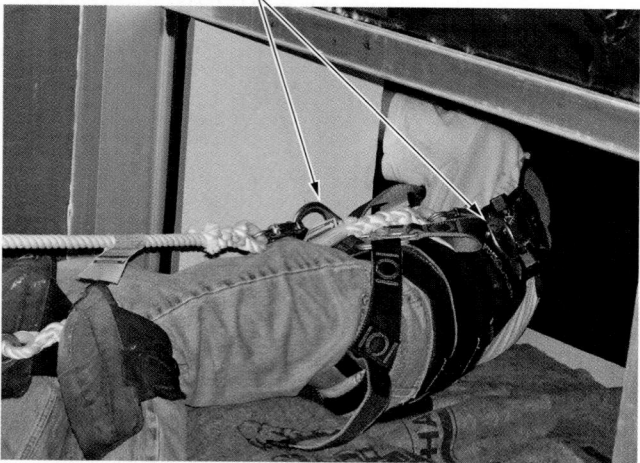

102F18.EPS

Figure 18 ◆ Electrician tied off to two lanyards.

should feel the grab around the thighs, chest, and buttocks. Jobs that require positioning must be accomplished using a full-body harness with side D-rings. Safety belts are not allowed.

Lanyards – Lanyards are used to connect the harness to the attachment point. As no employee can be exposed to a fall of more than 6', standard lanyards must be no longer than 6'. You can be exposed to 1,800 lbs of force in a properly worn harness. The use of shock absorbing lanyards or retractable lanyards can reduce that force to as low as 400 to 600 lbs. Shock absorbers work by slowing the employee to a stop by ripping stitches while elongating up to 42". All lanyards must have locking snap hooks. Never attach two locking snap hooks to the same D-ring as they can foul each other, causing the relatively weak gates to break. *Figure 18* shows an electrician making an adjustment to an outdoor wallwasher lighting fixture while hanging out of a 15th-story window. For extra protection, he is attached to two lanyards on two separate D-rings.

A twin-tailed lanyard is required when climbing. While climbing, you cannot unhook your lanyard to move it to another anchorage and still have 100% fall protection. Thus, with two lanyards, you can "walk" to where you are working.

Retractable lanyards come in a variety of sizes from 10' to over 150'. *Figure 19* shows a worker

RETRACTABLE LANYARD

102F19.EPS

Figure 19 ◆ Proper fall protection on a boom lift.

tied off to a retractable lanyard when working on a boom lift. Retractable lanyards are used where movement or close proximity with the ground will render standard lanyards ineffective or inefficient. Retractable lanyards can be longer than 6' because when a fall occurs they grab hold within 2'. This quick reaction also eliminates the need for a shock absorber.

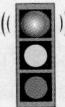

WARNING!

Do not put a shock absorber in line with a retractable lanyard because it may interfere with the quick response of the retractable lanyard.

Anchorage devices – Anchorage devices and points are the interface between the PFAS and the structure to which they are attached. This point must hold 5,000 lbs (this is the equivalent of a full-size extended cab pickup truck).

A lanyard cannot be wrapped around an anchorage and then attached to itself unless it is specially designed with cross arm straps. Cross arm straps are made of webbing, two inches wide, of any necessary length, with two different size D-rings. The cross arm straps are passed over whatever object you are going to attach them to and wrapped around to reduce the length of the lanyard. Beam clamps, wire hangers, trolleys, and other manufactured devices are also used for specific applications.

What's wrong with this picture?

102SA14.EPS

Equipment inspection process – All PFAS must be inspected when received and before each use. Check the manufacturer's tag for the manufacturer's inspection date. If the fall equipment has no date, it should be disposed of immediately. Carefully look over the webbing. If you observe any burns, ripped stitches, color marker threads, distorted grommets, bent or cracked buckle tongues, distorted D-rings, or bent, cracked, or pulled fabrics, remove the PFAS from service. Retractable lanyards must undergo the same inspection process as other PFAS. In addition, you must also pull out the entire lanyard for inspection, let it back in, and then pull out 2' to 4' of lanyard and give it a swift tug to see if it properly engages. If any of these inspections fail, the equipment must be destroyed immediately or tagged DO NOT USE and returned to the shop.

WARNING!

All fall protection equipment that is involved in a fall must be taken out of service and destroyed. Any employees involved in a fall must receive medical attention, even if they do not feel they have been injured. Falls can cause internal injuries that are not readily apparent to the victim.

Rescue – Never pull anyone up by their fall protection; always rescue them with ladders or equipment from below. If the standard equipment is not available to provide rescue, a plan must be created before work can proceed. Rescues must be accomplished from below using ladders, lifts, and/or scaffolds.

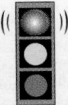

WARNING!

Unless a person is in immediate danger, never attempt to lift them up by their lanyard. This could cause an additional drop for the fallen worker and/or injure the rescuers.

Immediately summon the fire department to assist in the rescue effort unless you can rescue the person without assistance. Rescue must take place as quickly as possible, as hanging from a harness presents additional hazards. If you fall, continue to move your limbs while awaiting rescue. This will help maintain circulation in your lower extremities.

Putting It All Together

THINK ABOUT IT

This module has described a professional approach to electrical safety. How does this professional outlook differ from an everyday attitude? What do you think are the key features of a professional philosophy of safety?

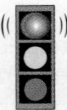

WARNING!

Any employee whose weight exceeds 310 pounds, including their tools, cannot wear fall protection, as that is the maximum weight for which it is designed.

PFAS selection – The type of system selected depends on the fall hazards associated with the work to be performed. First, a hazard analysis must be conducted by the job site supervisor prior to the start of work. Based on the hazard analysis, the job site supervisor and project manager, in consultation with the safety manager, will select the appropriate fall protection system. All employees must be instructed in the use of the fall protection system before starting work.

16.1.3 Controlled Access Zones

There are times when a guardrail cannot be attached to the building. In these cases, a controlled access zone must be installed. A controlled access zone may consist of guards, barricades, badge systems, or other administrative measures.

Figure 20 shows a controlled access zone on a roof. Note that the barricade is located 6' from the edge. This way, even if someone trips over the barricade, they won't fall off the roof.

SIX FEET FROM EDGE

102F20.EPS

Figure 20 ◆ Controlled access zone.

1. The most life-threatening hazards on a construction site typically include all of the following *except* _____.
   a. falls
   b. electric shock
   c. being crushed or struck by falling or flying objects
   d. chemical burns

2. If a person's heart begins to fibrillate due to an electrical shock, the solution is to _____.
   a. leave the person alone until the fibrillation stops
   b. immerse the person in ice water
   c. use the Heimlich maneuver
   d. have a qualified person use emergency defibrillation equipment

3. Low-voltage conductors rarely cause injuries.
   a. True
   b. False

4. Class 00 rubber gloves are used when working with voltages less than _____.
   a. 500 volts
   b. 1,000 volts
   c. 5,000 volts
   d. 7,500 volts

5. An important use of a hot stick is to _____.
   a. replace busbars
   b. test for voltage
   c. replace fuses
   d. test for continuity

6. Which of these statements correctly describes a double-insulated power tool?
   a. There is twice as much insulation on the power cord.
   b. It can safely be used in place of a grounded tool.
   c. It is made entirely of plastic or other nonconducting material.
   d. The entire tool is covered in rubber.

7. Which of the following applies in a lockout/tagout procedure?
   a. Only the supervisor can install lockout/tagout devices.
   b. If several employees are involved, the lockout/tagout equipment is applied only by the first employee to arrive at the disconnect.
   c. Lockout/tagout devices applied by one employee can be removed by another employee as long as it can be verified that the first employee has left for the day.
   d. Lockout/tagout devices are installed by every authorized employee involved in the work.

8. The *NEC*® provides requirements for safe electrical installations, while *NFPA 70E* provides guidance for establishing safe electrical workplaces.
   a. True
   b. False

9. What is the proper distance from the feet of a straight ladder to the wall?
   a. one-fourth the working height of the ladder
   b. one-half the height of the ladder
   c. three feet
   d. one-fourth of the square root of the height of the ladder

10. What are the minimum and maximum distances (in inches) that a scaffold plank can extend beyond its end support?
    a. 4; 8
    b. 6; 10
    c. 6; 12
    d. 8; 12

11. All of the following are considered confined spaces *except* a _____.
    a. 3 ft trench
    b. sewer
    c. pipeline
    d. manhole

12. The best way to protect yourself from solvent hazards is to _____.
    a. always wear vinyl gloves and a paper filter mask
    b. ask your supervisor or a co-worker
    c. ask the supplier
    d. read and follow all instructions on the product's MSDS

13. Asbestos was banned in _____; therefore, you must assume that any facility constructed before _____ has asbestos in it.
    a. 1943; 1945
    b. 1964; 1966
    c. 1978; 1980
    d. 1988; 1990

14. You may throw vapor lamps out with regular trash as long as you wrap them carefully to avoid breakage.
    a. True
    b. False

15. A PFAS anchorage point must be able to hold _____ lbs.
    a. 250
    b. 500
    c. 1,000
    d. 5,000

Summary

Safety must be your concern at all times so that you do not become either the victim of an accident or the cause of one. Safety requirements and safe work practices are provided by OSHA and your employer. It is essential that you adhere to all safety requirements and follow your employer's safe work practices and procedures. Also, you must be able to identify the potential safety hazards of your job site. The consequences of unsafe job site conduct can often be expensive, painful, or even deadly. Report any unsafe act or condition immediately to your supervisor. You should also report all work-related accidents, injuries, and illnesses to your supervisor immediately. Remember, proper construction techniques, common sense, and a good safety attitude will help to prevent accidents, injuries, and fatalities.

Notes

Trade Terms Quiz

1. A life-threatening condition of the heart in which the muscle fibers contract irregularly is called _____.

2. Any tool that has a case made of nonconductive material and that has been constructed so that the case is insulated from electrical energy is a _____.

3. _____ are chemicals often found in liquids that are used to cool certain types of large transformers and capacitors.

4. A _____ will de-energize a circuit or a portion of it if the current to ground exceeds some predetermined value.

5. A _____ has a three-prong plug at the end of its power cord or some other means to ensure that stray current travels to ground without passing through the body of the operator.

Trade Terms

Double-insulated/ungrounded tool
Fibrillation
Grounded tool
Ground fault circuit interrupter (GFCI)
Polychlorinated biphenyls (PCBs)

Michael J. Powers

Tri-City Electrical Contractors, Inc.

How did you choose a career in the electrical field?
My father was an electrician and after I "burned out" with a career in fast-food management, I decided to choose a completely different field.

Tell us about your apprenticeship experience.
It was excellent! I worked under several very knowledgeable electricians and had a pretty good selection of teachers. Over my four-year apprenticeship, I was able to work on a variety of jobs, from photomats to kennels to colleges.

What positions have you held and how did they help you to get where you are now?
I have been an electrical apprentice, a licensed electrician, a job-site superintendent, a master electrician, and am currently a corporate safety and training director. The knowledge I acquired in electrical theory in apprenticeship school and preparing for my licensing exams, as well as the practical on-the-job experience over thirty years in the trade, were wonderful training for my current position.

I also serve on the authoring team for NCCER's Electrical Curricula, which has provided me with not only the opportunity to share what I have learned, but is also a great way to keep current in other areas by meeting with electricians from a variety of disciplines (we have commercial, residential, and industrial electricians on the team, as well as instructors).

What would you say was the single greatest factor that contributed to your success?
Choosing a company that recognized and rewarded competent, hard workers and provided them with the support and guidance to allow them to develop and succeed in the industry.

What does your current job entail?
I am responsible for safe work practices and procedures through the job-site management team at Tri-City Electrical Contractors, which currently has a workforce of over 1,100. I also assist in developing, delivering, and administering the training program, from apprenticeship to in-house to outsourced training.

What advice do you have for trainees?
Training in all its aspects is the key to your success and advancement in the industry. Any time you are given a training opportunity, take it, even if it might not appear relevant at the time. Eventually, all knowledge can be applied to some situation.

Most importantly, have fun! The construction industry is composed of good people. I firmly believe that construction workers, as a group, are much more honest and direct than any other comparable group. Wait—did I say comparable group? That's a misstatement—there is no comparable group. Construction workers build America!

Trade Terms
Introduced in This Module

Double-insulated/ungrounded tool: An electrical tool that is constructed so that the case is insulated from electrical energy. The case is made of a nonconductive material.

Fibrillation: Very rapid irregular contractions of the muscle fibers of the heart that result in the heartbeat and pulse going out of rhythm with each other.

Grounded tool: An electrical tool with a three-prong plug at the end of its power cord or some other means to ensure that stray current travels to ground without passing through the body of the user. The ground plug is bonded to the conductive frame of the tool.

Ground fault circuit interrupter (GFCI): A protective device that functions to de-energize a circuit or portion thereof within an established period of time when a current to ground exceeds some predetermined value. This value is less than that required to operate the overcurrent protective device of the supply circuit.

Polychlorinated biphenyls (PCBs): Toxic chemicals that may be contained in liquids used to cool certain types of large transformers and capacitors.

Additional Resources

This module is intended to present thorough resources for task training. The following reference works are suggested for further study. These are optional materials for continued education rather than for task training.

29 CFR Parts 1900–1910, Standards for General Industry. Occupational Safety and Health Administration, U.S. Department of Labor.

29 CFR Part 1926, Standards for the Construction Industry. Occupational Safety and Health Administration, U.S. Department of Labor.

National Electrical Code® Handbook, Latest Edition. Quincy, MA: National Fire Protection Association.

Standard for Electrical Safety in the Workplace, Latest Edition. Quincy, MA: National Fire Protection Association.

CONTREN® LEARNING SERIES – USER UPDATE

NCCER makes every effort to keep these textbooks up-to-date and free of technical errors. We appreciate your help in this process. If you have an idea for improving this textbook, or if you find an error, a typographical mistake, or an inaccuracy in NCCER's Contren® textbooks, please write us, using this form or a photocopy. Be sure to include the exact module number, page number, a detailed description, and the correction, if applicable. Your input will be brought to the attention of the Technical Review Committee. Thank you for your assistance.

Instructors – If you found that additional materials were necessary in order to teach this module effectively, please let us know so that we may include them in the Equipment/Materials list in the Annotated Instructor's Guide.

Write: Product Development and Revision
National Center for Construction Education and Research
3600 NW 43rd St., Bldg. G, Gainesville, FL 32606

Fax: 352-334-0932

E-mail: curriculum@nccer.org

Craft _____   Module Name _____

Copyright Date _____   Module Number _____   Page Number(s) _____

Description _____

_____

_____

_____

(Optional) Correction _____

_____

_____

(Optional) Your Name and Address _____

_____

_____

Residential
Electrical Services

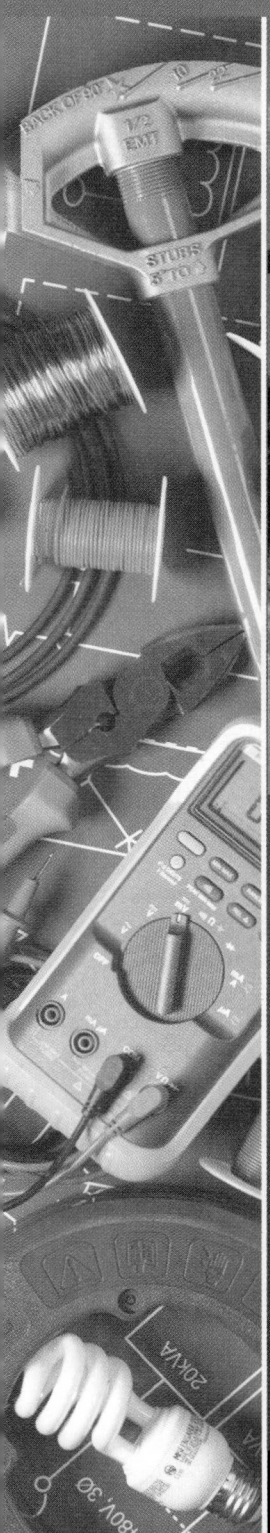

Phoenix Fire Station No. 50

Phoenix's Fire Station No. 50 sports many environmentally protective measures in its construction, such as a roof made of recycled aluminum cans and terra-cotta colored terrazzo flooring made by grinding down the concrete structural slab. Recycled countertops are used in the kitchen and more than 80 percent of the lighting takes advantage of natural sources to save energy. The landscape is a xeriscape design that will require no irrigation after two years.

68113-09

68113-09
Residential Electrical Services

Topics to be presented in this module include:

1.0.0	Introduction	13.2
2.0.0	Sizing the Electrical Service	13.2
3.0.0	Sizing Residential Neutral Conductors	13.8
4.0.0	Sizing the Load Center	13.9
5.0.0	Grounding	13.15
6.0.0	Installing the Service Entrance	13.24
7.0.0	Panelboard Location	13.27
8.0.0	Wiring Methods	13.29
9.0.0	Equipment Grounding System	13.34
10.0.0	Branch Circuit Layout for Power	13.37
11.0.0	Branch Circuit Layout for Lighting	13.45
12.0.0	Outlet Boxes	13.47
13.0.0	Wiring Devices	13.51
14.0.0	Lighting Control	13.52
15.0.0	Electric Heating	13.61
16.0.0	Residential Swimming Pools, Spas, and Hot Tubs	13.64

Overview

The first step in residential wiring is a complete review of the floor plan layout. Electrical floor plans show approximate locations of panels, switches, receptacles, lighting, and other outlets. They do not show, however, the routing of the wiring that interconnects these devices, as this task is generally left up to the electrician.

If the design engineer has not performed the load calculations, the electrician must determine the connected load for the residence, and then size the electrical service accordingly. In order to figure total connected load, certain formulas must be applied based on livable square footage of the house and other factors. Residential electricians must know how to perform load calculations accurately.

Specific wiring methods, grounding requirements, and ground fault circuit interrupting techniques for residences are strictly regulated by the *National Electrical Code®* because building occupants are constantly exposed to the hazards associated with electrical systems and devices. If the *NEC®* and local codes are not strictly followed, the inspector will not sign off on the installation. This will cause construction delays and rework expenses.

Note: *National Electrical Code®* and *NEC®* are registered trademarks of the National Fire Protection Association, Inc., Quincy, MA 02269. All *National Electrical Code®* and *NEC®* references in this module refer to the 2008 edition of the *National Electrical Code®*.

Objectives

When you have completed this module, you will be able to do the following:

1. Explain the role of the *National Electrical Code®* in residential wiring and describe how to determine electric service requirements for dwellings.
2. Explain the grounding requirements of a residential electric service.
3. Calculate and select service-entrance equipment.
4. Select the proper wiring methods for various types of residences.
5. Compute branch circuit loads and explain their installation requirements.
6. Explain the types and purposes of equipment grounding conductors.
7. Explain the purpose of ground fault circuit interrupters and tell where they must be installed.
8. Size outlet boxes and select the proper type for different wiring methods.
9. Describe rules for installing electric space heating and HVAC equipment.
10. Describe the installation rules for electrical systems around swimming pools, spas, and hot tubs.
11. Explain how wiring devices are selected and installed.
12. Describe the installation and control of lighting fixtures.

Trade Terms

Appliance
Bonding bushing
Bonding jumper
Branch circuit
Feeder
Load center
Metal-clad (MC) cable
Nonmetallic-sheathed (Type NM) cable
Romex®
Roughing in
Service drop
Service entrance
Service-entrance conductors
Service-entrance equipment
Service lateral
Switch
Switch leg

Required Trainee Materials

1. Paper and pencil
2. Copy of the latest edition of the *National Electrical Code®*
3. Appropriate personal protective equipment

Prerequisites

Before you begin this module, it is recommended that you successfully complete *Core Curriculum*, and *Construction Technology*, Modules 68101-09 through 68112-09.

This course map shows all of the modules in *Construction Technology*. The suggested training order begins at the bottom and proceeds up. Skill levels increase as you advance on the course map. The local Training Program Sponsor may adjust the training order.

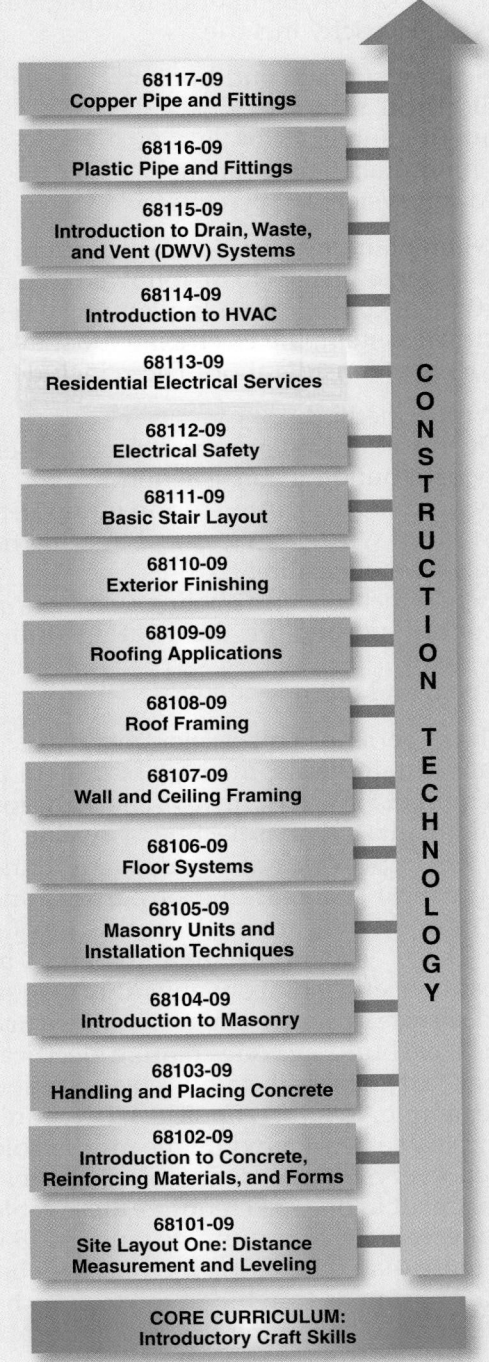

68117-09
Copper Pipe and Fittings

68116-09
Plastic Pipe and Fittings

68115-09
Introduction to Drain, Waste, and Vent (DWV) Systems

68114-09
Introduction to HVAC

68113-09
Residential Electrical Services

68112-09
Electrical Safety

68111-09
Basic Stair Layout

68110-09
Exterior Finishing

68109-09
Roofing Applications

68108-09
Roof Framing

68107-09
Wall and Ceiling Framing

68106-09
Floor Systems

68105-09
Masonry Units and Installation Techniques

68104-09
Introduction to Masonry

68103-09
Handling and Placing Concrete

68102-09
Introduction to Concrete, Reinforcing Materials, and Forms

68101-09
Site Layout One: Distance Measurement and Leveling

CORE CURRICULUM:
Introductory Craft Skills

CONSTRUCTION TECHNOLOGY

113CMAP.EPS

1.0.0 ◆ INTRODUCTION

The use of electricity in houses began shortly after the opening of the California Electric Light Company in 1879 and Thomas Edison's Pearl Street Station in New York City in 1882. These two companies were the first to enter the business of producing and selling electric service to the public. In 1886, the Westinghouse Electric Company secured patents that resulted in the development and introduction of alternating current; this paved the way for rapid acceleration in the use of electricity.

The primary use of early home electrical systems was to provide interior lighting, but today's uses of electricity include:

- Heating and air conditioning
- Electrical appliances
- Interior and exterior lighting
- Communications systems
- Alarm systems

When planning any electrical system, there are certain general steps to be followed, regardless of the type of construction. In planning a residential electrical system, the electrician must take certain factors into consideration. These include:

- Wiring method
- Overhead or underground electrical service
- Type of building construction
- Type of service entrance and equipment
- Grade of wiring devices and lighting fixtures
- Selection of lighting fixtures
- Type of heating and cooling system
- Control wiring for the heating and cooling system
- Signal and alarm systems

The experienced electrician readily recognizes, within certain limits, the type of system that will be required. However, always check the local code requirements when selecting a wiring method. The *NEC*® provides minimum requirements for the practical safeguarding of persons and property from hazards arising from the use of electricity. These minimum requirements are not necessarily efficient, convenient, or adequate for good service or future expansion of electrical use. Some local building codes require electrical installations that surpass the requirements of the *NEC*®. For example, *NEC Section 230.51(A)* requires that service cable be secured by means of cable straps placed every 30 inches. The electrical inspection department in one area requires these cable straps to be placed at a minimum distance of 18 inches.

If more than one wiring method may be practical, a decision as to which type to use should be made prior to beginning the installation.

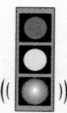

NOTE

See the *Appendix* for other codes and electrical standards that apply to residential electrical installations.

In a residential occupancy, the electrician should know that a 120/240-volt (V), single-phase service entrance will invariably be provided by the utility company. The electrician knows that the service and feeders will be three-wire, that the branch circuits will be either two- or three-wire, and that the safety switches, service equipment, and panelboards will be three-wire, solid neutral. On each project, however, the electrician must consult with the local utility to determine the point of attachment for overhead connections and the location of the metering equipment.

2.0.0 ◆ SIZING THE ELECTRICAL SERVICE

It may be difficult to decide at times which comes first, the layout of the outlets or the sizing of the electric service. In many cases, the service (main disconnect, panelboard, service conductors, etc.) can be sized using the *NEC*® before the outlets are actually located. In other cases, the outlets will have to be laid out first. However, in either case, the service entrance and panelboard locations will have to be determined before the circuits can be installed—so the electrician will know in which direction (and to what points) the circuit homeruns will terminate. In this module, an actual residence will be used as a model to size the electric service according to the latest edition of the *NEC*®.

2.1.0 Floor Plans

A floor plan is a drawing that shows the length and width of a building and the rooms that it contains. A separate plan is made for each floor.

Figure 1 shows how a floor plan is developed. An imaginary cut is made through the building as shown in the view on the left. The top half of this cut is removed (bottom view), and the resulting floor plan is what the remaining structure looks like when viewed directly from above.

The floor plan for a small residence is shown in *Figure 2*. This building is constructed on a concrete slab with no basement or crawl space. There is an unfinished attic above the living area and an open carport just outside the kitchen entrance. Appliances include a 12 kilovolt-ampere (kVA) electric

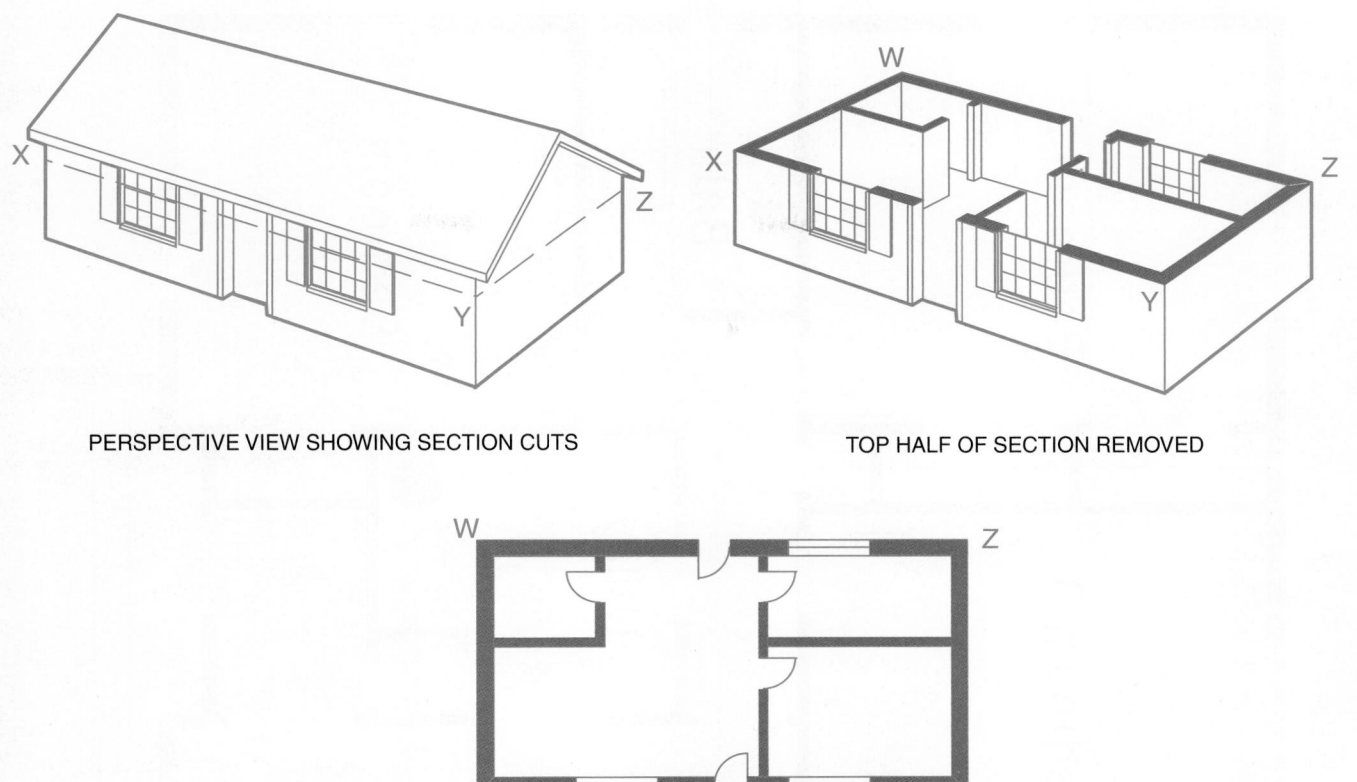

PERSPECTIVE VIEW SHOWING SECTION CUTS

TOP HALF OF SECTION REMOVED

RESULTING FLOOR PLAN IS WHAT THE REMAINING
STRUCTURE LOOKS LIKE WHEN VIEWED FROM ABOVE

111F01.EPS

Figure 1 ◆ Principles of floor plan layout.

range, a 4.5kVA water heater, a ⅓hp 120V disposal, and a 1.5kVA dishwasher.

There is also a washer/dryer (rated at 5.5kVA) in the utility room. A gas furnace with a ⅓hp 120V blower supplies the heating. In this module, the electrical requirements of this example building will be computed.

2.2.0 General Lighting Loads

General lighting loads are calculated on the basis of *NEC Table 220.12.* For residential occupancies, three volt-amperes (watts) per square foot of living space is the figure to use. This includes non-appliance duplex receptacles into which lamps, televisions, etc. may be connected. Therefore, the area of the building must be calculated first. If the building is under construction, the dimensions can be determined by scaling the working drawings used by the builder. If the residence is an existing building with no drawings, actual measurements will have to be made on the site.

Using the floor plan of the residence in *Figure 2* as a guide, an architect's scale is used to measure the longest width of the building (using outside dimensions). It is determined to be 33 feet. The longest length of the building is 48 feet. These two measurements multiplied together give $33 \times 48 = 1,584$ square feet of living area. However, there is an open carport on the lower left of the drawing. This carport area will have to be calculated and then deducted from 1,584 to give the true amount of living space. This open area (carport) is 12 feet wide by 19.5 feet long: $12 \times 19.5 = 234$ square feet. Subtract the carport area from 1,584 square feet: $1,584 - 234 = 1,350$ square feet of living area.

When using the square-foot method to determine lighting loads for buildings, *NEC Section 220.12* requires the floor area for each floor to be computed from the outside dimensions. When calculating lighting loads for residences, the computed floor area must not include open porches, carports, garages, or unused or unfinished spaces that are not adaptable to future use.

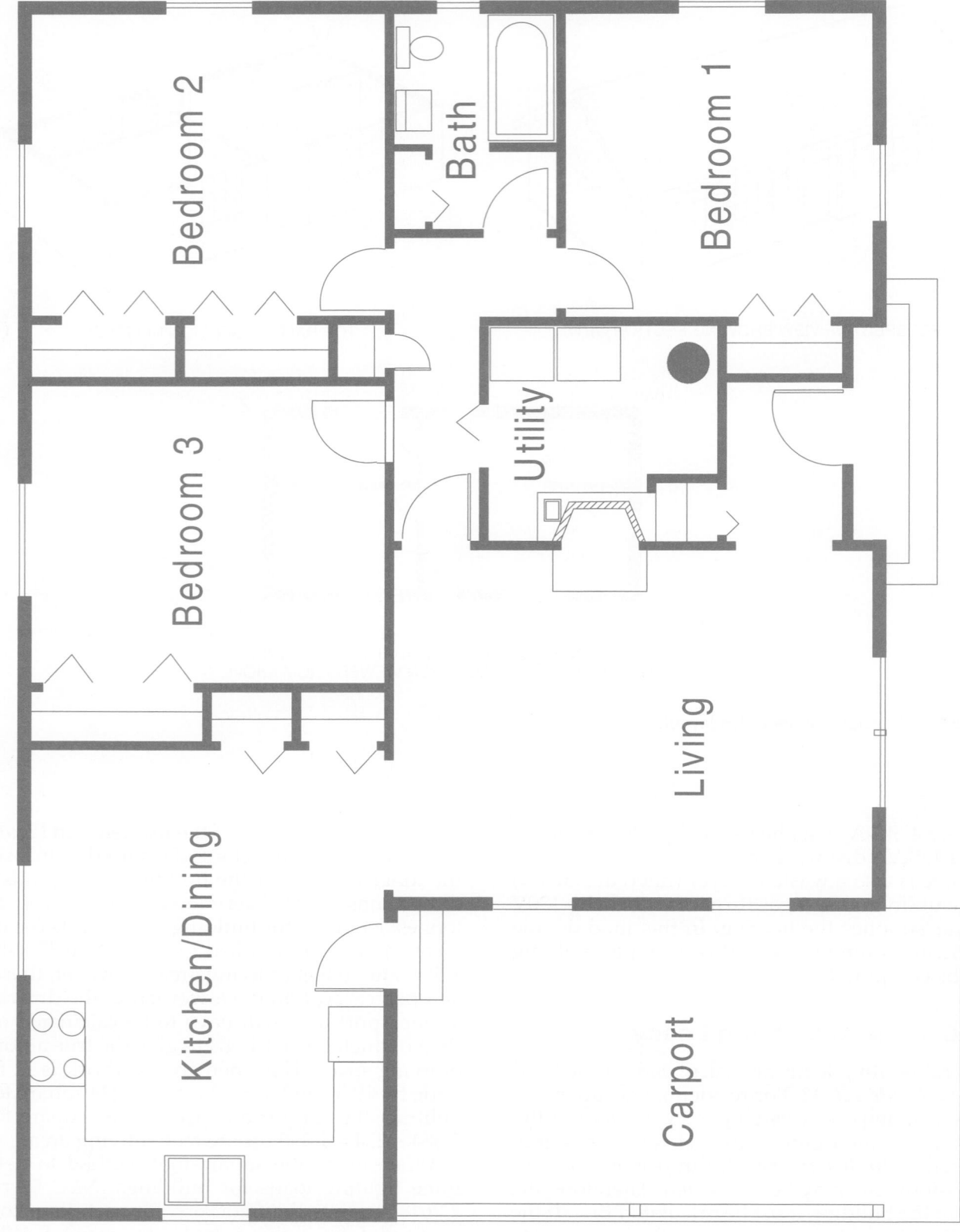

Figure 2 ◆ Floor plan of a typical residence.

111F02.EPS

2.3.0 Calculating the Electric Service Load

Figure 3 shows a standard calculation worksheet for a single-family dwelling. This form contains numbered blank spaces to be filled in while making the service calculation. Using this worksheet as a guide, the total area of our sample dwelling has been previously determined to be 1,350 square feet of living space. This figure is entered in the appropriate space (Box 1) on the form and multiplied by 3 volt-amperes (VA) for a total general lighting load of 4,050VA (Box 2).

2.3.1 Small Appliance Loads

NEC Section 210.11(C)(1) requires at least two 120V, 20A small appliance branch circuits to be installed for the small appliance loads in each kitchen area of a dwelling. Kitchen areas include the dining area, breakfast nook, pantry, and similar areas where small appliances will be used. *NEC Section 220.52* gives further requirements for residential small appliance circuits; that is, the load for those circuits is to be computed at 1,500VA each. Since our example dwelling has only one kitchen area, the number 2 is entered in Box 3 for the number of required kitchen small appliance branch circuits. Multiply the number of these circuits by 1,500 and enter the result in Box 4.

2.3.2 Laundry Circuit

NEC Section 210.11(C)(2) requires an additional 20A branch circuit to be provided for the exclusive use of the laundry area (Box 5). This circuit must not have any other outlets connected except for the laundry receptacle(s). Therefore, enter 1,500VA in Box 6 on the form.

General Lighting Load							Phase	Neutral
Square footage of the dwelling	[1]	1350	× 3VA =	[2]	4050			
Kitchen small appliance circuits	[3]	2	× 1500 =	[4]	3000			
Laundry branch circuit	[5]	1	× 1500 =	[6]	1500			
Subtotal of gen. lighting loads per *NEC Section 220.52* =				[7]	8550			
Subtract 1st 3000VA per *NEC Table 220.42*				[8]	3000	× 100% =	[9] 3000	
Remaining VA times 35% per *NEC Table 220.42*				[10]	5550	× 35% =	[11] 1943	
Total demand for general lighting loads =							[12] 4943	[13]

Fixed Appliance Loads (Nameplate or NEC FLA of motors) per *NEC Section 220.53*				
Hot water tank, 4.5kVA, 240V	[14]	4500		
Dishwasher 1.5kVA, 120V	[15]	1500		
Disposal 1/2HP, 120V per *NEC Table 430.248* = 9.8A	[16]	1176		
Blower 1/3HP, 120V per *NEC Table 430.248* = 7.2A	[17]	864		
	[18]			
	[19]			
Subtotal of fixed appliances	[20]	8040		
If 3 or less fixed appliances take @ 100% =			[21]	[22]
If 4 or more fixed appliances take @ 75% =			[23] 6030	[24]

Other Loads per *NEC Section 220.14*			
Electric Range per *NEC Section 220.55* [neutral @ 70% per *NEC Section 220.61(B)*]		[25] 8000	[26]
Electric Dryer per *NEC Section 220.54* [neutral @ 70% per *NEC Section 220.61(B)*]		[27] 5500	[28]
Electric Heat per *NEC Section 220.51*			
Air Conditioning *NEC Section 220.82(C)*	omit smaller load per *NEC Section 220.60*	[29]	[30]
Largest Motor = 1176	× 25% (per *NEC Section 430.24*) =	[31] 294	[32]
Total VA Demand =		[33] 24767	[34]
(VA divided by 240 volts) **Amps** =		[35] **103**	[36]
Service OCD and minimum size grounding electrode conductor		[37] 125	[38]
AWG per *NEC Sections 310.15(B)(6)* and *Table 310.16* for neutral		[39]	[40]

111F03.EPS

Figure 3 ◆ Calculation worksheet for residential requirements.

So far, there is enough information to complete the first portion of the service calculation form:

- General lighting 4,050VA (Box 2)
- Small appliance load 3,000VA (Box 4)
- Laundry load 1,500VA (Box 6)
- Total general lighting and appliance loads 8,550VA (Box 7)

2.3.3 Lighting Demand Factors

All residential electrical outlets are never used at one time. There may be a rare instance when all the lighting may be on for a short time every night, but even so, all the small appliances and receptacles throughout the house will never be used simultaneously. Knowing this, *NEC Section 220.42* allows a diversity or demand factor to be used when computing the general lighting load for services. Our calculation continues as follows:

- The first 3,000VA is rated at 100% 3,000VA (Box 8)
- The remaining 5,550VA (Box 10) may be rated at 35% (the allowable demand factor) Therefore, 5,550 × 0.35 = 1,943VA (Box 11)
- Net general lighting and small appliance load (rounded off) 4,943VA (Box 12)

2.3.4 Fixed Appliances

NEC Section 220.53 permits the loads for fixed appliances to be computed at 75% as long as they are not electric heating, air conditioning, electric cooking, or electric clothes dryer loads. To compute the load of the fixed appliances in this dwelling, list all the fixed appliances that meet *NEC Section 220.53.* Enter the nameplate rating of the appliance or VA for motors by using *NEC Table 430.248* to find the FLA of each motor. *NEC Section 220.5(A)* tells us to use 120V (not 115V) for calculation purposes. The fixed appliances would be as follows:

- Hot water tank 4,500VA (Box 14)
- Dishwasher 1,500VA (Box 15)
- ½hp 120V disposal (9.8A × 120V) 1,176VA (Box 16)
- Gas furnace blower (7.2A × 120V) 864VA (Box 17)
- Add the loads for the fixed appliances 8,040VA (Box 20)
- Since there are four or more fixed appliances, multiply the total in Box 20 by 75% 6,030VA (Box 23)

2.3.5 Other Loads

The remaining loads of the dwelling are now computed in the Other Loads section in *Figure 3. NEC Section 220.14* allows electric dryers to be computed as permitted in *NEC Table 220.54* and electric cooking appliances to be computed per *NEC Table 220.55.* For a single range rated over 8.75kVA, but not over 12kVA, Column C of *NEC Table 220.55* permits a demand of 8kVA for the range in this dwelling. Enter 8,000VA in Box 25.

The electric dryer must be computed at 5,000VA or the nameplate, whichever is greater, according to *NEC Section 220.54.* Up to four electric dryers must be taken at 100%. Enter 5,500VA in Box 27.

If this dwelling had electric space heating and/or air conditioning, it would be computed in this section using the larger of the two loads. Since they are typical noncoincidental loads, *NEC Section 220.60* permits the smaller of those loads to be omitted. There are no demand factors for either electric heating or air conditioning; therefore, the larger of the two loads would be computed at 100%.

The final step in this calculation is to add in 25% of the largest motor in the dwelling. This dwelling unit has two motors: the disposal at 9.8A and the blower at 7.2A. (See *NEC Section 430.17.*) In this case, the larger motor is the disposal; therefore, we must add 25% of the rating to meet the requirements of *NEC Section 430.24.* Enter 294VA (1,176 × 25%) in Box 31. Adding together the individual loads as computed, we have a minimum demand of 24,767VA (Box 33) for the phase conductors.

2.3.6 Required Service Size

The conventional electric service for residential use is 120/240V, three-wire, single-phase. Services are sized in amperes, and when the volt-amperes are known on single-phase services, amperes may be found by dividing the highest voltage into the total volt-amperes. For example:

24,767VA ÷ 240V = 103A (Box 35)

The **service-entrance conductors** have now been calculated and must be rated at a minimum of 110A, which is a standard rating for overcurrent protection. However, this is not a typical trade size; therefore, we will use the more common rating of 125A as the size of our service.

If the demand for our dwelling unit had resulted in a load of less than 100A, *NEC Section 230.79(C)* would have required that the minimum rating of the service disconnect be 100A. *NEC Section 230.42(B)* would have required the ampacity of the service conductors to be equal to the rating of the 100A disconnect as well.

2.4.0 Demand Factors

NEC Article 220, Part III provides the rules regarding the application of demand factors to certain types of loads. Recall that a demand factor is the maximum amount of volt-amp load expected at any given time compared to the total connected load of the circuit. The maximum demand of a feeder circuit is equal to the connected load times the demand factor. The loads to which demand factors apply can be found in the *NEC*® as follows:

- Lighting loads   *NEC Table 220.42*
- Receptacle loads   *NEC Table 220.44*
- Dryer loads   *NEC Table 220.54*
- Range loads   *NEC Table 220.55*
- Kitchen equipment loads   *NEC Table 220.56*

In addition to those demand factors listed in *NEC Article 220, Part III*, alternative (optional) methods for computing loads can be found in *NEC Article 220, Part IV*. They include the following:

- Dwelling unit loads   *NEC Section 220.82*
- Existing dwelling unit loads   *NEC Section 220.83*
- Multi-family dwelling unit loads   *NEC Section 220.84*

2.5.0 General Lighting and Receptacle Load Demand Factors

NEC Table 220.42 provides the demand factors allowed for various types of lighting situations.

2.6.0 Appliance Loads

NEC Section 210.11(C) provides the number of branch circuits required for small appliances and laundry loads. Demand factors for dryers and ranges are found in *NEC Tables 220.54 and 220.55*.

2.6.1 Small Appliance Loads

The small appliance branch circuits required by *NEC Section 210.11(C)(1)* for small appliances supplied by 15A or 20A receptacles on 20A branch circuits for each kitchen area served are calculated at 1,500VA. If a dwelling has more than one kitchen area, the *NEC*® will require two small appliance branch circuits computed at 1,500VA for each kitchen area served. Where a dwelling with only one kitchen area has more than the required two small appliance branch circuits installed to serve a single kitchen area, only the first two required circuits need be computed. Additional circuits for countertops or refrigeration provide a separation of load, not additional loads. If a dwelling has two kitchen areas, then the total small appliance branch circuits required would be four at 1,500VA each. These loads are permitted to be included with the general lighting load and subjected to the demand factors of *NEC Table 220.42*.

2.6.2 Laundry Circuit Load

A 1,500VA feeder load is added to load calculations for each two-wire laundry branch circuit installed in a home. The branch circuit is required by *NEC Section 210.11(C)(2)*. This load may also be added to the general lighting load and subjected to the same demand factors provided in *NEC Section 220.42*.

2.6.3 Dryer Load

The dryer load for each electric clothes dryer is 5,000VA or the actual nameplate value of the dryer, whichever is larger. Demand factors listed in *NEC Table 220.54* may be applied for more than one dryer in the same dwelling. If two or more single-phase dryers are supplied by a three-phase, four-wire feeder, the total load is computed by using twice the maximum number connected between any two phases.

2.6.4 Range Load

Range loads and other cooking appliances are covered under *NEC Section 220.55*. The feeder demand loads for household electric ranges, wall-mounted ovens, countertop cooking units, and other similar household appliances individually rated over 1¾kW are permitted to be computed in accordance with *NEC Table 220.55*. If two or more single-phase ranges are supplied by a three-phase, four-wire feeder, the total load is computed by using twice the maximum number connected between any two phases.

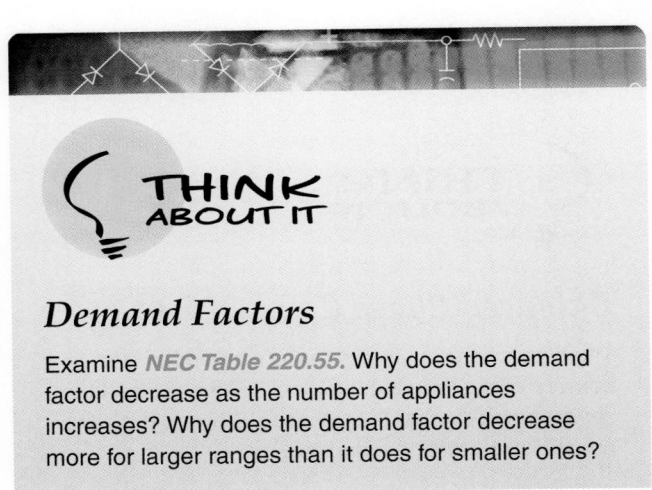

THINK ABOUT IT

Demand Factors

Examine *NEC Table 220.55*. Why does the demand factor decrease as the number of appliances increases? Why does the demand factor decrease more for larger ranges than it does for smaller ones?

2.6.5 Demand Loads for Electric Ranges

Ranges can be computed in various ways that depend on which part of *NEC Article 220* you are using and the occupancy type for the ranges involved. Note the demand factors permitted for the following occupancy types:

- Dwelling units per Part III — *NEC Section 220.55*
- Dwelling units per Part IV — *NEC Section 220.82*
- Additions to existing dwellings per Part IV — *NEC Section 220.83*
- Multi-family dwellings per Part III — *NEC Section 220.55*
- Multi-family dwellings per Part IV — *NEC Section 220.84*
- Restaurant loads per Part III — *NEC Section 220.56*
- Restaurant loads per Part IV — *NEC Section 220.88*

2.7.0 Demand Factors for Neutral Conductors

The neutral conductor of electrical systems generally carries only the maximum current imbalance of the phase conductors. For example, in a single-phase feeder circuit with one phase conductor carrying 50A and the other carrying 40A, the neutral conductor would carry 10A. Since the neutral in many cases will never be required to carry as much current as the phase conductors, the *NEC®* allows us to apply a demand factor. (See *NEC Section 220.61*.) Note that in certain circumstances such as electrical discharge lighting, data processing equipment, and other similar equipment, a demand factor cannot be applied to the neutral conductors because these types of equipment produce harmonic currents that increase the heating effect in the neutral conductor.

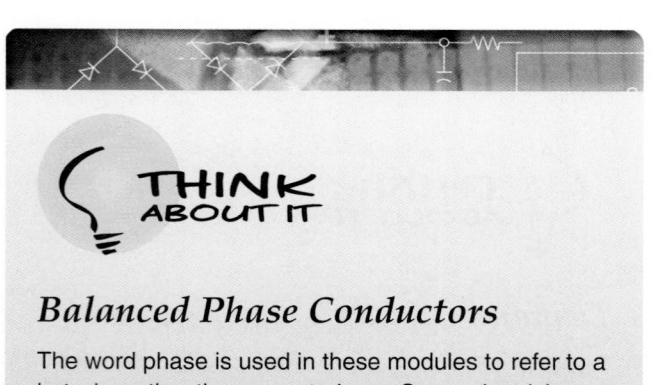

Balanced Phase Conductors

The word phase is used in these modules to refer to a hot wire rather than a neutral one. Some electricians call these legs rather than phases. Why must the two phase conductors be balanced?

3.0.0 ◆ SIZING RESIDENTIAL NEUTRAL CONDUCTORS

The neutral conductor in a three-wire, single-phase service carries only the unbalanced load between the two ungrounded (hot) wires or legs. Since there are several 240V loads in the above calculations, these 240V loads will be balanced and therefore reduce the load on the service neutral conductor. Consequently, in most cases, the service neutral does not have to be as large as the ungrounded (hot) conductors.

In the previous example, the water heater does not have to be included in the neutral conductor calculation, since it is strictly 240V with no 120V loads. This takes the total number of fixed appliances on the neutral conductor down to three appliances. Therefore, each of the fixed appliance loads on the neutral must be computed at 100% (dishwasher at 1,500VA, plus disposal at 1,176VA, plus the blower at 864VA). The neutral loads of the electric range and clothes dryer are permitted by *NEC Section 220.61* to be computed at 70% of the demand for the phase conductors since these appliances have both 120V and 240V loads. In this case, the largest motor is the same for the neutral conductors as it is for the phase conductors; therefore, it is computed in the same manner. Using this information, the neutral conductor may be sized accordingly:

- Net general lighting and small appliance load — 4,943VA (Box 13)
- Fixed appliance loads — 3,540VA (Box 22)
- Electric range (8,000VA × 0.70) — 5,600VA (Box 26)
- Clothes dryer (5,500VA × 0.70) — 3,850VA (Box 28)
- Largest motor — 294VA (Box 32)
- Total — 18,227VA (Box 34)

To find the total phase-to-phase amperes, divide the total volt-amperes by the voltage between phases:

$$18{,}227\text{VA} \div 240\text{V} = 75.9\text{A or } 76\text{A}$$

The service-entrance conductors have now been calculated and are rated at 125A with a neutral conductor rated for at least 76A. See *Figure 4* for a completed calculation form for the example residence.

In *NEC Section 310.15(B)(6)*, special consideration is given to 120/240V, single-phase residential services and feeders. Conductor sizes are shown in *NEC Table 310.15(B)(6)*. Reference to this table shows that the *NEC®* allows a No. 2 AWG copper or a 1/0 AWG aluminum conductor for a 125A service. The neutral conductor is sized per *NEC*

General Lighting Load								Phase	Neutral
Square footage of the dwelling	[1]	1350	× 3VA =	[2]	4050				
Kitchen small appliance circuits	[3]	2	× 1500 =	[4]	3000				
Laundry branch circuit	[5]	1	× 1500 =	[6]	1500				
Subtotal of gen. lighting loads per *NEC Section 220.52* =				[7]	8550				
Subtract 1st 3000VA per *NEC Table 220.42*				[8]	3000	× 100% =	[9]	3000	
Remaining VA times 35% per *NEC Table 220.42*				[10]	5550	× 35% =	[11]	1943	
Total demand for general lighting loads =							[12]	4943	[13] 4943

Fixed Appliance Loads (Nameplate or NEC FLA of motors) per *NEC Section 220.53*				
Hot water tank, 4.5kVA, 240V	[14]	4500		
Dishwasher 1.5kVA, 120V	[15]	1500		
Disposal 1/2HP, 120V per *NEC Table 430.248* = 9.8A	[16]	1176		
Blower 1/3HP, 120V per *NEC Table 430.248* = 7.2A	[17]	864		
	[18]			
	[19]			
Subtotal of fixed appliances	[20]	8040		
If 3 or less fixed appliances take @ 100% =			[21]	[22] 3540
If 4 or more fixed appliances take @ 75% =			[23] 6030	[24]

Other Loads per *NEC Section 220.14*			
Electric Range per *NEC Section 220.55* [neutral @ 70% per *NEC Section 220.61(B)*]		[25] 8000	[26] 5600
Electric Dryer per *NEC Section 220.54* [neutral @ 70% per *NEC Section 220.61(B)*]		[27] 5500	[28] 3850
Electric Heat per *NEC Section 220.51*	omit smaller load per *NEC Section 220.60*	[29]	[30]
Air Conditioning *NEC Section 220.82(C)*			
Largest Motor = 1176	× 25% (per *NEC Section 430.24*) =	[31] 294	[32] 294
Total VA Demand =		[33] 24767	[34] 18227
(VA divided by 240 volts) **Amps** =		[35] **103**	[36] **76**
Service OCD and minimum size grounding electrode conductor		[37] 125	[38] 8 AWG
AWG per *NEC Sections 310.15(B)(6) and Table 310.16* for neutral		[39] 2 AWG	[40] 4 AWG

111F04.EPS

Figure 4 ◆ Completed calculation form.

Tables 310.15(B)(6) or 310.16 using the appropriate column for the markings on the service equipment per *NEC Section 110.14(C)*. Assuming our service panel is marked as suitable for use with 75°C-rated conductors, the minimum size of the neutral would be a No. 4 AWG copper or No. 2 AWG aluminum.

When sizing the grounded conductor for services, the provisions stated in *NEC Sections 215.2, 220.61, and 230.42* must be met, along with other applicable sections.

4.0.0 ◆ SIZING THE LOAD CENTER

Each ungrounded conductor in all circuits must be provided with overcurrent protection in the form of either fuses or circuit breakers. If more than six such devices are used, a means of disconnecting the entire service must be provided using either a main disconnect switch or a main circuit breaker.

To calculate the number of fuse holders or circuit breakers required in the sample residence, look at the general lighting load first. The total general lighting load of 4,050VA can be divided by 120V to find the amperage:

$$4,050VA \div 120V = 33.75A$$

Either 15A or 20A circuits may be used for the lighting load. Two 20A circuits (2 × 20) equal 40A, so two 20A circuits would be adequate for the lighting. However, two 15A circuits total only 30A and 33.75A are needed. Therefore, if 15A circuits are used, three will be required for the total lighting load. In this example, three 15A circuits will be used.

In addition to the lighting circuits, the sample residence will require a minimum of two 20A

circuits for the small appliance load and one 20A circuit for the laundry. So far, the following branch circuits can be counted:

• General lighting load	Three 15A circuits
• Small appliance load	Two 20A circuits
• Laundry load	One 20A circuit
• Total	Six branch circuits

Most **load centers** and panelboards are provided with an even number of circuit breaker spaces or fuse holders (for example, four, six, eight, or ten). But before the panelboard can be selected, space must be provided for the remaining loads. Each 240V load will require two spaces. In some existing installations, you might find a two-pole fuse block containing two cartridge fuses being used to feed a residential electric range. Each 120V load will require one space each. Thus, the remaining number of circuits for this example is as follows:

• Hot water heater	One two-pole breaker
• Dishwasher	One single-pole breaker
• Disposal	One single-pole breaker
• Blower	One single-pole breaker
• Electric range	One two-pole breaker
• Electric dryer	One two-pole breaker

These additional appliances will therefore require an additional nine spaces in the load center or panelboard. *NEC Section 210.11(C)(3)* requires that a separate 20A branch circuit be provided for the bathroom receptacles. While this circuit requires extra space within a load center, it does not add to the demand on the service for a dwelling unit. Adding the nine spaces for the other loads in the dwelling, plus one for a bathroom circuit, to the six required for the general lighting and small appliance loads requires at least a 16-space load center to handle the circuits.

4.1.0 Ground Fault Circuit Interrupters

Under certain conditions, the amount of current it takes to open an overcurrent protective device can be critical. You should remember from the *Electrical Safety* module that when persons are subject to very low current values (less than one full ampere), it can be fatal. The overcurrent protection installed on services, feeders, and branch circuits protects only the conductors and equipment.

Because of this fact, the *NEC*® requires ground fault circuit interrupter (GFCI) protection for receptacle outlets and/or equipment in many locations and occupancies. The *NEC*® defines a GFCI as "a device intended for the protection of personnel that functions to de-energize a circuit or portion thereof within an established period of time when a current to ground exceeds the values established for a Class A device." Class A GFCIs trip when the current to ground has a value in the range of 4mA to 6mA.

For dwelling units, the majority of requirements to provide protection for 15A or 20A, 125V-rated receptacles can be found in *NEC Section 210.8(A).* Further requirements for GFCI protection at dwelling units can be found in other *NEC*® articles such as *NEC Article 590* for temporary construction sites; *NEC Article 620* for special equipment such as elevators; or in *NEC Article 680* for special equipment such as swimming pools, hot tubs, and hydromassage tubs. These articles may also expand the requirements for GFCI protection to include circuits rated at more than 20A or operating at 240V.

According to *NEC Section 210.8(A),* the 15A and 20A, 125V-rated receptacles in our dwelling that require GFCI protection will be those receptacles located in the following areas:

- Bathrooms
- Outdoor receptacles (except those provided on dedicated circuits for snow melting and de-icing equipment)
- Receptacles that serve the countertops in kitchens

Further requirements for GFCI protection at dwelling units are as follows:

- Receptacles within garages and accessory buildings, such as storage sheds or workshops, or similar uses that have a floor located at or below grade level
- Receptacles in unfinished basements
- Crawl spaces at or below grade level
- Receptacles that serve countertops and are within 6' of wet bar sinks, utility, or laundry sinks
- Boathouses

One way to provide this GFCI protection is through the use of a GFCI circuit breaker. GFCI circuit breakers require the same mounting space as standard single-pole circuit breakers and provide the same branch circuit wiring protection as standard circuit breakers. They also provide Class A ground fault protection.

Listed GFCI circuit breakers are available in single- and two-pole construction; 15A, 20A, 25A, and 30A, 50A, and 60A ratings; and have a 10,000A interrupting capacity. Single-pole units are rated at 120VAC; two-pole units are rated at 120/240VAC.

GFCI breakers can be used not only in load centers and panelboards, but they are also available factory-installed in meter pedestals and power outlet panels for recreational vehicle (RV) parks and construction sites.

The GFCI sensor continuously monitors the current balance in the ungrounded or energized (hot) load conductor and the neutral load conductor. If the current in the neutral load wire becomes less than the current in the hot load wire, then a ground fault exists, since a portion of the current is returning to the source by some means other than the neutral load wire. When a current imbalance occurs, the sensor, which is a differential current transformer, sends a signal to the solid-state circuit, which activates the ground trip solenoid mechanism and breaks the hot load connection (*Figure 5*). A current imbalance as low as four milliamps (4mA) will cause the circuit breaker to interrupt the circuit. This is indicated by the trip indicator on the front of the device.

The two-pole GFCI breaker (*Figure 6*) continuously monitors the current balance between the two hot conductors and the neutral conductor. As long as the sum of these three currents is zero, the device will not trip; that is, if the A load wire is carrying 10A of current, the neutral is carrying 5A, and the B load wire is carrying 5A, then the sensor is balanced and will not produce a signal. A current imbalance from a ground fault condition as low as 4mA will cause the sensor to produce a signal of sufficient magnitude to trip the device.

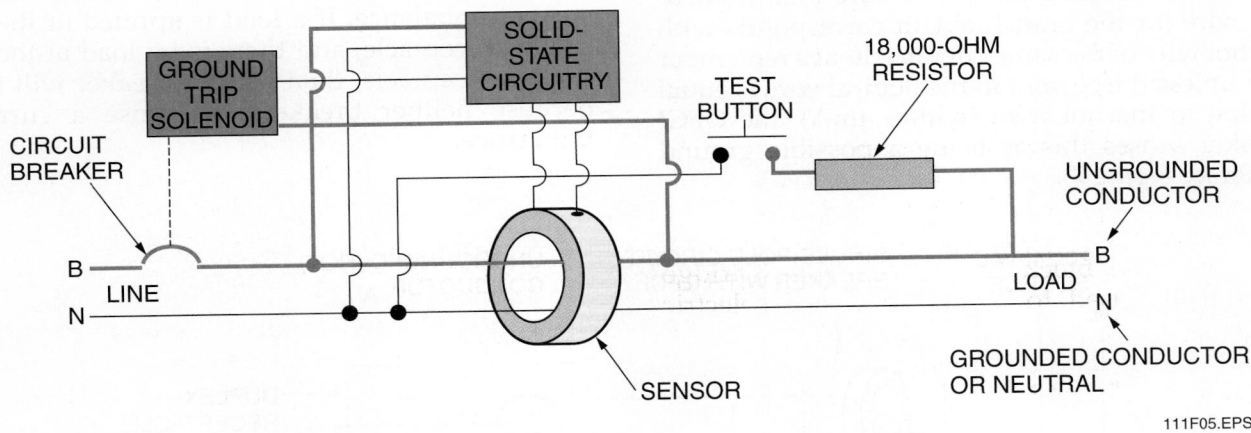

Figure 5 ◆ Operating circuitry of a typical GFCI.

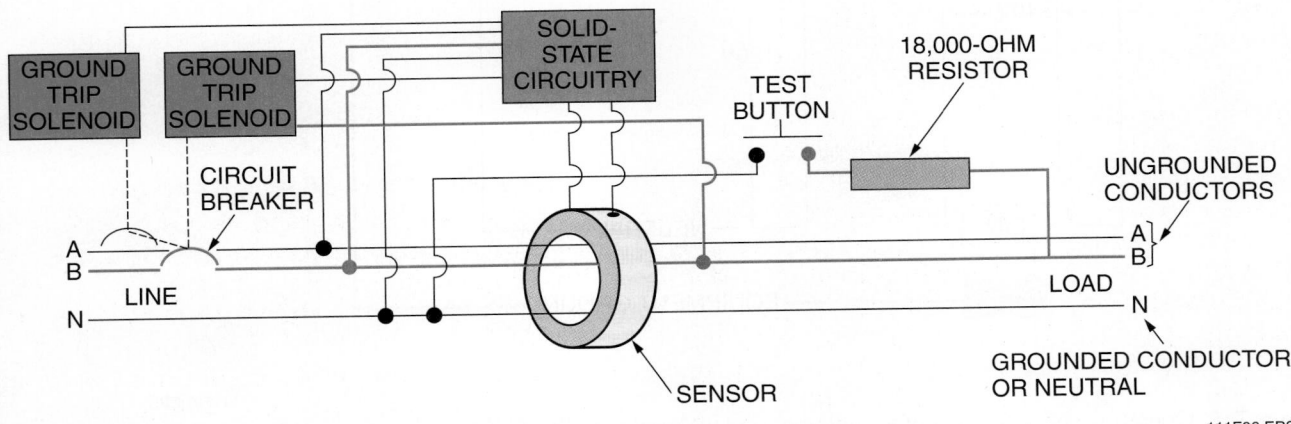

Figure 6 ◆ Operating characteristics of a two-pole GFCI.

4.1.1 Single-Pole GFCI Circuit Breakers

The single-pole GFCI breaker has two load lugs and a white wire pigtail in addition to the line side plug-on or bolt-on connector. The line side hot connection is made by installing the GFCI breaker in the panel just as any other circuit breaker is installed. The white wire pigtail is attached to the panel neutral (S/N) assembly. Both the neutral and hot wires of the branch circuit being protected are terminated in the GFCI breaker. These two load lugs are clearly marked LOAD POWER and LOAD NEUTRAL in the breaker case. Also in the case is the identifying marking for the pigtail, PANEL NEUTRAL.

NOTE

Single-pole GFCI circuit breakers must be installed on independent circuits. They cannot be used on multi-wire circuits.

Care should be exercised when installing GFCI breakers in existing panels. Be sure that the neutral wire for the branch circuit corresponds with the hot wire of the same circuit. Always remember that unless the current in the neutral wire is equal to that in the hot wire (within 4mA), the GFCI breaker senses this as being a possible ground fault (see *Figure 7*).

4.1.2 Two-Pole GFCI Circuit Breakers

A two-pole GFCI circuit breaker can be installed on a 120/240VAC single-phase, three-wire system; the 120/240VAC portion of a 120/240VAC three-phase, four-wire system; or the two phases and neutral of a 120/208VAC three-phase, four-wire system. Regardless of the application, the installation of the breaker is the same—connections are made to two hot buses and the panel neutral assembly. When installed on these systems, protection is provided for two-wire 240VAC or 208VAC circuits, three-wire 120/240VAC or 120/208VAC circuits, and 120VAC multiwire circuits.

The circuit in *Figure 8* illustrates the problems that are encountered when a common load neutral is used for two single-pole GFCI breakers. Either or both breakers will trip when a load is applied at the #2 duplex receptacle. The neutral current from the #2 duplex receptacle flows through breaker #1; this increase in neutral current through breaker #1 causes an imbalance in its sensor, thus causing it to produce a fault signal. At the same time, there is no neutral current flowing through breaker #2; therefore, it also senses a current imbalance. If a load is applied at the #1 duplex receptacle, and there is no load at the #2 duplex receptacle, then neither breaker will trip because neither breaker will sense a current imbalance.

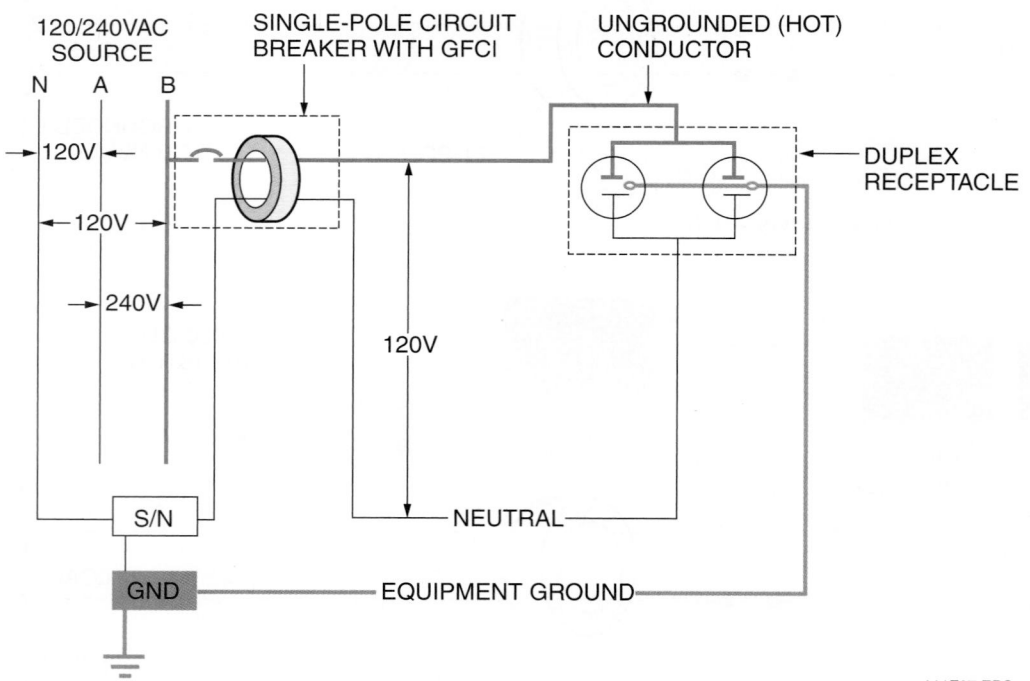

111F07.EPS

Figure 7 ◆ Operating characteristics of a single-pole circuit breaker with a GFCI.

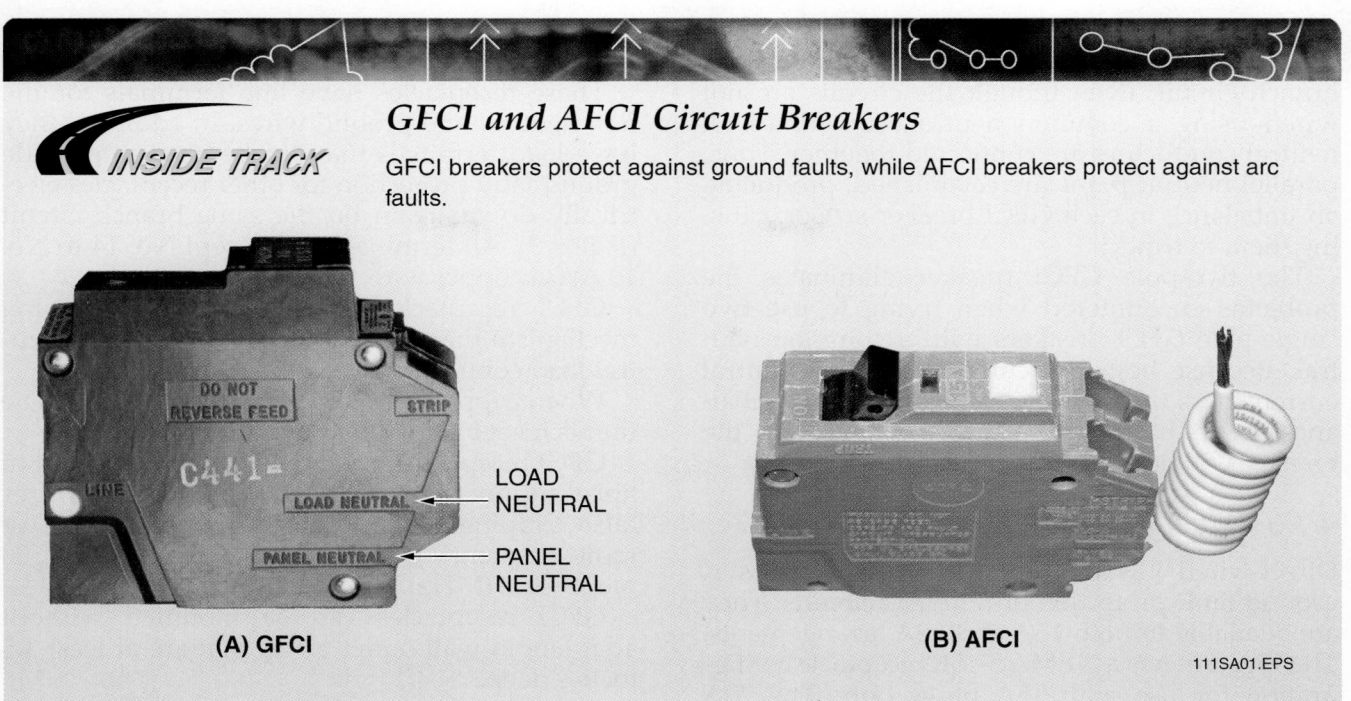

GFCI and AFCI Circuit Breakers

GFCI breakers protect against ground faults, while AFCI breakers protect against arc faults.

LOAD NEUTRAL

PANEL NEUTRAL

(A) GFCI

(B) AFCI

111SA01.EPS

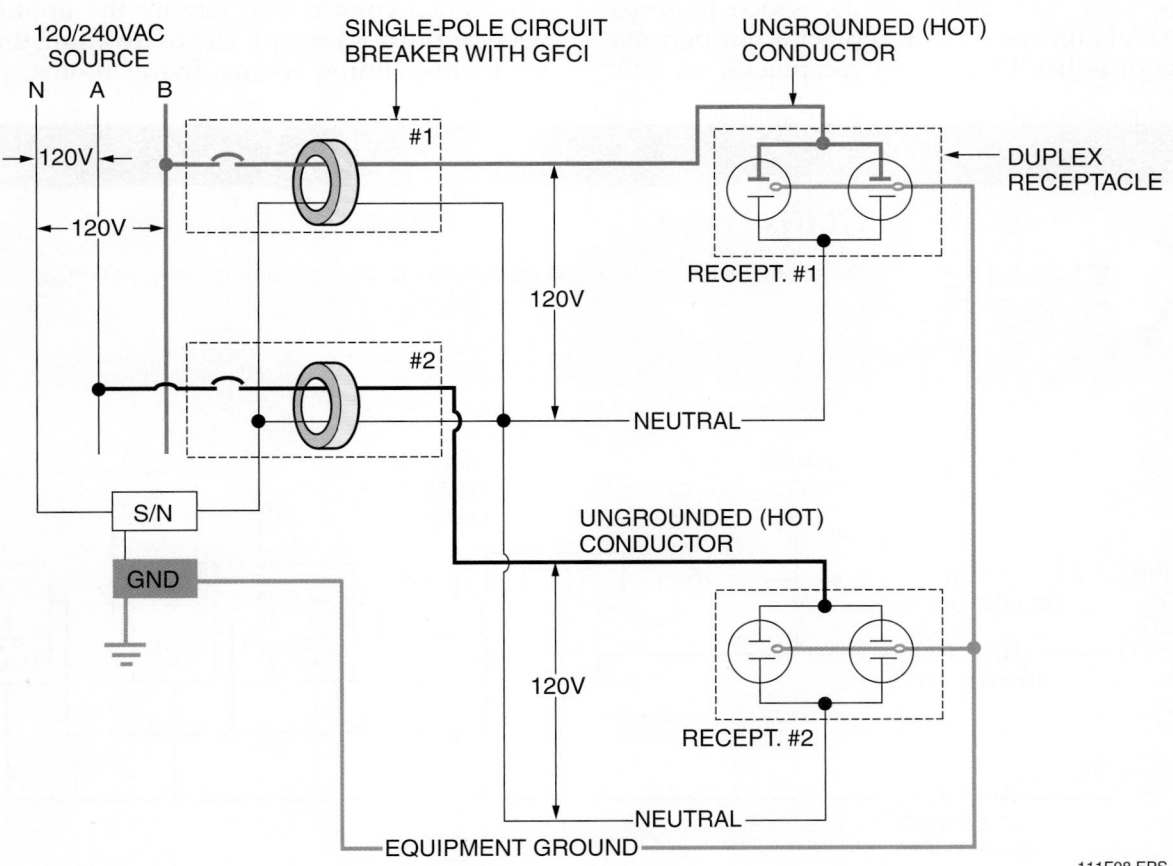

Figure 8 ◆ Circuit depicting the common load neutral.

111F08.EPS

Junction boxes can also present problems when they are used to provide taps for more than one branch circuit. Even though the circuits are not wired using a common neutral, sometimes all neutral conductors are connected together. Thus, parallel neutral paths are established, producing an imbalance in each GFCI breaker sensor, causing them to trip.

The two-pole GFCI breaker eliminates the problems encountered when trying to use two single-pole GFCI breakers with a common neutral. Because both hot currents and the neutral current pass through the same sensor, no imbalance occurs between the three currents, and the breaker will not trip.

4.1.3 Direct-Wired GFCI Receptacles

Direct-wired GFCI receptacles provide Class A ground fault protection on 120VAC circuits. They are available in both 15A and 20A arrangements. The 15A unit has a NEMA 5-15R receptacle configuration for use with 15A plugs only. The 20A device has a NEMA 5-20R receptacle configuration for use with 15A or 20A plugs. Both 15A and 20A units have a 120VAC, 20A circuit rating. This is to comply with *NEC Table 210.24,* which requires that 15A circuits use 15A receptacles but permits the use of either 15A or 20A receptacles on 20A

circuits. Therefore, GFCI receptacle units that contain a 15A receptacle may be used on 20A circuits.

These receptacles have line terminals for the hot, neutral, and ground wires. In addition, they have load terminals that can be used to provide ground fault protection for other receptacles electrically downstream on the same branch circuit (*Figure 9*). All terminals will accept No. 14 to No. 10 AWG copper wire.

GFCI receptacles have a two-pole tripping mechanism that breaks both the hot and the neutral load connections.

When tripped, the RESET button pops out. The unit is reset by pushing the button back in.

GFCI receptacles have the additional benefit of noise suppression. Noise suppression minimizes false tripping due to spurious line voltages or radio frequency (RF) signals between 10 and 500 megahertz (MHz).

GFCI receptacles can be mounted without adapters in wall outlet boxes that are at least 1.5 inches deep.

4.2.0 Arc Fault Circuit Interrupters

All branch circuits that supply the lighting and general-purpose receptacles in dwelling unit family rooms, dining rooms, living rooms, parlors,

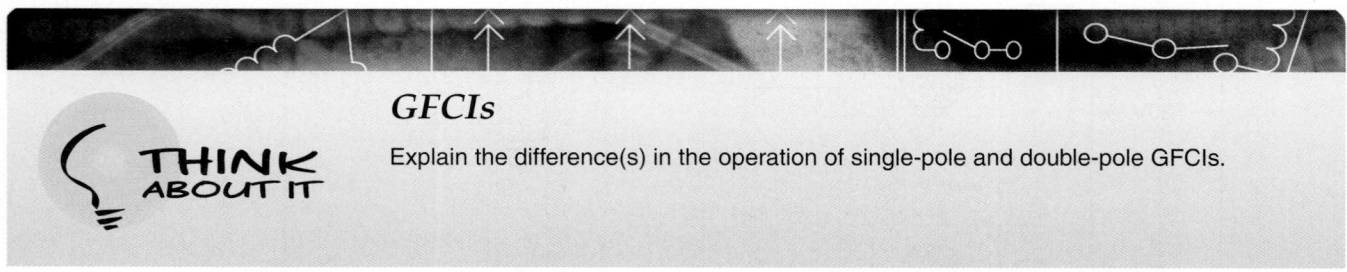

GFCIs

Explain the difference(s) in the operation of single-pole and double-pole GFCIs.

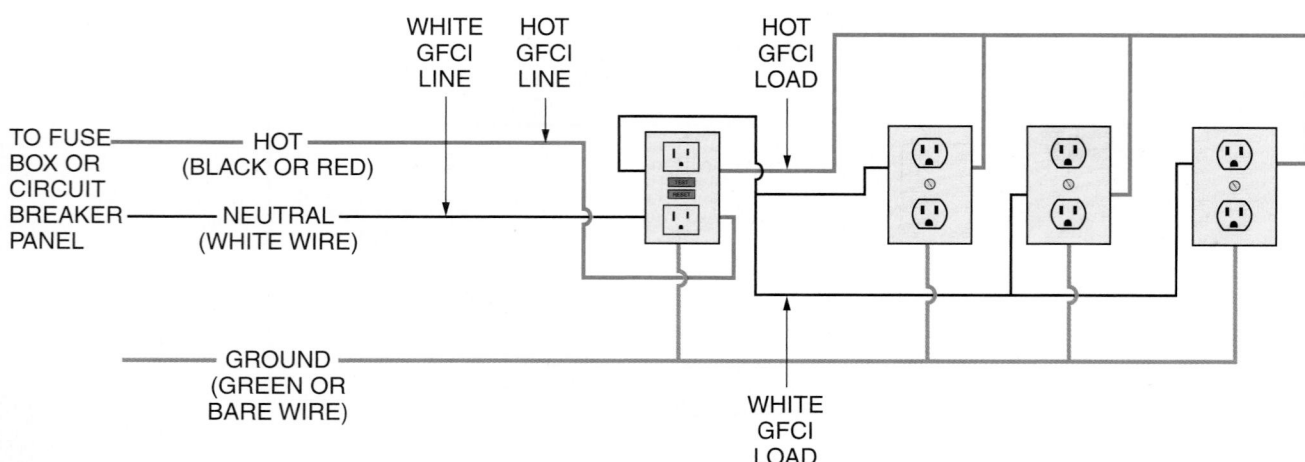

Figure 9 ◆ GFCI receptacle used to protect other outlets on the same circuit.

libraries, dens, bedrooms, sunrooms, recreation rooms, closets, hallways, or similar rooms or areas must have arc fault circuit interrupter protection to comply with *NEC Section 210.12.*

5.0.0 ◆ GROUNDING

NEC Section 250.4(A) provides the general requirements for grounding and bonding of grounded electrical systems. In order to ensure systems are properly grounded and bonded, the prescriptive requirements of *NEC Article 250* must be followed.

The grounding system is a major part of the electrical system. Its purpose is to protect people and equipment against the various electrical faults that can occur. It is sometimes possible for higher-than-normal voltages to appear at certain points in an electrical system or in the electrical equipment connected to the system. Proper grounding ensures that the electrical charges that cause these higher voltages are channeled to the earth or ground and that an effective ground fault path is provided throughout the system so that overcurrent devices will open before people are endangered or equipment is damaged.

The word ground refers to ground potential or earth ground. If a conductor is connected to the earth or some conducting body that serves in place of the earth, such as a driven ground rod (electrode), the conductor is said to be grounded. The neutral conductor in a three- or four-wire service, for example, is intentionally grounded, and therefore becomes a grounded conductor. This is the path back to the source of supply for all ground faults in an electrical system. This conductor is intended not only to carry the unbalanced loads of an installation, but also to provide the low-impedance path back to the source so that enough current will flow in the system to open the overcurrent devices. A wire that is used to connect this neutral conductor to a grounding electrode or electrodes is referred to as a grounding electrode

conductor (GEC). Note the difference in the two meanings: one is grounded, while the other provides a means for grounding.

There are two general classifications of protective grounding:

- System grounding
- Equipment grounding

The system ground relates to the **service-entrance equipment** and its interrelated and bonded components; that is, the system and circuit conductors are grounded to limit voltages due to lightning, line surges, or unintentional contact with higher voltage and to stabilize the voltage to ground during normal operation per *NEC Sections 250.4(A)(1) and (2).*

The noncurrent-carrying conductive parts of materials enclosing electrical conductors or equipment, or forming a part of such equipment, and electrically conductive materials that are likely to become energized are all connected together to the supply source in a manner that establishes an effective ground fault path per *NEC Sections 250.4(A)(3) and (4).*

NEC Section 250.4(A)(5) defines the requirements for an effective ground path. It requires that electrical equipment and wiring and other electrically conductive materials likely to become energized shall be installed in a manner that creates a permanent, low-impedance circuit capable of safely carrying the maximum ground fault current likely to be imposed on it from any point on the wiring system where a ground fault may occur to the electrical supply source. The earth shall not be used as the sole equipment grounding conductor or effective ground fault current path.

To better understand a complete grounding system, a conventional residential system will be examined, beginning at the power company's high-voltage lines and transformer, as shown in *Figure 10.* The pole-mounted transformer is fed with a two-wire, single-phase 7,200V system,

Grounding

Systematic grounding wasn't required by the *NEC*® until the mid-1950s; even then, electricians commonly grounded an outlet by wrapping an uninsulated wire around a cold-water pipe and taping it. Three-hole receptacles with grounding terminals became common in the 1960s, but in many older houses, you cannot assume that receptacles are grounded, even when you see a three-hole receptacle. Sometimes, new receptacles have simply been screwed onto old boxes where there is no equipment grounding conductor.

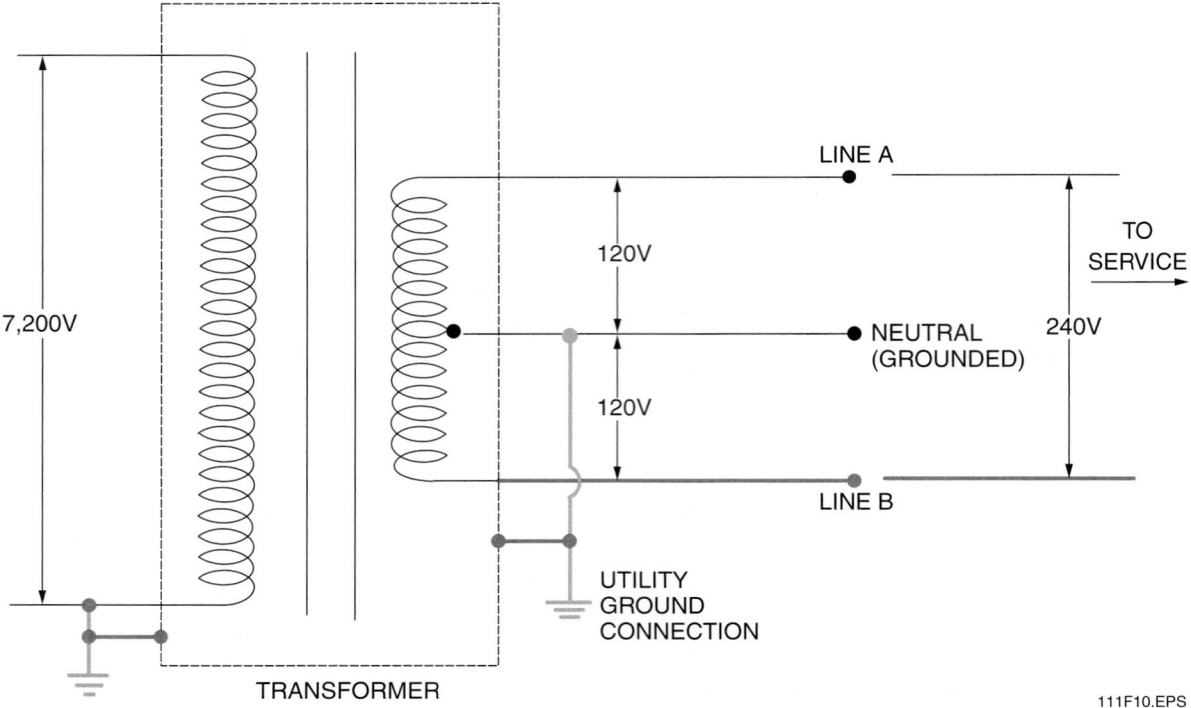

Figure 10 ◆ Wiring diagram of a 7,200V to 120/240V, single-phase transformer connection.

which is transformed and stepped down to a three-wired, 120/240V, single-phase electric service suitable for residential use. Note that the voltage between line A and line B is 240V. However, by connecting a third (neutral) wire on the secondary winding of the transformer—between the other two—the 240V is split in half, providing 120V between either line A or line B and the neutral conductor. Consequently, 240V is available for household appliances such as ranges, hot water heaters, and clothes dryers, while 120V is available for lights and small appliances.

Referring again to *Figure 10*, conductors A and B are ungrounded conductors, while the neutral is a grounded conductor. If only 240V loads were connected, the neutral (grounded conductor) would carry no current. In this instance, the neutral would be used to carry any ground fault currents from the load side of the service back to the utility instead of depending on the earth as the path back to the source. However, since 120V loads are present, the neutral will carry the unbalanced load and become a current-carrying conductor. For example, if line A carries 60A and line B carries 50A, the neutral would carry only 10A (60A − 50A = 10A). This is why the *NEC*® allows the neutral conductor in an electric service to be smaller than the ungrounded conductors. However, *NEC Section 250.24(C)(1)* requires that it must be sufficient to carry fault currents back to the source and, therefore, must not be less than the

required grounding electrode conductor using *NEC Table 250.66* for service conductors up to 1,100 kcmil and not less than 12.5% of the area of the service-entrance conductors (or equivalent) larger than 1,100 kcmil. The typical pole-mounted service drop conductors are normally routed by a messenger cable from a point on the pole to a point on the building being served, terminating at the point where service-entrance conductors exit a weatherhead. Service-entrance conductors are then typically routed through metering equipment into the service disconnecting means. This is the point where most services are grounded. See *Figure 11*. *NEC Section 250.24(A)(1)* requires that the grounding electrode for the structure connection to the neutral (grounded conductor) be at any accessible point from the load end of the service drop or **service lateral** to and including the terminal or bus to which the neutral (grounded service conductor) is connected to the service disconnecting means.

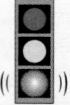

NOTE

Effectively grounded means intentionally connected to earth through one or more ground connection(s) of sufficiently low impedance and having sufficient current-carrying capacity to prevent the buildup of voltages that may result in a hazard to people or connected equipment.

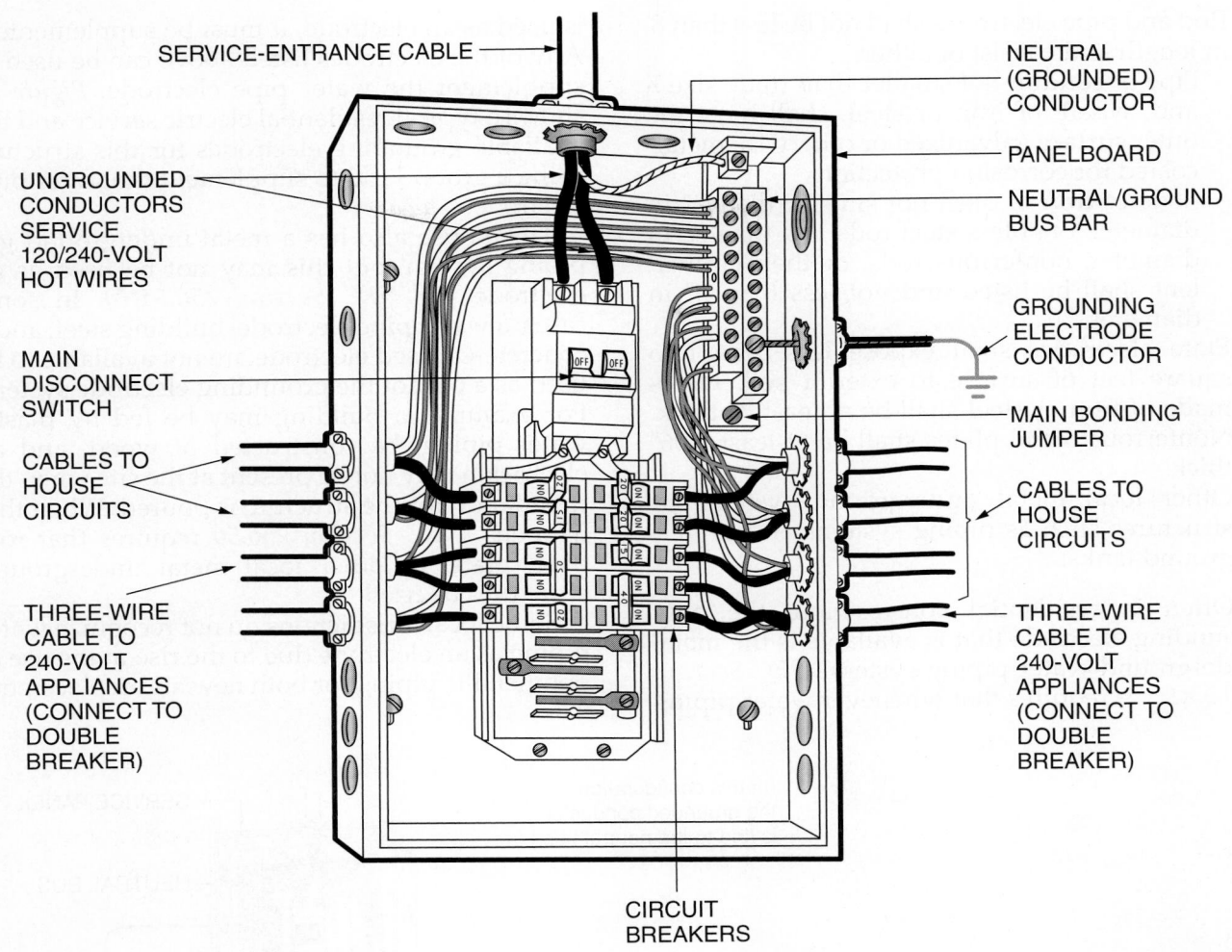

SERVICE-ENTRANCE CABLE

NEUTRAL (GROUNDED) CONDUCTOR

UNGROUNDED CONDUCTORS SERVICE 120/240-VOLT HOT WIRES

PANELBOARD

NEUTRAL/GROUND BUS BAR

MAIN DISCONNECT SWITCH

GROUNDING ELECTRODE CONDUCTOR

CABLES TO HOUSE CIRCUITS

MAIN BONDING JUMPER

THREE-WIRE CABLE TO 240-VOLT APPLIANCES (CONNECT TO DOUBLE BREAKER)

CABLES TO HOUSE CIRCUITS

THREE-WIRE CABLE TO 240-VOLT APPLIANCES (CONNECT TO DOUBLE BREAKER)

CIRCUIT BREAKERS

111F11.EPS

Figure 11 ◆ Interior view of panelboard showing connections.

5.1.0 Grounding Electrodes

NEC Article 250, Part III provides the requirements for connecting an electric service to the grounding electrode system of a building or structure. *NEC Section 250.50* requires, in general, that all of the electrodes described in *NEC Section 250.52(A)* be used (if present), and they must be bonded together to form the grounding electrode system. The electrodes listed in *NEC Section 250.52(A)* are as follows:

• Metal underground water pipe in direct contact with the earth for 10' or more and electrically continuous (or made electrically continuous by bonding around insulating joints or insulating pipe) to the points of connection of the grounding electrode conductor and the bonding conductors. Interior metal water piping located more than 5' from the point of entrance to

the building shall not be used as part of the grounding electrode system or as a conductor to interconnect electrodes that are part of the grounding electrode system.

• Metal frame of the building or structure that complies with *NEC Section 250.52(A)(2)*.

• An electrode encased by at least 2" of concrete may be used if it is located within and near the bottom of a concrete foundation or footing that is in direct contact with the earth. The electrode must be at least 20' long and must be made of electrically conductive coated steel reinforcing bars or rods of not less than ½" in diameter, or consisting of at least 20' of bare copper conductor not smaller than No. 4 AWG wire size.

• A ground ring encircling the building or structure, in direct contact with the earth, consisting of at least 20' of bare copper conductor not smaller than No. 2 AWG.

- Rod and pipe electrodes shall not be less than 8' in length and consist of either:
  - Pipe or conduit not smaller than trade size ¾ and, where of iron or steel, shall have the outer surface galvanized or otherwise metal-coated for corrosion protection.
  - Rods of iron or steel not smaller than ⅝" in diameter. Stainless steel rods less than ⅝" in diameter, nonferrous rods, or their equivalent shall be listed and not less than ½" in diameter.
- Plate electrodes shall expose less than two square feet of surface to exterior soil. Plates made of iron or steel shall be at least ¼" thick. Nonferrous metal plates shall be at least 0.06" thick.
- Other local metal underground systems or structures such as piping systems and underground tanks.

Often in residential construction, the only grounding electrode that is available is the metal underground water piping system. *NEC Section 250.53(D)(2)* requires that whenever water piping is used as an electrode, it must be supplemented. Any of the electrodes listed above can be used to supplement the water pipe electrode. *Figure 12* shows a typical residential electric service and the available grounding electrodes for this structure using a ground rod to supplement the water pipe electrode.

This house also has a metal underground gas piping system, but this may not be used as an electrode per *NEC Section 250.52(B).* In some cases, a water pipe electrode, building steel, and a concrete-encased electrode are not available to be used as a part of the grounding electrode system. For example, a building may be fed by plastic water piping, be constructed of wood, and an electrician may not be present at the site when the foundation for the structure is poured. When that happens, *NEC Section 250.50* requires that rod, pipe, plate, or other local metal underground structures be used.

Some local jurisdictions do not recognize water piping as an electrode due to the rise in the use of nonmetallic piping for both new and replacement

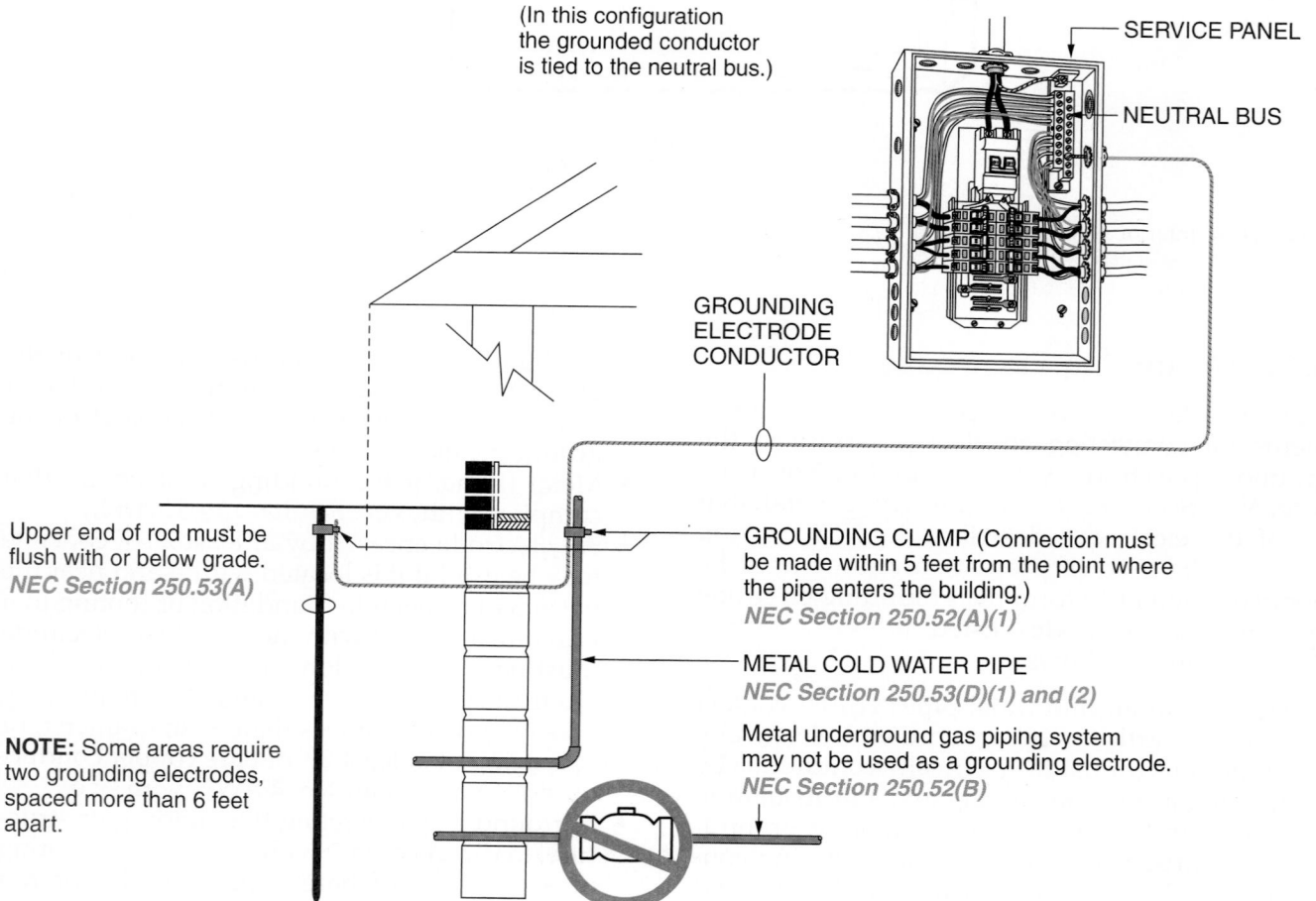

(In this configuration the grounded conductor is tied to the neutral bus.)

SERVICE PANEL

NEUTRAL BUS

GROUNDING ELECTRODE CONDUCTOR

Upper end of rod must be flush with or below grade. *NEC Section 250.53(A)*

GROUNDING CLAMP (Connection must be made within 5 feet from the point where the pipe enters the building.) *NEC Section 250.52(A)(1)*

METAL COLD WATER PIPE *NEC Section 250.53(D)(1) and (2)*

Metal underground gas piping system may not be used as a grounding electrode. *NEC Section 250.52(B)*

NOTE: Some areas require two grounding electrodes, spaced more than 6 feet apart.

111F12.EPS

Figure 12 ◆ Components of a residential grounding system.

water systems. They do not want to rely on the maintained viability of an existing metallic water service and therefore require the use of other electrodes such as the concrete-encased or rod electrodes. This means electricians must be involved with the construction prior to the foundation being poured in order to utilize concrete-encased electrodes.

In most cases, the supplemental electrode used for a water pipe electrode will consist of either a driven rod or pipe electrode, the specifications for which are shown in *Figure 13*.

> **WARNING!**
> A metal underground gas piping system must never be used as a grounding electrode.

5.1.1 Grounding Electrode Installations

NEC Section 250.53(A) requires that rod, pipe, and plate electrodes, where practical, be buried below the permanent moisture level and that they are free from any nonconductive coatings, such as paint or enamel. This section also requires that each electrode system used for a structure be at least 6' from other electrode systems, such as those for lightning protection.

NEC Section 250.53(G) permits a rod or pipe electrode to be driven at a 45-degree angle if rock bottom is encountered and prevents the rod or pipe from being driven vertically for at least 8'. Where driving a rod or pipe electrode at a 45-degree angle will not work, it is permitted to lay a rod or pipe horizontally in a trench that is at least 30" deep. For rod or pipe electrodes longer than 8', it is permitted to have the upper end above ground level if a suitable means of protection is provided for the grounding electrode conductor attachment; otherwise, the upper end must be flush with the earth surface.

NEC Section 250.56 requires that a single rod, pipe, or plate electrode that does not have a resistance to ground of 25Ω or less shall be augmented by one additional electrode of any of the types specified by *NEC Sections 250.52(A)(2) through (7)*. In fact, many local jurisdictions require two electrodes regardless of the resistance to ground.

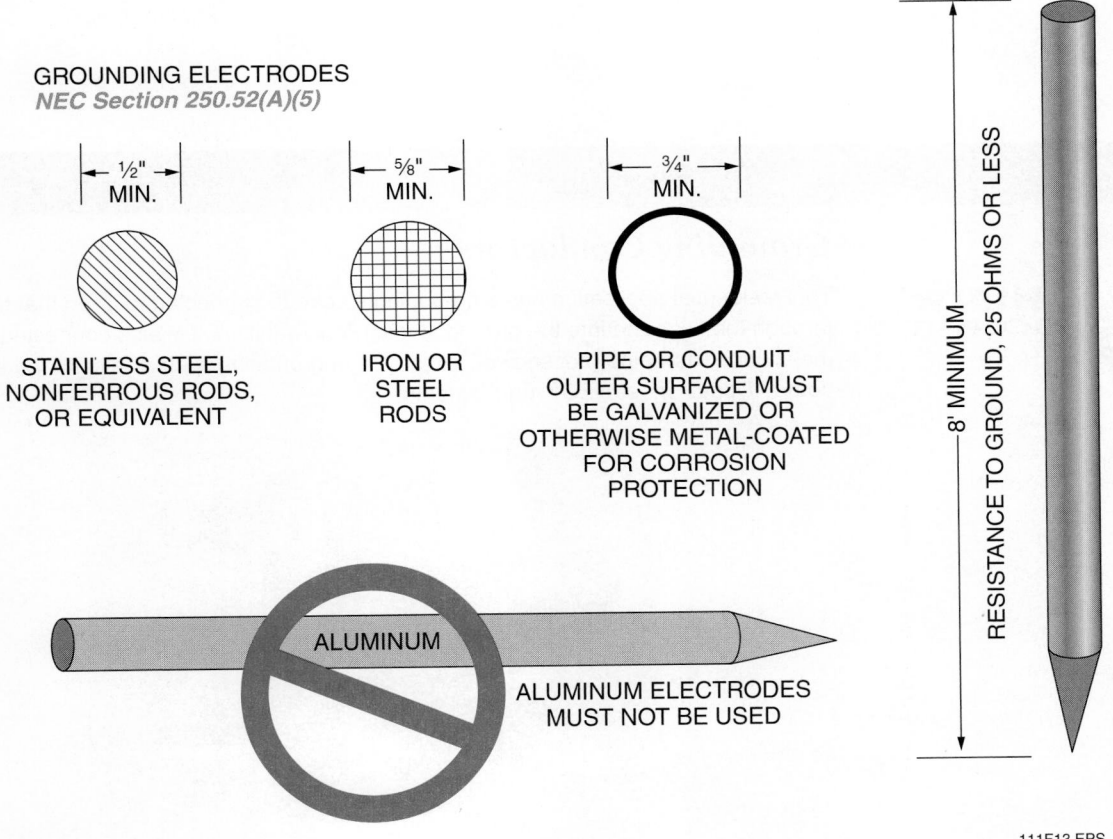

GROUNDING ELECTRODES
NEC Section 250.52(A)(5)

½" MIN.

⅝" MIN.

¾" MIN.

STAINLESS STEEL, NONFERROUS RODS, OR EQUIVALENT

IRON OR STEEL RODS

PIPE OR CONDUIT OUTER SURFACE MUST BE GALVANIZED OR OTHERWISE METAL-COATED FOR CORROSION PROTECTION

8' MINIMUM

RESISTANCE TO GROUND, 25 OHMS OR LESS

ALUMINUM

ALUMINUM ELECTRODES MUST NOT BE USED

111F13.EPS

Figure 13 ◆ Specifications for rod and pipe grounding electrodes.

Always check with the local inspection authority, including the local utility, for rules that surpass the requirements of the *NEC*®.

- Where multiple rod, pipe, or plate electrodes are installed to meet the requirements of this section, they shall not be less than 6' apart.
- Plate electrodes must be buried at least 30" below the surface of the earth, according to *NEC Section 250.53(H)*.
- Where two or more electrodes are effectively bonded together, they are treated as a single electrode system.

5.1.2 Grounding Electrode Conductors (GECs)

The grounding electrode conductor (GEC) connecting the neutral (grounded conductor of the service) at the panelboard neutral bus to the grounding electrodes must meet the requirements of *NEC Section 250.62*. This requires that it be made of copper, aluminum, or copper-clad aluminum. The material selected must be suitably protected against corrosion. The GEC may be either solid or stranded, covered or bare. Note that the GEC is not an equipment grounding conductor, and thus is not required to be identified by the use of the color green or green with yellow stripes, if insulated.

5.1.3 Installation of GECs

NEC Section 250.64 provides the installation requirements for GECs and does not permit bare aluminum or copper-clad aluminum grounding conductors to be used where in direct contact with masonry, the earth, where subject to corrosive conditions, or where used outside within 18" of the earth at the termination point. Other *NEC*® requirements include the following:

- A GEC or its enclosure is required to be securely fastened to the surface on which it is carried.
- A No. 4 AWG or larger copper or aluminum GEC is required to be protected where it will be exposed to severe physical damage.
- A No. 6 AWG or larger GEC that is free from exposure to physical damage is permitted to be run along the surface of the building without metal covering or protection where it is securely fastened to the building. Otherwise, it must be installed in RMC, IMC, RNC, EMT, or a cable armor.
- GECs smaller than No. 6 AWG must be protected by RMC, IMC, RNC, EMT, or cable armor.

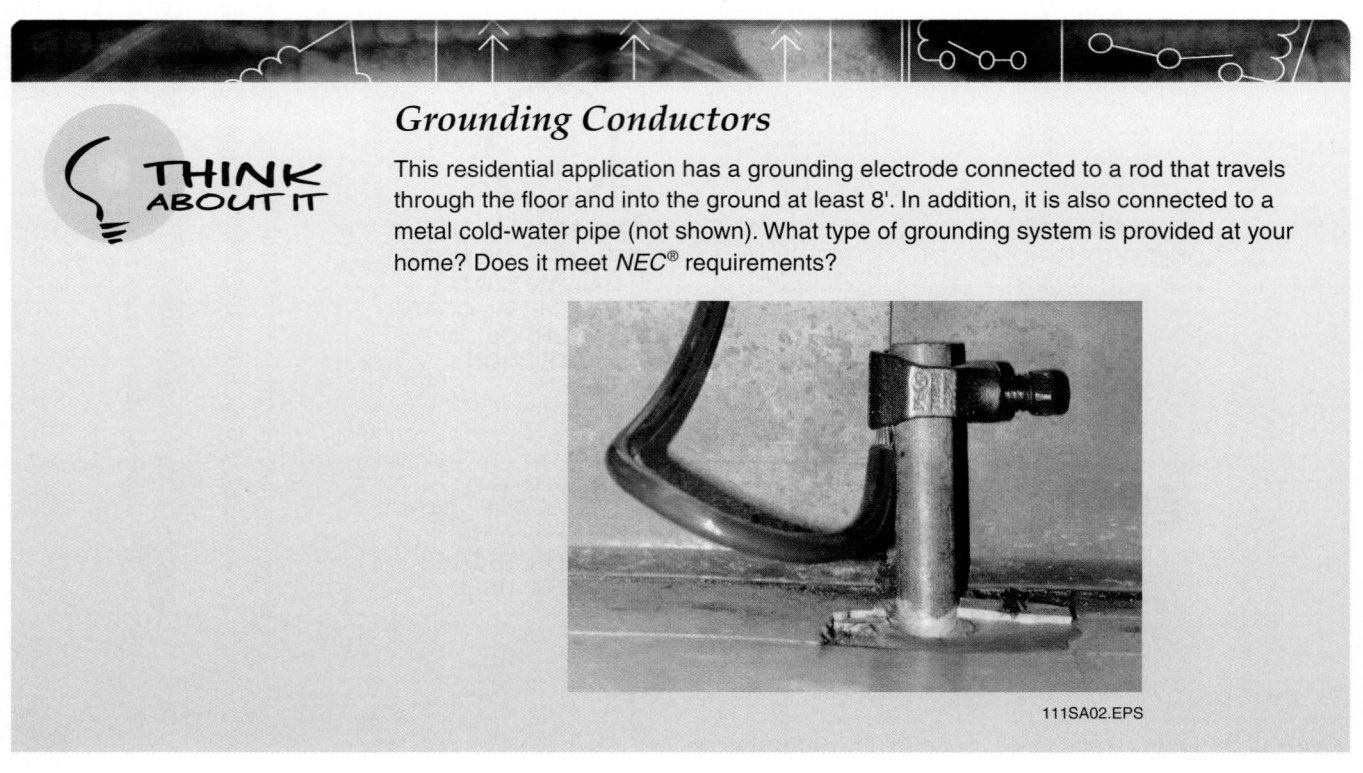

Grounding Conductors

This residential application has a grounding electrode connected to a rod that travels through the floor and into the ground at least 8'. In addition, it is also connected to a metal cold-water pipe (not shown). What type of grounding system is provided at your home? Does it meet *NEC*® requirements?

111SA02.EPS

Grounding Electrode Conductors

THINK ABOUT IT

Which *NEC®* table would you use to size the minimum GEC required for a typical residential service?

- The GEC shall be installed in one continuous length without a splice or joint, unless spliced only by irreversible compression-type connectors listed for the purpose or by an exothermic welding process. Connecting sections of busbars together to form a GEC is not considered to be a splice.
- Where a service consists of more than a single enclosure, it is permissible to connect taps to the GEC, provided each tap extends all the way into the inside of each such enclosure. The tap conductors shall be connected to the GEC in such a manner that the GEC remains without a splice.
- Ferrous metal enclosures for the GEC are required to be electrically continuous from the point of attachment to metal cabinets or metallic equipment enclosures to the GEC. They must also be securely fastened to the ground clamp or fitting.
- Ferrous metal enclosures for the GEC that are not physically continuous from a metal cabinet or metallic equipment enclosure to the grounding electrode must be made electrically continuous by bonding each to the enclosed GEC.
- GECs may be run to any convenient grounding electrode available in the grounding electrode system, or to one or more grounding electrode(s) individually. The GEC shall be sized for the largest grounding electrode conductor required among all the electrodes connected together.

5.1.4 Methods of Connecting GECs

NEC Section 250.70 requires the GEC to be connected to electrodes using exothermic welding, listed pressure connectors, listed clamps, listed lugs, or other listed means. Connections that depend on solder must never be used. To prevent corrosion, the ground clamp must be listed for the material of the grounding electrode and the GEC.

Where used on a pipe, ground rod, or other buried electrodes, the fitting must be listed for direct soil burial or concrete encasement. More than one conductor is not permitted to be connected to the grounding electrode using a single clamp or fitting unless the clamp or fitting is specifically listed for the connection of more than one conductor.

For the connection to an electrode, you must use one of the following:

- A listed, bolted clamp of cast bronze or brass, or plain or malleable iron
- A pipe fitting, pipe plug, or other approved device that is screwed into a pipe or pipe fitting
- For indoor telecommunication purposes only, a listed sheet metal strap-type ground clamp with a rigid metal base that seats on the electrode with a strap that will not stretch during or after installation
- An equally substantial approved means

The connection of a GEC or a **bonding jumper** to a grounding electrode must be accessible unless that connection is to the concrete-encased or buried grounding electrodes permitted in *NEC Section 250.68.* Where it is necessary to ensure the grounding path for metal piping used as a grounding electrode, effective bonding shall be provided around insulated joints and around any equipment likely to be disconnected for repairs or replacement. Bonding conductors shall be of sufficient length to permit removal of such equipment while retaining the integrity of the bond. Coatings on metal piping systems must be removed to ensure that a permanent and effective grounding path is provided.

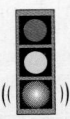

NOTE

The UL listing states that "strap-type ground clamps are not suitable for attachment of the grounding electrode conductor of an interior wiring system to a grounding electrode."

For the example house, the point of connection to the water piping is shown in *Figure 12* and would be required to be accessible after any wall coverings are installed. Any nonconductive coatings on the water piping would also have been scraped off or removed prior to installing the clamp on the water pipe.

5.1.5 Sizing GECs

Grounding electrode conductors must be sized per *NEC Section 250.66,* which uses the area of the largest service-entrance conductor (or equivalent area for paralleled conductors). Except as noted below, *NEC Table 250.66* will provide the minimum size GEC and any bonding jumpers used to interconnect grounding electrodes used.

- Where connected to rod, pipe, or plate electrodes, that portion of the GEC that is the sole connection to the grounding electrode shall not be required to be larger than No. 6 AWG copper or No. 4 AWG aluminum.
- Where connected to a concrete-encased electrode, that portion of the GEC that is the sole connection to the grounding electrode shall not be required to be larger than No. 4 AWG copper wire.
- Where the GEC is connected to a ground ring, that portion of the conductor that is the sole connection to the grounding electrode shall not be required to be larger than the conductor used for the ground ring.
- Where multiple sets of service-entrance conductors are used as permitted in *NEC Section 230.40, Exception 2,* the equivalent size of the largest service-entrance conductor is required to be determined by the largest sum of the areas of the corresponding conductors of each set.
- Where there are no service-entrance conductors, the GEC size is required to be determined by the equivalent size of the largest service-entrance conductor required for the load to be served.

For our sample dwelling unit, the size of the service-entrance conductors is No. 2 AWG. Using *NEC Table 250.66,* we can determine that the size of the conductor coming from the service panel to the water pipe (the GEC) must be at least No. 8 AWG copper. This No. 8 AWG may continue on without a splice to the ground rod, as shown in *Figure 12,* or a separate No. 6 AWG could be installed for the ground rod(s). This conductor would have to be connected to the service panel as an individual run or with a separate connector to any portion of the No. 8 AWG. It may not be connected directly to the water piping.

5.1.6 Air Terminals

Air terminal conductors and driven pipes, rods, or plate electrodes used for grounding air terminals are not permitted to be used in lieu of the grounding electrodes covered in *NEC Section 250.50* for grounding wiring systems and equipment. However, *NEC Section 250.106* requires that they be bonded to the wiring and equipment grounding electrode system for the structure.

5.2.0 Main Bonding Jumper

NEC Section 250.24(B) requires that an unspliced main bonding jumper (MBJ) shall be used to connect the equipment grounding conductor(s) and the service disconnect enclosure to the grounded conductor (neutral) of the system within the enclosure of each service disconnect.

The MBJ must be of copper or other corrosion-resistant material. An MBJ may be in the form of a wire, bus, screw, or similar suitable conductor.

Where an MBJ is in the form of a screw, it is required to be identified with a green finish so that the head of the screw is visible for inspection. An MBJ must be attached using exothermic welding, a listed pressure connector, listed clamp, or other listed means.

The MBJ cannot be smaller than the sizes given in *NEC Table 250.66* for grounding electrode conductors. See *NEC Section 250.28(D)* for service conductors that exceed 1,100 kcmil.

The MBJ is the means by which any ground fault in the branch circuits and feeders of the electrical system travels back to the source of supply at the utility. A ground fault will travel along the equipment conductors of the circuits back to the service disconnecting means. Where metallic

INSIDE TRACK

What Does a Lightning Rod Do?

An interesting fact about grounding is that a lightning rod (air terminal) isn't meant to bring a bolt of lightning to ground. To do this, its conductors would have to be several feet in diameter. The purpose of the rod is to dissipate the negative static charge that would cause the positive lightning charge to strike the house.

raceways are used as equipment grounding conductors, there will be no connection to the grounded conductor at the service. Without the MBJ, the path back to the source would be through the grounding electrode system and the earth. This does not provide a low-impedance path, and thus will not allow enough current to flow in the circuit to let the overcurrent devices open.

For example, suppose a phase conductor makes contact with the metallic housing of a 120V, 15A appliance that was wired using EMT. Further suppose that the total combined resistance of the EMT being used as the equipment grounding conductor connected to the appliance and the resistance of the metal water piping and our ground rod in the sample house is 20Ω (less than the 25Ω permitted in *NEC Section 250.56*). The amount of current that could flow back to the utility source would be 120V ÷ 20 = 6A. The smallest overcurrent device in our electrical system is 15A, and would not trip. With the MBJ installed, the path back to the utility source is through the MBJ to the grounded conductor of the service. This resistance will be much less than 1Ω, and thus would allow enough current to flow to open up the overcurrent devices within the system. The MBJ provides the path back to the source for faults that occur within the service disconnect means.

5.2.1 Bonding at the Service

Electrical continuity is required at the service per *NEC Section 250.92(A)*, which states that all of the following must be bonded:

- The service raceways, auxiliary gutters, or service cable armor or sheaths, except for underground metallic sheaths of continuously underground cables as noted in *NEC Section 250.84*
- All service enclosures containing service-entrance conductors, including meter fittings, boxes, or the like interposed in the service raceway or armor
- Any metallic raceway or armor enclosing a grounding electrode conductor as specified in *NEC Section 250.64(E)*

Bonding shall apply at each end and to all intervening raceways, boxes, and enclosures between the service equipment and the grounding electrode.

The items that typically require bonding include the mast and weatherhead, the meter enclosure, the armor of the SE cable (if it has armor), and the service disconnect.

5.2.2 Methods of Bonding at the Service

The electrical continuity of the service equipment, raceways, and enclosures will be ensured per *NEC Section 250.92(B)* through the use of the following methods:

- Bonding equipment to the grounded service conductor in a manner provided in *NEC Section 250.8*
- Connections utilizing threaded couplings or threaded bosses on enclosures where made up wrenchtight
- Threadless couplings and connectors where made up tight for metal raceways and metal-clad cables
- Other approved devices, such as bonding-type locknuts and **bonding bushings**

Bonding jumpers must be used around concentric or eccentric knockouts that are punched or otherwise formed so as to impair the electrical connection to ground. Standard locknuts or bushings shall not be the sole means for bonding.

5.2.3 Bonding and Grounding Requirements for Other Systems

An accessible means external to the service equipment enclosure is required for connecting intersystem bonding and grounding conductors and connections for the communications, radio and television (TV), community antenna television (CATV), and network-powered broadband communication system. The intersystem bonding termination must consist of at least three terminals and be installed in accordance with *NEC Section 250.94*. Any one of the following can be used:

- A set of listed terminals securely mounted to and electrically connected to the meter enclosure
- A bonding bar near the service-entrance enclosure, meter enclosure, or raceway for service conductors connected to an equipment grounding conductor in the enclosure or raceway with a minimum No. 6 AWG copper conductor
- A bonding bar near the grounding electrode conductor connected with a minimum No. 6 AWG copper conductor

5.2.4 Bonding of Water Piping Systems

Metallic water piping systems in or on a structure must be bonded as required by *NEC Section 250.104(A)*. The metallic water piping system(s) must be bonded by means of a bonding jumper sized in accordance with *NEC Table 250.66* and connected to one of the following:

- The service-entrance enclosures
- The grounded (neutral) conductor at the service
- The grounding electrode conductor where of sufficient size
- The grounding electrode(s) used

The points of attachment of the bonding jumper(s) shall be accessible. It shall be installed in accordance with *NEC Section 250.64(A), (B), and (E).* Note that while this conductor is sized in the same manner as if the water piping system is a grounding electrode, the point of attachment to the water piping is permitted to be at any convenient point on the water piping system and not just within the first 5' of where the water enters the building.

NEC Section 250.104(A)(2) states that in multifamily dwelling units (or other multiple occupancy buildings) where the metal water piping system(s) installed in or attached to a building or structure for the individual occupancies is metallically isolated from all other occupancies by use of nonmetallic water piping, the metal water piping system(s) for each occupancy shall be permitted to be bonded to the equipment grounding terminal of the panelboard or switchboard enclosure (other than service equipment) supplying that occupancy. The bonding jumper shall be sized in accordance with *NEC Table 250.122.*

5.2.5 Bonding of Other Piping Systems

NEC Section 250.104(B) requires that other piping systems, where installed in or attached to a building or structure, including gas piping, that may become energized shall be bonded to one of the following:

- The service equipment enclosure
- The grounded conductor at the service
- The grounding electrode conductor where of sufficient size
- One or more grounding electrodes used

The bonding jumper(s) shall be sized in accordance with *NEC Table 250.122* using the rating of the circuit that may energize the piping system(s). The equipment grounding conductor for the circuit that may energize the piping shall be permitted to serve as the bonding means. The points of attachment of the bonding jumper(s) shall be accessible.

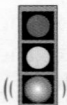

NOTE

Bonding all piping and metal air ducts within the premises will provide additional safety.

6.0.0 ◆ INSTALLING THE SERVICE ENTRANCE

In practical applications, the electric service is normally one of the last components of an electrical system to be installed. However, it is one of the first considerations when laying out a residential electrical system. For instance:

- The electrician must know in which direction and to what location to route the circuit home-runs while **roughing in** the electrical wiring.
- Provisions must be made for sleeves through footings and foundations in cases where underground systems (service laterals) are used.
- The local power company must be notified as to the approximate size of service required so they may plan the best way to furnish a **service drop** to the property.

6.1.0 Service Drop Locations

The location of the service drop, electric meter, and load center should be considered first. It is always wise to consult the local power company to obtain their requirements; where you want the service drop and where they want it may not coincide. A brief meeting with the power company about the location of the service drop can prevent problems later on.

The service drop must be routed so that the service drop conductors have a clearance of not less than 3' horizontally and below windows that open, doors, porches, fire escapes, or similar locations. In addition, they must have a 10' vertical clearance that extends 3' horizontally from porches, fire escapes, balconies, and so forth, as required in *NEC Section 230.9.* Where service drop conductors pass over rooftops, driveways, yards, and so forth, they must have clearances as specified in *NEC Section 230.24.*

A plot plan (also called a site plan) is often available for new construction. The plot plan shows the entire property, with the building or buildings drawn in their proper location on the plot of land. It also shows sidewalks, driveways, streets, and existing utilities—both overhead and underground.

A plot plan of the sample residence is shown in *Figure 14*. In reviewing this drawing, you can see that the closest power pole is located across a public street from the house. By consulting with the local power company, it is learned that the service will be brought to the house from this pole by triplex cable, which will connect to the residence at a point on its left (west) end. The steel uninsulated conductor of triplex cable acts as both the grounded conductor (neutral) and as a support for the insulated (ungrounded) conductors. It is also suitable for overhead use.

When service-entrance cable is used, it will run directly from the point of attachment and service head to the meter base. However, since the carport is located on the west side of the building, a service mast *(Figure 15)* will have to be installed.

The *NEC*® requires a clearance of not less than 8' over rooftops, unless the roof has a slope of 4" in 12" or greater, in which case the clearance may be reduced to 3'. Where the service drop conductors pass over only the overhang (eaves) of a roof, the clearance may be reduced to 18" as long as no more than 6' of the conductors travel over no more than 4' of the overhang (eave). This minimum height requirement extends beyond the roof for a distance of not less than 3' in all directions, except the final portion of the span where the service drop conductors attach to the sides of a building.

6.2.0 Vertical Clearances of Service Drop

NEC Section 230.24(B) specifies the distances by which service drop conductors must clear the ground. These distances vary according to the surrounding conditions.

In general, the *NEC*® states that the vertical clearances of all service drop conductors that carry 600V or under are based on a conductor temperature of 60°F (15°C) with no wind and with the final unloaded sag in the wire, conductor, or cable. Service drop conductors must be at least 10' above the ground or other accessible surfaces at all times. More distance is required under most conditions. For example, if the service conductors pass over residential property and driveways or commercial property that is not subject to truck traffic, the conductors must be at least 15' above the ground. However, this distance may be reduced to 12' when the voltage is limited to 300V to ground.

In other areas, such as public streets, alleys, roads, parking areas subject to truck traffic, driveways on other-than-residential property, the minimum vertical distance is 18'. The conditions of the sample residence are shown in *Figure 16*.

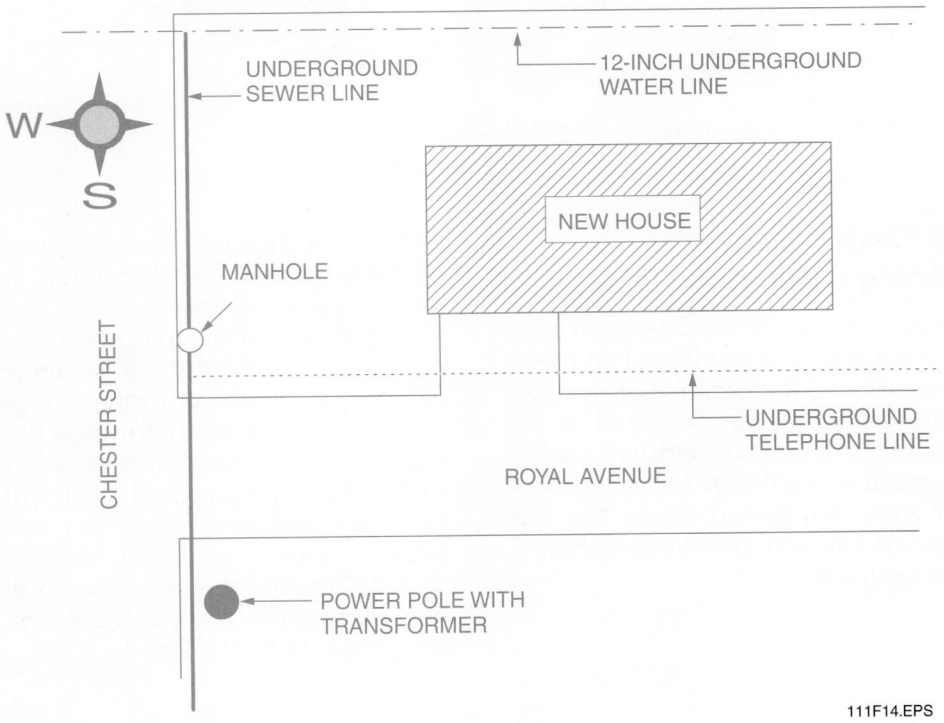

Figure 14 ◆ Plot plan of the sample residence.

111F14.EPS

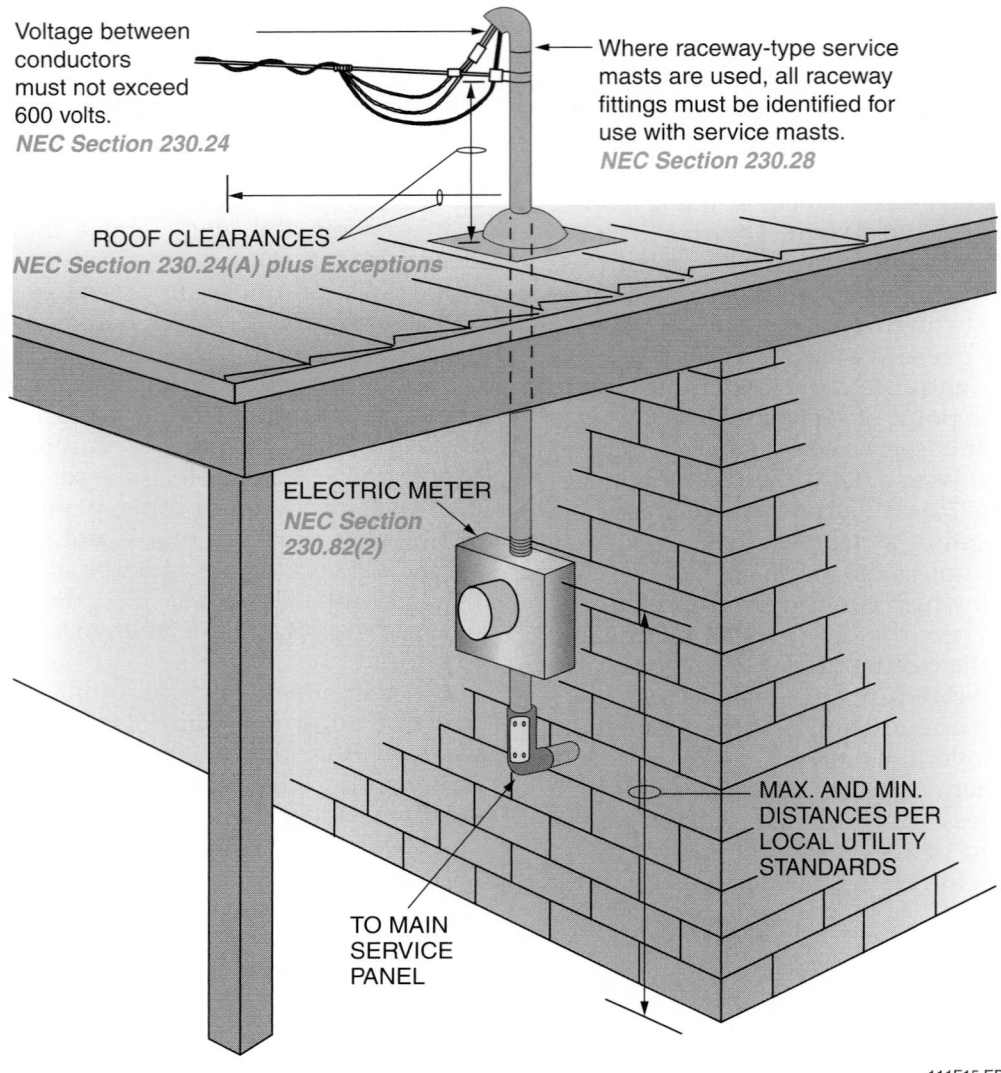

Voltage between conductors must not exceed 600 volts.
NEC Section 230.24

Where raceway-type service masts are used, all raceway fittings must be identified for use with service masts.
NEC Section 230.28

ROOF CLEARANCES
NEC Section 230.24(A) plus Exceptions

ELECTRIC METER
NEC Section 230.82(2)

MAX. AND MIN. DISTANCES PER LOCAL UTILITY STANDARDS

TO MAIN SERVICE PANEL

111F15.EPS

Figure 15 ◆ *NEC*® sections governing service mast installations.

6.3.0 Service Drop Clearances for Building Openings

Service conductors that are installed as open conductors or multiconductor cable without an overall outer jacket must have a clearance of not less than 3' from windows that are designed to be opened, doors, porches, balconies, ladders, stairs, fire escapes, or similar locations (*NEC Section 230.9*). However, conductors run above the top level of a window are permitted to be less than 3' from the window opening.

The 3' of clearance is not applicable to raceways or cable assemblies that have an overall outer jacket approved for use as a service conductor. The intention of this requirement is to protect the conductors from physical damage and/or physical contact with unprotected personnel when evacuating a structure through the window opening. The exception allows service conductors, including drip loops and service drop conductors, to be located just above the window openings because they would not interfere with ladders leaning against the structure to the right, left, or below the window opening when used to evacuate people from the building.

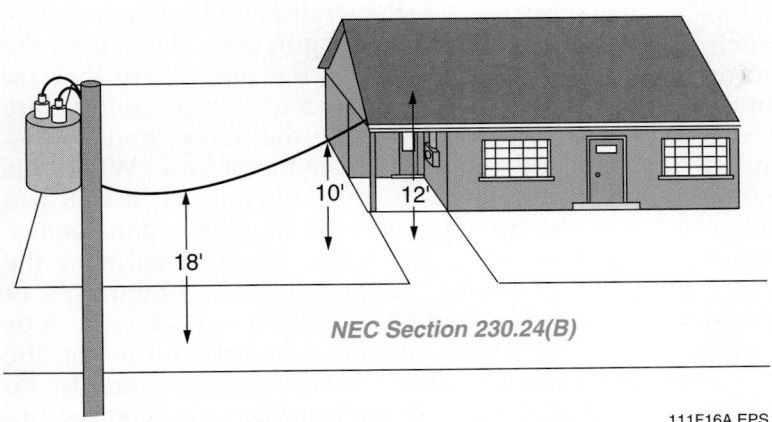

NEC Section 230.24(B)

111F16A.EPS

111F16B.EPS

Figure 16 ◆ Vertical clearances for service drop conductors.

Failure to De-energize Panelboard

A 31-year-old electrician was finishing the installation of an outdoor floodlight on a new home. He borrowed an aluminum ladder from another contractor and then proceeded with his task. He did not verify that power was removed at the panelboard, and when he used his wire strippers to remove the conductor insulation, his right thumb and index finger contacted a 110V circuit. He received a fatal shock.

The Bottom Line: Always de-energize circuits at the panelboard before beginning any electrical task, and never use aluminum ladders while working in or near electrical devices. In this case, the ladder provided a path to ground.

7.0.0 ◆ PANELBOARD LOCATION

The main service disconnect or panelboard is normally located in a portion of an unfinished basement or utility room on an outside wall so that the service cable coming from the electric meter can terminate immediately into the switch or panelboard when the cable enters the building. In the example home, however, there is no basement and the utility room is located in the center of the house with no outside walls. Consequently, a somewhat different arrangement will have to be used. A load center is a type of panelboard that is normally located at the service entrance of a residential installation. The load center usually contains a main circuit breaker, which is the main disconnect. Circuit breakers are provided for equipment such as electric water heaters, ranges, dryers, air conditioning and heating units, and breakers that feed subpanels such as lighting panels.

NEC Section 230.70 requires that the service disconnecting means be installed in a readily accessible location—either outside or inside the building. If located inside the building, it must be located nearest the point of entrance of the service conductors. In the sample home, there are at least two methods of installing the panelboard in the utility room that will comply with this *NEC*® regulation, as well as the requirements in *NEC Sections 110.26 and 240.24.*

The first method utilizes a weatherproof 100A disconnect (safety switch or circuit breaker enclosure) mounted next to the meter base on the outside of the building. With this method, service conductors are provided with overcurrent protection; the neutral conductor is also grounded at this point, as this becomes the main disconnect switch. Three-wire cable with an additional grounding wire is then routed from this main disconnect to the panelboard in the utility room. All three current-carrying conductors (two ungrounded and one neutral) must be insulated with this arrangement; the equipment ground, however, may be bare. The panelboard containing overcurrent protection devices for the branch circuits, which is located in the utility room, now becomes a subpanel. See *Figure 17.*

An alternate method utilizes conduit from the meter base that is routed under the concrete slab and then up to a main panelboard located in the utility room. *NEC Section 230.6* considers conductors to be outside of a building when they are installed under not less than 2" of concrete beneath a building or installed in a conduit not less than 18" deep beneath a building. The sample residence has a 4"-thick reinforced concrete slab—well within the *NEC*® regulations. Therefore, the service conductors from the meter base that are installed under the concrete slab in conduit are considered to be outside the house, and no disconnect is required at the meter base. When this conduit emerges in the utility room, it will run straight up into the bottom of the panelboard, again meeting the *NEC*® requirement that the panel be located nearest the point of entrance of the service conductors. Always check with your local authority having jurisdiction for specific requirements about where services are to be located. Details of this service arrangement are shown in *Figure 18.*

NOTE

Local ordinances in some areas may require a disconnect at the meter base, making the panel in the utility room a subpanel.

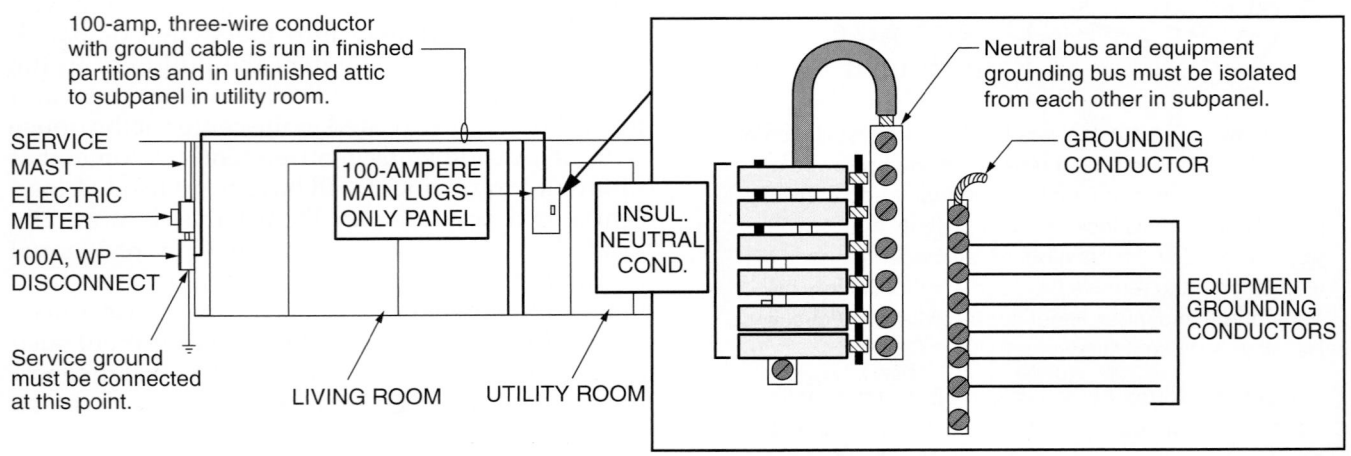

111F17.EPS

Figure 17 ◆ One method of wiring a panelboard for the sample residence.

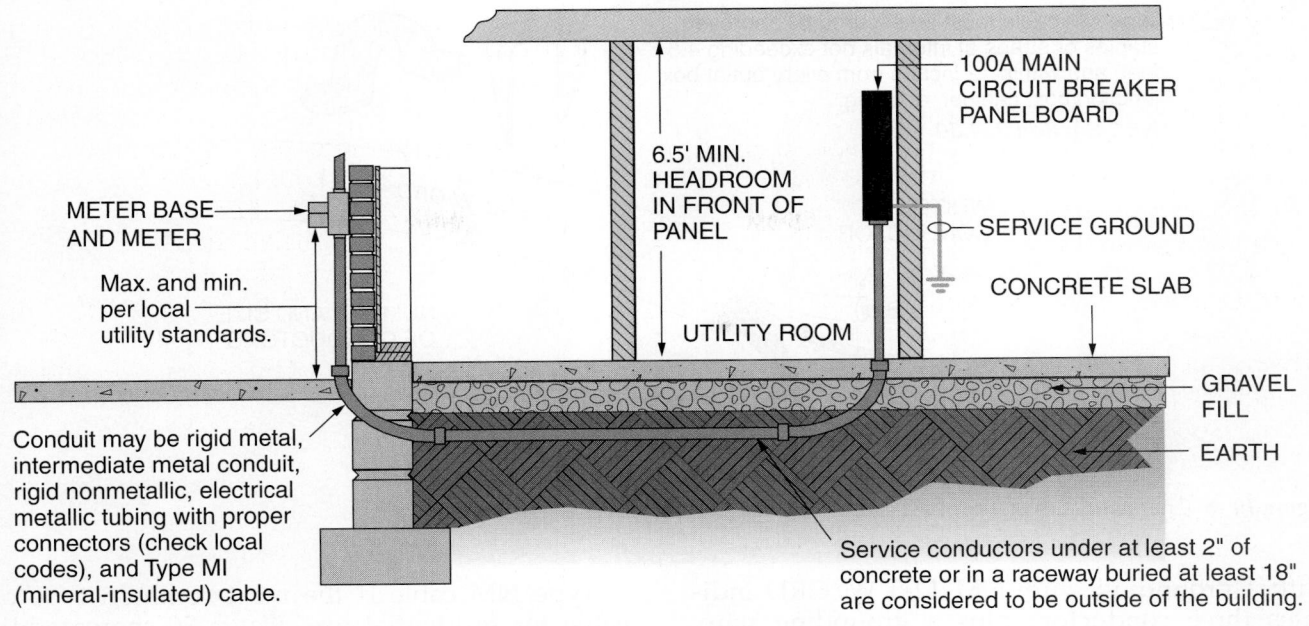

METER BASE AND METER

Max. and min. per local utility standards.

Conduit may be rigid metal, intermediate metal conduit, rigid nonmetallic, electrical metallic tubing with proper connectors (check local codes), and Type MI (mineral-insulated) cable.

6.5' MIN. HEADROOM IN FRONT OF PANEL

UTILITY ROOM

100A MAIN CIRCUIT BREAKER PANELBOARD

SERVICE GROUND

CONCRETE SLAB

GRAVEL FILL

EARTH

Service conductors under at least 2" of concrete or in a raceway buried at least 18" are considered to be outside of the building.

111F18.EPS

Figure 18 ◆ Alternate method of service installation for the sample residence.

8.0.0 ◆ WIRING METHODS

Branch circuits and feeders are used in residential construction to provide power wiring to operate components and equipment, and control wiring to regulate the equipment. Wiring may be further subdivided into either open or concealed wiring.

In open wiring systems, the cable and/or raceways are installed on the surface of the walls, ceilings, columns, and other areas where they are in view and are readily accessible. Open wiring is often used in areas where appearance is not important, such as in unfinished basements, attics, and garages.

Concealed wiring systems are installed inside walls, partitions, ceilings, columns, and behind baseboards or moldings where they are out of view and are not readily accessible. This type of wiring is generally used in all new construction with finished interior walls, ceilings, and floors, and it is the preferred type of wiring where appearance is important.

In general, there are two basic wiring methods used in the majority of modern residential electrical systems. They are:

• Sheathed cables of two or more conductors
• Raceway (conduit) systems

The method used on a given job is determined by the requirements of the *NEC*®, any amendments made by local authorities, the type of build-

ing construction, and the location of the wiring in the building. In most applications, either of the two methods may be used, and both methods are frequently used in combination.

8.1.0 Cable Systems

Several types of cable are used in wiring systems to feed or supply power to equipment. These include nonmetallic-sheathed cable, **metal-clad (MC) cable**, underground feeder cable, and service-entrance cable.

8.1.1 *Nonmetallic-Sheathed Cable*

Nonmetallic-sheathed (Type NM) cable *(NEC Article 334)* is manufactured in two- or three-wire configurations with varying sizes of conductors. In both two- and three-wire cables, conductors are color-coded: one conductor is black while the other is white in two-wire cable; in three-wire cable, the additional conductor is red. Both types also have a grounding conductor, which is usually bare, but it is sometimes covered with green plastic insulation, depending upon the manufacturer. The jacket or covering consists of rubber, plastic, or fiber. Most also have markings on this jacket giving the manufacturer's name or trademark, wire size, and number of conductors (see *Figure 19*). For example, NM 12-2 W/GRD indicates that the jacket contains two No. 12 AWG conductors along

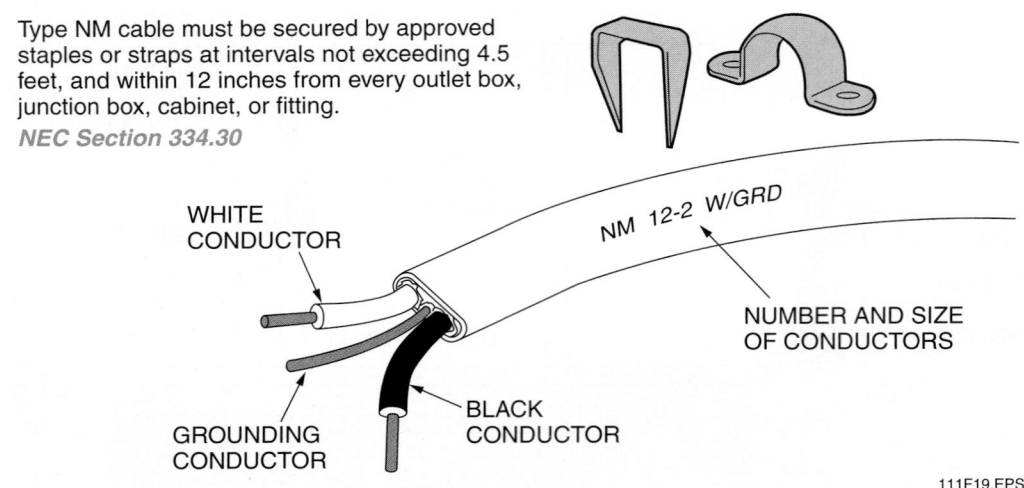

Type NM cable must be secured by approved staples or straps at intervals not exceeding 4.5 feet, and within 12 inches from every outlet box, junction box, cabinet, or fitting.
NEC Section 334.30

WHITE CONDUCTOR

NM 12-2 W/GRD

NUMBER AND SIZE OF CONDUCTORS

GROUNDING CONDUCTOR

BLACK CONDUCTOR

111F19.EPS

Figure 19 ◆ Characteristics of Type NM cable.

with a grounding wire; NM 12-3 W/GRD indicates three conductors plus a grounding wire. Type NM cable is often referred to as **Romex**®.

NEC Section 334.10 permits Type NM cable to be used in the following applications:

- One- and two-family dwelling units
- Multi-family dwellings when they are of Types III, IV, and V construction
- Other structures if concealed behind a 15-minute finish barrier and are of Types III, IV, and V construction

NEC Section 334.12 prohibits the use of Type NM cable in the following applications:

- As open runs in dropped or suspended ceilings in other than dwelling units
- As service-entrance cable
- In commercial garages that have hazardous (classified) areas
- In theaters and similar locations, except as permitted by *NEC Section 518.4*
- In motion picture studios
- In storage battery rooms
- In hoistways or on elevators or escalators
- Embedded in poured cement, concrete, or aggregate
- In hazardous (classified) areas
- Where exposed to corrosive fumes or vapors
- Embedded in masonry, adobe, fill, or plaster
- In a shallow chase in masonry, concrete, or adobe and covered with plaster, adobe, or similar finish
- Where exposed or subject to excessive moisture or dampness

Type NM cable is the most common type of cable for residential use. *Figure 20* shows additional *NEC*® regulations pertaining to the installation of Type NM cable.

8.1.2 Metal-Clad Cable

Metal-clad (MC) cable is manufactured in two-, three-, and four-wire assemblies with varying sizes of conductors, and is used in locations similar to those where Type NM cable is allowed. Unlike Type NM, it can also be used as service-entrance cable and in other locations permitted by *NEC Section 330.10*.

The metallic spiral covering on Type MC cable offers a greater degree of mechanical protection than Type NM cable and also provides a continuous grounding bond without the need for additional grounding conductors.

Type MC cable may be embedded in plaster finish, brick, or other masonry, except in damp or wet locations. It may also be run in the air voids of masonry block or tile walls, except where such walls are exposed or subject to excessive moisture or dampness. It may be used in wet locations if the conditions of *NEC Section 330.10(A)(11)* are met. It may not be used where subject to physical damage. See *Figures 21* and *22*.

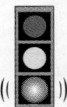

NOTE

In the past, armored cable (Type AC), also called BX® cable, was commonly used in residential applications. Today, Type MC cable is used because it has a plastic wrapping to protect the conductors and does not require an insulating bushing at cable terminations.

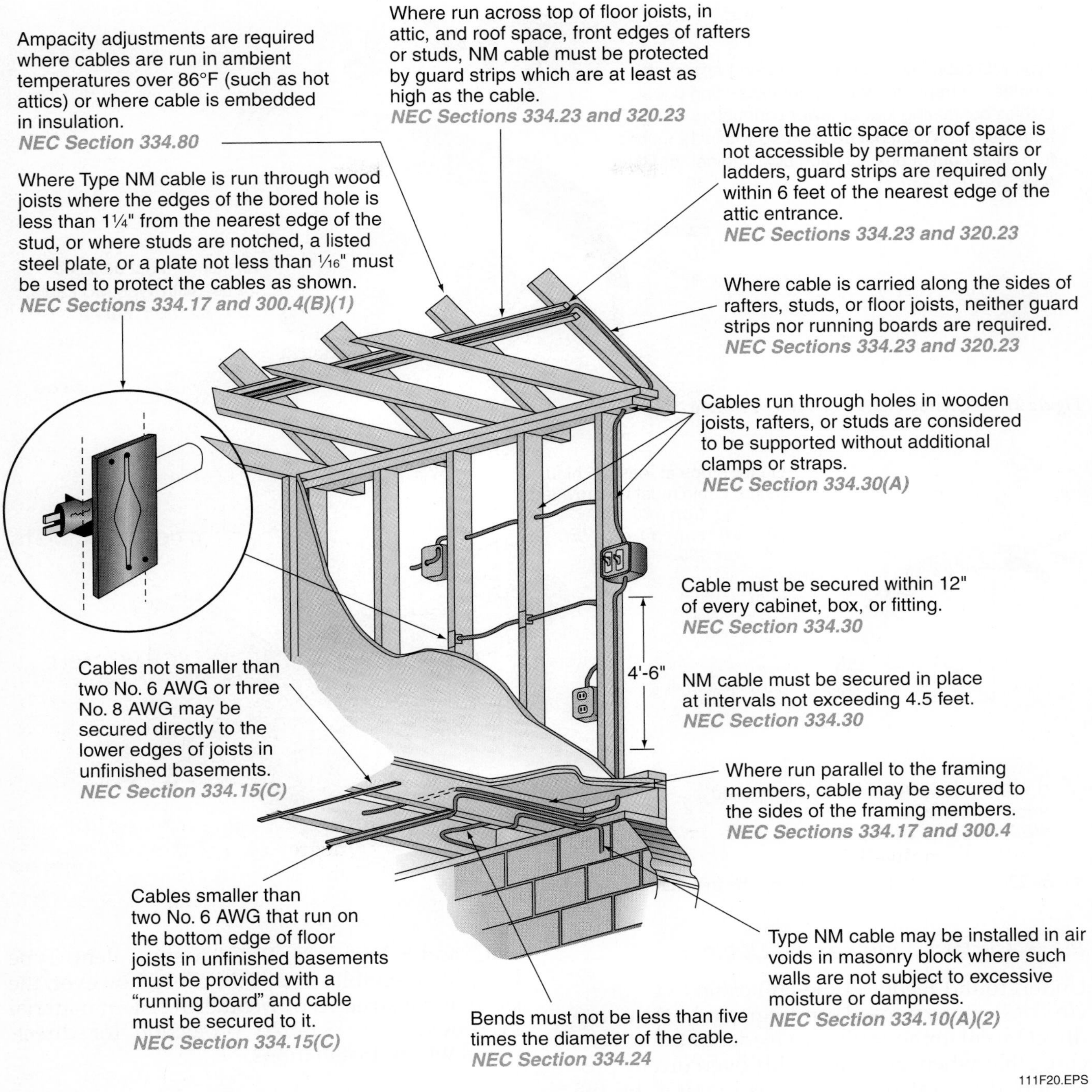

Ampacity adjustments are required where cables are run in ambient temperatures over 86°F (such as hot attics) or where cable is embedded in insulation.
NEC Section 334.80

Where Type NM cable is run through wood joists where the edges of the bored hole is less than 1¼" from the nearest edge of the stud, or where studs are notched, a listed steel plate, or a plate not less than ¹⁄₁₆" must be used to protect the cables as shown.
NEC Sections 334.17 and 300.4(B)(1)

Where run across top of floor joists, in attic, and roof space, front edges of rafters or studs, NM cable must be protected by guard strips which are at least as high as the cable.
NEC Sections 334.23 and 320.23

Where the attic space or roof space is not accessible by permanent stairs or ladders, guard strips are required only within 6 feet of the nearest edge of the attic entrance.
NEC Sections 334.23 and 320.23

Where cable is carried along the sides of rafters, studs, or floor joists, neither guard strips nor running boards are required.
NEC Sections 334.23 and 320.23

Cables run through holes in wooden joists, rafters, or studs are considered to be supported without additional clamps or straps.
NEC Section 334.30(A)

Cable must be secured within 12" of every cabinet, box, or fitting.
NEC Section 334.30

4'-6"

NM cable must be secured in place at intervals not exceeding 4.5 feet.
NEC Section 334.30

Cables not smaller than two No. 6 AWG or three No. 8 AWG may be secured directly to the lower edges of joists in unfinished basements.
NEC Section 334.15(C)

Where run parallel to the framing members, cable may be secured to the sides of the framing members.
NEC Sections 334.17 and 300.4

Cables smaller than two No. 6 AWG that run on the bottom edge of floor joists in unfinished basements must be provided with a "running board" and cable must be secured to it.
NEC Section 334.15(C)

Bends must not be less than five times the diameter of the cable.
NEC Section 334.24

Type NM cable may be installed in air voids in masonry block where such walls are not subject to excessive moisture or dampness.
NEC Section 334.10(A)(2)

111F20.EPS

Figure 20 ◆ *NEC® sections governing the installation of Type NM cable.*

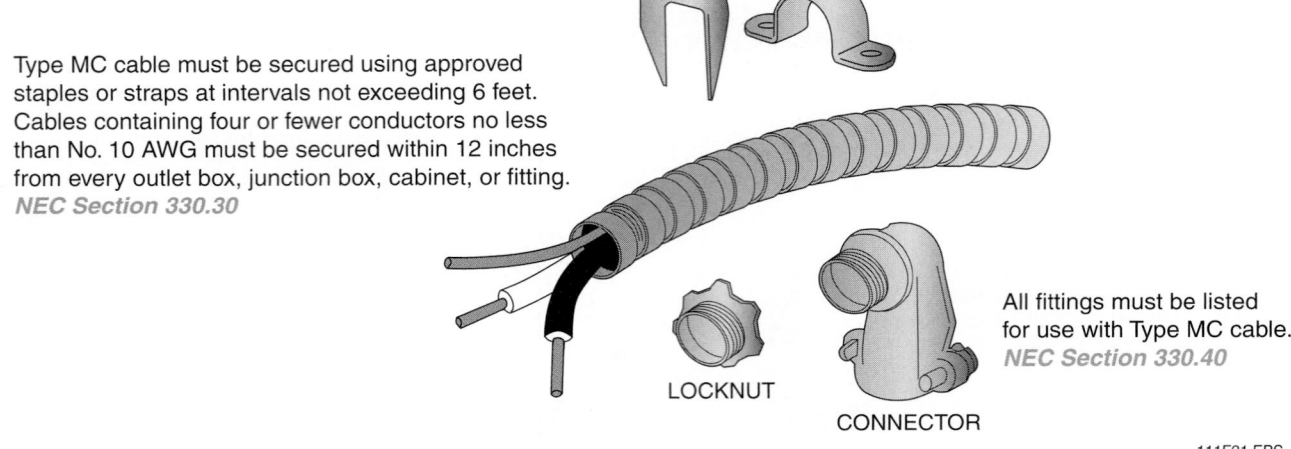

Type MC cable must be secured using approved staples or straps at intervals not exceeding 6 feet. Cables containing four or fewer conductors no less than No. 10 AWG must be secured within 12 inches from every outlet box, junction box, cabinet, or fitting. *NEC Section 330.30*

All fittings must be listed for use with Type MC cable. *NEC Section 330.40*

LOCKNUT

CONNECTOR

111F21.EPS

Figure 21 ◆ Characteristics of Type MC cable.

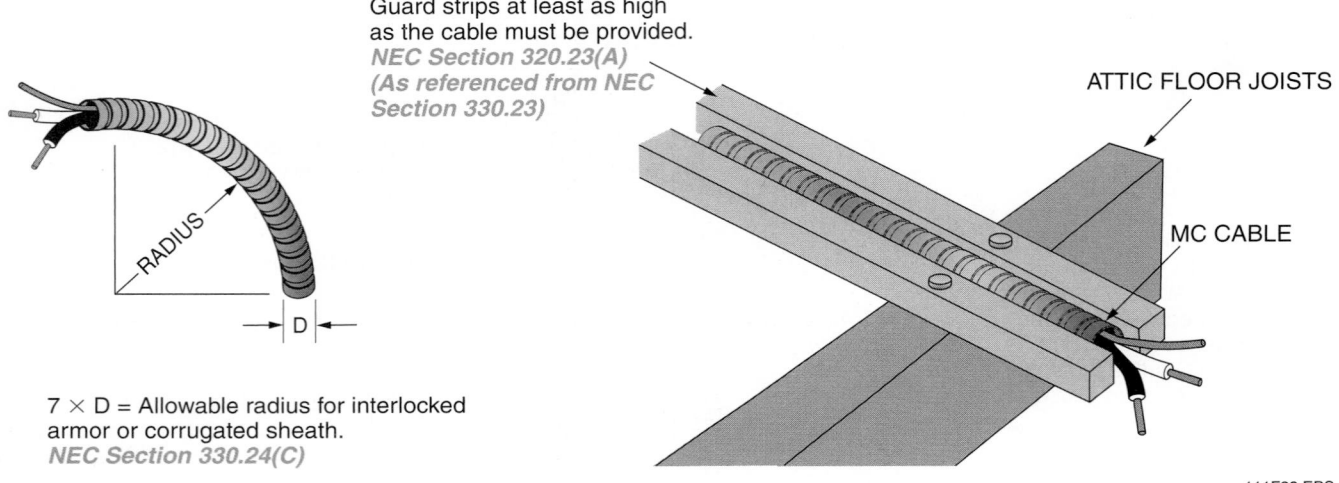

Guard strips at least as high as the cable must be provided. *NEC Section 320.23(A) (As referenced from NEC Section 330.23)*

ATTIC FLOOR JOISTS

MC CABLE

RADIUS

D

7 × D = Allowable radius for interlocked armor or corrugated sheath. *NEC Section 330.24(C)*

111F22.EPS

Figure 22 ◆ *NEC®* sections governing the installation of Type MC cable.

8.1.3 Underground Feeder Cable

Underground feeder (Type UF) cable *(NEC Article 340)* may be used underground, including direct burial in the earth, as a feeder or branch circuit cable when provided with overcurrent protection at the rated ampacity as required by the *NEC®*. When Type UF cable is used above grade where it will come in direct contact with the rays of the sun, its outer covering must be sun-resistant. Furthermore, where Type UF cable emerges from the ground, some means of mechanical protection must be provided. This protection may be in the form of conduit or guard strips. *NEC Section 300.5(D)(1)* requires that the protection extend from the minimum burial depth below grade to a point at least 8' above grade. *NEC Section 300.5(D)(4)* states that if conduit is used as protection, the permitted types are RMC,

IMC, and Schedule 80 PVC, or equivalent. Type UF cable resembles Type NM cable; however, the jacket is constructed of weather-resistant material to provide the required protection for direct-burial wiring installations.

8.1.4 Service-Entrance Cable

Service-entrance (Type SE) and underground service-entrance (Type USE) cable, when used for electrical services, must be installed as specified in *NEC Articles 230 and 338*. Service-entrance cable is available with the grounded conductor bare for outside service conductors, and also with an insulated grounded conductor for interior wiring systems.

Type SE cable is permitted for use on branch circuits or feeders provided that all current-carrying conductors are insulated; this includes

Cable Stripping

Special strippers are used to remove the jackets from Type NM and Type MC cable.

(A) NM CABLE RIPPER

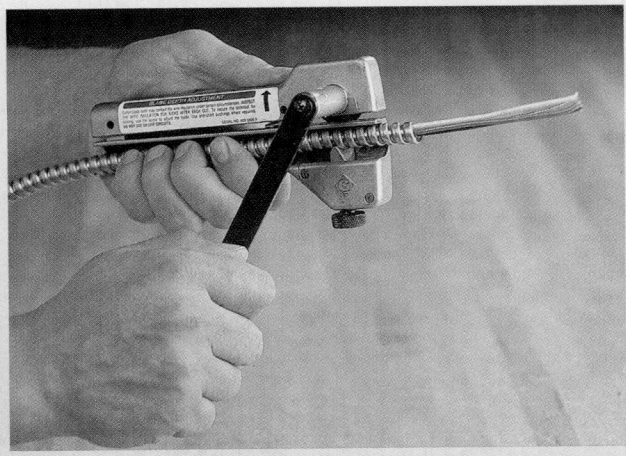

(B) MC CABLE CUTTER

111SA03.EPS

the grounded or neutral conductor. Where a conductor in the cable is not insulated, it is only permitted to be used as an equipment grounding conductor for branch circuits or feeders. Where used as an interior wiring method, the installation requirements of *NEC Article 334* must be followed, except for determining the ampacity of the cable. Where installed as exterior wiring, the requirements of *NEC Article 225* must be met, with the supports for the cable in accordance with *NEC Section 334.30*.

SE Style R (SER) cable is used in residential applications for subfeeds for ranges, and it is also used for service laterals in multi-family dwellings.

Figure 23 summarizes the installation rules for Type SE cable for both exterior and interior wiring.

8.2.0 Raceways

A raceway is any channel that is designed and used solely for the purpose of holding wires, cables, or busbars. Types of raceways include rigid metal conduit, intermediate metal conduit, rigid nonmetallic conduit, flexible metallic conduit, electrical metallic tubing, and auxiliary gutters. Raceways are constructed of either metal or insulating material, such as polyvinyl chloride or PVC (plastic). Metal raceways are joined using threaded, compression, or setscrew couplings; nonmetallic raceways are joined using cement-coated couplings. Where a raceway terminates in an outlet box, junction box, or other enclosure, an approved connector must be used.

Raceways provide mechanical protection for the conductors that run in them and also prevent accidental damage to insulation and the conducting material. They also protect conductors from corrosive atmospheres and prevent fire hazards to life and property by confining arcs and flames that may occur due to faults in the wiring system. Conduits or raceways are used in residential applications for service masts, underground wiring embedded in concrete, and sometimes in unfinished basements, shops, or garage areas.

Another function of metal raceways is to provide a continuous equipment grounding system throughout the electrical system. To maintain this feature, it is extremely important that all raceway systems be securely bonded together into a continuous conductive path and properly connected to the system ground. The following section explains how this is accomplished.

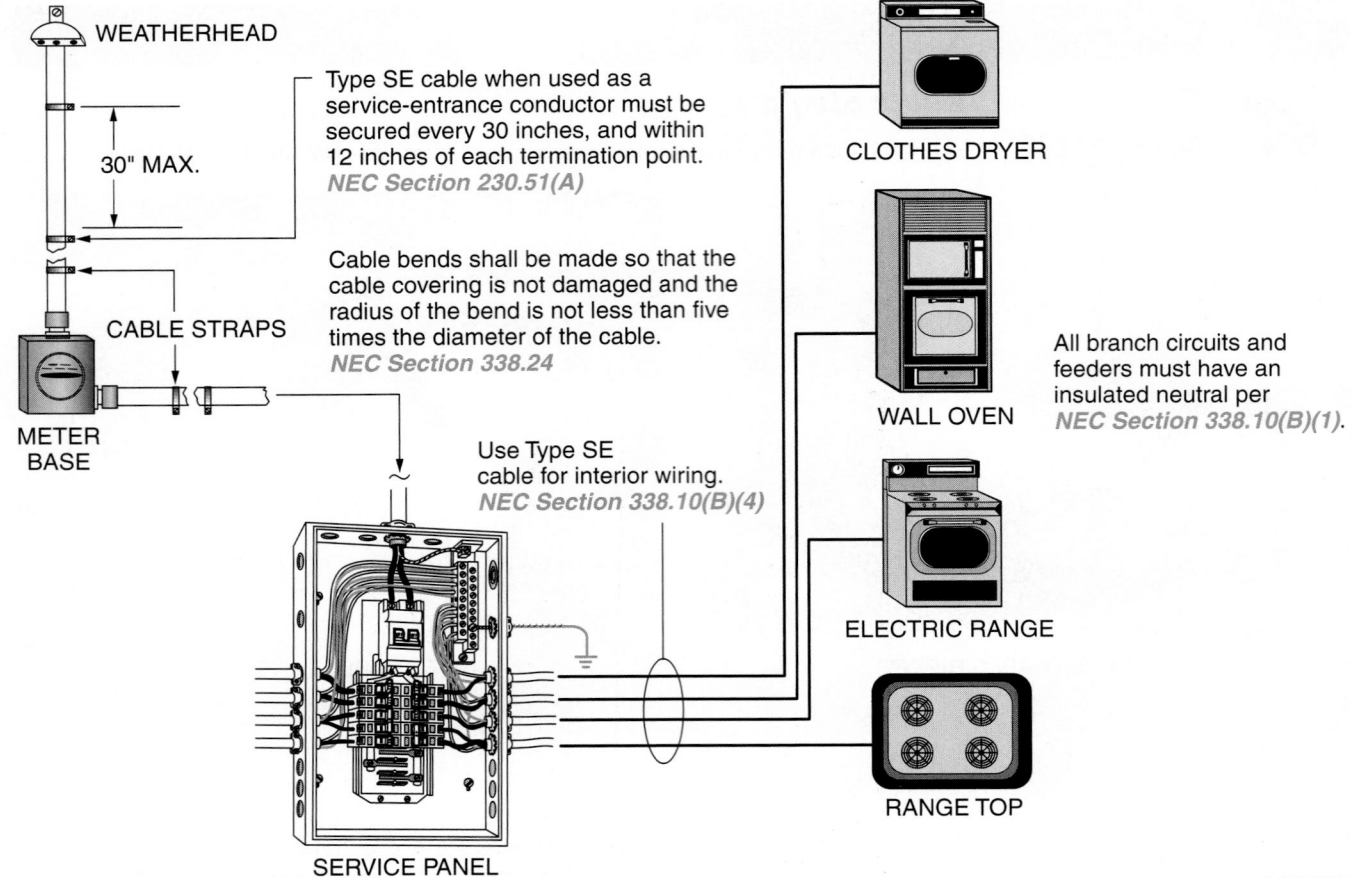

WEATHERHEAD

30" MAX.

CABLE STRAPS

METER BASE

Type SE cable when used as a service-entrance conductor must be secured every 30 inches, and within 12 inches of each termination point. *NEC Section 230.51(A)*

Cable bends shall be made so that the cable covering is not damaged and the radius of the bend is not less than five times the diameter of the cable. *NEC Section 338.24*

Use Type SE cable for interior wiring. *NEC Section 338.10(B)(4)*

SERVICE PANEL

CLOTHES DRYER

WALL OVEN

ELECTRIC RANGE

RANGE TOP

All branch circuits and feeders must have an insulated neutral per *NEC Section 338.10(B)(1).*

111F23.EPS

Figure 23 ◆ *NEC®* sections governing Type SE cable.

9.0.0 ◆ EQUIPMENT GROUNDING SYSTEM

NEC Article 250, Part IV generally requires that all metallic enclosures, raceways, and cable armor be grounded. The exceptions in *NEC Sections 250.80 and 250.86* allow metal enclosures or short sections of raceways that are used to provide support or physical protection to be ungrounded under specific conditions.

NEC Article 250, Part VI covers equipment grounding and equipment grounding conductors. This section generally requires that the exposed noncurrent-carrying metal parts of fixed equipment likely to become energized be grounded under the following conditions:

- Where within 8' vertically or 5' horizontally of ground or grounded metal objects and subject to contact by occupants or others
- Where located in wet or damp locations

- Where in electrical contact with metal
- Where in hazardous (classified) locations as covered by *NEC Articles 500 through 517*
- Where supplied by a metal-clad, metal-sheathed, or metal raceway, or other wiring method that provides an equipment ground
- Where equipment operates with any terminal at over 150V to ground

Specific equipment that is required to be grounded regardless of the voltage is listed in *NEC Section 250.112* and includes equipment such as motors, motor controllers, and light fixtures. Types of cord- and plug-connected equipment in dwelling units that are required to be grounded are found in *NEC Section 250.114* and include equipment such as refrigerators, freezers, air conditioners, information technology equipment (computers), clothes washers, clothes dryers, and dishwashing machines.

The types of equipment grounding conductors that are acceptable to be used are found in *NEC Section 250.118*. Note that among the list of wiring methods approved for use as equipment grounding conductors, both flexible metal conduit (FMC) and liquidtight flexible metal conduit (LFMC) are permitted to be used. Listed FMC is permitted to be used as an equipment grounding conductor only when the following conditions are met:

- The conduit is terminated in fittings listed for grounding.
- The circuit conductors contained in the conduit are protected by overcurrent devices rated at 20A or less.
- The combined length of FMC, FMT, and LFMC in the same ground return path does not exceed 6'.
- The conduit is not installed for flexibility.

Type LFMC is also used in dwelling units and has slightly different requirements when used as an equipment grounding conductor:

- The conduit is terminated in fittings listed for grounding.
- For trade sizes ⅜ through ½, the circuit conductors contained in the conduit are protected by overcurrent devices rated at 20A or less.
- For trade sizes ¾ through 1¼, the circuit conductors contained in the conduit are protected by overcurrent devices rated at 60A or less and there is no FMC, FMT, or LFMC in trade sizes ⅜ through ½ in the grounding path.

- The combined length of FMC, FMT, and LFMC in the same ground return path does not exceed 6'.
- The conduit is not used for flexibility.

Where external bonding jumpers are used to provide the continuity of the fault current path, *NEC Section 250.102(E)* limits the length to not more than 6', except at outside pole locations for the purposes of bonding or grounding the isolated sections of metal raceways or elbows installed in exposed risers at those pole locations. When installing an equipment grounding conductor in a raceway, *NEC Table 250.122* is used to determine the size of the equipment grounding conductor. It is permitted to install one equipment grounding conductor in a raceway that has several circuits. In that case, the size of the equipment grounding conductor is based on the rating of the largest overcurrent device protecting the circuits contained in the raceway.

NEC Section 250.148 requires that where circuit conductors are spliced within a box, or terminated on equipment within or supported by a box, separate equipment grounding conductors associated with those circuit conductors shall be spliced or joined within the box or to the box with devices suitable for the use. *Figure 24* shows several types of fittings that are suitable for this purpose.

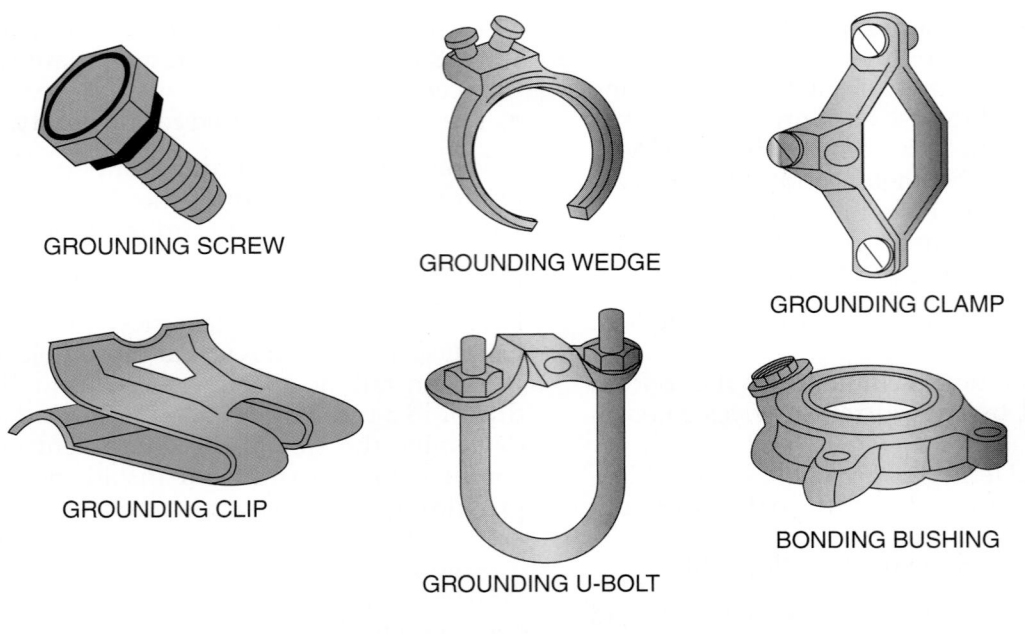

GROUNDING SCREW

GROUNDING WEDGE

GROUNDING CLAMP

GROUNDING CLIP

GROUNDING U-BOLT

BONDING BUSHING

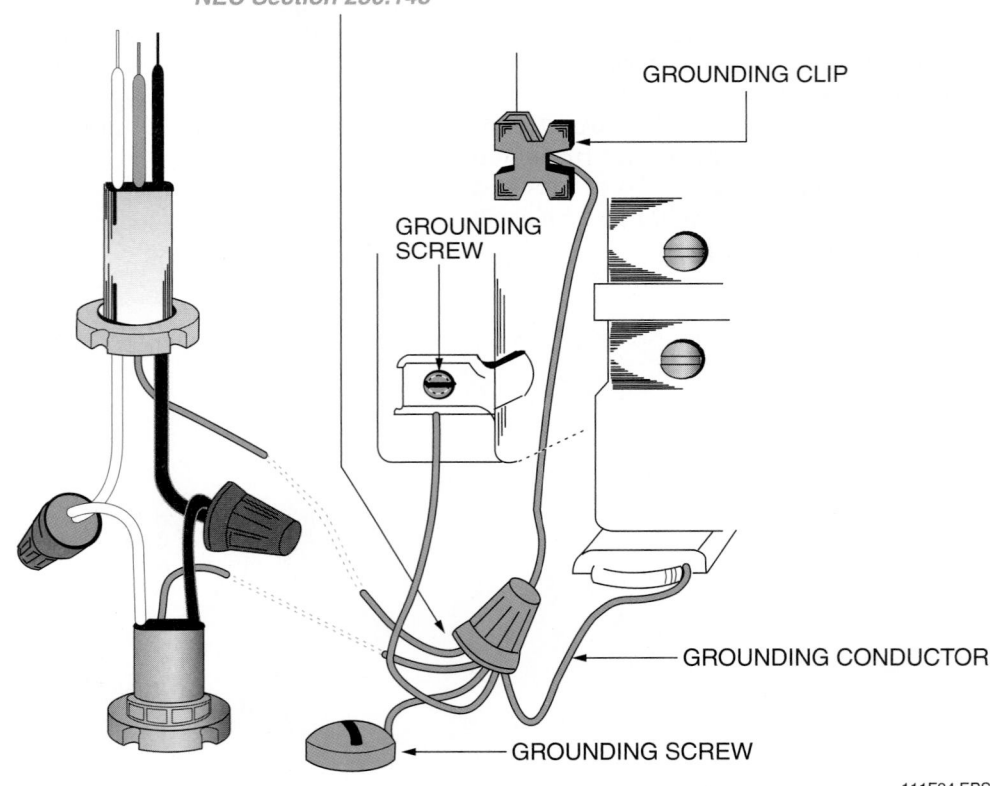

Where splices are made in a junction box, the grounding conductors must be spliced to the metal junction box.
NEC Section 250.148

GROUNDING CLIP

GROUNDING SCREW

GROUNDING CONDUCTOR

GROUNDING SCREW

111F24.EPS

Figure 24 ◆ Equipment grounding methods.

10.0.0 ◆ BRANCH CIRCUIT LAYOUT FOR POWER

The point at which electrical equipment is connected to the wiring system is commonly called an outlet. There are many classifications of outlets: lighting, receptacle, motor, appliance, and so forth. This section, however, deals with the power outlets normally found in residential electrical wiring systems.

When viewing an electrical drawing, outlets are indicated by symbols (usually a small circle with appropriate markings to indicate the type of outlet). The most common symbols for receptacles are shown in *Figure 25*.

10.1.0 Branch Circuits and Feeders

The conductors that extend from the panelboard to the various outlets are called branch circuits and are defined by the *NEC*® as the point of a wiring system that extends beyond the final overcurrent device protecting the circuit. See *Figure 26*.

A feeder consists of all conductors between the service equipment and the final overcurrent device. See *Figure 27*.

In general, the size of the branch circuit conductors varies depending upon the load requirements of the electrically operated equipment connected to the outlet. For residential use, most branch circuits consist of either No. 14 AWG, No. 12 AWG, No. 10 AWG, or No. 8 AWG conductors.

The basic branch circuit requires two wires or conductors to provide a continuous path for the flow of electric current, plus a third wire for equipment grounding. The usual receptacle branch circuit operates at 120V.

Fractional horsepower motors and small electric heaters usually operate at 120V and are connected to 120V branch circuits by means of a receptacle, junction box, or direct connection.

With the exception of very large residences and tract-development houses, the size of the average residential electrical system of the past has not been large enough to justify the expense of preparing complete electrical working drawings and specifications. Such electrical systems were usually laid out by the architect in the form of a sketchy outlet arrangement or else laid out by the electrician on the job, often only as the work progressed. However, many technical developments in residential electrical use—such as electric heat

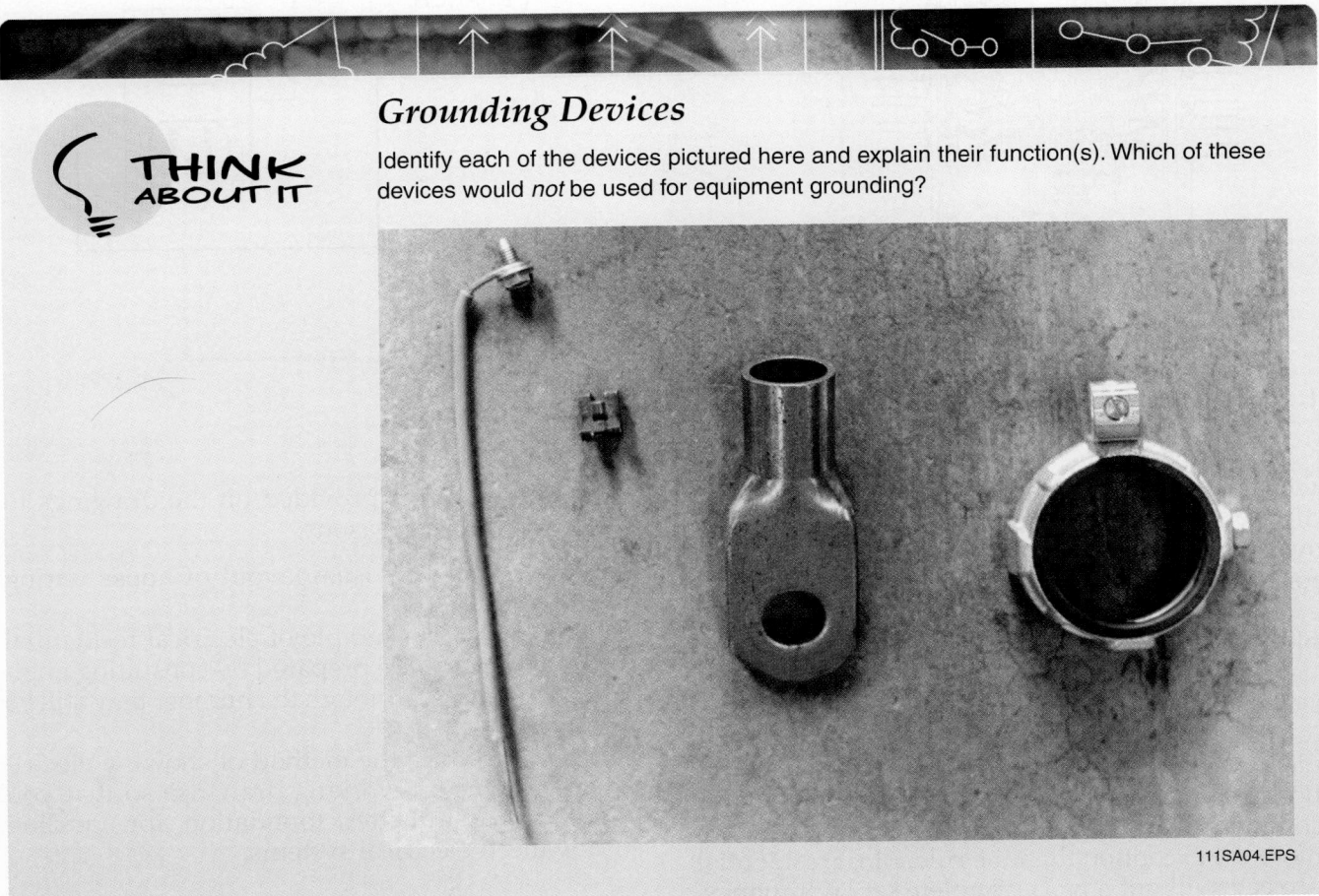

THINK ABOUT IT

Grounding Devices

Identify each of the devices pictured here and explain their function(s). Which of these devices would *not* be used for equipment grounding?

111SA04.EPS

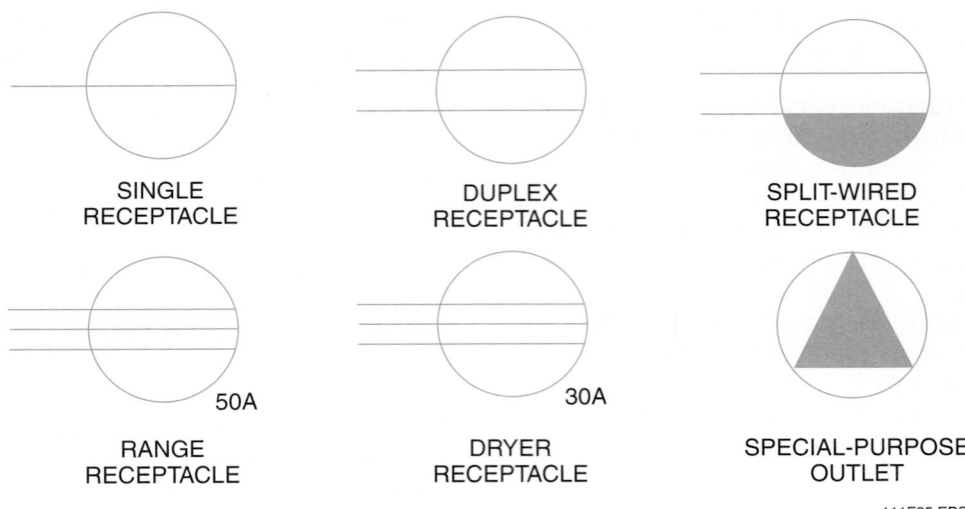

SINGLE
RECEPTACLE

DUPLEX
RECEPTACLE

SPLIT-WIRED
RECEPTACLE

RANGE
RECEPTACLE
50A

DRYER
RECEPTACLE
30A

SPECIAL-PURPOSE
OUTLET

111F25.EPS

Figure 25 ◆ Typical outlet symbols appearing in electrical drawings.

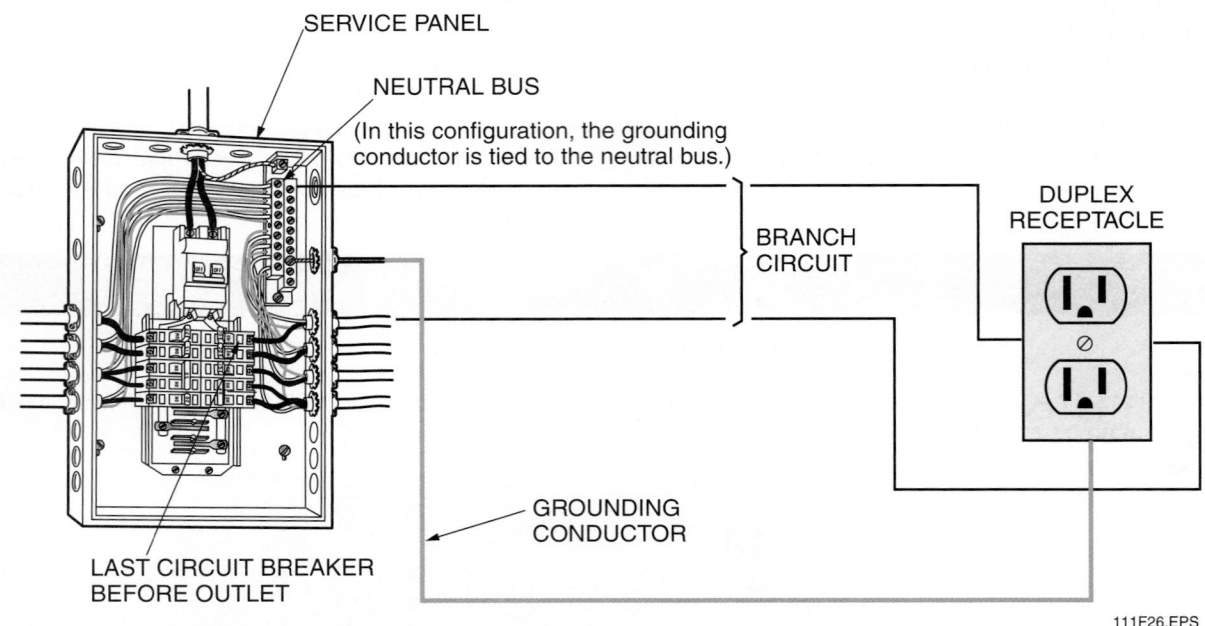

SERVICE PANEL

NEUTRAL BUS

(In this configuration, the grounding
conductor is tied to the neutral bus.)

BRANCH
CIRCUIT

DUPLEX
RECEPTACLE

GROUNDING
CONDUCTOR

LAST CIRCUIT BREAKER
BEFORE OUTLET

111F26.EPS

Figure 26 ◆ Components of a duplex receptacle branch circuit.

with sophisticated control wiring, increased use of electrical appliances, various electronic alarm systems, new lighting techniques, and the need for energy conservation techniques—have greatly expanded the demand and extended the complexity of today's residential electrical systems.

Each year, the number of homes with electrical systems designed by consulting engineering firms increases. Such homes are provided with complete electrical working drawings and specifications, similar to those frequently provided for commercial and industrial projects. Still, these are more the exception than the rule. Most residential projects will not have a complete set of drawings.

Circuit layout is provided on the drawings to follow for several reasons:

• They provide a visual layout of house wiring circuitry.
• They provide a sample of electrical residential drawings that are prepared by consulting engineering firms, although the number may still be limited.
• They introduce the method of showing electrical systems on working drawings so that you will have a better foundation for tackling advanced electrical systems.

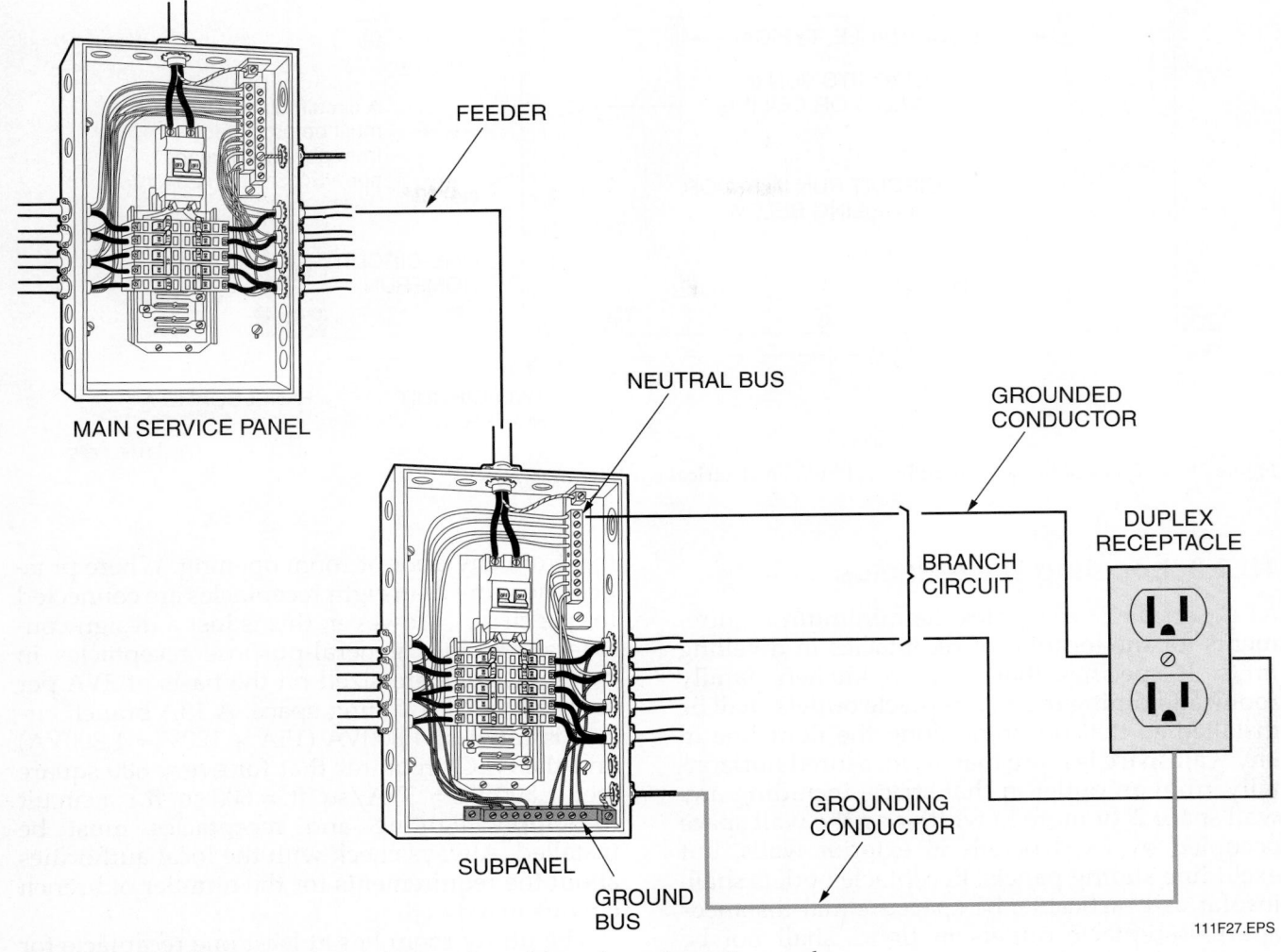

FEEDER

NEUTRAL BUS

GROUNDED CONDUCTOR

GROUNDING CONDUCTOR

MAIN SERVICE PANEL

DUPLEX RECEPTACLE

BRANCH CIRCUIT

SUBPANEL

GROUND BUS

111F27.EPS

Figure 27 ◆ A feeder being used to feed a subpanel from the main service panel.

Branch circuits are shown on electrical drawings by means of a single line drawn from the panelboard (or by homerun arrowheads indicating that the circuit goes to the panelboard) to the outlet or from outlet to outlet where there is more than one outlet on the circuit.

The lines indicating branch circuits can be solid to show that the conductors are to be run concealed in the ceiling or wall, dashed to show that the conductors are to be run in the floor or ceiling below, or dotted to show that the wiring is to be run exposed. *Figure 28* shows examples of these three types of branch circuit lines.

In *Figure 28*, No. 12 indicates the wire size. The slash marks shown through the circuits in *Figure 28* indicate the number of current-carrying conductors in the circuit. Although two slash marks are shown, in actual practice, a branch circuit containing only two conductors usually contains no slash marks; that is, any circuit with no slash marks is assumed to have two conductors. However, three or more conductors are always indicated on electrical working drawings—either by slash marks for each conductor, or else by a note.

Never assume that you know the meaning of any electrical symbol. Although great efforts have been made in recent years to standardize drawing symbols, architects, consulting engineers, and electrical drafters still modify existing symbols or devise new ones to meet their own needs. Always consult the symbol list or legend on electrical working drawings for an exact interpretation of the symbols used.

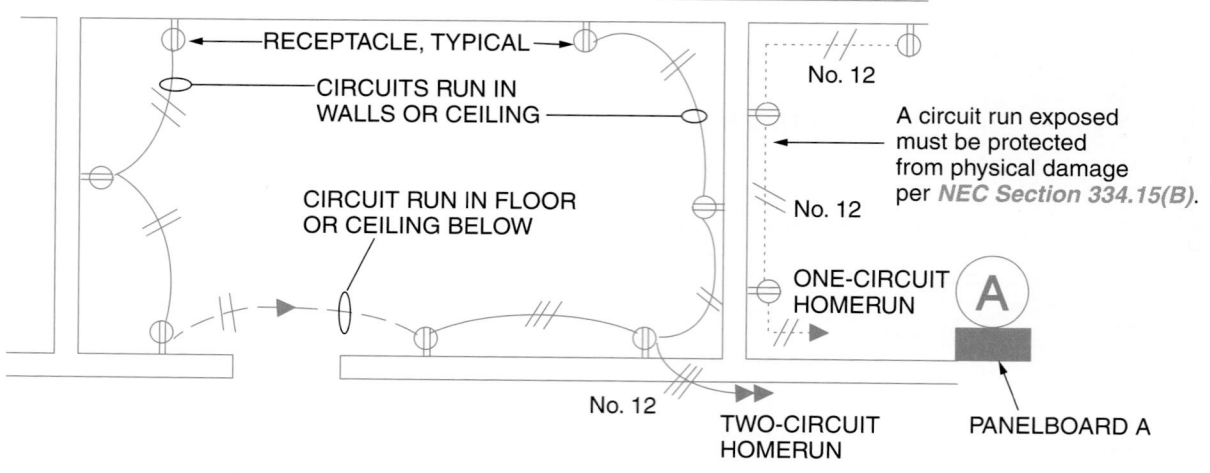

RECEPTACLE, TYPICAL

CIRCUITS RUN IN WALLS OR CEILING

CIRCUIT RUN IN FLOOR OR CEILING BELOW

No. 12

A circuit run exposed must be protected from physical damage per *NEC Section 334.15(B).*

No. 12

ONE-CIRCUIT HOMERUN

A

No. 12

TWO-CIRCUIT HOMERUN

PANELBOARD A

111F28.EPS

Figure 28 ◆ Types of branch circuit lines shown on electrical working drawings.

10.2.0 Locating Receptacles

NEC Section 210.52 states the minimum requirements for the location of receptacles in dwelling units. It specifies that in each kitchen, family room, and dining room, receptacle outlets shall be installed so that no point along the floor line in any wall space is more than 6', measured horizontally, from an outlet in that space, including any wall space 2' or more in width and the wall space occupied by fixed panels in exterior walls, but excluding sliding panels. Receptacle outlets shall, insofar as practicable, be spaced equal distances apart. Receptacle outlets in floors shall not be counted as part of the required number of receptacle outlets unless located within 18" of the wall.

The *NEC*® defines wall space as a wall that is unbroken along the floor line by doorways, fireplaces, or similar openings. Each wall space that is two feet or more in width must be treated individually and separately from other wall spaces within the room.

The purpose of *NEC Section 210.52* is to minimize the use of cords across doorways, fireplaces, and similar openings.

With this *NEC*® requirement in mind, outlets for our sample residence will be laid out (see *Figure 29*). In laying out these receptacle outlets, the floor line of the wall is measured (also around corners), but not across doorways, fireplaces, passageways, or other spaces where a flexible cord extended across the space would be unsuitable.

In general, duplex receptacle outlets must be no more than 12' apart. When spaced in this manner, a 6' extension cord will reach a receptacle from any point along the wall line.

Note that at no point along the wall line are any receptacles more than 12' apart or more than six

feet from any door or room opening. Where practical, no more than eight receptacles are connected to one circuit. However, this is just a design consideration since general-purpose receptacles in dwelling units are sized on the basis of 3VA per square foot of dwelling space. A 15A branch circuit is rated at 1,800VA (15A × 120V = 1,800VA) and the *NEC*® requires that for every 600 square feet (1,800VA ÷ 3VA/sq. ft. = 600 sq. ft.), a circuit to supply lighting and receptacles must be installed. Always check with the local authorities about the requirements for the number of branch circuits in a dwelling.

The utility room has at least one receptacle for the laundry on a separate circuit in order to comply with *NEC Sections 210.11(C)(2) and 210.52(F).*

One duplex receptacle is located in the vestibule for cleaning purposes, such as feeding a portable vacuum cleaner or similar appliance. It is connected to the living room circuit. An additional duplex receptacle is required per *NEC Section 210.52(H)* in hallways of 10' or more.

Although this is not shown in the figure, the living room outlets could be split-wired (the lower half of each duplex receptacle is energized all the time, while the upper half can be switched on or off). The reason for this is that a great deal of the illumination for this area will be provided by portable table lamps, and the split-wired receptacles provide a means to control these lamps from several locations, such as at each entry to the living room, if desired. Split receptacles are discussed in more detail in the next section.

To comply with *NEC Sections 210.11(C)(1) and 210.52(B),* the kitchen receptacles are laid out as follows. In addition to the number of branch circuits determined previously, two or more 20A small appliance branch circuits must be provided

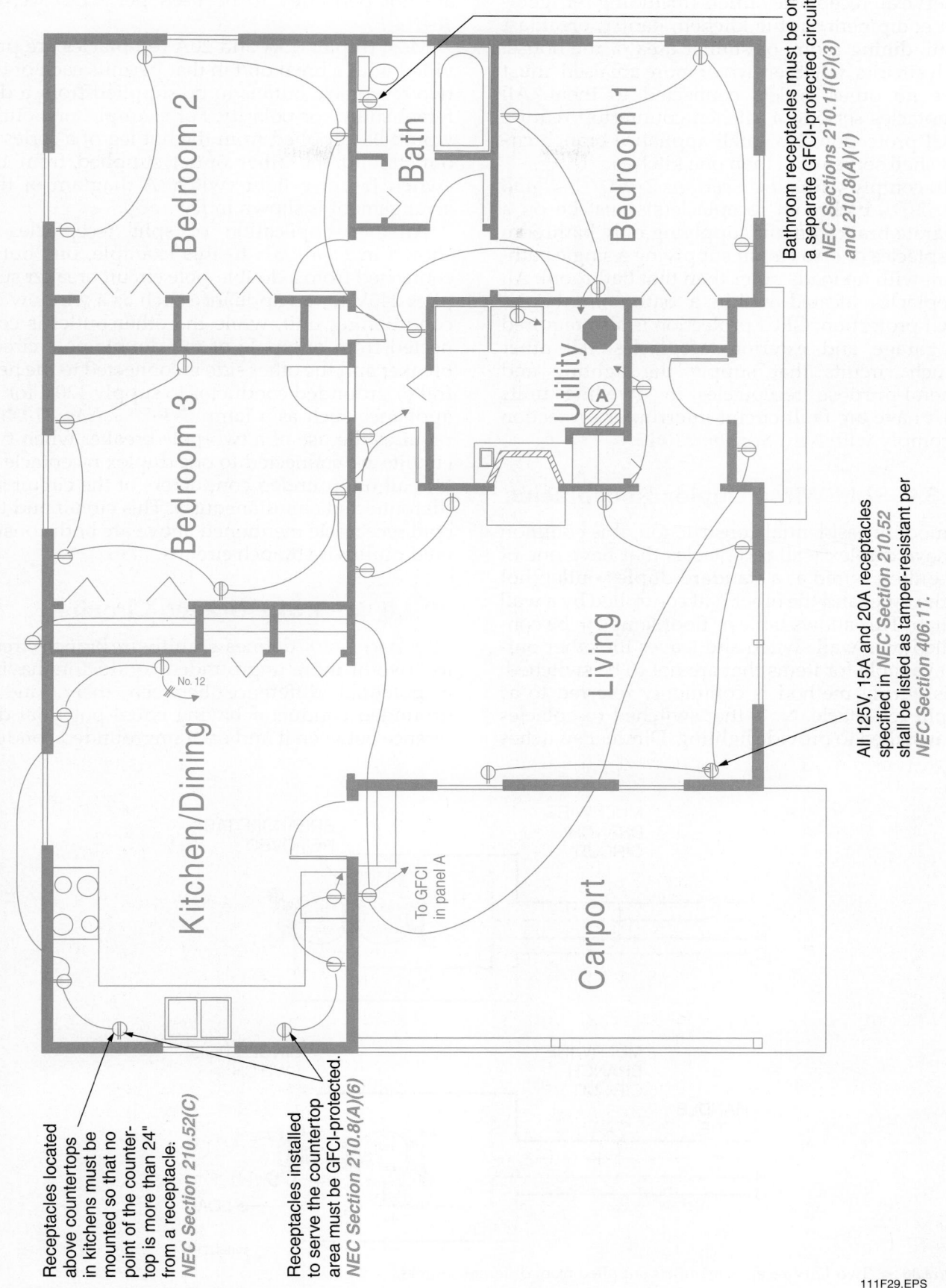

Bathroom receptacles must be on a separate GFCI-protected circuit. *NEC Sections 210.11(C)(3) and 210.8(A)(1)*

All 125V, 15A and 20A receptacles specified in *NEC Section 210.52* shall be listed as tamper-resistant per *NEC Section 406.11.*

Bedroom 2

Bath

Bedroom 1

Bedroom 3

Utility

Ⓐ

No. 12

Living

Kitchen/Dining

To GFCI in panel A

Carport

Receptacles located above countertops in kitchens must be mounted so that no point of the countertop is more than 24" from a receptacle. *NEC Section 210.52(C)*

Receptacles installed to serve the countertop area must be GFCI-protected. *NEC Section 210.8(A)(6)*

111F29.EPS

Figure 29 ◆ Floor plan of the sample residence.

to serve all receptacle outlets (including refrigeration equipment) in the kitchen, pantry, breakfast room, dining room, or similar area of the house. Such circuits, whether two or more are used, must have no other outlets connected to them. All receptacles serving a kitchen countertop require GFCI protection. No small appliance branch circuit shall serve more than one kitchen.

To comply with *NEC Sections 210.11(C)(3) and 210.52(D),* bathroom receptacle(s) must be on a separate branch circuit supplying only bathroom receptacles or on a circuit supplying a single bathroom with no loads other than that bathroom. All receptacles located within a bathroom require GFCI protection. GFCI protection is also required on garage and exterior receptacles. All other branch circuits that supply the lighting and general-purpose receptacles in dwelling units must have arc fault circuit interrupter protection to comply with *NEC Section 210.12.*

10.3.0 Split-Wired Duplex Receptacles

In modern residential construction, it is common to have duplex wall receptacles that have one of the outlets wired as a standard duplex outlet (hot all the time) and the other half controlled by a wall switch. This allows table or floor lamps to be controlled by a wall switch and leaves the other outlet available for items that are not to be switched. This wiring method is commonly referred to as a split receptacle. Note that switched receptacles are installed to provide lighting. Dimmer switches

are not permitted to be used per *NEC Section 404.14(E).*

Most duplex 15A and 20A receptacles are provided with a breakoff tab that permits each of the two receptacle outlets to be supplied from a different source or polarity. For example, one outlet would be supplied from the hot leg of a series of outlets and the other outlet supplied from the **switch leg** of a light switch. A diagram of this arrangement is shown in *Figure 30.*

Another application of split receptacles is shown in *Figure 31.* In this example, one outlet connected from a double-pole circuit breaker supplies 240V for an appliance such as a window air conditioning unit, while the other outlet is connected from one pole of the double-pole circuit breaker and the other side is connected to the neutral or grounded conductor to supply 120V for an appliance such as a lamp. *NEC Section 210.4(B)* requires the use of a two-pole breaker when two circuits are connected to one duplex receptacle so that all ungrounded conductors of the circuit are disconnected simultaneously. This circuit and the split receptacle mentioned above are both considered multiwire branch circuits.

10.4.0 Multiwire Branch Circuits

NEC Article 100 defines a multiwire branch circuit as "two or more ungrounded conductors having a potential difference between them, and a grounded conductor having equal potential difference between it and each ungrounded conduc-

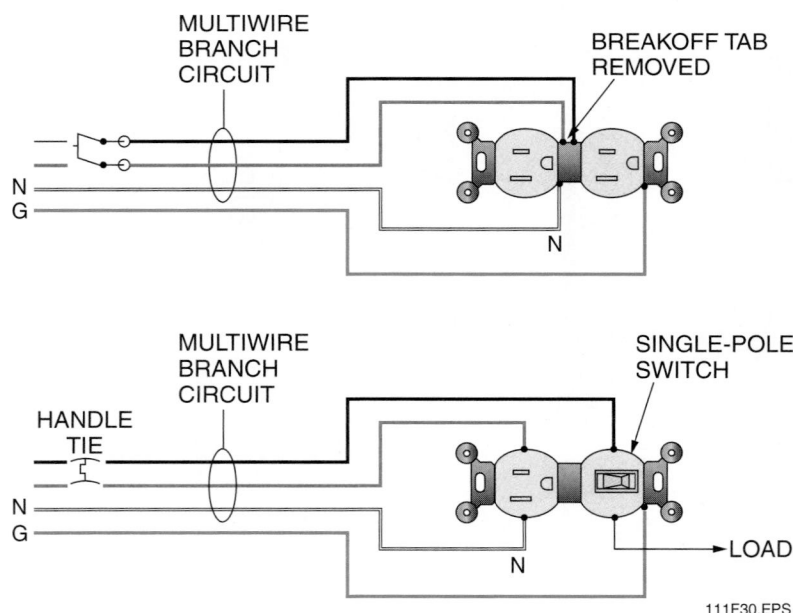

Figure 30 ◆ Two 120V receptacle outlets supplied from different sources.

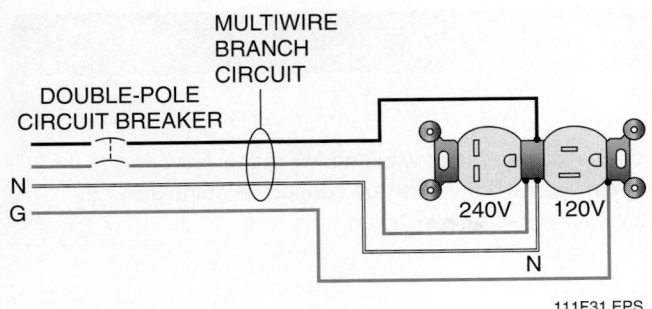

Figure 31 ◆ Combination receptacle.

tor of the circuit and that is connected to the neutral conductor of the system."

Multiwire branch circuits have many advantages, such as three wires doing the work of four (in place of two two-wire branch circuits), less raceway fill, easier balancing and phasing of a system, and less voltage drop. See *NEC Section 210.4(B)*.

10.5.0 240-Volt Circuits

The electric range, clothes dryer, and water heater in the sample residence all operate at 240VAC. Each will be fed by a separate circuit and connected to a two-pole circuit breaker of the appropriate rating in the panelboard. To determine the conductor size and overcurrent protection for the range, proceed as follows:

Step 1 Find the nameplate rating of the electric range. This has previously been determined to be 12kVA.

Step 2 Refer to *NEC Table 220.55.* Since Column A of this table applies to ranges rated at 12kVA (12kW) and under, this will be the column to use in this example.

Step 3 Under the Number of Appliances column, locate the appropriate number of appliances (one in this case), and find the maximum demand given for it in Column A. Column A states that the circuit should be sized for 8kVA (not the nameplate rating of 12kVA).

Step 4 Calculate the required conductor ampacity as follows:

$$\frac{8{,}000VA}{240V} = 33.33A$$

The minimum branch circuit must be rated at 40A since common residential circuit breakers are rated in steps of 15A, 20A, 30A, 40A, and so forth. A 30A circuit breaker is too small, so a 40A circuit breaker is selected. The conductors must have a current-carrying capacity that is equal to or greater than the overcurrent protection. Therefore, No. 8 AWG conductors will be used.

If a cooktop and wall oven were used instead of the electric range, the circuit would be sized similarly. The *NEC*® specifies that a branch circuit for a counter-mounted cooking unit and not more than two wall-mounted ovens, all supplied from a single branch circuit and located in the same room, is computed by adding the nameplate ratings of the individual appliances and treating this total as equivalent to one range. Therefore, two appliances of 6kVA each may be treated as a single range with a 12kVA nameplate rating.

Figure 32 shows how the electric range circuit may appear on an electrical drawing. The connection may be made directly to the range junction box, but more often a 50A range receptacle is mounted at the range location and a range cord-and-plug set is used to make the connection. This facilitates moving the appliance later for maintenance or cleaning.

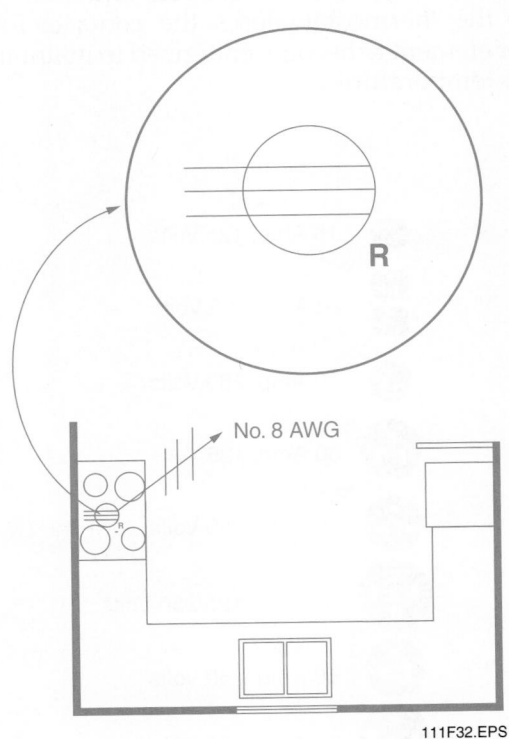

Figure 32 ◆ Range circuit shown on an electrical drawing.

240V Circuits

Calculate the ampacity required for a kitchen range with an 8kW rating. Now design the practical wiring in a labeled diagram. How will the wires be connected at the service panel and at the appliance? How will the cable be installed?

THINK
ABOUT IT

Figure 33 shows several types of receptacle configurations used in residential wiring applications. You will eventually recognize these configurations at a glance.

The branch circuit for the water heater in the sample residence must be sized for its full capacity because there is no diversity or demand factor for this appliance. Since the nameplate rating on the water heater indicates two heating elements of 4,500W each, the first inclination would be to size the circuit for a total load of 9,000W (volt-amperes). However, only one of the two elements operates at a time. See *Figure 34*. Note that each element is controlled by a separate thermostat. The lower element becomes energized when the thermostat calls for heat, and at the same time, the thermostat opens a set of contacts to prevent the upper element from operating. When the lower element's thermostat is satisfied, the lower contacts open, and at the same time, the thermostat closes the contacts for the upper element to become energized to maintain the water temperature.

With this information in hand, the circuit for the water heater may be sized as follows:

$$\frac{4,500VA}{240V} = 18.75A \times 1.25 = 23.44A$$

NEC Section 422.13 requires that the branch circuits that supply storage type water heaters having a capacity of 120 gallons or less be rated not less than 125% of the nameplate rating of the water heater. Our calculation shows this to be not

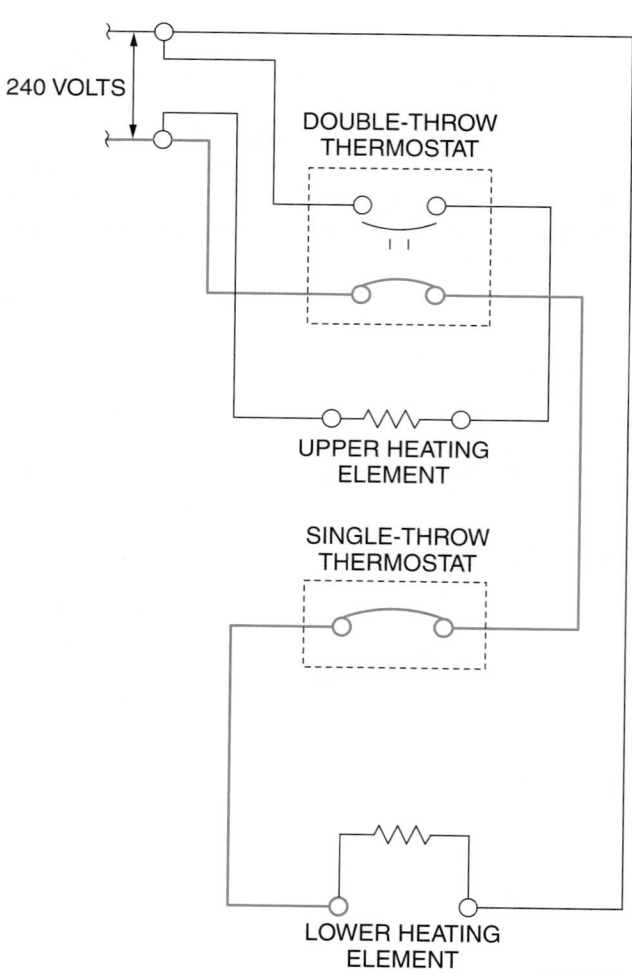

Figure 33 ◆ Residential receptacle configurations.

	15 Amp, 125 Volts
	20 Amp, 125 Volts
	20 Amp, 250 Volts
	30 Amp, 125 Volts
	30 Amp, 250 Volts
	30 Amp, 125/250 Volts
	50 Amp, 250 Volts
	50 Amp, 125/250 Volts

111F33.EPS

Figure 34 ◆ Wiring diagram of water heater controls.

111F34.EPS

less than 23A. Normally, this would require a maximum rating for the branch circuit to be not more than 25A. (See standard ratings of over-current devices in *NEC Section 240.6*.) However, *NEC Section 422.11(E)* permits a single nonmotor-operated appliance to be protected by overcurrent devices rated up to 150% of the nameplate rating of the appliance. In this case, 4,500VA ÷ 240V = 18.75A × 150% = 28.125A. Since the next standard rating is 30A, the water heater will be wired with No. 10 AWG conductors protected by a 30A over-current device.

The *NEC*® specifies that electric clothes dryers must be rated at 5kVA or the nameplate rating, whichever is greater. In this case, the dryer is rated at 5.5kVA, and the conductor current-carrying capacity is calculated as follows:

$$\frac{5,500VA}{240V} = 22.92A$$

A three-wire, 30A circuit will be provided (No. 10 AWG wire). It is protected by a 30A circuit breaker. The dryer may be connected directly, but a 30A dryer receptacle is normally provided for the same reasons as mentioned for the electric range.

Large appliance outlets rated at 240V are frequently shown on electrical drawings using lines and symbols to indicate the outlets and circuits. In some cases, no drawings are provided.

11.0.0 ◆ BRANCH CIRCUIT LAYOUT FOR LIGHTING

A simple lighting branch circuit requires two conductors to provide a continuous path for current flow. The usual lighting branch circuit operates at 120V; the white (grounded) circuit conductor is therefore connected to the neutral bus in the panelboard, while the black (ungrounded) circuit conductor is connected to an overcurrent protection device.

Lighting branch circuits and outlets are shown on electrical drawings by means of lines and symbols; that is, a single line is drawn from outlet to outlet and then terminated with an arrowhead to indicate a homerun to the panelboard. Several methods are used to indicate the number and size of conductors, but the most common is to indicate the number of conductors in the circuit by using slash marks through the circuit lines and then indicate the wire size by a notation adjacent to these slash marks. For example, two slash marks indicate two conductors; three slash marks indicate three conductors. Some electrical designers omit slash marks for two-conductor circuits. In this case, the conductor size is usually indicated in the symbol list or legend.

The circuits used to feed residential lighting must conform to standards established by the *NEC*® as well as by local and state ordinances. Most of the lighting circuits should be calculated to include the total load, although at times this is not possible because the electrician cannot be certain of the exact wattage that might be used by the homeowner. For example, an electrician may install four porcelain lampholders for the unfinished basement area, each to contain one 100-watt (100W) incandescent lamp. However, the homeowners may eventually replace the original lamps with others rated at 150W or even 200W. Thus, if the electrician initially loads the lighting circuit to full capacity, the circuit will probably become overloaded in the future.

It is recommended that no residential branch circuit be loaded to more than 80% of its rated capacity. Since most circuits used for lighting are rated at 15A, the total ampacity (in volt-amperes) for the circuit is as follows:

$$15A \times 120V = 1,800VA$$

Therefore, if the circuit is to be loaded to only 80% of its rated capacity, the maximum initial connected load should be no more than 1,440VA.

Figure 35 shows one possible lighting arrangement for the sample residence. All lighting fixtures are shown in their approximate physical location as they should be installed.

Electrical symbols are used to show the fixture types. Switches and lighting branch circuits are also shown by appropriate lines and symbols. The meanings of the symbols used on this drawing are explained in the symbol list in *Figure 36*.

In actual practice, the location of lighting fixtures and their related switches will probably be the extent of the information shown on working drawings. The circuits shown in *Figure 35* are meant to illustrate how lighting circuits are routed, not to imply that such drawings are typical for residential construction. If fixtures are used in a closet, they must meet the requirements of *NEC Section 410.16*.

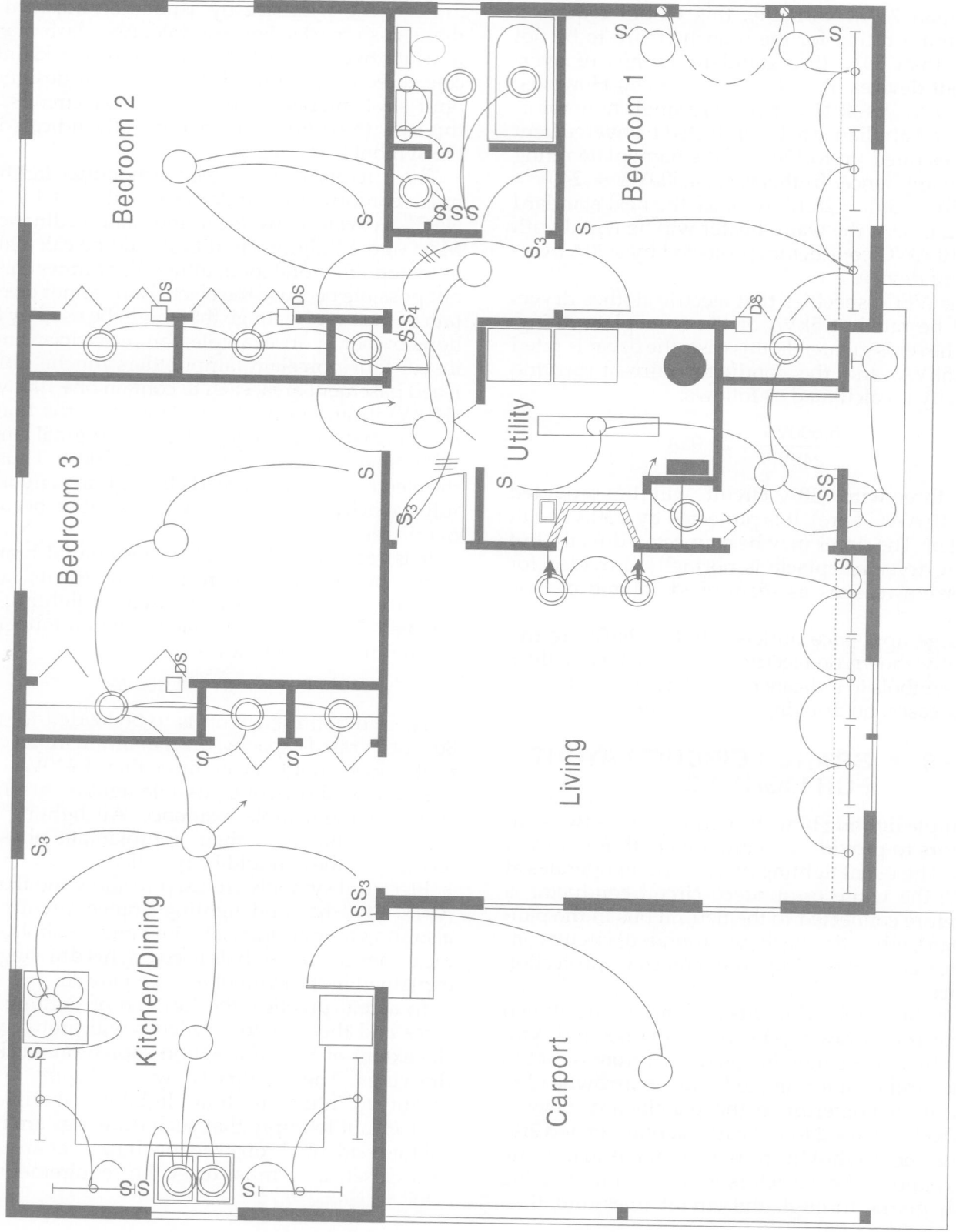

Figure 35 ◆ Lighting layout of the sample residence.

111F35.EPS

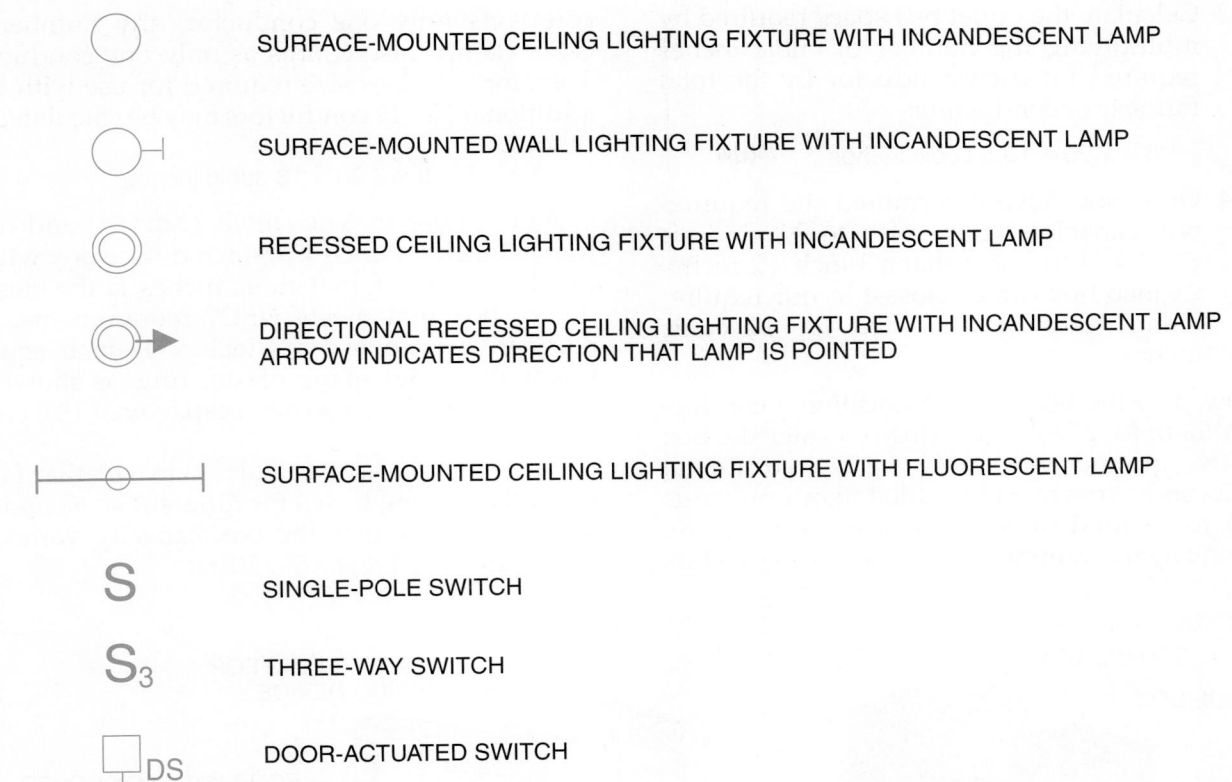

SURFACE-MOUNTED CEILING LIGHTING FIXTURE WITH INCANDESCENT LAMP

SURFACE-MOUNTED WALL LIGHTING FIXTURE WITH INCANDESCENT LAMP

RECESSED CEILING LIGHTING FIXTURE WITH INCANDESCENT LAMP

DIRECTIONAL RECESSED CEILING LIGHTING FIXTURE WITH INCANDESCENT LAMP
ARROW INDICATES DIRECTION THAT LAMP IS POINTED

SURFACE-MOUNTED CEILING LIGHTING FIXTURE WITH FLUORESCENT LAMP

S SINGLE-POLE SWITCH

S₃ THREE-WAY SWITCH

DS DOOR-ACTUATED SWITCH

111F36.EPS

Figure 36 ◆ Symbols.

12.0.0 ◆ OUTLET BOXES

Electricians installing residential electrical systems must be familiar with outlet box capacities, means of supporting outlet boxes, and other requirements of the *NEC*®. Boxes were discussed in detail in an earlier module, but a general review of the rules and necessary calculations is provided here.

The maximum numbers of conductors of the same size permitted in standard outlet boxes are listed in *NEC Table 314.16(A).* These figures apply where no fittings or devices such as fixture studs, cable clamps, switches, or receptacles are contained in the box and where no grounding conductors are part of the wiring within the box. Obviously, in all modern residential wiring systems there will be one or more of these items contained in every outlet box installed. Therefore, where one or more of the above-mentioned items are present, the total number of conductors will be less than that shown in the table. Also, if the box contains a looped, unbroken conductor 12" or more in length, it must be counted twice.

For example, a deduction of two conductors must be made for each strap containing a wiring device entering the box (based on the largest size conductor connected to the device) such as a switch or duplex receptacle; a further deduction of one conductor must be made for one or more equipment grounding conductors entering the box (based on the largest size grounding conductor). For instance, a 3-inch × 2-inch × 2¾-inch box is listed in the table as containing a maximum of six No. 12 wires. If the box contains cable clamps and a duplex receptacle, three wires will have to be deducted from the total of six, providing for only three No. 12 wires. If a ground wire is used, which is always the case in residential wiring, only two No. 12 wires may be used.

For example, to size a metallic outlet box for two No. 12 AWG conductors with a ground wire, cable clamp, and receptacle, proceed as follows:

Step 1 Calculate the total number of conductors and equivalents [*NEC Section 314.16(B)*]. One ground wire plus one cable clamp plus one receptacle (two wires) plus two No. 12 conductors equals a total of six No. 12 conductors.

Step 2 Determine the amount of space required for each conductor. *NEC Table 314.16(B)* gives the box volume required for each conductor. No. 12 AWG equals 2.25 cubic inches.

Step 3 Calculate the outlet box space required by multiplying the number of cubic inches required for each conductor by the total number of conductors:

$$6 \times 2.25 = 13.5 \text{ cubic inches}$$

Step 4 Once you have determined the required box capacity, again refer to *NEC Table 314.16(A)* and note that a 3-inch × 2-inch × 2¾-inch box comes closest to our requirements. This box is rated for 14 cubic inches.

Now, size the box for two additional conductors. Where four No. 12 conductors enter the box with two ground wires, only the two additional No. 12 conductors must be added to our previous count for a total of 8 conductors (6 + 2 = 8). Remember, any number of ground wires in a box counts as only one conductor; any number of cable clamps also counts as only one conductor. Therefore, the box size required for use with two additional No. 12 conductors may be calculated as follows:

$$8 \times 2.25 = 18 \text{ cubic inches}$$

Again, refer to *NEC Table 314.16(A)* and note that a 3-inch × 2-inch × 3½-inch device box with a rated capacity of 18.0 cubic inches is the closest device box that meets *NEC*® requirements. An alternative is to use a 4-inch × 1¼-inch square box with a single-gang plaster ring, as shown in *Figure 37*. This box also has a capacity of 18.0 cubic inches.

Other box sizes are calculated in a similar fashion. When sizing boxes for different size conductors, remember that the box capacity varies as shown in *NEC Table 314.16(B)*.

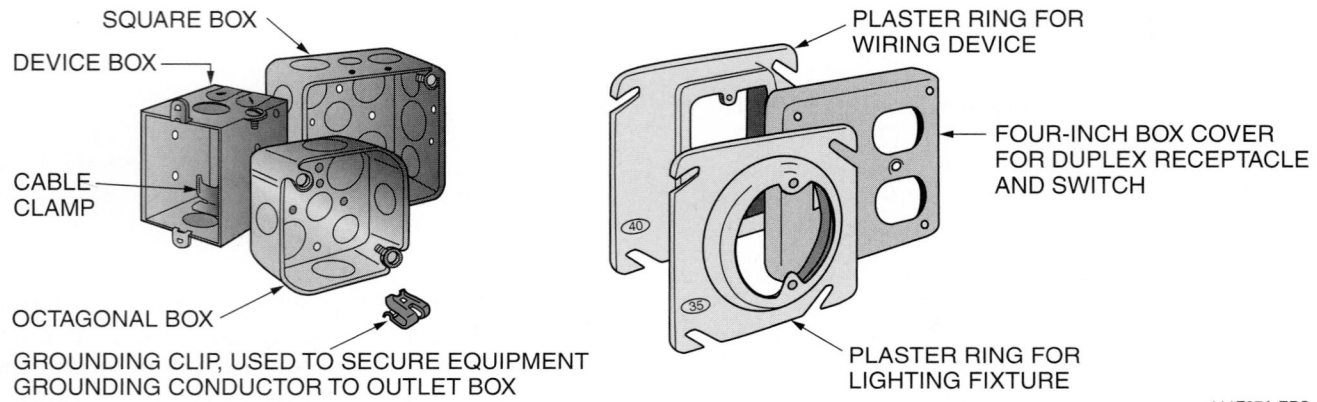

SQUARE BOX

DEVICE BOX

CABLE CLAMP

OCTAGONAL BOX

GROUNDING CLIP, USED TO SECURE EQUIPMENT GROUNDING CONDUCTOR TO OUTLET BOX

PLASTER RING FOR WIRING DEVICE

FOUR-INCH BOX COVER FOR DUPLEX RECEPTACLE AND SWITCH

PLASTER RING FOR LIGHTING FIXTURE

111F37A.EPS

111F37B.EPS

Figure 37 ◆ Typical metallic outlet boxes with extension (plaster) rings.

Calculating Conductors

In a 4" × 4" × 1½" metal box, one 14/3 cable with ground feeds three 14/2 cables with ground wires. The red wire of the 14/3 cable feeds a receptacle, and the black wire feeds the 14/2 black wires. All of the white wires are spliced together, with one brought out to the receptacle terminal. The ground wires are all spliced, with one brought out to the grounding terminal on the receptacle and one to the ground clip on the box. All four cables are connected with box connectors rather than internal clamps. Using *NEC Section 314.16,* decide whether this wiring violates the code.

111SA05.EPS

12.1.0 Mounting Outlet Boxes

Outlet box configurations are almost endless, and if you research the various methods of mounting these boxes, you will be astonished. In this section, some common outlet boxes and their mounting considerations will be reviewed.

The conventional metallic device box, which is used for residential duplex receptacles and switches for lighting control, may be mounted to wall studs using 16d (penny) nails placed through the round mounting holes passing through the interior of the box. The nails are then driven into the wall stud. When nails are used for mounting outlet boxes in this manner, the nails must be located within ¼" of the back or ends of the enclosure.

Nonmetallic boxes normally have mounting nails fitted to the box for mounting. Other boxes have mounting brackets. When mounting outlet boxes with brackets, use either wide-head roofing nails or box nails about 1¼" in length. *Figure 38* shows various methods of mounting outlet boxes.

Before mounting any boxes during the rough wiring process, first find out what type and thickness of finish will be used on the walls. This will dictate the depth to which the boxes must be mounted to comply with *NEC®* regulations. For example, the finish on plastered walls or ceilings is normally ½" thick; gypsum board or drywall is either ½" or ⅝" thick; and wood paneling is normally only ¼" thick. (Some tongue-and-groove wood paneling is ½" to ⅝" thick.)

The *NEC®* specifies the amount of space permitted from the edge of the outlet box to the finished wall. When a noncombustible wall finish (such as plaster, masonry, or tile) is used, the box may be recessed ¼". However, when combustible finishes are used (such as wood paneling), the box must be flush (even) with the finished wall or ceiling. See *Figure 39* and *NEC Section 314.20.*

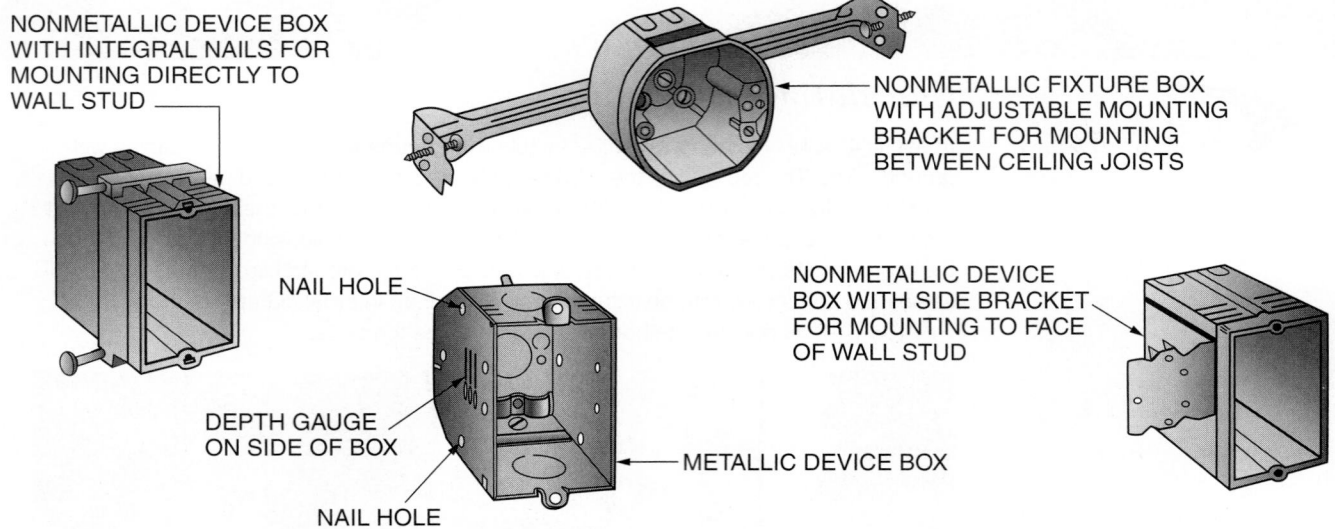

NONMETALLIC DEVICE BOX WITH INTEGRAL NAILS FOR MOUNTING DIRECTLY TO WALL STUD

NONMETALLIC FIXTURE BOX WITH ADJUSTABLE MOUNTING BRACKET FOR MOUNTING BETWEEN CEILING JOISTS

NAIL HOLE

DEPTH GAUGE ON SIDE OF BOX

NAIL HOLE

METALLIC DEVICE BOX

NONMETALLIC DEVICE BOX WITH SIDE BRACKET FOR MOUNTING TO FACE OF WALL STUD

111F38.EPS

Figure 38 ◆ Several methods of mounting outlet boxes.

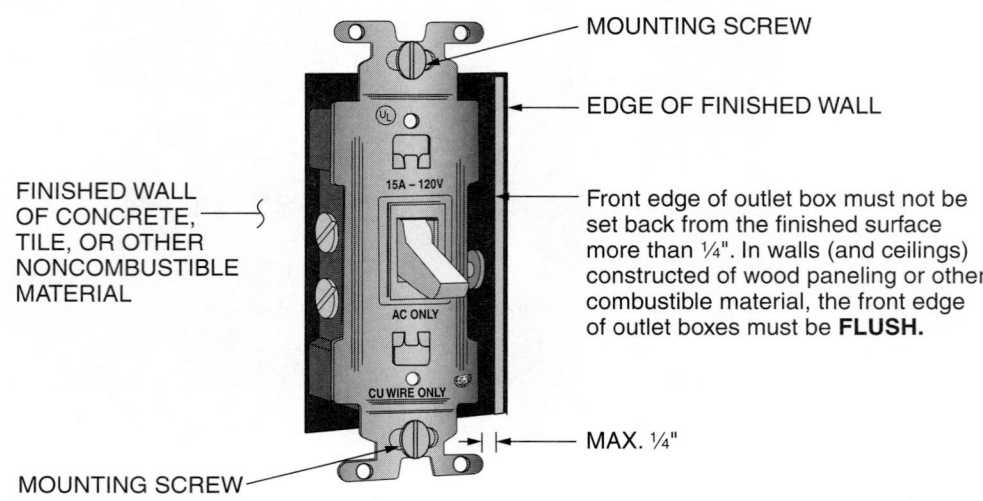

MOUNTING SCREW

EDGE OF FINISHED WALL

FINISHED WALL OF CONCRETE, TILE, OR OTHER NONCOMBUSTIBLE MATERIAL

Front edge of outlet box must not be set back from the finished surface more than ¼". In walls (and ceilings) constructed of wood paneling or other combustible material, the front edge of outlet boxes must be **FLUSH.**

MAX. ¼"

MOUNTING SCREW

111F39.EPS

Figure 39 ◆ Outlet box installation.

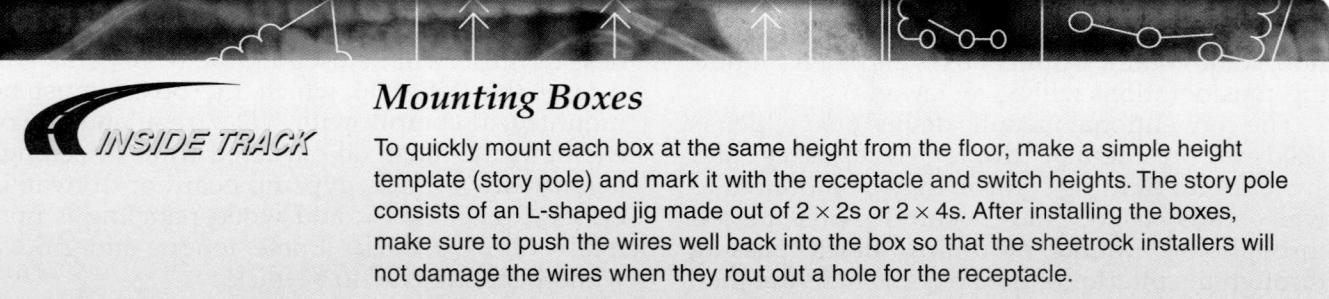

Mounting Boxes

To quickly mount each box at the same height from the floor, make a simple height template (story pole) and mark it with the receptacle and switch heights. The story pole consists of an L-shaped jig made out of 2 × 2s or 2 × 4s. After installing the boxes, make sure to push the wires well back into the box so that the sheetrock installers will not damage the wires when they rout out a hole for the receptacle.

When Type NM cable is used in either metallic or nonmetallic outlet boxes, the cable assembly, including the sheath, must extend into the box by not less than ¼" [*NEC Section 314.17(C)*]. In all instances, all permitted wiring methods must be secured to the boxes by means of either cable clamps or approved connectors. The one exception to this rule is where Type NM cable is used with 2¼-inch × 4-inch (or smaller) nonmetallic boxes where the cable is fastened within eight inches of the box. In this case, the cable does not have to be secured to the box. See *NEC Section 314.17(C), Exception.*

13.0.0 ◆ WIRING DEVICES

Wiring devices include various types of receptacles and switches, the latter being used for lighting control. Switches are covered in *NEC Article 404*, while regulations for receptacles may be found in *NEC Article 406*.

13.1.0 Receptacles

Receptacles are rated by voltage and amperage capacity. *NEC Section 406.3* requires that receptacles connected to a 15A or 20A circuit have the correct voltage and current rating for the application, and be of the grounding type. *NEC Section 406.11* requires that all 15A and 20A, 125V receptacles installed in dwelling units be listed as tamper-resistant.

Where there is only one outlet on a circuit, the receptacle's rating must be equal to or greater than the capacity of the conductors feeding it per *NEC Section 210.21(B)(1)*. For example, if one receptacle is connected to a 20A residential laundry circuit, the receptacle must be rated at 20A or more. When more than one outlet is on a circuit, the total connected load must be equal to or less than the capacity of the branch circuit conductors feeding the receptacles.

Refer to *Figure 40* for some of the characteristics of a standard 125V, 15A duplex receptacle. Note that the terminals are color coded as follows:

- *Green* – Connection for the equipment grounding conductor
- *Silver* – Connection for the neutral or grounded conductor
- *Brass* – Connection for the ungrounded conductor

A standard 125V, 15A receptacle is also typically imprinted with the following symbols:

- *UL* – Underwriters Laboratories, Inc., listing
- *CSA* – Canadian Standards Association
- *CO/ALR* – Designed for use with both copper and aluminum wire
- *15A* – Receptacle rated for a maximum of 15A
- *125V* – Receptacle rated for a maximum of 125V

The UL label means that the receptacle has undergone testing by Underwriters Laboratories, Inc., and meets minimum safety requirements. Underwriters Laboratories, Inc., was created by the National Board of Fire Underwriters to test electrical devices and materials. The UL label is a safety rating only and does not mean that the device or equipment meets any type of quality standard. The CSA label means that the receptacle is approved by the Canadian Standards Association, the Canadian equivalent to Underwriters Laboratories, Inc. The CSA label means that the receptacle is acceptable for use in Canada.

The CO/ALR symbol means that the device is suitable for use with copper, aluminum, or copper-clad aluminum wire. The CO in the symbol stands for copper while ALR stands for aluminum revised. The CO/ALR symbol replaces the earlier CU/AL mark, which appeared on

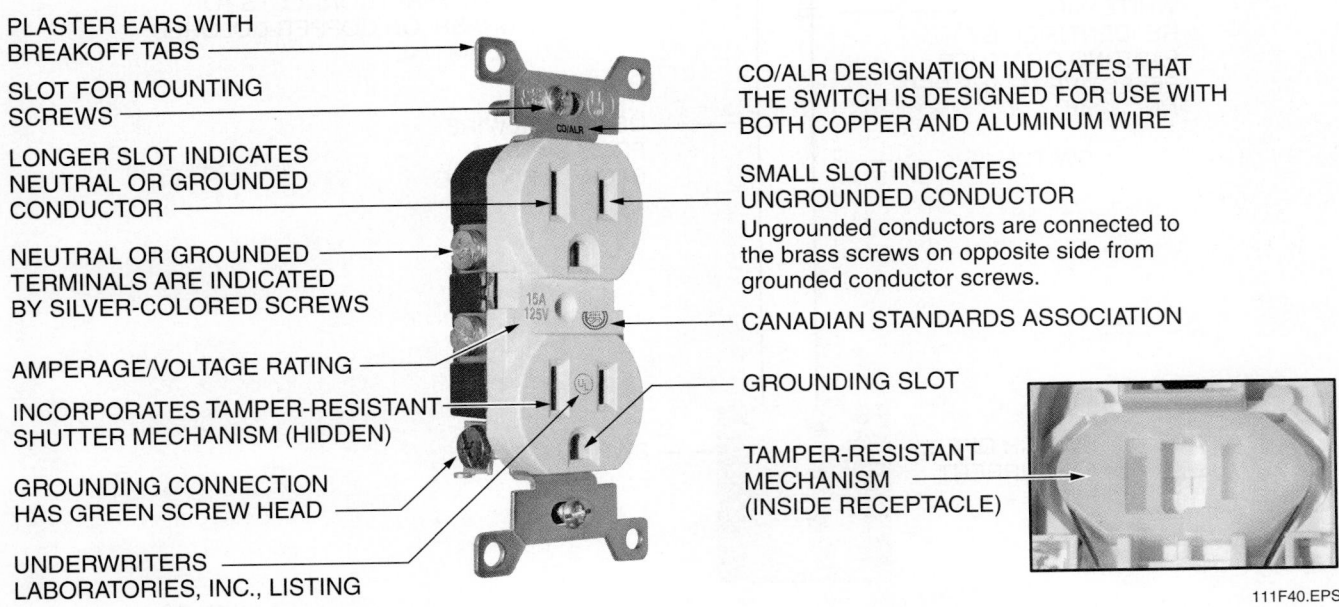

Figure 40 ◆ Standard 125V, 15A duplex receptacle.

wiring devices that were later found to be inadequate for use with aluminum wire in the 15A to 20A range. Therefore, any receptacle or wall switch marked with the CU/AL configuration or anything other than CO/ALR should be used only for copper wire.

These same configurations also apply to wall switches used for lighting control. These will be discussed next.

14.0.0 ◆ LIGHTING CONTROL

There are many types of lighting control devices. These devices have been designed to make the best use of the lighting equipment provided by the lighting industry. They include:

- Automatic timing devices for outdoor lighting
- Dimmers for residential lighting
- Common single-pole, three-way, and four-way switches

For the purposes of this module, a switch is defined as a device that is used on branch circuits to control lighting. Switches fall into the following basic categories:

- Snap-action switches
- Quiet switches

A single-pole snap-action switch consists of a device containing two stationary current-carrying elements, a moving current-carrying element, a toggle handle, a spring, and a housing. When the contacts are open, as shown in *Figure 41*, the circuit is broken and no current flows. When the moving element is closed by manually flipping the toggle handle, the contacts complete the circuit and the lamp is energized. See *Figure 42*.

The quiet switch (*Figure 43*) is the most common switch for use in lighting applications. Its operation is much quieter than the snap-action switch.

The quiet switch consists of a stationary contact and a moving contact that are close together when the switch is open. Only a short, gentle movement is required to open and close the switch, producing very little noise. This type of switch may be used only on alternating current.

Quiet switches are common for loads from 10A to 20A, and are available in single-pole, three-way, and four-way configurations.

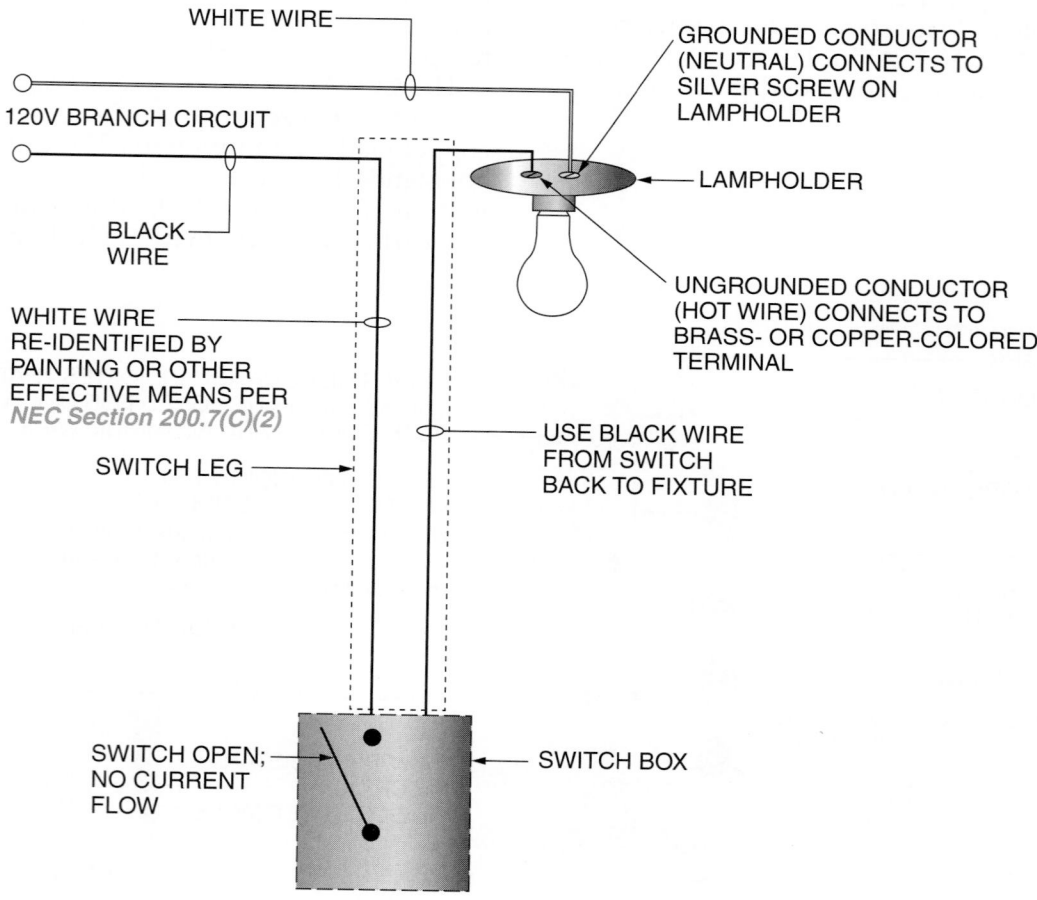

Figure 41 ◆ Switch operation, contacts open.

111F41.EPS

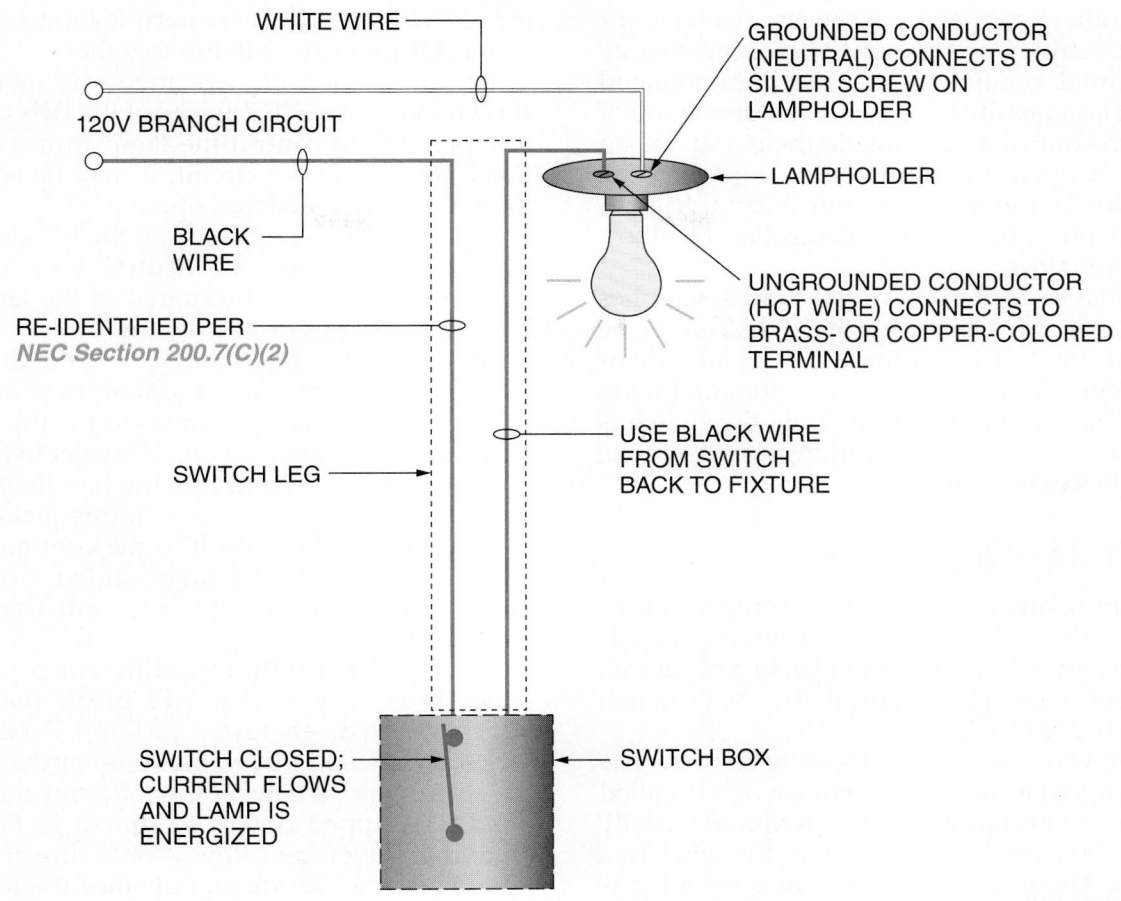

WHITE WIRE

GROUNDED CONDUCTOR (NEUTRAL) CONNECTS TO SILVER SCREW ON LAMPHOLDER

120V BRANCH CIRCUIT

LAMPHOLDER

BLACK WIRE

UNGROUNDED CONDUCTOR (HOT WIRE) CONNECTS TO BRASS- OR COPPER-COLORED TERMINAL

RE-IDENTIFIED PER *NEC Section 200.7(C)(2)*

USE BLACK WIRE FROM SWITCH BACK TO FIXTURE

SWITCH LEG

SWITCH CLOSED; CURRENT FLOWS AND LAMP IS ENERGIZED

SWITCH BOX

111F42.EPS

Figure 42 ◆ Switch operation, contacts closed.

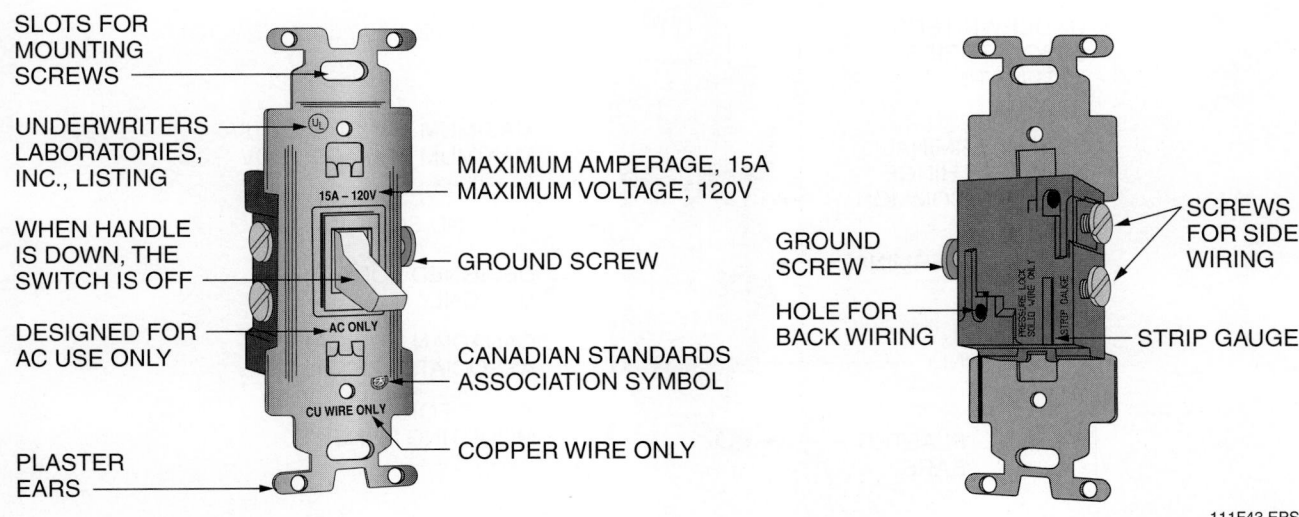

SLOTS FOR MOUNTING SCREWS

UNDERWRITERS LABORATORIES, INC., LISTING

MAXIMUM AMPERAGE, 15A
MAXIMUM VOLTAGE, 120V

15A – 120V

WHEN HANDLE IS DOWN, THE SWITCH IS OFF

GROUND SCREW

AC ONLY

DESIGNED FOR AC USE ONLY

CANADIAN STANDARDS ASSOCIATION SYMBOL

CU WIRE ONLY

COPPER WIRE ONLY

PLASTER EARS

GROUND SCREW

SCREWS FOR SIDE WIRING

HOLE FOR BACK WIRING

STRIP GAUGE

111F43.EPS

Figure 43 ◆ Characteristics of a single-pole quiet switch.

Many other types of switches are available for lighting control. One type of switch used mainly in residential occupancies is the door-actuated switch. It is generally installed in the door jamb of a closet to control a light inside the closet. When the door is open, the light comes on; when the door is closed, the light goes out. Most refrigerator and oven lights are also controlled by door-actuated switches.

Combination switch/indicator light assemblies are available for use where the light cannot be seen from the switch location, such as an attic or garage. Switches are also made with small neon lamps in the handle that light when the switch is off. These low-current-consuming lamps make the switches easy to find in the dark.

14.1.0 Three-Way Switches

Three-way switches are used to control one or more lamps from two different locations, such as at the top and bottom of stairways, in a room that has two entrances, etc. A typical three-way switch is shown in *Figure 44*.

A three-way switch has three terminals. The single terminal at one end of the switch is called the common or hinge point. This terminal is easily identified because it is darker than the other two terminals. The feeder (hot wire) or switch leg is always connected to the common dark or black terminal. The two remaining terminals are called traveler terminals. These terminals are used to connect three-way switches together.

The connection of two three-way switches is shown in *Figure 45*. By means of the two switches, it is possible to control the lamp from two locations. By tracing the circuit, it may be seen how these three-way switches operate.

A 120V circuit emerges from the left side of the drawing. The white or neutral wire connects directly to the neutral terminal of the lamp. The hot wire carries current, in the direction of the arrows, to the common terminal of the three-way switch on the left. Since the handle is in the up position, the current continues to the top traveler terminal and is carried by this traveler to the other three-way switch. Note that the handle is also in the up position on this switch; this picks up the current flow and carries it to the common point, which continues to the ungrounded terminal of the lamp to make a complete circuit. The lamp is energized.

Moving the handle to a different position on either three-way switch will break the circuit, which in turn de-energizes the lamp. For example, let's say a person leaves the room at the point of the three-way switch on the left, and the switch handle is flipped down, as shown in *Figure 46*. Note that the current flow is now directed to the bottom traveler terminal, but since the handle of the three-way switch on the right is still in the up position, no current will flow to the lamp.

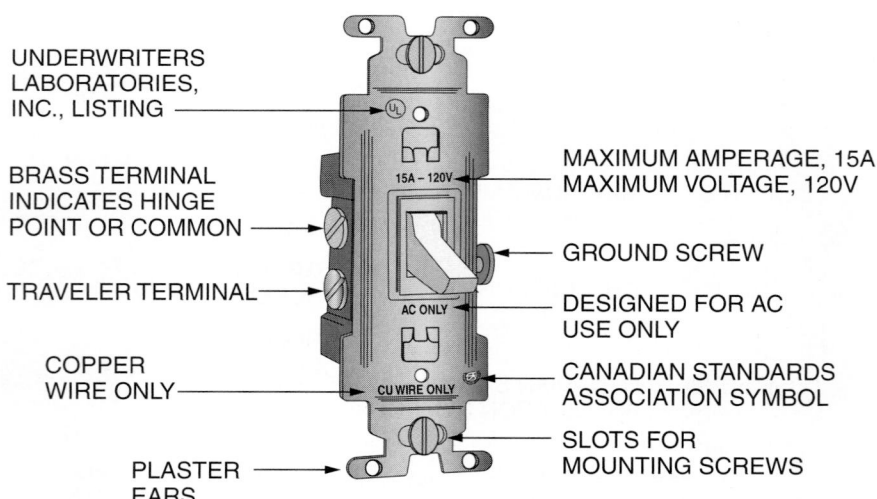

UNDERWRITERS LABORATORIES, INC., LISTING

BRASS TERMINAL INDICATES HINGE POINT OR COMMON

TRAVELER TERMINAL

COPPER WIRE ONLY

PLASTER EARS

MAXIMUM AMPERAGE, 15A
MAXIMUM VOLTAGE, 120V

GROUND SCREW

DESIGNED FOR AC USE ONLY

CANADIAN STANDARDS ASSOCIATION SYMBOL

SLOTS FOR MOUNTING SCREWS

111F44.EPS

Figure 44 ◆ Typical three-way switch.

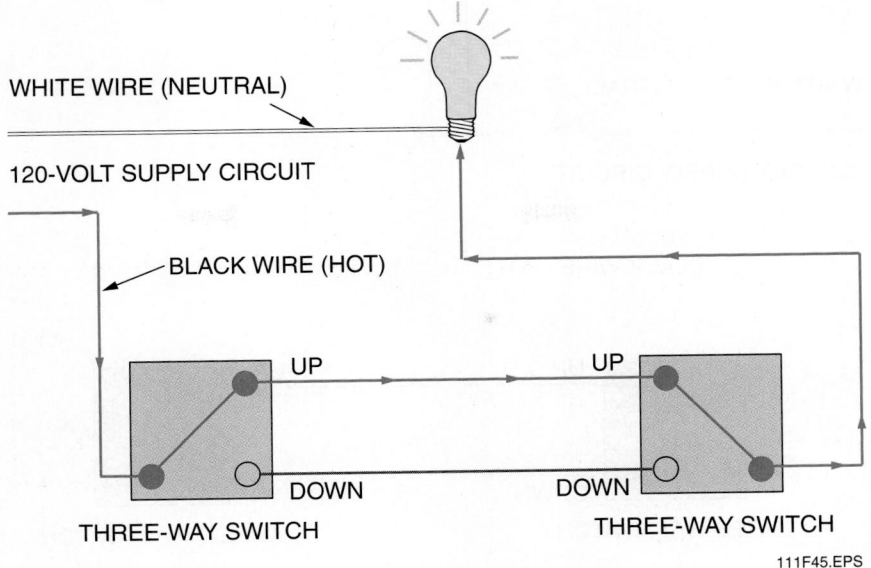

WHITE WIRE (NEUTRAL)

120-VOLT SUPPLY CIRCUIT

BLACK WIRE (HOT)

UP   UP

DOWN   DOWN

THREE-WAY SWITCH   THREE-WAY SWITCH

111F45.EPS

Figure 45 ◆ Three-way switches in the ON position; both handles are up.

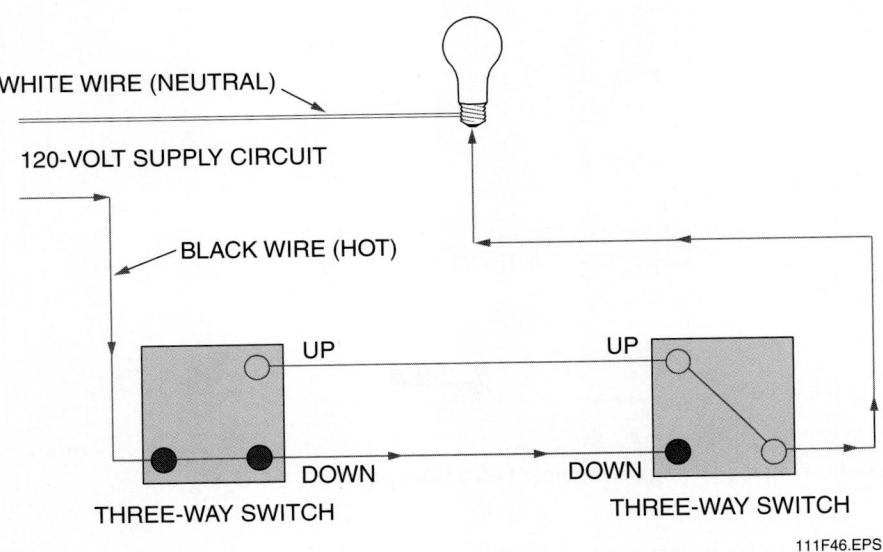

WHITE WIRE (NEUTRAL)

120-VOLT SUPPLY CIRCUIT

BLACK WIRE (HOT)

UP   UP

DOWN   DOWN

THREE-WAY SWITCH   THREE-WAY SWITCH

111F46.EPS

Figure 46 ◆ Three-way switches in the OFF position; one handle is down, one handle is up.

If another person enters the room at the location of the three-way switch on the right, and the handle is flipped downward, as shown in *Figure 47*, this change provides a complete circuit to the lamp, which causes it to be energized. In this example, current flow is on the bottom traveler. Again, changing the position of the switch handle (pivot point) on either three-way switch will de-energize the lamp.

In actual practice, the exact wiring of the two three-way switches to control the operation of a lamp will be slightly different from the routing shown in these three diagrams. There are several ways that two three-way switches may be connected. One solution is shown in *Figure 48*. In this case, two-wire, Type NM cable is fed to the three-way switch on the left.

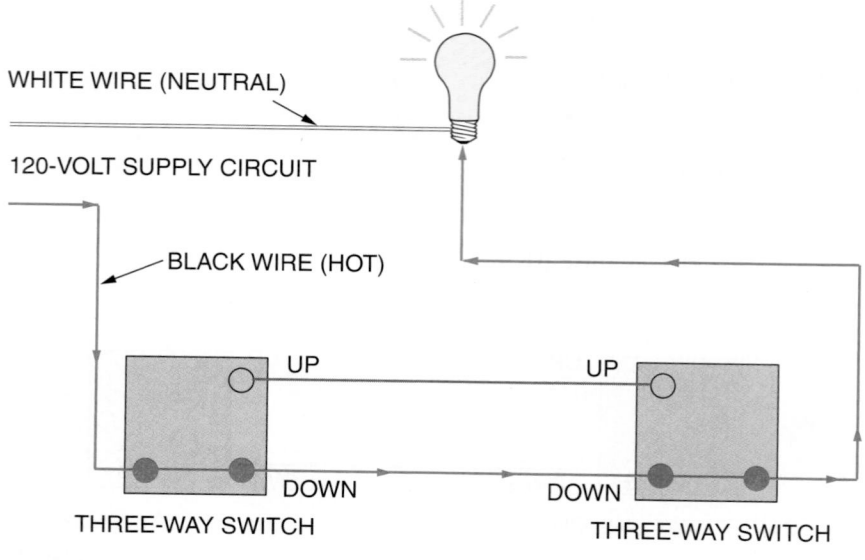

WHITE WIRE (NEUTRAL)

120-VOLT SUPPLY CIRCUIT

BLACK WIRE (HOT)

UP   UP

DOWN   DOWN

THREE-WAY SWITCH   THREE-WAY SWITCH

111F47.EPS

Figure 47 ◆ Three-way switches with both handles down; the light is energized.

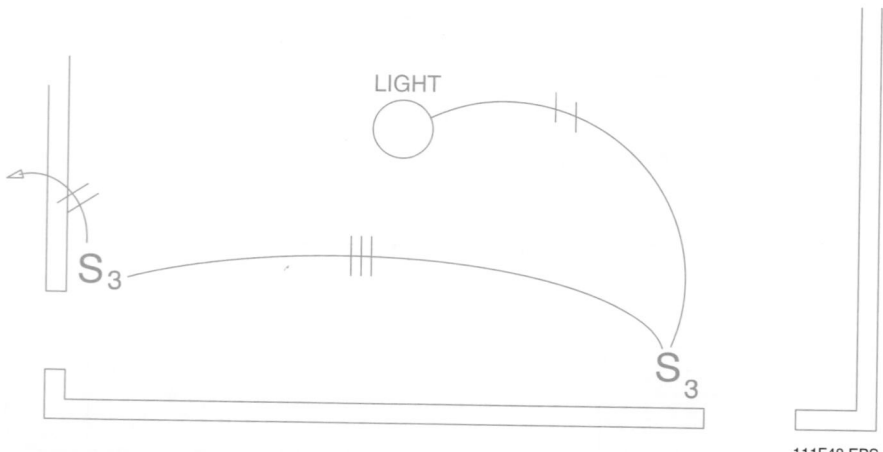

LIGHT

S₃

S₃

111F48.EPS

Figure 48 ◆ Method of showing the wiring arrangement on a floor plan.

The black or hot conductor is connected to the common terminal on the switch, while the white or neutral conductor is spliced to the white conductor of the three-wire, Type NM cable leaving the switch. This three-wire cable is necessary to carry the two travelers plus the neutral to the three-way switch on the right. At this point, the black and red wires connect to the two traveler terminals, respectively. The white or neutral wire is again spliced—this time to the white wire of another two-wire, Type NM cable. The neutral wire is never connected to the switch itself. The black wire of the two-wire, Type NM cable connects to the common terminal on the three-way switch. This cable, carrying the hot and neutral conductors, is routed to the lighting fixture outlet for connection to the fixture.

Another solution is to feed the lighting fixture outlet with two-wire cable. Run another two-wire cable carrying the hot and neutral conductors to one of the three-way switches. A three-wire cable is pulled between the two three-way switches, and then another two-wire cable is routed from the other three-way switch to the lighting fixture outlet.

Some electricians use a shortcut method that eliminates one of the two-wire cables in the preceding method. In this case, a two-wire cable is run from the lighting fixture outlet to one three-way switch. Three-wire cable is pulled between the two three-way switches—two of the wires for travelers and the third for the common point return. This method is shown in *Figure 49*.

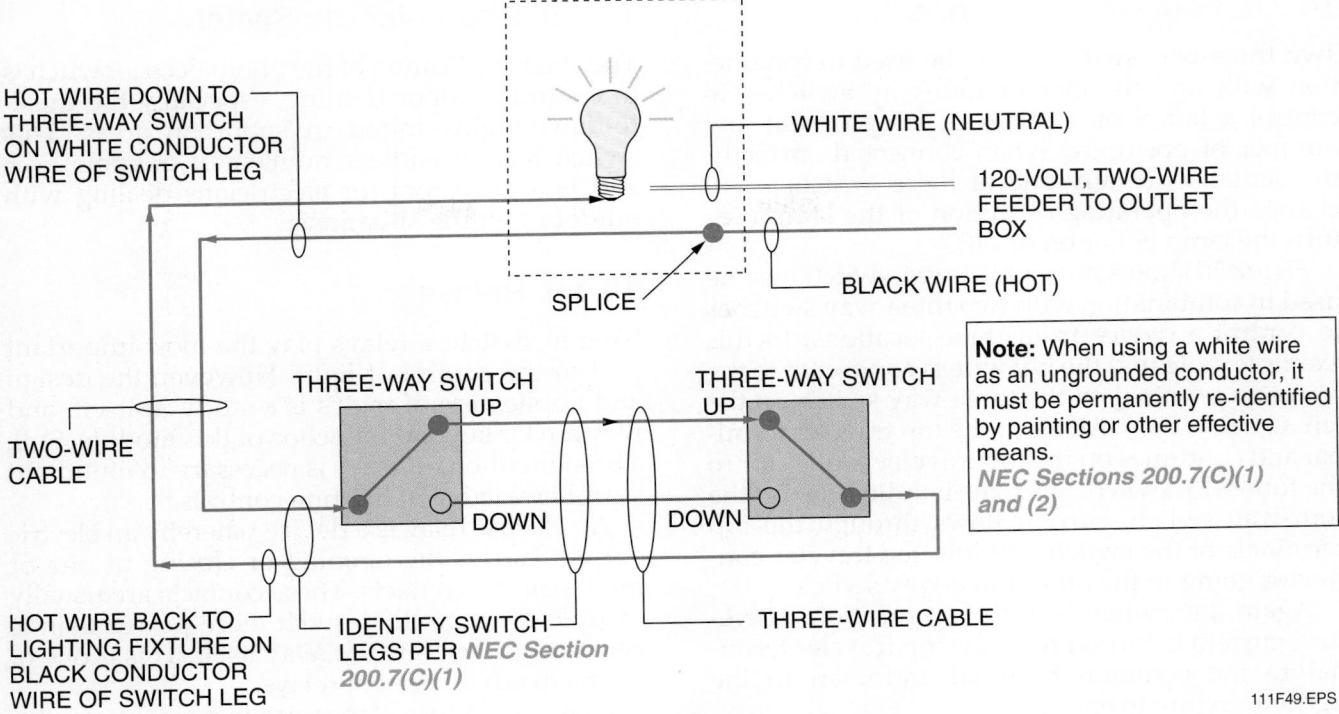

HOT WIRE DOWN TO
THREE-WAY SWITCH
ON WHITE CONDUCTOR
WIRE OF SWITCH LEG

WHITE WIRE (NEUTRAL)

120-VOLT, TWO-WIRE
FEEDER TO OUTLET
BOX

SPLICE

BLACK WIRE (HOT)

Note: When using a white wire as an ungrounded conductor, it must be permanently re-identified by painting or other effective means. *NEC Sections 200.7(C)(1) and (2)*

THREE-WAY SWITCH
UP

THREE-WAY SWITCH
UP

TWO-WIRE
CABLE

DOWN

DOWN

HOT WIRE BACK TO
LIGHTING FIXTURE ON
BLACK CONDUCTOR
WIRE OF SWITCH LEG

IDENTIFY SWITCH
LEGS PER *NEC Section
200.7(C)(1)*

THREE-WIRE CABLE

111F49.EPS

Figure 49 ◆ One way to connect a pair of three-way switches to control one lighting fixture.

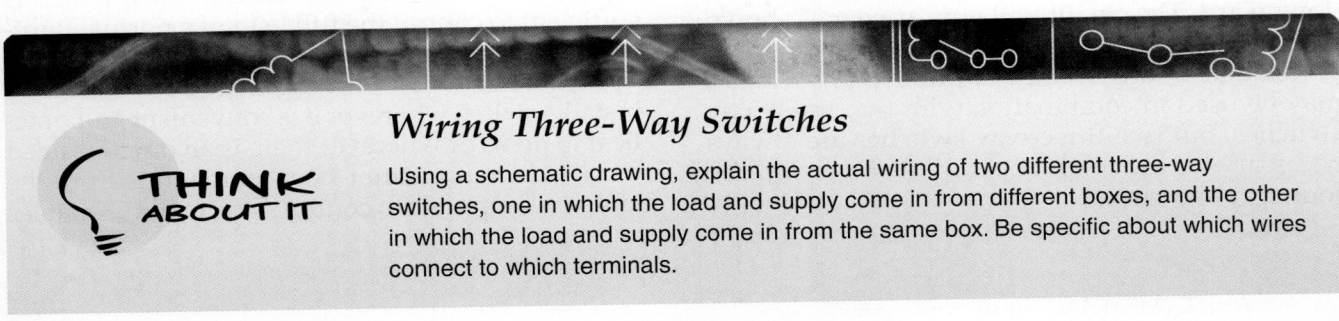

Wiring Three-Way Switches

Using a schematic drawing, explain the actual wiring of two different three-way switches, one in which the load and supply come in from different boxes, and the other in which the load and supply come in from the same box. Be specific about which wires connect to which terminals.

14.2.0 Four-Way Switches

Two three-way switches may be used in conjunction with any number of four-way switches to control a lamp, or a series of lamps, from any number of positions. When connected correctly, the actuation of any one of these switches will change the operating condition of the lamp (i.e., turn the lamp either on or off).

Figure 50 shows how a four-way switch may be used in combination with two three-way switches to control a device from three locations. In this example, note that the hot wire is connected to the common terminal on the three-way switch on the left. Current then travels to the top traveler terminal and continues on the top traveler conductor to the four-way switch. Since the handle is up on the four-way switch, current flows through the top terminals of the switch and into the traveler conductor going to the other three-way switch.

Again, the switch is in the up position. Therefore, current is carried from the top traveler terminal to the common terminal and then to the lighting fixture to energize it.

If the position of any one of the three switch handles is changed, the circuit will be broken and no current will flow to the lamp. For example, assume that the four-way switch handle is flipped downward. The circuit will now appear as shown in *Figure 51*, and the light will be out.

Remember, any number of four-way switches may be used in combination with two three-way switches, but two three-way switches are always necessary for the correct operation of one or more four-way switches.

14.3.0 Photoelectric Switch

The chief application of the photoelectric switch is to control outdoor lighting, especially the dusk-to-dawn lights found in suburban areas. This switch has an endless number of possible uses and is a great tool for electricians dealing with outdoor lighting situations.

14.4.0 Relays

Next to switches, relays play the most important part in the control of light. However, the design and application of relays is a study in itself, and they are far beyond the scope of this module. Still, a brief mention of relays is necessary to round out your knowledge of lighting controls.

An electric relay is a device whereby an electric current causes the opening or closing of one or more pairs of contacts. These contacts are usually capable of controlling much more power than is necessary to operate the relay itself. This is one of the main advantages of relays.

One popular use of the relay in residential lighting systems is that of remote control lighting. In this type of system, all relays are designed to operate on a 24V circuit and are used to control 120V lighting circuits. They are rated at 20A, which is sufficient to control the full load of a normal lighting branch circuit, if desired.

Remote control switching makes it possible to install a switch wherever it is convenient and practical to do so or wherever there is an obvious need for a switch, no matter how remote it is from the lamp or lamps it is to control. This method enables

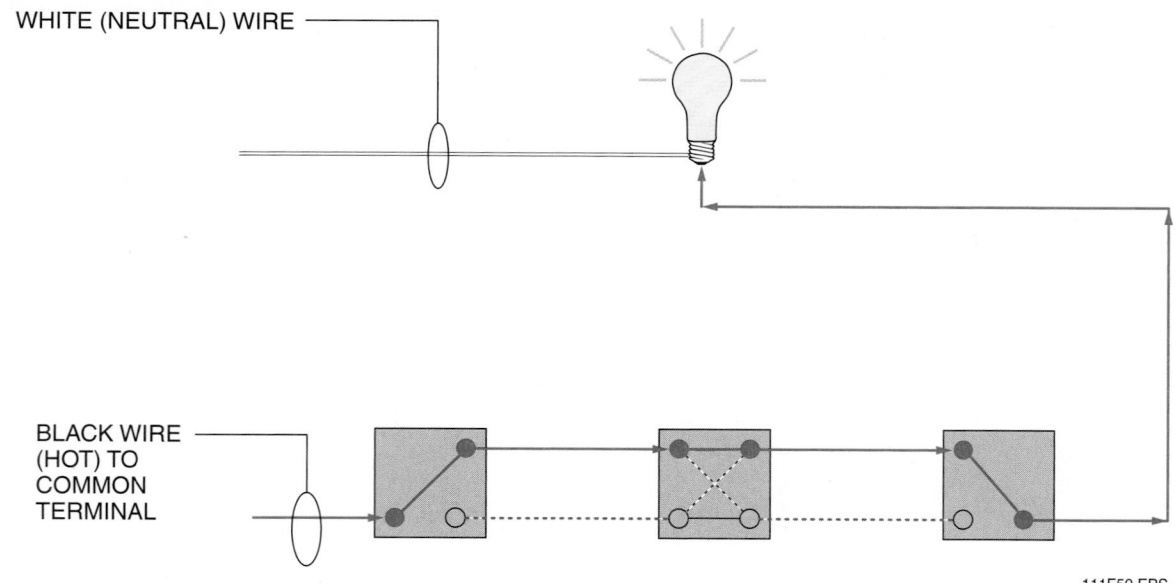

WHITE (NEUTRAL) WIRE

BLACK WIRE (HOT) TO COMMON TERMINAL

111F50.EPS

Figure 50 ◆ Three- and four-way switches used in combination; the light is on.

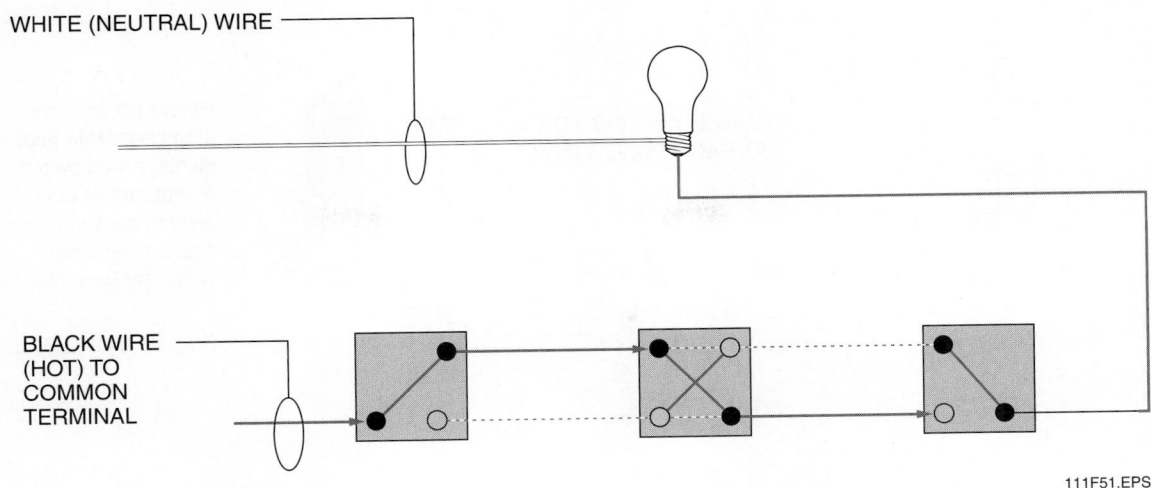

WHITE (NEUTRAL) WIRE

BLACK WIRE
(HOT) TO
COMMON
TERMINAL

111F51.EPS

Figure 51 ◆ Three- and four-way switches used in combination; the light is off.

lighting designs to achieve new advances in lighting control convenience at a reasonable cost. Remote control switching is also ideal for rewiring existing homes with finished walls and ceilings.

One relay is required for each fixture or each group of fixtures that are controlled together. Switch locations for remote control follow the same rules as for conventional direct switching. However, since it is easy to add switches to control a given relay, no opportunities should be overlooked for adding a switch to improve the convenience of control.

Remote control lighting also has the advantage of using selector switches at central locations. For example, selector switches located in the master bedroom or in the kitchen of a home enable the owner to control every lighting fixture on the property from this location. For example, the selector switch may be used to control outside or basement lights that might otherwise be left on inadvertently.

14.5.0 Dimmers

Dimming a lighting system provides control of the quantity of illumination. It may be done to create certain moods or to blend the lighting from different sources for various lighting effects.

For example, in homes with formal dining rooms, a chandelier mounted directly above the dining table and controlled by a dimmer switch becomes the centerpiece of the room while providing general illumination. The dimmer adds versatility since it can set the mood for the activity—low brilliance (candlelight effect) for formal dining or bright for an evening of playing cards. When chandeliers with exposed lamps are used, the dimmer

is essential to avoid a garish and uncomfortable atmosphere. The chandelier should be sized in proportion to the dining area.

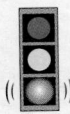

NOTE
It is very important that dimmers be matched to the wattage of the application. Check the manufacturer's data.

14.6.0 Switch Locations

Although the location of wall switches is usually provided for convenience, the *NEC*® also stipulates certain mandatory locations for lighting fixtures and wall switches. See *NEC Section 210.70(A)* for specific switch locations in dwelling units. These locations are deemed necessary for added safety in the home for both the occupants and service personnel.

For example, the *NEC*® requires adequate light in areas where heating, ventilating, and air conditioning (HVAC) equipment is placed. Furthermore, these lights must be conveniently controlled so that homeowners and service personnel do not have to enter a dark area where they might come in contact with dangerous equipment. Three-way switches are required under certain conditions. The *NEC*® also specifies regulations governing lighting fixtures in clothes closets, along with those governing lighting fixtures that may be mounted directly to the outlet box without further support. *Figure 52* summarizes some of the *NEC*® requirements for light and switch placement in the home. For further details, refer to the appropriate sections in the *NEC*®.

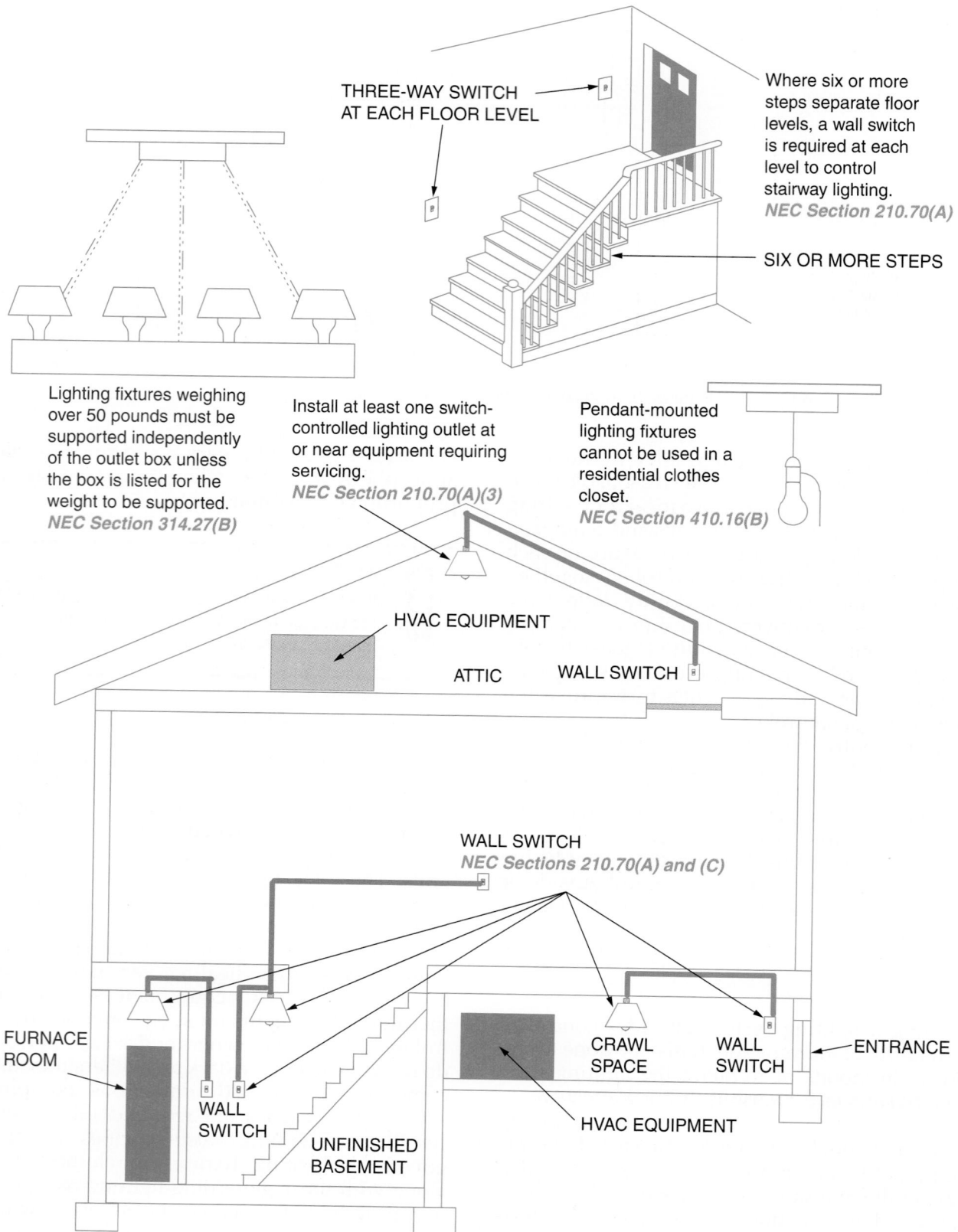

THREE-WAY SWITCH AT EACH FLOOR LEVEL

Where six or more steps separate floor levels, a wall switch is required at each level to control stairway lighting.
NEC Section 210.70(A)

SIX OR MORE STEPS

Lighting fixtures weighing over 50 pounds must be supported independently of the outlet box unless the box is listed for the weight to be supported.
NEC Section 314.27(B)

Install at least one switch-controlled lighting outlet at or near equipment requiring servicing.
NEC Section 210.70(A)(3)

Pendant-mounted lighting fixtures cannot be used in a residential clothes closet.
NEC Section 410.16(B)

HVAC EQUIPMENT

ATTIC

WALL SWITCH

WALL SWITCH
NEC Sections 210.70(A) and (C)

FURNACE ROOM

WALL SWITCH

UNFINISHED BASEMENT

HVAC EQUIPMENT

CRAWL SPACE

WALL SWITCH

ENTRANCE

Figure 52 ◆ *NEC®* requirements for light and switch placement.

111F52.EPS

14.7.0 Low-Voltage Electrical Systems

Conventional lighting systems operate and are controlled by the same system voltage, generally 120V in residential lighting circuits. The *NEC®* permits the use of low-voltage systems to control lighting circuits. There are some advantages to low-voltage systems. One advantage is that the control of lighting from several different locations is more easily accomplished, such as with the remote control system discussed earlier. For example, outside flood lighting can be controlled from several different rooms in a house. The cost of the control wiring is less in that it is rated for a lower voltage and only carries a minimum amount of current compared to a standard lighting system. When extensive or complex lighting control is required, low-voltage systems are preferred. Also, since these circuits are low-energy circuits, circuit protection is not required.

14.7.1 NEC® Requirements for Low-Voltage Systems

NEC Article 725 governs the installation of low-voltage system wiring. These provisions apply to remote control circuits, low-voltage relay switching, low-energy power circuits, and low-voltage circuits. The *NEC®* divides these circuits into three categories:

- Remote control
- Signaling
- Power-limited circuits

As mentioned earlier, circuit protection of the low-voltage circuit is not required; however, the high-voltage side of the transformer that supplies the low-voltage system must be protected. *NEC Chapter 9, Tables 11(A) and 11(B)* cover circuits that are inherently limited in power output and therefore require no overcurrent protection or are limited by a combination of power source and overcurrent protection.

There are a number of requirements of the power systems described in *NEC Chapter 9, Tables 11(A) and 11(B)* and the notes preceding the tables. You should read and study all applicable portions of the *NEC®* before installing low-voltage power systems.

Low-voltage systems are described in more detail in later modules.

15.0.0 ◆ ELECTRIC HEATING

The use of electric heating in residential occupancies has risen tremendously over the past decade or so, and the practice will no doubt continue. This is due to the following advantages of electric heat over most other heating systems:

- Electric heat is noncombustible and is therefore safer than combustible fuels.
- It requires no storage space, fuel tanks, or chimneys.
- It requires little maintenance.
- The initial installation cost is relatively inexpensive when compared to other types of heating systems.
- The comfort level may be improved since each room may be controlled separately by its own thermostat.

There are also some disadvantages to using electric baseboard heat, especially in northern climates. Some of these disadvantages include:

- Electric heat is often more expensive to operate than other types of fuels.
- Receptacles must not be installed above electric baseboard heaters.
- Electric baseboard heaters tend to discolor the wall area immediately above the heater, especially if there are smokers in the home.

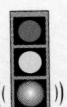

NOTE

It is very important to calculate the extra electric load of an electric heater installation (especially in an add-on situation). Ensure that the extra load does not exceed the maximum amperage draw of either the circuit or the panel.

The type of electric heating system used for a given residence will usually depend on the structural conditions, the kind of room, and the activities for which the room will be used. The homeowner's preference will also enter into the final decision.

Electric heating equipment is available in baseboard, wall, ceiling, kick space, and floor units; in resistance cable embedded in the ceiling or concrete floor; in forced-air duct systems similar to conventional oil- or gas-fired hot air systems; and in electric boilers for hot water baseboard heat.

Electric heat pumps have also become popular for HVAC systems in certain parts of the country. The term heat pump, as applied to a year-round air conditioning system, commonly denotes a system in which refrigeration equipment is used in such a manner that heat is taken from a heat source and transferred to the conditioned space when heating is desired; heat is removed from the space and discharged to a heat sink when cooling and dehumidification are desired.

A heat pump has the unique ability to furnish more energy than it consumes. This is due to the fact that under certain outdoor conditions, electrical energy is required only to move the refrigerant and run the fan; thus, a heat pump can attain a heating efficiency of two or more to one; that is, it will put out an equivalent of two or three watts of heat for every watt consumed. For this reason, its use is highly desirable for the conservation of energy.

In general, electric baseboard heating equipment should be located on the outside wall near the areas where the greatest heat loss will occur, such as under windows, etc. The controls for wall-mounted thermostats should be located on an interior wall, about 50 inches above the floor to sense the average room temperature. *Figure 53* shows an electric heating arrangement for the sample residence. *NEC*® regulations governing the installation of these units are also noted.

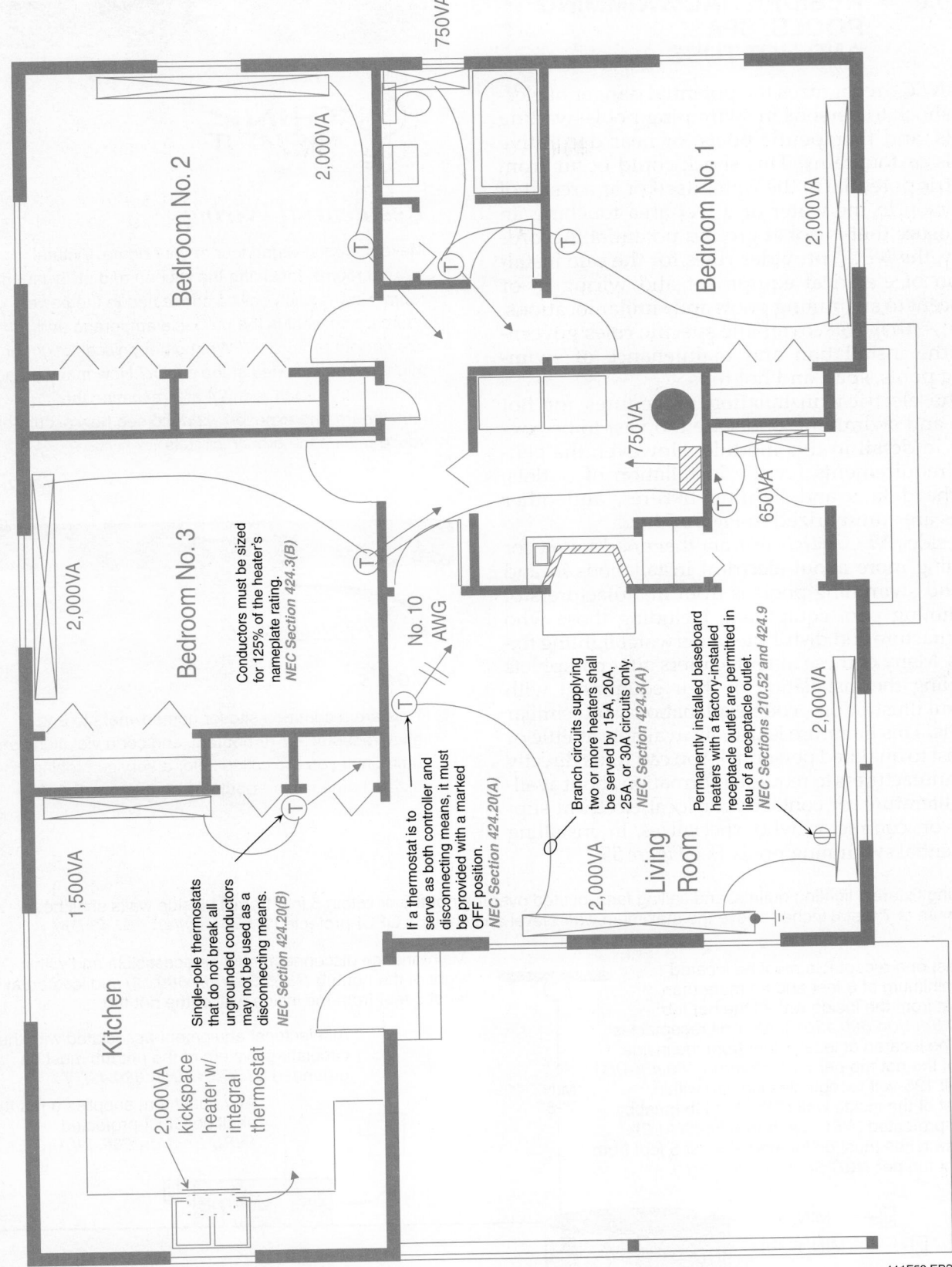

750VA

2,000VA

Bedroom No. 2

Bedroom No. 1

2,000VA

750VA

650VA

Bedroom No. 3

2,000VA

Conductors must be sized for 125% of the heater's nameplate rating.
NEC Section 424.3(B)

No. 10 AWG

2,000VA

Single-pole thermostats that do not break all ungrounded conductors may not be used as a disconnecting means.
NEC Section 424.20(B)

If a thermostat is to serve as both controller and disconnecting means, it must be provided with a marked OFF position.
NEC Section 424.20(A)

Branch circuits supplying two or more heaters shall be served by 15A, 20A, 25A, or 30A circuits only.
NEC Section 424.3(A)

Permanently installed baseboard heaters with a factory-installed receptacle outlet are permitted in lieu of a receptacle outlet.
NEC Sections 210.52 and 424.9

Living Room

2,000VA

1,500VA

Kitchen

2,000VA kickspace heater w/ integral thermostat

111F53.EPS

Figure 53 ◆ Electric heating arrangement for the sample residence.

16.0.0 ◆ RESIDENTIAL SWIMMING POOLS, SPAS, AND HOT TUBS

The *NEC*® recognizes the potential danger of electric shock to persons in swimming pools, wading pools, and therapeutic pools, or near decorative pools or fountains. This shock could occur from electric potential in the water itself or as a result of a person in the water or a wet area touching an enclosure that is not at ground potential. Accordingly, the *NEC*® provides rules for the safe installation of electrical equipment and wiring in or adjacent to swimming pools and similar locations. *NEC Article 680* covers the specific rules governing the installation and maintenance of swimming pools, spas, and hot tubs.

The electrical installation procedures for hot tubs and swimming pools are too vast to be covered in detail in this module. However, the general requirements for the installation of outlets, overhead fans and lighting fixtures, and other items are summarized in *Figure 54*.

Besides *NEC Article 680,* another good source for learning more about electrical installations in and around swimming pools is from manufacturers of swimming pool equipment, including those who manufacture and distribute underwater lighting fixtures. Many of these manufacturers offer pamphlets detailing the installation of their equipment with helpful illustrations, code explanations, and similar details. This literature is usually available at little or no cost to qualified personnel. You can write directly to manufacturers to request information about available literature, or contact your local electrical supplier or contractor who specializes in installing residential swimming pools. See *Figure 55*.

Residential Wiring

Make a mental wiring tour of your home. Picture several rooms, including the kitchen and utility/laundry room. How is each device connected to the power source, and what is the probable amperage and overcurrent protection? What other devices might or might not be included in the circuit? How many branch circuits serve each room? Later, examine the panelboard and exposed wiring to see how accurately you identified the branch circuits.

Pools

Pools are a common site for homeowners to add lights, receptacles, or heaters, and code violations are common. If you are called in for a service problem, ask the homeowner about any do-it-yourself wiring.

Lighting fixtures, lighting outlets, and ceiling fans located over the hot tub or within 5 feet from its inside walls shall be a minimum of 7 feet 6 inches above the maximum water level and shall be GFCI-protected [*NEC Section 680.43(B)*].

At least one receptacle must be located at a minimum of 6 feet and no more than 10 feet from the inside wall of the hot tub [*NEC Section 680.43(A)*]. Also, all receptacles must be located at least 6 feet from the inside wall of the hot tub per *NEC Section 680.43(A)(1)* and all 125-volt receptacles located within 10 feet of the inside wall of the hot tub must be GFCI-protected [*NEC Section 680.43(A)(2)*]. Wall switches must be located at least 5 feet from the hot tub per *NEC Section 680.43(C)*.

Maintenance disconnect must be accessible and within sight of the hot tub (*NEC Section 680.12*) and located at least 5 feet from the inside wall of the hot tub.

All electrical equipment associated with the circulating system of the hot tub must be grounded [*NEC Section 680.43(F)*].

Any outlet that supplies a hot tub shall be GFCI-protected [*NEC Section 680.43(A)(3)*].

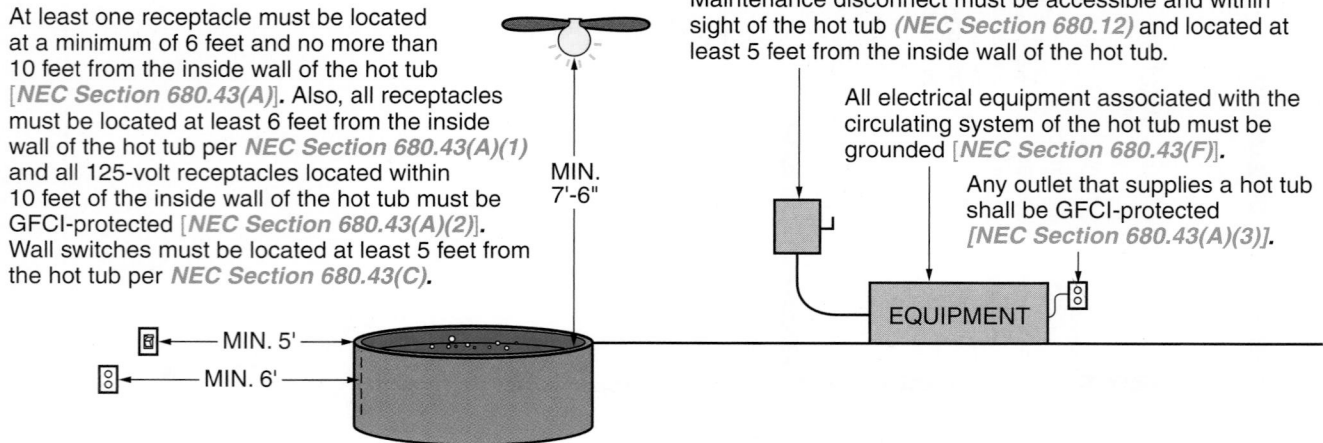

Figure 54 ◆ *NEC*® requirements for packaged indoor hot tubs.

111F54.EPS

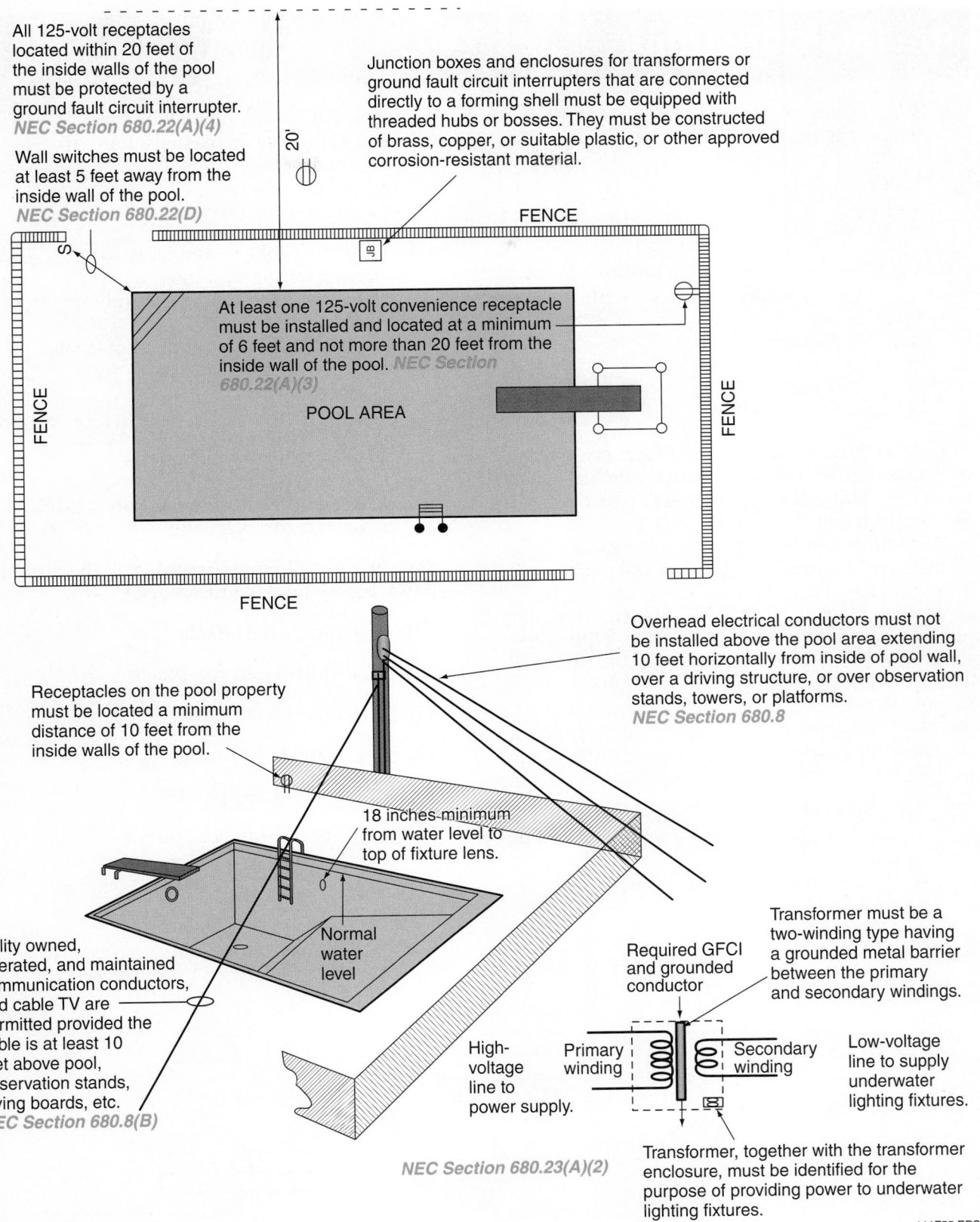

All 125-volt receptacles located within 20 feet of the inside walls of the pool must be protected by a ground fault circuit interrupter. *NEC Section 680.22(A)(4)*

Wall switches must be located at least 5 feet away from the inside wall of the pool. *NEC Section 680.22(D)*

Junction boxes and enclosures for transformers or ground fault circuit interrupters that are connected directly to a forming shell must be equipped with threaded hubs or bosses. They must be constructed of brass, copper, or suitable plastic, or other approved corrosion-resistant material.

20'

FENCE

JB

FENCE

At least one 125-volt convenience receptacle must be installed and located at a minimum of 6 feet and not more than 20 feet from the inside wall of the pool. *NEC Section 680.22(A)(3)*

POOL AREA

FENCE

FENCE

Receptacles on the pool property must be located a minimum distance of 10 feet from the inside walls of the pool.

Overhead electrical conductors must not be installed above the pool area extending 10 feet horizontally from inside of pool wall, over a driving structure, or over observation stands, towers, or platforms. *NEC Section 680.8*

18 inches minimum from water level to top of fixture lens.

Normal water level

Utility owned, operated, and maintained communication conductors, and cable TV are permitted provided the cable is at least 10 feet above pool, observation stands, diving boards, etc. *NEC Section 680.8(B)*

Required GFCI and grounded conductor

Transformer must be a two-winding type having a grounded metal barrier between the primary and secondary windings.

High-voltage line to power supply.

Primary winding

Secondary winding

Low-voltage line to supply underwater lighting fixtures.

NEC Section 680.23(A)(2)

Transformer, together with the transformer enclosure, must be identified for the purpose of providing power to underwater lighting fixtures.

111F55.EPS

Figure 55 ◆ NEC® requirements for typical swimming pool installations.

1. When sizing electrical services, at what percentage is the first 3,000VA rated?
   a. 20%
   b. 45%
   c. 80%
   d. 100%

2. What section of the *NEC*® requires that fittings be identified for use with service masts?
   a. *NEC Section 230.28*
   b. *NEC Section 230.40*
   c. *NEC Section 250.46*
   d. *NEC Section 250.83*

3. A service conductor without an overall jacket must have a clearance of not less than _____ above a window that can be opened.
   a. two feet
   b. three feet
   c. eight feet
   d. ten feet

4. *NEC Section 230.6* considers conductors installed under at least _____ inch(es) of concrete to be outside the building.
   a. one
   b. two
   c. four
   d. five

5. Type NM cable may *not* be used in _____.
   a. shallow chases of masonry, concrete, or adobe
   b. the framework of a building
   c. protective strips
   d. attic spaces

6. Type MC cable may *not* be used _____.
   a. in concrete or plaster where dry
   b. in dry masonry
   c. in attic spaces
   d. where subject to physical damage

7. Type SE cable is available with _____ for interior wiring systems.
   a. a non-insulated ground or neutral conductor
   b. an insulated grounded conductor
   c. no ground conductor
   d. guard strips

8. Type SER cable may be used _____.
   a. in overhead applications
   b. underground
   c. as a subfeed under certain conditions
   d. in hazardous locations

Identify the following receptacles by numbers shown below for each receptacle.

9. _____ Single receptacle

10. _____ Split-wired receptacle

11. _____ Dryer receptacle

12. _____ Range receptacle

13. _____ Duplex receptacle

14. _____ Special-purpose outlet

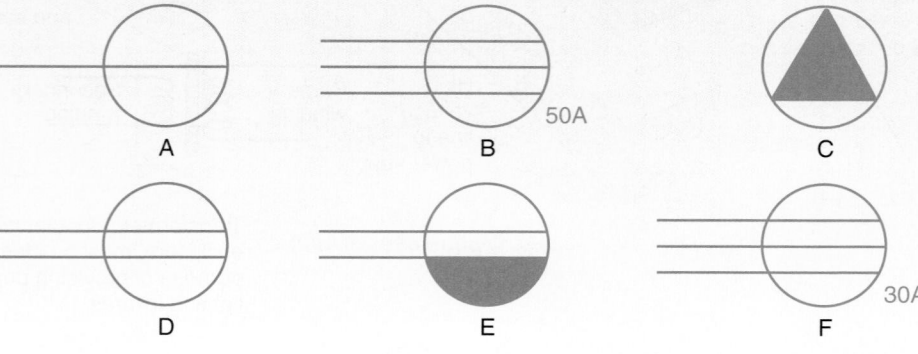

111RQ01.EPS

15. Using *NEC Table 314.16(A),* calculate the cubic inches required for the receptacle outlets shown in the table below. Then, indicate the size of the metallic box that should be used.

Number and Size of Conductors in Box	Free Space within Box for Each Conductor	Total Cubic Inches of Box Space Required	What Size Metallic Box May Be Used?
A. Six No. 12 conductors and three ground wires	2.25	_____	_____
B. Seven No. 12 conductors and three ground wires with one receptacle	2.25	_____	_____
C. Two No. 14 conductors and one ground wire	2.0	_____	_____
D. Four No. 14 conductors and two ground wires	2.0	_____	_____
E. Six No. 14 conductors and three ground wires with one receptacle	2.0	_____	_____

Summary

This module covered the basics of residential wiring, including load calculations and wiring devices.

Residential electrical system design begins with the floor plan. The square footage of the home is used to calculate the lighting loads. The lighting loads, along with small appliance loads, fixed appliance loads, and other loads must be calculated to determine the service size. Demand factors are applied and the neutral conductor is sized per *NEC Article 220*. GFCI and AFCI breakers must be installed in all required locations, and the system must be grounded and bonded per *NEC Article 250*. Most residences use Type NM or MC cable for branch circuits. Switches must be installed in all locations required by the *NEC®*. Special consideration must be given to electric heating units and circuits supplying swimming pools and hot tubs.

A thorough knowledge of the *NEC®* is essential to the safe and successful installation of residential wiring systems.

Notes

1. _____ consists of wires with a spiral-wound, flexible steel outer jacketing.

2. A type of cable that is popular in residential and small commercial wiring systems, _____ may be used for both exposed and concealed work in normally dry locations; this type of cable is also referred to as _____.

3. A(n) _____ is a piece of equipment that has been designed for a particular purpose.

4. A(n) _____ is used for turning an electrical circuit on and off.

5. The circuit that is routed to a switch box for controlling electric lights is known as a(n) _____.

6. A(n) _____ is equipped with a conductor terminal to take a bonding jumper.

7. A(n) _____ is a bare or green insulated conductor used to ensure conductivity between metal parts that are required to be electrically connected.

8. The _____ is comprised of the conductors that extend from the last power company pole to the point of connection at the service facilities.

9. The _____ is the point where power is supplied to a building.

10. _____ lie between the point of termination of the overhead service drop or underground service lateral and the main disconnecting device in the building.

11. _____ mainly provides overcurrent protection to the feeder and service conductors.

12. A(n) _____ is comprised of the underground conductors through which service is supplied between the power company's distribution facilities and their first point of connection to the building.

13. The portion of a wiring system that extends beyond the final overcurrent device is the _____.

14. A(n) _____ is a circuit that carries current from the service equipment to a subpanel or a branch circuit panel or to some point in the wiring system.

15. Normally located at the service entrance of a residential installation, a(n) _____ usually contains the main disconnect.

16. Raceway, cable, wires, boxes, and other equipment are installed during _____.

Trade Terms

Appliance
Bonding bushing
Bonding jumper
Branch circuit
Feeder
Load center

Metal-clad (Type MC) cable
Nonmetallic-sheathed (Type NM) cable
Romex®

Roughing in
Service drop
Service entrance
Service-entrance conductors

Service-entrance equipment
Service lateral
Switch
Switch leg

Dan Lamphear

Associated Builders and Contractors, Inc.

Like many other people, Dan Lamphear just fell into his career as an electrician. But once he discovered the electrical trade, he knew he had found a home. Since then, he has progressed from a helper to an apprentice, a journeyman, an independent contractor, an inventor, and finally, a teacher.

It was as much luck as anything else that led Dan toward a career as a professional electrician more than two decades ago. He wasn't sure what he wanted to do with his life after graduating from high school. However, after watching an electrician perform a commercial wiring job at a friend's business—and providing a helping hand—he was hooked. "It seemed like a challenging career, and I was curious to learn more about how electricity works," he recalls.

Dan was hired by that same electrician, under whom he apprenticed for several years before hearing about the NCCER program. He jumped at the chance to further his skills through the program. "Like they say, knowledge is money," he smiles.

After graduating from the program, Dan struck out on his own as an independent electrician,

specializing in plant maintenance, industrial, and commercial work. His ability to diagnose and repair electrical problems in factory machinery soon made him a valuable contractor in Milwaukee's industrial sector.

He also discovered his knack for invention, and he has designed and built specialized machinery for a company that hired him as its full-time electrical maintenance supervisor. "Knowing the electrical side of machinery allowed me to understand how they operate mechanically," he says of his work as an inventor.

Dan later returned to the Associated Builders and Contractors (ABC) chapter, which trains out of a local community college, to repay the favor that helped him embark on his career. He teaches Electrical Level 2 courses for students who represent the next generation of professional electricians.

"Knowing how to use test instruments is perhaps the most important aspect of the job," he notes. "I still have some of the same meters I started out with."

David Lewis

Instructor
Putnam Career & Technical Center

David Lewis started his career working in coal mines. After a few years he opened his own electrical business. Now he is an electrical instructor and works with the State Department of Education on curriculum development. He also serves on the NCCER revision team for the Electrical curriculum.

How did you first get interested in the field?
After graduating from high school in 1972 and attending college for a while, I decided that I wanted to work in the coal mines. Electricity interested me, so I became a maintenance foreman/electrician. Eventually, I started my own business.

What kind of training have you been through?
While working in the coal mines, I attended several electrical training classes and obtained my underground electrical license. While in business for myself I got my Master Electrician license and attended many update classes. Since I have started teaching, I have attended classes on PLCs and other topics. I also went back to college and obtained a bachelor of science degree in Career and Technical Education.

What work have you done in your career?
In the coal mines I worked on all types of mining equipment. After starting my own business, I worked mainly in residential and light commercial wiring.

Tell us about your present job and what you like about it.
I enjoy being an instructor in Electrical Technology. I work mostly with high school students and during the two years they are with me, it is great to see them grasp the knowledge of electricity.

What factors have contributed most to your success?
Hard work and the willingness to learn from experienced electricians.

What advice would you give to those new to the field?
Try to learn all you can. Work with an experienced electrician and learn from them. Attend any training or classes that you can. There is always something new to learn.

Trade Terms
Introduced in This Module

Appliance: Equipment designed for a particular purpose (for example, using electricity to produce heat, light, or mechanical motion). Appliances are usually self-contained, are generally available for applications other than industrial use, and are normally produced in standard sizes or types.

Bonding bushing: A special conduit bushing equipped with a conductor terminal to take a bonding jumper. It also has a screw or other sharp device to bite into the enclosure wall to bond the conduit to the enclosure without a jumper when there are no concentric knockouts left in the wall of the enclosure.

Bonding jumper: A bare or green insulated conductor used to ensure the required electrical conductivity between metal parts required to be electrically connected. Bonding jumpers are frequently used from a bonding bushing to the service-equipment enclosure to provide a path around concentric knockouts in an enclosure wall, and they may also be used to bond one raceway to another.

Branch circuit: The portion of a wiring system extending beyond the final overcurrent device protecting a circuit.

Feeder: A circuit, such as conductors in conduit or a cable run, that carries current from the service equipment to a subpanel or a branch circuit panel or to some point in the wiring system.

Load center: A type of panelboard that is normally located at the service entrance of a residential installation. It usually contains the main disconnect.

Metal-clad (Type MC) cable: A type of cable with a metal jacket that is popular for use in residential and small commercial wiring systems. In general, it may be used for both exposed and concealed work in normally dry locations. It may also be used as service-entrance cable and in other locations as permitted by the *NEC®*.

Nonmetallic-sheathed (Type NM) cable: A type of cable that is popular for use in residential and small commercial wiring systems. In general, it may be used for both exposed and concealed work in normally dry locations. *See Romex®.*

Romex®: General Cable's trade name for Type NM cable; however, it is often used generically to refer to any nonmetallic-sheathed cable.

Roughing in: The first stage of an electrical installation, when the raceway, cable, wires, boxes, and other equipment are installed. This is the electrical work that must be done before any finishing work can be done.

Service drop: The overhead conductors, through which electrical service is supplied, between the last power company pole and the point of their connection to the service facilities located at the building.

Service entrance: The point where power is supplied to a building (including the equipment used for this purpose). The service entrance includes the service main switch or panelboard, metering devices, overcurrent protective devices, and conductors/raceways for connecting to the power company's conductors.

Service-entrance conductors: The conductors between the point of termination of the overhead service drop or underground service lateral and the main disconnecting device in the building.

Service-entrance equipment: Equipment that provides overcurrent protection to the feeder and service conductors, a means of disconnecting the feeders from energized service conductors, and a means of measuring the energy used.

Service lateral: The underground conductors through which service is supplied between the power company's distribution facilities and the first point of their connection to the building or area service facilities located at the building.

Switch: A mechanical device used for turning an electrical circuit on and off.

Switch leg: A circuit routed to a switch box for controlling electric lights.

Other Codes and Standards That Apply to Electrical Installations

Until 2000, there were three model building codes. These included the following:

- *Standard Building Code (SBC)* – Published by the Southern Building Code Congress International.
- *BOCA National Building Code (NBC)* – Published by the Building Officials and Code Administrators.
- *Uniform Building Code (UBC)* – Published by the International Conference of Building Officials.

The three code writing groups, SBCCI, BOCA, and UBC, combined into one organization called the International Code Council with the purpose of writing one nationally accepted family of building and fire codes. The first edition of the *International Building Code* was published in 2000 and the second edition in 2003, and updated again in 2006. It is intended to continue on a three-year cycle.

The International Residential Code (IRC) is adopted as part of the electrical code requirements in many areas of the country. The IRC covers one- and two-family dwellings of three stories or less. The IRC includes requirements for such things as ventilating fans for bathrooms, requirements for smoke detectors, and other items not specified by the *NEC®*.

The IRC covers all trades, including building, plumbing, mechanical, gas, energy, and electrical.

In 2002, the NFPA published its own building code, *NFPA 5000*. There are now two nationally recognized codes competing for adoption by the 50 states.

To be thoroughly competent in the electrical trade, you should become familiar with the contents of these codes and the terminology used in them.

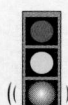

NOTE

Always refer to the latest editions of codes in effect in your area.

This module is intended to present thorough resources for task training. The following reference work is suggested for further study. This is optional material for continued education rather than for task training.

National Electrical Code® Handbook, Latest Edition. Quincy, MA: National Fire Protection Association.

CONTREN® LEARNING SERIES – USER UPDATE

NCCER makes every effort to keep these textbooks up-to-date and free of technical errors. We appreciate your help in this process. If you have an idea for improving this textbook, or if you find an error, a typographical mistake, or an inaccuracy in NCCER's Contren® textbooks, please write us, using this form or a photocopy. Be sure to include the exact module number, page number, a detailed description, and the correction, if applicable. Your input will be brought to the attention of the Technical Review Committee. Thank you for your assistance.

Instructors – If you found that additional materials were necessary in order to teach this module effectively, please let us know so that we may include them in the Equipment/Materials list in the Annotated Instructor's Guide.

Write: Product Development and Revision
National Center for Construction Education and Research
3600 NW 43rd St., Bldg. G, Gainesville, FL 32606

Fax: 352-334-0932

E-mail: curriculum@nccer.org

Craft   Module Name _____

Copyright Date   Module Number   Page Number(s) _____

Description _____

_____

_____

_____

(Optional) Correction _____

_____

_____

(Optional) Your Name and Address _____

_____

_____

Introduction to HVAC
68114-09

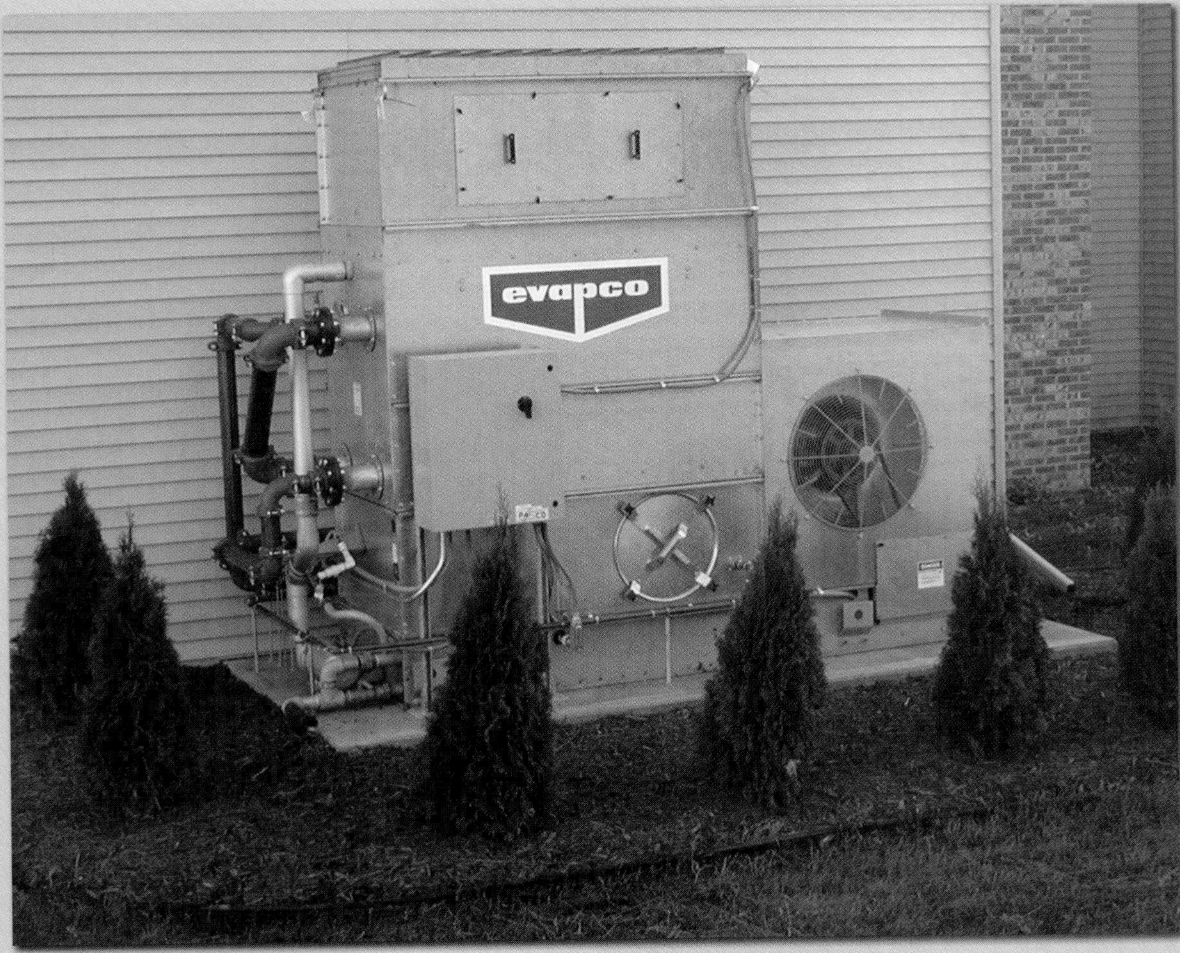

68114-09
Introduction to HVAC

Topics to be presented in this module include:

1.0.0 Introduction 14.2
2.0.0 Heating 14.2
3.0.0 Ventilation 14.6
4.0.0 Air Conditioning 14.6
5.0.0 Blueprints, Codes, and Specifications 14.9
6.0.0 Careers in HVAC 14.13
7.0.0 Types of Training Programs 14.15
8.0.0 The HVAC Technician and the Environment 14.18

Overview

This program will pave the way for you to begin a career in one of America's most dynamic industries. Virtually every one of the tens of millions of homes and businesses in the United States has a heating system, and a large percentage have comfort cooling systems as well. Workers trained in this industry have the opportunity to install systems in new construction, service equipment in existing construction, and replace aging systems.

Working in the HVAC industry is challenging and rewarding because environmental technology is constantly changing. Technical advances in HVAC systems are made every day in advanced computerized controls, greater operating efficiency, and improved packaging.

The HVAC industry offers many opportunities for advancement. The training you are receiving can qualify you to become an installer, troubleshooter, sales technician, system design specialist, and eventually even the owner of your own HVAC service business.

Objectives

When you have completed this module, you will be able to do the following:

1. Explain the basic principles of heating, ventilating, and air conditioning.
2. Identify career opportunities available to people in the HVAC trade.
3. Explain the purpose and objectives of an apprentice training program.
4. Describe how certified apprentice training can start in high school.
5. Describe what the Clean Air Act means to the HVAC trade.
6. Describe the types of regulatory codes encountered in the HVAC trade.
7. Identify the types of schedules/drawings used in the HVAC trade.

Trade Terms

Chiller
Chlorofluorocarbon (CFC) refrigerant
Compressor
Condenser
Evaporator
Expansion device
Heat transfer
HVAC plan
Hydrochlorofluorocarbon (HCFC) refrigerant
International Building Code
Mechanical refrigeration
Noxious
Reclamation
Recovery
Recycling
Refrigeration cycle
Takeoffs
Toxic

Required Trainee Materials

1. Paper and pencil
2. Appropriate personal protective equipment

Prerequisites

Before you begin this module it is recommended that you successfully complete *Core Curriculum*, and *Construction Technology* modules 68101-09 through 68113-09.

This course map shows all of the modules in *Construction Technology*. The suggested training order begins at the bottom and proceeds up. Skill levels increase as you advance on the course map. The local Training Program Sponsor may adjust the training order.

CONSTRUCTION TECHNOLOGY

68117-09
Copper Pipe and Fittings

68116-09
Plastic Pipe and Fittings

68115-09
Introduction to Drain, Waste, and Vent (DWV) Systems

68114-09
Introduction to HVAC

68113-09
Residential Electrical Services

68112-09
Electrical Safety

68111-09
Basic Stair Layout

68110-09
Exterior Finishing

68109-09
Roofing Applications

68108-09
Roof Framing

68107-09
Wall and Ceiling Framing

68106-09
Floor Systems

68105-09
Masonry Units and Installation Techniques

68104-09
Introduction to Masonry

68103-09
Handling and Placing Concrete

68102-09
Introduction to Concrete, Reinforcing Materials, and Forms

68101-09
Site Layout One: Distance Measurement and Leveling

CORE CURRICULUM:
Introductory Craft Skills

114CMAP.EPS

1.0.0 ◆ INTRODUCTION

Since ancient times, people have sought ways to make the buildings in which they live, work, and play more comfortable. Today, the Heating, Ventilating, and Air Conditioning (HVAC) industry provides the means to control the temperature, humidity, and even the cleanliness of the air in our homes, schools, offices, and factories. The members of the HVAC trade are skilled workers who install, maintain, and repair the equipment that makes this possible.

2.0.0 ◆ HEATING

Early humans burned fuel as a source of heat. That hasn't changed; what's different between then and now is the way it's done. We no longer need to huddle around a wood fire to keep warm. Instead, a central heating source such as a furnace or boiler does the job using the **heat transfer** principle; that is, heat is created in one place and carried to another place by means of air or water.

For example, in a common household furnace, fuel oil, natural gas, or propane/butane gas is burned to create heat, which warms metal plates known as heat exchangers (*Figure 1*). Air from living spaces is circulated over the heat exchangers and returned to the living spaces as heated air. This type of system is known as a forced-air system, and it is the most common type of central-heating system used in the United States.

Water is also used as a heat exchange medium. The water is heated in a boiler (*Figure 2*), then pumped through pipes to heat exchangers where the heat it contains is transferred to the surrounding air. The heat exchangers are usually baseboard heating elements located in the space to be heated. This type of system is known as a hydronic heating system, and it is more common in the Northeast and Midwest than in other parts of the country.

Natural gas and fuel oil are, by far, the most widely used heating fuels. Natural gas is currently the most popular fuel. Oil heat is more common in the Northeast than in other parts of the country. Propane/butane gas is used in many parts of the country in place of oil or natural gas. Oil and propane/butane fuels are primarily used in rural areas of the country where natural gas pipelines are not readily available.

Electricity is also used as a heat source. In an electric heating system, electricity flows through coils of heavy wire, causing the coils to become hot. Air from the conditioned space is passed over the coils and the heat from the coils is transferred to the air. Because electricity is so expensive, total electric heat is no longer common in cold climates. Total electric heat is more likely to be used in warm climates where heat is seldom required. Heat pumps are used, and in most cases are preferred over total electric heat. In some areas of the country, a main or supplementary heating system may be fueled by wood or coal. Due to increases in the cost of fossil fuels, such as oil and gas, over the last several decades, and the fact that wood is a relatively cheap and renewable resource, wood-burning stoves and furnaces have remained quite popular.

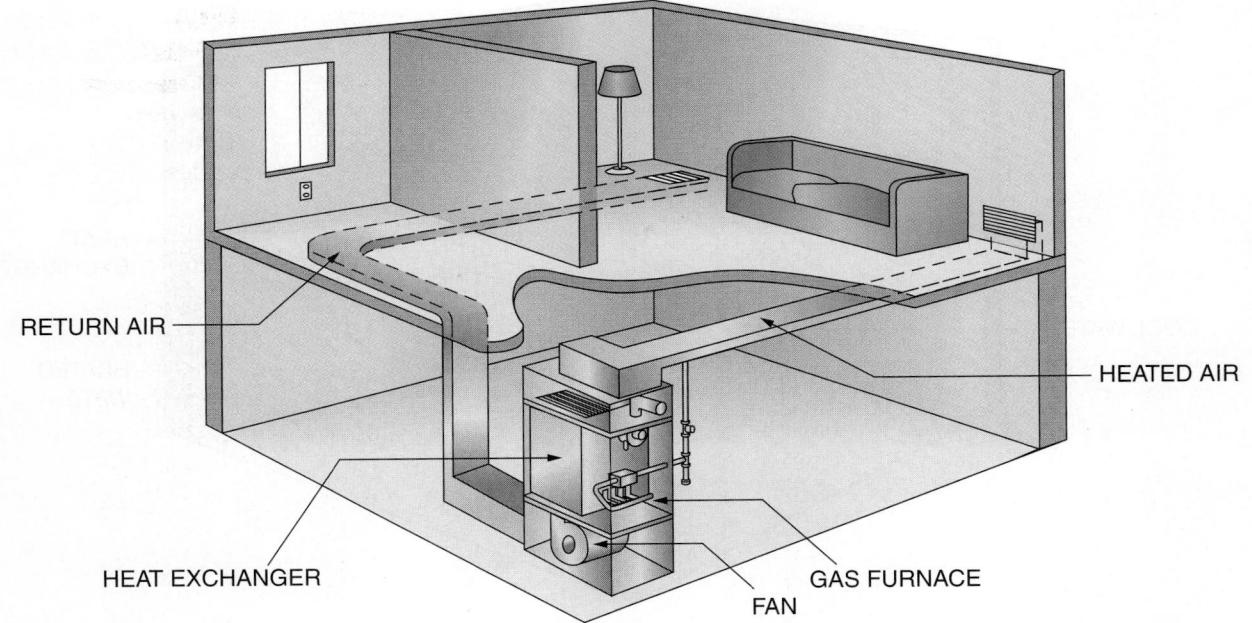

RETURN AIR

HEATED AIR

HEAT EXCHANGER

FAN

GAS FURNACE

BASEMENT INSTALLATION

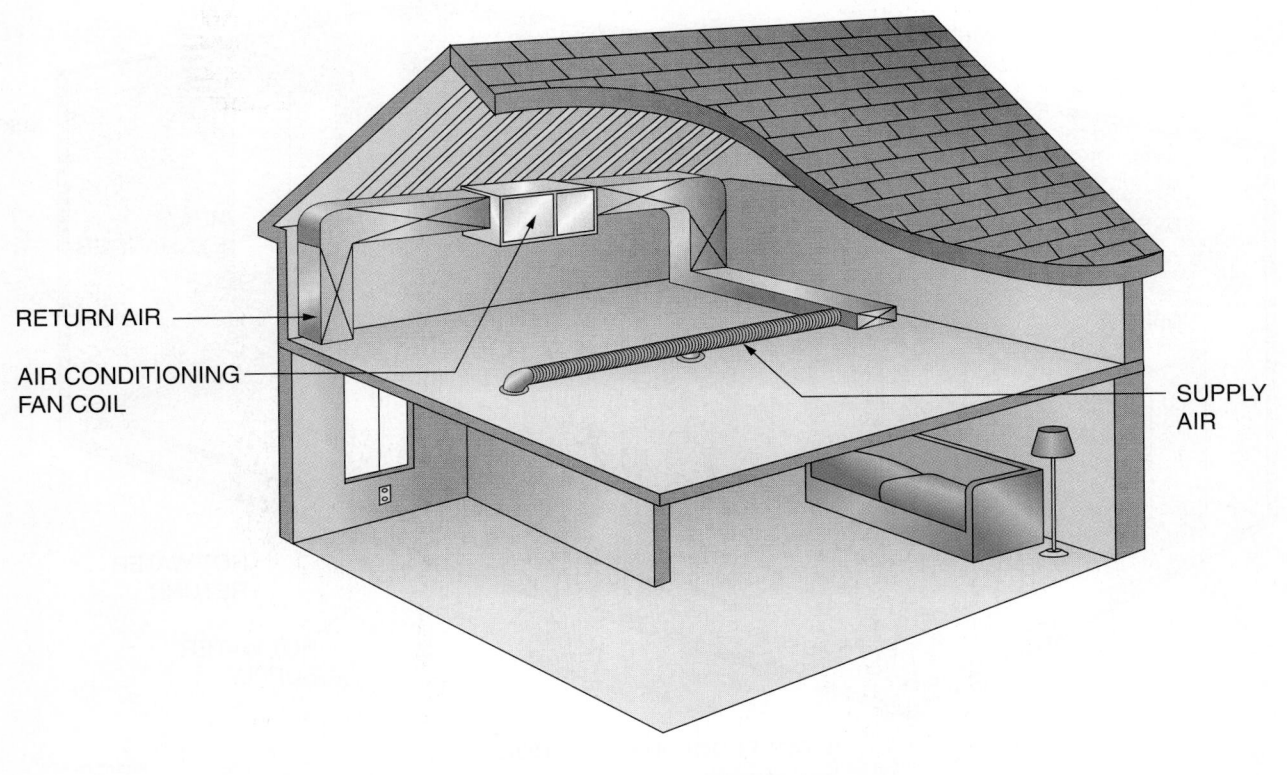

RETURN AIR

AIR CONDITIONING
FAN COIL

SUPPLY
AIR

ATTIC INSTALLATION

101F01.EPS

Figure 1 ◆ Forced-air heating.

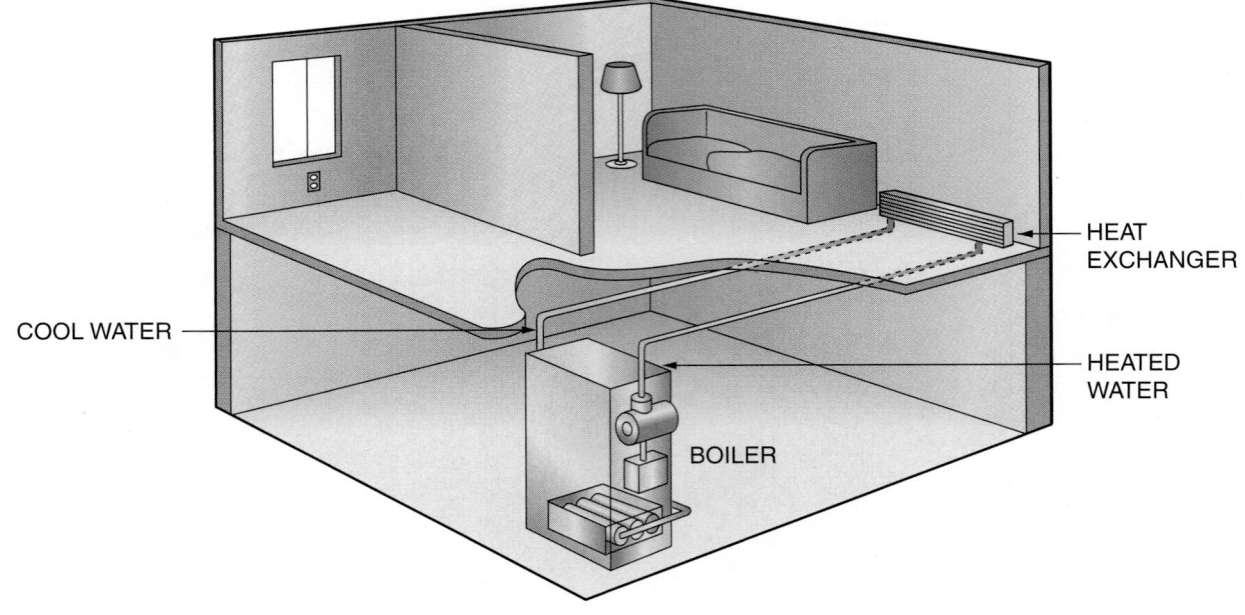

HEAT
EXCHANGER

COOL WATER

HEATED
WATER

BOILER

BASEMENT INSTALLATION

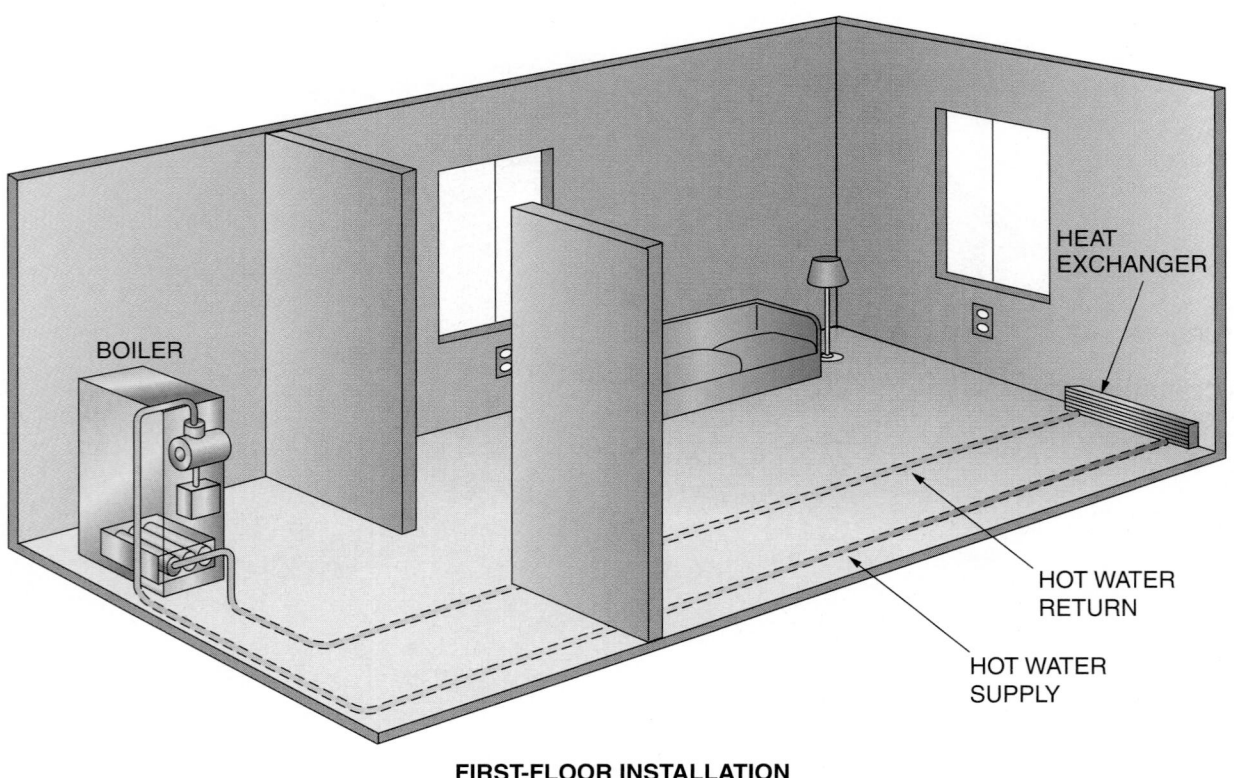

BOILER

HEAT
EXCHANGER

HOT WATER
RETURN

HOT WATER
SUPPLY

FIRST-FLOOR INSTALLATION

101F02.EPS

Figure 2 ◆ Hot water heating.

High-Efficiency Furnaces and Boilers

Many of the newer furnaces and boilers available today are high-efficiency units with efficiency ratings ranging from 90 percent to 96.6 percent for gas. This compares to ratings of approximately 70 percent for old-style gas-burning units and 65 percent for standard oil-burning units. Although new or replacement high-efficiency equipment costs more initially, the payback savings in energy costs for a customer in the northern states could occur in as little as five years when compared to less efficient equipment. Moreover, high-efficiency units used for replacement purposes are often eligible for energy-saving cash incentives from local utilities.

A significant difference between a standard furnace and a condensing furnace is that in a condensing furnace, polyvinylchloride (PVC) piping is used to supply outdoor air for combustion and to exhaust the combustion products to the outdoors. An exhaust-vent blower is used to maintain combustion airflow through the burner section of the furnace. Because the primary and secondary heat exchangers for these furnaces are so efficient at removing heat, the exhaust gases are relatively cool and plastic exhaust-vent pipe can be used instead of metal pipe. The low temperature of the exhaust gases causes moisture to condense out of the gases in the secondary heat exchanger. The condensates are drained from the secondary heat exchanger and are usually pumped to a drain by a condensate pump.

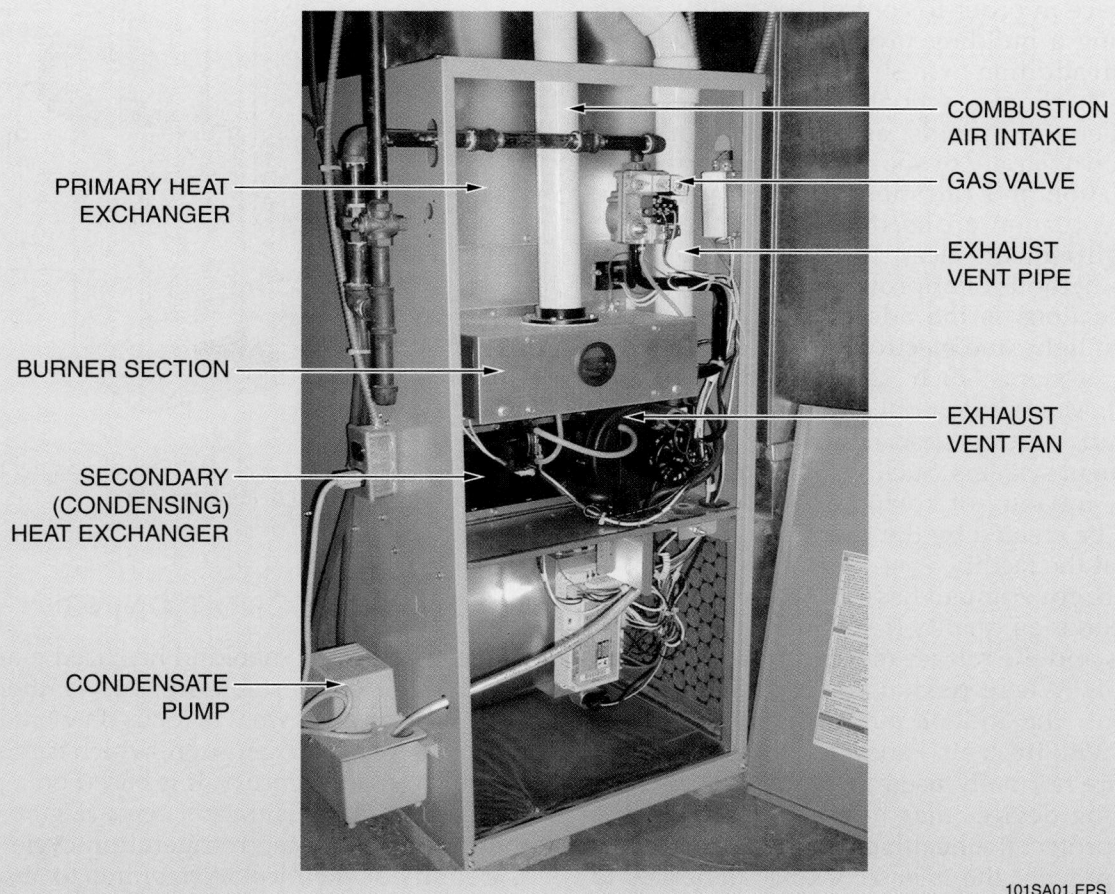

PRIMARY HEAT EXCHANGER

BURNER SECTION

SECONDARY (CONDENSING) HEAT EXCHANGER

CONDENSATE PUMP

COMBUSTION AIR INTAKE

GAS VALVE

EXHAUST VENT PIPE

EXHAUST VENT FAN

101SA01.EPS

Indoor Environmental Quality

The energy-efficient buildings constructed between the mid-1970s and early 1990s actually contributed to indoor air quality problems. Building construction became so tight that the interior was robbed of the fresh air that would normally infiltrate around windows and doors and tiny openings in the structure. In the 1990s, indoor air quality became a major concern and the term sick building was coined to represent structures in which the air being breathed by the occupants lacked sufficient fresh air and contained excessive amounts of dust, germs, molds, and other pollutants. The recognition of this problem has led to the much wider use of electronic air cleaners, fresh air ventilation, duct cleaning, and other measures aimed at improving the indoor air we breathe. The problem has also created more opportunities in the HVAC industry to manufacture, sell, install, and service equipment designed to improve indoor air quality.

3.0.0 ◆ VENTILATION

Ventilation is the introduction of fresh air into a closed space in order to control air quality. Fresh air entering a building provides the oxygen we breathe. In addition to fresh air, we want clean air. The air in our homes, schools, and offices contains dust, pollen, and molds, as well as vapors and odors from a variety of sources. Relatively simple air circulation and filtration methods, including natural ventilation, are used to help keep the air in these environments clean and fresh. Among the common methods of improving the air in residential applications is the addition of humidifiers, ultraviolet light, and electronic air cleaners to air-handling systems such as forced-air furnaces (*Figure 3*). Many industrial environments, on the other hand, require special ventilation and air-management systems. Such systems are needed to eliminate noxious or toxic particles and fumes that may be created by the processes and materials used at the facility.

The U.S. government has strict regulations governing indoor air quality (IAQ) in industrial environments and the release of toxic materials to the outside air. Where noxious or toxic fumes may be present, the indoor air must be constantly replaced with fresh air. Fans and other ventilating devices are normally used for this purpose. Special filtering devices may also be required; these not only protect the health of building occupants, but also prevent the release of toxic materials to the outside air.

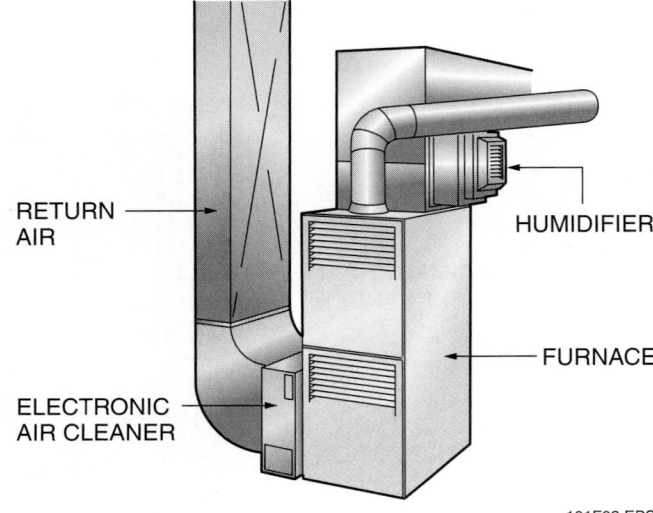

101F03.EPS

Figure 3 ◆ Humidifier and electronic air cleaner in a residential heating system.

4.0.0 ◆ AIR CONDITIONING

Through the ages, mankind has used many methods to stay comfortable in hot weather. In this module, however, we will focus on what is known as mechanical refrigeration, which came into use in the twentieth century. It is based on a principle known as the refrigeration cycle (*Figure 4*).

Simply stated, the refrigeration cycle relies on the ability of chemical refrigerants to absorb heat. If a cold refrigerant flows through a warm space,

The Mechanical Refrigeration Cycle

Many people think that an air conditioner adds cool air to an indoor space. In reality, the basic principle of air conditioning and the mechanical refrigeration cycle is that heat is extracted from the indoor air and transferred to another location (the outdoors) by the refrigerant that flows through the cycle.

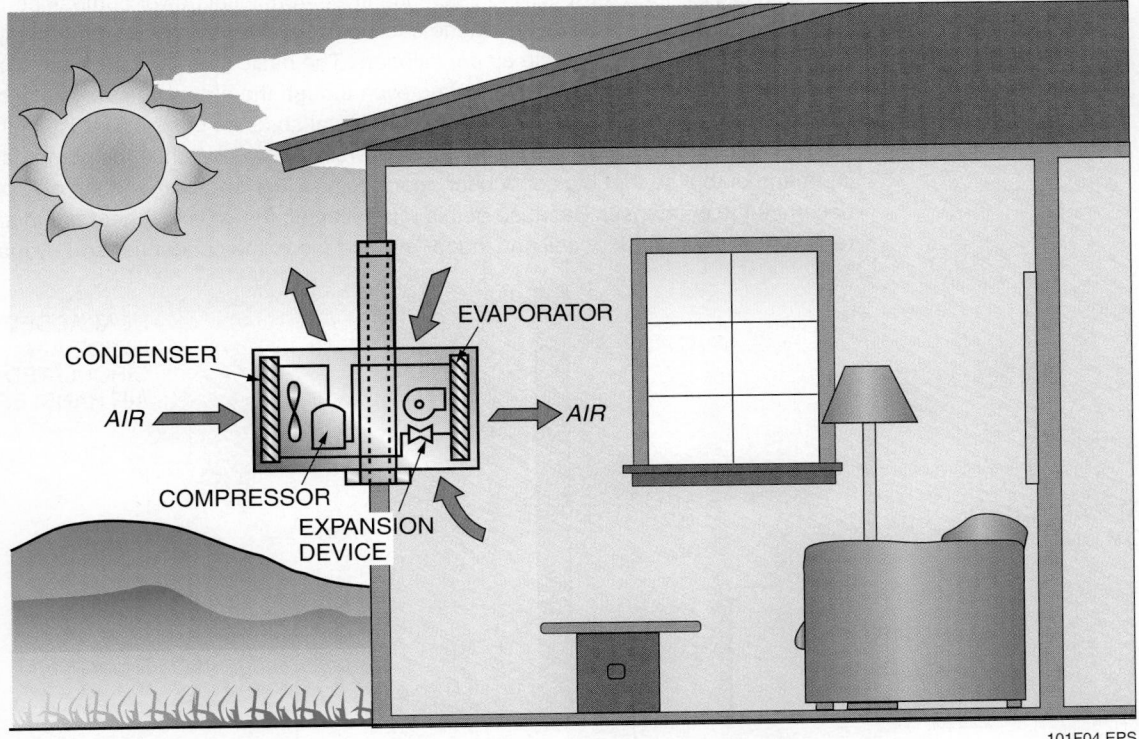

101F04.EPS

Figure 4 ◆ Basic refrigeration cycle for an air conditioner.

it will absorb heat from the space. Having given up heat to the refrigerant, the space becomes cooler. The colder the refrigerant, the more heat it will absorb, and the cooler the space will become. If the super-hot refrigerant flows to a cooler location, the outdoors for example, the refrigerant will give up the heat it absorbed from the indoors and become cool again.

A mechanical refrigeration system is a sealed system operating under pressure. The main elements of a mechanical refrigeration system are the:

- **Compressor** – Provides the force that circulates the refrigerant and creates the pressure differential necessary for the refrigeration cycle to work. A special refrigerant compressor is used.
- **Evaporator** – A heat exchanger where the heat in the warm indoor air is transferred to the cold refrigerant.
- **Condenser** – Also a heat exchanger. In the condenser, the heat absorbed by the refrigerant is transferred to relatively cooler outdoor air.
- **Expansion device** – Provides a pressure drop that lowers the boiling point and pressure of the refrigerant as it enters the evaporator. This allows the refrigerant to become a cold liquid/gas mixture and absorb heat in the evaporator.

Heat Pumps

A special type of air conditioner known as a heat pump is widely used in moderate climates to provide both cooling and heating. Heat pumps are extremely efficient; however, they are most effective in climates where the temperature generally does not fall below 25°F or 30°F. In colder parts of the country, heat pumps can be combined with a gas- or oil-fired, forced-air furnace. In such arrangements, the furnace automatically takes over heating duties when the outdoor temperature falls below the efficient range of the heat pump. Like high-efficiency furnaces and boilers, heat pumps are often eligible for energy-saving cash incentives from local power companies.

Heat pumps operate by reversing the cooling cycle. For this reason, heat pumps are sometimes called reverse-cycle air conditioners. The basic operating principle of a heat pump is that there is some heat in the air, even though the air may be very cold. In fact, the temperature would have to be −460°F for a total absence of heat to exist. In the heating mode, a special valve, known as a reversing valve, switches the compressor input and output so that the condenser operates as the evaporator and the evaporator becomes the condenser. Because of this role reversal, the coils in a heat pump are referred to as the outdoor coil and indoor coil instead of the condenser and evaporator.

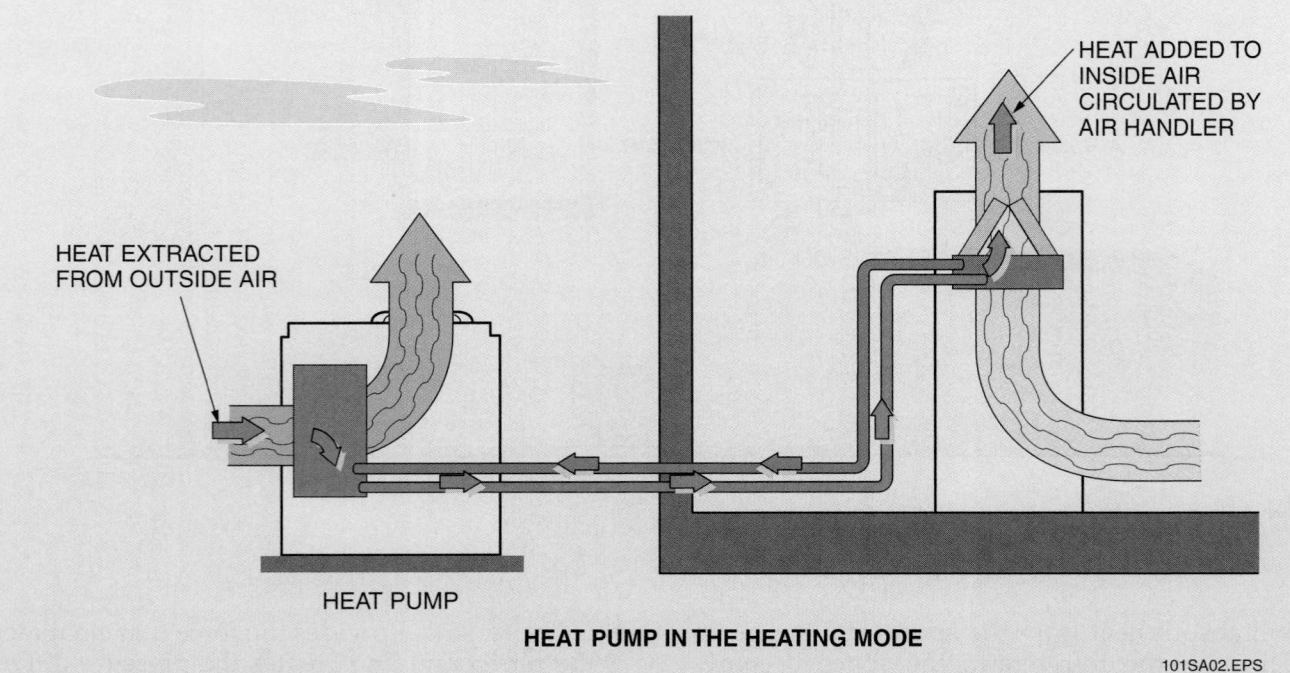

HEAT ADDED TO INSIDE AIR CIRCULATED BY AIR HANDLER

HEAT EXTRACTED FROM OUTSIDE AIR

HEAT PUMP

HEAT PUMP IN THE HEATING MODE

101SA02.EPS

The relationship between temperature and pressure is critical to mechanical refrigeration. As you study the process, you will learn that the same refrigerant can be very cold at one point in the system (the evaporator input) and very hot at another (the condenser input). These two points are often only inches apart. This is possible because of pressure changes caused by the compressor and expansion device. In addition to the circulation of refrigerant, air must also circulate. Fans at the condenser and evaporator move air across the condenser and evaporator coils.

This is a simple explanation of the refrigeration cycle. It is meant to give you a basic idea of how an air conditioner works. Later in the training program, you will explore this subject in greater detail. The relationship between temperature and pressure will also be studied in depth. It is the key

Hermetic and Semi-Hermetic (Servicible) Compressors

Hermetically sealed compressors are typically used in residential and light commercial air conditioners and heat pumps. Semi-hermetic compressors, also known as servicible compressors, are used in large-capacity refrigeration or air conditioning chiller units. Semi-hermetic compressors can be partially disassembled for repair in the field. Hermetic compressors are sealed and cannot be repaired in the field.

HERMETIC RECIPROCATING COMPRESSOR

HERMETIC ROTARY COMPRESSOR

101SA03.EPS

SEMI-HERMETIC (SERVICIBLE) COMPRESSOR

101SA04.EPS

to understanding and troubleshooting mechanical refrigeration systems.

The refrigeration cycle is the same in all refrigeration equipment, from the small air conditioner in your car to the huge system that cools the largest office building. The difference is in the size and construction of the components and piping and the amount and type of refrigerant.

5.0.0 ◆ BLUEPRINTS, CODES, AND SPECIFICATIONS

Commercial construction is a complex process involving the work of many different trades, as well as the participation of people from local government. As an HVAC technician, it is important that you be aware of the HVAC drawings that you will encounter and recognize that building codes will affect the work you do. Just as important are the specifications that answer the many questions that arise on any job site.

5.1.0 Blueprints

For a commercial project, a complete set of blueprints will contain several types of drawings or plans. HVAC drawings are generally part of the mechanical plans. Mechanical plans show what is required to install building systems such as hot water or chilled-water distribution, air distribution, refrigeration, sprinkler, and control systems. For more complex jobs, a separate HVAC plan is added to the set of plans.

Information about the HVAC system equipment, ductwork, and wiring can be found on the architectural, electrical, and mechanical plans of the construction drawings.

The power supplied to the various HVAC systems is typically shown on an electrical drawing. HVAC drawings (*Figure 5*) show piping for water supplies and returns, air handling equipment, AC systems, HVAC component diagrams and schematics, and more.

In addition to the detail work of installation, the HVAC drawings are also used for developing takeoffs, the process of itemizing and counting all materials and equipment needed for the HVAC installation.

Changes that occur during the course of a project must be marked on the drawings. Offsets in ductwork and piping are often not specified on the drawings. These offsets must be documented during the installation, in order to have an accurate set of drawings that reflects how the system was actually installed. At the end of the project, the marked-up drawings represent the as-built configuration of the building. For that reason, the as-builts are often considered to be the most valuable document for representing the completed project.

5.1.1   HVAC Schedules

Schedules are not drawings. They are tables shown on HVAC drawings and on other drawings in the overall drawing set. The schedule provides details you will need, for example, on the mechanical components and equipment shown on the mechanical/HVAC plans. Portions of a typical mechanical equipment schedule are shown in *Figure 6*.

5.1.2   Shop Drawings

Shop drawings show how a specific portion of the work is to be done, such as the fabrication and installation of duct runs. For large commercial jobs, a drafter creates shop drawings based on a design drafted by an engineer. On smaller jobs, the drafter may work from freehand sketches based on field measurements. Shop drawings, like section and detail drawings, are drawn to a larger scale than the engineer's design drawing.

5.1.3   Piping

When reading and using the project drawings, you will encounter various types of pipe. Knowing which type of piping is used on the systems you are working with is important.

The materials commonly used in piping systems include steel (coated and uncoated), black iron, copper (soft and hard), alloy steel and stainless steel, and thermoplastic such as PVC.

Steel pipe, produced with wall thicknesses that are identified by schedule and weight, is used for gas piping and, occasionally, hot water heating. Steel and stainless steel may occasionally be used for refrigerant piping.

Copper tubing is seamless and is manufactured in ¼" to 12" sizes. It is classified by wall thickness. Copper tubing is generally used in plumbing, heating, and refrigeration applications. It should not be used in ammonia systems.

PVC and CPVC are widely used types of plastic pipe. PVC is a rigid material, resistant to chemicals and corrosion and used for many types of drain, waste, and vent (DWV) systems. CPVC, a slightly different formula than PVC, adds high-temperature performance and improved impact resistance over PVC. PEX (cross-linked polyethylene) tubing is a flexible tubing that can withstand heat and high pressure. It is commonly used to carry liquid in radiant floor heating systems.

The piping material selected should be checked for design, temperature, and pressure ratings and must conform to code requirements.

5.2.0 Building Codes

As noted earlier, people from local government are involved in construction projects, primarily enforcement of building codes. The objective of a building code is to regulate the health and safety aspects of building construction in a community. Building codes regulate new construction, for example, by establishing limits on the height and

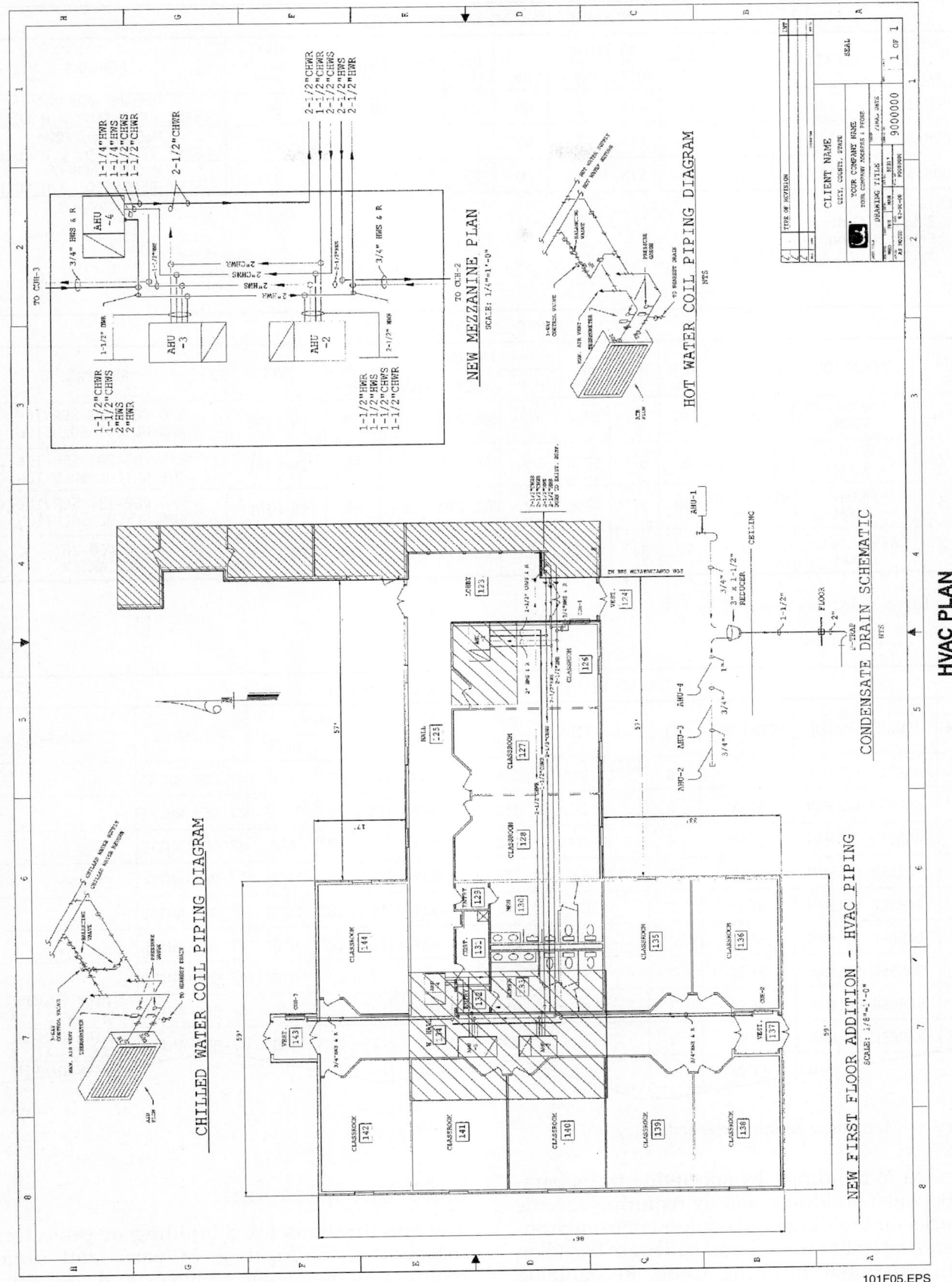

HVAC PLAN

Figure 5 ◆ HVAC drawing.

101F05.EPS

CABINET UNIT HEATER SCHEDULE

UNIT HEATER NO.	LOCATION	C.F.M.	FAN MOTOR				MBH	GPM	EWT	EAT	MAX. WATER P.D.	REMARKS
			H.P.	VOLTS	PHASE	Hz						
CUH-1	124	400	1/12	115	1	60	23	2.3	180°F	60°F	2.7	McQUAY #CHF004 SEMI-RECESSED, R.H COIL
CUH-2	137	400	1/12	115	1	60	23	2.3	180°F	60°F	2.7	McQUAY #CHF004 SEMI-RECESSED, L.H COIL
CUH-3	143	400	1/12	115	1	60	23	2.3	180°F	60°F	2.7	McQUAY #CHF004 SEMI-RECESSED, R.H COIL

NOTES:
1. 3 Speed Control
2. Front Discharge
3. With Return Air Filters

PUMP SCHEDULE

UNIT NO.	LOCATION	SERVICE.	GPM	MBH	MOTOR					TYPE	REMARKS
					RPM	H.P.	VOLTS	PHASE	Hz		
P-1	MECH. ROOM	NAVE	45	41'	1750	1 1/2	208/230	3	60	IN-LINE	B/G #60-20T SERVICE 40% GLYCOL SOLUTION
P-2	MECH. ROOM	CHW TO AHU 1-4	65	37'	1750	1 1/2	208/230	3	60	IN-LINE	B/G #60-20T SERVICE 40% GLYCOL SOLUTION
P-3	MECH. ROOM	RECIR. TO TANK	40	17'	1750	1/2	208/230	3	60	IN-LINE	B/G #60-13T SERVICE 40% GLYCOL SOLUTION
P-4	EXIST. MECH. ROOM	HW	73	31'	1750	1 1/2	208/230	3	60	IN-LINE	B/G #60-20T HOT WATER

NOTES:
1. Starters And Disconnects By E.C.

GRILLE, REGISTER AND DIFFUSER SCHEDULE

ITEM	MANUFACTURER	MODEL NO.	QTY.	LOCATION	CFM EACH	AIR PATTERN	SIZE		FINISHES	REMARKS
							FRAME	NECK		
A	BARBER COLMAN	SFSV	8	126,127,128 144	245	4-WAY	12"× 12"	8"Ø	#7 OFF-WHITE	
B	BARBER COLMAN	SFSV	2	142	275	4-WAY	18"× 18"	10"Ø	#7 OFF-WHITE	
C	BARBER COLMAN	SFSV	4	140, 141	240	4-WAY	12"× 12"	8"Ø	#7 OFF-WHITE	
D	BARBER COLMAN	SFSV	2	139	270	4-WAY	18"× 18"	10"Ø	#7 OFF-WHITE	
E	BARBER COLMAN	SFSV	2	138	280	4-WAY	18"× 18"	10"Ø	#7 OFF-WHITE	
F	BARBER COLMAN	SFSV	2	136	250	4-WAY	12"× 12"	6"Ø	#7 OFF-WHITE	
G	BARBER COLMAN	SFSV	2	135	235	4-WAY	12"× 12"	6"Ø	#7 OFF-WHITE	
H	BARBER COLMAN	SFSV	3	134	100	4-WAY	12"× 12"	6"Ø	#7 OFF-WHITE	FIRE DAMPER SEE DETAIL A
I	BARBER COLMAN	SFSV	1	134	190	4-WAY	12"× 12"	8"Ø	#7 OFF-WHITE	FIRE DAMPER SEE DETAIL A
	COLMAN	SFSV	1	134		WAY	12"× 12"			FIRE DAMPER

101F06.EPS

Figure 6 ◆ Mechanical equipment schedules.

floor area of buildings, by specifying the separation between buildings, and by requiring specific set-backs for buildings and equipment from property and easement lines. Local codes are based on the **International Building Code**. Recognizing that building codes directly affect HVAC work is crucial to HVAC workers.

5.3.0 Specifications

The specifications for a building or project are the written descriptions of work and duties required by the owner, architect, and consulting engineer. Together with the working drawings, these specifications form the basis of the contract requirements for the construction of the building or project. Those who use the construction drawings and specifications must always be alert to discrepancies between the working drawings and the written specifications. These are some situations where discrepancies may occur:

- Architects or engineers use standard or prototype specifications and attempt to apply them without any modification to specific working drawings.
- Previously prepared standard drawings are changed or amended by reference in the specifications only and the drawings themselves are not changed.
- Items are duplicated in both the drawings and specifications, but an item is subsequently amended in one and overlooked in the other contract document.

In such instances, the person in charge of the project has the responsibility to ascertain whether the drawings or the specifications take precedence. Such questions must be resolved, preferably before the work begins, to avoid added costs to the owner, architect/engineer, or contractor.

5.4.0 Commissioning

System commissioning is a process by which a formal and organized approach is taken to obtaining, verifying, and documenting the installation and performance of a particular system or systems. The goal of commissioning is to make sure that a system operates as intended and at optimum efficiency. Commissioning is normally required for newly installed or retrofitted commercial and industrial systems. Because each building and its systems are different, the specific elements of the commissioning process must be tailored to fit each situation. However, the objectives for performing system commissioning are the same:

- Verify that the system design meets the functional requirements of the owner.
- Verify that all systems are properly installed in accordance with the design and specifications.
- Verify that all systems and components meet required local, state, and other required codes.
- Verify and document the proper operation of all equipment, systems, and software.
- Verify that all documentation for the system is accurate and complete.
- Train building operator and maintenance personnel to efficiently operate and maintain the installed equipment and systems.

6.0.0 ◆ CAREERS IN HVAC

Career opportunities in the HVAC trade are many and varied. There is a large existing base of HVAC systems that needs service, repair, and replacement. In addition, every time a new residential, commercial, or industrial building is constructed, it contains one or more HVAC system elements.

To get an idea of how vast the HVAC trade is, picture the town in which you live. Then think about the fact that almost every building in town contains some form of equipment to provide heating and cooling, as well as air circulation and purification. Expand that view to include the entire country and you realize that there are tens of millions of heating, air conditioning, and air-management systems. New ones are being added every day; old ones are wearing out and being repaired or replaced. From that perspective, the opportunities in the trade appear limitless.

Figure 7 provides an overview of career opportunities in the HVAC trade. For the purposes of this discussion, it is convenient to view the HVAC industry as having three segments:

- *Community-based* – Companies that sell, install, and service residential and light commercial equipment and systems such as furnaces and packaged air conditioners.
- *Commercial/industrial* – Companies that install and maintain systems for large office buildings, factories, apartment complexes, shopping malls, and so forth.
- *Manufacturing* – Companies that build and market HVAC systems and equipment.

6.1.0 Community-Based

At the community level, you might find anything from a one-person installation and service business to a firm with a hundred or more employees, including heating and air conditioning specialists, installers, sheet metal workers, and sales engineers. In such businesses, HVAC specialists may work alone or with a single partner. They typically respond to service calls from homes or small businesses. They may also install furnaces and air conditioning equipment sold by their firm's sales engineer. In other firms, one group may do installations while another group handles troubleshooting and maintenance. At this level, a technician is expected to work with a wide variety of products from many different manufacturers. Systems can consist of anything from a window air conditioner to a heating/air-conditioning system containing two or three major components.

Local or regional distributors provide equipment, parts, special tools, and other services for

COMMUNITY-BASED

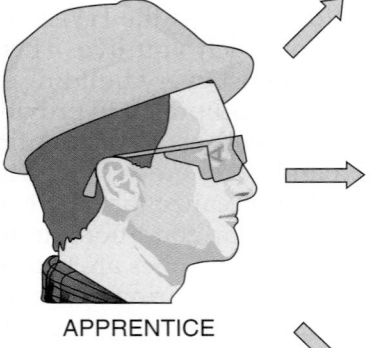

• Installer • Service Technician • Distributor Sales	• Service Supervisor • Sales Engineer • Owner • Inspector • Code Enforcement Officer • Estimator

COMMERCIAL/INDUSTRIAL

• Installer • Air System Technician • Service Technician	• Foreman • Technical Specialist • Training Specialist • Operating Engineer • Project Manager • Estimator

MANUFACTURING

• Test Technician • Engineering Assistant • Assembler	• Supervisor • Technical Specialist • Trainer • Field Service Engineer

APPRENTICE

101F07.EPS

Figure 7 ◆ Career opportunities in the HVAC trade.

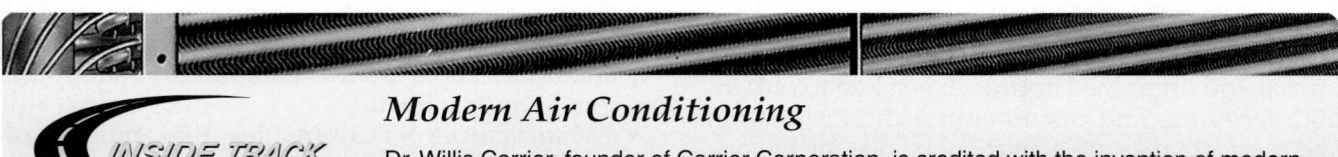

Modern Air Conditioning

Dr. Willis Carrier, founder of Carrier Corporation, is credited with the invention of modern air conditioning. In 1902, he developed a system that could control both humidity and temperature using a non-toxic, non-flammable refrigerant. Air conditioning started out as a means of solving a problem in a printing facility where heat and humidity were causing paper shrinkage. Later systems served similar purposes in textile plants. The concept wasn't applied to comfort air conditioning until about 20 years later when Carrier's centrifugal chillers began to be installed in department stores and movie theaters.

High-Efficiency Air Conditioners and Heat Pumps

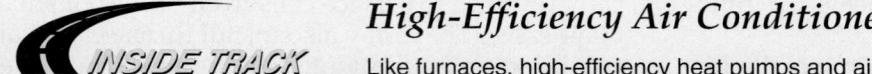

Like furnaces, high-efficiency heat pumps and air conditioners are also available for both commercial and residential installations. A number of manufacturers have voluntarily partnered with the Environmental Protection Agency (EPA) in an ENERGY STAR® program to market units that exceed minimum efficiency standards. Building designs may also qualify for this program through the use of high-efficiency HVAC systems combined with thermal storage systems. For example, to reduce the energy costs of air conditioning a building, ice is generated by refrigeration systems at night using low-cost power. The ice is stored in insulated tanks and is used to aid the building's air-conditioning system during the daytime. This reduces energy consumption during peak usage hours and makes power generation by the utility company more efficient.

INSIDE TRACK

Large Commercial Chiller Unit

As the name implies, chillers use chilled water as a cooling medium. The chiller acts as the evaporator. Chillers are often combined with a cooling tower that acts as the condensing unit. Water flowing over the cooling tower absorbs the heat extracted from the indoor air. This photo shows a large commercial chiller unit.

101SA05.EPS

the firms that sell, install, and service HVAC equipment. The distributor needs salespeople who know HVAC equipment. The distributor may also provide engineering support and service training for its dealers. Distributorships are often affiliated with a single manufacturer.

NOTE

Many community-based firms are subsidiaries of nationwide firms known as consolidators. A consolidator is an umbrella organization that provides centralized management, purchasing, training, and other functions that give small, local companies the power of a large, national company.

6.2.0 Commercial/Industrial

Large commercial and industrial systems have many components and may require thousands of feet of ductwork and piping. Such systems are designed by engineers and architects. Many

HVAC trade people are required for these projects and an individual is more likely to specialize. For example, where residential systems are usually controlled by a single wall thermostat, a large commercial system will often have central computer controls that are installed and serviced by a system control specialist. Others may specialize in working with large steam boilers; still others may choose to become experts in installing and servicing high-volume cooling units known as chillers.

Companies that install such equipment are often large construction firms that may work anywhere in the world. They are likely to do only the installation and let the building owner contract with another firm for maintenance. Many large facilities employ their own HVAC maintenance people.

6.3.0 Manufacturing

There are thousands of HVAC manufacturers. Some of the larger ones cover the entire HVAC spectrum, while others focus on a particular product or market. For example, one may make window air conditioners, another gas furnaces, and yet another might make only heavy commercial equipment. Regardless of their market, they employ HVAC specialists such as:

- Test technicians
- Engineering assistants
- Training specialists
- Instruction book writers
- Field service technicians

Working as an HVAC journeyman is satisfying and financially rewarding. If you want to go beyond the journeyman level, there are many opportunities in supervision and executive management with dealerships, construction companies, and manufacturers. Companies that install and service HVAC equipment are usually founded by someone who started as a service technician. For those who want to teach, there are opportunities to work in vocational schools and training programs such as the one in which you are now participating.

7.0.0 ◆ TYPES OF TRAINING PROGRAMS

There are two basic forms of training programs that most employers consider. The primary one is on-the-job training (OJT) to improve the competence of their employees in order to provide better customer service and for the continuity and growth of the company. The second is formal apprenticeship training, which provides the same type of training, but also conforms to federal and state requirements under the *Code of Federal Regulations (CFR), Titles 29:29* and *29:30*.

7.1.0 Standardized Training by the NCCER

The National Center for Construction Education and Research (NCCER) is a not-for-profit education foundation established by the nation's leading construction companies. NCCER was created to provide the industry with standardized construction education materials, the Contren® Learning Series (the HVAC modules are part of this series), and a system for tracking and recognizing students' training accomplishments—NCCER's National Registry.

NCCER also offers accreditation, instructor certification, and skills assessments. NCCER is committed to developing and maintaining a training process that is internationally recognized, standardized, portable, and competency-based.

Working in partnership with industry and academia, NCCER has developed a system for program accreditation that is similar to those found in institutions of higher learning. NCCER's accreditation process ensures that students receive quality training based on uniform standards and criteria. These standards are outlined in NCCER's *Accreditation Guidelines* and must be adhered to by NCCER Accredited Training Sponsors.

More than 550 training and assessment centers across the U.S. and eight other countries are proud to be NCCER Accredited Training Sponsors. Millions of craft professionals and construction managers have received quality construction education through NCCER's network of Accredited Training Sponsors and the thousands of Training Units associated with the Sponsors. Every year the number of NCCER Accredited Training Sponsors increases significantly.

A craft instructor is a journeyman craft professional or career and technical educator trained and certified to teach NCCER's Contren® Learning Series. This network of certified instructors ensures that NCCER training programs will meet the standards of instruction set by the industry. There are more than 3,350 master trainers and 33,000 craft instructors within the NCCER instructor network. More information is available at **www.nccer.org**.

The basic idea of the NCCER is to replace governmental control and credentialing of the construction workforce with industry-driven training and education programs. The NCCER departs from traditional classroom or distance learning by offering a competency-based training regimen. Competency-based training means that instead of simply attaining required hours of classroom training and set hours of OJT, you have to prove that you know what is required and successfully demonstrate specific skills. All completion information on every trainee is sent to the NCCER and kept within the National Registry. The NCCER can confirm training and skills for workers as they move from company to company, state to state, or to different offices in the same company. These are portable credentials and are recognized nationally.

In an effort to provide industry credentials and ensure national portability of skills, more than three million module completions have been

INSIDE TRACK

Technician Certification

A knowledgeable HVAC installation technician or service technician can obtain nationally recognized certification by taking the Air Conditioning Excellence (ACE) exams. North American Technical Excellence, Inc. (NATE) administers the exams. Although other certification programs are available, the NATE certification has become the most widely accepted in the industry.

NATE tests are divided into tests for service technicians and tests for installation technicians. NATE certification is for either the service path or the installation path, with each specialty having a core test plus a test in that specialty. Specialty tests are specific to the path you choose.

A technician who wants to obtain certification can take the core exam and any one of the specialty exams to be certified for that specialty. A technician may want to be certified in several specialties. For example, a technician living in the Middle Atlantic states may want to obtain certification in all five specialties, whereas a person living in the desert Southwest might not have a need to be certified for gas or oil heating.

delivered to students and craft professionals nationwide. The National Registry provides transcripts, certificates, and wallet cards for students of the Contren® Learning Series when training is delivered through an NCCER Accredited Training Sponsor. These valuable industry credentials benefit students as they seek employment and build their careers.

7.1.1 Apprenticeship Training

As stated earlier, formal apprenticeship programs conform to federal and state requirements under *CFR Titles 29:29* and *29:30*. All approved apprenticeship programs provide OJT as well as classroom instruction. The related training requirement is fulfilled by all NCCER craft training programs. The main difference between NCCER training and registered apprenticeship programs is that apprenticeship has specific time limits in which the training must be completed. Apprenticeship standards set guidelines for recruiting and outreach, and a specific time limit for each of a variety of OJT tasks. Additionally, there are reporting requirements and audits to ensure adherence to the apprenticeship standards. Companies and employer associations register their individual apprenticeship programs with the Office of Apprenticeship within the U.S. Department of Labor, and in some instances, with state apprenticeship councils (SAC). OJT of 2,000 hours per year and a minimum of 144 hours of classroom-related training are required. Apprenticeship programs vary in length from 2,000 hours to 10,000 hours.

7.1.2 Youth Training and Apprenticeship Programs

Youth apprenticeship programs are available that allow students to begin their apprenticeship or craft training while still in high school. A student entering the program in the 11th grade may complete as much as one year of the NCCER training program by high school graduation. In addition, programs (in cooperation with local construction industry employers) allow students to work in the craft and earn money while still in school. Upon graduation, students can enter the industry at a higher level and with more pay than someone just starting in a training program.

Students participating in the NCCER or youth apprenticeship training are recognized through official transcripts and can enter the second level or year of the program wherever it is offered. They may also have the option of applying credits at two-year or four-year colleges that offer degree or certificate programs in their selected field of study.

8.0.0 ◆ THE HVAC TECHNICIAN AND THE ENVIRONMENT

Scientific studies have shown that the chlorine in chlorofluorocarbon (CFC) and hydrochlorofluorocarbon (HCFC) refrigerants can damage the ozone layer that protects the Earth and its inhabitants from the sun's ultraviolet rays. One of the effects is an increased rate of skin cancer. Many refrigerants used in air conditioning equipment are CFCs and HCFCs.

In addition to the toxic effect of chlorine, there is evidence that the release of refrigerant to the atmosphere contributes to global warming, a condition known as the greenhouse effect. The heat trapped in the atmosphere leads to a gradual increase in the Earth's temperature. Refrigerants, especially CFCs, are viewed as major contributors to global warming.

In the 1980s the nations of the world made a commitment to phase out these chemicals. They agreed to take steps in the interim to prevent the discharge of refrigerants into the atmosphere. In 1990, the U.S. Congress passed the *Clean Air Act,* which calls for early phaseout of the most toxic refrigerants, eventual elimination of all CFCs and HCFCs, and strict control and labeling of refrigerants. As a result of the requirement to phase out CFCs and HCFCs, a number of new, environmentally friendly refrigerants have been developed. The U.S. Environmental Protection Agency (EPA) is responsible for implementing and enforcing this law, which has a significant impact on the HVAC trade. For example, it has imposed the following restrictions on refrigerants, regardless of the chlorine content:

- Anyone releasing these refrigerants to the atmosphere is subject to a stiff fine and possibly a prison term.
- Anyone handling these refrigerants must have EPA-sanctioned certification. Without it, you cannot even buy refrigerants. The training needed to obtain EPA certification is included in this curriculum.
- Records must be kept on all transactions involving these refrigerants. These include purchase, use, reprocessing, and disposal.

Before a sealed refrigeration system can be opened for repair, the refrigerant it contains must be identified. With few exceptions, all refrigerants must be recovered and stored in approved containers. When the repair is complete and the system is resealed, the same refrigerant may be returned to the system. It may also be used in another system belonging to the same owner. Recovered refrigerant should, however, be recycled before reuse. This removes moisture and impurities that could damage the system. Some refrigerant recovery units have a built-in recycling capability.

If the refrigerant is badly contaminated or no longer needed, it can be reclaimed. This is done at remanufacturing centers where the refrigerant is returned to the standards of purity that govern new refrigerants. Reclaimed refrigerant can be resold on the open market, but cannot be classified as "new" refrigerant.

In addition to federal regulations, there may be state regulations that apply to refrigerants. State regulations may be stricter. It is critical that everyone in the HVAC industry understands and follows the EPA regulations regarding the handling, storage, and labeling of refrigerants. Failure to do so could be very costly to both you and your employer.

Summary

The HVAC trade involves equipment used for heating, cooling, and purifying indoor air. It covers everything from the window air conditioners and furnaces used in our homes to the giant heating and cooling systems used in large office buildings and industrial complexes.

A construction project is a complex process involving many people who represent the various construction trades and local government. Without a clear set of blueprints and specifications, assigning and organizing the work efficiently would be impossible. Blueprints and specifications spell out every detail of the construction job, from preparing the site to putting on the finishing touches.

As an HVAC trainee, you must learn the basics of blueprint reading so that you can interpret the drawings that show how to accurately install an HVAC system and its accessories. However, you must also be able to read the prints used by other trades because you and your co-workers will have to coordinate your work with them.

Because of the widespread use of HVAC equipment, there are many career opportunities in the trade. Jobs are available with small local firms, large industrial and commercial contractors, and the manufacturing firms that build and market HVAC equipment. The apprentice program provides an opportunity to learn the trade through a combination of hands-on training and related classroom learning. The Youth Apprentice Program allows students to begin their training while still in high school.

One of the most serious issues affecting the trade is the damage that can be done to the Earth's ozone layer by the improper release of refrigerants into the atmosphere. There are severe penalties for improper refrigerant disposal.

Notes

Review Questions

1. In a common household forced-air furnace, heat is transferred from the _____ .
   a. heat exchangers to the air
   b. natural gas or oil to the conditioned space
   c. air to the heat exchangers
   d. refrigerant to the outside air

2. Water is often used in heating systems as a _____ .
   a. fuel
   b. refrigerant
   c. heat exchange medium
   d. heat exchanger

3. Ventilation is concerned with _____ .
   a. air circulation
   b. air temperature
   c. the circulation and cleanliness of indoor air
   d. the circulation and cleanliness of indoor air and air discharged to the outdoors

4. The term refrigeration cycle refers to _____.
   a. the process by which circulating refrigerant absorbs heat in one location and moves it to another location
   b. the process by which refrigerant moves through a compressor
   c. mobile refrigeration units
   d. the process by which refrigerant is recycled to remove impurities

5. Heat is transferred from the indoor air to the refrigerant at the _____ .
   a. compressor
   b. furnace
   c. evaporator
   d. condenser

6. The expansion device _____ .
   a. raises the boiling point of the refrigerant entering the evaporator
   b. lowers the boiling point of the refrigerant entering the evaporator
   c. gets bigger as the temperature increases
   d. lowers the pressure at the evaporator outlet

7. The layout of the HVAC system ductwork is shown in the _____ plans.
   a. site
   b. structural
   c. civil
   d. mechanical

8. The primary purpose of a building code is to regulate the quality of all mechanical installation on a commercial site.
   a. True
   b. False

9. In order to be resold on the open market, a refrigerant must have been _____ .
   a. reclaimed
   b. recycled
   c. recovered
   d. refurbished

10. Releasing CFC and HCFC refrigerants or certain substitutes to the atmosphere is _____ .
    a. okay, provided that the refrigerant has been recycled
    b. okay, provided that the refrigerant has been reclaimed
    c. prohibited by federal law
    d. prohibited in some states

Trade Terms Quiz

1. Anything _____ is considered poisonous.

2. _____ is the remanufacturing of used refrigerant to bring it up to the standards required of new refrigerant.

3. A class of refrigerants that contains hydrogen, chlorine, fluorine, and carbon is called _____.

4. A(n) _____ is a heat exchanger that transfers heat from the air flowing over it to the cooler refrigerant flowing through it.

5. A class of refrigerants that contains chlorine, fluorine, and carbon is called _____.

6. The removal and temporary storage of refrigerant in containers approved for that purpose is called _____.

7. A liquid metering device is also known as a(n) _____.

8. A(n) _____ is a heat exchanger that transfers heat from the refrigerant flowing inside it to the air or water flowing over it.

9. The process by which a circulating refrigerant absorbs heat from one location and transfers it to another location is called a(n) _____.

10. _____ is the transfer of heat from a warmer substance to a cooler substance.

11. _____ is circulating recovered refrigerant through filtering devices that remove moisture, acid, and other contaminants.

12. The use of machinery to provide cooling is called _____.

13. A substance harmful to your health is said to be _____.

14. A(n) _____ is added to the mechanical plans for complex jobs that require separate heating, ventilating, and air conditioning systems.

15. A high-volume cooling unit is called a(n) _____.

16. A(n) _____ is the process of itemizing and counting all material and equipment needed for the HVAC installation.

17. The _____ is a series of model construction codes that set standards that apply across the country.

18. The mechanical device that converts low-pressure, low-temperature refrigerant gas into high-temperature, high-pressure refrigerant gas in a refrigeration system is a(n) _____.

Trade Terms

Chiller
Chlorofluorocarbon (CFC) refrigerant
Compressor
Condenser
Evaporator

Expansion device
Heat transfer
HVAC plan
Hydrochlorofluorocarbon (HCFC) refrigerant

International Building Code
Mechanical refrigeration
Noxious
Reclamation
Recovery

Recycling
Refrigeration cycle
Takeoffs
Toxic

John Tianen

HVAC Training Consultant
Tucson, AZ

John Tianen's career is a good example of how a commitment to learning can help an individual succeed. John started his HVAC career working on the assembly line for an air conditioning manufacturer. By applying himself to learning HVAC technology, he became an installation and service technician, then an instructor, and finally a designer and developer of HVAC training. Along the way, he earned bachelor's and master's degrees by taking college classes on his own time.

Following high school, John spent four years in the U.S. Air Force as an electronic systems technician. After completing military service, he worked on a production line in a factory. After a year he was hired by a major manufacturer of air conditioning equipment as a maintenance technician on the production line. His duties included maintaining the equipment used to evacuate (empty) and charge (fill) room air conditioners as they moved down the assembly line. John studied HVAC technology and soon had a job in the laboratory where central air conditioners were tested. This experience led him to establish a business installing and servicing residential air conditioners and furnaces. This

additional experience enabled him to obtain a position as a service engineer in which he had worldwide service support responsibility for all split-system air conditioners and heat pumps. A subsequent promotion moved John into the technical training department, where he produced the installation and service training programs for all residential heating and cooling products. During that period, he took advantage of his company's generous tuition refund program and obtained college degrees, including a master of science in education. Now semi-retired, John has formed his own consulting company, which provides training support to the HVAC industry.

John's advice to someone just entering the trade is to make the most of apprentice training. It gives the foundation you need for a successful career. Also, take advantage of any other learning opportunities that come your way. The industry is constantly changing, and the only way to keep up with it is to make a lifelong commitment to learning. Although this sometimes requires taking classes, it can also be done by subscribing to trade journals and by participating in local, regional, and national trade organizations.

Trade Terms Introduced in This Module

Chiller: A high-volume cooling unit. The chiller acts as an evaporator.

Chlorofluorocarbon (CFC) refrigerant: A class of refrigerants that contains chlorine, fluorine, and carbon. CFC refrigerants have a very adverse effect on the environment.

Compressor: In a refrigeration system, the mechanical device that converts low-pressure, low-temperature refrigerant gas into high-temperature, high-pressure refrigerant gas.

Condenser: A heat exchanger that transfers heat from the refrigerant flowing inside it to the air or water flowing over it.

Evaporator: A heat exchanger that transfers heat from the air flowing over it to the cooler refrigerant flowing through it.

Expansion device: Also known as the liquid metering device or metering device. Provides a pressure drop that converts the high-temperature, high-pressure liquid refrigerant from the condenser into the low-temperature, low-pressure liquid refrigerant entering the evaporator.

Heat transfer: The transfer of heat from a warmer substance to a cooler substance.

HVAC plan: Added to the mechanical plans for complex jobs that require separate heating, ventilating, and air-conditioning systems.

Hydrochlorofluorocarbon (HCFC) refrigerant: A class of refrigerants that contains hydrogen, chlorine, fluorine, and carbon. Although not as high in chlorine as chlorofluorocarbon (CFC) refrigerants, HCFCs are still considered hazardous to the environment.

International Building Code: A series of model construction codes. These codes set standards that apply across the country. This is an ongoing process led by the International Code Council (ICC).

Mechanical refrigeration: The use of machinery to provide cooling.

Noxious: Harmful to health.

Reclamation: The remanufacturing of used refrigerant to bring it up to the standards required of new refrigerant.

Recovery: The removal and temporary storage of refrigerant in containers approved for that purpose.

Recycling: Circulating recovered refrigerant through filtering devices that remove moisture, acid, and other contaminants.

Refrigeration cycle: The process by which a circulating refrigerant absorbs heat from one location and transfers it to another location.

Takeoffs: The process of itemizing and counting all the material and equipment needed for an installation.

Toxic: Poisonous.

This module is intended to present thorough resources for task training. The following reference work is suggested for further study. This is optional material for continued education rather than for task training.

Career Opportunities in Heating, Air Conditioning, and Refrigeration, Latest Edition. Fairfax, VA: Air Conditioning and Refrigeration Institute (ARI).

CONTREN® LEARNING SERIES — USER UPDATE

NCCER makes every effort to keep these textbooks up-to-date and free of technical errors. We appreciate your help in this process. If you have an idea for improving this textbook, or if you find an error, a typographical mistake, or an inaccuracy in NCCER's Contren® textbooks, please write us, using this form or a photocopy. Be sure to include the exact module number, page number, a detailed description, and the correction, if applicable. Your input will be brought to the attention of the Technical Review Committee. Thank you for your assistance.

Instructors – If you found that additional materials were necessary in order to teach this module effectively, please let us know so that we may include them in the Equipment/Materials list in the Annotated Instructor's Guide.

Write: Product Development and Revision
National Center for Construction Education and Research
3600 NW 43rd St, Bldg G, Gainesville, FL 32606

Fax: 352-334-0932

E-mail: curriculum@nccer.org

Craft _____ Module Name _____

Copyright Date _____ Module Number _____ Page Number(s) _____

Description _____

(Optional) Correction _____

(Optional) Your Name and Address _____

Introduction to Drain, Waste, and Vent (DWV) Systems
68115-09

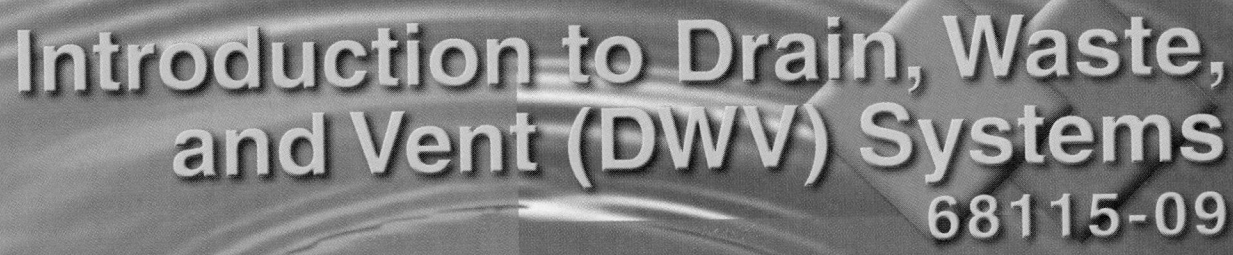

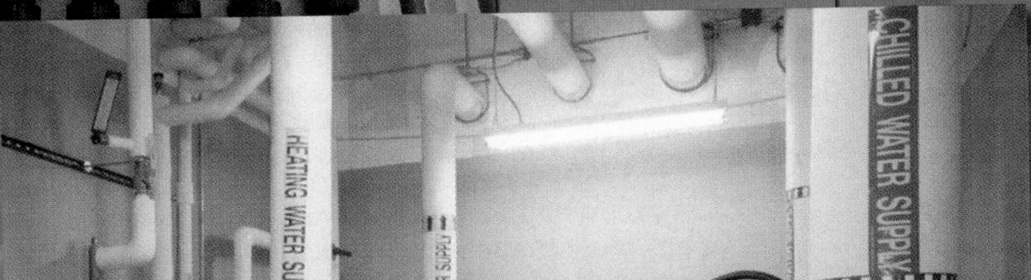

68115-09

Introduction to Drain, Waste, and Vent (DWV) Systems

Topics to be presented in this module include:

1.0.0	Introduction	15.2
2.0.0	DWV Systems	15.2
3.0.0	Fixture Drains	15.2
4.0.0	Traps	15.2
5.0.0	Vents	15.12
6.0.0	Sizing Drains and Vents	15.13
7.0.0	Fittings and Their Applications	15.14
8.0.0	Grade	15.21
9.0.0	Building Drain	15.22
10.0.0	Building Sewer	15.22
11.0.0	Sewer Main	15.22
12.0.0	Waste Treatment	15.22
13.0.0	Code and Health Issues	15.24

Overview

To design, install, and maintain drain, waste, and vent (DWV) systems, plumbers must be familiar with the factors that affect them. Sanitary drainage systems include the pipes inside the building, the drain pipe buried outside the building, and the public sewer. Knowing how drains, fittings, vents, and pipe move waste out of a building enables plumbers to prevent system malfunctions.

Fixture drains connect fixtures to the building's DWV piping system. Traps and vents protect people from pathogens, odors, and sewer gases. While codes govern specific requirements for these drains and traps, plumbers must understand the installation requirements, including dimensions, and they must be able to troubleshoot trap failure. Plumbers install vents to provide a free flow of air and to maintain equalized pressure throughout the drainage system. To properly size drains and vents, plumbers must thoroughly understand the mechanics of fluid flow in the pipes.

Drainage systems are categorized as storm water drains and building drains. Because these systems rely on gravity to move solid and liquid waste, plumbers install them on a slope, or grade. The local municipality installs, maintains, and controls the public or municipal sewer systems. These systems collect waste, treat it at a sewage plant, and then discharge it back into the ecosystem. Properly designed, installed, and maintained DWV systems are essential to public safety.

▣ Focus Statement

The goal of the plumber is to protect the health, safety, and comfort of the nation job by job.

▣ Code Note

Codes vary among jurisdictions. Because of the variations in code, consult the applicable code whenever regulations are in question. Referencing an incorrect set of codes can cause as much trouble as failing to reference codes altogether. Obtain, review, and familiarize yourself with your local adopted code.

Objectives

When you have completed this module, you will be able to do the following:

1. Explain how waste moves from a fixture through the drain system to the environment.
2. Identify the major components of a drainage system and describe their functions.
3. Identify the different types of traps and their components, explain the importance of traps, and identify the ways that traps can lose their seals.
4. Identify the various types of drain, waste, and vent (DWV) fittings and describe their applications.
5. Identify significant code and health issues, violations, and consequences related to DWV systems.

Trade Terms

Adapter	Pipe scale
Back pressure	Run
Backpressure backflow	S-trap
Branch interval	Sanitary combination
Building sewer	Sanitary fitting
Capillary attraction	Sanitary increaser
Cleanout	Sanitary upright wye
Crown weir/trap weir	Sanitary wye
Double ¼ bend	Short sweep ¼ bend
Double trapping	Side inlet
Drainage fitting	Siphonage
DWV system	Slope (percent of grade)
Elevation	Sludge
Evaporation	Stack
Fall	Test tee
Fixture drain	Velocity
Grade	Vent branch (branch
Hydraulic gradient	vent)
Interceptor	Vent ell
Inverted wye	Vent tee
Long sweep ¼ bend	Weir (trap or crown)
P-trap	

Required Trainee Materials

1. Appropriate personal protective equipment
2. Pencils and paper
3. Copy of local adopted code

Prerequisites

Before you begin this module, it is recommended that you successfully complete the following: *Core Curriculum*, and *Construction Technology*, Modules 68101-09 through 68114-09.

This course map shows all of the modules *Construction Technology*. The suggested training order begins at the bottom and proceeds up. Skill levels increase as you advance on the course map. The local Training Program Sponsor may adjust the training order.

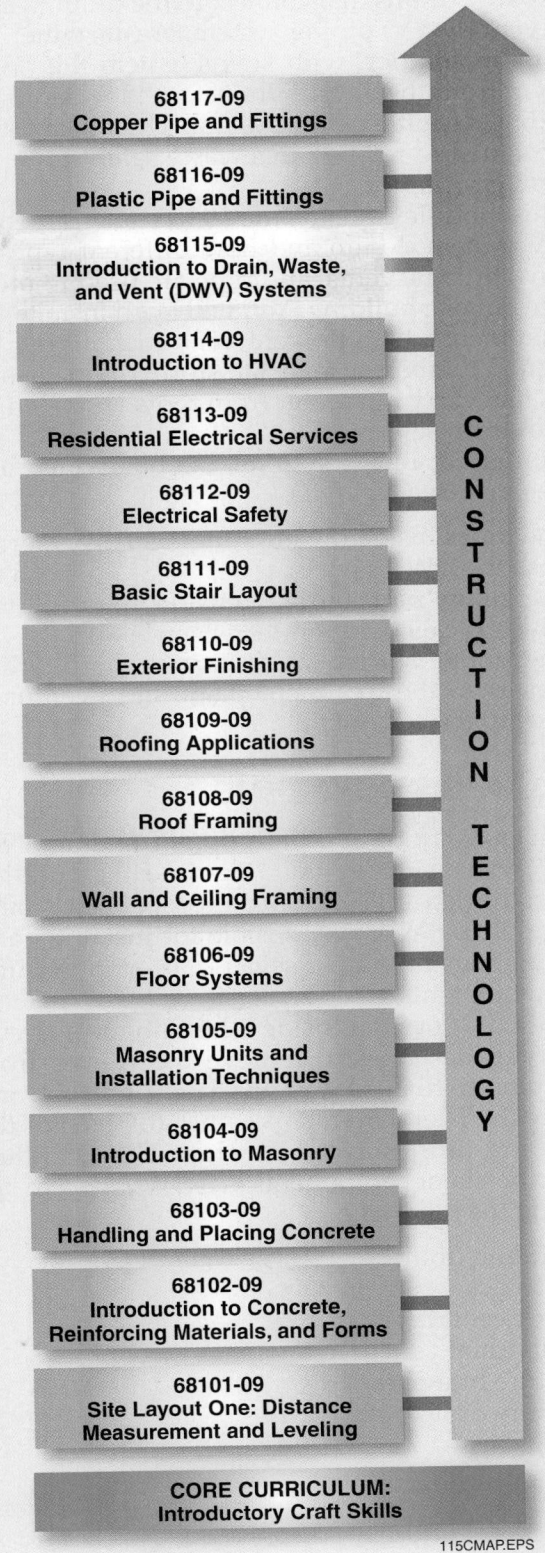

68117-09
Copper Pipe and Fittings

68116-09
Plastic Pipe and Fittings

68115-09
Introduction to Drain, Waste, and Vent (DWV) Systems

68114-09
Introduction to HVAC

68113-09
Residential Electrical Services

68112-09
Electrical Safety

68111-09
Basic Stair Layout

68110-09
Exterior Finishing

68109-09
Roofing Applications

68108-09
Roof Framing

68107-09
Wall and Ceiling Framing

68106-09
Floor Systems

68105-09
Masonry Units and Installation Techniques

68104-09
Introduction to Masonry

68103-09
Handling and Placing Concrete

68102-09
Introduction to Concrete, Reinforcing Materials, and Forms

68101-09
Site Layout One: Distance Measurement and Leveling

CORE CURRICULUM:
Introductory Craft Skills

CONSTRUCTION TECHNOLOGY

115CMAP.EPS

1.0.0 ◆ INTRODUCTION

As a plumber, you need to understand how drainage systems work. This module describes the flow of waste products from a building, to the treatment facilities, and back into the ecosystem (streams, rivers, and lakes). This cycle begins at the fixture drains that connect to the drain, waste, and vent (DWV) piping system. Waste, either liquid or in solution with solids, enters the **DWV system** from the fixture drains and flows into the building's sanitary pipe system. The pipe system is designed to remove this waste safely from the building's interior.

This module explains the factors that influence DWV system design and how different types of drains, fittings, vents, and pipe are used to move waste out of a building. You will learn installation requirements that prevent malfunctions in the system. Plumbers also install storm drainage systems that carry rainwater from roofs and open areas to storm sewers.

Sanitary drainage systems can be divided into three parts (see *Figure 1*):

- The pipes inside the building, usually referred to as the DWV system
- The drain pipe buried outside the building, which is called the **building sewer**
- The public sewer, which carries the building wastes to the treatment plant and eventually back to the ecosystem

2.0.0 ◆ DWV SYSTEMS

Plumbers may design, install, and maintain the DWV systems inside buildings and the building sewers buried outside on the property. Usually, the municipality is responsible for installing and maintaining the public sewers, lift stations, and treatment plants.

The DWV system inside a building is a circuit of piping designed to remove the wastes from plumbing fixtures and drains safely, reliably, and efficiently. There are many names for each of the pipes and fittings in this network. *Figure 2* illustrates the major components of a DWV system, including the following:

- Building drain
- Soil stack
- Stack vent
- Individual vents
- Fixture branches
- Fixture drain or trap arm
- Traps

- Bends or elbows
- Tees
- Wyes
- Couplings, reducers, and adapters

3.0.0 ◆ FIXTURE DRAINS

Fixture drains connect fixtures to the building's DWV piping system. Many fixtures have drains that strain the wastewater before it enters the drainage piping. Examples of fixture drains include a basket strainer for a kitchen sink, PO (pop-up) plugs for lavatories, and other strainers for bidets and showers (see *Figure 3*).

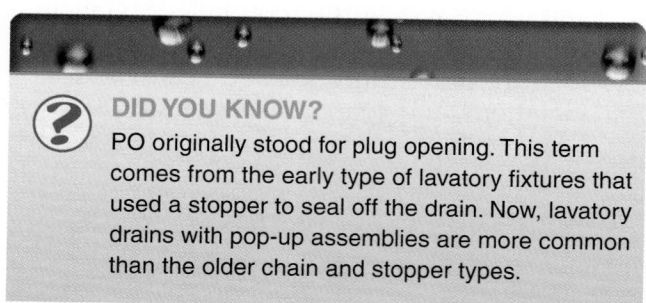

DID YOU KNOW?

PO originally stood for plug opening. This term comes from the early type of lavatory fixtures that used a stopper to seal off the drain. Now, lavatory drains with pop-up assemblies are more common than the older chain and stopper types.

4.0.0 ◆ TRAPS

Traps and vents protect the safety of homes and other buildings. The water seal in a trap protects people from airborne pathogens (germs), foul odors, and potentially explosive sewer gases.

Traps are important components of the DWV piping system. A trap is a fitting or device that provides a liquid seal of 2 to 4 inches. This seal prevents sewer gases from leaking back into the building but should not affect the flow of sewage or wastewater through the drain. As will be explained later in this module, trap seals are protected by vents.

Modern codes require every plumbing fixture to have a trap that protects the fixture and users from the sanitary drain system. Some fixtures, such as water closets and many urinals, have integral or built-in traps.

A fixture trap is a vital part of any DWV system. To function properly, a fixture trap must flush completely, be self-cleaning, have a smooth interior waterway, and be accessible for cleanout. The depth of the seal and the amount of water normally held in the trap are important factors in trap design. The fixture trap, which creates a water seal, requires a vent system to protect it from **siphonage, back pressure**, wind, and aspiration. Back pressure is often referred to as **backpressure backflow**.

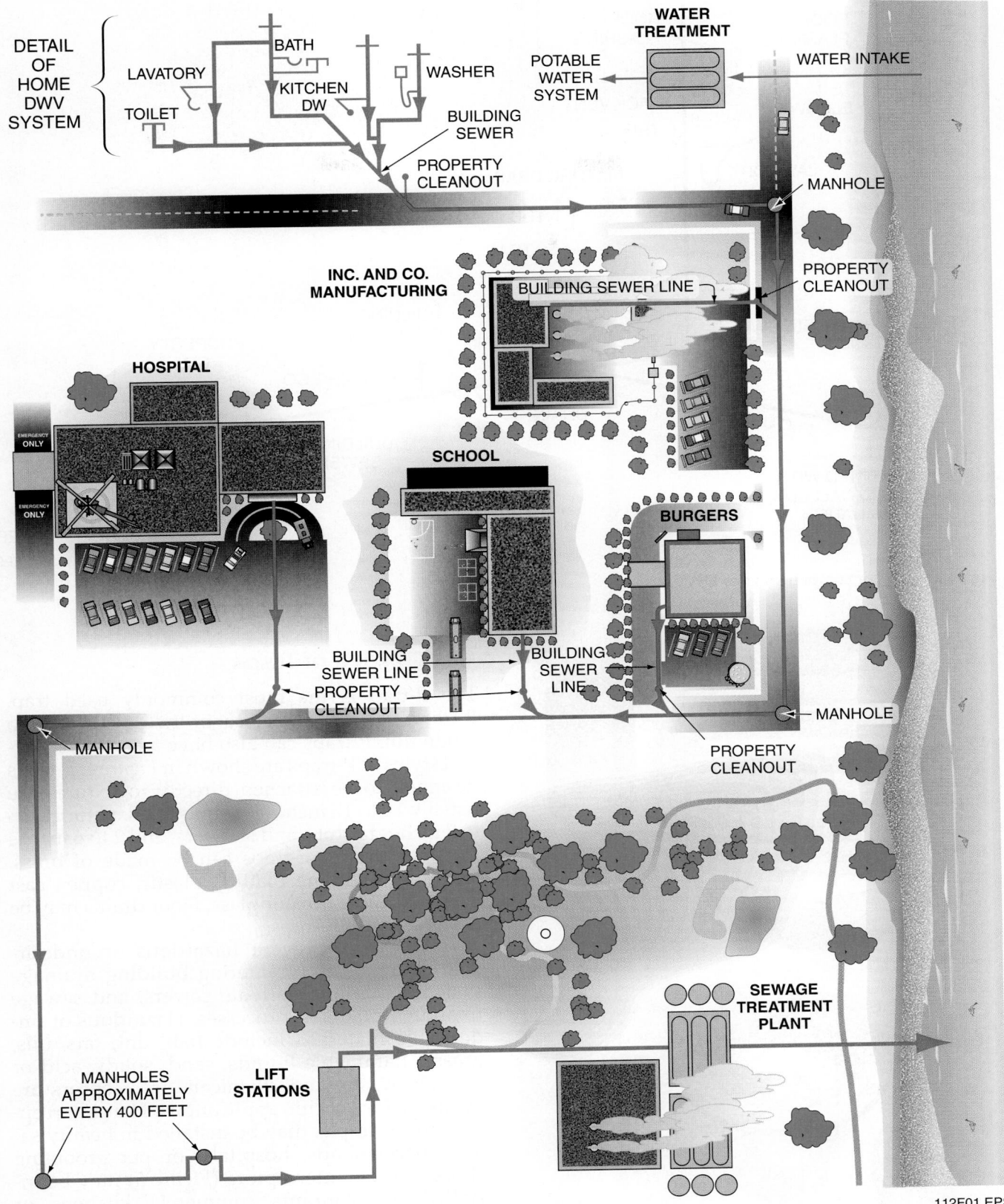

DETAIL OF HOME DWV SYSTEM

LAVATORY
BATH
TOILET
KITCHEN DW
WASHER

WATER TREATMENT

POTABLE WATER SYSTEM
WATER INTAKE

BUILDING SEWER
PROPERTY CLEANOUT

MANHOLE

PROPERTY CLEANOUT

INC. AND CO. MANUFACTURING
BUILDING SEWER LINE

HOSPITAL
EMERGENCY ONLY
EMERGENCY ONLY

SCHOOL

BURGERS

BUILDING SEWER LINE
PROPERTY CLEANOUT
BUILDING SEWER LINE

MANHOLE

PROPERTY CLEANOUT

MANHOLE

MANHOLES APPROXIMATELY EVERY 400 FEET

LIFT STATIONS

SEWAGE TREATMENT PLANT

112F01.EPS

Figure 1 ◆ Overview of a typical community sewer system.

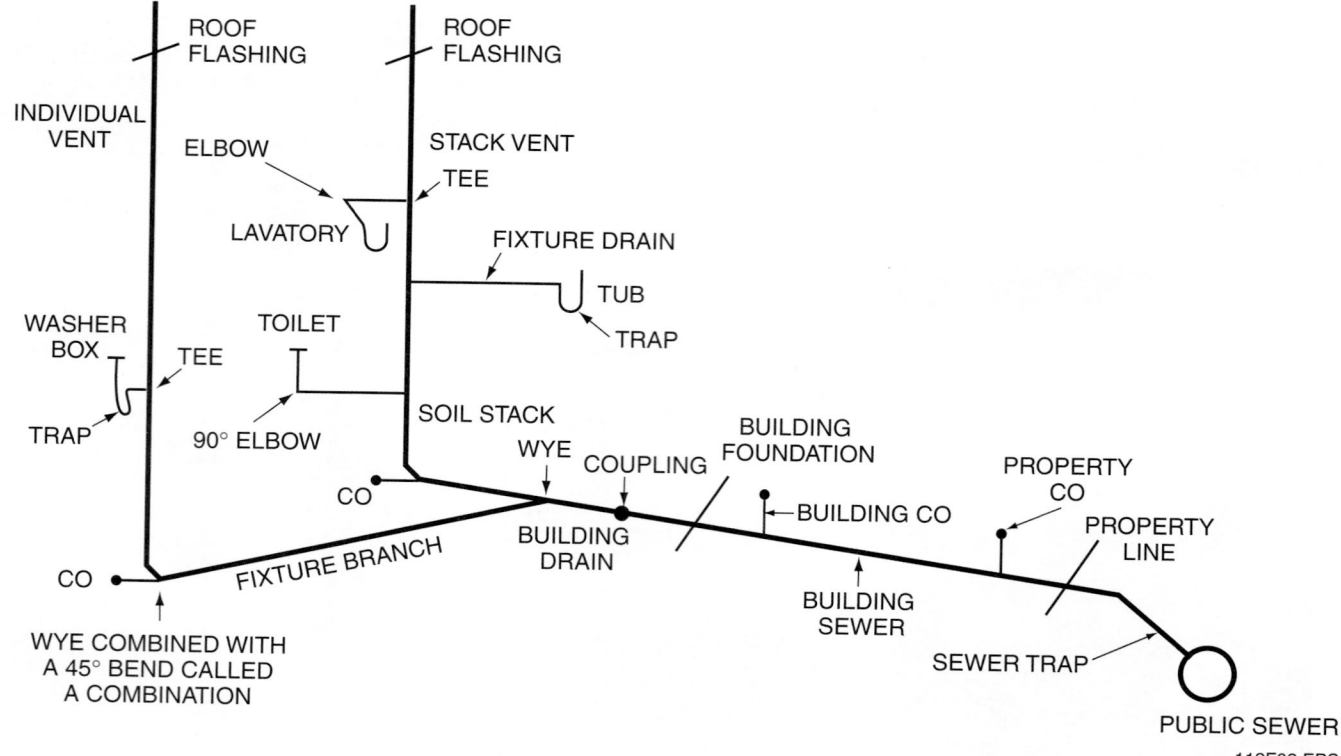

Figure 2 ◆ Major components of a DWV system.

112F02.EPS

Figure 3 ◆ Fixture drains.

112F03.TIF

4.1.0 Types of Traps

The P-trap is the most commonly used trap. **P-traps** can be one-piece or two-piece with a union nut. P-traps can also have a **cleanout**. Several styles of P-traps are shown in *Figure 4*. P-traps designed to be attached directly to fixtures are usually 1¼ to 1½ inches for sinks and lavatories, 1½ to 2 inches for tubs and showers, and 2 to 4 inches for floor drains. P-traps can be made of brass, brass with chrome plating, plastic, copper, cast iron, malleable iron, or glass. Floor drains may be designed with P-traps.

Interceptors prevent hazardous or undesirable materials from entering building drainage systems, public or private sewers, and sewage treatment plants or processes. Hazardous or undesirable materials include hair, lint, fats, oils, grease, flammable liquids, sand, solids, acid or alkaline waste, and chemicals. Interceptors are available for specific applications. Hair interceptors, for example, may be installed in beauty salons, barbershops, hospitals, or pet grooming shops. Grease interceptors (*Figure 5*) may be installed in restaurants, commercial kitchens, or auto repair shops. Although they differ in design, all interceptors operate on similar principles.

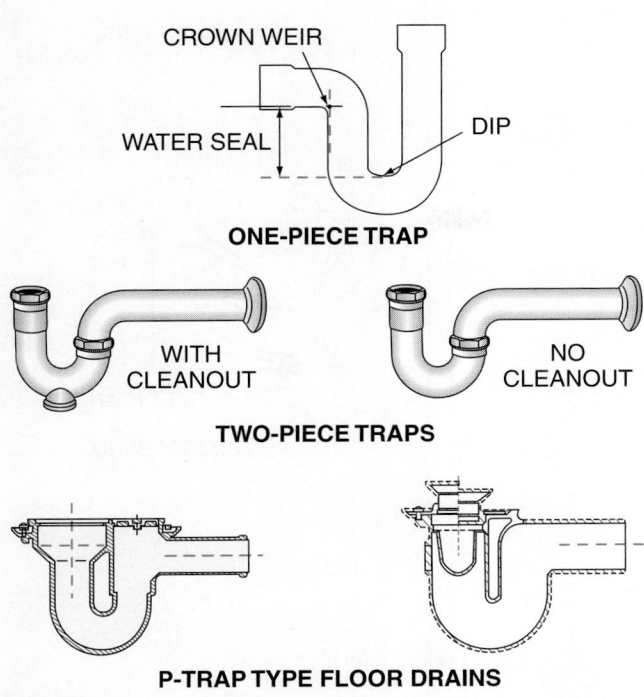

CROWN WEIR

WATER SEAL

DIP

ONE-PIECE TRAP

WITH CLEANOUT

NO CLEANOUT

TWO-PIECE TRAPS

P-TRAP TYPE FLOOR DRAINS

112F04.EPS

Figure 4 ◆ P-traps.

Wastewater flows through a chamber where harmful materials are separated before the wastewater flows out again. For example, in interceptors designed to capture solid wastes such as grease, wastewater flows into a chamber through screens. Because the solids are heavier than the wastewater, gravity causes them to fall to the bottom of the chamber, where they are retained until the chamber is cleaned out.

Current plumbing codes prohibit the use of **S-traps** (*Figure 6*). P-traps are more efficient and effective. The long downstream leg of the S-trap tends to promote siphonage. Plumbers could make the trap more effective by increasing the depth of the seal, but this led to other problems. With greater depth, there was a greater chance that solids would stay in the trap. Fungus growth was also a problem in traps that were too deep. Two types of S-traps were used: the full S-trap and the ¾ S-trap. Many older homes still have S-traps. Plumbing supply stores stock these traps for service but not for installation in new construction.

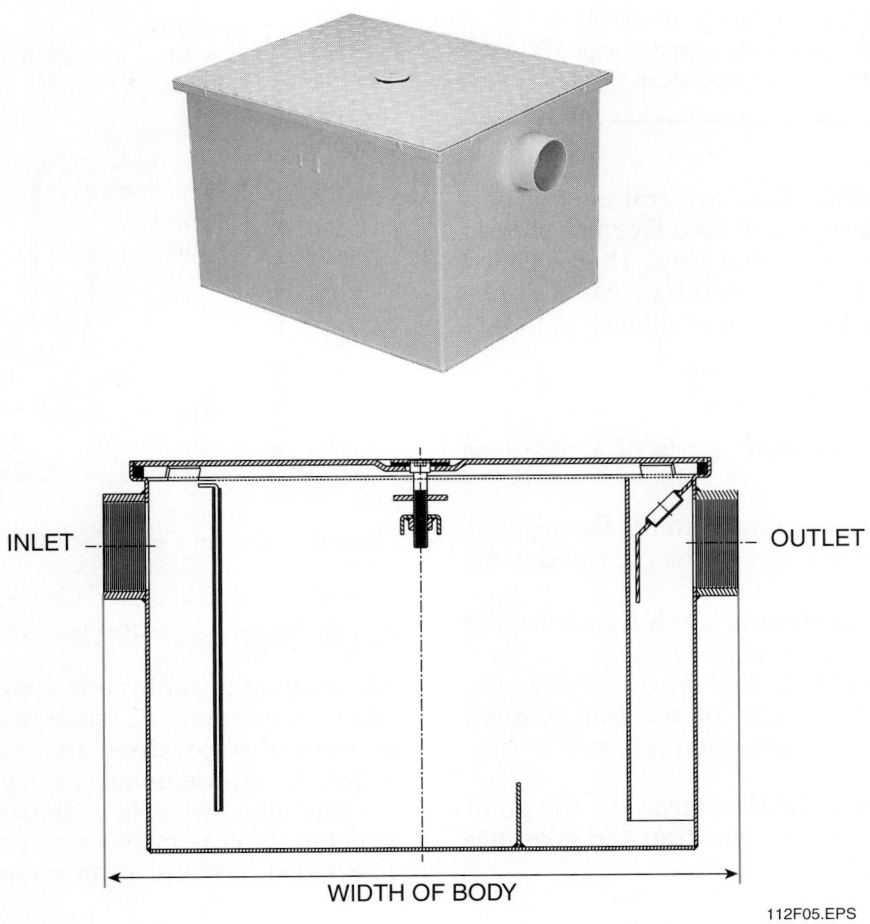

INLET

OUTLET

WIDTH OF BODY

112F05.EPS

Figure 5 ◆ Grease interceptor.

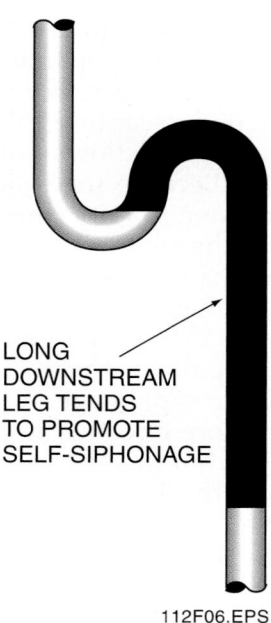

LONG
DOWNSTREAM
LEG TENDS
TO PROMOTE
SELF-SIPHONAGE

112F06.EPS

Figure 6 ◆ S-trap.

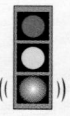

NOTE

Nonsiphon traps are available with deeper trap seals. These are used where the plumbing system is subjected to abnormal changes in pressure or to a lot of evaporation.

Water closets (toilets) have integral traps as part of their design. Water closets (see *Figure 7*) should never be attached to another trap. This is called **double trapping**. Double trapping creates a pressure that will stop the flow of drainage.

4.2.0 Parts of Traps

The following are the basic parts of a trap (see *Figure 8*):

- *Inlet* – where water enters from the fixture
- *Top dip* – the inside curve of the pipe under the inlet
- *Bottom dip* – the bottom of the lowest curve of the pipe beneath the inlet
- *Crown weir (sometimes called the trap weir)* – the highest point in the seal of the trap. Crown weirs and trap weirs are often referred to simply as **weirs**.
- *Fixture drain (also called the trap arm)* – the point where wastewater leaves the trap and goes into the drainage piping

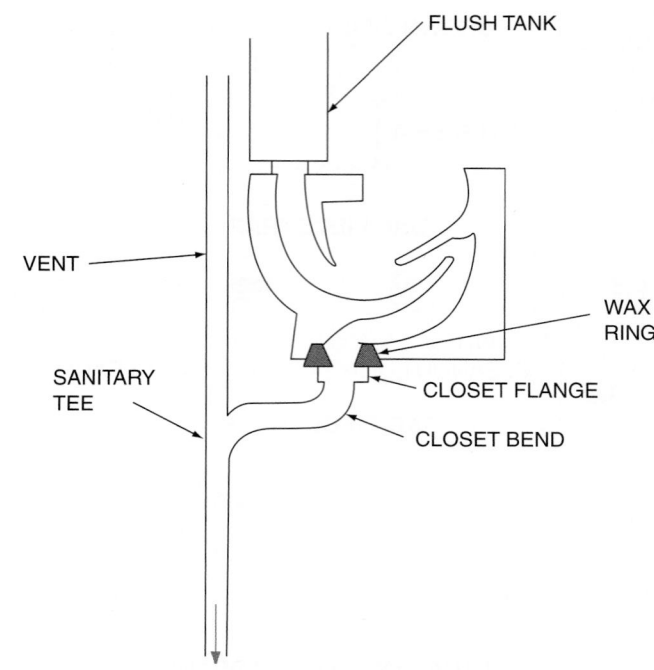

FLUSH TANK

VENT

SANITARY TEE

WAX RING

CLOSET FLANGE

CLOSET BEND

112F07.EPS

Figure 7 ◆ Integral or built-in trap.

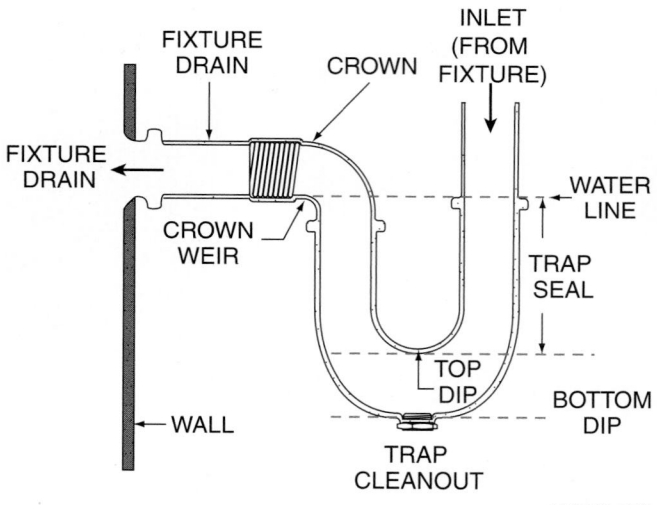

FIXTURE DRAIN

CROWN

INLET (FROM FIXTURE)

FIXTURE DRAIN

WATER LINE

CROWN WEIR

TRAP SEAL

TOP DIP

BOTTOM DIP

WALL

TRAP CLEANOUT

112F08.EPS

Figure 8 ◆ Parts of a trap.

4.3.0 Trap Installation Requirements

Although applicable code governs specific installation requirements, such as dimensions and location of traps, there are some typical trap installation requirements (see *Figure 9*).

Generally, the vertical distance from the fixture outlet to the crown weir may not exceed 24 inches. The second critical dimension is the horizontal

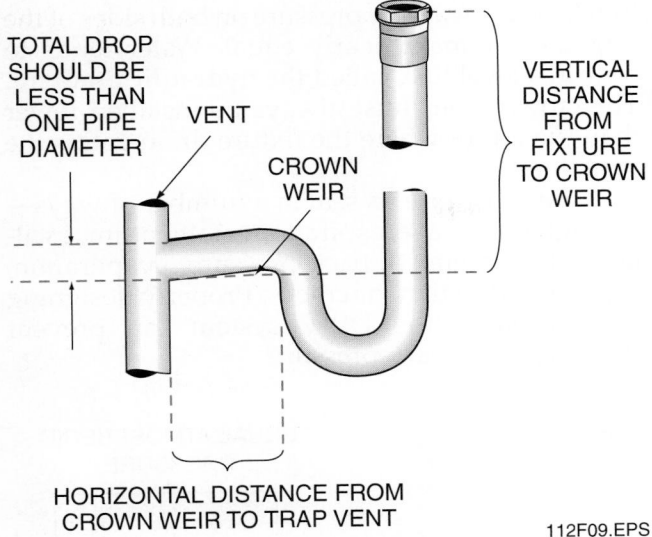

TOTAL DROP SHOULD BE LESS THAN ONE PIPE DIAMETER

VENT

CROWN WEIR

VERTICAL DISTANCE FROM FIXTURE TO CROWN WEIR

HORIZONTAL DISTANCE FROM CROWN WEIR TO TRAP VENT

112F09.EPS

Figure 9 ◆ Critical dimensions of a trap vent.

distance from the crown weir to the trap vent. This distance varies depending on the diameter of the trap (see *Table 1*). The third important dimension is the total drop (or **fall**) in the horizontal pipe from the crown weir to the vent. It may not exceed one pipe diameter. Usually, the fall is ¼ inch per foot. If the horizontal leg is installed with greater drop, the trap is likely to siphon.

Many P-traps are manufactured in two pieces. The J-bend piece joins with the fixture tailpiece and is secured with a gasket and compression nut called a slip-joint washer and a slip-joint nut. The outlet end, called the trap arm, joins the horizontal drain, which connects to the vent.

Generally, traps should be installed for each plumbing fixture that does not have a built-in trap. Some codes permit one trap to serve more than one fixture. For example, three lavatories

 DID YOU KNOW?

In the earliest home plumbing fixtures (in the 1850s), the main safeguards against odors and sewer gases were handmade traps that the plumber installed in the drains of individual fixtures. These traps often lost their water seals because of siphonage and back pressure (described in this section) and became ineffective. Efforts to prevent seal loss failed, because the principle of venting fixture drains (covered later in this module) was not known at the time.

In the early 1900s, the problems with fixture traps led health officials to require a secondary safeguard: the installation of building traps on each sanitary or combined building sewer. Without this additional safeguard, rats were able to travel freely from one building to another. Building traps became the second line of defense against rats in the sewer systems.

This requirement was a big advance at the time. However, since the development of modern collection, drainage, and venting systems, most model codes don't require building traps. In fact, many codes actually prohibit building traps. The only exceptions are in areas where sewer gases are extremely corrosive or where the sewer gases contain high explosive gas content, creating a risk of explosions in the public sewer system that might, for instance, blow off manhole covers and cause considerable damage.

that are 30 inches or less apart may be connected to one trap. Sinks containing two or three compartments also may be connected to one trap.

Table 1 Horizontal Distance of Fixture Trap from Vent at Slope of ¼ Inch per Foot

		Distance from Trap		
		Standard Plumbing Code (1994) International Plumbing Code® (1997)	Uniform Plumbing Code (1997)	National Standard Plumbing Code
Size of Fixture Drain (1996)	Size of Trap			
1¼"	1¼"	3'6"	2'6"	3'6"
1½"	1½"	5'0"	3'6"	5'0"
2"	2"	6'0"	5'0"	8'0"
3"	3"	10'0"	6'0"	10'0"
4" and larger*	4"	12'0"	10'0"	12'0"

*Size of fixture drain specification is 4" only (not 4" and larger) for *Standard Plumbing Code* (1994) and *International Plumbing Code* (1997).

112T01.EPS

4.4.0 Why a Trap Loses Its Seal

A thorough understanding of how traps function will help you understand how important it is to install traps correctly. Also, you can use this knowledge to determine what causes a trap to malfunction.

When a trap functions properly, waste from the fixture flows into and through the trap (see *Figure 10*). The trap is refilled with the last of the wastewater to leave the fixture. This water provides the necessary liquid seal. For the trap to function this way, the pressure on both sides of the trap must remain nearly equal. Water tends to flow in a level line, called the **hydraulic gradient.** The crown weir must always be installed lower than the top of where the fixture drain enters the vent line.

A trap may lose its seal in a number of ways—through siphonage, aspiration, momentum, oscillation (wind effect), back pressure, evaporation, capillary attraction, or cracks. Properly designing and installing the DWV system can prevent siphonage and back pressure.

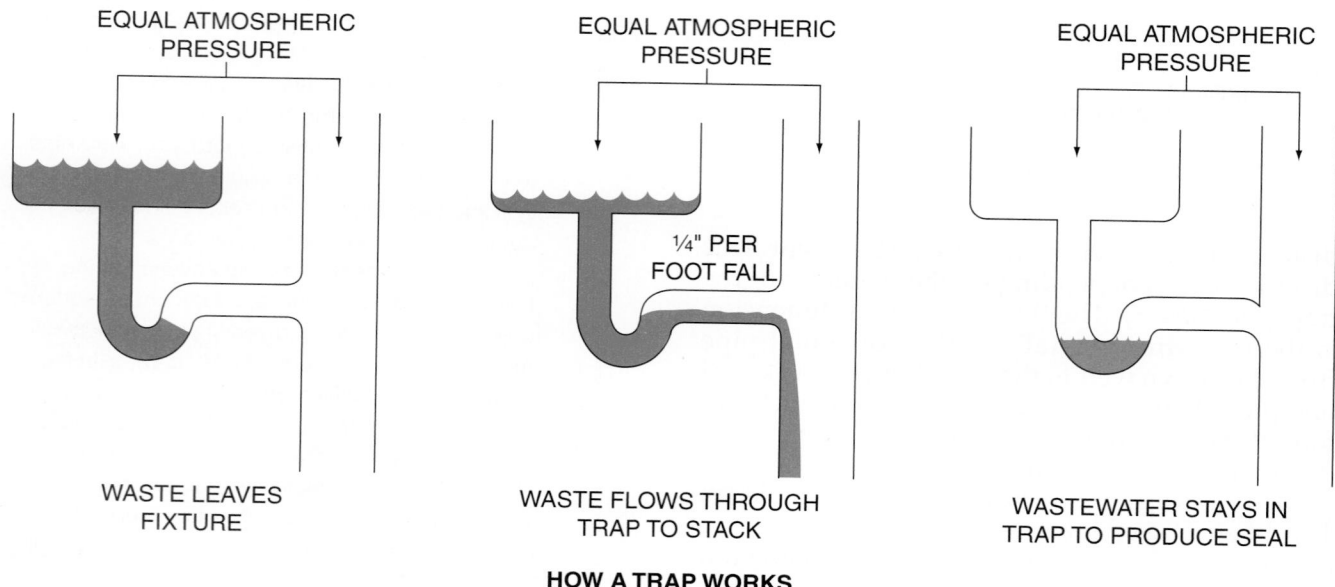

EQUAL ATMOSPHERIC PRESSURE

WASTE LEAVES FIXTURE

EQUAL ATMOSPHERIC PRESSURE

¼" PER FOOT FALL

WASTE FLOWS THROUGH TRAP TO STACK

EQUAL ATMOSPHERIC PRESSURE

WASTEWATER STAYS IN TRAP TO PRODUCE SEAL

HOW A TRAP WORKS

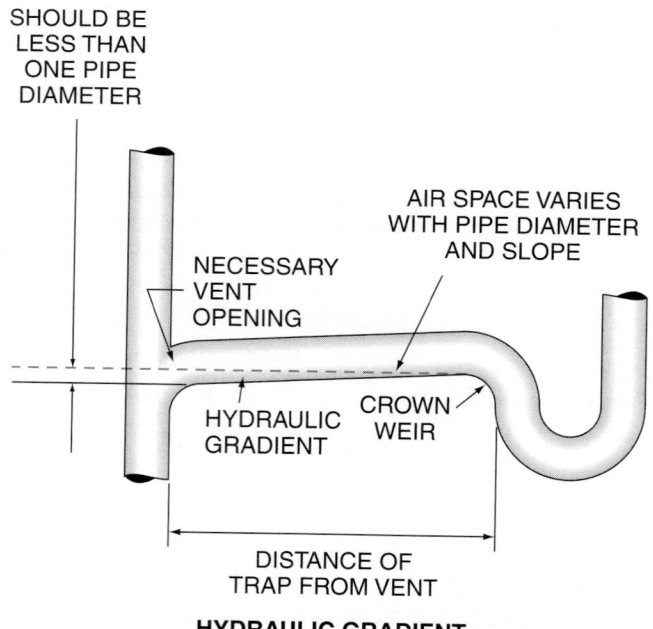

SHOULD BE LESS THAN ONE PIPE DIAMETER

NECESSARY VENT OPENING

AIR SPACE VARIES WITH PIPE DIAMETER AND SLOPE

HYDRAULIC GRADIENT

CROWN WEIR

DISTANCE OF TRAP FROM VENT

HYDRAULIC GRADIENT

112F10.EPS

Figure 10 ◆ How a trap works.

4.4.1 Siphonage

If the trap is not properly vented, it is likely to siphon. Siphonage occurs when there is negative pressure inside the DWV piping. This pressure difference pushes the water that is normally held in the trap into the DWV piping system. Generally, siphonage occurs when the DWV piping is improperly vented or the vent is blocked (see *Figure 11*). As the waste leaves the trap, an area of reduced pressure is created in the drainage piping. Because of the difference in pressure, the water is forced from the trap. This destroys the trap seal. Siphonage happens when there is too much fall and the crown weir is higher than the top of where the fixture drain enters the vent.

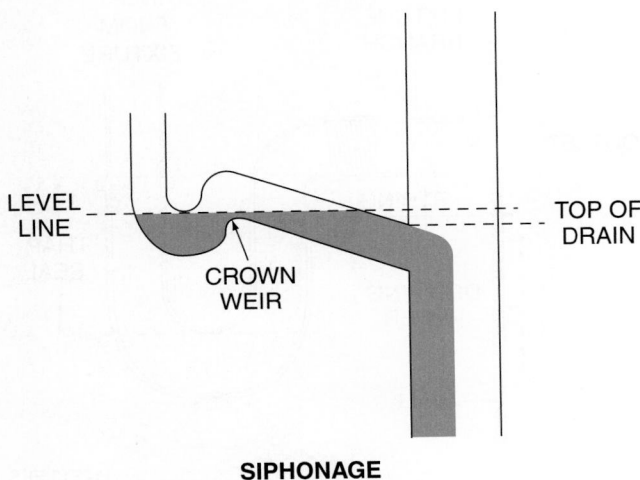

SIPHONAGE

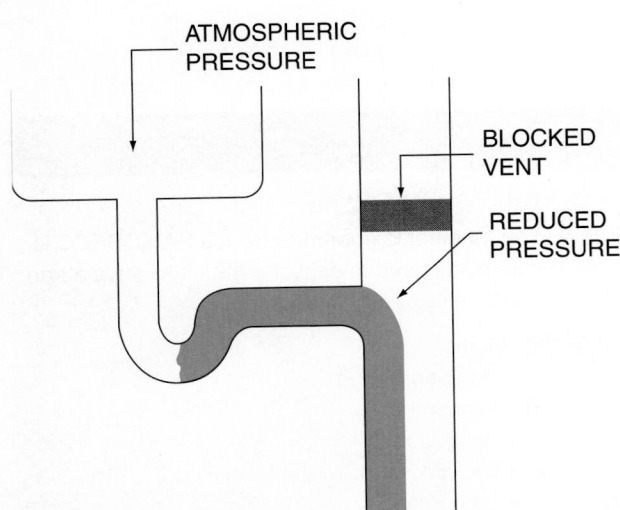

SIPHONAGE CAUSED BY REDUCED PRESSURE ZONE

112F11.EPS

Figure 11 ◆ Siphonage.

4.4.2 Aspiration

The term aspiration means the drawing in, out, or up of something, usually a fluid. In piping, aspiration takes place when a large volume of water flows near the trap, creating negative pressure. This negative pressure draws the water from the trap and causes the seal to fail. Because of aspiration, trap seals can absorb gases and odors. When the trap seal is saturated with the gas or odor, it will emit the same, often unpleasant, odor into the building.

Fixtures with S-traps are the most vulnerable to aspiration. Although modern plumbing codes prohibit S-traps, you may still encounter them in older buildings. Where an S-trap is installed, you can detect a failed seal by the smell of sewer gases or by a gurgling sound in the pipe. Drainage systems must provide adequate circulation of air in the piping to prevent siphonage and aspiration and to protect trap seals.

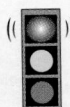 **WARNING!**
Many toxic gases do not emit an odor. Always test for gas before working in an area where toxic gases may be present. Use a smoke, water, or peppermint test to locate gas that may be entering the system through a leak in the piping. Use a toxic gas detector to monitor potentially dangerous levels of gas before working in an exposed area.

4.4.3 Momentum

The momentum, or speed, of water rushing through a pipe can force the standing water out of a trap and empty it, thus breaking the seal. Water can gain enough speed to empty a trap when the vertical distance between the fixture outlet and the trap is too long. In most cases, that distance should be limited to around 12 inches, although longer vertical distances may be required for certain types of standpipes. Refer to applicable codes to determine the correct distance.

4.4.4 Oscillation

Oscillation, or wind effect, is one of the least likely ways a trap can lose its seal. Where there are strong upward or downward air currents, the pressure or suction of the moving air may cause the water in the trap to rise or fall. If it rises enough to spill over into the waste pipe, less water remains in the trap and the seal is weakened. Lower than normal back pressures could break the seal.

4.4.5 Back Pressure

Back pressure (*Figure 12*) can cause a trap seal to break. Back pressure is pressure inside the DWV piping that is greater than atmospheric pressure. If enough wastewater from a fixture enters the **stack** so that a slug of water forms a moving plug, the air in the stack below the plug is compressed. This excess pressure tries to escape through the trap in fixture B illustrated in *Figure 12*. To prevent normal back pressure from destroying the trap seal, the stack must be properly sized, and the trap must be properly sized and protected by a vent. Also, the seal must be deep enough; generally a trap seal of 2 to 4 inches is required.

4.4.6 Evaporation

A trap may lose its seal as a result of **evaporation.** This is most likely to happen in traps that are seldom used. The water evaporates, causing the seal to break. If the DWV piping is properly designed, evaporation will become a problem only during long periods of nonuse. When sewer gas enters a structure, unused floor drains are often the cause. If you anticipate long periods of nonuse, you can install extra-deep traps. Many codes require trap primers where evaporation of the trap seal is likely. Trap primers are connected to a regularly used water line. Water flows through a small tube into the trap so that it can keep its seal.

4.4.7 Capillary Attraction

Capillary attraction (*Figure 13*) may cause a trap seal to break if a porous material, such as string or

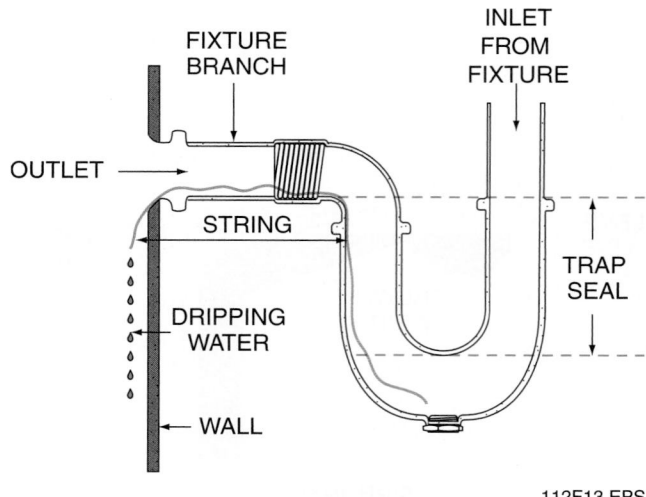

Figure 13 ◆ Capillary attraction.

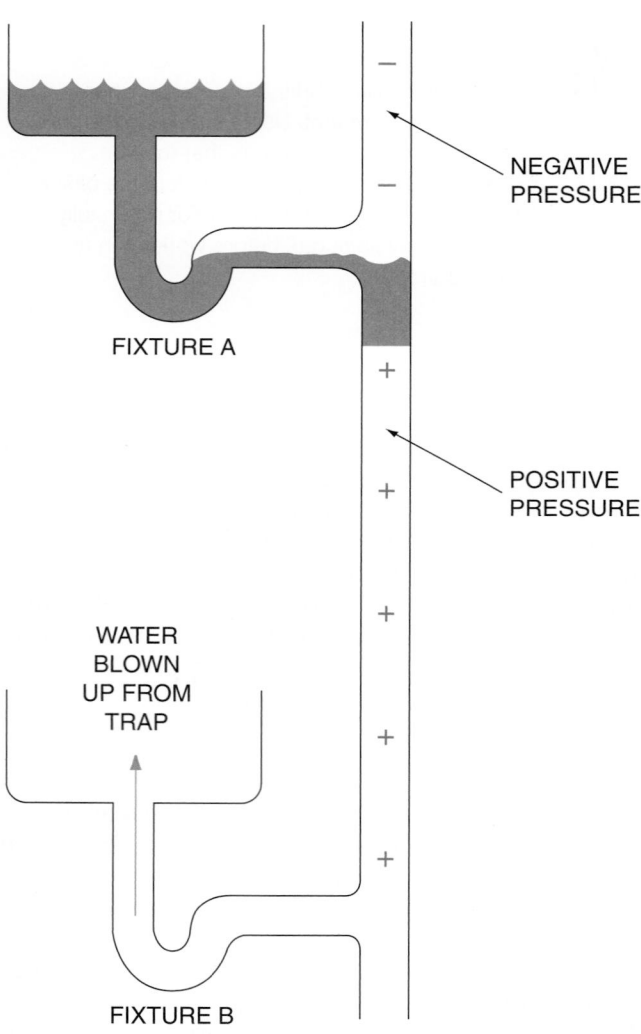

Figure 12 ◆ Back pressure.

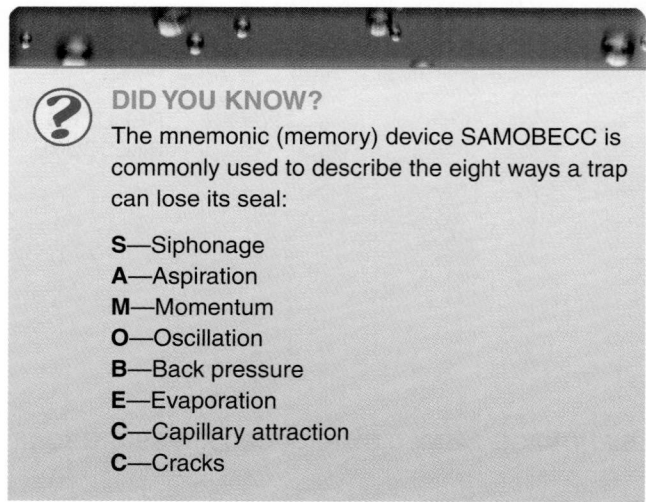

DID YOU KNOW?

The mnemonic (memory) device SAMOBECC is commonly used to describe the eight ways a trap can lose its seal:

S—Siphonage
A—Aspiration
M—Momentum
O—Oscillation
B—Back pressure
E—Evaporation
C—Capillary attraction
C—Cracks

paper, is caught in the trap. The porous material acts as a wick and draws the water out of the trap by capillary action. Cleaning the trap will solve this problem.

4.4.8 Cracks

A more common cause of waste and sewer gas leaking into a building is a crack in the trap. Cracks can be caused by worn washers, or by a broken nut, solder joint, or glue joint.

Review Questions

Sections 1.0.0–4.0.0

1. Sanitary drainage systems can be divided into three parts: the pipes inside the building, the drain pipe buried outside the building, and the _____.
   a. building sewer
   b. public sewer
   c. treatment plant
   d. lift station

2. Usually the _____ is responsible for installing and maintaining the public sewers, lift stations, and treatment plants.
   a. municipality
   b. federal government
   c. contractor
   d. land owner

3. The _____ inside a building is a circuit of piping designed to remove the wastes from plumbing fixtures and drains safely, reliably, and efficiently.
   a. drainage system
   b. sewage-disposal system
   c. DWV system
   d. vent system

4. Many fixtures have _____ that strain the wastewater before it enters the drainage piping.
   a. drains
   b. seals
   c. traps
   d. vents

5. The _____ seal in a trap protects people from airborne pathogens (germs), foul odors, and potentially explosive sewer gases.
   a. gas
   b. air
   c. pressure
   d. water

6. A fixture trap must _____, be self-cleaning, have a smooth interior waterway, and be accessible for cleanout.
   a. flush completely
   b. contain an antibacterial filter
   c. be constructed of at least four separate pieces
   d. have a nonrusting hinge

7. To avoid double trapping, _____ should never be attached to another trap.
   a. sinks
   b. tubs
   c. water closets
   d. shower assemblies

8. The highest point in the seal of the trap is the _____.
   a. crown weir
   b. inlet
   c. bottom dip
   d. fixture

9. Siphonage occurs when there _____.
   a. is negative pressure outside the DWV piping
   b. is negative pressure inside the DWV piping
   c. are not enough bends in the DWV piping
   d. are two or more vents in a vertical piping installation

10. A properly sized stack will prevent _____ from breaking the trap seal.
    a. wind effect
    b. back pressure
    c. backflow
    d. capillary attraction

5.0.0 ◆ VENTS

Every trap requires a vent of some type. Vent pipes are critical for plumbing fixtures to function correctly as part of the sanitary drainage system. Venting prevents back pressure or siphonage from breaking the water trap seals that serve the fixtures. All the vent pipes of a building create the vent system and are connected to the drain pipes. The system may include one or more pipes. Vents are installed to provide a free flow of air and to maintain equalized pressure throughout the drainage system. There are many types of vents, one of which is shown in *Figure 14*.

5.1.0 Distance from Trap to Vent

If a vent is placed too far from the trap, the vent opening will fall below the crown weir because of the fall of the waste pipe. This situation could cause the trap to self-siphon and lose its seal. To prevent this, most codes publish tables based on pipe fall and pipe diameter. These tables give the maximum distances allowed between the pipe and the vent. Check applicable code for details.

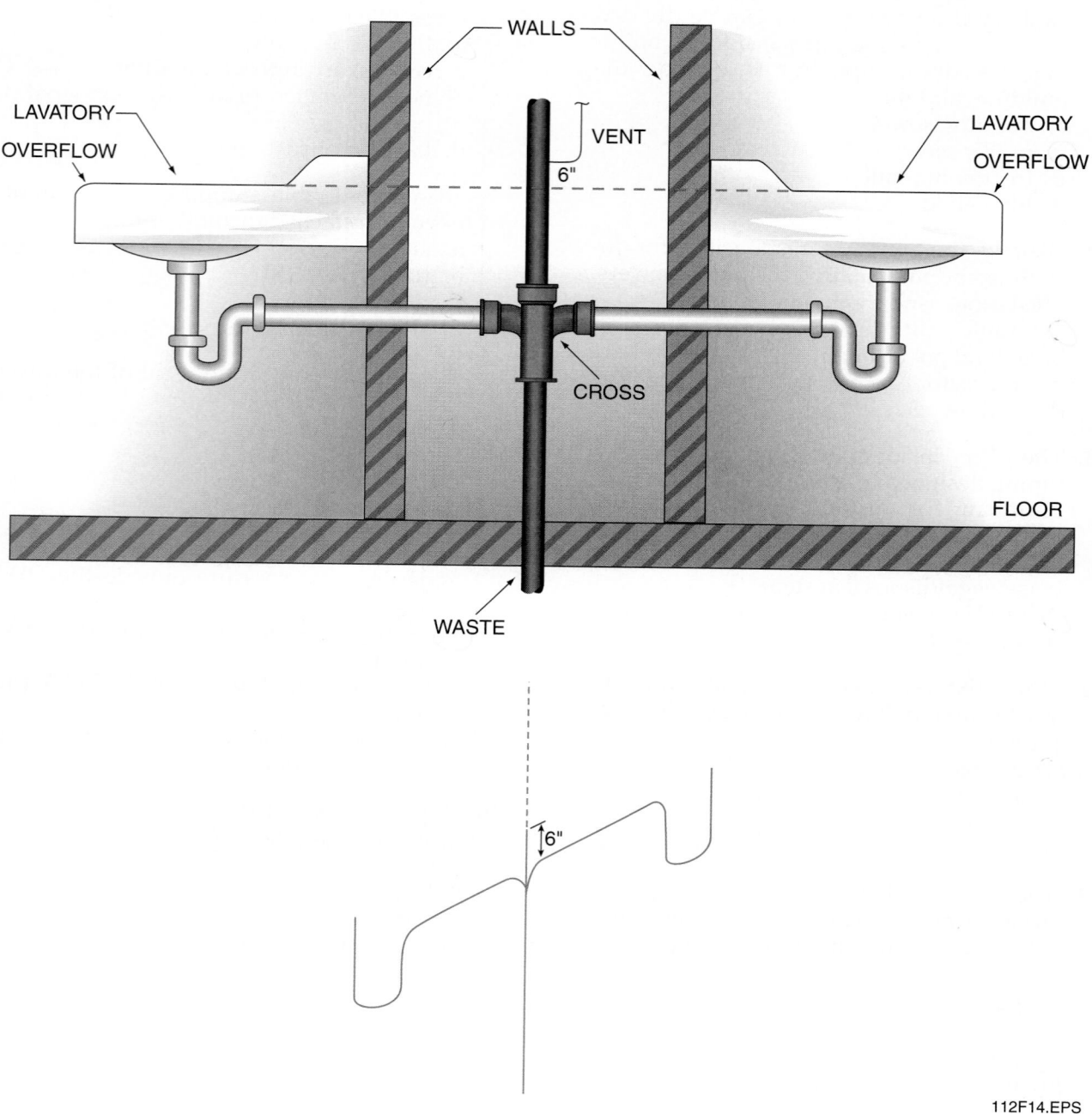

Figure 14 ◆ Vent.

112F14.EPS

6.0.0 ◆ SIZING DRAINS AND VENTS

Drainage systems fall into two major categories: storm water drains and building drains. Storm water drains collect storm water from roofs and pavement. The water is either held for on-site disposal or is disposed of at a certain rate into a storm sewer system. The sizing of storm drainage systems is based on expected rainfall. Building drainage systems must be appropriately sized based on the expected water use in the building. Plumbers must understand the mechanics of fluid flow in pipes to properly size drains and vents.

Vents are used in plumbing systems to balance pressure in the piping network. This balance of pressure is necessary to prevent the fixture traps from losing their seals. For the vents to function properly, plumbers must size the vents correctly according to how many fixtures are connected, the number of drainage fixture units (how much water discharges into the drain per minute), and the length of the vent pipe. If inadequately sized vents are installed, the plumbing system will not work properly.

Review Questions

Sections 5.0.0–6.0.0

1. _____ balance the pressure within the drainage system to maintain the water seals in fixture traps.
   a. Fixture drains
   b. Vents
   c. P-traps
   d. Cleanouts

2. Vents use _____ to maintain equalized pressure throughout the drainage system.
   a. air
   b. sewage gas
   c. liquid wastes
   d. water

3. Tables based on fall and pipe diameter give the maximum _____ allowed between the crown weir and the vent.
   a. sizes
   b. distances
   c. fittings
   d. drains

4. If the total drop between the crown weir and the fixture vent exceeds one pipe diameter, the trap is likely to _____.
   a. evaporate
   b. crack
   c. siphon
   d. pressurize

5. Plumbers must size the vents correctly according to how many _____ are connected, the number of drainage fixture units, and the length of the vent pipe.
   a. fittings
   b. fixtures
   c. drains
   d. cleanouts

7.0.0 ◆ FITTINGS AND THEIR APPLICATIONS

Fittings are devices used to connect pipe. Those used in DWV systems are called **drainage fittings.** This section describes the various types of drainage fittings and their uses within the DWV piping system.

7.1.0 DWV Fitting Materials

DWV fittings are made from many different materials, including copper, brass, lead, steel, cast iron, clay, glass, and various types of plastic. Cast iron and plastic are the most commonly used materials (see *Figure 15*). Some codes may prohibit the use of certain materials for certain applications, so be sure to check applicable code requirements before installing a DWV piping system. Not all fittings are available in all materials.

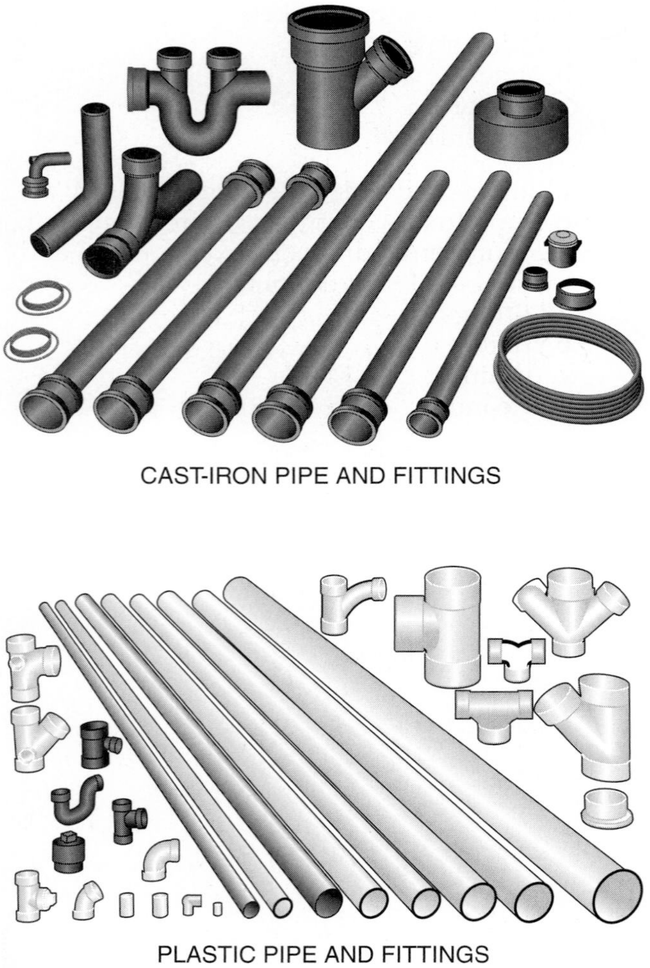

CAST-IRON PIPE AND FITTINGS
PLASTIC PIPE AND FITTINGS
112F15.EPS

Figure 15 ◆ DWV fittings.

Although the fitting materials may vary, fittings of the same design have the same names. For example, a plastic sanitary tee and a cast-iron sanitary tee are basically identical, even though they are made from different materials.

A number of fittings are available for copper pipe for DWV purposes. Those fittings include 90-degree ells, 45-degree ells, 22½-degree ells, male adapters, tees, cleanout tees, reducing tees, and reducers.

7.2.0 General DWV Fitting Requirements

Most codes state that drainage fittings may not slow or block the flow of materials in the pipe. Because of this, sanitary drainage fittings are made with a sweeping design (*Figure 16*) to allow for the smooth flow of material in the system.

Another code requirement is that the direction of hub-type fittings should not go against the flow of the system—that is, the wastes should flow from the bell to the spigot end of the pipe.

7.3.0 DWV Fittings

Sanitary fittings are used to connect DWV branches to the main DWV system. The branch inlets of these fittings may be reducing (going from a larger pipe to a smaller pipe). If so, they can be joined to the system without reducers.

7.3.1 Bends

The term bend is often used in reference to cast-iron fittings. With other types of fittings, the term elbow is more common. Bends are used to change the direction of a **run** of pipe. A run is one or more lengths of pipe in a straight line. Bends are available in different sizes. *Figure 17* shows 1⁄16, 1⁄8, 1⁄6, 1⁄5, and 1⁄4 bends. The 1⁄5 bend is available only in cast-iron pipe.

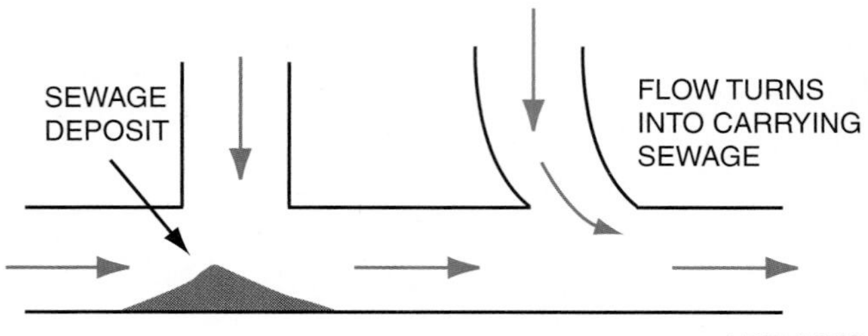

SEWAGE
DEPOSIT
FLOW TURNS
INTO CARRYING
SEWAGE
112F16.EPS

Figure 16 ◆ Sweeping design.

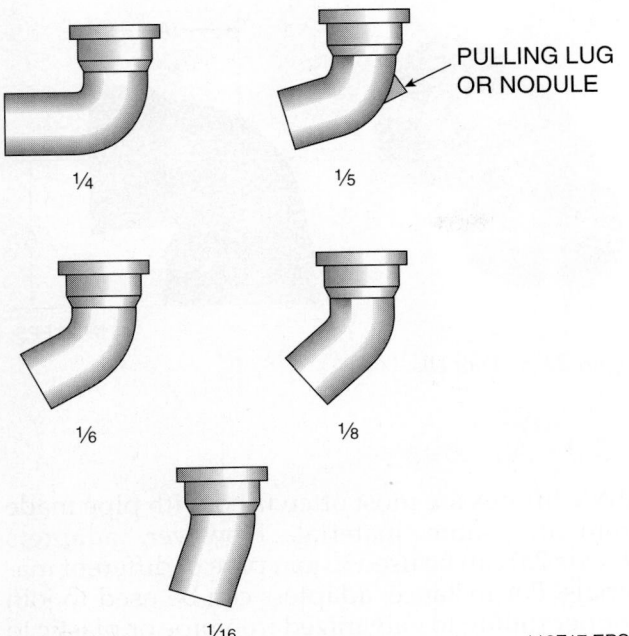

Figure 17 ◆ Bends.

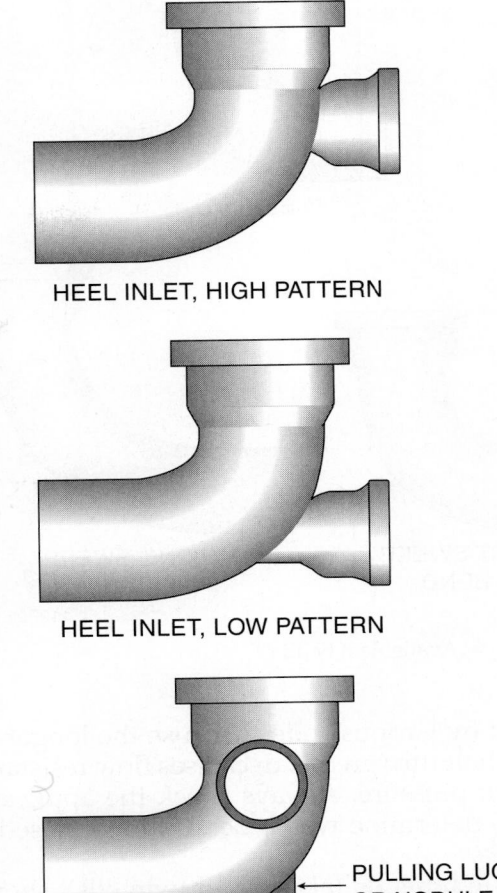

HEEL INLET, HIGH PATTERN

HEEL INLET, LOW PATTERN

SIDE INLET

112F18.EPS

Figure 18 ◆ Bends with heel and side inlets.

The plumbing design tells you which bends are used and where they are placed in the system. Bends are expressed as fractions of a complete circle. A circle contains 360 degrees. You can easily determine the number of degrees a given bend turns by multiplying the bend fraction by 360 degrees twice. For example, to determine the bend angle of a ¼ bend, multiply the bend type (¼) by 360 (¼ × 360 = 0.025). Next, multiply that answer by 360 to determine the number of degrees in a ¼ bend (0.025 × 360 = 90 degrees). Bends are available with heel inlets and side inlets to allow smaller lines to be connected to the bend (see Figure 18). Be sure to check applicable code requirements. It is important to note that a high- or low-pattern side inlet bend cannot be used as a vent if the inlet is horizontal.

Bends with side inlets are available with single and double side inlets. To determine whether the inlets are right or left inlets, place the spigot of the bend down and look through the hub (or bell) end (this is the same direction water would be flowing down the drain) (see Figure 19). Left inlets will be on the left side, right inlets on the right side.

Three patterns of ¼ bends are available (see Figure 20). The basic fitting is simply called a ¼ bend.

Short sweep ¼ bends and long sweep ¼ bends may be used at the base of DWV stacks. They are

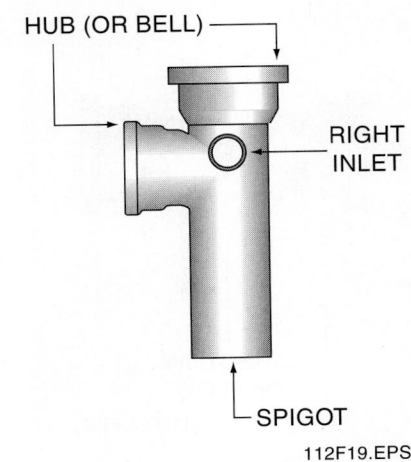

Figure 19 ◆ Determining left or right side inlet.

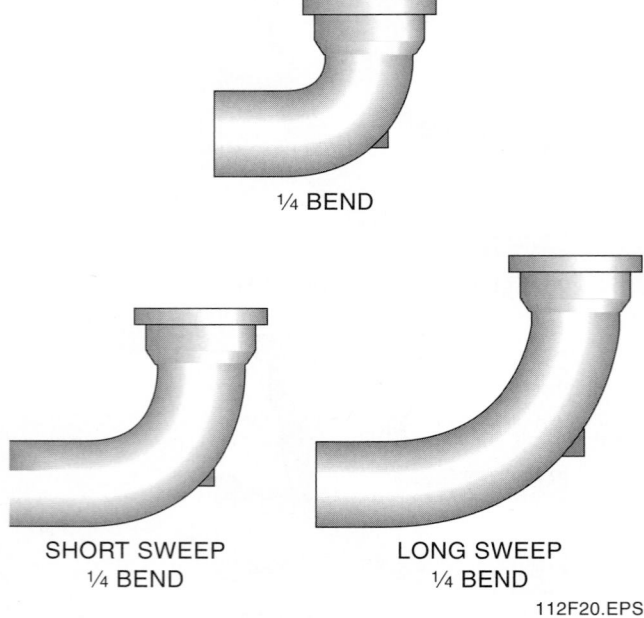

¼ BEND

SHORT SWEEP
¼ BEND

LONG SWEEP
¼ BEND

112F20.EPS

Figure 20 ◆ Available ¼ bends.

required by various codes because the longer radius of their turn greatly decreases flow resistance and back pressure. Always check the applicable codes to determine which bend must be used in your area.

Double ¼ bends (also called twin ells), shown in *Figure 21*, are used to collect and combine the flow from two opposite runs into a single run of pipe. Note the direction of flow in *Figure 21*. Be sure to check the applicable code requirements.

Codes allow the use of **vent ells** (*Figure 22*) only in vent lines. If they were placed in drainage or waste lines, their sharp turn radius would severely restrict the flow of materials. They are available only in plastic fittings.

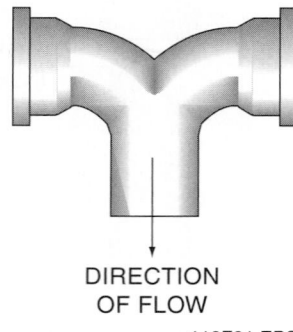

DIRECTION
OF FLOW

112F21.EPS

Figure 21 ◆ Double ¼ bend.

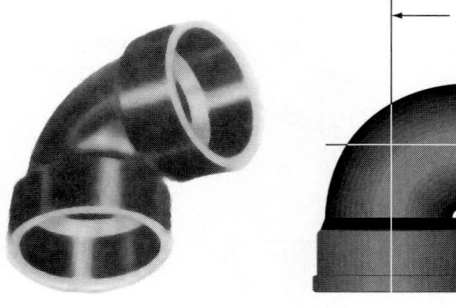

112F22.EPS

Figure 22 ◆ Vent ells.

7.3.2 Adapters

DWV fittings are most often used with pipe made from the same material. However, **adapters** (*Figure 23*) can be used to join pipe of different materials. For instance, adapters can be used to join copper tubing to galvanized iron pipe or plastic to cast-iron pipe.

7.3.3 Cleanouts

Fittings are also available for cleanouts. Cleanout fittings have internal threads on the branch fitting to accept a threaded cleanout plug. A cleanout adapter (*Figure 24*) may be installed in one end or branch of a fitting to provide a cleanout access. In both cases, a cleanout plug is provided to permit access to the DWV piping system to remove blockage and to prevent leakage.

Sanitary fittings are available in single and double patterns. The double pattern is used to connect two branch lines entering the system from opposite directions. This allows for central placement of horizontal runs and vertical runs. Sanitary branch fittings consist of tees, wyes, and combinations of the two.

7.3.4 Tees

Sanitary tees, shown in *Figure 25*, are used for branches that run from horizontal to vertical. Model codes restrict their use to sanitary drainage systems where the flow of material is from the horizontal to the vertical.

Sanitary tees are available with side inlets. The side inlet allows smaller drains to be connected from the right or left.

Most codes restrict the use of **vent tees** (*Figure 26*) and they may not be used to vent lines or as cleanout fittings. They are prohibited for use in drainage systems because their design restricts the flow of material. This design restriction may allow wastes and **pipe scale** to collect within the fitting.

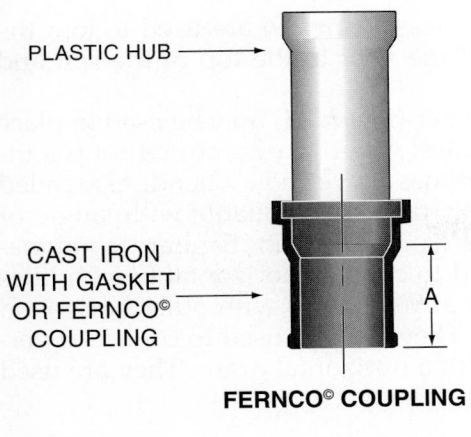

PLASTIC HUB

CAST IRON
WITH GASKET
OR FERNCO©
COUPLING

A

FERNCO© COUPLING

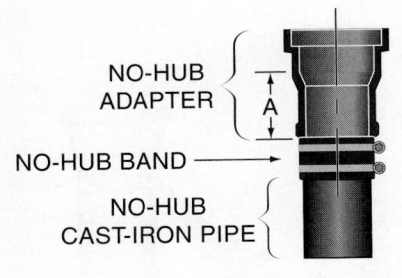

NO-HUB
ADAPTER

A

NO-HUB BAND

NO-HUB
CAST-IRON PIPE

NO-HUB ADAPTER

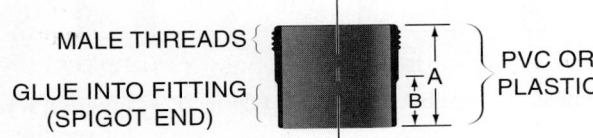

MALE THREADS

GLUE INTO FITTING
(SPIGOT END)

A
B

PVC OR
PLASTIC

PVC MALE ADAPTER WITH SPIGOT

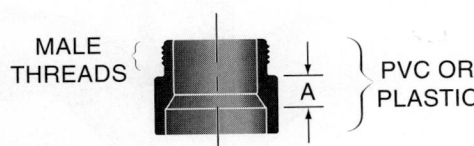

MALE
THREADS

A

PVC OR
PLASTIC

PVC MALE ADAPTER WITH HUB

112F23.EPS

Figure 23 ◆ Adapters.

FEMALE ADAPTER

MALE ADAPTER

112F24.EPS

Figure 24 ◆ Cleanout adapters.

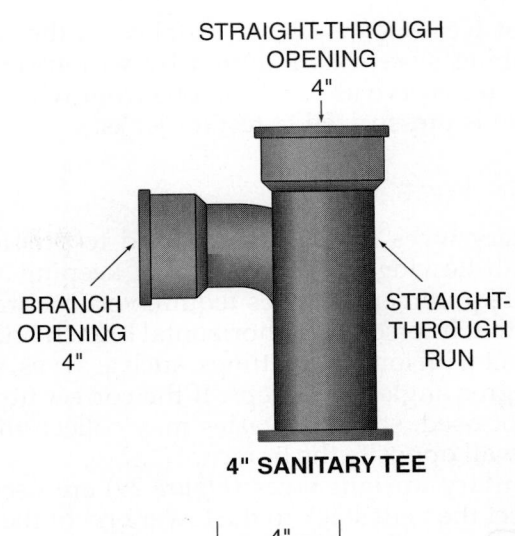

STRAIGHT-THROUGH
OPENING
4"

BRANCH
OPENING
4"

STRAIGHT-
THROUGH
RUN

4" SANITARY TEE

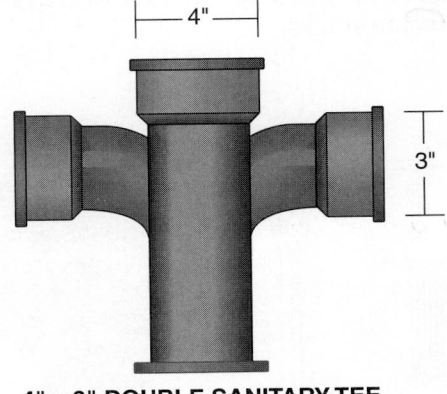

4"

3"

**4" × 3" DOUBLE SANITARY TEE
(SANITARY CROSS)**

112F25.EPS

Figure 25 ◆ Sanitary tee and cross.

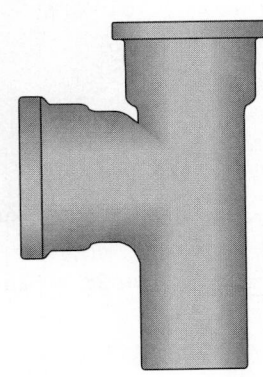

112F26.EPS

Figure 26 ◆ Vent tee.

Test tees (*Figure 27*) are installed in the DWV plumbing system as required by various codes. These tees serve as test locations from which the system is pressurized to test for leaks.

7.3.5 Wyes

Sanitary wyes (*Figure 28*) are used to provide a smooth-flowing DWV system, in keeping with code requirements. Codes require vertical branch lines that intersect with horizontal branch lines to connect with long turn fittings, such as wyes, with 45-degree angles or sweeps. If the correct fittings are not used, soil and wastes may collect on the pipe wall opposite the branch.

Sanitary upright wyes (*Figure 29*) are used to connect the vent stack to the lower end of the soil and waste stacks.

Vent branches (*Figure 30*) are used to join the upper end of the vent to the top of the soil and waste stacks.

Inverted wyes (*Figure 31*) may be used in place of vent branches. A **sanitary combination** is a fitting that combines a wye and a ⅛ bend. Also called a tee-wye, this fitting is available with single or double inlets (see *Figure 32*). Sanitary combinations are used to connect horizontal branch lines that intersect at 90 degrees with other horizontal branch lines. They also are used to connect a vertical stack with a horizontal drain. They are used

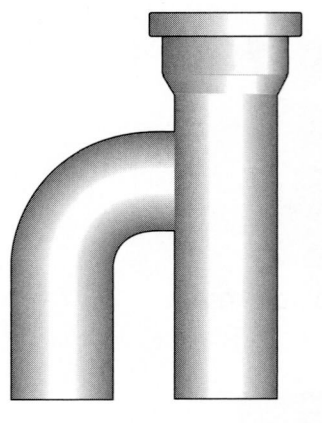

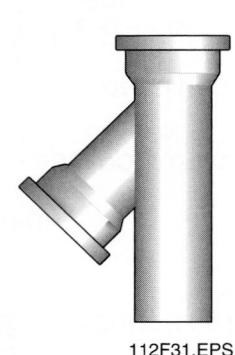

112F30.EPS

112F31.EPS

Figure 30 ◆ Vent branch.　　***Figure 31*** ◆ Inverted wye for venting.

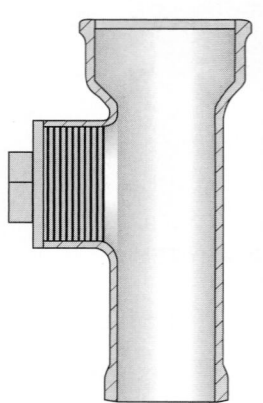

112F27.EPS

Figure 27 ◆ Test tee.

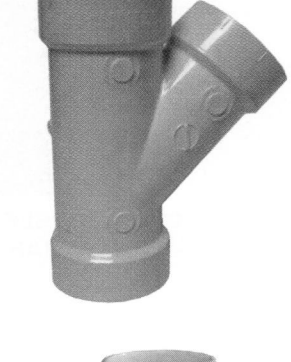

112F28.EPS

Figure 28 ◆ Sanitary wyes.

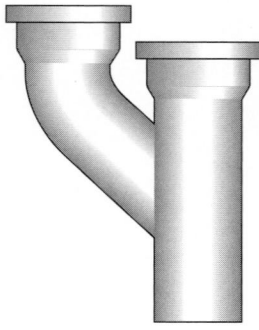

112F29.EPS

Figure 29 ◆ Sanitary upright wye.

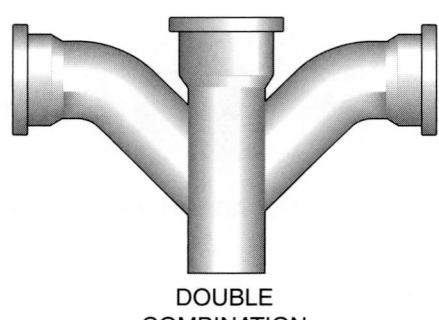

DOUBLE COMBINATION

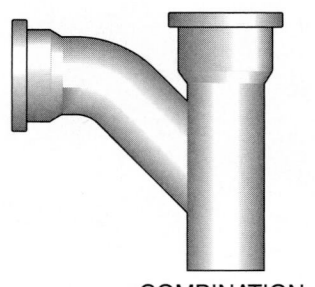

COMBINATION

112F32.EPS

Figure 32 ◆ Sanitary combinations.

in place of sanitary tees whenever space permits because they offer less resistance to the flow of materials than sanitary tees do. Combination fittings also reduce the number of fittings needed. *Figure 33* shows that a sanitary wye and a ⅛ bend would be needed to do the same job as a combination fitting. Reducing the number of fittings also reduces the number of joints to be made and the amount of time needed to make them.

7.3.6   Miscellaneous Fittings

Sanitary increasers (*Figure 34*) are used to enlarge the diameter of vent stacks and are usually placed at least 1 foot below the stack's intersection with the roof, although this distance may vary by code. Sanitary increasers are necessary in cold climates to keep condensing water vapor from freezing and gradually closing the vent opening. The loss of the vent could cause the loss of the trap seals, allowing sewer gas to enter the building.

As you have already learned, offsets (*Figure 35*) are used to change the path of the pipe to avoid obstruction. They can offset the run of the pipe from 2 to 12 inches.

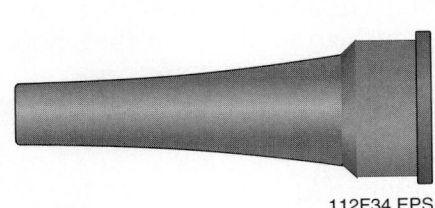

112F34.EPS

Figure 34 ◆ Sanitary increaser.

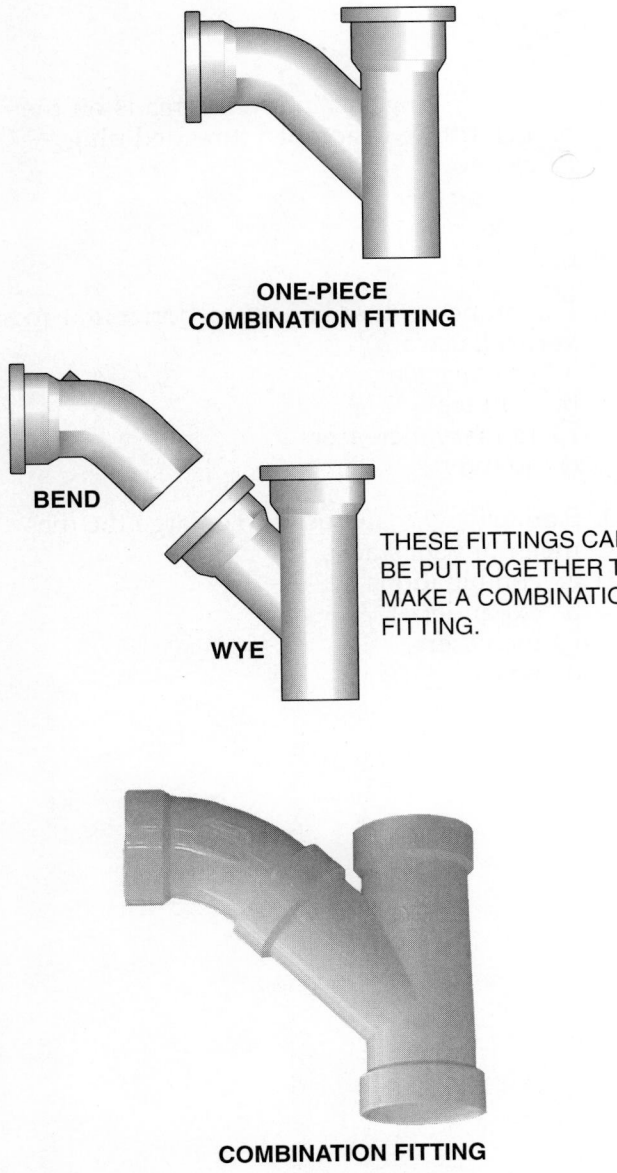

ONE-PIECE
COMBINATION FITTING

BEND

WYE

THESE FITTINGS CAN BE PUT TOGETHER TO MAKE A COMBINATION FITTING.

COMBINATION FITTING
112F33.EPS

Figure 33 ◆ Combination fittings.

112F35.EPS

Figure 35 ◆ Offset.

Section 7.0.0

1. _____ and plastic are the most commonly used materials for DWV fittings.
   a. Brass
   (b) Cast iron
   c. Clay
   d. Glass

2. Fittings of the same design will have the same name unless they are made of different materials.
   a. True
   (b) False

3. Sanitary drainage fittings are made with a _____ design to allow for the smooth flow of material in the system.
   a. circular
   b. straight
   (c) sweeping
   d. downward

4. The direction of hub-type fittings should flow from the _____ of the pipe.
   (a.) bell to the spigot end
   b. spigot end to the bell
   c. elbow to the spigot end
   d. bell to the elbow

5. If the branch inlets of sanitary fittings are _____, they can be joined to the DWV system without reducers.
   (a) reducing
   b. expanding
   c. increasing
   d. decreasing

6. A ¼ bend turns _____ of a complete circle.
   a. 45 degrees
   (b) 90 degrees
   c. 30 degrees
   d. 22½ degrees

7. Use a(n) _____ to join pipes of different materials in the DWV piping system.
   a. gasket
   b. increaser
   (c) adapter
   d. tee

8. A _____ fitting has internal threads on the branch fitting to accept a threaded plug.
   (a) cleanout
   b. sanitary
   c. wye
   d. reducer

9. For branches that run from horizontal to vertical, use a(n) _____.
   a. sanitary tee
   b. vent tee
   (c) sanitary increaser
   d. adapter

10. Sanitary _____ are used to enlarge the diameter of vent stacks.
    a. combinations
    b. expanders
    (c) increasers
    d. wyes

8.0.0 ◆ GRADE

Drainage and waste systems (see *Figure 36*) rely on gravity to move solid and liquid wastes, so these piping systems must be installed at a slope toward the point of disposal. In the plumbing industry, this slope is called **grade.** Grade is also often referred to as **slope** or percent of grade. Drainage and waste piping systems are designed with the grade engineered into the system. Grade determines the velocity of the liquid waste flowing through the piping. If the grade is too shallow, the liquid waste moves too slowly and will not scour the pipe and remove solid wastes. If the grade is too steep, the liquid waste will flow too fast and leave solids behind in the piping. The architects and engineers who design the piping system normally determine the grade. However, in some cases, such as in residential plumbing, the plumber selects the grade according to applicable code.

8.1.0 The Importance of Grade

In a system with the proper grade, the liquid wastes will flow at the right **velocity,** or speed, to scour the insides of the pipe, and the solids will be carried away. Proper grade is essential. If too much grade is used, the liquid wastes may flow too fast, leaving the solids behind. If too little grade is used, the liquid wastes will not flow fast enough to scour the pipe and remove the solid wastes. If the grade of a pipe does not remain constant, the velocity of the liquid wastes will change at the point where the grade changes. In any of these cases, the pipe will soon become blocked with solid wastes.

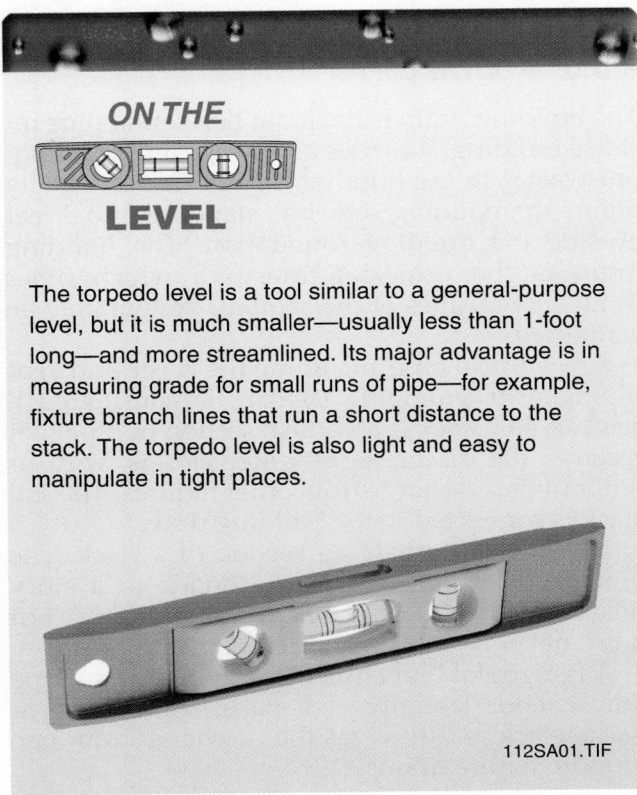

ON THE LEVEL

The torpedo level is a tool similar to a general-purpose level, but it is much smaller—usually less than 1-foot long—and more streamlined. Its major advantage is in measuring grade for small runs of pipe—for example, fixture branch lines that run a short distance to the stack. The torpedo level is also light and easy to manipulate in tight places.

112SA01.TIF

← TO WASTE TREATMENT PLANT

GRADE (SLOPE)

112F36.EPS

Figure 36 ◆ Grade.

Plumbers must determine the grade before they begin their work. This information may be in the local plumbing codes, in the specifications (specs) for the structure, or in the construction drawings. If it is not in any of these places, contact the local plumbing inspector for a decision.

9.0.0 ◆ BUILDING DRAIN

The building drain is the main horizontal pipe inside a building. It carries all sewage and other liquid wastes to the building sewer. Codes usually define the building sewer as standing 2 to 3 feet outside the building foundation. The building drain is the principal artery to which other drainage branches of the sanitary system may be connected.

Any vertical pipe, including the waste and vent piping of a plumbing system, is considered a stack. A soil stack is a vertical section of pipe that receives the discharge of water closets, with or without the discharge from other fixtures. The soil stack is connected to the building drain.

A **branch interval** is a section of a stack. The branch interval usually corresponds to a story height (the height of one floor in a building), but it can never be less than 8 feet long.

A horizontal branch is the part of a drain pipe that extends laterally (sideways) from a soil or waste stack and receives the discharge from one or more fixture drains.

9.1.0 Cleanouts

Cleanouts (*Figure 37*) are fittings with removable plugs. The plugs provide access to the inside of drainage and waste piping systems so that

PVC DWV TWISTLOK™ PLUG

112F37.TIF

Figure 37 ◆ Cleanout adapter with recessed hubs.

blockages can be removed. Cleanout adapters may be used to convert other fittings so they can be used as cleanouts.

10.0.0 ◆ BUILDING SEWER

The building sewer or house sewer is the drainage piping that runs from the building's foundation to the sewer main or septic tank (private waste disposal system). It normally starts approximately 2 to 3 feet outside the building.

Building sewers are commonly made using ABS (acrylonitrile-butadiene-styrene) or PVC (polyvinyl chloride) plastic, cast-iron, or vitrified clay (a hard, nonporous clay) pipe. When sewers are laid, the ground must be tamped to keep the pipes from settling and losing their grade. Sometimes the pipes are installed over a bed of gravel that supports them. In some parts of the country, plumbers lay the entire sewer from the foundation wall to the sewer main. In other areas, the municipal sewer crew may be in charge of the installation from the property line to the sewer main. All operations associated with building sewers are regulated by code. Always check the applicable codes when you are working on a sewer system.

Manholes must be provided for underground piping that is 8 inches or larger in diameter. They should be located at intervals not more than 400 feet apart and at every major change in direction, grade, **elevation**, or pipe size. To meet applicable codes for traffic and loading conditions, the manholes must have metal covers of sufficient weight and strength.

11.0.0 ◆ SEWER MAIN

A public or municipal sewer is installed, maintained, and controlled by the local municipality or town. The sewer main is usually located in a street or alleyway or within an easement on privately owned land. Sewer mains carry waste to the treatment plant.

Municipalities or towns usually install a 6-inch sewer laterally from the sewer main to the edge of each building lot. This lateral pipe connects the building sewer to the public sewage system.

12.0.0 ◆ WASTE TREATMENT

Many municipalities have sewer systems in which wastes are collected and treated at a sewage plant, then discharged back into the ecosystem. Other municipalities require households to treat their waste individually in private waste disposal systems.

12.1.0 Municipal Waste Treatment Systems

Municipal waste treatment plants (see *Figure 38*) are highly sophisticated facilities, as public health and safety depend on their operation.

These systems are designed to handle thousands of gallons of sewage each day. The treatment facilities receive sewage into huge holding tanks, where heavier substances settle to the bottom and lighter substances float to the top. The heavier layer is called **sludge.** Both the sludge and the wastewater are then treated.

12.2.0 Private Waste Disposal Systems

Private waste disposal systems are designed to meet the needs of individual households and the requirements of applicable codes and health departments. As in municipal waste treatment systems, but on a much smaller scale, the waste flows into a holding tank, where the sludge settles out and is digested by bacteria. Liquid waste flows through a distribution box into a leachfield, where it seeps into the earth in a natural purification cycle.

Plumbers must know about different types of private disposal systems and the advantages and disadvantages of each. Plumbers also must be able to install private waste disposal systems correctly and according to code requirements. The following is a description of one basic component of a private waste disposal system—the conventional septic tank system.

A conventional septic system consists of a septic tank, a distribution box, a leachfield, and piping between those parts (see *Figure 39*). A septic tank system provides partial treatment of raw wastewater. It protects the soil absorption system from becoming clogged by solids that are suspended in the raw wastewater. Applicable codes strictly regulate the use of these systems.

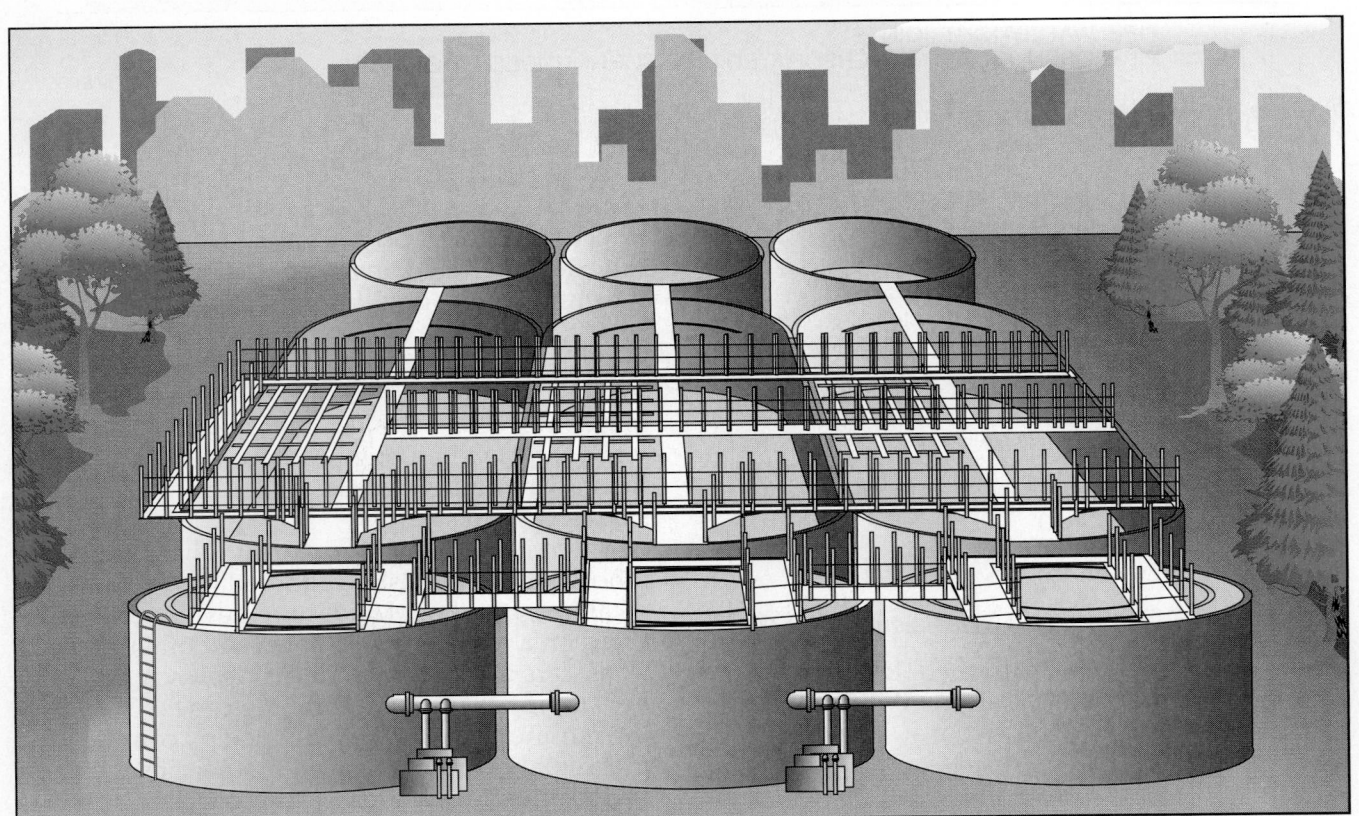

112F38.EPS

Figure 38 ◆ Municipal sewage treatment plant.

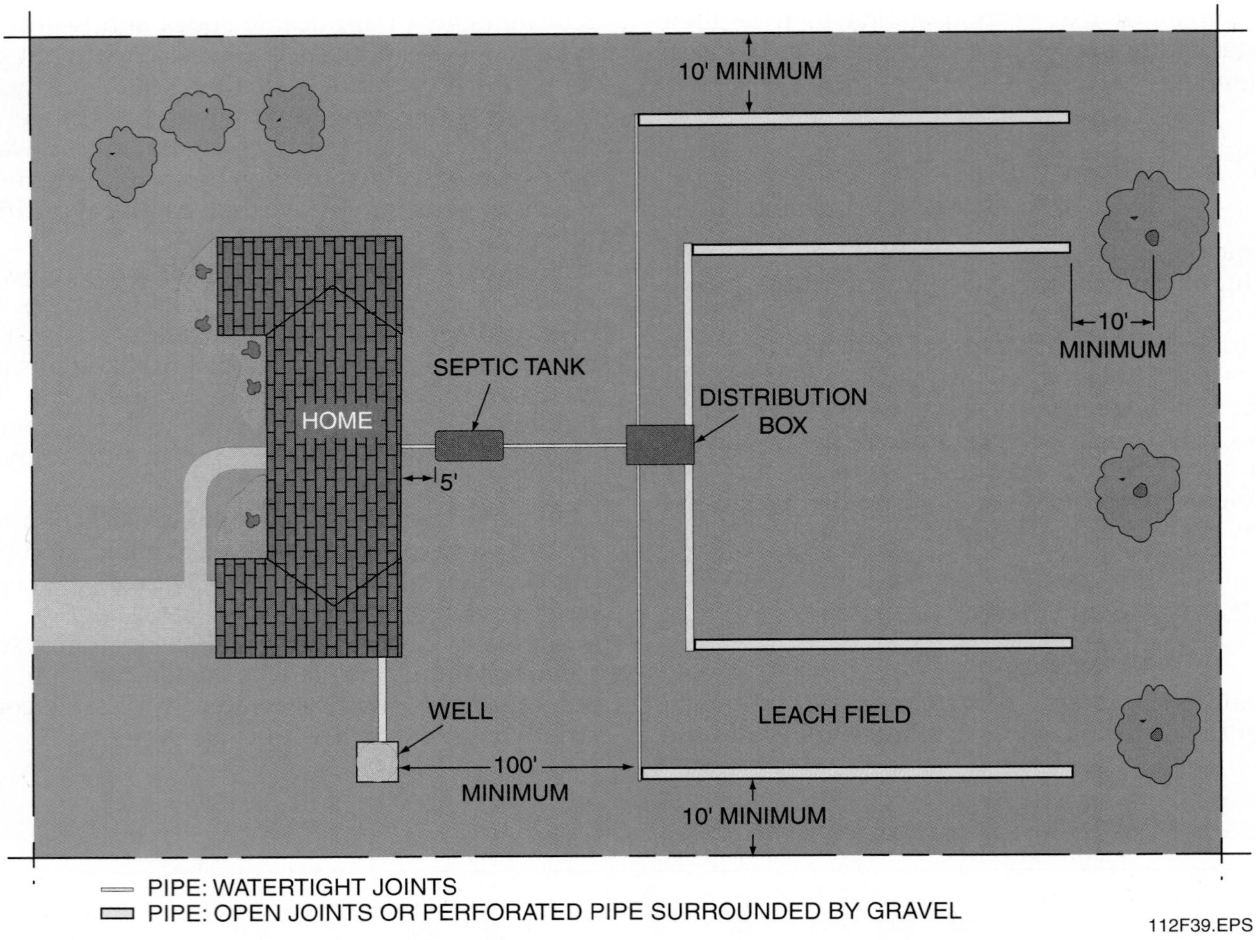

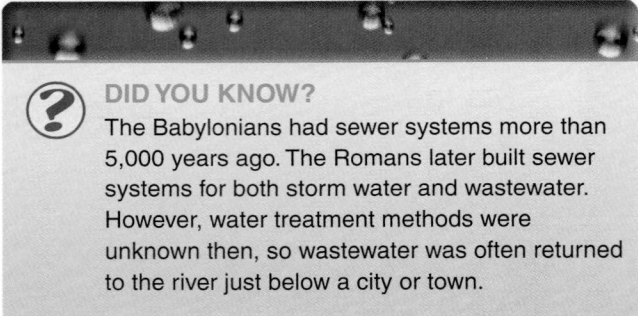

10' MINIMUM

SEPTIC TANK

HOME

5'

DISTRIBUTION BOX

10' MINIMUM

WELL

100' MINIMUM

LEACH FIELD

10' MINIMUM

PIPE: WATERTIGHT JOINTS
PIPE: OPEN JOINTS OR PERFORATED PIPE SURROUNDED BY GRAVEL

112F39.EPS

Figure 39 ◆ Typical layout for a septic tank system.

> **DID YOU KNOW?**
> The Babylonians had sewer systems more than 5,000 years ago. The Romans later built sewer systems for both storm water and wastewater. However, water treatment methods were unknown then, so wastewater was often returned to the river just below a city or town.

13.0.0 ◆ CODE AND HEALTH ISSUES

Properly designed and installed DWV systems are essential to public safety. Without DWV systems, the public would be at great risk of waste-borne illness and disease. The plumbing profession has improved public health, safety, and comfort over the past 150 years. Many serious health risks have been dramatically reduced as a direct result of good plumbing, especially properly designed and installed DWV systems, and the enforcement of plumbing codes.

13.1.0 DWV Plumbing Codes

Plumbing codes protect the safety, health, and welfare of the public. Although code requirements vary, all codes are based on principles of sanitation and safety.

There is no single national plumbing code whose requirements are adopted by all states and localities, but model codes are developed and revised on a regular basis. States and other jurisdictions can use these model codes as a basis for developing their own plumbing codes. DWV piping requirements are also defined by counties and municipalities and enforced by their inspectors.

You must always check the codes in the area where you are working. Model codes, for example, vary widely on how far vents have to be from traps and on the size of fixture drains. Consult applicable code before you start a job. Some variations in model codes are shown in *Table 1*.

Almost all codes require you to install cleanouts to provide access to all parts of the drainage system so that obstructions can be removed. Cleanouts range from removable plugs in horizontal drainage piping to manhole covers in building sewers. Codes vary widely on the specifics, so you must check applicable code requirements.

Under most codes, you may not use a cleanout plug opening to install new fixtures, unless there is another cleanout of equal accessibility and capacity and you have written approval from the plumbing inspector. On pipes that are 4 inches or less in diameter, many codes specify that cleanouts must be the same size as the pipe they serve. For larger pipes, the size of the cleanout must be at least 4 inches in diameter.

The various plumbing codes require that cleanout fittings be installed at specified locations within the DWV piping system and that the fittings be accessible. Most codes have requirements for cleanout locations in horizontal runs of pipe. In horizontal drain lines 4 inches in diameter or less, cleanouts generally must be installed no more than 50 feet apart. For larger lines, cleanouts cannot be more than 100 feet apart. Local codes may differ regarding these distances. Always refer to the applicable code.

Some codes also require cleanouts at or near the foot of each waste or soil stack and near the junction of the building drain and the building sewer.

13.2.0 Health Issues

The health issues related to the improper design, installation, and maintenance of DWV systems are significant. In the past, diseases such as cholera, typhoid fever, typhus, and dysentery have been traced to failures in DWV systems. These diseases can spread rapidly when bacteria from the sewer system enter buildings through damaged or improperly installed vents and traps.

Another health issue related to DWV systems is the accumulation of toxic sewer gases. Explosions, fires, and suffocation can occur when sewer gases are not properly released through good ventilation. Control of sewage through DWV systems can eliminate many public sanitation and health problems.

? DID YOU KNOW?

Good plumbing saves lives. Thus, good plumbing systems are extremely important to the health of our nation. In 2003, the World Health Organization commissioned an investigation to find out if plumbing systems contributed to the outbreak of severe acute respiratory syndrome (SARS) in Hong Kong. The investigation stated that poor installation, operation, and improper plumbing applications were likely contributors. Virus-rich droplets re-entered apartments through sewage and drainage systems where there were strong upward airflows, inadequate traps, and nonfunctioning water seals. Investigators also noted that plumbing, when properly designed, installed, and maintained, is an important tool in stopping transmission of disease through a building's drainage system.

Source: 2003 World Health Organization Press Release, www.who.int/mediacentre/releases/2003/pr70/en/print.html.

Sections 8.0.0–13.0.0

1. Drainage and waste systems depend on _____ to move solid and liquid wastes through the pipes.
   a. seepage
   b. gravity
   c. specifications
   d. adapters

2. The main horizontal drain pipe located inside a building is called a _____.
   a. DWV stack
   b. building drain
   c. soil stack
   d. branch interval

3. Fittings that work as removable plugs used to provide access to the inside of DWV piping are called _____.
   a. ports
   b. cleanouts
   c. inlets
   d. outlets

4. The _____ is the drainage piping that goes from the building's foundation to the sewer main or septic tank (private waste disposal system).
   a. horizontal branch
   b. building sewer
   c. building drain
   d. vent stack

5. In a septic tank, _____ settles out of the raw wastewater.
   a. seepage
   b. bacteria
   c. sewage
   d. sludge

6. Plumbing codes do not vary among jurisdictions.
   a. True
   b. False

Summary

The DWV system of a building is part of a plumbing system designed to protect the health and safety of the people who use its facilities. The system carries wastewater out of the building, treats it, and returns it to the ecosystem.

Drain and waste piping removes wastewater from a building. The type and size of the piping system selected depends on critical factors such as the amount of fluids expected to flow through the piping, the types of fluids that will be carried, and the grade. Drainage and waste systems rely on gravity to move solid and liquid wastes, so the plumber must install piping at the correct grade toward the building sewer where the wastes leave the building and enter the public or private waste disposal system. Grade determines the velocity of the liquid waste flowing through the piping. If the grade is too shallow, the liquid waste moves too slowly and will not scour the pipe and remove solid wastes. If the grade is too steep, the liquid waste will flow too fast and leave solids behind in the piping.

The vent system is an important part of the overall DWV system. Vent piping provides for the free flow of air in the drainage system. This air equalizes the atmospheric pressure inside the pipes and prevents back pressure or siphoning from destroying the water trap seals in the fixtures. The water trap seals keep sewer gas and odors out of the building. Traps provide a way for the wastewater or sewage to flow through the fixture and into the piping system while protecting the occupants of the building from bacteria and potentially explosive gases.

The DWV piping system relies on different kinds of fittings to join the lengths of pipe. The fittings may be made of copper, brass, lead, steel, cast iron, clay, glass, or plastic. The type of installation determines the best piping and fitting material to use. The fittings are designed so they do not block or slow the flow of materials in the pipe. The sweeping design of DWV fittings allows for the smooth flow of material within the system. Different fitting shapes serve specific purposes, depending on whether a pipe runs horizontally or vertically. Correct selection and installation of fittings are vital for the system to function properly.

Notes

Trade Terms Quiz

Fill in the blank with the correct trade term that you learned from your study of this module.

1. DWV branches are connected to the main DWV system using _____.

2. _____ describes the attachment of one trap to another trap in a water closet.

3. The substance that settles on the bottom of holding tanks is _____.

4. The underground drain pipe that carries waste from the building to the public sewer is called a(n)_____.

5. _____ allow pipes made from different materials to be connected.

6. The flow from two opposite runs into a single run of pipe are collected and combined in _____.

7. _____ are bend fittings with a short radius that are used at the base of a DWV stack.

8. In cold climates, _____ can prevent vent openings from closing as a result of frozen condensation in pipes.

9. _____ may not be used in drainage systems because their design can cause waste to collect in the fitting.

10. The sharp turning radius of _____ severely restricts the flow of materials, making their use allowable only in vent lines.

11. _____ are designed to keep undesirable or hazardous materials from entering a building drainage system, a public or private sewer, or sewage treatment plant or process.

12. Vertical branch lines that intersect with horizontal branch lines must be joined at an angle with _____.

13. _____ serve as locations to conduct leakage tests.

14. _____ collects on fittings, especially on iron or steel, as a result of metal corrosion.

15. Horizontal branch lines that intersect with other horizontal branch lines at a 90-degree angle are connected using _____.

16. _____ is one of four factors that must be considered when determining the intervals for manholes.

17. The _____, also called _____, of the piping system works with gravity to move solid and liquid waste through DWV systems.

18. The discharge overflow of the trap outlet is the _____, which is also simply known as a(n) _____.

19. _____ strain the wastewater before it enters the drainage piping.

20. A(n) _____ is one or more lengths of pipe that continue in a straight line.

21. The pressure inside the DWV piping that is greater than atmospheric pressure is called _____, or _____.

22. A(n) _____ is the access point to all parts of the drainage system for the removal of blockages.

23. _____ allow smaller lines to be connected to a bend.

24. A(n) _____ is the section of stack that connects branch pipes to the main DWV stack.

25. An imbalance in pressure between the inside and outside DWV piping can cause _____.

26. Code generally requires the use of _____ at the base of stacks.

27. _____ is a general term for most vertical line including offsets of soil, waste, vent, or inside conductor piping.

28. A one-piece or two-piece trap with a union nut is called a(n) _____.

29. A(n) _____ consists of a circuit of piping inside a building.

30. Improvements in plumbing have replaced the _____ with the P-trap.

31. Porous material caught in a trap can cause a seal to be broken by _____.

32. The infrequent use of a trap can cause _____ of the seal.

33. _____ are devices used to connect pipe in DWV systems.

34. The upper ends of vents are joined to the top of soil and waste stacks by _____.

35. Vent stacks are attached to the lower end of the soil and waste stacks by _____.

36. Connecters that may be used in place of vent branches are called _____.

37. The level line in which water tends to flow through the trap is the _____.

38. _____ is the speed at which waste flows through pipes.

39. The _____ in a horizontal pipe from crown weir to vent must not exceed one pipe diameter.

Trade Terms

Adapter	DWV system	Pipe scale	Slope (percent of grade)
Back pressure	Elevation	Run	Sludge
Backpressure backflow	Evaporation	S-trap	Stack
Branch interval	Fall	Sanitary combination	Test tee
Building sewer	Fixture drain	Sanitary fitting	Velocity
Capillary attraction	Grade	Sanitary increaser	Vent branch (branch vent)
Cleanout	Hydraulic gradient	Sanitary upright wye	Vent ell
Crown weir/trap weir	Interceptor	Sanitary wye	Vent tee
Double ¼ bend	Inverted wye	Short sweep ¼ bend	Weir (trap or crown)
Double trapping	Long sweep ¼ bend	Side inlet	
Drainage fitting	P-trap	Siphonage	

Jonathan Byrd

Jonathan Byrd was born in Fort Collins, Colorado, and grew up in the nearby Boulder area. After moving to Georgia in 1992, his dissatisfaction with the retail industry led him to a job in the plumbing trade, where he worked for the same plumber for several years.

Jonathan then earned his journeyman classification and joined Kosciusko, Mississippi-based Ivey Mechanical Company. Taking advantage of the company's training programs, he enrolled in a prep class that led to his Unrestricted Master's License, a Georgia license that allows a plumber to perform any type of plumbing work within the state.

Throughout his career, Jonathan has worked on dozens of projects in the Southeast, from sports arenas to broadcasting studios. Most notably, he participated in the restoration of the Georgia State Capitol Building. Built in the 1890s, the capitol building was in desperate need of new fixtures and plumbing. Jonathan assisted in the removal of the existing plumbing and the installation of the new plumbing and fixtures, often through terra cotta walls that were several feet thick.

While at Ivey Mechanical, Jonathan was offered a position teaching the apprenticeship program for the Construction Education Foundation of Georgia (CEFGA). It was there that he discovered his love of teaching. He taught the apprenticeship program at

CEFGA part time for two years before moving to a full-time position. Eventually, he returned to a full-time position at Ivey, where he now teaches the prep class for the journeyman's license and arranges training opportunities for Ivey employees. As well as teaching continuing-education classes for the Plumbing and Mechanical Association of Georgia, Jonathan continues to teach CEFGA's apprenticeship program part time.

According to Jonathan, the plumbing trade offers more freedom today than it did years ago. He says, "Today you can specialize. Back in the day, a plumber had to do a little of everything in order to survive. There's a lot of things a person can do."

Beyond the training opportunities that companies offer to their employees, craftspeople can go to associations to continue their training. "Associations are the ones who are going to assure that craft training is done right," Jonathan says. "Associations have a vested interest in training. Some are just for continuing education rather than apprenticeship, but almost every association is going to offer some training."

Jonathan advises, "Take the time to learn the trade. That might sound trite, but some people out there believe that they can be a helper for three months and then start calling themselves a plumber. I was a helper for years. I might have cut that time in half if I had some formal training."

He believes that textbook learning is also important. "The things you learn in a textbook you might not need for several years, but when you finally come across different materials on a job site, you'll know what to do." Jonathan has been in the trade long enough to see too many untrained people make some serious mistakes. "They'll try to take shortcuts, but somebody down the road will end up having to fix them. That just makes the property owner upset. Plumbing has to be done right the first time. Eighty-five percent of the work a plumber does is behind a wall. So, if a mistake has been made, someone has to go back, cut the wall, fix the mistake, and patch up the wall. It's time-consuming."

When asked what he likes most about his job, Jonathan replied, "Although I do the same thing every day, I'm always doing it somewhere new. So it's the same, and it's not the same. Once you know a trade, there's going to be enough to keep it interesting."

Trade Terms
Introduced in This Module

Adapter: A fitting that joins pipes of different sizes or materials, such as copper and galvanized pipe or cast-iron and plastic pipe.

Back pressure: A condition which may occur in the potable water distribution system, whereby a higher pressure than the supply pressure is created causing a reversal of flow into the potable water piping. Also referred to as back-pressure backflow.

Backpressure backflow: Another term for back pressure.

Branch interval: A distance along a soil or waste stack corresponding, in general, to a story height, but in no case less than 8 feet within which the horizontal branches from one floor or story of a building are connected to the stack.

Building sewer: That part of the drainage system which extends from the end of the building drain and conveys its discharge to a public sewer, private sewer, individual sewage-disposal system, or other point of disposal.

Capillary attraction: The tendency of water to be drawn by porous material. If a piece of paper or string is stuck in a pipe, the water is drawn out of the trap and the seal is broken.

Cleanout: An access point to all parts of the drainage system for the removal of blockages.

Crown weir/trap weir: Discharge overflow of the trap outlet. Also referred to as weir, trap, or crown.

Double ¼ bend: A fitting used to collect and combine the flow from two opposite runs into a single run of pipe.

Double trapping: A situation in which one trap is attached to another, creating negative pressure that will stop the flow of drainage.

Drainage fitting: Any of a variety of fittings used in the DWV system to remove waste from a building.

DWV system: An acronym for "drain-waste-vent" referring to the combined sanitary drainage and venting systems. This term is technically equivalent to "soil-waste-vent" (SWV).

Elevation: The height above an established reference point, such as a grade reference point on a construction drawing.

Evaporation: Loss of water, especially in a drainage system, into the atmosphere.

Fall: The amount of slope given to horizontal runs of pipe.

Fixture drain: The drain from the trap of a fixture to the junction of that drain with any other drain pipe.

Grade: The slope of a horizontal run of pipe. Also referred to as slope (percent of grade).

Hydraulic gradient: The level line in which water tends to flow in a pipe.

Interceptor: A device designed and installed so as to separate and retain deleterious, hazardous, or undesirable matter from normal waste while permitting normal sewage or liquid wastes to discharge into the drainage system by gravity.

Inverted wye: A fitting used to join the upper end of the vent to the top of the soil and waste stacks. It looks like an upside-down letter Y.

Long sweep ¼ bend: A bend used at the base of DWV stacks because the longer radius of its design greatly decreases flow resistance and back pressure. Use is regulated by code.

P-trap: A P-shaped trap that provides a water seal in a waste or soil pipe, used mostly at sinks and lavatories.

Pipe scale: Flaky material resulting from corrosion of metals, especially iron or steel. Also, a heavy oxide coating on copper or copper alloys resulting from exposure to high temperatures and an oxidizer.

Run: One or more lengths of pipe that continue in a straight line.

S-trap: A trap with a long downstream leg, which tends to promote siphonage. S-traps are no longer installed but are still found in older buildings.

Sanitary combination: A fitting that combines a wye and ⅛ bend. It is used to connect horizontal branch lines that intersect other horizontal branch lines. It offers less resistance to the flow of material than a sanitary tee. Also called a tee-wye.

Sanitary fitting: A fitting used to connect DWV branches to the main DWV system and to serve as a cleanout.

Sanitary increaser: A fitting used to enlarge the diameter of the vent stack. It is usually placed at least 1 foot below the intersection of the stack and the roof. Use is regulated by code.

Sanitary upright wye: A fitting used to connect the vent stack to the lower end of the soil and waste stacks.

Sanitary wye: A drainage fitting, shaped like the letter Y, that joins the main run of pipe at an angle.

Short sweep ¼ bend: A bend fitting with a short radius used at the base of a DWV stack. Use is regulated by code.

Side inlet: An opening in an ell or tee fitting at right angles to the line of the run, used to connect smaller lines to the main line.

Siphonage: Loss of water in a trap seal caused by unequal pressure inside and outside DWV piping.

Slope (percent of grade): Another term for grade.

Sludge: Semi-liquid matter that settles out in a holding tank during the waste treatment process.

Stack: A general term for any vertical line including offsets of soil, waste, vent, or inside conductor piping. This does not include vertical fixture and vent branches that do not extend through the roof or that pass through not more than two stories before being reconnected to the vent stack or stack vent.

Test tee: A tee installed as a test location for pressurizing the system to test for leaks.

Velocity: Speed of motion, such as the speed of sewage or wastewater through the drainage piping.

Vent branch (branch vent): A vent connecting one or more individual vents with a vent stack or stack vent.

Vent ell: A plastic fitting with a sharp turn radius, used only in vent piping systems. Use is regulated by code.

Vent tee: A fitting used in venting systems or as a cleanout. It may not be used in the drainage system because it restricts the flow of material. Use is regulated by code.

Weir (trap or crown): Also referred to as crown weir or trap weir.

Additional Resources and References

ADDITIONAL RESOURCES

This module is intended to present thorough resources for task training. The following reference works are suggested for further study. These are optional materials for continued education rather than for task training.

Basic Plumbing with Illustrations, Revised Edition, 1994. Howard C. Massey. Carlsbad, CA: Craftsman Book Company.

Plumbing Systems: Analysis, Design and Construction, 1996. Tim Wentz. Upper Saddle River, NJ: Prentice Hall.

REFERENCE

2003 National Standard Plumbing Code. Chapter 1, Definitions (definitions of back pressure, branch interval, building sewer, crown weir or trap, DWV system, fixture drain, interceptor, stack, and vent branch). Falls Church, VA: Plumbing-Heating-Cooling Contractors–National Association.

NCCER makes every effort to keep these textbooks up-to-date and free of technical errors. We appreciate your help in this process. If you have an idea for improving this textbook, or if you find an error, a typographical mistake, or an inaccuracy in NCCER's Contren® textbooks, please write us, using this form or a photocopy. Be sure to include the exact module number, page number, a detailed description, and the correction, if applicable. Your input will be brought to the attention of the Technical Review Committee. Thank you for your assistance.

Instructors – If you found that additional materials were necessary in order to teach this module effectively, please let us know so that we may include them in the Equipment/Materials list in the Annotated Instructor's Guide.

Write: Product Development and Revision
National Center for Construction Education and Research
3600 NW 43rd St, Bldg G, Gainesville, FL 32606

Fax: 352-334-0932

E-mail: curriculum@nccer.org

Craft _____ Module Name _____

Copyright Date _____ Module Number _____ Page Number(s) _____

Description _____

_____

_____

_____

(Optional) Correction _____

_____

_____

(Optional) Your Name and Address _____

_____

_____

Plastic Pipe and Fittings
68116-09

68116-09
Plastic Pipe and Fittings

Topics to be presented in this module include:

1.0.0   Introduction 16.2

2.0.0   Plastic Pipe 16.2

3.0.0   Fittings 16.8

4.0.0   Measuring, Cutting, and Joining Plastic
          Pipe and Fittings 16.10

5.0.0   Pipe Supports 16.19

6.0.0   Pressure Testing 16.22

Overview

Plastic has changed the way plumbers work. Plastic pipe is strong, durable, and requires little maintenance. Although it is easier to use than metal pipe, it does have some disadvantages. It can be affected by temperatures, give off toxic fumes, and become flammable. Plumbers must learn how to safely and properly handle plastic pipe and fittings.

Plumbers must be familiar with several types and sizes of plastic pipe, each with its own properties. Each type of plastic pipe is designed for different applications and has different installation requirements. All plastic pipe must be handled and stored properly. A pipe's application determines the type of fittings and joints that are compatible. Specific fittings are used for water supply and drain, waste, and vent (DWV) systems. Plumbers use special tools to measure, cut, and join plastic pipe and fittings. While measuring and cutting techniques are standard, different types of plastic pipe have their own methods for joining.

Plastic pipe can be supported using a variety of methods. The material, size, applications, and installation affect the type of support. Hangers are used for horizontal support; pipe riser clamps provide vertical support. Plans and specifications dictate codes that must be followed. After completing an installation, the system must be hydrostatically pressure-tested.

Focus Statement

The goal of the plumber is to protect the health, safety, and comfort of the nation job by job.

Code Note

Codes vary among jurisdictions. Because of the variations in code, consult the applicable code whenever regulations are in question. Referencing an incorrect set of codes can cause as much trouble as failing to reference codes altogether. Obtain, review, and familiarize yourself with your local adopted code.

Objectives

When you have completed this module, you will be able to do the following:

1. Identify types of materials and schedules of plastic piping.
2. Identify proper and improper applications of plastic piping.
3. Identify types of fittings and valves used with plastic piping.
4. Identify and determine the kinds of hangers and supports needed for plastic piping.
5. Identify the various techniques used in hanging and supporting plastic piping.
6. Properly measure, cut, and join plastic piping.
7. Explain proper procedures for the handling, storage, and protection of plastic pipes.

Trade Terms

ABS
Bell-and-spigot pipe
Cellular core wall
Chamfer
Chemically inert
Compression collar
CPVC
Elastomeric
Fusion fitting
Hydronic
Hydrostatically pressure-test
Inside diameter
Interference fit
Outside diameter
PB
PE
PEX
Pipe riser clamp
Pressure rating
psi
PVC
Ring-tight gasket fitting
Schedule
Size dimension ratio
Solid wall
Solvent weld
Thermoplastic pipe
Thermosetting pipe
Transition fitting
Water hammer

Required Trainee Materials

1. Appropriate personal protective equipment
2. Pencils and paper
3. Copy of local adopted code

Prerequisites

Before you begin this module, it is recommended that you successfully complete the following: *Core Curriculum*, and *Construction Technology*, Modules 68101-09 through 68115-09.

This course map shows all of the modules in *Construction Technology*. The suggested training order begins at the bottom and proceeds up. Skill levels increase as you advance on the course map. The local Training Program Sponsor may adjust the training order.

68117-09
Copper Pipe and Fittings

68116-09
Plastic Pipe and Fittings

68115-09
Introduction to Drain, Waste, and Vent (DWV) Systems

68114-09
Introduction to HVAC

68113-09
Residential Electrical Services

68112-09
Electrical Safety

68111-09
Basic Stair Layout

68110-09
Exterior Finishing

68109-09
Roofing Applications

68108-09
Roof Framing

68107-09
Wall and Ceiling Framing

68106-09
Floor Systems

68105-09
Masonry Units and Installation Techniques

68104-09
Introduction to Masonry

68103-09
Handling and Placing Concrete

68102-09
Introduction to Concrete, Reinforcing Materials, and Forms

68101-09
Site Layout One: Distance Measurement and Leveling

CORE CURRICULUM:
Introductory Craft Skills

CONSTRUCTION TECHNOLOGY

116CMAP.EPS

1.0.0 ◆ INTRODUCTION

Plastic has revolutionized plumbing, a trade that was based for hundreds of years on metal pipe and fittings. Not widely used by the plumbing industry until the 1950s, plastic is the newest of plumbing materials and has changed the way plumbers work. Plumbers use plastic today in a wide variety of ways, including drain, waste, and vent (DWV) piping; water distribution; chemical waste systems; and fuel gas piping.

2.0.0 ◆ PLASTIC PIPE

Plastic pipe has both advantages and disadvantages that you need to know. In addition, you need to understand the different properties of plastic pipe and how the industry measures plastic piping. The following sections discuss these characteristics and properties, as well as the standard labeling practice for plastic piping and the most common plastic pipe and fitting manufacturers. You will also learn about the specific types of plastic pipe that you will use in plumbing applications.

2.1.0 Advantages and Disadvantages of Using Plastic Pipe

Most plastic pipe and fittings are **chemically inert**, meaning they do not react with other substances or materials, so they can withstand most chemicals that are used in the home, office, or factory. Strong and durable, plastic pipe and fittings also require little maintenance because they are generally resistant to corrosion and do not pit or scale. In addition, they are easier to handle than metals because of their relative light weight, which can make them more cost-effective to install.

The use of plastic pipe helps to eliminate one of the most common hazards associated with installation—fire. During the installation process, no welding or soldering is involved. Therefore, it is less likely that fires will occur. In the past, when plumbers joined pipe using molten lead, accidents involving spilled lead and tipped-over heating furnaces were a constant concern.

Plastic does have some drawbacks. Temperature changes cause it to expand and contract more than metal. Plastic pipe can also be flammable and give off fumes when it burns. In addition, fumes from solvent chemicals used to join plastic pipes are harmful; plumbers must be familiar with the types of safety hazards each variety of pipe and chemical solvent presents. It is your responsibility to know how to protect yourself on the job site. Always consult the manufacturer's specifications regarding the properties of the pipe you are using.

All plastic pipe is affected by ultraviolet (UV) radiation to some degree. That means that you should store plastic piping under cover during a construction project to protect it from the sun. In general, short-term sun exposure will not harm the piping, but long-term exposure will degrade the plastic material. Always consult the manufacturer's instructions regarding the proper way to select and protect the particular piping you are using.

2.2.0 Properties of Plastic Pipe

Plastics contain polymers, long chains of molecules that manufacturers can mold, cast, or force through a die into desired shapes. Manufacturers can heat and shape resins in the form of pellets, powders, and solutions in various combinations to give the finished product the desired properties.

Each type of plastic pipe has its own unique properties. These properties determine where and how you can install it, how you can join it to other piping, and how you should support it. There are two general categories of plastic: thermoplastic and thermosetting. **Thermoplastic pipe** can be repeatedly softened by heating and hardened by cooling. When softened, thermoplastics can be molded into desired shapes. **Thermosetting pipe** changes chemically when heated, so that once hardened by heat or chemicals, it is hardened permanently.

In addition, plastic pipe can be either rigid or flexible. Rigid pipe is straight and maintains its shape. Flexible pipe bends, so although it requires fewer fittings and joints, it requires more support. Flexible pipe is often referred to as tubing. Manufacturers sell flexible pipe in coils.

The construction of plastic piping can vary as well. Pipes can be **solid wall** or **cellular core wall** construction. Solid wall does not contain trapped air. A pipe with cellular core construction, also called foam core construction, has walls that contain trapped air, so it is lighter weight. Cellular core wall piping consists of three layers of plastic: solid inner and outer layers and a foam middle layer. Because it consists of less material, it is also less expensive than solid wall pipe. Plumbers use solid wall and cellular core wall pipe interchangeably for DWV applications.

> **NOTE**
>
> When choosing plastic piping, always consult the specifications for your project. Always adhere to applicable local codes.

2.3.0 Plastic Pipe Sizing

Plastic pipe is manufactured with various wall thicknesses, commonly called schedules or size dimension ratios (SDRs). SDR relates wall thickness to the diameter of the pipe, using a set ratio of wall thickness to diameter. As the pipe diameter gets bigger, the wall gets thicker. An SDR number, such as SDR 35, designates this ratio. The ratio between a pipe's outside diameter (OD), or the distance between the outer walls of a pipe, and its wall thickness is constant for each pipe size. The larger the SDR number, the thinner the wall versus the diameter. For example, a 2-inch pipe (OD of 2.375 inches) with an SDR of 11 has a wall thickness of 0.216 inches (see *Table 1*). The same pipe with an SDR of 17 has a wall thickness of 0.14 inches. Be aware that pipe can also be measured by its inside diameter (ID), the distance between its inner walls.

In an effort to simplify and standardize the use of plastic pipe and fittings, manufacturers designated two SDR ratios as Schedule 40 (thin wall), which is the most commonly used, and Schedule 80 (thick wall). Most plumbing codes designate Schedule 40 as the minimum size for use in or under buildings.

2.4.0 Labeling (Markings)

Plumbers must use plastic pipe and fittings that are labeled and approved for use in plumbing systems. The label on a plastic pipe will include the following:

- Nominal pipe diameter, copper tube size (CTS) or iron pipe size (IPS)
- Schedule or SDR
- Type of material, PVC (polyvinyl chloride), CPVC (chlorinated polyvinyl chloride), and so on
- Pressure rating, usually expressed as psi (pounds per square inch) at a certain temperature
- Relevant standard for the pipe, for example, *American Society for Testing and Materials (ASTM) D-3309*
- Listing body or laboratory, for example, the National Sanitation Foundation
- Name of manufacturer or brand name
- Country of origin

This label ensures that a recognized third party—the listing body or laboratory—has tested the material and approved it for the use intended (see *Figure 1*). For example, the ASTM numbers on cellular core piping are different from those on Schedule 40 solid core piping. Solvent cements must also be listed and labeled for their specific use.

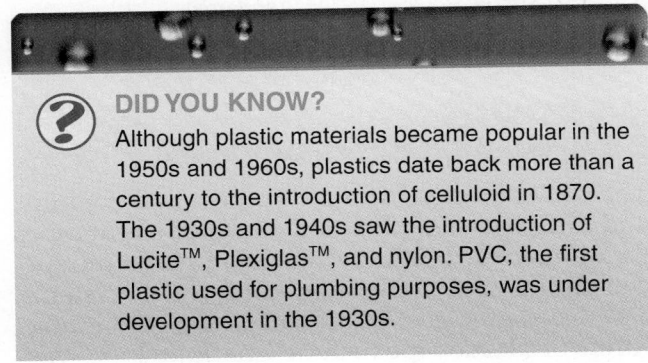

DID YOU KNOW?

Although plastic materials became popular in the 1950s and 1960s, plastics date back more than a century to the introduction of celluloid in 1870. The 1930s and 1940s saw the introduction of Lucite™, Plexiglas™, and nylon. PVC, the first plastic used for plumbing purposes, was under development in the 1930s.

Table 1   Minimum Wall Thickness of 2-Inch Pipe

Plastic Pipe and Fittings Association Minimum Wall Thicknesses of 2-Inch Pipe Based on SDR/SIDR			
IPS-OD SDR (Outside Diameter = 2.375 inches) **SDR (D 3035)**	**Wall**	**IPS-ID SIDR** (Inside Diameter = 2.067 inches) **SIDR (D 2239)**	**Wall**
7	0.339	5.3	0.390
9	0.264	7	0.295
11	0.216	9	0.230
13.5	0.176	11.5	0.180
17	0.140	15	0.138
21	0.113	19	0.109
26	0.091	24	0.089
32.5	0.073	30.5	0.071

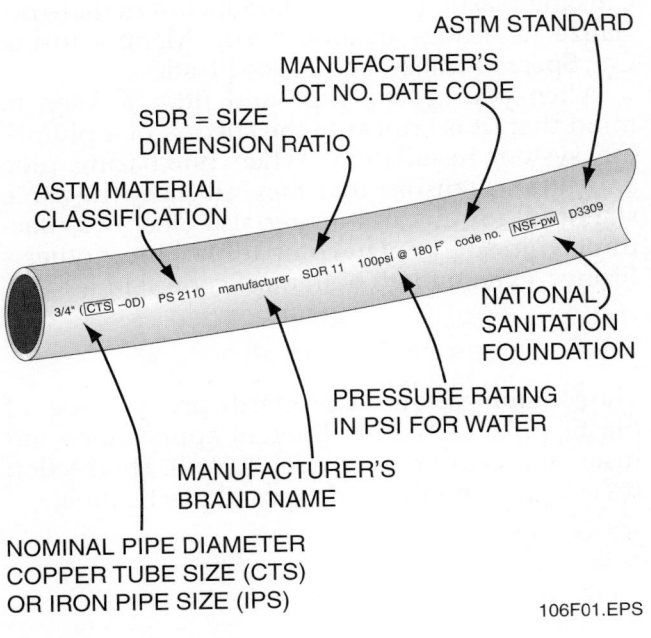

ASTM STANDARD

MANUFACTURER'S LOT NO. DATE CODE

SDR = SIZE DIMENSION RATIO

ASTM MATERIAL CLASSIFICATION

3/4" (CTS) –OD) PS 2110 manufacturer SDR 11 100psi @ 180 F° code no. NSF-pw D3309

NATIONAL SANITATION FOUNDATION

PRESSURE RATING IN PSI FOR WATER

MANUFACTURER'S BRAND NAME

NOMINAL PIPE DIAMETER COPPER TUBE SIZE (CTS) OR IRON PIPE SIZE (IPS)

106F01.EPS

Figure 1   ◆   Typical pipe label.

ON THE LEVEL **Restrictions on Plastics**

Consult your local codes for restrictions on the use of plastics in commercial fire-rated buildings and in multistory applications.

Table 2 Types of Plastic Pipe and Their Applications

Pipe	Applications
ABS	DWV sanitary systems, corrosive waste
PVC	DWV sanitary systems, cold water service
CPVC	Hot and cold water distribution
PE	Cold water service, corrosive waste, gas service
PEX	Hot and cold water distribution
PB	Hot and cold water distribution

2.5.0 Manufacturers of Plastic Pipe

There are hundreds of plastic pipe manufacturers, and different manufacturers make different pipes for different applications. For example, some specialize in pipes for hot or cold water systems, DWV systems, gas, or radiant heating. Manufacturers of plastic pipe include Charlotte Pipe, Vanguard, Bristol Pipe, J-M Manufacturing, and Cresline Plastic Pipe Co. Manufacturers that specialize in fittings include Nibco Manufacturing Co., Spears Plastics, and Lasco Plastics.

When you specify pipe and fittings, keep in mind that fit is critical to the success of a plumbing system installation. When purchasing pipe and fittings, ensure that they all meet the same standard for fit and materials. One manufacturer's pipe may not be compatible with another's fittings, even though both meet code standards.

2.6.0 Types of Plastic Pipe

The plumbing industry primarily uses six types of plastic pipe, each with different applications and installation requirements (see *Table 2*). This section reviews commonly used types of plastic pipe.

2.6.1 ABS

ABS (acrylonitrile-butadiene-styrene) pipe and fittings (*Figure 2*) are made from a thermoplastic resin. ABS is the standard material for many types of DWV systems. ABS pipe is available in diameters ranging from 1½ inches to 6 inches. ABS pipe wall comes in both solid wall and cellular core wall construction, which are interchangeable in plumbing applications.

Because ABS is light, it is easy to handle and install. A 3-inch-diameter, 10-foot-long section weighs less than 10 pounds. ABS performs well at extreme temperatures because it absorbs heat and cold slowly, an important feature for a system that handles both hot and cold wastes.

ABS is highly resistant to household chemicals. In tests, it showed no effect from such common products as detergent, bleach, and household drain cleaners. Sewage treatment plants use ABS because it stands up to the highly corrosive and abrasive liquids commonly found in such systems.

ABS is strong and long-lasting. In 1959, ABS pipe was used in an experimental residence. Twenty-five years later, an independent research firm dug up and analyzed a section of the pipe and found no evidence of rot, rust, or corrosion. The pipe also withstands earth loads, slab foundations, and high surface loads without collapsing.

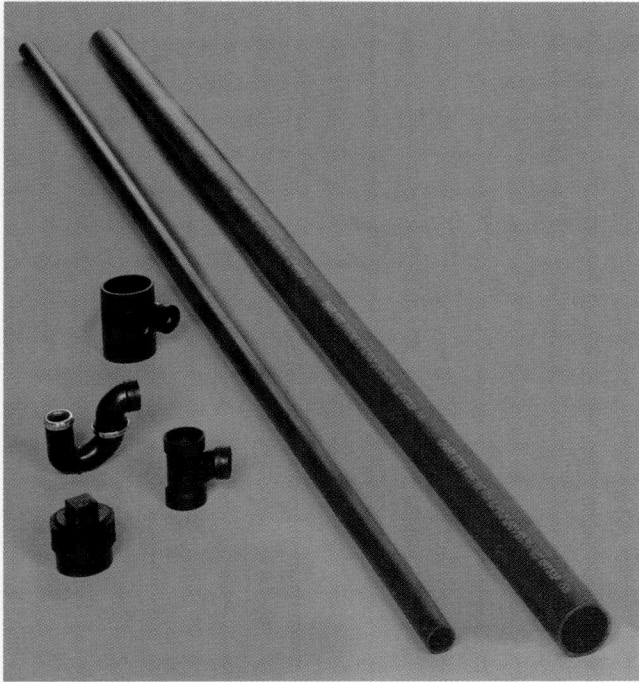

Figure 2 ◆ ABS pipe and fittings.

2.6.2 PVC

PVC is a rigid pipe with high-impact strength that is manufactured from a thermoplastic material (see *Figure 3*). The material has an indefinite life span under most conditions. Plumbers frequently use PVC in cold water systems. PVC is interchangeable with ABS pipe for DWV systems. It is also used to transport many chemicals because of its chemical-resistant properties. PVC is available in solid wall and cellular core wall construction.

Solid core PVC pipe can be used in high-pressure systems, but only to carry low-temperature water. You must protect it from sunlight because ultraviolet light degrades the thermoplastic materials. It is lightweight, easy to handle and install, has joint flexibility that handles ground movement without leaking, and lasts a long time with no maintenance as long as it is protected from sunlight.

2.6.3 CPVC

CPVC pipe and fittings (*Figure 4*) are made from an engineered vinyl polymer. Plumbers use CPVC in hot and cold water distribution systems. Improvements made to its parent polymer, polyvinyl chloride, added high-temperature performance and improved impact resistance to this material. CPVC is produced in standard CTSs from ½ inch to 2 inches, with a full line of fittings.

CPVC is acceptable under many model codes for indoor use. Its molecular structure practically eliminates condensation in the summer and heat loss in the winter, decreasing the likelihood of costly drip damage to walls or structures. CPVC pipe's smooth, friction-free interior surfaces result in lower pressure loss and higher flow rates and provide less opportunity for bacteria growth. CPVC does not break down in the presence of aggressive, or chemically reactive, water. Like other types of plastic, it does not rust, pit, or scale.

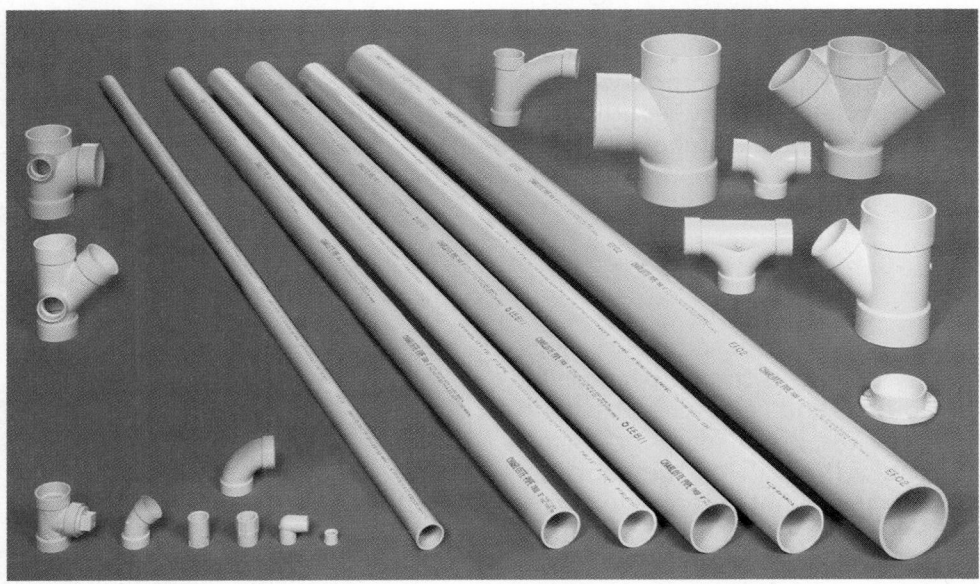

106F03.TIF

Figure 3 ◆ PVC pipe and fittings.

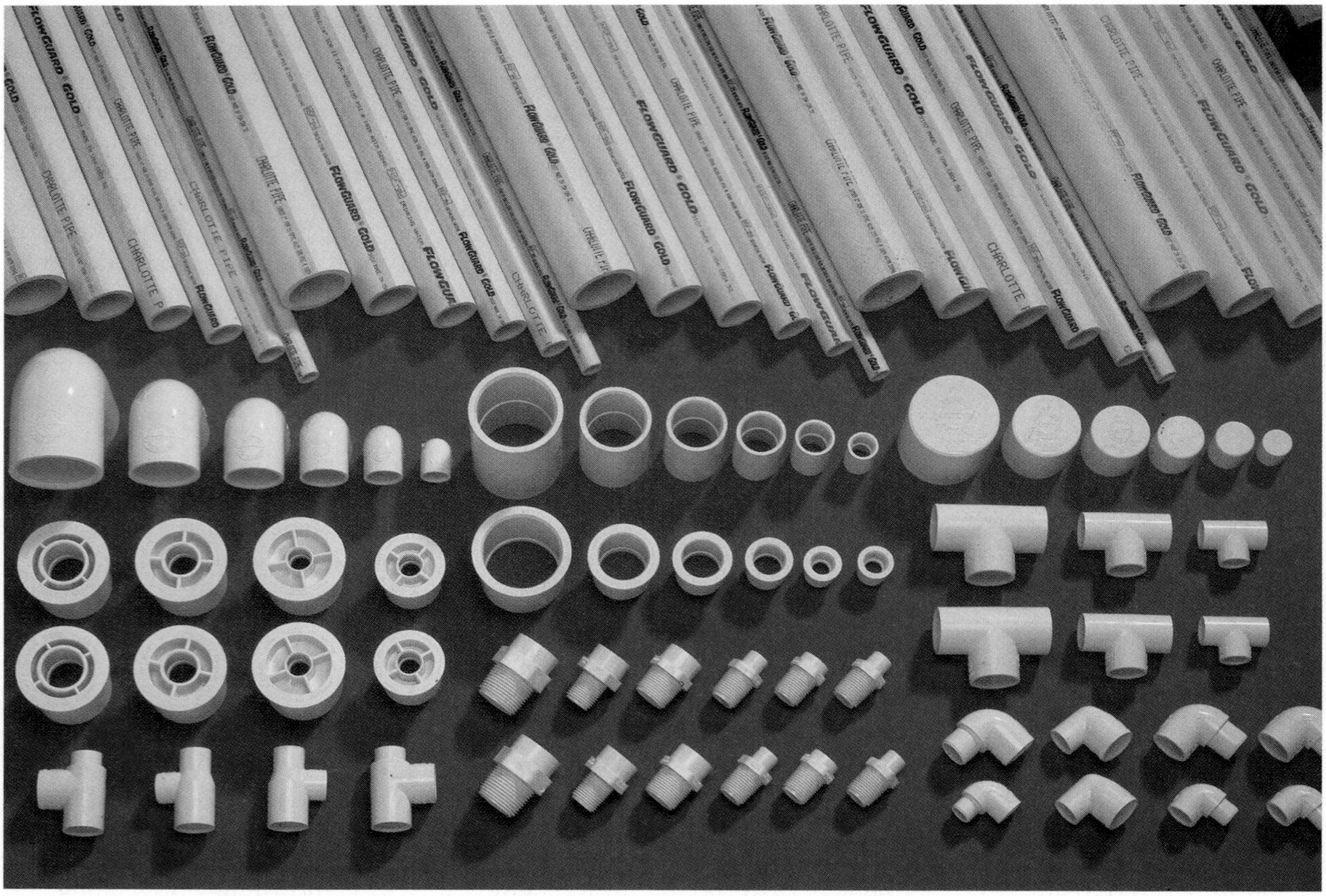

Figure 4 ◆ CPVC pipe and fittings.

106F04.TIF

CPVC is lightweight and easy to install. Recent improvements to CPVC have made the pipe stronger and more durable during installation. The strength of CPVC is a clear advantage for plumbers working in cold-weather states. In laboratory tests down to 20°F, CPVC pipe withstood a **water hammer** drop that substantially damaged copper piping under the same conditions. Water hammer is an extreme change in water pressure within a pipe that can cause a loud, banging sound.

2.6.4 PE

PE (polyethylene) is a thermosetting plastic. Plumbers commonly use PE as tubing because of its strength, flexibility, and chemical-resistant properties (see *Figure 5*). PE is also corrosion-resistant, which makes it ideal for transporting chemical compounds. It will not deteriorate when exposed to ultraviolet light, so you can install it outdoors without a protective coating. PE is used for cold water and underground gas service lines (outside buildings).

Figure 5 ◆ PE tubing.

106F05.EPS

DID YOU KNOW?
PE is the most inert of all plastic pipe materials, meaning that it is unlikely to react with other substances. That is why it is used for hospital tubing and in soda fountains.

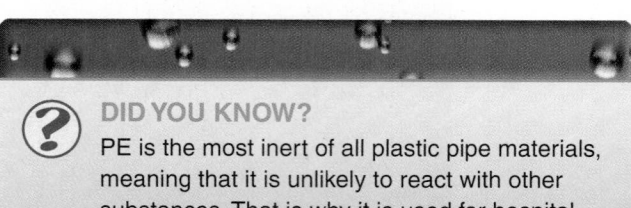

2.6.5 PEX

PEX (cross-linked polyethylene) tubing (*Figure 6*) is formed when high-density polyethylene is subjected to heat and high pressure. Because it resists high temperatures, pressure, and chemicals, it is ideal for potable (free from impurities in amounts sufficient to cause disease or harmful effects) hot and cold water systems, **hydronic** radiant floor systems, baseboard and radiator connections, and snow-melt systems. Plumbers commonly use PEX for manifold plumbing distribution systems because of its flexibility.

2.6.6 PB

PB (polybutylene) is a thermoplastic pipe that plumbers used extensively for water supply piping from the late 1970s to the mid-1990s. In many cases, PB piping has become weak and failed without warning. Experts believe this may be because of exposure to oxidants, such as chlorine, in public water systems. For this reason, PB is no longer available, so it is very unlikely that you will install PB piping. Many buildings still contain PB piping, though, and repair fittings are available, so you need to be able to identify it (see *Figure 7*).

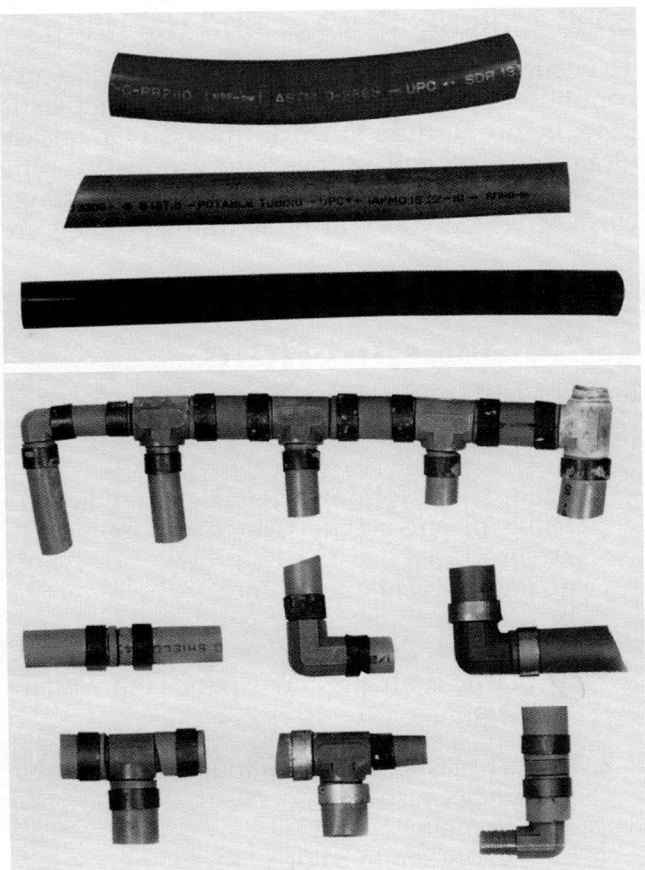

106F07.EPS

Figure 7 ◆ PB piping and fittings.

106F06.TIF

Figure 6 ◆ PEX tubing.

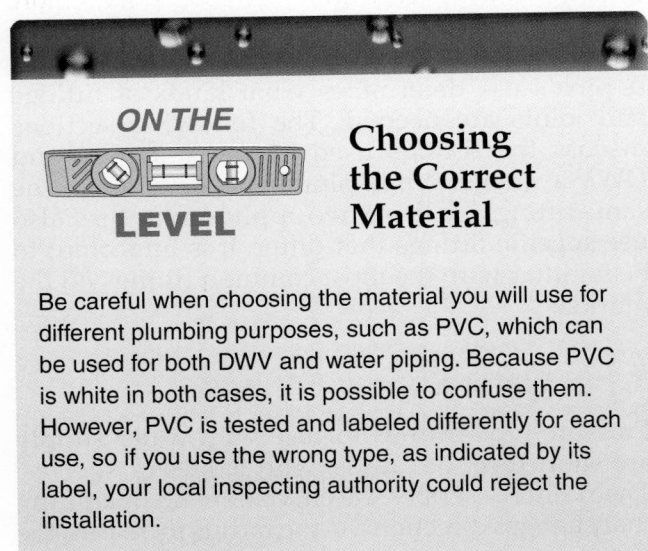

ON THE LEVEL

Choosing the Correct Material

Be careful when choosing the material you will use for different plumbing purposes, such as PVC, which can be used for both DWV and water piping. Because PVC is white in both cases, it is possible to confuse them. However, PVC is tested and labeled differently for each use, so if you use the wrong type, as indicated by its label, your local inspecting authority could reject the installation.

2.7.0 Material Storage and Handling

To ensure maximum productivity on the job, use common sense when storing and handling plastic pipe. Before starting to work, always plan ahead. Organize pipe and fittings into groups by pipe size and type. In addition, store pipe and fittings close to where you will be working so they are easy to access. Remember that you must protect many types of plastic pipe from ultraviolet light. When handling pipe and fittings, be careful not to bend or damage them. Pipe that is bent or otherwise damaged not only costs money, but also makes you less productive.

3.0.0 ◆ FITTINGS

A pipe's use determines what kinds of fittings and joints are needed. The following sections discuss the fittings used in water supply and DWV systems. While plumbers use some of the same fittings for these two applications, they also use specific fittings that differ. It is important to be familiar with the most common fittings on the market.

3.1.0 Water Supply Fittings

Pressure-type fittings for use with water supply are short turn, or radius, with ledges or shelves. The radius describes the curve or bend of a fitting that changes direction. As the radius increases, the change in flow direction becomes smoother. This is particularly important for waste drain systems, where the smoother flow keeps organic waste from being deposited along the pipeline.

The most common types of water supply fittings include those listed in *Table 3* and shown in *Figure 8*.

3.2.0 DWV Fittings

DWV fittings have smooth interior passages with no ledges. They also have a longer radius that makes directional changes smoother and less likely to collect solids (see *Table 4* and *Figure 9*). When selecting DWV fittings, always consult all applicable codes and manufacturers' installation recommendations.

Table 3 Water Supply Fittings and Their Uses

Fitting	Use
Union	Mechanically connects two pipes, usually at a termination downstream of a service valve
Reducer	Connects pipes of different sizes
Elbow	Changes the direction of rigid pipe by either 90 degrees or 45 degrees
Tee	Provides an opening to connect a branch pipe at 90 degrees to the main pipe run
Coupling	Joins two lengths of the same pipe size when making a straight run
Cap	Plugs water outlets when testing the system or creates an air chamber to eliminate water hammer
Plug	Closes openings in other fittings or seals the end of a pipe
Manifold	Runs several water supply lines from the main supply to different fixtures

Table 4 DWV Fittings and Their Uses

2003 International Plumbing Code®
Table 706.3, Fittings for Change in Direction

Type of Fitting Pattern	Change in Direction		
	Horizontal to Vertical	Vertical to Horizontal	Horizontal to Horizontal
Sixteenth bend	X	X	X
Eighth bend	X	X	X
Sixth bend	X	X	X
Quarter bend	X	X[a]	X[a]
Short sweep	X	X[a,b]	X[a]
Long sweep	X	X	X
Sanitary tee	X[c]		
Wye	X	X	X
Combination wye and eighth bend	X	X	X

For SI: 1 inch = 25.4 mm

a. The fittings shall only be permitted for a 2-inch or smaller fixture drain.

b. Three inches or larger.

c. For a limitation on double sanitary tees, see *Section 706.3*.

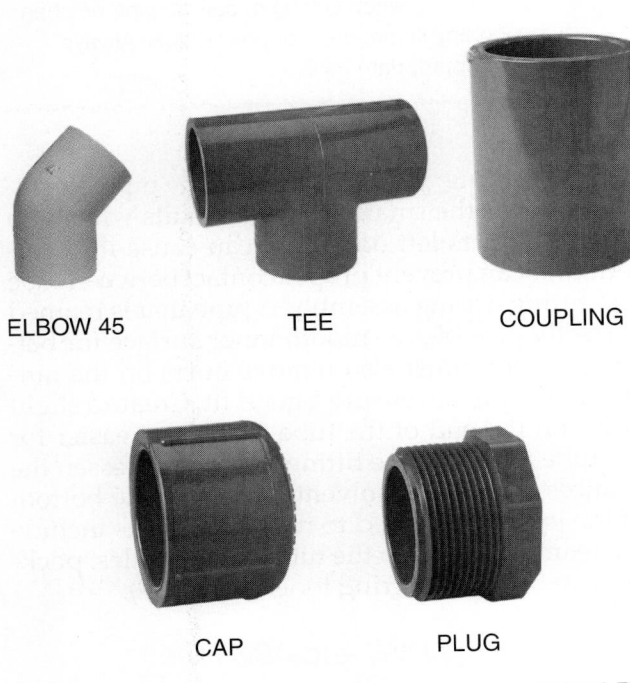

UNION   REDUCER   ELBOW

ELBOW 45   TEE   COUPLING

CAP   PLUG

106F08A.EPS

Figure 8 ◆ Water supply fittings.

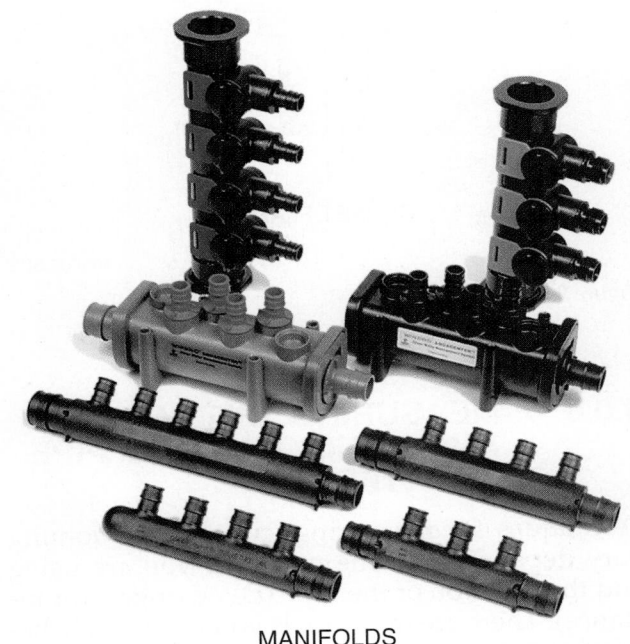

MANIFOLDS

106F08B.EPS

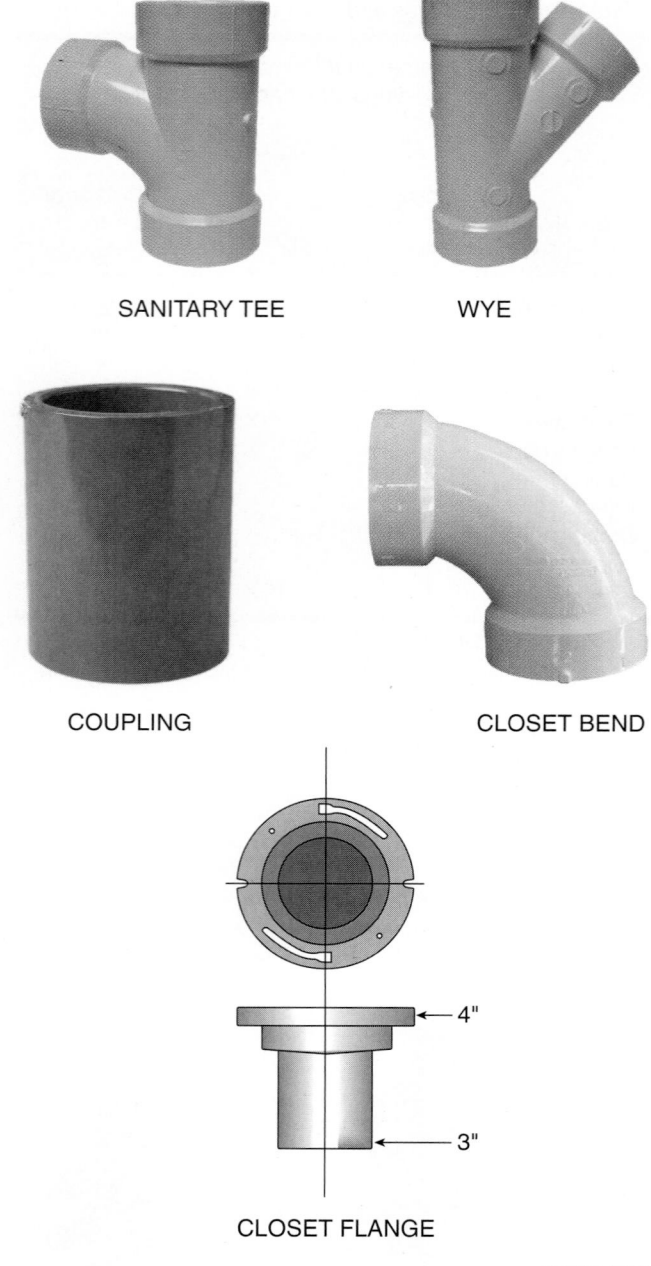

SANITARY TEE   WYE

COUPLING   CLOSET BEND

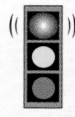

4"

3"

CLOSET FLANGE

106F09.EPS

Figure 9 ◆ DWV fittings.

4.0.0 ◆ MEASURING, CUTTING, AND JOINING PLASTIC PIPE AND FITTINGS

Techniques for measuring, cutting, and joining vary depending on the materials you are using and the function of the pipe (DWV or gas, for instance). There are many tools you will need to become familiar with in order to measure, cut, and join materials properly. The following sections explain methods for working with ABS, PVC, CPVC, PEX, and PE pipe used in DWV and water systems.

4.1.0 Measuring

It is important to plan ahead when you are installing ABS and PVC pipe and fittings in DWV systems. These systems have a built-in slope, also called a pitch or fall, and you must lay them accurately, with pipes cut to exact lengths. You cannot fix mistakes later with heat or hammers.

When you are measuring any pipe, be sure to allow for depth of joints. Take measurements to the full depth of the socket, not with pipe partly inserted into the socket. This is especially important when you use solvent cement to join piping. With this method, you must dry fit the installation, then mark alignment for fittings before you make the joint.

4.2.0 Cutting

Plastic pipe requires a square cut for good joint integrity. Cut tubing as squarely as possible to create the best bonding area within a joint. If you see any indication of damage or cracking at the tubing end, cut off at least 2 inches beyond any visible crack.

> ((◉)) **WARNING!**
> Follow all manufacturer-recommended precautions when cutting or sawing pipe or when using any flame, heat, or power tools. Always wear appropriate PPE.

After cutting, you should ream the pipe. Reaming removes the small burr that results when you cut pipe. Burrs left on piping can cause it to corrode and can prevent proper contact between tube and fitting during assembly. A pipe that is reamed correctly provides a smooth inner surface for better flow. You must also remove burrs on the outside of the pipe to ensure a good fit. Create a slight bevel on the end of the tube to make it easier for the tube to fit into the fitting socket and lessen the chances of pushing solvent cement to the bottom of the joint. Tools used to ream pipe ends include the reaming blade on the tubing cutter, files, pocketknives, and deburring tools (*Figure 10*).

4.2.1 Cutting PVC and ABS Pipe

You can cut PVC and ABS pipe with appropriate pipe cutters, a handsaw, or a power saw

DEBURRING TOOLS

INSIDE OUTSIDE REAMER

106F10.EPS

Figure 10 ◆ Deburring tools.

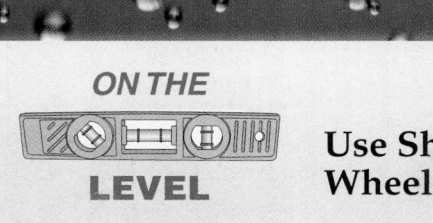

equipped with a carbide tip or abrasive blade. Plastic pipe cutters have one to four cutting wheels that can be rotated around a pipe to cut it. Wheels are available to fit standard cutters. You can also use ratchet shears or lightweight, quick-adjusting cutters designed exclusively for plastic piping (*Figure 11*).

To make sure you get a square cut, use a power saw on large jobs and a miter box on small jobs (see *Figure 12*). Ensure that you use the proper plastic saw for cutting PVC pipe. If these are not available, scribe the pipe and cut to the mark. After cutting the pipe, ream it inside and **chamfer** the edge to remove burrs, shoulders, and ragged spots. Chamfering involves beveling the edge of the pipe to a 45-degree angle. Chamfering is good practice because it provides for a more secure joint. When the socket and pipe fit tighter, you reduce the possibility of leakage from the pipes.

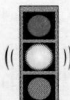

CAUTION

Mark carefully! Proper alignment in the final assembly is critical. To ensure alignment, carefully mark the positions of any fittings that will be rolled or otherwise aligned.

4.3.0 Joining

Typically, water supply and DWV fittings are joined using the same installation techniques.

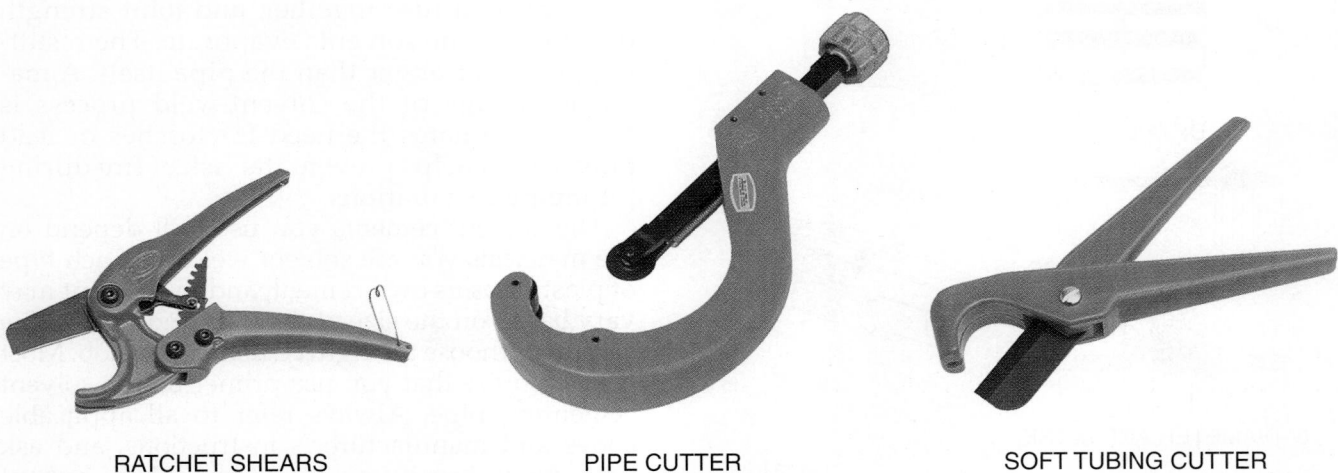

RATCHET SHEARS   PIPE CUTTER   SOFT TUBING CUTTER

106F11.EPS

Figure 11 ◆ Cutting tools used for plastic pipe.

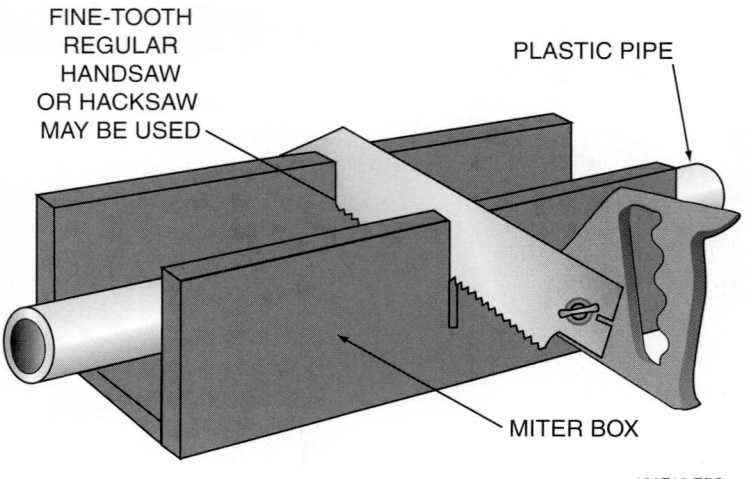

FINE-TOOTH
REGULAR
HANDSAW
OR HACKSAW
MAY BE USED

PLASTIC PIPE

MITER BOX

106F12.EPS

Figure 12 ◆ Using a miter box on small jobs.

Among the installation methods are mechanical fittings, such as **ring-tight gaskets, bell-and-spigot,** threaded joints, heat fusion, and **solvent weld.**

Ring-tight gasket fittings have a rubber O-ring or gasket in the socket. You must bevel the pipe at the leading edge to allow it to pass the gasket that forms the seal (see *Figure 13*). The bell-and-spigot pipe has a bell on one end with an internal **elastomeric** seal. The spigot (straight end) of the next pipe is fitted into the bell end to form a fluid-tight joint.

Plumbers can also connect some plastic pipe to dissimilar pipe with a **transition fitting.** Refer to your local applicable code and manufacturer's specifications to ensure that you are using the proper transition fittings.

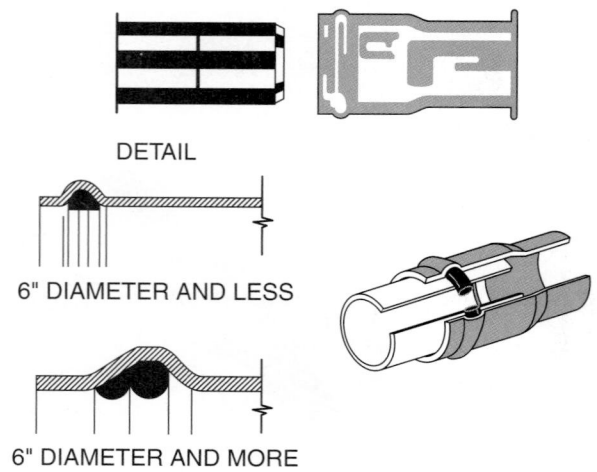

DETAIL

6" DIAMETER AND LESS

6" DIAMETER AND MORE

106F13.EPS

Figure 13 ◆ Ring-tight gasket fitting.

There are many different types of **fusion fittings,** so always consult your local applicable code and the manufacturer's specifications for proper joining and installation methods (see *Figure 14*).

4.3.1 *Solvent Welding (Solvent Cementing)*

Plumbers often join ABS, PVC, and CPVC plastic pipe and fittings with solvent cement to form a solvent weld. The process of solvent welding is also referred to as solvent cementing. Solvent-weld fittings have sockets that the pipe fits into (see *Figure 15*). Plumbers apply the solvent cement to the pipe end and the inside of the fitting end, which temporarily softens the joining surfaces. This brief softening period lets you seat the pipe into the socket's **interference fit**—that is, the fit gets tighter as the pipe is pushed into the socket. The softened surfaces then fuse together, and joint strength develops as the solvents evaporate. The resulting joint is stronger than the pipe itself. A major advantage of the solvent-weld process is that it eliminates the need for torches or lead pots, which helps prevent the risk of fire during plumbing installations.

The solvent cements you use will depend on the materials you are solvent welding. Each type of plastic has its own cement, and the cement may vary based on the size of the pipe and fittings. Be careful to choose the right cement for the job. Most codes require that you use primer before solvent cementing pipe. Always refer to all applicable codes and manufacturer's instructions, and ask your supervisor if you are unsure which cement to choose.

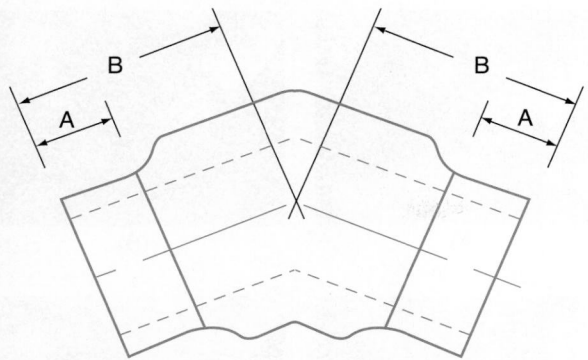

45-DEGREE ELBOW — BUTT FUSION — MOLDED

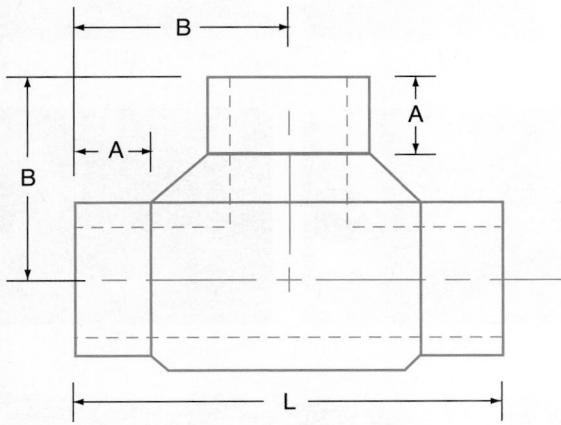

TEE — BUTT FUSION — MOLDED

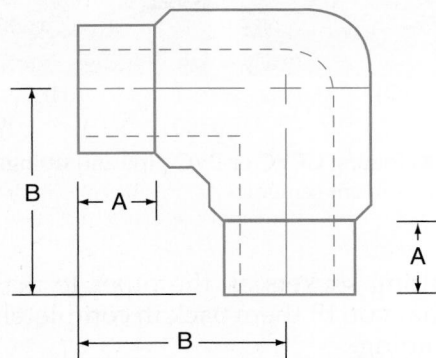

90-DEGREE ELBOW — BUTT FUSION — MOLDED

106F14.EPS

Figure 14 ◆ Fusion fittings.

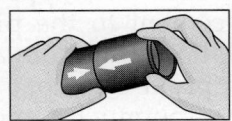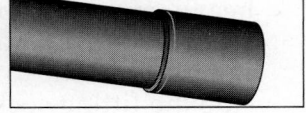

106F15.EPS

Figure 15 ◆ Solvent-welded fitting.

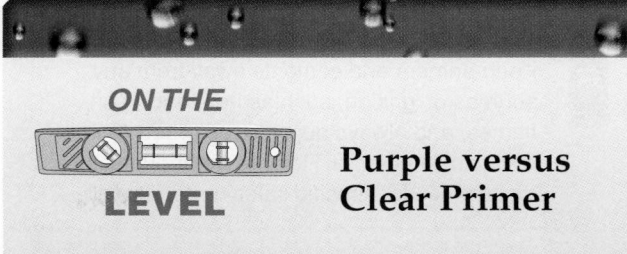

ON THE LEVEL

Purple versus Clear Primer

Two types of plastic pipe primer are available: purple and clear. Use purple primer; code inspectors will look specifically for the purple stain as proof that the pipes were primed before the solvent cement was applied. In some cases, with permission, you may use a clear primer on trim-out work.

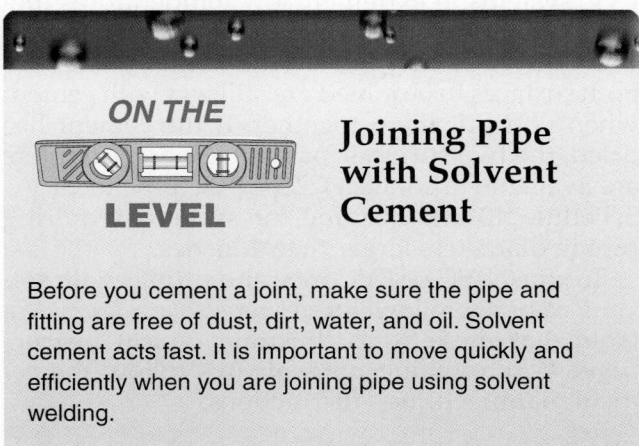

ON THE LEVEL

Joining Pipe with Solvent Cement

Before you cement a joint, make sure the pipe and fitting are free of dust, dirt, water, and oil. Solvent cement acts fast. It is important to move quickly and efficiently when you are joining pipe using solvent welding.

Because the cement hardens fast, you must move quickly and efficiently when joining pipe. The manufacturer's instructions will show minimum cure times for different size tubes at different temperatures before you can pressure-test the joint. Solvent cement and cure times also depend on relative humidity. Cure time is shorter for drier environments, smaller sizes, and higher temperatures.

 **WARNING!**

Avoid unnecessary skin or eye contact with primers and cements. If contact occurs, wash immediately. Use protective eyewear and gloves during any solvent-weld procedure and consult the MSDS for the cement you are using. An MSDS should be available to anyone on the job site.

4.3.2 Joining CPVC or PVC Pipe and Fittings

When you join CPVC or PVC pipe and fittings, use only CPVC or PVC cement or all-purpose cement conforming to *ASTM F-493* standards, or the joint may fail.

Use special care when assembling CPVC and PVC systems in extremely low temperatures (below 40°F) or extremely high temperatures (above 100°F). In extremely hot environments, make sure both surfaces to be joined are still wet with cement when you put them together. If the cement has dried, the two surfaces may not adhere. Adapters are available to connect CTS CPVC pipe to CPVC Schedule 40 and 80 pipe for systems requiring piping diameters larger than 2 inches.

To join CPVC or PVC pipe and fittings with solvent cement, follow these steps (see *Figure 16*). Note that these steps illustrate typical instructions. When joining pipe, always follow the cement manufacturer's instructions.

Step 1 Cut the pipes to the lengths you need, then test fit the pipes and fittings. The pipes should fit tightly against the bottom of the hubbed sockets on the fittings.

Step 2 Clean the surfaces you are joining by lightly scouring the ends of the pipe with emery paper, then wiping with a clean cloth. This ensures that the primer you apply later will soften the plastic before you apply solvent cement.

Step 3 Clean the socket interior and the spigot area of all dirt with a rag or brush.

Step 4 Mark the pipe and fitting with a felt-tipped pen to show the proper position for alignment. Also mark the depth of the

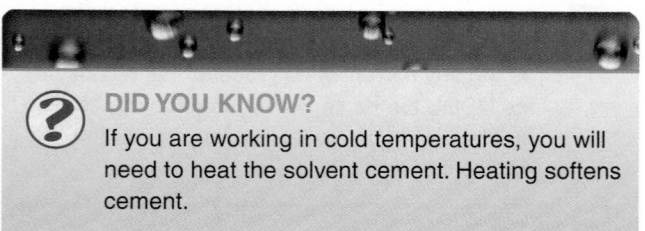

DID YOU KNOW?

If you are working in cold temperatures, you will need to heat the solvent cement. Heating softens cement.

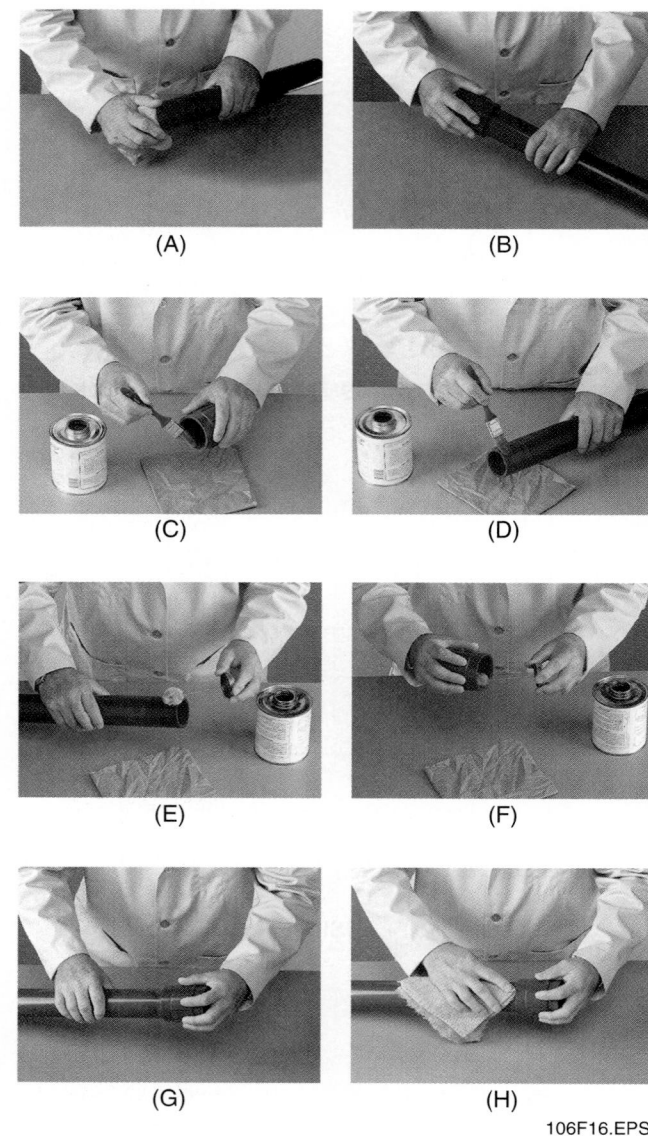

(A) (B) (C) (D) (E) (F) (G) (H)

106F16.EPS

Figure 16 ◆ Joining CPVC or PVC pipe and fittings with solvent cement.

fitting sockets on the pipes to make sure that you fit them back in completely when joining.

Step 5 Apply the primer to the surfaces you are joining. The primer will soften the plastic in preparation for the solvent-weld process.

Step 6 While the primer is still wet, apply a heavy, even coat of cement to the pipe end. Use the same applicator without additional cement to apply a thin coat inside the fitting socket. Too much cement can clog waterways. Do not allow excess cement to puddle in the fitting and pipe assembly.

Step 7 Immediately insert the tubing into the fitting socket, rotating the tube one-quarter to one-half turn while inserting. This motion ensures an even distribution of cement in the joint. Properly align the fitting.

Step 8 Hold the assembly firmly until the joint sets up. An even bead of cement should be visible around the joint. If this bead does not appear all the way around the socket edge, you may not have applied enough cement. In this case, remake the joint to avoid the possibility of leaks. Wipe excess cement from the tubing and fitting surfaces. Always follow the cement manufacturer's instructions.

4.3.3 Installing PVC Bell-and-Spigot Pipe

PVC bell-and-spigot pipe is generally used outdoors for gravity sewers. These outdoor systems are typically installed to connect with municipal utilities. To install PVC bell-and-spigot pipe, follow these steps (see *Figure 17*):

Step 1 Prepare the inner surface of the bell according to the manufacturer's instructions. Ensure that the groove is free of dirt and other particles.

Step 2 Fold the gasket into a heart shape with the nose or rounded part of its cross section facing out of the mouth of the bell.

Step 3 Insert the gasket into the bell and work it into its groove until it is smooth and free from waves. You may have to snap the gasket, or wet it with clean water or a wet rag, to make sure it goes in place completely. Mark the pipe, creating a memory mark to show the proper position for alignment.

Step 4 After you place the gasket in the bell, thoroughly coat its exposed surface with lubricant. Then apply the lubricant to the entire surface of the spigot end up to the memory mark. Make especially sure that the tapered portion of the spigot is thoroughly coated. When you have finished lubricating, the pipe is ready to be joined.

Step 5 Line up the spigot with the bell and insert it straight into the bell. The spigot end of the pipe has a mark to indicate the proper depth of insertion. This mark will be about flush with the end of the bell when the joint is fully assembled. The memory mark must never be more than ⅜ inch from the end of the bell after assembly.

CAUTION

When you are installing a ring-tight PVC gasket, you must assemble the pipe either by hand or by using a bar or block. Never swing or stab the pipe to join it.

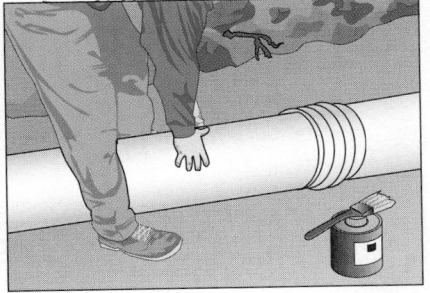

106F17.EPS

Figure 17 ◆ Installing PVC gravity sewer pipe.

4.3.4 Joining PEX Tubing

Because PEX tubing resists high temperature and chemicals, you cannot join it with solvent cement or heat fusion. The most common method is to use an insert and a crimp-ring system. The other tools used to join PEX tubing include the tubing cutter, hand-crimping tool, and go-no-go gauge (see *Figure 18*).

> **? DID YOU KNOW?**
> Crimping tools must be calibrated regularly. Consult and follow the manufacturer's specifications for the correct calibration intervals.

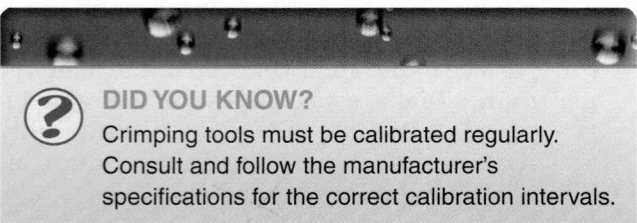

CRIMPSERT INSERT FITTINGS AND CRIMP RINGS

HAND-CRIMPING TOOLS

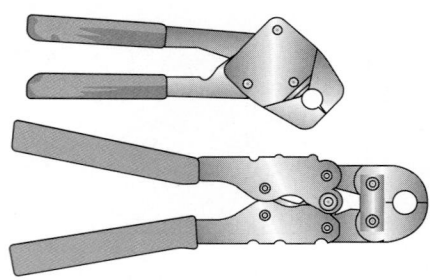

TUBING CUTTERS   **GO-NO-GO CRIMP MEASURING GAUGE**

106F18.EPS

Figure 18 ◆ Tools for joining PEX tubing.

To join PEX tubing, follow these steps (see *Figure 19*):

Step 1 Square cut the tubing perpendicular to the length of the tubing, using a cutter designed for plastic tubing. Remove all excess material or burrs that might affect the fitting connection.

Step 2 Slide a PEX ring over the end of the tube, and extend it no more than ¹⁄₁₆ inch.

Step 3 Open the handles of an expander tool and insert the tool's expansion head into the end of the tubing until it stops. Be sure you

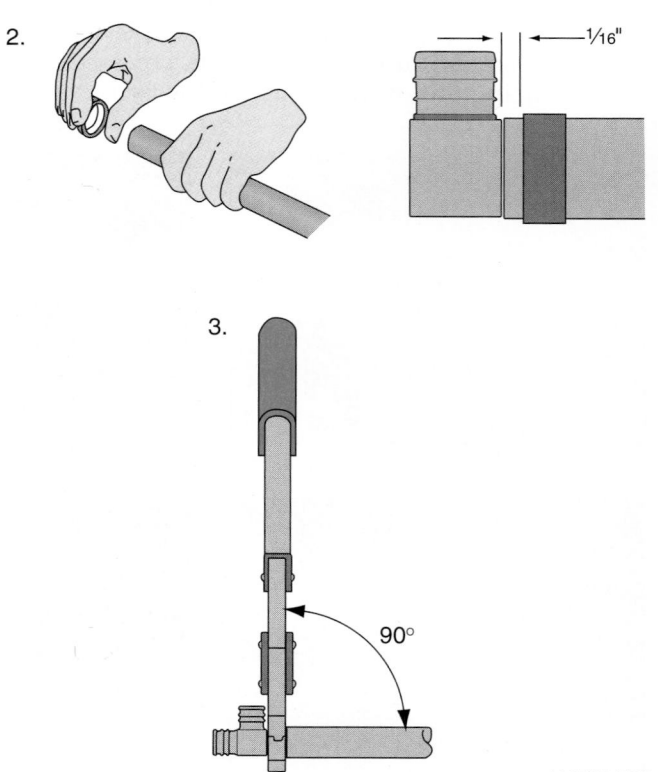

106F19.EPS

Figure 19 ◆ Joining PEX tubing.

have the correct size expander head in the tool. Place the free handle of the tool against your hip, or place one hand on each handle when necessary.

Step 4 Fully separate the handles and bring them together. Repeat this process until the tubing and ring are snug against the shoulder on the expansion head. Before the final expansion, withdraw the tubing from the tool and rotate the tool one-eighth of a turn. This prevents the tool from forming ridges in the tubing.

Step 5 Expand the tubing one final time. Immediately remove the tool and slide the expanded tubing over the fitting until the tubing reaches the stop on the fitting. Hold the fitting in place for two or three seconds until the tubing shrinks onto the fitting so that it holds the fitting firmly. For a proper connection, the tubing and PEX ring must be snug against the stop of the shoulder fitting. If there is more than 1/16 inch between the ring and the fitting, square cut the tubing 2 inches away from the fitting and make another connection using a new PEX ring.

4.3.5 Joining PE Tubing

Because PE tubing is resistant to chemicals, you must join it by heat fusion or with mechanical joints and clamps. PE joined by fusion is similar to a weld on steel—the materials of the joined parts merge so they are indistinguishable from each other. This process gives the joint the same positive characteristics as the pipe itself.

PE fusion often requires special training and certification. Manufacturers of PE products and joining equipment often provide this training and certification free of charge. New techniques that involve **compression collars** for joining PE are becoming popular because they require less training.

Some of the tools that plumbers use in the fusion process include a temperature indicator stick, a heating tool, a fusion timer, a socket face, and a cold ring (see *Figure 20*). Plumbers use the temperature indicator stick to make sure that piping has reached the required temperature for successful fusion. They use the stick to mark a particular area on the pipe. When the pipe reaches the desired temperature, the mark will melt. During heat fusion, the surface of the socket face comes in direct contact with the pipe or fitting.

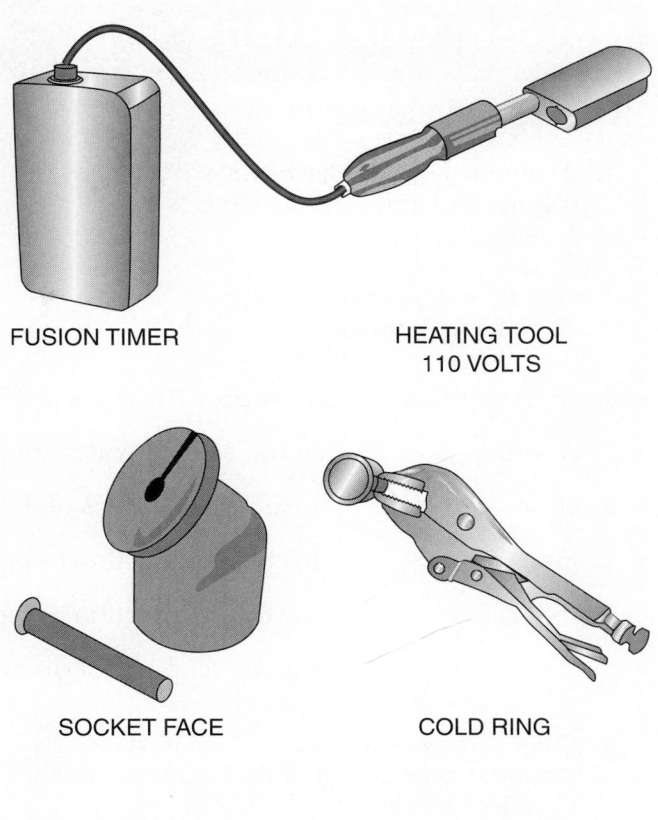

FUSION TIMER   HEATING TOOL
110 VOLTS

SOCKET FACE   COLD RING

TEMPERATURE INDICATOR STICK

106F20.EPS

Figure 20 ◆ PE fusion tools.

One of the most common methods for joining PE tubing is the butt-fusion method. To join PE tubing using the butt-fusion method, follow these steps:

Step 1 Cut the ends of the tubing square with a tubing cutter.

Step 2 Mark the tubing with the proper temperature indicator stick and heat the tubing ends with a heating tool.

Step 3 When the tubing reaches the required temperature, remove the heating tool.

Step 4 Press the tubing ends together to form a tight seal at the joint.

Step 5 Allow the joint to cool before applying force.

Sections 3.0.0–4.0.0

1. A pipe's _____ determines what kinds of fittings and joints are needed.
   a. length
   b. use ←
   c. inside diameter
   d. outside diameter

Match the following water supply fittings with their intended use.

2. coupling   a. to run several water supply lines from the main supply to different fixtures

3. elbow   b. to join two lengths of the same pipe size when making a straight run

4. manifold   c. to change direction of pipe by 22.5 or 60 degrees

5. reducer   d. to change direction of pipe by 45 or 90 degrees

      e. to connect pipes of different sizes

6. DWV fittings have a _____ that makes directional changes smoother and less likely to collect solids.
   a. smaller circumference
   b. longer radius ←
   c. shorter diameter
   d. greater perimeter

7. In DWV systems, a sanitary tee can be used to change the direction of a pipe from _____.
   a. horizontal to vertical ←
   b. vertical to horizontal
   c. horizontal to horizontal
   d. vertical to vertical

8. When you are measuring any pipe that needs to be cut and joined, be sure to calculate the depth of the _____.
   a. weld
   b. fitting
   c. joint ←
   d. socket

9. If you see any indication of damage or cracking at the tubing end, cut off at least _____ inch(es) beyond any visible crack.
   a. 1
   b. 2 ←
   c. 3
   d. 4

10. The bell-and-spigot pipe has a bell on one end with an internal _____ seal.
    a. pressure
    b. air
    c. water
    d. elastomeric ←

11. Plumbers can connect some plastic pipe to dissimilar pipe with a _____ fitting.
    a. solvent weld
    b. heat fusion
    c. transition ←
    d. threaded joint

12. A(n) _____ temporarily softens the pipe and the fitting materials, allowing them to be fitted together and fused.
    a. expander tool
    b. adapter
    c. solvent cement ←
    d. polymer

13. Because the cement hardens fast, you must move slowly and methodically when joining pipe to avoid making mistakes.
    a. True
    b. False ←

14. Use special care when assembling CPVC and PVC systems in temperatures _____.
    a. below 40°F
    b. below 40°F and above 100°F
    c. between 40°F and 100°F
    d. above 100°F

15. _____ bell-and-spigot pipe is generally used outdoors for gravity sewers, which are typically installed to connect with municipal utilities.
    a. PVC
    b. PEX
    c. CPVC
    d. PE

16. Because it resists high temperature and chemicals, you cannot join _____ tubing with solvent cement or heat fusion.
    a. ABS
    b. PEX
    c. CPVC
    d. PE

17. Because _____ is resistant to chemicals, you must join it by heat fusion or with mechanical joints and clamps.
    a. ABS
    b. PEX
    c. CPVC
    d. PE

18. Plumbers use a _____ to make sure that piping has reached the required temperature for successful fusion.
    a. mercury thermometer
    b. temperature indicator stick
    c. heating tool
    d. fusion timer

5.0.0 ◆ PIPE SUPPORTS

Plastic pipe can be supported using several different methods. The type of support you use depends on the pipe material, its size, its use, and whether the pipe is installed in a horizontal or vertical position, as well as the system specifications and applicable plumbing codes.

When architects and engineering consultants design plumbing installations, they provide plans and specifications that completely describe the proposed system. Specifications are based on codes or ordinances and must be followed. A specification for pipe hangers, for example, may read, "All piping shall be supported with hangers spaced no more than 10 feet apart (on center). Hangers shall be the malleable iron split-ring type and shall be as manufactured by XYZ Hangers,

Inc., or other approved vendor." You should always refer to applicable codes and the manufacturer's installation instructions for specific types and intervals appropriate for your particular installation.

If you are installing pipe in a seismically active area—that is, where earthquakes are a possibility—local codes will require seismic restraints. The purpose of these additional requirements is to ensure that the pipe is securely fastened to the structure in the event of excessive vibration. For example, some codes require hangers and supports to be used at closer intervals than in non-seismically active areas. In addition, they may require that you leave extra spacing for pipes where they meet walls and floors to allow for anticipated movement.

Table S–1

2003 International Plumbing Code®
Table 308.5, Hanger Spacing

Piping Material	Maximum Horizontal Spacing (feet)	Maximum Vertical Spacing (feet)
ABS pipe	4	10[b]
Aluminum tubing	10	15
Brass pipe	10	10
Cast-iron pipe	5[a]	15
Copper or copper-alloy pipe	12	10
Copper or copper-alloy tubing, 1¼-inch diameter and smaller	6	10
Copper or copper-alloy tubing, 1½-inch diameter and larger	10	10
PEX pipe	2.67 (32 in)	10[b]
PEX/aluminum/PEX (PEX-AL-PEX) pipe	2.67 (32 in)	4[b]
CPVC pipe or tubing, 1 inch or smaller	3	10[b]
CPVC pipe or tubing, 1¼ inches or larger	4	10[b]
Steel pipe	12	15
Lead pipe	Continuous	4
PB pipe or tubing	2.67 (32 in)	4
PE/aluminum/PE (PE-AL-PE) pipe	2.67 (32 in)	4[b]
PVC pipe	4	10[b]
Stainless steel drainage systems	10	10[b]

For SI: 1 inch = 25.4 mm, 1 foot = 304.8 mm

a. The maximum horizontal spacing of cast-iron pipe hangers shall be increased to 10 feet where 10-foot lengths of pipe are installed.

b. Mid-story guide for sizes 2 inches and smaller.

5.1.0 Hangers

Plumbers use hangers (*Figure 21*) for horizontal support of pipes and piping. The main purpose of hangers and brackets is to keep the piping in alignment and to prevent it from bending or distorting, but they can also prevent pipes from vibrating. You can attach horizontal hangers to wooden structures with lag screws or large nails. If vibration is a concern, follow the manufacturer's specifications to determine what type of hanger to use for your installation.

Hangers should be strong enough to support the weight of the pipe and its contents and maintain its alignment without sagging. In general, you should support piping at intervals of 4 feet or less, as well as at branches and changes of direction and when using large fittings. Although supports should provide free movement, they must prevent lateral runs from moving up, which could create a reverse grade (also known as slope) on branch piping and back up the system. Avoid hangers that may cut or squeeze pipe and tight clamps or straps that prevent pipe from moving or expanding. Size any holes made for pipe through framing members to allow for free movement. When working with piping in the ground, lay it on a firm bed for its entire length. Always consult applicable codes for specific requirements.

5.2.0 Fasteners

Plumbers often use beam clamps, C-clamps, and suspension clamps to fasten pipes to beams and other metal structures (see *Figure 22*). *Figure 23* shows other examples of horizontal support clamps and brackets. Vertical hangers, also called **pipe riser clamps**, provide vertical support for pipes and tubing (see *Figure 24*). These hangers consist of a friction clamp that you can attach to structural site components to support the vertical load of the pipe. You must use specific fasteners to attach hangers to masonry, concrete, or steel.

CAUTION

The area where the support is to be fastened should be smooth so that the item will have solid footing. Uneven footing might cause the support being fastened to twist, warp, or not tighten properly.

CAUTION

Make sure the fastener is straight after working it around in the hole and installing the washer. The washer centers the fastener and holds it in place until the grout or epoxy hardens. If the grout or epoxy sets and the fastener is not straight, the fastener will be unusable and will have to be removed and the installation repeated.

Use supports for vertical piping at each floor level or as required by the installation design. Mid-story guides can provide greater stability for vertical pipes that run up through the building.

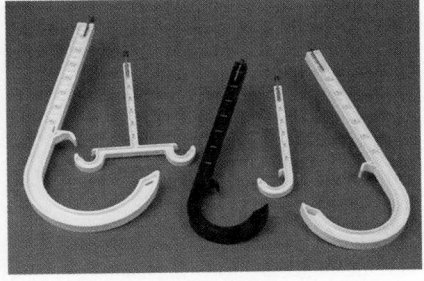

J-HOOKS

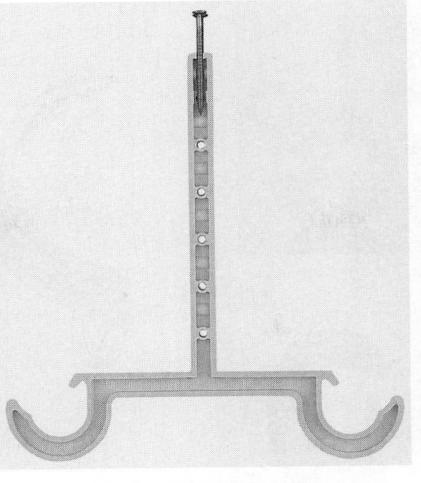

DOUBLE-J HOOKS

DOUBLE-J HOOKS INSTALLED

LOCKING TUBE STRAPS

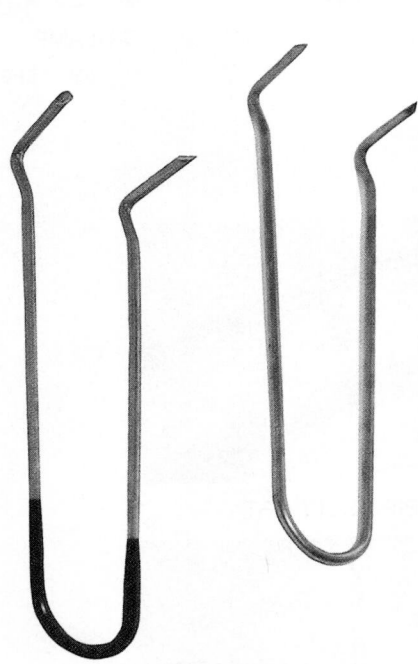

PIPE HOOKS

106F21.EPS

Figure 21 ◆ Pipe hangers.

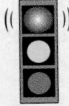

WARNING!

Do not use metal hangers on any plastic pipe. Always use hangers that are made of the same material as the pipe itself. For more information, refer to the *MSS40* hanger standards established by the Manufacturers Standardization Society.

WARNING!

Testing PVC pipe with air can increase the chance of an explosion. Serious injury or death can result if you use too much air, fail to vent trapped air, or fail to depressurize the system.

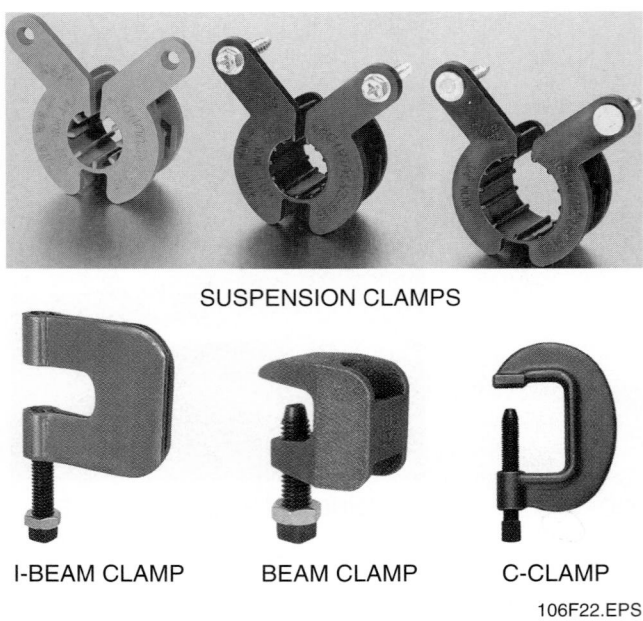

SUSPENSION CLAMPS

I-BEAM CLAMP   BEAM CLAMP   C-CLAMP

106F22.EPS

Figure 22 ◆ Clamps.

PIPE SUPPORT

SUPPORT BRACKET

SUPPORT BRACKET INSTALLED

106F23.EPS

Figure 23 ◆ Horizontal support clamps and brackets.

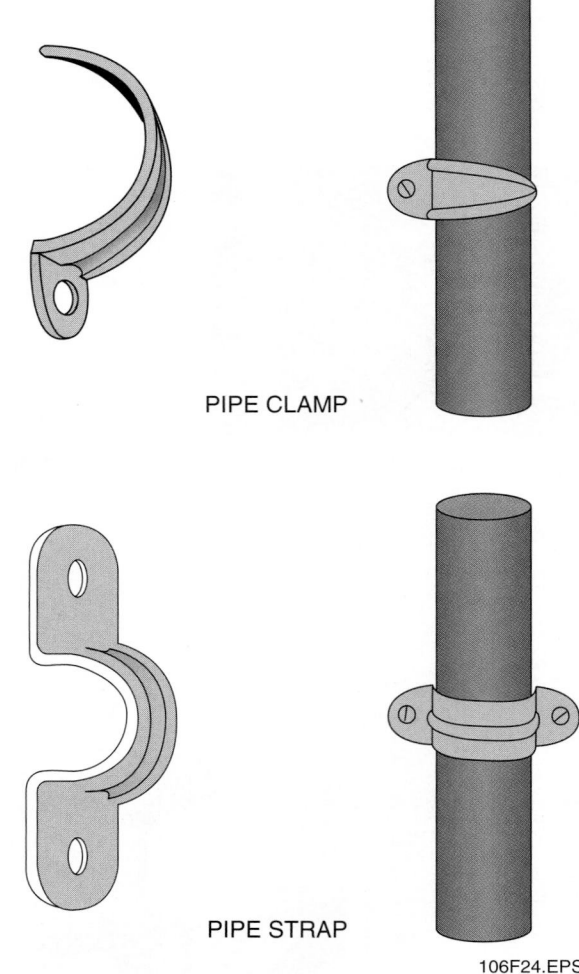

PIPE CLAMP

PIPE STRAP

106F24.EPS

Figure 24 ◆ Vertical hangers.

6.0.0 ◆ PRESSURE TESTING

Once you have completed and cured an installation, you must **hydrostatically pressure-test** the system in accordance with applicable code requirements. This process involves filling the system with water and bleeding all air out from the highest and farthest points in the run. If you find a leak, you must remove and replace it. You can install a new section using couplings. During subfreezing temperatures, you should blow water out of the lines after testing to avoid possible damage to the pipes from freezing.

CAUTION

Never pressure-test a connection until the manufacturer's recommended cure times have been met. After testing a connection, thoroughly flush the system for at least 10 minutes to remove any remaining trace amounts of solvent cement.

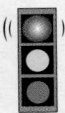

WARNING!

Air testing is not recommended for plastic pipe, although it may be necessary under certain temperatures. Never use pure oxygen for pressure tests. Never test PVC pipe with air, as it will explode. Use air testing for other plastic pipe only when hydrostatic testing is not practical. Always use extreme caution, because air under pressure is explosive. Never test with more than 5 pounds of air pressure. If you perform an air test, you must notify all site personnel of the test, use protective eyewear, and take precautions to prevent impact damage to the system during the test.

Review Questions

Sections 5.0.0–6.0.0

1. The main purpose of pipe hangers is to _____.
   a. ensure adequate pipe vibration
   b. allow pipes to bend easily
   c. maintain horizontal alignment of pipes
   d. maintain vertical alignment of pipes

2. In general, you should support piping at intervals of _____ feet or less, as well as at branches and changes of direction and when using large fittings.
   a. 3
   b. 4
   c. 5
   d. 8

3. _____ are used as vertical support for pipes and tubing.
   a. Beam clamps
   b. C-clamps
   c. Pipe riser clamps
   d. Special fasteners

4. If you find a leak when testing a connection, you must _____.
   a. solvent weld the connection
   b. remove and replace it
   c. perform the hydrostatic test again
   d. air test the connection

5. During subfreezing temperatures, you should blow water out of pipe lines _____ to avoid possible damage to the pipes from freezing.
   a. before installation
   b. during installation
   c. before testing
   d. after testing

Summary

The use of plastics in the plumbing trade has greatly increased the effectiveness and safety of indoor and outdoor plumbing systems. The introduction of different types of plastic pipe and tubing means that today's plumbers must know about new techniques, tools, and applications. Plastic may seem like the answer to all plumbing problems, but exactly which type of plastic you use makes a big difference. You must learn the special techniques for cutting and joining a variety of plastic pipe types. Recognizing the common types of materials used on a job site is a basic skill.

Developing technology affects plumbers now more than ever. Manufacturers are constantly researching and producing new products. To be competitive in the modern plumbing industry, you must keep up with innovations and improvements in installation techniques and materials.

Notes

Trade Terms Quiz

Fill in the blank with the correct trade term that you learned from your study of this module.

1. A(n) _____ allows threaded plastic pipe to be connected to steel pipe.

2. The _____ is a measurement of a pipe's wall thickness.

3. Plastic is _____; it does not react with other substances or materials.

4. A(n) _____ means that the fit tightens as the pipe is pushed into the socket.

5. The _____ relates wall thickness to the diameter of the pipe, using a set ratio of wall thickness to diameter.

6. Plumbers commonly use _____ for manifold plumbing distribution systems because of its flexibility.

7. _____ changes chemically when heated, so that once hardened by heat or chemicals, it is hardened permanently.

8. After reaming cut pipe, _____ the edge to remove burrs, shoulders, and ragged spots.

9. The pressure that a pipe can withstand continuously, called the _____, is required on the pipe's label.

10. When softened, _____ can be molded into desired shapes.

11. Vertical hangers, also called _____, provide vertical support for pipes and tubing.

12. Pipe with _____ construction consists of a single layer of plastic that does not contain trapped air.

13. _____ pipe will not deteriorate when exposed to ultraviolet light, so you can install it outdoors without a protective coating.

14. Techniques for joining PE that use _____ are becoming popular because they require less training than heat fusion.

15. A(n) _____ system heats and cools by circulating water or steam through a closed piping system.

16. After completing an installation, you must _____ the system by filling the system with water and bleeding all air out from the highest and farthest points in the pipe run.

17. An extreme change in water pressure that can cause a loud, banging sound within a pipe is called _____.

18. Because it has been known to become weak and fail without warning in previous installations, _____ pipe is no longer available for installation today.

19. _____ pipe has a bell on one end with an internal _____ seal.

20. Pipe can be measured by its _____, which is the distance between its inner walls.

21. Pipe with _____ construction is more lightweight and less expensive than solid wall pipe.

22. _____ fittings have a rubber O-ring or gasket in the socket.

23. The ratio between a pipe's _____ and its wall thickness is constant for each pipe size.

24. _____ pipe performs well at extreme temperatures because it absorbs heat and cold slowly.

25. The molecular structure of _____ pipe practically eliminates condensation in the summer and heat loss in the winter, decreasing the likelihood of costly drip damage to walls or structure.

26. To _____ joints, plumbers apply solvent cement to the pipe end and the inside of the fitting end, which temporarily softens the joining surfaces.

27. A pipe's pressure rating is measured in _____.

28. To form a joint using a(n) _____, you must heat the accessories to the manufacturer's specifications and press them together.

29. _____ is a rigid pipe with high-impact strength that is manufactured from a thermoplastic material and has an indefinite life span under most conditions.

Trade Terms

ABS (acrylonitrile-butadiene-styrene)
Bell-and-spigot pipe
Cellular core wall
Chamfer
Chemically inert
Compression collar
CPVC (chlorinated polyvinyl chloride)

Elastomeric
Fusion fitting
Hydronic
Hydrostatically pressure-test
Inside diameter (ID)
Interference fit
Outside diameter (OD)
PB (polybutylene)

PE (polyethylene)
PEX (cross-linked polyethylene)
Pipe riser clamp
Pressure rating
psi (pounds per square inch)
PVC (polyvinyl chloride)
Ring-tight gasket fitting

Schedule
Size dimension ratio (SDR)
Solid wall
Solvent weld
Thermoplastic pipe
Thermosetting pipe
Transition fitting
Water hammer

Bob Muller

John J. Muller Plumbing & Heating Inc.
President/Chief Executive Officer
Matawan, New Jersey

Bob Muller was born in Matawan, New Jersey, and attended Matawan Regional High School. He joined the U.S. Air Force and attended technical school at the Air Force School of Applied Sciences, where he studied aerospace ground equipment, which, along with weapons loading, covered hydraulics, pneumatics, electrical generation and control, and air conditioning. After he returned from Vietnam, Bob left the service and took extension courses in math and science at the University of Southern California, Victor Valley College. He spent three years in college and worked for several companies in the industry but always felt that something was missing. He finally found the missing piece when he returned to his family's business. Today, he is president and CEO of John J. Muller Plumbing & Heating Inc., a third-generation business his grandfather started in 1927.

How did you become interested in this industry?
I always had an interest in mechanical things; I enjoyed the challenge of using my brain and hands and figuring out how systems worked. During high school, I worked as a plumber's helper, digging, threading pipe, running for material, and helping the other mechanics during summers. I had several different jobs before going back to the family business, but there was always something missing. I went back to the family business in 1979 when my grandfather became ill. The tugging feeling was gone once I went back to my family business. I love the industry; it's a great deal of fun.

What path did you take to your current position?
I came from an old German family. There was no free ride, no allowance without work. My family taught me a good work ethic. My father was harder on me than on most of the employees, so I worked my tail off. I had to prove myself. While I was going to school, I worked as a plumber's helper, and then I was an apprentice, journeyman, and master plumber. I worked in the family business until my father retired, and then I was thrown into the business of management.

I would not take the position without proper training, so I attended vocational school. I completed a four-year apprenticeship program and one year of journeyman training. Then I applied for my master plumber's license and hit it on the first try. I now hold certifications for the International Boiler and Radiator Manufacturers' Association and for various manufacturer-sponsored training programs. I renew my licenses every two years. As part of the renewal process, our state requires that every plumber take at least five hours of continuing education prior to renewal. In addition, I continue my education because there are many changes in our industry in material, code issues, and installation practices, which occur almost daily. As the saying goes, "If ya' snooze, ya' lose."

What does it take to be successful in your trade?
If it's one thing, it's that you have to have a good work ethic. And you have to develop this work ethic; it's not God-given. You have to study math and the sciences. With geometry, for example, you must be able to do the numbers, then apply them to something tangible. Reading and retention are also important. You have to be able to read and understand plans and apply what you have read. Written and oral communication is essential. You have to be able to communicate both in writing and speech. Understanding the latest technology, such as computers, is a necessity.

What are some things you do on the job?

It's not just a nine-to-five job. Sometimes it's 24 hours a day, seven days a week. You may try to shut off, but it's hard not to take work with you. So manage your time, stick to deadlines, set short- and long-term goals, and prioritize.

You wear a lot of hats in this business. I came from a military background, but nothing compared to what I am doing now. I took college courses in management and made connections with the Plumbing-Heating-Cooling Contractors—National Association (PHCC—NA). The PHCC—NA provided a lot of help in offering seminars in management, profitability, safety, employee training, and apprenticeship training. But even more important, my affiliation with the PHCC—NA has allowed me a connection with plumbers from all over the country and around the world.

My father was standing by for help, and he is still always ready to help. As a manager, I still go into the field with my employees. I start digging just like everyone else. We have seven employees, and we are adding and growing all the time. We currently affiliate ourselves with the Local Union #9, and we are working with their training programs. Our company offers residential, commercial, and light industrial work.

Now I am also an instructor for continuing education. After I received my master plumbing license, I joined the New Jersey PHCC—NA. I worked through the committees and was president of the association from 1995 to 1996. I am a certified instructor in the apprenticeship program for the Middlesex County Vocational and Technical School. I teach all four years; currently, I'm teaching fourth-year plumbing. I've been doing this for 15 to 20 years now. I've been recognized by the New Jersey General Assembly and the Board of Education of Middlesex County for my contributions to the Apprenticeship Training Program.

What is the most interesting aspect of your profession? What makes your trade stand out from others?

We are in the business of health. We live by the credo, "The plumber protects the health and safety of the nation." Basic sanitation cannot be taken for granted; it is our job to maintain it and improve upon it daily. All you have to do is look at the news and you'll see that, without basic sanitation, thousands of people die daily.

We all have a role. Everyone has a place in the community and we all have a part to make it work. I am not sure I can answer specifically what makes our trade better than another. All the trades are important to the community; plumbing is important, indeed, but what's the point if the carpenter didn't build the structure and the electrician didn't install the lighting? We are part of the overall big picture.

What do you like most—and least—about your job?

I love it all! Bringing young individuals in and making sure that they achieve what they set out to achieve is rewarding. I reached my goals and I enjoy helping others reach theirs. Nothing, absolutely nothing, pleases me more than when a student, former employee, colleague, or friend in the business comes up to me and thanks me for some success they have had in the business.

Probably what I like the least is when I hear of a business or personal failure because somebody didn't apply the education they have received.

What would you say to someone entering the trade today?

You have to be physically strong and mentally focused. Education, education, education. Then, application. Keep yourself in good physical condition. It is an industry where physical and mental conditioning are important. Stress is a killer in this business. A lot of good people have ruined their lives because they turned to drugs and alcohol to relieve the stress. This job requires a positive mental attitude and the ability to communicate with others. Develop a good mental attitude and a good work ethic. If you don't know what that is, just find a good plumber!

What can an apprentice expect to earn in his/her first years on the job in your area? What can he/she expect to earn after 10-plus years in the industry?

New Jersey has higher wages but also high overhead and living expenses. The base wage for plumber pipe fitter Local #9 journeymen is $22.08. This includes the surety fund, welfare fund, and almost 100 percent pension, as well as two weeks of vacation time.

The overtime rate is time-and-a-half for weekends, and holidays get double time. There are a lot of opportunities for people to work the weekends and get those rates if they can withstand the work.

The apprentice salary is based on a percentage of the journeyman rate: 50 percent of the journeyman wage, plus full benefit package, pension fund, and so on, which is managed by the union. Apprentices make 60 percent of the wage in their second year, 70 percent in the third, and 80 percent in the fourth. In their fifth year they begin to earn the full journeyman wage. It's a nonbinding contract between employer and apprentice. Under the bureau of apprenticeships, the formula is the same, but the rate is negotiated.

Trade Terms
Introduced in This Module

ABS (acrylonitrile-butadiene-styrene): Plastic pipe and fittings used extensively in drain, waste, and vent (DWV) systems.

Bell-and-spigot pipe: Pipe that has a bell, or enlargement, also called a hub, at one end of the pipe and a spigot, or smooth end, at the other end. The bell and spigot of two different pipes slide together to form a joint. Also called hub-and-spigot pipe.

Cellular core wall: Plastic pipe wall that is low-density, lightweight plastic containing entrained (trapped) air.

Chamfer: To bevel the edge of construction material to a 45-degree angle.

Chemically inert: Does not react with other chemicals.

Compression collar: A piece of hardware that uses compression force to connect sections of polyethylene piping.

CPVC (chlorinated polyvinyl chloride): Plastic pipe and fittings used extensively in hot and cold water distribution systems.

Elastomeric: Rubberized. Made of an elastic substance such as a polyvinyl elastomer.

Fusion fitting: A fitting with a butt that has the same outside diameter and inside diameter as the pipe. It is usually joined to a pipe by heat.

Hydronic: A system that heats and cools by circulating water or steam through a closed piping system.

Hydrostatically pressure-test: To fill a pipe with water and bleed all air out from the highest and farthest points in the run.

Inside diameter (ID): The distance between the inner walls of a pipe; the standard measure of piping used in heating and plumbing.

Interference fit: Fit that tightens as the pipe is pushed into the socket.

Outside diameter (OD): The distance between the outer walls of a pipe.

PB (polybutylene): Plastic piping that was formerly used for plumbing pipe; it is no longer used but is still found in some residences.

PE (polyethylene): Flexible plastic pipe, tubing, and fittings, usually used for water distribution, that do not deteriorate when exposed to sunlight.

PEX (cross-linked polyethylene): Tubing and fittings made with heat and high pressure that resist high temperatures, pressure, and chemicals.

Pipe riser clamp: A vertical extension of pipe hanger that provides support for pipe and tubing.

Pressure rating: The maximum pressure at which a component or system may be operated continuously.

psi (pounds per square inch): A measurement of pressure.

PVC (polyvinyl chloride): Plastic pipe and fittings used for cold water distribution and for industrial water and chemicals, as well as for drain, waste, and vent (DWV) systems.

Ring-tight gasket fitting: Fitting with a rubber O-ring or gasket in the socket.

Schedule: A measurement that describes pipe wall thickness.

Size dimension ratio (SDR): A measurement of pipe size that relates pipe wall thickness to pipe diameter.

Solid wall: Plastic pipe wall that does not contain trapped air.

Solvent weld: A joint created by joining two pipes using solvent cement that softens the material's surface.

Thermoplastic pipe: Pipe that can be repeatedly softened by heating and hardened by cooling. When softened, thermoplastic pipe can be molded into desired shapes.

Thermosetting pipe: Pipe that changes chemically when heated, so that once hardened by heat or chemicals, it is hardened permanently.

Transition fitting: A special fitting used to connect plastic pipe to pipe of a dissimilar material, as specified by applicable code.

Water hammer: An extreme change in water pressure within a pipe that can cause a loud, banging sound and even damage the system.

ADDITIONAL RESOURCES

This module is intended to present thorough resources for task training. The following reference works are suggested for further study. These are optional materials for continued education rather than for task training.

Basic Plumbing with Illustrations, Revised, 1994. Howard C. Massey. Carlsbad, CA: Craftsman Book Company.

Pipefitting Level 2, 2006. NCCER. Upper Saddle River, NJ: Prentice Hall.

Plumber's Handbook, Revised Edition, 1998. Howard C. Massey. Carlsbad, CA: Craftsman Book Company.

Plumbing: Design and Installation, Second Edition, 2002. L. V. Ripka. Homewood, IL: American Technical Publishers.

REFERENCES

J&L Supply website, www.jlsupply.com, *J&L Supply Instruments,* www.jlsupply.com/testing/testinginstr.htm#TEMPILSTIK_TEMPERATURE_INDICATORS, reviewed July 2003.

National Standard Plumbing Code, 2003. Falls Church, VA: Plumbing-Heating-Cooling Contractors—National Association.

Plastic Pipe and Fittings Association. Table 3-c Minimum Wall Thicknesses of 2-Inch Pipe Based on SDR/SIDR (page 20).

Plumbing Apprentice Training Manual for Plastic Piping Systems, 2002. Glen Ellyn, IL: Plastic Pipe and Fittings Association and Plastic Piping Educational Foundation.

Ridgid Tool Company website, www.ridgid.com/Tools/Reamer-and-Deburring-Tools/, reviewed March 8, 2004.

2003 International Plumbing Code. Table 706.3 Fittings for Change in Direction. Falls Church, VA: International Code Council.

2003 International Plumbing Code. Table 308.5 Hanger Spacing. Falls Church, VA: International Code Council.

Wirsbo Systems website, www.wirsbo.com, *Comfort Heating Frequently Asked Questions,* www.wirsbo.com/main.php?pm=1&mm=1&sm=4&pc=homeowner/ho_mm1sm4.php, reviewed July 2003.

NCCER makes every effort to keep these textbooks up-to-date and free of technical errors. We appreciate your help in this process. If you have an idea for improving this textbook, or if you find an error, a typographical mistake, or an inaccuracy in NCCER's Contren® textbooks, please write us, using this form or a photocopy. Be sure to include the exact module number, page number, a detailed description, and the correction, if applicable. Your input will be brought to the attention of the Technical Review Committee. Thank you for your assistance.

Instructors – If you found that additional materials were necessary in order to teach this module effectively, please let us know so that we may include them in the Equipment/Materials list in the Annotated Instructor's Guide.

Write: Product Development and Revision
National Center for Construction Education and Research
3600 NW 43rd St, Bldg G, Gainesville, FL 32606

Fax: 352-334-0932

E-mail: curriculum@nccer.org

Craft _____ Module Name _____

Copyright Date _____ Module Number _____ Page Number(s) _____

Description _____

_____

_____

_____

(Optional) Correction _____

_____

_____

(Optional) Your Name and Address _____

_____

_____

_____

Copper Pipe and Fittings
68117-09

68117-09
Copper Pipe and Fittings

Topics to be presented in this module include:

1.0.0   Introduction 17.2
2.0.0   Copper Pipe 17.2
3.0.0   Fittings and Valves 17.4
4.0.0   Measuring, Cutting, Bending, Joining, and Grooving . . .17.8
5.0.0   Installing Pipe Hangers and Supports 17.16
6.0.0   Insulating Pipes 17.21
7.0.0   Pressure Testing 17.21

Overview

Copper pipe and fittings are used in a variety of plumbing applications. Copper is expensive, but because it is reliable and easy to use, it is also cost-effective. Copper pipe comes in a series of sizes and different wall thicknesses. Annealed copper is soft and flexible, while drawn copper is rigid and hard. Plumbers must use copper pipe and fittings that have been labeled and approved by the manufacturer.

Based on a pipe's application, plumbers select appropriate fittings, which change the direction or size of a pipe run, and valves, which control flow. Plumbers use specific methods to measure, cut, bend, join, and groove copper pipe. These methods depend on the type of pipe and its function, and the various special tools that are needed. Copper pipe can be joined by soldering a sweat joint, creating a compression joint, or making a flare joint. Plumbers must learn to perform each method safely and use all tools properly.

Hangers and supports secure horizontal and vertical runs of copper pipe and are used to prevent leaks and damage. Plumbers select the type of hanger depending on job specifications. Applicable codes dictate appropriate pipe attachments and connectors for each installation, as well as whether the pipe needs to be insulated. After completing an installation, plumbers pressure-test the system for leaks and secure connections.

◾ Focus Statement

The goal of the plumber is to protect the health, safety, and comfort of the nation job by job.

◾ Code Note

Codes vary among jurisdictions. Because of the variations in code, consult the applicable code whenever regulations are in question. Referencing an incorrect set of codes can cause as much trouble as failing to reference codes altogether. Obtain, review, and familiarize yourself with your local adopted code.

Objectives

When you have completed this module, you will be able to do the following:

1. Identify the types of materials and schedules used with copper piping.
2. Identify the material properties, storage, and handling requirements of copper piping.
3. Identify the types of fittings and valves used with copper piping.
4. Identify the techniques used in hanging and supporting copper piping.
5. Properly measure, ream, cut, and join copper piping.
6. Identify the hazards and safety precautions associated with copper piping.

Key Trade Terms

ACR
ACR tubing
Annealing
Bullhead tee
Capillary action
Clevis
Compression joint
Drawn copper
Drop forged
Ferrule

Flare joint
Formability
Head
Insulation
Nominal size
Pressure drop
Roll grooved
Sizing tool
Sweat joint

Required Trainee Materials

1. Appropriate personal protective equipment
2. Pencils and paper
3. Copy of local adopted code

Prerequisites

Before you begin this module, it is recommended that you successfully complete the following: *Core Curriculum*, and *Construction Technology*, Modules 68101-09 through 68116-09.

This course map shows all of the modules in *Construction Technology*. The suggested training order begins at the bottom and proceeds up. Skill levels increase as you advance on the course map. The local Training Program Sponsor may adjust the training order.

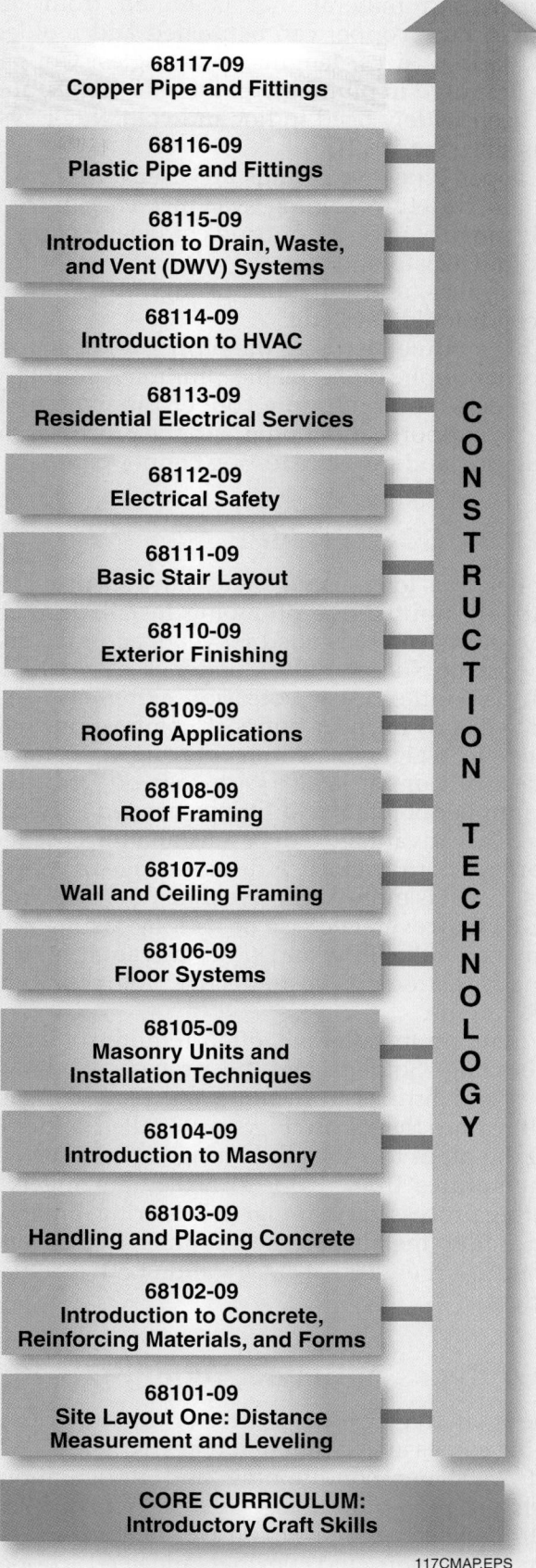

68117-09
Copper Pipe and Fittings

68116-09
Plastic Pipe and Fittings

68115-09
Introduction to Drain, Waste, and Vent (DWV) Systems

68114-09
Introduction to HVAC

68113-09
Residential Electrical Services

68112-09
Electrical Safety

68111-09
Basic Stair Layout

68110-09
Exterior Finishing

68109-09
Roofing Applications

68108-09
Roof Framing

68107-09
Wall and Ceiling Framing

68106-09
Floor Systems

68105-09
Masonry Units and Installation Techniques

68104-09
Introduction to Masonry

68103-09
Handling and Placing Concrete

68102-09
Introduction to Concrete, Reinforcing Materials, and Forms

68101-09
Site Layout One: Distance Measurement and Leveling

CONSTRUCTION TECHNOLOGY

CORE CURRICULUM:
Introductory Craft Skills

117CMAP.EPS

1.0.0 ◆ INTRODUCTION

Copper is a mineral that is mined from the ground. Pure copper can be melted and molded into various sizes, lengths, and angles. Copper was first used in plumbing in the early 1800s, and has been widely used in hot and cold water systems since the 1930s.

Copper pipe and fittings are used in a nearly endless variety of piping systems. They have a wide range of uses: for hot and cold water supply; for drain, waste, and vent (DWV) systems; for fuel gas supplies; and for transporting refrigerant in air conditioning systems.

This module discusses the properties of copper pipe, its applications in the plumbing industry, the processes used to join copper pipe, the tools used to support copper pipe, and the processes for insulating and pressure testing copper pipe.

2.0.0 ◆ COPPER PIPE

Copper is a long-lasting material that provides relatively trouble-free plumbing installations. In fact, copper pipe installed 60 years ago may still be working successfully. Copper is easy to join and dismantle, resists corrosion extremely well, and will not burn or support combustion. It is lighter in weight than iron, making it easier to transport. Copper is also easy to bend, reducing the number of joints and fittings needed to install pipe. This advantage reduces installation cost and improves performance. A disadvantage of copper is its cost; it is more expensive than other pipe materials. However, because of its long-lasting performance, reliability, and other advantages, it is actually extremely cost-effective for plumbing applications.

Plumbing installations often require soldering or brazing. Soldering is the process of joining pipes and fittings by heating the materials and melting into the joint a filler metal called solder to seal the pipes together. Soldering is performed at temperatures from 350°F to 550°F. Brazing is a heating process very similar to soldering, but it requires filler metals that melt at much higher temperatures (between 1,100°F and 1,500°F) than solders.

2.1.0 Types of Copper Tubing

Pipe is often referred to as tube or tubing. Copper tubing comes in five different types: K, L, M, DWV, and ACR (air conditioning and refrigeration). Each type represents a series of sizes with different wall thicknesses. The tubing can be hard drawn copper, or soft and flexible annealed copper.

Drawn copper is produced by pulling the tube through dies to reduce its diameter. Drawing hardens the copper and makes it very rigid. Annealed copper is produced by the annealing process, in which the material is heated and slowly cooled to relieve internal stress. This process reduces brittleness and increases toughness. All types of tube are available in a hard form. Hard forms come in 12- to 20-foot lengths and in diameters ranging from ¼ to 12 inches. Drawn copper tubing is widely used in commercial refrigeration and air conditioning systems. Types K, L, and ACR also are available in soft coils in 40- to 100-foot lengths in diameters ranging from ⅛ to 2 inches.

Manufacturers fill lengths of drawn tubing with nitrogen and plug them at each end to maintain a clean, moisture-free internal condition. This tubing is intended for use with formed fittings to make the necessary bends or changes in direction. It is more self-supporting than annealed copper tubing, such as tubing for air conditioning and refrigeration systems (ACR tubing); therefore, it needs fewer supports.

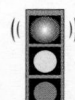

WARNING!

Always check for electrical grounding when working with copper pipe. Copper water supply lines are often used as the electrical grounding. Always shut off electrical power if you break the grounding. When you shut off electrical power, always use approved lockout/tagout procedures to avoid electrical shock. Check with your immediate supervisor before proceeding whenever electrical power is applied.

2.2.0 Copper Pipe Sizing

Pipe sizing varies depending on the type of copper pipe. Types K, L, M, and DWV use nominal, or standard, sizing. This means that the outside diameter (OD) is always ⅛ inch larger than the nominal size (see *Figure 1*). The nominal size is the approximate measurement in inches of the inside diameter (ID) for most copper pipes. For example, the OD of a ¾-inch nominal type M pipe measures ⅞ inch. This allows the same fittings to be used with the different wall thicknesses and IDs of different types of copper pipe. The ID is usually close to the stated size. A pipe with a ½-inch ID has approximately ½ inch between the inside walls of the pipe.

Type ACR pipe uses actual OD sizing. A ⅞-inch OD ACR copper pipe is actually the same OD as ¾-inch K, L, or M copper pipe. This means that you can use the same fittings with these pipes.

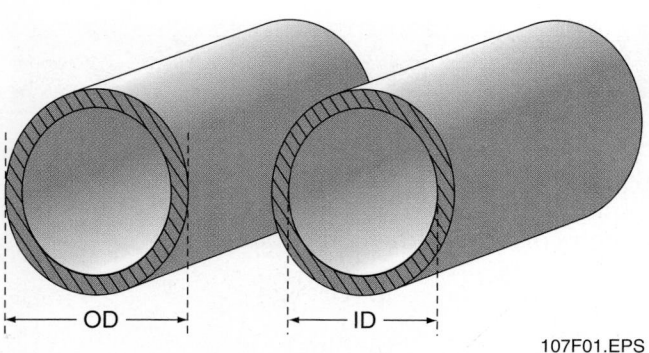

Figure 1 ◆ Copper pipe sizing.

107F01.EPS

Table 1 Copper Tubing Color Codes

Type	Color Code	Application
K	Green	_____
L	Blue	_____
M	Red	_____
DWV	Yellow	_____
ACR	Blue	_____

2.3.0 Labeling (Markings)

Plumbers must use copper pipe and fittings that are labeled and approved for use in plumbing systems. The labels on copper pipe contain very important information. Manufacturers must permanently mark Types K, L, M, and DWV to show the tube type, the name or trademark of the manufacturer, and the country of origin. Manufacturers also print this information on the hard tubes in a color that identifies the tube type. Manufacturers label most soft copper by stamping the information into the product. *Table 1* shows five types of copper tubing and the corresponding color code for labeling.

Exercise: Using *Table 1*, fill in the blanks to indicate the applications that correspond to each type of copper tubing based on applicable codes in your area.

2.4.0 Copper Pipe Applications

Consider strength and **formability** (ease of bending) when selecting copper pipe. Consult your local plumbing and mechanical codes when selecting pipe; these codes govern which types of pipe you can use in particular applications.

2.5.0 Material Storage and Handling

To ensure maximum productivity on the job, it is important to use common sense when storing and handling copper pipe. Before starting to work, always plan ahead. Always store copper pipe in a secure area. Store pipe and fittings near where you will be working so they are easy to access. Organize pipe and fittings into groups by pipe size and type. When handling pipe and fittings, be careful not to bend or damage them. Pipe that is bent or otherwise damaged not only costs money, but also makes you less productive. Take extra care when working in cold temperatures. Cold copper pipe can stick to bare hands.

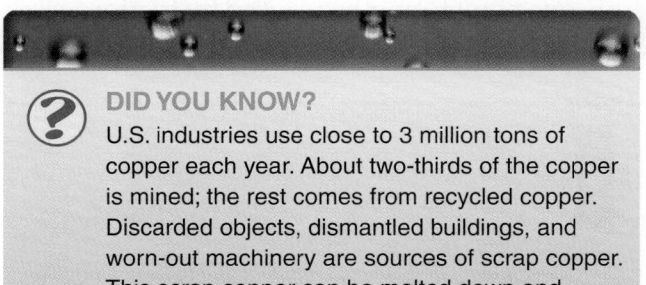

Sections 1.0.0–2.0.0

1. Copper pipe is generally less expensive than other types of pipe.
   a. True
   (b.) False

2. Pipes that are measured in nominal sizing have an OD that is _____ larger than the nominal size.
   (a.) ⅛ inch
   b. ¼ inch
   c. ½ inch
   d. 1 inch

3. Types K, L, M, and DWV copper tubing are measured using _____.
   a. ID sizing
   b. OD sizing
   (c.) nominal sizing
   d. copper tube sizing

Match the following types of copper pipe with their corresponding color code.

4. DWV   a. Orange
5. K   b. Blue
6. M   c. Yellow
7. L   d. Green
              e. Red

8. Take extra care when working with copper pipe in cold temperatures because it can _____.
   a. break
   b. shrink
   (c.) harm your hands
   d. corrode

3.0.0 ◆ FITTINGS AND VALVES

The types of fittings and valves used with copper pipe depend on how the pipe is used. Fittings allow you to join pipes to each other and change the direction or size of a pipe run, for example. Fittings are designed so they do not block or slow the flow of materials in the pipe. This section discusses water supply fittings, water supply valves, and DWV fittings, as well as relatively new alternatives.

3.1.0 Water Supply Fittings

Individual fitting shapes serve specific purposes, depending on whether a pipe runs horizontally or vertically. Use as few fittings as possible. Fewer fittings mean fewer chances for leaks and **pressure drops**, which are decreases in pressure from one point to another caused by friction losses in a water system. Common copper fittings for use with water supply systems include those listed in *Table 2* and shown in *Figure 2*.

When brazing, use brazing fittings, not standard fittings. Brazing fittings have sockets that are half the depth of standard fittings.

Solder fittings, also called sweat fittings, are special copper or brass fittings that are used for soldering or brazing copper pipe. Solder fittings are made slightly larger than the pipes to be joined, leaving only enough room for solder to flow into the joint. Adapter fittings allow copper tubing to be joined with threaded pipe on one end while the other end is soldered.

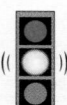

CAUTION

Never use dielectric unions in hot water systems. The unions cannot handle the temperature changes and they may leak.

Source: University of California–Santa Barbara Design, Construction, and Physical Facilities website. http://facilities.ucsb.edu/Standards/15000Plumbing.pdf, reviewed March 9, 2004.

DID YOU KNOW?

Two kinds of solder fittings are available with copper tubing. The first is a wrought copper fitting, which is made from copper tubing that is shaped into different types of fittings. Wrought copper fittings are generally lightweight, are smooth on the outside, and have thin walls. The second type is a cast solder fitting. This type of fitting is made using a mold. The first cast fittings had holes in the sockets to put solder in. Today, the heated copper is poured into the mold and allowed to cool. Cast solder fittings have a rough exterior and come in a wide variety of shapes. They are heavier than wrought solder fittings.

Table 2 Common Copper Fittings and Their Descriptions

Fitting	Description
90-degree ell and 45-degree ell	Elbow used to change the direction of the pipeline by 90 or 45 degrees.
Drop ear ell	An elbow that allows you to attach the pipeline to the building frame; frequently used at the last joint before the pipe comes through the wall to be attached to a fixture.
Street 90	An elbow with a male end and a female end that is used to change the direction of the pipeline by 90 degrees.
Street 45	An elbow with a male end and a female end that is used to change the direction of the pipeline by 45 degrees.
Tee	A fitting with three openings used to make branches at 90-degree angles to the main pipe. Reducing tees are used to make branches at 90-degree angles from the main pipe to smaller outlet pipes.
Coupling	A fitting used to connect lengths of pipe on straight runs.
Sweat cap	A fitting used to close the end of a copper pipe.
Female and male adapters	Fittings soldered onto the end of a length of copper tubing to provide a threaded end for attaching a pipe to another threaded pipe.
Sweat-to-compression adapter	A fitting used to adapt a soldered copper tube to a compression joint by means of a ferrule, which is a brass compression ring used for joining. A compression joint consists of a threaded nut that has been squeezed over a compression ring to seal the joint.
Sweat flange	A fitting used to adapt soldered copper tube to iron pipe flange.
Reducer coupling and reducer bushing	A fitting used to connect pipe to pipe or pipe to fitting with different sizes of pipe.
Sweat union	A fitting soldered to copper tubes, allowing the tubes to be joined to male and female half unions by a threaded shell nut.
Flare fittings	Fittings with a flared end that can be joined with a male cone-shaped tubing end or union. Flare fittings used in refrigeration and air conditioning include a variety of elbows, tees, and unions. They are drop-forged brass and are accurately machined to form the 45-degree flare face. The fittings used are based on the size of the tubing. Flare nuts are hexagon-shaped—they have six sides. An adjustable, or open-end, wrench is used with these fittings.
Grooved copper	A mechanical coupling material for rigidly connecting copper tubing that has been roll-grooved. Grooved copper fittings are made for connecting copper tubing in sizes from 2 to 6 inches. These fittings have grooved ends, so they can be installed using a wrench. This eliminates the need for soldering or brazing.

ON THE LEVEL

Dielectric Unions

If you attach unlike metals such as copper and galvanized pipe, it results in a process called electrolysis. Transition fittings, such as dielectric unions, prevent electrolysis. Dielectric unions isolate the different materials by using a mixture of brass, plastic, or rubber. However, they can cause other problems. For example, calcium deposits can build up, which eventually will lead to blockages. Therefore, you should never use dielectric unions unless they are plastic lined.

Source: Peace of Mind Home Inspection Services, LLC, website. www.getpeace.com/plumbing.htm, reviewed March 9, 2004.

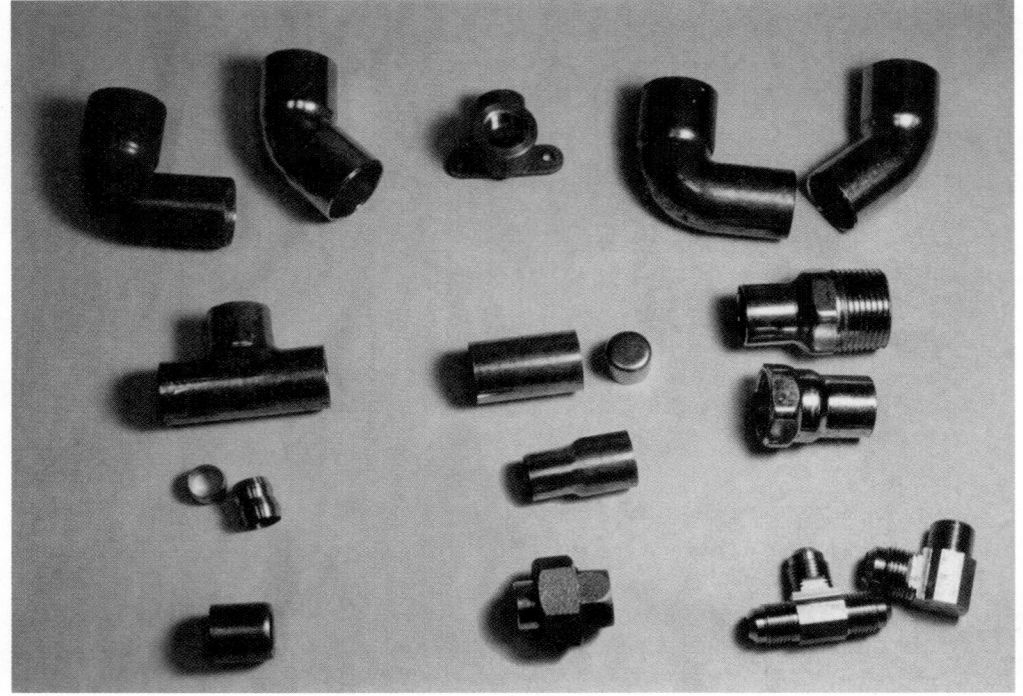

107F02A.EPS   107F02B.TIF

Figure 2 ◆ Fittings for copper tubing.

3.2.0 Water Supply Valves

Valves regulate the flow of liquid. They may provide on/off service or prevent flow reversal through a line. Valves use one or more of the following methods to control flow through a piping system:

- Move a disc or plug into or against a passageway.
- Slide a flat cylindrical or spherical surface across a passageway.
- Rotate a disc or ellipse around a shaft extending across the diameter of a pipe.
- Move a flexible material into the passageway.

Common valves for water supply systems include the following (see *Figure 3*):

- *Gate* – A valve with a wedge-shaped or tapered metal disc that fits into a smooth-ground surface or seat with the same shape, allowing a straight-line flow with little obstruction. It is a good choice for lines that will remain either completely open or completely closed most of the time.
- *Globe* – A valve that controls the flow of liquid with a movable spindle, which lowers to restrict flow through the valve opening. Because this valve is reliable and easy to repair, it is often used in water supply lines inside buildings.
- *Ball* – A valve that consists of a ball with a hole bored through its diameter, mounted on a spindle. When the valve is closed, the hole is at 90 degrees to the valve body so no flow can take place. When the valve is turned a quarter turn

GATE VALVE   GLOBE VALVE

107F03A.EPS

STOP AND WASTE VALVE   CHECK VALVE

107F03B.EPS

Figure 3 ◆ Common valves for water supply systems.

or opened completely, water flows through the hole. This valve is commonly used at the inlets and outlets of heat exchangers in HVAC (heating, ventilating, and air conditioning) systems and in systems where quick shutoffs may be necessary for in-line maintenance.

- *Compression, stop, and waste* – A valve that is opened or closed by raising or lowering a horizontal disc using a threaded stem. An elastomeric, or rubberized, washer on the end of the stem seals the valve seat, closing off water flow. This valve is most commonly used for draining and freeze protection above ground.
- *Check* – An automatic valve that permits the flow of liquid in one direction only. It prevents reverse flow. This valve is commonly used on domestic and irrigation wells.
- *Stop* – A valve that controls flow of liquids or gases between a building and supply source. It is also called a ground-key valve.

3.3.0 DWV Fittings

DWV fittings are designed to allow liquids and other materials to flow smoothly through them. Common fittings used in DWV systems are elbows (90 degrees, 45 degrees, 22½ degrees), male adapters, ferrule adapters, sanitary tees, cleanout tees, reducing tees, wyes, and reducers.

3.4.0 Alternative Fittings

In addition to the common fittings discussed above, you can use some alternative fittings to join copper pipe. These alternatives include press fittings and mechanically formed tee connections.

Press fittings, also called press-connect fittings, are mechanical fittings that connect pipe by means of a cold press fit system. First introduced in the 1970s in Germany and more recently in the U.S., they are used primarily for potable water systems. Press fittings are created by an electric press tool or a hand pressing tool (*Figure 4*). The tool crimps the fitting around the pipe against an O-ring inside the fitting. This ensures that the connection is strong and leakproof.

A **bullhead tee** is a tee fitting used on branches that are longer than the main line. The term is also used to describe a tee fitting in which the outlet is larger than the opening of the straight run. If tee fittings are not properly installed, they can cause a condition known as bullheading, which results from the larger outlet opening. When the flow of liquid hits the back wall of the tee, it causes turbulence. This adds to the pressure drop caused by liquid moving from a smaller pipe to a larger pipe. This turbulence may also cause a banging in the line. If more than one tee is installed in the line, a straight piece of pipe with a length between tees of 10 times the pipe's diameter is recommended to reduce turbulence. For example, a pipe that is 4 inches in diameter should have 40 inches of pipe between each tee (10 inches × 4 inches = 40 inches).

Mechanically formed tee connections are joints created by a tee-pulling tool. The tool allows you to drill into a section of pipe and create a tee connection

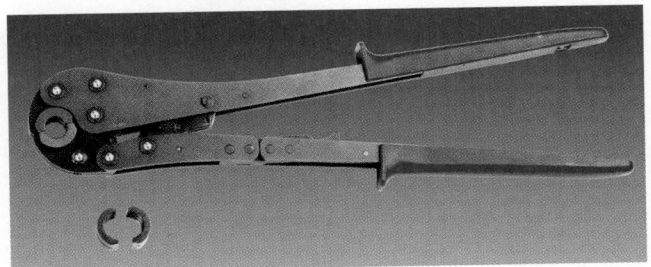

107F04.TIF

Figure 4 ◆ Hand pressing tool.

(see *Figure 5*). You must use brazing to join a branch line to the pipe. This method increases productivity because you create only one brazed joint rather than three soldered joints to form the tee connection (see *Figure 6*). This method is commonly used to create manifolds as well as copper fire sprinkler installations. Some codes do not permit tee-pulling tools for drainage because of the possibility of leakage. Follow the manufacturer's instructions and always consult applicable codes.

WARNING!

Always work under the direct supervision of your instructor or foreman, and always practice safety precautions to protect yourself and your co-workers from injury and to prevent equipment damage. Do not try to do any work that you have not been specifically trained to do. For example, do not operate valves unless you know exactly what the result will be. Before you work independently on a system, you must know the temperature and pressure conditions that exist at every point in the system, and how those conditions can be affected by malfunctions or by changes in valve positions.

107F06.TIF

Figure 6 ◆ Mechanical tee connection.

4.0.0 ◆ MEASURING, CUTTING, BENDING, JOINING, AND GROOVING

Some measuring, cutting, reaming, bending, joining, and grooving techniques are related specifically to copper. Different techniques may be used depending on the type and function of the copper pipe you are using. You will need to become familiar with many tools to install copper tubing properly. The following sections explain methods for measuring, cutting, bending, joining, and grooving copper tubing.

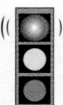

WARNING!

Before you begin a job, think through the potential hazards and wear the appropriate PPE, including, but not limited to, safety glasses, a hard hat, and gloves. In addition, beware of hair creams and shave creams; they can be flammable. Always make sure you have quick access to a fire extinguisher on the job.

4.1.0 Measuring

It is extremely important to measure pipe carefully. Several methods of measuring copper pipe are described in the following list (see *Figure 7*):

- *End-to-end* – Measure the full length of the pipe.
- *End-to-center* – Use for pipe that has a fitting joined on one end only; pipe length is equal to the measurement minus the end-to-center dimension of the fitting.
- *Center-to-center* – Use with a length of pipe that has fittings joined on both ends; pipe length is equal to the measurement minus the sum of the end-to-center dimensions of the fittings.

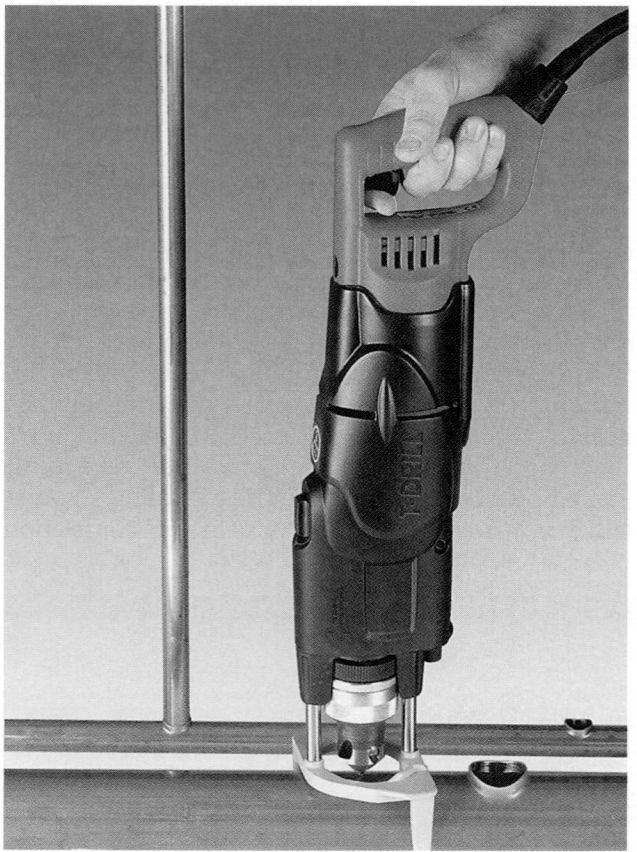

107F05.TIF

Figure 5 ◆ Tee-pulling tool.

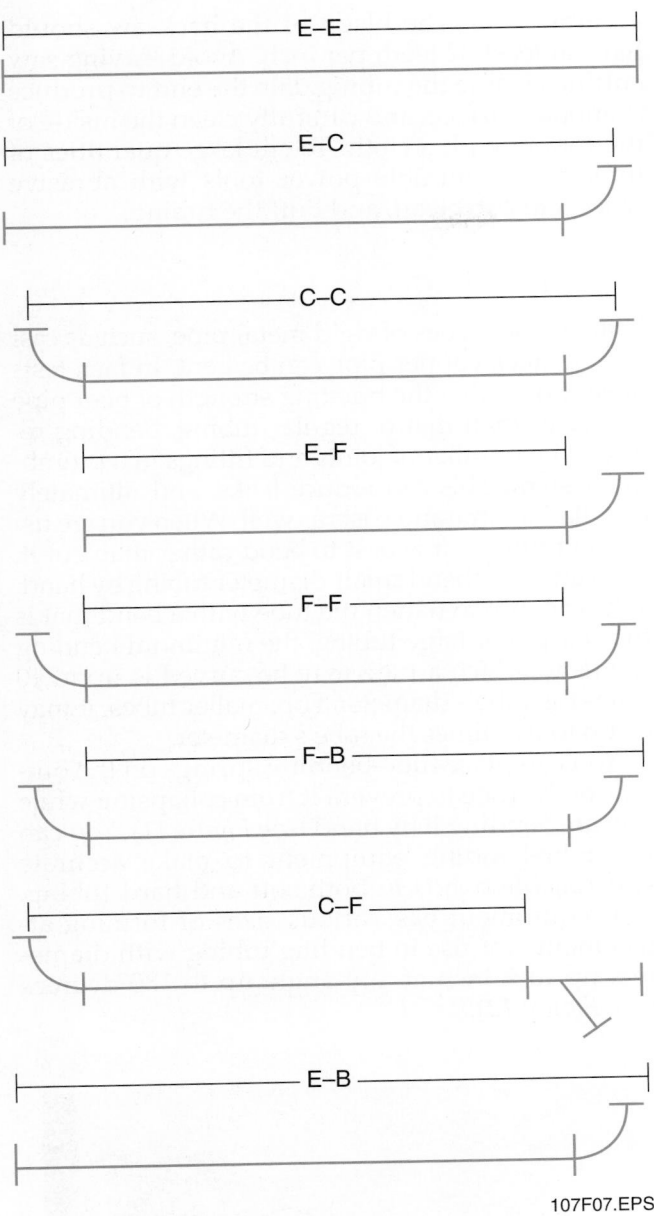

Figure 7 ◆ Measuring copper pipe.

- *End-to-face* – Use for pipe that has a fitting joined on one end only; pipe length is equal to the measurement.
- *Face-to-face* – Use for same situation as center-to-center measurement; pipe length is equal to the measurement.
- *Face-to-back* – Use with a length of pipe that has fittings joined on both ends; pipe length is equal to the measurement plus the distance from the face to the back of one sweated-on fitting.
- *Center-to-face* – Use with a length of pipe that has fittings on both ends; pipe length is equal to the measurement from the center of one of the

fittings to the face of the opposite fitting, plus twice the insertion length.
- *End-to-back* – Use for pipe that has a fitting joined on one end only; pipe length is equal to the measurement plus the length of the sweated-on fitting.

4.2.0 Cutting

You can cut copper tubing with a handheld tube cutter, a hacksaw, or a midget cutter. The handheld tube cutter (*Figure 8*) is preferred because it makes a cleaner joint and leaves no metal particles. Use a tube cutter that is the right size for the copper you are cutting, and make sure that the proper cutting wheel is in place. Plumbers use internal tube cutters (*Figure 9*) for trimming extended ends of installed water closet bowl and shower waste lines below the level of the flange.

107F08.TIF

Figure 8 ◆ Handheld tube cutter.

107F09.TIF

Figure 9 ◆ Internal tube cutter.

After cutting, ream all cut tube ends to the full inside diameter of the tube. Reaming removes the small burr (rough inside edge) created when you cut the pipe. Burrs left on tubing can cause the pipe to corrode. A tube that is reamed correctly provides a smooth inner surface for better flow. You must also remove burrs on the outside of the tube to ensure a good fit. Tools used to ream tube ends include the reaming blade on the tube cutter, files (round or half-round), a pocketknife, and a deburring tool. If your pipe becomes deformed, you can use a *sizing tool*, which consists of a plug and a sizing ring, to bring the pipe back to roundness. Refer to *Figure 8* for an example of a deburring blade on a tube cutter. A variety of models are available, and the cutting sizes range from ⅛- to 4⅛-inches OD.

To use the handheld tube cutter, follow these steps (see *Figure 10*):

Step 1 Place the tube cutter on the tube at the point where you want to cut. Tighten the knob, forcing the cutting wheel against the tube.

Step 2 Make the cut by rotating the cutter around the tube under constant pressure.

Step 3 Use the built-in deburring blade to remove any burrs from inside the tube.

To cut larger size drawn tubing, you can use a hacksaw and a vise. A vise is a gripping tool that secures an object while you work on it. Using a vise helps you square the ends and allows more accurate cuts. The blade of the hacksaw should have at least 32 teeth per inch. Avoid leaving saw cuttings inside the tubing. File the end to produce a smooth surface, and carefully clean the inside of the tubing with a cloth. To cut large quantities of tubing, use portable power tools with abrasive wheels to cut, clean, and buff the tubing.

4.3.0 Bending

Unlike other types of rigid metal pipe, such as cast iron or steel, copper pipe can be bent. In fact, tests have shown that the bursting strength of bent pipe is greater than that of regular tubing. Bending reduces the number of joints and fittings in a plumbing system. This can reduce leaks, and ultimately installation time and costs as well. When you are using soft tubing, it is best to bend rather than cut it. You can easily bend small-diameter tubing by hand. Take care not to flatten the tube with a bend that is too sharp. For large tubing, the minimum bending radius to which a tube may be curved is up to 10 times the tube's diameter. For smaller tubes, it may be up to five times the tube's diameter.

You can place tube-bending springs on the outside of the tube to prevent it from collapsing while you are bending it by hand (see *Figure 11*). You can use tube-bending equipment to make accurate and reliable bends in both soft and hard tubing. This equipment has various sizes of forming attachments for use in bending tubing with diameters up to ⅞ inch at any angle up to 180 degrees (see *Figure 12*).

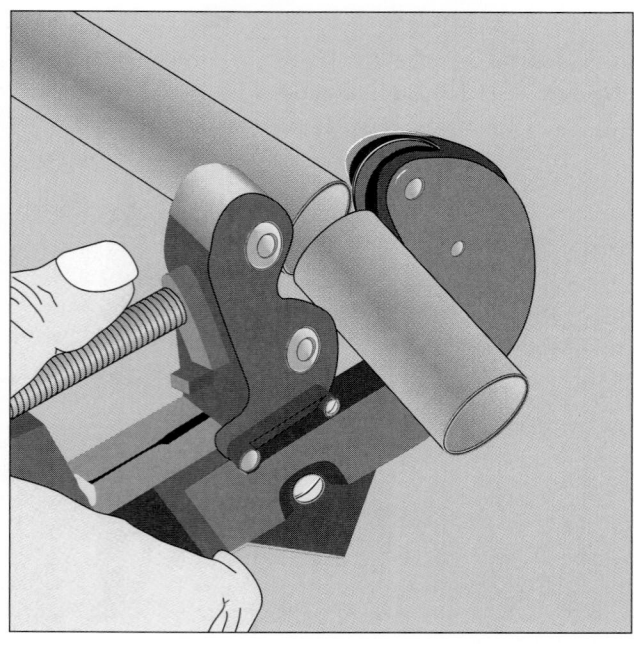

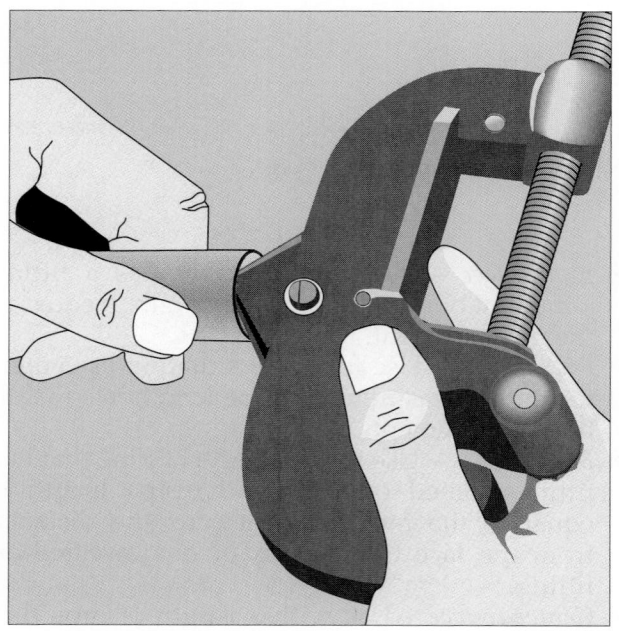

107F10.EPS

Figure 10 ◆ Using a handheld tube cutter.

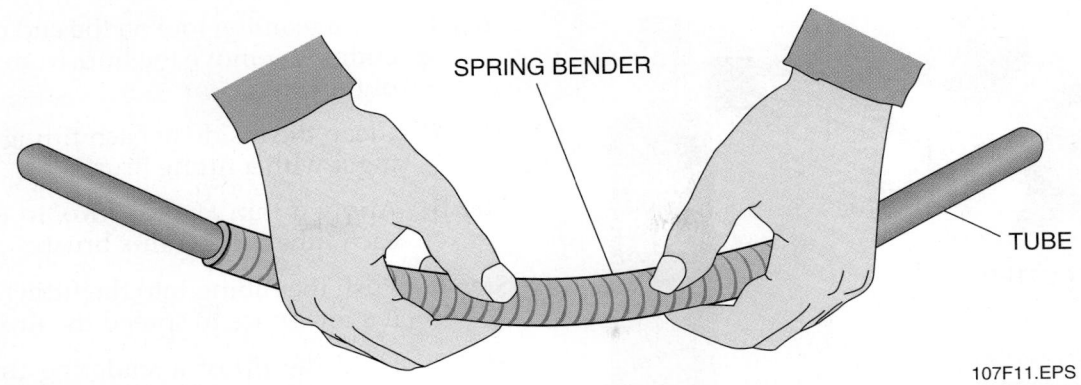

SPRING BENDER

TUBE

107F11.EPS

Figure 11 ◆ Tube-bending spring.

107F12.EPS

Figure 12 ◆ Tube-bending equipment.

4.4.0 Joining

Copper can be joined in many ways, including by a **sweat joint**, a compression joint, or a **flare joint**. The method depends on the plumbing application and environmental factors. Sweat joints, for example, require a heating process. You should not use this method when a fire hazard exists at the job site. As always, follow all applicable codes and manufacturer's instructions when joining copper pipe.

4.4.1 Soldering

Soldering is a type of heat bonding in which copper pipe is joined when a soft filler metal is melted in the joint between the two pipes. Solder fittings are made slightly larger than the pipes to be joined, leaving only enough space for solder to flow into the joint. This space is called the capillary action gap. Solder melts and flows into the gap in a process known as **capillary action.** Capillary

action occurs regardless of whether the molten solder is flowing up, down, or horizontally. Adapter fittings allow copper tubing to be joined with threaded pipe on one end while the other end is soldered.

A sweat joint is made by measuring, cutting, reaming, cleaning, and applying flux to the copper pipe, then adding solder and heating to a certain temperature until the solder flows into the joint. Always use lead-free solder. Flux is a paste that acts as a wetting and cleaning agent and aids the soldering process by preventing oxidation of the joint, which would damage the copper. Flux should be lead-free and water-soluble. The heat required for soldering (350°F to 550°F) can be generated by electricity or various kinds of gases. Use sweat joints with hard or soft copper with nominal sizes from ⅛ to 4 inches.

Tools used to solder copper tubing to fittings include a tube cutter, fitting brush, solder, flux brush, and soldering torch (see *Figure 13*). In place

TUBE CUTTER

SOLDERING TORCH

WIRE SOLDER

FITTING BRUSH

FLUX BRUSH AND SOLDER PASTE

107F13.EPS

Figure 13 ◆ Tools for soldering copper tubing to a fitting.

of a fitting brush, you can use any abrasive other than steel wool. Steel wool contains oil, which contaminates the pipe.

To solder copper tubing to a fitting, follow these steps (see *Figure 14*):

Step 1 Measure and mark tubing to length, including the fitting allowance for the portion of the tube that will extend inside fittings.

Step 2 Use a tube cutter to cut the tubing.

Step 3 Use a reaming tool on the end of the tube cutter to remove the burr from the inside of the tubing.

Step 4 Clean the inside of each fitting by scouring it with a fitting brush.

Step 5 Apply a thin layer of flux to the end of each tube using a flux brush.

Step 6 Push the tubing into the fitting, and turn it a few times to spread the flux evenly.

Step 7 Hold the tip of a soldering torch flame against the fitting. When the flux starts to sizzle, move the flame around to the other side of the fitting to heat it evenly. Some plumbers use a special soldering torch fired from a 20-pound tank of gas called a B-tank.

Step 8 When the flux starts to bubble, remove the torch, and touch the wire solder to the point where the tubing enters the fitting. The solder will melt and be drawn into the joint. Once a line of solder shows completely around the joint, the connection is filled with solder.

Step 9 Allow the joint to cool until the solder solidifies.

Step 10 Once the solder has solidified, wipe it clean with a soft, wet cloth to remove any flux.

Brazing is a heating process very similar to soldering, but it requires filler metals that melt at much higher temperatures (between 1,100°F and 1,500°F) than solders. Plumbers use brazing when joints must be very strong. For example, brazing is often required to join refrigeration pipe.

4.4.2 *Creating Compression Joints*

A compression joint (*Figure 15*) is a mechanical joint that is made by measuring, cutting, and reaming the pipes and using compression fittings. With this method, you tighten a threaded nut to squeeze a compression ring to seal the joint. This

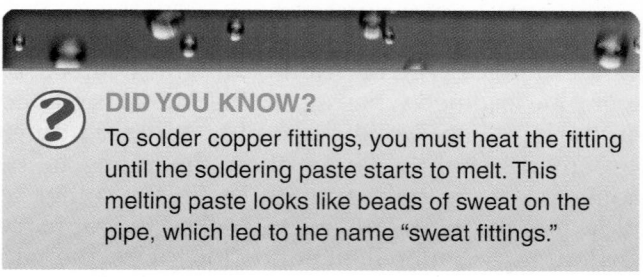

DID YOU KNOW?

To solder copper fittings, you must heat the fitting until the soldering paste starts to melt. This melting paste looks like beads of sweat on the pipe, which led to the name "sweat fittings."

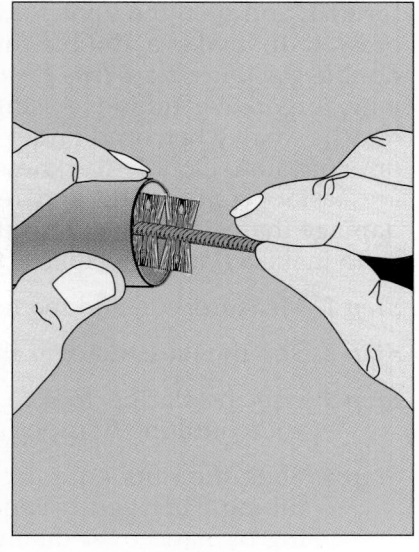

CLEANING

HEATING

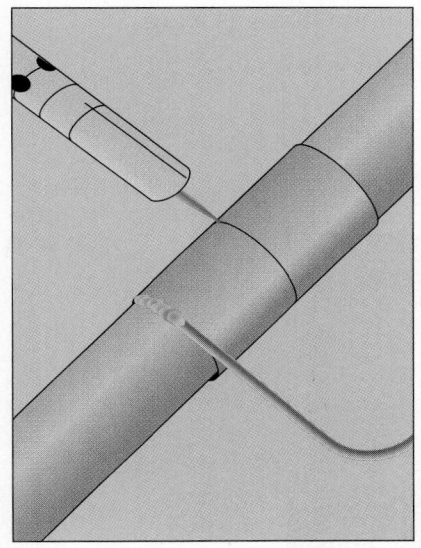

MELTING THE SOLDER

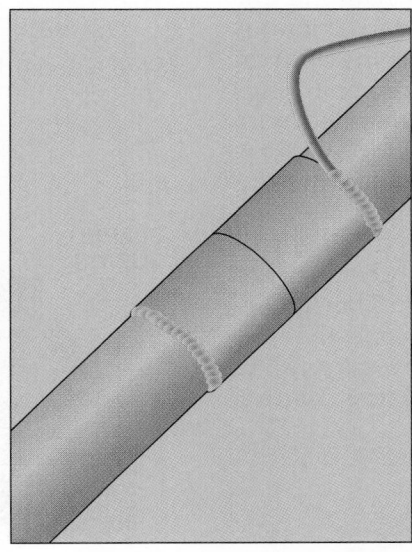

FILLING THE JOINTS

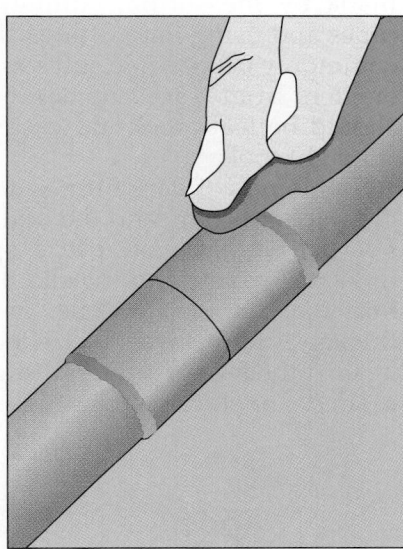

WIPING REMAINING FLUX
AFTER SOLDER COOLS

107F14.EPS

Figure 14 ◆ Soldering copper tubing to a fitting.

kind of joint is often used for joining refrigerant tubing. It is a popular method because it uses a threaded fitting and takes less time than making soldered or flared joints.

To create a compression joint, follow these steps:

Step 1 Measure, cut, and ream the tube. Make sure the tube is cut square and is free of burrs.

Step 2 Slip the nut over the tube, and slide the compression ring on the tube with the teeth facing the tube's end.

Step 3 Install the cone with the convex surface toward the end of the tube. To be sure the fitting goes together completely, make sure that ¼ inch of the tube extends beyond the cone when working with a ½-inch tube, and that ½ inch of the tube extends beyond the cone when working with a ¾-inch tube.

Step 4 Push the nut onto the fitting, and tighten. When the fitting squeaks, turn the nut one more full turn.

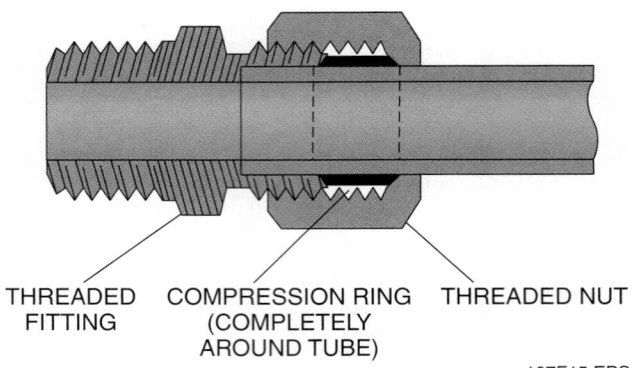

THREADED FITTING   COMPRESSION RING (COMPLETELY AROUND TUBE)   THREADED NUT

107F15.EPS

Figure 15 ◆ Compression joint.

4.4.3 Creating Flare Joints

Flare joints may be required in an installation in which a fire hazard exists and a torch for soldering or brazing is not permitted. A flare joint is made by measuring, cutting, and reaming the pipes and using flare fittings. This kind of joint is commonly used to join soft copper tubing with diameters from ¼ to 2 inches. Soft copper fittings should be leakproof and easily dismantled with the right tools.

Two kinds of flare fittings are popular: the single-thickness flare and the double-thickness flare. For both types, you use a special flaring tool (*Figure 16*) to expand the end of the tube outward into the shape of a cone, or flare. The single-thickness flare forms a 45-degree cone that fits against the face of a flare fitting (see *Figure 17*). In a single operation, the single-thickness flare is

formed, and then the lip is folded back and compressed to make a double-thickness flare. The double-thickness flare (see *Figure 18*) is preferable with larger-size tubing. A single-thickness flare may be weak when used under excessive pressure or expansion. Double-thickness flare connections are easier to dismantle and reassemble without damage than single-thickness flare connections.

To make a flare joint, follow these steps:

Step 1   Measure, cut, and ream the tubing.

Step 2   Slip the flare nut over the tubing.

Step 3   Use the flaring tool to flare the tubing's ends until the fit is perfect.

Step 4   Slide the nuts onto each end of the flare tubing. Then gently bend or shape the tubing by hand over the male thread of each fitting.

Step 5   Use a smooth-jaw adjustable wrench to tighten until the fitting is snug.

Step 6   Test the joint for leaks.

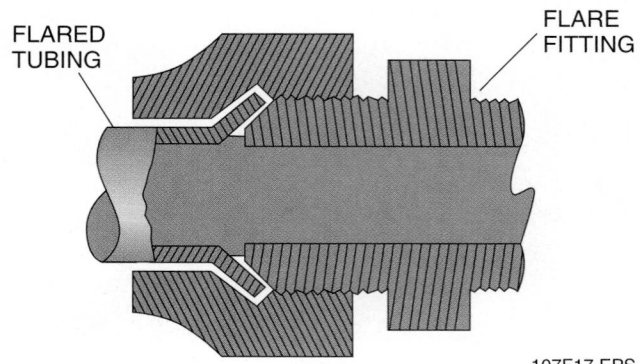

FLARED TUBING   FLARE FITTING

107F17.EPS

Figure 17 ◆ Single-thickness flare.

107F16.TIF

Figure 16 ◆ Flaring tool.

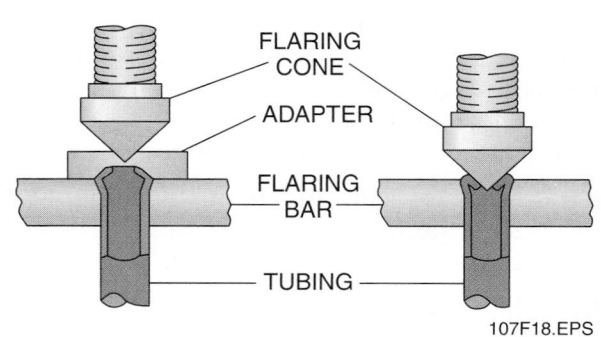

FLARING CONE
ADAPTER
FLARING BAR
TUBING

107F18.EPS

Figure 18 ◆ Double-thickness flare.

4.5.0 Grooving

You can groove copper pipe in two ways: by rolling or by cutting. Roll grooving involves cold forming pipe—it does not remove any metal from the pipe. Cut grooving removes metal from the OD of the pipe. You will learn more about grooving pipe elsewhere in this curriculum. Always follow applicable code and manufacturer's instructions when grooving copper pipe.

Review Questions

Sections 3.0.0–4.0.0

1. _____ are used to make branches in a pipeline.
   a. Tees
   b. Adapters
   c. Couplings
   d. Flanges

2. Grooved copper fittings are made for connecting copper tubing in sizes from _____.
   a. 0 to 4 inches
   b. 2 to 6 inches
   c. 4 to 8 inches
   d. 6 to 10 inches

3. The valve most often used for draining and freeze protection above ground is the _____ valve.
   a. ball
   b. compression, stop, and waste
   c. globe
   d. gate

4. The automatic valve that permits the flow of liquid in one direction only and prevents reverse flow is the _____ valve.
   a. globe
   b. gate
   c. check
   d. stop

5. A _____ tee is a fitting that is used on branches that are longer than the main line.
   a. bullhead
   b. sanitary
   c. reducing
   d. cleanout

6. To measure a length of pipe that has fittings joined on both ends, use the _____ method.
   a. end-to-end
   b. end-to-center
   c. center-to-center
   d. end-to-face

7. When cutting copper tubing, the _____ is preferred because it makes a cleaner joint and leaves no metal particles.
   a. handheld tube cutter
   b. hacksaw
   c. midget cutter
   d. power saw

8. A _____ helps prevent the tube from collapsing while you are bending it by hand.
   a. tube-reaming tool
   b. handheld burring blade
   c. tube-shielding cover
   d. tube-bending spring

9. A(n) _____ joint is commonly used to join soft copper tubing from ¼ to 2 inches.
   a. compression
   b. flare
   c. sweat
   d. angle

10. You can _____ copper pipe in two ways: by rolling or by cutting.
    a. bend
    b. groove
    c. join
    d. measure

5.0.0 ◆ INSTALLING PIPE HANGERS AND SUPPORTS

Hangers and supports are designed to hold and support pipe in either a horizontal or a vertical position. All pipes must be installed and supported so that both the pipe and its joints remain leakproof. Improper support can cause the piping system to sag. This causes stress on the pipe and fittings and, over time, increases the chance of breaks or leaks in the piping system. Without proper support, the drainage pipe can shift from its proper angle and form traps. These traps fill with liquid and solid wastes that block the pipeline. Copper pipe is supported in a variety of ways and with various sizes of hangers and supports. Be sure to follow the manufacturer's instructions for specific installation instructions, and consult applicable codes.

If you are installing pipe in a seismically active area—that is, where earthquakes are a possibility—local codes will require seismic restraints. The purpose of these restraints is to ensure that the pipe is securely fastened to the structure in the event of excessive vibration. For example, some codes require hangers and supports to be used at closer intervals than in non-seismically active areas. In addition, they may require you to leave extra spacing for pipes where they meet walls and floors to allow for anticipated movement.

5.1.0 Types of Pipe Hangers and Supports

Which hanger to use depends on the job specifications, the documents that describe the quality of the materials and work required. Specifications for hangers and supports are determined by the following factors:

- The combined weight of the pipe fittings and valves
- The maximum weight of the contents that the pipe might carry
- The material (wood, concrete, steel) the hanger will be attached to
- The distance from the anchor point to the pipe
- The potential for corrosion between the hanger and the pipes, fittings, or valves
- The expansion and contraction of the piping system
- The vibration of equipment attached to the piping system

Different types of hangers and supports are designed to hold and support pipe in either a horizontal or vertical position. They are manufactured in various materials, including carbon steel, malleable iron, cast iron, and plastics. They are available with different finishes, including copper plate, black, galvanized, chrome, and brass. To reduce corrosion, hangers and supports should be made of the same material as the pipe. If another material is used, the pipe must be shielded.

The basic components of hangers and supports can be placed into the following three major categories:

- Pipe attachments
- Connectors
- Structural attachments

5.1.1 Pipe Attachments

A pipe attachment is the part of the hanger that touches or connects directly to the pipe. It may be designed for either heavy duty or light duty, for covered (insulated) pipe or plain pipe. Applicable fire codes may restrict the use of hangers made of material other than copper. In some cases, you can use plastic-coated hangers. For specific requirements, consult applicable codes. Examples of pipe attachments are shown in *Figure 19*.

Hangers are used for horizontal or vertical support of pipes. The main purpose of hangers and brackets is to keep the piping in alignment and prevent it from bending or distorting. Examples of hangers are shown in *Figure 20*.

Pipe can be supported on wood-frame construction with several styles of pipe attachments. These include pipe hooks, J-hooks, tube straps, plumber's tape (also called strap iron or band

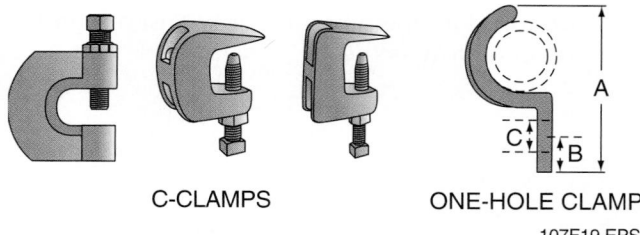

C-CLAMPS   ONE-HOLE CLAMP

107F19.EPS

Figure 19 ◆ Pipe attachments.

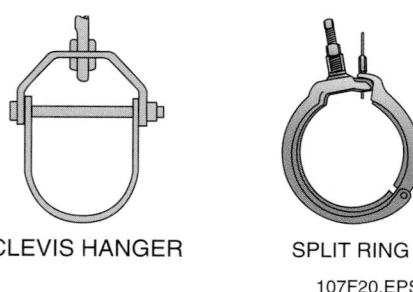

CLEVIS HANGER   SPLIT RING

107F20.EPS

Figure 20 ◆ Pipe hangers.

iron), and pipe straps (see *Figure 21*). In these examples, the pipe attachment, connector, and structural attachment are all one unit. Note that plumber's tape may not be allowed in your area. When choosing hangers of any type, always consult applicable codes.

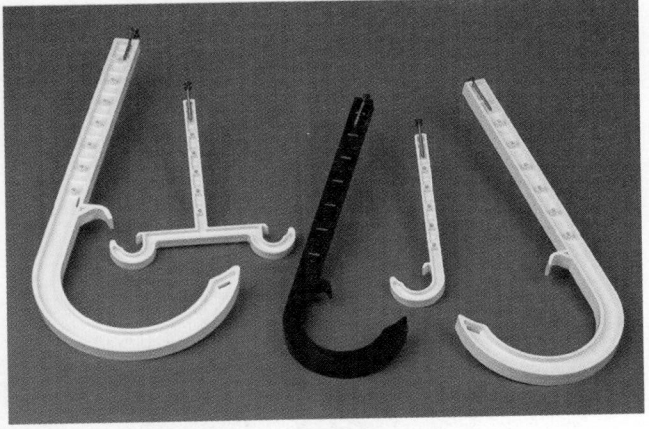

J-HOOKS

Pipe hangers are used mainly to support pipe, but they can also be used as a support for vibration isolators (see *Figure 22*). If no vibration problems are expected, use commonly accepted plumbing practices. If vibration may be a problem, the specifications should tell you what materials to use.

To fasten pipes to beams and other metal structures, one-hole clamps, steel brackets, beam clamps, or C-clamps are used. Vertical hangers, also called pipe riser clamps, consist of a friction clamp that can be attached to structural site components to support the vertical load of the pipe (see *Figure 23*). Special fasteners are used to attach hangers to masonry, concrete, or steel.

Other pipe attachments include universal pipe clamps and standard 1⅜-inch or 1½-inch channels (see *Figure 24*). The notched steel clamps are inserted by twisting them into position along the slotted side of the channel. The pipes can be aligned as close to one another as the couplings allow.

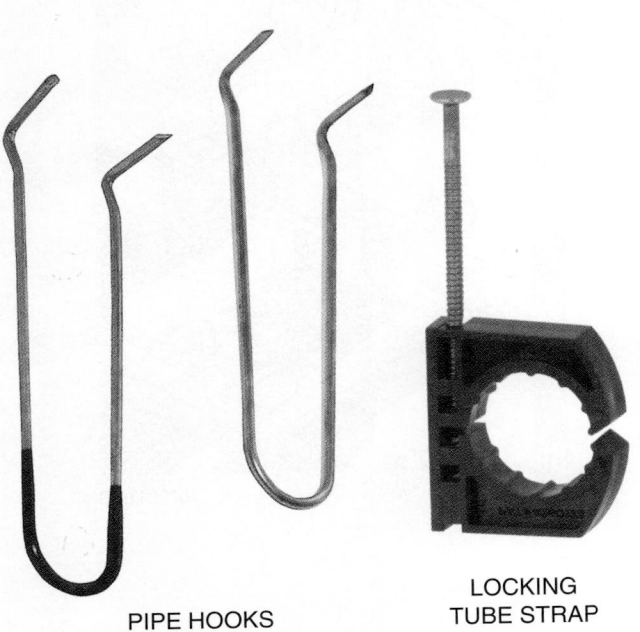

PIPE HOOKS

LOCKING TUBE STRAP

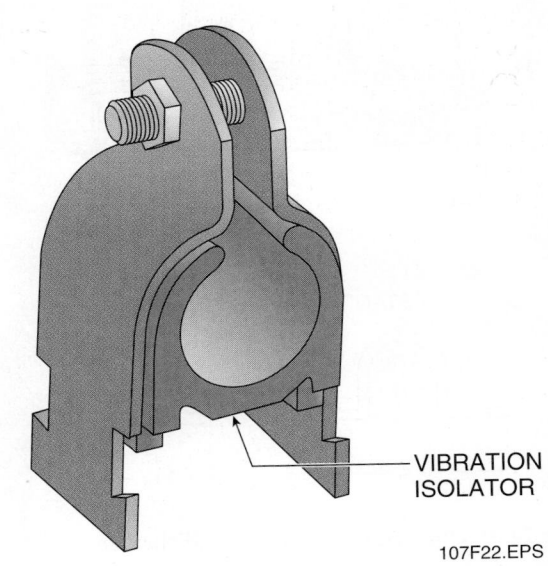

VIBRATION ISOLATOR

107F22.EPS

Figure 22 ◆ Vibration isolator installed in hanger.

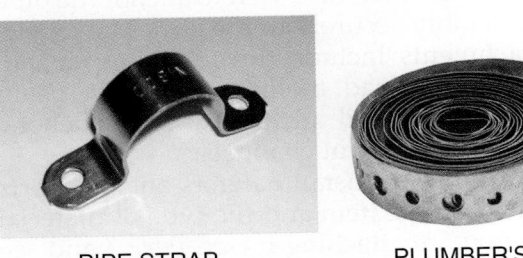

PIPE STRAP

PLUMBER'S TAPE

107F21.EPS

Figure 21 ◆ Pipe hangers for wood-frame construction.

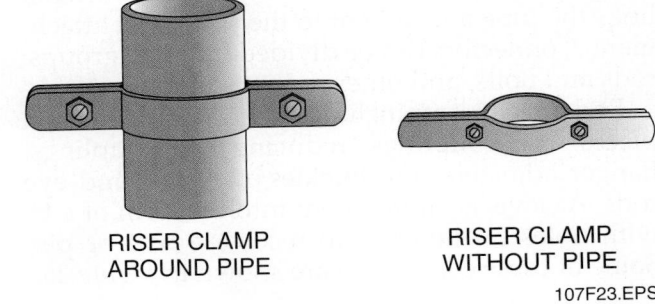

RISER CLAMP AROUND PIPE

RISER CLAMP WITHOUT PIPE

107F23.EPS

Figure 23 ◆ Vertical hangers.

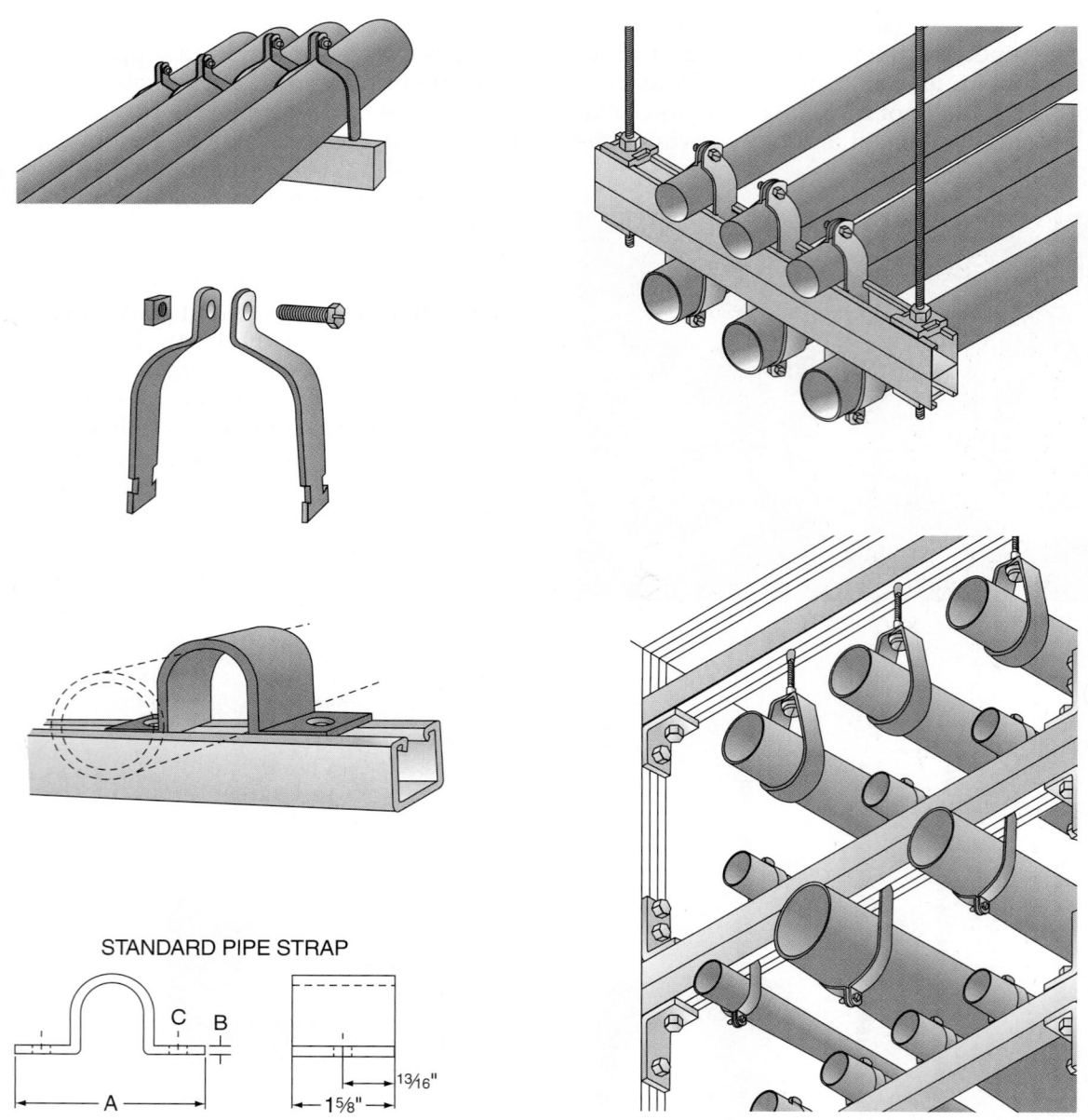

STANDARD PIPE STRAP

Figure 24 ◆ Universal pipe clamps and channels.

107F24.EPS

5.1.2 Connectors

The connector section of the hanger is the part that links the pipe attachment to the structural attachment. Connectors can be divided into two groups: rods and bolts, and other rod attachments.

Rod attachments include eye sockets, extension pieces, rod couplings, reducing rod couplings, hanger adjusters, turnbuckles, **clevises**, and eye rods. A clevis is an iron bent into the form of a U, with holes in the ends to receive a bolt or pin. Some of these connectors are shown in *Figure 25*.

5.1.3 Structural Attachments

Structural attachments are used to anchor the pipe hanger assembly securely to the structure. Structural attachments include threaded drop-in anchors, plastic or lead mollies, toggles, C-beam clamps, pound-in nail anchors, wedge anchors, threaded rods, and strut channels.

What you use to install hangers and supports will depend on the item and the type of material to which you are attaching it (see *Table 3* and see *Figure 26*).

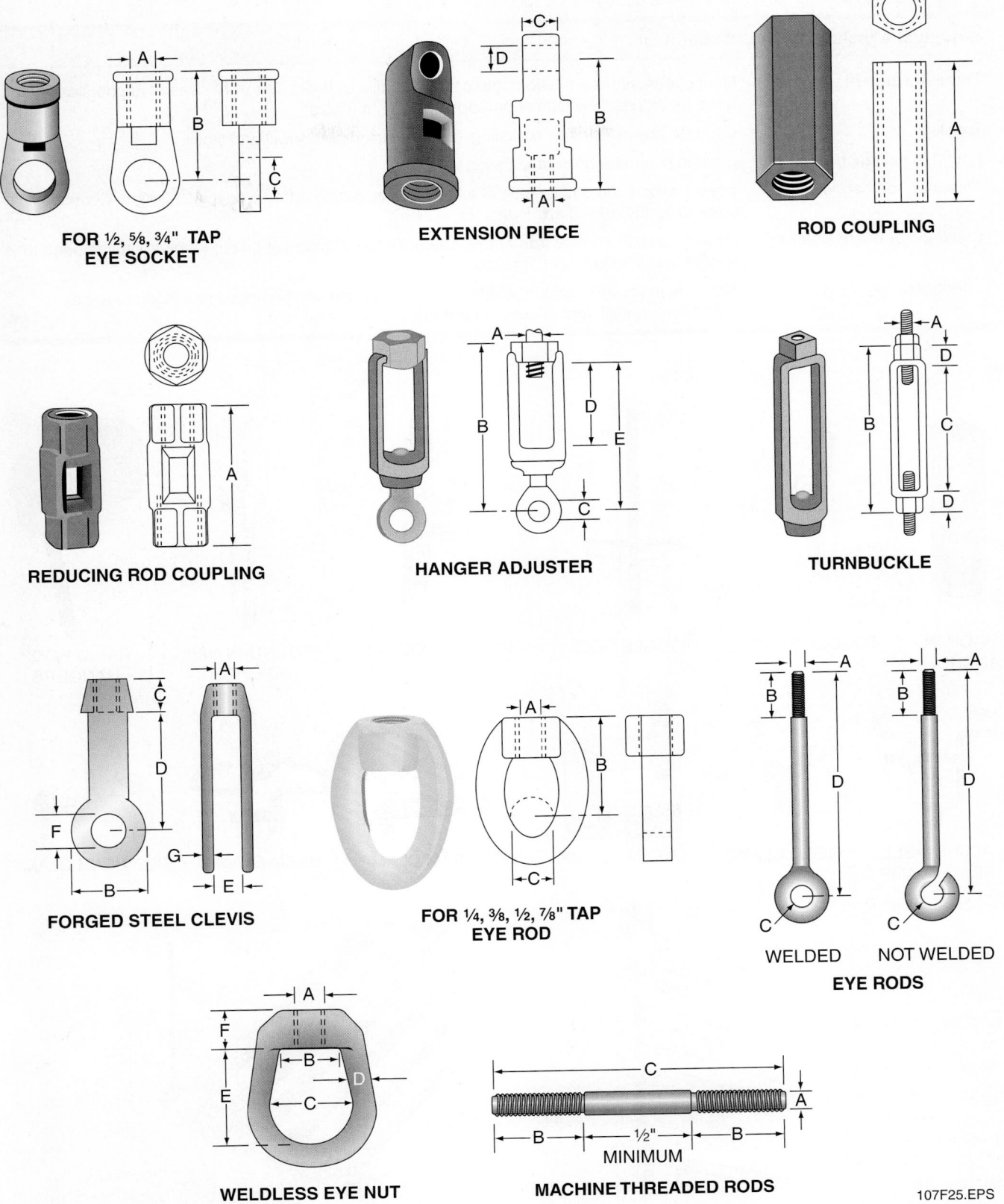

FOR ½, ⅝, ¾" TAP
EYE SOCKET

EXTENSION PIECE

ROD COUPLING

REDUCING ROD COUPLING

HANGER ADJUSTER

TURNBUCKLE

FORGED STEEL CLEVIS

FOR ¼, ⅜, ½, ⅞" TAP
EYE ROD

WELDED   NOT WELDED
EYE RODS

WELDLESS EYE NUT

MACHINE THREADED RODS

½"
MINIMUM

107F25.EPS

Figure 25 ◆ Rod attachment connectors.

Table 3 Types of Structural Attachments and Their Descriptions

Structural Attachment	Description
Threaded drop-in anchors	Most commonly used in concrete ceilings, walls, or floors. Different sizes of anchors are rated for different weights or spreads between anchors.
Toggles	Used for any hollow wall or ceiling support, including drywall and block.
Pound-in nail anchors	Used in concrete, brick, and stone.
Threaded rod hangers	Used in wood. Various lengths of all-thread rod can be cut to a specified length. Common sizes of all-thread rod are $\frac{3}{8}$ inch to $\frac{1}{2}$ inch.
C-clamps or beam clamps	Usually used to hook to iron or to a beam. These include set-bolts that must be tightened to ensure that they will not come loose.
Channel (strut)	Available in various lengths, widths, and heights and can be bolted or welded together. Various floor, wall, and overhead strut brackets are available.

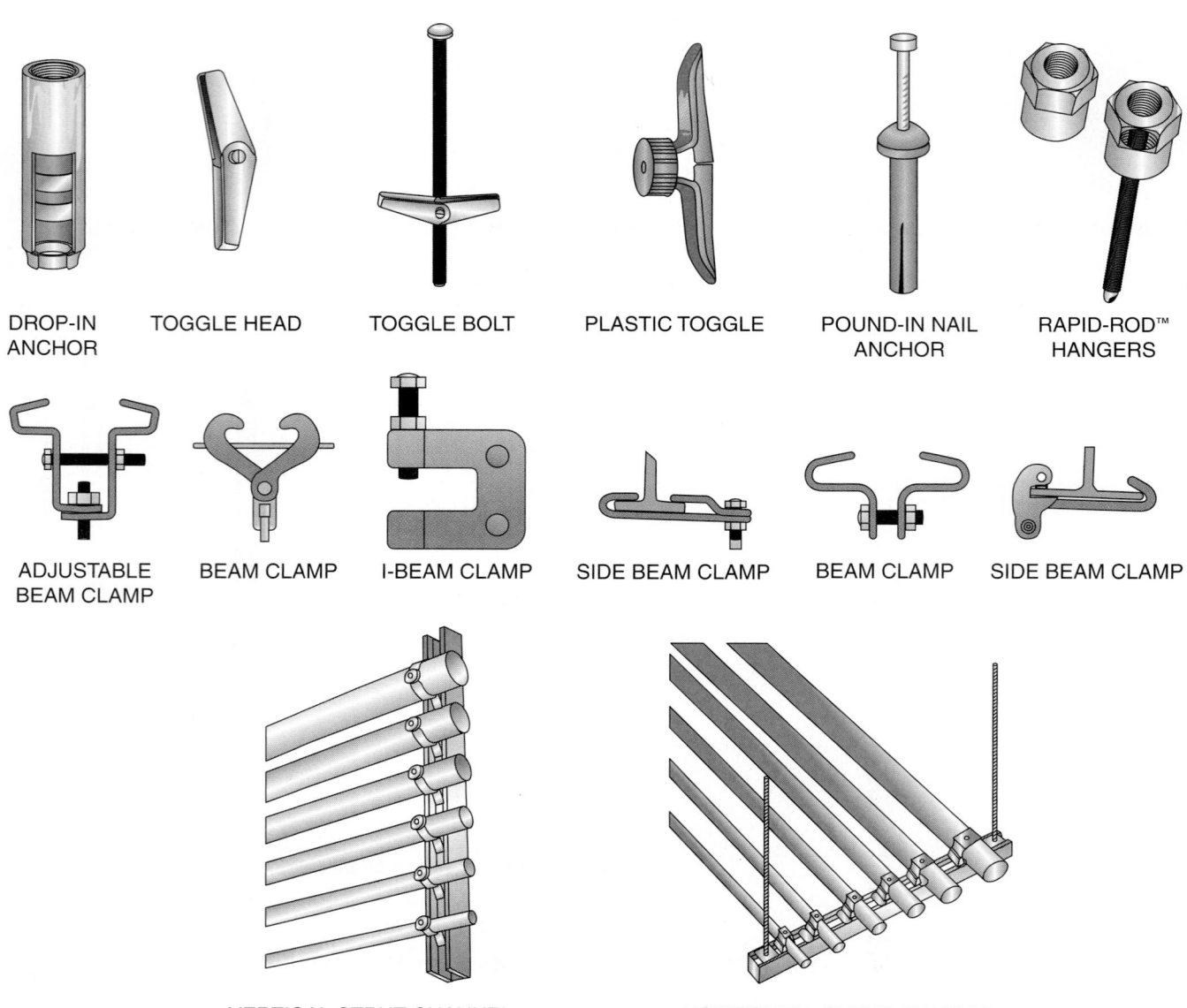

DROP-IN ANCHOR   TOGGLE HEAD   TOGGLE BOLT   PLASTIC TOGGLE   POUND-IN NAIL ANCHOR   RAPID-ROD™ HANGERS

ADJUSTABLE BEAM CLAMP   BEAM CLAMP   I-BEAM CLAMP   SIDE BEAM CLAMP   BEAM CLAMP   SIDE BEAM CLAMP

VERTICAL STRUT CHANNEL   HORIZONTAL STRUT CHANNEL

107F26.EPS

Figure 26 ◆ Structural attachments.

6.0.0 ◆ INSULATING PIPES

Under certain temperature and humidity conditions, condensation will form on cold water pipe and may drip into equipment or occupied areas. Similarly, heat can escape from hot water piping. To prevent these occurrences, some pipe is insulated (*Figure 27*).

Insulation is a material that prevents the transfer of heat. Cork, glass fibers, mineral wool, and polyurethane foams are examples of insulating materials. Insulation should be fire-resistant, moisture-resistant, and vermin-proof.

If the insulation cannot be installed before the tubing is connected, it must be split lengthwise to fit onto the pipe. Split seams and connecting seams must then be sealed. Insulation should not be stretched, because its effectiveness will be reduced. You generally do not put the insulation on the pipe until after pressure testing is performed. However, if you must do so before testing takes place, you must keep the joints exposed for the testers.

Some pipes are always insulated, and others are insulated only under certain conditions. Local building codes and job specifications will describe insulation requirements.

7.0.0 ◆ PRESSURE TESTING

Once an installation is complete, you must pressure test the system for leaks to be sure that all connections are secure. The system must also be inspected to make sure that the work has been done according to the job specifications and applicable codes.

You can pressure test copper pipe with water or air. You are required to supply all the equipment needed for testing, including test plugs, caps, plugs, and a source of compressed air (if air testing is required). Water testing can be done with test plugs and a hose. To test, close all outlets to the piping system with test plugs, caps, and plugs. Fill the system with water or air under pressure and look for leaks.

To water test a DWV system, you must have a minimum of 10 feet of water head. Consult all applicable codes to determine the appropriate pressure for testing. To do this, fill the vent stack completely with water. Once you have filled the DWV piping, inspect the whole piping system for leaks. This inspection is required before any part of the system can be covered. The plumbing inspector will also check for cross-connections, defective or inferior materials, and poor work.

To water test the water supply system, close all outlets and fill the system with water from the main. Inspect the system for leaks or other problems. Where codes specify water testing at pressures higher than those available from the normal water supply, you can use a hydraulic test pump (*Figure 28*) to produce the required pressure. Refer to code for the correct pressures to use.

To air test a system, the plumber or inspector fills the piping with compressed air. By using a pressure gauge at the test plug, the plumber or inspector can determine if any pressure is lost somewhere in the piping. Another useful method for identifying leaks is to brush soapsuds onto joints. If there are leaks in a joint, bubbles will form there.

WARNING!

Never use oxygen, acetylene, or other gases to pressure test a refrigerant system. Oxygen will cause an explosion if it comes into contact with refrigerant oil. Acetylene is highly flammable. The only gas other than refrigerant that should be introduced into a system is nitrogen.

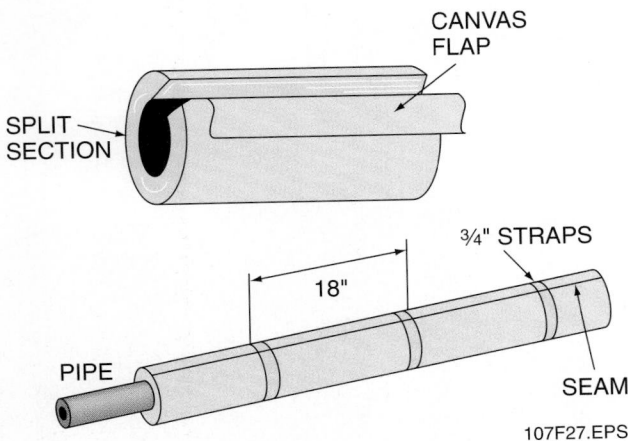

Figure 27 ◆ Insulated pipe.

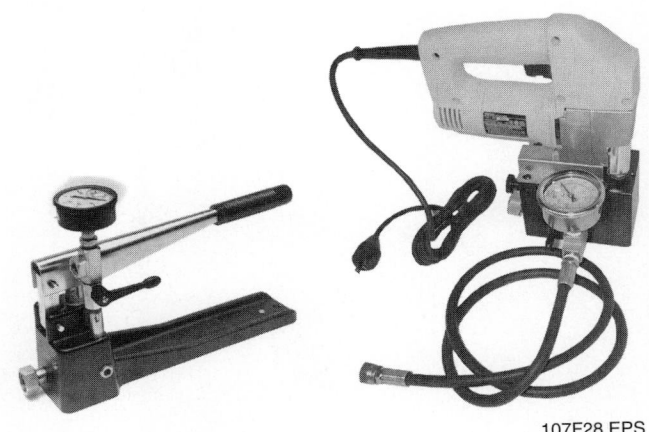

Figure 28 ◆ Hydraulic test pumps.

Sections 5.0.0–7.0.0

1. Pipe hangers are primarily used to support pipe, but they can also be used to support _____.
   a. pressure checks
   b. channel struts
   c. vibration isolators
   d. insulation support

2. _____ can be attached to structural site components to support the vertical load of the pipe.
   a. J-hooks
   b. Eye sockets
   c. Wedge anchors
   d. Pipe riser clamps

3. A connector is the part of the hanger that links the pipe itself to the structural attachment.
   a. True
   b. False

4. Each of the following is an example of an insulating material for copper pipe *except* _____.
   a. cork
   b. polyethylene foam
   c. glass fibers
   d. mineral wool

5. In addition to checking for leaks, a plumbing inspector checks for _____ when water testing the completed installation of a water system.
   a. inadequate water pressure
   b. defective materials
   c. improper insulation
   d. inadequate pipe supports and hangers

Summary

A plumber must be able to work with several kinds of pipe, including copper. This module introduced you to soft and hard copper pipe and its uses. You have also learned about the various fittings and valves that you can use to join copper pipe for water distribution systems and DWV systems. You now know that pipe must be properly supported and insulated, and you know the different kinds of hangers and supports that you can use. The requirements for pipe hangers and insulation are usually specified in the job specifications or by local building codes.

Copper can be joined in many ways, including sweat joints, compression joints, or flare joints. This module has introduced you to all three methods of joining copper pipe, and how to measure, cut, ream, and bend pipe that you are going to join.

A completed piping installation must be inspected and tested. At this point, you will not be doing the testing, but you should be familiar with the procedures. Throughout this module you have read about special safety practices to prevent injury to yourself and others or damage to equipment. Follow these practices as you go into the field and work with copper pipe and fittings.

Notes

Trade Terms Quiz

Fill in the blank with the correct trade term that you learned from your study of this module.

1. Grooved copper is a mechanical coupling material for rigidly connecting copper tubing that has been _____.

2. _____ is the approximate measurement in inches of the inside diameter of pipe for most copper pipes. The exception is the measurement for ACR tubing, which is based on the outside diameter.

3. A brass compression ring used for joining is called a(n) _____.

4. A(n) _____ is commonly used to join soft copper tubing with diameters from ¼ to 2 inches.

5. Copper tubing that is described as soft and flexible is created through a process called _____.

6. Flare fittings are made of _____ brass.

7. If pipe becomes deformed, use a(n) _____ to work the pipe back to roundness.

8. _____, or air conditioning and refrigeration systems, is one of many plumbing applications in which copper tubing is used.

9. The use of a(n) _____, a fixture in which the main line is smaller than the branch, can prevent turbulence in the system.

10. A(n) _____ is a rod attachment made of iron that is commonly in the form of a U.

11. To prevent the transfer of heat, install _____. Materials used for this purpose include cork, glass fibers, mineral wool, and polyurethane foams.

12. Annealed copper tubing that is manufactured specifically for use in air conditioning and refrigeration work is called _____.

13. A(n) _____ consists of a threaded nut that has been squeezed over a compression ring to join pipes.

14. To water test a DWV system, you must have at least 10 feet of water _____ in the system.

15. _____, or hard, copper tubing is widely used in commercial refrigeration and air conditioning systems.

16. A(n) _____ is created by soldering.

17. Using fewer fittings reduces the chance for _____, or decreases in pressure from one point to another caused by friction losses in a water system.

18. The process that occurs during soldering in which solder melts and flows into the gap between the pipe and the fitting is known as _____.

19. You should consider strength and _____, or ease of bending, when selecting copper pipe.

Trade Terms

ACR	Clevis	Flare joint	Pressure drop
ACR tubing	Compression joint	Formability	Roll grooved
Annealing	Drawn copper	Head	Sizing tool
Bullhead tee	Drop forged	Insulation	Sweat joint
Capillary action	Ferrule	Nominal size	

Charles Robbins

M. Davis and Sons
Training Director
Wilmington, Delaware

Charles Robbins was born in Cook County, Illinois. He attended Upper Darby High School in Pennsylvania and began college not knowing exactly what he wanted to do. Charles began his career in construction as a shipfitter, but a friend gave him the opportunity to get into an accelerated apprenticeship program. When shipping took a downturn, he went to work for a plumbing contractor, where he was offered his plumbing apprenticeship. The blueprint reading, shipbuilding, and basic measuring skills that he learned as a shipfitter served him well in plumbing. For the past 19 years, Charles has been with M. Davis and Sons, the oldest privately owned company in Delaware.

How did you become interested in this industry?
I enjoyed plumbing more than shipfitting. The man who hired me actually had me drive to North Philadelphia just to dig ditches. I worked all day for $15. One day he asked, "Can I call you if I need you?" I thought about it and said yes. He had a one-man shop, so when he needed help, he called me. I started learning more and more, and soon he would leave me alone to do the work on my own.

What path did you take to your current position?
I was working for a man who ran a job shop—bathrooms and additions—but he wasn't keeping me busy. I had the opportunity to go into new construction when I was still an apprentice. At a plumbing class I ran into a man that I knew when I was a Boy Scout. His father owned a plumbing shop and he invited me to come work for them. They remembered me from when I was a kid. I finished the last two years of my apprenticeship with Charles H. Fischer, Inc.

The normal apprenticeship was a five-year program. I finished in four years because of the accelerated program at Sun Ship. I had taken welding and blueprint reading and they credited that amount of time to me. I was a journeyman plumber when I left Charles H. Fischer for M. Davis and Sons.

After I began with M. Davis and Sons, I took a year off from school. I got my master plumber's license and began a heating, ventilating, and air conditioning (HVAC) apprenticeship program. I got my HVAC journeyman's card and then my master's

HVAC license. Today I hold master plumber's licenses in New Jersey; Delaware; Maryland; New Castle County, Delaware (upper portion of the state); the city of Wilmington; and the city of Philadelphia. In Delaware I have an HVAC master's license, in Maryland I hold a limited electrical license, and I am a certified member of the Refrigeration, Service, and Engineers Society.

M. Davis and Sons is an industrial contractor. I do a lot of work in the petrochemical arena. I found that very interesting, so that's where I ended up. I started working in the field in industrial maintenance. I took on more responsibility and started estimating. Then I became a foreman and took the route to superintendent and division manager. I oversaw $9 million worth of work and 110 employees.

I went from division manager to training when the CEO asked me what I wanted to do. I always enjoyed training. I was always sent to classes, because I would ask, "Can I take this class, or that class?" I thought that I might as well start training other people. I have enough gray hair! And it's a pleasant opportunity to help others.

What does it take to be successful in your trade?
I started as a plunger-carrying plumber and worked my way up. I took on more responsibility. It takes hard work. Always look for opportunities to learn and stay on top. Be open-minded to new things. I find it exciting to learn something new. I wanted to see new things and see how they work.

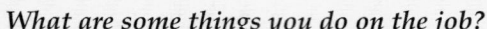

What are some things you do on the job?
I conduct small-group training. On some assignments I go to the job site and grab the new guys. I show them basic knot tying, how to pick out the right tips, how to use the right number of teeth for the specific material they are cutting, how to select the right glue. I conduct on-site training and classroom training. I am really hands-on. I teach them how to use equipment and machinery. Adult learners need instant gratification. My trainees feel good about learning new methods.

I also work with people who are switching careers. I focus on an area they are interested in and help them. I do a lot with high schools, talking about careers in construction. A lot of schools invite me to be on panels.

I am active in the Plumbing-Heating-Cooling Contractors–National Association (PHCC–NA). I am past president of PHCC of Delaware, past president of Refrigeration Service Engineers Society (RSES), and treasurer of PHCC–NA. I am also active in Associated Building Contractors and serve as the educational chair. As a master trainer for NCCER, I am currently teaching the second-year apprenticeship program in Delaware.

What is the most interesting aspect of your profession? What makes your trade stand out from others?
A plumber protects the health of the nation. A house isn't livable without plumbing. I truly believe that.

I have worked in all aspects of the trade. It's great to go from start to finish, and it is exciting to see something come out—to see the end product. A tradesperson wants to show off his or her accomplishments. Take a bridge builder, for example; some people in the trade are just waiting on quitting time, while another guy will say, "I'm doing this so my family can cross the water and go to the beach."

What do you like most—and least—about your job?
I most like interaction with other people and seeing something completed. I like to see something come together, something get built, the end product completed professionally and in an appealing way.

I least like the people who are along for the ride, but who do not have an interest in what they are doing. They say, "That's good enough," but that's not the right attitude.

What would you say to someone entering the trade today?
Training and education is forever. It is never ending.

What can an apprentice expect to earn in his/her first years on the job in your area? What can he/she expect to earn after 10-plus years in the industry?
Rates are geared toward geography. In this area, apprentices can start at $10.00 per hour. After completing their apprenticeship, they can earn around $18.00 to $19.00 per hour. After about 10 years, they earn around $25.00 per hour.

Trade Terms
Introduced in This Module

ACR: Air conditioning and refrigeration system.

ACR tubing: Annealed copper tubing that is manufactured specifically for use in air conditioning and refrigeration work.

Annealing: Heating a material and slowly cooling it to relieve internal stress. This process reduces brittleness and increases toughness.

Bullhead tee: A tee fitting used on a branch that is longer than the main line, or that has one outlet larger than the run openings.

Capillary action: The process during soldering in which the molten solder flows into the narrow gap between the pipe and the joint, regardless of whether the solder is flowing up, down, or horizontally.

Clevis: An iron, or link in a chain, bent into the form of a horseshoe, stirrup, or letter U, with holes in the ends to receive a bolt or pin.

Compression joint: A method of connection in which tightening a threaded nut squeezes a compression ring to seal the joint.

Drawn copper: Tubing produced by pulling the tube through dies to reduce its diameter. The drawing process hardens the copper and makes it very rigid.

Drop forged: A characteristic of a product made when heated metal is pounded or shaped between dies with a drop hammer or press.

Ferrule: A brass compression ring used for joining.

Flare joint: A fitting in which one end of each tube to be joined is flared outward using a special tool. The flared tube ends mate with the threaded flare fitting and are secured to the fitting with flare nuts.

Formability: The ease with which a material can bend.

Head: The height of a water column, measured in feet. One foot of head is equal to 0.433 pounds per square inch (psi).

Insulation: A substance that retards the flow of heat.

Nominal size: Approximate measurement in inches of the inside diameter (ID) of pipe for most copper pipes. However, nominal size of ACR tubing is based on the outside diameter (OD).

Pressure drop: A decrease in pressure from one point to another caused by friction losses in a fluid system, such as a water system.

Roll grooved: A type of copper piping that is compatible with grooved copper fittings.

Sizing tool: A tool consisting of a plug and a sizing ring that is used to reshape a deformed pipe back to roundness.

Sweat joint: A pipe joint made by applying solder to the joint and heating it until it flows into the joint.

Additional Resources and References

ADDITIONAL RESOURCES

This module is intended to present thorough resources for task training. The following reference works are suggested for further study. These are optional materials for continued education rather than for task training.

The Copper Tube Handbook, 1995. New York: Copper Development Association.

Engineering Lab, Inc. website, www.engineeringlab.com, *Flux, Solder, and Cleaning* (Online Course), www.engineeringlab.com/fluxsolder.html?source=goto, reviewed July 2003.

Pipefitter's Handbook, Third Edition, 1967. Forrest R. Lindsey. New York: Industrial Press, Inc.

Plumbing and Mechanical website, www.pmmag.com, *Throw Away Your Torches,* Julius Ballanco, P.E. http://www.pmmag.com/pm/cda/articleinformation/features/bnp__features__item/0,,5020,00+en-uss_01dbc.html, publication date: June 2000, reviewed July 2003.

REFERENCES

Canadian Copper and Brass Development Association website, www.ccbda.org, *Copper Tube and Fittings Manual, Publication No. 28E,* www.ccbda.org/publications/pub28E/28e-publicationp1.html, reviewed July 2003.

The Copper Tube Handbook, 1995. New York: Copper Development Association.

Efficient Building Design, Volume 3: Water and Plumbing, 2000. Ifte Choudhury and J. Trost. Upper Saddle River, NJ: Prentice Hall.

NSF International website, www.nsf.org, *Certification Underway for Press-Connect Fittings,* www.nsf.org/newsletters/plumbing01-1/page4.html, reviewed July 2003.

NCCER makes every effort to keep these textbooks up-to-date and free of technical errors. We appreciate your help in this process. If you have an idea for improving this textbook, or if you find an error, a typographical mistake, or an inaccuracy in NCCER's Contren® textbooks, please write us, using this form or a photocopy. Be sure to include the exact module number, page number, a detailed description, and the correction, if applicable. Your input will be brought to the attention of the Technical Review Committee. Thank you for your assistance.

Instructors – If you found that additional materials were necessary in order to teach this module effectively, please let us know so that we may include them in the Equipment/Materials list in the Annotated Instructor's Guide.

Write: Product Development and Revision
National Center for Construction Education and Research
3600 NW 43rd St, Bldg G, Gainesville, FL 32606

Fax: 352-334-0932

E-mail: curriculum@nccer.org

Craft _____ Module Name _____

Copyright Date _____ Module Number _____ Page Number(s) _____

Description _____

_____

_____

_____

(Optional) Correction _____

_____

_____

(Optional) Your Name and Address _____

_____

_____

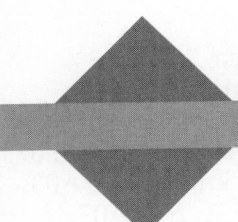

Photo Credits

Module 68101-09

John Hoerlein, 202P0201, 202P0202, 202F02, 202F04, 202F22
Sokkia Corporation, 202F21, 202F25
David White, 202F20, 202F23

Module 68102-09

Associated Builders and Contractors, Inc., module divider
Benner-Nawman, Inc., 108SA05
Portland Cement Association, 108SA01, 108SA03, 108SA04, 108F03
Rosamond Gifford Zoo, 108SA02, 108SA08, 108SA09
Topaz Publications, Inc., module overview photo, 108SA06, 108SA07

Module 68103-09

Baker Concrete Construction, Inc., 305F10, 305F24, 305F25, 305F28, 305F30, 305F35, 305F36, 305F39
Bon Tool Company, 305SA02, 305SA03, 305F38
Chapin International, 305F44
Photo courtesy of constructionphotographs.com, 305F07, 305F16
CS Unitec, www.csunitec.com, 305F31, 305F45 (middle)
Granite City Tools, 305F17
Metal Forms Corporation, 305F19
MK Diamond, 305F32, 305F45 (left, right)
Multiquip Inc., 305F20 (roller screed)
Portland Cement Association, 305F05, 305F06, 305F14, 305F15
REED Concrete Pumps, 305SA01, 305F11
Soff-Cut International, 305F33
Somero Enterprises, 305F21
Superchute, 305F12 (photo)
Topaz Publications, Inc., 305F08, 305F26
Wacker Corporation, 305F20 (steel truss screed)

Module 68104-09

Copyright © 2004, Corel Corporation, 101F01, 101SA01
Topaz Publications, Inc., 101SA02–101SA04, 101F15, 101F16, 101F21, 101F24, 101F26, 101F27, 101F36
Hanson Brick, 101F03–101F05, 101F07
Used with permission of the Brick Industry Association, Reston, Virginia, www.gobrick.com, 101F06, 101F08, 101F10, 101F11
Portland Cement Association, 101F12–101F14, 101T01
Associated General Contractors, 101F18, 101F33
Easi-Set Industries, 101SA05
Bon Tool Company, 101F19
www.freefoto.com, 101F28–101F31
Coleman Cable, Inc., 101SA06
Construction Safety Association of Ontario, 101F40A
Fall Protection Systems, Inc., 101F40B, 101F40C
Protecta International, Inc., 101F42–101F48
Sellick Equipment Limited, 101F49, 101F51
Manitou North America, Inc., 101F52

Module 68105-09

Concrete Masonry Association, 105F01, 105F03, 105F33
Topaz Publications, Inc., 105F02, 105SA03, 105F10, 105SA04–105SA06, 105F13, 105F15, 105F22, 105F39, 105F42, 105F44, 105F45, 105F47, 105F48, 105SA09, 105F54, 105F55, 105F59, 105SA10, 105F62–105F64
Used with permission of the Brick Industry Association, Reston, Virginia, www.gobrick.com, 105F04, 105F05, 105F12, 105F16, 105F26, 105F27, 105F29–105F32, 105F34, 105F38, 105F67, 105T01
Portland Cement Association, 105SA01,105F09, 105F11, 105F20, 105F24, 105SA08, 105F49–105F53, 105F61, 105F65, 105F66

ChemGrout, Inc., 105SA02
ICD Corporation, www.selecticd.com, 105F06
Air Vol Block, Inc., 105F08
Indiana Limestone Institute of America, Inc.,
105F19
Granite City Tool Company, 105F40, 105F41
Bon Tool Company, 105F43

Module 68106-09

APA—The Engineered Wood Association, Table 1
Associated Builders and Contractors, Inc.,
module divider
Topaz Publications, Inc., module overview
photo, 105SA02, 105SA04, 105SA05, 105SA06,
105SA07
Trus Joist, A Weyerhaeuser Business, 105SA03,
105F20

Module 68107-09

Associated Builders and Contractors, Inc.,
module divider
Topaz Publications, Inc., module overview
photo, 106SA01, 106SA02, 106SA03, 106F31,
106F32

Module 68108-09

SkillsUSA, module divider
Southern Forest Products Association, 107F30
The Stanley Works, 107F19
Topaz Publications, Inc., module overview
photo, 107SA01, 107SA02, 107SA03, 107F27,
107F28, 107SA05, 107F31, 107F32
Reprinted with permission from the Wood Truss
Council of America (WTCA). For more
information, visit www.woodtruss.com,
107F24, 107SA04

Module 68109-09

Alum-a-Pole Corporation, 202SA05
ATAS International, Inc., 202SA11 (top)
Cedar Shake & Shingle Bureau, 202F05, 202F90,
Copyright © 2009, All Rights Reserved.
CertainTeed Corporation, 202SA01, 202F02,
202F40. Copyright © 2006. Used with
permission.
Cornell Corporation, 202SA14
DaVinci Roofscapes LLC, 202SA02, 202SA09.
For more information about DaVinci's family
of products, call 1-800-328-4624 or visit
www.davinciroofscapes.com.
DBI/SALA & Protecta, 202F24
Follansbee Steel, 202F10
Hilltop Slate, Inc., 202F109

Johns Manville, 202F13
Maruhachi Ceramics of America, 202SA13
Reimann and Georger Corporation, 202SA06
The Stanley Works, 202F19 (hook knife)
Topaz Publications, Inc., 202F01, 202F04, 202F07,
202F12, 202F19 (except hook knife), 202SA03,
202F20, 202F44, 202F48, 202F51, 202SA07,
202F66, 202F69 (photo), 202SA10, 202SA11
(bottom), 202SA12, 202F130–202F137, 202F140,
202F145, 202F148, 202F149 (photo), 202F150
(photos), 202SA15

Module 68110-09

Alum-a-Pole Corporation, 204SA03
Cedar Valley Shingle Systems, 204SA13
CertainTeed Corporation, 204SA15. Copyright ©
2006. Used with permission.
Cummins Industrial Tools, 204SA06
James Hardie Building Products, 204SA14,
204F45, 204F46
Milwaukee Electric Tool Corp., 204F44 (top)
Tapco Integrated Tool Systems, 204F50
Topaz Publications, Inc., 204SA01, 204SA02,
204SA05, 204F26, 204SA08, 204SA09, 204SA11,
204F57, 204F59
Zircon Corporation, 204F56

Module 68111-09

Classic Stair Works, Ltd., module overview
photo, 110SA01
SkillsUSA, module divider
Topaz Publications, Inc., 110F19, 110SA02

Module 68112-09

TIC, The Industrial Company, module divider
Topaz Publications, Inc., 102F02, 102F05,
102F06A
Panduit Corp., 102F06B
Mike Powers, 102F08–102F10, 102F14,
102F16–102F20, 102SA01, 102SA03–102SA10,
102SA12–102SA14
The Stanley Works, 102F12A, 102F13
Greenlee Textron, Inc., a subsidiary of Textron
Inc., 102F12B, 102F12D
RIDGID®, 102F12C
Tim Ely, 102SA02
Datum Filing Systems, Inc., 102SA11
Occupational Safety and Health Administration,
102T01, 102T02

Module 68113-09

Associated Builders and Contractors, Inc., module divider

Veronica Westfall, module overview

John Traister, 111F02, 111F05–111F08, 111F10–111F13, 111F19–111F23, 111F25–111F29, 111F35, 111F43, 111F44, 111F53, 111F55

Topaz Publications, Inc., 111F16B, 111F37B, 111SA01A, 111SA02, 111SA04, 111SA05

Tim Dean, 111F40, 111SA01B

Greenlee Textron, Inc., a subsidiary of Textron Inc., 111SA03

Module 68114-09

Topaz Publications, Inc., 101SA01, 101SA03, 101SA04, 101SA05, 101SA06

Module 68115-09

Charlotte Pipe and Foundry Company, 112F24, 112F28 (top)

Fernco®, Inc., 112F23

Genova Products, Inc., 112F03 (top), 112F28 (bottom), 112F37

Ivey Mechanical Company, module divider, TOC art

The Stanley Works, 112SA01

Watts Regulator Company, 112F03 (bottom)

Zurn Plumbing Products Group, 112F05

Module 68116-09

Adjustable Clamp Company, 106F22 (C-clamp)

Charlotte Pipe and Foundry Company, 106F02, 106F03, 106F04, 106F09 (sanitary tee, wye), 106F16

Consumer Plumbing Recovery Center, Inc., 106F07

Genova Products, Inc., 106F09 (closet bend)

Hudson Extrusions, Inc., 106F05

International Code Council, Inc., Table 4, Table S-1. International Plumbing Code 2003, Falls Church, VA: International Code Council, Inc. Reproduced with permission.

Ivey Mechanical Company, module divider, TOC art

LASCO Fittings, Inc., 106F08 (union, elbow 90, elbow 45, tee, coupling, cap, plug), 106F09 (coupling)

Plastic Pipe and Fittings Association, Table 1

Reed Manufacturing Company, 106F11

Courtesy of Ridge Tool Company, 106F10

Sioux Chief Manufacturing Co., Inc., 106F21, 106F22 (suspension, I-beam, beam clamps), 106F23

Tempil, Inc., An Illinois Tool Works, Inc., Company, 106F20 (temperature indicator stick)

Uponor Wirsbo AB, 106F08 (manifold)

Vanguard Piping Systems, Inc., 106F18 (crimpsert)

Zurn Plumbing Products Group, 106F06

Module 68117-09

Adobe® Image Library, module divider

BernzOmatic, 107F13 (soldering torch)

LeDuc & Dexter Plumbing, TOC art

Mueller/B&K Industries, Inc., 107F03 (stop and waste valve)

NIBCO, 107F21 (pipe strap)

Oatey Co., 107F13 (flux brush and solder paste, wire solder)

Courtesy of Ridge Tool Company, 107F08, 107F09, 107F12, 107F13 (fitting brush), 107F16

Sioux Chief Manufacturing Co., Inc., 107F21 (pipe hooks, J-hooks, locking tube strap, plumber's tape)

Becky Swinehart, 107F02 (top)

T-Drill Industries, Inc., 107F05, 107F06

Victaulic Company of America, 107F02 (bottom)

Watts Regulator Company, 107F03 (gate valve, globe valve, check valve)

Weil-McLain, 107F04

Wheeler Rex Pipe Tools, 107F28

Glossary

ABS (acrylonitrile-butadiene-styrene): Plastic pipe and fittings used extensively in drain, waste, and vent (DWV) systems.

ACI: American Concrete Institute.

ACR: Air conditioning and refrigeration system.

ACR tubing: Annealed copper tubing that is manufactured specifically for use in air conditioning and refrigeration work.

Adapter: A fitting that joins pipes of different sizes or materials, such as copper and galvanized pipe or cast-iron and plastic pipe.

Admixture: A chemical or mineral other than water, cement, or aggregate added to mortar immediately before or during mixing to change its setting time or curing time, to reduce water requirements, or to change the overall properties of the mortar.

Adobe: Sun-dried, molded clay brick.

Aggregates: Materials used as filler in concrete; may include mixtures of sand, gravel, crushed stone, crushed gravel, or blast-furnace slag.

Alkaline: Bitter, slippery, or caustic.

American Society for Testing and Materials (ASTM) International: The publisher of masonry standards.

Annealing: Heating a material and slowly cooling it to relieve internal stress. This process reduces brittleness and increases toughness.

Appliance: Equipment designed for a particular purpose (for example, using electricity to produce heat, light, or mechanical motion). Appliances are usually self-contained, are generally available for applications other than industrial use, and are normally produced in standard sizes or types.

Ashlar: A squared or rectangular cut stone masonry unit; or, a flat-faced surface having sawed or dressed bed and joint surfaces.

Asphalt roofing cement: An adhesive that is used to seal down the free tabs of strip shingles. This plastic asphalt cement is mainly used in open valley construction and other flashing areas where necessary for protection against the weather.

Autoclave: A pressurized, steam-heated tank used for sterilizing and cooking.

Axle-steel: Deformed reinforcing bars that are rolled from carbon-steel axles used on railroad cars.

Back pressure: A condition which may occur in the potable water distribution system, whereby a higher pressure than the supply pressure is created causing a reversal of flow into the potable water piping. Also referred to as back-pressure backflow.

Backpressure backflow: Another term for back pressure.

Backsight (BS): A reading taken on a leveling rod held on a point of known elevation to determine the height of the leveling instrument.

Baluster: A supporting column or member; a support for a railing, particularly one of the upright columns of a balustrade.

Balustrade: A stair rail assembly consisting of a handrail, balusters, and posts.

Barge rafter: A gable end roof member that extends beyond the gable to support a decorative end piece. Also known as a fly rafter.

Base flashing: The protective sealing material placed next to areas vulnerable to leaks, such as chimneys.

Bell-and-spigot pipe: Pipe that has a bell, or enlargement, also called a hub, at one end of the pipe and a spigot, or smooth end, at the other end. The bell and spigot of two different pipes slide together to form a joint. Also called hub-and-spigot pipe.

Bleed: A condition in concrete in which the solids settle and the water moves to the top.

Blocking: A wood block used as a filler piece and a support between framing members.

Board-and-batten: A type of vertical siding consisting of wide boards with the joint covered by narrow strips known as battens.

Bonding bushing: A special conduit bushing equipped with a conductor terminal to take a bonding jumper. It also has a screw or other sharp device to bite into the enclosure wall to bond the conduit to the enclosure without a jumper when there are no concentric knockouts left in the wall of the enclosure.

Bonding jumper: A bare or green insulated conductor used to ensure the required electrical conductivity between metal parts required to be electrically connected. Bonding jumpers are frequently used from a bonding bushing to the service-equipment enclosure to provide a path around concentric knockouts in an enclosure wall, and they may also be used to bond one raceway to another.

Brace: A diagonal supporting member used to reinforce a form against the weight of the concrete.

Branch circuit: The portion of a wiring system extending beyond the final overcurrent device protecting a circuit.

Branch interval: A distance along a soil or waste stack corresponding, in general, to a story height, but in no case less than 8 feet within which the horizontal branches from one floor or story of a building are connected to the stack.

Breaking the tape: Making measurements using a portion of a full tape's length in a series of steps.

Brown coat: A coat of plaster with a rough face on which a finish coat will be placed.

Building paper: A heavy paper used for construction work. It assists in weatherproofing the walls and prevents wind infiltration. Building paper is made of various materials and is not a vapor barrier.

Building sewer: That part of the drainage system which extends from the end of the building drain and conveys its discharge to a public sewer, private sewer, individual sewage-disposal system, or other point of disposal.

Bullhead tee: A tee fitting used on a branch that is longer than the main line, or that has one outlet larger than the run openings.

Bundle: A package containing a specified number of shingles or shakes. The number is related to square foot coverage and varies with the product.

Butter: Apply mortar to the end of a masonry unit.

Cap flashing: The protective sealing material that overlaps the base and is embedded in the mortar joints of vulnerable areas of a roof, such as a chimney.

Capillary action: The process during soldering in which the molten solder flows into the narrow gap between the pipe and the joint, regardless of whether the solder is flowing up, down, or horizontally.

Capillary attraction: The tendency of water to be drawn by porous material. If a piece of paper or string is stuck in a pipe, the water is drawn out of the trap and the seal is broken.

Capital: The top part of an architectural column.

Cellular core wall: Plastic pipe wall that is low-density, lightweight plastic containing entrained (trapped) air.

Chamfer: To bevel the edge of construction material to a 45-degree angle.

Chemically inert: Does not react with other chemicals.

Chiller: A high-volume cooling unit. The chiller acts as an evaporator.

Chlorofluorocarbon (CFC) refrigerant: A class of refrigerants that contains chlorine, fluorine, and carbon. CFC refrigerants have a very adverse effect on the environment.

Cleanout: An access point to all parts of the drainage system for the removal of blockages.

Clevis: An iron, or link in a chain, bent into the form of a horseshoe, stirrup, or letter U, with holes in the ends to receive a bolt or pin.

Closed stairway: A stairway that has solid walls on each side.

Closure unit: The last brick or block to fill a course.

Compression collar: A piece of hardware that uses compression force to connect sections of polyethylene piping.

Compression joint: A method of connection in which tightening a threaded nut squeezes a compression ring to seal the joint.

Compressor: In a refrigeration system, the mechanical device that converts low-pressure, low-temperature refrigerant gas into high-temperature, high-pressure refrigerant gas.

Concrete masonry unit (CMU): A hollow or solid block made from portland cement and aggregates.

Condenser: A heat exchanger that transfers heat from the refrigerant flowing inside it to the air or water flowing over it.

Consolidating concrete: Working freshly placed concrete so that each layer is compacted with the layer below and voids caused by water or air pockets are eliminated.

Control points: A series of horizontal and/or vertical points established in the field to serve as a known framework for all points on the site.

Cornice: The horizontal projection crowning the wall of a building; the construction under the eaves where the roof and side walls meet.

Course: One row of brick, block, or siding as it is placed in the wall; a row or horizontal layer of masonry units.

CPVC (chlorinated polyvinyl chloride): Plastic pipe and fittings used extensively in hot and cold water distribution systems.

Cripple stud: In wall framing, a short framing stud that fills the space between a header and a top plate or between the sill and the soleplate.

Crosshairs: A set of lines, typically horizontal and vertical, placed in a telescope used for sighting purposes.

Crowding the line: A person touching the mason's line, or a masonry unit too close to the line.

Crown: The high point of the crooked edge of a framing member.

Crown weir/trap weir: Discharge overflow of the trap outlet. Also referred to as weir, trap, or crown.

Cube: A strapped bundle of approximately 500 standard bricks, or 90 standard blocks, usually palletized. The number of units in a cube will vary according to the manufacturer.

Cured concrete: Concrete that has hardened and gained its structural strength.

Cut: Removing soil or rock on site to achieve a required elevation.

Dead load: The weight of permanent, stationary construction and equipment included in a building.

Differential leveling: A method of leveling used to determine the difference in elevation between two points.

Double ¼ bend: A fitting used to collect and combine the flow from two opposite runs into a single run of pipe.

Double top plate: A plate made of two members to provide better stiffening of a wall. It is also used for connecting splices, corners, and partitions that are at right angles (perpendicular) to the wall.

Double trapping: A situation in which one trap is attached to another, creating negative pressure that will stop the flow of drainage.

Double-insulated/ungrounded tool: An electrical tool that is constructed so that the case is insulated from electrical energy. The case is made of a nonconductive material.

Drainage fitting: Any of a variety of fittings used in the DWV system to remove waste from a building.

Drawn copper: Tubing produced by pulling the tube through dies to reduce its diameter. The drawing process hardens the copper and makes it very rigid.

Drop forged: A characteristic of a product made when heated metal is pounded or shaped between dies with a drop hammer or press.

Dry bond: Laying out masonry units without mortar to establish spacing.

Drying-in: Applying sheathing, windows, and exterior doors to a framed building.

DWV system: An acronym for "drain-waste-vent" referring to the combined sanitary drainage and venting systems. This term is technically equivalent to "soil-waste-vent" (SWV).

Earthwork: All construction operations connected with excavating (cutting) or filling earth.

Eave: The lower part of a roof, which projects over the side wall.

Elastomeric: A material having the properties of excellent flexibility and elongation; rubberized, made of an elastic substance such as a polyvinyl elastomer.

Elevation: The height above an established reference point, such as a grade reference point on a construction drawing.

Evaporation: Loss of water, especially in a drainage system, into the atmosphere.

Evaporator: A heat exchanger that transfers heat from the air flowing over it to the cooler refrigerant flowing through it.

Expansion device: Also known as the liquid metering device or metering device. Provides a pressure drop that converts the high-temperature, high-pressure liquid refrigerant from the condenser into the low-temperature, low-pressure liquid refrigerant entering the evaporator.

Exposure: The distance (in inches) between the exposed edges of overlapping shingles.

Facing: That part of a masonry unit or wall that shows after construction; the finished side of a masonry unit.

Fall: The amount of slope given to horizontal runs of pipe.

Fascia: The exterior finish member of a cornice on which the rain gutter is usually hung.

False fascia: The board that is attached to the tails of the rafters to straighten and space the rafters and provide a nailer for the fascia. Also called sub fascia and rough fascia.

Feeder: A circuit, such as conductors in conduit or a cable run, that carries current from the service equipment to a subpanel or a branch circuit panel or to some point in the wiring system.

Felt paper: An asphalt-impregnated paper.

Ferrule: A brass compression ring used for joining.

Fibrillation: Very rapid irregular contractions of the muscle fibers of the heart that result in the heartbeat and pulse going out of rhythm with each other.

Field notes: A permanent record of field measurement data and related information.

Fill: Adding soil or rock on site to achieve a required elevation.

Finish coat: The final coat of plaster or paint.

Firestop: An approved material used to fill air passages in a frame to retard the spread of fire.

Fixture drain: The drain from the trap of a fixture to the junction of that drain with any other drain pipe.

Flare joint: A fitting in which one end of each tube to be joined is flared outward using a special tool. The flared tube ends mate with the threaded flare fitting and are secured to the fitting with flare nuts.

Flatwork: Work connected with concrete slabs used for walks, driveways, patios, and floors.

Footing: The base of a foundation system for a wall, column, and chimney. It bears directly on the undisturbed soil and is made wider than the object it supports to distribute the weight over a greater area. Also, the base for a masonry unit wall, or concrete foundation, that distributes the weight of the structural member resting on it.

Foresight (FS): A reading taken on a leveling rod held on a point in order to determine a new elevation.

Formability: The ease with which a material can bend.

Forms: Wood or metal structures built to contain plastic concrete until it hardens.

Foundation: The supporting portion of a structure, including the footings.

Frieze board: A horizontal finish member connecting the top of the sidewall, usually abutting the soffit. Its bottom edge usually serves as a termination point for various types of siding materials.

Furring strip: Narrow wood strips nailed to a wall or ceiling as a nailing base for finish material.

Furrowing: Making an indentation with a trowel point along the center of the mortar bed joint.

Fusion fitting: A fitting with a butt that has the same outside diameter and inside diameter as the pipe. It is usually joined to a pipe by heat.

Gable: The triangular wall enclosed by the sloping ends of a ridged roof.

Gable roof: A roof with two slopes that meet at a center ridge.

Geometrical stair: A winding stairway built around a well. Examples include circular, elliptical, and spiral stairs.

Grade: The slope of a horizontal run of pipe. Also referred to as slope (percent of grade).

Green concrete: Concrete that has hardened but has not yet gained its structural strength.

Ground fault circuit interrupter (GFCI): A protective device that functions to de-energize a circuit or portion thereof within an established period of time when a current to ground exceeds some predetermined value. This value is less than that required to operate the overcurrent protective device of the supply circuit.

Grounded tool: An electrical tool with a three-prong plug at the end of its power cord or some other means to ensure that stray current travels to ground without passing through the body of the user. The ground plug is bonded to the conductive frame of the tool.

Grout: A mixture of portland cement, lime, and water, with or without fine aggregate, with a high enough water content that it can be poured into spaces between masonry units and voids in a wall.

Grout: A thin mortar.

Guardrail: A rail secured to uprights and erected along the exposed sides and ends of platform stairs, etc.

Handrail: A member supported on brackets from a wall or partition to furnish a handhold.

Head: The height of a water column, measured in feet. One foot of head is equal to 0.433 pounds per square inch (psi).

Head joint: A vertical joint between two masonry units.

Head lap: The distance between the top of the bottom shingle and the bottom edge of the one covering it.

Header: A horizontal structural member that supports the load over a window or door.

Header joist: A framing member used in platform framing into which the common joists are fitted, forming the box sill. Header joists are also used to support the free ends of joists when framing openings in a floor.

Headroom: The vertical and clear space in height between a stair tread and the ceiling or stairs above.

Heat transfer: The transfer of heat from a warmer substance to a cooler substance.

Height of instrument (HI): The elevation of the line of sight of the telescope relative to a known elevation. It is determined by adding the backsight elevation to the known elevation.

Hip roof: A roof with four sides or slopes running toward the center.

Housed stringer: A stair stringer with horizontal and vertical grooves cut (mortised) on the inside to receive the ends of the risers and treads. Wedges covered with glue are often used to hold the risers and treads in place in the grooves.

HVAC plan: Added to the mechanical plans for complex jobs that require separate heating, ventilating, and air-conditioning systems.

Hydration: The catalytic action water has in transforming the chemicals in portland cement into a hard solid. The water interacts with the chemicals to form calcium silicate hydrate gel.

Hydraulic gradient: The level line in which water tends to flow in a pipe.

Hydrochlorofluorocarbon (HCFC) refrigerant: A class of refrigerants that contains hydrogen, chlorine, fluorine, and carbon. Although not as high in chlorine as chlorofluorocarbon (CFC) refrigerants, HCFCs are still considered hazardous to the environment.

Hydronic: A system that heats and cools by circulating water or steam through a closed piping system.

Hydrostatically pressure-test: To fill a pipe with water and bleed all air out from the highest and farthest points in the run.

Hygroscopic: The tendency of a substance to absorb moisture.

Inside diameter (ID): The distance between the inner walls of a pipe; the standard measure of piping used in heating and plumbing.

Insulation: A substance that retards the flow of heat.

Interceptor: A device designed and installed so as to separate and retain deleterious, hazardous, or undesirable matter from normal waste while permitting normal sewage or liquid wastes to discharge into the drainage system by gravity.

Interference fit: Fit that tightens as the pipe is pushed into the socket.

International Building Code: A series of model construction codes. These codes set standards that apply across the country. This is an ongoing process led by the International Code Council (ICC).

Inverted wye: A fitting used to join the upper end of the vent to the top of the soil and waste stacks. It looks like an upside-down letter Y.

Jamb: The top (head jamb) and side members of a door or window frame that come into contact with the door or window.

Joints: The area between each brick or block that is filled with mortar.

Joist hanger: A metal stirrup secured to the face of a structural member, such as a girder, to support and align the ends of joists flush with the member.

Kip: An informal unit of force that equals one thousand (kilo) pounds.

Landing: A horizontal area at the end of a flight of stairs or between two flights of stairs.

Ledger: A board to which the lookouts are attached and which is placed against the outside wall of the structure. It is also used as a nailing edge for the soffit material.

Let-in: Any type of notch in a stud, joist, etc., which holds another piece. The item that is supported by the notch is said to be let in.

Lintel: The support beam over an opening such as a window or door. Also called a header.

Live load: The total of all moving and variable loads that may be placed upon a building.

Load center: A type of panelboard that is normally located at the service entrance of a residential installation. It usually contains the main disconnect.

Long sweep ¼ bend: A bend used at the base of DWV stacks because the longer radius of its design greatly decreases flow resistance and back pressure. Use is regulated by code.

Lookout: A member used to support the overhanging portion of a roof or frame an overhang.

Louver: A slatted opening used for ventilation, usually in a gable end or a soffit.

Mason: A person who assembles masonry units by hand, using mortar, dry stacking, or mechanical connectors.

Masonry unit: Any building block made of brick, cement, ashlar, clay, adobe, rubble, glass, tile, or any other material, that can be assembled into a structural unit.

Mechanical refrigeration: The use of machinery to provide cooling.

Metal-clad (Type MC) cable: A type of cable with a metal jacket that is popular for use in residential and small commercial wiring systems. In general, it may be used for both exposed and concealed work in normally dry locations. It may also be used as service-entrance cable and in other locations as permitted by the *NEC*®.

Monolithic slab (monolithic pour): Concrete placed in forms in a continuous pour without construction joints.

Mortar: A mixture of portland cement, lime, fine aggregate, and water, plastic or stiff enough to hold its shape between masonry units.

Newel post: An upright post supporting the handrail at the top and bottom of a stairway or at the turn of a landing; also, the main post about which a circular staircase winds or a stone column carrying the inner ends of the treads of a spiral stone staircase.

Nominal size: Approximate measurement, in inches, of the inside diameter (ID) of pipe for most copper pipes. However, nominal size of ACR tubing is based on the outside diameter (OD).

Nonmetallic-sheathed (Type NM) cable: A type of cable that is popular for use in residential and small commercial wiring systems. In general, it may be used for both exposed and concealed work in normally dry locations. See *Romex*®.

Nonstructural: Not bearing weight other than its own.

Nosing: The portion of the stair tread that extends beyond the face of the riser.

Noxious: Harmful to health.

Open stairway: A stairway that is open on at least one side.

Outside diameter (OD): The distance between the outer walls of a pipe.

Overhang: The part that extends beyond the building line. The amount of overhang is always given as a projection from the building line on a horizontal plane.

Parallax: The apparent movement of the crosshairs in a surveying instrument caused by movement of the eyes.

Parapet: A low wall or railing.

Pascal: A metric measurement of pressure.

PB (polybutylene): Plastic piping that was formerly used for plumbing pipe; it is no longer used but is still found in some residences.

PE (polyethylene): Flexible plastic pipe, tubing, and fittings, usually used for water distribution, that do not deteriorate when exposed to sunlight.

Peg test: A procedure used to check for an out-of-adjustment bubble vial on levels and other instruments.

PEX (cross-linked polyethylene): Tubing and fittings made with heat and high pressure that resist high temperatures, pressure, and chemicals.

Pier: A column of masonry used to support other structural members, typically girders or beams.

Pilaster: A square or rectangular pillar projecting from a wall.

Piles: Column-like structural members that penetrate through unstable, nonbearing soil to lower levels of loadbearing soil. They provide support for grade beams or columns that carry the structural load of a building.

Pipe riser clamp: A vertical extension of pipe hanger that provides support for pipe and tubing.

Pipe scale: Flaky material resulting from corrosion of metals, especially iron or steel. Also, a heavy oxide coating on copper or copper alloys resulting from exposure to high temperatures and an oxidizer.

Pitch: The ratio of the rise to the span indicated as a fraction. For example, a roof with a 6' rise and a 24' span will have a ¼ pitch.

Pitch board: A board that serves as a pattern for marking cuts for stairs. The shortest side is the height of the riser cut, and the next longer side is the width of the tread. This is used mainly when there is great repetition such as in production housing.

Plancier: The same as a soffit, but the member is usually fastened to the underside of a rafter rather than the lookout.

Plastic concrete: Concrete when it is first mixed and is in a semiliquid and moldable state.

Plyform: American Plywood Association's tradename for a reusable material for constructing concrete forms.

Polychlorinated biphenyls (PCBs): Toxic chemicals that may be contained in liquids used to cool certain types of large transformers and capacitors.

Pozzolan: The name given by the ancient Romans to describe the volcanic ash they used as a type of cement. Today, the term is used for natural or calcined materials (including fly ash and silica fume) or air-cooled blast furnace slag.

Pressure drop: A decrease in pressure from one point to another caused by friction losses in a fluid system, such as a water system.

Pressure rating: The maximum pressure at which a component or system may be operated continuously.

psi (pounds per square inch): A measurement of pressure.

P-trap: A P-shaped trap that provides a water seal in a waste or soil pipe, used mostly at sinks and lavatories.

Purlin: A horizontal roof support member parallel to the plate and installed between the plate and the ridgeboard.

PVC (polyvinyl chloride): Plastic pipe and fittings used for cold water distribution and for industrial water and chemicals, as well as for drain, waste, and vent (DWV) systems.

Rabbet: A groove cut in the edge of a board so as to receive another board.

Rackback: A lead or other structure built with each course of masonry shorter than the course below it.

Racking: Shortening each course of masonry by one unit so it is shorter than the course below it, resulting in a pyramid shape.

Rafter plate: The top or bottom horizontal member at the top of a wall.

Rail-steel: Deformed reinforcing bars that are rolled from selected used railroad rails.

Rake: The slope or pitch of the cornice that parallels the roof rafters on the gable end.

Ranging: Aligning a corner by using a line. Corners can be ranged around themselves or from one corner to another.

Rebars: Abbreviation for reinforcing bars. Also called rerod.

Reclamation: The remanufacturing of used refrigerant to bring it up to the standards required of new refrigerant.

Recovery: The removal and temporary storage of refrigerant in containers approved for that purpose.

Recycling: Circulating recovered refrigerant through filtering devices that remove moisture, acid, and other contaminants.

Refrigeration cycle: The process by which a circulating refrigerant absorbs heat from one location and transfers it to another location.

Return: A corner in a structure or lead.

Ribband: A 1 × 4 nailed to the ceiling joists at the center of the span to prevent twisting and bowing of the joists.

Ridge: The horizontal line formed by the two rafters of a sloping roof that have been nailed together. The ridge is the highest point at the top of the roof where the roof slopes meet.

Ring-tight gasket fitting: Fitting with a rubber O-ring or gasket in the socket.

Rise: The vertical dimension of a set of stairs. Also called the total rise.

Rise and run: A term used to indicate the degree of incline.

Riser: A vertical board under the tread of a stair step; in other words, a board set on edge for connecting the treads of a stairway.

Roll grooved: A type of copper piping that is compatible with grooved copper fittings.

Romex®: General Cable's trade name for Type NM cable; however, it is often used generically to refer to any nonmetallic-sheathed cable.

Roof sheathing: Usually 4 × 8 sheets of plywood, but can also be 1 × 8 or 1 × 12 roof boards, or other new products approved by local building codes. Also referred to as decking.

Roughing in: The first stage of an electrical installation, when the raceway, cable, wires, boxes, and other equipment are installed. This is the electrical work that must be done before any finishing work can be done.

Run: One or more lengths of pipe that continue in a straight line.

Run: The horizontal distance from the face of the first or upper riser to the face of the last or lower riser. Also called the total run.

R-value: A numerical designation given to a material based on its insulating ability, such as R-19.

Saddle: An auxiliary roof deck that is built above the chimney to divert water to either side. It is a structure with a ridge sloping in two directions that is placed between the back side of a chimney and the roof sloping toward it. Also referred to as a cricket.

Sanitary combination: A fitting that combines a wye and ⅛ bend. It is used to connect horizontal branch lines that intersect other horizontal branch lines. It offers less resistance to the flow of material than a sanitary tee. Also called a tee-wye.

Sanitary fitting: A fitting used to connect DWV branches to the main DWV system and to serve as a cleanout.

Sanitary increaser: A fitting used to enlarge the diameter of the vent stack. It is usually placed at least 1 foot below the intersection of the stack and the roof. Use is regulated by code.

Sanitary upright wye: A fitting used to connect the vent stack to the lower end of the soil and waste stacks.

Sanitary wye: A drainage fitting, shaped like the letter Y, that joins the main run of pipe at an angle.

Scab: A length of lumber applied over a joint to strengthen it.

Scarf: To join the ends of stock together with a sloping lap joint so there appears to be a single piece.

Schedule: A measurement that describes pipe wall thickness.

Scratch coat: The first coat of cement plaster consisting of a fine aggregate that is applied through a diamond mesh reinforcement or on a masonry surface.

Screeding: Leveling newly placed concrete to an established grade. Also called striking off.

Scrim: A loosely knit fabric.

Segregation: The separation of sand-cement ingredients from the gravel due to the improper placement of concrete.

Selvage: The section of a composition roofing roll or shingle that is not covered with an aggregate.

Service drop: The overhead conductors, through which electrical service is supplied, between the last power company pole and the point of their connection to the service facilities located at the building.

Service entrance: The point where power is supplied to a building (including the equipment used for this purpose). The service entrance includes the service main switch or panelboard, metering devices, overcurrent protective devices, and conductors/raceways for connecting to the power company's conductors.

Service-entrance conductors: The conductors between the point of termination of the overhead service drop or underground service lateral and the main disconnecting device in the building.

Service-entrance equipment: Equipment that provides overcurrent protection to the feeder and service conductors, a means of disconnecting the feeders from energized service conductors, and a means of measuring the energy used.

Service lateral: The underground conductors through which service is supplied between the power company's distribution facilities and the first point of their connection to the building or area service facilities located at the building.

Set: The hardening of concrete.

Shakes: Hand- or machine-split wood shingles.

Shoring: Temporary bracing used to support above-grade concrete slabs while they harden.

Short sweep ¼ bend: A bend fitting with a short radius used at the base of a DWV stack. Use is regulated by code.

Side inlet: An opening in an ell or tee fitting at right angles to the line of the run, used to connect smaller lines to the main line.

Side lap: The distance between adjacent shingles that overlap, measured in inches.

Sill: The lower framing member attached to the top of the lower cripple studs to form the base of a rough opening for a window.

Siphonage: Loss of water in a trap seal caused by unequal pressure inside and outside DWV piping.

Size dimension ratio (SDR): A measurement of pipe size that relates pipe wall thickness to pipe diameter.

Sizing tool: A tool consisting of a plug and a sizing ring that is used to reshape a deformed pipe back to roundness.

Skirtboard: A baseboard or finishing board at the junction of the interior wall and floor. Also called a finished stringer.

Slab-on-grade (slab-at-grade): A ground supported concrete slab 3½" or thicker that is used as a foundation system. It combines concrete foundation walls with a concrete floor slab that rests directly on an approved base that has been placed over the ground.

Slack to the line: Masonry units set too far away from the mason's line.

Slag: The ash produced during the reduction of iron ore to iron in a blast furnace.

Slope (percent of grade): Another term for grade.

Slope: The ratio of rise to run. The rise in inches is indicated for every foot of run.

Sludge: Semi-liquid matter that settles out in a holding tank during the waste treatment process.

Slump: The distance a standard-sized cone made of freshly mixed concrete will sag. This is known as a slump test.

Soffit: The underside of a roof overhang.

Soleplate: The bottom horizontal member of a wall frame.

Solid wall: Plastic pipe wall that does not contain trapped air.

Solvent weld: A joint created by joining two pipes using solvent cement that softens the material's surface.

Spalling: The condition of concrete breakup, chipping, splitting, or crumbling.

Span: The distance between structural supports such as walls, columns, piers, beams, or girders.

Spread: A row of mortar placed into a bed joint.

Square: The amount of shingles needed to cover 100 square feet of roof surface. For example, square means 10' square or 10' × 10'.

Stack: A general term for any vertical line including offsets of soil, waste, vent, or inside conductor piping. This does not include vertical fixture and vent branches that do not extend through the roof or that pass through not more than two stories before being reconnected to the vent stack or stack vent.

Stairwell: A compartment extending vertically through a building into which stairs are placed.

Station: Instrument setting locations in differential leveling.

S-trap: A trap with a long downstream leg, which tends to promote siphonage. S-traps are no longer installed but are still found in older buildings.

Stringer: The inclined member that supports the treads and risers of a stairway.

Stringing: Spreading mortar with a trowel on a wall or footing for a bed joint.

Strongback: An L-shaped arrangement of lumber used to support ceiling joists and keep them in alignment.

Structural: Bearing weight in addition to its own.

Studs: Vertical members of a form panel used to support sheathing.

Subgrade: Soil prepared and compacted to support a structure or pavement system.

Sweat joint: A pipe joint made by applying solder to the joint and heating it until it flows into the joint.

Switch: A mechanical device used for turning an electrical circuit on and off.

Switch leg: A circuit routed to a switch box for controlling electric lights.

Tail: To check the spacing of head joints by checking the diagonal edges of the courses on a lead or corner.

Tail joist: Short joists that run from an opening to a bearing.

Takeoffs: The process of itemizing and counting all the material and equipment needed for an installation.

Temporary benchmark: A point of known (reference) elevation determined from benchmarks through leveling, and permanent enough to last for the duration of a project.

Test tee: A tee installed as a test location for pressurizing the system to test for leaks.

Thermoplastic pipe: Pipe that can be repeatedly softened by heating and hardened by cooling. When softened, thermoplastic pipe can be molded into desired shapes.

Thermosetting pipe: Pipe that changes chemically when heated, so that once hardened by heat or chemicals, it is hardened permanently.

Top lap: The distance, measured in inches, between the lower edge of an overlapping shingle and the upper edge of the lapping shingle.

Top plate: The upper horizontal framing member of a wall used to carry the roof trusses or rafters.

Toxic: Poisonous.

Transition fitting: A special fitting used to connect plastic pipe to pipe of a dissimilar material, as specified by applicable code.

Tread: The horizontal member of a step.

Trimmer joist: A full-length joist that reinforces a rough opening in the floor.

Trimmer stud: A vertical framing member that forms the sides of rough openings for doors and windows. It provides stiffening for the frame and supports the weight of the header.

Truss: An engineered assembly made of wood, or wood and metal members, that is used to support floors and roofs.

Tuckpointing: Filling fresh mortar into cutout or defective joints in masonry.

Turning point (TP): A temporary point within an open or closed differential leveling circuit whose elevation is determined by differential leveling. It is normally the leveling rod location. Its elevation is determined by subtracting the foresight elevation from the height of the instrument elevation.

Underlayment: A material, such as particleboard or plywood, laid on top of the subfloor to provide a smoother surface for the finished flooring.

Underlayment: Asphalt-saturated felt protection for sheathing; 15-lb roofer's felt is commonly used. The roll size is 3' × 144' or a little over four squares.

Unit rise: The vertical distance from the top of one stair tread to the top of the next one above it; also called the stair rise.

Unit run: The horizontal distance from the face of one riser to the face of the next riser.

Valley flashing: Watertight protection at a roof intersection. Various metals and asphalt products are used; however, materials vary based on local building codes.

Valley: The internal part of the angle formed by the meeting of two roofs.

Velocity: Speed of motion, such as the speed of sewage or wastewater through the drainage piping.

Veneer: A brick face applied to the surface of a frame structure.

Vent: A small opening to allow the passage of air.

Vent branch (branch vent): A vent connecting one or more individual vents with a vent stack or stack vent.

Vent ell: A plastic fitting with a sharp turn radius, used only in vent piping systems. Use is regulated by code.

Vent stack flashing: Flanges that are used to tightly seal pipe projections through the roof. They are usually prefabricated.

Vent tee: A fitting used in venting systems or as a cleanout. It may not be used in the drainage system because it restricts the flow of material. Use is regulated by code.

Walers: Horizontal pieces placed on the outsides of the form walls to strengthen and stiffen the walls. The form ties are also fastened to the walers.

Wall flashing: A form of metal shingle that can be shaped into a protective seal interlacing where the roof line joins an exterior wall. Also referred to as step flashing.

Water hammer: An extreme change in water pressure within a pipe that can cause a loud, banging sound and even damage the system.

Water-cement ratio: The ratio of water to cement, usually by weight (water weight divided by cement weight), in a concrete mix. The water-cement ratio includes all cementitious components of the concrete, including fly ash and pozzolans, as well as portland cement.

Weephole: A small opening in mortar joints or faces to allow the escape of moisture.

Weir (trap or crown): Also referred to as crown weir or trap weir.

Winding stairway: A type of geometrical staircase that changes direction by means of winders or a landing and winders. The stair opening is relatively wide, and the balustrade follows the curve with only a newel post at the bottom.

Wythe: A continuous section of masonry wall, one masonry unit in thickness, or that part of a wall which is one masonry unit in thickness.

Index

Index

Figures are indicated by *f*.

A

Abbreviations
    of construction terms, 1.47–1.48
    on control points, 1.9
ABS (acrylonitrile-butadiene-styrene) pipe, 16.4, 16.5*f*
    defined, 16.30
Absorption testing, 5.11
Accidents
    costs of, 4.28–4.29, 4.28*f*
    electrical accidents, 12.2, 12.3–12.4
    forklifts and, 4.44–4.45
    tools and, 5.1
Accuracy
    of estimates, 7.25
    taping and, 1.13
ACE (Air Conditioning Excellence), 14.16
ACI. *See* American Concrete Institute (ACI)
Acids, safety, 12.33
Acid washes, 5.44
Acoustic tile, 4.7
ACR (air conditioning and refrigeration), defined, 17.27
ACR tubing, 17.2
    defined, 17.27
Acrylonitrile-butadiene-styrene (ABS) pipe, 16.4, 16.5*f*
    defined, 16.30
Adaptability, 4.21
Adapter fittings, 17.4, 17.5
Adapters, 15.16, 15.17*f*
    defined, 15.32
Adjustments, to bricks already laid, 5.32
Admixtures, 2.2, 2.4–2.5, 4.9
    defined, 2.38, 4.57
Adobe, 4.1
    defined, 4.57
Aerated concrete blocks, 4.11, 5.9
Affected employee, defined, 12.17
Aggregates, 2.2, 2.4, 4.9
    defined, 2.38, 4.57
Air conditioning. *See also* Mechanical refrigeration
    high-efficiency and, 14.14
    introduction to, 14.6–14.9, 14.7*f*
    loads and, 13.6
Air Conditioning Excellence (ACE), technician certification,
    14.16

Air infiltration control, 10.4
Air pressure testing, 16.21, 16.23, 17.21
Air-purifying respirators, 12.32
Air slots, ridge vent installation, 9.89–9.90, 9.89*f*, 9.90*f*
Air terminals, 13.22
Air testing
    plastic pipe installations, 16.21, 16.23
    rubber gloves, 12.8, 12.9, 12.9*f*
Aligning walls, 7.16–7.19, 7.17*f*, 7.18*f*
Alignment
    distance measurement and, 1.14, 1.14*f*
    marking pipe before assembly, 16.11
Alkaline, 4.31
    defined, 4.57
Aluminum and vinyl fascia and soffits, 10.16, 10.16*f*, 10.17*f*
Aluminum floats, 3.22, 3.22*f*
Aluminum foil underlayment, 10.52
Aluminum nails, 10.54, 10.55, 10.56, 10.58
Aluminum siding corner caps, 10.60, 10.60*f*
Aluminum wire, grounding electrode conductors, 13.20
American bonds, 5.23, 5.23*f*
American Concrete Institute (ACI)
    defined, 2.28
    floor flatness and, 3.18
    purpose of, 2.6
    tables of concrete mix, 2.5
American National Standards Institute (ANSI)
    as national consensus standard, 12.20
    portland cement standards, 2.2
    rubber protective equipment classification system, 12.7
American Plywood Association (APA), 6.26, 7.18–7.19
American Society for Testing and Materials (ASTM)
    brick standards, 4.4, 4.5
    clay masonry units standards, 5.10
    concrete masonry units (CMUs) standards, 4.9, 5.2, 5.5,
      5.8, 5.9, 5.12
    concrete reinforcement standards, 2.12–2.13
    defined, 4.57
    floor flatness specifications, 3.18
    mortar performance specifications, 4.14
    as national consensus standard, 12.20
    plastic pipe, 16.3, 16.4
    portland cement standards, 2.2
    rubber protective equipment classification system, 12.7
Anchoring devices, personal fall-arrest systems, 4.42, 4.42*f*,
    9.25, 12.37

Anchor points, safety in selecting, 9.26–9.27
Anchors
    drop-in, 17.18, 17.20, 17.20*f*
    for girder supports, 6.17–6.18, 6.17*f*
    masonry walls and, 5.14, 5.15*f*
Annealed copper, 17.2
Annealing, 17.2
    defined, 17.27
ANSI. *See* American National Standards Institute (ANSI)
APA (American Plywood Association), 6.26, 7.18–7.19
APP (atactic polypropylene), 9.9
Appliances
    defined, 13.72
    equipment grounding system and, 13.34
    fixed appliance circuits, 13.6
    neutral conductors and, 13.8
    planning residential electrical systems, 13.2
    small appliance loads, 13.5, 13.7–13.8
Appointed authorized employee, defined, 12.17
Apprentice masons, 4.4, 4.18–4.19, 4.19*f*, 4.20
Apprenticeship training, 14.15–14.16, 14.17
Approach boundaries for electrical conductors, 12.6
Arc blast, 12.5, 12.6, 12.20
Arc burns, 12.5
Arc fault circuit interrupters, 13.13, 13.14–13.15
Arc flash, 12.20
Arches, brick architecture, 4.2, 4.3*f*, 4.4*f*
Architect's rods, 1.23–1.24, 1.24*f*, 1.25
Architectural drawings. *See also* Drawing sets
    floor systems, 6.7–6.8, 6.8*f*, 6.9*f*, 6.10
Architectural shingles
    cap rows and, 9.49, 9.49*f*
    description of, 9.2–9.3, 9.3*f*
Architectural terra-cotta, 4.7, 5.10
Area, formulas for finding, 2.40, 10.20–10.21
Armored cable (Type AC), 13.30
Arresting forces, 4.39, 9.23
Asbestos
    safety and, 12.32–12.33
    shingles and, 9.2
Asbestosis, 12.32
Asbestos warning signs, 12.33
Ashlar stone, 4.1, 4.13, 5.11–5.12
    defined, 4.57
Aspdin, Joseph, 2.3, 4.14
Asphalt roofing cements, 9.13
    defined, 9.98
Asphalt-saturated felt, 9.12
Asphalt shingles. *See* Composition shingles
Aspiration, 15.2, 15.8, 15.9
Assured equipment grounding conductor programs, 12.15
ASTM. *See* American Society for Testing and Materials (ASTM)
Atactic polypropylene (APP), 9.9
Attachment plugs, receptacles for, 12.15
Attic ventilation, 9.87, 9.87*f*
Attitudes about work, 4.21, 12.13
Authorized employee, defined, 12.17
Authorized supervisor, defined, 12.17
Autoclaves, 5.9
    defined, 5.49
Automatic leveling instruments, 1.21, 1.21*f*
Automatic timing devices, 13.52
Aviation shears, 10.57
Axle-steel, 2.13
    defined, 2.38

B
Babylonian plumbing, 15.24
Back injuries, 12.24, 12.25
Back pressure, 15.2, 15.8, 15.9, 15.9*f*
    defined, 15.32
Backpressure backflow, 15.2
    defined, 15.32
Backsights (BS)
    defined, 1.46
    differential leveling and, 1.32
    line-of-sight check and, 1.29
Balanced phase conductors, 13.8
Ballasted membrane roofs, 9.10, 9.11*f*
Balloon-frame construction, 6.4, 6.5*f*, 7.4
Ball valves, 17.6–17.7
Balusters, 11.6, 11.7*f*
    defined, 11.26
Balustrades, 11.6
    defined, 11.26
Band iron, 17.16–17.17, 17.17*f*
Band joists, 6.20
Barge rafters, 8.8, 8.14
    defined, 8.35
Bar joists, 8.28, 8.28*f*
Barrel-style tiles, 9.77*f*, 9.80, 9.81*f*, 9.82–9.83, 9.82*f*, 9.83*f*
Barrows, 3.7
Bar supports, 2.2, 2.16, 2.16*f*
Base flashing, 9.45–9.46, 9.45*f*, 9.46*f*, 9.47*f*
    defined, 9.98
Batch plants, 3.5, 3.5*f*
Bathroom receptacles, 13.10
Battens, clay, ceramic, and concrete tile installation, 9.78, 9.78*f*, 9.79
Batter boards, 1.38–1.40, 1.39*f*
Batteries, 12.33
BCSI (Building Component Safety Information) series, 8.25
Beam blocks, 4.10*f*, 5.6*f*
Beam clamps, 16.20, 16.22*f*, 17.17
Beams. *See also* Girders; Wood I-beams
    grade beams, 2.20, 2.21, 2.22*f*
    lumber grades for, 6.16
Bell-and-spigot pipes
    defined, 16.30
    described, 16.12
    joining, 16.15, 16.15*f*
Belt conveyors, 3.7, 3.13
Benchmarks (BM), 1.7, 1.32
Bending, copper pipe and fittings, 17.8, 17.10, 17.11*f*
Bends, 15.14–15.16, 15.15*f*, 15.16*f*
Beveled rake, wood shingles and shakes, 9.61
Beveled wood siding
    around windows, 10.25, 10.25*f*
    corners and, 10.25, 10.26*f*, 10.27*f*
    installation of, 10.22–10.23, 10.22*f*, 10.23*f*, 10.24, 10.24*f*, 10.25, 10.25*f*, 10.26*f*, 10.27*f*
    as style, 10.18*f*, 10.21, 10.21*f*
Bird's mouths, 8.7, 8.7*f*, 8.10, 8.11*f*, 8.39, 8.39*f*
Black iron pipe, 14.10
Bleed, 3.16, 3.21
    defined, 3.38
Bleed water, 3.28
Blind corners, and forklifts, 4.47
Block bed joints, 5.19–5.20, 5.20*f*
Block bond patterns, 5.23, 5.24*f*
Block cleaning solutions, 5.46
Blocking
    defined, 7.35
    wall framing and, 7.2

Blowers, and furnaces, 14.5
Blueprints
    floor systems and, 6.11–6.12
    HVAC and, 14.9, 14.10, 14.11*f*
Board-and-batten wood siding, 10.25, 10.26, 10.27–10.28, 10.27*f*, 10.28*f*, 10.29*f*
    defined, 10.71
Board-on-board wood siding, 10.28, 10.28*f*, 10.29*f*
Boards, picking up mortar from, 4.24, 4.24*f*
Board subfloors, 6.26–6.27
BOCA National Building Code, 13.73
Body belts, 4.40, 4.41
Body harnesses
    description of, 9.23–9.24, 9.24*f*
    personal positioning systems and, 4.40, 4.40*f*
    wearing, 9.26, 9.26*f*
Body resistance, 12.4–12.5, 12.4*f*
Boilers
    heating and, 14.2, 14.4*f*
    high-efficiency, 14.5
    hydronic heating systems and, 14.2
Bond beam blocks, 5.9*f*
Bond beam cut blocks, 5.26, 5.26*f*
Bonding
    intersystem, 13.23
    piping systems and, 13.23–13.24
Bonding at the service, 13.23
Bonding bushings, 13.23
    defined, 13.72
Bonding jumpers, 13.21, 13.35
    defined, 13.72
Bonding masonry units
    block bond patterns, 5.23, 5.24*f*
    bond patterns, 4.5, 4.5*f*
    mechanical bonds, 5.20–5.21
    pattern bonds, 5.21–5.22, 5.21*f*, 5.22*f*
    structural bonds, 5.22–5.23, 5.22*f*, 5.23*f*, 5.24*f*
    structural pattern bonds, 5.22–5.23, 5.22*f*, 5.23*f*, 5.24*f*
Bond patterns, 4.5, 4.5*f*
Boston hips, 9.76
Bottom dips, 15.6, 15.6*f*
Bottom plates, as wall components, 2.25, 2.25*f*
Bows, in lumber, 7.15
Box beam headers, 7.6, 7.7*f*
Box cornices, 10.2, 10.5, 10.7, 10.8–10.12*f*, 10.13, 10.13*f*
Boxes, outlet boxes, 13.47–13.50, 13.48*f*, 13.50*f*
Box framing, 6.2. *See also* Platform framing
Box sills, 6.20
Box vents, 9.87, 9.87*f*, 9.88–9.89, 9.88*f*
Braced frames, 6.2, 6.4
Braces
    defined, 2.38
    for wall forms, 2.25, 2.25*f*, 2.26*f*
Bracing
    diagonal bracing, 7.23*f*, 7.25–7.26
    for long roof spans, 8.11, 8.12*f*
    metal wall bracing, 7.18, 7.18*f*
    permanent bracing, 7.18, 7.18*f*, 8.23, 8.24, 8.24*f*
    safety and, 8.25
    temporary bracing, 7.18, 8.23, 8.24, 8.24*f*, 8.25
    for trusses, 8.23, 8.24, 8.24*f*, 8.25
Branch circuits
    defined, 13.72
    feeders and, 13.37–13.39, 13.38*f*, 13.39*f*, 13.40*f*
    layout for lighting, 13.45, 13.46*f*, 13.47*f*
    layout for power, 13.37–13.40, 13.38*f*, 13.39*f*, 13.40*f*, 13.41*f*, 13.42–13.45, 13.42*f*, 13.43*f*, 13.44*f*
    residential electrical service planning, 13.2

residential wiring methods for, 13.29–13.30, 13.30*f*, 13.31*f*, 13.32–13.33, 13.32*f*, 13.34*f*
Branch intervals, 15.22
    defined, 15.32
Brazing, 17.2, 17.12
Break-back spreader ties, 2.26
Breaking the tape, 1.16, 1.16*f*
    defined, 1.46
Break-off line ties, 2.26*f*
Brick and stone veneers, 10.65
Brick cleaning solutions, 5.44
Brick Institute of America, cleaning methods for brick, 5.44, 5.45
Bricklaying basics
    buttering joints, 4.26–4.27, 4.27*f*
    cutting or edging mortar, 4.25–4.26, 4.26*f*
    furrowing mortar, 4.26, 4.26*f*, 5.19
    general rules, 4.27–4.28
    picking up mortar, 4.23–4.25, 4.23*f*, 4.24*f*
    preparing mortar, 4.22, 4.23*f*
    spreading mortar, 4.23, 4.25, 4.25*f*
Brick on block walls, 5.21
Bricks
    ASTM standards for, 4.4, 4.5
    classification of, 4.7
    cleaning, 5.44–5.45
    cutting, 5.25, 5.25*f*, 5.26–5.29, 5.27*f*, 5.28*f*
    installation positions of, 5.10–5.11, 5.11*f*
    laying to the line, 5.34–5.36, 5.36*f*
    manufacture of, 4.4–4.5
    mud, 4.1, 4.2
    Norman, 4.4
    Roman, 4.3
    size relationship to blocks, 5.5
    solid masonry units, 4.5, 4.5*f*, 4.6*f*, 5.10–5.11
    special shaped bricks, 4.5, 4.6*f*
    standard sizes of, 4.4–4.5, 4.5*f*
    working stacks of, 4.33–4.34, 4.34*f*
Brick trowels, 4.17
Bridging
    estimating material for, 6.41
    floor systems and, 6.24–6.25, 6.25*f*
    installing, 6.35–6.36, 6.36*f*
Brooming, 3.27, 3.27*f*
Brown coats
    defined, 10.71
    stucco (cement) finishes and, 10.64–10.65, 10.65*f*
Bucket cleaning brick, 5.44–5.45
Bucket trowels, 4.17*f*
Buggies, 3.7
Builder's levels, 1.21, 1.21*f*. *See also* Leveling instruments
Building brick, 4.5, 4.7
Building codes
    HVAC industry and, 14.9–14.10, 14.12
    stairs and, 11.10
Building Component Safety Information (BCSI) series, 8.25
Building drains, 15.22
Building layout points, 1.8, 1.8*f*
Building paper, 10.2
    defined, 10.71
Building plan drawings, 1.2–1.3, 1.4*f*
Building sewers, 15.2, 15.22
    defined, 15.32
Building traps, 15.7
Building wraps, exterior finishing, 10.3
Built-in-place forms, 2.28
Built-up headers, 7.6, 7.6*f*
Built-up lumber girders, 6.15–6.16, 6.15*f*

Built-up roofing (BUR) membrane, 9.8, 9.9f, 9.10
Bullfloats, 3.21–3.22, 3.21f
Bull headers, 4.7, 4.8f
Bullhead tees, 17.7
    defined, 17.27
Bull nose blocks, 4.10f, 5.6f
Bull's eye rod levels, 1.25, 1.25f
Bull stretchers, 4.7, 4.8f
Bundles
    composition shingles and, 9.4
    defined, 9.98
    scattering, 9.32
Bungalow wood siding, 10.18f, 10.21, 10.21f
BUR (built-up roofing) membrane, 9.8, 9.9f, 9.10
Burns
    chemical burns, 2.2, 4.23, 4.31
    from electrical shock, 12.3, 12.5
Burrs, 17.10
Bush hammers, 3.29, 3.31f
Butane, as heating fuel, 14.2
Butted splices, 7.21, 7.21f
Butter, 4.17
    defined, 4.57
Buttering joints
    in blocks, 5.18–5.19, 5.19f, 5.20f
    in bricks, 4.26–4.27, 4.27f
Butt joints, in hardboard and particleboard siding, 10.41,
    10.43f
Bypass framing. See Balloon-frame construction
Byrd, Jonathan, profile in success, 15.30–15.31

C
Cable strippers, 13.33
Cable systems
    armored cable (Type AC), 13.30
    metal-clad cable (MC), 13.30, 13.32f
    nonmetallic-sheathed cable (Type NM), 13.29–13.30,
        13.30f, 13.31f, 13.55–13.56
    residential wiring methods and, 13.29–13.33, 13.30f,
        13.31f, 13.32f, 13.34f
Calcium silicate units, 5.2, 5.9, 5.9f
Calculating
    electrical loads, 13.5–13.6, 13.5f
    jack rafter lengths, 8.42
    rafter length, 8.4–8.5, 8.6f, 8.9–8.10, 8.9f
    rise and pitch, 8.8
    stud length, 7.12–7.14, 7.13–7.14f, 7.15
    volume calculations for concrete, 2.9–2.10, 2.9f, 2.10f, 2.11,
        2.11f, 2.12, 2.12f
Calculators, decimal conversions, 1.17
Caldwell, Berry, profile in success, 7.34
Calibration
    crimping tools and, 16.16
    distance measurement and, 1.13, 1.14, 1.14f, 1.18, 1.22
    leveling instruments and, 1.28–1.30, 1.28f, 1.29f
California Electric Light Company, 13.2
Canadian Standards Association (CSA), 13.51
Cantilever applications, and joists, 6.38
Cantilevered floors, installing joists for, 6.38, 6.39f, 6.40
Capacitors, discharging, 12.11
Capacity of outlet boxes, 13.47–13.48, 13.49
Cap blocks, 5.9f
Cap flashing
    composition shingle installation and, 9.42, 9.48, 9.48f,
        9.49f
    defined, 9.98
    preformed, 9.50

Capillary action, 17.11
    defined, 17.27
Capillary attraction, 15.8, 15.10–15.11, 15.10f
    defined, 15.32
Capitals, 4.7
    defined, 4.57
Cap row shingles, 9.49, 9.49f, 9.51, 9.51f, 9.52f
Carborundum rubbing stones, 3.31, 3.31f
Cardiopulmonary resuscitation (CPR), 12.3, 12.30
Careers
    in HVAC, 14.13, 14.14, 14.14f, 14.15
    in masonry, 4.18–4.20, 4.19f
Carrier, Willis, 14.14
Carrier Corporation, 14.14
Cast-iron fittings, bends, 15.14
Cast solder fittings, 17.4
Cast stone, 4.13, 5.12, 5.12f
Catch basin concrete units, 4.12, 5.2, 5.10, 5.11f
Caulking, vinyl and metal siding, 10.64
Cavity walls, 4.8–4.9, 4.8f, 4.15, 4.15f
C-beam clamps, 17.18, 17.20, 17.20f
C-clamps, 16.20, 16.22f, 17.16f, 17.17
C-clips, 9.71–9.72, 9.72f
Ceiling joists
    cutting and installing, 7.21, 7.22f
    estimating material for, 7.23f, 7.26
    layout, 7.19–7.21, 7.20f, 7.21f
    reinforcing, 7.21, 7.22f
    splicing, 7.20–7.21, 7.21f
Ceilings
    framing, 7.19–7.21, 7.22f
    framing spacing and, 7.19
    laying out, 7.19–7.21, 7.22f
Cells, concrete blocks, 4.11f, 5.4f
Cellular core wall plastic pipe, 16.2
    defined, 16.30
Celotex Thermax, 10.4
Celotex Tuff-R, 10.4
Cement dermatitis, 3.32, 3.33, 4.31
Cement dust, safety, 2.2, 4.31, 4.32
Cement-lime mortars, 4.14
Cements
    asphalt roofing cements, 9.13, 9.98
    cold temperatures and, 16.14
    portland cement, 2.2, 2.3, 4.14
    solvent cements, 16.12–16.13
Center of gravity, 4.49, 4.49f, 4.50f
Center-to-center measurements, 7.4
Center-to-center pipe measuring, 17.8, 17.9f
Center-to-face pipe measuring, 17.9, 17.9f
Central mix plants, 3.5
Ceramic glazed brick, 4.5
Certification
    powder actuated tools and, 4.35, 12.29
    of technicians, 14.16, 14.17
CFCs (Chlorofluorocarbons), 14.18, 14.23
CFR (Code of Federal Regulations), 12.5, 14.16, 14.17
Chaining, 1.11. See also Taping
Chaining pins, 1.11, 1.11f, 1.13
Chalklines, 7.9
Chamfering, 16.11
    defined, 16.30
Channel blocks, 5.7f
Channel rustic wood siding, 10.19f, 10.31, 10.31f
Channel strips, 9.72, 9.73f
Check valves, 17.7, 17.7f
Cheek cuts, 8.38, 8.38f, 8.39
Chemical aprons, 12.30

Chemical burns
  concrete and, 2.2
  mortar and, 4.23, 4.31
Chemical cartridge respirators, 12.30
Chemical face shields, 12.30
Chemically inert, 16.2
  defined, 16.30
Chemicals, safety, 12.32
Chicago rods, 1.23
Chillers
  defined, 14.23
  as evaporators, 14.15
Chimneys
  composition shingle installation, 9.45–9.46, 9.45f, 9.46f,
    9.47f, 9.48, 9.48f, 9.49f
  flashing and, 5.14, 9.46, 9.46f
Chipping hammers, 3.29, 3.31f
Chisels, cutting masonry with, 5.26–5.27, 5.27f
Chlorine, and ozone layer, 14.18
Chlorofluorocarbons (CFCs)
  defined, 14.23
  environment and, 14.18
Chutes, 3.7, 3.11
Cinders, in concrete blocks, 4.9
Circuit breakers, and lockout, 12.14
Circuits
  de-energizing circuits, 12.8, 12.11, 12.13–12.14, 12.19, 13.27
  laundry circuits, 13.5–13.6, 13.7
  multiwire branch circuits, 13.42–13.43, 13.42–13.43f
  water heater circuits, 13.44–13.45, 13.44f
Civil plans, 1.2–1.3, 1.3f, 1.4f, 1.5f
Classification systems, for rubber protective equipment, 12.7
Clay, ceramic, and concrete tile installation
  barrel-style tiles, 9.77f, 9.80, 9.81f, 9.82–9.83, 9.82f, 9.83f
  battens, 9.78, 9.78f, 9.79
  flat tiles, 9.79, 9.79f, 9.80, 9.80f
  hold-down brackets for, 9.76, 9.76f, 9.77f
  nailing tiles, 9.7, 9.76
  roof deck preparation, 9.78–9.79, 9.78f
  substitutes for clay tiles, 9.77
Clay masonry units, 5.3, 5.10–5.11, 5.11f
Clay products, 4.4–4.7, 4.5f, 4.6f, 4.8f, 5.3, 5.10–5.11, 5.11f
Clean Air Act, and refrigerants, 14.18
Cleaning
  concrete blocks, 5.46
  duct cleaning, 14.6
  masonry units, 5.44–5.46
  vinyl and metal siding, 10.64
Cleanout adapters, 15.16, 15.17f, 15.22, 15.22f
Cleanout plugs, 15.16, 15.22
Cleanouts
  building drain and, 15.22, 15.22f
  codes and, 15.25
  defined, 15.32
  DWV fittings and, 15.16, 15.17f
  P-traps and, 15.4, 15.5f
Clearances
  OSHA standards, 12.6
  for service drops near building openings, 13.26
  vertical clearances for service drops, 13.25, 13.27f
Cleated stringers, 11.5, 11.6f, 11.20, 11.20f
Clevises, 17.18
  defined, 17.27
Clevis hangers, 17.16f
Clinker bricks, 4.7
Closed cornices, 10.5, 10.6–10.7, 10.6f
Closed-cut valleys, 9.38–9.39, 9.40f

Closed finish stringers, 11.7f, 11.8
Closed loops, differential leveling, 1.32
Closed stairways, 11.2, 11.2f
  defined, 11.26
Closed-woven valleys, 9.38, 9.39f
Closing springs, discharging, 12.14
Closure units
  defined, 5.49
  laying, 5.31–5.32, 5.31f
Clothing
  for cold weather, 4.36
  OSHA standards, 12.19
  personal, 12.10, 12.19
  for personal protection, 4.29–4.30, 4.29f
CMUs. See Concrete masonry units (CMUs)
Coal, as heat source, 14.2
CO/ALR symbol, 13.51
Code of Federal Regulations (CFR), 12.5, 14.16, 14.17
Codes
  DEFS and EIFS cladding and, 10.67
  drain, waste, and vent (DWV) systems and, 15.24–15.25
  drainage fittings and, 15.14, 15.14f
  exterior finish carpentry and, 10.2
  International Plumbing Code (IPC), 16.20
  model codes, 15.7, 15.24–15.25, 17.6
  seismic codes, 16.19
  for supports and fasteners, 16.20
Cofferdam concrete seal courses, 3.13
Coil ties, 2.27f
Cold exposure, 4.36
Cold temperatures
  cements and, 16.14
  footings and, 2.19
  mortar and, 5.16
  safety and, 4.36–4.37
Cold weather
  clothing for, 4.36
  consolidating concrete and, 3.16
Colonial wood siding, 10.21
Color-coding
  of control points, 1.9
  copper pipe, 17.3
  of main bonding jumpers, 13.22
  of powder charges, 4.34–4.35
  of receptacle terminals, 13.51
  of switches, 13.51
Colored concrete, 2.8
Columns
  beam/girder supports and, 6.17–6.18, 6.17f, 6.18f,
    6.30–6.31, 6.31f
  consolidating concrete and, 3.16
Column spacing, 6.18, 6.18f
Combed ridges, 9.75–9.76, 9.75f
Combination fittings, 15.19
Combination tools, 3.29, 3.30f
Combined center of gravity, 4.49, 4.50f
Combustible liquids, 4.52
Commercial handrail requirements, 11.6, 11.7f
Commercial stairways, 11.11
Common bonds, 5.23, 5.23f
Common load neutrals, 13.12, 13.13f, 13.14
Common rafters
  described, 8.3, 8.4f
  laying out, 8.7–8.8, 8.7f, 8.10, 8.11f, 8.18, 8.18f, 8.19f, 8.20
Common studs, measuring and cutting, 7.12–7.13, 7.13–7.14f
Composite masonry walls, 4.15

Composition shingle installation
    chimneys and, 9.45–9.46, 9.45*f*, 9.46*f*, 9.47*f*, 9.48, 9.48*f*, 9.49*f*
    dormer roof valleys, 9.42, 9.43, 9.43*f*, 9.44, 9.44*f*
    gable roofs and, 9.32–9.34, 9.34*f*, 9.36, 9.36*f*, 9.37*f*, 9.43
    hip roofs and, 9.36, 9.37*f*, 9.49, 9.49*f*, 9.51, 9.51*f*, 9.52*f*
    roof projections and flashing, 9.40*f*, 9.41*f*, 9.42*f*, 9.43, 9.43*f*, 9.44–9.46, 9.44*f*, 9.45*f*, 9.46*f*, 9.47*f*, 9.48–9.49, 9.48*f*, 9.49*f*, 9.50, 9.51, 9.51*f*, 9.52*f*
    soil stacks and, 9.39–9.40, 9.41*f*
    valleys and, 9.37–9.39, 9.38*f*, 9.39*f*, 9.40*f*, 9.43
    vertical wall flashing, 9.42*f*, 9.43*f*
Composition shingles
    characteristics of, 9.2–9.4, 9.3*f*, 9.30, 9.31*f*
    nailing, 9.34, 9.35, 9.35*f*, 9.51, 9.51*f*
    nailing points, 9.32, 9.32*f*
    terminology for placing, 9.30, 9.32, 9.32*f*
    top lap and, 9.31*f*, 9.32*f*
    types of, 9.30, 9.31*f*
    underlayment and, 9.3–9.4, 9.12
Compressed gas, air testing gloves, 12.9
Compression, stop, and waste valves, 17.7, 17.7*f*
Compression collars, 16.17
    defined, 16.30
Compression joints, 17.12–17.13, 17.14*f*
    defined, 17.27
Compressive strength, 2.12, 5.2
Compressors
    defined, 14.23
    hermetic rotary, 14.9
    hermetic (welded hermetic), 14.9
    mechanical refrigeration and, 14.7
    reciprocating, 14.9
    semi-hermetic (serviceable hermetic), 14.9
Concave ends, concrete blocks, 4.11*f*, 5.4*f*
Concrete. *See also* Concrete placement; Finishing concrete; Forms for concrete
    admixtures, 2.2, 2.4–2.5, 4.9
    aggregates, 2.2, 2.4, 4.9
    consolidating concrete, 3.2, 3.16–3.17, 3.17*f*
    curing of, 2.8, 3.27–3.28
    estimating volume, 2.9–2.10, 2.9*f*, 2.10*f*, 2.11, 2.11*f*, 2.12, 2.12*f*, 3.15
    joint sealants, 3.28–3.29
    joints in concrete structures, 3.2–3.4, 3.3*f*, 3.4*f*
    methods of conveying, 3.7–3.8
    mix proportions and measurements, 2.5–2.6, 2.5*f*
    moving and handling concrete, 3.5–3.6, 3.5*f*, 3.6*f*, 3.7–3.8, 3.9–3.14, 3.9*f*, 3.10*f*, 3.12*f*, 3.13*f*
    portland cement and, 2.2, 2.3
    reinforcement materials, 2.12–2.16, 2.14*f*, 2.16*f*, 2.17*f*
    removing forms, 2.26, 2.31, 3.29
    safety and, 2.2, 3.9, 3.10, 3.24, 3.32–3.33, 4.31–4.32
    slump testing, 2.8, 2.9*f*
    special types, 2.6, 2.7*f*
    tools for working concrete, 3.29, 3.30*f*, 3.31, 3.31*f*, 3.32*f*
    water and, 2.4, 2.8
Concrete blocks
    block bed joints, 5.19–5.20, 5.20*f*
    buttering, 5.18–5.19, 5.19*f*, 5.20*f*
    characteristics of, 4.11, 5.3, 5.4*f*, 5.5, 5.6*f*, 5.7, 5.7*f*, 5.8
    cleaning, 5.46
    common sizes of, 4.9, 4.10*f*
    cutting, 5.25–5.29, 5.25*f*, 5.26*f*, 5.27*f*, 5.28*f*
    development of, 4.8–4.9
    furrowed joints and, 5.19
    general rules for laying, 5.20
    head joints and, 5.18–5.20

laying to the line, 5.36
    rebars and, 5.13
Concrete bond, 3.2–3.3
Concrete bricks, 4.11, 4.12*f*, 5.2, 5.8, 5.8*f*
Concrete buckets, 3.6, 3.7, 3.9, 3.9*f*
Concrete calculators, 2.9, 2.9*f*
Concrete carts, 3.12–3.13, 3.12*f*
Concrete masonry units (CMUs)
    ASTM standards for, 4.9, 5.2, 5.5, 5.8, 5.9
    contraction and expansion joints, 5.2–5.3, 5.3*f*, 5.4*f*
    defined, 4.57
    development of, 4.8–4.9
    mason's use of, 4.1
    moisture content and absorption of, 5.2, 5.11
Concrete placement
    placing concrete in forms, 3.14–3.16, 3.15*f*
    segregation of concrete materials and, 3.6, 3.12, 3.15
Concrete power saws, 3.24, 3.25*f*
Concrete products
    cavity walls and, 4.8–4.9, 4.8*f*, 4.15, 4.15*f*
    concrete blocks, 4.8–4.9, 4.10*f*, 4.11, 4.11*f*
    concrete bricks, 4.11, 4.12*f*, 5.2, 5.8, 5.8*f*
    pre-faced concrete units, 4.12, 4.12*f*, 5.2, 5.8, 5.9, 5.9*f*, 5.10
Concrete pumps, 3.8, 3.10–3.11, 3.10*f*
Concrete strength, cured concrete, 2.2, 3.27–3.28
Concrete tables, 2.9, 2.10*f*
Concrete volume estimation, rectangular volume calculations, 2.9–2.10, 2.9*f*, 2.10*f*, 2.11, 2.11*f*, 3.15
Condensers
    defined, 14.23
    mechanical refrigeration and, 14.7
Condensing furnaces, 14.5
Conductors
    calculating for outlet boxes, 13.47–13.48, 13.49
    electrical injuries from contact with, 12.3
    residential neutral conductors, 13.8–13.9, 13.9*f*
Cone snap-in form ties, 2.20*f*
Confined spaces
    entry procedures, 12.29–12.30, 12.29*f*
    general guidelines, 12.29, 12.29*f*
    hazard reviews, 12.29–12.30
    OSHA standards, 12.29–12.30
Connectors, copper pipe hangers and supports, 17.18, 17.19*f*
Consolidating concrete, 3.2, 3.16–3.17, 3.17*f*
    defined, 3.38
Consolidators, 14.15
Construction joints, 3.2–3.3, 3.3*f*, 3.4
Construction terms, abbreviations of, 1.47–1.48
Continuing education, for masons, 4.21
Continuous flashing, 9.42, 9.42*f*, 9.43*f*, 9.46, 9.47*f*
Continuous footings, 2.19, 2.20, 2.20*f*, 2.21*f*, 2.24
Continuously adjustable aluminum scaffolding, 10.3
Continuous roof lines, 9.89, 9.89*f*
Contour lines, 1.5, 1.5*f*, 1.6*f*
Contraction joints
    concrete masonry units, 5.2–5.3, 5.3*f*, 5.4*f*
    as control joints, 3.4, 3.4*f*
Contractors, 4.20, 5.14–5.15
Control joints
    finishing concrete and, 3.22–3.24, 3.24*f*
    shrinkage cracking and, 5.2–5.3, 5.3*f*
    types of, 3.4, 3.4*f*
Controlled access zones, 12.38, 12.38*f*
Control points
    abbreviations and, 1.9
    carpenter's laying out, 1.2
    color-coding of, 1.9
    defined, 1.46

information on, 1.9
placement of, 1.2, 1.8, 1.9, 1.9f
types of, 1.7–1.8, 1.7f, 1.8f
Conversions
decimal conversion tables, 2.39
feet and inches to decimal feet, 1.17
metric conversion charts, 2.39, 2.41
Cooking appliances, 13.7–13.8
Copper, and tile roofing, 9.7, 9.76
Copper fittings and valves
alternative fittings, 17.7–17.8, 17.7f, 17.8f
descriptions, 17.5
for drain, waste, and vent (DWV), 17.7
water supply fittings, 17.4, 17.5, 17.6f
water supply valves, 17.6–17.7, 17.7f
Copper pipe
grounding and, 17.2
labeling, 17.3
sizing, 17.2, 17.3f
storage and handling, 17.3
types of, 17.2
Copper pipe and fittings
bending, 17.8, 17.10, 17.11f
cutting, 17.8, 17.9–17.10, 17.9f, 17.10f
fittings and valves, 17.4–17.8, 17.6f, 17.7f, 17.8f
grooving, 17.5, 17.8, 17.15
insulating, 17.21, 17.21f
joining, 17.8, 17.11–17.14, 17.12f, 17.13f, 17.14f
measuring, 17.8–17.9, 17.9f
pipe hangers and supports, 17.16–17.18, 17.16f, 17.17f, 17.18f, 17.19f, 17.20f
pressure testing, 17.21, 17.21f
Copper pipe hangers and supports
connectors, 17.18, 17.19f
earthquakes and, 17.16
pipe attachments, 17.16–17.17, 17.16f, 17.17f, 17.18f
structural attachments, 17.18, 17.20, 17.20f
Copper tubing, uses in HVAC, 14.10
Cored bricks, 4.7
Cores, concrete blocks, 4.11f, 5.4f
Corner blocks, 4.10f, 5.6f, 5.7f
Corner boards, beveled siding, 10.25, 10.27f
Corner-bullnose blocks, 5.9f
Corner caps
for vinyl and metal siding, 10.58–10.60, 10.59f, 10.60f
for wood siding, 10.25, 10.26, 10.26f
Corner flashing, 10.26
Corner layout
batter boards and, 1.39–1.40, 1.39f
dry bonding and, 5.18, 5.19f
Corner poles, mason's lines, 5.33
Corners, wall framing, 7.3, 7.3f, 7.4f
Cornices
aluminum or vinyl fascia and soffits, 10.16, 10.16f, 10.17f
box cornices, 10.2, 10.5, 10.7, 10.8–10.12f, 10.13, 10.13f
closed cornices, 10.5, 10.6–10.7, 10.6f
cornice returns, 10.14, 10.15f, 10.16
defined, 4.57, 10.71
fascia and, 10.7, 10.7f
installation, 10.13–10.14, 10.13f, 10.14f, 10.15f
open cornices, 10.5, 10.7, 10.7f
ornamental terra-cotta for, 4.7
tail rafter cuts and, 10.7, 10.7f
Corona cutting, 12.7
Corrections
distance measurement and, 1.14, 1.14f
tape measurements and, 1.18–1.19

Corrugated metal roofing, 9.68–9.69, 9.69f
Corrugated roofing light panels, 9.70
Couplings
copper, 17.5
plastic, 16.22
Coursed ashlar block bonds, 5.24f
Courses, 4.7, 4.8f, 10.6
defined, 4.57, 10.71
Cove base blocks, 5.9f
Cox-comb ridges, 9.76
CPR (cardiopulmonary resuscitation), 12.3, 12.30
CPVC (chlorinated polyvinyl chloride), 16.3
defined, 16.30
CPVC (chlorinated polyvinyl chloride) pipe
HVAC and, 14.10
joining, 16.14–16.15, 16.14f
as type of plastic pipe, 16.5–16.6, 16.6f
Cracking
construction joints and, 3.2
curing concrete and, 3.27
Cracks, and loss of trap seal, 15.8, 15.11
Crane deliveries, 8.16
Cranes
erecting walls and, 7.16
hammerhead cranes, 3.6, 3.9f
moving and handling concrete, 3.6, 3.7, 3.9, 3.9f
safety and, 12.23–12.24
safety working near, 4.31, 4.31f
tower cranes, 3.6, 3.9f
Cribbing, mobile boom-type concrete pumps, 3.11
Crickets, 9.45, 9.45f, 9.46, 9.47f, 9.48
Cripple studs, 7.2, 7.2f, 7.14
defined, 7.35
Crosshairs, 1.27, 1.28f
defined, 1.46
Cross-linked polyethylene. See PEX (cross-linked polyethylene)
Cross-section leveling, 1.38
Cross webs, concrete blocks, 4.11f, 5.4f
Crowding the line, 5.35
defined, 5.49
Crowns, 6.15, 7.15
defined, 6.50
Crown weirs, 15.6, 15.6f
defined, 15.32
Cubes
of brick, 4.4, 5.8
defined, 4.57
Cured concrete
concrete strength and, 2.2, 3.27–3.28
defined, 2.38
Cure times, 16.13, 16.22
Curing blankets, 2.8
Curing compounds, 2.8
Curing concrete, 2.8, 3.27–3.28
Curing mats, 2.8
Curing paper, 2.8
Curl, roll roofing, 9.53
Curtain walls, 4.16
Cut, defined, 1.46
Cut grooving, 17.15
Cutout stringers, 11.5, 11.5f
Cutting
bricks, 5.25, 5.25f, 5.26–5.29, 5.27f, 5.28f
concrete blocks, 5.25–5.29, 5.25f, 5.26f, 5.27f, 5.28f
copper pipe and fittings, 17.8, 17.9–17.10, 17.9f, 17.10f

Cutting (*continued*)
    king studs, 7.12–7.13, 7.13–7.14f
    masonry units, 5.25–5.29, 5.25f, 5.26f, 5.27f, 5.28f
    mortar, 4.25–4.26, 4.26f
    overcutting, 11.17, 11.19f
    plastic plumbing materials, 16.10–16.11, 16.11f, 16.12f
    PVC (polyvinyl chloride) pipe, 16.10–16.11, 16.11f
    slate, 9.75
    stringers, 11.15–11.17, 11.15f, 11.16f, 11.18, 11.18f, 11.19,
      11.19f, 11.20, 11.20f
    studs, 7.12–7.14, 7.13f, 7.14f
    vinyl and metal siding, 10.57, 10.57f, 10.58f
Cutting patterns, 10.63
Cutting tables, for siding, 10.51

D
Dadoed stringers, 11.5, 11.6f, 11.20, 11.20f
Damaged equipment, safety, 4.43
Danger tags, 12.17, 12.18f
Darby floats, 3.21, 3.21f
Dead loads, 6.20
    defined, 6.50
Dead time, and mobile boom-type concrete pumps, 3.10
Deburr tools, 16.10, 16.11f, 17.10
Deceleration devices, 4.39, 4.41, 4.41f, 9.23, 9.24–9.25, 9.25f
Deceleration distance, 4.39, 9.23
Decimal conversion tables, 2.39
Decimal feet, converting to feet and inches, 1.17
Decorative joints, 3.4, 3.4f
Deeds, Adam, profile in success, 6.49
De-energizing circuits, 12.8, 12.11, 12.13–12.14, 12.19, 13.27
Defects, in lumber, 7.15
Deflection, of floor joists, 6.20
Deformed bars, 2.13
DEFS (Direct-applied exterior finish systems), 10.66–10.67,
    10.66f
Demand factors, 13.7, 13.8, 13.9f
Dependability, 4.21
Detail drawings, floor systems, 6.8
Detergents, cleaning blocks, 5.46
Diagonal block bonds, 5.24f
Diagonal bracing, estimating material for, 7.23f, 7.25–7.26
Diagonal measurements
    batter boards and, 1.40
    squaring sill plates and, 6.28
Diamond wheel hand grinders, 9.15, 9.16f
Dielectric unions
    electrolysis and, 17.5
    hot water systems and, 17.4
Differential leveling
    defined, 1.46
    procedure for, 1.32, 1.33f, 1.34–1.35, 1.35f
    process of, 1.2, 1.30, 1.31f
    terminology used in, 1.30–1.32
    tools and equipment for, 1.20–1.30, 1.21f, 1.22f, 1.23f, 1.24f,
      1.25f, 1.26f, 1.27f, 1.28f, 1.29f
Dimensions versus scale, in drawings, 6.11
Dimmer switches, 13.52, 13.59
Direct-applied exterior finish systems (DEFS), 10.66–10.67,
    10.66f
Direct-wired ground-fault circuit interrupters (GFCI), 13.14,
    13.14f
Disconnect switches
    lockout and, 12.14
    safety marking, 12.14, 12.15
Distance measurement
    electronic distance measurements, 1.19–1.20, 1.20f
    estimation by pacing, 1.19

by taping, 1.11, 1.13–1.19, 1.14f, 1.15f, 1.16f
    tools and equipment for, 1.11–1.13, 1.11f
Dock plates, 4.51
Docks, and forklift operation, 4.51
Dolly Varden wood siding, 10.18f, 10.21, 10.21f
Domes, brick architecture, 4.2
Door-actuated switches, 13.54
Doors
    furring out around, 10.52
    installing vinyl and metal siding around, 10.60–10.61,
      10.62f
    installing wood siding around, 10.25, 10.25f, 10.32, 10.33
    in masonry walls, 7.28
Dormer roof valleys, composition shingle installation, 9.42,
    9.43, 9.43f, 9.44, 9.44f
Dormers, 8.26, 8.27f
Double basket weave block bonds, 5.24f
Double bends, 15.16, 15.16f
    defined, 15.32
Double brick, 4.12f
Double corner blocks, 4.10f, 5.6f
Double coursing, 10.35, 10.35f, 10.36
Double headed (duplex) nails, 2.21
Double-insulated/ungrounded tools, 12.3, 12.15, 12.15f
    defined, 12.44
Double-J hooks, 16.21f
Double joists, 6.20, 6.32, 6.33, 6.33f
Double-L stairways, 11.3f
Double stretcher garden wall patterns, 5.23, 5.24f
Double-thickness flares, 17.14, 17.14f
Double top plates, 7.2, 7.2f
    defined, 7.35
Double trapping, 15.6
    defined, 15.32
Dovetail garden wall patterns, 5.23, 5.24f
Dow Styrofoam, 10.4
Drain, waste, and vent (DWV) systems
    building drains, 15.22
    building sewers, 15.2, 15.22
    copper fittings for, 17.7
    defined, 15.32
    described, 15.2, 15.3f, 15.4f
    fittings and applications, 15.14–15.16, 15.14f, 15.15f,
      15.16f, 15.17f, 15.18–15.19, 15.18f, 15.19f
    fixture drains, 15.2, 15.4f, 15.6, 15.6f
    grade and, 15.21–15.22, 15.21f
    health issues, 15.25
    municipal waste treatment systems, 15.22–15.23, 15.23f
    piping drawings and, 14.10
    plastic fittings for, 16.8, 16.9, 16.10f
    plumbing codes, 15.24–15.25
    private waste control systems, 15.22, 15.23, 15.24f
    sewer mains, 15.22
    sizing, 15.13
    traps, 15.2–15.11, 15.5f, 15.6f, 15.7f, 15.8f, 15.9f, 15.10f
    vents, 15.12, 15.12f
    waste treatment, 15.22–15.23, 15.24, 15.24f
Drainage fittings
    bends, 15.14–15.16, 15.15f, 15.16f
    code requirements for, 15.14, 15.14f
    defined, 15.32
    materials for, 15.14, 15.14f
    wyes, 15.18–15.19, 15.18f, 15.19f
Drawings
    building plan drawings, 1.2–1.3, 1.4f
    dimensions versus scale in, 6.11
    electrical drawings, 13.37–13.39, 13.40f
    HVAC drawings, 14.9–14.10, 14.11f

identifying construction and control joints on, 3.4
plan view drawings, foundation plans, 5.17, 5.17f, 6.7
working drawings, 6.7–6.8, 6.9f, 6.10–6.12
Drawing sets
blueprints, 6.11–6.12, 14.9, 14.10, 14.11f
detail drawings and, 6.8
floor systems and, 6.7–6.8, 6.8f, 6.9f, 6.10
plumbing, mechanical, and electrical plans, 6.11
schedules and, 7.9, 7.10f
section drawings, 6.8, 6.9f
structural drawings, 6.10
Drawn copper, 17.2
defined, 17.27
Dress, for masonry work, 4.29–4.30, 4.29f
Drilling wooden joists, 6.22, 6.23f
D-rings, 4.40, 4.40f, 4.42
Drip edges, 8.15, 8.16f, 9.10–9.12, 9.11f, 9.12f
Drop chutes, 3.8, 3.12, 3.12f, 3.15
Drop ear ells, 17.5
Drop-forged, 17.5
defined, 17.27
Dropping loads, and forklifts, 4.50
Drops, and traps, 15.7, 15.7f
Drop testing, of safety nets, 4.44, 9.27
Drop wood siding, 10.19f, 10.29, 10.29f, 10.31, 10.31f
Dry bonding, 5.16, 5.18, 5.18f, 5.19f
Dry bonds, defined, 5.49
Dryer loads, 13.6, 13.7
Drying-in, 7.19
defined, 7.35
Drying time, stucco, 10.65
Duck bill trowels, 4.17f
Duct cleaning, 14.6
Duplex or scaffold (doublehead) nails, 2.21
Duplex receptacles
duplex receptacle branch circuit, 13.38f
grounding-type duplex receptacles, 13.51, 13.51f
split-wired duplex receptacles, 13.42, 13.42f, 13.43f
Dutch bonds, 5.23, 5.23f
DWV systems. See Drain, waste, and vent (DWV) systems

E
Ears, and concrete blocks, 4.11f, 5.4f
Earthquakes
copper pipe hangers and supports, 17.16
footings and, 2.19, 2.22
plastic pipe supports and, 16.19
post-and-beam framing and, 6.7
Earthwork, 1.38
defined, 1.46
Eaves
defined, 10.71
frieze boards and, 10.6, 10.7, 10.7f
installing vinyl and metal siding under, 10.63–10.64,
10.63f, 10.64f
Edge forms, 2.27–2.28, 2.29f, 2.30f, 2.31
Edgers, 3.22, 3.23f
Edging, and finishing concrete, 3.22, 3.22f, 3.23f
Edging mortar, 4.25–4.26, 4.26f
Edison, Thomas, 13.2
EDMIs (Electronic distance measurement instruments),
1.19–1.20, 1.20f
Efficiency ratings, for furnaces, 14.5
EIFS (Exterior insulation and finish systems), 10.66–10.67,
10.66f
Elastomeric, 3.28
defined, 3.38

Elastomeric seals, 16.12
defined, 16.30
Elbows, 15.14, 17.5
Electrical accidents, 12.2, 12.3–12.4
Electrical burns, 12.3, 12.5
Electrical current
effects on human body, 12.2–12.5, 12.4f
safety and, 12.3, 12.4, 12.5
Electrical drawings, 13.37–13.39, 13.40f
Electrical hand tools, electrical injuries, 12.3
Electrical plans, and floor systems, 6.11
Electrical service
calculating load, 13.5–13.6, 13.5f
grounding and, 13.15–13.24, 13.16f, 13.17f, 13.18f, 13.19f
sizing, 13.2–13.3, 13.3f, 13.4f, 13.5–13.8, 13.5f
Electrical shock
body resistance and, 12.4–12.5, 12.4f
burns from, 12.3, 12.5
as cause of electrical injuries, 12.3–12.4
electrical current and, 12.2–12.3, 12.4
factors affecting severity of, 12.5
Electrical symbols, 13.45, 13.47f
Electric heating, 13.6, 13.61–13.62, 13.63f, 14.2
Electric heating loads, 13.6, 13.61
Electricity, as heat source, 14.2
Electric range circuits, 13.43, 13.43f
Electrocution
construction sites and, 12.3
GFCIs and, 4.30, 12.15
grounding prongs and, 12.28
mobile boom-type concrete pumps and, 3.11
scaffolding clearance and, 12.23
Electrolysis, and dielectric unions, 17.5
Electronic air cleaners, 14.6, 14.6f
Electronic distance measurement (EDM), 1.19–1.20, 1.20f
Electronic distance measurement instruments (EDMIs),
1.19–1.20, 1.20f
Electro-optical EDMIs, 1.19–1.20
Elephant trunks (tremies), 3.8, 3.12, 3.12f
Elevated loads, placing with forklifts, 4.49
Elevations
civil plans and, 1.2–1.3, 1.3f, 1.4f, 1.5f
defined, 15.32
differential leveling and, 1.31
manholes and, 15.22
transferring up structures, 1.37–1.38, 1.37f
Ells, 15.14, 17.5
Emergency hand signals, 4.31, 4.31f
Emergency removal authorization, 12.18
Emergency responses
fall rescues, 4.44, 9.27, 12.37
for overheating injuries, 4.37
Employees
lockout/tagout procedures and, 12.17
responsibilities of, 12.12
safety obligations of, 12.12
Employers
responsibilities of, 12.12
safety obligations of, 12.12
Ending newel posts, 11.6, 11.7f
End nosing, 11.7f, 11.8
Ends, and concrete blocks, 4.11f, 5.4f
End-to-back pipe measuring, 17.9, 17.9f
End-to-center pipe measuring, 17.8, 17.9f
End-to-end pipe measuring, 17.8, 17.9f
End-to-face pipe measuring, 17.9, 17.9f
Energy-efficient buildings, indoor environmental quality,
14.6

Energy isolation device, defined, 12.17
Energy Star® program, 14.14
Engineered lumber girders, 6.15, 6.15*f*, 6.16
Engineered wood products, 6.15, 6.15*f*, 6.16, 7.6, 7.7
Engineer's levels, 15.36
Engineer's rods, 1.23–1.24, 1.24*f*, 1.25
English bonds, 5.23, 5.23*f*
English cross bonds, 5.23, 5.23*f*
Entry procedures, for confined spaces, 12.29–12.30, 12.29*f*
Environmental Protection Agency (EPA)
    Clean Air Act and, 14.18
    Energy Star® program, 14.14
    PCB disposal standards, 12.33
    refrigerant transition and recovery certification, 14.17
Environmental safety, and HVAC technicians, 14.18
Equipment grounding, 13.15–13.16, 13.34–13.35, 13.36*f*, 13.37
Erecting trusses, 8.21, 8.21*f*
Erecting walls, 7.16–7.19, 7.16*f*, 7.17*f*, 7.18*f*
Errors, in distance measurement, 1.13, 1.17, 1.18
Escape routes, 4.52
Estimates, 7.25
Estimating material
    for bridging, 6.41
    for ceiling joists, 7.23*f*, 7.26
    for diagonal bracing, 7.23*f*, 7.25–7.26
    for floor joists, 6.40
    floor plans and, 7.23
    for floor systems, 6.40–6.41, 6.40*f*
    for gable roofs, 8.26
    for girders, 6.40
    for headers, 7.23*f*, 7.24–7.25
    for panel and board siding, 10.20–10.21
    for rafters, 8.26
    for ridgeboards, 8.26
    for roof framing, 8.25–8.26
    for roofing, 9.27–9.28, 9.28*f*
    for sheathing, 8.26
    for sills, 6.40
    for soleplates, 7.23*f*, 7.24
    for studs, 7.23*f*, 7.24
    for subflooring, 6.41
    for top plates, 7.23*f*, 7.24
    for wall framing, 1.23–1.26, 1.23*f*
Estimating volume of concrete, 2.9–2.10, 2.9*f*, 2.10*f*, 2.11, 2.11*f*, 2.12, 2.12*f*
Ethical principles, 4.22
Evaporation, 15.8, 15.10
    defined, 15.32
Evaporators
    chillers as, 14.15
    defined, 14.23
    mechanical refrigeration and, 14.7
Excavations, and mobile boom-type concrete pumps, 3.11
Expanded polystyrene fiberboard sheathing, 10.4
Expansion devices
    defined, 14.23
    mechanical refrigeration and, 14.7
Expansion joints, 3.3, 3.3*f*, 5.2–5.3, 5.3*f*, 5.4*f*
Exposed lamps, and safety, 12.15
Exposed rebar, 2.16
Exposure
    defined, 9.98
    wood shingles and shakes and, 9.5, 9.59, 9.59*f*
    wood siding and, 10.22, 10.23, 10.23*f*
Extension cords, and safety, 12.15
Extension ladders, and safety, 12.21, 12.21*f*
Extension rings, 13.48, 13.48*f*

Exterior finishing. *See also* Cornices; Flashing; Wood siding
    direct-applied exterior finish systems, 10.66–10.67, 10.66*f*
    exterior insulation and finish systems, 10.66–10.67, 10.66*f*
    fiber-cement siding, 10.44, 10.45, 10.46–10.47*f*, 10.48*f*
    insulation and, 10.4
    stucco (cement) finishes, 10.64–10.65, 10.65*f*
    veneers, 10.4, 10.65, 10.71
Exterior insulation and finish systems (EIFS), 10.66–10.67, 10.66*f*
External vibrators, 3.16, 3.17
Eye bolts, 4.42, 4.42*f*
Eye injuries, 4.29
Eye protection, 12.11

F
Fabrication, steel stud walls, 7.28, 7.29*f*
Face cuts, 5.25, 5.25*f*
Face protection, 12.11
Face shell mortar bedding, compressive strength, 5.2
Face shells, concrete blocks, 4.11*f*, 5.4*f*
Face shields, 12.9, 12.10, 12.19
Face-to-back pipe measuring, 17.9, 17.9*f*
Face-to-face pipe measuring, 17.9, 17.9*f*
Facial hair, and respirators, 12.32
Facing bricks, 4.5, 4.7
    defined, 4.57
Fall. *See also* Grade; Slope
    defined, 15.32
    traps and, 15.7, 15.7*f*
Fall-arrest systems. *See* Fall protection; Personal fall-arrest systems (PFAS)
Falling objects, 4.30–4.31, 4.31*f*
Fall protection. *See also* Personal fall-arrest systems (PFAS); Personal protective equipment (PPE); Safety
    equipment, 9.23–9.27, 9.24*f*, 9.25*f*, 9.26*f*
    erecting roofs and, 8.11
    guardrails and, 4.38, 4.38*f*, 7.20
    improper use of, 4.37, 4.38*f*
    masonry work and, 4.37–4.44, 4.38*f*, 4.39*f*, 4.40*f*, 4.41*f*, 4.42*f*
    procedures, 12.34–12.38, 12.34*f*, 12.36*f*, 12.38*f*
Fall rescues, 4.44, 9.27, 12.37
False fascia, 8.7
    defined, 8.35
Fascia
    cornices and, 10.7, 10.7*f*
    defined, 10.71
    false fascia, 8.7, 8.35
    safety and, 10.4
Fatal shock, 12.3–12.4
Feeders
    branch circuits and, 13.37–13.39, 13.38*f*, 13.39*f*, 13.40*f*
    defined, 13.72
    residential electrical service planning, 13.2
    wiring methods for, 13.29–13.33, 13.30*f*, 13.31*f*, 13.32*f*, 13.34*f*
Felt installation, 8.17
Felt paper, 10.39
    defined, 10.71
Female adapters, 17.5
Fernco couplings, 15.17*f*
Ferrous metal piping, steel pipe, 14.10
Ferrules, 17.5
    defined, 17.27
Fiber-cement siding, 10.44, 10.45, 10.46–10.47*f*, 10.48*f*
Fibrillation, 12.3, 12.4
    defined, 12.44

Field-fabricated housed stringers, 11.19
Field-marked flash protection signs, 12.10
Field notes, 1.7, 1.36–1.37, 1.36f
    defined, 1.46
Fill, defined, 1.46
Finish coats
    defined, 10.71
    stucco (cement) finishes and, 10.64–10.65, 10.65f
Finished floors, and riser height, 11.18
Finished stringers, 11.7f, 11.8
Finishing concrete
    brooming, 3.27, 3.27f
    control joints and, 3.22–3.24, 3.24f
    edging and, 3.22, 3.22f, 3.23f
    floating, 3.25–3.26, 3.25f, 3.26f
    leveling, 3.21–3.22, 3.21f
    screeding, 3.18–3.19, 3.19f, 3.20, 3.20f, 3.21
    troweling, 3.26–3.27, 3.26f, 3.27f
Finish-to-finish measurements, 7.4
Fire and explosion hazards, and forklifts, 4.51–4.52
Fire bricks, 4.5, 4.7
Fireman's rule, for ladders, 12.22
Firestops, 6.4, 7.15, 7.16f
    defined, 6.50
Fire suits, 12.9
First aid, 12.30
Fittings
    for drain, waste, and vent (DWV) systems, 15.14–15.16,
        15.14f, 15.15f, 15.16f, 15.17f, 15.18–15.19, 15.18f, 15.19f
    for plastic pipe, 16.8, 16.9, 16.9f, 16.10f
Fixed appliance circuits, 13.6
Fixture drains, 15.2, 15.4f, 15.6, 15.6f
    defined, 15.32
Fixture traps, 15.2
Flagstone, 4.13
Flammable gases, and fire fighting, 4.52
Flammable liquids, and safety, 4.32, 4.52
Flare fittings, 17.5, 17.14, 17.14f
Flare joints, 17.11, 17.14, 17.14f
    defined, 17.27
Flaring tools, 17.14, 17.14f
Flashing. See also Roof projections and flashing; Step flash-
        ing; Valley flashing; Wall flashing
    chimneys and, 5.14, 9.46, 9.46f
    corner flashing, 10.26
    exterior finishing and, 10.4–10.5, 10.6f
    masonry walls, 5.13, 5.13f
    ridge flashing, 9.72, 9.73f
    vent stack flashing, 9.40, 9.40f, 9.98
    windows and, 10.5, 10.6f
Flash protection boundaries, 12.20
Flash suits, 12.10
Flat roofs, 9.52
Flat tile roofs, 9.79, 9.79f, 9.80, 9.80f
Flatwork, 2.27
    defined, 2.38
Flemish bonds, 5.23, 5.23f
Flexible cords, and safety, 12.15
Flexible metal conduit (FMC), 13.35
Floating, and finishing concrete, 3.25–3.26, 3.25f, 3.26f
Floor blocks, 4.10f, 5.6f
Floor joists
    described, 6.20, 6.20f, 6.21f, 6.22
    estimating material for, 6.40
    installing, 6.33, 6.34f
    notching and drilling, 6.22, 6.23f
    sizing, 6.43, 6.43f
    trusses as, 6.24, 6.24f
    wood I-beams, 6.22, 6.23, 6.23f

Floor joist transitions, 6.37
Floor loads, 6.41, 6.42f, 6.43
Floor plans
    estimating material and, 7.23
    floor systems and, 6.7, 6.10
    laying out wall openings and, 7.9, 7.9f
    locating receptacles and, 13.40, 13.41f
    sizing electrical service and, 13.2–13.3, 13.3f, 13.4f
Floors
    board subfloors, 6.26–6.27
    cantilevered floors, 6.38, 6.39f, 6.40
    finished floors, 11.18
    flatness and, 3.18
    framing openings in, 6.35, 6.35f
    plywood subfloors, 6.26
Floor systems
    bridging, 6.24–6.25, 6.25f
    cantilevered floors and, 6.38, 6.39f, 6.40
    construction of, 6.33, 6.34f, 6.35–6.38, 6.35f, 6.36f, 6.37f,
        6.38f
    drawings and specifications and, 6.7–6.8, 6.8f, 6.9f,
        6.10–6.12
    estimating material quantities, 6.40–6.41, 6.40f
    floor joists, 6.20, 6.20f, 6.21, 6.21f, 6.22, 6.23f, 6.24, 6.24f
    girders and supports, 6.15–6.18, 6.15f, 6.17f, 6.18f, 6.19f
    house framing methods and, 6.2, 6.3f, 6.4, 6.5f, 6.6f, 6.7
    layout of, 6.27–6.32, 6.28f, 6.29f, 6.30f, 6.31f, 6.32f
    sills, 6.12, 6.13f, 6.14, 6.14f, 6.15
    sizing girders and joists, 6.41, 6.42f, 6.43, 6.43f
    subflooring, 6.25–6.27, 6.26f
Floor-type concrete saws, 3.24, 3.25f
Florida rods, 1.23
Flush stringer attachment, 11.17, 11.18f
Flush top surfaces, 6.21
Fluxes, 17.11
F-numbers, 3.18
Foam core construction. See Cellular core wall plastic pipe
Footing forms, 2.21–2.22, 2.23f, 2.24f
Footings
    components of, 2.21–2.22, 2.23f, 2.24f
    concrete forms for, 2.21–2.22, 2.23f, 2.24f
    defined, 2.38, 4.57
    as foundation elements, 2.18–2.21, 2.20f, 2.21f
    laying out, 2.24
    masonry and, 4.17
    size and shape of, 2.18–2.19
    types of, 2.20–2.21, 2.20f, 2.21f, 2.22f
Forced-air heating, 14.2, 14.3f
Foresights (FS)
    defined, 1.46
    differential leveling and, 1.32
    line-of-sight check and, 1.29
Forklifts
    accidents involving, 4.44–4.45
    daily inspections of, 4.45, 4.46f
    fire and explosion hazards and, 4.51–4.52
    general safety precautions for operating, 4.45, 4.47
    load handling, 4.48–4.50, 4.49f, 4.50f, 4.51
    pedestrian safety and, 4.47, 4.52–4.53
    training and certification for operating, 4.45
    traveling, 4.47–4.48, 4.47f
Forks, in low position for travel, 4.48, 4.48f
Formability, 17.3
    defined, 17.27
Form oil, 3.14
Form pressures, 2.17, 2.18
Form-release compounds, 3.14

Forms
    checking of, 3.14
    for concrete stairs, 11.21, 11.21*f*
    defined, 2.38
    hardening of concrete and, 2.2
Forms for concrete
    edge forms, 2.27–2.28, 2.29*f*, 2.30*f*, 2.31
    footings, 2.21–2.22, 2.23*f*, 2.24*f*
    placing concrete and, 3.14–3.16, 3.15*f*
    preparing for concrete placement, 3.14, 3.15
    removing, 2.26, 2.31, 3.29
    safety and, 2.17–2.18
    slabs, 2.27–2.28, 2.29*f*, 2.30*f*, 2.31
    wall forms, 2.24–2.27, 2.25*f*, 2.26*f*, 2.27*f*
Form spreaders, 2.25, 2.25*f*
Form ties, 2.25, 2.25*f*, 2.26, 2.26*f*
Formulas
    for finding areas, 2.40, 10.20–10.21
    for finding volumes, 2.9, 2.10, 2.12, 2.40
Foundation plans
    floor systems and, 6.7
    locating masonry walls and, 5.17, 5.17*f*
Foundations
    checking for squareness, 6.14, 6.27–6.28, 6.28*f*
    defined, 6.50
    footings and, 2.18–2.21, 2.20*f*, 2.21*f*
Foundation walls, as girder supports, 6.18, 6.19*f*
Four-way switches, residential wiring, 13.52, 13.58, 13.58*f*, 13.59*f*
Framing. *See also* Roof framing; Wall framing
    balloon-frame construction, 6.4, 6.5*f*, 7.4
    braced frames, 6.2, 6.4
    ceilings, 7.19–7.21, 7.22*f*
    gable ends, 8.12, 8.13*f*
    house framing methods, 6.2, 6.3*f*, 6.4, 6.5*f*, 6.6*f*, 6.7
    of joist at girder, 6.20, 6.21*f*
    plank-and-beam framing, 8.27–8.28, 8.28*f*
    post-and-beam framing, 6.4, 6.6*f*, 6.7, 8.27–8.28, 8.28*f*, 8.36*f*
    spacing of, 7.2
    stairs, 11.8–11.9, 11.9*f*
    western platform framing, 6.2, 7.4
Framing openings
    in floors, 6.35, 6.35*f*
    in roofs, 8.13, 8.14, 8.15*f*, 8.16
    for stairwells, 11.13–11.14
    in steel stud walls, 7.28, 7.29*f*
Framing squares
    cutting stringers and, 11.15–11.17, 11.15*f*, 11.16*f*, 11.18*f*
    hip rafter length and, 8.36, 8.38–8.39, 8.38*f*
    rafter framing squares, 8.4–8.5, 8.5*f*, 8.6*f*
    rafter length and, 8.4–8.5, 8.6*f*, 8.9–8.10, 8.9*f*
Free-fall distances, 4.39, 9.23
Fresh air ventilation, 14.6
Frieze boards, 10.6, 10.7, 10.7*f*
    defined, 10.71
Frogged brick concrete blocks, 4.10*f*, 5.6*f*, 5.8, 5.8*f*
Frogs, 4.12*f*, 5.8, 5.8*f*
Frost, and footings, 2.19
Frostbite, 4.36–4.37
FS (foresights), 1.29, 1.32, 1.46
Fuel oils, 14.2
Fuels, and heating, 14.2
Fullback cut blocks, 5.26, 5.26*f*
Full-body harnesses, 9.23–9.24, 9.24*f*, 9.26, 9.26*f*, 12.35–12.36
Full-cut header blocks, 4.10*f*
Full mortar bedding, and compressive strength, 5.2
Fumes, and liquid-fueled tools, 4.35

Furnaces
    condensing furnaces, 14.5
    heat exchangers and, 14.2, 14.3*f*, 14.4*f*, 14.5
    high-efficiency furnaces, 14.5
Furring out windows and doors, 10.52
Furring strips
    defined, 7.35
    metal and vinyl siding and, 10.51–10.52, 10.52*f*, 10.53*f*
    vertical boards and, 10.28
    wall framing in masonry and, 7.26, 7.27*f*, 7.28, 7.28*f*
Furrowing, defined, 4.57
Furrowing mortar, 4.26, 4.26*f*, 5.19
Fuse handling equipment, 12.10, 12.15
Fuse pullers, 12.10, 12.15
Fuse replacements, 12.14–12.15
Fusion fittings, 16.12, 16.13*f*
    defined, 16.30
Fusion joining, 16.17, 16.17*f*
Fusion tools, 16.17, 16.17*f*

G
Gable and valley roofs, 8.2, 8.2*f*, 8.36, 8.36*f*, 8.37*f*
Gable dormers, 8.26, 8.27*f*
Gable ends
    framing, 8.12, 8.13*f*
    siding installation and, 10.63, 10.63*f*
Gable end trim, vinyl and metal siding, 10.56, 10.57*f*
Gable end vents, 8.12, 8.13*f*
Gable overhangs, 8.13, 8.14*f*
Gable roofs
    ceiling layout and, 7.19
    composition shingle installation and, 9.32–9.34, 9.34*f*, 9.36, 9.36*f*, 9.37*f*, 9.43
    defined, 7.35
    erecting, 8.11–8.13, 8.11*f*, 8.12*f*, 8.13*f*, 8.14*f*, 8.15*f*
    estimating materials and, 8.26
    as roof type, 8.2, 8.2*f*
    wall framing and, 7.19
Gables, 8.2
    defined, 8.35
Gable studs, 8.12, 8.13*f*
Gambrel roofs, 8.2, 8.2*f*
Gammon reels, 1.11, 1.11*f*, 1.12
Garden wall bonds, 5.23, 5.24*f*
Gas forced-air furnaces, high-efficiency, 14.5
Gasoline, safety, 4.32
Gasoline-powered tools, safety, 4.34
Gas-removing respirators, 12.32
Gate valves, 17.6, 17.7*f*
Gauge blocks, aligning walls, 7.17–7.18, 7.17*f*
Genie lifts, 12.24
Geometrical stairs, 11.4
    defined, 11.26
GFCI. *See* Ground-fault circuit interrupters (GFCI)
Gibbs, Joseph, 4.8
Girder hangers, 6.18, 6.19*f*
Girder pockets, 6.18, 6.19*f*
Girders. *See also* Beams
    estimating material for, 6.40
    floor systems and, 6.15–6.18, 6.15*f*, 6.17*f*, 6.18*f*, 6.19*f*
    installing, 6.30–6.31, 6.31*f*
    laying out, 6.31–6.32, 6.32*f*
    sizing of, 6.41, 6.42*f*, 6.43, 6.43*f*
    support columns and, 6.17–6.18, 6.17*f*, 6.18*f*, 6.30–6.31, 6.31*f*
    trusses and, 8.21, 8.23
Girder supports, 6.17–6.18, 6.17*f*, 6.18*f*, 6.30–6.31, 6.31*f*
Glazed bricks, 4.7

Globe valves, 17.6, 17.7f
Gloves, and safety, 3.33
Glue-laminated lumber (glulam), 6.15f, 6.16
Glue-laminated lumber (glulam) headers, 7.6
Go-no-go crimp measuring gauges, 16.16, 16.16f
Goosenecks, 11.6, 11.7f
Grade. See also Fall; Slope
     defined, 15.32
     drain, waste, and vent systems and, 15.21–15.22, 15.21f
Grade beams, 2.20, 2.21, 2.22f
Grading, of rebar, 2.12–2.13, 2.14, 2.14f
Grease interceptors, 15.4, 15.5f
Green concrete, 2.2
     defined, 2.38
Grid leveling, 1.38
Grooved copper, 17.5
Grounded tools, 12.15
     defined, 12.44
Ground-fault circuit interrupters (GFCI)
     circuit breakers, 13.11–13.12, 13.11f, 13.13, 13.13f, 13.14, 13.14f
     defined, 12.44
     direct-wired ground-fault circuit interrupters, 13.14, 13.14f
     kitchen countertop receptacles and, 13.42
     noise suppression and, 13.14
     operating circuitry of, 13.11, 13.11f
     power tool safety and, 4.30, 12.3, 12.15, 12.16, 12.16f
     safety and, 4.30, 12.3, 12.15, 12.16, 12.16f, 13.10
     single-pole ground-fault circuit interrupters, 13.11–13.12, 13.12f
     sizing residential load centers and, 13.10–13.12, 13.13, 13.13f, 13.14, 13.14f
     two-pole ground-fault circuit interrupters, 13.11, 13.12, 13.13f, 13.14
Grounding
     equipment grounding, 13.15–13.16, 13.34–13.35, 13.36f, 13.37
     with flexible metal conduit (FMC), 13.35
     with liquidtight flexible metal conduit (LFMC), 13.35
     NEC® standards, 13.15–13.24, 13.18f, 13.19f
     residential grounding system and, 13.15–13.24, 13.16f, 13.17f, 13.18f, 13.19f
     system grounding, 13.15–13.16
Grounding devices, 13.37
Grounding electrode conductors (GEC), 13.15, 13.20–13.22
Grounding electrodes, 13.17–13.22, 13.18f, 13.19f
Grounding prongs, and safety, 12.28
Grounding-type duplex receptacles, 13.51, 13.51f
Ground rings, 13.17
Grout
     architectural terra-cotta and, 4.7
     characteristics of, 4.14
     concrete blocks and, 4.11
     in construction joints, 3.2
     defined, 3.38, 4.57
     pumping, 5.5
Guardrails
     defined, 11.26
     safety and, 4.38, 4.39f, 7.20, 9.23
     stairways and, 11.6, 11.7f
     temporary guardrails, 11.10, 12.34, 12.34f
Guardrail systems, 12.34–12.35, 12.34f
Guilds, 4.4
Gunite, 3.14
Gutters, and exterior finishing, 10.2, 10.64, 10.64f

H
Hacked bricks, 5.31f
Hacksaws, 17.9, 17.10
Halfback cut blocks, 5.26, 5.26f
Half-cut header blocks, 4.10f
Half-high rib blocks, 5.25
Half-laced valleys. See Closed-cut valleys
Half weave valleys. See Closed-cut valleys
Hammerhead cranes, 3.6, 3.9f
Hammers, cutting masonry with, 5.26–5.27, 5.27f
Hand-crimping tools, 16.16, 16.16f
Hand floats, 3.25, 3.25f
Hand grinders with diamond wheels, 9.15, 9.16f
Handheld vibrating screeds, 3.17, 3.20, 3.21
Hand jointers (groovers), 3.23–3.24, 3.24f
Handrails, 11.6, 11.7f, 11.9
     defined, 11.26
Hand signals, 1.9, 1.10f, 1.11, 4.31, 4.31f
Hand tools
     cutting masonry and, 5.26–5.28, 5.27f
     safety and, 12.26, 12.27f
Hand trowels, 3.26, 3.26f, 3.27f
Hardboard and particleboard siding, 10.39–10.41, 10.40f, 10.41f, 10.42f, 10.43f
Hardboard shingle/shake panels, 9.67, 9.67f, 9.68f
Hard hats, 12.19
Hardipanel, 10.45, 10.48f
Hardscape, 5.7, 5.7f
Hazard analysis, 12.20
Hazards. See also Safety
     confined spaces and, 12.29–12.30
     of construction sites, 4.30, 12.2, 12.3
     fire and explosion, 4.51–4.52
     grounding and, 13.16
     residential wiring and, 13.2
HCFC (hydrochlorofluorocarbon) refrigerant, 14.18, 14.23
H-clips, 8.15, 8.15f
Header blocks, 6f.9f
Header bricks, 5.11f, 5.21, 5.22f
Header joists, 6.20, 6.20f
     defined, 6.50
Headers
     bricks as, 4.7, 4.8f
     defined, 7.35
     estimating material for, 7.23f, 7.24–7.25
     types of, 7.6, 7.6f, 7.7f
     as wall components, 7.2, 7.2f, 7.4, 7.6, 7.6f, 7.7, 7.7f
Head joints, 4.17, 5.18–5.20
     defined, 4.57
Head laps
     defined, 9.98
     slate roofing and, 9.74
Headroom
     clearances, 11.7f, 11.8
     defined, 11.26
Heads
     defined, 17.27
     of water, 17.21
Heart rhythms, and electrical shock, 12.3, 12.4
Heat cramps, 4.37
Heat exchangers, and furnaces, 14.2, 14.3f, 14.4f, 14.5
Heat exhaustion, 4.37
Heating
     electric heating, 13.6, 13.61–13.62, 13.63f, 14.2
     forced-air heating, 14.2, 14.3f
     fuels and, 14.2
     hot water heating, 14.2, 14.4f, 17.4
     introduction to, 14.2

Heating, Ventilating, and Air Conditioning (HVAC) drawings, 14.9–14.10, 14.11f
Heating, Ventilating, and Air Conditioning (HVAC) industry
    blueprints, 14.9, 14.10, 14.11f
    building codes, 14.9–14.10, 14.12
    careers in, 14.13, 14.14, 14.14f, 14.15
    environmental concerns, 14.18
    indoor environmental quality and, 14.6
    introduction, 14.2
    specifications, 14.12–14.13
    training programs, 14.15–14.18
Heating, Ventilating, and Air Conditioning (HVAC) plans, 14.10, 14.23
Heating, Ventilating, and Air Conditioning (HVAC) schedules, 14.10, 14.12f
Heat pumps
    described, 14.8
    electric heating and, 13.62
    high-efficiency, 14.14
    use of, 14.2
Heat strokes, 4.37
Heat transfer
    defined, 14.23
    principle of, 14.2
Heavy rollers, 9.15, 9.15f
Heavyweight (high-density) concrete, 2.6
Heel inlets, 15.15, 15.15f
Height, checking when laying masonry units, 5.30, 5.30f
Height of instrument (HI)
    defined, 1.46
    differential leveling and, 1.32
Helpers, 4.18
Hermetic rotary compressors, 14.9
Hermetic (welded hermetic) compressors, 14.9
Herringbone block bonds, 5.24f
Herringbone patterns, 4.2, 4.2f
High-efficiency air conditioning, 14.14
High-efficiency boilers, 14.5
High-efficiency furnaces, condensing furnaces, 14.5
High-efficiency gas forced-air furnaces, 14.5
High-efficiency heat pumps, 14.14
High-pressure water washing, 5.45
High-strength concrete, 2.6
High-voltage, and fatal accidents, 12.3
High-voltage boots, 12.9
High-voltage sleeves, 12.9
High wind areas. See also Wind
    composition shingles and, 9.35
    fastening systems for tile roofs, 9.76, 9.76f, 9.77f
    high-windload vinyl siding, 10.59
    siding seam laps and, 10.58
Hip and valley roofs, 8.2, 8.2f, 8.36, 8.36f, 8.37f
Hip caps. See Ridge and hip caps
Hip rafter position, 8.38, 8.38f
Hip rafters
    described, 8.3, 8.4f
    laying out with speed square, 8.42–8.43, 8.43f
Hip roofs, 7.19, 8.2, 8.2f
    defined, 7.35
Hip rows
    composition shingle installation and, 9.49, 9.49f, 9.51, 9.51f, 9.52f
    slate roofing and, 9.74, 9.74f
    wood shingles and, 9.63, 9.64f
Hoists, and safety, 12.23–12.24
Hold-down systems, clay, ceramic, and concrete tile installation, 9.76, 9.76f, 9.77f
Hollow brick, 4.5, 4.12f

Hollow concrete blocks, 4.11, 5.3
Hollow masonry units/tiles, 4.6–4.7, 5.10
Hollow masonry walls, 4.15, 4.15f
Hoppers, 3.12, 3.12f
Horizontal crosshair tests, 1.28, 1.28f
Horizontal lifelines, 4.42, 4.42f
Horizontal pipe supports, 16.20, 16.22f
Horizontal scored blocks, 5.9f
Horseplay, forklifts and, 4.48
Hot sticks, 12.10
Hot tubs, residential, 13.65, 13.65f
Hot water heating, 14.2, 14.4f, 17.4
Hot weather
    placing and finishing concrete and, 3.16
    safety and, 4.37
Housed stringers, 11.5, 11.6f, 11.19
    defined, 11.26
House framing methods, 6.2, 6.3f, 6.4, 6.5f, 6.6f, 6.7
Hub stakes, 1.7, 1.7f, 1.8, 1.8f
Humidifiers, 14.6, 14.6f
Hurricane-prone areas
    balloon framing and, 6.4
    fastening systems, 9.76, 9.77f
HVAC drawings, 14.9–14.10, 14.11f
HVAC industry. See Heating, Ventilating, and Air Conditioning (HVAC) industry
HVAC plans, 14.10
    defined, 14.23
HVAC schedules, 14.10, 14.12f
Hydration
    as chemical reaction, 2.2
    curing concrete and, 3.27–3.28
    defined, 2.38
Hydraulic gradients, 15.8, 15.8f
    defined, 15.32
Hydraulic test pumps, 17.21, 17.21f
Hydrochlorofluorocarbon (HCFC) refrigerant
    defined, 14.23
    environment and, 14.18
Hydrofluorocarbons (HFCs), 14.18
Hydronic, defined, 16.30
Hydronic heating systems, boilers, 14.2
Hydronic radiant floor heating, 16.7
Hydrostatically pressure-test, 16.22
    defined, 16.30
Hydrostatic testing, 16.22, 16.23
Hygroscopic, 4.31
    defined, 4.57

I
IBC. See International Building Code (IBC)
Ice dams, 9.30, 9.30f, 9.91, 9.94
Ice edging, 9.91, 9.92f, 9.93, 9.94
Icy conditions, and safety, 4.41
Improvised safety equipment, 12.32
Indoor air quality (IAQ), 14.6
Indoor coils, 14.8
Indoor environmental quality and, 14.6
Injuries
    back injuries, 12.24, 12.25
    electrical injuries, 12.3–12.4
    eye injuries, 4.29
    overheating injuries, 4.37
Inlets, 15.6, 15.6f
Inside corner posts, 10.54, 10.54f
Inside corner strips, and wood siding, 10.20, 10.20f
Inside diameters (ID), 16.3
    defined, 16.30

Inspections
    commercial formwork and, 3.14, 3.15
    of fall arrest systems, 12.37
    of rubber gloves, 12.8–12.9, 12.9f
    of rubber protective equipment, 12.7–12.8
Installation. *See also* Composition shingle installation
    of beveled siding, 10.22–10.23, 10.22f, 10.23f, 10.24, 10.24f, 10.25, 10.25f, 10.26f, 10.27f
    of board-and-batten siding, 10.27–10.28, 10.27f, 10.28f, 10.29f
    of bridging, 6.35–6.36, 6.36f
    of ceiling joists, 7.21, 7.22f
    of clay, ceramic, and concrete tile, 9.76, 9.76f, 9.77, 9.77f, 9.78–9.80, 9.78f, 9.79f, 9.80f, 9.81f, 9.82–9.83, 9.82f, 9.83f
    of cornices, 10.13–10.14, 10.13f, 10.14f, 10.15f
    of direct-applied exterior finish systems (DEFS), 10.67
    of exterior insulation and finish systems, 10.67
    of felt, 8.17
    of fiber-cement siding, 10.44, 10.46–10.47f, 10.48f
    of floor joists, 6.33, 6.34f
    of girders, 6.30–6.31, 6.31f
    of hardboard and particleboard siding, 10.39–10.41, 10.40f, 10.41f, 10.42f, 10.43f
    of joists for cantilevered floors, 6.38, 6.39f, 6.40
    of panelboards, 13.27–13.28, 13.28f, 13.29f
    of plywood siding, 10.39
    of rafters, 8.11, 8.11f, 8.12f
    of roll roofing, 9.52–9.57, 9.53f, 9.54f, 9.55f, 9.56f, 9.57f
    of service entrances, 13.24–13.26, 13.25f, 13.26f, 13.27f, 13.29f
    of sheathing, 8.15–8.16, 8.15f, 8.16f, 8.17
    of shingle or shake siding, 10.34–10.35, 10.35f, 10.36, 10.37
    of shiplap siding, 10.31, 10.32, 10.33
    of sills, 6.28–6.30, 6.29f, 6.30f
    of slate or synthetic slate, 9.73–9.76, 9.74f, 9.75f
    of subflooring, 6.36–6.38, 6.37f, 6.38f
    of tongue-and-groove siding, 10.29, 10.30, 10.32, 10.33
    of traps, 15.6–15.7, 15.7f
    of trusses, 8.21, 8.22f, 8.23–8.24, 8.23f, 8.24f, 8.25
    of valley flashing, 9.28
    of vinyl and metal siding, 10.58–10.61, 10.59f, 10.60f, 10.61f, 10.62f, 10.63–10.64, 10.63f, 10.64f
Installation Safety Requirements, 12.20
Insulating blankets, 12.9
Insulation
    for copper pipe and fittings, 17.21, 17.21f
    defined, 17.27
    exterior finishing and, 10.4
Integral traps, 15.2, 15.6, 15.6f
Interceptors, 15.4–15.5
    defined, 15.32
Interference fits, 16.12
    defined, 16.30
Internal tube cutters, 17.9, 17.9f
Internal vibrators, 3.16, 3.17, 3.17f
International Building Code (IBC)
    defined, 14.23
    development of, 13.73
    local codes and, 14.12
    stair requirements, 11.8
International Plumbing Code (IPC), 16.20
International Residential Code (IRC), 13.73
Intersections
    forklifts and, 4.47
    in partitions, 7.4, 7.5f, 7.6f
Inverted wyes, 15.18, 15.18f
    defined, 15.32
IPC (*International Plumbing Code*), 16.20
IRC (*International Residential Code*), 13.73

Isolation joints, 3.3, 3.3f

J
Jack rafters
    described, 8.3, 8.4f
    layout, 8.36f, 8.39–8.40, 8.39f, 8.40f, 8.41f, 8.42
Jackscrews, 6.17f, 6.18
Jamb blocks, 4.10f, 5.6f, 5.7f
Jambs, 7.10
    defined, 7.35
J-channel, 10.55, 10.56, 10.57f
Jericho, 4.2
Jewelry, safety, 12.11
J-hooks, 16.21f, 17.16, 17.17f
Johnson, Isaac Charles, 2.3
Joining
    bell-and-spigot pipe, 16.15, 16.15f
    copper pipe and fittings, 17.8, 17.11–17.14, 17.12f, 17.13f, 17.14f
    CPVC (chlorinated polyvinyl chloride) pipe, 16.14–16.15, 16.14f
    PE (polyethylene) pipe, 16.17, 16.17f
    PEX (cross-linked polyethylene) tubing, 16.16–16.17, 16.16f
    plastic plumbing materials, 16.11–16.17, 16.12f, 16.13f, 16.14f, 16.15f, 16.16f, 16.17f
    PVC (polyvinyl chloride) pipe, 16.14–16.15, 16.14f
Joint fillers, 5.14, 5.14f
Joint rakers, 5.41, 5.42f
Joint reinforcement ties, 5.13, 5.13f
Joints
    classes of, 3.2–3.4
    construction joints, 3.2–3.3, 3.3f, 3.4
    control joints, 3.4, 3.4f, 3.22–3.24, 3.24f, 5.2–5.3, 5.3f
    decorative joints, 3.4, 3.4f
    defined, 4.57
    isolation joints, 3.3, 3.3f
    joint finishes, 5.40–5.41, 5.40f
    mortar joints, 4.5, 5.40–5.42, 5.40f, 5.41f, 5.42f
    striking of, 5.41–5.42, 5.41f, 5.42f
Joint sealants, 3.28–3.29
Joint hangers, 6.20, 6.21f
    defined, 6.50
Joist headers, 6.33, 6.34f, 6.40
Joist layout, 6.31–6.32, 6.32f, 6.33f, 7.19–7.21, 7.20f, 7.21f
Joist nails, 6.22
Joists. *See* Ceiling joists; Floor joists; Girders; Header joists
Journeyman masons, 4.4, 4.19–4.20

K
Keratin, 12.4
Keyways, 2.22, 2.23f
King closure bricks, 5.25, 5.25f
King studs
    described, 7.2, 7.2f
    measuring and cutting, 7.12–7.13, 7.13–7.14f
Kips, 2.13
    defined, 2.38
Kirklocks, 12.14
Knowledge, of masons, 4.20–4.21
Kobe, Japan, earthquake in, 6.7

L
Labeling
    copper pipe, 17.3
    plastic pipe, 16.3, 16.3f
Ladder conveyors, 9.22

Ladder jacks, 9.19, 9.19f
Ladders, and safety, 9.19, 9.20, 9.21, 9.21f, 12.20–12.22, 12.21f
Lally columns, 6.17f, 6.18
Laminated veneer lumber (LVL), 6.15f, 6.16
Laminated veneer lumber (LVL) headers, 7.6
Landing newel posts, 11.6, 11.7f
Landings, 11.2, 11.7f, 11.8, 11.11
    defined, 11.26
Lanyards, 4.40, 4.41, 4.41f, 9.24, 9.24f, 12.36–12.37, 12.36f
Laser leveling instruments
    described, 1.22, 1.22f
    wood siding and, 10.22
Laundry circuits, 13.5–13.6, 13.7
Laying masonry units
    cleaning, 5.44–5.46
    corners, 5.36, 5.37–5.40, 5.37f, 5.38f
    laying brick in place, 5.29–5.32, 5.29f, 5.30f, 5.31f
    laying to the line, 5.33–5.36, 5.33f, 5.34f, 5.35f, 5.36f
    leads, 5.36–5.39, 5.37f, 5.38f
    mortar joints, 5.40–5.42, 5.40f, 5.41f, 5.42f
    patching mortar, 5.42–5.43, 5.43f
    placing blocks, 5.32–5.33, 5.32f
Laying out
    ceilings, 7.19–7.21, 7.22f
    continuous footing forms, 2.24
    floor systems, 6.27–6.32, 6.28f, 6.29f, 6.30f, 6.31f, 6.32f
    hips and valleys, 8.36, 8.36f, 8.37f, 8.38–8.40, 8.38f, 8.39f, 8.40f, 8.41f, 8.42–8.43, 8.43f
    joist locations, 6.32, 6.33f
    masonry jobs, 5.16–5.18, 5.16f, 5.17f, 5.18f, 5.19f
    rafters, 8.7–8.8, 8.7f, 8.18, 8.18f, 8.19f, 8.20, 8.39, 8.39f, 8.40f, 8.42–8.43, 8.43f
    roof framing, 8.3–8.5, 8.4f, 8.5f, 8.6f, 8.7–8.14, 8.7f, 8.9f, 8.11f, 8.12f, 8.13f, 8.14f, 8.15f
    sills and girders for floor joists, 6.31–6.32, 6.32f
    soleplates, 7.8, 7.8f
    stairways, 11.10–11.17, 11.11f, 11.13f, 11.15f, 11.16f, 11.18, 11.18f, 11.19, 11.19f, 11.20, 11.20f, 11.21f
    stringers, 11.15–11.20, 11.15f, 11.16f, 11.18f, 11.19f, 11.20f
    studs, 7.8, 7.8f
    top plates, 7.8, 7.8f
    wall framing, 7.8–7.11, 7.8f, 7.9f, 7.10f, 7.11f, 7.12f
    wall openings, 7.9–7.11, 7.9f, 7.10f, 7.11f, 7.12f
Lead-free solder, 17.11
Lead mollies, 17.18
Leads, laying masonry units, 5.36–5.39, 5.37f, 5.38f
Leather glove covers, 12.6, 12.8
Leave-behind she-bolts, 2.26f
Ledgers
    concrete floors and, 11.21
    cornice installation and, 10.13, 10.13f
    defined, 10.71
    framing joists and, 6.21, 6.21f
Left-thumb rule, 1.27
Let-ins, 6.4
    defined, 6.50
Leveling
    differential leveling process and procedures, 1.2, 1.30–1.32, 1.31f, 1.33f, 1.34–1.35, 1.35f
    differential leveling tools and equipment, 1.20–1.30, 1.21f, 1.22f, 1.23f, 1.24f, 1.25f, 1.26f, 1.27f, 1.28f, 1.29f
    finishing concrete and, 3.21–3.22, 3.21f
    laying masonry units, 5.30, 5.30f
    leveling applications, 1.37–1.38, 1.37f
Leveling instruments
    initial setup and adjustment, 1.26–1.27, 1.26f, 1.27f, 1.28f
    testing calibration of, 1.28–1.30, 1.28f, 1.29f
    types of, 1.21–1.22, 1.21f, 1.22f

Leveling rods, 1.23–1.26, 1.23f, 1.24f, 1.25f
Levels, line-of-sight checks for, 1.28–1.30, 1.29f
Lewis, David, profile in success, 13.71
Lifting, and safety, 12.24, 12.25, 12.25f, 12.26
Lifts
    placing concrete and, 3.16, 3.17
    safety and, 12.23–12.24
Lighting
    layout of branch circuits for, 13.45, 13.46f, 13.47f
    lighting demand factors, 13.6, 13.7
Lighting control
    dimmer switches and, 13.52, 13.59
    four-way switches, 13.52, 13.58, 13.58f, 13.59f
    low-voltage electrical systems, 13.61
    photoelectric switches, 13.58
    switch locations, 13.59, 13.60f
    three-way switches and, 13.52, 13.54–13.56, 13.54f, 13.55f, 13.56f, 13.57f, 13.58f
    types of devices, 13.52, 13.54
Lighting loads, sizing electrical service, 13.3, 13.5
Lightning rods, 13.22
Lightweight concrete, 2.6, 2.7f
Lightweight concrete blocks, 4.11, 5.3
Lime, in mortar, 4.2
Line blocks and stretchers, 5.33–5.34, 5.33f
Line-of-sight checks
    of levels and transits, 1.28–1.30, 1.29f
    of theodolites and total stations, 1.30
Line pins, 5.34, 5.35f
Line trigs, 5.34, 5.35f
Lintel blocks, 4.10f, 5.6f, 5.9f
Lintels, 5.13, 5.14, 5.26
    defined, 5.49
Lion Gate (Istanbul), 4.2, 4.2f
Liquid-fueled tools, and safety, 4.35
Liquid propane (LP) gas, and forklifts, 4.51–4.52
Liquidtight flexible metal conduit (LFMC), 13.35
Live-dead-live test, 12.11
Live loads, 6.20
    defined, 6.50
Loadbearing and nonbearing partitions
    estimating and, 7.24
    joist layout and, 6.32, 6.33f
Loadbearing concrete blocks, 4.11, 5.2, 5.3, 5.7, 5.7f
Load centers
    defined, 13.72
    sizing of, 13.9–13.12, 13.13, 13.13f, 13.14–13.15
    as type of panelboard, 13.27
Load handling, and forklifts, 4.48–4.50, 4.49f, 4.50f, 4.51
Locking tube straps, 16.21f, 17.16, 17.17f
Lockout/tagout
    company procedures for, 12.16
    NEC standards, 12.16
    OSHA general safety precautions for, 12.14
    OSHA rule for, 12.11, 12.16–12.18, 12.18f, 12.19
    procedures for, 12.16–12.18
Lockout/tagout devices, 12.17, 12.18f
Lockout/tagout hardware, defined, 12.17
Log cabin wood siding, 10.19f, 10.31, 10.31f
Long loads, traveling with, 4.48–4.49
Long-L stairways, 11.3f
Long sweep bends, 15.15–15.16, 15.16f
    defined, 15.32
Lookouts, 8.14, 8.14f, 10.13, 10.14f
    defined, 8.35, 10.71
Louvers, 10.63
    defined, 10.71
Low-voltage electrical systems, 13.61

Low voltages, attitudes about, 12.3
Lumber
    defects in, 7.15
    engineered wood products, 6.15, 6.15f, 6.16, 7.6, 7.7
    warped lumber, 7.25
Lung cancer, and asbestos, 12.32

M

Machine-made concrete blocks, 4.9
Magnesium bullfloats, 3.22
Magnesium screeds, 3.19
Main bonding jumpers (MBJ), 13.22–13.24
Maintenance
    measuring tools and, 1.13, 1.14, 1.14f, 1.22
    NFPA 70E requirements for, 12.20
Male adapters, 17.5
Manhole brick, 4.5
Manhole concrete units, 4.12, 4.13f, 5.2, 5.10, 5.11f
Manholes
    elevations and, 15.22
    safety and, 12.29–12.30
Mansard roofs, 8.2, 8.2f, 8.3
Manual pipe cutters, 16.10–16.11, 16.11f
Manual screeds, 3.18–3.19, 3.19f
Manufactured board panel subfloors, 6.26
Manufacturers Standardization Society hanger standards, 16.21
Manufacturing, and careers in HVAC, 14.13, 14.14f, 14.15
Margin trowels, 4.17f
Mason-Dixon line, 1.6
Masonic Temples, 4.4
Masonry
    as art and craft, 4.17–4.18
    basic bricklaying, 4.22–4.28, 4.23f, 4.24f, 4.25f, 4.26f, 4.27f
    brick masonry terms, 4.7
    as career, 4.18–4.20, 4.19f
    clay products, 4.4–4.7, 4.5f, 4.6f, 4.8f, 5.3, 5.10–5.11, 5.11f
    concrete products, 4.8–4.9, 4.8f, 4.10f, 4.11–4.12, 4.11f, 4.12f, 4.13f
    fall protection and, 4.37–4.44, 4.38f, 4.39f, 4.40f, 4.41f, 4.42f
    forklifts and, 4.44–4.45, 4.44f, 4.46f, 4.47–4.53, 4.47f, 4.48f, 4.49f, 4.50f
    history of, 4.2–4.4, 4.2f, 4.3f, 4.4f
    introduction to, 4.1
    laying out jobs, 5.16–5.18, 5.16f, 5.17f
    modern work environment, 4.17–4.18
    mortars and grouts, 4.13–4.14
    qualities of good masons, 4.20–4.22
    safety practices, 4.28–4.37, 4.28f, 4.29f, 4.31f, 4.33f, 4.34f, 5.1–5.2
    setting up jobs, 5.14–5.16
    stone, 4.13, 4.13f, 5.11–5.12, 10.65
Masonry cement mortars, 4.14
Masonry hammers, cutting masonry with, 5.27, 5.27f
Masonry saws, cutting masonry with, 5.28, 5.28f
Masonry units. See also Bricks; Concrete blocks
    bonding, 5.20–5.23, 5.21f, 5.22f, 5.23f, 5.24f
    characteristics of, 4.1
    checking before cutting, 5.29
    clay masonry units, 5.3, 5.10–5.11, 5.11f
    cleaning, 5.44–5.46
    cutting, 5.25–5.29, 5.25f, 5.26f, 5.27f, 5.28f
    defined, 4.57
    laying, 5.29–5.40, 5.29f, 5.30f, 5.31f, 5.32f, 5.33f, 5.34f, 5.35f, 5.36f, 5.37f, 5.38f
    mortar joints, 5.40–5.42, 5.40f, 5.41f, 5.42f
    patching mortar, 5.42–5.43, 5.43f

Masonry walls
    construction techniques, 4.14–4.15, 4.15f, 4.16, 4.16f
    foundation plans and, 5.17, 5.17f
    framing, 7.26, 7.27f, 7.28, 7.28f
    locating, 5.17–5.18, 5.17f
    reinforced, 4.15, 4.16f, 5.3, 5.3f, 5.5
Masons, 4.1
    defined, 4.57
Mason's lines
    setting up using corner poles, 5.33
    setting up using line pins, 5.34, 5.35f
    setting up using line trigs, 5.34, 5.35f
    setting up with line blocks and stretchers, 5.33–5.34, 5.33f, 5.34f
Mason's rules, 5.30, 5.30f
Mass concrete, 2.6
Materials. See also Plastic plumbing materials; Roofing materials
    material movement and handling safety, 4.33, 9.21, 9.22
    stacking building materials, 4.33–4.34, 4.34f, 5.1
Material Safety Data Sheets (MSDS)
    block cleaning solutions and, 5.46
    brick cleaning solutions and, 5.44
    pipe cement and, 16.13
    siding materials and, 10.4
    solvents and, 12.30, 12.31, 12.31f, 12.32
Mathematical corrections, distance measurement, 1.14, 1.14f
Maximum weight limits, for fall protection, 12.38
McLawhorn, Curtis, profile in success, 11.25
Mean sea level (MSL), 1.31
Measurements
    concrete mix proportions and, 2.5–2.6, 2.5f
    copper pipe and fittings, 17.8–17.9, 17.9f
    floor systems and, 6.27–6.28, 6.28f, 6.32
    measure twice cut once, 7.13
    measuring and cutting studs, 7.12–7.14, 7.13–7.14f
    plastic pipe, 16.10
Mechanical bonds, 5.20–5.21
Mechanical energy, and safety, 12.16
Mechanically adhered membrane roofs, 9.10, 9.11f
Mechanically formed tee connections, 17.7–17.8, 17.8f
Mechanical plans, 6.11
Mechanical refrigeration
    compressors, 14.7
    condensers, 14.7
    defined, 14.23
    evaporators, 14.7
    expansion devices and, 14.7
    metering devices, 14.7
    refrigeration cycle, 14.6–14.9, 14.7f
Mesothelioma, and asbestos, 12.32–12.33
Metal building frames, as grounding electrodes, 13.17
Metal-clad (Type MC) cable, 13.30, 13.32f
    defined, 13.72
Metal drip caps and edges, and cornices, 10.9
Metal framing, and roofs, 8.28, 8.28f
Metal roofing
    corrugated metal roofing, 9.68–9.69, 9.69f
    description of, 9.7–9.8, 9.8f
    simulated standing-seam metal roofing, 9.69–9.72, 9.71f, 9.72f, 9.73f
    snug-rib system, 9.73
Metal roof trusses, 8.28, 8.28f
Metal studs. See Steel studs
Metal tiedowns, 8.23, 8.23f
Metal ties, 5.13, 5.13f
Metal tiles, 9.77

Metal underground gas piping, as unsuitable for grounding electrodes, 13.18, 13.19
Metal underground water pipes, as grounding electrodes, 13.17–13.19, 13.18f
Metal wall bracing, 7.18, 7.18f
Metering devices, and mechanical refrigeration, 14.7
Metric conversion charts, 2.39, 2.41
Metric conversion factors, 2.39
Metric measurements, 2.13, 2.14f
Metric rods, 1.23
Metric shingles, 9.2
Microwave EDMIs, 1.19
Midget tubing cutters, 17.9
Minimum coverage of steel reinforcement, 2.15
Minimum safe clearance distances, 12.23
Minimum stairway widths, 11.8, 11.9f
Mission tiles. *See* Barrel-style tiles
Miter boxes, 16.11, 16.12f
Mitered corners, and beveled siding, 10.25, 10.26f
Mitered hips, slate floors, 9.74, 9.74f
Mixing concrete
    off-site equipment for, 3.5, 3.5f
    on-site equipment for, 3.5–3.6, 3.5f, 3.6f, 3.7–3.8, 3.9–3.14, 3.9f, 3.10f, 3.12f, 3.13f
Mixing mortar, 4.22, 4.23f
Mobile boom-type concrete pumps, 3.10–3.11, 3.10f
Mobile concrete batch plants, 3.6, 3.6f
Mobile continuous mixers, 3.7
Model codes, 15.7, 15.24–15.25, 17.6
Modified bitumen membrane roofing systems, 9.9, 9.9f, 9.10
Moisture content and absorption, of concrete masonry units (CMUs), 5.2, 5.11
Moisture control, siding and, 10.23, 10.27
Moisture infiltration
    cavity walls and, 4.8
    masonry walls and, 4.15
Momentum, 15.8, 15.9
Monolithic slabs (monolithic pours), 2.28
    defined, 2.38
Monsanto Fome-Cor, 10.4
Monuments, 1.7
Mortar
    bonding concrete masonry units with, 4.1
    buttering joints, 4.26–4.27, 4.27f
    cleaning from masonry units, 5.44–5.46
    cutting, 4.25–4.26, 4.26f
    defined, 4.57
    furrowing, 4.26, 4.26f, 5.19
    joints, 4.5, 5.40–5.42, 5.40f, 5.41f, 5.42f
    mixing, 4.22, 4.23f
    mud as, 4.2, 4.13
    patching, 5.42–5.43, 5.43f
    performance specifications for, 4.14
    picking up, 4.23–4.25, 4.23f, 4.24f
    portland cement as, 4.14
    safety and, 4.31–4.32, 5.1–5.2
    spreading, 4.23, 4.25, 4.25f
Mortar bonds, 5.19–5.20
Mortar burrs, 5.42
MSDS. *See* Material Safety Data Sheets (MSDS)
Mud, as mortar, 4.2, 4.13
Mud bricks, 4.1, 4.2
Muller, Bob, profile in success, 16.27–16.29
Multiple lockout/tagout devices, 12.17, 12.18f
Multiple-rotary finishing machines, 3.26, 3.26f
Multiwire branch circuits, 13.42–13.43, 13.42–13.43f
Municipal waste treatment systems, 15.22–15.23, 15.23f

N
Nailing fiber-cement siding, 10.44, 10.46–10.47f, 10.48f
Nailing flanges, 10.54, 10.55f
Nailing partitions, 7.4, 7.5f
Nailing shingles
    for composition shingle roof installation, 9.34, 9.35, 9.35f, 9.51, 9.51f
    for wood shingle and shake roof installation, 9.60–9.61, 9.60f, 9.61, 9.66
    for wood siding, 10.34, 10.35, 10.35f
Nailing slate roofing, 9.75, 9.75f
Nailing step flashing, 9.41, 9.42, 9.42f
Nailing tiles, 9.76
Nailing wood siding
    beveled siding, 10.18f, 10.21, 10.21f, 10.22f, 10.25
    board-and-batten, 10.25, 10.27–10.28, 10.27f, 10.28f, 10.29f
    plywood, 10.39
    shingles or shakes, 10.34, 10.35, 10.35f
    shiplap, 10.19f, 10.31, 10.31f, 10.32
    tongue-and-groove, 10.18f, 10.29, 10.29f
Nail rippers, 9.14, 9.15, 9.15f
Nails
    aluminum nails, 10.54, 10.55, 10.56, 10.58
    protruding nails, 9.30
    roofing nails, 9.13, 9.13f, 9.14f, 9.60
    stainless steel nails, 10.13
    wood siding and, 10.17, 10.20f
Nail spacing, closed soffits, 10.13
Narrow-U stairways, 11.3f
NATE (North American Technical Excellence), 14.16
National Board of Fire Underwriters, 13.51
National Center for Construction Education and Research (NCCER), standardized training, 4.20, 14.16–14.18
National consensus standards, 12.20
National Electric Code® (NEC®) standards
    for 240-volt circuits, 13.43, 13.45
    for air terminals, 13.22
    assured equipment grounding conductor programs and, 12.15
    for bathroom receptacles, 13.10
    for bonding at the service, 13.23
    for bonding of piping systems, 13.23–13.24
    for branch circuits for lighting, 13.45
    for building opening service drop clearances, 13.26
    for calculating loads, 13.6
    for demand factors, 13.7
    for dryer loads, 13.7
    for electric heating, 13.62, 13.63f
    for equipment grounding systems, 13.34–13.35, 13.36f
    for fixed appliances, 13.6
    for flash protection signs, 12.10
    for general lighting and receptacle demand factors, 13.7
    for general lighting loads, 13.3
    for ground-fault protection, 12.15, 13.10
    for grounding electrical services, 13.15–13.24, 13.18f, 13.19f
    for grounding electrode conductors (GEC), 13.20–13.22
    for grounding electrodes, 13.17–13.22, 13.18f, 13.19f
    for indoor hot tubs, 13.65, 13.65f
    for intersystem bonding, 13.23
    for laundry circuits, 13.5, 13.7
    for lighting demand factors, 13.6
    for locating light fixtures, 13.59, 13.60f
    for locating receptacles, 13.40, 13.41f, 13.42
    for locating switches, 13.59, 13.60f
    for lockout, 12.16
    for low-voltage systems, 13.61
    for main bonding jumpers (MBJ), 13.22–13.24

for metal-clad (TYPE MC) cable, 13.30, 13.32f
for multiwire branch circuits, 13.42–13.43
for neutral conductor demand factors, 13.8
for nonmetallic-sheathed cable (Type NM), 13.29–13.30, 13.30f, 13.31f, 13.50
OSHA standards and, 12.15
for outlet boxes, 13.47–13.50
for panelboard location, 13.27–13.28
for range loads, 13.7–13.8
for rating receptacles, 13.51
for service drop locations, 13.24–13.25, 13.25f, 13.26f
for service drop vertical clearances, 13.25, 13.27f
for service masts, 13.25, 13.26f
for sizing load centers, 13.10
for sizing residential neutral conductors, 13.8–13.9
for small appliance circuits, 13.5, 13.7
for split-wired duplex receptacles, 13.42
for swimming pools, 13.64, 13.65f
for switches, 13.51, 13.59, 13.60f
for three-way switches, 13.57f
for Type SE (service entrance) cable, 13.32–13.33, 13.34f
for underground feeder (Type UF) cables, 13.32
for wiring low-voltage systems, 13.62, 13.63f
National Fire Protection Association (NFPA)
    Annexes A through M, 12.20
    eye and face protection standards, 12.11
    Installation Safety Requirements, 12.20
    NFPA 5000 building code, 13.73
    Safety-Related Maintenance Requirements, 12.20
    Safety-Related Work Practices, 12.20
    Safety Requirements for Special Equipment, 12.20
    Standard for Electrical Safety in the Workplace (NFPA 70E), 12.11, 12.19–12.20
National Institute for Occupational Safety and Health (NIOSH)
    construction industry electrocutions and, 12.3
    reports of spontaneous fires, 4.53
Natural gas, as heating fuel, 14.2
NEC®. See National Electric Code® (NEC®) standards
Neutral conductor demand factors, 13.8
Neutral conductors, residential, 13.8–13.9, 13.9f
Newel posts, 11.6, 11.7f
    defined, 11.26
NFPA. See National Fire Protection Association (NFPA)
Nibbling tools, 9.15, 9.16f
NIOSH (National Institute for Occupational Safety and Health), 4.53, 12.3
No-hub cast-iron pipe, 15.17f
Noise suppression, and ground-fault circuit interrupters, 13.14
Nominal sizes, 17.2, 17.3f
    defined, 17.27
Nonagitating trucks, 3.7
Nonbearing concrete blocks, 5.2, 5.3
Nonconductive eyewear, 12.10
Nonconductive helmets, 12.9
Nonmetallic-sheathed cable (Type NM), 13.29–13.30, 13.30f, 13.31f, 13.55–13.56
    defined, 13.72
Nonpermit confined spaces, and safety, 12.30
Nonsiphon traps, 15.6
Nonstructural, defined, 4.57
Nonstructural concrete blocks, 4.11, 5.8
Normal weight concrete blocks, 4.11, 5.3
Norman bricks, 4.4
North American Technical Excellence (NATE), 14.16
Nosing, 11.7f, 11.8
    defined, 11.26
Nosing returns, 11.7f, 11.8

Notching wooden joists, 6.22, 6.23f
Notre Dame Cathedral, 4.1, 4.2f
Noxious, defined, 14.23
Noxious particles and fumes, 14.6

O
Obstructed views, and forklifts, 4.50, 4.51
Occupational Safety and Health Act of 1970!
Occupational Safety and Health Administration (OSHA)
    clothing requirements, 12.19
    Code of Federal Regulations (CFR), 12.5, 14.16, 14.17
    concrete equipment and tools requirements, 2.18
    concrete tools requirements, 3.32
    construction industry requirements, 12.5
    de-energizing circuits requirements, 12.13–12.14
    electrical regulations, 12.15–12.18, 12.19
    employee responsibilities and, 12.12
    employer responsibilities and, 12.12
    exposed rebar requirements, 2.16
    exterior finishing requirements, 10.2, 10.4
    eye protection requirements, 12.11
    fall protection equipment requirements, 9.23, 9.24, 9.25, 9.26
    fall protection requirements, 12.34
    first aid requirements, 12.30
    gasoline-powered equipment requirements, 4.34
    general industry requirements, 12.5
    general protective equipment and tools requirements, 12.6, 12.7
    general requirements, 12.12
    guidelines for testing fall protection systems, 9.27
    hard hat areas requirements, 12.19
    lockout/tagout requirements, 12.14
    lockout/tagout rule, 12.11, 12.16–12.18, 12.18f, 12.19
    minimum safe clearance distances requirements, 12.6
    National Electric Code (NEC) standards and, 12.15
    personal protective equipment requirements, 4.29, 12.6–12.7
    powder actuated tool certifications requirements, 4.35
    roofing brackets requirements, 9.22
    rubber protective equipment requirements, 12.7
    safety glasses requirements, 9.22, 12.19
    safety philosophy, 12.13–12.15
    safety shoes requirements, 12.19
    safety standards, 12.13
    scaffolding requirements, 10.2, 10.3, 10.51, 12.22–12.23, 12.23f
    scaffold safety rules, 12.22–12.23
    standards for confined spaces, 12.29–12.30
    storage and handling of materials, 4.33
    temporary guardrails requirements, 11.10
    training requirements, 12.17
    uncovered openings requirements, 12.13
    vaults and manholes requirements, 12.29–12.30
    work practices requirements, 12.13–12.15, 12.20
Odorless toxic gases, 15.9
OD (outside diameters), 16.3, 16.30
Offsets, 15.19, 15.19f
One-hole clamps, 17.16f, 17.17
On-the-job training (OJT), 14.15, 14.17
Open/closed stairways, 11.2, 11.2f
Open cornices, 10.5, 10.7, 10.7f
Open finished stringers, 11.7f, 11.8
Open-flame heat welding, 9.10
Open sheathing, wood shingles and shakes, 9.59, 9.59f
Open stairways, 11.2, 11.2f
    defined, 11.26
Open valleys, 9.37–9.38, 9.38f

Operator's compartments, and forklift safety, 4.47
Orders of Freemasonry, 4.4
Oriented strand board (OSB), 6.16, 6.22, 6.27
Oscillation, 15.8, 15.9
OSHA. *See* Occupational Safety and Health Administration
    (OSHA)
Outdoor coils, 14.8
Outlet boxes
    calculating conductors for, 13.47–13.48, 13.49
    metallic outlet boxes, 13.48*f*
    mounting, 13.49–13.50, 13.50*f*
Outlets
    classifications of, 13.37
    symbols for, 13.37, 13.38*f*
Out of plumb bricks, 5.30, 5.31*f*
Outside corner posts, 10.54, 10.55*f*
Outside diameters (OD), 16.3
    defined, 16.30
Overcutting, 11.17, 11.19*f*
Overhangs, 9.27, 9.28*f*
    defined, 9.98
Overheating injuries, 4.37
Overlapped splices.20-21, 7.21*f*
Ozone, 12.7
Ozone resistant (Type 2) rubber protective equipment, 12.7

P
Pacing, estimation by, 1.19
Padlocks, 12.17, 12.18*f*
Palletized bricks, 4.33, 4.33*f*
Palmer, Harmon, 4.9
Panelboards
    interior view of, 13.17*f*
    locating and installing, 13.27–13.28, 13.28*f*, 13.29*f*
Panelized shingles/shakes
    applying, 9.66, 9.66*f*, 9.67*f*
    wood shingles and shakes compared to, 9.5, 9.5*f*
    wood siding and, 10.37, 10.38
Panelized walls, 7.19
Panel pinch, 6.38
Pans, picking up mortar from, 4.24–4.25, 4.24*f*
Paper filter masks, 12.30
Parallax, 1.27
    defined, 1.46
Parallel strand lumber (PSL), 7.6
Parallel strand lumber (PSL) headers, 7.6
Parapets, 4.14
    defined, 4.57
Parging trowels, 4.17*f*
Partially adhered membrane roofs, 9.10, 9.11*f*
Particleboard siding, 10.39–10.41, 10.42*f*, 10.43*f*
Particulate-removing respirators, 12.32
Partition blocks, 4.10*f*, 5.6*f*, 5.7*f*
Partitions
    laying out joist locations for, 6.32, 6.33*f*
    loadbearing and nonbearing partitions, 7.24
    masonry construction and, 4.10*f*, 4.11
    nailing surfaces for, 7.4, 7.5*f*
    partition backing, 7.26, 7.27*f*
    partition intersections, 7.4, 7.5*f*, 7.6*f*
    as wall components, 7.2, 7.2*f*
Pascals, 2.6
    defined, 2.38
Passengers, and forklifts, 4.47
Pattern bonds, 5.21–5.22, 5.21*f*, 5.22*f*
Patterned concrete, 2.8
Patterns, of bricklaying, 4.2, 4.2*f*
Paving brick, 4.5

Payne, John, profile in success, 2.37
PB (polybutylene) pipe, 16.7, 16.7*f*
    defined, 16.30
Pedestrian safety, and forklifts, 4.47, 4.52–4.53
Peg tests, 1.28–1.30, 1.29*f*
    defined, 1.46
PE (polyethylene) pipe, 16.6, 16.6*f*
    defined, 16.30
    joining, 16.17, 16.17*f*
Permanent bracing
    roof trusses and, 8.23, 8.24, 8.24*f*
    wall framing and, 7.18, 7.18*f*
Permit-required confined spaces, 12.30
Permit-required confined spaces controlled by ventilation,
    12.30
Personal clothing, 12.10, 12.19
Personal fall-arrest systems (PFAS)
    anchoring devices, 4.42, 4.42*f*, 9.25, 12.37
    anchoring devices and connectors, 4.42, 4.42*f*, 9.25, 12.37
    body harnesses, 4.40, 4.40*f*, 9.23–9.24, 9.24*f*, 9.26, 9.26*f*
    deceleration devices, 4.39, 4.41, 4.41*f*, 9.23, 9.24–9.25, 9.25*f*
    described, 4.39
    description of, 9.23–9.25
    lanyards, 4.40, 4.41, 4.41*f*, 9.24, 9.24*f*, 12.36–12.37, 12.36*f*
    lifelines, 4.41–4.42, 4.41*f*, 4.42*f*, 9.25, 9.25*f*
    proper use of, 4.43
    purpose of, 4.39
    safety and, 9.26–9.27, 12.35–12.38, 12.36*f*
    short-distance falls and, 4.44
    tying-off to, 4.42–4.43
Personal positioning systems, and body harnesses, 4.40, 4.40*f*
Personal protection, 3.33
Personal protective equipment (PPE). *See also* Fall protection;
    Personal fall-arrest systems (PFAS); Safety
    high voltage and, 12.6
    masonry work and, 4.29–4.30, 4.29*f*
    OSHA standards for, 4.29, 12.6–12.7
PEX (cross-linked polyethylene)
    defined, 16.30
    functions of, 16.7, 16.7*f*
PEX (cross-linked polyethylene) tubing
    functions of, 16.7, 16.7*f*
    HVAC and, 14.10
    joining, 16.16–16.17, 16.16*f*
Phasing testers, 12.11
Philadelphia rods, 1.23
Photoelectric switches, 13.58
Picking up loads, and forklifts, 4.48
Pier blocks, 4.10*f*, 5.6*f*
Pier forms, 2.20, 2.21*f*, 2.22, 2.24*f*
Piers
    defined, 6.50
    as girder supports, 6.17–6.18, 6.17*f*, 6.19*f*
    post-and-beam framing and, 6.4
    as type of footings, 2.20, 2.21*f*
Pigtails, 2.16, 2.16*f*
Pilasters, 4.25
    defined, 4.57
Piles, 2.21
    defined, 2.38
Pipe. *See* Copper pipe; Plastic pipe
Pipe cement, 16.12–16.13
Pipe channels, 17.17, 17.18*f*
Pipe clamps, 16.22*f*
Pipe cutters, 16.10–16.11, 16.11*f*
Pipe fasteners
    for plastic pipe, 16.20, 16.22*f*
    for wooden structures, 16.20

Pipe hooks, 16.21f, 17.16, 17.17f
Pipeline length and layout, and mobile boom-type concrete pumps, 3.11
Pipe reamers, 16.10, 16.11f, 17.10
Pipe riser clamps, 16.20, 17.17, 17.17f
    defined, 16.30
Pipe scale, 15.16
    defined, 15.32
Pipe straps, 17.16–17.17, 17.17f
Piping drawings, 14.10
Pitch
    defined, 9.98
    flat roofs and, 9.52
    roofing materials and, 9.2
    roof layout and, 8.3, 8.4f, 8.5
    unequally pitched roof intersections, 9.38
Pitch (asphalt), in mortar, 4.2
Pitch boards
    defined, 11.26
    laying out stringers and, 11.15, 11.15f, 11.19f, 11.20
Placing blocks, 5.32–5.33, 5.32f
Placing bricks, 5.29–5.30, 5.29f
Placing loads, and forklifts, 4.49
Plain ends, concrete blocks, 4.11f, 5.4f
Plain wood siding, 10.18f, 10.21, 10.21f, 10.29, 10.29f, 10.31, 10.31f
Planciers, 10.7
    defined, 10.71
Plank-and-beam framing, 8.27–8.28, 8.28f. See also Post-and-beam framing
Plan view drawings, foundation plans, 5.17, 5.17f, 6.7
Plastic concrete
    central mix plants and, 3.5
    defined, 2.38
    properties of, 2.2
Plastic fittings
    for drain, waste, and vent (DWV), 16.8, 16.9, 16.9f, 16.10f
    for water supply, 16.8, 16.9, 16.9f
Plastic mollies, 17.18
Plastic pipe
    advantages and disadvantages of, 16.2
    labeling, 16.3, 16.3f
    manufacturers, 16.4
    pipe supports, 16.19–16.20, 16.21f, 16.22f
    properties of, 16.2
    sizing, 16.3
    storage and handling, 16.8
    types of, 16.4–16.7, 16.5f, 16.6f, 16.7f
Plastic pipe hangers, 16.20, 16.21f
Plastic pipe supports, 16.19–16.20, 16.21f, 16.22f
Plastic plumbing materials
    measuring cutting and joining, 16.10–16.17, 16.11f, 16.12f, 16.13f, 16.14f, 16.15f, 16.16f, 16.17f
    pipe supports, 16.19–16.20, 16.21f, 16.22f
    plastic pipe, 16.2–16.8, 16.3f, 16.5f, 16.6f, 16.7f
    pressure testing, 16.21, 16.22, 16.23
Plastics, 16.3
Plastic sheeting, and curing concrete, 2.8
Plate electrodes, 13.18–13.20
Plates, 2.25, 2.25f, 8.3, 8.4f
Plate vibrators, 3.17
Platform framing, 6.12, 6.13f
Platforms. See Landings
Platform stairways, 11.4
Plot plans, 1.2–1.3, 1.3f, 1.4f, 1.5f
Plug openings (PO), 15.2, 15.2f
Plumb, checking when laying masonry units, 5.30, 5.31f
Plumb bobs, 1.11, 1.11f, 1.12

Plumber's levels, 15.36
Plumber's tape, 17.16–17.17, 17.17f
Plumbing and aligning walls, 7.16–7.19, 7.17f, 7.18f
Plumbing plans, floor systems, 6.11
Plyform, 2.23, 2.25
    defined, 2.38
Plywood forms, 2.22, 2.23
Plywood siding, 10.37–10.39, 10.37f, 10.38f
Plywood subfloors, 6.26
Pneumatic guns, 3.8
Pneumatic nail guns, and safety, 12.28, 12.28f
Pneumatic roofing nailers, 9.14, 9.17, 9.17f, 9.22–9.23
Pointing, 5.42–5.43
Pointing and margin trowels, 3.29, 3.31f
Pointing trowels, 4.17f
Pollutants, indoor air quality, 14.6
Polybutylene pipe. See PB (polybutylene) pipe
Polychlorinated biphenyls (PCBs)
    defined, 12.44
    safety and, 12.33–12.34
Polyethylene pipe. See PE (polyethylene) pipe
Polyvinyl chloride. See PVC (polyvinyl chloride)
PO (plug opening), 15.2, 15.2f
Portable brakes, 9.15, 9.16f, 10.51, 10.51f
Portland cement, 2.2, 2.3, 4.14
Positioning stringers for layout, 11.17, 11.18f
Post anchors, 6.17, 6.17f
Post-and-beam framing, 6.4, 6.6f, 6.7, 8.27–8.28, 8.28f, 8.36f
Post caps, 6.17f, 6.18
Posts, 6.17–6.18, 6.17f, 6.19f
Pound-in nail anchors, 17.18, 17.20, 17.20f
Powder-actuated tools, and safety, 4.34–4.35, 4.34f, 12.29
Power, layout of branch circuits for, 13.37–13.40, 13.38f, 13.39f, 13.40f, 13.41f, 13.42–13.45, 13.42f, 13.43f, 13.44f
Power buggies, 3.12–3.13, 3.12f
Powered finishing machines, 3.25–3.26, 3.26f
Powered screeds, 3.19, 3.20f
Power grinders, 3.31, 3.32f
Power lines, and mobile boom-type concrete pumps, 3.11
Power nailers
    roofing and, 9.17, 9.17f
    safety and, 9.22–9.23
Powers, Michael J., profile in success, 12.43
Power shears, 10.44, 10.44f
Power tools, and safety, 12.26–12.29, 12.28f
Pozzolan, 2.5
    defined, 2.38
PPE. See Personal protective equipment (PPE)
Precast reinforced concrete girders, 6.15
Precut studs, 7.12
Prefabricated hip and ridge units, wood shingles and shakes, 9.63, 9.64f
Pre-faced concrete units, 4.12, 4.12f, 5.2, 5.8, 5.9, 5.9f, 5.10
Pre-framed stairwells, 11.14–11.15
Premanufactured membrane roofing, 9.8–9.10, 9.9f, 9.10f
Preplaced aggregate (prepacked) concrete, 2.6
Prescription specifications, for concrete, 2.5
Pressed bricks, 4.7
Press fittings, 17.7, 17.7f
Pressure, pressure/temperature relationships, 14.8–14.9
Pressure drops, 17.4
    defined, 17.27
Pressure ratings, 16.3
    defined, 16.30
Pressure testing
    copper pipe and fittings, 17.21, 17.21f
    plastic plumbing materials, 16.21, 16.22, 16.23
Pressure tools, and safety, 4.35–4.36

Pressure-treated furring and furring substitutes, 10.62
Pressure-treated lumber, 6.12
Pressure waves, from arc blast, 12.5
Pride, 4.21
Primary control points, 1.7, 1.7*f*
Prisms, and EDMIs, 1.19–1.20, 1.20*f*
Private waste control systems, 15.22, 15.23, 15.24*f*
Profile leveling, 1.38
Projections, installing joists for, 6.38, 6.39*f*, 6.40
Propane, as heating fuel, 14.2
Propane torches, torch-down roofing, 9.85, 9.85*f*
Proportions, for concrete, 2.5–2.6, 2.5*f*
Protective apparel, 12.9–12.10
Protruding nails, roof deck preparation, 9.30
ProWrap, 10.3
Psi (pounds per square inch), 16.3
    defined, 16.30
P-traps, 15.4, 15.5*f*, 15.7
    defined, 15.32
Public health, and drain, waste, and vent (DWV) systems, 15.25
Pump jacks, 9.19, 9.20*f*
Pump pipeline diameters, and mobile boom-type concrete pumps, 3.11
Pumps, concrete pumps, 3.8, 3.10–3.11, 3.10*f*
Pump truck stabilization, 3.11
Punching slate, 9.75
Purlins, 8.11
    defined, 8.35
Purple versus clear primers, 16.13
PVC (polyvinyl chloride), 16.3
    defined, 16.30
PVC (polyvinyl chloride) membranes, 9.10
PVC (polyvinyl chloride) pipe
    cutting, 16.10–16.11, 16.11*f*
    described, 16.5, 16.5*f*, 16.7
    HVAC and, 14.10
    joining, 16.14–16.15, 16.14*f*
PVC (polyvinyl chloride) saws, 16.10–16.11
Pythagorean theorem, 1.40, 11.17

Q
Qualified individuals, 12.13, 12.30
Quality in work, 4.22
Quarterback cut blocks, 5.25–5.26, 5.26*f*
Queen closure bricks, 5.25, 5.25*f*
Quiet switches, 13.52, 13.53*f*

R
Rabbeted beveled siding, 10.21, 10.21*f*
Rabbets, 10.6
    defined, 10.71
Raceways, residential wiring methods, 13.33
Rackback, defined, 5.49
Rackback corners
    block and, 5.39–5.40
    brick and, 5.37–5.39, 5.37*f*, 5.38*f*
Rackback leads, 5.37, 5.37*f*
Racking, 5.37
    defined, 5.49
Rafter framing squares, 8.4–8.5, 8.5*f*, 8.6*f*
Rafter marks, 7.19, 7.20*f*, 8.12
Rafter pitch, cutting ceiling joists to match, 7.21, 7.22*f*
Rafter plates, 6.4
    defined, 6.50
Rafters
    estimating material for, 8.26

    installing, 8.11, 8.11*f*, 8.12*f*
    laying out hip rafters, 8.42–8.43, 8.43*f*
    laying out jack rafters, 8.36*f*, 8.39–8.40, 8.39*f*, 8.40*f*, 8.41*f*, 8.42
    laying out locations of, 8.7–8.8, 8.7*f*
    laying out valley rafters, 8.39, 8.39*f*, 8.40*f*
    laying out with speed square, 8.18, 8.18*f*, 8.19*f*, 8.20
Rafter tables, 8.4–8.5, 8.5*f*
Rafter tails, 8.16
Rail-steel, 2.13
    defined, 2.38
Rain, and curing concrete, 3.28
Raising walls, 7.16, 7.16*f*, 7.17
Rake edges
    flat tile roofs and, 9.80, 9.80*f*
    simulated standing-seam metal roofing and, 9.72, 9.72*f*
Rakes, 10.2
    defined, 10.71
Ramps, and forklift operation, 4.51
Range loads, 13.7–13.8
Range poles, 1.11, 1.11*f*, 1.12
Rangers. *See* Walers (wales)
Ranging, 5.38
    defined, 5.49
Ratchet shears, 16.11, 16.11*f*
RCC (roller-compacted concrete), 2.6, 2.7
Reaming, 16.10, 16.11*f*, 17.10, 17.10*f*
Rebar cutters and benders, 2.15
Rebars
    bar supports for, 2.2, 2.16, 2.16*f*
    concrete blocks and, 5.13
    concrete masonry units and, 5.3, 5.3*f*
    as concrete reinforcement material, 2.12–2.13, 2.14, 2.14*f*, 2.15
    defined, 2.38
    identifying rebar, 2.13, 2.14*f*, 2.15
    spacing of, 2.4
    splicing, 2.16
Rebar stakes, 1.38
Receptacles
    for attachment plugs, 12.15
    demand factors, 13.7
    direct-wired ground-fault circuit interrupters and, 13.14, 13.14*f*
    grounding-type duplex receptacles, 13.51, 13.51*f*
    locating, 13.40, 13.41*f*, 13.42
    NEC standards, 13.7, 13.10, 13.40, 13.41*f*, 13.42, 13.51
    safety and, 12.15
    split-wired duplex receptacles, 13.42, 13.42*f*, 13.43*f*
    symbols for, 13.37, 13.38*f*
    as wiring devices, 13.51–13.52, 13.51*f*
Reciprocating compressors
    hermetically sealed, 14.9
    semi-hermetic, 14.9
Reclamation
    defined, 14.23
    refrigerant and, 14.18
Recording measurements, 1.14, 1.14*f*
Recovery
    defined, 14.23
    of refrigerants, 14.18
Rectangular volumes
    calculations for finding, 2.9–2.10, 2.9*f*, 2.10*f*, 2.11, 2.11*f*
    estimating, 3.15
Recycling
    defined, 14.23
    of refrigerants, 14.18
Reducer bushings, 17.5

Reducer couplings, 17.5
Reducers, 17.5
Reference lines, vinyl and metal siding, 10.53–10.54, 10.53f, 10.54f
Reflectors, and EDMIs, 1.19–1.20, 1.20f
Refrigerants
    Clean Air Act and, 14.18
    refrigeration cycle and, 14.6–14.7, 14.8
Refrigeration. *See* Mechanical refrigeration
Refrigeration cycle, 14.6–14.9, 14.7f
    defined, 14.23
Reinert, Zachary, profile in success, 5.48
Reinforced cutout stringers, 11.20, 11.20f, 11.21f
Reinforced footings, 2.19
Reinforced masonry walls, 4.15, 4.16f, 5.3, 5.3f, 5.5
Reinforced stringer mounting, 11.20, 11.21f
Reinforcement bars. *See* Rebars
Reinforcement materials for concrete, 2.12–2.16, 2.14f, 2.16f, 2.17f
Reinforcing bars. *See* Rebars
Reinforcing ceiling joists, 7.21, 7.22f
Relays, and lighting control, 13.58–13.59
Remote control switching, 13.58–13.59
Removing concrete forms, 2.26, 2.31, 3.29
Repeating measurements, 1.14, 1.14f
Required service size, 13.6
Rerods. *See* Rebars
Rescue plans, 4.44, 9.27
Rescues, after falls, 4.44, 9.27, 12.37
Residential grounding systems
    components of, 13.15–13.16, 13.16f, 13.17f, 13.18f
    grounding electrodes, 13.17–13.22, 13.18f, 13.19f
    main bonding jumper, 13.22–13.24
Residential receptacle configurations, 13.44, 13.44f
Residential secondary roof systems, 9.93
Residential wiring. *See also* Branch circuits
    electric heating, 13.6, 13.61–13.62, 13.63f
    equipment grounding system, 13.15–13.16, 13.34–13.35, 13.36f, 13.37
    grounding electric services, 13.15–13.24, 13.16f, 13.17f, 13.18f, 13.19f
    installing service entrance, 13.24–13.26, 13.25f, 13.26f, 13.27f, 13.29f
    introduction, 13.2
    lighting control, 13.52, 13.52–13.62, 13.52f, 13.53f, 13.54–13.56, 13.54f, 13.55f, 13.56f, 13.57, 13.57f, 13.58–13.59, 13.58f, 13.59f, 13.60f, 13.61
    outlet boxes, 13.47–13.50, 13.48f, 13.50f
    panelboard location, 13.27–13.28
    sizing electrical service, 13.2–13.3, 13.3f, 13.4f, 13.5–13.8, 13.5f
    sizing load center, 13.9–13.12, 13.13, 13.13f, 13.14–13.15
    sizing residential neutral conductors, 13.8–13.9, 13.9f
    swimming pools, spas, and hot tubs, 13.64, 13.64f, 13.65f
    wiring devices, 13.51–13.52, 13.51f
    wiring methods, 13.29–13.30, 13.30f, 13.31f, 13.32–13.33, 13.32f, 13.34f
Resistance, body resistance, 12.4–12.5, 12.4f
Respiratory protection
    exterior finish materials and, 10.4
    solvents and toxic vapors and, 12.30, 12.32
Responsibilities
    of employees and employers, 12.12
    of masons, 4.21
Retractable lanyards, 12.36–12.37, 12.36f
Retractable lifelines, 4.41, 4.41f
Returns, 5.37
    defined, 5.49

Reverse batten (board-on-batten) wood siding, 10.28, 10.28f
Ribbands, 7.21, 7.22f
    defined, 7.35
Ribbon courses, 9.36, 9.37f
Ribs. *See* Walers (wales)
Ridge and hip caps
    alternate end treatments of, 9.90
    composite shingle installation and, 9.36, 9.37f, 9.51, 9.51f, 9.52f
    flat tile roofs and, 9.80, 9.80f
    roof ventilation and, 9.90–9.91, 9.90f, 9.91f, 9.92f
    slate roofs and, 9.75–9.76, 9.75f
    wood shingles and shakes, 9.63, 9.64f
Ridgeboards
    described, 8.3, 8.4f
    estimating material for, 8.26
Ridge flashing, simulated standing-seam metal roofing, 9.72, 9.73f
Ridge rows, 9.49, 9.51, 9.51f, 9.52f
Ridges. *See also* Ridgeboards
    defined, 9.98
    ridge tiles, 9.7
    roll roofing installation and, 9.54, 9.56, 9.56f, 9.57
Ridge vent installation, 9.87, 9.87f, 9.88, 9.88f, 9.89–9.91, 9.89f, 9.90f, 9.91f
Rigging loads, forklifts and, 4.50
Rigging trusses, 8.21, 8.21f, 8.23
Rigid vent material, roof ventilation, 9.90, 9.91f
Rim joists, 6.20
Ring-tight gasket fittings, defined, 16.30
Ring-tight gaskets, 16.12, 16.12f
Rip blocks, 5.25, 5.25f
Rise
    defined, 11.26
    roof layout and, 8.3, 8.4f
    stairs and, 11.10, 11.11
Rise and run, 11.9
    defined, 11.26
Riser heights, 11.8, 11.10–11.12
Risers, 11.2, 11.5, 11.8
    defined, 11.26
Riser/tread combinations, 11.14
Risk reduction, 12.5–12.12, 12.6f, 12.9f
Robbins, Charles, profile in success, 17.25–17.26
Rod and pipe electrodes, 13.18–13.20, 13.19f
Rod attachments, 17.18, 17.19f, 17.20
Rods and bolts, 17.18
Rollerbug tampers, 3.22, 3.22f
Roller-compacted concrete (RCC), 2.6, 2.7
Roll-grooved, 17.5, 17.15
    defined, 17.27
Roll roofing
    concealed nail single-coverage installation, 9.55–9.56, 9.55f, 9.56f
    description of, 9.4, 9.4f
    double-coverage installation, 9.56–9.57, 9.56f, 9.57f
    exposed nail single-coverage installation, 9.53–9.54, 9.53f, 9.54f, 9.55f
    installation, 9.52–9.57, 9.53f, 9.54f, 9.55f, 9.56f, 9.57f
    top lap and, 9.3f, 9.52
Roll vent material, 9.90, 9.91f
Roman bricks, 4.3
Roman masonry, 4.3, 4.4f
Roman mortar, 4.14
Roman plumbing, 15.24
Romex cable, defined, 13.72
Roof cut. *See* Slope
Roof deck preparation, 9.28, 9.29f

Roof framing
    dormers, 8.26, 8.27*f*
    estimating materials for, 8.25–8.26
    installing sheathing, 8.15–8.16, 8.15*f*, 8.16*f*, 8.17
    laying out hips and valleys, 8.36, 8.36*f*, 8.37*f*, 8.38–8.40,
      8.38*f*, 8.39*f*, 8.40*f*, 8.41*f*, 8.42–8.43, 8.43*f*
    layout of roofs, 8.3–8.5, 8.4*f*, 8.5*f*, 8.6*f*, 8.7–8.14, 8.7*f*, 8.9*f*,
      8.11*f*, 8.12*f*, 8.13*f*, 8.14*f*, 8.15*f*
    metal roof framing, 8.28, 8.28*f*
    plank-and-beam framing, 8.27–8.28, 8.28*f*
    rafter layout using speed square, 8.18, 8.18*f*, 8.19*f*, 8.20
    rafter locations, 8.7–8.8, 8.7*f*
    truss construction, 8.20–8.21, 8.20*f*, 8.21*f*, 8.22*f*, 8.23–8.24,
      8.23*f*, 8.24*f*, 8.25
    types of, 8.2, 8.2*f*
Roofing applications. *See also* Composition shingle installa-
    tion; Composition shingles; Wood shingles and shakes
    ice edging, 9.91, 9.92*f*, 9.93, 9.94
    metal roofing, 9.7–9.8, 9.8*f*, 9.68–9.73, 9.69*f*, 9.71*f*, 9.72*f*,
      9.73*f*
    preparation for application, 9.28–9.30
    roll roofing installation, 9.3*f*, 9.4, 9.4*f*, 9.52–9.57, 9.53*f*,
      9.54*f*, 9.55*f*, 9.56*f*, 9.57*f*
    roofing materials, 9.2–9.13, 9.3*f*, 9.4*f*, 9.5*f*, 9.6*f*, 9.8*f*, 9.9*f*,
      9.10*f*, 9.11*f*, 9.12*f*, 9.13*f*
    roof ventilation, 9.87–9.91, 9.87*f*, 9.88*f*, 9.89*f*, 9.90*f*, 9.91*f*
    safety and, 9.17–9.19, 9.19*f*, 9.20, 9.20*f*, 9.21–9.27, 9.21*f*,
      9.23*f*, 9.24*f*, 9.25*f*, 9.26*f*
    single-ply roofing, 9.9–9.10, 9.10*f*, 9.11*f*, 9.83–9.85, 9.84*f*,
      9.85*f*
    slate roofing, 9.5–9.6, 9.6*f*, 9.73–9.76, 9.74*f*, 9.75*f*
    tile roofing, 9.6–9.7, 9.6*f*, 9.76, 9.76*f*, 9.77, 9.77*f*, 9.78–9.80,
      9.78*f*, 9.79*f*, 9.80*f*, 9.81*f*, 9.82–9.83, 9.82*f*, 9.83*f*
    tools, 9.14–9.15, 9.15*f*, 9.16*f*
    torch-down roofing, 9.9, 9.9*f*, 9.85–9.87, 9.85*f*, 9.86*f*, 9.87*f*
Roofing brackets, safety, 9.22, 9.23*f*
Roofing materials
    asphalt roofing cements, 9.13
    built-up roofing membrane, 9.8, 9.9*f*, 9.10
    composition shingle roofing, 9.2–9.4, 9.3*f*
    drip edge and flashing, 9.10–9.12, 9.11*f*, 9.12*f*
    estimating, 9.27–9.28, 9.28*f*
    metal roofing, 9.7–9.8, 9.8*f*
    premanufactured membrane roofing, 9.8–9.10, 9.9*f*, 9.10*f*
    roll roofing, 9.3*f*, 9.4, 9.4*f*
    roofing nails, 9.13, 9.13*f*, 9.14*f*, 9.60
    slate roofing, 9.5–9.6, 9.6*f*
    tile roofing, 9.6–9.7, 9.6*f*
    underlayment, 9.12
    waterproof membranes, 9.12–9.13
    wood shingle and shake roofing, 9.4–9.5, 9.5*f*
Roofing nails, 9.13, 9.13*f*, 9.14*f*, 9.60
Roof openings, 8.13, 8.14, 8.15*f*, 8.16
Roof projections and flashing
    composition shingles and, 9.40*f*, 9.41*f*, 9.42*f*, 9.43, 9.43*f*,
      9.44–9.46, 9.44*f*, 9.45*f*, 9.46*f*, 9.47*f*, 9.48–9.49, 9.48*f*, 9.49*f*,
      9.50, 9.51, 9.51*f*, 9.52*f*
    flat tile roofs and, 9.79, 9.79*f*
Roofs. *See also* Roof framing; Roofing applications; *and spe-
    cific types of roofs*
    types of, 8.2, 8.2*f*
Roof sheathing, 9.22
    defined, 9.98
Roof ventilation
    attic ventilation, 9.87, 9.87*f*
    box vent installation, 9.87, 9.87*f*, 9.88–9.89, 9.88*f*
    ridge vent installation, 9.87, 9.87*f*, 9.88, 9.88*f*, 9.89–9.91,
      9.89*f*, 9.90*f*, 9.91*f*
    ventilation requirements, 9.88

Roof windows, 8.13, 8.15*f*
Rope grabs, 4.41, 4.41*f*, 9.24, 9.25*f*
Roughing in, defined, 13.72
Rough openings, 7.2, 7.2*f*, 7.9*f*, 7.11
Rough sills, 7.2, 7.2*f*
Rowlocks, 4.7, 4.8*f*, 5.11*f*, 5.22, 5.22*f*
Rowlock stretcher bricks, 5.22, 5.22*f*
Rubber blankets, 12.9
Rubber gloves, 12.6, 12.6*f*, 12.7, 12.8–12.9, 12.9*f*
Rubber protective equipment, 12.7–12.9, 12.9*f*
Rubber sleeves, 12.9
Rubble stone, 4.13, 5.11–5.12
Rules and regulations. *See also Code of Federal Regulations
    (CFR); National Electric Code (NEC) standards; National
    Fire Protection Association (NFPA); Occupational Safety
    and Health Administration (OSHA)*
    employee responsibilities for, 12.12
Run, 8.3, 8.4*f*, 11.10
    defined, 11.26
Running bonds, 5.22, 5.22*f*, 5.24*f*
Runs of pipe, 15.14
    defined, 15.32
Rust, on rebar, 2.15
R-values
    defined, 10.71
    exterior finishing and, 10.4

S
Saddle ridges, 9.75–9.76, 9.75*f*
Saddles. *See also* Crickets
    defined, 9.98
Safety. *See also* Fall protection; Hazards; Personal protective
    equipment (PPE)
    acids and, 12.33
    anchor points and, 9.26–9.27
    asbestos and, 12.32–12.33
    batteries and, 12.33
    before starting jobs, 17.8
    bracing and, 8.25
    cement and, 2.2, 4.31, 4.32
    chemical cleaning solutions and, 5.44
    chemicals and, 12.32
    concrete and, 2.2, 3.9, 3.10, 3.24, 3.32–3.33, 4.31–4.32
    concrete forms and, 2.17–2.18
    copper pipe grounding and, 17.2
    cranes and, 12.23–12.24
    cubes of brick and, 5.8
    cutting masonry units and, 5.2, 5.26, 5.28
    cutting pipe and, 16.10
    cutting ridge vents and, 9.90
    cutting siding materials and, 10.4
    disconnect switch markings and, 12.14, 12.15
    electrical current and, 12.3, 12.4, 12.5
    electrical shock and, 12.2–12.5, 12.4*f*
    employer and employee obligations, 12.12
    energized circuits and, 12.8, 12.11, 12.13–12.14
    erecting roofs and, 8.11
    exterior finishing and, 10.2, 10.4
    falling objects and, 4.30–4.31, 4.31*f*
    fall protection and, 4.37–4.44, 4.38*f*, 4.39*f*, 4.40*f*, 4.41*f*,
      4.42*f*, 7.20, 8.11, 12.34–12.38, 12.34*f*, 12.36*f*, 12.38*f*
    fall protection equipment, 9.23–9.27, 9.24*f*, 9.25*f*, 9.26*f*
    fiber-cement siding and, 10.44
    flammable liquids and, 4.32, 4.52
    floor openings and, 7.20
    forklifts and, 4.44–4.45, 4.44*f*, 4.46*f*, 4.47–4.53, 4.47*f*, 4.48*f*,
      4.49*f*, 4.50*f*
    fuse installation or removal and, 12.7, 12.10, 12.14–12.15

gasoline and, 4.32
gasoline-powered tools and, 4.34
general precautions, 12.11–12.12
GFCIs with power tools and, 4.30, 12.3, 12.15, 12.16, 12.16*f*
ground-fault circuit interrupters (GFCI) and, 12.15, 12.16, 12.16*f*, 13.10
grounding and, 13.16
guardrails and, 4.38, 4.39*f*, 7.20, 9.23
handling and placing concrete and, 3.32–3.33
hand tools and, 12.26, 12.27*f*
hoists and, 12.23–12.24
hot weather and, 4.37
ice edging and, 9.93, 9.94
jewelry and, 12.11
joist installation and, 7.20
ladders and, 9.19, 9.20, 9.21, 9.21*f*, 12.20–12.22, 12.21*f*
lifting and, 12.24, 12.25, 12.25*f*, 12.26
lifts and, 12.23–12.24
liquid-fueled tools and, 4.35
masonry and, 4.28–4.37, 4.28*f*, 4.29*f*, 4.31*f*, 4.33*f*, 4.34*f*, 5.1–5.2
material movement and handling and, 4.33, 9.21, 9.22
mechanical energy and, 12.16
mobile boom-type concrete pumps and, 3.11
mortar and concrete and, 4.31–4.32, 5.1–5.2
odorless toxic gases and, 15.9
open-flame heat welding and, 9.10
OSHA electrical regulations, 12.15–12.18, 12.19
personal fall arrest systems and, 9.26–9.27, 12.35–12.38, 12.36*f*
pneumatic nail guns and, 12.28, 12.28*f*
polychlorinated biphenyls and, 12.33–12.34
powder-actuated tools and, 4.34–4.35, 4.34*f*, 12.29
power nailers and, 9.22–9.23
power tools and, 12.26–12.29, 12.27*f*, 12.28*f*
pressure testing and, 16.21, 16.23, 17.21
pressure tools and, 4.35–4.36
primers and cements and, 16.13, 16.14
rebar and, 2.16
receptacles and, 12.15
risk reduction, 12.5–12.12, 12.6*f*, 12.9*f*
roofing and, 9.17–9.19, 9.19*f*, 9.20, 9.20*f*, 9.21–9.27, 9.21*f*, 9.23*f*, 9.24*f*, 9.25*f*, 9.26*f*
roofing safety summary, 9.17
roof openings and, 8.16
scaffolds and, 4.41, 9.18–9.19, 9.19*f*, 9.20*f*, 9.32
sheathing and, 8.17
shock-absorbing lanyards and, 4.41, 9.24, 12.36, 12.37
slate roofs and, 9.74
soffits and, 10.4
solvents and, 12.30, 12.31, 12.31*f*, 12.32
stairways and, 11.4
static electricity and, 4.32, 4.53
torch-down roofing and, 9.85
training and, 4.34, 4.35, 4.52
trenches and, 2.19
trowels and, 4.23
trusses and, 8.21, 8.23
uncovered openings and, 12.12, 12.13
valves and, 17.8
weather hazards and, 4.36–4.37
wheel chocks and, 4.51
wind and, 10.4
wood shingles and shakes and, 9.61
working stacks and, 4.33–4.34, 4.34*f*, 5.1
worn or damaged equipment and, 4.43
Safety glasses, 9.22, 12.19

Safety locks, 12.14
Safety nets, 4.44, 4.45, 9.23, 9.27
Safety-Related Maintenance Requirements, 12.20
Safety-Related Work Practices, 12.20
Safety Requirements for Special Equipment, 12.20
Safety shoes, 12.19
Safety switches, 13.2
Sailors, 4.8*f*, 5.11*f*
SAMOBECC, and loss of trap seal, 15.10
Sand, and concrete, 2.4
Sandblasting, 5.45
Sand-lime bricks, 5.9
San Francisco rods, 1.23, 1.23*f*
Sanitary combinations, 15.18–15.19, 15.18*f*, 15.19*f*
    defined, 15.32–15.33
Sanitary drainage systems, 15.2
Sanitary fittings
    bends, 15.14–15.16, 15.15*f*, 15.16*f*
    defined, 15.33
    DWV fittings, 15.14–15.16, 15.14*f*
    tees, 15.16, 15.17*f*, 15.18
    wyes, 15.18–15.19, 15.18*f*, 15.19*f*
Sanitary increasers, 15.19, 15.19*f*
    defined, 15.33
Sanitary tees, 15.16, 15.17*f*, 15.18
Sanitary upright wyes, 15.18, 15.18*f*
    defined, 15.33
Sanitary wyes, 15.18, 15.18*f*
    defined, 15.33
SARS (severe acute respiratory syndrome), 15.25
Saw cut joints, 3.24
Saws
    concrete power saws, 3.24, 3.25*f*
    floor-type concrete saws, 3.24, 3.25*f*
    hacksaws, 17.9, 17.10
    masonry saws, 5.28, 5.28*f*
    PVC saws, 16.10–16.11
    soft-cut concrete saws, 3.24, 3.25*f*
    wet saws, 9.15, 9.16*f*
SBC (Standard Building Code), 6.22, 13.73
SBS (styrene butadiene styrene), 9.9
Scabs, 6.33
    defined, 6.50
Scaffolding
    continuously adjustable aluminum scaffolding, 10.3
    OSHA regulations and, 10.2, 10.3, 10.51, 12.22–12.23, 12.23*f*
    safety and, 4.41, 9.18–9.19, 9.19*f*, 9.20*f*, 9.32
    for vinyl and metal siding application, 10.51
    wood shingle and shake installation and, 9.60
Scale
    drawings and, 1.2
    pipe scale, 15.16, 15.32
    versus dimensions in drawings, 6.11
Scarf, 6.33
    defined, 6.50
Scarifiers, 3.31, 3.32*f*
Scattering shingle bundles, 9.32
Schedules. *See also* Size dimension ratios (SDR)
    defined, 16.30
    drawing sets and, 7.9, 7.10*f*
    laying out wall openings and, 7.9, 7.10*f*
Scratch coats
    defined, 10.71
    stucco (cement) finishes and, 10.64–10.65, 10.65*f*
Screeding
    defined, 2.38
    and finishing concrete, 3.18–3.19, 3.19*f*, 3.20, 3.20*f*, 3.21

Screeds
    handheld vibrating screeds, 3.17, 3.20, 3.21
    manual screeds, 3.18–3.19, 3.19f
    self-propelled laser screeds, 3.19, 3.20f, 3.21
Screw spreaders, 3.8
Scrims
    defined, 9.98
    single-ply membrane roofing and, 9.10
SCUBA (self-contained breathing apparatus), 12.30
SDR (size dimension ratios), 16.3, 16.3f, 16.30
Seal, loss of in traps, 15.8–15.11, 15.8f, 15.9f, 15.10f
Seat cuts. See Bird's mouths
Secondary control points, 1.7, 1.7f
Section drawings, floor systems, 6.8, 6.9f
Segregation
    concrete placement and, 3.6, 3.12, 3.15
    defined, 3.38
    discharging concrete from bucket and, 3.9
    internal vibrators and, 3.17
    transporting concrete in wheelbarrows and, 3.13
Seismic codes, 16.19
Seismic restraints, 16.19, 17.16. See also Earthquakes
Seismic risk zones, and footings, 2.19, 2.22
Self-contained breathing apparatus (SCUBA), 12.30
Self-propelled laser screeds, 3.19, 3.20f, 3.21
Self-retracting lifelines, 9.25, 9.25f
Selvage, 9.4, 9.4f
    defined, 9.98
Semi-hermetic (serviceable hermetic) compressors, 14.9
Service drops
    clearances for building openings, 13.26
    defined, 13.72
    locating, 13.24–13.25, 13.25f, 13.26f
    vertical clearances and, 13.25, 13.27f
Service entrance
    defined, 13.72
    residential electrical system planning and, 13.2
    residential installation of, 13.24–13.26, 13.25f, 13.26f, 13.27f, 13.29f
Service-entrance conductors, 13.6, 13.8
    defined, 13.72
Service-entrance equipment, 13.15
    defined, 13.72
Service entrance (Type SE) cable, 13.32–13.33, 13.34f
Service laterals, 13.16
    defined, 13.72
Service masts, 13.25, 13.26f
SE style R (SER) cable, 13.33
Set, 3.3
    defined, 3.38
Setting up masonry jobs
    contractors and, 5.14–5.15
    preliminary checklist, 5.15–5.16
Settling
    balloon framing and, 6.4
    platform framing and, 6.2
Severe acute respiratory syndrome (SARS), 15.25
Sewer brick, 4.5
Sewer mains, 15.22
Shakes. See also Wood shingles and shakes
    defined, 10.71
    as wood siding, 10.4, 10.34–10.35, 10.35f, 10.36, 10.37
Sheathing
    estimating material for, 8.26
    installing, 8.15–8.16, 8.15f, 8.16f, 8.17
    safety and, 8.17
    wall framing and, 2.25, 2.25f, 7.16, 7.18–7.19
Shed dormers, 8.26, 8.27f

Shed roofs, 8.2, 8.2f
Shiner bricks. See Rowlock stretcher bricks
Shingle alignment methods, 9.35
Shingle hatchets, 9.14, 9.15f, 9.59, 9.60, 9.60f
Shingle knives, 9.14, 9.15f
Shingle or shake siding, 10.4, 10.34–10.35, 10.35f, 10.36, 10.37
Shingles. See Architectural shingles; Composition shingles; Panelized shingles/shakes; Wood shingles and shakes
Shiplap wood siding, 10.19f, 10.31, 10.31f, 10.32, 10.33
Shock-absorbing lanyards, 4.40, 4.41, 4.41f, 9.24, 9.24f, 12.36, 12.37
Shop drawings, 14.10
Shoring, 2.18
    defined, 2.38
Short distance sighting, 1.25
Shorting probes, 12.11
Short sweep bends, 15.15–15.16, 15.16f
    defined, 15.33
Shotcrete, 3.13–3.14, 3.13f
Shrinkage cracking, and control joints, 5.2–5.3, 5.3f
Shrink mixing, 3.5
Shueck, Arnold, profile in success, 4.55
Sick buildings, 14.6
Side cuts, 8.38, 8.38f, 8.39
Side inlets, 15.15, 15.15f
    defined, 15.33
Side lap
    chimney flashing and, 5.14
    defined, 9.98
    roll roofing and, 9.52
    tile roofing and, 9.76f, 9.77f
    underlayment and, 8.15, 8.16f, 9.29f
Side shots, 1.34
Siding. See Fiber-cement siding; Vinyl and metal siding; Wood siding
Siding gauges, 10.23, 10.23f, 10.24
Siding junctures, 10.23, 10.24f
Siding reference marks, 10.23
Sight levels, 1.11, 1.11f, 1.12–1.13
Silicosis, 2.2, 5.28, 10.44
Sill blocks, 5.9f
Sill extensions, cutting to install siding, 10.51, 10.51f
Sill plates, and diagonal measurements, 6.28
Sills
    defined, 7.35
    estimating material for, 6.40
    floor systems and, 6.12, 6.13f, 6.14, 6.14f, 6.15
    installing, 6.28–6.30, 6.29f, 6.30f
    laying out for joists, 6.31–6.32, 6.32f
Sill sealers, 6.40
Simulated board-and-batten wood siding, 10.29
Simulated standing-seam metal roofing, 9.69–9.72, 9.71f, 9.72f, 9.73f
Single basket weave block bonds, 5.24f
Single coursing, 10.35, 10.35f, 10.36
Single-ply membrane roofing systems
    anchoring to roof structure, 9.10, 9.11f
    applying single-ply roofing, 9.83–9.85, 9.84f, 9.85f
    description of, 9.9–9.10, 9.10f
Single-pole ground-fault circuit interrupters (GFCI), 13.11–13.12, 13.12f
Single-pole switches, 13.52, 13.52f, 13.53f
Single-rotary finishing machines, 3.26, 3.26f
Single-thickness flares, 17.14, 17.14f
Siphonage, 15.2, 15.8, 15.9, 15.9f
    defined, 15.33
Site layout
    batter boards, 1.38–1.40, 1.39f

building plan drawings, 1.2–1.3, 1.4*f*

contour lines, 1.5, 1.5*f*, 1.6*f*

control points, 1.2, 1.7–1.9, 1.7*f*, 1.8*f*, 1.9*f*

differential leveling process and procedures, 1.2, 1.30–1.32, 1.31*f*, 1.33*f*, 1.34–1.35, 1.35*f*

differential leveling tools and equipment, 1.20–1.30, 1.21*f*, 1.22*f*, 1.23*f*, 1.24*f*, 1.25*f*, 1.26*f*, 1.27*f*, 1.28*f*, 1.29*f*

distance estimation by pacing, 1.19

distance measurement by taping, 1.11, 1.13–1.19, 1.14*f*, 1.15*f*, 1.16*f*

distance measurement tools and equipment, 1.11–1.13, 1.11*f*

electronic distance measurements, 1.19–1.20, 1.20*f*

field notes, 1.7, 1.36–1.37, 1.36*f*

hand signal communication, 1.9, 1.10*f*, 1.11

leveling applications, 1.37–1.38, 1.37*f*

3-4-5 rule, 1.40, 1.40*f*

Site plans, 1.2–1.3, 1.3*f*, 1.4*f*, 1.5*f*

Six-foot rule, 12.34

Six inch pattern shingle layout, 9.33–9.34, 9.34*f*, 9.36

Size dimension ratios (SDR), 16.3, 16.3*f. See also* Schedules

    defined, 16.30

Sizing

copper pipe, 17.2, 17.3*f*

drain, waste, and vent (DWV) systems, 15.13

floor joists, 6.43, 6.43*f*

girders and joists, 6.41, 6.42*f*, 6.43, 6.43*f*

plastic pipe, 16.3

Sizing electrical service

calculating electric service load, 13.5–13.6, 13.5*f*

demand factors and, 13.7, 13.8

floor plans and, 13.2–13.3, 13.3*f*, 13.4*f*

general lighting loads and, 13.3, 13.5

lighting demand factors, 13.6, 13.7

load centers and, 13.9–13.10

residential neutral conductors and, 13.8–13.9, 13.9*f*

Sizing tools, 17.10

    defined, 17.27

Skate rakers, 5.41, 5.42*f*

Skin, and electrical resistance, 12.4

Skirtboards, 11.7*f*, 11.8

    defined, 11.26

Skylights, 8.16

Slab-on-grade (slab-at-grade), 2.27, 2.28

    defined, 2.38

Slabs

concrete forms for, 2.27–2.28, 2.29*f*, 2.30*f*, 2.31

concrete placement and, 3.15

curing concrete and, 3.28

with foundations, 2.28, 2.29*f*

with thickened edges, 2.28, 2.29*f*

Slack to the line, 5.36

    defined, 5.49

Slag, 2.2

    defined, 2.38

Slate cutters, 9.14, 9.15*f*, 9.75, 9.75*f*

Slate roofing

description of, 9.5–9.6, 9.6*f*

slate or synthetic slate installation, 9.73–9.76, 9.74*f*, 9.75*f*

Slater's hammer, 9.14, 9.15*f*

Slope. *See also* Fall; Grade

composition shingles and, 9.3–9.4, 9.3*f*

defined, 9.98, 15.33

distance measurement and, 1.14, 1.14*f*, 1.15–1.17, 1.16*f*, 1.18–1.19

DWV systems and, 15.21

roof framing and, 8.3, 8.4*f*, 8.5

Sludge, 15.23

    defined, 15.33

Slump

defined, 2.38

plant-mixed concrete and, 3.5

testing for, 2.2, 2.8, 2.9*f*

Slump brick, 4.11, 4.12*f*, 5.8, 5.9*f*

Slump testing, 2.2, 2.8, 2.9*f*

Small appliance loads, 13.5, 13.7–13.8

Smeaton, John, 2.3

Snap-action switches, 13.52, 13.52*f*, 13.53*f*

Snaphooks, 4.42, 4.42*f*

Snug-rib system, 9.73

Soffit floor blocks, 4.10*f*, 5.6*f*

Soffits

aluminum and vinyl, 10.16, 10.16*f*, 10.17*f*

cornices and, 10.16, 10.16*f*, 10.17*f*

defined, 10.71

safety and, 10.4

Soffit ventilation, 9.87, 9.87*f*

Soft-cut concrete saws, 3.24, 3.25*f*

Soil stacks

building drain and, 15.22

composition shingle installation and, 9.39–9.40, 9.41*f*

flashing, 9.40*f*, 9.41*f*

Solder, 17.11, 17.12*f*

Solder fittings, 17.4

Soldering, 17.2, 17.11–17.12, 17.12*f*, 17.13*f*

Soldering tools, 17.11–17.12, 17.12*f*

Soldiers, 4.7, 4.8*f*, 5.11*f*, 5.22, 5.22*f*

Soleplates

defined, 6.50

estimating material for, 7.23*f*, 7.24

floor systems and, 6.2

layout, 7.8, 7.8*f*

as wall components, 7.2, 7.2*f*

Solid brick concrete blocks, 4.10*f*, 5.6*f*, 5.8, 5.8*f*

Solid concrete blocks, 4.10*f*, 4.11, 5.3, 5.6*f*

Solid lumber girders, 6.15, 6.15*f*

Solid masonry units, 4.5, 4.5*f*, 4.6*f*, 5.10–5.11

Solid masonry walls, 4.15, 4.15*f*

Solid top blocks, 4.10*f*, 5.6*f*

Solid wall plastic pipe, 16.2

Solid walls, defined, 16.30

Solid-web trusses, 6.22, 6.23, 6.23*f*

Solvent cements, 16.12–16.13

Solvents, and safety, 12.30, 12.31, 12.31*f*, 12.32

Solvent welding, 16.12–16.13, 16.13*f*

Solvent welds, 16.12, 16.13*f*

    defined, 16.30

Spacing jigs, 5.18

Spacing requirements, plastic pipe supports, 16.19–16.20, 16.21*f*, 16.22*f*

Spalling, 3.22

    defined, 3.38

Spanish tiles. *See* Barrel-style tiles

Spans

defined, 6.50

post-and-beam framing and, 6.4

roof layout and, 8.3, 8.4*f*

Spas, residential, 13.64, 13.64*f*

Special shaped bricks, 4.5, 4.6*f*

Special shingling effects, 9.62, 9.62*f*, 9.63

Specifications

described, 14.12–14.13

floor systems and, 6.12

Speed squares

and hip rafter layout, 8.42–8.43, 8.43*f*

and rafter layout, 8.18, 8.18*f*, 8.19*f*, 8.20

Spindles, 11.6, 11.7*f*

Splicing ceiling joists, 7.20–7.21, 7.21*f*

Splicing rebar, 2.16
Split blocks, 5.25, 5.25*f*
Split brick, 4.12*f*
Splitters, cutting masonry with, 5.28, 5.28*f*
Split-wired duplex receptacles, 13.42, 13.42*f*, 13.43*f*
Spontaneous fires, 4.53
Sprayers, 3.31, 3.32*f*
Spread footings, 2.20, 2.20*f*, 2.21*f*
Spreading mortar, 4.23, 4.25, 4.25*f*
Spreads, 4.25
        defined, 4.57
Spring benders, 17.10, 17.11*f*
Squareness, and foundations, 6.14, 6.27–6.28, 6.28*f*
Squares
        composition shingles and, 9.4
        defined, 9.98
Squaring brick corners, 5.39
Stack bonds, 5.21, 5.21*f*, 5.24*f*
Stacking building materials, 4.33–4.34, 4.34*f*, 5.1
Stacks, 15.10, 15.22
        defined, 15.33
Staggered joints, and siding installation, 10.58–10.59, 10.59*f*
Staggered shingles and shakes, 9.63
Staging, safety, 9.18–9.19
Stainless steel nails, 10.13
Stair gauges, 11.15, 11.15*f*
Stairs
        components of, 11.5–11.6, 11.5*f*, 11.6*f*, 11.7*f*, 11.8
        forms for concrete stairs, 11.21, 11.21*f*
        framing, 11.8–11.9, 11.9*f*
        International Building Code requirements for, 11.8
        laying out and cutting stringers, 11.15–11.17, 11.15*f*,
                11.16*f*, 11.18, 11.18*f*, 11.19, 11.19*f*, 11.20, 11.20*f*
        reinforced cutout stringers, 11.20, 11.20*f*, 11.21*f*
        stairway design and layout, 11.10–11.17, 11.11*f*, 11.13*f*,
                11.15*f*, 11.16*f*, 11.18, 11.18*f*, 11.19, 11.19*f*, 11.20, 11.20*f*,
                11.21*f*
        stairwells, 11.2, 11.9, 11.9*f*, 11.13–11.15
        terms on plans, 11.6, 11.7*f*, 11.8
        types of, 11.2, 11.2*f*, 11.3*f*, 11.4
Stairway design and layout
        pre-framed stairwells and, 11.13*f*, 11.14–11.15
        riser height and, 11.10–11.12, 11.11*f*
        stairwell openings and, 11.13–11.14, 11.13*f*
        stringers and, 11.15–11.17, 11.15*f*, 11.16*f*, 11.18, 11.18*f*,
                11.19, 11.19*f*, 11.20, 11.20*f*
        total stair run and, 11.12–11.13, 11.13*f*
Stairwells, 11.2, 11.9, 11.9*f*, 11.13–11.15
        defined, 11.26
Standard Building Code (SBC), 6.22, 13.73
Standard for Electrical Safety in the Workplace (NFPA 70E),
        12.11, 12.19–12.20
Standing-seam valley flashing, 9.11–9.12, 9.12*f*
Starter courses, slate roofs, 9.74, 9.74*f*
Starter strips, 9.33, 9.55, 9.55*f*, 10.55, 10.55*f*
Starting newel posts, 11.6, 11.7*f*
Static electricity, and safety, 4.32, 4.53
Stations, 1.30
        defined, 1.46
Steam cleaning, 5.45
Steel, in rebar, 2.12–2.13, 2.14*f*, 2.15
Steel I-beam girders, 6.15, 6.15*f*, 6.16
Steel I-beam headers, 7.6
Steel pipe, uses in HVAC, 14.10
Steel reinforcing rods, as grounding electrodes, 13.17–13.18
Steel stud channels, 7.28, 7.29*f*
Steel studs, 6.4, 7.28, 7.29*f*
Steel tapes, 1.11, 1.11*f*, 1.12

Steep slopes, and composition shingles, 9.4
Step flashing. *See also* Wall flashing
        chimneys and, 9.46, 9.46*f*
        nailing, 9.41, 9.42, 9.42*f*
Step ladders, and safety, 12.21–12.22
Stepped continuous footings, 2.20, 2.21*f*
Stone, 4.13, 4.13*f*, 5.11–5.12, 10.65
Stop valves, 17.7
Storage and handling
        construction material, 4.33
        copper pipe, 17.3
        plastic pipe, 16.8
Storm drains, 15.2
Story poles
        mounting boxes and, 13.50*f*
        wood siding and, 10.22
Straightedge, for shake/shingle application, 9.59, 9.60*f*,
        10.36
Straight-flight stairways, 11.4
Straight ladders, and safety, 12.21, 12.21*f*
Straightness, checking when laying masonry units, 5.31,
        5.31*f*
Straight reference lines, 10.53–10.54, 10.53*f*, 10.54*f*
Straight-run stairways, 11.2, 11.2*f*, 11.4
Strap iron, 17.16–17.17, 17.17*f*
S-traps, 15.5, 15.6*f*, 15.9
        defined, 15.32
Straub, Francis, 4.9
Street 45s, 17.5
Street 90s, 17.5
Strength, of concrete, 2.4, 2.5–2.6, 2.5*f*
Stretcher blocks, 4.10*f*, 4.11*f*, 5.4*f*, 5.5, 5.6*f*, 5.7*f*, 5.9*f*
Stretcher bricks, 5.11*f*, 5.21, 5.22*f*
Stretchers
        bull stretchers, 4.7, 4.8*f*
        mason's lines and, 5.34, 5.34*f*
Striking joints
        striking, 5.41–5.42, 5.41*f*, 5.42*f*
        testing mortar for, 5.41
Stringer mounting attachments, 11.17, 11.18*f*
Stringers
        cleated stringers, 11.5, 11.6*f*, 11.20, 11.20*f*
        closed finish stringers, 11.7*f*, 11.8
        cutout stringers, 11.5, 11.5*f*
        dadoed stringers, 11.5, 11.6*f*, 11.20, 11.20*f*
        defined, 11.26
        field-fabricated housed stringers, 11.19
        housed stringers, 11.5, 11.6*f*, 11.26
        laying out and cutting, 11.15–11.17, 11.15*f*, 11.16*f*, 11.18,
                11.18*f*, 11.19, 11.19*f*, 11.20, 11.20*f*
        open finished stringers, 11.7*f*, 11.8
        reinforced cutout stringers, 11.20, 11.20*f*, 11.21*f*
Stringing
        defined, 4.57
        mortar, 4.25
Strip-saddle ridges, 9.76
Strongbacks
        defined, 7.35
        joists and, 7.21, 7.22*f*
        and stringers, 11.20, 11.20*f*
        wall forms and, 2.25, 2.25*f*
Structural, defined, 4.57
Structural attachments, copper pipe hangers and supports,
        17.18, 17.20, 17.20*f*
Structural bonds, 5.22–5.23, 5.22*f*, 5.23*f*, 5.24*f*
Structural clay facing tile, 4.6, 4.7
Structural clay products, 4.4
Structural clay tile, 4.6

Structural drawings, 6.10
Structural pattern bonds, 5.22–5.23, 5.22f, 5.23f, 5.24f
Strut channels, 17.18, 17.20, 17.20f
Stub nails, 6.22
Stucco (cement) finishes, 10.64–10.65, 10.65f
Stucco locks, 10.65, 10.65f
Studs
    defined, 2.38
    estimating material for, 7.23f, 7.24
    layout, 7.8, 7.8f
    measuring and cutting, 7.12–7.14, 7.13–7.14f
    as wall components, 2.25, 7.2, 7.2f
    for wall forms, 2.25, 2.25f
Stud-to-stud measurements, 7.4
Styrene butadiene styrene (SBS), 9.9
Subflooring
    board subfloors, 6.26–6.27
    estimating material for, 6.41
    floor systems and, 6.25–6.27, 6.26f
    installing, 6.36–6.38, 6.37f, 6.38f
Subgrades, 2.18
    defined, 2.38
Subpanels, 13.28, 13.28f
Superintendents, 4.20
Supervisors, 4.20
Surface preparation, for vinyl and metal siding, 10.51, 10.51f
Surface treatments, for blocks, 5.7
Surfactants, cleaning blocks and, 5.46
Survey plans, 1.2–1.3, 1.3f, 1.4f, 1.5f
Suspension clamps, 16.20, 16.22f
Sweat caps, 17.5
Sweat fittings, 17.4
Sweat flanges, 17.5
Sweat joints, 17.11–17.12
    defined, 17.27
Sweat-to-compression adapters, 17.5
Sweat unions, 17.5
Swimming pools, residential, 13.64, 13.65f
Switchboard blankets, 12.10
Switchboard matting, 12.9
Switches
    color-coding of, 13.51
    defined, 13.72
    disconnect switches, 12.14, 12.15
    four-way switches, 13.52, 13.58, 13.58f, 13.59f
    locating, 13.59, 13.60f
    NEC standards, 13.51, 13.59, 13.60f
    quiet switches, 13.52, 13.53f
    snap-action switches, 13.52, 13.52f, 13.53f
    symbols for, 13.45, 13.47f
    three-way switches, 13.52, 13.54–13.56, 13.54f, 13.55f, 13.56f, 13.57f, 13.58f
Switchgear operations, safety precautions, 12.14
Switch/indicator light assemblies, 13.54
Switching operations, protective apparel, 12.10, 12.14
Switch legs, defined, 13.72
Symbols
    for circuit elements, 13.45, 13.47f
    electrical symbols, 13.45, 13.47f
    for outlets, 13.37, 13.38f
    receptacle symbols, 13.37, 13.38f
Synthetic-fiber clothing, safety and, 12.10
Synthetic roofing materials, 9.6, 9.7
Synthetic stone, 10.65
System commissioning, 14.13
System grounding, 13.15–13.16

Tagging, scaffolding, 9.19
Tailing
    defined, 5.49
    of diagonals, 5.37, 5.38f
Tail joists, 6.20, 6.20f
    defined, 6.50
Tail rafter cuts, and cornices, 10.7, 10.7f
Takeoffs, 14.10
    defined, 14.23
Tampers, 3.22, 3.22f
Tape clamps, 1.13
Tape length, corrections for, 1.18
Taper form ties, 2.26, 2.26f, 2.27, 2.27f
Tape sag, 1.18
Taping
    accuracy and tolerances and, 1.13
    conversion between measurement systems and, 1.17
    corrections of measurements, 1.18–1.19
    guidelines for, 1.14, 1.14f
    procedure for, 1.14–1.17, 1.15f, 1.16f
    tools and equipment for, 1.11–1.13, 1.11f
Targets, as rod accessories, 1.25, 1.25f
T-clips, 9.71–9.72, 9.72f
Technician certification
    Air Conditioning Excellence (ACE) exams and, 14.16
    EPA refrigerant transition and recovery certification, 14.17
Tee-pulling tools, 17.7–17.8, 17.8f
Tees
    copper pipe and, 17.5
    drain, waste, and vent (DWV) systems and, 15.16, 15.17f, 15.18
    sanitary tees, 15.16, 15.17f
Temperature
    cements and, 16.14
    curing concrete and, 3.27–3.28
    footings and, 2.19
    mortar and, 5.16
    placing and consolidating concrete and, 3.16
    pressure/temperature relationships, 14.8–14.9
    safety and, 4.36–4.37
    tape measurements and, 1.13, 1.18
Temperature indicator sticks, 16.17
Temporary benchmarks
    defined, 1.46
    transferring elevations and, 1.37–1.38, 1.37f
Temporary bracing
    roof trusses and, 8.23, 8.24, 8.24f, 8.25
    wall framing and, 7.18
Temporary guardrails, 11.10, 12.34, 12.34f
Temporary lighting, OSHA standards for, 12.15
Temporary supports, for girders, 6.18
Tension, on measuring tapes, 1.14, 1.14f, 1.18
Tension springs, and taping, 1.13, 1.18
Termite shields, 6.14, 6.40
Testing fall protection systems, 9.27
Testing installations, plastic plumbing materials, 16.22, 16.23
Test tees, 15.18, 15.18f
    defined, 15.33
T&G (tongue-and-groove) wood siding, 10.18f, 10.29, 10.29f, 10.30, 10.32, 10.33
Theodolites, line-of-sight checks of, 1.30
Thermal contact burns, 12.5
Thermometers, and taping, 1.13
Thermoplastic pipe, 16.2
    defined, 16.30
Thermoplastic (plastic polymer) roofing systems, 9.9, 9.10
Thermoplastic polyolefin (TPO) membranes, 9.10

Thermo-ply foil-faced paper board, 10.4
Thermoset (rubber polymer) roofing systems, 9.9–9.10
Thermosetting pipe, 16.2
    defined, 16.30
Thin veneer brick, 4.5
Threaded drop-in anchors, 17.18, 17.20, 17.20f
Threaded rod hangers, 17.18, 17.20
3-4-5 rule, 1.40, 1.40f
Three-way switches, residential lighting control, 13.52,
    13.54–13.56, 13.54f, 13.55f, 13.56f, 13.57f, 13.58f
Tianen, John, 14.22
Tie-off points. See Anchoring devices
Tie wire, 2.16, 2.16f
Tile cutters, 9.15, 9.16f
Tile roofing
    barrel-style tiles, 9.77f, 9.80, 9.81f, 9.82–9.83, 9.82f, 9.83f
    clay, ceramic, and concrete tile installation, 9.76, 9.76f,
        9.77, 9.77f, 9.78–9.80, 9.78f, 9.79f, 9.80f, 9.81f, 9.82–9.83,
        9.82f, 9.83f
    description of, 9.6–9.7, 9.6f
    flat tiles and, 9.78f, 9.79, 9.79f
    synthetic tiles, 9.6, 9.7
Tiles, hollow masonry units/tiles, 4.6–4.7, 5.10
Tilesetting trowels, 4.17
Tin snips, 10.57, 10.57f
Tipping, and forklifts, 4.49, 4.49f
Toed bricks, 5.31f
Toggles, 17.18, 17.20, 17.20f
Tolerances, and distance measurement, 1.13
Tongue-and-groove (T&G) wood siding, 10.18f, 10.29,
    10.29f, 10.30, 10.32, 10.33
Tools
    accidents and, 5.1
    concrete working tools, 3.29, 3.30f, 3.31, 3.31f, 3.32f
    deburr tools, 16.10, 16.11f, 17.10
    for differential leveling, 1.20–1.30, 1.21f, 1.22f, 1.23f, 1.24f,
        1.25f, 1.26f, 1.27f, 1.28f, 1.29f
    for distance measurement, 1.11–1.13, 1.11f
    double-insulated/ungrounded tools, 12.3, 12.15, 12.15f
    flaring tools, 17.14, 17.14f
    fusion tools, 16.17, 16.17f
    gasoline-powered tool safety, 4.34
    grounded tools, 12.15
    hand-crimping tools, 16.16, 16.16f
    for joining PEX tubing, 16.16, 16.16f
    measuring tools, 1.13, 1.14, 1.14f, 1.22
    powder-actuated tools, 4.34–4.35, 4.34f, 12.29
    pressure tools, 4.35–4.36
    roofing and, 9.14–9.15, 9.15–9.16f
    rules for care and safe use, 3.32–3.33
    safety and, 12.26–12.29, 12.27f, 12.28f
    sizing tools, 17.10
    soldering tools, 17.11–17.12, 17.12f
    for taping, 1.11–1.13, 1.11f
    tee-pulling tool, 17.7–17.8, 17.8f
    tube cutters, 16.10–16.11, 16.11f, 17.9, 17.9f
    vinyl and metal siding and, 10.50, 10.50f
    vises, 17.10
Top dips, 15.6, 15.6f
Top lap
    composition shingles and, 9.31f, 9.32f
    defined, 9.98
    roll roofing and, 9.3f, 9.52
    underlayment and, 8.15, 8.16f, 9.12, 9.29f
Topographical features, in civil plans, 1.2–1.3, 1.4f, 1.5f
Top plates
    defined, 7.35
    estimating material for, 7.23f, 7.24

layout, 7.8, 7.8f
    as wall components, 2.25, 2.25f, 7.2, 7.2f
Torch-down roofing, 9.9, 9.9f, 9.85–9.87, 9.85f, 9.86f, 9.87f
Torpedo levels, 15.36
Total stair run, 11.12–11.13, 11.13f, 11.15
Total stairway rise, 11.11, 11.11f
Total station telescopes, line-of-sight checks for, 1.30
Tower cranes, 3.6, 3.9f
Toxic, defined, 14.23
Toxic particles and fumes, 14.6
Toxic vapors, and safety, 12.30, 12.31f, 12.32
TPO (thermoplastic polyolefin) membranes, 9.10
TP (turning points), 1.30, 1.32, 1.46
Traffic, and curing concrete, 3.28
Training
    apprenticeship training, 14.15–14.16, 14.17
    asbestos work and, 12.33
    confined space entry and, 12.30
    fall protection and, 12.34
    laser leveling instruments and, 1.22
    lockout/tagout rules and, 12.17
    Occupational Safety and Health Administration (OSHA)
        standards, 12.17
    safety and, 4.34, 4.35, 4.52
Training and certification, for operating forklifts, 4.45
Training programs, 14.15–14.18
    Heating, Ventilating, and Air Conditioning (HVAC)
        industry and, 14.15–14.18
Transformers
    fuse replacements and, 12.14–12.15
    residential grounding systems and, 13.16, 13.16f
Transition fittings, 16.12
    defined, 16.30
Transits
    line-of-sight checks for, 1.28–1.30, 1.29f
    and wood siding, 10.22
Trap arms, 15.6
Trap primers, 15.10
Traps
    functioning of, 15.2, 15.8, 15.8f
    installation requirements, 15.6–15.7, 15.7f
    loss of seal and, 15.8–15.11, 15.8f, 15.9f, 15.10f
    parts of, 15.6, 15.6f
    types of, 15.4–15.6, 15.5f, 15.6f
Trap-to-vent distances, 15.6–15.7, 15.7f, 15.12
Trap weirs. See Crown weirs
Traveling, and forklifts, 4.47–4.49, 4.47f
Tread run, 11.12–11.13, 11.13f
Treads
    defined, 11.26
    as stairway component, 11.2, 11.5, 11.8
Tread thickness, and stringers, 11.17, 11.19f
Tread widths, 11.10
Trenches, and safety, 2.19
Trenching and excavation, and mobile boom-type concrete
    pumps, 3.11
Trim, vinyl and metal siding, 10.55, 10.56, 10.56f, 10.57f
Trimmer joists, 6.20, 6.20f
    defined, 6.50
Trimmer studs, 7.2, 7.2f, 7.14
    defined, 7.35
Tripods, 1.22–1.23, 1.23f
Troubleshooting, pressure/temperature relationship,
    14.8–14.9
Trough blocks, 5.7f
Troweling, and finishing concrete, 3.26–3.27, 3.26f, 3.27f
Trowels
    buttering joints, 4.26–4.27, 4.27f

cutting masonry with, 5.27–5.28
holding, 4.23–4.24, 4.23f
picking up mortar, 4.23–4.25, 4.23f, 4.24f
spreading, cutting, and furrowing mortar, 4.25–4.26, 4.25f, 4.26f
striking mortar with, 5.41
as tools of masons, 4.17, 4.17f
Truck agitators, 3.7
Truck mixers, 3.5, 3.5f, 3.7
Trusses
    bracing for, 8.23, 8.24, 8.24f, 8.25
    components of, 8.20, 8.20f
    defined, 6.50
    erecting trusses, 8.21, 8.21f
    floor systems and, 6.24, 6.24f
    installation of, 8.21, 8.22f, 8.23–8.24, 8.23f, 8.24f, 8.25
    metal roof trusses, 8.28, 8.28f
    truss placement diagrams, 8.21, 8.22f
    types of, 8.20–8.21, 8.20f
Truss headers, 7.6, 7.7f
Truss rigging, 8.21, 8.21f, 8.23
Truss storage, 8.21, 8.23
Tube bending equipment, 17.10, 17.11f
Tube cutters, 16.10–16.11, 16.11f, 17.9, 17.9f
Tuckpointer trowels, 4.17f
Tuckpointing, 4.18, 5.43, 5.43f
    defined, 4.57
Turbine ventilation, 9.87, 9.87f, 9.88
Turning points (TP)
    defined, 1.46
    described, 1.30
    differential leveling and, 1.32
Twin ells. See Double bends
Twin-tailed lanyards, 12.36
240-volt circuits, 13.43–13.45
Two-pole ground-fault circuit interrupters (GFCI), 13.11, 13.12, 13.13f, 13.14
Tying-off
    for fall protection equipment, 9.26–9.27
    personal fall-arrest systems and, 4.42–4.43
Type K mortar, 4.14
Type M mortar, 4.14
Type NM cable, 13.29–13.30, 13.30f, 13.31f, 13.55–13.56
Type N mortar, 4.14
Type O mortar, 4.14
Type SE (service entrance) cable, 13.32–13.33, 13.34f
Type S mortar, 4.14
Type USE cable, 13.32
Tyvek, 10.3

U

UBC (Uniform Building Code), 6.30, 13.73
Ultraviolet light, and ventilation, 14.6
Ultraviolet protection, torch-down roofing, 9.86, 9.87, 9.87f
Ultraviolet (UV) radiation, and plastic pipe, 16.2, 16.5
Uncovered openings, 12.12, 12.13
Under-eave furring, 10.52, 10.53f
Underground feeder (Type UF) cables, 13.32
Underlayment. See also Waterproof membranes
    aluminum foil underlayment, 10.52
    composition shingles and, 9.3–9.4, 9.12
    defined, 6.50, 9.98
    roof deck preparation and, 9.28, 9.29f
    sheathing installation and, 8.15–8.16, 8.16f
    single-ply membrane roofing and, 9.10
    subflooring and, 6.25, 8.16, 8.16f, 8.17
    torch-down roofing and, 9.85, 9.85f
    wood shakes and, 9.64, 9.65, 9.65f

Undersill furring, 10.52, 10.52f
Underwriters Laboratories, Inc. (UL), and receptacles, 13.51
Unequal roof pitch, 9.38
Uniform Building Code (UBC), 6.30, 13.73
Unit rise
    defined, 11.26
    roof layout and, 8.3–8.4
    stairs and, 11.10, 11.11f, 11.14–11.15
Unit run, 11.10, 11.14–11.15
    defined, 11.26
Universal pipe clamps, 17.17, 17.18f
Utility locations, and civil plans, 1.3

V

Valley flashing
    defined, 9.98
    installing, 9.28
    roll roofing and, 9.4, 9.54, 9.55f, 9.56
    simulated standing-seam metal roofing and, 9.72, 9.73f
    tile roofing and, 9.80, 9.82f
    types of, 9.11–9.12, 9.12f
Valley jacks, 8.39–8.40, 8.39f, 8.40f, 8.41
Valley rafters
    described, 8.3, 8.4f
    layout, 8.39, 8.39f, 8.40f
Valleys
    composition shingle installation and, 9.37–9.39, 9.38f, 9.39f, 9.40f, 9.43
    defined, 9.98
    flashing and, 9.4, 9.11–9.12, 9.28, 9.54, 9.55f, 9.56, 9.72, 9.73f, 9.80, 9.82f
    roll roofing installation and, 9.4, 9.54, 9.55f, 9.56
    simulated standing-seam metal roofing and, 9.72, 9.73f
    wood shake installation, 9.66
    wood shingle installation and, 9.63, 9.63f
Van Holten, Alan, profile in success, 8.34
Vapor lamps, disposal of, 12.34
Vapor-removing respirators, 12.32
Vaults, and safety, 12.29
Vault units, 4.12, 4.13f, 5.10, 5.11f
V-edged siding, 10.29, 10.29f, 10.30
Velocity, 15.36
    defined, 15.33
Veneer masonry walls, 4.15, 4.15f
Veneer moisture barriers, 10.65
Veneers
    brick and stone veneers, 10.65
    defined, 10.71
    exterior finishing and, 10.4
Veneer stone walls, 4.13, 4.13f
Veneer ties, 5.13, 5.13f
Vent branches, 15.18, 15.18f
    defined, 15.33
Vent ells, 15.16, 15.16f
    defined, 15.33
Ventilation
    confined spaces and, 12.29, 12.30
    introduction to, 14.6
    solvents and, 12.30
Vents. See also Drain, waste, and vent (DWV) systems
    defined, 10.71
    gable end vents, 8.12, 8.13f
    plumbing, 15.12, 15.12f
    in siding, 10.63
Vent stack flashing, 9.40, 9.40f
    defined, 9.98
Vent systems, 15.12, 15.12f

Vent tees, 15.16, 15.17*f*
    defined, 15.33
Vent-to-trap distances, 15.6–15.7, 15.7*f*, 15.12
Vertical lifelines, 4.41–4.42, 4.42*f*
Vertical pipe hangers, 16.20, 16.22*f*
Vertical scored blocks, 5.9*f*
Vertical siding, 10.18*f*, 10.19*f*, 10.25, 10.29, 10.31, 10.32, 10.33
Vertical towers, 12.24
Vertical wall flashing, 9.42*f*, 9.43*f*
Vibration, and consolidating concrete, 3.16, 3.17, 3.17*f*
Vibration isolators, 17.17, 17.17*f*
Vibratory hand floats, 3.17
Vibratory hand trowels, 3.17
Vinyl and metal siding
    caulking and cleanup, 10.64
    cutting procedures, 10.57, 10.57*f*, 10.58*f*
    furring and insulation techniques, 10.51–10.52, 10.52*f*,
       10.53*f*
    gable end trim, 10.56, 10.57*f*
    inside corner posts and, 10.54, 10.54*f*
    installing around windows and doors, 10.60–10.61, 10.61*f*,
       10.62*f*
    installing at gable ends, 10.63, 10.63*f*
    installing under eaves, 10.63–10.64, 10.63*f*, 10.64*f*
    materials and components, 10.49–10.50, 10.49*f*
    outside corner posts, 10.54, 10.55*f*
    reference lines and, 10.53–10.54, 10.53*f*, 10.54*f*
    required equipment, 10.51, 10.51*f*
    siding and corner cap installation, 10.58–10.60, 10.59*f*,
       10.60*f*
    starter strips and, 10.55, 10.55*f*
    surface preparation, 10.51, 10.51*f*
    tools, 10.50, 10.50*f*
    window and door trim, 10.55, 10.56*f*, 10.57*f*
Vises, 17.10
Voltage ratings, rubber protective equipment, 12.7, 12.8
Volume calculations for concrete
    circular, 2.12, 2.12*f*
    rectangular, 2.9–2.10, 2.9*f*, 2.10*f*, 2.11, 2.11*f*, 3.15
Volumes, formulas for finding, 2.9, 2.10, 2.12, 2.40
Volumetric mixers, 3.6, 3.6*f*

W
Walers (wales), 2.22, 2.25, 2.25*f*
    defined, 2.38
Wall construction techniques, 4.14–4.15, 4.15*f*, 4.16, 4.16*f*
Wall flashing. *See also* Step flashing
    defined, 9.98
    tile roofing and, 9.80, 9.82*f*
    use of, 9.41
    vertical wall flashing, 9.40–9.42, 9.42*f*, 9.43*f*
Wall forms
    components of, 2.25–2.27, 2.25*f*, 2.26*f*, 2.27*f*
    concrete placement and, 3.16
    described, 2.24
Wall framing
    assembling walls, 7.14–7.15, 7.16*f*
    components of, 2.25, 2.25*f*, 7.2–7.4, 7.2*f*, 7.3*f*, 7.4*f*, 7.5*f*, 7.6,
       7.6*f*, 7.7, 7.7*f*
    erecting walls, 7.16–7.19, 7.16*f*, 7.17*f*, 7.18*f*
    estimating material for, 1.23–1.26, 1.23*f*
    laying out, 7.8–7.11, 7.8*f*, 7.9*f*, 7.10*f*, 7.11*f*, 7.12*f*
    masonry wall framing, 7.26, 7.27*f*, 7.28, 7.28*f*
    measuring and cutting studs, 7.12–7.14, 7.13–7.14*f*
    sheathing and, 2.25, 2.25*f*, 7.16, 7.18–7.19
    steel studs, 6.4, 7.28, 7.29*f*
Wall lifting jacks, 7.16, 7.16*f*

Wall openings, layout, 7.9–7.11, 7.9*f*, 7.10*f*, 7.11*f*, 7.12*f*
Walls. *See also* Masonry walls; Wall framing
    aligning of, 7.16–7.19, 7.17*f*, 7.18*f*
    assembly of, 7.14–7.15, 7.16*f*
    components of, 2.25, 2.25*f*, 7.2–7.4, 7.2*f*, 7.3*f*, 7.4*f*, 7.5*f*, 7.6,
       7.6*f*, 7.7, 7.7*f*
    curtain walls, 4.16
    erecting of, 7.16–7.19, 7.16*f*, 7.17*f*, 7.18*f*
    plumbing and aligning, 7.16–7.19, 7.17*f*, 7.18*f*
Wall ties, 2.27, 2.27*f*
Warped lumber, and estimating material, 7.25
Wash stations, and safety, 12.33
Waste. *See also* Drain, waste, and vent (DWV) systems
    and calculating concrete volume, 2.11, 2.12
Waste allowances, for siding, 10.21
Waste treatment, 15.22–15.23, 15.24, 15.24*f*
Water
    cofferdam concrete seal course and, 3.13
    and concrete, 2.4, 2.8
    construction joints and, 3.2
    curing concrete and, 3.28
    as heat exchange medium, 14.2
Water-cement ratio, 2.5
    defined, 2.38
Water hammer, 16.6
    defined, 16.30
Water heater circuits, 13.44–13.45, 13.44*f*
Water heater controls, 13.44, 13.44*f*
Water leveling, and reference lines, 10.54, 10.54*f*
Water levels, in batteries, 12.33
Waterproof membranes. *See also* Underlayment
    for roofing, 9.12–9.13, 9.28, 9.29*f*, 9.40, 9.52
Waterproof mortar, 4.3
Water supply, plastic fittings for, 16.8, 16.9, 16.9*f*
Water testing, 17.21. *See also* Hydrostatic testing
Weather
    curing of concrete and, 3.28
    placing and finishing concrete and, 3.16
    safety and, 4.36–4.37
Wedge anchors, 17.18
Weepholes, 4.15
    defined, 4.57
Weight, of concrete, 3.9
Weirs, 15.6
    defined, 15.33
Welded-wire fabric, 2.16, 2.17, 2.17*f*
Western platform framing, 6.2, 7.4. *See also* Platform framing
Western red cedar, as allergen, 10.4
Westinghouse Electric Company, 13.2
Wet-mix plants, 3.5
Wet saws, 9.15, 9.16*f*
Wheelbarrows
    mixing mortar in, 4.22, 4.23*f*
    transporting concrete and, 3.7, 3.12–3.13, 3.12*f*
Wheel chocks, 4.51
Wide-L stairways, 11.3*f*
Wide-U stairways, 11.3*f*
Wind. *See also* High wind areas
    curing concrete and, 3.28
    safety working in, 10.4
    siding installation and, 10.58, 10.59
Wind effect. *See* Oscillation
Winding stairways, 11.3*f*, 11.4
    defined, 11.26
Window flashing systems, and exterior finishing, 10.5, 10.6*f*

Windows
    furring out around, 10.52
    trim installation, 10.55, 10.56f, 10.57f
    vinyl and metal siding installed around, 10.60–10.61, 10.61f, 10.62f
    wood siding installed around, 10.25, 10.25f, 10.32, 10.33
Window sills, boxing in, 10.55, 10.56f
Wire ties, 2.16, 2.16f
Wiring. *See* Residential wiring
W-metal flashing, 9.11–9.12, 9.12f
Wood, as heat source, 14.2
Wood I-beams
    as floor joists, 6.22, 6.23, 6.23f
    floor systems and, 6.16
    headers and, 7.6
Wood shingles and shakes
    applying hardboard shingle/shake panels, 9.67, 9.67f, 9.68f
    applying wood shakes, 9.64, 9.65f, 9.66
    applying wood shingles, 9.59–9.63, 9.60f, 9.61f, 9.62f, 9.63f, 9.64f
    description of, 9.4–9.5, 9.5f
    ridge and hip caps, 9.63, 9.64f
    roof exposure and, 9.5, 9.59, 9.59f
    safety and, 9.61
    special effects, 9.62, 9.62f, 9.63
    types and grades of, 9.57–9.59, 9.64, 9.64f
    valleys and, 9.63, 9.63f, 9.66
Wood siding
    beveled siding, 10.21–10.23, 10.21f, 10.22f, 10.23f, 10.24, 10.24f, 10.25f, 10.26f, 10.27f
    board-and-batten siding, 10.25, 10.26, 10.27–10.28, 10.27f, 10.28f, 10.29f
    estimating panel and board siding, 10.20–10.21

    hardboard and particleboard siding, 10.39–10.41, 10.40f, 10.41f, 10.42f, 10.43f
    plywood siding, 10.37–10.39, 10.37f, 10.38f
    shingle siding or shakes, 10.4, 10.34–10.35, 10.35f, 10.36, 10.37
    shiplap wood siding, 10.19f, 10.31, 10.31f, 10.32, 10.33
    styles, 10.17, 10.18–10.19f, 10.20
    tongue-and-groove wood siding, 10.18f, 10.29, 10.29f, 10.30, 10.32, 10.33
Work attitudes, 4.21, 12.13
Working control points, 1.8, 1.8f
Working drawings, 6.7–6.8, 6.9f, 6.10–6.12
Working life, of mortar, 4.28
Working stacks, and safety, 4.33–4.34, 4.34f, 5.1
World Health Organization, 15.25
Worn equipment, and safety, 4.43
Wrought copper fittings, 17.4
Wyes, 15.18–15.19, 15.18f, 15.19f
Wythe
    bricks and, 4.7, 4.8
    defined, 4.57
    hollow masonry units/tiles as, 4.6
    pattern bonds and, 5.21
    structural and structural pattern bonds and, 5.22–5.23

Y
Youth training and apprenticeship programs, 14.17–14.18

Z
Zero, finding on tapes, 1.14, 1.14f
Z-strips, 9.72, 9.73f